ÁIST

Encyclopedia of Clinical Pharmacy

Encyclopedia of Clinical Pharmacy

edited by

Joseph T. DiPiro
Panoz Professor of Pharmacy
University of Georgia College of Pharmacy
Athens, Georgia, U.S.A.
and
Clinical Professor of Surgery
Medical College of Georgia
Augusta, Georgia, U.S.A.

MARCEL DEKKER, INC. NEW YORK • BASEL

ISBN
Print: 0-8247-0752-4
Online: 0-8247-0608-0

This book is printed on acid-free paper.

Headquarters
Marcel Dekker, Inc.
270 Madison Avenue, New York, NY 10016
tel: 212-696-9000; fax: 212-685-4540

Eastern Hemisphere Distribution
Marcel Dekker AG
Hutgasse 4, Postfach 812, CH-4001 Basel, Switzerland
tel: 41-61-260-6300; fax: 41-61-260-6333

World Wide Web
http://www.dekker.com

The publisher offers discounts on this book when ordered in bulk quantities. For more information, write to Special Sales/Professional Marketing at the headquarters address above.

Current printing (last digit):
10 9 8 7 6 5 4 3 2 1

PRINTED IN THE UNITED STATES OF AMERICA

Encyclopedia of Clinical Pharmacy

Editorial Advisory Board

Thomas Hardin, Pharm.D., FCCP
Ortho-McNeil Pharmaceutical, Inc., San Antonio, Texas, U.S.A.

Mary Anne Koda-Kimble, Pharm.D.
University of California at San Francisco, San Francisco, California, U.S.A.

Milap C. Nahata, Pharm.D.
The Ohio State University, Columbus, Ohio, U.S.A.

Atsuhiko Nishitani, Ph.D.
Juntendo University Hospital, Tokyo, JAPAN

Cynthia Raehl, Pharm.D.
Texas Tech University HSC at Amarillo, Amarillo, Texas, U.S.A.

Myrella Roy, Pharm.D., FCCP
Consultant, Ottawa, Ontario, CANADA

Barbara Wells, Pharm.D., FCCP, FASHP, BCPP
University of Mississippi, University, Mississippi, U.S.A.

Georges Zelger, Ph.D., PD
Pharmacie des hôpitaux, du Nord Vaudois et de la Broye, Yverdon-les-Bains, SWITZERLAND

List of Contributors

Marwan S. Abouljoud / *Henry Ford Hospital, Detroit, Michigan, U.S.A.*
Azucena Aldaz / *Clinica Universitaria de Navarra, Pamplona, Spain*
Christopher P. Alderman / *Repatriation General Hospital, Adelaide, Australia*
Kathryn T. Andrusko-Furphy / *Atascadero, California, U.S.A.*
Joint Commission on Accreditation of Healthcare Organizations (JCAHO)Illinois, U.S.A.
Calvin J. Anthony / *National Community Pharmacy Association, Arlington, Virginia, U.S.A.*
Edward P. Armstrong / *University of Arizona, Tucson, Arizona, U.S.A.*
Carolyn H. Asbury / *University of Pennsylvania, Philadelphia, Pennsylvania, U.S.A.*
Iman E. Bajjoka / *Henry Ford Hospital, Detroit, Michigan, U.S.A.*
Teresa Bassons Boncompte / *Pharmaceutical Association of Catalonia, Barcelona, Spain*
Jerry L. Bauman / *University of Illinois, Chicago, Illinois, U.S.A.*
Diane E. Beck / *Auburn University, Auburn, Alabama, U.S.A.*
William Benda / *Big Sur, California, U.S.A.*
John D. Benz / *American College of Clinical Pharmacy, Kansas City, Missouri, U.S.A.*
Deena Bernholtz-Goldman / *Fujisawa Healthcare, Inc., Deerfield, Illinois, U.S.A.*
Richard J. Bertin / *Council on Credentialing in Pharmacy, Washington, D.C., U.S.A.*
Theresa M. Bianco / *Oregon Health Sciences University, Portland, Oregon, U.S.A.*
Beverly L. Black / *American Society of Health-System Pharmacists, Bethesda, Maryland, U.S.A.*
James L. Blackburn / *University of Saskatchewan, Saskatoon, Canada*
Deborah E. Boatwright / *Veterans Affairs Medical Center, San Francisco, California, U.S.A.*
Joaquin Bonal de Falgas / *Pharmaceutical Care Spain Foundation, Barcelona, Spain*
Christine M. Bond / *University of Aberdeen, Aberdeen, U.K.*
Joseph K. Bonnarens / *Applied Health Outcomes, Tampa, Florida, U.S.A.*
Brent M. Booker / *University at Buffalo, Buffalo, New York, U.S.A.*
Lynn Bosco / *Agency for Healthcare Research and Quality, Rockville, Maryland, U.S.A.*
James Buchanan / *Tularik, Inc., South San Francisco, California, U.S.A.*
Marcia L. Buck / *University of Virginia Medical Center, Charlottesville, Virginia, U.S.A.*
Nanette C. Bultemeier / *Oregon State University, Portland, Oregon, U.S.A.*
Kathleen M. Bungay / *The Health Institute, Boston, Massachusetts, U.S.A.*
Naomi Burgess / *Society of Hospital Pharmacists of Australia, South Melbourne, Australia*
Judith A. Cahill / *Academy of Managed Care Pharmacy, Alexandria, Virginia, U.S.A.*
Karim Anton Calis / *National Institutes of Health, Bethesda, Maryland, U.S.A.*
William Campbell / *University of North Carolina, Chapel Hill, North Carolina, U.S.A.*
Alex A. Cardoni / *The Institute of Living, Hartford, Connecticut, U.S.A.*
Jannet M. Carmichael / *VA Sierra Nevada Health Care System, Reno, Nevada, U.S.A.*
Allen Cato / *Cato Research Ltd., Durham, North Carolina, U.S.A.*
Allen Cato III / *Cato Research Ltd., San Diego, California, U.S.A.*
Christi Cawood Marsh / *University Health Care System, Augusta, Georgia, U.S.A.*
Jack J. Chen / *Western University of Health Sciences, Pomona, California, U.S.A.*

John Jackson / *Integrated Pharmacy Services, Victoria, Australia*
Judith Jacobi / *Methodist Hospital/Clarian Health, Indianapolis, Indiana, U.S.A.*
Kevin M. Jarvis / *Biocentric, Inc., Berkley, Michigan, U.S.A.*
Tommy Johnson / *University of Georgia College of Pharmacy, Athens, Georgia, U.S.A.*
Joan Kapusnik-Uner / *First DataBank, Inc., San Bruno, California, U.S.A.*
Yasmin Khaliq / *Rochester, Minnesota, U.S.A.*
Arthur H. Kibbe / *Wilkes University, Wilkes-Barre, Pennsylvania, U.S.A.*
Daren L. Knoell / *The Ohio State University, Columbus, Ohio, U.S.A.*
Jill M. Kolesar / *University of Wisconsin, Madison, Wisconsin, U.S.A.*
Sheldon X. Kong / *Merck & Co. Inc., Whitehouse Station, New Jersey, U.S.A.*
Joan K. Korth-Bradley / *JB Ashtin Group Inc., Canton, Michigan, U.S.A.*
Edward P. Krenzelok / *Children's Hospital of Pittsburgh, Pittsburgh, Pennsylvania, U.S.A.*
Patricia Dowley Kroboth / *University of Pittsburgh, Pittsburgh, Pennsylvania, U.S.A.*
Suzan E. Kucukarslan / *Henry Ford Hospital, Detroit, Michigan, U.S.A.*
Peggy G. Kuehl / *American College of Clinical Pharmacy, Kansas City, Missouri, U.S.A.*
Mae Kwong / *American Society of Health-System Pharmacists, Bethesda, Maryland, U.S.A.*
Carlos Lacasa / *Clinica Universitaria de Navarra, Pamplona, Spain*
Y.W. Francis Lam / *University of Texas Health Science Center at San Antonio, San Antonio, Texas, U.S.A.*
Grace D. Lamsam / *University of Pittsburgh School of Pharmacy, Pittsburgh, Pennsylvania, U.S.A.*
Kimberly J. La Pointe / *Children's Hospital of The King's Daughters, Norfolk, Virginia, U.S.A.*
Beth A. Lesher / *Hamilton House, Virginia Beach, Virginia, U.S.A.*
Donald E. Letendre / *American Society of Health-System Pharmacists, Bethesda, Maryland, U.S.A.*
Arthur G. Lipman / *University of Utah, Salt Lake City, Utah, U.S.A.*
Elizabeth Tracie Long / *ISPOR—International Society for Pharmacoeconomics and Outcomes Research, Lawrenceville, New Jersey, U.S.A.*
Neil J. MacKinnon / *Dalhousie University, Halifax, Nova Scotia, Canada*
Robert L. Maher, Jr. / *Duquesne University, Pittsburgh, Pennsylvania, U.S.A.*
Robert A. Malone / *National Community Pharmacy Association, Arlington, Virginia, U.S.A.*
Patrick M. Malone / *Creighton University, Omaha, Nebraska, U.S.A.*
Henri R. Manasse, Jr. / *American Society of Health-System Pharmacists, Bethesda, Maryland, U.S.A.*
Henar Martinez Sanz / *Sociedad Española de Farmacia Hospitalaria, Madrid, Spain*
Pilar Mas Lombarte / *Fundación Hospital-Asil de Granollers, Granollers, Spain*
Gary R. Matzke / *University of Pittsburgh, Pittsburgh, Pennsylvania, U.S.A.*
J. Russell May / *Medical College of Georgia Hospitals & Clinics, Augusta, Georgia, U.S.A.*
Theresa A. Mays / *Cancer Therapy and Research Center, San Antonio, Texas, U.S.A.*
William McLean / *Les Consultants Pharmaceutiques de l'Outaouais, Cantley, Quebec, Canada*
William A. Miller / *University of Iowa, Iowa City, Iowa, U.S.A.*
David I. Min / *University of Iowa, Iowa City, Iowa, U.S.A.*
Josi-Bruno Montoro-Ronsano / *Hospital General Vall d'Hebron, Barcelona, Spain*
Phylliss Moret / *American Society of Consultant Pharmacists, Alexandria, Virginia, U.S.A.*
Walter J. Morrison / *Provider Payment Solution, Inc., Little Rock, Arkansas, U.S.A.*
Gene D. Morse / *University at Buffalo, Buffalo, New York, U.S.A.*
Pat Murray / *Royal Edinburgh Hospital, Edinburgh, U.K.*
Milap C. Nahata / *The Ohio State University, Columbus, Ohio, U.S.A.*
Gloria J. Nichols-English / *University of Georgia College of Pharmacy, Athens, Georgia, U.S.A.*
Edward H. O'Neil / *Western University, Pomona, California, U.S.A.*
Ana Ortega / *Clinica Universitaria de Navarra, Pamplona, Spain*
María-José Otero / *Hospital Universitario de Salamanca, Salamanca, Spain*
Thomas W. Paton / *University of Toronto, Toronto, Ontario, Canada*
Stephen H. Paul / *Temple University, Philadelphia, Pennsylvania, U.S.A.*

Richard P. Penna / *American Association of Colleges of Pharmacy, Alexandria, Virginia, U.S.A.*
Carmen Permanyer Munné / *Pharmaceutical Association of Catalonia, Barcelona, Spain*
Matthew Perri III / *University of Georgia College of Pharmacy, Athens, Georgia, U.S.A.*
Vanita K. Pindolia / *Henry Ford Health System, Detroit, Michigan, U.S.A.*
Stephen C. Piscitelli / *Virco Lab Inc., Rockville, Maryland, U.S.A.*
Sylvie Poirier / *Regie Regionale de la Sante et des Services Sociaux Monteregie, Quebec, Canada*
Therese I. Poirier / *Duquesne University, Pittsburgh, Pennsylvania, U.S.A.*
Samuel M. Poloyac / *University of Pittsburgh, Pittsburgh, Pennsylvania, U.S.A.*
Jeff Poston / *Canadian Pharmacists Association, Ottawa, Ontario, Canada*
Leigh Ann Ramsey / *University of Mississippi, Jackson, Mississippi, U.S.A.*
Teresa Requena Caturla / *Hospital "Principe de Asturias", Madrid, Spain*
Renee F. Robinson / *The Ohio State University, Columbus, Ohio, U.S.A.*
Chip Robison / *BeneScript Services, Inc., Norcross, Georgia, U.S.A.*
David S. Roffman / *University of Maryland, Baltimore, Maryland, U.S.A.*
Brendan S. Ross / *Department of Veterans Affairs Medical Center, Jackson, Mississippi, U.S.A.*
Simone Rossi / *Australian Medicines Handbook Pty Ltd., Adelaide, Australia*
Myrella T. Roy / *The Ottawa Hospital, Ottawa, Ontario, Canada*
John Russo, Jr. / *The Medical Communications Resource, Mahwah, New Jersey, U.S.A.*
Melody Ryan / *University of Kentucky, Lexington, Kentucky, U.S.A.*
Rosalie Sagraves / *University of Illinois, Chicago, Illinois, U.S.A.*
Richard T. Scheife / *American College of Clinical Pharmacy, Boston, Massachusetts, U.S.A.*
Lauren Schlesselman / *Niantic, Connecticut, U.S.A.*
Patricia H. Schoch / *Denver VA Medical Center, Denver, Colorado, U.S.A.*
Jon C. Schommer / *University of Minnesota, Minneapolis, Minnesota, U.S.A.*
Terry L. Schwinghammer / *University of Pittsburgh School of Pharmacy, Pittsburgh, Pennsylvania, U.S.A.*
Gayle Nicholas Scott / *Medical Communications and Consulting, Chesapeake, Virginia, U.S.A.*
Leon Shargel / *Eon Labs Manufacturing, Inc., Laurelton, New York, U.S.A.*
Marjorie A Shaw Phillips / *Medical College of Georgia Hospitals and Clinics, Augusta, and University of Georgia College of Pharmacy, Athens, Georgia, U.S.A.*
Kenneth I. Shine / *Institute of Medicine, Washington, District of Columbia, U.S.A.*
Ingrid S. Sketris / *Dalhousie University, Halifax, Nova Scotia, Canada*
Betsy L. Sleath / *University of North Carolina, Chapel Hill, North Carolina, U.S.A.*
William E. Smith / *Virginia Commonwealth University, Richmond, Virginia, U.S.A.*
Patrick F. Smith / *University at Buffalo, Buffalo, New York, U.S.A.*
Mary Ross Southworth / *University of Illinois, Chicago, Illinois, U.S.A.*
Lara E. Storms / *Virginia Commonwealth University School of Pharmacy, Richmond, Virginia, U.S.A.*
Lynda Sutton / *Cato Research Ltd., Durham, North Carolina, U.S.A.*
C. Richard Talley / *American Society of Health-System Pharmacists, Bethesda, Maryland, U.S.A.*
Carl E. Trinca / *Western University, Pomona, California, U.S.A.*
Patrice Trouiller / *University Hospital of Grenoble, Grenoble, France*
Carl J. Tullio / *Pfizer, Inc., Yorktown, Virginia, U.S.A.*
Kimberly Vernachio / *Canton, Georgia, U.S.A.*
Peter H. Vlasses / *American Council on Pharmaceutical Education, Chicago, Illinois, U.S.A.*
Jeffrey W. Wadelin / *American Council on Pharmaceutical Education, Chicago, Illinois, U.S.A.*
Margaret C. Watson / *University of Aberdeen, Aberdeen, U.K.*
Timothy Webster / *American Society of Consultant Pharmacists, Alexandria, Virginia, U.S.A.*
Barbara G. Wells / *University of Mississippi, University, Mississippi, U.S.A.*
Albert I. Wertheimer / *Temple University, Philadelphia, Pennsylvania, U.S.A.*
Roger L. Williams / *U.S. Pharmacopeia, Rockville, Maryland, U.S.A.*
Harold H. Wolf / *University of Utah, Salt Lake City, Utah, U.S.A.*

Carol Wolfe / *American Society of Health-System Pharmacists, Bethesda, Maryland, U.S.A.*
Andrew B.C. Yu / *U.S. Food and Drug Administration, Rockville, Maryland, U.S.A.*
Woodie M. Zachry III / *University of Arizona, Tucson, Arizona, U.S.A.*
Dawn G. Zarembski / *American Council on Pharmaceutical Education, Chicago, Illinois, U.S.A.*
Barbara Zarowitz / *Henry Ford Health System, Bingham Farms, Michigan, U.S.A.*
David S. Ziska / *Applied Health Outcomes, Tampa, Florida, U.S.A.*

Contents

Foreword

Everyone reading this foreword already knows that clinical pharmacy is evidence-based. Those in the field rely heavily on therapeutics textbooks, drug information compendia, journal articles, and the World Wide Web as critical sources of information and knowledge to guide their patient care and research decisions. So what can a resource like the *Encyclopedia of Clinical Pharmacy* add to the growing (and some might say already overcrowded) library of professional literature?

This encyclopedia is not intended to be a textbook of therapeutics or a compendium of drug information. Rather, its goal is to document information about key people, events, publications, legislation, regulations, and the myriad of things other than therapeutics per se that shape what clinical pharmacists do, why they do it, and how they do it. It was this opportunity that convinced the American College of Clinical Pharmacy and the American Society of Health-System Pharmacists to partner with Marcel Dekker, Inc. in this unique project.

Although efforts to publish the *Encyclopedia of Clinical Pharmacy* began in early 1999, its true origins lie in the pioneering work of visionary pharmacists like Donald Brodie, Donald Francke, Paul Parker, Harvey A. K. Whitney, and others in the 1950s, '60s, and '70s. They foresaw the increasing complexity of pharmacotherapy, the problems of medication-related morbidity and mortality, and the impact that clinically empowered pharmacists have on assuring safe and effective pharmaceutical care for patients. The *Encyclopedia of Clinical Pharmacy*—in print and online—is designed to be a dynamic resource that will expand as future events unfurl and as time allows a more complete documentation of important past contributions and contributors. Thus, as clinical pharmacy continues to evolve, so will the *Encyclopedia*.

Robert M. Elenbaas, Pharm.D., FCCP
Executive Director
American College of Clinical Pharmacy
Kansas City, Missouri, U.S.A.

Henri R. Manasse, Jr., Ph.D., Sc.D.
Executive Vice President and Chief Executive Officer
American Society of Health-System Pharmacists
Bethesda, Maryland, U.S.A.

Preface

The term "clinical pharmacy" has come to describe a wide range of pharmacy practices that occur in a variety of settings, including health-systems, community pharmacies, clinics, pharmaceutical industry, and government agencies. Clinical Pharmacy incorporates the patient-oriented practices of pharmaceutical care as well as drug policy management, research, education, and many other aspects within the field. As the scope of clinical pharmacy has grown, it has been less easy to capture in a simple definition. The range of topics included in the *Encyclopedia of Clinical Pharmacy* attest to the complexity and expansion of the clinical pharmacy practice.

The *Encyclopedia of Clinical Pharmacy* is a valuable resource for today's clinical pharmacist and pharmacotherapist. Practitioners require a large set of information on diverse topics to effectively conduct their practices. While some of this information can be obtained from drug information compendia, therapeutics textbooks, and primary literature reports, these sources do not thoroughly address the different dimensions of knowledge and information required by clinical pharmacists. The *Encyclopedia of Clinical Pharmacy* assembles information for practicing clinical pharmacists and students not found in disease-oriented or drug-oriented resources, and provides information and insights to topics and issues that relate to clinical pharmacy practice.

All entries in the *Encyclopedia* have been written by experts in their fields and reviewed by appropriate subject matter authorities. Categories of clinical pharmacy-related topics included in the *Encyclopedia* are:

- Important reports, position papers, and consensus statements.
- Clinical pharmacy practices and models of health service delivery.
- State-of-the-art practices in selected, rapidly changing areas of pharmaceutical and medical sciences, or allied fields.
- Federal and state agencies and pharmacy-related organizations.
- Relevant legal issues and court decisions.
- Processes, methods, and guidelines impacting clinical researchers.
- Biographies of leaders and innovators in clinical pharmacy.
- Educational and training programs.

The *Encyclopedia* is available in a printed text and an online version. The online version includes everything in the print version while also providing the convenience of a keyword search engine. New articles and revised articles will be digitally posted quarterly and available to all subscribers of the electronic version as soon as available. Although the content will initially parallel the print version, unique electronic enhancements will be available on the online version.

The intended audience for the *Encyclopedia* is clinical pharmacists and pharmacy students throughout the world. The *Encyclopedia* should become an essential resource for libraries and drug information centers as well as for personal libraries. The *Encyclopedia* will also be of interest to other health care practitioners and students who wish to learn about clinical pharmacy practice. The *Encyclopedia's* international Editorial Advisory Board represents the United States, Canada, Australia, the United Kingdom, Germany, Spain, Switzerland, and Japan.

This project was begun with the intent of representing clinical practice expertise emanating from two major organizations, the American College of Clinical Pharmacy (ACCP) and the American Society of Health-System Pharmacists (ASHP). Robert Elenbaas from ACCP and Richard Talley from ASHP have been particularly helpful in marshalling the expertise within their organizations to contribute to the *Encyclopedia*. The completion of this project was only possible through the sound advice and contributions of the *Encyclopedia of Clinical Pharmacy's* international Editorial Advisory Board, contributors, and the competent and diligent efforts of the staff at Marcel Dekker, Inc. In particular, the efforts of Carolyn Hall, Ellen Lichtenstein, and Alison Cohen have been much appreciated.

Joseph T. DiPiro

Encyclopedia of Clinical Pharmacy

Academia, Clinical Pharmacy Careers in

Diane E. Beck
Auburn University, Auburn, Alabama, U.S.A.

INTRODUCTION

If you are contemplating a pharmacy practice career in academia, this may stem from the intellectual and cultural stimulation and the variety of interesting and eager people that you encountered while in college. However, there are many aspects of this career path that are not readily apparent. The goal of this article is to expand your perspectives about clinical pharmacy academia so that you are more informed about it as a career opportunity. Specifically, this article provides an overview of the clinical pharmacy careers in academia, insight into important issues in academia, and recommendations for making informed decisions about career options.

OPTIONS AND ISSUES

Academic positions are often cited as being either a *tenure-track* or a *nontenure-track* and are often available to clinical pharmacy faculty at either a pharmacy or a medical school. A tenure-track position is one that requires an individual to successfully demonstrate scholarly accomplishments within the initial six years of employment, and in return, the university guarantees lifetime employment unless the institution falls into financial exigency or the faculty member exhibits misconduct. With a nontenure-track position, the faculty member must also demonstrate some level of scholarly accomplishment, but there is more flexibility in the window of time during which this occurs. Although the university does not commit itself to lifetime employment of individuals in nontenure positions, it does provide a contract that conveys a commitment of employment for a given period of time and the contract is usually renewable at the end of each period. Individuals who enjoy teaching but who do not desire scholarship/research responsibilities are encouraged to seek an adjunct or affiliate clinical faculty appointment with a nearby pharmacy and/or medical school.

Activities and Overview of Career Options

Most full-time clinical pharmacy faculty positions require the individual to balance and accomplish responsibilities related to four missions: 1) teaching, 2) scholarship/research, 3) service, and 4) patient care/practice. Although all faculty must make some level of contributions to each of these responsibilities, many institutions expect faculty to excel in only one or two of these four missions. There are several important career options that must be considered in order to successfully balance and achieve these responsibilities. First, an individual pursuing an academic career must decide upon either a tenure- or nontenure-track option. Second, because it is difficult to be accomplished in all four of these areas, there are primarily two types of tenure-track clinical faculty position models now offered by pharmacy schools.[1,2] Individuals who assume the practitioner-educator model have equal responsibility for patient care/teaching and research/scholarship. Those who fulfill the researcher/educator model spend the majority of their time in research and teaching with much less time devoted to patient care. A nontenure-track position is similar to the practitioner-educator model but, at some institutions they have less time devoted to research. The following sections provide insight into issues and factors to consider when contemplating either a full-time, college-based tenure- or nontenure-track position.

Academia in Transition

Higher education is currently undergoing rigorous financial restriction, external demands for accountability, and self-scrutiny. Therefore, any future academic career will likely be different than that of the faculty you had in college. For example, those entering academic clinical

Encyclopedia of Clinical Pharmacy
DOI: 10.1081/E-ECP 120006174

pharmacy today are encountering changes in the tenure system, a different reward system, and continued evolution of academic clinical pharmacy.

Changes in the tenure system

In the early 1900s the tenure system was established so that faculty would have the academic freedom to express their thoughts without fear of dismissal.[3] Due to statutory repeal of the mandatory retirement age and financial restrictions in higher education, many institutions are finding tenure limits their flexibility in responding to fluctuations in admissions to different degree programs.[4] Advocates believe tenure is necessary since it provides a healthy environment that encourages new ideas; individuals can say what they want and be protected from dismissal.[5] Opponents believe it protects the incompetent and reduces the flexibility institutions need to be responsive to new demands such as limited financial resources and accountability.

At most institutions, the tenure system provides new assistant professors a time span of approximately six years in which to demonstrate achievement in teaching, service, practice, and scholarship/research.[5] After this window of time, the faculty decides whether to grant tenure. The granting of tenure implies a guaranteed position within the university unless there is financial exigency or the faculty member is found guilty of misconduct.

When contemplating the tenure- versus nontenure-track alternatives, the individual should weigh the benefit of academic freedom with the difficulty, stress, and anxiety that many academicians encounter when trying to accomplish promotion and tenure criteria within a six-year window. Because clinical pharmacy faculty members have to establish a practice in addition to teaching, scholarship/research, and service, these time constraints can be particularly stressful and inflexible.[6] To facilitate long-term success, some institutions are offering clinical faculty members the opportunity to begin on a nontenure-track and move to the tenure track once their practice and research abilities are established.

Changes in the reward system

In the early 1600s when the first colleges were established in the United States, the primary college mission of the faculty was teaching.[7] However, since then our universities have added service as a second mission in order to meet a need of society for practical assistance in everyday living, and research as a third mission due to the influence of higher education from Germany. This research mission requires faculty to generate new knowledge rather than just convey knowledge.[7] Institutions with medical and health-related profession educational programs also have practice/patient care as an extension of their service or outreach mission.

Unfortunately, by the mid-1900s the faculty reward system, even in institutions with no graduate programs, had evolved to one in which decisions about faculty promotions were largely based only on the individuals research and publication record.[7] The effectiveness of one's teaching was given little, if any consideration at the time of promotions.

In 1990, Boyer called for academia to move beyond this problem.[7] He stated that each institution should clearly establish its unique missions and measure itself by these values rather than the traditional research reputation. To accomplish this, Boyer asserted that higher education must establish and adopt a new way of defining and rewarding *scholarship*. Boyer proposed that the term scholarship must take on a broader meaning than just original research in order to characterize the full scope of duties an academician is expected to accomplish in today's higher-education institutions. He further characterized the scholarship expected among a faculty as encompassing four distinct functions: 1) the scholarship of discovery (i.e., original research), 2) the scholarship of application, 3) the scholarship of integration, and 4) the scholarship of teaching. Glassick and colleagues[8] have since facilitated acceptance of this concept by establishing criteria by which scholarship can be measured. These criteria have helped distinguish the difference between achievement of excellence and scholarship, and they are enabling institutions to place equal value on all four types of scholarship.[9] Although all four types of scholarship are recognized by most institutions, most institutions require faculty to excel in only one or two of these four options.

During the last 11 years, these pivotal reports have led most institutions to reevaluate how faculty members prioritize their time and the faculty reward system. Therefore, when contemplating an academic position, an individual should clearly understand the institution's mission and faculty reward system. The faculty candidate should also ascertain whether the assigned duties can be accomplished according to the projected allocation of time and effort and that they are consistent with the faculty reward system. Individuals who select either a college-based tenure- or nontenure-track position should clearly understand that success in academia requires achievement of not only excellence in completion of assigned duties, but also scholarship.[1,9]

Evolution of clinical pharmacy academicians

Similar to the evolution of other clinical disciplines, clinical pharmacy practice grew from the commitment of a cadre of clinicians who contributed significant time in practice and teaching.[10] Because they had little time for research and scholarship and often had limited skills in these areas, early clinical pharmacy academicians were sometimes viewed as quasi-faculty members. In the last 30 years, the discipline has overcome this stigma by establishing peer-reviewed discipline-specific journals, becoming accepted authors to journals of other academic disciplines, and developing specialty residency and fellowship programs that prepare future clinical pharmacy faculty members for academic careers.

The discipline now also has a cadre of clinical scientists who are fulfilling the typical academic scientist role; but in order to accomplish this, these individuals have had to focus primarily on teaching and research with minimal activities in practice. Since the fundamental role of the discipline is to prepare students for actual practice, a cadre of clinical faculty whose primary duties are practice and teaching are therefore also essential.

Because a broader definition of scholarship is now accepted in higher education and it is realized that faculty members cannot effectively accomplish all four missions, two faculty models are most frequently used today. Individuals who assume the practitioner-educator model are enabled to accomplish either the scholarship of integration, application, and/or teaching because of their focus on teaching and practice. Individuals who fulfill the researcher-educator model are able to pursue the scholarship of discovery because they have minimal practice expectations.

Although clinical pharmacy has established itself in academia, it is still in its infancy and there are unresolved needs. Individuals who are planning for an academic clinical pharmacy career or who are pursuing an academic position should talk with their mentors and do further reading about these needs. For example, fellowship-training programs are providing fellows with too little time devoted to development of research abilities and too much time devoted to teaching and practice.[11] It has been proposed that clinical pharmacy training programs at the Ph.D. level may better prepare clinical faculty to compete for NIH funds and to study the pharmacotherapeutic and practice issues that need to be addressed in today's healthcare environment.[12] Other needs that pharmacy schools are addressing include development of promotion and tenure guidelines that are equitable and consistent with the assigned duties or faculty model, strategies to ensure that both faculty models are equally respected and valued, and the relative proportions of each faculty model that are needed at a pharmacy school in order to achieve the institutional mission.[1,2,6,10]

Faculty Roles and Responsibilities

As noted above, a faculty member's roles and specific responsibilities are largely determined by the institution's mission and the assigned duties. Although there are differences in the percentage of time that is assigned to the areas of teaching, practice, scholarship/research, and service, all college-based faculty members are expected to demonstrate some level of scholarship.[1,9]

Because faculty members have significant autonomy in accomplishing university assignments, their responsibilities are not always clearly defined. Kennedy[13] has concluded that in accepting academic freedom, academicians must also realize their academic duty to the institution. Kennedy further vows that academic duty is accomplished by meeting a set of responsibilities. He notes that the primary responsibility of academia is to meet the needs of society since it nurtures our existence. Faculty members can help accomplish this by meeting specific responsibilities such as: 1) accomplishing all four missions/responsibilities with excellence, 2) demonstrating commitment to students by mentoring and being well-prepared for classes, 3) maintaining high standards of individual scholarship, 4) working "collegially" with other faculty members so that all academic missions are accomplished, and 5) completing all assignments ethically. It should be emphasized that collegiality requires the faculty member to be a team member by actively contributing to the departmental and school/college work and actively participating in decision-making; collegiality does not infer that the faculty member is just friendly to everyone. Kennedy also notes that academicians have a responsibility of commitment such that outside activities (e.g., consulting) do not interfere with one's responsibilities to the university.

DESCRIPTION OF ACADEMIC SITES AND SETTINGS

At the present time, there are 84 pharmacy schools in the United States and each has clinical pharmacy faculty members. Because the number of new pharmacy schools continues to increase and a number of senior faculty members are likely to retire in upcoming years, it has been predicted that we may encounter a shortage of academic faculty members.[14,15] Since the current shortage of pharmacy practitioners is expected to continue in the foreseeable future, the need for new clinical pharmacy faculty members will likely persist.

Both public and private institutions serve as the settings for these pharmacy schools. Therefore, an individual pursuing a faculty position should learn about the characteristics of working in each of these settings. Other considerations when selecting the institutional setting for your faculty position include whether the institution has its own medical center and whether your practice will be located at a distant site away from the pharmacy school. These factors will greatly impact opportunities for research/scholarship, access to mentors, and ability to develop collegial relationships.

USUAL DEGREE, TRAINING, AND EXPERIENCE REQUIRED

Today, most academic clinical pharmacy faculty positions require a Pharm.D. degree although individuals with another advanced degree may be considered at some institutions. Completion of post-doctoral training programs such as specialized residencies and fellowships are also required. Because specialized residencies focus on development of practice skills and teaching, they can appropriately prepare an individual for a tenure-track practitioner-educator or a nontenure-track faculty position. Individuals who desire a researcher-educator faculty position should complete a fellowship that emphasizes development of research skills. Postgraduate coursework such as biostatistics and research design would also facilitate success in a researcher-educator position and a fellowship should provide opportunity to complete such coursework. Many specialized residency and fellowship training programs require completion of a general practice pharmacy residency as a prerequisite. The requirement for specialty faculty to become board certified is increasing and individuals planning to enter academic clinical pharmacy are encouraged to obtain this credential.[16]

Some individuals may elect to gain several years of experience as a practitioner before pursuing either post-doctoral training or a faculty position. Although this is not required, the experience can certainly be beneficial.

CAREER LADDER AND GROWTH

Most entry-level faculty positions involve appointments at the level of assistant professor. The successful junior faculty member usually achieves the criteria established for promotion to the level of associate professor after a period of 5–7 years.[5] At most institutions, this requires demonstration of scholarship/research and excellence in accomplishing assigned duties. Most faculty members are able to achieve the rank of full professor approximately 5–10 years after promotion to the associate professor level. Promotion to this rank usually requires development of an established scholarship/research theme and recognition by peers at a national level.[5] The American Association of Colleges of Pharmacy (AACP) surveys pharmacy schools on an annual basis about the salaries of each academic rank and publishes the results. Individuals interested in an academic career should review these data to gain insight into the financial aspects of the academic pharmacy career ladder and growth. These data are published annually and may be obtained by contacting either an AACP faculty member or the senior vice president at AACP.

Once either the associate professor or full professor rank is achieved, the academician may also opt for an academic administration career.[5] An administrative position requires an additional set of knowledge and skills that emphasize leadership and management. These attributes may be gained by pursuing postgraduate degrees and, attending workshops and/or fellowships in higher education. Most individuals begin this track by serving as a department chair or assistant/associate dean. Success in one of these positions can enable the individual to become a dean. After several years of experience in any of these positions, an individual may also pursue administrative positions in higher education that are outside of pharmacy schools.

CONCLUSIONS

Although the need for clinical pharmacy faculty members will likely continue, the transformation that is occurring in higher education will make the expectations of faculty in the future different than what they have been in the past. Individuals contemplating an academic pharmacy career must have a clear understanding about the current issues and needs in order to make informed career decisions.

REFERENCES

1. Carter, B.L. Chair report of the section of teachers of pharmacy practice task force on scholarship definition and evaluation. Am. J. Pharm. Educ. **1994**, *58* (2), 220–227.
2. Carter, B.L. Chair report of the AACP section of teachers

of pharmacy practice task force on faculty models. Am. J. Pharm. Educ. **1992**, *56* (2), 195–201.

3. De George, R.T. *Academic Freedom and Tenure: Ethical Issues*; Rowman & Littlefield Publishers, Inc.: Lanham, Massachusetts, 1997; 1–28.
4. McGhan, W.F. Tenure and academic freedom: Vital ingredients for advancing practice, science, and society. Am. J. Pharm. Educ. **1996**, *60* (1), 94–97.
5. Heiberger, M.M.; Vick, J.M. *The Academic Job Search Handbook,* 2nd Ed.; University of Pennsylvania Press: Philadelphia, Pennsylvania, 1996; 161–165.
6. Bradberry, J.C. Some thoughts on promotion and tenure. Am. J. Pharm. Educ. **1990**, *54* (3), 321–322.
7. Boyer, E.L. *Scholarship Reconsidered. Priorities of the Professoriate*; Princeton University Press: Princeton, New Jersey, 1990; 1–27.
8. Glassick, C.E.; Huber, M.T.; Maeroff, G.I. *Scholarship Assessed: Evolution of the Professoriate*; Jossey-Bass: San Francisco, California, 1995; 1–25.
9. Hutchings, P.; Shulman, L.S. The scholarship of teaching: New elaborations, new developments. Change **1999**, 11–15, September/October.
10. Amerson, A.B. The evolution of scholarship in pharmacy practice: Examining our direction. Am. J. Pharm. Educ. **1992**, *56* (4), 421–424.
11. Rhoney, D.H.; Brooks, V.G.; Patterson, J.H.; Piper, J.A. Pharmacy fellowship programs in the United States: Perceptions from fellows and preceptors. Am. J. Pharm. Educ. **1998**, *62* (3), 290–296.
12. Cohen, J.L. The need to nurture the clinical pharmaceutical sciences. Am. J. Pharm. Educ. **1998**, *62* (4), 471.
13. Kennedy, D. *Academic Duty*; Harvard University Press: Cambridge, Massachusetts, 1997; 1–58.
14. Penna, R.P. Academic pharmacy's own workforce crisis. Am. J. Pharm. Educ. **1999**, *63* (4), 453–454.
15. Broedel-Zaugg, K.; Henderson, M.L. Factors utilized by pharmacy faculty in selecting their first academic position. Am. J. Pharm. Educ. **1997**, *61* (4), 384–387.
16. Wells, B.G. Board certification and clinical faculty. Am. J. Pharm. Educ. **1999**, *63* (2), 251.

Academy of Managed Care Pharmacy

Judith A. Cahill
Academy of Managed Care Pharmacy, Alexandria, Virginia, U.S.A.

INTRODUCTION

The Academy of Managed Care Pharmacy (AMCP)[a] was founded in 1989 as a professional society for pharmacists practicing in managed care settings and for their associates who subscribe to the principles underlying managed care pharmacy. It has grown steadily and substantially since its founding and is today's voice for managed care pharmacy.

ORGANIZATIONAL STRUCTURE AND GOVERNANCE

An nine-member Board of Directors, eight of whom are elected by the active membership, governs the Academy of Managed Care Pharmacy; the ninth member is the employed Executive Director. Thirteen committees aid the Board in its governance, with jurisdiction over various segments of the organization's activities.

Committees are comprised of member volunteers who develop policy recommendations for the Board. All policy-making authority is vested in the Board of Directors. Professional staff handles the implementation of those policies.

The Academy is comprised of over 4800 individual members nationally who are part of more than 600 healthcare organizations that provide comprehensive coverage and services to the 170 million Americans served by managed care organizations. Its active members are those pharmacists who have responsibility and accountability for the design and implementation of managed care pharmacy benefits, and those individuals who aspire to that status.

[a] The AMCP is located at 100 N. Pitt Street, Alexandria, Virginia 22314; phone: (703) 683-8416; fax: (703) 683-8417; www.AMCP.org.

MISSION STATEMENT

The AMCP is a professional association of pharmacists and associates who serve patients and the public through the promotion of wellness and rational drug therapy by the application of managed care principles.

The mission of AMCP is to serve as an organization through which the membership pursues its common goals: to provide leadership and support for its members; to represent its members before private and public agencies and healthcare professional organizations; and to advance pharmacy practice in managed healthcare systems.

VISION STATEMENT

By 2005, the AMCP will be:

- Recognized as the primary national professional association for pharmacists and associates who practice in managed healthcare systems.
- The principal source of knowledge regarding pharmaceutical care in managed healthcare systems.
- An association whose members value and promote application and advancement of pharmaceutical care principles in managed care pharmacy.
- An effective voice for the principles and practices of managed care pharmacy.
- An effective and credible public policy advocate.

Encyclopedia of Clinical Pharmacy
DOI: 10.1081/E-ECP 120006282

- Effective at maintaining a dynamic organizational structure that allows the association to meet its goals through the responsible management of human, financial, technological, and other resources.

CURRENT MAJOR INITIATIVES

- Serve AMCP's core constituency.
- Establish a model defining the role of pharmacy within managed care.
- Define the value of managed care pharmacy.
- Be public policy advocates.
- Develop a professional policy digest.
- Encourage the professional development of members; develop and improve the governance and leadership of the organization.
- Define AMCP's relationship with the pharmaceutical industry.
- Define AMCP's customers and audiences.
- Identify AMCP's leadership on quality issues.

MAJOR MEETINGS

In the spring of each year, the AMCP holds its Annual Meeting. In the fall of each year, an Educational Conference is held.

ACPE Standards 2000

Jeffrey W. Wadelin
Peter H. Vlasses
American Council on Pharmaceutical Education, Chicago, Illinois, U.S.A.

INTRODUCTION

The American Council on Pharmaceutical Education (ACPE) is the national agency for accreditation of professional degree programs in pharmacy and for approval of providers of continuing pharmaceutical education. The ACPE was established in 1932 for accreditation of preservice education. In 1975, its scope of activity was broadened to include continuing pharmaceutical education. The ACPE is an autonomous and independent agency whose Board of Directors (the decision and policy-making body) includes pharmacy educators, pharmacy practitioners, state board of pharmacy members/executives, and public representation. A public interest panel having at least two members also provides public perspectives in the policy and decision-making processes of accreditation.

ACCREDITATION

Accreditation is the public recognition accorded a professional program in pharmacy that is judged by ACPE to meet established standards through initial and subsequent periodic evaluations. The values of accreditation are several and the ACPE accreditation process serves several constituencies concurrently including the general public, students and prospective students, licensing bodies, colleges and schools of pharmacy and their parent institutions, and the profession. Graduates of accredited professional programs in pharmacy should be educationally prepared for practice and should satisfy educational requirements for licensure. However, decisions concerning eligibility for licensure reside with the respective licensing bodies in accordance with their state statutes and administrative rules and regulations.

ACCREDITATION STANDARDS

Accreditation standards reflect professional and educational qualities identified by ACPE as essential to quality professional programs of Colleges and Schools of Pharmacy and serve as the basis for program evaluation. Standards are set by the ACPE in accordance with a procedure that provides adequate time and opportunity for all parties significantly affected by the accreditation process to comment on such standards prior to their adoption. Advance notice is given whenever revision of standards is proposed by ACPE. The initial standards were published in 1937 and revisions have been effected, on the average, every seven years in keeping with changes in pharmaceutical education and practice. (The standards and guidelines in use prior to those presented herein were adopted in July 1984 and became effective in January 1985.) These standards and guidelines are presented in the *ACPE Accreditation Manual*, 9th Edition, September, 2000.

"STANDARDS 2000"

New accreditation standards and guidelines were adopted June 14, 1997. This occurred following a nearly 50-year consensus building process, often fraught with controversy. The revision process leading to *Accreditation Standards and Guidelines for the Professional Program in Pharmacy Leading to the Doctor of Pharmacy Degree* was initiated in September 1989 and conducted in accord with the *Procedure and Schedule for the Revision of Accreditation Standards and Guidelines*, issued January 7, 1990. This *Procedure and Schedule* involved a stepwise, decade-long process. The early years were devoted to study and formulation of proposed revisions and the later years provided for two comment periods, each affording open hearings and opportunities to submit written comments. Final consideration of the last iteration of proposed revisions, *Proposed Revision, January 15, 1996*, was given during the June 1997 meeting of the ACPE. The *Accreditation Standards and Guidelines for the Professional Program in Pharmacy Leading to the Doctor of Pharmacy Degree*, as adopted June 14, 1997, will be contained in the next edition of the *ACPE Accreditation Manual*. Copies of the new standards and guidelines may be obtained by writing the ACPE office (311 West Superior Street, Chicago, Illinois 60610, U.S.A.), and may be found on the ACPE web site (www.acpe-accredit.org).

Encyclopedia of Clinical Pharmacy
DOI: 10.1081/E-ECP 120006267

"Standards 2000" reflects broad input from the profession, and sets forth expectations for quality in Doctor of Pharmacy programs offered by colleges and schools of pharmacy. It is expected that colleges and schools of pharmacy maintain a fundamental commitment to the preparation of students for the general practice of pharmacy with provision of the professional competencies necessary to the delivery of pharmaceutical care. For these purposes, pharmaceutical care is defined as the responsible provision of drug therapy for the purpose of achieving definite outcomes that improve a patient's quality of life. These outcomes are: 1) cure of a disease; 2) elimination or reduction of a patient's symptomatology; 3) arresting or slowing of a disease process; or 4) preventing a disease or symptomatology.

Pharmaceutical care involves the process through which a pharmacist cooperates with a patient and other professionals in designing, implementing, and monitoring a therapeutic plan that will produce specific therapeutic outcomes for the patient. This in turn involves three major functions: 1) identifying potential and actual drug-related problems; 2) resolving actual drug-related problems; and 3) preventing drug-related problems.

Pharmaceutical care is a necessary element of healthcare, and should be integrated with other elements. Pharmaceutical care is, however, provided for the direct benefit of the patient, and the pharmacist is responsible directly to the patient for the quality of that care. The fundamental relationship in pharmaceutical care is a mutually beneficial exchange in which the patient grants authority to the provider, and the provider gives competence and commitment (accepts responsibility) to the patient. The fundamental goals, processes, and relationships of pharmaceutical care exist regardless of practice setting.

"Standards 2000" sets forth 18 professional competencies that should be achieved through the college or school of pharmacy's curriculum. additionally, "Standards 2000":

1. Emphasizes pharmaceutical care, as considered in the professional literature and as presented in the Position Paper of the AACP Commission to Implement Change in Pharmaceutical Education, as a part of the mission statement of a college or school of pharmacy, and as an organizing principle for curricular development.
2. Reflects new competencies and outcome expectations for the preparation of a generalist practitioner, which are requisite to the rendering of pharmaceutical care in a variety of practice settings.
3. Encourages the development of non-traditional curricular pathways and innovative program delivery modes (e.g., external degrees) to address the needs of baccalaureate-degreed practitioners already in practice.
4. Encourages increased practitioner involvement in pharmaceutical education as volunteer faculty and in the affairs of colleges and schools of pharmacy.
5. Places emphasis upon the importance of developing good problem-solving, decision-making, critical-thinking, and communication skills.
6. Does not distinguish between externships and clerkships; rather, it is expected that experiential education will be incorporated as a curricular continuum throughout the professional program, as both introductory and advanced practice experiences, and that experiences will begin earlier in the educational process.
7. Increases expectations regarding quality control in the pharmacy practice experience component of the curriculum (introductory and advanced practice experiences).
8. Encourages innovation in the development and innovation of new tactics for teaching and learning, with particular emphasis upon increasing student involvement as active learners.
9. Encourages the development and implementation of new and innovative methods for student evaluation and assessment which measure learning at a variety of levels beyond the memorization and reiteration of facts.
10. Incorporates expectations that the leadership of colleges and schools of pharmacy will undergo formal evaluations in a regular and systematic manner.
11. Expects that curricular management and editing processes will strive to assure that the addition of material will be counterpoised with the elimination of outdated and/or unnecessary material, so as to avoid unnecessary and undesirable overlap.
12. Incorporates Total Quality Management (TQM) principles throughout the standards and guidelines.
13. Expects particular emphasis to be placed upon the professionalization (professional development) of students.
14. Recognizes the broad range of responsibilities of pharmacy faculty, including teaching, research and scholarly activities, professional practice, service, and administration.
15. Expects that colleges and schools will develop and utilize admission criteria, policies, and procedures that consider not only academic qualifications but also other factors which may impact upon success in the professional program (e.g., communication skills, etc.).

Adherence to Pharmaceutical Care

Gloria J. Nichols-English
University of Georgia College of Pharmacy, Athens, Georgia, U.S.A.

Sylvie Poirier
Regie Regionale de la Sante et des Services Sociaux Monteregie, Quebec, Canada

INTRODUCTION

Nonadherence to medication regimens remains a major problem in health care. The National Council on Patient Information and Education (NCPIE) has termed noncompliance ''America's other drug problem.''[a] Pharmacists are in an ideal position to assess and treat adherence-related problems that can adversely affect patients' health outcomes. Strategies to monitor and improve adherence are key components of pharmaceutical care plans, especially for patients with chronic diseases, such as hypertension, diabetes, and atherosclerotic heart disease. Nonadherence is a behavioral disorder that can be assessed and managed through a carefully devised pharmaceutical care plan.

NONADHERENCE: DEFINITION AND SCOPE OF THE PROBLEM

Medication nonadherence is most simply defined as the number of doses not taken or taken incorrectly that jeopardizes the patient's therapeutic outcome.[2] NCPIE[1] has noted that nonadherence can take a variety of forms, including not having a prescription filled, taking an incorrect dose, taking a medication at the wrong time, forgetting to take doses, or stopping therapy too soon. In this article, we use the term ''adherence'' instead of compliance, because the former connotes an interactive, collaborative relationship between pharmacist and patient. Compliance originates from a practitioner-centered paradigm and is more control oriented. It relies on patient obedience and sometimes stigmatizes the patient as engaging in deviant behavior if another course of action is chosen.[3,4] A patient-centered approach is one in which the pharmacist engages patients to become more active in the continuum of decision making about their treatment and the consequent health outcomes.

Although medication nonadherence is the primary focus of this article, it is only one form of nonadherence. Poorer health outcomes may also result when a patient does not adhere to recommended lifestyle changes, such as exercise or smoking cessation, or to prescribed nonpharmacologic interventions, such as physical therapy or dietary plans. Pharmacists who counsel patients with chronic diseases, such as asthma, hypertension, or diabetes, need to assess and promote adherence to these nonpharmacologic treatments as well.

Medication nonadherence is a major public health problem that has been called an ''invisible epidemic.''[5,6] Nonadherence to pharmacotherapy has been reported to range from 13% to 93%, with an average rate of 40%.[7] The problem encompasses all ages and ethnic groups. It has been estimated that 43% of the general population, 55% of the elderly, and 54% of children and teenagers are nonadherent.[8] A host of individual characteristics also influence adherence, such as the patient's religion, health beliefs, social support system, and ethnicity.

Rates of nonadherence vary with different disease states. For example, the nonadherence rate for hypertension is reported to be 40%, while that for arthritis has been found to range between 55% and 70%.[9] Nonadherence rates are especially high among patients with chronic diseases.[10] These patients, who typically require long-term, if not lifelong, medications to control symptoms and prevent complications, often must make significant behavioral changes to adhere with pharmacotherapy. Such changes can be difficult to integrate into everyday life.

Nonadherence to pharmacotherapy has been shown to decrease productivity and increase disease morbidity, physician office visits, admissions to nursing homes, and

[a]Copyright © 2000 by the American Pharmaceutical Association. Originally published in the *Journal of the American Pharmaceutical Association*; adapted with permission.

Encyclopedia of Clinical Pharmacy
DOI: 10.1081/E-ECP 120006252

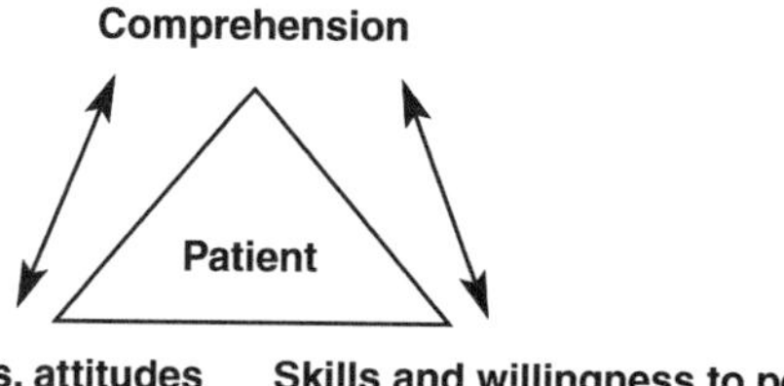

Fig. 1 Patient-centered adherence paradigm. In the patient-centered adherence paradigm, the pharmacist integrates information about a patient's medication use from three perspectives: the patient's knowledge of the medication (comprehension); the patient's beliefs and attitudes toward his or her illness and its treatment (beliefs, values, and attitudes); and the patient's ability and motivation to follow the regimen (skills and willingness to perform).

death.[1,9,11] For example, an estimated 125,000 deaths per year have been attributed to nonadherence to treatment for cardiovascular disease.[11] Many studies have documented poorer health outcomes due to nonadherence, especially in patients with chronic diseases such as hypertension, diabetes, and epilepsy.[5,6,12,13]

Finally, nonadherence places a huge burden on the United States' economy. Its direct and indirect costs have been estimated to be $100 billion per year in this country alone.[12] Pharmacies also lose revenue because patients often fail to refill prescription medications, especially for chronic diseases.[14] According to The Task Force for Compliance,[9] only 25% of prescriptions for chronic conditions are refilled after 1 year.

For pharmacists, the message is clear: To improve adherence to pharmacotherapy, and hence to improve health outcomes, we must assess each patient individually, then provide targeted interventions that are responsive to his or her unique risk factors and needs (see Fig. 1). Research, such as the American Pharmaceutical Association Foundation's Project ImPACT: Hyperlipidemia,[15] has clearly documented the value of pharmacist-led patient care in fostering better adherence and outcomes.

NONADHERENCE AS A BEHAVIORAL DISORDER

Nonadherence has been studied widely by behavioral scientists whose models, such as the Health Belief Model and the Theory of Reasoned Action, attempt to explain and predict nonadherence.[16] However, despite the numerous articles that have been published on this topic, nonadherence remains a problem of epidemic proportions. An alternative model that can be useful for understanding and treating nonadherence is to view the problem as a disorder—a behavioral disorder.[3] Although not a true physiological disease, nonadherence shares many of the same characteristics as a medical disorder. For example:

- *Numerous risk factors for nonadherence have been identified.* Clearly, nonadherence is a multifactorial problem, and a host of contributing social, economic, medical, and behavioral factors have been identified.[5,6,9,17–19] As shown in Table 1, some risk factors for nonadherence relate to the disease (e.g., a chronic or asymptomatic illness), others relate to the patient (forgetfulness, sensory impairment, and economic problems), and still others relate to the drug regimen (concerns about cost, real or perceived adverse effects, or dosing schedule).
- *Nonadherence can be assessed and monitored.* A variety of direct and indirect methods are available to assess the presence and severity of nonadherence. As pharmacotherapy specialists, pharmacists may be the best suited of health providers to evaluate adherence problems on an ongoing basis.
- *Effective interventions are available to treat nonadherence.* Many cases of nonadherence can be treated with carefully selected interventions. However, other cases may not be resolvable, despite the best efforts of health care providers.[5]
- *Nonadherence frequently leads to increased morbidity and mortality.* Just as untreated medical disorders often progress to serious complications, nonadherence has a well-documented adverse impact on health outcomes.[17,20,21]
- *Nonadherence tends to have a variable course.* Nonadherence is not a stable condition, but tends to progress or change over time in a given patient.[7] Just as most chronic medical conditions require periodic reevaluation and therapeutic adjustments, patients with adherence problems should also be reassessed on a regular basis.

Table 1 Major risk factors for nonadherence

Asymptomatic conditions
Chronic conditions
Cognitive impairments, especially forgetfulness
Complex regimens
Multiple daily doses
Patient fears and concerns related to medication effects
Poor communication between patients and practitioners
Psychiatric illness

ASSESSING ADHERENCE

Before effective strategies can be devised to improve adherence, pharmacists need to evaluate how well a patient is adhering to pharmacotherapy and identify risk factors that may predispose the individual to nonadherence. Both direct and indirect methods are available to assess adherence.

Direct Methods

Direct and objective methods of assessing adherence include blood-level monitoring and urine assay for the measurement of drug metabolites or marker compounds. Collecting blood or urine samples can be expensive and inconvenient for patients and, moreover, only a limited number of drugs can be monitored in this way. The bioavailability and completeness of absorption of various drugs, as well as the rate of metabolism and excretion, are factors that make it difficult to correlate drug levels in blood or urine with adherence. The ability of direct methods to identify nonadherence also depends on the accuracy of the test and the degree to which the patient was nonadherent before the urine or blood sample was taken.

Indirect Methods

Indirect methods of assessing adherence include patient interviews, pill counts, refill records, and measurement of health outcomes. In one study, the use of patient interviews identified 80% of nonadherent patients, as verified by pill counts.[22] The interview method is inexpensive and allows the pharmacist to show concern for the patient and provide immediate feedback. A drawback of this method is that it can overestimate adherence, and its accuracy depends on the patient's cognitive abilities and the honesty of their replies, as well as the interviewer's correct interpretation of responses. Pill counts provide an objective measure of the quantity of drug taken over a given time period. However, this method is time consuming and assumes that medication not in the container was consumed. The refill record provides an objective measure of quantities obtained at given intervals, but assumes that the patient obtained the medication only from the recorded source.

Pharmacists can generally obtain reliable information on medication-taking behaviors from the patient or a family member or caregiver. The interview should be systematic and include specific questions on forgetfulness, the patient's understanding of medication instructions, and the conditions for which therapy has been prescribed. The patient's health beliefs and the degree of support available from friends and family should also be assessed.[4]

Interviewing patients to detect nonadherence is most effective when indirect probes are used. For instance, the probe "Most people have trouble remembering to take their medications. Do you have any trouble remembering to take yours?" will solicit more reliable information than asking: "Are you taking your medications as prescribed?" Table 2 gives examples of specific probes that the pharmacist can use to assess whether a patient has been or is likely to be adherent.

Table 2 Probes pharmacists can use to assess adherence

Assessing the patient's medication knowledge or medication-taking behavior
What is the reason you are taking this drug?
How do you take this medication?
Are you taking the medication with food or fluid?
Where did you receive information about this medication?
Are you taking nonprescription drugs while on this medication?
Do you use any memory aids to help you remember to take your medication?
Do you depend on anyone to help you remember to take your medication or to assist you in taking it?
Assessing attitudes, values, and beliefs regarding medication-taking behaviors
What results do you expect to receive from this medication?
What are the chief problems that you feel your illness has caused you?
Do you have any concerns about your illness and its treatment?
Are you satisfied with your current treatment plan?
How well do you usually follow a treatment plan?
What is the main concern you have about your medication?
Do you feel comfortable asking your physician or pharmacist questions about your medications?
Assessing whether the patient has the proper skills and is motivated or willing to follow through on the therapy plan
Have you encountered any problems with your medication- or pill-taking procedure?
Are you confident that you can follow your treatment plan?
What might prevent you from following the recommended treatment plan?
How likely is it that you will ask your physician or pharmacist about your medications?
Can you explain how you remind yourself to take your medication on schedule?
Do you normally write down questions to ask your physician or pharmacist before an appointment?

Pharmacy computerized prescription records provide perhaps the most practical and least intrusive method for assessing adherence. This method allows the pharmacist to review and monitor prescription records to determine whether the patient is refilling medications in a timely manner. Computer algorithms can be incorporated into the pharmacy computer software system as a tool for monitoring adherence and measuring the timeliness of prescription refills.[23] This method also has the potential to flag potential adherence problems that may develop over the course of several refills. One disadvantage of this method is that it does not assess actual medication-taking behaviors (e.g., this method would not detect a patient who was swallowing a sublingual tablet or improperly inhaling an asthma medication from a metered-dose inhaler).

Factors that have a negative or positive influence on medication adherence are shown in Table 3. This table may be used both to identify factors that contribute to nonadherence and to develop interventions to address adherence problems.

Table 3 Factors that affect medication adherence

Factors that promote adherence
Disease-related factors
Perceived or actual severity of illness
Perceived susceptibility to the disease or developing complications
Treatment-related factors
Perceived benefits of therapy
Written and verbal instructions
Convenience of treatment
Medication provides symptomatic relief
Patient-related factors
Good communication and satisfactory relationship with physician
Participation in devising the treatment plan
Confidence in the physician, the diagnosis, and the treatment
Support of family members and friends
Knowledge about the illness
Factors that reduce adherence
Disease-related factors
Chronic disease
Lack of symptoms
Treatment-related factors
Treatment requires significant behavioral changes
Actual or perceived unpleasant side effects
Regimen complexity and duration
Medication takes time to take effect
Patient-related factors
Sensory or cognitive impairments
Physical disability or lack of mobility
Lack of social support
Educational deficiencies (literacy problem or poor English fluency)
Failure to recognize the need for medication
Health is a low priority
Conflicting health beliefs
Economic problems
Negative expectations or attitudes toward treatment

From Refs. [3,56–58].

DESIGNING PATIENT-FOCUSED INTERVENTIONS FOR NONADHERENCE

Strategies to improve adherence should target the specific risk factors and causes identified during the patient assessment. Adherence aids may be used alone or in combination, but should be tailored to the individual patient. For example, a forgetful patient may benefit from a special package or container that provides a visual reminder that a medication was taken (e.g., blister packaging or a computer-aided compliance package). Forgetful patients also can be advised to take dosages in conjunction with other routine daily activities, such as at mealtimes or before tooth brushing. Refill reminders or automatic delivery to the home can also be valuable for the forgetful patient, as can simplification of the dosage schedule, such as changing to a once-daily prescription.

Once the initial adherence plan is implemented, follow-up is important to gauge how well the plan is working and whether changes are needed. Most studies have reported that almost all adherence strategies, regardless of their initial acceptability, will decline in responsiveness over time.[7] Therefore, the pharmaceutical care plan must include periodic reinforcement strategies for long-term success. The plan should also be reevaluated from time to time to assess its effectiveness and determine how well it meets patient expectations.

Identifying and measuring the outcomes of a pharmaceutical care adherence plan is also important. Objective measures of improved health status and/or reduced health care expenditures document success in a well-designed pharmaceutical care plan. Examples of measurable outcomes include a reduction in inappropriate use of the health care system (e.g., fewer emergency department visits for asthma exacerbations) or improved control of the patient's disease (e.g., HbA_{1c} levels below 7% in a patient with type 2 diabetes).

The results of Project ImPACT: Hyperlipidemia demonstrate that a pharmacist-oriented program to improve adherence can dramatically improve health outcomes.[15] Project ImPACT, which stands for *Im*prove *P*ersistence *A*nd *C*ompliance with *T*herapy, was conducted in 26 community-based ambulatory care pharmacies in 12 states. The program's objective was to demonstrate that pharmacists, working collaboratively with patients and physicians, could improve patients' adherence to prescribed therapy for dyslipidemia and help them achieve their National Cholesterol Education Program (NCEP) goals.

Remarkably, over an average of 24.6 months, 93.6% of Project ImPACT patients adhered to their prescribed therapy and 90.1% persisted with therapy through the study's end.[15] Among patients with existing coronary artery disease, 48% attained their NCEP goal, far better than in any previously published national study of patients with hyperlipidemia. The authors stated that collaboration between pharmacists, patients, and physicians, using pharmacy-based testing for blood lipids and pharmacist-led counseling, could reduce the risk of heart disease and stroke by one-third.

STRATEGIES FOR ENHANCING ADHERENCE TO PHARMACOTHERAPY

Although pharmaceutical care plans should be individualized, some adherence-promoting strategies tend to be helpful in the majority of patients. Whenever possible, the pharmacist should strive to

- *Promote self-efficacy.* Encourage patients to assume an active role in their own treatment plans. In general, the more confident people feel about their ability to manage a problem, the more likely they will be to take positive action to solve that problem. Involving patients in decisions about their care is important for promoting self-efficacy. For example, a study by Nessman and colleagues[24] showed that patients with hypertension who were highly involved in decisions about their therapy and trained to take their own blood pressure had significantly better health outcomes than patients who did not have these characteristics. The authors attributed the improved outcomes to the patients' ability to make choices about health care decisions and follow through on a monitoring plan.
- *Empower patients to become informed medication consumers.* A pharmaceutical care plan to enhance adherence should first focus on educating the patient and family members or caregivers about the patient's disease and medications. Pharmacists should provide both written and oral information to address such basic questions as: What is the disease? Which treatments have been prescribed or recommended and why? What is the patient's role in managing the disease? Which adverse effects may occur? Perhaps surprisingly, the amount of factual information that a patient has about his or her medication is *not* highly correlated with adherent behavior.[7] Instead, the patient's functional knowledge—that is, information that is directly useful and meaningful to the patient—and clear instructions for medication use are more significant.[25] Opportunities to impart functional knowledge begin with the physician and/or nurse at the time of the initial prescription, and should be reinforced by the pharmacist when the prescription is filled or refilled.
- *Avoid fear tactics.* Scaring patients or giving them dire warnings about the consequences of less-than-perfect adherence can backfire and may actually worsen adherence.[26] A more constructive approach is to help the patient focus on ways to integrate medication taking into their daily routine.[27]
- *Help the patient to develop a list of short- and long-term goals.* These goals should be realistic, achievable, and individualized. The pharmacist can also make "contractual" agreements with the patient to encourage development of constructive behaviors, such as getting more exercise or beginning a smoking cessation program.
- *Plan for regular follow-up.* The pharmacist should plan to interact with the patient at regular, usually brief intervals to reinforce the adherence plan. For example, brief appointments can be scheduled when patients visit the pharmacy for prescription refills. The plan should be adapted to the patient's lifestyle and be reevaluated from time to time to adjust for life changes, such as aging or a change in work or school schedules. If possible, the time for counseling on adherence should be separated from the dispensing and pick-up functions.
- *Implement a reward system.* Giving prescription coupons or specific product discounts for successfully reaching a goal in the treatment plan can help to increase adherence, particularly in patients with low motivation.

Considerations for Special Populations

Although the problem of nonadherence affects all ethnic and age groups, some populations are more vulnerable than others. Pharmacists should be especially alert for adherence problems in high-risk populations, such as the

Table 4 Resources for improving patient adherence

Organizations
National Council on Patient Information and Education (NCPIE) 4915 Saint Elmo Ave., Suite 505 Bethesda, MD 20814-6053 301-656-8653 www.talkaboutrx.org Among other resources, NCPIE publishes ''Prescription Medicines and You: A Consumer Guide,'' a large print brochure available in English, Spanish, and Asian languages.
United States Pharmacopeia (USP) 12601 Twinbrook Parkway Rockville, MD 20852 800-822-8772 www.usp.org USP's many resources include ''MedCoach'' patient information leaflets, which are available at two reading levels and may contain pictograms.
Resources for Special Populations
For low literacy patients
Responding to the Challenge of Health Literacy. The Pfizer Journal. Spring 1998;2(1):1-37. Available from: Impact Communications, Inc. 330 Madison Avenue, 21st Floor New York, NY 10017 212-490-2300
For older adults
The ElderCare Patient Education Series The Peter Lamy Center for Drug Therapy and Aging University of Maryland School of Pharmacy 506 West Fayette Street, Suite 101 Baltimore, MD 21201 http://gerontology.umaryland.edu/docs/lamy.html e-mail: lamycenter@rx.umaryland.edu
For children
The Pediatric Medication Text (Patient information for 200 commonly prescribed pediatric medications; available in English and Spanish) American College of Clinical Pharmacy 3101 Broadway, Suite 380 Kansas City, MO 64111 816-531-2177, ext. 20 www.accp.com/ped_medtxt.html
For ethnic minorities
Closing-the-Gap.com This online magazine provides resources for health care providers and consumers to promote minority health through culturally relevant care.

elderly, children, low-literacy individuals, and some ethnic minorities. Table 4 provides resources that can aid pharmacists in improving adherence in these high-risk groups.

The elderly

Although older Americans (ages 65 and older) account for less than 15% of the population, they consume about 33% of all prescription medications and 40% of nonprescription drugs.[28] Poor adherence in the elderly often leads to additional physician or emergency department visits, hospitalization, and uncontrolled chronic diseases. One study estimated that about 17% of elderly hospitalizations are due to adverse medication reactions—nearly six times the rate in the nonelderly population.[29]

A variety of often-interacting risk factors increase the risk of nonadherence among the elderly. Risk factors in this population include

- *Polypharmacy.* Elderly patients are more likely to take multiple medications, including both prescription and nonprescription products. Whenever possible, the medication regimen should be simplified. The pharmacist also should consider the extent to which the mode of drug delivery (e.g., pill, patch, or inhaler) may influence adherence.
- *Physical impairments.* Age-related physical disabilities, such as difficulty getting out of bed or a chair, may limit an elderly person's ability to take medication consistently. Traditional packaging of medication also may be an impediment to some elderly patients; for example, individuals with arthritis in their hands may have trouble opening containers. For these patients, consider options such as use of unit-of-use packaging, unit-dose packing, or blister packaging. The pharmacy environment should also be friendly to senior citizens. For example, elderly patients with hearing problems may need a quiet place to receive patient counseling so as not to be distracted by ambient noise. Written materials should be available in large type (14-point font size) for people with vision problems.[30]
- *Cognitive limitations.* Memory loss and other cognitive problems may interfere with adherence by causing patients to fail to understand or remember medication instructions.[30] For these patients, pharmacists may need to provide medication instructions several times and in different formats, such as both verbal and written information.
- *Limited access to or affordability of health care services.* Many elderly patients are on fixed incomes. A study conducted by the consumer advocacy group Families USA reported that over the past 5 years, the prices for the 50 prescription drugs most commonly used by the elderly have increased faster than inflation.[31] Elderly patients who are unable to afford

certain medications may be eligible for various forms of state or federal aid, or special discounts from pharmaceutical manufacturers.

Pharmacists should also consider how an elderly patient's relationship with other health care providers might influence adherence. For example, research shows that the elderly tend to favor partnership-type relationships with their physicians and that satisfying patient–provider relationships contribute to better adherence.[32] However, with the growing number of managed care and group practices, these relationships are often more difficult to develop. A good pharmaceutical care plan can help elderly patients relate more effectively with primary care providers by helping these patients understand the nature of their diseases and how to better communicate their needs to physicians.

The role of a patient's caregivers in helping or hindering medication adherence also should be considered. A motivated and well-informed caregiver can be essential for optimizing adherence in an elderly patient. However, caregivers can sometimes hinder adherence efforts. For example, a caregiver who is having trouble coping with an elderly patient's behavioral or cognitive problems may demand medications to sedate the patient. Pharmacists who serve communities with a large elderly population may want to hold special classes to teach caregivers about medication management, addressing topics such as medication administration and how to monitor and report adverse effects.

Low-literacy patients

Patients who read poorly or not at all are at high risk for poor adherence. According to the U.S. Department of Education National Adult Literacy Survey,[33] 40 million people in the United States are functionally illiterate and another 55 million are only marginally literate. Patients with low literacy skills are less likely to be adherent to their medication regimens and appointments, or to present for care early in the course of their disease.[34]

Inadequate health literacy skills have been shown to adversely affect the management of a number of chronic diseases, including diabetes and hypertension. For example, in a study of hospitalized patients, 49% of patients with hypertension and 44% of those with diabetes were found to have inadequate health literacy.[35] In that study, as many as 50% of patients did not understand how many times a prescription should be refilled. After examining a standard appointment slip, up to 33% could not describe when a follow-up appointment was scheduled, and as many as 50% could not determine whether they were eligible for financial assistance based on their income and number of children.[35]

People with low health literacy may not understand the health risks associated with errors in medication management. Shame or embarrassment about their low literacy may deter them from seeking help with medication instructions. Pharmacists can assess health literacy using nonobtrusive screening tests such as the Test of Functional Health Literacy in Adults (TOFHLA), which is available in English and Spanish versions.[36] This test includes items that assess the patient's ability to understand labeled prescription vials, blood glucose test results, clinic appointment slips, and financial information forms.

On a more practical level, pharmacists also should strive to provide patient educational materials that are written at a low literacy level. The National Work Group on Literacy and Health[37] recommends that materials should be at the fifth-grade level or lower, yet most patient education materials are written at the eleventh-grade level. Patient education materials should be short, simple, and contain culturally sensitive graphics. Easy-to-read written materials should be combined with verbal instructions, which ideally should be repeated on several different occasions to reinforce patient understanding. Involving family members in the patient education process also can promote adherence.

Many literacy organizations recommend that pictograms and warning stickers be affixed to prescription bottles and nonprescription product packages. A detailed list of pictograms and a summary of research on their usefulness for low-literacy populations are available from the United States Pharmacopeia (USP) at www.usp.org. In addition, multimedia computer-based educational programs are available that permit patients to choose to see or hear information about their particular medical condition.

Ethnic minorities

An extensive literature documents persistent differences in health outcomes between ethnic minorities and white Americans. These disparities include differences in health care access and utilization as well as health status and outcomes. Wolinsky[38] showed that differences in access and use of health services by various ethnic groups stems in part from their varying cultural traditions. Pharmacists can assist in closing this gap in health outcomes by providing culturally sensitive patient care. Information about patients' cultural health care beliefs and practices is essential for devising interven-

tions to improve adherence. To provide care that is responsive to cultural differences, pharmacists should strive to develop the following three skills:[37]

- *Communicate information that is both accurate and understandable to the patient.* This skill involves the use of interviewing techniques to assess the patient's literacy level, possible language barriers, and cultural health beliefs. Insufficient English language skills are a major barrier for some minority patients. Depending on the pharmacy's location and clientele, Spanish or other foreign language versions of patient education materials may be necessary.
- *Openly discuss racial or ethnic differences.* A patient's cultural health beliefs can contribute greatly to adherence problems. For example, a patient may believe that the body needs periodic rests from medications during long-term therapy or that daily medication use is dangerous because it can lead to addiction. Getting to know the patient and their beliefs requires time, but it fosters the development of a trusting relationship. The pharmacist should try to ascertain the answers to the following questions: Does the patient understand their diagnosis and the purpose of the medication? How do the patient's cultural health beliefs influence their understanding of the illness? Is the patient using any other therapies, such as complementary or alternative medicine, in addition to prescription medications? Does the patient have any religious beliefs that might affect the decision to adhere to the treatment plan?
- *Use community and other resources on behalf of the patient.*[37] A disproportionate number of patients in some minority groups have limited incomes, which can be a major barrier to obtaining medications. Patients with low or fixed incomes who do not qualify for medicare and medicaid often have difficulty in securing the appropriate supply of their medications. A number of programs are in place to provide free medication and counseling for low-income patients. For example, the volunteer-managed Crisis Control Pharmacy in North Carolina provides free medications that range from one-time-only prescriptions to long-term maintenance therapy. Each patient is evaluated on the basis of their financial need. Another example is the Medical Access Program (MAP), offered by the University of Georgia College of Pharmacy through the Carlos and Marguerite Mason Trust. The mission of MAP, which serves an ethnically diverse low-income population, is to increase medication access for organ transplant patients who live in Georgia.[39]

Children

With a growing number of prescription drugs being developed and marketed specifically for children and adolescents, nonadherence is becoming a significant problem in the pediatric population. According to NCPIE,[40] only one-third of children take medications as prescribed or recommended by physicians. In a study by Bush et al.[43] one-third of the children in grades 3 to 7 reported they had used one or more prescription or nonprescription medications in a 48-hour period. Another study of children 9 to 16 years old, who were attending summer camp, revealed that almost one-half had brought and used a supply of medications, many without the knowledge of camp personnel.[42] Adherence plans for children often require innovative approaches to teach them how to use their medications appropriately and to encourage active participation in caring for their own health.

The literature offers a number of recommendations that can help pharmacists to improve adherence in children. Some suggestions are as follows:

- *Teach children early in life to assume some responsibility for taking their medications.* According to the Children's Health Belief model developed by Bush and Ianotti,[43] children formulate health beliefs and expectations about medication use early in their development. The authors recommend that children, especially those with chronic illnesses, assume some responsibility at an early age for taking their medications. Young children who are taught to use medications wisely may be less likely in later life to engage in high-risk behaviors such as illicit drug use or medication abuse.[44] Such children may also be more discerning about the quality of information they receive about medications from their peers, and from television and other media.
- *Educate the parents, too—particularly the mother.* In young children, most risk factors for nonadherence reside in the parent. In most cultures, the mother plays an extremely important role in supervising the care of a sick child. For example, even though young children may have an aversion to the ''bad taste'' of the drug, they usually take their medications because their mothers tell them it is necessary to feel better. Research shows that children internalize parental beliefs, which greatly influence their attitudes and behaviors toward health problems as they mature into adults.[41]
- *Adapt the educational program to the child's cognitive level and stage of development.* Education should be based on the child's maturity and ability to grasp

essential concepts about the disease and medication. According to one study, physicians and pharmacists rarely talked with children about medications, yet most children wanted to know about their medicines and would ask their physicians or pharmacist if they could.[41] Children as young as 5 years of age knew there was a difference between medications for children versus those for adults.[41] They could grasp the concept that medications for adults would be "too powerful for a little body." Older children perceived the risk for adverse reactions better than the younger children did. Older children also could understand the "cost–benefit" of getting well despite the need to take a bad-tasting medicine. These children wanted to have more personal control and independence in making decisions about their medication use. Finally, although most children did not know how medications worked, they were very much interested in this topic.

Bush and her colleagues[45] developed a cognitive developmental model for educating children about medications that is based on Piaget's cognitive development theory. This model recommends teaching children about the therapeutic purpose of their medications and that medications can be both helpful and harmful (i.e., good drugs versus bad drugs, or poisons). For younger children, learning activities should be interactive and fun. For older children, education should correct earlier misconceptions and naive theories about medications that may have been learned earlier in their development. Older children may enjoy learning about medications through the use of computer games, videos, and reading materials.

- *Relate the need for medications to a child's past experiences with the illness.* For example, if child is being recalcitrant about receiving immunization against influenza, the pharmacist might use a probe such as, "Do you remember the yucky flu you had last year? Would you like to avoid that this year?" This approach can help the child remember previous bouts of the flu as an awful-feeling illness. The child then can understand the need to prevent the illness by receiving the flu vaccination.

Specific guidelines for developing interventions to address adherence problems in children can be found in the USP's *Ten Guiding Principles for Teaching Children and Adolescents about Medicines.* These principles were developed on recommendations from more than 100 health care professionals, educators, and consumer representatives who attended the USP's fall 1996 open conference, *Children and Medicines: Information Isn't Just for Grownups.* The proceedings of this conference and the recommendations can be accessed at www.usp.org/information/programs/children/principles.htm.

PATIENT-CENTERED ADHERENCE MANAGEMENT FOR CHRONIC DISEASES

Each chronic disease presents its own constellation of adherence problems. A brief overview of adherence strategies for two major public health problems—hypertension and type 2 diabetes—illustrates disease-specific risk factors for nonadherence and shows how pharmaceutical care services can enhance adherence.

Hypertension

Because hypertension is usually a silent disease, most patients do not experience symptoms that remind them of the need for taking medications. Without symptoms, it is more difficult to establish a link in the patient's mind between taking the medication and controlling hypertension and its complications. Because patients often do not feel or perceive the benefits of their treatment, the first step in enhancing adherence is to educate them about hypertension and its serious complications, such as coronary heart disease, stroke, and renal failure.

Pharmacists who want to maximize adherence to pharmaceutical care programs for hypertension should first read the *Sixth Report of the Joint National Committee on Prevention, Detection, Evaluation, and Treatment of High Blood Pressure.*[46] This report encourages a greater interdisciplinary role for pharmacists in monitoring medication use and providing patient information. Adherence to therapy is a key consideration for reaching the 2010 national goals for blood pressure control.[46] Only one-half of patients with hypertension still take their medications after the first year of treatment, and one-third of them do not take enough medications to keep their blood pressure under control.[7]

The primary goals of a pharmaceutical care plan for hypertension are to improve patient adherence, decrease the risk of developing complications, and reduce the cost of unnecessary emergency department visits and hospital stays. Simplified dosage regimens, such as once- or twice-daily dosing, have been shown to enhance adherence in hypertensive patients. In one study, adherence rates were 73% and 70% for once-or twice-daily regimens, respectively, versus 52% and 42% for three- and four-times-a-day regimens.[47] Improving adherence is particularly important with the newer regimens, because drug concentrations may be subtherapeutic when dosing delays or omissions occur.[48] Common adverse effects of antihy-

pertensive therapy, such as fatigue, impotence, and lightheadedness, also can adversely affect adherence.

Patients may need advice on how to incorporate medications and other antihypertensive treatments, such as exercise recommendations, into their daily activities and lifestyles. One useful strategy is to help patients establish cues that will serve as reminders to take medication, such as after breakfast, after brushing teeth, or just before bed.

As with other chronic diseases, education of caregivers and family members is crucial. In one study, 70% of patients wanted their family members to know more about hypertension. The patients reported that negative attitudes, insufficient family support, and lack of confidence in the management of their blood pressure were contributing factors to their long-term adherence problems.[49] Whenever possible, a family member or caregiver should be included in educational sessions to help the patient follow instructions and stay on track over time.

Social or group support can also help to boost the patient's confidence and sense of self-efficacy. Group social support may be available from a patient advocacy organization, such as a local chapter of the American Heart Association.

To promote adherence to long-term therapeutic interventions, the pharmacist and patient may agree on a "contract" that includes a series of mutually agreed-upon and realistic health goals. Once a target goal has been achieved, the pharmacist can provide the patient with a reward, such as a discount on a prescription, a coupon for store merchandise, or a colorful certificate announcing successful goal attainment. Rewards should be carefully staged so they serve as motivators and are not so ostentatious as to overpower the effect of personal satisfaction from a job well done. The pharmacist and patient also can collaboratively develop periodic reports about the patient's progress for the primary care physician.

The pharmaceutical care plan should include outcome measures to gauge the success of adherence strategies for hypertensive patients. Outcomes might include refill patterns for patients taking long-term medications and periodic measurement of blood pressure control over time. Quality-of-life measurements and patient satisfaction surveys are also appropriate outcome measures. The former are useful to monitor the progress or potential complications in patients receiving lifelong therapy for asymptomatic diseases such as hypertension.[50]

Type 2 Diabetes

Type 2 diabetes is reaching epidemic proportions in the United States, largely because of rising rates of obesity, physical inactivity, and an aging population. Studies have conclusively demonstrated that the complications of type 2 diabetes can be greatly reduced or delayed by intensive medical management.[51] However, it is estimated that only 7% of patients with diabetes adhere fully to all aspects of their regimen.[52] Adherence rates for insulin-injection regimens range from 20% to 80%, adherence to dietary recommendations is about 65%, and adherence to exercise regimens varies from 19% to 30%. Glucose-monitoring adherence rates range from 57% to 70%.[52]

Hsiao and Salmon[53] reported that patients' beliefs about the benefits of diabetes therapy are important in determining whether they obtain and use medication. In general, the more severe the patient's disease and the greater the perceived susceptibility to complications, the more likely the patient is to be adherent. Patients must be convinced of the seriousness of their disease and empowered to monitor themselves for diabetic complications. Patients with diabetes who were at high risk for nonadherence included older people, men, and those with low socioeconomic status.[53]

Pharmacist-led programs can be extremely effective in improving adherence to diabetes care, as two independent pharmacies in Richmond, VA, recently demonstrated in a year-long program. During the first 6 months of the program, enrolled patients experienced an average decrease in their morning glucose values from 178.6 mg/dL to 159.3 mg/dL.[54] Remarkably, over the 12-month study period, participants had an average adherence rate of 90% for their use of diabetes medications.

To help the pharmacists identify medication problems, a prescription record review was performed 6 months after the start of the study. In addition, a computerized "diabetes checklist" was generated and given to each patient to complete at every prescription refill. Along with other information, the checklist asked about any medication-related problems the patient had experienced since the last refill and assessed the patient's pattern of blood glucose self-monitoring. The program also included a systematic review of appropriate medication dosages, potential drug or disease interactions, and potential adverse drug reactions.

At each refill visit, the pharmacist reviewed the plan with the patient and provided reminders about the need for other preventive care, such as yearly eye exams and proper foot care. When appropriate, the physician was contacted, with the patient's consent, regarding specific treatment recommendations. In summary, this diabetes monitoring program showed the value of combining multiple interventions to improve adherence and outcomes.

TIME AND MONEY: PRACTICAL ASPECTS OF ADHERENCE SERVICES

Payment for Adherence Services

Considering that pharmacies lose nearly $8 billion yearly from unrefilled prescriptions, improving adherence is well worth the effort.[14] Huffman and Jackson[55] estimated that by increasing the number of refills by only 10%, a pharmacy could increase its annual sales by $55,000 and net profit by more than $8000. Adherence screening, monitoring, and implementation of interventions also take time, and pharmacists may seek compensation for the hours they spend in those activities. Third-party payers have begun to realize the value of adherence management, and some payers may be willing to pay for adherence-related services. Patients also may be willing to pay out of pocket for these services. To increase the likelihood of reimbursement, pharmacists should be sure to document their adherence-related activities, such as patient assessment, education, and counseling.

Pharmacists also can benefit from building professional relationships with a core network of physicians who can refer patients to the pharmacy for adherence-related services. Reimbursement for cognitive services or disease state management programs is often tied to provider referrals. Providers usually make referrals to other specialists based on trust and their expertise and professional competence. A physician is more likely to refer a patient to a pharmacy when they have confidence in the content of the services and the competence of the pharmacist administering the therapeutic plan. Accountability (i.e., having the name of an individual, rather than an organization, responsible for the services rendered) is also important.

Space Considerations

Assessment of and counseling on adherence is best done face to face. The use of a special counseling area is recommended, especially when counseling requires more time or privacy. Although extensive renovation of the pharmacy is usually not needed, the environment should be conductive to open communication, with enough privacy for patients to feel free to discuss personal matters.

Environmental barriers, such as a desk or prescription counter, may pose a physical barrier to communication and should be avoided, if possible. Adequate privacy is also important, especially when patients are discussing sensitive medical matters and others could overhear. Ideally, the counseling area should be free of distractions, such as ringing telephones or other conversations. The counseling area should have enough space for the pharmacist to demonstrate the use of medications or devices, to write instructions, and to store written materials for distribution. A chair should also be available for patients to sit during counseling sessions.

Making Time for Adherence Services

It can be challenging for pharmacists to find ways to incorporate adherence screening and monitoring into their current organizational structures. Use of pharmacy technicians to perform routine dispensing duties can free time for the pharmacist to provide cognitive services, such as assessment and counseling. Innovative scheduling methods may also free up time for patient education and counseling. For example, there may be a brief overlap of pharmacist coverage during the times immediately before and after work shifts. Another strategy is to schedule patient appointments during times when the pharmacy workload is lighter.

SUMMARY

Adherence to pharmacotherapy is essential to optimal therapeutic outcomes. The pivotal role of the pharmacist in optimizing adherence encompasses many actions: assessing the adherence problem, identifying predisposing factors, providing comprehensive counseling, and recommending specific adherence strategies targeted to the patient's needs. Patients who have chronic conditions, physical or cognitive impairments, or cultural backgrounds outside the mainstream may have special needs that should be addressed in the adherence plan. Pharmaceutical care plans also should take into account the patient's age, stage of life, and literacy level. Although a wide range of adherence aids and strategies are available, the key to success is to tailor the intervention to the individual patient and, when necessary, to combine interventions to optimize adherence.

REFERENCES

1. National Council on Patient Information and Education (NCPIE). *The Other Drug Problem: Statistics on Medicine Use and Compliance*; 1997, Bethesda, Maryland, Available at: www.talkaboutrx.org/compliance.html#problem. Accessed May 8, 2000.
2. Smith, D.L. *Patient Compliance: An Educational Mandate*; Norwich Eaton Pharmaceuticals, Inc. and Consumer Health Information Corp.: McLean, Virginia, 1989.
3. Poirier, S.; Jackson, R.A.; Perri, M., et al. Compliance en-

hancement: All have a stand, all stand to gain. Am. Pharm. **1999**, 31–42, June.

4. Felkey, B.G. Adherence screening and monitoring. Am. Pharm. **1995**, *NS35*, 42–51.
5. Smith, M.C. Predicting and Detecting Noncompliance. In *Social and Behavioral Aspects of Pharmaceutical Care*; Smith, M.C., Wertheimer, A.I., Eds.; Pharmaceutical Products Press, Inc.: New York, New York, 1996.
6. Fincham, J.E.; Wertheimer, A.I. Using the health belief model to predict initial drug therapy defaulting. Soc. Sci. Med. **1985**, *20* (1), 101–105.
7. Bond, W.S.; Hussar, D.A. Detection methods and strategies for improving medication compliance. Am. J. Hosp. Pharm. **1991**, *48*, 1978–1988.
8. Gladman, J. Pharmacists paid to improve drug compliance, persistency. Paym. Strat. Pharm. Care **1997**, 4–8, October.
9. *Noncompliance with Medications: An Economic Tragedy with Important Implications for Health Care Reform*; The Task Force for Compliance: Baltimore, Maryland, 1994; 1–39.
10. Blandford, L.; Dans, P.E.; Ober, J.D., et al. Analyzing variations in medication compliance related to individual drug, drug class, and prescribing physician. J. Manage. Care Pharm. **1999**, *5* (1), 47–51.
11. Burrell, C.D.; Levy, R.A. Therapeutic Consequences of Noncompliance. In *Improving Medication Compliance: Proceedings of a Symposium*; National Pharmaceutical Council: Washington, DC, 1984; 7–16.
12. *Noncompliance with Medication Regimens: An Economic Tragedy. Emerging Issues in Pharmaceutical Cost Containing*; National Pharmaceutical Council: Washington, DC, 1992; 1–16.
13. Cramer, J. Relationship between medication compliance and medical outcomes. Am. J. Hosp. Pharm. **1995**, *52* (suppl. 3), S27–S29.
14. Jackson, R.A.; Worthen, D.B.; Barnett, C.W. The Financial Aspects of Improved Refill Management. In *Practice Opportunities*; Proctor & Gamble Health Care: Cincinnati, Ohio, 1998; 1–15.
15. Bluml, B.M.; McKenney, J.M.; Cziraky, M.J. Pharmaceutical care services and results in Project ImPACT: Hyperlipidemia. J. Am. Pharm. Assoc. **2000**, *40*, 157–165.
16. Ried, L.D.; Christensen, D.B. A psychological perspective in the explanation of patients' drug taking behavior. Soc. Sci. Med. **1988**, *27* (3), 277–285.
17. Berg, J.S.; Dischler, J.; Wagner, D.J., et al. Medication compliance: A healthcare problem. Ann. Pharmacother. **1993**, *27* (suppl. 9), S1–S24.
18. Morris, L.S.; Schultz, R.M. Patient compliance and overview. J. Clin. Pharm. Ther. **1992**, *17*, 283–295.
19. Stephenson, B.J.; Rowe, B.H.; Haynes, R.B., et al. Is this patient taking the treatment as prescribed? JAMA, J. Am. Med. Assoc. **1993**, *269*, 2779–2781.
20. Hepler, C.D.; Strand, L.M. Opportunities and responsibilities in pharmaceutical care. Am. J. Hosp. Pharm. **1990**, *47*, 533–543.
21. Healthcare Compliance Packaging Council Noncompliance. The invisible epidemic. Drug Top. **1992**; 1–11, August 17.
22. Stewart, M. The validity of an interview to assess a patient's drug taking. Am. J. Prev. Med. **1987**, *3* (2), 95–100.
23. Christensen, D.B.; Williams, B.; Goldberg, H.I., et al. Assessing compliance to antihypertensive medications using computer-based pharmacy records. Med. Care **1997**, *35* (11), 1164–1170.
24. Nessman, D.G.; Carnahan, J.E.; Nugent, C.A. Increasing compliance. Patient-operated hypertension groups. Arch. Intern. Med. **1980**, *140*, 1427–1430.
25. Hulka, B.S.; Kupper, L.; Cassel, J.C., et al. Medication use and misuse: Physician-patient discrepancies. J. Chronic. Dis. **1975**, *28*, 7–21.
26. Rudd, P. Maximizing compliance with antihypertensive therapy. Drug Ther. **1992**, *22*, 25–32.
27. Mullen, P.O.; Green, L.W.; Pessinger, G.S. Clinical trials of patient education for chronic conditions: A comparative meta-analysis of intervention types. Prev. Med. **1985**, *14*, 753–757.
28. National Council on Patient Information and Education. *Medication Communication Needs of Select Population Groups*; Available at: www.talkaboutrx.org/select.html#old. Accessed May 8, 2000.
29. Nanada, C.; Fanale, J.; Kronholm, P. The role of medication noncompliance and adverse reactions in hospitalizations of the elderly. Arch. Intern. Med. **1990**, *150*, 841–846.
30. Mallet, L. Counseling in special populations: The elderly patient. Am. Pharm. **1992**, *NS32* (10), 835–843.
31. Families USA. Available at: www.familiesusa.org. Accessed May 8, 2000.
32. Stewart, R.B.; Caranasos, G.J. Medication Compliance in the elderly. Med. Clin. North Am. **1989**, *73*, (6), 1551–1563.
33. Kirsch, I.; Jungeblit, A.; Jenkins, L., et al. *Adult Literacy in America. US Department of Education. National Center for Educational Statistics. National Adult Literacy Survey. Princeton, NJ: Educational Testing Service*; 1993.
34. Malveaux, J.O.; Murphy, P.W.; Arnold, C., et al. Improving patient education for patients with low literacy skills. Am. Fam. Phys. **1996**, *53* (1), 205–211.
35. Williams, M.V.; Baker, D.W.; Parker, R.M., et al. Relationship of functional health literacy to patient's knowledge of their chronic disease: A study of patients with hypertension and diabetes. Arch. Intern. Med. **1998**, *158* (2), 166–172.
36. Nurss, J.R.; Parker, R.M.; Williams, M.V., et al. *Test of Functional Health Literacy in Adults (TOFHLA)*; Georgia State University and Emory University School of Medicine: Atlanta, Georgia, 1995.
37. Weiss, B.D. Communicating with patients who have limited literacy skills. Report of the National Work Group

on Literacy and Health. J. Fam. Pract. **1998**, *46* (2), 168–176.

38. Wolinsky, F.D. Racial differences in illness behavior. J. Commun. Health **1982**, *8*, 87–101.
39. Medical Access Program (MAP). The University of Georgia College of Pharmacy, Clinical Pharmacy Program at the Medical College of Georgia, Augusta, Georgia.
40. Children and America's Other Drug Problem: Guidelines for Improving Prescription Medicine Use Among Children and Teenagers. National Council of Patient Information and Education. Available at: www.talkaboutrx.org/select.html#child Accessed May 9, 2000.
41. Menacker, F.; Aramburuzabala, P.; Minian, N., et al. Children and medicines: What they want to know and how they want to learn. J. Soc. Adm. Pharm. **1999**, *16* (1), 38–51.
42. Rudolf, C.J.; Alaria, A.J.; Youth, B., et al. Self-medication in childhood: Observations at a residential summer camp. Pediatrics **1993**, *91*, 1182–1185.
43. Bush, P.J.; Ianotti, R.J. A children's health belief model. Med. Care **1990**, *28* (1), 69–83.
44. United States Pharmacopeia. USP recommends: Children and adolescents have a right to information and direct communications about medicines. Available at: www.usp/org/aboutusp/releases/pr_9819.htm. Accessed May 8, 2000.
45. *Children, Medicines and Culture*; Bush, P.J., Trakas, D.J., Sanz, E.J., et al., Eds.; Pharmaceutical Products Press, Inc.: New York, New York, 1996; Vol. 131; 263–270.
46. The Sixth Report of the Joint National Committee on Prevention, Detection, Evaluation, and Treatment of High Blood Pressure (JNC-VI). Arch. Intern. Med. Available at: www.nhlbi.nih.gov/guidelines/hypertension/jncintro.htm. Accessed May 8, 2000.
47. Greenberg, R.N. Overview of patient compliance with medication dosing: A literature review. Clin. Ther. **1993**, *6*, 590–599.
48. Rudd, P. Clinicians and patients with hypertension: Unsettled issues about compliance. Am. Heart J. **1995**, *130* (3), 573–579.
49. Becker, M.H.; Maiman, L.A. Strategies for enhancing patient compliance. J. Commun. Health **1980**, *6* (2), 113–130.
50. MacKeigan, L.D.; Pathak, D.S. Overview of health-related quality-of-life measures. Am. J. Hosp. Pharm. **1992**, *49*, 2236–2245.
51. Ohkubo, Y.; Kishikawa, H.; Araki, E., et al. Intensive insulin therapy prevents the progression of diabetic microvascular complications in Japanese patients with non-insulin dependent diabetes mellitus: A randomized prospective 6-year study. Diabetes Res. Clin. Pract. **1995**, *28*, 103–117.
52. McNabb, W.L. Adherence in diabetes: Can we define it and can we measure it? Diabetes Care **1997**, *20*, 215–218.
53. Hsiao, L.C.D.; Salmon, J.W. Predicting adherence to prescription medication purchase among HMO enrollees with diabetes. J. Manag. Care Pharm. **1999**, *5* (4), 336–341.
54. Berringer, R.; Shibley, M.C.; Cary, C., et al. Outcomes of a community pharmacy-based diabetes monitoring program. J. Am. Pharm. Assoc. **1999**, *39* (6), 791–797.
55. Huffman, D.C.; Jackson, R.A. The financial benefits of improved patient compliance. NARD J. **1995**, 108–111, October.
56. Lasagna, L.; Hutt, P.B. Health Care, Research, and Regulatory Impact on Noncompliance. In *Patient Compliance in Medical Practice and Clinical Trials*; Cramer, J.A., Spilker, B., Eds.; Raven Press, Ltd.: New York, New York, 1991.
57. Morrow, D.; Leirer, V.; Sheikh, J. Adherence and medication instructions. Review and recommendations. J. Am. Geriatr. Soc. **1988**, *36*, 1147–1160.
58. Horne, R. One to be taken as directed: Reflections on non-adherence (non-compliance). J. Soc. Adm. Pharm. **1993**, *10* (4), 150–156.

Adverse Drug Reactions

Therese I. Poirier
Robert L. Maher Jr.
Duquesne University, Pittsburgh, Pennsylvania, U.S.A.

INTRODUCTION

Adverse drug reactions (ADRs) are types of adverse drug events (ADEs) (1). ADEs include ADRs, medication errors, and other drug-related problems. ADEs are the negative consequences of drug misadventures. Henri Manasse defined drug misadventure as the iatrogenic hazard that is an inherent risk when drug therapy is indicated. This chapter will focus on ADRs.

DEFINITIONS

The World Health Organization's (WHO) and Karch and Lasagna's definitions of an ADR are quite similar. An ADR is any response to a drug that is noxious and unintended, and occurs at doses used for prophylaxis, diagnosis, or therapy, excluding failure to accomplish the intended purpose (2). The Food and Drug Administration (FDA) focuses on ADRs that have unexpected reactions and/or those of more significant morbidity. These ADRs would include those where the patient outcome is death, life-threatening, hospitalization, disability, congenital anomaly, or required intervention to prevent permanent impairment or damage (3). The Joint Commission on Accreditation of Healthcare Organizations (JCAHO) is concerned with the reporting of significant ADRs. Those that result in morbidity, require additional treatment, require an increased length of stay, temporarily or permanently cause disability, or cause death must be reported to the FDA (4). The American Society of Health-System Pharmacists (ASHP) defines significant ADRs as any unexpected, unintended, undesired, or excessive response to a drug that includes the following:

- Requires discontinuing the drug
- Requires changing the drug therapy
- Requires modifying the dose
- Necessitates admission to the hospital
- Prolongs stay in a health care facility
- Necessitates supportive treatment
- Significantly complicates diagnosis
- Negatively affects prognosis or results in temporary or permanent harm, disability, or death (5)

The ASHP definition does not include reactions due to drug withdrawal, drug abuse, poisoning, or drug complications.

Other terms that may be included as ADRs are side effects, drug intolerance, idiosyncratic reactions, toxic reactions, allergic reactions, or hypersensitivity reactions (6). *Side effects* are reactions that are unintended and unwanted but are known pharmacologic effects of the drug and occur with predictable frequency. *Drug intolerance* is a mild reaction to a drug that results in little or no change in patient management. *Idiosyncratic reaction* is an unexpected response that occurs with usual dose of a drug. *Toxic reaction* is a predictable response that results from greater than recommended drug dosages or drug concentration in the body. *Allergic or hypersensitivity reaction* is an unusual sensitivity to a drug of an immunologic nature.

CLASSIFICATION SYSTEMS

Four classification systems are used to describe ADRs (1, 7). ADRs can be classified according to thepharmacologic effect of the drug—Type A, B, C, and D reactions. Type A reactions are exaggerated but normal pharmacologic actions of a drug. They are predictable and dose dependent. Type B reactions are not predictable given the known pharmacologic action of a drug and are not dose related. Many of these Type B reactions are hypersensitivity or immune-based. These reactions can be further subdivided into type I (IgE-mediated reaction), II(IgG or IgM-mediated cytotoxic reaction), III (IgG-mediated immune complex reactions), and IV (cell-mediated immune reaction). Type C reactions are those due to long-term use of a drug. Type D reactions are delayed drug effects, such as due to carcinogenicity or teratogenicity.

ADRs can also be classified according to the dose relationship, i.e., dose-related and non-dose-related reactions. Another classification system is based on the causal relationship between the reaction and the drug. One of the

Encyclopedia of Clinical Pharmacy
DOI: 10.1081/E-ECP 120006420

most widely used causality classifications is based on Naranjo's descriptions. These categories include definite (drug is likely the true cause), probable (drug is the apparent cause), possible (drug appears to be associated), and remote (drug is not likely to be the cause). The fourth classification system is based on degree of injury or severity of reaction. There are mild reactions (temporary discomfort and tolerable), moderate (significant discomfort), and severe (potentially life threatening or causing permanent disability or death).

INCIDENCE

The frequency of ADRs in the general population is unknown. However, the reported rates of new occurrences for ADRs are noted for selected patient populations. A meta-analysis of 39 prospective studies reported an overall incidence of serious ADRs in hospitalized patients of 6.7% and of fatal ADRs of 0.32% (8). The fatality rate makes ADRs the fourth to sixth leading cause of death in the United States. Another meta-analysis of 36 studies indicated that approximately 5% of hospital admissions are due to ADRs (9). The costs of ADRs are estimated to be $1.56–$4 billion in direct hospital costs per year in the United States (10).

FACTORS PREDISPOSING TO ADRS

Two major factors predispose to adverse drug reactions: the drug itself and patient factors. Factors related to the drug include its dose, dosage form and delivery system, and interactions between drugs. Patient-related factors include age, disease states, genetics, gender, nutrition, multidrug therapy use, and use of herbal therapies.

Drug-Related Factors

Dose

ADRs may be the result of ingestion of increased amounts of a drug. Dosing issues are especially likely with narrow therapeutic index drugs. Examples of these types of drugs include digoxin, anticoagulants, anticonvulsants, antiarrhythmics, antineoplastic agents, bronchodilators, sedatives, and hypnotics (11).

Dosage form and delivery system

Many of the ADRs related to the dosage form and delivery system are the result of local irritation or hypersensitivity reactions (12). Local irritation to the gastrointestinal (GI) tract can occur with oral dosages. For example, toxicity resulting in mouth ulcerations is associated with antineoplastic drugs. In addition, the use of certain formulations, such as sustained release preparations, can increase esophageal injury if esophageal transit is delayed. For example, a controlled release wax matrix of potassium chloride has been associated with significant esophageal erosions. Factors identified to predispose to esophageal injury include large film-coated tablets, capsules, large sustained-release preparations, rapidly dissolving formulations, and ingestion of solid oral dosage forms before bed rest with very little water intake (12).

Localized tissue irritation can be seen from the intramuscular (IM) route. This is especially an issue when the formulation pH differs from the pH of the surrounding tissue or when precipitation of poorly soluble drugs occurs (12). Incorrect administration of IM injections is probably the most important factor that causes local adverse effects. Local skin irritation can also be seen with transdermal delivery systems due to the alcohols, nonionic surfactants, and adhesives.

Hypersensitivity reactions can occur due to the presence of contaminants or excipients in pharmaceutical dosage forms (e.g., outbreaks of eosinophilia-myalgia syndrome associated with oral tryptophan contaminants in various drugs) (12). Another example is the anaphylactoid reactions to the surfactant Cremaophor EL, which is used in paclitaxel (Taxol).

Direct toxicity effects related to use of preservatives also has been documented. For example, severe metabolic acidosis and death in infants was attributed to the presence of benzyl alcohol, a preservative used in bacterostatic normal saline that was used to flush catheters (12).

The use of specific intravenous (IV) delivery devices also can cause ADRs. For instance, use of plastic infusion sets for IV administration of nitroglycerin has resulted in subtherapeutic effects due to diffusion of the drug into the plastic tubes (12).

Formulation effects, such as bioavailability differences, can cause ADRs when patients are switched to generic products. For example, significant adverse effects have occurred with anticonvulsants and thyroid preparations (12).

Interactions between drugs

It has been estimated that 6.9% of ADRs are due to drug–drug interactions (6). The most likely reason for an adverse drug interaction is the pharmacokinetic changes that result in altered metabolism or excretion of drugs, or the pharmacodynamic changes that result in synergistic or additive effects of drugs.

Patient-Related Factors

Age, disease states, genetics, gender, nutrition, multidrug therapy use, and herbal therapies use are patient-related factors that influence the likelihood of adverse drug reactions.

Age—geriatrics

Age-related alterations in pharmacokinetics and pharmacodynamics may affect the response of elderly patients to certain medications, and may increase the susceptibility for ADRs among elderly patients (13–15) (Table 1). The risk of ADRs among elderly patients is probably not due to age alone. ADRs may be related more to the degree of frailty and medical conditions of the patient (15). On average, older persons have five or more coexisting diseases that may increase the risk of adverse events. Polypharmacy seems to be more of a common problem among the elderly. The average elderly patient takes 4.5 chronic medications and fills 13 prescriptions yearly (15). Elderly patients appear to have a decline in homeostatic mechanisms. The imbalance of homeostatic mechanisms and the decline in function reserves may put a patient at greater risk for ADEs due to decreased tolerance of medications and the ability to handle stressful situations (16).

Age—pediatrics

The two factors responsible for increasing risks of ADRs in children are pharmacokinetic changes and dose delivery issues. Age-related differences in pharmacokinetics in children are documented (17). However, the data on both efficacy and safety are often limited or not studied at all in this population. Thus, it is unclear whether an increased risk for ADRs exists in this group. However, there is a potential risk for increased ADRs if appropriate considerations are not taken into account in view of pharmacokinetic changes (18).

It is important to note that only one-fourth of the drugs approved by the FDA have indications specific for use in a pediatric population (17). Medications used in adults are often given to children without FDA safety and efficacy data. Compatibility and stability issues with dosage forms intended for adults that have been altered (e.g., dilution or reformulation) can increase risks for ADRs.

Information on pediatric age-related difference in neonates, children, and adolescents may aid in prevention of pediatric ADRs (18) (Table 2). Further studies of drug use in pediatrics are needed in order to prevent ADRs.

Concurrent diseases

Diseases such as hepatic or renal diseases can influence the incidence of ADRs by altering the pharmacokinetics of drugs, such as absorption, distribution, metabolism, or excretion (6).

Hepatic disease

Patients with liver disease have an increased susceptibility to certain drugs due to decreased hepatic clearance for drugs metabolized by the liver or due to enhanced sensitivity (6). For example, impaired hepatic metabolism can precipitate central nervous system (CNS) toxicity in patients on theophylline, phenytoin, or lidocaine; or ergot poisoning on ergotamine (19).

Increased sensitivity to drugs is also encountered in liver disease(19). The use of anticoagulants increases the risk of bleeding due to the reduced absorption of vitamin K or decreased production of vitamin K-dependent clotting factors. There is an enhanced risk for respiratory depression and hepatic encephalopathy due to morphine

Table 1 Geriatric age-related changes in pharmacokinetics

Pharmacokinetic phase	Pharmacokinetic parameters
Gastrointestinal absorption	Unchanged passive diffusion and no change in bioavailability for most drugs
	↓ Active transport and ↑ bioavailability for some drugs
	↓ First-pass effect and ↑ bioavailability
Distribution	↓ Volume of distribution and ↑ concentration of water soluble drugs
	↑ Volume of distribution and ↑ half-life for fat soluble drugs
	↑ or ↓ free fraction of highly plasma protein-bound drugs
	↓ Clearance and ↑ half-life for some Phase I
Oxidation drugs	↓ Clearance and ↑ half-life of drugs with high extraction ratio
Renal excretion	↓ Clearance and ↑ half-life of renally eliminated drugs

↓ = Decreased; ↑ = Increased.

Table 2 Pediatric age-related risk factors and causes of ADRs

Neonates:
Placental transfer of drug before birth
Differing drug action
Altered pharmacokinetics
Increased percutaneous absorption
Decreased renal/hepatic function
Decreased plasma protein binding
Use of multiple drugs
Limited information on drug action in critically ill and premature neonates

Children:
Paradoxical effect of medications (excitability rather than sedation from antihistamines)
Excipients of liquid dosage forms
Sugar as sweeteners
Propylene glycol as solvent
Large volume intravenous solutions
Treatment of viral infections with antibiotics
Disruption of neurologic and somatic development

Adolescents:
Autonomy seeking
Use and misuse of devices (e.g., tampons)
Use and misuse of prescription and nonprescription medications
Poor compliance with instructions
Use of multiple medications
Recreational use of alcohol and illicit drugs
Effects of changing hormone levels on drugs

(From Ref. 7.)

or barbiturates in patients with severe liver disease. Vigorous use of diuretics can precipitate hepatic coma due to potassium loss in liver disease. There is an increased risk of hypoglycemia with sulphonylurea antidiabetic drugs due to decreased glycogenesis in liver disease.

Liver disease can also cause hypoalbuminemia due to decreased liver synthesis of albumin. For drugs that are extensively bound to albumin, such as phenytoin, an enhanced risk of drug toxicity could occur because of the increase in free drug concentration.

There are no useful methods to quantify the degree of liver disease that can assist in dosage adjustment. A practical approach involves checking patients for elevated prothrombin time, rising bilirubin levels, and/or falling albumin levels. In such instances, drugs that have an altered response in liver disease or cause hepatotoxicity need to be avoided.

Renal disease

Impaired renal function increases the incidence of ADRs for drugs that depend on the kidney for their elimination. Unlike liver disease, use of pharmacokinetic dosing principles can minimize the risk for adverse effects.

Mechanisms responsible for enhanced ADRs in renal disease include delayed drug excretion, decreased protein binding due to hypoalbuminemia, and increased drug sensitivity (6). Delayed renal excretion is responsible for enhanced toxicity with drugs such as aminoglycosides, digoxin, vancomycin, chlorpropamide, H2-antagonists, allopurinol, lithium, insulin, and methotrexate (20). For some drugs, the accumulation of a toxic metabolite during renal failure is responsible for ADRs. This is the case with meperidine, where a toxic metabolite, normeperidine, accumulates in renal failure (20).

Patients with accumulation of uremic toxins have increased sensitivity to certain drugs. There may be an enhanced response to CNS depressants (such as barbiturates and benzodiazepines), hemorrhagic effects from aspirin or warfarin, and other bleeding effects from antibiotics that inhibit platelet aggregation, such as carbenicillin, ticarcillin, and piperacillin.

Other diseases

On theoretical grounds, other diseases associated with hypoalbuminemia could predispose patients to adverse reactions and to altered responses to drugs that are highly protein bound (21) (Table 3).

The presence of other diseases can influence the risk for ADRs. Many of these adverse effects are related to an extension of the pharmacologic effects of the drug in the presence of certain pathophysiology. Numerous examples are given in Table 4 (6).

Patients who have had a previous reaction to drugs are also more likely to experience an ADR (22). Patients with history of allergic diseases also have an increased risk due to a genetically related ability to form immunoglobulin E.

Genetic factors

Genetic factors account for some ADRs due to either altered pharmacokinetics or by altering tissue responsiveness. Altered metabolism of drugs occurs due to

Table 3 Conditions associated with hypoalbuminemia

Aging	Liver disease
Burns	Nephrotic syndrome
Cancer	Nutritional deficiency
Cardiac failure	Pregnancy
Protein-losing enteropathy	Renal failure
Inflammatory diseases	Sepsis
Injury	Stress
Immobilization	Surgery

Table 4 Influence of diseases on adverse drug reactions

Disease	Drug	Adverse reactions
Gastrointestinal		
Peptic ulcer	Aspirin, corticosteroids, nonsteroidal antiinflammatory drugs	Risk of bleeding or perforation of ulcer
Cardiovascular		
Heart failure	β-Blockers	Aggravate or precipitate heart failure
	Lidocaine, theophylline	Enhanced toxicity—seizures
Myocardial ischemia	Tricyclic antidepressants	Disturbances of cardiac rate, rhythm, and conduction
	Digoxin	Arrhythmias
Bradycardia	β-Blockers	Cardiac standstill
	Quinidine	
Hypertension	Oral contraceptives, vasoconstrictors	Increased blood pressure
	Phenothiazines, nitrates	Decreased blood pressure
	Tricyclic antidepressants	
Hematologic		
Bleeding disorders—hemophilia	Aspirin	Increased risk of hemorrhage
Neurological disorders		
Myasthenia gravis	Aminoglycosides	Aggravate muscle weakness
	Quinidine, quinine	Paralysis
Epilepsy	Phenothiazines	Lower seizure threshold
	Tricyclic antidepressants	
Cerebrovascular	Ergotamine	Ischemic episodes
Rheumatic		
Systemic lupus	Drugs	Increased incidence of drug reactions in general
Hyperuricemia	Thiazide diuretics, furosemide	Gouty attack
Respiratory		
Asthma	β-Blockers	Acute bronchospasms
Respiratory insufficiency	Narcotic analgesics	Hypoventilation, respiratory arrest
Endocrine disorders		
Diabetes mellitus	Thiazide diuretics, furosemide, corticosteroids, oral contraceptives	Hyperglycemia; aggravates diabetic control
Hypothyroidism	Digoxin	Enhanced response
Hyperthyroidism	Oral anticoagulants	Enhanced response
	Digoxin	Decreased response
Ocular		
Narrow-angle glaucoma	Anticholinergics	Glaucoma attack

differences in hydrolysis, acetylation, and hepatic oxidation of drugs. Altered pharmacodynamic reactions could be either an exaggerated response or a qualitative response. These types of reactions are unpredictable. Examples of altered drug response due to genetic factors are found in Table 5 (6).

Gender

A higher incidence of ADRs has been reported for women in comparison to men (6). One reason for this observation is that women take more drugs than men. Yet, no sex-linked differences in drug pharmacokinetics have been documented. Other reports have not supported a higher incidence of ADRs in women as compared to men. Thus, sex alone is unlikely to be a major determinant of ADRs.

Nutrition

Nutritional factors are also responsible for ADRs. These factors include the interaction of drugs and nutrients, and altered pharmacokinetics related to nutritional status.

One study reported a very low incidence (0.4%) of clinically significant drug–nutrient interactions in a teaching hospital (23). Three mechanisms postulated for drug–nutrient interactions are interference with drug absorption, alteration of drug excretion, and affecting drug activity. For example, the absorption of tetracycline is reduced by chelation with iron, calcium, and magnesium. Foods that acidify or alkalinize the urine can affect drug excretion. Foods that contain a large amount of vitamin K can inhibit the activity of warfarin. A listing of important drug–nutrient interactions is found in Table 6 (23). A review article on drug–food interactions in clinical practice is found in Ref. 24.

Drug–nutrient interactions may be more highly significant in renal failure patients. A review article of drug–nutrient interactions in renal failure has been published (25).

Nutritional status can affect drug pharmacokinetics. Malnutrition states can cause the following: 1) the liver and kidneys changes affect drug elimination; 2) GI system changes affect drug absorption; 3) changes in the heart affect blood flow; 4) hormone changes affect metabolic enzymes and drug binding proteins; 5) plasma, tissue proteins, and body composition changes affect protein binding and elimination; 6) mineral and electrolyte changes affect drug metabolism and protein binding; and 7) tissue changes affect uptake of drugs and drug–receptor interactions (26).

Multidrug use

According to several epidemiological studies, multiple drug use has a strong association in the causality of ADRs. It has been suggested that the more medications used, the higher the risk for ADRs (27). Consistent drug regimen reviews by healthcare providers in order to reduce polypharmacy may decrease the risk of ADRs.

Herbal therapies use

The use of herbal therapies increased dramatically during the 1990s. Herbal therapy sales are estimated to be $4 billion a year, with sales increasing at 20% per year since the early 1990s (28). Patients often mistakenly believe that since these products are natural, they do not possess the potential harm as in prescription medications. Since herbal medications are sold and marketed without stringent FDA approval and guidelines, limited evidence-based data on efficacy, adverse effects, and drug interactions exist. Recently, two review articles examined available data on ADRs for the most common herbal medications (28, 29). Many of these available reports fall short on documentation of temporal relationship with the specific ADR and the herbal drug.

For most conditions, herbal products are not a replacement for proven prescription or nonprescription drugs. Patients should be aware that health care practitioners cannot guarantee the safety and consistency of herbal products. Patients should start with the recommended effective doses and report any unusual side effects to their health care practitioner. Patients should always consult with their pharmacist for possible drug–herbal interactions. Side effects and possible drug interactions for the ten most commonly used herbals are listed in Table 7.

ADVERSE DRUG REACTION REPORTING SYSTEMS

The WHO, the FDA, the JCAHO, and the Health Care Financing Administration (HCFA) have all addressed and mandated the need for health care institutions to implement an ADE detection and reporting system. Detection systems are instrumental in postmarketing surveillance of ADRs. The JCAHO requires all accredited health care institutions to have an ongoing drug surveillance program (4). The goals of ADR detecting and reporting systems are to aid in postmarketing surveillance of FDA approved medications and to identify ways to decrease ADR risks. The main focus of all of these reporting systems is to aid in promoting improvements in the medication use process.

Table 5 Genetic factors and altered drug responses

Genetic mechanism	Drug(s)	Adverse drug response
Pharmacokinetic		
Low plasma pseudocholinesterase	Succinylcholine	Prolonged neuromuscular blockade leading to apnea
Slow acetylator	Isoniazid	Increased incidence of peripheral neuropathy; SLE-like syndrome; and more prone to phenytoin toxicity
	Hydralazine, procainamide	Increased incidence of SLE-like syndrome
	Phenelzine, sulfasalazine	More prone to side effects
Rapid acetylator	Isoniazid	More prone to hepatitis
Deficiency of epoxide hydrolase	Phenytoin, carbamazepine, phenobarbital	Life threatening hypersensitivity syndrome due to accumulation of toxic intermediates
Pharmacodynamic		
Glucose 6-phosphate dehydrogenase deficiency (G-6-PD)	Aspirin, BAL (dimercaprol), chloroquine, chloramphenicol, dapsone hydroxychloroquine, nalidixic acid, nitrofurantoin, primaquine, probenecid, quinine, quinidine, sulfonamides	Hemolytic anemia
Methemoglobin reductase deficiency	Acetaminophen, anesthetics, topical, benzocaine, chloroquine, dapsone, nitrites, primaquine, sulfonamides	Methemoglobinemia
Abnormality of calcium regulation	Anesthetics, general, (halothane), muscle relaxants (succinylcholine)	Malignant hyperpyrexia

Table 6 Important drug-nutrient interactions

Drug	Nutrient	Interaction
Phenytoin	Alcohol	Enhanced metabolism of phenytoin
	Enteral feedings	Decreased phenytoin absorption
Tetracycline	Dairy products	Impaired drug absorption
Theophylline	Caffeine	Potential for toxic effects
Warfarin	Foods high in vitamin K	Decreases anticoagulant response
Chlorpropamide, tolbutamide, tolazamide, acetohexamide, metronidazole	Alcohol	Disulfiram-like reaction
Trancylcypromide	Foods high in tyramine	Hypertensive crisis
Disulfiram	Alcohol	Nausea, blurred vision, chest pain, dizziness, fainting
Spironolactone	Foods high in potassium	Hyperkalemia

(Adapted from Ref. 23.)

ADR Screening Methods

The best methodology for screening for ADRs has not been determined. However, several screening methods have been proposed. In particular, the literature has highlighted five screening methods using clinical data (30–34). The five include screening for: 1) "tracer drugs," e.g., antidotes such as vitamin K and diphenhydramine; 2) "narrow therapeutic range drugs," e.g., follow-up of computer lab values for warfarin and digoxin; 3) change in medications, e.g., documentation of discontinued medications or decreased dose; 4) diagnosed ADRs documented in the medical record, e.g., chart review or reviewing ICD-9 CM (International Classification of Diseases, Ninth Revision, Clinical Modification) codes; and 5) ADR computer report tracking systems. Although each of these ADR screening methods has been described in detail, limited data are available on the productivity of these screens.

Systems for Pharmaco-epidemiologic Studies

Pharmacoepidemiology is used to detect ADRs (35, 36). Several types of systems use pharmacoepidemiologic methods. These include spontaneous reporting, studies of therapeutic classes, and studies of specific medical syndromes.

Spontaneous reporting

Spontaneous reporting is currently the major backbone for the detection of ADRs (37). It occurs in one of three ways:

1. Reporting to the FDA as part of clinical trials;
2. Reporting by practitioners to medical journals; or
3. Patients' self-reporting to either manufacturers or the FDA (38).

Clinical trials in new drug development cannot detect all the possibilities for drug safety. Limitations in Phase III clinical trials include a relatively small sample size, short duration of the trial, restricted populations (e.g., geriatrics and pediatrics), uncomplicated patients, (e.g., limited disease states), and limited power for adverse drug reaction detection (30). Thus, the FDA relies heavily on spontaneous reporting of suspected ADRs (39). Spontaneous reporting is important in early market history of the drug to determine previously unidentified drug reactions. This has been particularly true in the last few years because of numerous new medications that have entered the market and now carry a black box warning. For example, Rezulin® and Trovan® are associated with hepatotoxicity and carry black box warnings.

Additional advantages of spontaneous reporting systems include the detection of extremely rare ADRs and ability to identify at-risk subgroups. In order to enhance the spontaneous reporting system approach, the FDA developed the MedWatch form. This form can be faxed to the agency (1-800-FDA-1078) or called in (1-800-FDA-1088) (40). The forms also can be obtained by the "MedWatch Online" internet-based website (http://www.fda.gov/medwatch/).

Limitations of FDA spontaneous reporting include both under-reporting and over-reporting.

An example of over-reporting occurs with recently approved drugs. This is partly due to enhanced publicity about these drugs.

Table 7 ADRs for the top ten herbal medicines

Herbal	Common use	Side effects and interactions
Echinacea	Treatment and prevention of upper respiratory infections, common cold	Rash, pruritis, dizziness, unclear long-term effects on the immune system.
St. John's wort	Mild to moderate depression	Gastrointestinal upset, photo-sensitivity. Mild serotonin syndrome with the following medications: paroxetine, trazodone, sertraline, and nefazodone. May decrease digoxin levels. May decrease cyclosporine serum concentrations. Combined oral contraceptives—breakthrough bleeding.
Gingko biloba	Dementia	Mild gastrointestinal distress, headache, may affect warfarin (increase INR). Interaction with aspirin (spontaneous hyphema)
Garlic	Hypertension, hypercholesterolemia	Gastrointestinal upset, gas, reflux, nausea, allergic reactions, and antiplatelet effects. May effect warfarin (increase INR)
Saw palmetto	Benign prostatic hyperplasia	Uncommon
Ginseng	General health promotion, sexual function, athletic ability, energy, fertility	High doses may cause diarrhea, hypertension, insomnia, nervousness, may affect warfarin (decreased INR)
Goldenseal	Upper respiratory infections, common cold	Diarrhea, hypertension, vasoconstriction
Aloe	Topical application for dermatitis, herpes, wound healing, and psoriasis, orally for constipation	May delay wound healing after topical application. Diarrhea, and hypokalemia with oral use
Siberian ginseng	Similar to ginseng	May raise digoxin levels. May affect warfarin (increased INR)
Valerian	Insomnia, anxiety	Fatigue, tremor, headache, paradoxical insomnia (not advised with other sedative-hypnotics)

Studies of therapeutic classes

Observational cohort or case control designs have been used to determine ADR relationships with specific therapeutic classes (36, 41). Medical claims data are often used in these studies and caution should be warranted due to lack of definite confirmation of drug exposure and the potential for confounding variables (38). However, these studies have been beneficial in determining risk of ADRs with specific classes (e.g., NSAIDs and the risk of peptic ulcer disease) (42).

Studies of specific medical syndromes

Observational cohort or case control designs can also be useful to study possible causality relationships of specific medical conditions or syndromes due to drug exposure (36, 41). These types of studies have been particularly useful in examining ADRs in a specific population, such as geriatric or pediatric patients. These groups of patients are often excluded in Phase III trials. However, a disadvantage of these studies is that they also often use administrative data. These data can warrant risk of problems in determining causality due to potential confounding variables (38).

Assessing Adverse Drug Reactions

After detection of a possible ADE, causality assessment needs to be performed. It is important to be able to rank the likelihood of an ADR as unlikely, possible, probable, or definite. A major problem with determining causality is that confounding variables can contribute to the complexity of causality assessment (43). In order to determine causality, several important points of data are required. These include the nature of the adverse event, name of the putative drug, other potential causes, and the temporal relationship

Table 8 ADR Naranjo causality algorithm

	Yes	No	Do not know	Score
1. Are there previous conclusive reports on this reaction?	+1	0	0	
2. Did the adverse event appear after the suspected drug was administered?	+2	−1	0	
3. Did the adverse reaction improve when the drug was discontinued, or a specific antagonist was administered?	+1	0	0	
4. Did the adverse reaction reappear when the drug was readministered?	+2	−1	0	
5. Are there alternative causes (other than drug) that could on their own caused this reaction?	−1	+2	0	
6. Did the reaction reappear when a placebo was given?	−1	+1	0	
7. Was the drug detected in the blood (or other fluids) in concentrations known to be toxic?	+1	0	0	
8. Was the reaction more severe when the dose was increased, or less severe when the dose was decreased?	+1	0	0	
9. Did the patient have a similar reaction to the same or similar drugs in any previous exposure?	+1	0	0	
10. Was the adverse event confirmed by any objective evidence?	+1	0	0	
			Total score	

Probability category scores: Definite ≥ 9; Probable 5–8; Possible 1–4; Doubtful ≤ 0.

between the drug and adverse event. Potential causes are obtained by examining the medical history, physical examination findings, and directed diagnostic tests.

Identification of causality can be performed simply by using a health care provider's clinical reasoning and judgment. The main disadvantage to this approach is a low inter-rater and intra-rater agreement for ADR causality (44, 45).

An ADR causality algorithm addresses the issue of inter-rater and intra-rater reliability with a series of clinical questions. For example, the Naranjo algorithm consists of a series of clinical questions that focus on temporal and dose–response relationships, consistency of the ADR with previous clinical reports or patient experiences, placebo response, drug dechallenge and rechallenge, toxic blood drug concentrations, alternative causes of the reaction, and whether the event was confirmed by objective evidence (44) (Table 8). Numerous health care institutions and the FDA use some type of causality algorithm to minimize disagreement among different evaluators and improve inter-rater and intra-rate agreement.

PREVENTING ADVERSE DRUG REACTIONS

ADRs are problematic in that they cause significant morbidity and mortality. Almost 95% of ADRs are Type A (predictable) reactions, and thus with quality improvement measures, ADRs can be avoided and prevented (46). Knowledge of causative factors and an increase in patient education may help prevent ADRs. Improvements in the documentation of allergic reactions (e.g., via computer tracking), development of tools to enhance compliance, and application of tools to improve prescribing and administration of drugs are other preventative approaches to ADRs.

In 1994, the ASHP, the American Medical Association (AMA), and the American Nurses Association (ANA) generated the following system of recommendations to prevent ADRs in health care systems:

1. Health care systems should establish processes in which prescribers enter medication orders directly into computer systems.
2. Health care systems should evaluate the use of machine-readable coding (e.g., bar coding) in their medication use processes.
3. Health care systems should develop better systems for monitoring and reporting adverse drug events.
4. Health care systems should use unit dose medication distribution and pharmacy-based intravenous medication admixture systems.
5. Health care systems should assign pharmacists to work in patient care areas in direct collaboration with prescribers and those administering medications.

6. Health care systems should approach medication errors as system failures and seek system solutions in preventing them.
7. Health care systems should ensure that medication orders are routinely reviewed by the pharmacist before first doses and should ensure that prescribers, pharmacists, nurses, and other workers seek resolution whenever there is any question of safety with respect to medication use (47).

SUMMARY

Adverse drug reactions are of significant concern in the pharmaceutical technology arena. Various drug and patient factors that predispose to ADRs have been identified. Reporting systems used to screen and assess ADRs facilitate the understanding of risk factors and contribute to the development of systematic improvement in the prevention of ADRs.

REFERENCES

1. Schumock, G.; Guenette, A. Adverse Drug Events. *Pharmacotherapy Self-Assessment Program*; Carter, B., Ed.; ACCP: Kansas City, MO, 1999; 5, 103–130.
2. Karch, F.E.; Lasagna, L. Adverse Drug Reactions: A Critical Review. JAMA **1975**, *234*, 1236–1241.
3. Rossi, A.; Knapp, D. Discovery of New Adverse Drug Reactions. A Review of the Food and Drug Administration's Spontaneous Reporting System. JAMA **1984**, *252*, 1030–1033.
4. *Joint Commission on Accreditation of Health Care Organizations. 1998 Comprehensive Accreditation Manual for Hospitals*; A.M.H: Oakbrook Terrace, IL, 1997.
5. American Society of Health-System Pharmacists. Suggested Definitions and Relationships Among Medication Misadventures, Medication Errors, Adverse Drug Events and Adverse Drug Reactions. Am. J. Health-Syst. Pharm. **1998**, *55*, 165–166.
6. Edwards, I.R. Pharmacological Basis of Adverse Drug Reactions. *Avery's Drug Treatment*; Speight, T., Holford, N., Eds.; ADIS International LTD: Auckland, New Zealand, 1997; 261–299.
7. Young, L.R.; Wurtzbacher, J.D.; Blankenship, C.S. Adverse Drug Reactions: A Review for Healthcare Practitioners. Am. J. Managed Care **1997**, *12*, 1884–1906.
8. Lazarou, J.; Pomeranz, B.; Corey, P. Incidence of Adverse Drug Reactions in Hospitalized Patients. JAMA **1998**, *279* (15), 1200–1205.
9. Einarson, T. Drug-Related Hospital Admissions. Ann. Pharmacother. **1993**, *27*, 832–840.
10. Clasen, D.C.; Pestonik, S.L.; Evans, R.S.; Lloyd, J.F.; Burke, J.P. Adverse Drug Events in Hospitalized Patients: Excess Length of Stay, Extra Costs, and Attributable Mortality. JAMA **1997**, *277*, 301–306.
11. *Applied Biopharmaceutics and Pharmacokinetics*; Shargel, L., Yu, A., Eds.; Appleton and Lange: Stamford, CT, 1999.
12. Uchegbu, I.; Florence, A. Adverse Drug Events Related to Dosage Forms and Delivery Systems. Drug Safety Concepts **1996**, *14* (1), 39–67.
13. Swift, C.G. Pharmacodynamics: Changes in Homeostatic Mechanisms, Receptor and Target Organ Sensitivity in the Elderly. Br. Med. Bull. **1990**, *46*, 36–52.
14. Parker, B.M.; Cusack, B.J.; Vestal, R.E. Pharmacokinetic Optimisation of Drug Therapy in Elderly Patients. Drugs Aging **1995**, *7*, 10–18.
15. Gurwitz, J.H.; Avorn, J. The Ambiguous Relation Between Aging and Adverse Drug Reactions. Ann. Intern. Med. **1991**, *114*, 956–966.
16. Taffet, G.E. Age-Related Physiologic Changes. *Geriatric Review Syllabus: A Core Curriculum in Geriatric Medicine*; Reuben, D.B., Yoshikawa, T.T., Besdine, R.W., Eds.; Kendall/Hunt for the American Geriatric Society: Dubuque, IA, 1996; 11–24.
17. Nahata, M.C. Pediatrics. *Pharmacotherapy A Pathophysiologic Approach*; DiPiro, J.T., Talbert, R., Yee, G.C., Matzke, G.R., Wells, B.G., Posey, M.L., Eds.; Appleton & Lange: Stanford, CT, 1999; 44–51.
18. Gupta, A.; Waldhauser, L.K. Adverse Drug Reactions from Birth to Early Childhood. Pediatr. Clin. North. Am. **1997**, *44*, 79–92.
19. Piper, D.W.; deCarle, D.J.; Talley, N.J.; Gallagher, N.D.; Wilson, J.S.; Powell, L.W.; Crawford, D.; Gibson, P.R.; Sorrell, T.C.; Kellow, J.E.; Roberts, R.K. Gastrointestinal and Hepatic Diseases. *Avery's Drug Treatment*; Speight, T., Holford, N., Eds.; ADIS International LTD: Auckland, New Zealand, 1997; 1010–1012.
20. Critchley, J.A.; Chan, T.Y.K.; Cumming, A.D. Renal Diseases. *Avery's Drug Treatment*; Speight, T., Holford, N., Eds.; ADIS International LTD: Auckland, New Zealand, 1997; 1107–1109.
21. Tillement, J.P.; Lhoste, F.; Giudicelli, J.F. Diseases and Drug Protein Binding. Clin. Pharmacokinet **1978**, *3*, 144–154.
22. Smith, J.W.; Seidl, L.G.; Cluff, L.E. Studies on the Epidemiology and Adverse Drug Reactions. Ann. Intern. Med. **1966**, *65*, 629–640.
23. Franse, V.; Stark, N.; Powers, T. Drug-Nutrient Interactions in a Veterans Administration Medical Center Teaching Hospital. Nutrition in Clinical Practice **1988**, *3* (4), 145–147.
24. Yamreudeewong, W.; Henann, N.; Fazio, A.; Lower, D.; Cassidy, T. Drug-Food Interactions in Clinical Practice. J. Fam. Pract. **1995**, *40* (4), 376–384.
25. Mason, N.; Boyd, S. Drug-Nutrient Interactions in Renal Failure. J. Renal Nutrition **1995**, *5* (4), 214–222.
26. Krishnaswamy, K. Drug Metabolism and Pharmacokinetics in Malnutrition. Clin. Pharmacokinet **1978**, *3*, 216–240.
27. Grymonpre, R.E.; Mitenko, P.A.; Sitar, D.S.; Aoki, F.Y.; Montgomery, P.R. Drug-Associated Hospital Admissions in Older Medical Patients. J. Am. Geriatr. Soc. **1988**, *36*, 1092–1098.

28. Mar, C.; Bent, S. An Evidence-Based Review of the 10 Most Commonly Used Herbs. West. J. Med. **1999**, *171*, 168–171.
29. Fugh-Berman, A. Herb-Drug Interactions. Lancet **2000**, *355*, 134–38.
30. American Society of Health-System Pharmacists. ASHP Guidelines on Adverse Drug Reaction Monitoring and Reporting. Am. J. Health-Syst. Pharm. **1995**, *52*, 417–419.
31. Johnston, P.E.; Morrow, J.D.; Branch, R. Use of a Database Computer Program to Identify Trends in Reporting Adverse Drug Reactions. Am. J. Hosp. Pharm. **1990**, *47*, 1321–1327.
32. Koch, K.E. Use of Standard Screening Procedures to Identify Adverse Drug Reactions. Am. J. Hosp. Pharm. **1990**, *47*, 1314–1320.
33. O'Neil, A.C.; Petersen, L.A.; Cook, E.F.; Bates, D.Q.; Lee, T.H.; Berman, T.A. Physician Reporting Compared With Medical-record Review to Identify Adverse Medical Events. Ann. Intern. Med. **1993**, *119*, 370–376.
34. Prosser, T.R.; Kamysz, P.L. Multidisciplinary Adverse Drug Reaction Surveillance Program. Am. J. Hosp. Pharm. **1990**, *47*, 1334–1339.
35. Porta, M.S.; Hartzema, A.G. The Contribution of Epidemiology to the Study of Drugs. Drug. Intell. Clin. Pharm. **1987**, *21*, 41–47.
36. Strom, B.L., Ed. *Pharmacoepidemiology*, 2nd Ed.; John Wiley: Chichester, 1994.
37. *The Detection of New Adverse Drug Reactions*; Stephens, M.D.B., Ed.; The MacMillan Press: Basingstoke, 1988.
38. Hanlon, J.T.; Schmader, K.; Lewis, I. Adverse Drug Reactions. *Therapeutics in the Elderly*; Delafuente, J.C., Stewart, R.B., Eds.; Harvey Whitney Books: Cincinnati, 1995; 212–227.
39. Kennedy, D.L.; McGinnis, T. Monitoring Adverse Rrug Reactions: The FDA's New MedWatch Program. Pharmacol. Ther. **1993**, *833–834*, 839–842.
40. Kessler, D.A. Introducing MEDWatch: New Approach to Reporting Medication and Device Adverse Effects and Product Problems. J.A.M.A. **1993**, *269*, 2765–2768.
41. *Pharmacoepidemiology: An Introduction*; Hartzema, A.G., Porta, M.S., Tilson, H.H., Eds.; Harvey Whitney Books: Cincinnati, 1991.
42. Guess, H.A.; West, R.; Strand, L.M.; Helston, D.; Lydick, E.G.; Bergman, C.A.; Wolski, K. Fatal Upper Gastrointestinal Hemorrhage of Perforation Among Users and Nonusers of Nonsteroidal Anti-Inflammatory Drugs in Saskatchewan, Canada 1983. J. Clin. Epidemiol. **1988**, *41*, 35–45.
43. Meyboom, R.H.B.; Hekster, Y.A.; Antoine, C.G.; Gribnau, F.W.J.; Edward, I.R. Causal or Casual? The Role of Causality Assessment in Pharmacovigilance. Drug Saf. **1997**, *6*, 374–389.
44. Naranjo, C.A.; Busto, U.; Sellers, E.M.; Sandor, P.; Ruiz, I.; Roberts, E.A.; Janecek, E.; Domecq, C.; Greenblatt, D.J. A Method for Estimating the Probability of Adverse Drug Reactions. Clin. Pharmacol. Ther. **1981**, *30*, 239–245.
45. Karch, F.E.; Lasagna, L. Adverse Drug Reactions: A Critical Review. JAMA **1975**, *234*, 1236–1241.
46. Rawlins, M.D. Adverse Reactions to Drugs. Br. Med. J. **1981**, *82*, 974–976.
47. American Society of Health System Pharmacists. Top-Priority Actions for Preventing Adverse Drug Events in Hospitals: "Recommendations of An Expert Panel.". Am. J. Health-Syst. Pharm. **1996**, *53*, 747–751.

Agency for Healthcare Research and Quality

William Campbell
Betsy L. Sleath
University of North Carolina, Chapel Hill, North Carolina, U.S.A.

Lynn Bosco
Agency for Healthcare Research and Quality, Rockville, Maryland, U.S.A.

INTRODUCTION

As early as 1965, when Medicare and Medicaid became law, it was recognized that research and evaluation would be necessary to guide the progress of these programs. Because Medicare and Medicaid were born as amendments to the Social Security Act, the Office of Research in the Social Security Administration provided modest support for research to address important issues. Under the rubric "health services research" the effort was expanded, first as the National Center for Health Services Research (NCHSR), and later as the Agency for Healthcare Policy and Research (AHCPR). The legislature that established AHCPR described its mission as:

> The purpose of the Agency is to enhance the quality, appropriateness, and effectiveness of healthcare services, and access to such services, through the establishment of a broad base of scientific research and through the promotion of improvements in clinical practice and in the organization, financing, and delivery of healthcare services.[1]

In 1999 the name was changed to the Agency for Healthcare Research and Quality, removing the word "policy" from the name. The new name clarified the mission of AHRQ by indicating it is to conduct and disseminate research that may be used by policymakers, but does not itself determine Federal healthcare policies and regulations. It is a scientific research organization; it is not a policy-setting organization. Further, adding the word "quality" to the name established that AHRQ is the lead federal agency on quality of healthcare research. This responsibility includes coordinating all federal quality improvement efforts and health services.

MISSION

The agency supports, conducts, and disseminates research that improves access to care and outcomes of care, as well as the cost, quality, and utilization of healthcare services. Succinctly stated, AHRQ's mandate is to sponsor research that provides better information and enables better decisions about healthcare. In order to accomplish this mission, the agency has three strategic goals:[2]

1. Support improvements in health outcomes.
2. Strengthen quality measurement and improvement.
3. Identify strategies to improve access, foster appropriate use, and reduce unnecessary expenditures.

These goals are pursued and measured using the following definitions:

Health outcomes research examines the end results of the structures and processes employed in delivery of care. An important consideration in this research is the patient's perspective, as well as the public and private-sector policymakers who are concerned with the impact of their investment in healthcare.

Quality measurement and improvement requires developing and testing quality measures and investigating the best ways to collect, compare, and communicate these data so they are useful to decision makers. AHRQ's research emphasizes studies of the most effective ways to implement these measures and strategies in order to improve patient safety and healthcare quality.

Improving access, fostering appropriate use, and reducing unnecessary expenditures continues to be a challenge for the poor, the uninsured, minority groups, rural and inner city residents, and other priority populations. The agency supports studies of access, healthcare utilization, and expenditures to identify whether particular approaches to healthcare delivery and payment alter behaviors in ways that promote access and/or economize on healthcare resource use.

Four specific areas of research were mandated by Congress in 1999 and have been adopted by AHRQ as

Encyclopedia of Clinical Pharmacy
DOI: 10.1081/E-ECP 120006178

current research priorities. Research grants or contracts awarded to investigators in the foreseeable future will likely fall within these areas.[3]

Improve the Quality of Healthcare

AHRQ is to coordinate, conduct, and support research, demonstrations, and evaluations related to the measurement and improvement of healthcare quality. AHRQ is also to disseminate scientific findings about what works best and facilitate public access to information on the quality of, and consumer satisfaction with, healthcare.

Promote Safety and Reduce Medical Errors

AHRQ will develop research and build partnerships with heath care practitioners and healthcare systems, and establish a permanent program of Centers for Education and Research in Therapeutics (CERTs). These initiatives will help address concerns raised in a 1999 report by the Institute of Medicine (IOM) that estimates as many as 98,000 patients die as a result of medical errors in hospitals each year.[4]

Advance the Use of Information Technology for Coordinating Patient Care and Conducting Quality and Outcomes Research

AHRQ will promote the use of information systems to develop and disseminate performance measures, create effective linkages between health information sources to enhance healthcare delivery and coordinate evidence-based healthcare services, and promote protection of individually identifiable patient information used in health services research and healthcare quality improvement.

Establish an Office of Priority Populations

The needs of low-income groups, minorities, women, children, the elderly, and individuals with special healthcare needs will be addressed through the agency's intramural and extramural research portfolio.

SPECIAL INITIATIVES

AHRQ has several initiatives that should be of particular interest to the clinical pharmacy community. These include the Centers for Education and Research in Therapeutics (CERTs), the Evidence Based Practice Centers (EPCs), National Guidelines Clearinghouse (NGC), and a coordinated set of activities with the goal of Translating Research Into Practice (TRIP).

Centers for Education and Research in Therapeutics (CERTs)

In 1994 Woosley raised concern about the quality and quantity of prescription drug information available to physicians and other practitioners.[5] He contrasted the billions of dollars available from commercial interests to promote prescribing and use of (primarily new) drugs, with the limited funds available to help practitioners select cost-effective therapeuties. What was missing was a balance between commercially driven information and nonproprietary information, a vacuum that Woosley proposed would be filled by CERTs. The conceptual basis for CERTs was that of an academic entity capable of striking a balance between the relative abundance of pharmaceutical industry-generated information, and the relative paucity of NIH- or FDA-generated information available to practitioners.

Congress recognized the importance of the CERTs concept and directed the formation of the CERTs in Section 409 of the Food and Drug Modernization Act (FDAMA) of 1997 and authorized AHRQ to establish CERTs as a demonstration effort. Congress intended CERTs to have a dual mission of conducting essential research not otherwise performed by the pharmaceutical industry, and to communicate to practitioners information concerning the most effective, safest, and least-expensive therapies. Each Center was to have a focus based upon a therapeutic area of interest and a defined population, leading to a national network with complementary resources and interests.

In October, 1999 AHRQ awarded funds to four CERTs, and soon followed with awards for three additional CERTs. The seven CERTs and their areas of interest are:

Duke University	Improving Prescribing for Cardiovascular Illness
Georgetown University	Preventing Drug–Drug Interactions in Women
Harvard University	Demonstration of Implementation of Improved Prescribing Practices in an Integrated Network of HMOs
University of Alabama at Birmingham	Improving Drug Therapy for Musculo-skeletal Disorders
University of North Carolina at Chapel Hill	Rational Drug Therapy for the Pediatric Population
University of Pennsylvania	Applying Pharmaco-epidemiological Methods to Improved Prescribing
Vanderbilt University	Comparison of Therapeutic Effectiveness of Selected Drugs in the TennCare System

The CERTs legislation has been transferred from FDAMA to AHRQ, and is now a permanently authorized program. The CERTs network is expected to expand, both through addition of new Centers as well as establishment of collaborations with investigators and practitioners throughout the country. Additional information is available at http://www.certs.hhs.gov.

Evidence-Based Practice Centers (EPCs)

The philosophy of evidence-based practice is widely accepted, although operational and implementation issues represent major barriers. One of the significant barriers is a shortage of evidence reports on topics of critical interest, and the lack of a national infrastructure to prepare such reports. In response to this need, AHRQ has funded 12 Evidence-based Practice Centers to conduct systematic, comprehensive analyses and syntheses of the scientific literature to develop evidence reports and technology assessments on clinical topics that are common, expensive, and present challenges to decision makers. Since December 1998, 11 evidence reports have been released on topics that include sleep apnea, traumatic brain injury, alcohol dependence, cervical cytology, urinary tract infection, depression, dysphasia, sinusitis, stable angina, testosterone suppression, and attention deficit hyperactivity disorder.

Pharmacotherapy is a significant interest within the EPCs, and AHRQ welcomes partners such as specialty societies and health systems to submit topics for evidence reports, participate with the EPC's in preparing reports, and most importantly to use the findings of EPC's to develop tools and materials that will improve the quality of care.

National Guidelines Clearinghouse (NGC)

Developed in partnership with the American Medical Association and the American Association of Health Plans, the NGC is a Web-based resource for information on evidence-based clinical practice guidelines. The NGC began providing online access to guidelines at http://www.guideline.gov in 1998. Since becoming fully operational, the site receives over 100,000 visits each month. The site provides information to help healthcare professionals and health system leaders select appropriate treatment recommendations by providing full text or an abstract of the recommendations, by comparing and evaluating different recommendations, and by describing how they were developed. Because almost all guidelines include some consideration of pharmacotherapy, and the NGP should be regarded as an invaluable resource to clinical practitioners, clinical investigators, educators, and others from the pharmacy community.

Translating Research Into Practice (TRIP)

One of the most pressing challenges in healthcare is to apply the knowledge that is currently available; in other words, to close the gap between knowledge and practice. The first round of TRIP initiatives supported development and implementation of evidence-based tools into diverse healthcare settings. Translational efforts included cost-effective approaches to implement smoking cessation, chlamydia screening of adolescents, diabetes care in medically underserved areas, and treatment of respiratory distress syndrome in preterm infants. The second round of TRIP initiatives (funded in 2000) focused on continued development of partnerships between researcher and healthcare systems and organizations (e.g., integrated health service delivery systems, academic health systems, purchaser groups, managed care programs including health maintenance organizations, practice networks, worksite clinics) to help accelerate and magnify the impact of practice-based, patient outcome research in applied settings.

Databases

AHRQ achieves its mission through a combination of efforts, described as a ''research pipeline.'' This pipeline of activities builds the infrastructure, tools, and knowledge for improvements in the American healthcare system. An important part of that pipeline for pharmacy is the maintenance of public use databases that can help identify problems and formulate solutions to improve pharmacotherapy. One database of particular interest is the Medical Expenditure Panel Survey (MEPS) which provides up-to-date, highly detailed information on how Americans as a group, as well as segments of the population, use and pay for healthcare. This ongoing survey of about 10,000 households and 24,000 individuals also studies insurance coverage and other factors related to access to healthcare. AHRQ encourages investigators to write applications that analyze the MEPS data.

CONCLUSION

Today's AHRQ has an annual budget of $270 million for fiscal year 2001, with approximately 80% awarded as grants and contracts to researchers at universities and other institutions across the country. The remaining 20% is allocated to intramural research and administrative support. The agency is administratively located within the

Department of Health and Human Services (HHS), where it reports to the Secretary of HHS through the Undersecretary for Health. Virtually every topic of interest to AHRQ has a pharmacotherapy component, and virtually every topic of interest to the pharmacotherapy community fits within a research priority of AHRQ.

New research investigators might first consider applying for a small grant which funds up to $100,000 total costs. Investigators who have a clinical degree or a research doctoral degree and who are no more than five years out of their latest research training experience might consider applying for an Independent Scientist Award (K02). Individuals with a clinical doctoral degree, who have identified a mentor with extensive research experience, and are willing to spend a minimum of 75% of full-time professional effort conducting research and developing a research career during the award might consider applying for a Mentored Clinical Scientist Award (K08). The agency also sponsors dissertation grants for students working on their doctoral degrees.

Opportunities are numerous for collaboration between AHRQ and pharmacotherapy investigators, educators, practitioners, and administrators. Potential collaborators are encouraged to contact AHRQ to initiate discussions on topics of interest. Two agency publications are particularly useful in describing researchable questions and methodologies: ''The Outcome of Outcomes Research at AHCPR'' and ''Greatest Hits of Outcomes Research at AHCPR.''[6,7] The agency is committed to a research agenda that is ''user driven'' and welcomes contacts from the pharmacotherapy community on topics of healthcare quality, cost, and effectiveness.

REFERENCES

1. *The Omnibus Budget Reconciliation Act of 1989 (Public Law 101–239), Part A, Section 901 (b).*
2. Eisenberg, J.M. Health services research in a market-oriented healthcare system. Health Aff. **1998**, *17* (1), 98–108.
3. Kohn, L.T.; Corrigan, J.M.; Donaldson, M.S. *To Err is Human: Building a Safer Health System*; Kohn, L.T., Corrigan, J.M., Davidson, M.S., Eds.; Institute of Medicine, National Academy Press: Washington, DC, 1999.
4. The President's Advisory Commission on Consumer Protection and Quality in the Healthcare Industry. In *Quality First: Better Healthcare for All Americans. Final Report to the President of the United States*; U.S. Government Printing Office: Washington, DC, March 1998 (GPO 017-012-00396-6).
5. Woosley, R.L. Centers for education and research in therapeutics. Clin. Pharmacol. Ther. **1994**, *55*, 249–255.
6. Tunis, S.; Stryer, D. *The Outcome of Outcomes Research at AHCPR*; March 1999 (AHCPR Publication # 99-R044).
7. AHRQ. *Selected ''Greatest Hits'' of Outcomes Research at AHCPR*; March, 1999 (AHCPR Publication #99-R043).

Ambulatory Care/Primary Care, Clinical Pharmacy Careers in

Dave Hachey
Idaho State University, Pocatello, Idaho, U.S.A.

INTRODUCTION

Community pharmacy practitioners have provided ambulatory care services to their customers for years. However, more students who graduated with advanced degrees since the 1980s have moved from the traditional dispensing role to providing direct ambulatory care patient services. Pursuing this patient care role in ambulatory care and primary care settings has increased job opportunities, positioned pharmacists in patient care areas, and changed the expectations and duties of pharmacists. This growth in clinical pharmacy careers has also pushed recent graduates and those seeking employment in this arena to pursue further training and education.

Integration of pharmacists with various disciplines of medicine offers many benefits to the health care system and the patient, including lower costs and improved health outcomes.[1–3] Pharmacists have also ventured away from the team approach to become more independent, which has lead to the same benefits of improved health outcomes and cost savings.[2] Whether integrated into team approach, or operating independently, pharmacists working in ambulatory and primary care have evolved slowly. This chapter discusses these various careers, including typical work settings and job activities. The type of degree, training, salary, and experience, in addition to long-term growth potential, is also discussed. Finally, to give more insight as to what these clinical sites might be like, descriptions of various sites are given.

JOB ACTIVITIES, RANGE OF CAREERS, AND WORK SETTINGS

The job activities, range of careers, and work settings for ambulatory care pharmacists vary as much as disease states. However, most duties required of ambulatory care pharmacists incorporate three to four components. These include clinical, distributive, and administrative duties, and sometimes a teaching or academic role.[4]

To expand on these job activities, a survey of pharmacists who work in ambulatory care positions was conducted, and it was reported that 45% of the pharmacist's time was spent performing distributive functions, while 30% clinical and 21% of the pharmacist's time was spent performing administrative activities. Distributive functions of these pharmacists may be defined as filling and dispensing prescriptions, as well preparing intravenous medications. The clinical portion of ambulatory care pharmacists includes a variety of activities, such as monitoring patient outcomes and compliance, conducting specialized clinics, providing therapeutic drug monitoring services and, in some settings, practicing independently with prescriptive authority. Pharmacists who have been included on multidisciplinary care teams or work at teaching hospitals may have a responsibility to teach students, residents, and fellows about various aspects of drug and disease state management.

Responses of this survey also indicated that approximately one-half of the pharmacists worked in conjunction with a physician or a nurse on an interdisciplinary ambulatory care team and that the medical staff and senior management were very supportive of having pharmacists on these teams.

The aforementioned specialty clinics allow pharmacists to have an enormous impact on patient care and a variety of career opportunities. Pharmacists may be expected to participate on a team in the multidisciplinary approach to patient care, or may need to act independently in disease-specific clinics. The Veterans Affairs Medical Centers (VAMCs) have been leaders in the area of a multidisciplinary approach to health care, and pharmacists have been involved on these teams for a long time. In a study conducted to determine the in-

Encyclopedia of Clinical Pharmacy
DOI: 10.1081/E-ECP 120006261

volvement of clinical pharmacy services in 50 of the VAMCs' specialty ambulatory clinics, it was found that 310 of the 401 (77%) specialty clinics were staffed with a clinical pharmacist and 144 (36%) were managed by pharmacists. These clinics covered a variety of disease states, including congestive heart failure, anticoagulation, lipid management, geriatrics, diabetes, and therapeutic drug monitoring.[2,5]

The work settings for pharmacists in ambulatory and primary care clinics also vary. More recently, for example, pharmacists can be found in private physician offices or large teaching hospitals. Anywhere ambulatory care is being provided by physicians, nurse practitioners, or physician assistants, there is opportunity for pharmacist involvement.

DEGREE, TRAINING, EXPERIENCE, AND SALARY RANGE

Since the push and support for an entry-level Doctor of Pharmacy degree by the American Association of Colleges of Pharmacy in 1992, the majority of colleges of pharmacy in the United States have been converting from the Bachelor of Science degree. One of the goals in switching to the Doctor of Pharmacy degree is for the colleges of pharmacy to produce patient care providers rather than medication dispensers. Because providing patient care is one of the major activities of ambulatory care pharmacists, the majority of graduates have come from an entry-level Doctor of Pharmacy program. Others, however, have returned to school for additional education to gain the knowledge needed to move into the clinical setting and manage the diversity of disease states.

For many ambulatory care pharmacists, training does not end at graduation with acceptance of the degree. Additional training in residency or fellowship for 1 to 2 years is sometimes completed to obtain more clinical experience in patient care as well as to develop a deeper knowledge base. In a 1995 survey of pharmacists practicing in an ambulatory care setting, 67% of the 99 respondents indicated that they had residency training and 21% had fellowship training. Forty-six percent of respondents also specified that they had received board certification.[6]

New graduates who select a career in ambulatory care pharmacy may decide to complete a 1-year general pharmacy practice residency program or to choose a specialized residency in ambulatory care or primary care. The invaluable experience gained in residency programs provides guidance and practical training to pharmacists who are seeking more education and skills to provide patient care. These programs are offered in a variety of work settings from VAMCs and large teaching hospitals to smaller family medicine groups and community pharmacies.

Although the majority of ambulatory care pharmacists have chosen the route of the Doctor of Pharmacy degree and residency, it is not the only course to becoming an ambulatory care pharmacist. Some pharmacists who have been practicing several years have grown and established positions in ambulatory care without residency or fellowship training. However, many institutions require that they have continuing education to practice in an ambulatory care setting with a team or independently, and one means of continuing education is through certificate programs. Many certificate programs that are available will teach specific disease state management such as anticoagulation, diabetes, or asthma care. However, other certificate programs may be more inclusive, covering a broader spectrum of ambulatory care.[7]

Salary range for pharmacists practicing in an ambulatory care setting varies depending on geographic region, years in the work force, and board certification status. However, the median salary in 1995 was $53,500 (average, $55,861; range, $35,000–$90,000), with a higher salary reflective of more years employed.[6]

GROWTH AND LONG-TERM OPPORTUNITIES

Since the 1990s, clinicians in the fields of ambulatory care and primary care have embraced pharmacists as colleagues and as an invaluable source of information. This acceptance has lead to an increase in demand of pharmacists in the ambulatory care arena in several capacities. First, the educational system has experienced the need to increase the education of students in this area, thus producing more students that choose paths in ambulatory care. Second, pharmacists have been dedicating themselves to improving patient outcomes in primary and ambulatory care, which has lead to a tremendous growth and need for pharmacists in this area.

As it is evident that this field is growing, the question arises as to the longevity of these positions. Many pharmacists that are practicing in ambulatory care have created their own positions. Since the mid-1980s and early 1990s, many of these pharmacists have moved from dispensing to the clinical role. Therefore, pharmacists

who have been in ambulatory care several years are some of the first to experience this long-term job stability. However, as the goals for health care continue to move toward more cost-effective ways to administer better health care, pharmacists continue to prove themselves to fit this equation. Thus, pharmacists will most likely continue to be in these positions for quite some time.

SITE DESCRIPTION

One benefit of practicing as an ambulatory care pharmacist is that there are a variety of practice settings. These practice sites vary from physician office buildings to physician residency training programs, as well as large hospitals and retail pharmacies.

One example is that of a private physician's office. Studies have been performed to determine the impact of having a pharmacist providing pharmaceutical care in a physician's office.[8] In this scenario, pharmacists usually have unlimited access to patient information and may have their own office or exam room to evaluate and educate patients. The pharmacist may see these patients independently of the physician or evaluate the patient for pharmaceutical issues before the physician sees them. Other services may be available at these offices such as a laboratory or radiological services, depending on size and specialty of the office.

Another model that may be used to integrate pharmacists into ambulatory care settings is that of a university-based family practice center or residency training program.[9] In this setting, the pharmacist typically works at a larger physician training program and teaching clinic. The pharmacist usually has patient care duties such as specialty clinics, medication refill services, or other consultative services. However, in this environment, the pharmacists also have obligations to teach and evaluate the medical residents in both didactic and clinical situations. This type of position is sometimes affiliated with higher academic institutions, and clinical duties may need to be balanced with administrative, research, or teaching obligations.

Other traditional sites are changing the way that pharmacists see and educate patients. More hospitals are moving to outpatient treatment programs and becoming involved in the multidisciplinary approach to ambulatory care patients, and practice sites for pharmacists are moving from the central pharmacy to walk-in or ambulatory care clinics and even home care teams. These opportunities have allowed pharmacists who traditionally process orders and mix intravenous medications to become involved in the treatment decisions for patients.

Retail pharmacy is also making an effort to get pharmacists out from behind the counter by establishing a variety of clinics in the retail setting. Some pharmacists are providing pharmaceutical care, such as working in conjunction with physician offices to counsel newly diagnosed diabetic patients on the proper use of glucometers and insulin injections, whereas others work more independently, offering services in durable medical equipment, home infusion, and home oxygen.

The progress in ambulatory care has not always been a clear road. Many barriers have risen along the way that have prevented pharmacists from being accepted in the clinical community as a patient care provider. Some clinicians still view pharmacists as dispensers of medications and believe patient care is not within the scope of a pharmacists' practice. In some cases, this barrier has been surpassed in settings such as VAMCs and HMOs, where a capitated health care system is practiced. Pharmacists have saved these institutions money as well as improved health outcomes, acting not as dispensers of medication, but as clinicians. In addition, the pharmacy profession itself has in some respects inhibited its own growth. Large retail pharmacy chains whose salaries are significantly more than that of an ambulatory care pharmacist, absorb a large portion of graduates that may desire to pursue a career in ambulatory care but are attracted to a higher salary. However, the future for ambulatory care pharmacists appears brighter as legislative change is looking to recognize pharmacists as providers and reimburse them for providing pharmaceutical care.

CONCLUSION

As the need for change in the profession of pharmacy has evolved since the 1990s, pharmacy schools have responded by producing a well-rounded practitioner and provider of pharmaceutical care. More graduates are choosing to gain patient care skills and training in ambulatory care residency or fellowship programs, allowing them to be focused practitioners and teachers. The future for pharmacists in the area of ambulatory care looks bright as pharmacists lobby to be recognized as providers and to be reimbursed for providing pharmaceutical care. Finally, as pharmacists continue to demonstrate that their clinical services improve patient outcomes and decrease overall health care costs, jobs in the ambulatory care setting will continue to expand to other venues, allowing pharmacists to accomplish what they were trained to do.

REFERENCES

1. McMullin, S.T.; Hennenfent, J.A.; Rithie, D.J.; Huey, W.Y.; Lonergan, T.P.; Schaiff, R.A.; Tonn, M.E.; Bailey, T.C. A prospective, randomized trial to assess the cost impact of pharmacist-initiated interventions. Arch. Intern. Med. **1999**, *159*, 2306–2309.
2. Gattis, W.A.; Hasselblad, V.; Whellan, D.J.; O'Connor, C.M. Reduction in heart failure events by the addition of a clinical pharmacist to the heart failure management team: Results of the Pharmacist in Heart Failure Assessment Recommendation and Monitoring (PHARM) Study. Arch. Intern. Med. **1999**, *159*, 1939–1945.
3. Chiquette, E.; Amato, M.G.; Bussey, H.I. Comparison of an antiocogulation clinic with usual care. Arch. Intern. Med. **1998**, *158*, 1641–1647.
4. Reeder, C.E.; Kozma, C.M.; O'Malley, C. ASHP Survey of ambulatory care responsibilities of pharmacists in integrated health systems-1997. Am. J. Health-Syst. Pharm. **1998**, *55*, 35–43.
5. Carter, B.L. Clinical pharmacy in disease specific clinics. Pharmacotherapy **2000**, *20* (10 Pt 2), 273s–277s.
6. Anastasio, G.D.; Shaughnessy, A.F. Salary survey of ambulatory care clinical pharmacists. Pharmacotherapy **1997**, *17* (3), 565–568.
7. Jannsen, R.K.; Murphay, C.M.; Kendzierski, D.L.; Brown, D.H.; Carter, B.L.; Furmaga, E.M.; Schoen, M.D.; Woker, D.R. Ambulatory care certificate program for pharmacists. Am. J. Health-Syst. Pharm. **1996**, *53*, 1018–1023.
8. Campbell, R.K.; Saulie, B.A. Providing pharmaceutical care in a physician office. J. Am. Pharm. Assoc. **1998**, *38*, 495–499.
9. Lilley, S.H.; Cummings, D.M.; Whitley, C.W.; Pippin, H.J. Intergration of a pharmacotherapy clinic in a university-based family practice center. Hosp. Pharm. **1998**, *33*, 1105–1110.

American College of Clinical Pharmacy (ACCP)

Robert M. Elenbaas
*American College of Clinical Pharmacy,
Kansas City, Missouri, U.S.A.*

INTRODUCTION

The American College of Clinical Pharmacy (ACCP) was founded in 1979 when 29 clinical pharmacists—organized largely by Donald C. McLeod, M.S.—gathered in Kansas City, Missouri, with a common goal: to promote the rational use of medications in society by forming an organization dedicated to and focused on advancing the cutting-edge of clinical pharmacy practice and research. Those ideals held by ACCP's founding members still find themselves in the College's mission:

> ACCP is a professional and scientific society that provides leadership, education, advocacy, and other resources enabling clinical pharmacists to achieve excellence in practice and research.

ORGANIZATION

Only two years after its founding, ACCP created its Research Institute in 1981 to advance pharmacotherapy through support and promotion of research, training, and educational programs. Through 2000, this has largely taken the form of a number of Research Awards that support specific research projects conducted by College members in a variety of therapeutic areas, and Fellowships that provide for the stipends of postgraduate clinical pharmacists in an intensive research training experience. Both types of programs are available to ACCP members on a competitive basis.

Also in 1981, Russell R. Miller, Ph.D., founded the journal *Pharmacotherapy* as a publication dedicated to human pharmacology and drug therapy. When first established, *Pharmacotherapy* was not affiliated with any medical or pharmacy associations. In 1988, ACCP adopted *Pharmacotherapy* as its official journal, and in 1994, ACCP acquired the journal. Now a monthly publication, *Pharmacotherapy* publishes a complementary array of original clinical research and evidence-based reviews in the broad field of pharmacotherapy and clinical pharmacology.

Membership

As of the end of 2001, ACCP had approximately 7000 members, located mostly in the United States and Canada. ACCP members can be found in all practice venues, including ambulatory clinics and community pharmacies, community hospitals, the pharmaceutical industry, pharmacy and medical school faculties, university hospitals, and VA and military hospitals. More than 80% of ACCP members hold the Pharm.D. degree, 70% have completed a postgraduate residency, and 25% have completed a research fellowship training program. Approximately 25% of ACCP members are certified in one or more of the specialty practice areas recognized by the Board of Pharmaceutical Specialties (BPS; i.e., Nuclear Pharmacy, Nutrition Support, Oncology, Pharmacotherapy, Psychiatry). Consistent with one of ACCP's founding tenets—to promote the rational use of medications in society—College members directly assume responsibility for the drug therapy of individual patients through collaborative practice agreements with physicians; regularly consult with and advise physicians, other health professionals, and patients regarding drug therapy; serve on key institutional or other committees that oversee the medication use process; and teach pharmacy or other health profession students. In addition, many ACCP members are responsible for conducting basic, clinical, health services, economic, or other applied research. This research, for example, may involve clinical trials of new drug entities, pharmacokinetic and pharmacodynamic studies in normal volunteers and patients, pharmacoeconomic evaluations of drug therapies, and health services research to examine the impact of pharmacy services.

The practice and research interests of ACCP members span the broad array of pharmacotherapy. As one way to provide for the unique needs of clinical pharmacists with diverse interests, ACCP currently includes approximately 20 Practice and Research Networks (PRNs). The PRNs form special interest groups within the College in areas ranging from Ambulatory Care to Infectious Diseases to Women's Health.

Encyclopedia of Clinical Pharmacy
DOI: 10.1081/E-ECP 120006306

MAJOR PROGRAMS

Education

ACCP holds three major scientific and/or educational meetings each year. The ACCP Annual Meeting, held in late-October or early-November, and the Spring Practice and Research Forum, held in April, include a variety of educational symposia as well as poster or platform presentations of original research. Both meetings include educational and networking sessions conducted by the College's PRNs. The ACCP Recruitment Forum takes place at the Annual Meeting and provides an opportunity for employers and prospective applicants to interview. Recruitment On-Line, a year-round job listing service, is available on the College's web site.

Each year, ACCP also conducts its "Updates in Therapeutics," designed as both a comprehensive review of therapeutics and as a preparatory course for clinical pharmacists planning to sit for BPS specialty certification in Pharmacotherapy, Nutrition Support, Oncology, or Psychiatry. ACCP is expanding its use of technology to facilitate distance learning. ACCP educational programs will be increasingly available through the College's web site at www.accp.com.

In April 1999, ACCP partnered with the European Society of Clinical Pharmacy (ESCP) to co-host the first International Congress on Clinical Pharmacy in Orlando, Florida. With a theme of "Documenting the Value of Clinical Pharmacy Services," the Congress was attended by more than 1300 pharmacists from 51 countries.[1] ACCP and ESCP plan to organize a second International Congress in 2004.

Publications

In addition to *Pharmacotherapy*, publications produced by ACCP include the *Pharmacotherapy Self-Assessment Program* (PSAP) and the College's annual *Directory of Residencies and Fellowships*. In addition to its use as a general professional development tool, PSAP is approved by BPS for use by Board Certified Pharmacotherapy Specialists (BCPS) in obtaining their required recertification. With publication of its fourth edition (PSAP-IV) in 2001, this modular-based program is available in both hardcopy and Internet versions. The ACCP *Directory of Residencies and Fellowships* provides a comprehensive index and description of postgraduate training opportunities offered by ACCP members. It is published in the fall of each year to assist students and residents in their career development. Other publications available from ACCP are described on the College's web site.

Professional Leadership and Advocacy

ACCP participates in several coalitions with other national organizations, including the Council on Credentialing in Pharmacy, the Joint Commission of Pharmacy Practitioners, the Alliance for Pharmaceutical Care, and the Pharmaceutical Sciences Consortium. In general, ACCP's advocacy efforts are focused on the federal government, with the overall goal of better enabling clinical pharmacists to provide patient care and perform research.

A bibliography and reprints of ACCP white papers, position statements, and guidelines are available on the College's web site. These include two comprehensive reviews of published literature that document the value of clinical pharmacy services.[2,3]

GOVERNANCE

ACCP is governed by an 11-person Board of Regents, elected from and by the College's members. The President of the College serves as chair of the Board of Regents. Members of the 2001 Board of Regents include:

- President: Barry L. Carter, Pharm.D., FCCP, BCPS
- President-Elect: Bradley A. Boucher, Pharm.D., FCCP, BCPS
- Past President: Thomas C. Hardin, Pharm.D., FCCP, BCPS
- Secretary: J. Herbert Patterson, Pharm.D., FCCP, BCPS
- Treasurer: Marsha A. Raebel, Pharm.D., FCCP, BCPS
- Regents: Betty J. Dong, Pharm.D.; Julie A. Johnson, Pharm.D., FCCP, BCPS; Mary Lee, Pharm.D., FCCP, BCPS; Michael Maddux, Pharm.D., FCCP; Ralph H. Raasch, Pharm.D., FCCP, BCPS; and David R. Rush, Pharm.D., BCPS

REFERENCES

1. Proceedings of the First International Congress on Clinical Pharmacy. Pharmacotherapy **2000**, *20*, 233S–346S.
2. Willett, M.S.; Bertch, K.E.; Rich, D.S.; Ereshefsky, L. Prospectus on the economic value of clinical pharmacy services. Pharmacotherapy **1989**, *9*, 45–56.
3. Schumock, G.T.; Meek, P.D.; Ploetz, P.A.; Vermeulen, L.C. Economic evaluations of clinical pharmacy services—1988–1995. Pharmacotherapy **1996**, *16*, 1188–1208.

PROFESSIONAL ORGANIZATIONS

American Council on Pharmaceutical Education

A

Dawn G. Zarembski
Peter H. Vlasses
American Council on Pharmaceutical Education,
Chicago, Illinois, U.S.A.

INTRODUCTION

The American Council on Pharmaceutical Education (ACPE), located at 311 W. Superior Street, Suite 512, Chicago, Illinois 60610-3537, is the national agency for accreditation of professional degree programs in pharmacy and providers of continuing pharmaceutical education including certificate programs in pharmacy. The ACPE was established in 1932 for accreditation of preservice (entry-level) pharmacy education. Accreditation standards reflect professional and educational qualities identified by ACPE as essential to quality professional programs at colleges and schools of pharmacy. Standards are established through a comprehensive and broadly based procedure that provides opportunities for contributions from the community of interests affected by the accreditation process.

OVERVIEW

Accreditation of Professional Degree Programs

The first ACPE standards, developed between 1932 and 1937, called for sweeping changes in pharmaceutical education. These standards required the completion of a four-year course of study in order to attain the baccalaureate degree in pharmacy. The standards were published in 1937 and were subsequently revised during the 1940s and early 1950s. In the 1960s, revision of the standards led to the incorporation of a five-year baccalaureate in pharmacy program and a doctor of pharmacy program which involved a four-year professional program that was preceded by two years of preprofessional studies. In the mid-1970s, standards for two separate entry-level, professional programs, the baccalaureate in pharmacy program and the doctor of pharmacy program were developed. In addition, professional practice experiences were incorporated into the curriculum for the first time. The revisions of the 1980s expanded upon curricular expectations for the doctor of pharmacy program leading to greater emphasis on curricular and programmatic outcomes. In 1989, ACPE issued its intention to propose new standards that would merge the two programmatic standards with a focus on a doctor of pharmacy program. The new accreditation standards and guidelines for the professional program in pharmacy leading to the Doctor of Pharmacy degree (Standards 2000) were adopted June 14, 1997. Implementation Procedures for these new accreditation standards became effective on July 1, 2000, in accord with a stated transition period.

Continuing Education Accreditation Program

Requirements for continued pharmaceutical education began in the early 1970s, when State Boards of Pharmacy began requiring licensed pharmacists to participate in continuing pharmaceutical education activities. Between 1972 and 1974, the American Association of Colleges of Pharmacy and the American Pharmaceutical Association convened a task force to discuss issues related to the continued competence in pharmacy practice. In 1974, the Board of the American Pharmaceutical Association recommended that ACPE initiate accreditation of continuing pharmaceutical education. Consequently, ACPE began the accreditation of providers of continuing pharmaceutical education in 1975. Participation in continuing education programs earned through an ACPE-accredited provider is accepted for licensure renewal by all state boards of pharmacy requiring continuing education. The symbol used by the ACPE to designate that a continuing education provider is accredited is

In 1998, the profession charged ACPE to develop standards for certificate programs. A certificate program is a structured continuing education experience that is narrower in focus and shorter in duration than a degree

DOI: 10.1081/E-ECP 120006199

program. Certificate programs are designed to instill, expand, or enhance practice competencies through the systematic acquisition of specified knowledge, skills, attitudes, and behaviors. In June 1999, Standards and Quality Assurance Procedures for Providers Offering Certificate Programs in Pharmacy were adopted. As of Fall 2000, 30 providers had been accredited to provider certificate programs. The symbol used by the ACPE to designate that a certificate training program is provided by an accredited provider is

Of note, the standards originally developed for the accreditation of providers of continuing pharmaceutical education were revalidated during development of the standards pertaining to providers of certificate programs. In addition, the new term ''statements of credit'' should be when documenting completion of a continuing education activity provided by an ACPE-accredited provider. The term ''certificates of credit'' should be reserved for use in conjunction with ACPE-accredited certificate programs only.

Annually, or more frequently if necessary, the ACPE publishes the Directory of Accredited Doctor of Pharmacy Programs of Colleges and Schools of Pharmacy and the Directory of Accredited Providers of Continuing Pharmaceutical Education.

The Pharmacists' Learning Assistance Network (P.L.A.N.®) is an information service developed by the ACPE to allow pharmacists to access information on continuing education programs. The P.L.A.N. service, which is operated by ACPE, maintains a database on all continuing pharmaceutical education programs offered by ACPE-approved providers. Pharmacists may request a computer search of continuing pharmaceutical education programs to suit their learning needs or conduct their own search on ACPE's web site.

ORGANIZATIONAL STRUCTURE AND GOVERNANCE

The Council is an autonomous and independent agency whose 10-member Board of Directors is derived through the American Association of Colleges of Pharmacy, the American Pharmaceutical Association, the National Association of Boards of Pharmacy (three appointments each), and the American Council on Education (one appointment). These organizations are not members of the ACPE, and appointees to the Board of Directors are not delegates of these organizations. The organizational structure of ACPE assures the integrity of the accreditation program through responsive, responsible, and independent operation. The Board of Directors has authority for management of corporate affairs and is responsible for establishing policies and procedures, setting standards for accreditation of professional programs of colleges and schools of pharmacy, establishing standards for accreditation of providers of continuing education, including certificate programs in pharmacy, and taking actions concerning accreditation. A Public Interest Panel serves in an advisory capacity. The ACE appointee and the Public Interest Panel assure a public perspective in policy- and decision-making processes.

Dr. Daniel A. Nona served as the Executive Director from 1975–2000. Dr. Peter H. Vlasses assumed the Executive Director position effective January 1, 2000.

MISSION

ACPE is organized for the purpose of promoting and encouraging educational, research, and scientific activities. The ACPE formulates educational, research, and scientific standards which an accredited professional program of a college or school of pharmacy or an accredited provider of continuing education will be expected to meet and maintain. The essential purpose of the professional degree program accreditation process is to provide a professional judgment of the quality of a college or school of pharmacy's professional program(s) and to encourage continued improvement thereof. Accreditation concerns itself with quality assurance and quality enhancement. The responsibilities of the ACPE's professional degree accreditation program are as follows:

- To advance the standards of pharmaceutical education in the United States and associated commonwealths.
- To formulate the educational, scientific, and professional principles and standards for professional programs in pharmacy which a college or school of pharmacy is expected to meet and maintain for accreditation of its programs, and to revise these principles and standards when deemed necessary or advisable.
- To formulate policies and procedures for the accreditation process.
- To evaluate the professional program(s) of any college or school of pharmacy within or beyond its national geographic scope that requests accreditation of its program(s).
- To publish a directory of accredited professional programs of colleges and schools of pharmacy for the use of state boards of pharmacy or appropriate

state licensing agencies in pharmacy, other interested agencies, and the public, and to revise such directory annually or as frequently as deemed desirable.
- To provide assurances to constituencies that the professional programs that have been accredited continue to comply with standards, and therefore, to conduct periodic evaluations in a manner similar to that for original accreditation.
- To assist the advancement and improvement of pharmacy education as well as prerequisites and procedures for licensure and to provide a basis for interinstitutional relationships.

The ACPE's Continuing Education Provider Accreditation Program is designed to assure pharmacists, boards of pharmacy, and other members of pharmacy's community of interests, of the quality of continuing pharmaceutical education programs. The purposes of the Continuing Education Provider Accreditation Program are to:

- Assure and advance the quality of continuing pharmaceutical education, including Certificate Programs in Pharmacy, thereby assisting in the advancement of the practice of pharmacy.
- Establish standards for accredited providers of continuing pharmaceutical education, including accredited providers offering Certificate Programs in Pharmacy.
- Provide pharmacists with a dependable basis for selecting accredited continuing education experiences.
- Provide a basis for uniform acceptance of continuing education credits among states.
- Provide feedback to providers about their continuing education programs through periodic comprehensive reviews and ongoing monitoring activities with a need toward continuous improvement and strengthening.

CURRENT INITIATIVES

The new accreditation standards and guidelines for the professional program in pharmacy leading to the Doctor of Pharmacy degree (Standards 2000) were adopted June 14, 1997. Implementation procedures for these new accreditation standards became effective on July 1, 2000, in accord with a stated transition period.

ACPE is a member of the Council on Credentialing in Pharmacy, a coalition consisting of 11 national pharmacy organizations, which was founded in 1999 to provide leadership, standards, public information, and coordination for voluntary professional credentialing programs in pharmacy.

In April 2000, ACPE developed a Web-based survey as part of a strategic planning initiative. The purpose of the Web-based survey was to:

- Assess current awareness of ACPE within the profession.
- Assess current effectiveness of ACPE activities.
- Identify opportunities for ACPE process improvement.
- Receive feedback and opinions on how else ACPE can serve pharmacy.

Data obtained from the Web-based survey will be used to develop a new strategic plan targeted for review by the profession and for approval by January 30, 2001.

MEETINGS

The Board of Directors meets twice a year. A regular annual meeting of the Board of Directors is held in January, while a second annual meeting is held every June.

American Journal of Health-System Pharmacy (ASHP)

C. Richard Talley
American Society of Health-System Pharmacists, Bethesda, Maryland, U.S.A.

INTRODUCTION

The *American Journal of Health-System Pharmacy* (*AJHP*) is the official publication of the American Society of Health-System Pharmacists (ASHP). *AJHP*'s mission is to facilitate communication among members and subscribers and to create an archive of publications that both reflect and lead contemporary pharmacy practice.

OBJECTIVES

In addition to publishing a substantial body of peer-reviewed scientific papers, *AJHP* provides extensive, timely, in-depth news coverage of pharmacy issues. Numerous columns feature advice on management and therapeutic problems. Current content priorities include contemporary drug therapy issues; descriptions of practice innovations in acute care, long-term care, ambulatory care, home care, and managed care; and outcomes research, including pharmacoeconomics. Content is also guided by ASHP's Leadership Agenda, which currently emphasizes creating fail-safe medication use in health systems, advancing the pharmacist's role in patient care, fostering pharmacy practice leadership, accelerating the adoption of high-level pharmaceutical services for patients across the continuum of care, and assisting pharmacists in applying advances in electronic technology and science to the care of patients.

HISTORY

Leo Mossman edited the first issue, published in June 1943. The periodical was originally entitled the *Official Bulletin of the American Society of Hospital Pharmacists* and was mimeographed on beige paper. Harvey A.K. Whitney became coeditor in September 1943, and Mossman and Whitney continued monthly publication through December 1943. Donald E. Francke and Mossman coedited two issues, distributed in January and July of 1944, and Francke then assumed sole editorship, a responsibility he would hold for 22 years.

Because of financial difficulties in funding the production and mailing of the *Bulletin*, a decision was made to accept paid advertising to offset expenses. The first paid advertisement appeared in the January–February 1950 issue. The influx of advertising revenue allowed the use of commercial offset printing with this issue. Over the years, advertising income has far exceeded the production cost of *AJHP*. This margin has contributed to ASHP's ability to provide member services in excess of those possible through membership dues alone. Most other pharmacy association periodicals have been subsidized through other income sources.

By 1955 the *Bulletin* had achieved worldwide distribution. In 1958 the publication began monthly distribution and was renamed the *American Journal of Hospital Pharmacy*. When Francke resigned from ASHP in 1966 to pursue other publishing interests, George P. Provost succeeded him as editor and continued in that post until 1974. William A. Zellmer succeeded Provost as *AJHP*'s editor and provided a voice for pharmacy for more than 18 years in his widely read editorials. In 1992, when Zellmer's other ASHP obligations had expanded to the point where it was difficult for him to fulfill the responsibilities of *AJHP*'s editorship, C. Richard Talley succeeded to the post.

In 1982, in response to the growth of clinical pharmacy practice in hospitals and the increasing number of clinical papers being published in *AJHP*, ASHP created a new periodical entitled *Clinical Pharmacy*. This journal began as a bimonthly publication and attracted 7000 subscribers in its first year. It expanded to monthly distribution in 1986. ASHP's creation of this periodical is widely believed to have enhanced the growth of clinical pharmacy.

As the practice of clinical pharmacy became increasingly mainstream among ASHP members, more of them asked ASHP to provide the content of *Clinical Pharmacy* as a benefit to all members, not just subscribers. Through the persistent application of new publishing technology,

DOI: 10.1081/E-ECP 120006272

the *AJHP* and *Clinical Pharmacy* staff was able to reduce production costs dramatically. That enabled ASHP to merge *Clinical Pharmacy* into *AJHP* in 1994, creating pharmacy's only peer-reviewed scientific journal to be published 24 times each year. *AJHP* has published more than 2000 pages annually for the past 20 years.

AJHP (Coden: AHSPEK; ISSN: 1079-2082) is published by the American Society of Health-System Pharmacists twice a month, on the 1st and 15th. Circulation in 2001 was 37,870; the 2001 subscription rate was $195 for nonmembers (USA). A subscription is included as a benefit to ASHP members. Editorial offices are located at 7272 Wisconsin Avenue, Bethesda, Maryland 20184, U.S.A. (Telephone: 301-657-3000, ext. 1200; Fax: 301-664-8857; E-mail: ajhp@ashp.org).

AJHP is abstracted and indexed by all the major secondary sources (e.g., *International Pharmaceutical Abstracts*, *Biological Abstracts*, *Chemical Abstracts*, *Cumulative Index to Nursing and Allied Health Literature*, *Current Contents: Clinical Medicine*, *Current Contents: Life Sciences*, *Excerpta Medica*, *Index Medicus*, and the Iowa Drug Information Service.

In 1997 some components of *AJHP* began appearing on ASHP's Web site. Starting in September 1999, the full text of *AJHP* was posted. Members and other subscribers can now read and print type-quality copies of *AJHP* content about 10 days before an issue is mailed. Furthermore, important scientific findings that affect patient safety can be conveyed to the public and the media the instant they have finished undergoing the traditional peer-review, editing, and composition steps, saving weeks or months of delay in many cases. No other pharmacy organization in the world is currently doing this.

Enhancements of *AJHP*'s Web-based distribution are in the works. The searchability of past issues is being improved, and new procedures and mechanisms for authors and reviewers to use in electronically submitting manuscripts, letters, and other communications are being developed.

BIBLIOGRAPHY

www.ashp.org.

PROFESSIONAL RESOURCES

American Journal of Pharmaceutical Education

George H. Cocolas
University of North Carolina,
Chapel Hill, North Carolina, U.S.A.

INTRODUCTION

The *American Journal of Pharmaceutical Education* is a service publication for the community of pharmacy educators. It publishes articles as a means of disseminating information, methods, and techniques concerning pharmacy education.

AIMS AND SCOPE

The Journal chronicles the numerous changes that have occurred in pharmacy education since its launch issue in 1937. During the most recent 25 years, many of changes that have occurred in pharmacy education have been described in this Journal. Pharmacy programs have incorporated the advent of physical pharmacy, biopharmaceutics, and pharmacokinetics courses in their curriculums. Changes in the pharmaceutical sciences of social and administrative sciences as well as medicinal chemistry have impacted the teaching of pharmacotherapy to pharmacy students. In addition, the role of the pharmacist as a health professional in the management of disease states through drug therapy has seen the development and teaching of clinical pharmacy in the pharmacy curriculum. The inclusion of an experiential component to pharmacy education has created a vast emphasis on the application of the science of pharmacy to the delivery of care to the patient. Current issues of the Journal continue to describe pharmacy education and also include a section based on the Innovations in Teaching Competition sponsored by the AACP Council of Faculties and a section called ''Teachers' Topics'' that features course content presented by the best instructors selected by their school or college.

The Journal is published quarterly, from its office at 1426 Prince Street, Alexandria, Virginia 22314–2841 (phone: 703-739-2330; fax: 703-839-8982; www.aacp.org) under the editorship of George H. Cocolas, School of Pharmacy, Chapel Hill, North Carolina.

HISTORY

The Journal is the official publication of the American Association of Colleges of Pharmacy (AACP). Its purpose is to document pharmacy education and to advance it. The Journal originated in 1937 from the efforts of Rufus A. Lyman, dean of the College of Pharmacy at the University of Nebraska. AACP had published a *Proceedings* since 1900 but these once-a-year volumes were not a useful mechanism to energize pharmaceutical education. With the support of the Executive Committee of AACP, the Journal began its publication as a quarterly to describe pharmaceutical education in the schools and colleges in the United States. In 2002, the Journal reached its 66th year and remains a quarterly. Since 1991 the Journal has added a winter supplement to archive committee reports and minutes of meetings in one single issue.

FUTURE DIRECTIONS

Pharmacy education is ever changing. The mission of the Journal is to continue to document current teaching methodologies and studies about pharmacy education in schools and colleges of pharmacy.

The technology of the Web is becoming more evident in the publication of printed journals. This Journal, in the near future, will be offering its publication online.

Encyclopedia of Clinical Pharmacy
DOI: 10.1081/E-ECP 120006273

American Pharmaceutical Association

John A. Gans
American Pharmaceutical Association, Washington, D.C., U.S.A.

INTRODUCTION

The American Pharmaceutical Association (APhA) is the national professional society of pharmacists and is located at 2215 Constitution Avenue, NW, Washington, D.C. The main phone number is (202) 628-4410; the main fax line is (202) 783-2351; the primary web address is www.aphanet.org. APhA also hosts a site for consumers with the address www.pharmacyandyou.org. Since its founding, APhA has been a leader in the professional and scientific advancement of pharmacy, and in safeguarding the well-being of the individual patient.

HISTORY

The American Pharmaceutical Association was founded in Philadelphia, Pennsylvania, in 1852 by a group of pharmacists from across the young nation who were concerned about the quality of medicinal products and the standards of practice of those engaged in the apothecary, or pharmacy, trade.

The first permanent home for APhA was established through an act of Congress in 1932. Land on the national mall adjacent to the Lincoln Memorial in Washington, D.C., was identified as appropriate to be occupied by a national organization dedicated to the advancement of science and practice of pharmacy. To this day, APhA is the only nongovernmental organization with its operations on the national mall. The building itself was designed by noted architect John Russell Pope who, among other prominent buildings, designed the Jefferson Memorial and National Archives Building.

ORGANIZATIONAL STRUCTURE AND GOVERNANCE

A 19-member Board of Trustees elected by the membership governs APhA. The chief executive officer is John A. Gans, PharmD, who assumed his position as Executive Vice-President and a member of the APhA Board in 1989. The term of APhA President is three years (President-Elect, President, and Immediate Past-President). Nine trustees are elected for 3-year terms (three in each election cycle). The Board elects a Treasurer for a term of 3 years, and this individual serves with the presidential officers and another sitting Trustee on the APhA Executive Committee. Each of the three APhA Academy presidents serves a 1-year term as Trustee. The other two Trustees are the Speaker and Speaker-Elect of the APhA House. These officers are elected at the APhA Annual Meeting by delegates to the APhA House and serve for 2-year terms.

Policy for the Association, on issues important to the profession and the public health, is established by a nearly 400-member House of Delegates—the largest representative body of pharmacists in the United States. Eleven national pharmacy organizations have secured voting representation in the APhA House of Delegates and collaborate on the formation of policy for the Association and the profession. The Board of Trustees is empowered, if necessary, to make policy in the interim period between meetings of the House of Delegates and determines implementation strategies for House-adopted policies.

The membership of APhA exceeds 50,000 pharmacists, pharmacy students, pharmaceutical scientists, and pharmacy technicians. An active member must be licensed to practice pharmacy in the United States or hold a degree from a U.S.-accredited school or college of pharmacy. Nonpharmacists with an interest in the mission of APhA may become associate members.

Each member is served by one of three APhA Academies: The Academy of Pharmacy Practice and Management (APhA–APPM), the Academy of Pharmaceutical Research and Science (APhA–APRS), or the Academy of Students of Pharmacy (APhA–ASP). Each Academy annually elects officers to direct the development of programs, products, and services to meet the needs of members.

Encyclopedia of Clinical Pharmacy
DOI: 10.1081/E-ECP 120006280

MISSION AND KEY OBJECTIVES

APhA is dedicated to improving public health by assisting members and enhancing the profession of pharmacy. It accomplishes this mission by pursuing activities consistent with the following goals:

- Expand access to and promote the value of pharmacists' caregiving services in obtaining positive health outcomes through optimal use of medications.
- Equip pharmacists and others allied to the profession with information and education resources to support provision of patient care.
- Become the primary membership organization of choice for America's pharmacists and others allied to the profession.
- Develop the resource base necessary for successful achievement of the Association's mission.

CURRENT MAJOR INITIATIVES

APhA programs, products, and services are built in recognition of the dual roles of contemporary pharmacists: 1) to insure that the public has access to a safe, efficient, accurate, and patient-sensitive drug distribution system, and 2) that patients achieve optimal outcomes from medication use (prescription, nonprescription, and nontraditional therapies) with the assistance of pharmacists. Pharmacists may engage in one or both aspects of these roles, and APhA members practice in a wide variety of different settings, either directly or indirectly affecting patient care.

Publications and education are important elements of APhA program development. New product bulletins and special reports are aimed at equipping practitioners with timely and unbiased information for their practice. APhA periodical publications include *Journal of the American Pharmaceutical Association*, *Pharmacy Today*, *Journal of Pharmaceutical Sciences*, *Pharmacy Student*, and *Drug-Info Line*, the newest monthly newsletter which provides concise summaries of key pharmacotherapeutic issues.

APhA reference books include the *Handbook of Nonprescription Drugs*, 13th Edition, *Medication Errors*, *Handbook of Pharmaceutical Excipients*, *APhA Drug Treatment Protocols*, and several texts to guide those engaged in compounding practice.

APhA advocacy efforts aim principally at earning deserved recognition for pharmacists as practitioners involved in direct patient care and securing appropriate compensation for those services. This includes educating and lobbying legislative and regulatory officials at the state and national levels, in collaboration with other state and national pharmacy organizations. Private sector advocacy on these same issues extends to business leaders, colleagues in other health professions, insurers, and other decision makers involved with healthcare and the medication use process.

APhA has several affiliate organizations, and through the work of its Foundation and credentialing organizations, the Association has made a major commitment to research, quality measurement, and accountability. The APhA Foundation has sponsored and directed several significant research and demonstration projects to contribute to the body of evidence that pharmacists' services enhance patient health-seeking behavior and improve outcomes. Research on quality measurement and the development of tools to reduce medication use problems and errors is a priority of the Foundation.

Three credentialing organizations with APhA affiliations help pharmacists and technicians secure meaningful credentials to advance their careers. The Board of Pharmaceutical Specialties is a 25-year-old certification agency awarding specialty recognition to pharmacists in five board-recognized areas. The Pharmacy Technician Certification Board, founded by APhA, the American Society of Health-System Pharmacists, the Michigan Pharmacists Association, and the Illinois Council of Health-System Pharmacists in 1995, has certified over 75,000 pharmacy technicians. A disease-specific certification agency, the National Institute for Standards in Pharmacist Credentialing, was founded in 1997. APhA shares responsibility with the National Association of Boards of Pharmacy, the National Association of Chain Drug Stores, and the National Community Pharmacists Association for overseeing this credentialing program for pharmacists. Four disease states are currently certified by computer-based examination through NISPC.

MAJOR APhA MEETINGS

APhA hosts one national meeting annually. The meeting provides continuing education for pharmacists, pharmacy technicians, scientists, and numerous affiliated organizations. The Federal pharmacists hold meetings in conjunction with APhA, as do the American Institute on the History of Pharmacy, the American Society of Pharmacy Law, and several professional fraternities and honor societies. Dates for meetings in upcoming years are as follows:

- APhA2003—New Orleans, Louisiana, March 28–April 1.
- APhA2004—Seattle, Washington, March 26–30.
- APhA2005—Orlando, Florida, April 1–5.
- APhA2006—Phoenix, Arizona, March 24–28.

PROFESSIONAL ORGANIZATIONS

American Society of Consultant Pharmacists

A

Timothy Webster
Phylliss Moret
American Society of Consultant Pharmacists, Alexandria, Virginia, U.S.A.

INTRODUCTION

The American Society of Consultant Pharmacists (ASCP) is the international professional association that provides leadership, education, advocacy, and resources enabling senior care pharmacists to enhance quality of care and quality of life for older individuals through the provision of pharmaceutical care and the promotion of healthy aging. Consultant pharmacists specializing in senior care pharmacy are essential participants in the healthcare system, recognized and valued for the practice of pharmaceutical care for the senior population and people with chronic illness.

For millions of senior citizens and individuals with chronic illnesses, consultant pharmacists play a vital role in ensuring optimal drug therapy. In their role as medication therapy experts, consultant pharmacists take responsibility for their patients' medication-related needs; ensure that their patients' medications are the most appropriate, the most effective, the safest possible, and are used correctly; and identify, resolve, and prevent medication-related problems that may interfere with the goals of therapy. Consultant pharmacists manage and improve drug therapy and improve the quality of life of the senior population and other individuals residing in a variety of environments, including hospitals, nursing facilities, subacute care and assisted living facilities, psychiatric hospitals, hospice, and home- and community-based care.

SENIOR CARE PHARMACY

While medications are probably the single most important factor in improving the quality of life for older Americans, the nation's seniors are especially at risk for medication-related problems due to age-related physiological changes, higher incidence of multiple chronic diseases and conditions, and greater consumption of prescription and over-the-counter medications.

The economic impact of medication-related problems in persons over the age of 65 now rivals that of Alzheimer's disease, cancer, cardiovascular disease, and diabetes. Medication-related problems are estimated to be one of the top five causes of death in that age group, and a major cause of confusion, depression, falls, disability, and loss of independence.

For more than a generation, consultant pharmacists have dedicated themselves to protecting the health of some of our most vulnerable citizens—residents of nursing facilities. Today, the senior care pharmacists ASCP represents are patient advocates for all of our nation's seniors, wherever they reside.

CONSULTANT PHARMACY PRACTICE

Consultant pharmacists are committed to caring for the well-being of each individual, taking into account the complex interrelationships between disease states, nutrition, medications, and other variables. They are essential players on the healthcare team, and influential decision makers in all aspects of drug therapy. Consultant pharmacists counsel patients, provide information and recommendations to prescribers and caregivers, review patients' drug regimens, present in-service educational programs, and oversee medication distribution services.

In addition to these basic responsibilities, consultant pharmacists provide a wide range of other primary care services to the nation's seniors, including pain management counseling, pharmacokinetic dosing services, intravenous therapy, nutrition assessment and support, and durable medical equipment services.

A clear picture of the enormous impact being made by consultant pharmacists in achieving optimal therapeutic outcomes and reducing medication-related problems is emerging from the Fleetwood Project—ASCP Foundation's landmark three-phase study to document the value of pharmacists' services. The Fleetwood Phase I study found that consultant pharmacists' drug regimen

Encyclopedia of Clinical Pharmacy
DOI: 10.1081/E-ECP 120006284

review services in the nation's nursing facilities improve the frequency of optimal drug therapy outcomes by 43% and save as much as $3.6 billion annually in costs associated with medication-related problems.[11] In Fleetwood Phase II, the Fleetwood model was developed and tested for feasibility. The Fleetwood model includes prospective drug regimen review, direct communication with prescribers to resolve therapeutic issues, patient assessment, and formalized pharmaceutical care planning for geriatric patients at highest risk for medication-related problems. The results of the Phase II pilot study were published in the October 2000 issue of *The Consultant Pharmacist*. The Fleetwood model will be further refined in Fleetwood Phase III, currently underway, by identifying and validating ''pharmacist-sensitive outcomes''—those clinical outcomes most sensitive to pharmacist intervention for older patients at high risk for medication-related problems.

ASCP: SERVING THE NEEDS OF A DYNAMIC PROFESSION

The ASCP was founded in 1969 to represent the interests of its members and promote safe and effective medication therapy for the residents of nursing facilities—mostly frail elderly patients. The term ''consultant pharmacists'' is rooted in federal regulations, which requires a pharmacist to provide drug regimen reviews for nursing facility residents. The organization has grown dramatically over the past quarter century and its membership continues to diversify and expand their services to people who need them the most—America's seniors, wherever they reside (Tables 1 and 2).

Table 1 Percentage of ASCP members that provide the following services

Drug regimen review	64%
IV therapy	34%
Drug utilization review	34%
Pharmacokinetic monitoring	27%
Drug formulary management	23%
Pain management	20%
Drug research and studies	20%
Nutrition	18%
Compliance packaging	15%
Home care	13%
DME/surgical appliances	11%
Services for fees	7%
Laboratory testing	5%

Table 2 Percentage of ASCP pharmacists that provide services to the following sites

Nursing homes	76%
Counseling to long-term care facilities (LTCF)	70%
Dispensing to LTCF	54%
Administrative responsibility to LTCF	42%
Residential	41%
Subacute	31%
Hospice	30%
Mental health	28%
Home care	25%
Retail dispensing	19%
Acute care	15%
Correctional facilities	11%
Hospital LTCF	8%

The Executive Director of ASCP is Tim Webster; elected leadership consists of President (Mark Sey), President-elect (Stephen Feldman), Vice President (Ross Buckley), Secretary/Treasurer (Herb Langsam), and the Immediate Past President and Chairman of the Board. These officers and ten directors comprise the Board of Directors. The Board of Directors has full administrative authority in all Society matters, except as otherwise provided in ASCP bylaws.

ASCP has chapters in 20 states and Canada, 30 state affiliates, over 600 pharmacy student members, and hundreds of international members in 18 countries. As consultant pharmacists' practice activities expand and diversify, so does their need for innovative programs, information, and resources. ASCP is strongly committed to meeting these needs.

Education

ASCP offers many opportunities for ACPE-accredited continuing education at its annual meeting, midyear conference, and other regional and chapter-sponsored meetings, seminars, and workshops. ASCP also enables pharmacists to gain geriatric pharmacy knowledge thought its web-based education sites, which include geriatricpharmacyreview.com and scoup.net. (SCOUP is the acronym for Senior Care Online University for Professionals.)

The ASCP Research and Education Foundation also funds, coordinates, and conducts a wide range of traineeships and research programs in long-term care and geriatric healthcare.

Advocacy

ASCP protects the interests of consultant pharmacists and their patients in lobbying and congressional testimony on

Capitol Hill, with federal regulatory agencies, and with state legislatures. The Society tracks and analyzes hundreds of legislative and regulatory developments nationwide, and maintains an effective political presence through the ASCP-PAC (political action committee) and the Capitol Fund, a legislative lobbying fund.

Practice Resources

To help consultant pharmacists succeed in a demanding and changing healthcare environment, ASCP offers a broad array of manuals, texts, videotapes, and software programs. These include the widely used texts: *Drug Regimen Review: A Process Guide for Pharmacists*, *100% Immunization Campaign Resource Manual*, *The Medication Policy and Procedure Manual for Assisted Living*, and *Nursing Home Survey Procedures and Interpretive Guidelines: A Resource for the Consultant Pharmacist*. A multitude of resource directories are also available at ascp.com, which include *Medication-Related Problems in Older Adults*, *Geriatrics Resource Page*, and *Fact Sheet on Medication Use in Nursing Facilities*. Other ASCP-related web sites include ccgp.com, immunizeseniors.org, ascpfoundation.org, geriatricpharmacyreview.com, and scoup.net.

Publications

ASCP members receive several publications including *The Consultant Pharmacist*, the Society's award-winning monthly journal presenting peer-reviewed clinical research, news, and practice management information; *ASCP Update*, a monthly newsletter focusing on pharmacy news, ASCP programs and initiatives, and state and federal legislative and regulatory developments; and *Clinical Consult*, continuing education newsletter, providing in-depth information on a wide range of clinical topics.

A

MISSION

ASCP is the international professional association that provides leadership, education, advocacy, and resources enabling senior care pharmacists to enhance quality of care and quality of life for older individuals through the provision of pharmaceutical care and the promotion of healthy aging.

ASCP's vision include:

- The senior population realizes improved quality of care and quality of life through the provision of pharmaceutical care.
- Senior care pharmacists are recognized and valued for their care of patients.
- Senior care pharmacists are professionals, essential in healthcare systems.
- ASCP is the acknowledged leader in Senior Care Pharmacy practice.

For more information, contact the American Society of Consultant Pharmacists, 1321 Duke Street, Alexandria, Virginia 22314-3563; Tel: 703-739-1300; Fax: 703-739-1321; e-mail: info@ascp.com; www.ascp.com.

REFERENCE

1. Bootman, J.L.; Harrison, D.L.; Cox, E. The healthcare cost of drug-related morbidity and mortality in nursing facilities. Arch. Intern. Med. **1997**, *157* (18), 2089–2096.

American Society of Health-System Pharmacists (ASHP)

C. Richard Talley
American Society of Health-System Pharmacists, Bethesda, Maryland, U.S.A.

INTRODUCTION

The American Society of Health-System Pharmacists (ASHP) is the 30,000-member national professional association representing pharmacists who practice in hospitals, health maintenance organizations, long-term care facilities, home care, and other components of the health-care system. ASHP believes that the mission of pharmacists is to help people make the best use of medicines, and assisting pharmacists in fulfilling this mission is ASHP's primary objective. The Society has extensive publishing and educational programs designed to help members improve their delivery of pharmaceutical care and is the national accrediting organization for pharmacy residency and pharmacy technician training programs. Among pharmacy associations, ASHP stands out as having the largest staff (more than 200), the largest budget (more than $34 million in 2001), the largest educational meeting (more that 21,000 attendees in 2001), and the largest variety of products and educational opportunities.

HISTORY AND ACHIEVEMENTS

Among ASHP's most important achievements are its successes in building practitioner consensus on the societal role of pharmacy. One of the landmark events in the emergence of clinical pharmacy was the ASHP invitational conference on ''Directions for Clinical Practice in Pharmacy,'' held in 1985 in Hilton Head, South Carolina. In similar fashion, ASHP's involvement with the four ''Pharmacy in the 21st Century'' conferences (the second of four such meetings, conceived and presented through the Joint Commission of Pharmacy Practitioners), helped establish the clinical roles of pharmacists.

ASHP further established new direction for clinical pharmacy by being one of the creators—along with the American Pharmaceutical Association, the Michigan Pharmacists Association, and the Illinois Council of Health-System Pharmacist—of the Pharmacy Technician Certification Board (PTCB). It is widely believed that standardized education for pharmacy technicians, enabled by PTCB, will facilitate pharmacists in delegating more distributive roles to technicians and in expanding clinical roles for pharmacists.

ASHP provides services and products to members in ten practice domains: acute care, ambulatory care, clinical specialist, home care, long-term and chronic care, managed care, new practitioner, pharmacy practice management, student, and technician. Pharmacy students receive all member services plus the special services of the ASHP Student Forum—all at greatly reduced rates.

MEMBERSHIP

Membership in ASHP includes a wide variety of benefits. For example, members have full access to all components of ashp.org, ASHP's content-rich, interactive Web site that provides information and e-commerce opportunities 24 hours a day, 7 days a week. Membership includes subscriptions to two periodicals: the *American Journal of Health-System Pharmacy* (*AJHP*), a peer-reviewed, scientific journal published 24 times each year, and ASHP *News and Views*, a newsletter addressing practice issues and upcoming events that is published 12 times each year. Members can attend ASHP's major continuing education meetings—The ASHP Annual Meeting and the ASHP Midyear Clinical Meeting—at rates discounted by an amount equal to ASHP's annual dues. Members can take advantage of more than 150 self-study courses, publications, videos, and software at an average saving of 20% compared with nonmember rates. In addition to the networking opportunities available at ASHP meetings and seminars and similar events conducted by ASHP's Affiliated State Chapters, members can enroll in the ASHP Practice Advancement Links (PALs) program to expand their interconnections with other practitioners. ASHP members can achieve recognition as Fellows of ASHP by applying

Encyclopedia of Clinical Pharmacy
DOI: 10.1081/E-ECP 120006370

to the Practitioner Recognition Program. National and international recognition can be achieved through authorship in *AJHP* and through presentations at ASHP's educational meetings. Members qualify for financial benefits through ASHP's MemberCard program, available at low rates and with no annual fee. There is also an ASHP Member loan program, and members are eligible for insurance benefits, including group insurance plans for family term life, short-term medical, catastrophic major medical, accident, and disability income protection and in-hospital care.

PUBLICATIONS AND EDUCATIONAL RESOURCES

ASHP is the leading publisher of pharmacy information. The best-known products include:

- *American Journal of Health-System Pharmacy.*
- *AHFS Drug Information*, the print product; *AHFS first*, the electronic database, and eBookman, the hand-held multimedia content player version.
- *International Pharmaceutical Abstracts.*
- *Handbook on Injectable Drugs.*
- *Clinical Skills Programs*, for acute care and ambulatory care.
- *Medication Teaching Manual*, the print product; MedTeach, the customizable electronic database; and safemedication.com, the Web-based consumer medication guide.

Over several decades, ASHP has worked with members to develop *Best Practices for Health-System Pharmacy*, a compilation of statements, guidelines, therapeutic position statements, and residency accreditation standards. In addition to its ongoing creation of practice standards, the Office of Professional Practice and Scientific Affairs at ASHP monitors professional practice needs, works with other major health organizations, works toward the prevention of medication misadventures, and communicates with federal and state regulatory bodies that define pharmacy practice in hospitals and other components of health systems.

ASHP is the sole accrediting body for postgraduate residency training programs and pharmacy technician training programs. In 2001 there were 536 ASHP-accredited programs for pharmacists and 83 ASHP-accredited programs for technicians throughout the United States.

ASHP's Government Affairs Division staff provides substantial advocacy for public policy on behalf of ASHP members before the U.S. Congress, federal agencies (e.g., FDA and HCFA), state legislators, and boards of pharmacy.

The ASHP Section of Clinical Specialists focuses on "bringing science to practice" through 17 specialty networks, dedicated Web-site content, and section-specific electronic listservs and an online membership directory available only to section members.

The ASHP Section of Home Care Practitioners provides home infusion providers in alternative sites with special programming at the Midyear Clinical Meeting, advocacy on reimbursement and JCAHO issues, dedicated Web-site content, section-specific listserv news services, and an online membership directory available only to section members.

ASHP's Center on Pharmacy Practice Management monitors, analyzes, and reports on trends in pharmacy practice management. It conducts and publishes an annual national survey of pharmacy practice in health systems, conducts a leadership conference on pharmacy practice management, and coordinates other educational sessions at ASHP meetings.

ASHP's Center on Managed Care Pharmacy monitors, analyzes, and reports on trends in managed care pharmacy, creates specialized programming at ASHP's national meetings, conducts conferences and workshops, conducts surveys, monitors and influences quality-related measures, and coordinates networking opportunities.

ASHP's Center on Patient Safety helps pharmacists lead implementation of proven medication-use safety practices, fosters best practices, identifies training opportunities, promotes pharmacy's role, facilitates alliances, and collaborates with the ASHP Research and Education Foundation to achieve its goals.

To better support the success of its members, ASHP works with other pharmacy organizations, such as the Joint Commission of Pharmacy Practitioners, the Pharmacy Technician Certification Board, the Board of Pharmaceutical Specialties, the Institute for Safe Medication Practices, and the International Pharmaceutical Federation.

On a broader scale, ASHP's Public Relations Division works to influence the image of pharmacy with the U.S. Congress and regulatory bodies, other healthcare associations, hospital and health-system organizations, groups concerned with scientific issues, accrediting and licensing bodies, groups concerned with consumer and patient safety issues, key health-system decision-makers, and the news media.

PHARMACY PRACTICE ISSUES

Antibiotic Rotation

Steven Gelone
Temple University, Philadelphia, Pennsylvania, U.S.A.

INTRODUCTION

Antimicrobial resistance is secondary to a variety of variables (Fig. 1).[1] The link between drug use and the development of resistance is one that has been explored by many investigators resulting in common themes. The Society of Healthcare Epidemiology of America (SHEA) and the Infectious Diseases Society of America (IDSA) have authored a joint publication.[2] Their observations include the following: 1) changes in antimicrobial use are paralleled by changes in the prevalence of resistance; 2) antimicrobial resistance is more prevalent in nosocomial bacterial strains than in those from community-acquired infections; 3) during outbreaks of nosocomial infection, patients infected with resistant strains are more likely than control patients to have received antimicrobials previously; 4) areas within hospitals that have the highest rates of antimicrobial resistance also have the highest rates of antimicrobial use; and 5) increasing duration of patient exposure to antimicrobials increases the likelihood of colonization with resistant organisms.

As a result of these observations, a variety of methods have been utilized to modify antimicrobial use in an effort to combat resistance. These include the use of an antimicrobial formulary, restriction of agents, requirement of prior approval to obtain specific agents, multidisciplinary antibiotic management teams, and the use of computerized support systems. The impact of these programs on resistance is reviewed elsewhere,[3,4] and one or more of these methods is being employed in many institutions.

ANTIMICROBIAL ROTATION

One method that has seen only limited use is antimicrobial rotation. Defined as the prospective and purposeful altering of antimicrobial selection in an effort to prevent or delay the emergence and spread of bacterial resistance, this technique has evolved over the last several decades. Although antimicrobial rotation is simply a variation of antimicrobial control, its motive is focused on resistance prevention and reduction and not specifically on decreasing expenditures related to antimicrobial use.

It is implicit in this type of strategy that one uses a drug for a defined period of time, changes to another agent, and then reuses the original agent. In addition, what is *not* implied is a reactive alteration in antimicrobial selection due to an established epidemic of resistance. These definitions are quite important in evaluating the impact that antibiotic rotation may have, as many individuals have described the later scenario, and these reports do not meet the criteria for an evaluation of rotating drugs.

Several assumptions must be made in order for rotation to be a potentially effective strategy: 1) resistance to Drug A is independent from Drug B; 2) resistant organisms are "less fit" and will go away when selective pressure is decreased or removed (Fig. 2);[5] and 3) the environment in which rotation is being employed is a closed system. In addition, it must be understood that several pathways have been described for the appearance or spread of resistance (Table 1).

Given these assumptions and pathways, clearly, changes in antimicrobial use in a facility or part of a facility will only have potential impact on selection and to some extent on the likelihood of mutations to occur.

ESSENTIAL ELEMENTS OF AN OPTIMAL ANTIMICROBIAL ROTATION PROGRAM

Implementation of an antimicrobial rotation program into any setting is a daunting task. There are many practical issues that currently remain unanswered, including what agents should be included in the rotation, how frequently should rotation occur, what part(s) of an institution would be most likely to benefit from rotation and how to achieve buy-in from the medical staff so as to result in a high level of compliance with the program. Resolutions for these issues that will be broadly applicable to a variety of health care settings are unlikely. It is vital to prospectively answer these questions in a variety of institutions in order

Encyclopedia of Clinical Pharmacy
DOI: 10.1081/E-ECP 120006403

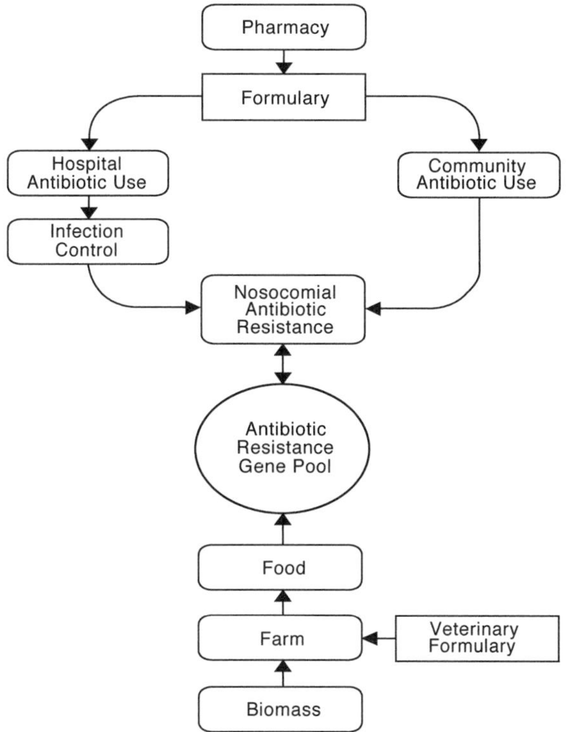

Fig. 1 Variables involved in antimicrobial resistance.

to be able to draw any conclusions about the success or failure of antibiotic rotation.

The impact of such a system will be measured in large part based on the changes in bacterial susceptibility, so

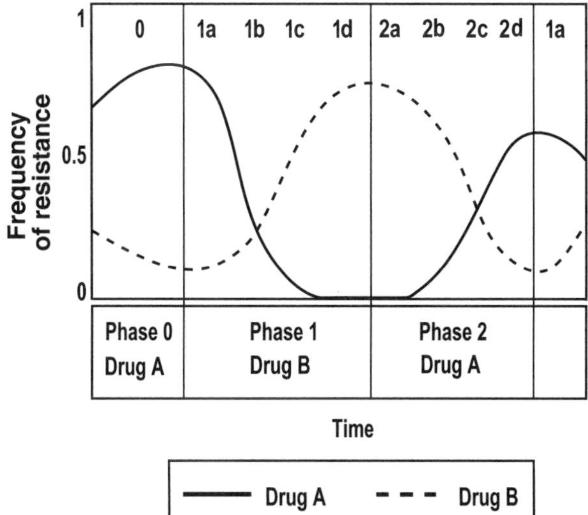

Fig. 2 Resistance assumptions.

Table 1 Pathways for the appearance or spread of resistance

Pathway	Issue
Introduction of a newly resistant organism	Patient, health care worker Other facility (hospital, etc.)
Mutation	Reservoir of high density of organisms with a high likelihood for random selection of resistance mutation
Selection of resistant organism	Selective pressure of antimicrobial use
Dissemination	Poor infection control

(From Ref. [6].)

one must prospectively define what organisms are of interest and what ''resistance'' is for each key organism. A system must be in place to monitor the presence or lack of resistance. This should include both clinical as well as environmental isolates. Organisms need to be able to be stratified based on site of infection, location within the hospital, and whether they are community or nosocomially acquired. As has been outlined above, antimicrobial agents are only one of the variables related to antimicrobial resistance, and antibiotic rotation should be seen as one of several modalities that will achieve antibiotic stewardship. A stable and effective infection control program must be in place, as well as a mechanism to optimize antimicrobial prescribing and track all antimicrobial use. This needs to include the ability to type organisms, identify genetic markers of resistance, and measure antimicrobial use in each area of an institution per unit of time. Lastly, an evaluation of clinical outcomes needs to be performed to insure that such a strategy does no harm to patients.

EVIDENCE-BASED REVIEW OF THE IMPACT OF ANTIMICROBIAL ROTATION

The first description of rotating antibiotics was conducted by Gerding et al.[7] In an effort to improve susceptibility to gentamicin, a shift to amikacin as the aminoglycoside of choice was conducted. During each period of amikacin use (on average 26 months duration), the susceptibility to gentamicin improved over baseline. Unfortunately, this project was reactive in that the switch was performed as a result of poor susceptibility to gentamicin. The duration of the cycles was not predetermined, there was no differentiation between nosocomial and community isolates, there was no specific infection control documentation, nonaminoglycoside antimicrobial use was not monitored, and clinical outcomes were not assessed.

Various investigators have conducted several other studies involving switches within the same class or a cessation of the use of one agent within a class.[8,9] Most have demonstrated improved antimicrobial susceptibility that was maintained as long as the original agent with which poor susceptibility had been seen, was not reintroduced into the environment. A recent study by Seppala et al. in Finland showed that by reducing the use of erythromycin, the susceptibility of Group A streptococci to macrolides significantly improved.[10] Unfortunately, as newer macrolides penetrated the marketplace in Finland and began to be utilized, this trend was quickly reversed (Fig. 3).[10]

Kollef et al. conducted a single switch study where they treated patients in a cardiothoracic intensive care unit empirically for 6 months with ceftazidime and then in the second 6-month period, treated patients with ciprofloxacin.[11] They showed a significant reduction in ventilator-associated pneumonia (VAP), mostly due to a decrease in the number of patients infected with resistant Gram-negative bacteria. This study did not employ true rotation in that a second 6-month rotation of each therapy was not conducted and, as outlined above, did not meet most of the criteria for an ideal rotation program.

Gruson et al. conducted a study in a 16-bed medical intensive care unit in patients mechanically ventilated for greater than 2 days.[12] During the first phase of the study, ciprofloxacin and ceftazidime were used empirically to treat Gram-negative infections. During the second phase of the study, these agents were restricted, and other antibiotics (including other beta-lactams with or without an aminoglycoside) were utilized. The authors noted a significant reduction in all VAP, but an increase in VAP caused by methicillin-sensitive *Staphylococcus aureus*. Susceptibility patterns for *Pseudomonas aeruginosa* and *Burkholderia cepacia* were evaluated and showed improvements over the baseline period.

Raymond et al. reported on a rotation study in a surgical intensive care unit with a different twist.[13] Patients were stratified as either having sepsis/peritonitis or pneumonia, and empiric therapy was cycled every 3 months by syndrome. Fourteen hundred fifty-six admissions and 540 infections were treated over a 2-year period. With similar severity of illness during the before and after periods (mean APACHE II = 19), the authors demonstrated a reduction of length of stay from a mean of 62 days to 39 days, a reduction of vancomycin-resistant enterococcal and methicillin-resistant staphylococcal infection from 14 per 100 admissions to 8 per 100 admissions and death due to any cause dropped from 25 in the before period to 18 in the rotation period. Antimicrobial susceptibility and several other key parameters needed to evaluate the effectiveness of this program were not reported.

Gelone et al. conducted a 3-year prospective study of antimicrobial rotation in three intensive care units at Temple University (the START trial).[14] This study, like those described above, used a before and after approach. Daily rounds were performed, and the following were assessed on every patient: 1) demographics; 2) antibiotic regimen and dosing; 3) the presence of infection (based on predefined criteria); and 4) organ-

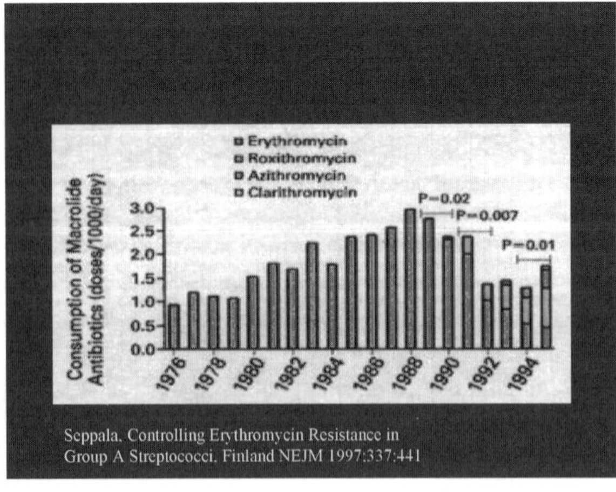

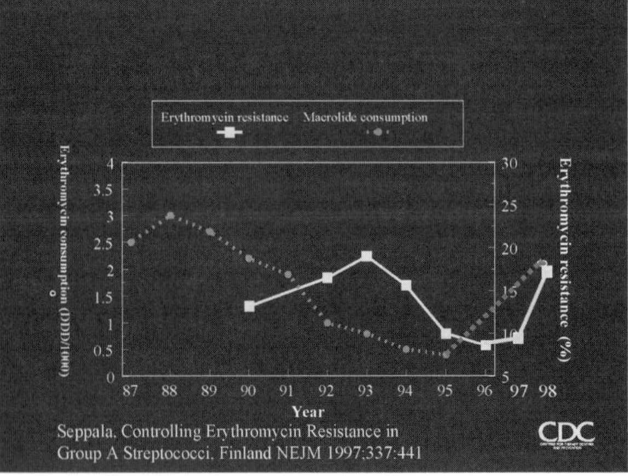

Fig. 3 Effect of erythromycin on Group A streptococci.

ism susceptibility. All definitions were developed prospectively. Criteria for bloodstream, skin and skin-structure, and urinary tract infection were defined based on modified Center's for Diseases Control and Prevention (CDC) criteria. Pneumonia was defined using the American Thoracic Society's criteria. Resistant organisms/infections were defined as resistance to two or more antibiotics typically used to empirically treat the above infection in these units (including aminoglycosides, ceftazidime, ciprofloxacin, imipenem–cilastatin, and piperacillin). Baseline data were collected for a 1-year period, and subsequently, a 2-year rotation period was carried out. Four regimens for empiric therapy of Gram-negative infection were rotated on a monthly basis (cefepime, imipenem–cilastatin, piperacillin–tazobactam, and ciprofloxacin). The addition of an aminoglycoside was left to the discretion of the primary care physician. Susceptibility to studied antimicrobials improved significantly for *Klebsiella pneumoniae*, *Enterobacter species*, *Pseudomonas aeruginosa*, and *Enterococcus species*. Susceptibility remained stable for *Staphylococcus aureus*. Susceptibility significantly decreased for *Acinetobacter species* to imipenem–cialstatin. Resistant bloodstream infection decreased significantly (11% versus 4%) as did resistant pneumonias (15% versus 5.4%). Overall, antimicrobial expenditures remained stable comparing before and after periods ($1.5 before and $1.3 million dollars after). Significant increases were seen in piperacillin–tazobactam (23%), imipenem–cilastatin (21%), and ciprofloxacin (18%) use, while cefepime use decreased by 34%. Review of crude mortality rates for bloodstream infections and pneumonia cases showed no significant differences.

The CDC has funded a 2-$^1/_2$ year, three-center study to evaluate antimicrobial rotation in adult intensive care units.[15] This trial will utilize 4-month cycles of the following drugs: cefepime, a fluoroquinolones, imipenem–cilastatin or meropenem, and piperacillin–tazobactam. Investigators will evaluate the acquisition of antibiotic-resistant Gram-negative organisms as gastrointestinal tract colonizers, those associated with clinical infection, changes in organism susceptibility over time, and adverse events and death. At the time of this publication, this project was just underway, and results were unavailable.

ROLE OF ANTIBIOTIC ROTATION

As stated above, antibiotic rotation should be viewed as one method of antibiotic control. As with most methods of antibiotic control, definitive evidence of its impact in a variety of health care settings is lacking. The available evidence does suggest that antibiotic rotation is associated with improvements in bacterial susceptibility and a decrease in the incidence of resistant infections. Many questions remain unanswered, including which agents to rotate, how long to rotate, and what settings would most benefit from this strategy.

The development of new or novel agents active against resistant pathogens is time consuming and at times lags behind new microbial threats. Strategies that enable the use of currently available agents for extended periods of time are exciting and necessary. Although antibiotic rotation is promising, there are currently gaps in knowledge regarding this method of antibiotic control. In addition, the data generated to date may not be generalizable across various health care settings. Caution should be exercised in establishing such a program. Individuals are encouraged to apply as many of the essential elements noted above in an effort to assess the impact of this strategy. Documentation of results (both positive and negative) is essential to define the ultimate role of antibiotic rotation.

REFERENCES

1. John, J.F., Jr.; Rice, L.B. The microbial genetics of antibiotic cycling. Infect. Control Hosp. Epidemiol. **2000**, *21*, S22–S31.
2. Shales, D.M.; Gerding, D.N.; John, J.F., Jr.; Craig, W.A.; Bornstein, D.L.; Duncan, R.A.; Eckman, M.R.; Farrer, W.E.; Greene, W.H.; Lorian, V.; Levy, S.; McGowan, J.E., Jr.; Paul, S.M.; Ruskin, J.; Tenover, F.C.; Watanakunakorn, C. Society for Hospital Epidemiology of America and the Infectious Diseases Society of America joint committee on the prevention of antibiotic resistnace: Guidelines for the prevention of antibiotic resistance in hospitals. Clin. Infect. Dis. **1997**, *25*, 584–599.
3. John, J.F., Jr.; Fishman, N.O. Programmatic role of the infectious diseases physician in controlling antibiotic costs in hospitals. Clin. Infect. Dis. **1997**, *24*, 471–485.
4. O'Donnell, J.A.; Gelone, S.P.; Levison, M.E. Antibiotic Control Systems. In *Saunders Infection Control Reference Service*, 2nd Ed.; Abrutyn, E., Ed.; W.B. Saunders Co., Orlando, 2000; 52–58.
5. McGowan, J.E., Jr., Unpublished.
6. McGowan, J.E., Jr. Strategies for the study of the role of cycling on antibiotic use and resistance. Infect. Control Hosp. Epidemiol. **2000**, *21*, S36–S43.
7. Gerding, D.N.; Larson, T.A.; Hughes, R.A.; Weiler, M.; Shanholtzer, C.; Peterson, L.R. Aminoglysocide resistance and aminoglycoside use: Ten years of experience in one

hospital. Antimicrob. Agents Chemother. **1991**, *35*, 1284–1290.

8. McGowan, J.E., Jr. Minimizing antibiotic resistance in hospital bacteria: Can switching or cycling drugs help? Infect. Control Hosp. Epidemiol. **1986**, *7*, 573–576.
9. Bonhoeffer, S.; Lipsitch, M.; Levin, B.R. Evaluation of treatment protocols to prevent antibiotic resistance. Proc. Natl. Acad. Sci. U. S. A. **1997**, *94*, 12106–12111.
10. Seppala, H.; Klankka, T.; Vuopio-Vakila, J.; Muotiala, A.; Helenius, H.; Lager, K.; Huovinen, P. The effect of changes in consumption macrolide antibiotics on erythromycin resistance in Findland. N. Engl. J. Med. **1997**, *337*, 441–446.
11. Kollef, M.H.; Vlasnik, J.; Sharpless, L.; Pasque, C.; Murphy, D.; Fraser, V. Schedule changes of antibiotic classes: A strategy to decrease the incidence of ventilator associated pneumonia. Am. J. Respir. Crit. Care Med. **1997**, *156* (4 Pt. 1), 1040–1048.
12. Gruson, D.; Hilbert, G.; Vargas, F.; Valentino, R.; Bebear, C; Allery, A.; Bebear, C.; Gbikpi-Benissan, G.; Cardinaud, J.P. Rotation and restricted use of antibiotics in a medical intensive care unit. Impact on the incidence of ventilator-associated pneumonia caused by antibiotic-resistant gram-negative bacteria. Am. J. Respir. Crit. Care Med. **2000**, *162* (3 Pt. 1), 837–843.
13. Raymond, D.P.; Pelletier, S.J.; Crabtree, T.G.; Gleason, T.G.; Hamm, L.L.; Pruett, T.L.; Sawyer, R.G. Impact of a rotating empiric antibiotic schedule on infectious mortality in an intensive care unit. Crit. Care Med. **2001**, *29*, 1101–1108.
14. Gelone, S.P.; Lorber, B.; St. John, K.; Badelino, M.; Axelrod, P.; Criner, G. Prospective evaluation of antibiotic rotation on three intensive care units at a tertiary care university hospital. Pharmacotherapy **2000**, *20*, abstract #107.
15. Centers for Disease Control and Prevention Program Announcement #99149.

Anticoagulation Clinical Pharmacy Practice

Patricia H. Schoch
Denver VA Medical Center, Denver, Colorado, U.S.A.

INTRODUCTION

Anticoagulation clinics serve the purpose of a coordinated clinic that oversees the management of oral anticoagulants (blood thinners), specifically, the medication warfarin. These anticoagulation clinics may also employ the use of intravenous or subcutaneous administration and monitoring of unfractionated heparin and low-molecular-weight heparin (LMWH). Anticoagulants in anticoagulation clinics are used for treatment and prevention of blood clots in the deep venous system and lungs, prevention of systemic blood clots in patients with atrial fibrillation, and prosthetic mechanical heart valves. They are also used in peripheral vascular disease, prior strokes, congestive heart failure, heart attacks, and hypercoagulable states.

Anticoagulation clinics are based in the inpatient or outpatient units of a hospital or managed-care organization, or in a community setting. They have been in existence since at least 1968, when they were managed solely by physicians.[1] Pharmacists have been involved with such clinics since the 1970s, both in the United States and in Europe.[2–7] The role of the pharmacist in an anticoagulation clinic varies widely. The pharmacist may be part of a multidisciplinary team, including physicians, physician assistants, and/or nurses, or the pharmacist may be directly responsible for the management of the clinic. In the winter of 1999, a survey by the Anticoagulation Forum elicited 233 responses out of 531 anticoagulation clinics contacted.[8] Anticoagulation clinics had a patient panel size ranging from 1 to 1000+ patients, with a median of 250 patients per clinic.[8] The majority of these clinics received referrals from cardiologists and general internists.[8] A little more than 50% of the staffing consisted of pharmacists, another 40% consisted of nurses.[8]

A variety of factors influence the safety of anticoagulation therapy, given the narrow therapeutic range of anticoagulation therapy and the risk of an adverse event versus the benefit of therapy. These factors include the following:

- Patient understanding of the medications.
- Patient compliance with medications and appointments.
- The need for frequent International Normalizing Ratio (INR) testing.
- Medications, diet, and illnesses.
- The need for careful assessment of results.
- The need for increased communication with patients.
- The need for coordination of care with other providers and services, such as laboratories.[9]

The majority of patients who are placed on oral anticoagulation are followed by their personal primary-care provider.[9] These providers often have an inadequate system for tracking patients and ensuring follow-up, and lack the time for thorough patient counseling.[9] Often practitioners are not aware of the multitude of drug interactions. These factors have deprived many patients of proper follow-up, or many times patients simply are not put on oral anticoagulation because of the complexity of monitoring.[9] Increasingly, patients are being placed on therapy with oral anticoagulation medications when there is a coordinated system in place to monitor them and thus improve overall patient care.[9] Anticoagulation clinics allow for a systemized approach to monitoring, evaluating, and adjusting appropriate anticoagulation therapy.[9] Pharmacists have been able to develop their role as an anticoagulation practitioner due to the significant role they play in counseling and educating patients about their medications. Pharmacists also possess the advanced drug knowledge and education in pharmacology and pharmacokinetics required in managing these patients.

The prothrombin time (PT) test is the most common method for monitoring oral anticoagulation.[10] Standardization of this test using the INR has improved reliability of warfarin monitoring in North America.[10] Findings from two pharmacist-managed anticoagulation clinics

Encyclopedia of Clinical Pharmacy
DOI: 10.1081/E-ECP 120006223

showed how unstandardized PT tests could lead to significant error and misjudgment.[11] The adoption of the INR is considered one of the most significant advances in anticoagulation therapy.

CLINICAL PHARMACY OPPORTUNITIES

Anticoagulation Clinics in the United States

There have been many articles published concerning pharmacist involvement with anticoagulation clinics. Typically, such anticoagulation clinics are in hospital-based, outpatient settings. In the United States, the majority of these clinics are in the outpatient units of the Department of Veterans Affairs Medical Centers (VAMC) and other teaching hospitals.[2–4,6,7,12–27] These anticoagulation clinics are usually responsible for the management and coordination of warfarin therapy. The first report of a pharmacist-managed anticoagulant clinic was published in 1979.[2] This report described a pharmacist-managed anticoagulation clinic at McGuire VAMC in Richmond, VA, that began in the Department of Cardiology. Although the majority of anticoagulation clinics exist in VAMC or university-affiliated teaching hospitals, there have been descriptions of an outpatient anticoagulation clinic in a private community hospital[28] and in managed-care settings.[29–31] There is a growing network of private practice-based, for-profit clinics.[32] Anticoagulation clinics can also be found in the inpatient setting.[33–35]

Outpatient Treatment of Deep Vein Thrombosis with LMWH

With the advent of LMWH, anticoagulation clinics are now using LMWH, by use of which patients can avoid being admitted into a hospital or shorten the length of hospitalization.[22–24,30,31] Patients can have their injections done at home. Patients with a confirmed diagnosis of deep vein thrombosis (DVT) are evaluated to determine eligibility to receive LMWH via outpatient. Once the patient is eligible, a nurse typically administers the first dose, and the pharmacist begins the counseling and evaluates the patient's ability to self-inject. The anticoagulation clinic then follows up with the patient until determining that anticoagulation therapy is no longer required.

Anticoagulation Clinics in the United Kingdom

In the United Kingdom, pharmacists have also managed anticoagulation clinics.[5,36–40] They can be found in hospital clinics, general practitioner surgery clinics, and community pharmacy-based clinics. In a year 2000 survey, approximately 20% of the 250 National Health Service hospitals had a pharmacist-led anticoagulation clinic; these anticoagulation clinics are now working together to develop a master protocol for the initiation of DVT treatment, management of anticoagulation treatment in outpatients, training of anticoagulation personnel, and research opportunities.[41] Pharmacists have also been involved in a pharmacist-led outpatient DVT clinic.[38] At Neath General Hospital, it was determined that there was a potential savings of 268 inpatient-bed days annually by employing the outpatient DVT clinic.[38] A former senior general practitioner at Downfield surgery in Dundee explained how working with a team, specifically with pharmacists, led to improved patient care, innovation, and development.[42] He noted that pharmacists are playing a major role in implementing national evidence-based guidelines to improve the quality of patient care.[42]

Description of Anticoagulation Clinics

There may not be such a thing as a typical anticoagulation clinic. Generally, however, a physician will serve as a director or consultant for the clinic. The physician may directly oversee each plan the pharmacist formulates, or the pharmacist may follow some type of protocol without being directly supervised by the physician. In many clinics, the pharmacist has responsibility for designing an appropriate anticoagulant regime for the patient. The pharmacist is normally responsible for the day-to-day operations of the clinic. Pharmacists are also responsible for enrolling patients in the anticoagulation clinic by obtaining medication and medical histories, and by assessing factors that may affect the control of the therapy. The pharmacist is responsible for counseling patients about the signs and symptoms of bleeding and clotting, compliance with medications and follow-up appointments, drug interactions, food interactions, and how health status affects warfarin. Generally, the patient presents for a blood draw either via venous puncture or via point-of-care testing. Point-of-care testing is currently being used in a few clinic sites and is done via portable machine. It measures the INR from a fingerstick sample of whole blood and provides results within minutes. If the patient has blood drawn from a venous puncture, the patient will have to wait for results, or some clinics will call or mail them to the patient. With modern technology, an efficient lab can provide INR results within 15 minutes of a blood draw. If blood is drawn from a finger stick, the patient will receive the results immediately and will be given instructions on dosing of warfarin and follow-up instructions for returning to the clinic. Regardless of the means of testing, patients are assessed for compliance

with the anticoagulant, signs or symptoms of bleeding or clotting, recent changes in diet, appetite, medications, alcohol consumption, and illnesses. Based on the patient's INR and assessment of the previous factors, the anticoagulant may be continued or adjusted. A follow-up appointment is then made.

Safety

Table 1 summarizes nonrandomized, mainly retrospective studies that compare frequency of hemorrhage and thromboembolism prior to being enrolled into a pharmacy-managed anticoagulation clinic and after enrollment. These studies demonstrate a decrease in adverse events when patients are followed in a pharmacist-managed anticoagulation clinic. Many of these reports compare adverse events before there was an anticoagulation clinic versus after an anticoagulation clinic was instituted. The study by Chiquette et al. was unique in that it compared results with three different inception groups.[17] All the patients were newly started on warfarin, two groups were in a pharmacy-managed anticoagulation clinic, and one group was managed by routine medical care.[17]

Cost Effectiveness

The cost effectiveness of pharmacy-managed anticoagulation clinics has been addressed in a few studies.[13,15,17] Usually the benefits are determined from decreased adverse events and decreased hospitalization and emergency visits. In one study, hospitalizations and emergency room visits were reduced by 50 to 80%.[17] Savings have been estimated between $860 and $4072 per patient per year of therapy for patients on oral anticoagulation therapy.[13,15,17] Anticoagulation clinics that have used LMWH for DVT treatment in the outpatient setting have estimated cost-avoidance savings between $1800 and $2470 per patient treated.[24,31] These savings not only reflect improved anticoagulation management, but also reflect pharmacists identifying and intervening with other medical conditions. Chiquette et al. described $300 savings per patient per year in regards to other interventions besides anticoagulation management.[17]

Patient and Physician Satisfaction

Three surveys have been published on patients' perceptions of a pharmacy-managed anticoagulation clinic.[19,25,29] Generally, patients found pharmacists to be caring and competent.[19] Patients perceived that they were at a decreased risk of having problems with warfarin and blood clotting due to pharmacist involvement, and they believed that frequent monitoring of their warfarin would mean less chance of bleeding or clotting.[25] Overall, patients were highly satisfied with the care they received from a pharmacist-managed anticoagulation clinic.[29] A survey of physicians published in the United States elicited 21 out of 41 responses and showed that physicians were positive about the care their patients were receiving from a pharmacy-managed anticoagulation clinic.[29]

Future

The future for anticoagulation clinics is bright. Every year there is evidence of increased interest in this area. The

Table 1 Frequency of hemorrhage and thromboembolism with routine medical care versus pharmacist-managed anticoagulation clinics

Study[b]	Type of care	No. of patients	Patient years	Major hemorrhage[c]	Minor hemorrhage[c]	Thromboembolism[c]
Cohen et al. 1985[6]	RMC	17	NA[a]	9.0[d]	NA[a]	NA[a]
	AC	18		6.9[d]	NA[a]	NA[a]
Garbedia-Ruffalo et al. 1985[12]	RMC	26	64.3	12.4	NA[a]	6.2
	AC	26	41.9	2.4	NA[a]	0
Wilt et al. 1995[15]	RMC	NA[a]	28	28.6	14.3	48.6
	AC		60	0	13.7	0
Chiquette et al. 1998[17]	RMC	142	102	3.9	62.8	11.8
	AC	176	123	1.6	26.1	3.3

AC, Pharmacist-managed anticoagulation clinic; RMC, routine medical care.
[a]NA, not available.
[b]Mix indications for warfarin (venous and arterial disease).
[c]Results expressed as percent per patient year of therapy.
[d]Combined major and minor hemorrhage.
Adapted from Ref. [9].

biggest problem facing anticoagulation clinics today is reimbursement. Pharmacists currently can only bill for a minimal visit, whereas other practitioners such as nurse practitioners and physician's assistants can charge three to five times as much per visit due to their provider status. Oral anticoagulation monitoring may become more common in the community setting as pharmacists learn more about anticoagulation and as more states allow pharmacists to adjust medications. Patients may find it to be more convenient due to proximity of local pharmacies, and with the availability of portable point-of-care testing, patients can receive results quickly.[43] Many patients may ultimately perform testing at home with portable machines.[44] Home testing can be performed in a couple of ways. Either the patient tests the blood at home and calls in the results to the anticoagulation clinic, or the patient follows a protocol to adjust the warfarin dosage at home.[44] There will always be some patients not capable of performing their own testing. Even if the patient is monitoring his or her blood at home, pharmacists will still need to be involved by providing education and making dosage change recommendations, especially when confounding factors exist, such as drug interactions and illnesses. Currently, the disadvantages of the portable machines are that they are expensive and that follow-up management is not billable. Recently, Roche Diagnostics withdrew its point-of-care testing machine for patient's home use due to insurance companies, hospitals, and clinics not willing to pay for such machines. Results of portable machine monitors versus routine lab results may diverge at high INRs; however, this also can occur between two different routine laboratories. Two studies reported that the portable monitor was more reliable, less variable, more reproducible, and less likely to give clinically misleading and erroneously high INR results than the laboratory of a major medical center.[45]

CERTIFICATION, TRAINING, AND CREDENTIALING

Certification

Pharmacists in anticoagulation clinics should be able to demonstrate advanced knowledge of anticoagulation. More recently, the importance of credentialing anticoagulation providers has come to the forefront. This may become more important in the future for pharmacy practitioners who want reimbursement for their services. The National Certification Board for Anticoagulation Providers (NCBAP) has a mission to optimize patient care through a multidisciplinary national certification process for registered nurses, advanced practice nurses, pharmacists, physician's assistants, or physicians. Any credentialed individual who is a Certified Anticoagulant Care Provider (CACP) should possess advanced antithrombotic/anticoagulant knowledge. Practitioners are required to submit evidence of their practice experience and obtain a passing score on a comprehensive examination. CACP providers must possess a valid U.S. professional license, as well as the knowledge and skills to provide high-quality care to patients. The certification and examination began in 1999; as of the spring of 2000, there were 68 CACP providers, 44 of these being pharmacists.[46] (To obtain more information, contact the NCBAP through www.acforum.org or write to NCBAP, c/o Anticoagulation Forum, Boston University Medical Center, Room E-113, 88 E. Newton Street, Boston, Massachusetts 02118-2395, phone 716-638-7265.) The National Institute for Standards in Pharmacists Credentialing (NISPC) was formed in 1998 by the American Pharmaceutical Association, the National Association of Boards of Pharmacy, the National Association of Chain Drug Stores, and the National Community Pharmacists Association. The NISPC provides a nationally recognized credentialing process that establishes appropriate standards of care and facilitates recognition of the value of disease state management services such as anticoagulation provided by pharmacists. (To obtain more information, contact the NISPC Testing Center through www.nispcnet.org, or write to NISPC Testing Center, 700 Busse Highway, Park Ridge, Illinois 60068-2402, phone 847-698-6227.)

Training

The Anticoagulant Therapy Management Certificate Program is an internet-delivered program developed by the University of Southern Indiana, School of Nursing and Health Professionals. This program is a collaborative effort among regional health care providers, members of the NCBAP, and the University of Southern Indiana School of Nursing and Health Professionals. Their goal is to prepare health professionals for monitoring and managing outpatient anticoagulation therapy. This program will also prepare health professionals for the National Certified Anticoagulation Care Provider Examination. (For information, visit the web site at http://healthusi.edu or call 1-877-874-4584.)

The American Society of Health-Systems Pharmacist (ASHP) Foundation provides a 5-day, experience-based certificate program called the Anticoagulation Management Service Traineeship Program. This program is designed to train pharmacy practitioners to establish and

maintain specialized services for the management of patients undergoing long-term anticoagulant therapy. The program is intended to provide individualized, intensive didactic and clinical training for selected candidates. Trainees will observe and participate in the activities of an established anticoagulation management service. This program provides 35 hours of continuing education. (Applications and additional information may be obtained by calling the ASHP Fax-on-demand system at 301-664-8888 and requesting documents 702 and 703, or by logging on to their web site at www.ashpfoundation.org.)

The American College of Clinical Pharmacy (ACCP) Research Institute, University of Texas (UT), and Anticoagulation Clinics of North America (ACNA) provides a minimum 4-week, intensive traineeship that includes a structured didactic component, extensive clinical experience in several ACNA practice sites in San Antonio, TX, and surrounding communities; and participation in ongoing clinic research. The program is targeted primarily to PharmD students in their final year of professional study, but pharmacy residents and fellows as well as practicing pharmacists are encouraged. Applicants have also come from other countries. Arrangements can be made with ACNA/UT faculty for students to receive academic credit from their home institution for the experience. (Applications can be obtained through www.accp.com/ClinNet/Research.html or by calling 816-531-2177.)

The University of Illinois at Chicago offers an Antithrombosis Management Service Certificate Program via the Internet. It is a 9-week certificate program and offers 40 contact hours. This program covers a wide array of topics, including pharmacotherapy, patient assessment, protocol development, and business planning. (Applications can be obtained through conted@uic.edu or by calling 866-742-7623.)

Credentialing

A 1996 survey conducted nationwide in the United States elicited 110 responses out of 177 pharmacist-managed anticoagulation clinics contacted.[47] As per the results, 23% offered some type of anticoagulation training program and 29% had at least one pharmacist who completed the ASHP Research and Education Foundation's Anticoagulation Service Traineeship.[47] At the McClellan Memorial VAMC in Little Rock, Arkansas, the anticoagulation clinic has specific guidelines on how their pharmacists should be trained and credentialed.[21] Pharmacists are required to have the following:

- Advanced knowledge of the pharmacology and pharmacokinetics of anticoagulants.
- Advanced knowledge of the pathophysiology of thromboembolic disease states.
- Experience with physical assessment and interviewing patients.
- Experience preparing and providing in-service education to other health care professionals and patients.
- A working knowledge of basic hospital and clinic policies and quality assurance practices.[21]

These requirements were met through reading review and research articles, observing patient interviews and conducting patient interviews under direct observation of a privileged anticoagulation clinic pharmacist, presenting an in-service, and passing a credentialing examination.[21]

RESOURCES

There are a few critical papers or publications that a practitioner should have available in the clinic:

- *Consensus Conferences on Antithrombotic Therapy* sponsored by the American College of Chest Physicians.[48] This provides a comprehensive review and recommendations performed by international experts on antithrombotic therapy.
- *British Committee for Standards in Haematology Guidelines.*[49] This document tends to be the guideline used in Europe. It contains information on indications for oral anticoagulation and management of an anticoagulation service.
- *The Consensus Guidelines for Coordinated Outpatient Oral Anticoagulation Therapy Management.*[50] This guideline contains information on the organization and management of anticoagulation clinics.
- *Managing Oral Anticoagulation Therapy, Clinical and Operational Guidelines.*[9] Written by a multidisciplinary group of health care providers. This is an excellent resource that covers the development and implementation of an anticoagulation management service and management of patients receiving oral anticoagulants. The chapters contain examples of actual policies, procedures, guidelines, algorithms, charts, and flow sheets used in anticoagulation clinics across the United States.
- The NCBAP puts forth competency statements for certified anticoagulation care providers. There are five domains:
    - Applied physiology and pathophysiology of thromboembolic disease.
    - Patient assessment and management.

- Patient education.
- Applied pharmacology of antithrombotic agents.
- Operation (administrative) procedures.
- Extensive recommendations on resources and references.

Pharmacy organizations or web sites that practitioners may find helpful are as follows:

- The ACCP's Practice and Research Networks (PRNs) are for members with common practice and research interests. An interactive e-mail group allows members to exchange information on a daily basis.
- ASHP:
  - ASHP Foundation provides an example of an anticoagulation clinic protocol.
  - Drug-use evaluation criteria that have been put forth by ASHP.[51] These criteria serve as guidelines for quality assurance.
- The web site of the Anticoagulation Forum (www.acforum.org) has links to continuing education, news and events, newsletters, and clinic locations.[52]
- The Department of Health in the United Kingdom provides an anticoagulant booklet, which provides patients with information on anticoagulation [obtained from DHSS Stores, No. 2 site, Manchester Road, Heywood, Lancashire OL10 2PZ, or SHHD (Div. IIID), Room 9, St. Andrews House, Edinburgh EH1 3DE].[49]

PROFESSIONAL NETWORKING OPPORTUNITIES

The Anticoagulation Forum, founded in 1991, is an excellent avenue for professional networking. It brings together three health care disciplines—medicine, nursing, and pharmacy. This organization has global membership and is interested in anticoagulation management in the setting of a coordinated anticoagulation management service. It is supported by the pharmaceutical and diagnostics industry. Currently, the organization's web site contains information about news and events, articles, meetings, and continuing education.[52] It can be accessed through the following web site address: www.acforum.org.[52] There are currently more than 2300 members representing 25 countries and more than 800 anticoagulation clinics.[52] These countries include the United States, Canada, Panama, Netherlands, Scotland, Germany, Brazil, Slovenia, Denmark, Holland, Sweden, Switzerland, Argentina, Iran, Uruguay, France, Italy, Korea, Saudi Arabia, Australia, United Kingdom, Singapore, Israel, Spain, and China.[53] This forum holds a conference biannually.

Another networking opportunity is the University of Wisconsin–Madison-sponsored Pharmacy Invitational Conference on Anticoagulation Therapy. This conference is held yearly, immediately preceding the ASHP Midyear Clinical Meeting in December. This event includes a full day of anticoagulation topics and qualifies for continuing education credit.

LEGAL ISSUES

Although important, only limited information is available in the literature concerning the legal issues of operating an anticoagulation clinic. Legal issues are mentioned briefly in the 2001 Chest supplement.[54] It describes strength in unanimity among anticoagulation clinics.[54] As the number of anticoagulation clinics increases and as more studies show that anticoagulation clinics improve patient care, anticoagulation clinics are becoming the standard of care. Other means of managing anticoagulated patients may have to demonstrate services that are equal or superior to an anticoagulation clinic.[54]

Another issue that may impact pharmacist-managed anticoagulation clinics will be collaborative drug therapy management (CDTM) based on legislated statute. Anticoagulation clinics are a good example of CDTM.[55] It is officially recognized in 25 states and by the federal government (i.e., U.S. armed forces, VAMCs, Indian Health Service).[55] Typically, the physician delegates management authority to the pharmacist with the terms of a formal agreement.[55] It allows pharmacists to order laboratory tests, assess patients, initiate and modify drug therapy, monitor patients, and administer medications.[55] States without CDTM may limit the pharmacist's role in an anticoagulation clinic.[55] To avoid litigation, pharmacists should be well trained, act within protocol framework, document thoroughly, and always be sure patients are aware that a pharmacist is providing care.[55]

REFERENCES

1. Davis, F.B.; Estruch, M.T.; Samson-Corvera, E.B.; Voigt, G.C.; Tobin, J.D. Management of anticoagulation in outpatients: Experience with an anticoagulation service in a municipal hospital setting. Arch. Intern. Med. **1977**, *137* (2), 197–202.
2. Conte, R.R.; Kehoe, W.A.; Nielson, N.; Lodhia, H. Nine-

year experience with a pharmacist-managed anticoagulation clinic. Am. J. Hosp. Pharm. **1986**, *43* (10), 2460–2464.
3. Reinders, T.P.; Steinke, W.E. Pharmacist management of anticoagulant therapy in ambulant patients. Am. J. Hosp. Pharm. **1979**, *36* (5), 645–648.
4. Davis, F.B.; Sczupak, C.A. Outpatient oral anticoagulation: Guidelines for long-term management. Postgrad. Med. **1979**, *66* (1), 100–109.
5. Pegg, M.; Bourne, J.; Mackay, A.D.; Lawton, W.A.; Cole, R.B. The role of the pharmacist in the anticoagulant clinic. J. R. Coll. Physicians London **1985**, *19* (1), 39–44.
6. Cohen, I.A.; Hutchison, T.A.; Kirking, D.M.; Shue, M.E. Evaluation of a pharmacist-managed anticoagulation clinic. J. Clin. Hosp. Pharm. **1985**, *10* (2), 167–175.
7. Bussey, H.I.; Rospond, R.M.; Quandt, C.M.; Clark, G.M. The safety and effectiveness of long-term warfarin therapy in an anticoagulation clinic. Pharmacotherapy **1989**, *9* (4), 214–219.
8. The management of warfarin therapy by anticoagulation clinics: Results of a forum survey. Anticoagulation Forum Newsl. **2000**, *5* (2).
9. *Managing Oral Anticoagulation Therapy: Clinical and Operational Guidelines,* 2nd Ed.; Ansell, J.E., Oertel, L.B., Wittowsky, A.K., Eds.; Aspen Publishers, Inc.: Gaithersburg, Maryland, 1998; 1A–2:4.
10. Hirsh, J.; Dalen, J.E.; Anderson, D.R.; Poller, L.; Bussey, H.; Ansell, J.; Deykin, D. Oral anticoagulants: Mechanism of action, clinical effectiveness, and optimal therapeutic range. Chest **2001**, *119* (1), 8S–21S, (Suppl.).
11. Bussey, H.I.; Force, R.W.; Bianco, T.M.; Leonard, A.D. Reliance on prothrombin time ratios causes significant errors in anticoagulation therapy. Arch. Int. Med. **1992**, *152*, 278–282.
12. Garabedian-Ruffalo, S.M.; Gray, D.R.; Sax, M.J.; Ruffalo, R.L. Retrospective evaluation of a pharmacist-managed warfarin anticoagulation clinic. Am. J. Hosp. Pharm. **1985**, *42* (2), 304–308.
13. Gray, D.R.; Garabedian-Ruffalo, S.M.; Chretien, S.D. Cost-justification of a clinical pharmacist-managed anticoagulation clinic. Drug Intell. Clin. Pharm. **1985**, *19* (7–8), 575–580.
14. Krokosky, N.J.; Vanscoy, G.J. Running an anticoagulation clinic. Am. J. Nurs. **1989**, *89* (10), 1304–1306.
15. Wilt, V.M.; Gums, J.G.; Ahmed, O.I.; Moore, L.M. Outcome analysis of a pharmacist-managed anticoagulation service. Pharmacotherapy **1995**, *15* (6), 732–739.
16. Lee, Y.-P.; Schommmer, J.C. Effect of a pharmacist-managed anticoagulation clinic on warfarin-related hospital admissions. Am. J. Health-Syst. Pharm. **1996**, *53* (13), 1580–1583.
17. Chiquette, E.; Amato, M.G.; Bussey, H.I. Comparison of an anticoagulant clinic with usual medical care. Arch. Intern. Med. **1998**, *158* (15), 1641–1647.
18. Kroner, B.A. Anticoagulation clinic in the VA Pittsburgh healthcare system. Pharm. Pract. Manage. Q. **1998**, *18* (3), 17–33.
19. Lewis, S.M.; Kroner, B.A. Patient survey of a pharmacist-managed anticoagulation clinic. Managed Care Interface **1997**, *10* (11), 66–70.
20. Foss, M.; Schoch, P.H.; Sintek, C. Efficient operation of a high-volume anticoagulation clinic. Am. J. Health-Syst. Pharm. **1998**, *56* (5), 443–449.
21. Santiago, M.E.; Rickman, H.S.; Hutchison, L.C. Training and credentialing for pharmacists in an anticoagulation clinic. Fed. Pract. **1997**, *14* (9), 35–44.
22. Pubentz, M.J.; Calcagno, D.E.; Teeters, J.L. Improving warfarin anticoagulation therapy in a community health system. Pharm. Pract. Manage. Q. **1998**, *18* (3), 1–16.
23. Wieland, K.A.; Ewy, G.A.; Wise, M. Quality assessment and improvement in a university-based anticoagulation management service. Pharm. Pract. Manage. Q. **1998**, *18* (3), 56–57.
24. Groce, J.B., III. Patient outcomes and cost analysis associated with an outpatient deep venous thrombosis treatment program. Pharmacotherapy **1998**, *18* (6 Pt. 3), 175S–180S.
25. Nau, D.P.; Ried, D.L.; Lipowski, E.E.; Kimberlin, C.; Pendergast, J.; Spivey-Miller, S. Patients' perceptions of the benefits of pharmaceutical care. J. Am. Pharm. Assoc. **2000**, *40* (1), 36–40.
26. Moherman, L.J.; Kolar, M.M. Complication rates for a telephone-based anticoagulation service. Am. J. Health-Syst. Pharm. **1999**, *56*, 1540–1542.
27. Yuen, M.L.; Shashoua, T.A.; Han, H.; Heng, M.K. Efficacy of pharmacist-managed anticoagulation clinic in a veteran population. Veterans Health Syst. J. **2000**, *5* (8), 37–41.
28. Norton, J.L.; Gibson, D.L. Establishing an outpatient anticoagulation clinic in a community hospital. Am. J. Health-Syst. Pharm. **1996**, *53* (10), 1151–1157.
29. Lodwick, A.; Sajbel, T.A. Patient and physician satisfaction with a pharmacist-managed anticoagulation clinic: Implications for managed care organizations. Managed Care **2000**, *9* (2), 47–50.
30. Dedden, P.; Change, B.; Nagel, D. Pharmacy-managed program for home treatment of deep vein thrombosis with enoxaparin. Am. J. Health-Syst. Pharm. **1997**, *54* (17), 1968–1972.
31. Witt, D.M.; Tillman, D.J. Clinical pharmacy anticoagulation services in a group model health maintenance organization. Pharm. Pract. Manage. Q. **1998**, *18* (3), 34–55.
32. 13 students, pharmacists complete anticoagulation traineeship. ACCP Rep. **2001**, *20* (3).
33. Chenella, F.C.; Klotz, T.A.; Gill, M.A.; Kern, J.W.; McGhan, W.F.; Paulson, Y.J.; Schuttenhelm, K.M.; Cheetham, T.C.; Noguchi, J.K.; McGehee, W.G. Comparison of physician and pharmacist management of anticoagulant therapy of inpatients. Am. J. Hosp. Pharm. **1983**, *40* (10), 1642–1645.
34. Ellis, R.F.; Stephens, M.A.; Sharp, G.B. Evaluation of a pharmacy-managed warfarin-monitoring service to coordinated inpatient and outpatient therapy. Am. J. Hosp. Pharm. **1992**, *49* (2), 387–394.

35. Mamdani, M.M.; Racine, E.; McCreadie, S.; Zimmerman, C.; O'Sullivan, T.L.; Jensen, G.; Ragatzki, P.; Stevenson, J.G. Clinical and economic effectiveness of an inpatient anticoagulation service. Pharmacotherapy **1999**, *19* (9), 1064–1074.
36. Radley, A.S.; Hall, J.; Farrow, M.; Carey, P.J. Evaluation of anticoagulant control in a pharmacist operated anticoagulant clinic. J. Clin. Pathol. **1995**, *48* (6), 545–547.
37. Macgregor, S.H.; Hamley, J.G.; Dunbar, J.A.; Dodd, T.R.P.; Cromarty, J.A. Evaluation of a primary care anticoagulant clinic managed by a pharmacist. BMJ **1996**, *312* (7030), 560.
38. Hughes, E.C.; John, N.R.; Swithenbank, P.J. Setting up and evaluating a pharmacist-led outpatient DVT clinic. Pharm. J. **1999**, *263* (7063), R66–R67.
39. Leach, R.H.; Calvert, P.S. Service developments at harrogate district hospital. Hosp. Pharm. **2000**, *7* (1), 20–23.
40. Radley, A.S.; Dixon, N.; Hall, J. Primary care group anticoagulant clinics. Prim. Care Pharm. **2000**, *1*, 70–72.
41. Dave.Roberts@uhw-tr.wales.nhs.uk, e-mail, September 25, 2000.
42. Macgregor, S. A general practitioner perspective. Prim. Care Pharm. **1999**, *1* (1), 18–19.
43. McCurdy, M. Oral anticoagulation monitoring in a community pharmacy. Am. Pharm. **1993**, *NS33* (10), 61–70.
44. Ansell, J.E.; Hughes, R. Evolving models of warfarin management: Anticoagulation clinics, patient self-monitoring, and patient self-management. Am. Heart J. **1996**, *132* (5), 1095–1100.
45. Bussey, H.I.; Chiquette, E.; Bianco, T.M.; Loweder-Bender, K.; Kraynak, M.A.; Linn, W.D.; Farnett, L.; Clark, G.M. A statistical and clinical evaluation of fingerstick and routine laboratory prothrombin time measurements. Pharmacotherapy **1997**, *17* (5), 861–866.
46. Certified Anticoagulant Care Provider (CACP) exam. Anticoagulation Forum Newsl. **2000**, *5* (1).
47. Mehlberg, J.; Wittowsky, A.K.; Possidente, C. National survey of training and credentialing methods in pharmacist-managed anticoagulation clinics. Am. J. Health-Syst. Pharm. **1998**, *55* (10), 1033–1036.
48. Sixth ACCP Consensus Conference on Antithrombotic Therapy. Dalen, J.E., Hirsh, J., Eds.; Chest, 2001; 119 (1); 1S–370S, (Suppl.).
49. Baglin, T.P.; Rose, P.E.; Walker, I.D.; Machin, S.; Baglin, T.P.; Barrowcliffe, T.W.; Colvin, B.T.; Greaves, M.; Ludlam, C.A.; Mackie, I.J.; Preston, F.E.; Rose, P.E. Guidelines on oral anticoagulation: Third edition. Brit. J. Haem. **1998**, *101*, 374–387.
50. Ansell, J.E.; Buttaro, M.L.; Thomas, O.V.; Knowlton, C.H. Anticoagulation guidelines task force. Consensus guidelines for coordinated outpatient oral anticoagulation therapy management. Ann. Pharmacother. **1997**, *31* (5), 1604–1615.
51. Hiatt, J.; Zablocki, C.J.; Wittowski, A.K. Criteria for use of warfarin in adult inpatients and outpatients. Clin. Pharm. **1993**, *12* (4), 307–313.
52. www.acforum.org (accessed September 2000).
53. Barbara.Ganick@bmc.org, e-mail, September 26, 2000.
54. McIntyre, K. Medicolegal implications of the consensus conference. Chest **2001**, *119* (1), 337S–343S, (Suppl.).
55. Koch, K.E. Trends in collaborative drug therapy management. Drug Benefit Trends **2000**, *12* (1), 45–54.

Association of Faculties of Pharmacy of Canada

James L. Blackburn
University of Saskatchewan, Saskatoon, Canada

INTRODUCTION

The Association of Faculties of Pharmacy of Canada (AFPC) was originally constituted as the Canadian Conference of Pharmaceutical Faculties in 1944. At that time there were seven pharmacy schools in Canada and this organization was formed to enhance pharmacy education in Canada. The name was changed to the Association of Faculties of Pharmacy of Canada in 1969 and the organization currently includes all pharmacy teaching faculty in the nine faculties of pharmacy in Canada.

ORGANIZATION STRUCTURE AND GOVERNANCE

The organization includes the nine faculties of pharmacy in Canada and all faculty members who have the equivalent of 20% FTE appointments within those institutions. There are currently 215 active members, 12 associate members, and 13 affiliate members within the organization.

The organization has a four person executive and a nine member council with each member representing one of the faculties of pharmacy. The executive director serves on a half time basis; the current executive and council members appear in the appendix.

SUMMARY OF MISSION AND KEY OBJECTIVES

The mission statement currently under review) is to develop and implement policies and programs which will provide a forum for exchange of ideas, ensure a liaison with other organizations, and foster and promote excellence in pharmaceutical education and research in Canada.

Primary Objectives

- To foster and promote progress in pharmaceutical education and research.
- To stimulate and provide opportunity for exchange of ideas and discussion among pharmaceutical educators with a view to improving curricula and teaching methods.
- To encourage high and uniform standards of education in pharmacy throughout Canada by assuming an advisory role for development of policies and standards used for the accreditation of programs of pharmaceutical education.
- To establish and maintain liaison with pharmacy and appropriate educational associations, other health professionals, government agencies and members of the pharmaceutical industry that may further the development, support, and improvement of pharmaceutical education, practice, and research.
- To represent, support, and protect the interests of members and to give recognition for achievement.

Current Initiatives

The AFPC has directed resources and energies to a number of activities that are designed to assist faculty become better teachers, to help schools with curricular planning, and to inform the academy of developments in education and learning. Educational outcomes have been established for both the baccalaureate and doctor of pharmacy programs. Recent annual conferences have focused on new teaching methods and means of evaluation of both students and the educational programs. The AFPC has also participated in several initiatives and position papers that were intended to serve the broader interests of the profession in the health care system. One very important issue—the human resource needs within the pharmacy profession—is currently being addressed. This not only includes the number of pharmacy personnel, but also the educational requirements for different levels of those personnel.

Major Meetings

The Association convenes an annual conference in late May or June of each year. The Executive and Council also

Encyclopedia of Clinical Pharmacy
DOI: 10.1081/E-ECP 120006279

hold a mid-year meeting in February with the primary purpose of conducting internal business and meeting with external pharmacy and related organizations.

APPENDIX: AFPC COUNCIL AND EXECUTIVE DIRECTORY (EFFECTIVE SEPTEMBER 12, 2000)

AFPC Executive

Dr. David Fielding, President, Faculty of Pharmaceutical Sciences, University of British Columbia, Vancouver, BC V6T 1Z3; Phone: (604) 822-5447; Fax: (604) 822-3035; E-mail: dwfield@unixg.ubc.ca

Dr. Fred Remillard, President Elect, College of Pharmacy and Nutrition, University of Saskatchewan, 110 Science Place, Saskatoon, SK S7N 5C9; Phone: (306) 966-6345; Fax: (306) 966-6377; E-mail: aj.remillard@usask.ca

Dr. David Hill, Past President, Faculty of Pharmaceutical Sciences, University of British Columbia, Vancouver, BC V6T 1Z3; Phone: (604) 822-4887; Fax: (604) 822-3035; E-mail: dhill@unixg.ubc.ca

Dr. Wayne Hindmarsh, ADPC Represenative, Faculty of Pharmacy, University of Toronto, Toronto, ON M5S 2S2; Phone: (416) 978-2880; Fax: (416) 978-8511; E-mail: wayne.hindmarsh@utoronto.ca

Dr. Jim Blackburn, Executive Director, Assoc. of Faculties of Pharmacy of Canada, 2609 Eastview, Saskatoon, SK S7J 3G7; Phone: (306) 374-6327; Fax: (306) 374-0555; E-mail: jblackburn@sk.sympatico.ca

AFPC Councillors

Simon Albon (2001), Faculty of Pharmaceutical Sciences, University of British Columbia, Vancouver, BC V6T 1Z3; Phone: (604) 822-2497; Fax: (604) 822-3035; E-mail: trout@unixg.ubc.ca

Dr. John Bachynsky (2001), Faculty of Pharmacy and Pharmaceutical Sciences, University of Alberta, Edmonton, AB T6G 2N8; Phone: (780) 492-0202; Fax: (780) 492-1217; E-mail: jbachynsky@pharmacy.ualberta.ca

Dr. Yvonne Shevchuk (2003), College of Pharmacy and Nutrition, University of Saskatchewan, 110 Science Place, Saskatoon, SK S7N 5C9; Phone: (306) 966-6330; Fax: (306) 966-6377; E-mail: shevchuk@duke.usask.ca

Dr. Lavern Vercaigne (2001), Faculty of Pharmacy, University of Manitoba, Winnipeg, MB R3T 2N2; Phone: (204) 474-6043; Fax: (204) 474-7617; E-mail: Lavern_Vercaigne@umanitoba.ca

Zubin Austin (2002), Faculty of Pharmacy, University of Toronto, Toronto, ON M5S 2S2; Phone: (416) 978-0186; Fax: (416) 978-8511; E-mail: zubin.austin@utoronto.ca

Dr. Sylvie Marleau (2003), Faculté de Pharmacie, Université de Montréal, C.P. 6128, Succursale Centre-Ville, Montréal, QC H3C 3J7; Phone: (514) 343-7110 (office); Phone: (514) 343-6110 (ext 3299-lab); Fax: (514) 343-2102; E-mail: sylvie.marleau@umontreal.ca

Dr. Pierre Bélanger (2002), Faculté de Pharmacie, Université Laval, Quebec, PQ G1K 7P4; Phone: (418) 656-2131; Fax: (418) 656-2305; E-mail: Pierre.Belanger@pha.ulaval.ca

Ms. Susan Mansour (2003), College of Pharmacy, Dalhousie University, Halifax NS B3H 3J5; Phone: (902) 494-3504; Fax: (902) 494-1396; E-mail: susan.mansour@dal.ca

Dr. Lili Wang (2002), School of Pharmacy, Memorial University, St. John's, NF A1B 3V6; Phone: (709) 737-7053; Fax: (709) 737-7044; E-mail lwang@morgan.ucs.mun.ca

BIBLIOGRAPHY

Association of Faculties of Pharmacy of Canada; www.afpc.info.

Australian Adverse Drug Reaction Advisory Committee

Christopher P. Alderman
Repatriation General Hospital, Adelaide, Australia

INTRODUCTION

In common with many other nations around the world, in Australia the Commonwealth Department of Health and Aged Care maintains an infrastructure that deals with issues relating to the safety and efficacy of pharmaceutical drug products. The Australian Drug Evaluation Committee (ADEC) was established in 1964 to address this role, and in 1970 a subcommittee known as the Adverse Drug Reactions Advisory Committee (ADRAC) was formed to facilitate the monitoring of medicinal drug safety in Australia. In this regard, ADRAC performs similar functions to those of the U.S. Food and Drug Administration (FDA) and the U.K. Committee on Safety of Medicines (CSM).

ADRAC MEMBERSHIP

The Executive Secretary of ADRAC is Dr. Ian Boyd (Ian.Boyd@health.gov.au), who administers the affairs of the committee. The Chair of the ADRAC committee is Dr. Timothy Mathew, a nephrologist based in Adelaide, South Australia. The current membership of the committee is constituted entirely of senior medical practitioners from locations around Australia. Each member has an extensive background in clinical and academic medicine, and all are highly respected by their peers. The Society of Hospital Pharmacists has recently lobbied without success for the addition of one or more senior clinical pharmacists to the committee. This request has been based on the sizable proportion of reports from the Australian hospital pharmacy sector and the unique skills and training that an appropriate pharmacist could bring to the committee.

REPORTING OF SUSPECTED ADVERSE DRUGS REACTIONS TO ADRAC

ADRAC utilizes a spontaneous, voluntary reporting system to identify suspected adverse reactions to drugs in hospitals and community-based settings. A standard reporting form (sometimes referred to as the ''blue card'') is widely promulgated by ADRAC by dissemination with commonly used Australian drug information resources such as the *Australian Medicines Handbook*, and it can also be downloaded from the Web site www.health.gov.au/tga/docs/html/adr.htm. The form is used to document information about all drug therapy at the time of the suspected adverse drug reaction (ADR), as well as basic demographic and clinical information such as age, gender, height, and weight. There is also space to record other relevant information, such as laboratory indices, relevant history, or previous exposure to the drug. Other information recorded on the form includes the details of any treatment administered and the outcome of the reaction (including sequelae). Completed forms are sent to the ADRAC secretariat in Australia's national capital city, Canberra, where the information is collated and analyzed in the ADRAC secretariat.

The last 10 years have seen substantial growth in the number of ADRs reported to ADRAC. Most recently, ADRAC has received some 13,000 reports per annum, and although the absolute number of reports may appear modest by international comparison, it is important to note that the proportionate rate of reporting (when adjusted for Australia's population) is higher than that of most other developed nations. Recent data suggests that approximately 1% of reports received have a fatal outcome. Approximately 50% of reports received by ADRAC now originate from the pharmaceutical industry, with a further 25% submitted by primary care physicians (also referred to in Australia as general medical practitioners). On the order of 20% of all submissions are received from Australian hospitals, and of these reports the majority are from hospital pharmacists. A small proportion of reports is also received from community-based pharmacists. Unlike the case in other parts of the world, submissions from nurses do not contribute significantly to the total number of reports received by ADRAC each year.

Encyclopedia of Clinical Pharmacy
DOI: 10.1081/E-ECP 120006371

Although ADRAC clearly encourages reporting of all suspected adverse drug reactions in Australia (including those to alternative medicines, including herbal and homeopathic products), the committee has provided guidance in relation to reactions of particular interest. Reactions that result in death, danger to life, admission to hospital, prolongation of hospitalization, absence from productive activity, or increased investigational or treatment costs have been identified as priority areas for ADRAC. In addition to these reports, the committee also requests the reporting of all suspected drug interactions, as well as all reactions thought to have been implicated as a cause of birth defects. Naturally, reaction reports are also sought for drugs that have been newly released onto the Australian market.

ANALYSIS AND USAGE OF ADR DATA BY ADRAC

The information received in ADR reports is considered by the ADRAC committee during its meetings, conducted eight times each year. After analyzing the clinical circumstances and ancillary data described in the report, assignment of a causality rating (in accordance with predefined criteria) is generally the next step. In a relatively small proportion of cases, the ADRAC secretariat may make further contact with the health practitioner who made the original report, seeking to gain additional clarifying information or to ascertain the outcome or sequelae of the suspected reaction. Once processed, the information is entered into a relational database that now stores details of many thousands of ADR reports. Health professionals such as clinical pharmacists can contact the ADRAC secretariat for information about possible reactions. Data available on request include the number of reactions (including reactions of a particular type) that have been received for any drug, as well as causality ratings assigned by the committee for individual reports. Upon further request, it is possible to access detailed clinical information about individual ADR reports. This service is provided in a very timely fashion, with the turnaround time for the feedback of information routinely less than 24 hours.

In addition to providing access to the information stored in the database, ADRAC also utilizes the information received in reports for a range of other purposes. The ADRAC bulletin is published four times each year and is widely distributed to medical practitioners and pharmacists free of charge. The bulletin, which is also available on the Internet (www.health.gov.au/tga/docs/html/aadrbidx.htm), summarizes details of common or important drug reactions and interactions and in this way serves an important educative function. Information from ADR reports is also summarized into reports that are published up to three times a year in the *Medical Journal of Australia*. ADR data from ADRAC are also forwarded to the Collaborating Centre for International Drug Monitoring of the World Health Organization in Uppsala, Sweden.

CONCLUSION

In summary, the ADRAC committee performs a unique and important public health role in Australia. The work of the committee and the secretariat provides vital support for Australian health care workers seeking information about the adverse effects of drugs. In return, ADRAC enjoys strong support from doctors, pharmacists, and the pharmaceutical industry in the form of voluntary, spontaneous reports of ADRs. In this way, ADRAC works with other stakeholders to make a positive contribution to the quality use of medicines in Australia.

BIBLIOGRAPHY

Roeser, H.P.; Rohan, A.P. Post-marketing surveillance of drugs. The spontaneous reporting scheme: Role of the Adverse Drug Reactions Advisory Committee. Med. J. Aust. **1990**, *153*, 720–726.

Australian Medicines Handbook

Simone Rossi
Australian Medicines Handbook Pty Ltd., Adelaide, Australia

INTRODUCTION

The Australian Medicines Handbook (AMH) is a compendium of drug and therapeutic information. It was developed to provide pharmacists, doctors, other health professionals, and their students with concise, independent and comparative information about drugs and quality drug use in Australia.

PHILOSOPHY

A compendium of drug and therapeutic information, the AMH is an initiative of Australia's National Health Policy in response to professional, consumer, and government concerns about the lack of independent drug information resources in Australia. AMH was developed to provide pharmacists, doctors, other health professionals, and their students with independent and comparative information about drugs.

AMH was initially modeled on the British National Formulary (BNF), and evolved to incorporate further comparative and therapeutic information. The best available evidence is used to support recommendations, thus discouraging drug use where evidence is lacking or poor.

AMH publications are designed as practice, teaching, and learning tools, and aim to promote Quality Use of Medicines (QUM) by providing readily accessible, concise, up-to-date, clinically relevant information that facilitates effective, rational, safe, and economical prescribing and dispensing.

AMH complements other independent Australian publications about drugs and therapeutics including Australian Prescriber, the Therapeutic Guideline series, and National Prescribing Service publications. It provides a different focus to drug information compendia based on government-approved Product Information.

KEY HISTORY

The concept of a "national formulary" modeled on the BNF was recommended in 1991 following a meeting convened by the Australian Society of Clinical and Experimental Pharmacologists and Toxicologists (ASCEPT) and the Consumers' Health Forum. In June 1995, money was granted by the Australian Government to develop a drug information database suitable for publishing printed and electronic versions of an Australian Medicines Handbook. The publishing phase was not initially funded. First staff were appointed in December 1995.

In May 1998, the first edition, AMH 1998, was published. The second edition, AMH 2000, was published in February 2000. CD-ROM and Internet products, based on second edition content, were released in May 2000, with upgrades released in April 2001. A set of third edition products is planned for release in early 2003.

Australian Medicines Handbook Pty Ltd., the business entity that owns the content and publishes the products, was established with Australian government funding. It received its final funding in March 1999, and now operates on a fully commercial basis. The three shareholders of the company are the Royal Australian College of General Practitioners, the Pharmaceutical Society of Australia, and the Australasian Society for Clinical and Experimental Pharmacologists and Toxicologists.

UPTAKE IN MAJOR TARGET GROUPS

The first edition was available only in print version (book) and sold approximately 9500 copies. The second edition, AMH 2000, is expected to at least match this figure in print version sales. CD-ROM and Internet versions based on the second edition have made a significant impact on uptake.

Encyclopedia of Clinical Pharmacy
DOI: 10.1081/E-ECP 120006366

Uptake by community and hospital pharmacists and undergraduate pharmacy training is very high. Most medical students in Australia now use AMH. In teaching hospitals, uptake by nurses and doctors is moderate to high but varies by state. Although uptake in general practice is moderate (about 10 to 20%), it appears to be steadily increasing (1).

CONTENT FEATURES

In Australia, the most commonly utilized and accessible information sources about prescription drugs are based on a drug's Product Information, the document approved by Australia's drug regulation authority, the Therapeutic Goods Administration (TGA).

AMH content is written after consideration of best available scientific evidence, the Product Information, standard international reference sources, and Australian evidence-based and consensus guidelines from government and nongovernment agencies. Information is concise and clinically relevant for Australian practice, providing comparisons between drugs and between drug classes, with a focus on comparative efficacy, safety, and cost. Key advice for patients about how to use medicines safely and effectively is included.

Types of Information Presented

Information in the AMH is organized into 20 chapters, according to organ-system (e.g., neurological drugs). Each chapter contains brief disease treatment summaries (e.g., hypertension), which incorporate discussion of various treatments, including nondrug and complementary therapies, and comparison of different classes of drugs. Treatment summaries also include advice for appropriate management for specific groups of patients or stages of disease.

Drug profiles are arranged by generic drug name and listed under drug class profiles (e.g., ACE inhibitors); this allows for comparison across a class to be made, and minimizes repetition of information common to all members of the class. Indications listed in monographs are generally those that are approved by TGA. Additional accepted clinical uses such as minor indications for which there are few or no other alternative drugs and there is evidence to support such use are also included. Dosage information is clear and concise; interactions information is limited to those likely to be important clinically, together with advice on management. Information and recommendations regarding use of a particular drug in pregnancy, breastfeeding, renal or hepatic impairment, children, and the elderly are included. AMH documents frequently include "Practice points," which include brief information, advice, and tips that are important for the safe and effective use of a particular drug or drug class.

General principles of drug use for groups of drugs, such as anti-infectives, antineoplastics, and ocular drugs, are also provided.

BENEFITS

Although passive provision of this high-quality targeted information alone is unlikely to change prescribing behavior and improve health outcomes, access to impartial drug information is recognized as a component of the WHO/INRUD drug use indicators (2) and is an identified strategy in the implementation plan of Australia's National Prescribing Service.

The perceived benefits of AMH are:

- It is a good starting point for drug-related questions, and can save time by providing reliable up-to-date information in a user-friendly format.
- It provides practical knowledge and advice that is independent of government and acts as a balance to promotional information provided by the pharmaceutical industry.
- It is written in minimally technical language and can be used as an aid to patient consultations.
- It is suited to both hospital and community use.

AMH has been credentialed by key health professional organizations such as the Australian Medical Association Council of General Practice, the Society of Hospital Pharmacists of Australia, the Australian Divisions of General Practice, and the National Prescribing Service. Consumer groups such as Consumers Health Forum and government advisory groups such as the Pharmaceutical Health and Rational Use of Medicines committee and the Australian Pharmaceutical Advisory Committee have also given support.

CONTENT

Creation

AMH content is derived in-house by a small team of highly skilled and experienced editors, most of whom are pharmacists. Editors possess postgraduate training and experience in areas such as drug information and critical appraisal, pharmacoepidemiology, pharmacoeconomics,

and editing. Work experiences are broad and include the pharmaceutical industry, academic detailing, hospital-based clinical pharmacy, and medical writing. In addition, a small number of paid external contributing writers help expand and update specific parts of the content. In addition to regular surveillance of published pharmacotherapeutics literature, spontaneous feedback from readers assists in developing and refining content.

Review

The review process for the scientific aspects of the content involves four steps. After initial drafting or updating by an AMH editor or external contributor, a second AMH editor will review material for adherence to house style and for scientific content. Content is modified based on this review. The draft material is then sent out in parallel to members of the Editorial Advisory Board, and a Review Panel (external reviewers) who are specifically recruited to look at particular sections. Information regarding drug use in renal impairment and pregnancy or breastfeeding is reviewed by external specialist clinical pharmacists. After consideration of all reviewers' comments, a final draft is created and sent to the Editorial Advisory Board for approval.

In this way, a combination of experts and end-users review AMH content. At present, a team of about 150 external reviewers assist in the Review Panel process. They include general practitioners and specialist physicians and surgeons, academics and researchers, hospital and community pharmacists, specialist nurses, and educators from organizations that support consumers with chronic illnesses. Each Review Panel contains expert specialist clinicians, a clinical pharmacologist with an interest in the specific area, hospital-based clinical pharmacists, community pharmacists, general practitioners (urban and rural), and allied health workers when relevant (e.g., diabetes and asthma nurses and educators). These reviewers are not remunerated, but they receive a complimentary copy of AMH, and their contribution is acknowledged in the publications.

AMH Editorial Advisory Board

The Editorial Advisory Board contains academics and practitioners in pharmacy, clinical pharmacology, and general practice. Some members have substantial experience in medical editing and publishing. They are asked to declare any potential conflicts of interest.

As well as assisting in the review process, the Editorial Advisory Board helps set editorial policy and approve plans for new content or major update.

FUTURE DEVELOPMENTS

Although a traditional publishing process was used in the creation of the first two editions of AMH books, the need to provide users with multiple formats has required the work environment to be re-engineered and editors now work in an SGML environment. This allows production of reports such as a print-ready file (from which the book is published); an HTML version that serves as the current CD-ROM and web-based products; and potentially other formats or subsets of AMH data, from one set of source files.

Future developments include adding a production pathway to allow a personal digital assistant version of AMH to be produced, and incorporation of an XML database into the work environment, which will allow for complex queries to be performed on the data set.

Best-Practices Documents (ASHP)

Joseph H. Deffenbaugh
*American Society of Health-System Pharmacists,
Bethesda, Maryland, U.S.A.*

INTRODUCTION

The American Society of Health-System Pharmacists (ASHP) develops and publishes professional Best-Practices documents that cover a wide range of clinical practice and therapeutic topics. There is a range of detail among types of Best-Practices documents: Statements documents express basic philosophy and Guidelines offer programmatic advice. Of the two types of therapeutic documents, Therapeutic Guidelines are thorough discussions of drug use, whereas Therapeutic Position Statements are concise responses to specific therapeutic issues.

TOPICS

The topics covered among the types of Best-Practices documents are varied and dynamic. Documents are continuously developed, reviewed, updated, and discontinued on the basis of assessments of relevance to contemporary clinical practices and trends. Currently, a sampling of active, clinically related documents includes the following topics summarized by document type:

- *ASHP Statement*: A declaration and explanation of basic philosophy or principle, as approved by the Board of Directors and the House of Delegates. ASHP Statements cover such topics as the pharmacist's role in clinical pharmacokinetic monitoring, infection control, and primary care.
- *ASHP Guideline*: Advice on the implementation or operation of pharmacy practice programs, as approved by the Board of Directors. ASHP Guidelines include, among other topics, pharmacists' activities in adverse drug-reaction monitoring and reporting, medication-use evaluation, patient education and counseling, development of clinical care plans, provision of medication information, and surgery and anesthesiology pharmaceutical services.
- *ASHP Therapeutic Guideline*: Thorough, systematically developed advice for healthcare professionals on appropriate use of medications for specific clinical circumstances, as approved by the Board of Directors. ASHP Therapeutic Guidelines cover surgical and nonsurgical antimicrobial prophylaxis, stress ulcer prophylaxis, pharmacologic management of nausea and vomiting caused by chemotherapy or radiation therapy, and the use of angiotensin-converting enzyme inhibitors in patients with left ventricular dysfunction.
- *ASHP Therapeutic Position Statement*: Concise statements that respond to specific therapeutic issues of concern to healthcare consumers and pharmacists, as approved by the Board of Directors. ASHP Therapeutic Position Statements include such topics as identifying and preventing pneumococcal resistance, preventing and treating multidrug-resistant tuberculosis, use of aspirin for prophylaxis of myocardial infarction, recognition and treatment of depression in older adults, smoking cessation, and optimizing treatment of hypertension.

PURPOSE OF ASHP BEST-PRACTICES DOCUMENTS

ASHP's Best-Practices documents represent a consensus of professional judgment, expert opinion, and documented evidence. They provide guidance and direction to ASHP members and pharmacy practitioners and to other audiences related to pharmacy practice. Their use may help with compliance to federal and state laws and regulations, to meet accreditation requirements, and to improve pharmacy practice and patient care. Best-Practices documents are written to establish goals that are progressive and challenging yet attainable in applicable health-system settings. They generally do not represent minimum levels of practice, unless titled as such, and should not be viewed as ASHP requirements. The use of ASHP's documents by members and other practitioners is strictly voluntary. Their content should be assessed and adapted to the needs of local health-system settings on the basis of independent judgment.

Encyclopedia of Clinical Pharmacy
DOI: 10.1081/E-ECP 120006405

DEVELOPMENT OF BEST-PRACTICES DOCUMENTS

The processes used to draft and review new or revised Best-Practices documents vary, depending on the body responsible for their development and on the type of document. These processes are described in the following. Once approved, documents become official ASHP policy and are published in the *American Journal of Health-System Pharmacy*, added to ASHP's Web site, and incorporated into the next edition of *Best-Practices for Health-System Pharmacy*.

ASHP Statements and Guidelines

Any of the ASHP policy-recommending bodies (councils and commissions) may initiate and oversee the development of ASHP Statements and Guidelines; however, most of them are initiated by the Council on Professional Affairs. The development of these documents generally includes the following steps:

- A team of up to five individuals is selected from volunteers on the basis of their demonstrated knowledge of the topic and practice settings. The team develops a preliminary draft. Drafters are usually ASHP members; however, depending on the subject, the team may include nonmember pharmacists and representatives of other healthcare disciplines.
- ASHP distributes drafts to reviewers who have interest and expertise in the topic. Reviewers consist of members, various ASHP bodies, and representatives of other health care disciplines and professional organizations. A draft may be presented at an open hearing or in a network forum during an ASHP Annual or Midyear Clinical Meeting, or may be posted on ASHP's Web site for comment.
- Drafts are revised on the basis of comments and a review of the literature, evaluated for content and quality, and submitted to the appropriate ASHP policy-recommending body for action. That body may suggest further revisions or recommend approval by the ASHP Board of Directors.

ASHP Therapeutic Guidelines and Position Statements

The Commission on Therapeutics has responsibility for the development of ASHP Therapeutic Guidelines and ASHP Therapeutic Position Statements.

ASHP therapeutic guidelines

The development of these documents generally includes the following steps:

- When the Commission on Therapeutics (COT) identifies a topic for Therapeutic Guideline development, ASHP formally solicits proposals for a contractual arrangement with an individual, group, or organization to draft the Guidelines document and coordinate its review. The contractor will work with a panel of six to ten experts appointed by ASHP who have diverse backgrounds relevant to the topic.
- A systematic analysis of the literature is performed, and scientific evidence is evaluated on the basis of predetermined criteria. Recommendations in the document rely on scientific evidence or expert consensus. When expert judgment must be used, the document indicates the scientific reasoning that influenced the decision. Scientific evidence takes precedence over expert judgment. Each recommendation is accompanied by projections of the relevant health and cost outcomes that could result.
- The expert panel and COT review every draft of the Guidelines document and provide comments. This process is repeated until the expert panel and COT are satisfied with the content.
- ASHP solicits multidisciplinary input on the final draft. Reviewers consist of members and selected individuals knowledgeable in the content area, representatives of various ASHP bodies, and other professional organizations.
- Once the above processes are completed, COT recommends that the ASHP Board of Directors approve the Guideline.

ASHP therapeutic position statements

The development of these documents generally includes the following steps:

- One or more experts on a given topic is assigned to draft the Therapeutic Position Statement. Drafters are selected on the basis of demonstrated knowledge of the topic and practice setting. Most often, the drafters are ASHP members.
- The proposed draft document is reviewed by COT, which may suggest modifications. This process is repeated until COT is satisfied with the content.
- ASHP solicits multidisciplinary input on the draft. Reviewers consist of members and selected indivi-

duals knowledgeable in the content area, representatives of various ASHP bodies, and other professional organizations.

- Once the above processes are completed, COT finalizes the draft and recommends that the ASHP Board of Directors approve it.

Timeliness

The goal is to have an approvable draft within one year of the initial decision to develop a new or revised practice standard. Development usually takes one to three years, depending on the availability of drafters, the strength of the evidence, the accumulated practice experience, and the extent of the reviews and revisions. ASHP Best-Practices documents are dynamic: Therapeutic documents are reviewed every three years, and Practice Statements and Guidelines every five years and are revised as needed.

ACCESS TO BEST-PRACTICES DOCUMENTS

Besides being published in *AJHP* and the *Best Practices for Health-System Pharmacy*, Best-Practice documents are available through ASHP's Fax On-Demand service by calling (301)664-8888 and on ASHP's Web site at www.ashp.org/bestpractices/index.html.

FURTHER READING

www.ashp.org/bestpractices/index.html.

Biopharmaceutics

Leon Shargel
Eon Labs Manufacturing, Inc., Laurelton, New York, U.S.A.

Andrew B.C. Yu[a]
U.S. Food and Drug Administration, Rockville, Maryland, U.S.A.

INTRODUCTION

Biopharmaceutics is the study of the interrelationship of the physicochemical properties of the drug [active pharmaceutical ingredient, (API)] and the drug product (dosage form in which the drug is fabricated) based on the biological performance of the drug (Table 1).

Biopharmaceutics also considers the impact of the various manufacturing methods and technologies on the intended performance of the drug product. Biopharmaceutics uses quantitative methods and theoretical models (1) to evaluate the effect of the drug substance, dosage form, and routes of drug administration on the therapeutic requirements of the drug and drug product in a physiological environment.

Bioavailability is often used as a measure of the biological performance of the drug and is defined as a measure of the rate and extent (amount) to which the active ingredient or active moiety becomes available at the site of action. Bioavailability is also a measure of the rate and extent of therapeutically active drug that is systemically absorbed.

Biopharmaceutics allows for rational design of drug products to deliver the drug at a specific rate to the body in order to optimize the therapeutic effect and minimize any adverse effects. As shown in Table 1, biopharmaceutics is based on the physicochemical characteristics of the active drug substance, the desired drug product, and considerations of the anatomy and physiology of the human body (1). Inherent in the design of a suitable drug product is knowledge of the pharmacodynamics of the drug, including the desired onset time, duration, and intensity of clinical response, and the pharmacokinetics of the drug including absorption, distribution, elimination, and target drug concentration.

Thus, biopharmaceutics involves factors that influence the: 1) protection and stability of the drug within the drug product; 2) the rate of drug release from the drug product; 3) the rate of dissolution of the drug at the absorption site; and 4) the availability of the drug at its site of action (Fig. 1).

BIOPHARMACEUTIC CONSIDERATIONS IN DRUG PRODUCT DESIGN

Drugs are generally given to a patient as a manufactured drug product (finished dosage form) that includes the active drug and selected ingredients (excipients) that make up the dosage form. Common pharmaceutical dosage forms include liquids, tablets, capsules, injections, suppositories, transdermal systems, and topical drug products. The formulation and manufacture of a drug product requires a thorough understanding of the biopharmaceutics.

Each route of drug application presents special biopharmaceutic considerations in drug product design (Table 2). Systemic drug absorption from an extravascular site is influenced by the anatomic and physiologic properties of the site and the physicochemical properties of the drug and the drug product. The anatomy, physiology, and the contents of the gastrointestinal tract (GI) are considered in the design of a drug product for oral administration. For example, considerations in the design of a vaginal tablet formulation for the treatment of a fungus infection include whether the ingredients are compatible with vaginal anatomy and physiology, whether the drug is systemically absorbed from the vagina and how the vaginal tablet is to be properly inserted and placed in the appropriate area for optimum efficacy. Requirements for an eye medication include pH, isotonicity, sterility, local irritation to the cornea, draining of the drug by tears, and concern for systemic drug absorption. An additional consideration might be the contact time of the medication with the cornea. Although, increased eye contact time might be achieved by an increase in viscosity of the ophthalmic solution, the patient may lose some visual acuity when a viscous product is administered. Biopharmaceutic considerations for a drugs administered

[a]The content in this article reflects the view of the authors and does not represent the view of FDA.

Encyclopedia of Clinical Pharmacy
DOI: 10.1081/E-ECP 120006419

Table 1 Biopharmaceutic considerations in drug product design

Active pharmaceutical ingredient (API)		
	Stability Solubility pH and pKa Crystalline form (polymorph) Excipient interaction and compatability	Impurities Salt form Particle size Complexation
Drug product		
	Type of drug product (capsule, tablet, solution, etc.) Immediate or modified release Dosage strength Bioavailability	Stability Excipients Manufacturing variables
Physiologic factors		
	Route of administration Permeation of drug across cell membranes Binding to macromolecules	Blood flow Surface area Biotransformation
Pharmacodynamic and pharmacokinetic considerations		
	Bioavailability Therapeutic objective Adverse reactions	Pharmacokinetics Dose Toxic effects
Manufacturing considerations		
	Production methodology and technology Quality control/quality assurance Specification of raw materials	Cost Stability testing
Patient considerations	Compliance, labeling, and product acceptance	Cost

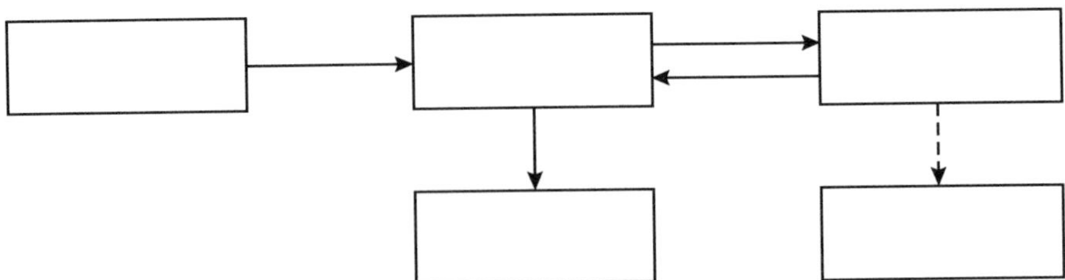

Fig. 1 Scheme demonstrating the dynamic relationships among the drug, the product, and pharmacologic effect. (From Ref. 1.)

by intramuscular injection include, local irritation, drug dissolution, and drug absorption from the injection site.

Biopharmaceutic studies may be performed using in vitro or in vivo methods (Table 3). In vitro methods are useful (2–6) to understand the physico-chemical properties of the drug and drug product and to evaluate the quality of the manufacturing process. Ultimately, the drug must be studied in vivo, in humans to assess drug efficacy, including the pharmacodynamic, pharmacokinetic, therapeutic and toxic profiles. Drug dissolution, absorption, metabolism, and potential interaction with food and other components in the GI tract are major biopharmaceutic topics for research and regulatory considerations in drug development.

A drug given by intravenous administration is considered complete or 100% bioavailable because the drug is placed directly into the systemic circulation. By carefully choosing the route of drug administration and proper design of the drug product, drug bioavailability can be varied from rapid and complete systemic drug absorption to a slow, sustained rate of absorption or even virtually no absorption, depending on the therapeutic objective. Once the drug is systemically absorbed, normal physiologic processes for distribution and elimination occur, which usually is not influenced by the specific formulation of the drug. The rate of drug release from the product, and the rate of drug absorption, are important in determining the onset, intensity, and duration of drug action of the drug.

RATE-LIMITING STEPS IN ORAL DRUG ABSORPTION

Systemic drug absorption from a drug product consists of a succession of rate processes (Fig. 2). For solid oral, immediate release drug products (e.g., tablet, capsule), the rate processes include 1) disintegration of the drug product and subsequent release of the drug; 2) dissolution of the drug in an aqueous environment; and 3) absorption across cell membranes into the systemic circulation. In the process of drug disintegration, dissolution, and absorption, the rate at which drug reaches the circulatory system is determined by the slowest step in the sequence.

The slowest step in a kinetic process is the rate-limiting step. Except for controlled release products, disintegration of a solid oral drug product is usually more rapid than drug dissolution and drug absorption. For drugs that have very poor aqueous solubility, the rate at which the drug dissolves (dissolution) is often the slowest step, and therefore exerts a rate-limiting effect on drug bioavailability. In contrast, for a drug that has a high aqueous solubility, the dissolution rate is rapid and the rate at which the drug crosses or permeates cell membranes is the slowest or rate-limiting step.

PHYSIOLOGIC FACTORS AFFECTING DRUG ABSORPTION

Passage of Drugs Across Cell Membranes

For systemic absorption, a drug must pass from the absorption site through or around one or more layers of cells to gain access into the general circulation. The permeability of a drug at the absorption site into the systemic circulation is intimately related to the molecular structure of the drug and the physical and biochemical properties of the cell membranes. For absorption into the cell, a drug must traverse the cell membrane. Transcellular absorption is the process of a drug movement across a cell. Some polar molecules may not be able to traverse the cell membrane, but instead, go through gaps or "tight junctions" between cells, a process known as paracellular drug absorption. Some drugs are probably absorbed by a mixed mechanism involving one or more processes.

Passive diffusion

Passive diffusion is the process by which molecules spontaneously diffuse from a region of higher concentration to a region of lower concentration. This process is passive because no external energy is expended. Drug

Table 2 Common routes of drug administration

Route	Bioavailability	Advantages	Disadvantages
Parenteral routes			
Intravenous bolus (IV)	Complete (100%) systemic drug absorption. Rate of bioavailability considered instantaneous.	Drug is given for immediate effect.	Increased chance for adverse reaction. Possible anaphylaxis.
Intravenous infusion (IV inf)	Complete (100%) systemic drug absorption.	Plasma drug levels more precisely controlled. May inject large fluid volumes.	Requires skill in insertion of infusion set.
	Rate of drug absorption controlled by infusion pump.	May use drugs with poor lipid solubility and/or irritating drugs.	Tissue damage at site of injection (infiltration, necrosis, or sterile abscess).
Intramuscular injection (IM)	Rapid from aqueous solution.	Easier to inject than intravenous injection.	Irritating drugs may be very painful.
	Slow absorption from nonaqueous (oil) solutions.	Larger volumes may be used compared to subcutaneous solution.	Different rates of absorption depending upon muscle group injected and blood flow.
Subcutaneous injection (SC)	Prompt from aqueous solution.	Generally, used for insulin injection.	Rate of drug absorption depends upon blood flow and injection volume.
	Slow absorption from repository formulations.		
Enteral Routes			
Buccal or sublingual (SL)	Rapid absorption from lipid-soluble drugs.	No “first-pass” effects.	Some drug may be swallowed. Not for most drugs or drugs with high doses.
Oral (PO)	Absorption may vary. Generally slower absorption rate compared to IV bolus or IM injection.	Safest and easiest route of drug administration. May use immediate-release and modified-release drug products.	Some drugs may have erratic absorption, be unstable in the gastointestinal tract, or be metabolized by liver prior to systemic absorption.
Rectal (PR)	Absorption may vary from suppository.	Useful when patient cannot swallow medication.	Absorption may be erratic. Suppository may migrate to different position.
	More reliable absorption from enema (solution).	Used for local and systemic effects.	Some patient discomfort.
Other routes			
Transdermal	Slow absorption, rate may vary.	Transdermal delivery system (patch) is easy to use.	Some irritation by patch or drug. Permeability of skin variable with condition, anatomic site, age, and gender.
	Increased absorption with occlusive dressing.	Used for lipid-soluble drugs with low dose and low MW.	Type of cream or ointment base affects drug release and absorption.
Inhalation	Rapid absorption. Total dose absorbed is variable.	May be used for local or systemic effects.	Particle size of drug determines anatomic placement in respiratory tract. May stimulate cough reflex. Some drug may be swallowed.

(From Ref. 1.)

Table 3 Examples of in vitro and in vivo biopharmaceutic studies

Biopharmaceutic studies (in vivo)	Bioavailability study	Measurement of drug in plasma, urine or other tissues
	Acute pharmacologic effect	Measurement of a pharmacodynamic effect, e.g., FEV_1, blood pressure, heart rate, skin blanching
	Clinical study	Measurement of drug efficacy
Biopharmaceutic studies (in vitro)	Drug release/dissolution	Measurement of the rate of drug dissolved under specified conditions
	Drug permeability	Use of CACO2 cells (an isolated colon cell line) are grown into membranes to study the intestinal permeability and gut metabolism of drugs.
	Drug biotransformation (metabolism)	Use of liver cells, homogenates or isolated cytochrome P450 isozymes to drug study biotransformation.

molecules move randomly forward and back across a membrane (Fig. 3). If the two regions have the same drug concentration, forward-moving drug molecules will be balanced by molecules moving back, resulting in no net transfer of drug. For a region that has a higher drug concentration, the number of forward-moving drug molecules will be higher than the number of backward-moving molecules, resulting in a transfer of molecules to the region with the lower drug concentration, as indicated by the big arrow. Flux is the rate of drug transfer and is represented by a vector to show its direction. Molecules tend to move randomly in all directions because molecules possess kinetic energy and constantly collide with each other in space. Only left and right molecule movements are shown in Fig. 3, because movement of molecules in other directions would not result in concentration changes because of the limitation of the container wall.

Passive diffusion is the major transmembrane process for most drugs. The driving force for passive diffusion is the difference in drug concentrations on either side of the cell membrane. According to Fick's Law of Diffusion, drug molecules diffuse from a region of high drug concentration to a region of low drug concentration

$$dQ/dt = \{DAK/h\}(C_{GI} - C_p)$$

where dQ/dt = rate of diffusion; D = diffusion coefficient; K = partition coefficient; A = surface area

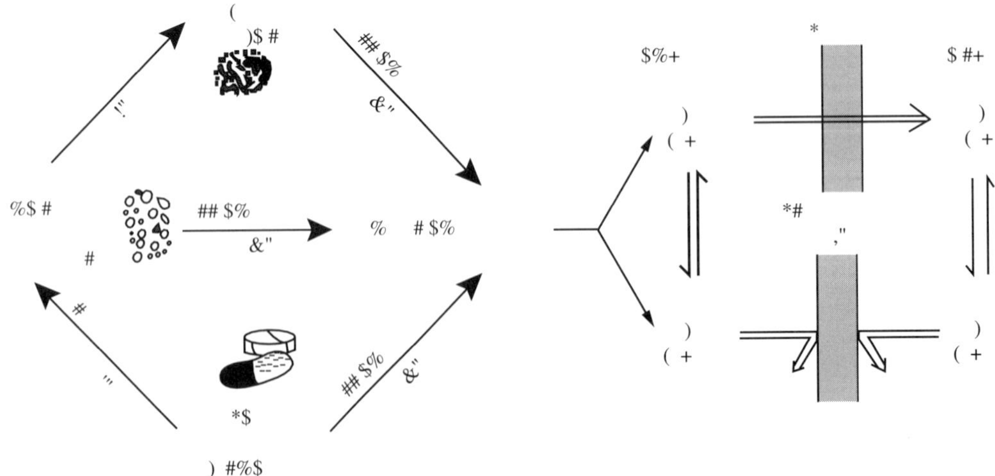

Fig. 2 Summary of processes involved following the oral administration of a drug in tablet or capsule form. (From Blanchard, J. Gastrointestinal absorption. II. Formulation factors affecting bioavailability. Am. J. Pharm. **1978**, *150*, 132–151.)

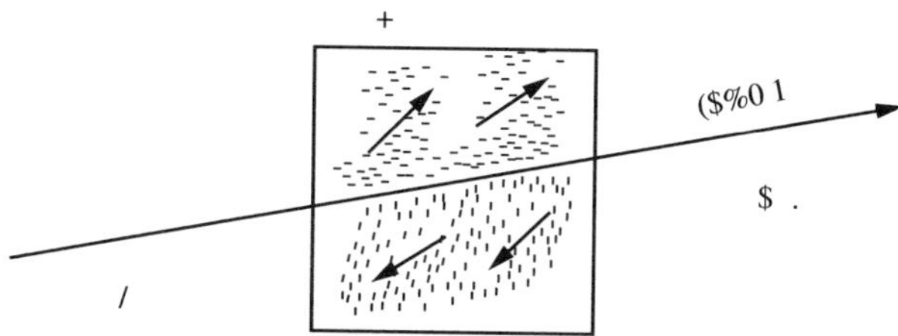

Fig. 3 Passive diffusion of molecules. Molecules in solution diffuse randomly in all directions. As molecules diffuse from left to right and vice versa (small arrows), a net diffusion from the high-concentration side to the low-concentration side results. This results in a net flux (J) to the right side. Flux is measured in mass per unit area (e.g., mg/cm^2). (From Ref. 1.)

of membrane; h = membrane thickness; and $C_{GI} - C_p$ = difference between the concentrations of drug in the GI tract and in the plasma.

Drug distributes rapidly into a large volume after entering the blood resulting in a very low plasma drug concentration with respect to the concentration at the site of drug administration. Drug is usually given in milligram doses, whereas plasma drug concentrations are often in the microgram per milliliter or nanogram per milliliter range. For drugs given orally, $C_{GI} \gg C_p$. A large concentration gradient is maintained driving drug molecules into the plasma from the GI tract.

As shown by Fick's Law of Diffusion, lipid solubility of the drug and the surface area and the thickness of the membrane influence the rate of passive diffusion of drugs. The partition coefficient, K, represents the lipid–water partitioning of a drug. More lipid soluble drugs have larger K values that theoretically increase the rate of systemic drug absorption. In practice, drug absorption is influenced by other physical factors of the drug, limiting its practical application of K. The surface area of the membrane through which the drug is absorbed directly influences the rate of drug absorption. Drugs may be absorbed from most areas of the GI tract. However, the duodenal area of the small intestine shows the most rapid drug absorption due to such anatomic features as villi and microvilli, which provide a large surface area. These villi are not found in such numbers in other areas of the GI tract.

The membrane thickness, h, is a constant at the absorption site but may be altered by disease. Drugs usually diffuse very rapidly into tissues through capillary cell membranes in the vascular compartments. In the brain, the capillaries are densely lined with glial cells creating a thicker lipid barrier (blood–brain barrier) causing a drug to diffuse more slowly into brain. In certain disease states (e.g., meningitis) the cell membranes may be disrupted or become more permeable to drug diffusion.

Many drugs have lipophilic and hydrophilic substituents. More lipid soluble drug molecules traverse cell membranes more easily than less lipid-soluble (i.e., more water-soluble) molecules. For weak electrolyte drugs (i.e, weak acids, bases), the extent of ionization influences drug solubility and the rate of drug transport. Ionized drugs are more water soluble than nonionized drugs which are more lipid soluble. The extent of ionization of a weak electrolyte depends on the pKa of the drug and the partition hypothesis (pH) of the medium in which the drug is dissolved. The Henderson and Hasselbalch equation describes the ratio of ionized (charged) to unionized form of the drug and is dependent on the pH conditions and the pKa of the drug:

For weak acids,

$$\text{Ratio} = -\frac{(\text{salt})}{(\text{acid})} = \frac{(\text{A}^-)}{(\text{HA})} = 10^{(\text{pH}-\text{pKa})}$$

For weak bases,

$$\text{Ratio} = -(\text{base})\,(\text{salt}) = (\text{RNH})2\,(\text{RNH}^{+3}) = 10^{(\text{pH}-\text{pKa})}$$

According to the pH, a weak acid (e.g., salicylic acid) should be rapidly absorbed from the stomach (pH 1.2) due to a favorable concentration gradient of the unionized (more lipid soluble) drug from the stomach to the blood, because practically all the drug in the blood compartment is dissociated (ionized) at pH 7.4. A weak base (e.g., quinidine) is highly ionized in acid pH and is poorly absorb from the stomach. Although many drugs obey by the pH, in practice, the major site of absorption of most drugs is usually in the small intestine (duodenum) due presence of a large surface area and high blood flow.

The drug concentration on either side of a membrane is also influenced by the affinity of the drug for a tissue component, which prevents the drug from freely moving back across the cell membrane. For example, drug that binds plasma or tissue proteins causes the drug to concentrate in that region. Dicumarol and sulfonamides strongly bind plasma proteins; whereas, chlordane, a lipid-soluble insecticide, partitions and concentrates into adipose (fat) tissue. Tetracycline forms a complex with calcium and concentrates in the bones and teeth. Drugs may concentrate in a tissue due to a specific uptake or active transport process. Such processes have been demonstrated for iodide in thyroid tissue, potassium in the intracellular water, and certain catecholamines in adrenergic storage sites.

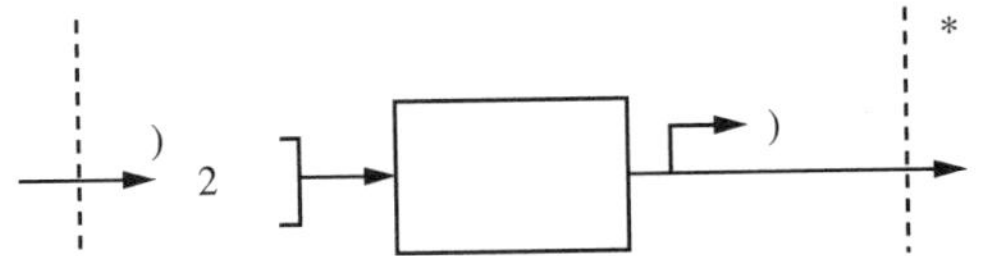

Fig. 4 Hypothetical carrier-mediated transport process. (From Ref. 1.)

Carrier-mediated transport

Theoretically, a lipophilic drug may pass through the cell or go around it. If drug has a low molecular weight and is lipophilic, the lipid cell membrane is not a barrier to drug diffusion and absorption. In the intestine, molecules smaller than 500 MW may be absorbed by paracellular drug absorption. Numerous specialized carrier-mediated transport systems are present in the body especially in the intestine for the absorption of ions and nutrients required by the body.

Active transport: Active transport is a carrier-mediated transmembrane process that is important for GI absorption of some drugs and also involved in the renal and biliary secretion of many drugs and metabolites. A carrier binds the drug to form a carrier–drug complex that shuttles the drug across the membrane and then dissociates the drug on the other side of the membrane (Fig. 4). Active transport is an energy-consuming system characterized by the transport of drug against a concentration gradient, that is, from regions of low drug concentrations to regions of high concentrations.

A drug may be actively transported, if the drug molecule structurally resembles a natural substrate that is actively transported. A few lipid-insoluble drugs that resemble natural physiologic metabolites (e.g., 5-fluorouracil) are absorbed from the GI tract by this process. Drugs of similar structure may compete for adsorption sites on the carrier. Because only a certain amount of carrier is available, the binding sites on the carrier may become saturated at high drug concentrations. In contrast, passive diffusion is not saturable.

Facilitated diffusion: Facilitated diffusion is a non-energy requiring, carrier-mediated transport system in which the drug moves along a concentration gradient (i.e., moves from a region of high drug concentration to a region of low drug concentration). Facilitated diffusion is saturable, structurally selective for the drug and shows competition kinetics for drugs of similar structure. Facilitated diffusion seems to play a very minor role in drug absorption.

Carrier-mediated intestinal transport: Various carrier mediated systems (transporters) are present at the intestinal brush border and basolateral membrane for the absorption of specific ions and nutrients essential for the body. Many drugs are absorbed by these carriers because of the structural similarity to natural substrates. An intestinal transmembrane protein, *P*-Glycoprotein (*P*-Gp) appears to reduce apparent intestinal epithelial cell permeability from lumen to blood for various lipophilic or cytotoxic drugs. Other transporters are present in the intestines. For example, many oral cephalosporins are absorbed through the amino acid transporter.

Vesicular transport

Vesicular transport is the process of engulfing particles or dissolved materials by the cell. Pinocytosis refers to the engulfment of small solutes or fluid, whereas phagocytosis refers to the engulfment of larger particles or macromolecules generally by macrophages. Endocytosis and exocytosis are the processes of moving macromolecules into and out of a cell, respectively.

During pinocytosis or phagocytosis, the cell membrane invaginates to surround the material, and then engulfs the material into the cell. Subsequently, the cell membrane containing the material forms a vesicle or vacuole within the cell. Vesicular transport is the proposed process for the absorption of orally administered sabin polio vaccine and various large proteins. An example of exocytosis is the transport of a protein such as insulin from insulin-producing cells of the pancreas into the extracellular space. The insulin molecules are first packaged into intracellular vesicles, which then fuse with the plasma membrane to release the insulin outside the cell.

ORAL DRUG ABSORPTION

Physiologic Considerations

Drugs may be administered by various routes of administration (Table 2). Except for intravenous drug administration, drugs are absorbed into the systemic circulation from the site of administration and are greatly affected by conditions at the administration site.

Oral administration is the most common route of drug administration. Major physiologic processes in the GI system include secretion, digestion, and absorption. Secretion includes the transport of fluid, electrolytes, peptides, and proteins into the lumen of the alimentary canal. Enzymes in saliva and pancreatic secretions are involved in the digestion of carbohydrates and proteins. Other secretions such as mucus protect the linings of the lumen of the GI tract. Digestion is the breakdown of food constituents into smaller structures in preparation for absorption. Both drug and food constituents are mostly absorbed in the proximal area (duodenum) of the small intestinal. The process of absorption is the entry of constituents from the lumen of the gut into the body. Absorption may be considered as the net result of both lumen-to-blood and blood-to-lumen transport movements.

Drugs administered orally pass through various parts of the enteral canal including the oral cavity, esophagus, and various parts of the GI tract. Residues eventually exit the body through the anus. Drugs may be absorbed by passive diffusion from all parts of the alimentary canal including sublingual, buccal, GI, and rectal absorption. For most drugs, the optimum site for drug absorption after oral administration is the upper portion of the small intestine or duodenum region. The unique anatomy of the duodenum provides an immense surface area for the drug to passively diffuse (Table 4). In addition, the duodenal region is highly perfused with a network of capillaries, which helps to maintain a concentration gradient from the intestinal lumen and plasma circulation.

The total transit time, including gastric emptying, small intestinal transit, and colonic transit ranges from 0.4 to 5 days. Small intestine transit time (SITT) ranges from 3 to 4 h for most healthy subjects. If absorption is not completed by the time a drug leaves the small intestine, drug absorption may be erratic or incomplete. The small intestine is normally filled with digestive juices and liquids, keeping the lumen contents fluid. In contrast, the fluid in the colon is reabsorbed, and the lumen content in the colon is either semisolid or solid, making further drug dissolution erratic and difficult.

Gastrointestinal motility

Once the drug is given orally, the exact location and/or environment of the drug product within the GI tract is difficult to discern. GI motility tends to move the drug through the alimentary canal so that it may not stay at the absorption site. For drugs given orally, an anatomic absorption window may exist within the GI tract in which the drug is efficiently absorbed. Drugs contained in a nonbiodegradable controlled-release dosage form must be completely released into this absorption window prior to the movement of the dosage form into the large bowel. The transit time of the drug in the GI tract depends upon the pharmacologic properties of the drug, type of dosage form, and various physiologic factors. Physiologic movement of the drug within the GI tract depends upon whether the alimentary canal contains recently ingested food (digestive or fed state) or is in the fasted or interdigestive state.

Gastric emptying time

After oral administration, the swallowed drug rapidly reaches the stomach. Because the duodenum has the greatest capacity for the absorption of drugs from the GI tract, a delay in the gastric emptying time will slow the rate and possibly the extent of drug absorption from the duodenum, thereby prolonging the onset time for the drug. Drugs, such as penicillin, that are unstable in acid, may decompose if stomach emptying is delayed. Other drugs, (e.g., aspirin) may irritate the gastric mucosa during prolonged contact.

Factors that tend to delay gastric emptying include consumption of meals high in fat, cold beverages, and anticholinergic drugs. Liquids and small particles less than 1 mm are generally not retained in the stomach. These small particles are believed to be emptied due to a slightly higher basal pressure in the stomach over the duodenum. Different constituents of a meal will empty from the stomach at different rates. For example, liquids are generally emptied faster than digested solids from the stomach. Large particles, including tablets and capsules, are delayed from emptying for 3–6 h by the presence of food in the stomach. Indigestible solids empty very slowly, probably during the interdigestive phase, a phase in which food is not present and the stomach is less motile but periodically empties its content due to housekeeper wave contraction.

Intestinal motility

Normal peristaltic movements mix the contents of the duodenum, bringing the drug particles into intimate contact with the intestinal mucosal cells. The drug must have a sufficient time (residence time) at the absorption site for optimum absorption. In the case of high motility in the intestinal tract, as in diarrhea, the drug has a very brief residence time and less opportunity for adequate absorption.

Blood perfusion of the gastrointestinal tract

The blood flow is important in carrying the absorbed drug from the absorption site to the systemic circulation. A large network of capillaries and lymphatic vessels perfuse the duodenal region and peritoneum. The splanchnic circulation receives about 28% of the cardiac output and is increased after meals. Drugs are absorbed from the small intestine into the mesenteric vessels which flows to the hepatic-portal vein and then to the liver prior to reaching the systemic circulation. Any decrease in mesenteric blood flow, as in the case of congestive heart failure, will decrease the rate of systemic drug absorption from the intestinal tract.

Table 4 Drug absorption in the gastrointestinal tract

Anatomic area	Function	Affect on drug absorption
Oral cavity	Saliva, pH 7, contains ptyalin (salivary amylase), digests starches. Mucin, a glycoprotein, lubricates food and may interact with drugs.	Buccal and sublingual absorption occurs for lipid-soluble drugs.
Esophagus	The esophagus connects the pharynx and the cardiac orifice of the stomach. The pH is 5–6. The lower part of the esophagus ends with the esophageal sphincter, which prevents acid reflux from the stomach.	Tablets or capsules may lodge in this area, causing local irritation. Very little drug dissolution occurs in the esophagus.
Stomach	The fasting stomach pH is about 2 to 6. In the fed state, the stomach pH is about 1.5 to 2, due to hydrochloric acid secreted by parietal cells. Stomach acid secretion is stimulated by gastrin and histamine. Mixing is intense and pressurized in the antral part of the stomach, a process of breaking down large food particles described as antral milling. Food and liquid are emptied by opening the pyloric sphincter into the duodenum.	Drugs are not efficiently absorbed in the stomach. Basic drugs are solubilized rapidly in acid. Stomach emptying influences the time for drug reaching the small intestine. The food content and osmolality influenced by stomach emptying. Fatty acids delay gastric emptying. High-density foods generally are emptied more slowly from the stomach.
Duodenum	A common duct from the pancreas and gall bladder enters the duodenum. Duodenal pH is 6 to 6.5 due to the presence of bicarbonate that neutralizes the acidic chyme emptied from the stomach. The pH is optimum for enzymatic digestion of protein and peptide food. Pancreatic juice containing enzymes is secreted into the duodenum from the bile duct. Trypsin, chymotrypsin, and carboxypeptidase are involved in the hydrolysis of proteins into amino acids. Amylase is involved in the digestion of carbohydrates. Pancreatic lipase secretion hydrolyzes fats into fatty acids.	The main site for drug absorption. An immense surface area for the passive diffusion of drug to due to the presence of villi and microvilli forming a brush border. A high blood perfusion maintains a drug concentration gradient from the intestinal lumen and plasma circulation. The complex fluid medium in the duodenum dissolves many drugs with limited aqueous solubility. Ester prodrugs are hydrolyzed during absorption. Proteolytic enzymes degrade many protein drugs in the duodenum, preventing adequate absorption. Acid drugs dissolve in the alkaline pH. Bile secretion helps to dissolve fats and hydrophobic drugs
Jejunum	The jejunum is the middle portion of the small intestine in between the duodenum and the ileum. Digestion of protein and carbohydrates continues after receiving pancreatic juice and bile in the duodenum, this portion of the small intestine generally has less contraction than the duodenum and is preferred for in vivo drug absorption studies.	Drugs generally absorbed by passive diffusion.
Ileum	The ileum, pH about 7, with the distal part as high as 8, is the terminal part of the small intestine and has fewer contractions than the duodenum. The ileocecal valve separates the small intestine with the colon.	Drugs generally absorbed by passive diffusion.
Colon	The colon, pH 5.5–7, is lined with mucin functioning as lubricant and protectant. The colon contains both aerobic and anaerobic micro-organisms that may metabolize some drugs. Crohn's disease affects the colon and thickens the bowel wall. The microflora may also become more anaerobic. Absorption of clindamycin and propranolol are increased, whereas other drugs have reduced absorption with this disease (Rubinstein et al. 1988).	Very limited drug absorption due to the lack of microvilli and the more viscous and semisolid nature of the lumen contents. A few drugs such as theophylline and metoprolol are absorbed in this region. Drugs that are absorbed well in this region are good candidates for an oral sustained-release dosage form.

(*Continued*)

Table 4 Drug absorption in the gastrointestinal tract (*Continued*)

Anatomic area	Function	Affect on drug absorption
Rectum	The rectum is about 15 cm long, ending at the anus. In the absence of fecal material, the rectum has a small amount of fluid, (about 2 m) with a pH about 7. The rectum is perfused by the superior, middle, and inferior hemorrhoidal veins. The inferior hemorrhoidal vein (closest to the anal sphincter) and the middle hemorrhoidal vein feed into the vena cava and back to the heart. The superior hemorrhoidal vein joins the mesenteric circulation, which feeds into the hepatic portal vein and then to the liver.	Drug absorption may be variable depending upon the placement of the suppository or drug solution within the rectum. A portion of the drug dose may be absorbed via the lower hemorrhoidal veins, from which the drug feeds directly into the systemic circulation; some drug may be absorbed via the superior hemorrhoidal veins, which feeds into the mesenteric veins to the hepatic portal vein to the liver, and metabolized prior to systemic absorption.

Some drugs may be absorbed into the lymphatic circulation through the lacteal or lymphatic vessels under the microvilli. Absorption of drugs through the lymphatic system bypasses the first-pass effect due to liver metabolism, because drug absorption through the hepatic portal vein is avoided. The lymphatics are important in the absorption of dietary lipids and may be partially responsible for the absorption for some lipophilic drugs such as bleomycin or aclarubicin which may dissolve in chylomicrons and be systemically absorbed via the lymphatic system.

Effect of food and other factors on GI drug absorption

Digested foods may affect intestinal pH and solubility of drugs. Food effects are not always predictable. The absorption of some antibiotics (e.g., penicillin, tetracycline) is decreased with food, whereas other drugs (e.g., griseofulvin) are better absorbed when given with food containing a high fat content. Food in the GI lumen stimulates the flow of bile. Bile contains bile acids. Bile acids are surfactants are involved in the digestion and solubilization of fats, and increases the solubility of fat-soluble drugs through micelle formation. For some basic drugs (e.g., cinnarizine) with limited aqueous solubility, the presence of food in the stomach stimulates hydrochloric acid secretion, which lowers the pH, causing more rapid dissolution of the drug and better absorption.

Generally, the bioavailability of drugs is better in patients in the fasted state and with a large volume of water (Fig. 5). However, to reduce GI mucosal irritation, drugs such as erythromycin, iron salts, aspirin, and nonsteroidal anti-inflammatory agents (NSAIDs) are given with food. The rate of absorption for these drugs may be reduced in the presence of food, but the extent of absorption may be the same.

The drug dosage form may also be affected by food. For example, enteric-coated tablets may stay in the stomach for a longer period of time because food delays stomach emptying. If the enteric-coated tablet does not reach the duodenum rapidly, drug release and subsequent systemic drug absorption are delayed. In contrast, enteric-coated beads or microparticles disperse in the stomach, are less affected by food, and demonstrate more consistent drug absorption from the duodenum.

Food may also affect the integrity of the dosage form, causing an alteration in the release rate of the drug. For example, theophylline bioavailability from Theo-24 controlled-release tablets is much more rapid (7) when given to a subject in the fed rather than fasted state (Fig. 6).

Some drugs, such as ranitidine, cimetidine, and dipyridamole, after oral administration produce a blood concentration curve consisting of two peaks. This double-peak phenomenon is generally observed after the administration of a single dose to fasted patients. The rationale for the double-peak phenomenon has been attributed to variability in stomach emptying, variable intestinal motility, presence of food, enterohepatic recycling, or failure of a tablet dosage form. For a drug with high water solubility, dissolution of the drug occurs in the stomach, and partial emptying of the drug into the duodenum will result in the first absorption peak. A delay in stomach emptying results in a second absorption peak as the remainder of the dose is emptied into the duodenum.

Diseases such as Crohn's disease that alter GI physiology and corrective surgery involving peptic ulcer, antrectomy with gastroduodenostomy and selective vagotomy may potentially affect drug absorption. Drug absorption may be unpredictable in many disease conditions. Drugs or nutrients or both may also affect the absorption of other drugs. For example, propantheline

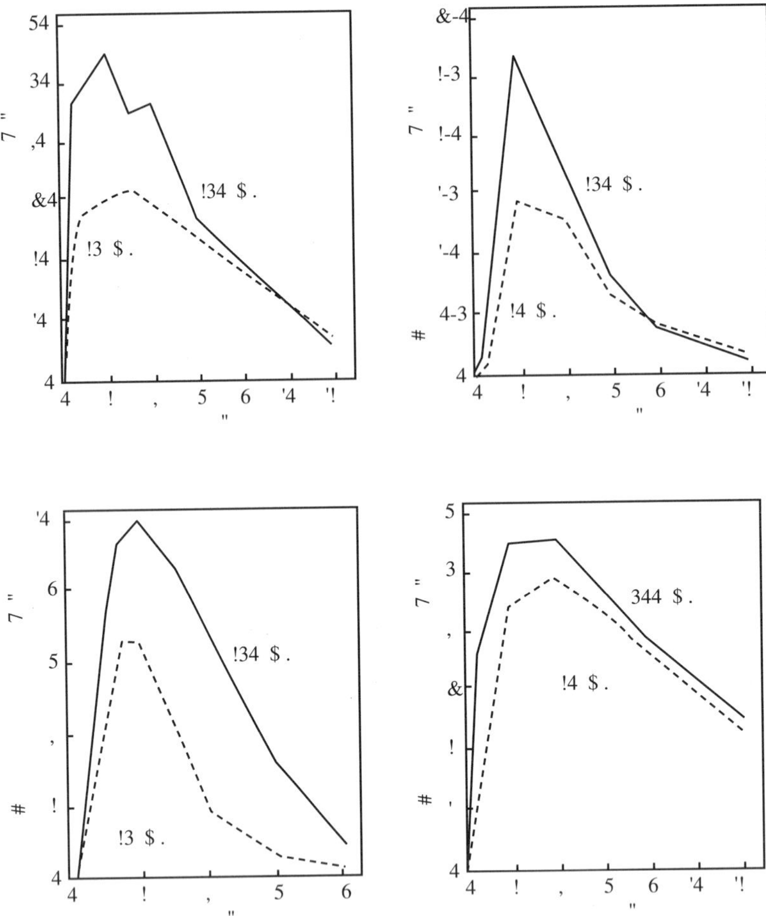

Fig. 5 Mean plasma or serum drug levels in healthy, fasting human volunteers ($n = 6$ in each case) who received single oral doses of aspirin (650 mg) tablets, erythromycin stearate (500 mg) tablets, amoxicillin (500 mg) capsules, and theophylline (260 mg) tablets, together with large. (From Welling P.G.; Drug Bioavailability and Its Clinical Significance. *Progress in Drug Metabolism*, Vol. 4; Bridges K.W.; Chassea, VD LF. Eds.; Wiley; London, 1980.)

bromide is an anticholinergic drug that slows stomach emptying and motility of the small intestine and may reduce stomach acid secretion. Grapefruit juice was found to increase the plasma level of many drugs due to inhibition of their metabolism in the liver.

PHARMACEUTICAL FACTORS AFFECTING DRUG BIOAVAILABILITY

Biopharmaceutic considerations in the design and manufacture of a drug product to deliver the active drug with the desired bioavailability characteristics include: 1) the type of drug product (e.g., solution, suspension, suppository), 2) the nature of the excipients in the drug product, 3) the physicochemical properties of the drug molecule, and 4) the route of drug administration.

Disintegration

Immediate release, solid oral drug products must rapidly disintegrate into small particles and release the drug. The United States Pharmacopoeia (USP) describes an official tablet disintegration test. The process of disintegration does not imply complete dissolution of the tablet and/or the drug. Complete disintegration is defined by the USP as "that state in which any residue of the tablet, except fragments of insoluble coating, remaining on the screen of

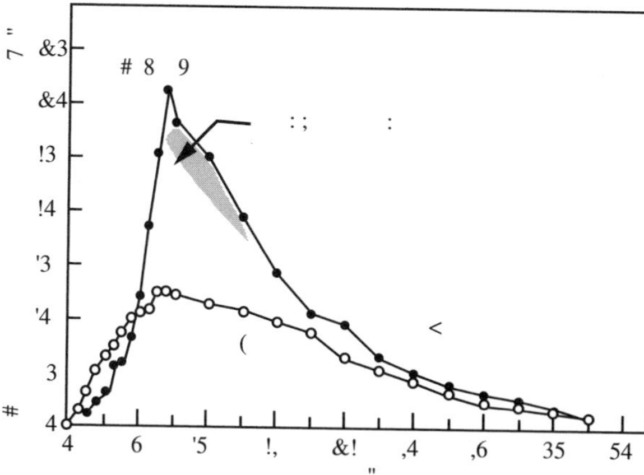

Fig. 6 Theophylline serum concentration in an individual subject after a single 1500 mg dose of Theo-24 taken during fasting, period during which this patient experienced nausea, repeated vomiting, or severe throbbing headache. The pattern of drug release during the food regimen is consistent with "dose-dumping." (From Ref. 7.)

the test apparatus in the soft mass have no palpably firm core." The USP provides specifications for uncoated tablets, plain coated tablets, enteric tablets, buccal tablets, and sublingual tablets. Exempted from USP disintegration tests are troches, tablets which are intended to be chewed, and drug products intended for sustained release or prolonged or repeat action.

Disintegration tests allow for precise measurement of the formation of fragments, granules, or aggregates from solid dosage forms, but do not provide information on the dissolution rate of the active drug. The disintegration test serves as a component in the overall quality control of tablet manufacture.

Dissolution

Dissolution is the process by which a chemical or drug becomes dissolved in a solvent. In biologic systems, drug dissolution in an aqueous medium is an important prior condition of systemic absorption. The rate at which drugs with poor aqueous solubility dissolve from an intact or disintegrated solid dosage form in the GI tract often controls the rate of systemic absorption of the drug. Thus, dissolution tests are discriminating of formulation factors that may affect drug bioavailability.

As the drug particle dissolves, a saturated solution (stagnant layer) is formed at the immediate surface around the particle. The dissolved drug in the saturated solution gradually diffuses to the surrounding regions. The overall rate of drug dissolution may be described by the Noyes–Whitney equation which models drug dissolution in terms of the rate of drug diffusion from the surface to the bulk of the solution. In general, drug concentration at the surface is assumed to be the highest possible, i.e., the solubility of the drug in the dissolution medium. The drug concentration C is the homogeneous concentration in the bulk solution which is generally lower than that in the stagnant layer immediate to the surface of the solid. The decrease in concentration across the stagnant layer is called the diffusion gradient

$$dC/dt = DA(CS - C)h$$

where, dC/dt = rate of drug dissolution, D = diffusion rate constant, A = surface area of the particle, CS = drug concentration in the stagnant layer, C = drug concentration in the bulk solvent, and h = thickness of the stagnant layer.

The rate of dissolution, $(dC/dt) \times (1/A)$, is the amount of drug dissolved per unit area per time (e.g., g/cm^2 per min).

The Noyes–Whitney equation shows that dissolution rate is influenced by the physicochemical characteristics of the drug, the formulation, and the solvent. In addition, the temperature of the medium also affects drug solubility and dissolution rate.

PHYSICOCHEMICAL NATURE OF THE DRUG

Solubility, pH, and Drug Absorption

The natural pH environment of the GI tract varies from acidic in the stomach to slightly alkaline in the small intestine. Drug solubility may be improved with the addition of acidic or basic excipients. Solubilization of aspirin, for example, may be increased by the addition of an alkaline buffer. Controlled release drug products are nondisintegrating dosage forms. Buffering agents may be added to slow or modify the release rate of a fast-dissolving drug in the formulation of a controlled release drug product. The buffering agent is released slowly rather than rapidly so that the drug does not dissolve immediately in the surrounding GI fluid. Intravenous drug solutions are difficult to prepare with drugs that have poor aqueous solubility. Drugs that are physically or chemically unstable may require special excipients, coating or manufacturing process to protect the drug from degradation.

Stability, pH, and Drug Absorption

The pH-stability profile is a plot of reaction rate constant for drug degradation versus pH and may help to predict if some of the drug will decompose in the GI tract. The stability of erythromycin is pH-dependent. In acidic medium, erythromycin decomposition occurs rapidly, whereas at neutral or alkaline pH the drug is relatively stable. Consequently, erythromycin tablets are enteric coated to protect against acid degradation in the stomach. In addition, less soluble erythromycin salts that are more stable in the stomach have been prepared.

Particle Size and Drug Absorption

The effective surface area of the drug is increased enormously by a reduction in the particle size. Because drug dissolution is thought to take place at the surface of the solute, the greater the surface area, the more rapid the rate of drug dissolution. The geometric shape of the drug particle also affects the surface area, and during dissolution the surface is constantly changing. In dissolution calculations, the solute particle is usually assumed to have retained its geometric shape.

Particle size and particle size distribution studies are important for drugs that have low water solubility. Particle size reduction by milling to a micronized form increased the absorption of low aqueous solubility drugs such as griseofulvin, nitrofurantoin, and many steroids. Smaller particle size results in an increase in the total surface area of the particles, enhances water penetration into the particles, and increases the dissolution rates. With poorly soluble drugs, a disintegrant may be added to the formulation to ensure rapid disintegration of the tablet and release of the particles.

Polymorphic Crystals, Solvates, and Drug Absorption

Polymorphism refers to the arrangement of a drug in various crystal forms (polymorphs). Polymorphs have the same chemical structure but different physical properties, such as solubility, density, hardness, and compression characteristics. Some polymorphic crystals may have much lower aqueous solubility than the amorphous forms, causing a product to be incompletely absorbed. Chloramphenicol (9), for example, has several crystal forms, and when given orally as a suspension, the drug concentration in the body depended on the percentage of β-Polymorph in the suspension. The β-form is more soluble and better absorbed (Fig. 7). In general, the crystal form that has the lowest free energy is the most stable polymorph. Polymorphs that are metastable may convert to a more stable form over time. A crystal form change may cause problems in manufacturing the product. For example, a change in crystal structure of the drug may cause cracking in a tablet or even prevent a granulation to be compressed into a tablet requiring reformulation of the product. Some drugs interact with solvent during preparation to form a crystal called solvate. Water may form a special crystal with drugs called hydrates, for example, erythromycin forms different hydrates (8) which may have quite different solubility compared to the anhydrous form of the drug (Fig. 8). Ampicillin trihydrate, for example, was reported

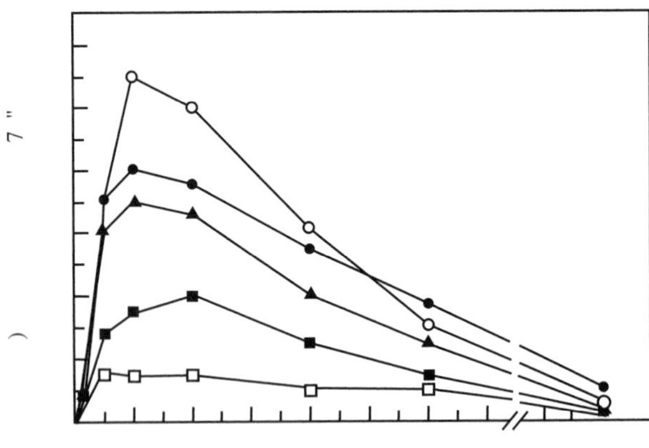

Fig. 7 Comparison of mean blood serum levels obtained with chloramphenicol palmitate suspensions containing varying ratios of α and β polymorphs, following single oral dose equivalent. (From Ref. 9.)

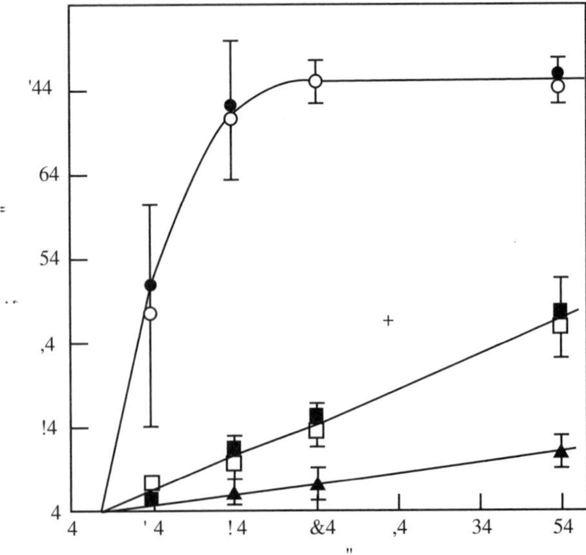

Fig. 8 Dissolution behavior of erythromycin dihydrate, monohydrate, and anhydrate in phosphate buffer (pH 7.5) at 37°C. (From Ref. 8.)

Table 5 Common excipients used in solid drug products

Excipient	Property in dosage form
Lactose	Diluent
Dibasic calcium phosphate	Diluent
Starch	Disintegrant, diluent
Microcrystalline cellulose	Disintegrant, diluent
Magnesium stearate	Lubricant
Stearic acid	Lubricant
Hydrogenated vegetable oil	Lubricant
Talc	Lubricant
Sucrose (solution)	Granulating agent
Polyvinyl pyrrolidone (solution)	Granulating agent
Hydroxypropylmethylcellulose	Tablet-coating agent
Titinium dioxide	Combined with dye as colored coating
Methylcellulose	Coating or granulating agent
Cellulose acetate phthalate	Enteric coating agent

(From Ref. 1.)

Table 6 Common excipients used in oral liquid drug products

Excipient	Property in dosage form
Sodium carboxymethylcellulose	Suspending agent
Tragacanth	Suspending agent
Sodium alginate	Suspending agent
Xanthan gum	Thixotropic suspending agent
Veegum	Thixotropic suspending agent
Sorbitol	Sweetener
Alcohol	Solubilizing agent, preservative
Propylene glycol	Solubilizing agent
Methyl propylparaben	Preservative
Sucrose	Sweetener
Polysorbates	Surfactant
Sesame oil	For emulsion vehicle
Corn oil	For emulsion vehicle

(From Ref. 1.)

to be less absorbed than the anhydrous form of ampicillin due to faster dissolution of the latter.

FORMULATION FACTORS AFFECTING DRUG DISSOLUTION

Excipients are pharmacodynamically inactive substances that are added to a formulation to provide certain functional properties to the drug and dosage form. Excipients may be added to improve the compressibility of the active drug, stabilize the drug from degradation, decrease gastric irritation, control the rate of drug absorption from the absorption site, increase drug bioavailability, etc. Some excipients used in the manufacture of solid and liquid drug products are listed in Tables 5 and 6. For solid oral dosage forms such as compressed tablets, excipients may include 1) diluent (e.g., lactose), 2) disintegrant (e.g., starch), 3) lubricant (e.g., magnesium stearate), and 4) other components such as binding and stabilizing agents. When improperly used in the formulation, excipients may alter drug bioavailability and possibly pharmacodynamic activity.

Excipients may affect the drug dissolution rate by altering the medium in which the drug is dissolving or by reacting with the drug itself. Some common manufacturing problems that affect drug dissolution and bioavailability are listed in Table 7. For example,

Table 7 Effect of excipients on the pharmacokinetic parameters of oral drug product[a]

Excipients	Example	k_a	t_{max}	AUC
Disintegrants	Avicel, Explotab	↑	←	↑/—
Lubricants	Talc, hydrogenated vegetable oil	←	↑	←/—
Coating agent	Hydroxypropylmethyl cellulose	—	—	—
Enteric coat	Cellulose acetate phthalate	←	↑	←/—
Sustained-release agents	Methylcellulose, ethylcellulose	←	↑	←/—
Sustained-release agents (waxy agents)	Castorwax, Carbowax	←	↑	←/—
Sustained-release agents (gum/viscous)	Veegum, Keltrol	←	↑	←/—

[a]This may be concentration and drug dependent.

↑ = Increase, ← = decrease, — = no effect. k_a = absorption rate constant, t_{max} = time for peak drug concentration in plasma, AUC = area under the plasma drug concentration time curve.

(From Ref. 1.)

suspending agents increase the viscosity of the drug vehicle, but may decrease the drug dissolution rate from the suspension. An excessive quantity of magnesium stearate (a hydrophobic lubricant) in the formulation may retard drug dissolution and slow the rate of drug absorption. The total amount of drug absorbed may also be reduced. To prevent this problem, the lubricant level should be decreased or a different lubricant selected. Sometimes, increasing the amount of disintegrant may overcome the retarding effect of lubricants on dissolution. However, with some poorly soluble drugs an increase in disintegrant level has little or no effect on drug dissolution because the fine drug particles are not wetted. The general influence of some common excipients on drug bioavailability parameters for typical oral drug products is summarized in Table 7.

Excipients may enhance or diminish the rate and extent of systemic drug absorption. Excipients that increase the aqueous solubility of the drug generally increase the rate of drug dissolution and absorption. For example, sodium bicarbonate in the formulation may change the pH of the medium surrounding the active drug substance. Aspirin, a weak acid, in an alkaline medium will form a water-soluble salt in which the drug rapidly dissolves. This process is known as dissolution in a reactive medium. The solid drug dissolves rapidly in the reactive solvent surrounding the solid particle. As the dissolved drug molecules diffuse outward into the bulk solvent, the drug may precipitate out of solution with a very fine particle size. The small particles have enormous collective surface area and disperse and redissolve readily for more rapid absorption on contact with the mucosal surface.

Excipients may interact directly with the drug to form a water-soluble or water-insoluble complex. If tetracycline is formulated with calcium carbonate, an insoluble complex of calcium tetracycline is formed that has a slow rate of dissolution and poor absorption.

Excipients may increase the retention time of the drug in the GI tract and therefore increase the amount of drug absorbed. Excipients may act as carriers to increase drug diffusion across the intestinal wall. The addition of surface-active agents may increase wetting as well as solubility of drugs. In contrast, many excipients may retard drug dissolution and thus reduce drug absorption.

Shellac used as a tablet coating, upon aging, can slow the drug dissolution rate. Surfactants may affect drug dissolution in an unpredictable fashion. Low concentrations of surfactants lower the surface tension and increase the rate of drug dissolution, whereas higher concentrations of surfactants tend to form micelles with the drug and thus decrease the dissolution rate. High tablet compression without sufficient disintegrant may cause poor disintegration in vivo of a compressed tablet.

IN VITRO DISSOLUTION TESTING

A dissolution test in vitro measures the rate and extent of dissolution of the drug in an aqueous medium in the presence of one or more excipients contained in the drug product. A potential bioavailability problem may be uncovered by a suitable dissolution method. The optimum dissolution testing conditions differ with each drug formulation. Different agitation rates, different medium (including different pH), and different dissolution apparatus should be tried to distinguish which dissolution method is optimum for the drug product and discriminating for drug formulation changes. The appropriate dissolution test condition for the drug product is then used to determine acceptable dissolution specifications.

The size and shape of the dissolution vessel may affect the rate and extent of dissolution. For example, the vessel may range in size from several milliliters to several liters. The shape may be round-bottomed or flat, so that the tablet might lie in a different position in different experiments. The amount of agitation and the nature of the stirrer affect the dissolution rate. Stirring rates must be controlled, and specifications differ between drug products. Low stirring rates (50–100 rpm) are more discriminating of formulation factors affecting dissolution than higher stirring rates. The temperature of the dissolution medium must be controlled and variations in temperature must be avoided. Most dissolution tests are performed at 37°C.

The nature of the dissolution medium, the solubility of the drug and the amount of drug in the dosage form will affect the dissolution test. The dissolution medium should not be saturated by the drug. Usually, a volume of medium larger than the amount of solvent needed to completely dissolve the drug is used in such tests. The usual volume of the medium is 500–1000 ml. Drugs that are not very water soluble may require use of a very-large-capacity vessel (up to 2000 ml) to observe significant dissolution. Sink conditions is a term referring to an excess volume of medium that allows the solid drug to continuously dissolve. If the drug solution becomes saturated, no further net drug dissolution will take place. According to the USP, "the quantity of medium used should be not less than three times that required to form a saturated solution of the drug substance."

Which medium is best is a matter of considerable controversy. The preferred dissolution medium in USP dissolution tests is deaerated water or if substantiated by the solubility characteristics of the drug or formulation, a buffered aqueous solution (typically pH 4–8) or dilute HCl may be used. The significance of dearation of the medium should be determined. Various investigators have used 0.1 *N* HCl, 0.01 *N* HCl, phosphate buffer, simulated gastric juice, water, and simulated intestinal juice, depending on the nature of the drug product and the location in the GI tract where the drug is expected to dissolve. No single apparatus and test can be used for all drug products. Each drug product must be tested individually with the dissolution test that best correlates to in vivo bioavailability.

The dissolution test usually states that a certain percentage of the labeled amount of drug in the drug product must dissolve within a specified period of time. In practice, the absolute amount of drug in the drug product may vary from tablet to tablet. Therefore, a number of tablets from each lot are usually tested to get a representative dissolution rate for the product. The USP provides several official (compendia) methods for carrying out dissolution tests of tablets, capsules and other special products such as transdermal preparations. The selection of a particular method for a drug is usually specified in the monograph for a particular drug product.

BIOAVAILABILITY AND BIOEQUIVALENCE

Bioavailability and bioequivalence may be determined directly using plasma drug concentration vs. time profiles, urinary drug excretion studies, measurements of an acute pharmacologic effect, clinical studies, or in vitro studies. Bioavailability studies are performed for both approved active drug ingredients or therapeutic moieties not yet approved for marketing by the FDA. New formulations of active drug ingredients or therapeutic moieties must be approved, prior to marketing, by the FDA. In approving a drug product for marketing, the FDA must ensure that the drug product is safe and effective for its labeled indications for use. To ensure that the drug product meets all applicable standards of identity, strength, quality, and purity, the FDA requires bioavailability/pharmacokinetic studies and where necessary bioequivalence studies for all drug products.

For unmarketed drugs which do not have full New Drug Application (NDA) approval by the FDA, in vivo bioavailability studies must be performed on the drug formulation proposed for marketing. Essential pharmacokinetic parameters of the active drug ingredient or therapeutic moiety is also characterized. Essential pharmacokinetic parameters include the rate and extent of systemic absorption, elimination half-life, and rates of excretion and metabolism should be established after single- and multiple-dose administration. Data from these in vivo bioavailability studies are important to establish recommended dosage regimens and to support drug labeling.

In vivo bioavailability studies are performed also for new formulations of active drug ingredients or therapeutic moieties that have full NDA approval and are approved for marketing. The purpose of these studies is to determine the bioavailability and characterize the pharmacokinetics of the new formulation, new dosage form, or new salt or ester relative to a reference formulation. After the bioavailability and essential pharmacokinetic parameters of the active ingredient or therapeutic moiety are established, dosage regimens may be recommended in support of drug labeling.

Bioequivalent Drug Products

Bioequivalent drug products are pharmaceutical equivalents whose bioavailability (i.e., rate and extent of systemic drug absorption) does not show a significant difference when administered at the same molar dose of the therapeutic moiety under similar experimental conditions, either single or multiple dose. Some pharmaceutical equivalents or may be equivalent in the extent of their absorption but not in their rate of absorption and yet may be considered bioequivalent because such differences in the rate of absorption are intentional and are reflected in the labeling, are not essential to the attainment of effective body drug concentrations on chronic use, or are considered medically insignificant for the particular drug product studied [21 CFR 320.1(e)].

Generic Drug Products

A generic drug product is considered bioequivalent to the reference listed drug product (generally the currently marketed, brand-name product with a full (NDA) approved by the FDA) if both products are pharmaceutical equivalents and its rate and extent of systemic drug absorption (bioavailability) do not show a statistically significant difference when administered in the same dose of the active ingredient, in the same chemical form, in a similar dosage form, by the same route of administration, and under the same experimental conditions.

Pharmaceutical equivalents are drug products that contain the same therapeutically active drug ingredient(s), same salt, ester, or chemical form; are of the same dosage form; and are identical in strength and concentration and route of administration. Pharmaceutical equivalents may differ in characteristics such as shape, scoring configuration, release mechanisms, packaging, and excipients (including colors, flavoring, preservatives).

Therapeutic equivalent drug products are pharmaceutical equivalents that can be expected to have the same clinical effect and safety profile when administered to patients under the same conditions specified in the labeling. Therapeutic equivalent drug products have the following criteria: 1) The products are safe and effective; 2) The products are pharmaceutical equivalents containing the same active drug ingredient in the same dosage form, given by the same route of administration, meet compendia or other applicable standards of strength, quality, purity, and identity and meet an acceptable in vitro standard; 3) The drug products are bioequivalent in that they do not present a known potential problem and are shown to meet an appropriate bioequivalence standard; 4) The drug products are adequately labeled; 5) The drug products are manufactured in compliance with current good manufacturing practice (GMP) regulations.

The generic drug product requires an abbreviated new drug application (ANDA) for approval by the FDA and may be marketed after patent expiration of the reference listed drug product. The generic drug product must be a therapeutic equivalent to the Reference drug product but may differ in certain characteristics including shape, scoring configuration, packaging, and excipients (includes colors, flavors, preservatives, expiration date, and minor aspects of labeling).

Pharmaceutical alternatives are drug products that contain same therapeutic moiety but are different salts, esters or complexes (e.g., tetracycline hydrochloride versus tetracycline phosphate) or are different dosage forms (e.g., tablet versus capsule; immediate release dosage form versus controlled release dosage form) or strengths.

In summary, clinical studies are useful in determining the safety and efficacy of the drug product. Bioavailability studies are used to define the affect of changes in the physico chemical properties of the drug substance and the affect of the drug product (dosage form) on the pharmacokinetics of the drug; whereas, bioequivalence studies are used to compare the bioavailability of the same drug (same salt or ester) from various drug products. If the drug products are bioequivalent and therapeutically equivalent, then the clinical efficacy and safety profile of these drug products are assumed to be similar and may be substituted for each other.

DRUG PRODUCT PERFORMANCE IN VITRO AS A MEASURE OF IN VIVO DRUG BIOAVAILABILITY

The best measure of a drug product's performance is to give the drug product to human volunteers or patients and then determine the in vivo bioavailability of the drug using a pharmacokinetic or clinical study. For some well characterized drug products and for certain drug products where bioavailability is self-evident (e.g., sterile solutions for injection), in vivo bioavailability studies may be unnecessary. In these cases, the performance of the drug product in vitro is used as a surrogate to predict the in vivo drug bioavailability. Because these products have predictable in vivo performance as judged by the in vitro characterization of the drug and drug product, the FDA may waive the requirement for performing an in vivo bioavailability study (Table 8).

Drug Products for which Bioavailability is Self-Evident

Drug bioavailability from a true solution is generally considered self-evident. Thus, sterile solutions, lyophilized powders for reconstitution, opthalmic solutions do not need bioequivalence studies but still must be manufactured according to current GMPs. However, highly viscous solutions may have bioavailability problems due to slow diffusion of the active drug.

In Vitro–In Vivo Correlation (IVIVC)

In vitro bioavailability data may be used to predict the performance of a dosage provided that the dissolution method selected is appropriate for the solid oral dosage form and prior information has been collected showing that the dissolution method will result in optimum drug absorption from the drug product. In general, IVIVC is best for well absorbed drugs for which the dissolution rate is the rate-limiting step. Some drugs are poorly absorbed and dissolution is not predictive of absorption (1). The objectives of IVIVC are to use rate of dissolution as a discriminating (i.e., sensitive to changes in formulation or manufacturing process), as an aid in setting dissolution specifications. When properly applied, IVIVC may be used to facilitate the evaluation

B

Table 8 Examples of drug products for which in vivo bioavailability studies may be waived

Condition	Example	Comment
Drug products for which bioavailability is self-evident	Drug solution (e.g., parenteral ophthalmic, oral solutions)	Drug bioavailability from a true solution is considered self-evident. However, highly viscous solutions may have bioavailability problems.
In vivo–in vitro correlation (IVIVC)	Modified release drug products	The dissolution of the drug from the drug product in vitro must be highly correlated to the in vivo bioavailability of the drug.
Biopharmaceutic classification (BCS) system	Immediate release solid oral drug products	Drug must be a highly soluble and highly permeable substance that is in a rapidly dissolving dosage form.
Biowaiver	Drug product containing a lower dose strength	Drug product is in the same dosage form, but lower strength and is proportionally similar in its active and inactive ingredients.

of drug products with manufacturing changes including minor changes in formulation, equipment, process, manufacturing site, and batch size. (see section on SUPAC) (2, 3, 10).

Three levels of IVIVC are generally recognized by the FDA (10). Level A correlation is usually estimated by deconvolution followed by comparison of the fraction of drug absorbed to the fraction of drug dissolved. A correlation of this type is the highest level of correlation and best predictor of bioavailability from the dosage form. A Level A correlation is generally linear and represents a point-to-point relationship between in vitro dissolution rate and the in vivo input rate. The Level A correlation should predict the entire in vivo time course from the in vitro dissolution data. Level B correlation utilizes the principles of statistical moment analysis, Various dissolution IVIVC methods were discussed by Shargel and Yu in 1985, 1993, 1999 (1). The mean in vitro dissolution time is compared to either the mean residence time or the mean in vivo dissolution time. Level B correlation, like Level A correlation, uses all of the in vitro and in vivo data but is not considered to be a point-to-point correlation and does not uniquely reflect the actual in vivo plasma level curve, since several different in vivo plasma level-time curves will produce similar residence times. A Level C correlation is the weakest IVIVC and establishes a single point relationship between a dissolution parameter (e.g., time for 50% of drug to dissolve, or percent drug dissolved in two hours, etc.) and a pharmacokinetic parameter (e.g., AUC, Cmax, Tmax). Level C correlation does not reflect the complete shape of the plasma drug concentration-time curve of dissolution profile.

BIOPHARMACEUTICS CLASSIFICATION SYSTEM (BCS)

The FDA may waive the requirement for performing an in vivo bioavailability or bioequivalence study for certain immediate release solid oral drug products that meets very specific criteria, namely, the permeability, solubility, and dissolution of the drug. These characteristics include the in vitro dissolution of the drug product in various media, drug permeability information, and assuming ideal behavior of the drug product, drug dissolution and absorption in the GI tract. For regulatory purpose, drugs are classified according to BCS in accordance the solubility, permeability and dissolution characteristics of the drug (FDA Draft Guidance for Industry, January, 1999, see FDA website for guidance) (11). Based on drug solubility and permeability, Amidon et al. (10, 12) recommended the following BCS in 1995 (Table 9).

This classification can be used as a basis for setting in vitro dissolution specifications and can also provide a basis for predicting the likelihood of achieving a successful in IVIVC. The solubility of a drug is determined by dissolving the highest unit dose of the drug in 250 ml of buffer adjusted between pH 1.0 and 8.0. A drug substance is considered highly soluble when the dose/solubility volume of solution are less than or equal to 250 ml. High-permeability drugs are generally those with an extent of absorption that is greater than 90%.

Solubility

An objective of the BCS approach is to determine the equilibrium solubility of a drug under approximate

Table 9 Biopharmaceutics classification system (BCS)

Condition	Comments
Solubility	A drug substance is considered highly soluble when the highest dose strength is soluble in 250 ml or less of water over a pH range of 1–8.
Dissolution	An immediate release (IR) drug product is considered rapidly dissolving when not less than 85% of the label amount of the drug substance dissolves within 30 min using the USP apparatus I at 100 rpm (or apparatus II at 50 rpm) in a volume of 900 ml or less.[a]
Permeability	A drug substance is considered highly permeable when the extent of absorption in humans is to be >90% of an administered dose based on mass balance determination.

[a]Media include: acidic media (e.g., 0.1 N HCl) or simulated gastric fluid, USP without enzymes, pH 4.5 buffer and pH 6.8 buffer of simulated intestinal fluid, USP without enzymes (From FDA Draft Guidance, Jan, 1999.)

physiological conditions. For this purpose, determination of pH-solubility profiles over a pH range of 1–8 is suggested. Preferably eight or more pH conditions should be evaluated. Buffers that react with the drug should not be used. An acid or base titration method can also be used for determining drug solubility. The solubility class is determined by calculating what volume of an aqueous media is sufficient to dissolve the highest anticipated dose strength. A drug substance is considered highly soluble when the highest dose strength is soluble in 250 ml or less of aqueous media over the pH range of 1–8. The volume estimate of 250 ml is derived from typical bioequivalence study protocols that prescribe administration of a drug product to fasting human volunteers with a glass (8 ounces) of water.

Solution stability of a test drug in selected buffers (or pH conditions) should be documented using a validated stability-indicating assay. Data collected on both pH-solubility and pH-stability should be submitted in the biowaiver application along with information on the ionization characteristics, such as pKa(s), of a drug.

Determining Permeability Class

Studies of the extent of absorption in humans, or intestinal permeability methods, can be used to determine the permeability class membership of a drug. To be classified as highly permeable, a test drug should have an extent of absorption >90% in humans. Supportive information on permeability characteristics of the drug substance should also be derived from its physical–chemical properties (e.g., octanol:water partition coefficient).

Some methods to determine the permeability of a drug from the GI tract include 1) in vivo intestinal perfusion studies in humans, 2) in vivo or in situ intestinal perfusion studies in animals, 3) in vitro permeation experiments using excised human or animal intestinal tissues, and 4) in vitro permeation experiments across a monolayer of cultured human intestinal cells. When using these methods, the experimental permeability data should correlate with the known extent-of-absorption data in humans.

Table 10 Postapproval change levels

Change level	Example	Comment
Level 1	Deletion or partial deletion of an ingredient to affect the color or flavor of the drug product	Level 1 changes are those that are unlikely to have any detectable impact on formulation quality and performance.
Level 2	Quantitative change in excipients greater that allowed in a Level 1 change.	Level 2 changes are those that could have a significant impact on formulation quality and performance
Level 3	Qualitative change in excipients	Level 3 changes are those that are likely to have a significant impact on formulation quality and performance. A Level 3 change may require in vivo bioequivalence testing.

Dissolution

The dissolution class is based on the in vitro dissolution rate of an immediate release drug product under specified test conditions and is intenended to indicate rapid in vivo dissolution in relation to the average rate of gastric emptying in humans under fasting conditions. An immediate release drug product is considered rapidly dissolving when not less than 85% of the label amount of drug substance dissolves within 30 min using the USP apparatus I at 100 rpm or apparatus II at 50 rpm in a voluume of 900 ml or less in each of the following media 1) acidic media such as 0.1 *N* HCl or Simulated Gastric Fluid USP without enzymes; 2) a pH 4.5 buffer; and 3) a pH 6.8 buffer or Simulated Intestinal Fluid USP without enzymes.

BIOWAIVERS

In addition to routine quality control tests, comparative dissolution tests have been used to waive bioequivalence requirements (biowaivers) for lower strengths of a dosage form. The drug products containing the lower dose strengths should be compositionally proportional or qualitatively the same as the higher dose strengths and have the same release mechanism. For biowaivers, a dissolution profile should be generated and evaluated using one of the methods described under Section V in this guidance, "Dissolution Profile Comparisons." Biowaivers are generally provided for multiple strengths after approval of a bioequivalence study performed on one strength, using the following criteria: For multiple strengths of IR products with linear kinetics, the bioequivalence study may be performed at the highest strength and waivers of in vivo studies may be granted on lower strengths, based on an adequate dissolution test, provided the lower strengths are proportionately similar in composition [21 CFR 320.22(d)(2)]. Similar may also be interpreted to mean that the different strengths of the products are within the scope of changes permitted under the category "Components and Composition," discussed in the SUPAC-IR guidance.

SCALE-UP AND POSTAPPROVAL CHANGES (SUPAC)

After a drug product is approved for marketing by the FDA, the manufacturer may want to make a manufacturing change. The pharmaceutical industry, academia and the FDA developed (2, 3, 5, 3, 10, 3, 12–17) a series of guidances for the industry that discuss scale-up and postapproval changes, generally termed, SUPAC guidances (11). The FDA SUPAC guidances are for manufacturers of approved drug products who want to change 1) a component and composition of the drug product; 2) the batch size; 3) the manufacturing site; 4) the manufacturing process or equipment; and/or 5) packaging. These guidances describe various levels of postapproval changes according to whether the change is likely to impact on the quality and performance of the drug product. The level of change as classified by the FDA as to the likelihood that a change in the drug product might affect the quality of the product (Table 10).

REFERENCES

1. Shargel, L.; Yu, A.B.C. *Applied Biopharmaceutics and Pharmacokinetics*; McGraw-Hill Medical Publishing: 1999.
2. SUPAC-MR: Modified Release Solid Oral Dosage Forms; Scale-Up and Post-Approval Changes: Chemistry, Manufacturing and Controls, In Vitro Dissolution Testing, and In Vivo Bioequivalence Documentation, FDA, Guidance for Industry, Sept. 1997.
3. Waiver of In Vivo Bioavailability and Bioequivalence Studies for Immediate Release Solid Oral Dosage Forms Containing Certain Active Moieties/Active Ingredients Based on a Biopharmaceutics Classification System, FDA Draft Guidance for Industry, Jan. 1999.
4. Skelly, J.P.; Shah, V.P.; Konecny, J.J.; Everett, R.L.; McCullouen, B.; Noorizadeh, A.C. Report of the Workshop on CR Dosage Forms: Issues and Controversies. Pharmaceutical Research **1987**, *4* (1), 75–78.
5. Shah, V.P.; Konecny, J.J.; Everett, R.L.; McCullouen, B.; Noorizadeh, A.C.; Shah, V.P. In Vitro Dissolution Profile of Water Insoluble Drug Dosage Forms in the Presence of Surfactants. Pharmaceutical Research **1989**, *6*, 612–618.
6. Moore, J.W.; Flanner, H.H. Mathematical Comparison of Dissolution Profiles. Pharmaceutical Technology **1996**, *20* (6), 64–74.
7. Hendeles, L.; Weinberger, M.; Milavetz, G.; Hill, M.; Vaughan, L. Food Induced Dumping from "Once-A-Day" Theophylline Product as Cause of Theophylline Toxicity. Chest **1985**, *87*, 758–785.
8. Allen, P.V.; Rahn, P.D.; Sarapu, A.C.; Vandewielen, A.J. Physical Characteristics of Erythromycin Anhydrate and Dihydrate Crystalline Solids. J. Pharm. Sci. **1978**, *67*, 1087–1093.
9. Aguiar, A.J.; Krc, J.; Kinkel, A.W.; Samyn, J.C. Effect of Polymorphism on the Absorption of Chloramphenical from Chloramphenical Palmitate. J. Pharm. Sci. **1967**, *56*, 847–853.
10. SUPAC-IR Immediate Release Solid Oral Dosage Forms. Scale-up and Post-Approval Changes: Chemistry, Manufacturing and Controls, In Vitro Dissolution Testing, and

In Vivo Bioequivalence Documentation, FDA Guidance for Industry, Nov. 1995.
11. FDA Regulatory Guidances FDA Website for Regulatory Guidances. www.fda.gov/cder/guidance/index.htm).
12. Amidon, G.L.; Lennernas, H.; Shah, V.P.; Crison, J.R. A Theoretical Basis For a Biopharmaceutic Drug Classification: The Correlation of In Vitro Drug Product Dissolution and In Vivo Bioavailability. Pharmaceutical Research **1995**, *12*, 413–420.
13. Skelly, J.P.; Amidon, G.L.; Barr, W.H.; Benet, L.Z.; Carter, J.E.; Robinson, J.R.; Shah, V.P.; Yacobi, A. In Vitro and In Vivo Testing and Correlation for Oral Controlled/Modified-Release Dosage Forms. Pharmaceutical Research **1990**, *7*, 975–982.
14. In Vitro-In Vivo Correlation for Extended Release Oral Dosage Forms, Pharmacopeial Forum Stimuli Article, United States Pharmacopeial Convention, Inc.: July 1988; 4160–4161.
15. In Vitro In Vivo Evaluation of Dosage Forms, U.S.P. XXIV<1088> United States Pharmacopeial Convention, Inc. 2051–2056.
16. Shah, V.P.; Skelly, J.P.; Barr, W.H.; Malinowski, H.; Amidon, G.H. Scale-up of Controlled Release Products — Preliminary Considerations. Pharmaceutical Technology **1992**, *16* (5), 35–40.
17. Skelly, J.P. Report of Workshop on In Vitro and In Vivo Testing and Correlation for Oral Controlled/Modified-Release Dosage Forms. Journal of Pharmaceutical Sciences **1990**, *79* (9), 849–854.
18. Cadwallader, D.E. *Biopharmaceutics and Drug Interactions*; Raven Press: New York, 1983.
19. Gibaldi, M. *Biopharmaceutics and Cinical Pharmacokinetics*; Lea & Febiger: Philadelphia, 1984.
20. Gibaldi, M.; Perrier, D. *Pharmacokinetics*; Marcel Dekker, Inc.: New York, 1982.
21. McGinity, J.W.; Stavchansky, S.A.; Martin, A. Bioavailability in Tablet Technology. *Pharmaceutical Dosage Forms: Tablets*; Lieberman, H.A., Lachman, L., Eds.; Marcel Dekker, Inc.: New York, 1981; 2.
22. Rowland, M.; Tozer, T.N. *Clinical Pharmacokinetics. Concepts and Applications*; Lea & Febiger: Philadelphia, 1995.

Board of Pharmaceutical Specialties

Barbara G. Wells
University of Mississippi, University, Mississippi, U.S.A.

Richard J. Bertin
Council on Credentialing in Pharmacy, Washington, D.C., U.S.A.

INTRODUCTION

The Board of Pharmaceutical Specialties (BPS) was established in 1976 under the auspices of the American Pharmaceutical Association (APhA). The overriding concern of BPS is to ensure that the public receives the highest possible quality pharmacy services, contributing toward outcomes that improve a patient's quality of life. The BPS has four primary responsibilities:

- To recognize specialties in pharmacy practice.
- To set standards for certification and recertification.
- To objectively evaluate individuals seeking certification and recertification.
- To serve as a source of information and coordinating agency for pharmacy specialties.

The BPS is located at 2215 Constitution Avenue, NW, Washington, D.C. 20037-2985, phone: (202) 429-7591; fax: (202) 429-6304; www.bpsweb.org.

OVERVIEW

Certification is a voluntary process by which a practitioner's education, experience, knowledge, and skills are confirmed by one's profession as meeting or surpassing a standard beyond that required for licensure. The standards and processes for certification (unlike those for licensure) are established by a professional, nongovernmental agency. BPS certification is at the specialty level and signifies that an individual has met a national professional standard and demonstrated mastery of a body of knowledge, skills, and abilities in an advanced level in a specialized area of practice.

Today, BPS functions as an agency of APhA with its own governing Board structure. The board is composed of six pharmacist members, two health care practitioners outside of pharmacy, and one public member. The chair of each Specialty Council and the BPS Executive Director serve as nonvoting members of the Board. The Executive Director of BPS is Richard J. Bertin, Ph.D., R.Ph. and the current Chair of BPS is Roger W. Anderson, R.Ph., DrPH, who is Director of Pharmacy at M.D. Anderson Cancer Center.

To date, five specialties have been recognized by BPS: 1) nuclear pharmacy; 2) nutrition support pharmacy; 3) oncology pharmacy; 4) pharmacotherapy; and 5) psychiatric pharmacy. As of August, 2002, 3414 pharmacists are certified as specialists in one or more of these specialties. Added Qualification is a process for providing recognition of pharmacists with further training and experience in areas of concentration within an existing specialty.

MISSION AND OBJECTIVES

The mission of BPS was refined in 1997 and is reviewed by the Board semiannually. The mission is to improve public health through recognition and promotion of specialized training, knowledge, and skills in pharmacy and certification of pharmacist specialists. The organization achieves its mission through accomplishment of six strategic objectives including 1) providing leadership for the profession of pharmacy in the discussion, evolution, direction, and recognition of specialties in pharmacy; 2) establishing the standards for identification and recognition of specialties in consultation with the profession; 3) establishing standards of training, knowledge, and skills as the basis for certification of individuals; 4) developing and administering objective and valid means to evaluate the knowledge and skills of pharmacist specialists; 5) evaluating areas of specialization for their value and

Encyclopedia of Clinical Pharmacy
DOI: 10.1081/E-ECP 120006201

viability; and 6) communicating the value of specialization and specialty certification in pharmacy.

THE SPECIALTIES

BPS has recognized five specialty practice areas. They are 1) nuclear pharmacy (1978); 2) nutrition support pharmacy (1988); 3) pharmacotherapy (1988); 4) psychiatric pharmacy (1992); and 5) oncology pharmacy (1996).

Nuclear pharmacy seeks to improve and promote public health through the safe and effective use of radioactive drugs for diagnosis and therapy. A nuclear pharmacist, as a member of the nuclear medicine team, specializes in procurement, compounding, quality assurance, dispensing, distribution, and development of radiopharmaceuticals. In addition, the nuclear pharmacist monitors patient outcomes and provides information and consultation regarding health and safety issues.

Nutrition support pharmacy addresses the care of patients receiving specialized parenteral or enteral nutrition. The nutrition support pharmacist is responsible for promoting restoration and maintenance of optimal nutritional status and designing and modifying treatment in accordance with patient needs. These specialists have responsibility for direct patient care and often function as members of multidisciplinary nutrition support teams.

Pharmacotherapy is the specialty responsible for ensuring the safe, appropriate, and economical use of drugs in patient care. The pharmacotherapy specialist has responsibility for direct patient care and often functions as a member of a multidisciplinary treatment team. These specialists may conduct clinical research and are frequently primary sources of drug information for other health care professionals.

Psychiatric pharmacy addresses the pharmaceutical care of patients with psychiatric disorders. As a member of a multidisciplinary treatment team, the psychiatric pharmacist specialist is often responsible for optimizing drug treatment and patient care by conducting patient assessments; recommending appropriate treatment plans; monitoring patient response; and preventing, identifying, and correcting drug-related problems.

Oncology pharmacy addresses the pharmaceutical care of patients with cancer. The oncology pharmacist specialist promotes optimal care of patients with various malignant diseases and their complications. These specialists are closely involved in recognition, management, and prevention of unique morbidities associated with cancer and cancer treatment; recognition of the balance between improved survival and quality of life as primary outcome indicators; and provision of safeguards against drug misadventures in a treatment area where novel and experimental drug therapies are frequently employed.

ADDED QUALIFICATIONS

Added Qualifications is the mechanism used by BPS to recognize further differentiation within a specialty which the Board has already recognized. This distinction may be granted to a BPS-certified specialist on the basis of a structured portfolio review process, administered by the Specialty Council responsible for the specialty. The first petition for Added Qualifications was in Infectious Diseases and was approved by the Pharmacotherapy Specialty Council and BPS in 1999. The first candidates were conferred the "Added Qualifications in Infectious Diseases" credential in 2000. A petition for Added Qualifications in Cardiology was approved in 2000, and the first candidates were conferred the Added Qualifications in Cardiology Pharmacotherapy credential in 2001.

THE CERTIFICATION PROCESS

When a group of interested pharmacists wishes to have a new specialty considered for recognition by the BPS, they submit a petition to the Board. The petition is evaluated against seven criteria: 1) need of the profession and the public for specifically trained practitioners in the specialty practice area to fulfill the responsibilities of the profession in improving the health and welfare of the public; 2) clear, significant demand for the specialty by the public and health care system; 3) presence of a reasonable number of pharmacist specialists practicing in and devoting significant time in the specialty area; 4) specialized knowledge of pharmaceutical sciences required by those practicing in the specialty area; 5) specialized functions provided by pharmacists in the specialty practice area that require education and training beyond the basic level attained by licensed pharmacists; 6) education and training in the specialty area provided by pharmacy colleges and other organizations; and 7) transmission of knowledge in the specialty practice area occurring through books, journals, symposia, professional meetings, and other media.

After a new specialty is recognized by BPS, a Specialty Council of content experts is appointed to work with the BPS and a professional testing firm to develop a psychometrically sound and legally defensible certification process. The Specialty Council is composed of six pharmacists practicing in the specialty area and three other pharmacists. Certification examinations consisting of 200 multiple choice questions are administered an-

nually at designated sites throughout the United States and in other countries. Each BPS-certified specialist must recertify every seven years. Approved professional development programs are available as an alternative to sitting for a 100-item recertification examination in nuclear pharmacy and pharmacotherapy. BPS continually evaluates and updates its certification and recertification processes. Approximately every five years, a new role delineation study is conducted for each specialty, and examination specifications are modified accordingly.

VALUE AND RECOGNITION OF CERTIFICATION

Specialty certification in pharmacy offers numerous potential benefits of significant value to patients, other health professionals, employers, health care systems, and the public. Specialty certification denotes that specialists are highly trained and skilled and have demonstrated the ability to identify, resolve, and prevent drug therapy problems. They have taken the initiative to seek additional education and experience in a specialized pharmacy field and exhibit a high level of commitment to patients and the profession. Certified pharmacist specialists function as valued members of treatment teams, optimizing and individualizing drug therapy. Employers can feel assured that the knowledge and skills of certified pharmacist specialists have been tested through a rigorous, objective, and peer-determined process.

Certification also provides a personal reward for pharmacist specialists. Specialty certification communicates to others that the specialist's educational and practice accomplishments differentiate the specialist from colleagues. Many specialists feel that they have a competitive edge in applying for positions, and some have received reimbursement from third-party payers, because their skills and knowledge have been validated through certification. Some pharmacist specialists have also reported increased salaries or one-time bonuses upon attaining BPS certification.

BPS certification has been formally recognized by the American Association of Colleges of Pharmacy, the American College of Clinical Pharmacy, the American Pharmaceutical Association, the American Society for Parenteral and Enteral Nutrition, the American Society of Health-System Pharmacists, the Ordre des Pharmaciens du Quebec, the Society of Infectious Diseases Pharmacists, and the Society of Hospital Pharmacists of Australia.

BPS-certified pharmacist specialists are recognized for their advanced level of knowledge, skills, and achievement by many government agencies and health care organizations. The following are examples of specific benefits that may be realized by BPS-certified pharmacist specialists:

- U.S. Nuclear Regulatory Commission: specialists may be licensed as Radiation Safety Officers and/or recognized as Authorized Users.
- U.S. Department of Defense: specialists may receive bonus pay.
- U.S. Department of Veterans Affairs: specialists may serve at higher pay steps.
- U.S. Public Health Service: specialists may receive bonus pay.
- New Mexico State Board of Pharmacy: specialists may apply for specified prescribing privileges.
- At least seven Colleges of Pharmacy may exempt BPS-certified specialists from some didactic courses in postbaccalaureate or nontraditional Pharm.D. programs. Other Colleges award advanced placement on an individual case basis and may recognize BPS certification in this process.

Many other national, regional, or local employers of BPS-certified pharmacists also recognize BPS certification in their hiring, salary, or privileging policies.

Bone Marrow Transplant Pharmacy Practice

Vanita K. Pindolia
Henry Ford Health System, Detroit, Michigan, U.S.A.

INTRODUCTION

The utilization of a bone marrow transplantation to treat hematologic malignancies, solid tumors, genetic disorders, metabolic diseases, immune deficiency disorders, and bone marrow failures has grown tremendously since the first successful transplant in 1968. If these diseases are not treated aggressively, they can be fatal. The seriousness of these diseases is reflected in the intense care provided to the patient during and after a bone marrow transplant. Whether the patient receives an autologous, syngeneic, or allogeneic bone marrow transplant, the patient needs to be followed very closely for the first several weeks to several years (depending on the type of bone marrow transplant and type of post-transplant complications) by the bone marrow transplant team. After an allogeneic bone marrow transplant, the patient needs to be seen in clinic 1 to 3 days per week for physical examinations, blood work, special microbiology testing, and medication adjustments. Once a bone marrow transplant patient is deemed to be stable, their outpatient visits will slowly decrease.

The medications used during a bone marrow transplant and post–bone marrow transplant carry extensive toxicity profiles and have numerous drug–drug/drug–food interactions, and the majority of medications are expensive. These characteristics in themselves justify the necessity of a pharmacist to be a key member of the bone marrow transplant team.

CLINICAL PHARMACY OPPORTUNITIES

The majority of bone marrow transplant centers (especially those performing allogeneic bone marrow transplants) in the United States have a pharmacist on the team. Currently, there are approximately 450 bone marrow transplant centers registered in 48 different countries with the National Marrow Donor Program.[1] In most cases, the pharmacist is employed by the Department of Pharmacy with partial or complete financial support from the Department of Medicine. With the stringent criteria developed for medical reimbursement by third-party payers (i.e., health maintenance organizations, preferred provider organizations, and medicaid/medicare), both the Departments of Pharmacy and Medicine have a strong interest in driving cost down. Since 11–37% of a bone marrow transplant cost is attributed to pharmaceuticals, the pharmacist plays a critical role in lowering the cost of a bone marrow transplant via close medication monitoring.[2] Similar to other clinical pharmacists, the bone marrow transplant pharmacist needs to monitor all drugs given to the patient for appropriate usage. However, the majority of medications used by a bone marrow transplant patient carry high levels of toxicities, narrow therapeutic windows, and life-threatening results if medication doses are forgotten or increased or decreased by the patient, thereby elevating the intensity of drug monitoring.

The bone marrow transplant team members primarily consists of a specially trained hematologist/oncologist, pharmacist, nurse practitioner/physician assistant, social worker, ± dietitian and/or total parenteral nutrition (TPN) personnel, and ± hematology/oncology medical fellow(s). The team members rely heavily on each other to ensure that appropriate, safe, and cost-effective care is given to each patient. The pharmacist–nurse practitioner/physician assistant relationship assists in combining physical assessment findings with drug outcomes. The pharmacist–social worker relationship is necessary for discharge planning to ensure medication affordability and appropriate home medication administration. The pharmacist–dietitian/TPN personnel relationship helps decrease drug–food interactions with mealtime plan changes and/or meal content changes and decrease cost by minimizing intravenous hyperalimentation usage. In some institutions, the pharmacist plays a lead role in dietary care, thereby negating the need for an additional dietary personnel. The pharmacist–hematology/oncology fellow relationship is primarily a teaching role for both parties. Finally, a sound pharmacist–physician relationship needs to be developed for the physicians to entrust patient care to a pharmacist. Once this bond has been established, a pharmacist's knowledge can be used for clinical advice on patient care, investigating innovative

Encyclopedia of Clinical Pharmacy
DOI: 10.1081/E-ECP 120006224

options for patient care, and developing research protocols to advance patient care. Therefore, in addition to a solid knowledge base in hematology-oncology, immunology, infectious diseases, fluid/electrolyte balance, and pain management, it is imperative for a bone marrow transplant pharmacist to possess excellent communication and people skills.

Because the patient's immune system is compromised for several months to several years secondary to the slow process of complete bone marrow recovery and/or long-term usage of immunosuppresives, the patient is vulnerable to numerous life-threatening infectious disease processes. In addition, the allogeneic bone marrow transplant patient always carries a certain risk (highest per-

Table 1 Pharmacist's responsibilities

Unstable patient	Stable patient
Adjust medication regimen(s) based on drug levels to enhance efficacy and/or prevent toxicity.[b]	Adjust medication regimen(s) based on drug levels to enhance efficacy and/or prevent toxicity.[a,b]
Monitor drug–drug and drug–food interactions to prevent toxicity.[c]	Monitor drug–drug and drug–food interactions to prevent toxicity.[a,c]
Educate the bone marrow transplant team about each medication's effect on the bone marrow.[d]	Educate the bone marrow transplant team about each medication's effect on the bone marrow.[a,d]
Offer advice on antibiotic choice(s) based on microbiology results, patient's immunologic state (i.e., neutropenic, type of underlying cancer), patient's infectious disease history (i.e., past infection(s), serology results), institution's microorganism–drug susceptibility record, and institution's microorganism resistance pattern.	
Adjust fluids and electrolytes based on daily laboratory values and medication changes.[e]	
Pain management. *Acute pain* secondary to mucositis needs persistent close monitoring and abrupt drug adjustments to attain near complete pain control within 4–6 hours.	Pain management. *Chronic pain* secondary to chronic GVHD needs close monitoring and drug adjustments to attain near complete pain control within 24–48 hours.
Monitor patients for *acute toxicities* secondary to conditioning regimen ± immunosuppressive agents. In addition, offer advice on preventive therapies for toxicities and treatment options for toxicities.	Monitor patients for *chronic toxicities* secondary to conditioning regimen ± immunosuppressive agents. In addition, offer advice on preventive therapies for toxicities and treatment options for toxicities.
Educate patient upon discharge on the importance of each medication by clearly writing the brand name and generic name of each medication, explaining the purpose of each medication, the time(s) of day each medication should be self-administered, the consequences of missing or doubling doses, and a contact name (preferably the pharmacist) and phone number for use if further questions arise at home.	Assure medication compliance by reviewing medication calendar. Once the patients are stable, they are more apt to develop their own regimens. These regimens may allow medications to be dosed too close together or too close to meal times, skip certain medications deemed unnecessary by the patient, and/or add certain medications (i.e., natural herbs, vitamins).
Conduct clinical research. Pharmacist-initiated research projects arise frequently from day-to-day unresolvable issues. Pharmaceutical industry-sponsored research projects are also available.[f]	Conduct outcome-based research. Because a pharmacist closely monitors the bone marrow transplant patients for a prolonged period of time, there is an abundance of data available for outcome-based research.

[a]Examples include, cyclosporine, tacrolimus, aminoglycosides, vancomycin.
[b]These tasks are performed at a lower intensity level than their counterparts.
[c]Examples include, choice of antihypertensive agent to be used while on cyclosporine or tacrolimus, timing medication ingestion around meal times.
[d]After identifying medications that are detrimental to the bone marrow, offer options for treatment that have no effect or minimal effect on the bone marrow.
[e]The pharmacist has the key role in preventing fluid and electrolyte abnormalities from occurring secondary to medications and aggressively correcting all fluid and electrolyte abnormalities.
[f]Pharmacists need to be aggressive in indentifying projects and seek funding for the projects.[3] Pharmacists should strive to be the primary investigator or coinvestigator on the projects.

centage if the donated bone marrow is from an unrelated individual without a 6/6 Human Lymphocyte Antigen match) for developing acute and/or chronic graft-versus-host disease (GVHD). If the patient develops GVHD, the immune system is even further compromised by not only the intense immunosuppressive agents needed for treatment, but also by the GVHD itself. Thus, an allogeneic bone marrow transplant patient may have numerous hospital readmissions for treatment of infectious disease processes, aggressive treatment, and close monitoring of moderate-to-severe GVHD or bone marrow failure (i.e., tumor relapse, bone marrow engraftment lost). The majority of patients are at highest risk for readmission during the first 100 days post-transplant. Bone marrow transplant recipients of mismatched or unrelated donors require more intense bone marrow immunosuppression for a longer period of time than their counterparts who receive matched or related bone marrow, thereby increasing their risk for hospital readmissions for a period greater than 100 days. Because the patient's medical needs can change drastically from day to day, the pharmacist needs to stay abreast of all new medication regimens required for patient care. It is imperative that the pharmacist has a good working relationship with the patient and patient's caregivers to ensure appropriate adherence to the evolving medication regimen.

Recently, there has been a surge of bone marrow transplant centers shifting inpatient care to the outpatient setting early in transplant (post bone marrow/peripheral blood stem cell infusion). The incentive for this trend has been to decrease the cost of bone marrow transplant and improve the patient's quality of life. Stringent, institution-specific criteria have been developed for patients to be outpatient bone marrow transplant candidates. The primary basis behind the criteria rely on the patient and a dedicated caregiver to be attentive to all their medical needs, including comprehension of appropriate medication administration guidelines. The patients are responsible for self-administration of scheduled and as needed oral, subcutaneous, and intravenous medications. By placing this level of responsibility on the patient and caregiver, the patient can be overcome with anxiety. A pharmacist plays a dominant role in alleviating any confusion or misunderstanding on medication self-administration. The bone marrow transplant pharmacist will need to thoroughly educate both the patient and caregiver on all the medications daily. Although the patient maybe medically stable in the outpatient setting, the initial amount of time the pharmacist needs to spend with the patient is equivalent to a complicated hospitalized bone marrow transplant patient.

There is a definite need for both an inpatient and outpatient bone marrow transplant pharmacist. Due to the patient's initial prolonged inpatient stay and high probability of multiple readmissions during the first several months post–bone marrow transplant, the distinction between an inpatient and outpatient bone marrow transplant pharmacist role becomes unclear. To help maintain continuity of care, usually one pharmacist (labeled as the inpatient pharmacist) will manage a patient's pharmaceutical needs both during the initial hospitalization and during the first several months of outpatient care. Once a patient's ambulatory visits decrease to at least once every 2 weeks, the outpatient pharmacist will attend to the patient's medication needs. If the institution predominantly performs autologous bone marrow transplants or if the number of bone marrow transplants (autologous combined with allogeneic) performed is low, then one pharmacist is sufficient to play both the inpatient and outpatient role.

The type and level of care provided by the pharmacist depends on the stability of the patient's health (Table 1). Generally, the patient is most unstable during the transplant and for the first several months post-transplant.

MODEL CLINICAL PRACTICES

The type of work a bone marrow transplant pharmacist performs on a daily basis depends on the goal of the employer. A pharmacist can be predominantly research based or clinically based.

Research-Based Practice

Most of the bone marrow transplant pharmacist positions with research emphasis are tenure-tracked or tenured with a teaching hospital. These pharmacists have minimal to no direct patient care duties assigned to them. Pharmacists are responsible for the following:

1. Developing research protocols
2. Applying for grants to fund the protocols
3. Screening and enrolling patients into study (when applicable)
4. Developing/running various assays
5. Publishing research results
6. Didactic, experiential university-based teaching

The majority of pharmacists will participate in pharmacy doctoral or postdoctoral programs or develop bone marrow transplant fellowships to attain reliable, hardworking assistance in the laboratory. In turn, the students/fellows will be closely mentored by the pharmacist. This symbiotic relationship will allow both parties to augment research productivity, increase the number of publications, and attain larger funding sources.

Clinical-Based Practice

Most of the bone marrow transplant pharmacist positions with clinical emphasis are nontenure-tracked. The pharmacist's primary job responsibilities revolve around direct patient care. Pharmacists are responsible for the following:

1. Reviewing patient's laboratory bloodwork, microbiology results, and medication profiles on a daily basis
2. Attending and actively participating in patient medical rounds and patient clinic visits
3. Reviewing medications used in bone marrow transplant patients for inpatient and outpatient formulary usage
4. Helping to standardize care by developing protocols and procedures for bone marrow transplant medication utilization
5. Publishing material related to bone marrow transplant patient care

Although a clinical pharmacist's emphasis is on direct patient care, many of the pharmacists do perform clinical research, seek for grants or awards to fund their research projects, publish research results, participate in university-based teaching (didactic ± experiential), and assist in mentoring pharmacy residents specializing in hematology-oncology. A clinical pharmacist is expected to remain abreast of the bone marrow transplant literature, especially in the following areas: GVHD, veno-occlusive disease, infectious disease processes, and chemotherapy-radiation-related toxicities. In addition to the pharmacist's clinical knowledge, the bone marrow transplant team heavily relies on the pharmacist for their knowledge of the practical aspects of pharmacy (i.e., compatibility issues, ability to compound products, maximum/minimum concentrations of intravenous medications, understanding of the institution's medication order entry and medication delivery processes). The clinical bone marrow transplant pharmacist's main goals are to provide safe, therapeutic, and cost-effective care to each patient, and to maintain continuity of patient care upon initial hospital discharge.

HEALTH OUTCOME AND ECONOMIC BENEFITS

Currently, there is no published literature documenting health outcome or economic benefits provided by a pharmacist to a bone marrow transplant patient. There are several review articles analyzing the economics of bone marrow transplant, peripheral blood stem cell transplant, and outpatient-based transplant.[2,4–7] The articles refer to various detailed cost-effective and cost-minimization studies. Unfortunately, they do not breakdown the cost analysis studies to note the impact a pharmacist has on the total cost of a bone marrow transplant procedure. The input provided by a pharmacist to the bone marrow transplant team is substantial and necessary for a cancer center to remain competitive and fiscally responsible. There is a strong need for bone marrow transplant pharmacists to generate outcome data and collectively or individually publish the benefits of retaining a pharmacist on the bone marrow transplant team.

NECESSARY TOOLS/MATERIALS

Medline

Many of the complications encountered during a bone marrow transplant have few (if any) standardized treatment protocols developed. Therefore, pharmacists need easy accessibility to a medline service during and after patient rounds to provide valuable information to the bone marrow transplant team in a timely manner. Although physicians may also perform their own research on the topic of discussion, it is important for pharmacists to critique and review the literature separately. This will allow the pharmacist to

1. Evaluate various innovative treatment options.
2. Choose the best option that complies with the hospital policies and procedures for drug attainment, drug compounding, and drug administration.
3. Resolve drug availablity issues (i.e., orphan drug status, length of time to receive drug in hospital).
4. Present the options to the bone marrow transplant physician in a timely manner.

Hematopoiesis Chart

The bone marrow transplant pharmacist needs to have a sound understanding of the maturation of the hematopoietic stem cell to form the three lineages. It is important to understand the relevance of immunomodulators at each step of cell maturation. Because ex-vivo cytokines are very expensive and many of them carry high toxicity profiles, it is important for a pharmacist to know which stem cell maturation step(s) will be influenced by the cytokine(s).

Other

Currently, there are no published documents providing guidelines or consensus statements on how medications should be administered during and following a bone marrow transplant. There are numerous review articles available on bone marrow transplant preparative regimens, prevention and treatment of graft versus host disease, infectious disease topics related to bone marrow transplant, and pain management. The chemotherapy/radiotherapy used as the preparative regimen for a bone marrow transplant varies from center to center, depending on the hematologist's past experience with the various regimens, the patient's eligibility for drug study enrollment, the patient's past chemotherapy/radiotherapy history, and the patient's past medical history. The medications and medication doses used to prevent and treat GVHD and to treat other bone marrow transplant related complications is also dependent on the physician's preference, patient's eligibility for drug study enrollment, and patient's medical history.

PROFESSIONAL NETWORKING OPPORTUNITIES

There are numerous bone marrow transplant web sites available; however, they are oncology center initiated to increase patient referral base or patient initiated to provide personal advice to other bone marrow transplant patients. The networking opportunities available for bone marrow transplant pharmacists are primarily in the following medical conferences/meetings:

1. American Society of Hematology (ASH)
2. American Society of Clinical Oncology (ASCO)

Unfortunately, a pharmacy conference/meeting specializing or subspecializing in bone marrow transplant has not been identified. There is a rising interest in forming a bone marrow transplant pharmacy network group; perhaps modeled after the infectious disease pharmacy group (Society of Infectious Disease Pharmacy meet annually at their medical counterpart conference, Interscience Conference on Antimicrobial Agents and Chemotherapy).

ACCP Oncology prn mainly targets the hematology/oncology pharmacists and ACCP Transplant prn mainly targets the solid organ transplant pharmacists. Thus, bone marrow transplant pharmacist members of ACCP are not strongly committed to any particular ACCP prn group.

REFERENCES

1. http://IBMTR.org/sitemap/sitemap/html (accessed September, 2000).
2. Bailey, E.M.; Pindolia, V.K. How to obtain funding for clinical research. Am. J. Hosp. Pharm. **1994**, *51*, 2858–2860.
3. Weeks, F.M.; Yee, G.C.; Bartfield, A.A.; Wingard, J.R. The true cost of bone marrow transplantation. Am. J. Med. Sci. **1997**, *314* (2), 101–112.
4. Waters, T.M.; Bennett, C.L.; Pajeau, T.S.; Sobocinski, K.A.; Klein, J.P.; Rowlings, P.A.; Horowitz, M.M. Economic analyses of bone marrow and blood stem cell transplantation for leukemias and lymphoma: What do we know? Bone Marrow Transplant. **1998**, *21*, 641–650.
5. Bennett, C.L.; Waters, T.M.; Stinson, T.J.; Almagor, O.; Pavletic, Z.S.; Tarantolo, S.R.; Bishop, M.R. Valuing clinical strategies early in development: A cost analyses of allogeneic peripheral blood stem cell transplantation. Bone Marrow Transplant. **1999**, *24*, 555–560.
6. Rizzo, J.D.; Vogelsang, G.B.; Krumm, S.; Frink, B.; Mock, V.; Bass, E.B. Outpatient-based bone marrow transplantation for hematologic malignancies: Cost savings or cost shifting? J. Clin. Oncol. **1999**, *17* (9), 2811–2818.
7. Barr, R.; Furlong, W.; Henwood, J.; Feeny, D.; Wegener, J.; Walker, I.; Brain, M. Economic evaluation of allogeneic bone marrow transplantation: A rudimentary model to generate estimates for the timely formulation of clinical policy. J. Clin. Oncol. **1996**, *14*, 1413–1440.

Canadian Hospital Pharmacy Residency Board

C

Thomas W. Paton
University of Toronto, Toronto, Ontario, Canada

INTRODUCTION

The mission of the Canadian Hospital Pharmacy Residency Board is to establish and apply standards for accreditation of pharmacy practice residency programs and to promote excellence in hospital pharmacy residency programs and practice. The key objectives are as follows:

1. To gain external recognition and support for pharmacy residency programs.
2. To provide support to residency program participants in their role through education, skill development and practice tools.
3. To foster an environment that facilitates the growth and development of pharmacy residency programs.

ORGANIZATIONAL STRUCTURE AND GOVERNANCE

The Canadian Hospital Pharmacy Residency Board is organized under the auspices of Canadian Society of Hospital Pharmacists. The Board consists of seven members. The terms of reference of the Board specifies that at least one of the members be from a recognized Faculty of Pharmacy. The members of the Board are selected by the Board itself and approved by CSHP Council. A chairperson and vice-chair are elected from the seven member Board, with a term of two years for each of the executives. The members themselves serve for two years, a term which is renewable twice for a total of six years.

The Canadian Hospital Pharmacy Residency Board currently conducts its work under the auspices of the Canadian Society of Hospital Pharmacists. As such, the Board is provided administrative support from CSHP (at 1145 Hunt Club Road, Suite 350, Ottawa, Ontario, K1B 0Y3; telephone 613-736-9733).

HISTORY

The Canadian Hospital Pharmacy Residency Board was established in the early 1960s. The assessment of the residency training programs was done following review of written documentation submitted to the Board. The on-site accreditation process and survey did not begin until the early 1980s, however. At the present time, the Board accredits residency training programs in pharmacy practice. Currently, there are 30 programs in Canada with 104 positions for prospective residents. The Board is not currently involved in the accreditation of specialty programs or pharmacy technician training programs.

CURRENT INITIATIVES

The current major initiatives consist of:

1. Consistent with the four-year cycle for accreditation, to update the standards of the Board for 2002.
2. To promote the use of the CHPRB-sponsored preceptor guidelines.
3. To evaluate the need for innovative specialty practice standards and, in particular, those to be used in an ambulatory setting.
4. To conduct a needs assessment of residents who have been in the residency training program over the past three years to determine future directions of residency training in Canada.

BIBLIOGRAPHY

www.cshp.ca.

Encyclopedia of Clinical Pharmacy
DOI: 10.1081/E-ECP 120006286

Canadian Pharmacists Association/Association des Pharmaciens du Canada

Jeff Poston
Canadian Pharmacists Association, Ottawa, Ontario, Canada

INTRODUCTION

The Canadian Pharmaceutical Association (CPhA) was founded in September 1907 as a national body for the profession of pharmacy in Canada. Its involvement in publishing began early in its history with the assumption of responsibility for the *Canadian Formulary* in 1929. The first edition of the *Compendium of Pharmaceuticals and Specialties (CPS)* was published in 1960 and continues today, along with a number of other well-respected health care publications including *Nonprescription Drug Reference for Health Professionals*, *Compendium of Nonprescription Products*, *Therapeutic Choices*, and *Herbs: Everyday Reference for Health Professionals*. The latter is published jointly with the Canadian Medical Association.

Throughout the late 1980s and early 1990s, CPhA scope of activities increased with staffing in the areas of professional development, research, and government and public affairs.

In 1995, CPhA began to change its structure from an organization representing provincial and national pharmacy organizations to an organization representing individual pharmacist members. This was followed in 1997 by a name change from Canadian Pharmaceutical Association to Canadian Pharmacists Association to better reflect the association's mandate.

ORGANIZATIONAL STRUCTURE AND GOVERNANCE

CPhA is an organization of approximately 9000 individual members. Members directly elect members of the Board of Directors to represent each province, pharmacy students, and the three practice specialties of hospital pharmacy, industrial pharmacy, and academia. The Board is responsible for managing the affairs of CPhA. The Board elects an Executive Committee comprised of president, president-elect, past president, and three vice presidents.

The executive director, who is a nonvoting member of the Executive Committee and Board, is responsible for the management and control of the affairs of the Association and general direction established by Board policy. A staff of approximately 50 reports to the executive director.

MISSION, VISION, AND KEY OBJECTIVES

The Canadian Pharmacists Association is the national voluntary organization of pharmacists committed to providing leadership for the profession of pharmacy.

The vision of CPhA is to establish the pharmacist as the health care professional whose practice, based on unique knowledge and skills, ensures optimal patient outcomes. CPhA will achieve its vision by serving its members through:

- Advocacy.
- Facilitation.
- Provision of knowledge.
- Participation in partnerships.
- Research and innovation.
- Education.
- Health promotion.

Strategic Plan

CPhA operates according to its strategic plan developed in 1999 and revised in 2001. The plan has five key result areas:

1. To represent the interests of a majority of pharmacists and to create cohesiveness within the profession on matters of practice, principle, and policy.

Encyclopedia of Clinical Pharmacy
DOI: 10.1081/E-ECP 120006202

2. To promote and facilitate the evolution of the pharmacy profession toward an expanded role in health care.
3. To foster public recognition of pharmacists as drug experts and as members of the health care team.
4. To secure appropriate reimbursement for pharmacists' professional services.
5. To effectively analyze and respond to the impact of advances in information technology on pharmacy practice.
6. To align resources to the key result areas.

CURRENT MAJOR INITIATIVES

Pharmacist Shortage

CPhA has taken the leadership position in addressing the shortage problem in the profession by initiating the development of a proposal for a labor market study to help pharmacy understand the current manpower shortage and its causes and develop tools to forecast future needs. Human Resources Development Canada (HRDC) commissioned an initial phase of the study, which is a literature search and key informant interviews to identify gaps in the available data. This study, "A Situational Analysis of Human Resource Issues in the Pharmacy Profession in Canada," is available on the CPhA web site.[1] The next step is development of a proposal for funding from HRDC for a comprehensive human resources study of the pharmacy profession in Canada to develop the foundation required to properly manage current and future pharmacy human resources.

Prescribing Authority

One of the key objectives of the strategic plan is a move to acquire prescribing authority for pharmacists. CPhA has developed a discussion paper to foster debate within the profession. The document has been distributed to solicit pharmacists' and other stakeholders' input.

Privacy Legislation

CPhA was pivotal in securing an amendment to the Personal Information Protection and Electronic Documents Act[2] which delayed its application to health care until January 2002. With the passage of the legislation, CPhA participated in a working group with six national health provider and consumer associations to examine the issue of privacy protection in Canada. The report of the Privacy Working Group focuses on the challenges of developing and implementing principles for privacy protection. It highlights the lack of consensus and the tension on this issue due to the disparate perspectives of the many stakeholders involved. CPhA continues to seek effective remedies to the shortcomings in the legislation, including presentations to federal officials.

Third-Party Payer Issues

It is recognized that pharmacists are spending an excessive amount of time dealing with claims reimbursement with third-party payers. This is having a significant impact on working conditions, and administrative burdens are proving an impediment to patient care. As a result, CPhA joined forced with the Canadian Association of Chain Drug Stores and the Ontario Pharmacists Association to sponsor a 2-day workshop to tackle these issues. Priorities for action include a standardized drug benefit card and PIN lists, patient awareness, benefit plan messages, and further collaboration with insurers and pharmacy software vendors.

Pharmacy Electronic Communications Standard (PECS)/National E-Claims Standard Initiative

Over the last decade, CPhA has been a leader in the development of pharmacy communication standards. CPhA's PECS Version 3.0 facilitates more than 98% of the electronic pharmacy claims in Canada. PECS Version 3.0 has undergone extensive revisions and now is being integrated into a National E-Claims Standard Initiative (NeCST)[3] designed to address the current need for a national electronic standard for health claims information.

E-Business and Enhanced Web Strategies

A major reengineering of the CPhA web site is underway. This revitalized site will offer Web services to benefit our members (e.g., electronic membership and order transactions, chat rooms, e-mentoring, e-broadcast).

In association with this initiative, an e-commerce advisory committee advises on e-commerce strategy and assists with visualizing and developing enhancements to CPhA's web site for member and nonmember pharmacists, other health care professionals, and consumers.

Advances in Publishing

CPhA, through its publishing program, provides pharmacists in every practice setting with accurate, current drug information and resource materials. However, on-line publishing of our drug information presents a new challenge. Work is underway on the *CPS* so that this publication can be easily accessible for print and electronic publishing. The *CPS* and our other publications are being repurposed for use on new e-media platforms.

MEETINGS

CPhA's annual meetings generally are held in May of each year in major locations in Canada.

REFERENCES

1. A Situational Analysis of Human Resource Issues in the Pharmacy Profession in Canada. http://www.cdnpharm.ca (accessed June 25, 2001).
2. Personal Information Protection and Electronic Documents Act. http://www.privcom.gc.ca (accessed June 22, 2001).
3. National e-Claims Standard Initiative. http://www.cihi.ca/eclaims/intro.shtml (accessed June 25, 2001).

PROFESSIONAL DEVELOPMENT

Cardiac Arrest/Emergency Pharmacy Services

C

David S. Roffman
University of Maryland, Baltimore, Maryland, U.S.A.

INTRODUCTION

Drug therapy plays a critical role in emergency medical care and, as a result, places the pharmacist in a position to have a significant impact on potentially life-saving therapeutic maneuvers. Pharmacists who practice in emergency medical care settings are often called upon to provide drug-related services and information without the luxury of time to retrieve information from external sources. This article reviews the role of pharmacy services in both the cardiac arrest setting and in the provision of other emergency medical services in which pharmacists play a central role.

CARDIAC ARREST

The spectrum of care required for patients in cardiac arrest ranges from providing basic life support (Cardiopulmonary resuscitation, or CPR) within or outside of an organized health care setting, to evaluating and treating complex cardiac arrhythmias in patients with multiple comorbidities receiving advanced cardiac life support. Pharmacists have been involved in the development of drug-related health services designed to support the care of such patients since the late 1960s.[1] As scientific and clinical outcome data has dramatically expanded since the 1960s and the efficacy of drug use in the cardiac arrest setting has been more critically evaluated, potential roles for the pharmacist have become more obvious. In addition, as hospital accrediting agencies begin to address quality improvement issues in the cardiac arrest setting, the evaluation and documentation of rational drug use will become a factor in setting standards of care.

CURRENT ROLES FOR PHARMACIST

The majority of information related to the activities of pharmacists participating in cardiac arrest teams in hospitals has been collected through questionnaires or surveys. Activities most frequently reported by pharmacists through such surveys are drug preparation, dosage and infusion rate calculation, drug use documentation, and the provision of drug information; very few pharmacists administer artificial respiration or chest compression.[2] Less frequently reported activities include setting up or operating infusion devices and administering medications.

According to data from the National Clinical Pharmacy Services study from 1992 to 1998, approximately 30% of 950–1600 hospitals surveyed had a pharmacist as an ''active'' member of the team attending most cardiac arrests when the CPR team pharmacist was in the hospital.[3–5] Despite the significant percentage of hospitals in which pharmacists are members of the cardiac arrest team, only 0.2%–0.3% of inpatients who experienced a cardiac arrest received resuscitation by a team that included a pharmacist. This disparity may be due to the fact that a CPR team pharmacist is not providing 24-hour, 7-day-per-week coverage, or that the CPR team pharmacist was assigned to provide service only in a specific area of the hospital. The national study also revealed that, of the hospitals with a pharmacist on the CPR team, approximately 65% routinely document pharmacists' involvement in patients' medical records. The average time commitment for pharmacists per arrest was 35 minutes. In the 1992 survey, this amount of time per encounter was more than the average amount of time for any other clinical service examined in the survey.

There is limited documentation of the value of pharmacists on cardiac arrest teams. What little data there are tend to be anecdotal in nature. As far back as 1972, Elenbaas responded to a review of the value of organized cardiac arrest teams in hospitals by noting the obvious absence of the inclusion of a pharmacist as a member of the team.[6] Given the extremely small number of actual patient arrests in which pharmacists participate, it would be difficult to accurately measure the actual or perceived value of pharmacist participation. Based on the small

Encyclopedia of Clinical Pharmacy
DOI: 10.1081/E-ECP 120006320

number of reports in the literature, the provision of drug information seems to evoke the most comments regarding the value of pharmacist participation. Because the role of physicians and nurses in the cardiac arrest setting has been so well established compared with the role that pharmacists currently fill and because there are currently no standards established for pharmacists' activities in this setting, the value of the pharmacist remains to be measured.

A well-delineated support role for the pharmacist in the provision of care in the cardiac arrest setting clearly exists. The pharmacy department in concert with the hospital Pharmacy and Therapeutics Committee has a defined and traditional role in maintenance of the emergency drug boxes (''crash carts'') located in various parts of the hospital. Assuring that emergency drug boxes are appropriately stocked and meet the needs of the institution, as mandated by the Pharmacy and Therapeutics Committee or other medical oversight committee has been a well-established support role for pharmacy services for many years. As Advanced Cardiac Life Support (ACLS) guidelines periodically change and new information related to pharmacotherapeutic efficacy of new and older drugs used in the cardiac arrest setting becomes available, the role of the pharmacist in updating the contents of the emergency drug box is critical to maintaining the standard of care for drug delivery in the cardiac arrest setting.

In addition to their role in drug delivery, the pharmacist's role of educator in the appropriate use of drug therapy in the cardiac arrest patient has been established in a number of hospitals across the United States. This educational role is important for several reasons. First, because cardiac arrests occur infrequently, especially in noncritical care areas of the hospital, the medical and nursing staff may not be as familiar with either the pharmacotherapeutic guidelines established by ACLS recommendations or the appropriate administration techniques for ACLS drug as are providers in critical care areas. Second, even in critical care units, implementing change in longstanding ACLS drug administration behaviors (modifying the role of lidocaine or sodium bicarbonate use) is often resisted by clinicians, but it is often essential to the provision of quality care. Third, despite the fact that changes in the official ACLS drug therapy guidelines occur only every 5–6 years, the results of landmark clinical trials often dictate changes in therapy prior to their incorporation into published guidelines. Pharmacists in many hospitals have taken active roles in teaching physicians, nurses, and affiliated health professionals the pharmacology and therapeutics of drugs used in the ACLS guidelines.

EDUCATION AND TRAINING

The education and training needs of pharmacists who serve as members of the cardiac arrest team vary from those that may be provided in some schools of pharmacy. All health care providers, as well as most laypersons, should be certified in Basic Life Support (BLS). In a study using questionnaires to determine attitudes toward and use of cardiopulmonary resuscitation training received in a school of pharmacy, 72% of responding graduates surveyed believed that a CPR-BLS program should be mandatory for graduation. Seventy percent of respondents believed that their training would be of value in their current practice, and 93% believed that such training would be of value in the future, despite the fact that only 5% had actually performed CPR since graduation.[7] Clearly, pharmacist members of a cardiac arrest team should possess BLS skills. In addition, since automated external defibrillators have been demonstrated to improve the chances of out-of-hospital cardiac arrest, training in the use of these devices is becoming an inherent part of BLS training both within hospitals and in routine BLS out-of-hospital training.[8,9] In addition to the prerequisite BLS training, cardiac arrest team pharmacist should be certified in ACLS, the procedure in which most drug therapy is instituted in the cardiac arrest setting. Even if the role of the pharmacist as a team member is limited to drug preparation or documentation, it is essential that the pharmacist understand the rationale, efficacy, and potential side effects associated with the use of the drug therapy being implemented. As in all other clinical pharmacy practices, the cardiac arrest team pharmacist should be trained to monitor for both the efficacy and side effects of the agents being used.

The need for the cardiac arrest team pharmacist to be certified in ACLS is further reinforced by the standards developed by the Joint Commission on Accreditation of Health Care Organizations (JCAHO) related to in hospital cardiac arrests.[10] Of the several standards adopted by the commission, several relate to activities in which the pharmacist may have a significant role. These standards include the development of ''appropriate policies, procedures, processes or protocols governing the provision of resuscitative services, appropriate data collection related to the process and outcomes of resuscitation,'' and ''ongoing review of outcomes, in the aggregate, to identify opportunities for improvement of resuscitative efforts.'' According to the Bethesda Conference on Cardiopulmonary Resuscitation, the majority of U.S. hospitals are deficient in one or more of the areas in which these new JCAHO standards have been established and will require significant restructuring of their resuscitative efforts.[11] It

is clear that a real opportunity exists for the pharmacist to influence those efforts related both to the improvement of drug use policies, and to the creation of quality improvement and feedback processes that can identify and improve hospital resuscitative efforts. Despite the assumption that hospitals function as self-contained emergency medical services (EMS) systems with respect to their management of cardiac arrest based on their abundance of health care providers in a defined environment, the Bethesda Conference report stated, ''the process of improving resuscitation in the hospital remains in its infancy.'' Such an assessment presents a unique opportunity for pharmacy to establish itself as a necessary component of a process that has become a part of required standards of hospital care.

EMERGENCY MEDICAL SERVICES

Pharmacists have provided clinical services in emergency medicine-related areas of hospitals since the late 1960s.[12] Documentation of clinical pharmacy services relate primarily to the provision of services in hospital emergency departments. Services vary from those provided fundamentally as support to the emergency department staff, to those provided directly to patients in specific disease management programs.[13,14] In a report of follow-up observations on 3787 pharmacotherapy consultations provided in an emergency department in a university hospital, 33% involved patients with pulmonary disease, 22% involved toxicology cases, 17% involved patients with seizure disorders, 11% involved cardiac cases, 7% were pharmacokinetic consultations, and 8% were miscellaneous consultations.[15] Consultations averaged 100 minutes each, and serum drug concentration determinations primarily involved theophylline, phenytoin, phenobarbital, and acetaminophen. An early questionnaire of the value of clinical pharmacy services in a medical center emergency department setting reported that clinical pharmacy services added benefit to both patient care and to educational programs in the department.[16] Eighty-seven percent of the responding physicians reported the pharmacist capable of providing primary care to specific patients once a physician-based diagnosis was established. Ninety-five percent of responders believed that clinical pharmacy services could be transferred to other emergency departments, and 83% were willing to have their patients charged for clinical services provided by the pharmacist. A more recent evaluation of the utility of clinical pharmacy services in the emergency department revealed that provision of a 24-hour consultative service by clinical pharmacy residents provided 3.1 consultations per 14-hour call period. Ninety percent of these consultations were completely followed by the recipient physicians.[17] Clinical pharmacy services involving a pediatric subspecialty emergency practice has been described in which a pharmacist is a member of a pediatric trauma team consisting of a pediatric surgeon, neurosurgeon, emergency physician, intensivist, radiology technician, and an intensive care unit nurse.[18]

Opportunities for clinical pharmacy service in the emergency department may expand because of the considerable effects resulting from changes in health insurance provision in the United States focused on reducing the number of hospital admissions. Perhaps the most prominent example of clinical pharmacy services in the emergency department setting is related to the management of asthma. The number of asthma-related deaths in the United States is increasing, especially among children. One potential cause for this increase may be inappropriate early discharge of patients from emergency departments. Pharmacists have participated in interdisciplinary efforts to maximize the urgent care of asthma patients in a number of environments, including the emergency department. In a university-affiliated urban teaching hospital, the number of emergency department visits for a group of asthma patients was significantly reduced after institution of a comprehensive program of asthma management.[19]

Finally, a number of nontraditional practice sites have been described in which pharmacist have participated in natural disaster relief or as part of a humanitarian effort relief team. These nontraditional practices have included a pharmacy consultative service for wilderness emergency drug planning, pharmacy involvement in emergency preparedness/response, the provision of pharmaceutical services at a medical site after Hurricane Andrew in Florida, and the experience of several pharmacists providing service in Bosnia-Herzegovina.[20–23]

In conclusion, there are numerous current emergency practice opportunities in which pharmacists play a significant role. Although the number of pharmacists who provide clinical services in these settings is relatively small, the critical nature of drug use in such setting suggests that the potential for direct pharmacotherapeutic intervention is large. Well-designed outcome evaluation of such service is sorely needed.

REFERENCES

1. Edwards, G.A.; Samuels, T.M. The role of the hospital pharmacist in emergency situations. Am. J. Hosp. Pharm. **1968**, *25*, 128–133.

2. Shimp, L.A.; Mason, N.A.; Toedter, N.M.; Atwater, C.B.; Gorenflow, D.W. Pharmacist participation in cardiopulmonary resuscitation. Am. J. Health-Syst. Pharm. **1995**, *52* (9), 980–984.
3. Bond, C.A.; Raehl, C.L.; Pitterle, M.E. 1992 National clinical pharmacy services survey. Pharmacotherapy **1994**, *14* (3), 282–304.
4. Bond, C.A.; Raehl, C.L.; Pitterle, M.E. 1992 National clinical pharmacy services survey. Pharmacotherapy **1998**, *18* (2), 302–326.
5. Bond, C.A.; Raehl, C.L.; Pitterle, M.E. 1992 National clinical pharmacy services survey. Pharmacotherapy **2000**, *20* (4), 436–460.
6. Elenbaas, R. Pharmacist on resuscitation team. N. Engl. J. Med. **1972**, *287* (3), 151.
7. Bond, C.A.; Reahl, C.L. Pharmacists' attitudes toward the use of cardiopulmonary resuscitation training received in pharmacy school. Am. J. Hosp. Pharm. **1989**, *46* (7), 1392–1394.
8. White, R.D.; Asplin, B.R.; Bugliosi, T.F.; Hankins, D.G. High discharge survival rate after out-of-hospital ventricular fibrillation with rapid defibrillation by police and paramedics. Ann. Emerg. Med. **1996**, *28*, 480–485.
9. White, R.D.; Hankins, D.G.; Bugliosi, T.F. Seven years' experience with early defibrillation by police and paramedics in an emergency medical services system. Resuscitation **1998**, *39*, 145–151.
10. *Comprehensive Accreditation Manual for Hospitals: The Official Handbook (CAMH)*, Joint Commission Resources, Inc.
11. Ewy, G.A.; Ornato, J.P. 31st Bethesda conference. Emergency cardiac care. Task force 1: Cardiac arrest. J. Am. Coll. Cardiol. **2000**, *35* (4), 832–846.
12. Edwards, G.A.; Samuels, T.M. The role of the hospital pharmacist in emergency situations. Am. J. Hosp. Pharm. **1968**, *25* (3), 128–133.
13. Culbertson, V.; Anderson, R.J. Pharmacist involvement in emergency room services. Contemp. Pharm. Pract. **1981**, *4* (3), 167–176.
14. Powell, M.F.; Solomon, D.K.; McEachen, R.A. Twenty-four hour emergency pharamaceutical services. Am. J. Hosp. Pharm. **1985**, *42* (4), 831–835.
15. Berry, N.S.; Folstad, J.E.; Bauman, J.L.; Leikin, J.B. Follow-up observations on 24-hour pharmacotherapy services in the emergency department. Ann. Pharmacother. **1992**, *26* (4), 476–480.
16. Elenbaas, R.M.; Waeckerle, J.F.; McNabney, W.K. The clinical pharmacist in emergency medicine. Am. J. Hosp. Pharm. **1977**, *34* (8), 843–846.
17. Kasuya, A.; Bauman, J.L.; Curtis, R.A.; Duarte, B.; Hutchinson, R.A. Clinical pharmacy on-call program in the emergency department. Am. J. Emerg. Med. **1986**, *4* (5), 464–467.
18. Vernon, D.D.; Furnival, R.A.; Hansen, K.W.; Diller, E.M.; Bolte, R.G.; Johnson, D.G.; Dean, J.M. Effect of a pediatric trauma response team on emergency department treatment time and mortality of pediatric trauma victims. Pediatrics **1999**, *103* (1), 20–24.
19. Pauuley, T.R.; Magee, M.J.; Cury, J.D. Pharmacist-managed, physician-directed asthma management program reduces emergency department visits. Ann. Pharmacother. **1995**, *29* (1), 5–9.
20. Closson, R.G. The pharmacist as consultant for wilderness emergency drug planning. J. Am. Pharm. Assoc. **1977**, *17* (12), 746–749.
21. Moore, S.R. Pharmacy involvement in emergency preparedness/response. J. R. Soc. Health **1998**, *118* (1), 28–30.
22. Nestor, A.; Aviles, A.I.; Kummerle, D.R.; Barclay, L.P.; Rey, J.A. Pharmaceutical services at a medical site after hurricane andrew. Am. J. Hosp. Pharm. **1993**, *50* (9), 1896–1898.
23. Bussieres, J.F.; St-Arnaud, C.; Schunck, C.; Lamarre, D.; Jouberton, F. The role of the pharmacist in humanitarian aid in Bosnia-Herzegovina: The experience of pharmaciens sans frontiers. Ann. Pharmacother. **2000**, *34* (1), 112–118.

PROFESSIONAL DEVELOPMENT

Cardiology, Clinical Pharmacy Practice in

C

Mary Ross Southworth
Jerry L. Bauman
University of Illinois, Chicago, Illinois, U.S.A.

INTRODUCTION

The roots of the clinical pharmacy movement are firmly imbedded in cardiology as a discipline. From the beginning, clinical pharmacists emphasized and specialized in cardiology pharmacotherapy. This is not surprising given several factors: the overall prevalence of heart disease in the United States, the prominence of cardiology as a discipline within medicine, and the fact that the cornerstone of the treatment of heart disease is drug therapy. Indeed, cardiac drugs tend to be complex, replete with drug interactions and narrow therapeutic margins and the need for close therapeutic monitoring, so that contributions by pharmacists to the care of patients with heart disease seem to be natural. The exact history of cardiology clinical pharmacy remains ill-defined, but programs with intensive training in cardiology therapeutics in the mid-1970s were the University of Missouri at Kansas City and the University of Texas at San Antonio. Of note, E. Grey Dimond, MD, a founding member of the American College of Cardiology, also initiated the clinical pharmacy program at Truman Medical Center and the University of Missouri at Kansas City School of Medicine. Strong training programs (specialized residencies and/or fellowships) subsequently developed in the early 1980s at the University of Illinois at Chicago, the University of Tennessee, and the University of Connecticut. Graduates of these programs in turn seeded many academic health centers and colleges of pharmacy throughout the United States. Today, many cardiology clinical pharmacists can trace their ancestry to a few of these programs.

Cardiology within pharmacy is not a stand-alone specialty. Rather, it is viewed as a subspecialty of sorts within the umbrella specialty of clinical pharmacy (or pharmacotherapy). In 2000, the Board of Pharmaceutical Specialties designated a process whereby a board-certified pharmacotherapy specialist may apply for "added qualifications" in cardiology pharmacotherapy practice. The applicant must submit a portfolio that documents training, clinical practice, educational efforts, and scholarly activities (among other details) specifically within cardiology. For this process to be approved by the Board of Pharmaceutical Specialties, a petition[1] was submitted (and subsequently approved) outlining the rationale, need, and demographics of the subspecialty, along with appropriate supporting information. This petition was prepared by the Cardiology Practice and Research Network (PRN) of the American College of Clinical Pharmacy (ACCP); within it are a number of facts that help to define the discipline:

1. In 2000, there were about 30 fellowships or specialized residencies in cardiology clinical pharmacy in the United States.
2. About 13% of all board-certified pharmacotherapy specialists list cardiology as the main emphasis of their practice.
3. The Cardiology PRN of ACCP has about 400 members, one of the largest subspecialties within this organization.
4. About 1100 members list cardiology practice as their primary emphasis on membership surveys of ACCP (750) and the American Society of Health-System Pharmacists (350).
5. From a survey of board-certified pharmacotherapy specialists performed for the BPS petition for added qualifications in cardiology, the following was listed as the respondent's practice area: cardiac intensive care (40%), stepdown/telemetry unit (26%), anticoagulation clinic (18%), lipid clinic (18%), managed care (7%), and other primary care clinic (35%). Of those responding, 24% had fellowship training, 13% had a specialized residency, and 19% had completed a certificate program.

Although the discipline of clinical pharmacy (or pharmacotherapy) within organized medicine is relatively young, specialized practice in cardiology is one of its more mature areas. It is not a stand-alone specialty because one uses the principles and skills of the specialty pharmacotherapy (as it is presently defined) simply applied to an area of knowledge (i.e., cardiology therapeutics). There are numerous citations documenting the role of and out-

Encyclopedia of Clinical Pharmacy
DOI: 10.1081/E-ECP 120006398

comes (clinical and economic) associated with clinical pharmacy practice in cardiology settings. These are summarized in the following sections. One will find that the role of pharmacists providing services to targeted areas (e.g., lipid clinic, anticoagulation clinic, smoking cessation) in ambulatory settings is much more frequently studied than the role of the pharmacist practicing in an acute care, inpatient setting where the clinical functions are more broad.

ACUTE CARE CARDIOLOGY PHARMACY PRACTICE

Twenty-five percent of Americans discharged from hospitals have a primary diagnosis of cardiovascular (CV) disease.[2] The pharmacist practicing in an acute care setting helps manage common cardiac disease states, including the spectrum of acute coronary syndromes (ACS), hypertensive emergencies and urgencies, acute heart failure, and cardiac arrhythmias, along with comorbid conditions. Decisions regarding optimal medication use in such patients are complex. Beginning with the initial choice of medication to treat a patient acutely, and through selection of appropriate chronic therapy and proper titration and monitoring, the acute care pharmacist is a vital component in the system of health care provision.

As part of the health care team, the acute care pharmacist works with attending physicians, physicians-in-training nurses, and other health care professionals to provide patient care. Daily activities are often centered around medical rounds, where the team reviews each inpatient's progress over the last day. Here, drug therapy decisions are made within the constructs of a team approach. Information shared during rounds includes results of lab tests, physical exams, diagnostic and therapeutic procedures, and symptomatology. Using this information, a pharmacist assists in evaluating patient response to medications, including assessing dose, route, and monitoring of each drug that the patient is receiving. When prospectively adding a medication to the patient's orders, the pharmacist recommends appropriate agents based on the clinical indication, dosing (initial and ''target''), and both efficacy and safety monitoring parameters. In providing such information, the pharmacist becomes a primary source of education regarding optimal medication use for all the members of the health care team. Other tasks the pharmacists might perform include obtaining medication histories from patients admitted to the hospital, patient medication education, and discharge counseling for patients discharged from the hospital on a new medication regimen.

An important role for the pharmacist is prevention of adverse drug events (ADEs), which significantly contribute to health care costs in numerous ways, including increases in lengths of stay, medication, and laboratory costs. Medications used in acute cardiac settings tend to have narrow therapeutic windows with substantial risk for toxicity and require close monitoring to optimize therapy (e.g., antithrombotics, antiarrythmic agents, intravenous inotropes, nitroprusside). Drug–drug interactions (also quite common with cardiac regimens), inappropriate dosing, and inappropriate drug selection are just a few examples of common ADEs where pharmacy intervention could have a tremendous impact. An important study by Leape et al. noted that the inclusion of a clinical pharmacist on a multidisciplinary team rounding in an intensive care setting reduced ADEs by 66%, through order clarification, provision of drug information, and recommendations of alternative therapy.[3]

A unique responsibility of a cardiology specialty pharmacist is the management of drug therapy of ACS, particularly those involving unstable angina and cardiac catheterization-associated procedures. Low-molecular-weight heparins and glycoprotein IIb/IIIa receptor antagonists are newer treatment modalities, but are considerably more expensive than older medications used for ACS. Newer thrombolytics used in treatment of acute myocardial infarction are easier to administer (in one or two bolus doses versus an infusion), yet are more expensive. Therefore, there is a need to develop cost-effective treatment strategies that encompass these newer agents. These strategies must take into account critical literature evaluation (i.e., are there superior outcomes between studies involving the newer agents?) and knowledge of patient characteristics (i.e., determining if the patient has an appropriate indication for use of a new therapy, identifying appropriate dosage adjustments in the face of renal insufficiency) when formulating guidelines. The cardiology specialty pharmacist may play a significant role in developing such guidelines for the institution, selecting individual patients for therapy, and selecting which therapy to use in particular ACS scenarios.

It is common to find a pharmacist as a member of the hospital cardiopulmonary resuscitation (CPR) team, which responds to emergent situations that may require immediate patient care. These scenarios usually involve a patient who suddenly becomes nonresponsive, ceases spontaneous respirations, and/or experiences a life-threatening cardiac arrhythmia. The CPR team responds to such patients by implementing advanced cardiac life support (ACLS), which involves quick provision of an airway and electrical (defibrillation) and/or pharmacologic interventions to sustain cardiac function. The pharmacist's role on such a team involves the preparation of intravenous infusions needed in an emergent situation, dose calculations, and consultation regarding appropriate medication use.

Participation by a pharmacist on a CPR team was associated with significantly lower hospital mortality rates in a study by Bond and colleagues.[4]

OUTPATIENT CARDIOLOGY SPECIALTY PRACTICE

In the outpatient setting. cardiology pharmacists frequently provide services in a wide array of clinic types, including general cardiology clinics, primary care/family medicine clinics, and disease management clinics. The impact of a cardiology pharmacist in these settings has been clearly documented in the medical literature. Generally, a pharmacist's knowledge of CV disease state pathophysiology, presentation, and course, coupled with extensive knowledge of drug therapy options and monitoring are invaluable insofar as enhancing comprehensive patient care. The following is a description of types of specialty care that a pharmacist might provide.

Hypertension

Some of the earliest published reports on the effects of provision of pharmaceutical care provided insight into the effects of a pharmacy program in the care of patients with hypertension. In an early study by McKenney and colleagues, the effects of clinical pharmacy services in a group of hypertensive patients were described.[5] Those patients who received pharmacy services in addition to standard care by their physician demonstrated an improvement in self-knowledge of their disease state, improved compliance, and better blood pressure control. Subsequent investigations have demonstrated a positive effect of pharmacy services on cost and quality of life in patients treated for hypertension.[6,7]

In this largely asymptomatic yet morbid disease, early identification and treatment are the mainstays for excellent patient care. The proper management of a hypertensive patient begins with selecting an appropriate goal blood pressure, recognizing other risk factors for CV disease, noting concomitant disease states, and selecting appropriate drug therapy for the patient. When selecting such therapy it is important to bear in mind compelling indications (as defined in the Sixth Report for the Joint National Committee on the Detection, Evaluation, and Treatment of High Blood Pressure,[8] which is the consensus guidelines for the treatment of hypertension), contraindications or cautions for using certain classes of medications, patient compliance, and cost. As a drug expert, pharmacists are in an ideal position to enhance care through the selection and monitoring of antihypertensive therapy. In addition, a pharmacist can provide patient education regarding the importance of compliance, adverse effects, and goals of therapy, thereby increasing the likelihood of successful control of patients' disease states.

Dyslipidemia

Dyslipidemia is a major risk factor for several CV diseases, including myocardial infarction, stable and unstable angina, and stroke. Control of cholesterol levels is important in reducing risk of both primary and secondary CV events. High cholesterol levels may be treated by altering diet and through pharmacologic intervention. Outpatient dyslipidemia clinic models that include intervention by a clinical pharmacist have demonstrated larger reductions in total cholesterol level, greater likelihood for achieving National Cholesterol Education Program[9] low-density lipoprotein goals, and better medication compliance.[10–12]

The decision to start cholesterol-lowering therapy can be complex and should involve patient assessment for concurrent risk factors (hypertension, diabetes), concomitant medications, diet, and social history (alcohol and cigarette use). The pharmacist can recommend and counsel regarding nonpharmacologic interventions such as reduction in body weight, dietary alterations, exercise, and cessation of cigarette smoking. It is also important to ensure that comorbid disease states (diabetes, hypertension) are adequately treated and monitored. If the decision is made to start a cholesterol-lowering agent, the pharmacist must ensure that the appropriate agent is selected because each agent may have a distinct effect on each lipoprotein component (low density, high density, triglycerides) of the lipid profile. In addition, prospective identification and avoidance of drug interactions when using cholesterol-lowering therapy is a salient pharmacist responsibility. For example, HMG CoA reductase inhibitors (some of the most commonly used cholesterol-reducing medications) are agents that inhibit a major metabolizing enzyme in the liver and may be a source of clinically significant drug interactions, including the occurrence of myositis, rhabdomyolysis, or renal dysfunction. Recommendation of pertinent monitoring parameters and patient education are additional contributions that the clinical pharmacist can make when caring for the dyslipidemic patient.

Chronic Heart Failure

There are many drug therapy-specific tasks unique to the care of a heart failure patient. A thorough review of the medication profile of a patient with heart failure should include ensuring the presence of appropriate medications

(ACE inhibitors, beta blockers) for reducing mortality related to this devastating disease. Cardiology pharmacists can optimize therapy by identifying and achieving goal doses (those doses achieved in clinical trials of heart failure) for each of these medications, depending on patient tolerability. Of equal importance is a check for medications that may potentially exacerbate heart failure or cause toxicity when given in conjunction with existing heart failure therapy.

Numerous trials document that many patients do not always receive drugs shown to decrease mortality in heart failure (e.g., ACE inhibitors) or do not receive the proper (''goal'') doses (see studies by Smith et al.[13] and Roe et al.[14]). Clearly, cardiology pharmacists have an impact on outcomes related to the use of these medications by ensuring appropriate dosing parameters. Such responsibilities may include recognition of an appropriate patient for ACE inhibitor or beta blocker therapy, proper uptitration of each agent, management of adverse effects related to therapy, and identification of true ACE or beta blocker intolerance.

Gattis and colleagues demonstrated the valuable contributions made by a clinical pharmacist in the care of patients with heart failure.[15] In this study, pharmacists made therapy recommendations (including ensuring attainment of goal doses of heart failure medications, avoidance of contraindicated medications), provided patient education regarding medical therapy, and monitored for adverse drug events. By providing intensive pharmacy services, the investigators were able to demonstrate a reduction in all cause mortality, attainment of higher ACE inhibitor dose, and greater use of alternate vasodilators in those patients intolerant to ACE inhibitors.

Antithrombosis Specialty Pharmacy Practice

Antithrombotic therapy is used in myriad CV diseases such as atrial fibrillation, heart failure, valve replacement, peripheral vascular disease, and stroke. This presents an ideal setting for a pharmacist-managed antithrombosis clinic. Today, many institutions have antithrombosis clinics managed by clinical pharmacists; this trend continues to grow.

In such a clinic, a pharmacist is responsible for the careful, periodic monitoring of prothrombin time (or international normalized ratio [INR]) to ensure safe and efficacious therapy with oral anticoagulants such as warfarin. In addition, the pharmacist emphasizes patient eduation regarding adverse effects, vitamin K-containing diets, and potentially interacting drugs. The patient's drug profile should be reviewed at every visit (for prescription and over-the-counter medication) to prevent possible drug interactions. Guidance is also provided to other health care providers regarding appropriate dosage changes and monitoring parameters.

Numerous studies have described clinic models and outcomes related to pharmacist-managed antithrombosis clinics. Chiquette and colleagues demonstrated fewer incidences of supratherapeutic levels of anticoagulation, more consistent maintenance of appropriate levels of anticoagulation, lower rates of bleeding complications, and thromboembolic events in a group of patients managed in a pharmacist-managed clinic versus those managed by usual medical care.[16] These investigators also showed lower rates of hospital admissions and emergency department visits due to warfarin therapy in those patients managed in the clinic. Similar superior care was noted in investigations published by Wilt and colleagues who also demonstrated a 20-fold increase in events (warfarin-related hospitalization, hemorrhagic or thrombotic events) in patients cared for in a family practice setting versus a pharmacist-managed clinic.[17] Both of these investigations translated these reduced event rates into significant cost savings; Chiquette estimated annual health care costs would be reduced by more than $130,000 per 100 patients, whereas Wilt attributed a $4000 cost avoidance per person-year of follow-up for those patients managed by a pharmacist.

Other antithrombotic therapies that are managed by pharmacists include the oral antiplatelet agents ticlopidine and clopidigrel, which are used for therapy for ischemic stroke and postcoronary stent placement. As with warfarin, therapy with these medications requires specific monitoring (especially of blood counts) and patient education. Another potential role for an antithrombosis pharmacist is management of patients on low-molecular-weight heparin (e.g., enoxaparin). As the use of these agents has expanded to the outpatient setting, particularly as a transition to oral anticoagulant therapy, the need exists for a skilled clinical pharmacist to maintain effective and safe treatment with the agents. Dedden and colleauges published a report on a pharmacist-managed program to treat proximal deep vein thrombosis.[18] Patients were treated at home with enoxaparin and warfarin until the patient's INR was therapeutic; all therapeutic monitoring was done by pharmacists. In treating 55 patients, a total of 294 patient hospital days were avoided, which could be translated into significant cost savings.

Other Clinic Types

Given the many clinical conditions related to or caused by cardiac conditions combined with the numerous medications used to treat such conditions, it is clear that the potential for pharmacist collaboration in the care of cardiology patients is endless. Other clinic types described in the literature include pharmacist-managed smoking cessation clinics[19–22] amiodarone monitoring clinics,[23] and cardiac medication assistance programs[24] (for those who cannot afford these medications).

NETWORKING OPPORTUNITIES

Many cardiology clinical pharmacists collaborate closely with their physician colleagues in patient care and scholarly matters. Presentations of the results of major clinical trials that will influence the daily practice of clinicians compel many to attend major medical cardiology meetings, such as the annual meetings of the American Heart Association or the American College of Cardiology, and follow cardiology specialty journals such as *Circulation*, *Journal of the American College of Cardiology*, *American Journal of Cardiology*, and *American Heart Journal*, among others. However, the primary forum for networking of cardiology clinical pharmacists is through the Cardiology PRN of the ACCP. This group meets at the annual meeting of ACCP, and maintains a useful and ac-

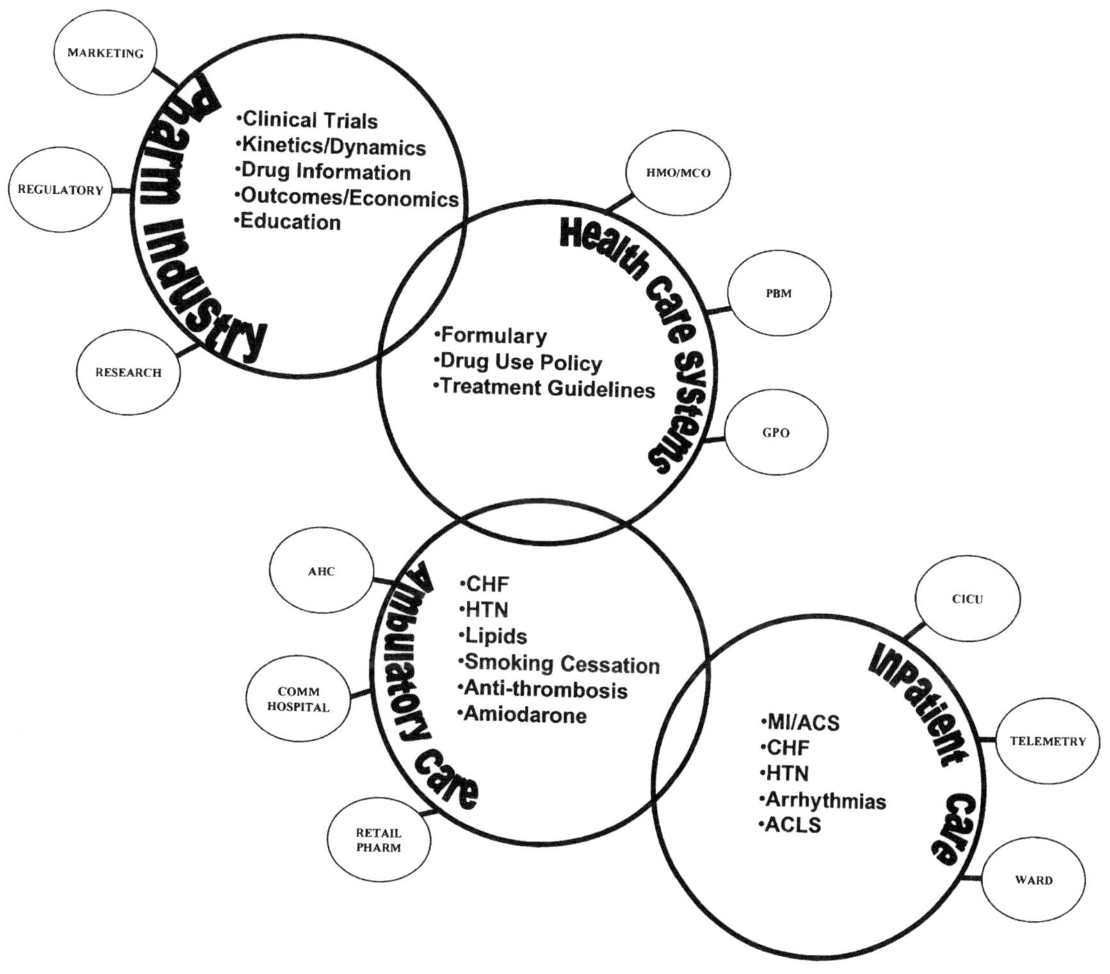

Fig. 1 Representation of the spectrum of cardiology clinical pharmacy practice. Clinical pharmacists may use their skills and knowledge of the drug treatment of heart disease in a variety of sites (large circles) such as the pharmaceutical industry, health care systems, ambulatory care, or inpatient settings. Specific duties are listed inside the large circles; they may include the direct care of patients (inpatient and ambulatory practice) or more global responsibilities for drug use (health care systems and industry), and overlap to some degree. The smaller circles represent more specific practice sites for each of the respective areas. Abbreviations: HMO, health maintenance organization; MCO, managed care organization; GPO, group purchasing organization; AHC, academic health center; Comm, community; Pharm, pharmacy or pharmaceutical; CICU, cardiac intensive care unit; CHF, congestive heart failure; HTN, hypertension; MI/ACS, myocardial infarction/acute coronary syndromes; ACLS, advanced cardiac life support (i.e., cardiac arrest team).

tive listserv for discussion on therapeutic problems or issues in clinical pharmacy practice.

PROFESSIONAL OPPORTUNITIES

There are numerous opportunities for cardiology clinical pharmacists, and these opportunities appear to be expanding (Fig. 1). Traditionally, positions with a predominantly clinical practice focus were concentrated in academic medical centers, often those affiliated with a college of pharmacy. Here, clinicians would practice—either collaboratively (particularly in an inpatient setting) within a health care team on cardiology units or independently (particularly in an ambulatory setting)—to manage specialized clinics. Typically, this type of clinician has teaching duties and some scholarly duties, in addition to clinical responsibilities. These roles have expanded to some degree into community hospitals as the clinical pharmacy movement grew. Further, because of the shift to managed health care, some clinical pharmacists with skills in cardiology may take responsibility for the drug use in a health care system, managing formularies and developing systemwide treatment guidelines and drug-use policies. Here, cardiology clinical pharmacists use their skills and knowledge to effect drug use in *populations* of patients with heart disease rather than select individuals.

For positions with a research focus, cardiology clinical pharmacists (usually with research fellowship training) have opportunities as clinical science faculty at research-intensive universities (colleges of pharmacy and/or medicine) or in the pharmaceutical industry. In industry positions, cardiology clinical pharmacists may coordinate clinical trials (phase III and IV) or work in drug disposition and pharmacokinetics. These types of positions remain, but other opportunities have arisen more recently. For instance, opportunities in the pharmaceutical industry for ''medical service managers'' or medical liaisons have expanded. These individuals may coordinate some smaller, single-site research projects (e.g., phase IV), have educational responsibilities to physicians and pharmacists, and also have a minor sales component within their duties (or combinations of all of the above, depending on the specific company). The industry is seeking sophisticated health care professionals who can represent the company and their products on a sophisticated level; thus, the cardiology clinical pharmacist seems well suited for such positions.

Last, there has been an expansion of cardiology clinical pharmacy into ambulatory settings. Due in part to prospective and fixed payment systems and the growing sophistication of pharmacists in therapeutic decision making, cardiology clinical pharmacists can find opportunities in managing disease state-specific clinics such as the ones previously reviewed (e.g., antithrombosis, smoking cessation, heart failure lipid management). Indeed, it has become very common (if not standard of care) for health care systems to employ cardiology pharmacists to manage outpatient antithrombosis treatment (e.g., warfarin, low-molecular-weight heparin). Usually, this is accomplished by establishing approved treatment protocols and collaborative drug therapy agreements with physician colleagues. It should be noted that a growing number of states have passed legislation to allow collaborative drug management by pharmacists (i.e., prescriptive authority under approved protocols and/or agreements with physicians).

The next frontier is cardiology clinical practice in community pharmacy settings. It is hoped that the progress made in ambulatory practice can be extrapolated into these environments. This possibility has been fueled by demonstration projects where pharmacists receive financial payments for cognitive services. Noteworthy is that some of these initial disease state management efforts (e.g., management of hypertension, lipid disorders, and thrombosis) require practice skills and specialized knowledge in cardiovascular pharmacotherapy.

APPENDIX

Clinical Pharmacy Guidelines, Consensus Statements, and Resources for Cardiology Specialty Pharmacy Practice

General

American Heart Association web site: www.americanheart.org.

Acute coronary syndromes

- Ryan TJ; Antman EM; Brooks NH; Califf RM; Hillis LD; Hiratzka LF; Rapaport E; Riegel B; Russell RO; Smith EE III; Weaver WD. 1999 update: ACC/AHA guidelines for the management of patients with acute myocardial infarction: executive summary and recommendations: a report of the American College of Cardiology/American Heart Association Task Force on Practice Guidelines (Committee on Management of Acute Myocardial Infarction). Circulation. **1999**, *100*, 1016–1030.
- Braunwald E; Antman EM; Beasley JW; Califf RM; Cheitlin MD; Hochman JS; Jones RH; Kereiakes D; Kupersmith J; Levin TN; Pepin CJ; Schaeffer JW;

Smith EE III; Steward DE; Theroux P. ACC/AHA 2002 Guideline update for the management of patients with unstable angina and non-ST-segment elevation myocardial infarction: a report of the American College of Cardiology/American Heart Association Task Force on Practice Guidelines (Committee on the Management of Patients with Unstable Angina). 2002. Available at: http://www.acc.org/clinical/guidelines/unstable/unstable.pdf.

Stable angina

- Gibbons RJ; Chatterjee K; Daley J; et al. ACC/AHA/ACP–ASIM guidelines for the management of ptients with chronic stable angina: executive summary and recommendations: a report of the American College of Cardiology/American Heart Association Task Force on Practice Guidelines (Committee on Management of Patients With Chronic Stable Angina). Circulation. **1999**, *99*, 2829–2848.

Hypertension

- The Sixth Report for the Joint National Committee on the Detection, Evaluation, and Treatment of High Blood Pressure. Arch Intern Med **1997**, *157*, 2413–2446.

Dyslipidemia

- Executive Summary of the Third Report of the National Cholesterol Education Program (NCEP) Expert Panel on Detection, Evaluation, and Treatment of High Blood Cholesterol in Adults (Adult Treatment Panel III), JAMA **2001**, *285*, 2486–2497.

Antithrombotic therapy

- Sixth ACCP Consensus conference on antithrombotic therapy. Chest **2001**, *119* (suppl), 1S–370S.

Atrial fibrillation

- Fuster V, Ryden LE, Asinger RW, Cannom DS, Crijns HJ, Frye RL, Halperin JL, Kay GN, Klein WW, Levy S, McNamara RL, Prystowsky EN, Wann LS, Wyse DG. ACC/AHA/ESC guidelines for the management of patients with atrial fibrillation: a report of the American College of Cardiology/American Heart Association Task Force on Practice Guidelines and Policy Conferences (Committee to Develop Guidelines for the Management of Patients With Atrial Fibrillation). J. Am Coll Cardiol **2001**, *38*, 1266i-lxx.

Heart failure

- Hunt SA, Baker DW, Chin MH, et al. ACC/AHA guidelines for the evaluation and mangement of chronic heart failure in the adult: a report of the American College of Cardiology/American Heart Association Task Force on Practice Guidelines (Committee to Revise the 1995 Guidelines for Evaluation and Management of Heart Failure. 2001. American College of Cardiology Web site. Available at: http://www.acc.org/clinical/guidelines/failure/hf_index.htm.

Advanced cardiac life support

- The American Heart Association in Collaboration with the International Liaison Committee on Resuscitation (ILCOR). Guidelines 2000 for Cardiopulmonary Resuscitation and Emergency Cardiovascular Care. Circulation **2000**, *102* (suppl I), I-1–I-384.

REFERENCES

1. The American College of Clinical Pharmacy Cardiology Practice and Research Network. *Executive Summary: Petition for Added Qualifications in Cardiology for Board Certified Pharmacotherapy Specialists*; 2000.
2. American Heart Association Web Page, www.americanheart.org, accessed November 2000.
3. Leape, L.L.; Cullen, D.J.; Clapp, M.D.; Burdick, E.; Demonaco, H.J.; Erickson, J.I.; Bates, D.W. Pharmacist participation on physician rounds and adverse drug events in the intensive care unit. JAMA **1999**, *282*, 267–270.
4. Bond, C.A.; Raehl, C.L.; Franke, T. Clinical pharmacy services and hospital mortality rates. Pharmacotherapy **1999**, *19*, 556–564.
5. McKenney, J.M.; Slining, J.S.; Henderson, H.R.; Devins, D.; Barr, M. The effect of a clinical pharmacy services on patients with essential hypertension. Circulation **1973**, *48*, 1104–1111.
6. Forstrom, M.J.; Ried, L.D.; Stergachis, A.S.; Corliss, D.A. Effect of a clinical pharmacist program on the cost of hypertension treatment in an HMO family practice clinic. DICP **1990**, *24*, 304–309.
7. Erickson, S.R.; Slaughter, R.; Halapy, H. Pharmacists ability to influence outcomes of hypertensive therapy. Pharmacotherapy **1997**, *17*, 140–147.
8. The sixth report for the Joint National Committee on the detection, evaluation, and treatment of high blood pressure. Arch. Intern. Med. **1997**, *157*, 2413–2446.
9. National cholesterol education program. Second report of the expert panel on detection, evaluation, and treatment of high blood cholesterol in adults. Circulation **1994**, *89*, 1333–1445.

10. Bogden, P.E.; Koontz, L.M.; Williamson, P.; Abbott, R.D. The physician and pharmacist team: An effective approach to cholesterol reduction. J. Gen. Intern. Med. **1997**, *12*, 158–164.
11. Shaffer, J.; Wexler, L.F. Reducing low-density lipoprotein cholesterol in an ambulatory care system. Results of a multidisciplinary collaborative practice lipid clinic compared with traditional physician-based care. Arch. Intern. Med. **1995**, *155*, 2330–2335.
12. Furmaga, E.M. Pharmacist management of a hyperlipidemia clinic. Am. J. Hosp. Pharm. **1993**, *50*, 91–95.
13. Smith, N.; Psaty, B.M.; Pitt, B.; Garg, R.; Gottdiener, J.S.; Heckbert, S.R. Temporal patterns in the medical treatment of congestive heart failure with angiotensin converting enzyme inhibitors in older adults, 1989 through 1995. Arch. Intern. Med. **1998**, *158*, 1074–1080.
14. Roe, C.M.; Motheral, B.R.; Teitelbaum, F.; Rich, M.W. Angiotensin converting enzyme inhibitor compliance and dosing among patients with heart failure. Am. Heart J. **1999**, *138*, 818–825.
15. Gattis, W.A.; Hasselblad, V.; Whellen, D.J.; O'Connor, C.M. Reduction in heart failure events by the addition of a clinical pharmacist to the heart failure management team. Arch. Intern. Med. **1999**, *159*, 1939–1945.
16. Chiquette, E.; Amato, M.G.; Bussey, H.I. Comparison of an anticoagulation clinic with usual medical care: Anticoagulation control, patient outcomes, and health care costs. Arch. Intern. Med. **1998**, *158*, 1641–1647.
17. Wilt, V.M.; Gums, J.G.; Ahmed, O.I. Outcome analysis of a pharmacist managed anticoagulation service. Pharmacotherapy **1995**, *15*, 732–739.
18. Dedden, P.; Chang, B.; Nagel, D. Pharmacy managed program for home treatment of deep venous thrombosis with enoxaparin. Am. J. Health-Syst. Pharm. **1997**, *54*, 1968–1972.
19. Tommasello, T. Two pharmacy-practice models for implementing the AHCPR smoking cessation guideline. Tob. Control **1997**, *6* (Suppl. 1), S36–S38.
20. Crealey, G.E.; McElnay, J.C.; Maguire, T.A.; O'Neill, C. Costs and effects associated with a community pharmacy-based smoking cessation programme. Pharmacoeconomics **1998**, *14*, 323–333.
21. Sinclair, H.K.; Bond, C.M.; Lennox, A.S.; Silcock, J.; Winfield, A.J.; Donnan, P.T. Training pharmacists and pharmacy assistants in the stage of change model of smoking cessation: A randomised controlled trial in Scotland. Tob. Control **1998**, *7*, 253–261.
22. Sanoski, C.A.; Schoen, M.D.; Gonzalez, R.C.; Avitall, B.; Bauman, J.L. Rationale, development, and clinical outcomes of a multidisciplinary amiodarone clinic. Pharmacotherapy **1998**, *18*, 146S–151S.
23. Carmichael, J.M.; O'Connell, M.B.; Devine, B.; Kelly, W.; Ereshefsky, L.; Linn, W.; Stimmel, G.L. ACCP position statement: Collaborative drug therapy management by pharmacists. Pharmacotherapy **1997**, *17*, 1050–1061.
24. Schoen, M.D.; DiDomenico, R.S.; Connor, S.E.; Dischler, J.E.; Bauman, J.L. Impact of the cost of prescription drugs on clinical outcomes in indigent patients with heart disease. Pharmacotherapy **2001**, *21*, 1455–1463.

PHARMACY PRACTICE ISSUES

Clinical Evaluation of Drugs

Allen Cato
Lynda Sutton
Cato Research Ltd., Durham, North Carolina, U.S.A.

Allen Cato III
Cato Research Ltd., San Diego, California, U.S.A.

INTRODUCTION

The process of developing a new drug, from the identification of a potential drug candidate to postmarketing surveillance, is extremely complex. The drug development process requires input from various members of a multidisciplinary team and the conduct of numerous studies. The time from drug discovery to marketing takes an average of 13 years. Once a chemical is identified as a new drug candidate, extensive preclinical analyses must be completed before the drug can be tested in humans. The pharmacology, toxicology, and preclinical pharmacokinetics must be characterized. The formulations of the drug product that were used in the preclinical studies may be different from the formulation of the final drug product, which may require that additional formulation work and pharmacokinetic analyses be performed. If the characteristics of the new drug candidate are acceptable for all of the preclinical assessments, it may then be tested in humans. The new drug candidate, at this point, enters the clinical research stage of drug development.

Clinical research represents a vital stage in the development process a stage that is no less daunting than the preclinical research stage. The data obtained from the first-time-in-human, Phase 1 pharmacokinetic studies, and initial safety evaluations in healthy volunteers can make or break the entire developmental program for a drug candidate. The sponsoring company, of course, hopes that the data collected in these initial studies will show minimal safety concerns over an adequate dose range. The pharmacokinetic data can then be used to help design future studies in which efficacy and long-term safety are assessed and additional pharmacokinetic and pharmacodynamic data are collected.

Although the basic designs of the initial single and multiple dose-escalating studies are generally straightforward (but the starting dose is often intensely debated), it is imperative that these studies and future studies be designed to address specific questions. The questions vary depending on numerous specific considerations, including the targeted disease characteristics (e.g., acute or chronic); desired safety, efficacy, and pharmacokinetic evaluations; and assessment of clinical pharmacology (e.g., dosage formulations or dose frequency). Basic study procedures must also be considered. Thus, the design, conduct, data reporting and analysis, and production of the final study reports can be completed only through the coordinated efforts of a multidisciplinary drug development team.

For every clinical study, input is required from multiple personnel with various areas of expertise. Members of a drug development team include physicians, scientists, pharmacists, project managers, statisticians, computer programers, study monitors, regulatory experts, and for some studies, a representative of the formulations group. While some team members may be able to perform multiple tasks, no one team member has the expertise or the time to do everything required to conduct a clinical study. In addition, some members may have overlapping abilities, but other members with particular expertise may be called upon. For example, pharmacokineticists are the experts in pharmacokinetics, but they may also be knowledgeable in pharmaceutics, biostatistics, and clinical care. However, scientists (PhDs) are trained primarily in basic research, while physicians (MDs) are trained in clinical medicine. Since a single drug development program is derived from both of these distinct disciplines, considerable overlap, cooperation, and coordination are necessary to take a drug successfully and efficiently from discovery to market.

Clinical drug development is generally divided into four phases: Phase 1 through Phase 4. For each study conducted within a particular phase, specific information is collected according to the requirements for individual drugs being developed. Collection of safety, efficacy, and pharmacokinetic data is the focus of most clinical trials. Although these topics appear to be distinct disciplines,

Encyclopedia of Clinical Pharmacy
DOI: 10.1081/E-ECP 120006416

they are intertwined and represent different ways of evaluating the intrinsic properties of a drug. While the safety, efficacy, and pharmacokinetics of a drug may be assessed in most studies, the team must establish the type and extent of information to be collected, which will vary based upon the specific objectives and designs of the studies.

A critical function of the drug development team is the development of the study protocol. The study protocol must clearly describe the study design and methodology that will be used to achieve the study objectives. Input from nonmedical and nonscientific members of the team, such as marketing and information technology experts, also is helpful in establishing development strategies and in designing and conducting of clinical studies. Finally, project planning efforts can synchronize team efforts, help contain the soaring costs of pharmaceutical research, and coordinate international development efforts.

The drug development team's primary goal is to gain approval to market the drug, which requires that a marketing application be submitted to a regulatory agency (e.g., a New Drug Application [NDA] is submitted to the Food and Drug Administration [FDA] in the United States and to the Health Products and Food Branch [HPFB] in Canada, while a Marketing Authorization Application [MAA] is submitted to European regulatory agencies). During the conduct of the studies and the compilation and analyses of the data, the team must consider and evaluate many issues, such as how to collect, categorize, and report adverse events. All of these decisions will affect the marketing application that is submitted and may ultimately define how the drug is to be administered.

Many of the decisions to be made by the team, and particularly by the investigators, pose ethical dilemmas. Legislation has been enacted to protect human research subjects. Recently, the most pressing ethical dilemma facing the clinical research scientist concerned biotechnology and genetic engineering research.

Frequent changes in the regulations and guidelines of various regulatory agencies, differences in interpretations of these rules, and special reporting mechanisms for adverse events represent only a few of the challenges facing a drug development team. Due to continuous advances in scientific information, understanding of disease processes, and gene therapy, change continues to be the rule in modern drug development. However, through the efficient application of sound scientific principles in an ethical manner and with a coordinated team effort, effective new therapies can continue to be developed and marketed.

ROLES OF THE DRUG DEVELOPMENT TEAM MEMBERS

Physicians

The physician's contribution to drug development and the physician's role on a drug development team have changed over the last few decades (1). Before the 1960s, medical departments of pharmaceutical companies were primarily composed of physicians who were routinely involved in responding to drug information requests rather than developing new drugs. The Kefauver–Harris Amendment, enacted in 1962, required pharmaceutical companies to demonstrate before marketing that a drug was efficacious, which necessitated that physicians increase their presence on drug development teams. Along with the advent of additional governmental regulations, the increase in complexity of medical knowledge has mandated that physicians become an integral member of any drug development team. In fact, because of the different roles of the physician within an organization, companies may now have various departments (e.g., a clinical research department and a clinical safety department) within the medical department.

Although physicians are trained in patient care, physicians who are typically employed by pharmaceutical companies have more training in scientific methodology than those in the past. The physician on the team is the one qualified to follow the progress of each patient enrolled in a clinical trial and to interpret the results. Some physicians continue to spend time treating patients at a university hospital or a specific clinic where their specialty can be utilized and practiced, which allows these physicians to maintain sharp diagnostic skills. Also, some may perform basic research in academic settings to develop or maintain their knowledge and skills in basic research.

However, much of today's clinical research is actually conducted by investigators who are not employed by the company sponsoring the development of the drug. The physician on the drug development team must help in the selection of appropriate investigators to conduct the clinical studies. Pharmaceutical physicians may rely on colleagues who are experts in their respective fields and who have appropriate patient populations and facilities for the targeted research project. The physician is also the expert who deals with emergency situations that may arise during the course of a clinical research project, such as an overdose or severe adverse experience (SAE) that might be experienced with the drug. Similarly, the physician assists investigators who are responsible for evaluating the severity of adverse experiences (AEs) and determining

the causal relationship of the AEs to the drug under development.

The physician's involvement in clinical research does not end with the completion of the clinical study. Medical reports, clinical study reports, and sections of NDAs must be written. Interactions with regulatory agencies that require the physician's input may occur frequently. Physicians in clinical research may also be called upon to promote new drugs in a scientific environment by organizing symposia and workshops and by reviewing journal advertisements and promotional material for medical validity and accuracy.

The role of the physician in a clinical drug development program has expanded and has been refined in the last 40 years. Physicians increasingly contribute clinical and scientific expertise and administrative skills. Many physicians on drug development teams today spend most of their time designing and implementing studies and interpreting and reporting data rather than being in direct contact with patients. An experienced clinician is an important member of any drug development team.

Scientists

While a drug development team may have only one primary physician, it may have multiple scientists. Pharmacokineticists, pharmacologists, toxicologists, and pharmaceutical scientists are all involved in the clinical development of drugs. The contributions of scientists to a drug development project are derived from their experience in both scientific methodology and basic research (1).

Although physicians are trained in patient care, scientists are trained in problem-solving skills related to scientific research. To obtain a doctoral degree, a scientist must conduct research and write a dissertation that covers a topic of sufficient scope and depth. During this process, the scientist learns how to solve problems from different perspectives. The scientist also collects extensive data and performs data analyses, thereby gaining valuable insight into the considerations necessary to determine the feasibility of collecting data in a clinical trial. Also, some scientists, such as pharmacokineticists with a pharmacy background, may receive some clinical experience during their training as a scientist.

Scientists help design major portions of study protocols and clinical case report forms (CRFs). The study protocol is the overall plan that the study follows, and it must contain certain types of information, including the following: 1) background data on the targeted disease; 2) the empirical and structural formula of the drug being studied; 3) preliminary pharmacology and toxicology of the drug (specific study objectives and designs); 4) the methods and materials to be used in the study; 5) information regarding drug packaging, labeling, dosage forms, and decoding procedures; 6) overdose management; 7) patient discontinuation procedures; 8) explanation of informed consent and provisions regarding institutional review board approval; and 9) any relevant references and appendices. The CRFs are the forms on which individual patient data are recorded during a clinical trial. From these data, clinical and statistical analyses are performed. All the information that is stipulated in the study protocol must be collected on the CRFs.

In conjunction with nonscientific personnel, scientists are responsible for ensuring that the CRFs will capture the appropriate information for each study subject according to the objectives, tests, and evaluations stipulated in the protocol. Careful attention must be given to the administration of special tests or collection of samples so that the timing of the assessments or sample collections do not conflict.

Experience in basic research enables the scientist to function as an important link between the basic research labs within the company and the drug development team. Departments specializing in drug metabolism, microbiology, pharmacology, and toxicology need feedback from early human safety and pharmacokinetic studies so they can continue to plan and conduct appropriate long-term animal studies. Thus, communication between the clinical scientist and the basic scientist is important throughout the progress of the drug development program.

Because clinical research has become increasingly more scientific, experts in the methodology of science are necessary for a complete research program. The drug development team's scientists may account for much of the scientific expertise, but the roles of the research team overlap to form a scientifically sound, medically astute cohesive group. In addition to scientific expertise, use of the scientist's administrative talents, such as organizational skills and familiarity with personnel practices, enables effective drug development. Thus, scientists with these skills are often employed in management positions in many organizations.

Pharmacists

The pharmacist's role on the drug development team has greatly expanded the professional opportunities of individuals with backgrounds in pharmacy. Pharmacists can provide valuable therapeutic insight into medical research. Training of pharmacists as clinical scientists with both clinical skills and scientific research skills continues to be an emphasis at many pharmacy schools. Several programs have been devised for the education and development of the pharmacist as clinical scientist (2). Pharmacists have a broad knowledge in both clinical medicine and pharmaceutics, and therefore are able to bridge the gap between the clinic and the laboratory.

Pharmacists' training focuses on drug therapies in disease states, whereas physicians' training focuses on the diagnosis of disease states. Studies regarding drug interaction, positive control, or drug comparison involve drugs that have been studied and marketed. Pharmacists can help in the design of such trials because of their knowledge of marketed drugs.

Additional roles of pharmacists appear in the areas of drug information and education and training. Pharmacists have the appropriate expertise in drug therapy to answer inquiries from physicians (and other health professionals) concerning both marketed and investigational drug products. Similarly, the clinic/laboratory bridge that the pharmacist builds makes this team member especially well suited to educate and train new employees in drug development. By offering both general and special skills, the research pharmacist blends clinical medicine with pharmaceutical science and is well qualified as an educator and drug information specialist.

Nonscientific Personnel

Drug development includes many tasks that may not require the specialized expertise of a physician or a scientist. Administrative skills, creativity, and excellent communication abilities, which are qualities not necessarily emphasized within traditional medical and scientific educational curricula, may be required for many of these tasks.

The administrative skills necessary for drug development include incorporating seemingly disparate but vitally linked concepts into a single overall plan. Integration planning may mean organizing study files into a logical sequence or helping to assemble the various parts of an NDA. In the first example, files must be set up in a way that can facilitate internal quality assurance audits and FDA inspections. In the second example, knowledge of the FDA's regulations and good abstracting capabilities are required.

Creativity is a quality that cannot be developed through formal training. Creativity requires bold conjecture and it expresses itself in newer, better ways to accomplish the same goals. An example of creativity in clinical drug research might involve the development of a variable report that could support all of the different research documents that are generated by drug research teams. With such a variable report, common information need not be recreated each time another document is generated.

Excellent communication skills may be the most important quality for individuals working in drug development, even for those with strong medical backgrounds. Clinical research requires extensive interactions with personnel within the organization and with outside vendors or clinical sites. The information flow must be both efficient and accurate. For example, marketing departments must communicate frequently with medical departments so that marketing studies, advertising, and package inserts can be planned and evaluated. Individuals who lack strong science backgrounds but who have excellent communication skills often act as liaisons in these situations.

One aspect of clinical research that requires extensive contribution by the drug development team personnel is study monitoring. Study monitors oversee the planning, initiation, conduct, and data processing of clinical studies (3). While monitoring studies, monitors must communicate frequently with investigators and help ensure the data are being collected properly, FDA regulations are being followed, and any administrative problems are resolved as quickly as possible. Although monitors traditionally have had a nonscientific background, many monitors today have training in the basic sciences, and some even have advanced degrees, which allows them to better understand the scientific aspects of the project. Effective study monitors have a wide range of talents.

The many facets of a clinical research program afford individuals with varying types of training, education, and experience, the opportunity to contribute to the drug development process. Although some tasks clearly require the clinical or scientific expertise of a physician or a scientist, other tasks are better suited to those individuals with less specialized and more general capabilities.

STAGES IN CLINICAL DRUG DEVELOPMENT

Before clinical drug development can begin, many years of preclinical development occur, millions of dollars are spent, and countless decisions are made. Basic research teams consisting of chemists, pharmacologists, biologists, and biochemists first identify promising therapeutic categories and classes of compounds. One or more compounds are selected for secondary pharmacology evaluations and for both acute and subchronic toxicology testing in animal models. A compound that is pharmacologically active and safe in at least two nonhuman species may then be selected for study in humans. Before the drug can be tested in humans, an Investigational New Drug (IND) application, which contains supporting preclinical information and the proposed clinical study designs, must be filed with an appropriate regulatory agency.

Clinical drug development follows a sequential process. By convention, development of a new drug in humans is divided into four phases: preapproval segments (Phases 1 through 3) and a postapproval segment (Phase 4) (1–4). The definitions of the three preapproval phases have relatively clear separations. However, the different phases refer to different types of studies rather than a specific time course of studies. For example, bioequivalence studies and drug–drug interaction studies are both Phase 1 studies, but they may be conducted after Phase 3 studies have been initiated. The generalized sequence of studies may be tailored to each new drug during development.

Phase 1

After the appropriate regulatory agency has approved a potential drug for testing in humans, Phase 1 of the clinical program begins. The primary goal of Phase 1 studies is to demonstrate safety in humans and to collect sufficient pharmacokinetic and pharmacological information to permit the determination of the dose strength and regimen for Phase 2 studies.

Phase 1 studies are closely monitored, are typically conducted in healthy adult subjects, and are designed to meet the primary goal (i.e., to obtain information on the safety, pharmacokinetics, and pharmacologic effects of the drug). In addition, the metabolic profile, adverse events associated with increasing dosages, and evidence of efficacy may be obtained. Because most compounds are available for initial studies as an oral formulation, the initial pharmacokinetic profile usually includes information about absorption. Additional studies, such as drug–drug interactions, assessment of bioequivalence of various formulations, or other studies that involve normal subjects, are included in Phase 1.

Generally, the first study in humans is a rising, single-dose tolerance study. The initial dose may be based on animal pharmacology or toxicology data, such as 10% of the no-effect dose. Doses are increased gradually according to a predetermined scheme, often some modification of the Fibonacci dose escalation scheme (5), until an adverse event is observed that satisfies the predetermined criteria of a maximum tolerated dose (MTD). Although the primary objective is the determination of acute safety in humans, the studies are designed to collect meaningful pharmacokinetic information. Efficacy information or surrogate efficacy measurements also may be collected. However, because a multitude of clinical measurements and tests must be performed to assess safety, measurements of efficacy parameters must not compromise the collection of safety and pharmacokinetic data.

Appropriate biological samples for pharmacokinetic assessment, typically blood and urine, should be collected at discrete time intervals based upon extrapolations from the pharmacokinetics of the drug in animals. Depending on the assay sensitivity, the half-life and other pharmacokinetic parameters in healthy volunteers should be able to be evaluated, particularly at the higher doses. The degree of exposure of the drug is an important factor in understanding the toxicologic results of the study. Pharmacokinetic linearity (dose linearity) or nonlinearity will be an important factor in the design of future studies.

Once the initial dose has been determined, a placebo-controlled, double-blind, escalating single-dose study is initiated. Generally, healthy male volunteers are recruited, although patients sometimes are used (e.g., when testing a potential anticancer drug that may be too toxic to administer to healthy volunteers). These studies may include two or three cohorts, with six or eight subjects receiving the active drug and two subjects receiving placebo. The groups may receive alternating dose levels, which allow assessment of dose linearity, intrasubject variability of pharmacokinetics, and dose-response (i.e., adverse events) relationship within individual subjects.

Participants in the first study are usually hospitalized or enrolled in a clinic so that clinical measurements can be performed under controlled conditions and any medical emergency can be handled in the most expeditious manner. This study is usually placebo-controlled and double-blinded so that the drug effects, such as drug-induced ataxia, can be distinguished from the nondrug effects, such as ataxia secondary to viral infection. The first study in humans is usually not considered successfully completed until an MTD has been reached. An MTD must be reached because the relationship between a clinical event (e.g., emesis) and a particular dose level observed under controlled conditions can provide information that will be extremely useful when designing future trials. Also, the dose range and route of administration should be established during Phase 1 studies.

A multiple-dose safety study typically is initiated once the first study in humans is completed. The primary goal of the second study is to define an MTD with multiple dosing before to initiating well-controlled efficacy testing. The study design of the multiple-dose safety study should simulate actual clinical conditions in as many ways as possible; however, scientific and statistical validity must be maintained. The inclusion of a placebo group is essential to allow the determination of drug-related versus nondrug-related events. The dosing schedule, which includes dosages, frequency, dose escalations, and dose tapering, should simulate the regimen to be followed in efficacy testing.

Typically, dosing in the second study lasts for 2 weeks. The length of the study may be increased depending on the pharmacokinetics of the drug so that both drug and metabolite concentrations reach steady state. Also, if the drug is to be used to treat a chronic condition, a 4-week study duration may be appropriate. To obtain information for six dose levels with six subjects receiving active drug and two receiving placebo for each of two cohorts, a minimum enrollment of 24 subjects should be anticipated. Similar to the first study in humans, these subjects would be hospitalized for the duration of the study.

Also similar to the first study, pharmacokinetic data must be obtained. These data will be used to help determine dosage in future efficacy trials. The new pharmacokinetic information that can be gathered includes the following: 1) determination regarding whether the pharmacokinetic parameters obtained in the previous acute safety study accurately predicted the multiple dose pharmacokinetic behavior of the drug; 2) verification of pharmacokinetic linearity (i.e., dose proportionality of C_{max} and AUC) observed in the acute study; 3) determination regarding whether the drug is subject to autoinduction of clearance upon multidosing; and 4) determination of the existence and accumulation of metabolites that could not be detected in the previous single-dose study. A number of experimental approaches can be used to gather this information, and all require frequent collection of blood and urine samples. The challenge to the clinical pharmacokineticist is to design an appropriate blood sample collection schedule that will maximize the pharmacokinetic information, yet can be gathered without biasing the primary objective—determination of clinical safety parameters.

Phase 2

After the initial introduction of a new drug into humans, Phase 2 studies are conducted. The focus of these Phase 2 studies is on efficacy, while the pharmacokinetic information obtained in Phase 1 studies is used to optimize the dosage regimen. Phase 2 studies are not as closely monitored as Phase 1 studies and are conducted in patients. These studies are designed to obtain information on the efficacy and pharmacologic effects of the drug, in addition to the pharmacokinetics. Additional pharmacokinetic and pharmacologic information collected in Phase 2 studies may help to optimize the dose strength and regimen and may provide additional information on the drug's safety profile (e.g., determine potential drug–drug interactions).

Efficacy trials should not to be initiated until the MTD has been defined. In addition, the availability of pharmacokinetic information in healthy volunteers is key to the design of successful efficacy trials. The clinical pharmacokineticist assists in the design and execution of these trials and analyzes the plasma drug concentration data upon completion of the efficacy studies.

During the planning stage of an efficacy trial, the focus is on the dosage regimen and its relationship to efficacy measurements. Plasma drug concentrations for various dosages can be simulated based upon the data collected in the first two studies in humans. The disease or physiological states of the test patients (e.g., organ dysfunction as a function of age), concurrent medications (e.g., enzyme inducers or inhibitors), and the safety data obtained earlier must be considered when choosing an optimal dosage regimen for the study. In addition, if the targeted site of the drug is in a tissue compartment, theoretical drug levels in this compartment can be simulated, which may help scientists determine the appropriate times for efficacy measurements.

On completion of the efficacy trial, a therapeutic window for plasma drug concentrations can be defined by reviewing the correlation between plasma drug concentrations and key safety and efficacy parameters. The goal is to improve efficacy and safety of the drug by individualizing the dosage based upon previous plasma drug concentration profiles in the same patient.

Phase 3

If the earlier clinical studies establish a drug's therapeutic, clinical pharmacologic, and toxicologic properties and if it is still considered to be a promising drug—Phase 3 clinical trials will be initiated. Phase 3 studies enroll many more patients and may be conducted both in a hospital or controlled setting and in general practice settings. The goals of Phase 3 studies are to confirm the therapeutic effect, establish dosage range and interval, and assess long-term safety and toxicity. Less common side effects and AEs that develop latently may be identified. In addition, studies targeted to evaluate and quantify specific effects of the drug, such as drowsiness or impaired coordination, are conducted during this phase.

Phase 3 studies are also used to identify the most appropriate population or subpopulation for the study drug and to establish a place for the drug in its therapeutic class. A drug may be developed in a therapeutic class that already has effective alternatives, but the investigative compound may have a better safety profile than its established competitors. A Phase 3 clinical study can be designed to assess relative safety profiles.

Closer inspection of drug interactions is warranted in Phase 3 clinical trials. In many disease states, the use of polytherapy is quite common, and the risk of drug–drug interactions is high, both from pharmacokinetic and pharmacodynamic perspectives. The likelihood of drug interactions and semiquantitative estimates of magnitude may be predicted from in vitro data (6). The potential for interactions needs to be evaluated from two perspectives: the potential that the new drug may affect the pharmacokinetics of other drugs, and the potential that other drugs may affect the pharmacokinetics of the new drug. The former generally depends on the ability of the new drug to affect various enzyme and carrier-mediated clearance processes. Most notably, this concerns the cytochrome P450 (CYP) isoforms but could also involve conjugative enzymes and transporters, such as p-glycoprotein. Drugs may be an effective inhibitor without being a substrate of a CYP isoform, as is the case for quinidine's inhibition of CYP2D6.

The potential for significant drug–drug interactions caused by other drugs requires knowledge of the components of clearance for the new drug and the likelihood that known inhibitors will be coadministered. For drugs with multiple pathways and a broad therapeutic index, the need for formal interaction studies may be limited. Population pharmacokinetic analyses of data obtained from Phase 3 studies may be used to help discover and quantify drug interactions due to classes of drugs often associated with inhibition (e.g., macrolides, systemic antifungals, calcium channel antagonists, fluoxetine, paroxetine) or induction (e.g., anticonvulsants, rifampin).

Most early clinical trials are conducted at university medical centers with physicians who specialize in a certain area of medicine. When study drugs are eventually marketed, however, general practitioners will be prescribing them as well. Therefore, it is important that family physicians are exposed to study drugs during this phase because they represent the segment of clinicians who will be writing most of the prescriptions. Similarly, to maximize the commercial return on drug development, a multi-indication strategy may be pursued (sometimes designated as Phase 5 if conducted postapproval). In addition, testing of the drug in foreign countries is appropriate during Phase 3; however, other countries may operate under different regulatory obligations than in the United States.

Phase 4

Whereas Phase 1, 2, and 3 studies are conducted prospectively using subjects or patients whose entrance into the study depends on strict inclusion and exclusion criteria, Phase 4 studies employ mainly observational, rather than exclusionary, study designs. Postmarketing surveillance and any additional studies requested by the regulatory agency as conditional approval of the NDA are conducted during Phase 4.

Data collection in premarketing clinical trials is an extensive, scientific exercise. Detailed blood work, special laboratory tests, and careful physiologic monitoring are typical in these studies. Postmarketing studies, however, are often targeted for much larger patient populations (5000–10,000 or more), which limits extensive data collection from each patient and emphasizes collection of safety information. These studies are complemented by reports of AEs from patients not enrolled in a study. The large numbers of patients in Phase 4 studies make it easier for researchers to determine rare AEs and can help identify patient populations that are at particular risk for certain AEs. For example, demographic trends toward side effects involving geographic locus, gender, or race may be determined from postmarketing surveillance data.

PROTOCOL CONSIDERATIONS

The task of designing a clinical study cannot be undertaken until the study objective of that trial has been rigorously defined. The objective should explicitly state what is being investigated and vague language should be avoided (7). Once an unbiased and specific objective has been developed, scientists can build the study design around it and then develop and write the protocol (8).

One of the main considerations when designing an investigational study concerns the type and number of comparative groups that will be involved. A control group of subjects may be evaluated in addition to the group taking the investigational drug. Sometimes more than one control group is used in a study. The control groups take either placebo or active medication and are compared with the group taking the investigational drug. This design is used to rule out the possibility of a placebo effect or to assess the efficacy and safety of the investigational drug relative to other drugs currently marketed.

Regulatory agencies frequently require the pivotal Phase 3 studies, which will be used to support an NDA, to be placebo-controlled studies. Placebo medication should be as similar as possible to the drug being investigated (e.g., same color, taste, and shape). No statistically significant difference in response between this group and the subjects taking the investigational drug is evidence against that drug having any real effectiveness.

Similar to the placebo considerations, active medication taken by the control group also should be as similar as possible to the drug being investigated (e.g., same color, taste, and shape). If the formulations cannot be made with similar appearances (e.g., tablet, suspension, etc.), a placebo of each formulation could be made so subjects would take one active formulation and the placebo of the other formulation to maintain the blind. No statistically significant difference in response in this group relative to the subjects taking the investigational drug is evidence that active medication has no advantage therapeutically over the existing therapy. However, a higher incidence of AEs in the control group and an equal rate of efficacy relative to the subjects taking the investigational drug are evidence of the new drug's advantage over the existing therapy.

In addition to determining the types and number of control groups that should be included in a study, the drug development team must decide between a parallel and a crossover design. For example, in a placebo-controlled clinical trial, a parallel design is one in which each study group takes the same medication (i.e., either placebo or active drug) throughout the study. With a crossover design, each study group eventually receives both placebo and active drug (e.g., one group may take placebo for a 6-week period and then cross over to receive active drug for the following 6-week period).

An advantage of the crossover design is that it allows each group to be its own control, thereby allowing a demonstration of efficacy to occur during the treatment with the drug. A disadvantage of the crossover design is that residual effects from one treatment period may carry over into the other treatment period. Absolute determination of efficacy and safety of the different treatments is difficult and sometimes impossible. One way to avoid the problem of residual effects on crossover studies is to have washout periods between the different treatment phases. During the washout period, the patient is either given a placebo or no treatment for several days or weeks so that any possible metabolite or effect of the drug is "washed out" of the patient before the next treatment phase begins.

An advantage of the parallel design is that it avoids the problems associated with possible residual effects of one treatment period influencing the other treatment period(s) because each treatment group is only exposed to one drug. Compared with a crossover study, more patients may be required for a parallel study so that statistical significance can be established between the study groups. In a parallel study, recruiting the required larger numbers of patients who fit the study criteria takes longer, but the duration of that study is usually shorter than the duration of a crossover study.

Crossover designs span greater periods of time because each group must sequentially take an active and a control medication over a period that is long enough to allow a treatment effect to emerge. When washout periods are added, the time required to conduct these studies becomes longer still, and more study subjects may drop out. These difficulties are often outweighed by the fact that statistical significance can be achieved with fewer patients in crossover studies.

Once the study design has been chosen, there are many other issues to consider when developing and writing clinical protocols. Among the topics to be considered are criteria for patient eligibility, efficacy and safety parameters, timing of the events, packaging and dispensing of the clinical trial material, and the informed consent form. Also, to be determined is how the study will be blinded. For most well-controlled studies, subjects are assigned to the various groups by using a randomization process so that biased selection is eliminated, the overall collection of the subjects' variables is comparable in each group, and statistical power is guaranteed (9). In these double-blind studies, neither the subject nor the investigating scientists know to which group the subject has been assigned. Thus, extensive input from the drug development team is required when designing studies and writing protocols.

DRUG DEVELOPMENT CONSIDERATIONS

Most drugs are tested in humans to treat a specific disease entity or some adverse clinical condition. Because the pathogenesis of diseases and the exact mechanisms of action of drugs are often poorly understood, the process of evaluating a drug's efficacy can be complicated. Upon treatment, a patient's adverse clinical condition may improve; however, for many diseases this occurrence can only be evaluated indirectly by clinical assessments (e.g., via blood pressure measurements in the treatment of hypertension). However, a drug's characteristics can also be measured directly. For example, measurement of blood concentrations of the drug enabling calculation of pharmacokinetic parameters is a direct evaluation of the drug.

Similar to efficacy assessments, evaluation of the safety of a drug may also involve indirect measurements. One of the primary methods of obtaining safety information in a

clinical trial is through a patient's reporting of AEs. Although the exact biochemical mechanisms responsible for many AEs cannot be evaluated directly, the indirect evaluation of the drug's adverse effect can be seen clinically. Because clinical assessments are indirect measures, AE reporting leads to several complex questions. The degree of drug-relatedness or causality, the effect of concomitant medication, the severity of the AE, the complications of the disease state, and the effects of other clinical conditions or diseases are usually difficult to determine, particularly early in the drug development program. Also, all reports of AEs in a clinical drug research program are recorded, tabulated, and cross-referenced to form a safety database, regardless of whether the AE is determined to be drug related. The information contained in this database is used to generate the package insert.

Although the clinical effect of a drug is perhaps the primary concern of drug development, an understanding of the drug's biochemical and physicochemical properties and mechanism of action is also desired. These direct measures are of equal concern in drug development as are the indirect evaluations of a drug's clinical effects. The primary tool used to study the intrinsic physicochemical properties of a drug is pharmacokinetics, which is a branch of biopharmaceutics. Pharmacokinetics describes the relationship between the processes of drug absorption, distribution, metabolism (biotransformation), and excretion (collectively abbreviated ADME) and the time course of therapeutic or adverse effects of drugs (10). Efficacy is determined by the drug concentration at the site of action, which generally is correlated with the drug concentration in the blood. The ultimate goal of pharmacokinetics is to characterize the sources of variability in the concentration–time profile, which may be correlated with variability in efficacy and adverse events.

Pharmacokinetics can be used to guide dosage regimen selection and thereby optimize pharmacologic effects and minimize toxicologic effects when a drug is administered to an individual patient. Thus, although the basic pharmacokinetic properties of a drug are identified during the earliest stage of clinical drug development, the many factors affecting the pharmacokinetics in the patient population must be identified throughout the drug development process to enable proper dose selection for individuals. Thus, both indirect and direct measures are used to evaluate a drug.

MARKETING INPUT

A successful pharmaceutical company has an appropriate blend of both research and marketing to enable a symbiotic, rather than antagonistic, relationship. Because an effective scientific and clinical research team often designs and executes experiments and clinical trials that involve costly overhead expenses, it is essential for marketing decisions to be geared toward company profitability being made allow the company profitable so these expenses can be met. Therefore, both medical and marketing input are necessary if a pharmaceutical company is to be successful.

By gathering data on all facets of the needs in the marketplace from clinicians and by maintaining a profile awareness of new products under development by competitors, marketing personnel are in an excellent position to advise their colleagues in the research arena who are responsible for the drug development program. Also, a marketing expert can help identify the problems other companies are having in selling their product and thereby avoid the same difficulties. For instance, sales problems may be related to ineffective advertising or faulty packaging; therefore, they do not concern clinical research. However, problems in sales can also be related to a drug's undesirable effects. An effective drug that does not lead to the AEs associated with an already approved drug would have a marketing advantage. Someone in marketing research may suggest conducting clinical studies that would evaluate the relative incidence of the AE with the hope that the data could be used to support effective advertising.

Thus, research and marketing are mutually benefical in a successful pharmaceutical company. Marketing groups help clinical research teams by supplying them with information about competing products, the needs of the marketplace, and suggestions for new formulations. Clinical research teams provide the data to support therapeutic and marketing claims and act as chief advisors to marketing personnel concerning drug research studies and promotional claims.

EFFECTIVE GLOBAL PLANNING

Because drugs are frequently marketed worldwide and the clinical development of drugs may involve studies that are conducted internationally effective global planning can present its own difficulties. Obviously, medical practice, regulatory guidelines, and the cultural environment may be different in various countries, but also the manner in which research is conceived can differ vastly between countries. Medical researchers in some countries may be more conservative than researchers in other countries, which could potentially lead to the underdosing of drugs.

These differences in research approaches actually stem from differences in ethical standards.

Another reason that international planning may be difficult in drug research concerns the way in which various countries view early clinical trials and drug safety. Some countries view volunteer subjects and patients differently from a regulatory perspective, making it easier to recruit and enroll subjects for Phase 1 studies than it is to recruit and enroll patients for Phase 2 or Phase 3 studies. In the United States, both patients and volunteers are viewed in the same way, and studies with patients and volunteers cannot be initiated until the FDA has authorized an IND.

In addition to regulatory guidelines, the regulatory process is still another aspect of clinical drug development that can differ widely between countries. In England, sponsoring research firms do not interact very much with the British drug regulatory agency the Committee on Safety of Medicines. This lack of direct interaction stems from the desire to keep commercial influence away from the objective evaluation of a pharmaceutical company's study data. This lack of communication results in British companies treating government guidelines for conducting clinical research as a routine checklist rather than an aid in forming the most appropriate development strategy.

In the United States, federal guidelines (Code of Federal Regulations, CFR) have been established by the FDA to help sponsoring research firms conduct good, consistent clinical studies. However, some of the items in these guidelines may not be appropriate for all clinical studies, and some items that may be appropriate to include in a clinical study may not have been incorporated into the federal guidelines. These variations occur because each drug and disease state is unique, and complete guidelines cannot be established for all cases. For these reasons, several meetings are held between clinical research teams and the FDA before an NDA submission to ensure that all appropriate methodology and experimentation is being incorporated into the overall drug development project.

Beginning in the early 1990s, the FDA participated in a collaborative effort to harmonize the technical procedures for development and regulatory approval of human pharmaceuticals internationally. Forces that led the agency in this direction included increased trade, the multinational nature of the pharmaceutical industry, trade agreements such as the North American Free Trade Agreement and the General Agreement on Tariffs and Trade by the World Trade Organization, European activism, and pressures on the industry to control costs (11). These pressures included intense competition and health care reimbursement controls. This harmonization effort is the work of the International Conference on Harmonisation (ICH) of Technical Requirements for Registration of Pharmaceuticals for Human Use. ICH has focused on achieving harmonization of technical requirements in three major regions of the world: the United States, the European Union, and Japan. Some of the earliest ICH guidelines addressed the format and content of the Investigator's Brochure (12), stability testing (13), and genotoxicity testing (14). The FDA also works with the World Health Organization and other international organizations to set standards for health care products (11).

Clinical drug research is a complicated, multidisciplinary task that may be conducted internationally. In fact, many pharmaceutical companies are multinational, with locations in several countries. Planning and coordination become even more complex for such global drug development programs. Despite the differences among countries in medical practice, regulation, and culture, international drug development and marketing are vital parts of many organizations. The successful multinational pharmaceutical company will plan its clinical research strategy according to any differences among nations before to implementing its international development plans.

ETHICAL CONSIDERATIONS

No topic in clinical drug development is more controversial and emotionally charged than the myriad ethical dilemmas that face physicians and scientists involved in clinical research. Given that clinical research has generally proved to have moral consequences through its direct and indirect influence on alleviating suffering, steps must be taken to ensure that abuses do not occur during the course of drug development. Therefore, guidelines for the protection of human subjects have been developed, proposed, and accepted worldwide (15).

Because of the atrocities committed by Nazi medical researchers in the 1930s, the Nuremberg Code (16) was written, and highlighted the importance of obtaining all research subjects' voluntary consent to their participation in clinical studies. The Declaration of Helsinki (17), which was published by the World Medical Association in 1964 and has been updated several times since, takes the informed consent issue one step further by giving only qualified medical scientists and physicians the right to conduct clinical research. However, similar concerns go back at least to the 1830s, when Dr. William Beaumont developed a contract with a patient, and in the late 1800s, when a leprosy worker experimented on a patient without her consent (18).

Legislation that ensures the protection of human research subjects in the United States includes the 1979 publication of the Belmont Report on the Ethical Principles and Guidelines for the Protection of Human Subjects of Research (19). This report concerns the fine line between biomedical research and the routine practice of medicine and explores the criteria that determine the risk-benefit ratio in the consideration of conducting clinical research. It also addresses basic guidelines for the proper selection of human research subjects and further defines the elements of informed consent.

Other important legislation in the United States includes the FDA's Guidance for Institutional Review Boards (IRBs) (20) for further guarantees of protection for human research subjects. IRBs are independent committees that review proposed clinical research projects before the commencement of the research. These committees decide whether the risk to research subjects outweighs the potential benefit of the research; they can suggest modifications in the research proposal or disapprove the project altogether. IRBs must consist of both men and women of varying professions. At least one member must have his or her primary concern in a nonscientific area (e.g., a lawyer or clergyperson), and at least one member must not be affiliated with the institution at which the research will be conducted.

Closely related to the rights of human research subjects are the rights of routine patients involved in nonresearch medical matters. In 1973, the American Hospital Association published the Patient's Bill of Rights (21), which requires that the acting physician give his patients complete information concerning their diagnosis, treatment, and prognosis; that the patient be given respectful care; that the patient be given the opportunity to refuse treatment; and that the patient's records, condition, and medical care be treated confidentially.

Another ethical issue facing clinical research scientists concerns study design, in particular, the placebo-controlled clinical trial. The reason placebo-controlled clinical trials are conducted is quite compelling from a scientific standpoint: to ensure that the evidence supporting the efficacy of an experimental drug is actually due to the properties of the drug and not to the psychologic properties of the study subjects. In other words, if a placebo effect from the experimental drug occurs rather than a true therapeutic effect, then a comparison of the drug group with the placebo group will show statistically similar response rates. It is a way to help separate actual drug responses from placebo responses, especially in studies investigating psychiatric compounds, but also in other therapeutic areas with a clearer "physiologic" or "biochemical" basis.

One defense for conducting placebo-controlled clinical trials is that the subjects chosen for the placebo group are randomly chosen, so that no malicious withholding occurs. Also, many study protocols have provisions of study extension that guarantee subjects in placebo groups have the opportunity to take the drug as an extension of the study after they complete the original part, or they are offered the chance to receive alternative therapy. Study subjects may be given monetary compensation for their participation in studies, in addition to free, thorough physical exams, lab work, and physician visits.

Interestingly, experimental drugs have unknown side effects that can cause serious biochemical and physiologic problems, whereas placebo medication does not. This fact makes possible the contrary argument and objection, on purely ethical grounds, to giving study subjects experimental and hence unproven drugs. Of course, informed consent and careful monitoring by trained medical personnel help to alleviate the ethical problems associated with giving subjects an active, investigational drug. The most important aspect of all studies is that the patient be completely informed of all study procedures and agree to willingly participate in the study.

The most recent pressing ethical dilemma facing the clinical research scientist surrounds the increasing amount of research that is being conducted in biotechnical and genetic engineering. Ethical issues will continue to play important parts in the medical and legal worlds. Whereas pure science is value-neutral, its application is always open to debate. Undesirable extremes are likely to exist at both ends of the spectrum.

CONCLUSIONS

To conduct a clinical study for the evaluation of a new drug, a vast array of personnel is required. Physicians are largely used because of their knowledge of clinical medicine and patient care, whereas scientists are used because of their knowledge of the methodology and the science. Pharmacists serve a bridging function due to their unique training in therapeutics and the pharmaceutical sciences. Nonscientific personnel are indispensable because of their ability to coordinate the many facets of a drug development project.

The clinical evaluation of drugs involves many different levels of scrutiny before a drug product can be marketed. These levels include Phase 1 for safety testing, Phase 2 for evaluating efficacy and determining the correct therapeutic dose, Phase 3 for large-scale studies and determination of drug interactions, and Phase 4 for postmarketing surveillance. Phase 1 studies are typically conducted in healthy volunteers, and Phase 2 through 4 studies are conducted in patients.

Study design plays a critical role in the clinical evaluation of drugs. A clinical study cannot be conducted without specifically outlined objectives and a definitive plan, which are vital components around which the study protocol is constructed. The use of placebo or active drug control groups in the study, and whether the design should be open, parallel, or crossover, must be determined. In most studies, patients are assigned to study groups randomly.

The developmental objectives facing the clinical research team include indirect evaluations of a drug's safety and efficacy, such as effects on vital signs or behavior, and direct evaluations of a drug's intrinsic properties, such as its pharmacokinetics and mode of action. Also, the marketing–medical liaison is important if research is to support future sales plans and advertising is to reflect study results. Finally, effective global planning is necessary because drugs are more frequently developed and marketed worldwide, and therefore involve differing patient populations and different government regulations. ICH guidelines have helped to standardize regulations worldwide.

The ethical dilemmas facing clinical research scientists affect much of the legislation that currently regulates the conduct of clinical trials. The goal of drug development research is to develop effective pharmacotherapy for mankind's ailments, and regulatory agencies have enacted legislation to prevent unethical research.

Although traditional medicines continue to be discovered and developed, the fields of biotechnology and gene therapy continue to advance. In addition, new methods to collect and evaluate clinical data on a real-time basis will help to speed the development process.

REFERENCES

1. Cato, A.; Cook, L. Clinical Research. *The Clinical Research Process in the Pharmaceutical Industry*; Matoren, G.M., Ed.; Marcel Dekker, Inc.: New York, 1984; 217–238.
2. Smith, R.V. Development of Clinical Scientists. Drug Intell. Clin. Pharm. **1987**, *21*, 101–103.
3. Spilker, B. Monitoring a Clinical Trial. *Guide to Clinical Trials*; Lippincott-Raven: Philadelphia, 1996; 430–448.
4. Burley, D.M.; Glynne, A. Clinical Trials. *Pharmaceutical Medicine*; Burley, D.M., Binns, T.B., Eds.; Edward Arnold: London, 1985; 18–38.
5. Edler, L. Statistical Requirements of Phase I studies. Onkologie **1990**, *13*, 90–95.
6. Bertz, R.J.; Granneman, G.R. Use of In Vitro and In Vivo Data to Estimate the Likelihood of Metabolic Pharmacokinetic Interations. Clin. Pharmacokinet. **1997**, *32* (3), 210–258.
7. Cato, A. Practical Insights on Designing the Correct Protocol. *Concepts and Strategies in New Drug Development*; Praeger Publishers: New York, 1983; 82–87.
8. Spilker, B. Part II: Developing and Writing Clinical Protocols. *Guide to Clinical Trials*; Lippincott-Raven: Philadelphia, 1996; 145–272.
9. Friedman, L.M.; Furberg, C.D. Basic Study Design. *Fundamentals of Clinical Trials*; PSG Publishing Co., Inc.: Littleton, MA, 1985; 35–38.
10. Gibaldi, M.; Levy, G. Pharmacokinetics in Clinical Practice I. Concepts. JAMA **1976**, *235*, 1864–1867.
11. Horton, L.R. Harmonization, Regulation, and Trade: Where Do We Go from Here? [editorial]. PDA J. Pharm. Sci. Technol. **1996**, *50*, 61–65.
12. Cocchetto, D.M. The Investigator's Brochure: A Comparison of the Draft International Conference on Harmonisation Guideline with Current Food and Drug Administration Requirements. Qual. Assur. **1995**, *4*, 240–246.
13. Haase, M. Stability Testing Requirements for Vaccines: Draft Guidelines of the International Conference on Harmonization. Dev. Biol. Stand. **1996**, *87*, 309–318.
14. Purves, D.; Harvey, C.; Tweats, D.; Lumley, C.E. Genotoxicity Testing: Current Practices and Strategies Used by the Pharmaceutical Industry. Mutagenesis **1995**, *10*, 297–312.
15. Levine, R.J. *Ethics and Regulation of Clinical Research*; Urban & Schwarzenberg: Baltimore, 1981.
16. The Nuremberg Code. JAMA **1997**, *276*, 1691.
17. *World Medical Association Declaration of Helsinki: Recommendations Guiding Physicians in Biomedical Research Involving Human Subjects*, Adopted by the 18th World Medical Assembly, Helsinki, Finland, June 1964 and Amended by the 29th World Medical Assembly, Tokyo, Japan, October 1975, the 35th World Medical Assembly, Venice, Italy, October 1983, the 41st World Medical Assembly, Hong Kong, September 1989, and the 48th General Assembly, Somerset West, Republic of South Africa, October 1996.
18. Lock, S. Research Ethics: A Brief Historical Review to 1965. J. Intern. Med. **1995**, *238* (6), 513–520.
19. Department of Health, Education, and Welfare. The Belmont Report. Ethical Principles and Guidlines for the Protection of Human Subjects of Research the National Commission for the Protection of Human Subjects of Biomedical and Behavioral Research April 18, 1979. Available at: http://ohsr.od.od.nih.gov/mpa/belmont.php3 (accessed March 2, 2001)
20. Food and Drug Administration Information Sheets. Guidance for Institutional Review Boards and Clinical Investigators; 1998 Update. Available at: http://www.fda.gov/oc/oha/IRB/toc.html (accessed March 2, 2001).
21. American Hospital Association. A Patient's Bill of Rights. First adopted by the AHA in 1973. revision approved October 21, 1992. Available at: http://www.aha.org/resource/pbillofrights.asp (accessed March 2, 2001).

PROFESSIONAL RESOURCES

Clinical Laboratory Improvement Amendments of 1988

Thomas P. Christensen
North Dakota State University, Fargo, North Dakota, U.S.A.

INTRODUCTION

Today, any facility that examines or tests material derived from the human body for patient care purposes is subject to federal regulation under the Clinical Laboratory Improvement Amendments of 1988 (CLIA 1988).[1,2] Historically, CLIA 1988 was passed in response to perceived and documented problems in cytology testing, but the law also expanded federal regulation to large numbers of previously unregulated laboratories, such as physician offices and health screening sites. Except for laboratories performing only the simplest of tests, CLIA provides a comprehensive regulatory framework including personnel, quality control and quality assurance standards. This is significant for pharmacists since advances in laboratory technology have created new opportunities for pharmacists to conduct testing services.[3,4] Although these services are not within the traditional pharmacist role, they are consistent with the scope of pharmacy practice under the definition of pharmaceutical care.[5,6] Since the role of the pharmacist is continually expanding, pharmacists should be familiar with the history and general regulatory framework of CLIA 1988.

HISTORY

Current federal regulation of clinical laboratories is found within the Clinical Laboratory Improvement Amendments of 1988 (CLIA 1988) and implementing regulations.[1,2] The law is quite comprehensive, but this wasn't always so. Prior to 1965, there was very little federal involvement in laboratory regulation.[7,8] In fact, only a few states regulated or licensed medical laboratories. Those states that did regulate laboratories did so mostly in conjunction with hospital licensing requirements.

After 1965, federal government involvement in clinical laboratory regulation increased, somewhat indirectly, through conditions of participation in Medicare and Medicaid.[7] The laboratory conditions of participation for Medicare and Medicaid were basically designed to ensure that the government and its beneficiaries received the services paid for and that such services met specified minimum standards. In general, the conditions of participation only affected hospital and larger independent laboratories participating in the Medicare and Medicaid programs.

In 1967, Congress took a direct role in expanding the federal regulation of clinical laboratories by passing the Clinical Laboratory Improvement Act of 1967.[7,9] Under CLIA 1967, private laboratories receiving specimens in interstate commerce had to be licensed. While this expanded the number of regulated laboratories, it was by no means comprehensive. Federal and state laboratories were exempt. In addition, any laboratory receiving fewer than 100 interstate specimens per year was exempt from licensure if it registered and applied for the exemption. Therefore, most physician office laboratories were exempt.

In addition to expanding the number of regulated laboratories, CLIA 1967 also increased federal quality standards. The new law required that regulated laboratories successfully participate in proficiency testing. Proficiency testing is the periodic testing of unknown samples by a laboratory under normal laboratory procedures and workload. The purpose of proficiency testing is to identify poorly performing laboratories.

While imperfect in scope, CLIA 1967 continued to govern the regulation of clinical laboratories for the next 20 years. In 1987, the *Wall Street Journal* published two articles suggesting that, at least for Pap smears, the quality of laboratory testing was not as good as expected.[10,11] In response to the articles, committee hearings were held in both the U.S. House of Representatives and the U.S. Senate. In addition, the Department of Health and Human Services began revising regulations implemented under CLIA 1967. On October 31, 1988, Congress enacted the Clinical Laboratory Improvement Amendments of 1988 (CLIA 1988).

Encyclopedia of Clinical Pharmacy
DOI: 10.1081/E-ECP 120006196

CLIA 1988 significantly increased the scope of federal laboratory regulation both in terms of the number of laboratories regulated and quality standards.[5] Under CLIA 1988, any facility that examines or tests material derived from the human body for patient care purposes is subject to federal regulation. Furthermore, the law provides for personnel, patient test management, quality control and quality assurance standards, as well as proficiency testing to identify poorly performing laboratories.

The legislation came under immediate criticism.[7] Initial criticism called the need for more comprehensive regulation into question. The *Wall Street Journal* articles that precipitated the legislation focused on "errors" that resulted in missed or delayed diagnosis of carcinoma of the uterine cervix. The problems described did exist to some degree but the articles were considered by many to be sensationalistic and misrepresentative. Furthermore, neither the *Wall Street Journal* article nor congressional testimony revealed that the CLIA 1967 regulations already applied to cytology and that all cited instances of unacceptable laboratory practice had been subject to enforcement under existing regulations.

Continuing criticism focused on the extent of the regulation.[12–14] Many laboratory professionals expressed concern that certain portions of the CLIA regulations were overly prescriptive and burdensome. While laboratory management welcomed the decreased stringency in personnel requirements as providing greater flexibility in staffing, laboratory professionals expressed concern that personnel changes were a dangerous weakening of standards.

Physicians operating office laboratories were also concerned. Much of the concern focused on the intrusiveness of quality standards, especially proficiency testing.[14] A chief criticism of proficiency testing was that the proficiency testing process is highly artificial with many operational, technical, and clerical variables that are absent from routine patient testing. Physicians argued that in an office setting significant inaccuracies were likely to be detected by the physician. Requiring physicians to comply with burdensome regulations might cause many physicians to eliminate office-based laboratory services. They argued quicker and more accessible results were more important for immediate patient care than highly accurate results after the patient has gone home.

In contrast to the concerns about overregulation were concerns about underregulation of persons engaged in relatively simple laboratory activities such as health screenings.[14] Under the CLIA 1988 regulations, persons collecting human specimens for patient care purposes using specified, relatively simple equipment and tests are required to register but are exempt from most quality standards. The exception raised concerns with regard to accuracy and interpretation of results in these settings.

RELATIONSHIP TO PHARMACY

Although relatively few pharmacies engaged in laboratory activities when CLIA 1988 was passed, the issues and concerns raised with the passage of CLIA 1988 are extremely important to pharmacists.[15] There is an increasing need for timely, accurate laboratory values in modern pharmacy practice. Objective laboratory measures help in determining dosages, evaluating medication efficacy, assessing for adverse drug reactions or toxicity, and monitoring adherence to therapy. Furthermore, in-pharmacy laboratory testing allows for the provision of screening services, a valuable public service in the detection of unrecognized disease.

Pharmacists share in the concerns about underregulated laboratories, especially with regard to accuracy and interpretation of results.[15] Since the results of pharmacist-conducted laboratory tests are used in making clinical decisions, the values must be accurate. Additionally, pharmacists must be knowledgeable about interpretations of laboratory results, the variables that can affect them, and their clinical implications. Finally, these concerns require that equipment be maintained and operated in full accordance with the manufacturer's instructions.

REGULATORY FRAMEWORK

Given the importance of laboratory data to pharmacy practice, pharmacists should be familiar with the regulatory framework of CLIA 1988. A major aim of expanding the reach of CLIA 1988 to virtually all clinical laboratories was to ensure that laboratory tests varied only by differences in methodology, equipment used, and training required for test performance.[16,17] In other words, the type of regulatory standards applied to a pharmacy, physician's office, or some other previously unregulated site would be determined only by the tests performed. This regulatory framework is known as a "complexity model."

Under the complexity model there are three categories of tests on which regulatory standards are based: waived tests, tests of moderate complexity (including a provider-performed microscopy subcategory), and tests of high complexity. The complexity of tests performed within the laboratory determine which personnel, proficiency test-

ing, patient management, and quality control and quality assurance standards will apply to the laboratory.

Waived Tests

Facilities that perform only simple procedures with an insignificant risk of an erroneous result may apply for waived status (Table 1).[18] A test may obtain waived status if it employs simple and accurate methodologies, the likelihood of erroneous results is negligible, and testing poses no reasonable risk of harm if performed incorrectly. Tests cleared by the FDA for home use are automatically waived.[19] A complete list of waived tests can be found at www.hcfa.gov/medicaid/clia/waivetbl.htm.

A laboratory performing only waived tests must register with Health Care Finance Administration (HCFA) and obtain a certificate of waiver. However, laboratories operating under a certificate of waiver are only required to permit inspections and follow manufacturers' instructions in performing tests. While proficiency testing, quality control, and personnel standards are not required, persons performing waived tests should adhere to basic tenets of quality control and quality assurance.[16]

Moderate- and High-Complexity Tests

Tests that are not specifically waived are categorized as either moderate- or high-complexity tests.[20] Categorization into high or moderate complexity depends on the knowledge and experience needed to perform the test, complexity of reagent materials and preparations used, characteristics of operational steps, characteristics of and availability of calibration, quality control, and proficiency testing materials; troubleshooting and maintenance required; and the degree of interpretation and judgement required in the testing process.

Using these criteria, virtually hundreds of tests have been classified as moderately complex, while fewer, highly

Table 1 Tests granted waived status under CLIA used in drug monitoring

Test name	Manufacturer	Use
Glucose monitoring devices cleared by the FDA for home use	Various	Monitoring of blood glucose levels
Bayer DCA 2000	Bayer	Measures the percent concentration of hemoglobin A1c in blood for monitoring long-term diabetic control
Metrika DRx HbA1c	Metrika, Inc.	
LXN Fructosamine Test System	LXN Corp.	Measures glucose/fructosamine to evaluate diabetic control over a 2–3 week period
LXN Duet Glucose Control Monitoring System		
LXN IN CHARGE Diabetes Control System		
ChemTrack AccuMeter	ChemTrak	Cholesterol monitoring
Advance Care	Johnson & Johnson	
Accu-Check Instant Plus Cholesterol	Boehringer Mannheim Corp.	
ENA.C.T Total Cholesterol Test	ActiMed Laboratories	
Lifestream Technologies Cholesterol Monitor	Lifestream Technologies	
MTM Bioscanner 1000 (for OTC use)	Polymer Technology Systems, Inc.	
PTS Bioscanner Test Strips Cholesterol	Polymer Technology Systems, Inc.	
Cholestech LDX	Cholestech	Measures total cholesterol, HDL cholesterol, triglycerides, and glucose levels
PTS Bioscanner (for OTC use)—for HDL	Polymer Technology Systems, Inc.	Measures HDL cholesterol in whole blood
PTS Bioscanner 2000 for Triglycerides	Polymer Technology Systems, Inc.	Measures triglycerides in whole blood
ITC Protime Microcoagulation System	International Technidyne Corp.	Evaluation of heparin, coumarin, or warfarin effect
CoaguChek PST	Boehringer Mannheim Corp.	
AvoSure Pro	Avocet Medical, Inc.	
Roche Diagnostics CoaguChek S Systems Test	Roche Diagnostics Corp.	

specialized tests (for example, cytogenetics, histopathology, histocompatibility, cytology, and other highly specialized tests) have been classified as high complexity.[16] The categorization of tests enables a laboratory to determine easily what level of regulation it will follow.[21,22] The intent is that testing environments not eligible for a certificate of waiver, and which do not conduct highly specialized testing, will be certified to perform moderate-complexity testing. The personnel, proficiency testing, patient test management, and quality control and assurance standards for moderate- and high-complexity tests are outlined below.

Personnel

Personnel standards are a significant differentiating element in the regulation of moderate- and high-complexity testing laboratories.[23] Both types of laboratories require a laboratory director, a technical consultant, a clinical consultant, and testing personnel. However, the education, training, and experience required for these positions differ. In addition, high-complexity laboratories are required to maintain technical supervisor and general supervisor positions.

Proficiency Testing

Each laboratory performing tests of moderate and high complexity must enroll in an approved proficiency testing program for each specialty or subspecialty for which it seeks certification.[24] In general, proficiency testing requires five challenges per testing and three testing events per year. Failure to attain an overall testing event score of at least 80% is unsatisfactory performance. Proficiency testing samples must be tested with the laboratory's regular patient workload, using routine testing methods, and by personnel who routinely perform testing.

Patient Test Management

Laboratories performing moderate-complexity or high-complexity testing must also employ and maintain a system that provides for proper patient preparation and proper specimen collection, identification, preservation, and processing.[25] This system must assure optimum patient specimen integrity and positive identification throughout the pretesting, testing, and posttesting processes and must meet the standards as they apply to the testing performed.

Quality Control

After December 31, 2000, laboratories performing either moderate- or high-complexity tests may, in general, meet quality control standards by following manufacturer's instructions when using a device cleared by the FDA as meeting CLIA requirements for quality control.[26] For other tests of moderate and high complexity, the regulations state specific quality control standards.

Quality Assurance

Finally, laboratories performing moderate-complexity or high-complexity testing must establish and follow written policies and procedures for a comprehensive quality assurance program that is designed to monitor and evaluate the ongoing and overall quality of the total testing process.[27] The laboratory's quality assurance program must evaluate the effectiveness of its policies and procedures; identify and correct problems; assure the accurate, reliable, and prompt reporting of tests results; and assure the adequacy and competency of the staff.

CONCLUSION

Currently, most pharmacy-based testing falls within the waived category and is exempt from extensive quality standards. However, given the importance of timely and accurate laboratory data in the provision of pharmaceutical care, it is likely that pharmacy-based laboratory testing will expand in breadth and scope.[28] Therefore, pharmacists need to be aware of the quality issues raised and addressed by CLIA 1988.

Although the detail is beyond the scope of this monograph, pharmacists should also be aware of other laws and regulations impacting pharmacy-based laboratory testing. Many states have their own laws regulating laboratory quality and professional competency.[15] In addition, there are Occupational Safety and Health Administration (OSHA) standards pertaining to laboratory operations.[29,30] Before engaging in laboratory testing pharmacists should be thoroughly familiar with all laboratory regulations.

REFERENCES

1. Clinical Laboratory Improvement Amendments of 1988 (CLIA), Pub. L. No. 100–578, 42 USC 201 (1988).
2. Laboratory Requirements. In *42 Code of Federal Regulations Parts 430 to End*; Revised as of October 1, 1999; 819–941.
3. Bluml, B.M.; McKenney, J.M.; Cziraky, M.J. Pharmaceutical care services and results in project ImPACT: Hyperlidemia. J. Am. Pharm. Assoc. **2000**, *40*, 157–165.
4. Bluml, B.M.; McKenney, J.M.; Cziraky, M.J.; Elswick,

R.K., Jr. Interim report from Project ImPACT: Hyperlidemia. J. Am. Pharm. Assoc. **1998**, *38*, 529–534.
5. Gore, M.J. Pharmacy-based laboratory testing: Adding another dimension to pharmaceutical care. J. Am. Pharm. Assoc. **1998**, *38*, 538–540.
6. Hepler, C.D.; Strand, L.M. Opportunities and responsibilities in pharmaceutical care. Am. J. Hosp. Pharm. **1990**, *47*, 533–542.
7. Bachner, P.; Hamlin, W. Federal regulation of clinical laboratories and the Clinical Laboratory Improvement Amendments of 1988—Part I. Clin. Lab. Med. **1993**, *13*, 739–752.
8. Bachner, P.; Hamlin, W. Federal regulation of clinical laboratories and the Clinical Laboratory Improvement Amendments of 1988—Part II. Clin. Lab. Med. **1993**, *13*, 987–994.
9. Clinical Laboratory Improvement Act of 1967 (CLIA), Pub. L. No. 90–174, 42 USC 216 (1967).
10. Bogdanich, W. Medical labs, trusted as largely error-free, are far from infallible. Wall St. J. **February 2, 1987,** 1.
11. Bogdanich, W. The Pap test misses much cervical cancer through labs' errors. Wall St. J. **November 2, 1987**, 1.
12. Neff, J.C.; Speicher, C.E. CLIA '88: More misguided regulation, or a promise of quality. Arch. Pathol. Lab. Med. **1992**, *116*, 679–680.
13. Hurst, J.; Nickel, K.; Hilborne, L.H. Are physicians' office laboratory results of comparable quality to those produced in other laboratory settings? JAMA, J. Am. Med. Assoc. **1998**, *279*, 468–471.
14. Bachner, P. Is it time to turn the page on CLIA 1988? JAMA, J. Am. Med. Assoc. **1998**, *279*, 473–475.
15. Rosenthal, W.M. Establishing a pharmacy-based laboratory service. J. Am. Pharm. Assoc. **2000**, *40*, 146–156.
16. Medicare, Medicaid and CLIA Programs: Regulations Implementing the Clinical Laboratory Improvement Amendments of 1988 (CLIA). Fed. Regist. **1992**, *57*, 7002–7186.
17. Stull, T.M.; Hearn, T.L.; Hancock, J.S.; Handsfield, J.H.; Collins, C.L. Variation in proficiency testing performance by testing site. JAMA, J. Am. Med. Assoc. **1998**, *279*, 463–467.
18. Subpart B—Certificate of Waiver. In *42 Code of Federal Regulations Section 493.35 to 493.39*; Revised as of October 1, 1999; 831–833.
19. Laboratories Performing Waived Tests. In *42 Code of Federal Regulations Section 493.15*; Revised as of October 1, 1999; 828.
20. Test Categorization. In *42 Code of Federal Regulations Section 493.17*; Revised as of October 1, 1999; 828–830.
21. Laboratories Performing Tests of Moderate Complexity. In *42 Code of Federal Regulations Section 493.20*; Revised as of October 1, 1999; 831.
22. Laboratories Performing Tests of High Complexity. In *42 Code of Federal Regulations Section 493.15*; Revised as of October 1, 1999; 831.
23. Subpart M—Personnel for Moderate Complexity (Including the Subcategory) and High Complexity Testing. In *42 Code of Federal Regulations Section 493.1351 to 493.1495*; Revised as of October 1, 1999; 900–925.
24. Subpart H—Participation in Proficiency Testing for Laboratories Performing Tests of Moderate Complexity (Including the Subcategory), High Complexity, or Any Combination of These Tests. In *42 Code of Federal Regulations Section 493.801 to 493.825*; Revised as of October 1, 1999; 853–856.
25. Subpart J—Patient Test Management for Moderate Complexity (Including the Subcategory), High Complexity, or Any Combination of These Tests. In *42 Code of Federal Regulations Section 493.1101 to 493.1111*; Revised as of October 1, 1999; 881–883.
26. Subpart K—Quality Control for Tests of Moderate Complexity (Including the Subcategory), High Complexity, or Any Combination of These Tests. In *42 Code of Federal Regulations Section 493.1201 to 493.1203*; Revised as of October 1, 1999; 883–885.
27. Subpart P—Quality Assurance for Moderate Complexity (Including the Subcategory) or High Complexity Testing, or Any Combination of These Tests. In *42 Code of Federal Regulations Section 493.1701 to 493.1717*; Revised as of October 1, 1999; 925–927.
28. Magarian, E.O.; Peterson, C.D.; McCullagh, M.E.; Kuzel, R.J. Role model ambulatory care clinical training site in a community-based pharmacy. Am. J. Pharm. Ed. **1993**, *57*, 1–9.
29. Blood-borne Pathogens. In *29 Code of Federal Regulations Section 1910.1030*; Revised as of July 1, 1999; 261–274.
30. Occupational Exposure to Hazardous Chemicals in the Laboratory. In *29 Code of Federal Regulations Section 1910.1450*; Revised as of July 1, 1999; 484–538.

Clinical Pharmacist as Principal Investigator (ACCP)

American College of Clinical Pharmacy
Kansas City, Missouri, U.S.A.

INTRODUCTION

Research is critical to advancing the practice of pharmacy. Indeed, the last three decades contain many examples where seminal research papers authored by clinical pharmacists have advanced pharmacy care and improved patient outcomes. Whereas many clinical pharmacists have established strong research programs, there must be continued efforts to increase the number of highly competent pharmacist researchers by reducing barriers into a research career and improving the quality of research training programs. The American College of Clinical Pharmacy (ACCP) continues to receive inquiries regarding the requisites for a pharmacist to serve as the principal investigator (PI) for industry-sponsored research. This article summarizes Food and Drug Administration (FDA) policies regarding the pharmacist's ability to serve as PI on clinical drug research; presents a brief view on this issue from the perspective of the pharmaceutical industry; describes FDA regulations governing clinical research, and the PI's responsibilities in meeting those regulations; and provides advice to clinical pharmacists who desire to serve as PIs on clinical drug research sponsored by the pharmaceutical industry.

HISTORY OF CLINICAL PHARMACIST AS PRINCIPAL INVESTIGATOR

The ACCP's role as an advocate for the clinical pharmacist as PI began in 1983. At that time, there was generally a great deal of reluctance within the pharmaceutical industry to allow pharmacists to function as PIs. Some clinical pharmacists were successfully participating as PIs in industry-sponsored and FDA-regulated clinical research, whereas others were being told by some industry sponsors that they interpreted FDA regulations to allow only physicians to be PIs. To clarify this matter then-ACCP president Peter H. Vlasses, Pharm.D., wrote then-FDA Commissioner Arthur Hull Hayes, MD, regarding FDA regulations on the issue. Stuart L. Nightingale, MD, then-FDA Associate Commissioner for Health Affairs, responded by writing, ''It has long been FDA policy to accept Doctors of Pharmacy as primary investigators of studies of investigational drugs within their areas of expertise.'' The FDA response noted that a person ''licensed to diagnose and treat disease be officially associated with the study'' as a coinvestigator (Fig. 1).[1] Subsequently, many clinical pharmacists used the correspondence to educate industry research sponsors and gain the ability to serve as PIs.

Some companies continued to have internal policies that required a physician to be the PI. Others interpreted the FDA letter to apply only to pharmacokinetic studies and not to clinical trials. To clarify the latter point, Dr. Vlasses again corresponded with the FDA. Dr. Nightingale responded, stating, ''Doctors of Pharmacy may serve as clinical investigators for both clinical pharmacology studies and clinical trials of a drug provided they do so in conjunction with a person licensed to diagnose and treat disease,'' again emphasizing that a physician must be a subinvestigator to assess the patient and make medical decisions.[2] In 1990, correspondence from the FDA to the American Association of Colleges of Pharmacy reiterated that ''pharmacists can serve as principal investigators in any clinical trial'' (Fig. 2).

Fifteen years later, many clinical pharmacists have served and continue to serve as PIs in human trials. Nonetheless, it is important to emphasize that one's academic degree alone—whether MD or Pharm.D.—does not qualify the individual to serve as PI. The person's overall training and experience, complemented by their research environment, are integral to their designation as PI. As noted by Dr. Nightingale, the FDA may ''reject any investigator as unsuitable as part of [the] review process'' if their overall portfolio is insufficient to demonstrate that they possess the knowledge, skills, and

From *Pharmacotherapy* 2000, 20(5):599–608, with permission of the American College of Clinical Pharmacy.

Encyclopedia of Clinical Pharmacy
DOI: 10.1081/E-ECP 120006414

C

DEPARTMENT OF HEALTH & HUMAN SERVICES

Public Health Service

Food and Drug Administration
Rockville MD 20857

MAY 10 1983

Peter H. Vlasses, Pharm.D.
Associate Director, Clinical Pharmacology Unit
Thomas Jefferson University Hospital
11th and Walnut Streets
Philadelphia, PA 19107

Dear Dr. Vlasses:

Your letter of February 24, 1983 to Dr. Hayes has been referred to me for response. In that letter you posed the question whether Doctors of Pharmacy (Pharm.D.s) may serve as investigators in clinical pharmacological studies of investigational drugs. You noted that you have received varying interpretations of our regulations on this point from different manufacturers who sponsor clinical pharmacological studies.

It has long been FDA policy to accept Doctors of Pharmacy as primary investigators of studies of investigational drugs within their areas of expertise. Because such studies may require the diagnosis of disease and the recognition and treatment of- adverse reactions or other medical incidents occurring during the course of the study, we have required that a person licensed to diagnose and treat disease be officially associated with the study to be performed. This is ordinarily done by naming such an individual in item 6(f) of the Form FD-1572 as being responsible to the principal investigator of record. Alternatively, both the Doctor of Pharmacy and the licensed individual may sign the Form FD-1572 as co-investigators, having equal responsibility in the performance of the study in question.

I trust that this clarifies FDA policy on the matter.

Sincerely yours,

Stuart L. Nightingale, M.D.
Associate Commissioner for
Health Affairs

Fig. 1 1983 Response letter from the FDA addressing participation of clinical pharmacists as PIs.

experience needed to assume the significant responsibilities of PI.[2]

CURRENT INDUSTRY PERSPECTIVE

To better understand current industry policies regarding pharmacists as PIs, a short, informal survey was administered by telephone to a small number of pharmaceutical companies. Information was obtained from five companies and suggests that most companies do not have specific policies that exclude pharmacists from functioning as PIs. However, the general consensus was that pharmacists were used primarily for pharmacokinetic studies and not for other types of clinical trials. In most instances, the program manager or study team leader

DEPARTMENT OF HEALTH & HUMAN SERVICES

Public Health Service

Food and Drug Administration
Rockville MD 20857

Original dated: April 3, 1990

American Association of Colleges of Pharmacy
Attention: Carl E. Trinca, Ph.D.
1426 Prince Street
Alexandria, Virginia 22314

Dear Dr. Trinca:

Your Letter of November 28, 1989, asked two questions about the qualifications for clinical investigators. Please excuse my delayed response.

You asked whether FDA would consider pharmacists, especially Pharm.D.'s with adequate training and experience eligible to be principal investigators in 1) pharmacokinetic studies and 2) clinical efficacy studies. This question has been addressed on a number occasions, beginning in 1980 with Dr. Marion Finkel, then director of the office of Scientific Evaluation (now called the Offices of Drug Evaluation I and II), and most recently in a memorandum (enclosed) from the Directors of the Offices of Drug Evaluation. In 1980 and at present, the conclusion is the same: pharmacists can serve as principal investigators in any clinical trial. Section 505(I) of the Food, Drug and Cosmetic Act requires that FDA assure that the investigational drug will be provided only to "experts qualified by training and experience to investigate" a new drug. Whether or not FDA will permit a particular pharmacist to be an investigator in a clinical study will be determined on a case by case basis, and may depend on the type of study (pharmacokinetic study, clinical efficacy study) proposed in the IND.

You also asked whether board certification is likely to become an important indicator of whether a person is qualified to be a principal investigator. Certainly, as boards become more widespread, and it becomes more and more probable that well-trained investigators will have then, lack of Boards will become more and more conspicuous. Nonetheless, the totality of the proposed investigator's experience will be considered and it is improbable that Boards will become necessary.

I hope this is helpful to you. If I can be of further assistance, please contact me.

Sincerely yours,

Carl Peck, M.D.
Director
Center for Drug Evaluation and Research
Food and Drug Administration

ENCLOSURE

Fig. 2 1989 and 1990 Response letters from the FDA reiterating that clinical pharmacists can serve as PIs in clinical trials.

MEMORANDUM

DEPARTMENT OF HEALTH AND HUMAN SERVICES
PUBLIC HEALTH SERVICE
FOOD AND DRUG ADMINISTRATION
CENTER FOR DRUG EVALUATION AND RESEARCH

DATE: October 4, 1989

FROM: Director, Office of Drug Evaluation I, HFD-100
Director, Office of Drug Evaluation II, HFD-500

SUBJECT: Non-M.D.'s as Clinical Investigators and Monitors

TO: Division Directors

At the June 1, 1989 Division Director's Policy Meeting, one of the topics was FDA policy on qualifications of principal investigators. It was agreed that FDA policy should continue to be as stated in the July 16, 1980 memorandum from Dr. Finkel, but that the memo should be updated to reflect the wording of the IND Rewrite as follows:

Clinical Investigators

Qualified individuals who are not M.D.'s can participate in clinical trials either as principal investigators or sub-investigators provided that an M.D. or D.O., (or D.O.S. depending upon the study) is either a sub-investigator or is listed in the IND as an individual who will be responsible for drug administration and evaluation of patient safety.

Clinical Monitors

Qualified individuals who are not M.D.'s can serve as monitors of clinical trials provided that an M.D., D.O., or D.O.S. is involved in the review and evaluation of the ensuing clinical data and the adverse reactions.

Please remind division staff of the above policy.

Robert Temple, M.D.

James Bilstad, M.D.

Fig. 2 1989 and 1990 Response letters from the FDA reiterating that clinical pharmacists can serve as PIs in clinical trials (*Continued*).

makes the decision regarding which investigators to approach as PIs based on their overall qualifications. If a pharmacist can demonstrate their expertise in a therapeutic area (as outlined later), and their ability to enroll the required number of qualified subjects into the clinical trial within an established time period, it would appear that most pharmaceutical companies will allow a pharmacist to serve as the PI for a study.

If the contact at a particular company states that a pharmacist may not serve as a PI, it may be worth inquiring about this policy. It could be that the decision maker is unaware of FDA's policy regarding pharmacists as PIs. Such information could influence the decision of the study manager, especially if there is no company policy that precludes pharmacists from functioning as PIs.

Given the importance of investigator experience and qualifications to their designation as PI, the remainder of this article centers on the capabilities and responsibilities (including FDA regulations) needed to serve in this capacity.

FDA POLICIES, REGULATIONS, AND GUIDELINES FOR INVESTIGATORS

FDA's Role and Responsibilities

Investigations of new chemical entities and dosage forms in the United States are regulated by the FDA. A qualified PI must have a comprehensive understanding of FDA policies and regulations regarding the conduct of clinical trials and the use of an investigational new drug. The PI must also demonstrate their ability to consistently abide by these policies and regulations. The Center for Drug Evaluation and Research (CDER) monitors the clinical development of drugs, whereas the Center for Biologics Evaluation and Research (CBER) supervises the clinical development of biologics, including most biotechnology products. Regulations are designed so clinical trials can proceed with new compounds, but with adequate safeguards to protect the health and safety of study subjects in particular and the well-being of the American people in general. Society benefits from the availability of new medicines in a timely, cost-efficient manner, but there must be protection from unanticipated adverse reactions or impure compounds. Information about CDER and CBER can be obtained from their respective web sites: www.fda.gov/cder and www.fda.gov/cber. An excellent review of the new drug development process is found in the *CDER Handbook*, available from CDER's web site.

The document, "Guidance for industry: E6—Good Clinical Practice: Consolidated Guidance," was developed by the Expert Working Group (Efficacy) of the International Conference on Harmonization of Technical Requirements for Registration of Pharmaceuticals for Human Use (ICH). The guidance encompasses the ethical tenets of the Declaration of Helsinki, as well as good clinical practices within the European Union, Japan, Australia, Canada, the Nordic countries, and the World Health Organization. It also represents the FDA's current thinking on good clinical practices. This Good Clinical Practice (GCP) guidance establishes standards for all aspects of the conduct of clinical studies; seeks to provide assurance that the data and results of these studies are credible and accurate; and assures that the rights, integrity, and confidentiality of trial subjects are protected.

Investigators are defined in the GCP guidance as the persons responsible for the conduct of a clinical trial at a given site. If that trial is conducted by a team of individuals, the PI is the responsible team leader. Subinvestigators are individual team members designated and supervised by the PI to perform critical trial-related procedures and/or make important study-related decisions. In the same document, a clinical trial or study is defined as any investigation in human subjects intended to discover or verify the clinical, pharmacological, and/or pharmacodynamic effects of an investigational product.

FDA Regulations and Policies for Investigators

The FDA's overwhelming principle of clinical investigation is that the rights, safety, and well-being of trial subjects should prevail over the interests of science and society. The medical care given to, and medical decisions made on behalf of, study subjects must always be the responsibility of a qualified physician (or when appropriate, a qualified dentist). However, responsibility for medical care is not the same thing as responsibility for conduct of the clinical trial.

All investigational new drug research must have an FDA-approved study protocol. In the case of research supported by the pharmaceutical industry, the sponsor (company) designs the trial to generate safety and efficacy data needed to support appropriate labeling claims for the new product. Frequently, sponsors consult with expert scientists to assure that the study has scientific and clinical merit. Investigators are frequently asked for suggestions to improve the protocol, but the decisions on study design rest with the sponsor and FDA. The sponsor identifies and recruits investigators, who then have the right to agree or decline to participate in the study. The PI must submit their credentials on an FDA Form 1572 for approval by the sponsor and FDA.

Before a trial is initiated, the FDA requires the protocol to be approved by an institutional review board (IRB) or independent ethics committee. The IRB must address the investigator's qualifications to conduct the proposed trial. It does so by reviewing a current curriculum vitae and other relevant documentation. Pharmacists seeking approval by an IRB must demonstrate experience in the conduct of clinical trials similar to the one for which approval is being sought. The investigator provides information to the IRB, but does not participate in the committee's deliberations.

The FDA requires each PI to comply with GCP and other applicable regulatory requirements, permit monitoring and audits by the sponsor and the FDA, and maintain a

list of qualified persons to whom the investigator has delegated significant trial-related duties. A qualified PI must show familiarity with the compound to be studied and demonstrate adequate resources to conduct the trial. This includes being able to recruit in a timely manner an adequate number of subjects that meet protocol-defined entry criteria, have an adequate number of qualified staff, and have necessary facilities available for the anticipated duration of the trial. The PI is responsible for assuring that all staff assisting in the trial are well informed about the protocol, the investigational drug, and their duties with regard to the trial.

RESPONSIBILITIES OF INVESTIGATOR

General Responsibilities

A PI is responsible for the conduct of a scientific investigation in accordance with study methodology, a signed investigator agreement or contract with the sponsor (if applicable), and any federal or state rules and regulations regarding performance of a study using human subjects. The underlying premise is to protect the rights, safety, and welfare of subjects under the investigator's care and to control the distribution and use of drugs under study.

As mentioned, prior to initiation of the trial, the investigator needs IRB approval, an approved subject informed consent form, and approved subject recruitment procedures. The investigator must provide the IRB with current copies of the Investigator's Brochure for each investigational drug before beginning the trial, and must provide updated copies if the Investigator's Brochure is revised during the conduct of the study. The PI is accountable for obtaining written informed consent from all subjects. In life-threatening situations necessitating the use of the study drug, the IRB may approve other methods of obtaining informed consent. Regardless, an investigator may only obtain consent after a subject or their legally authorized representative has had sufficient opportunity to consider the risks and benefits of participation without coercion or unreasonable influence.

During the clinical study, a qualified physician or dentist, as appropriate, who is at least a subinvestigator must be responsible for all trial-related medical (or dental) decisions. The PI and the institution must ensure that adequate medical care is provided to subjects for any adverse events related to the trial. Intercurrent illnesses that are detected during the course of the study must be noted and the subject informed. If agreeable to the study subject, the PI is also responsible for informing their primary physician about their participation in the trial. The investigator should make every effort to determine the reason for subject withdrawal from the trial, while being respectful of the subject's rights to withdraw without explanation.

During the performance of the study, the investigator must maintain adequate documentation of study activities and must promptly report any deviations from the protocol to both the IRB and study sponsor. Any changes to the protocol, all adverse drug reactions that are both serious and unexpected, and any new information that may affect adversely the safety of the subjects or their willingness to participate in the clinical trial must be promptly reported to the IRB. Substantive changes to the protocol must receive IRB approval. Investigators must also submit progress reports to the IRB at least annually; sponsors will usually request more frequent progress reports.

Investigator Record Keeping

The PI is responsible for maintaining records associated with the clinical study. These include case histories designed to record all observations or other pertinent data on each enrolled subject, independent of whether the subject received active treatment. Data for each trial subject is normally recorded on a case report form provided by the sponsor. The sponsor may also require study data to be recorded in a source document (patient chart). It is the responsibility of the PI to assure that the forms are filled out with the correct information and according to guidelines established by the sponsor.

Study drugs, whether investigational or commercially available, must be controlled and accounted for, and may only be administered to study subjects who have provided informed consent and who are under the supervised care of a trial investigator. Accordingly, a drug accountability log must be kept up to date and accurate. The log should record the disposition of the drug, including dates, quantity, and use by subjects. Study drugs must be secured in a locked storage area until destroyed, properly disposed of, or returned to the study sponsor.

Case histories, drug accountability logs, and all correspondence associated with the study must be secured and kept for 2 years after the marketing application is approved by the FDA. If no such application is filed, or if the application is not approved by FDA, these records must be retained for 2 years after the investigation is discontinued and the FDA is notified. It is advisable that records from clinical trials supported by grants from academic, government, or voluntary health organizations also be maintained for this same time period, unless otherwise specified by the agency. If the study is part of an international clinical trial, or if specified by the sponsor, the PI may be required to keep a patient list with study drug assignments for 15 years after the conclusion of the trial.

In the case of premature termination or suspension of the trial, the investigator must promptly inform trial subjects, assure appropriate follow up and treatment if required, and inform the IRB and regulatory authorities as appropriate. Similarly, when a trial is completed under normal conditions, the investigator must file final reports with the sponsor and inform the institution, the IRB, and regulatory authorities (as appropriate) that the trial is completed.

Finally, all records must be available on request from the FDA or study sponsor within a reasonable time. The authorized representative from the sponsor or FDA must be allowed to copy documents with patient identifiers omitted and must be provided sufficient time to verify records associated with the conduct of the study. Investigators are not required to provide the names of subjects unless the records of a specific individual require more intensive examination related to an adverse event, or there is suspicion that the records are falsified. Frequent or deliberate falsification of records may lead to disqualification of a PI to conduct future clinical studies, as well as potential criminal or civil prosecution.

ADVICE TO PHARMACIST INVESTIGATORS

Before agreeing to participate in a sponsored clinical study, the PI should answer a number of questions about the logistics of conducting the protocol at their site and

Study Logistics: Assessment and Plan

I. Administrative Plan

a. Institutional approvals

- Determine if contract and investigator agreement regarding responsibilities, intellectual property, confidentiality, indemnification, data ownership, and publication rights are acceptable to your institution.
- Is approval needed from other institutional regulatory committees? This might include radiation safety or biohazards committees.
- If the investigator is receiving funding for the study, how will these monies be handled within the institution? How much time is needed by the institution to set up the payment infrastructure?
- Does the institution have an overhead charge for government, voluntary organizations, or corporate studies? Will the PI have access to these excess monies after the study is complete?

b. IRB Logistics

- How often does the IRB meet?
- What are the requirements for submission of an IRB study packet?
- Take time to talk with the IRB personnel to understand the complete review process before submission. This will save time and prevent anxiety.

II. Operational Plan

a. Physical Resources

- Where will the study be conducted?
- Does the study require an ambulatory care clinic, a clinical research center, or will patients need to be hospitalized? Use of these facilities will need to be negotiated with the institution.
- What types of tests are necessary? Laboratory testing, radiography, nuclear medicine, physical therapy, or other services should be contacted about the proper method of sending samples, ordering tests, and paying for services. Some institutions may offer research discounts for tests.
- How will the test data be collected? If study data are recorded in patient's charts or institution computers, will the investigator have access to these charts after the patient is discharged or the study is complete? Verification of records may occur several years after a study is complete: therefore, access to records must be confirmed prior to starting the study.

b. Human Resources

- Determine need for coinvestigator(s) and outline their roles.
- Determine need for study coordinator and responsibilities.
- Assign responsibilities and establish deadlines and milestones for all personnel closely related to the study (e.g., study coordinator, coinvestigators, research fellows).
- Visit with the nursing staff of the faculty to avoid unnecessary delays in starting the protocol.

Fig. 3 Recommended guidelines for assessment and planning of study logistics before conducting a study protocol.

should create a sound plan for study implementation. To do so, the PI must understand the structure that governs research support and funding within their institution to avoid technical or financial problems during or after the study. Addressing the issues and questions identified in Fig. 3 before starting the protocol will help in conducting a timely and successful study.

Once the PI agrees to participate in a clinical study, its successful initiation requires that the PI have the following: 1) regulatory approval from the sponsor and FDA [if the study requires an Investigational New Drug application (IND)] through submission of FDA Form 1572; 2) legal approval by the investigator's institution of their contract and agreement with the study sponsor that details the scope of work, data ownership, intellectual property, publication rights, indemnification, and confidentiality; 3) budget approval from the institution's grants and contracts office; and 4) IRB approval of the study protocol and informed consent procedure. Investigators are encouraged to complete this process within 45 days to be competitive with other research service providers. Two major pitfalls of which investigators must be aware involve budgetary planning and ownership of data. Poor planning in either regard can jeopardize the PI's ability to complete the project and disseminate new scholarly information (data ownership and publication).

Budget

An essential component to any clinical trial is a well-planned budget that accounts for all resources and project costs. In many instances, the investigator will be competing with other sites for the research contract. Therefore, the budget must cover costs, provide a reasonable incentive to the investigative team, and be competitive in the marketplace. All direct and indirect costs must be identified and negotiated among the PI, their institution, and the sponsor before initiating the study. In doing so, the PI will assure that the study can be completed and that useful information will be provided to all parties. The most commonly made error on the part of the PI is to underestimate study costs because of failure to identify all resources required to complete the study or to underestimate their true costs. As a result, the study may not be completed and/or the investigator's research program will not benefit from residual funds remaining in the contract budget after study completion. These monies can usually be used to sustain the infrastructure of the clinical facility or laboratory.

To avoid these problems, it is recommended that the PI consult with internal support staff knowledgeable in budget design and institutional overhead. Most academic centers have a clinical trials office that can help construct a comprehensive budget and provide accurate estimates on all institutional costs. If such an office does not exist, one can contact the campus grants and contract office or a peer group with experience in conducting clinical studies in the investigator's setting. Many times, the sponsor will generate a proposed budget for a trial based on usual and customary costs. These budgets are typically very complete, although they can be modified through negotiation with the sponsor if the increased costs can be justified.

The investigator must remember that most clinical trial budgets derived in academic settings must conform to the Health Care Finance Administration's corporate compliance regulations. The regulations state that federal health care payers are responsible for covering only those resources that are medically necessary for the care of a patient. Taxpayer dollars (e.g., Medicare) cannot be used to subsidize purely research activities, or for experimental or unproven medical therapies.

Research-related costs are derived from the investigator's hospital or clinic, laboratory testing or analysis, or through contractual arrangements with other laboratories. In addition, there may be patient recruitment costs such as advertising, patient expense reimbursement (parking and transportation), and participation honoraria. Research resources usually include equipment, supplies, and salary support for the PI and associated personnel. Salary support for the PI and other study personnel generally has the greatest flexibility and provides an opportunity to generate residual funds to support the PI's overall research program.

From a financial perspective, the PI's primary objective when entering into a research contract is to complete the study successfully at or below the requested budget. The residual funds that result are often placed into a development fund on behalf of the PI and can be used at the investigator's discretion to support other research projects. Although this objective is perfectly reasonable, study expenditures and allocation of funds must be thoroughly documented throughout the course of the investigation. Accurate records are often requested by the sponsor, and occasionally by outside auditing agencies, and are essential to maintaining a productive clinical research program.

Publication Rights

A clear understanding of study responsibilities and publication rights should be negotiated among the PI, their collaborators, the study sponsor, and the investigator's institution prior to initiating the trial. It is best to involve all parties as early as possible and to negotiate all aspects before agreeing to the study contract. In this regard, most academic institutions have a liaison within their research office for business and industry contracts that can assist the investigator to organize and expedite this process. It should be the mission of the PI and liaison to ensure the investigator's freedom to publish the research findings. This includes the right to publish negative results. In doing so, the investigator has ownership of the data, materials, and documentation, but the sponsor receives copies and is given the right to use the materials for certain purposes. In general, sponsoring companies are aware of the research mission of academic institutions and are flexible in the negotiation process.

Any terms and conditions that restrict the publication rights of the investigator are usually reserved for the protection of the sponsor's patent rights that may arise from the contracted work. In this situation, the sponsor is granted a specified period of time (e.g., 30 days) to review the proposed publication before its submission and is provided the right to withhold publication for a specified time period (preferably no more than 90 days), pending submission of a patent application. It is always important to identify in the contract that the investigator and supporting institution will not allow any publication restriction in such a way as to impede the academic progress of a graduate student or research fellow if these individuals were integrally involved with the study. If any publication restrictions are accepted on behalf of the PI and students, they should be agreed to in contractual form before initiating the trial.

Finally, in single-center trials, the order in which authors are listed on the publication is usually the responsibility of the investigators involved with the study. Therefore, choose collaborators wisely. In multicenter trials, the publication rights of individual investigators and the expeditious publication of results become more complex. In this case, the investigator must accept the risk of sacrificing both ownership and timely publication of research findings despite a priori agreements with the study sponsor. Many sponsors choose authors for multicenter studies based on the reputation of those authors in the area of study or based on the number of subjects that their site enrolled in the trial. Thus, younger investigators, or investigators from smaller study sites, may find themselves excluded from the publication process.

QUALIFICATIONS OF A SUCCESSFUL PRINCIPAL INVESTIGATOR

It is important to emphasize that even though pharmacists may serve as PIs for industry-sponsored clinical research, the mere fact that a person is a pharmacist does not automatically qualify them to serve in this capacity. An investigator must establish their qualifications and credentials before they can reasonably expect an industry sponsor to trust them to serve as PI.

Experience

A PI typically evolves from a subinvestigator. Although we may consider our research trainees ready to assume a career as an independent scientist on completion of their program, 2–3 years of postdoctoral research training may not realistically provide them with sufficient experience. For example, a qualified PI should have relevant clinical experience in the proposed study population, usually gained from several years of experience in patient care. However, this patient care experience alone does not automatically qualify someone as an investigator. Understanding and demonstrating competence in clinical research practices, demonstrated compliance with research regulations and data management, and the ability to create and manage an investigational plan through task delegation must also be considered.

Local Resources

The PI must demonstrate that they have acceptable and adequate resources available to manage the trial, including access to or control over clinical space such as beds and clinics, if needed. The facilities must have appropriate staff and other resources to conduct the research and to protect the subjects. Specialized testing equipment needed for the experiments must be available, either in the form of general testing purchased from the health system or in the PI's own laboratory.

Command of Research Process

The PI must be knowledgeable about the multiple processes needed to manage a clinical study, including the medical records system, investigational drug pharmacy, clinical laboratory, other clinical departments necessary to support the project, the institutional review and approval processes, budget and financial management, and contract initiation.

Human Resources

The PI must generally demonstrate the existence of a qualified research and clinic staff, as appropriate. Very few studies can be managed by a single individual, and PIs have many other responsibilities. Thus, a successful PI must be able to delegate responsibility and hold other people accountable.

Access to Patients

An industry sponsor is very keen on knowing that the PI can enroll and complete the number of patients required or requested in a contractual agreement. The sponsor's research and business plans are entirely dependent on productive and enthusiastic PIs and study coordinators. Some pharmacists may be limited in identifying and enrolling study subjects by having insufficient direct professional responsibility for patient management, or by having access to patients only through physicians. To overcome this potential limitation, the PI should demonstrate the ability to recruit qualified patients through advertising or similar means.

Audits

Proof of competence is an increasing demand. Principal investigators must be able to show they have the ability to perform in compliance with research regulations through outside audits of process and product. The ability to demonstrate this competence through a portfolio of projects, meeting contractual obligations, and generating good data will serve as an effective reference for obtaining recognition as a PI.

Local Leadership and Environment

The leadership of local pharmacy organizations (e.g., colleges of pharmacy, departments of pharmacy, academic medical centers) must be supportive and encouraging, must remove barriers to successful research, and must encourage multidisciplinary collaboration permitting pharmacists to reach their potential. These local leaders can play a great role in promoting their new members into a career in clinical research and providing an environment conducive to this endeavor.

CONCLUSION

Clearly, an individual's academic degree alone does not imply that they possess adequate skills and experience to serve as a PI. The FDA and pharmaceutical industry are aware that pharmacists can be excellent investigators. However, as is the case with any PI, the pharmacist must have a proven track record that demonstrates successful clinical trial management. As outlined previously, serving as PI is an arduous task. There are many responsibilities that require the investigator to adhere to local institution, industry, regulatory, and human assurance guidelines and policies. Adherence to these guidelines requires extensive documentation. In addition, the investigator must enroll patients, execute the protocol, and meticulously collect and report patient data. Given these responsibilities, it is understandable why the industry sponsor needs to select PIs carefully. It is the opinion of the ACCP that the pharmaceutical industry should support policies and practices that use clinical pharmacists as PIs, as long as the pharmacist is an experienced clinical researcher with documented credentials, has the adequate institutional infrastructure and support, and has access to the required patient population.

ACKNOWLEDGMENTS

This document was prepared by the 1999 Research Affairs Committee: Myryam Bayat, Pharm.D.; Lisa Davis, Pharm.D, FCCP; Darren Knoell, Pharm.D.; Joan Korth-Bradley, Ph.D., Pharm.D., FCCP; Mark Munger, Pharm.D., FCCP; Mark Shaefer, Pharm.D., FCCP; Peter Vlasses, Pharm.D., FCCP; Daniel Wermeling, Pharm.D.; Chair: Michael Ujhelyi, Pharm.D., FCCP. Endorsed by the ACCP Board of Regents on January 17, 2000.

REFERENCES

1. American College of Clinical Pharmacy. Clinical pharmacists as principal investigators. Drug Intell. Clin. Pharm. **1983**, *17*, 675–676.
2. American College of Clinical Pharmacy. Clarification regarding clinical pharmacists as principal investigators. Drug Intell. Clin. Pharm. **1984**, *18*, 444.

Clinical Pharmacist, Evaluation of a (ACCP)

American College of Clinical Pharmacy
Kansas City, Missouri, U.S.A.

INTRODUCTION

As the focus of the standard of pharmacy practice moves from dispensing products to optimizing patient outcomes, so must the standards for evaluating pharmacists' performance. The goal of pharmacy practice is to deliver pharmaceutical care to patients. Opportunities are expanding for clinical pharmacists to have direct responsibility for patient well-being by ensuring optimum outcomes of therapy in various health care environments. In 1989 the Clinical Practice Affairs Committee of the American College of Clinical Pharmacy (ACCP) developed practice guidelines for pharmacotherapy specialists.[1] To complement the guidelines, the committee developed a template for evaluating clinical pharmacists, a tool for assessing the extent to which clinical pharmacists' performance meets predefined practice standards. The template can be adapted to meet site-specific requirements.

TEMPLATE DEVELOPMENT

The core materials for the development of this template (Appendix) were the ACCP practice guidelines for pharmacotherapy specialists,[1] the eight steps of the drug use process,[2] and the American Society of Hospital Pharmacists' technical assistance bulletin on assessment of departmental directions for clinical practice in pharmacy.[3] Initially, the standards, criteria for meeting the standards, and methods of assessment were developed for each drug use process step. These reflected important activities in drug distribution and outcomes for patient care. Subsequently, standards, criteria, and assessment methods for evaluating clinical activities outside of the drug use process were added.

The initial draft of the template was evaluated at four hospitals under the direction of clinical pharmacy administrators. The evaluators noted several advantages of the template; for example, it could lead to quality clinical pharmacy services universally, help to justify the development of clinical pharmacy programs, and improve efficiency by minimizing the need for different evaluation tools for every institution or practice site. In addition, the standards could be incorporated into policy and procedure manuals, if desired.

However, the first draft had several disadvantages, including its complexity, the time required to complete it, its general rather than specific focus, the assigned percentages for performance standards, and the lack of space for documenting findings or adding written comments. Opinions varied concerning appropriate predefined standards. Some believed that all standards should be met 100% of the time, whereas others believed such expectations were unrealistic. The latter group argued that standards must reflect pharmacists' level of education and training, as well as individual institutions' expectations and resources. In addition, certain sections appeared to be redundant. The evaluators recommended that sections on dispensing and administrative activities be deleted, and that the template evaluate either clinical pharmacists or clinical pharmacy services, but not both. Also, certain assumptions regarding pharmacists' activities were simply not true for every clinical pharmacist; for example, attending medical rounds and performing physical assessments.

Based on this feedback, a revised template that evaluates only clinical pharmacists was developed and tested at seven institutions. The revised version did not define standards, allowing it to be adapted to each institution's standards and to the expected performance of the clinical pharmacist. Response to the revised template was excellent; additional recommendations for revision as well as instructions for personalizing the tool were incorporated into the final template.

ASSESSMENT METHODS

Assessment methods describe how the evaluator will collect data to evaluate performance for each criterion. Examples from the template are review of selected monitoring forms, chart notations, or orders; review of adverse drug reaction or incidence reports; review of in-

From *Pharmacotherapy* 1993, 13(6):661–667, with permission of the American College of Clinical Pharmacy.

Encyclopedia of Clinical Pharmacy
DOI: 10.1081/E-ECP 120006413

service evaluations or formal evaluation of presentations; and comparison of monitoring forms with patient charts. Other methods that could be used are physician evaluations, formulary decisions, evaluations of patient profiles, acceptance of clinical pharmacotherapy suggestions, evaluation of patient response, peer review, documentation in patients' charts, patient consultation logs, drug use evaluation records, and patient satisfaction surveys.

Appendix. Template for the Evaluation of a Clinical Pharmacist[a]

The template for evaluating a clinical pharmacist is for use by clinical pharmacy managers. It should be revised to meet specific institutional requirements for clinical pharmacy practice prior to implementation. Specific numbers and types of patient interventions should be included and reviewed to reflect accurately the individual clinical pharmacist's practice responsibilities.

This template represents only part of the evaluation process. A letter or form should be submitted by an attending physician that addresses specific contributions to individual patient care by the practicing clinical pharmacist. Other health professionals (i.e., nurses, physician assistants, etc.) who interact daily with the individual clinical pharmacist should participate in the annual evaluation process.

Date ______________ Clinical Pharmacist ________________________ Supervisor ________________________

Instructions for Using the Template

The template for the evaluation of a clinical pharmacist was designed to be flexible and adaptable to a particular institution, clinical pharmacy service, and clinical pharmacist. The administrator (evaluator) and clinical pharmacist for whom it will be used should work together to modify and individualize the tool as necessary. Such communication is vital to its effective use. The pharmacist being evaluated must be an active participant in the process.

Step 1.

Review the performance appraisal requirements of the institution or practice site. Guidelines or other requirements for employee evaluations should be incorporated into the instrument.

Step 2.

Review the criteria within each section of the template and delete or add criteria as necessary to reflect activities performed at the practice site. The template can be tailored to an individual pharmacist's activities or to the patient care activities of the entire clinical staff.

Step 3.

Determine the standards (thresholds) for each criterion. These can reflect universal standards established for the institution's clinical program or expectations of an individual clinical pharmacist. The standards establish departmental expectations regarding the extent to which the individual performs each criterion. The expectations should be objective and expressed clearly. Some sites may decide not to use standards; however, they should be cautioned that objective evaluation may be difficult without standards.

Step 4.

Review the assessment methods for each criterion and tailor them to the practice site, the evaluator, and the pharmacist being evaluated. No matter who completes the performance assessment, the methods used must be discussed, documented, and understood by the evaluator and the person being evaluated. Methods such as chart review and direct observation are time consuming but yield useful information.

Step 5.

Use the individualized evaluation tool to identify areas for improvement and opportunities for professional growth. These can be noted in the "comments" section of the template.

I. Perception of the Need for a Drug

Criteria	Assessment Method	Standard	Meets Criteria	Comments
Interviews patient to obtain a complete list of prescription and OTC drug use, response, and toxicity.	Reliability testing between pharmacist and supervisor			
Actively participates in medical rounds to obtain pertinent information required to determine necessity of drug therapy.	Joint rounds with supervisor and evaluation by physicians			
Determines accurate problem list.	Comparison of monitoring form and medical chart			
Consults physicians about drugs without indications.	Review of selected patient-monitoring forms and medical charts			
Obtains pertinent information required to determine the necessity of drug therapy.	Discussion of selected patients and therapy with manager			

II. Selection of a Specific Drug

Criteria	Assessment Method	Standard	Meets Criteria	Comments
Assures the drug of choice for a particular patient condition is ordered.	Review of 25 patient-monitoring forms, chart notations, or orders			
Assures there are no contraindications for selected drug products (e.g., allergy, history of severe adverse reaction).	Review of selected monitoring forms and medical charts			
Selects drug products that are effective, are cost-beneficial, and promote patient compliance.	Review of selected monitoring forms and medical charts			
Participates in patient care rounds to provide input into drug selection.	Review of documentation on activity reports or productivity reports			
Actively participates in writing or evaluating drug therapy protocols.	Review of accuracy of standing protocols for drug therapy			
Suggests appropriate therapeutic alternatives for nonformulary drugs.	Order review with therapeutic and cost saving outcome evaluation			
Suggests nondrug therapy when appropriate.	Review of selected monitoring forms or medical charts			

C

III. Evaluation and Review of Drug Regimen

Criteria	Assessment Method	Standard	Meets Criteria	Comments
Determines drug therapy compliance with protocols, guidelines, or recommendations.	Review of selected patient-monitoring forms, chart notations, or orders			
Recommends drug discontinuation or dosage alteration when indicated.	Review of selected monitoring forms and medical charts			
Identifies potentially significant drug-drug, -food, -laboratory, and -disease interactions.	Review of selected monitoring forms and medical charts			
Communicates clinically relevant interactions with therapeutic alternatives to health care practitioners.	Review of selected monitoring forms and medical charts			
Obtains and uses clinical laboratory data to evaluate appropriateness of drug product selection and/or dosing regimen.	Review of selected monitoring forms, medical charts, and orders			
Adjusts drug therapy according to changes in concomitant therapy or the patient's condition.	Review of selected monitoring forms or selected patient charts for documentation			
Provides pharmacokinetic consultation for agents requiring such monitoring.	Review of selected patient-monitoring forms or selected patient charts for documentation			
Provides and evaluates drug therapy orders for appropriateness of dosage, route, interval, schedule, and duration throughout patient's hospital course.	Review of selected monitoring forms, medical charts, and orders			

IV. Monitoring Effects of Drug Therapy

Criteria	Assessment Method	Standard	Meets Criteria	Comments
Independently evaluates patient response to drug therapy.	Review of selected patient-monitoring forms, chart notations, or orders			
Participates in patient care rounds to provide input into the monitoring of drug therapy.	Review of documentation on activity reports or productivity reports			
Assures compliance with adverse drug reaction-reporting programs or medication error-reporting programs.	Review of adverse drug reaction or incident reports			
Records recommendations, interventions, or other appropriate activity in the medical record or appropriate activity report.	Review of selected medical charts or activity reports			

V. Education

Criteria	Assessment Method	Standard	Meets Criteria	Comments
Provides drug education and counsels patients on appropriate drug use and storage.	Review of selected patient-monitoring forms or chart notations			
Provides written information for appropriate drug products.	Review of selected patient-monitoring forms or chart notations			
Provides educational presentations to pharmacy staff, students, and other health care professionals.	Review of inservice presentations, evaluations, and/or scores on post-test evaluations			

VI. Evaluation of Drug Usage and Therapy

Criteria	Assessment Method	Standard	Meets Criteria	Comments
Employs drug usage evaluation information to alter therapy effectively.	Review of developed guidelines or drug therapy protocols			
Participates in drug use evaluation programs.	Review of selected patient-monitoring forms or chart notations			
Participates in research activities.	Review of selected protocols			

VII. Information Retrieval

Criteria	Assessment Method	Standard	Meets Criteria	Comments
Provides complete and accurate information.	Supervisor review; recipient evaluation			
Effectively communicates information in a timely manner to requester.	Supervisor review; recipient evaluation			

VIII. Committee Involvement

Criteria	Assessment Method	Standard	Meets Criteria	Comments
Contributes to department, hospital, or pharmacy committees.	Evaluation of meeting minutes; assessment of committee members			
Actively participates as a member or chairperson of hospital committees.	Evaluation of meeting minutes			

IX. Miscellaneous Activities

Criteria	Assessment Method	Standard	Meets Criteria	Comments
Participates in local, state, national, or international pharmacy organizations.	Organization membership, committee membership			

Comments______________________________

____

____

____

____

[a]This template may be photocopied and used for evaluating clinical pharmacists.

SUMMARY

Pharmacists are expected to deliver pharmaceutical care, that is, to accept responsibility for patients' well-being by ensuring optimum outcomes of drug therapy. Therefore, their performance must be evaluated based on this expectation. The template should be a useful tool for assessing the extent to which clinical pharmacists' performance meets predefined practice standards. Its adaptability will allow it to meet site-and pharmacist-specific requirements for performance appraisal. The evaluator and clinical pharmacist should work together to establish a priori percentage standards.

ACKNOWLEDGMENTS

The committee gratefully acknowledges the following institutions and persons who assisted in the review process: Patricia Hudgens, Medical University of South Carolina, Charleston, SC; Donald Kendzierski, University of Illinois at Chicago, Chicago, IL; Marsha A. Raebel, Scott and White Hospital, Temple, TX; Barbara Zarowitz, Henry Ford Hospital, Detroit, MI; Christine Rudd, Duke University Medical Center, Durham, NC; Beth Noer and Glen Schumock, University of Illinois at Chicago, Chicago, IL; Mary Anne Koda-Kimble, University of California, San Francisco, CA; Terri Graves Davidson, Emory University, Atlanta, GA; Louis Pagliaro, University of Alberta, Edmonton, Alberta, Canada; and Katherine Michael, Medical University of South Carolina, Charleston, SC.

This document was written by the following subcommittee of the 1989–1990, 1990–1991, and 1991–1992 ACCP Clinical Practice Affairs Committees: Mary Beth O'Connell, Pharm.D., FCCP, FASHP, Co-Chair 1991–1992; Christine Rudd, Pharm.D., FCCP, FASHP, Chair 1991–1992; Glen Schumock, Pharm.D.; Karen E. Bertch, Pharm.D.; Carl A. Hemstrom, Pharm.D.; and Marsha A. Raebel, PharmD, FCCP, BCPS, Board liaison. Other members of the 1989–1990 ACCP Clinical Practice Affairs Committee were Barry L. Carter, Pharm.D., FCCP, BCPS, Chair, 1989–1991; Michael W. Jann, Pharm.D.; W. Francis Lam, Pharm.D.; Milap C. Nahata, Pharm.D., FCCP; Douglas Anthony Powers, Pharm.D.; Anthony E. Ranno, Pharm.D.; and Nathan J. Schultz, Pharm.D.. Members of the 1990–1991 ACCP Clinical Practice Affairs Committee not mentioned above were Ryon Adams, Pharm.D.; Richard Berchou, Pharm.D.; G. Dennis Clifton, Pharm.D.; Joseph F. Dasta, MS, FCCP; Terri Graves Davidson, Pharm.D.; Donald Kendzierski, Pharm.D.; Edward Krenzelok, Pharm.D.; Bruce Kreter, Pharm.D.; Veronica Moriarty, Pharm.D.; Louis Pagliaro, Pharm.D.; Richard Ptachcinski, Pharm.D., FCCP; and Dominic Solimando Jr., Pharm.D. Members of the ACCP 1991–1992 Clinical Practice Affairs Committee not mentioned above were Daniel Buffington, Pharm.D.; Lea Ann Hansen, Pharm.D.; Kim Kelly, Pharm.D., FCCP; Gary Matzke, Pharm.D., FCP, FCCP; Katherine A. Michael, Pharm.D.; Beth Noer, Pharm.D.; Andrew T. Pennell, Pharm.D.; Dana Reid, Pharm.D.; and Joseph Tami, Pharm.D.. Staff editor: Toni Sumpter, Pharm.D. Approved by the Board of Regents on July 20, 1993.

REFERENCES

1. ACCP Clinical Practice Affairs Committee, 1989–1990. Practice guidelines for pharmacotherapy specialists. A position statement of the American College of Clinical Pharmacy. Pharmacotherapy **1990**, *10* (4), 308–311.
2. Hutchinson, R.A.; Vogel, D.P.; Witte, K.W. A model for inpatient clinical pharmacy practice and reimbursement. Drug Intell. Clin. Pharm. **1986**, *20*, 989–992.
3. ASHP. Technical assistance bulletin on assessment of departmental directions for clinical practice in pharmacy. Am. J. Hosp. Pharm. **1989**, *46*, 339–341.

PROFESSIONAL DEVELOPMENT

Clinical Pharmacokinetics Specialty Practice

C

Mary Ensom
BC's Children's & Women's Hospital, Vancouver, British Columbia, Canada

Y.W. Francis Lam
University of Texas Health Science Center at San Antonio, San Antonio, Texas, U.S.A.

INTRODUCTION

Clinical pharmacokinetics involves applying pharmacokinetic principles to determining optimal dosage regimens of specific drugs for specific patients to maximize pharmacotherapeutic effects and minimize toxic effects.[1] The birth of clinical pharmacokinetics as a discipline was spurred on by an increasing awareness of concentration–response relationships and knowledge of pharmacokinetic characteristics of various drugs, the advent of computerization, and advancements in analytical technology.[2] Therapeutic drug monitoring is an important aspect of clinical pharmacokinetics that has helped many pharmacists enter the clinical arena. Clinical pharmacokinetics emerged as a specialty pharmacy practice in the late 1960s and early 1970s to provide clinical pharmacokinetic consultation or dosing service.

Initially, many pharmacokinetics services were centralized programs with only specialists in pharmacokinetics providing these services.[3] The more recent advent of pharmaceutical care, however, has led to more integrated approaches such that clinical pharmacokinetic monitoring is becoming a fundamental responsibility of *all* pharmacists providing pharmaceutical care.[1,3] As such, most pharmacokinetics services are no longer centralized or stand-alone programs. However, various practice and research opportunities still exist for pharmacists with specialized education, training, or experience in clinical pharmacokinetics. These opportunities are outlined following a review of the specialty pharmacy practice since the 1980s.

CLINICAL PHARMACY EXPERIENCES AND OPPORTUNITIES

In 1994, Howard et al.[4] published the results of a survey mailed to Veterans Administration (VA) medical centers in the United States ($n = 160$, with 93% return rate) to assess the provision of pharmacokinetics services. Survey results indicated that 104 (70%) of the respondents provided pharmacokinetics services. Of the remaining 30% who did not, almost two-thirds planned to begin one in the future. Ninety-eight (94%) of the existing pharmacokinetics services were considered to be formally recognized in that they were approved by the Pharmacy and Therapeutics Committee, offered consultation on request, or had a contact person for serum drug concentration evaluations. Aminoglycosides, vancomycin, and theophylline were the most frequently monitored drugs. Selected characteristics (i.e., bed capacity, geographic region, pharmacy chief's highest degree, pharmacy residency, and teaching affiliation) were used to evaluate whether differences existed between VA medical centers with or without each characteristic. Medical centers with a pharmacy residency program were more likely to have a pharmacokinetics service compared with those without a residency program ($p<0.02$). Differences among geographic regions also were significant ($p<0.01$), with the following percentage of respondents having pharmacokinetics services: 51 (eastern); 77 (central); 82 (southern); 81 (western); 0 (unknown).[4]

In 1996, Murphy et al.[3] published the results of their national survey of hospital-based pharmacokinetics services. Altogether, 252 questionnaires were mailed to all respondents of the 1994 American Society of Health-System Pharmacists (ASHP) national survey of hospital-based pharmaceutical services[5] who indicated that their institution provided pharmacokinetics services. The response rate was 42.1% ($n=106$); however, only 98 surveys had complete data thus yielding a net response rate of 40.2%.[3] Aminoglycosides were the main focus (60.8±27.7% of total) of pharmacokinetic consultations, followed by vancomycin (21.1±18.3%), theophylline (4.6±7.4%), other (3.6±16.0%), warfarin (3.1±11.1%), digoxin (2.2±6.1%), phenytoin (1.2±3.0%), lithium

Encyclopedia of Clinical Pharmacy
DOI: 10.1081/E-ECP 120006225

(0.9±5.2%), carbamazepine (0.6±3.4%), valproic acid (0.5±3.0%), procainamide or N-acetylprocainamide (0.2±0.9%), phenobarbital (0.2±0.8%), caffeine (0.1±0.7%), quinidine (0.1±0.6%), cyclosporine (0.1±0.4%), ethosuximide (0.02±0.1%), and methotrexate (0.02±0.1). Pharmacokinetics services were paid primarily through the pharmacy budget (74.5±43.6%), although 15.8±36.5% billed the patient for the consultation via a pharmacy number. A disappointingly high number [i.e., 60 (61.2%)] of institutions providing pharmacokinetic consultations did not counsel patients receiving the service, although 12 (12.2%) provided educational materials to patients and 30 (30.6%) provided patients with individual instruction by the pharmacist or other health care professional. The survey showed that pharmacists, mostly staff pharmacists, spent on average of 19 hours per week providing pharmacokinetics services.[3]

Responses of 20 survey statements on attitudes toward pharmacokinetics services were all on the agreement side of neutral, with the exception of "Pharmacokinetic software is so sophisticated that very little judgment is needed for pharmacokinetic evaluation" (2.4±1.6 on a scale of 1 = strongly disagree and 7 = strongly agree, n = 92). Notably, the statements whose average responses were >6 included: "Pharmacokinetic reviews and/or consultations should be integrated into the duties of every hospital pharmacist" (6.3±1.3, n=94); "Provision of pharmacokinetic services is an important component of hospital pharmacy practice" (6.2±0.9, n=94); and "Our pharmacokinetic services are very successful in terms of acceptance and use by prescribers" (6.1±1.0, n=93). However, only about 6% of respondents indicated a potential for increased pharmacokinetic consultations or reviews in the future. The authors surmised that as the profession embraces the concept of pharmaceutical care, pharmacokinetic consultations are being integrated into the pharmaceutical care process, with the pharmacokinetics specialist in larger institutions acting as a consultant to the nonspecialists. Because the survey revealed that only a minority of pharmacokinetics service providers interact directly with patients or their advocates (12±22% of patients), the authors also encouraged greater direct involvement of pharmacists with patients.[3]

In 1998, Raehl et al.[6] published the results of the 1995 National Clinical Pharmacy Services study that determined the extent of hospital-based clinical pharmacy services in 1109 U.S. acute care, general, medical–surgical, and pediatric hospitals with 50 or more licensed beds. Pharmacokinetic consultations (i.e., pharmacist review of the serum drug concentration data and patient medical record with appropriate verbal or written follow up) were provided in 778 (70%) of hospitals surveyed, compared with only 54% in 1992. According to the 1995 survey, provision of pharmacokinetic consultations no longer varied by hospital ownership or geographic location as they did in 1992. Those hospitals with greater involvement in pharmacokinetic consultations included large hospitals, pharmacy teaching hospitals, hospitals in which the pharmacy director had a PharmD degree, and hospitals with decentralized pharmacists. Pharmacokinetic consultation was among the three services (the other two being adverse drug reaction management and drug history) that were more common in hospitals with a pharmaceutical care program than in those without one. In addition, pharmacokinetic consultations and drug therapy protocol management were the clinical services that experienced the most consistent and greatest growth since the first national survey in 1989. In hospitals providing pharmacokinetic consultations, 80.4±30.3% of inpatients on aminoglycoside therapy for greater than 48 hours had a measured serum aminoglycoside concentration. In the hospitals offering pharmacokinetic consultations, 74% routinely provided documentation in patient medical records.[6]

Detailed teaching affiliation data were available for 1102 of the 1109 surveyed hospitals in the 1995 national clinical pharmacy services study.[6,7] Hospitals affiliated with both PharmD and BS degree-granting Colleges of Pharmacy provided pharmacokinetic consultation as a core clinical service (defined as being offered in 50% or more of hospitals). However, the percentage was significantly greater for PharmD-affiliated hospitals [85% (PharmD), 66% (BS Pharmacy), 63% (nonpharmacy teaching), and 57% (nonteaching), $p<0.001$].[7]

In 1999, Ringold et al.[8] published the results of the 1998 ASHP national survey of pharmacy practice in acute care settings, which pertained to prescribing and transcribing practices. Of 1058 general and children's medical–surgical hospitals surveyed in the United States, 548 (51.8%) responded. Of the 536 hospitals or health systems for which information was available, 432 (80.6%) provided pharmacokinetic consultations. On average, 435.5±943.2 consultations were provided per year (n=362 hospitals or health systems). Of 411 total respondents, 71.2% indicated a >80% adoption rate for pharmacokinetic recommendations.[8]

In 2000, Raehl et al.[9] published the results of the 1998 National Clinical Pharmacy Services study that determined the extent of hospital-based clinical pharmacy services in 950 U.S. acute care, general, medical–surgical, and pediatric hospitals with 50 or more licensed beds. Eighty percent of hospitals offered pharmacokinetic consultations compared with 70% in 1995, 54% in 1992, and only 40% in 1989. Again, pharmacokinetic consultations and drug therapy protocol management were the clinical services that experienced the most consistent and greatest growth since the first national survey in 1989. According to the 1998 survey, those hospitals with greater involvement in pharmacokinetic consultations included pharmacy teaching hospitals, hospitals in which the pharmacy director had a PharmD degree, hospitals with decentralized pharmacists, and those in the Pacific region. In hospitals providing pharmacokinetic consultations, 68.9 ± 37.8% of inpatients on aminoglycoside therapy for greater than 48 hours had a measured serum aminoglycoside concentration. In the hospitals offering pharmacokinetic consultations, 79% routinely provided documentation in patient medical records.[9]

Clinicians in general agree that pharmacokinetic software cannot replace clinical judgment. Nevertheless, judicious use of software programs can facilitate optimization of patients' drug therapy. The USC*PACK PC program (University of Southern California Laboratory of Applied Pharmacokinetics, Los Angeles, CA, USA) is a software package developed by investigators at the University of Southern California. Within the package, individual drug programs enable the clinicians to fit a patient's dosing history and drug concentrations to a population model. This Bayseian approach has been shown to provide good prediction of concentrations and dosage regimens. Seminar courses regarding the scope and use of the USC*PACK PC programs are scheduled each year. DataKinetics (ASHP, Bethesda, MD, USA) is another dosing program based on one-compartment modeling. Most of the commonly monitored drugs are included in the program. With animation and drawing features, STELLA (High Performance Systems, Lyme, NH, USA) is a useful program for teaching. Programs primarily for analysis of pharmacokinetic data include PCNonlin (Scientific Consulting, Inc., Cary, NC, USA), RSTRIP (MicroMath, Salt Lake City, UT, USA), and NONMEM (University of California, San Francisco, CA, USA). Modeling of pharmacodynamic data is available with P-PHARM (Simed, Creteil, France) and MKMODEL (Biosoft, Ferguson, MO, USA), an National Institutes of Health and Prevention (NIH)-supported PROPHET program. A list of pharmacokinetic programs and other resources is also available on the world wide web at http://www.boomer.org/pkin/.[a]

Clinical pharmacokinetics specialists have opportunities as directors and/or consultants of clinical pharmacokinetics services[3–9] to assume the following responsibilities, as delineated in the ''ASHP Statement on the Pharmacist's Role in Clinical Pharmacokinetic Monitoring'': 1) design and conduct of clinical pharmacokinetic/dynamic research, explore concentration–response relationships for specific drugs, and evaluate and expand clinical pharmacokinetic monitoring as an integral part of pharmaceutical care; 2) develop and apply computer programs and point-of-care information systems; and 3) serve as an expert consultant to pharmacists with a general background in clinical pharmacokinetic monitoring.[1]

Aside from being directors and consultants of clinical pharmacokinetics services, clinical pharmacokinetic specialists also have professional practice opportunities as preceptors of pharmacokinetic residency and/or fellowship programs. The American College of Clinical Pharmacy's (ACCP's) *2000 Directory of Residencies and Fellowships* lists two residency preceptors (with a total of three positions) whose programs have ''pharmacokinetics'' as their primary specialty.[10] At least one of these residencies is accredited by ASHP.[10] Using the search engine ''Yahoo'' and the search terms ''Clinical Pharmacokinetic Service'' and ''Pharmacokinetic Practice,'' two more residencies in pharmacokinetics were identified.[11,12] All these residencies are designed to prepare the resident for a career in clinical practice or teaching with a focus on pharmacokinetics. The same directory lists 7 fellowship preceptors (with a total of 11 fellow positions) whose programs have ''pharmacokinetics'' as their primary specialty.[10] Of these 7 fellowship programs, 4 are ACCP recognized. These fellowship programs seek to provide fellows with clinical pharmacokinetic/dynamic research experience that involves, but is not limited to, study design, research methodology, study conduct, analytical methodology, data analysis, and scientific writing and research.[10]

Compared with the 1990s, the current numbers of specialized clinical pharmacokinetics pharmacy practices and fellowships are small. This likely reflects the general trend of integration of clinical pharmacokinetics into the concept of pharmaceutical care provided by pharmacists. In addition, concentration measurement is only an intermediate endpoint for optimal patient care. However, outside the traditional realm of therapeutic drug monitoring, there are other opportunities for pharmacists with indepth knowledge of clinical pharmacokinetics. This includes drug evaluation and regulatory review at the Food and Drug Administration (FDA), clinical research or drug development at pharmaceutical companies, the potential role of pharmacokinetic and monitoring for the increasing trend of home/community-based parenteral antibiotic therapy, and the relatively untested disciplines of toxicokinetics and clinical toxicology. Toxicokinetic studies are

[a]See the section on Professional Networking Opportunities later in this article.

not confined to preclinical drug development, and clinical toxicology presents a challenging opportunity for evaluating the adverse effects of drugs and poisons, as well as characterizing altered pharmacokinetics and response in overdose situations. Even within the current redefined practice of therapeutic drug monitoring, opportunities still exist. One good example is *optimizing* the use of once-daily aminoglycoside dosing.

DESCRIPTION OF MODEL CLINICAL PRACTICES

Following is a description of a model clinical practice in the specialty area of pharmacokinetics practice. This is an actual practice setting at the University of Kentucky Medical Center.

The mandate of the Clinical Pharmacokinetics Service at the University of Kentucky Medical Center is to ensure safe and efficacious dosage regimens through the application of pharmacokinetic/dynamic principles and the determination of serum drug concentrations.[13] The *Clinical Pharmacokinetics Service Policy and Procedure Manual* outlines standard dosing and monitoring guidelines (for aminoglycosides, carbamazepine, digoxin, fosphenytoin, lidocaine, lithium, phenobarbital, phenytoin, procainamide, quinidine, theophylline, valproic acid, and vancomycin) when providing clinical pharmacokinetic monitoring.[14] In addition, the Clinical Pharmacokinetics Service provides warfarin monitoring for patients on services or teams that do not have an assigned clinical pharmacist.[13]

At the University of Kentucky Medical Center, the primary pharmacist or pharmacy resident who attends rounds or precepts pharmacy students on a primary medical team is responsible for clinical pharmacokinetic monitoring of all patients on that team. The Clinical Pharmacokinetics Service oversees the pharmacokinetic monitoring process for all patients [i.e., those on teams with assigned pharmacists (''covered'')], as well as those who are on teams that do not have an assigned primary pharmacist or resident (''noncovered''). The Clinical Pharmacokinetics Service consists of a faculty member who serves as the director and of pharmacy practice residents and senior pharmacy students during their clinical pharmacokinetic rotations.[13]

Patients with serum drug concentrations on ''noncovered'' services are identified by two daily printouts provided by the Therapeutic Drug Monitoring (TDM) Laboratory (in the hospital's clinical laboratory, which analyzes all serum drug concentrations). Drug profiles also are reviewed at least thrice weekly for all patients on ''noncovered'' services to identify any patients who are prescribed ''monitorable'' drugs for which no serum drug concentrations have been ordered. Any physician also may request a pharmacist to provide a clinical pharmacokinetic evaluation by verbal or written communication.[13]

The TDM Laboratory notifies the primary pharmacist of any ''supratherapeutic'' drug concentrations, between 0800 and 1700 during the week; after 1700 and on weekends and holidays, the pharmacy resident on-call is notified. The TDM Laboratory also notifies the Clinical Pharmacokinetics Service of any supratherapeutic levels for any ''noncovered'' service. All other TDM issues are directed to the Director of the Clinical Pharmacokinetics Service.[13]

For every patient with a serum drug concentration ordered, the primary pharmacist writes a ''Clinical Pharmacokinetics'' note in the progress notes section of the patient's chart within 24 hours for normal or ''subtherapeutic'' concentrations. For ''supratherapeutic'' drug concentrations, the medical team is notified immediately if clinically warranted and a chart note written within 12 hours after the concentration was reported. The chart note contains all relevant patient information and pharmacokinetic parameters necessary to provide dosing and monitoring recommendations.[13]

At the University of Kentucky Medical Center, an average of approximately 800 serum drug concentrations are assessed every month, of which about 300 are directly evaluated through the Clinical Pharmacokinetics Service. On average, the Clinical Pharmacokinetics Service provides direct consultations for 100 to 125 patients monthly.[13]

Several other examples of clinical pharmacokinetics services are described in the literature. Briefly, Shevchuk and Poulin[15] described a pharmacist training program at Regina General Hospital, a 485-bed acute care facility in Saskatchewan, Canada, for an aminoglycoside monitoring service and a quality assurance program involving pharmacist certification. Ament and McGuire[16] described a pharmacokinetic dosing service at Latrobe Area Hospital, a 300-bed teaching-community hospital in Pennsylvania, wherein the pharmacist initiates and adjusts aminoglycoside and vancomycin regimens, and schedules serum drug concentration measurements and renal function lab tests without contacting a physician for verbal approval. Williams[17] also described a pharmacokinetic consult service at York Hospital, a 565-bed community teaching hospital in Pennyslvania, that expanded to include other clinical activities.

DOCUMENTATION OF THE BENEFITS OF THE PHARMACOKINETICS SPECIALTY

Numerous studies on clinical pharmacokinetic monitoring have demonstrated positive clinical outcomes. The reader is directed to a comprehensive review of the evidence to support such definitive outcomes.[2] Some examples of positive clinical outcomes for theophylline monitoring cited in the comprehensive review include decreased length of stay[18,19] and decreased toxicity.[18] For *traditional* aminoglycoside monitoring, examples include decreased length of treatment,[20,21] decreased length of hospital stay,[20–24] decreased febrile periods;[21] decreased duration to return to normal or baseline temperature,[22,25,26] decreased duration to stabilize heart rate,[26] decreased duration to stabilize respiratory rate,[22] increased patient survival,[23,27–29] and decreased changes in serum creatinine values from baseline.[23] For digoxin monitoring, examples include decreased length of hospital stay[30] and decreased toxicity.[30,31] Examples of beneficial clinical outcomes for anticonvulsants include decreased average number of readmissions per patient within 3 months of discharge[32] and decreased percentage of patients experiencing generalized tonic–clonic seizures.[33] Such examples for vancomycin include decreased length of hospital stay[34] and decreased toxicity.[34,35]

Several of these studies evaluating the impact of clinical pharmacokinetic monitoring on patient outcomes[21,22,24,26,32] were also cited in a 1996 landmark paper by Shumock et al.[36] that summarized and critiqued original economic assessments of clinical pharmacy services published from 1988 to 1995. Of the 104 literature articles that were identified, 13% fell under the category of pharmacokinetic monitoring services (defined as ''clinical pharmacy services that primarily involved evaluation of anticipated or actual serum drug concentrations and provision of subsequent dosing recommendations''). Two of the articles on pharmacokinetic monitoring calculated benefit : cost ratios of 75.84:1 (pharmacokinetic services for patients receiving aminoglycosides)[26] and 4.09:1 (computer-assisted aminoglycoside dosing),[24] respectively. Other notable findings of the pharmacoeconomic impact of pharmacokinetic monitoring services from individual institutions were as follows: charge avoidance of $500,000 annually;[37] despite an increased number of drug levels ordered, a decrease of $599 in hospital costs; increased rational ordering of serum drug concentration determinations leading to cost avoidance of up to $12,325;[38–40] decreased length of treatment;[22] decreased length of stay;[21,22,41,42] decreased febrile period;[21] decreased direct costs;[21] annual cost savings of $113,934;[22] reduction of $14,000 in drug costs associated with service;[41] savings of $3000;[39] decreased number of digoxin serum drug concentrations ordered;[43] overall cost savings of $100 after 1 year of the program;[32] equal cost of pharmacist monitoring and savings after 1 year;[44] $1311 savings per patient in the study group (computer-assisted aminoglycoside dosing);[24] and $490 savings per patient in the study group (pharmacist dosing of aminophylline).[42]

Several other examples of the value of pharmacokinetics services on patient or pharmacoeconomic outcomes are provided here.

Pharmacokinetic monitoring of cancer patients on methotrexate has been found to significantly reduce the incidence of serious toxicity and virtually eliminate death due to high-dose methotrexate.[45] Other examples for other drugs are discussed in a state-of-the-art paper on pharmacokinetic optimization of cancer chemotherapy and its effect on outcomes.[46]

An online pharmacy intervention program was developed to determine the value pharmacy brings to the medication use process to improve patient outcomes at the Memorial Sloan-Kettering Cancer Center, a 565-bed comprehensive cancer center in New York City. Of 2499 interventions within 1 year's time, the most common types were order clarification/change (18%), followed by pharmacokinetic consult (16%).[47] The authors concluded that ''pharmacy interventions elevated the standard of care and prevented major organ damage and potentially life-threatening events,'' thus demonstrating the major role that pharmacists play in improving patient outcomes.

In a cost-effectiveness analysis of the impact of goal-oriented and model-based clinical pharmacokinetic dosing of aminoglycosides on clinical outcomes, Van Lent-Evers et al.[48] found shorter length of hospital stay ($p=0.0450$), shorter duration of therapy ($p<0.001$), fewer dosage adjustments ($p=0.016$), and lower incidence of nephrotoxicity ($p=0.003$) in the group with active pharmacy-based clinical pharmacokinetic monitoring compared with a control group of patients not guided by therapeutic drug monitoring.[48]

Although not directly related to patient or pharmacoeconomic outcomes, deserving of mention is an exploratory study comparing survey data from 127 matched pairs of clinical pharmacists and physicians working together.[49] Interestingly, both pharmacists and physicans rated pharmacokinetic monitoring similarly as an activity in which pharmacists were highly competent. Pharmacy specialists perceived their influence on prescribing to be higher ($p = 0.015$) than that of generalists with respect to recommendations based on pharmacokinetics.[49]

In a study that determined associations among hospital characteristics, mortality rates, and staffing levels for professional health care workers in 3763 U.S. hospitals, Bond et al.[50] found that increased numbers of pharmacists in a hospital were associated with lower mortality rates. Although the reasons for this were unknown, the authors speculated that providing clinical pharmacy services such as pharmacokinetic dosing services, preventing and detecting adverse drug reactions, and admission drug histories may be responsible for improving patient care outcomes.[50]

Performing simple regression analysis on data from the 1992 National Clinical Pharmacy Services study,[51] Bond et al. found a significant ($p = 0.001$) association between provision of pharmacokinetic consultations and lower mortality rates. Further multiple regression analysis did not reveal a significant association ($p = 0.544$).[52] Subsequently, Bond et al.[53] evaluated direct relationships and associations among clinical pharmacy services, pharmacist staffing, and total cost of care in 1016 U.S. hospitals. Pharmacokinetic consultations were among seven clinical pharmacy services that were associated with lower cost of care ($p = 0.0001$), based on simple regression analysis. Further multiple regression analysis did not reveal a significant association ($p = 0.436$).[53]

MATERIALS USEFUL TO PRACTITIONERS

American Society of Health-System Pharmacists Statement on the Pharmacist's Role in Clinical Pharmacokinetic Monitoring:[1] This is a position statement that provides background information and lists responsibilities that should be a part of clinical pharmacokinetics services or monitoring conducted by all pharmacists. In addition, a list of responsibilities that should be assumed by pharmacists with specialized education, training, or experience in pharmacokinetics is provided.[1]

American Society of Health-System Pharmacists Supplemental Standard and Learning Objectives for Residency Training in Clinical Pharmacokinetics Practice:[54] This document provides comprehensive guidelines and objectives for residency training in clinical pharmacokinetics practice. The contents of the Preamble include sections on definition, purpose and philosophy, accreditation authority, qualifications of the program director, selection and qualifications of the resident, and content of the residency program. Both the Goal Statements and Associated Terminal and Enabling Objectives include sections on practice foundation skills, direct patient care, drug information, drug policy development, and practice management.[54]

PROFESSIONAL NETWORKING OPPORTUNITIES

A number of network opportunities are available within professional organizations. These include, but are not limited to, the following:

American Society of Health-System Pharmacists Section of Clinical Specialists Pharmacokinetics Specialty Network:[55] The Pharmacokinetics Specialty Network is available as one of 19 networking assemblies of ASHP's Section of Clinical Specialists, each of which is led by a facilitator. Modes of networking among pharmacokinetic specialists are via the listserv as well as at networking assemblies conducted during ASHP Midyear Clinical Meetings.

American College of Clinical Pharmacy Pharmacokinetics/Dynamics Practice Research Network (PRN):[56] The Pharmacokinetics/Dynamics PRN provides a mechanism for networking and collaboration, educational programming, and a forum in which ACCP members with similar interests can discuss pharmacokinetic and pharmacodynamic methods and research.

American Association of Pharmaceutical Scientists Pharmacokinetics, Pharmacodynamics and Drug Metabolism Section:[57] The Pharmacokinetics, Pharmacodynamics and Drug Metabolism Section provides an opportunity for AAPS members to present new developments and exchange ideas related to the field. The section is designed to bring together qualified individuals who are investigating or interested in pharmacokinetics, pharmacodynamics, and drug metabolism.

American Association of Pharmaceutical Scientists Population Pharmacokinetics and Pharmacodynamics Focus Group:[58] The AAPS Population Pharmacokinetics and Pharmacodynamics focus group is open to all individuals, not just members, who are interested in this focused area. The focus group is designed to serve as a vehicle for presentations, symposia, and other mechanisms of interchange of relevant ideas.

American Society for Clinical Pharmacology and Therapeutics Pharmacokinetics and Drug Metabolism

Section:[59] This is a new listserv (phk@lists.ascpt.org) established in June 2000 for Pharmacokinetics and Drug Metabolism Section members of the ASCPT. The goal is to facilitate communication for the section members.

Pharmacokinetic and Pharmacodynamic Resources:[60] The Pharmacokinetic and Pharmacodynamic Resources web site (http://www.boomer.org/pkin/) provides links to information about pharmacokinetics and pharmacodynamics. All individuals interested in discussing pharmacokinetics and pharmacodynamics with colleagues around the world are invited to subscribe to the listserv.

CONCLUSION

Although a number of studies have demonstrated significant positive outcomes from clinical pharmacokinetics services, the reader needs to be aware that several studies have shown clinical pharmacokinetic monitoring not to have a significant effect on specific patient outcomes. A few studies even found a negative effect on patient outcomes.[2] This equivocal result may reflect that conventional patient outcome indicators (e.g., length of hospital stay) may not be appropriate to evaluate the value of clinical pharmacokinetic service. In addition, specific outcome indicators are likely to vary from drug to drug. Thus, we need to define those patients who are most likely to benefit from clinical pharmacokinetic monitoring and incorporate this into our provision of pharmaceutical care, while minimizing the time and money spent on clinical pharmacokinetic monitoring that has limited value (e.g., monitoring of digoxin concentrations for efficacy assessment in patients with heart failure or atrial fibrillation). Specifically, we should provide clinical pharmacokinetics services in a particular situation only when the results of the drug assay will make a significant difference in the clinical decision-making process and provide more information than sound clinical judgment alone. The reader is directed to a decision-making algorithm for clinical pharmacokinetic monitoring in the twenty-first century.[2]

Historically, clinical pharmacokinetics as a specialty practice has focused on interpretation of measured drug concentration, and modification of dosage regimen when appropriate.

The practice is also mostly concentrated on a limited number of drugs that meet certain criteria for concentration monitoring. Although such criteria are necessary to ensure appropriate use of resources, they may not work in the real world setting (e.g., monitoring of digoxin concentrations mentioned previously). In addition, such criteria for therapeutic drug monitoring limit the practice and application of clinical pharmacokinetics to a small number of drugs, as well as imply that the specialty is purely a "concentration exercise." The authors would argue that clinical pharmacokinetics practice should expand beyond the traditional realm of therapeutic drug monitoring, in terms of concept as well as scope of service. For example, the fluoroquinolones are not drugs that typically require concentration monitoring. However, with the recent knowledge of Cmax to MIC and AUC to MIC ratios being important determinants of response, the fluoroquinolones may provide a good potential for this expanded role. There are obvious technical problems regarding assay availability and dose-limiting central nervous system toxicity, as well as the need for more evidence to apply these pharmacokinetic determinants of response in the clinical setting. However, the potential for characterizing drug efficacy (and toxicities) using pharmacokinetic parameters should be explored to expand the role of the clinical pharmacokinetic specialty beyond that of therapeutic drug monitoring.

The explosion of pharmacogenetic and pharmacogenomic research has been fueled by the tremendous amount of genetic data generated by the Human Genome Project. In the future, instead of targeting a patient's drug concentrations within a therapeutic range as in traditional clinical pharmacokinetic monitoring, we predict that pharmacists will likely be making dosage recommendations of certain drugs based on an individual patient's genotype. Given the waning use of traditional aminoglycoside therapy that has been the primary focus of pharmacokinetic monitoring over the years, pharmacogenetics-oriented monitoring of other drugs may well become the therapeutic drug monitoring of the future.

REFERENCES

1. Ensom, M.H.H.; Burton, M.E.; Coleman, R.W.; Ambrose, P.J.; McAuley, D.C. ASHP statement on the pharmacist's role in clinical pharmacokinetic monitoring. Am. J. Health-Syst. Pharm. **1998**, *55*, 1726–1727.
2. Ensom, M.H.H.; Davis, G.A.; Cropp, C.D.; Ensom, R.J. Clinical pharmacokinetics in the 21st century: Does the evidence support definitive outcomes? Clin. Pharmacokinet. **1998**, *34* (4), 265–279.
3. Murphy, J.E.; Slack, M.K.; Campbell, S. National survey of hospital-based pharmacokinetic services. Am. J. Health-Syst. Pharm. **1996**, *53*, 2840–2847.
4. Howard, C.E.; Capers, C.C.; Bess, D.T.; Anderson, R.J.

Pharmacokinetics services in Department of Veterans Affairs medical centers. Am. J. Hosp. Pharm. **1994**, *51*, 1672–1675.
5. Santell, J.P. ASHP national survey of hospital-based pharmaceutical services—1994. Am. J. Health-Syst. Pharm. **1995**, *52*, 1179–1198.
6. Raehl, C.L.; Bond, C.A.; Pitterle, M.E. 1995 National Clinical Pharmacy Services study. Pharmacotherapy **1998**, *18* (2), 302–326.
7. Raehl, C.L.; Bond, C.A.; Pitterle, M.E. Clinical pharmacy services in hospitals educating pharmacy students. Pharmacotherapy **1998**, *18* (5), 1093–1102.
8. Ringold, D.J.; Santell, J.P.; Schneider, P.J.; Arenberg, S. ASHP national survey of pharmacy practice in acute care settings: Prescribing and transcribing—1998. Am. J. Health-Syst. Pharm. **1999**, *56*, 142–157.
9. Raehl, C.L.; Bond, C.A. 1998 national clinical pharmacy services study. Pharmacotherapy **2000**, *20* (4), 436–460.
10. American College of Clinical Pharmacy. *2000 Directory of Residencies and Fellowships*; American College of Clinical Pharmacy: Kansas City, Missouri, 2000; 175–179.
11. Millard Fillmore Health System. The Clinical Pharmacokinetics Laboratory. In *Clinical Residency in Pharmacokinetics/Internal Medicine/Infectious Disease*; Available from URL: http://pharmacy.buffalo.edu/cpl/pk-im-id.html (accessed July 2000).
12. Thomas Jefferson University. *Jefferson Hospital Pharmacy Residency Opportunities*; Available from URL: http://jeffline.tju.edu/CWIS/DEPT/Pharmacy/8residen.html (accessed July 2000).
13. University of Kentucky Hospital Chandler Medical Center. Clinical Pharmacokinetics Service Policy/Procedures. In *Department of Pharmacy Policy*; Univ. of KY Med Ctr.: Lexington, Kentucky, 2000; policy number: PH-02-05.
14. *Clinical Pharmacokinetics Service Policy and Procedure Manual*, 23rd Ed.; Davis, G.A., Ed.; Univ. of KY Med Ctr.: Lexington, Kentucky, 2000.
15. Shevchuk, Y.M.; Poulin, S. Quality assurance and certification program for an aminoglycoside monitoring service. Can. J. Hosp. Pharm. **1990**, *43* (2), 49–55.
16. Ament, P.W. Setting up an automatic pharmacist-initiated pharmacokinetic dosing service. Hosp. Formul. **1993**, *28* (6), 589–592.
17. Williams, L.E. Benefits of clinical pharmacy services in a community hospital. Hosp. Pharm. (Saskatoon, Sask.) **1993**, *28* (8), 759–763.
18. Mungall, D.; Marshall, J.; Penn, D.; Robinson, A.; Scott, J.; Williams, R.; Hurst, D. Individualized theophylline therapy: The impact of clinical pharmacokinetics on patient outcomes. Ther. Drug Monit. **1983**, *5* (1), 95–101.
19. Hurley, S.F.; Dziukas, L.J.; McNeill, J.J.; Brignell, M.J. A randomized controlled clinical trial of pharmacokinetic theophylline dosing. Am. Rev. Respir. Dis. **1986**, *134*, 1219–1224.
20. Crist, K.D.; Nahata, M.C.; Ety, J. Positive impact of a therapeutic drug-monitoring program on total aminoglycoside dose and cost of hospitalization. Ther. Drug Monit. **1987**, *9* (3), 306–310.
21. Destache, C.J.; Meyer, S.K.; Rowley, K.M. Does accepting pharmacokinetic recommendations impact hospitalization? A cost–benefit analysis. Ther. Drug Monit. **1990**, *12* (5), 427–433.
22. Destache, C.J.; Meyer, S.K.; Padomek, M.T.; Ortmeier, B.G. Impact of a clinical pharmacokinetic service on patients treated with aminoglycosides for gram-negative infections. Drug Intell. Clin. Pharm. **1989**, *23*, 33–38.
23. Sveska, K.J.; Roffe, B.D.; Solomon, D.K.; Hoffmann, R.P. Outcome of patients treated by an aminoglycoside pharmacokinetic dosing service. Am. J. Hosp. Pharm. **1985**, *42*, 2472–2478.
24. Burton, M.E.; Ash, C.L.; Hill, D.P.; Handy, T.; Shepherd, M.D.; Vasko, M.R. A controlled trial of the cost benefit of computerized Bayesian aminoglycoside administration. Clin. Pharmacol. Ther. **1991**, *49* (6), 685–694.
25. Ho, K.K.L.; Thiessen, J.J.; Bryson, S.M.; Greenberg, M.L.; Einarson, T.R.; Leson, C.L. Challenges in comparing treatment outcome from a prospective with that of a retrospective study: Assessing the merit of gentamicin therapeutic drug monitoring in pediatric oncology. Ther. Drug Monit. **1994**, *16* (3), 238–247.
26. Destache, C.J.; Meyer, S.K.; Bittner, M.J.; Hermann, K.G. Impact of a clinical pharmacokinetic service on patients treated with aminoglycosides: A cost–benefit analysis. Ther. Drug Monit. **1990**, *12* (5), 419–426.
27. Bootman, J.L.; Wertheimer, A.I.; Zaske, D.; Rowland, C.; et al. Individualizing gentamicin dosage regimens in burn patients with gram-negative septicemia: A cost–benefit analysis. J. Pharm. Sci. **1979**, *68* (3), 267–272.
28. Zaske, D.E.; Bootman, J.L.; Solem, L.B.; Strate, R.G. Increased burn patient survival with individualized dosages of gentamicin. Surgery **1982**, *91* (2), 142–149.
29. Whipple, J.K.; Ausman, R.K.; Franson, T.; Quebbeman, E.J. Effect of individualized pharmacokinetic dosing on patient outcome. Crit. Care Med. **1991**, *19* (12), 1480–1485.
30. Horn, J.R.; Christensen, D.B.; deBlaquire, P.A. Evaluation of a digoxin pharmacokinetic monitoring service in a community hospital. Drug Intell. Clin. Pharm. **1985**, *19*, 45–52.
31. Duhme, D.W.; Greenblatt, D.J.; Koch-Weser, J. Reduction of digoxin toxicity associated with measurement of serum levels. Ann. Intern. Med. **1974**, *80* (4), 516–519.
32. Wing, D.S.; Duff, H.J. The impact of a therapeutic drug monitoring program for phenytoin. Ther. Drug Monit. **1989**, *11* (1), 32–37.
33. Ioannides-Demos, L.L.; Horne, M.K.; Tong, N.; Wodak, J.; Harrison, P.M.; McNeil, J.J.; Gilligan, B.S.; McLean, A.J. Impact of a pharmacokinetics consultation service on clinical outcomes in an ambulatory-care epilepsy clinic. Am. J. Hosp. Pharm. **1988**, *45*, 1549–1551.
34. Welty, T.E.; Copa, A.K. Impact of vancomycin therapeutic drug monitoring on patient care. Ann. Pharmacother. **1994**, *28*, 1335–1339.

35. De Gatta, del M.F.; Calvo, V.; Hernandez, J.M.; Caballero, D.; San Miguel, J.F.; Dominguez-Gil, A. Cost-effectiveness analysis of serum vancomycin concentration monitoring in patients with hematologic malignancies. Clin. Pharmacol. Ther. **1996**, *60* (3), 332–340.
36. Schumock, G.T.; Meek, P.D.; Ploetz, P.A.; Vermeulen, L.C.; Publications Committee of the American College of Clinical Pharmacy Economic evaluations of clinical pharmacy services—1988–1995. Pharmacotherapy **1996**, *16* (6), 1188–1208.
37. Ambrose, P.J.; Nitake, M.; Kildoo, C.W. Impact of pharmacist scheduling of blood-sampling times for therapeutic drug monitoring. Am. J. Hosp. Pharm. **1988**, *45* (2), 380–382.
38. D'Angio, R.G.; Stevenson, J.G.; Lively, B.T.; Morgan, J.E. Therapeutic drug monitoring improved performance through educational intervention. Ther. Drug Monit. **1990**, *12* (2), 173–181.
39. Klamerus, K.J.; Munger, M.A. Effect of clinical pharmacy services on appropriateness of serum digoxin concentration monitoring. Am. J. Hosp. Pharm. **1988**, *45*, 1887–1893.
40. Kraus, D.M.; Calligaro, I.L.S.; Hatoum, H.T. Multilevel model to assess appropriateness of pediatric serum drug concentrations. Am. J. Dis. Child. **1991**, *145* (10), 1171–1175.
41. Jorgenson, J.A.; Rewers, R.F. Justification and evaluation of an aminoglycoside pharmacokinetic dosing service. Hosp. Pharm. (Saskatoon, Sask.) **1991**, *26*, 605,609–611,615.
42. Cook, C.R.; Bush, N.D. Cost-saving impact of pharmacist-calculated IV aminophylline dosing schedules. Hosp. Formul. **1989**, *24* (12), 716–721.
43. Wade, W.E.; McCall, C.Y. Educational effort and CQI program improves ordering of serum drug levels. Hosp. Formul. **1994**, *29* (9), 657–659.
44. Wing, D.S.; Duff, H.J. Evaluation of therapeutic drug monitoring program for theophylline in a teaching hospital. Drug Intell. Clin. Pharm. **1987**, *21*, 702–706.
45. Relling, M.V.; Fairclough, D.; Ayers, D.; Crom, W.R.; Rodman, J.H.; Pui, C.-H.; Evans, W.E. Patient characteristics associated with high-risk methotrexate concentrations and toxicity. J. Clin. Oncol. **1994**, *12* (8), 1667–1672.
46. Masson, E.; Zamboni, W.C. Pharmacokinetic optimisation of cancer chemotherapy: Effect on outcomes. Clin. Pharmacokinet. **1997**, *32* (4), 324–343.
47. Chin, J.M.W.; Muller, R.J.; Lucarelli, C.D. A pharmacy intervention program: Recognizing pharmacy's contribution to improving patient care. Hosp. Pharm. (Saskatoon, Sask.) **1995**, *30* (2), 120, 123–126, 129–130.
48. Van Lent-Evers, N.A.E.M.; Mathot, R.A.A.; Geus, W.P.; van Hout, B.A.; Vinks, A.A.T.M.M. Impact of goal-oriented and model-based clinical pharmacokinetic dosing of aminoglycosides on clinical outcome: A cost–effectiveness analysis. Ther. Drug Monit. **1999**, *21* (1), 63–73.
49. Sulick, J.A.; Pathak, D.S. The perceived influence of clinical pharmacy services on physician prescribing behavior: A matched-pair comparison of pharmacists and physicians. Pharmacotherapy **1996**, *16* (6), 1133–1141.
50. Bond, C.A.; Raehl, C.L.; Pitterle, M.E.; Franke, T. Health care professional staffing, hospital characteristics and hospital mortality rates. Pharmacotherapy **1999**, *19* (2), 130–138.
51. Bond, C.A.; Raehl, C.L.; Pitterle, M.E. 1992 National clinical pharmacy services study. Pharmacotherapy **1994**, *14* (3), 282–304.
52. Bond, C.A.; Raehl, C.L.; Franke, T. Clinical pharmacy services and hospital mortality rates. Pharmacotherapy **1999**, *19* (5), 556–564.
53. Bond, C.A.; Raehl, C.L.; Franke, T. Clinical pharmacy services, pharmacy staffing and the total cost of care in United States hospitals. Pharmacotherapy **2000**, *20* (6), 609–621.
54. American Society of Health-System Pharmacists. *ASHP Supplemental Standard and Learning Objectives for Residency Training in Clinical Pharmacokinetics Practice*; Available from URL: http://www.ashp.org/public/trp/PKSUPPST.html (accessed July 2000).
55. Beringer, P.M.; American Society of Health-System Pharmacists. Pharmacokinetics. Available from URL: http://www.ashp.org/public/news/newsletters/specialist/1999/summer99/NewsBytes/pharmaco.html (accessed July 2000).
56. American College of Clinical Pharmacy. *ACCP Practice and Research Networks-Pharmacokinetics/Dynamics*; Available from URL: http://www.accp.com/ClinNet/prnkenetics.html (accessed Aug. 2000).
57. American Association of Pharmaceutical Scientists. *Pharmacokinetics, Pharmacodynamics and Drug Metabolism Section (PPDM)*; Available from URL: http://www.aaps.org/sections/ppdm/ (accessed Aug. 2000).
58. American Association of Pharmaceutical Scientists. *Population Pharmacokinetics and Pharmacodynamics Focus Group*; Available from URL: http://www.aaps.org/focus/poppk.html (accessed Aug. 2000).
59. American Society for Clinical Pharmacology and Therapeutics Pharmacokinetics and Drug Metabolism Section, Available from URL: http://www.ascpt.org/members/lists.
60. Bourne, D.W.A. Pharmacokinetic and pharmacodynamic resources, Available from URL: http://www.boomer.org/pkin/ (accessed Aug. 2000).

PROFESSIONAL RESOURCES

Clinical Pharmacy Practice Guidelines (Society of Hospital Pharmacists of Australia)

Michael Dooley
Peter MacCallum Cancer Institute, Victoria, Australia

INTRODUCTION

Clinical pharmacy services are provided in nearly all hospitals in Australia as an integral component of the pharmacy service.[1] The Society of Hospital Pharmacists of Australia (SHPA) Standards of Practice for Clinical Pharmacy, published in 1996, is the key reference point for the provision of clinical pharmacy services to hospitalized patients in Australia.[2]

The fundamental components of the Standard are the statement of objectives of clinical pharmacy and the documentation of procedures for selected clinical activities. In addition, the issues of boundaries of clinical practice, training, and education are addressed. These guidelines have been utilized in policy development at local and federal government levels, in the accreditation of clinical pharmacy services, and as a key standard for undergraduate and postgraduate teaching.

HISTORY

In Australia, approximately 90% of all hospitals and 100% of major government-funded hospitals provide clinical pharmacy services to admitted patients.[1]

SHPA is the primary professional body that represents pharmacists practicing in Australian hospitals and similar institutions. Founded in 1941, the Society today has a membership of more than 1700 pharmacists practicing in all states and territories of Australia.

The mission of the Society is to promote and develop the practice of pharmacy in hospitals and related areas. The Society promulgates practice standards, position statements, and other documents designed to provide professional guidance to members. Within the Society, the Division of Specialty Practice overseas a number of expert committees relating to selected areas of practice. These expert committees are known as Committee of Specialty Practice (COSP). The COSP in clinical pharmacy has representatives from a range of clinical practice backgrounds and membership is by invitation. One of the responsibilities of the committee is to develop standards of practice for clinical pharmacy for the Society and review these at least every five years. The current Standards are an expansion of previously published SHPA guidelines on selected clinical pharmacy activities.[3–8] The standards published by the Society are not legally binding as in the event of conflict or overlap with applicable legislation, the requirements of the legislation prevail.

The SHPA Standards of Practice for Clinical Pharmacy have been developed for all patient care settings and define the minimum requirements for service provision. Standards of practice in specialty areas such as drug usage evaluation, oncology, psychiatry, and other areas of pharmacy practice relevant to clinical pharmacy have been ratified and should be read in conjunction.[9–13] In addition, the SHPA Code of Ethics offers direction in relation to professional conduct.[14]

MAJOR STATEMENTS/RECOMMENDATIONS

The fundamental components of the Standard are the statement of objectives of clinical pharmacy, the documentation of procedures for selected clinical activities, the guidance on staffing levels, and the defining of an intervention. The issues of boundaries of clinical practice, resources required, training, education, quality assurance, and documentation requirements are addressed.

The objectives of clinical pharmacy are stated and defined and three key points are worth attention. The emphasis of the definition was to stress that patient care must be the focus of clinical practice. Reference was made to "quality use of medicines" as this is one of the four central objectives of the National Medicines Policy in Australia.[15] Educational and research activities were included in the definition to demonstrate that these are core clinical pharmacy activities.

Encyclopedia of Clinical Pharmacy
DOI: 10.1081/E-ECP 120006367

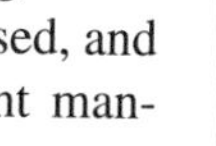

Leading on from this definition is a section that outlines the extent and operation of a clinical pharmacy service. Components that are discussed include the issue of the frequency of monitoring of patients' drug therapy. Clinical pharmacists need to be involved in activities that could improve patient clinical outcomes but are not necessarily focused on individual patients, for example, drug usage evaluations and the formulation of care plans. Research was addressed specifically by statements that involvement is an essential component of contemporary clinical pharmacy practice and should include a focus on optimizing drug therapy as well as research on the practice of clinical pharmacy. Education is also affirmed as a key core activity of clinical pharmacy practice. Specific details are not included in the document but rather motherhood statements that emphasize the need for involvement in undergraduate and postgraduate clinical teaching.

The third major section of the guidelines is the classification of clinical pharmacy services as a number of discrete activities. A goal for each activity is stated and then a recommended procedure outlines the major generic components of the activity. The activities were presented in an order that is somewhat reflective of the provision of these services. The Standard provides direction on which activities should be performed routinely. This is reflected in statements relating to medication history interview, the monitoring of drug therapy, and the provision of medication counseling.

The guidelines provide direction to the various resources recommended for the efficient provision of a clinical pharmacy service. There is limited detail as the primary aim was to briefly describe some selected components that should be considered rather than an attempt to quantify.

During the formulation and ratification of the guidelines, overwhelming input from members of the Society requested that the committee formulate a guide to staffing structure for the provision of a clinical service to particular clinical subspecialties. The document incorporates a guide to the ratio of pharmacists to patient bed numbers. These figures have not been substantiated by any objective measure or structured benchmarking study, but are based purely on consensus of the members of the Society.

Documentation of clinical service is featured and focuses on the very broad requirements of documenting actual activities performed as well as clinician documentation in the patient medical record. One of the key issues addressed is the defining of an intervention. The definition states that an intervention is "any action by a pharmacist that directly results in a change in patient management or therapy." By definition, the change must have occurred rather than have been merely proposed, and non-drug-related changes that impacted on patient management are included.

INFLUENCE AND RELEVANCE

These guidelines have been utilized in policy development at local and federal government levels, in the accreditation of provision of clinical pharmacy services, as a key standard for undergraduate and postgraduate teaching, and as a benchmark for practice.

The SHPA Standards of Practice for Clinical Pharmacy was one of the reference documents used in the formulation of the national guidelines to ensure continuity of medication management through hospital admission and treatment and postdischarge.[16] These guidelines were prepared by the Australian Pharmaceutical Advisory Council (APAC), which advises the Commonwealth Government of Australia on a wide range of pharmaceutical policy issues. The council includes representatives of major professional, industry, consumer, and media organizations as well as government members. The council published the guidelines in 1998, and they consist of broad principles on which standard procedures for individual institutions can be based, with the aim of ensuring continuity of medication management through hospital admission, treatment and postdischarge.

In another initiative by APAC, an integrated best-practice model for medication management in residential aged care facilities was developed.[17] One of the recommendations was that residents' medication be reviewed with cooperation between the prescriber and accredited pharmacists. Incorporated in the APAC model are guidelines for the performance of comprehensive medication review developed from the SHPA Standards of Practice for Clinical Pharmacy. Following on from this, the Commonwealth government agreed to reimburse accredited pharmacists to perform medication review services for nursing homes. Pharmacists can obtain accreditation through a number of mechanisms, including SHPA and also the Australian Association of Consultant Pharmacy (AACP). The AACP practice guidelines for the comprehensive medication review in residential care facilities utilize the Standard.

The SHPA Standards of Practice for Clinical Pharmacy are used as a reference point for benchmarking the provision of clinical pharmacy services in hospitals in Australia. An independent not-for-profit organization, the Australian Council of Healthcare Standards (ACHS), accredits healthcare organizations through a process evaluation and quality improvement program. The Standards have been used as a template for developing standards within the ACHS accreditation process for specifying the fundamental requirements for pharmacy services.

The structure of the definitions of the clinical activities has been incorporated within the national standards for classification in health in Australia. SHPA collaborated with the Australia National Centre for Classification in Health (NCCH) to develop pharmacy procedure codes within the International Statistical Classification of Diseases and Health Related Problems Australian Modification.[18] The classification has subsequently been piloted in 28 hospitals to develop a standard approach to the documentation of clinical services to individual patients.[19]

Other applications include the use of the intervention definition in projects incorporating evaluation of clinical pharmacists' activities. SHPA recently completed a prospective multicenter study of pharmacist-initiated changes to drug therapy and patient management in acute care government-funded hospitals in Australia. This was a prospective study performed in eight hospitals to examine resource implications of pharmacists' interventions and was assessed by an independent, multidisciplinary clinical panel. Results of the study are to be published soon.

The SHPA Standards of Practice for Clinical Pharmacy have been utilized extensively in a range of settings ranging from undergraduate and postgraduate training to experiential clinical teaching. Practical experience placements are incorporated into the various undergraduate curriculums, with students practicing individual clinical skills using the SHPA Standards of Practice for Clinical Pharmacy as the basis for these activities.

In addition to undergraduate teaching, the Standard has been utilized in developing training programs for pharmacy interns. In Australia, there is some variation in the requirement for obtaining Pharmacy Board registration. In most cases, this requires completion of the B Pharm and then a period of traineeship or internship followed by a Pharmacy Board examination. Guidelines have been developed for the hospital traineeships/internships that are based on the Standards. In 1999, a clinical residency program commenced in Victoria, the second-largest state in Australia. The program is structured around a more advanced level of clinical experiential teaching based on the Standard.

The current SHPA Standards of Practice for Clinical Pharmacy published in 1996 are under review. The revised document will be circulated extensively to the various members of SHPA for comment prior to adoption and publication. The new version will be available toward the end of 2002.

REFERENCES

1. McLennan, D.N.; Dooley, M.J. Documentation of clinical pharmacy activities. Aust. J. Hosp. Pharm. **2000**, *30*, 6–9.
2. The Society of Hospital Pharmacists of Australia Committee of Specialty Practice in Clinical Pharmacy. SHPA Standards of Practice for Clinical Pharmacy. In *Practice Standards and Definitions*; Johnstone, J.M., Vienet, M.D., Eds.; The Society of Hospital Pharmacists of Australia: Melbourne, 1996.
3. The Society of Hospital Pharmacists of Australia Committee of Specialty Practice in Clinical Pharmacy. SHPA policy guidelines for the practice of clinical pharmacy. Aust. J. Hosp. Pharm. **1984**, *14*, 7–8.
4. The Society of Hospital Pharmacists of Australia Committee of Specialty Practice in Clinical Pharmacy. SHPA guidelines for the practice of selected clinical pharmacy activities—Part 1. Aust. J. Hosp. Pharm. **1987**, *17*, 163–165.
5. The Society of Hospital Pharmacists of Australia Committee of Specialty Practice in Clinical Pharmacy. SHPA guidelines for the practice of selected clinical pharmacy activities—Part 2. Aust. J. Hosp. Pharm. **1987**, *17*, 261–264.
6. The Society of Hospital Pharmacists of Australia Committee of Specialty Practice in Clinical Pharmacy. SHPA guidelines for the practice of selected clinical pharmacy activities—Part 3. Aust. J. Hosp. Pharm. **1988**, *18*, 145–147.
7. The Society of Hospital Pharmacists of Australia Committee of Specialty Practice in Clinical Pharmacy. SHPA guidelines for the practice of selected clinical pharmacy activities—Part 4. Aust. J. Hosp. Pharm. **1990**, *20*, 192–194.
8. The Society of Hospital Pharmacists of Australia Committee of Specialty Practice in Clinical Pharmacy. SHPA guidelines for the practice of selected clinical pharmacy activities—Part 5. Aust. J. Hosp. Pharm. **1990**, *20*, 248–249.
9. The Society of Hospital Pharmacists of Australia Committee of Specialty Practice in Rehabilitation. SHPA Standards of Practice for the Community Liaison Pharmacists. In *Practice Standards and Definitions*; Johnstone, J.M., Vienet, M.D., Eds.; The Society of Hospital Pharmacists of Australia: Melbourne, 1996.
10. The Society of Hospital Pharmacists of Australia Committee of Specialty Practice in Rehabilitation. SHPA Standards of Practice for the Community Liaison Pharmacists. In *Practice Standards and Definitions*; Johnstone, J.M., Vienet, M.D., Eds.; The Society of Hospital Pharmacists of Australia: Melbourne, 1996.
11. The Society of Hospital Pharmacists of Australia Committee of Specialty Practice in Psychiatric Pharmacy. SHPA standards of practice for psychiatric pharmacy. Aust. J. Hosp. Pharm. **2000**, *30*, 292–299.

12. The Society of Hospital Pharmacists of Australia Committee of Specialty Practice in Oncology. SHPA Standards of Practice for the Oncology Pharmacist. In *Practice Standards and Definitions*; Johnstone, J.M., Vienet, M.D., Eds.; The Society of Hospital Pharmacists of Australia: Melbourne, 1996.
13. The Society of Hospital Pharmacists of Australia Committee of Specialty Practice in Drug Usage Evaluation. SHPA Standards of Practice for Drug Usage Evaluation in Australian Hospitals. In *Practice Standards and Definitions*; Johnstone, J.M., Vienet, M.D., Eds.; The Society of Hospital Pharmacists of Australia: Melbourne, 1996.
14. The Society of Hospital Pharmacists of Australia Code of Ethics. *Practice Standards and Definitions*; Johnstone, J.M., Vienet, M.D., Eds.; The Society of Hospital Pharmacists of Australia: Melbourne, 1996.
15. *National Medicines Policy 2000*, ISBN 0642415684; Commonwealth Department of Health and Aged Care Canberra: Australia, 2000; 1–7.
16. Australian Pharmaceutical Advisory Council. *National Guidelines to Achieve the Continuum of Quality Use of Medicines Between Hospital and Community*, ISBN 0642272646; Commonwealth Department of Health and Family Services Canberra, Australia 1998; 1–11.
17. Australian Pharmaceutical Advisory Council. *Integrated Best Practice Model for Medication Management in Residential Aged Care Facilities*, ISBN 0642415390; Commonwealth Department of Health and Aged Care. Canberra, Australia 2000; 1–17.
18. *ICD-10-AM Tabular List of Procedures (MBS-Extended). National Center for Classification in Health. Sydney, Australia*; Faculty of Health Sciences, University of Sydney: Australia, 1998.
19. Dooley, M.J.; Galbraith, K.; Burgess, N.; McLennan, D.N. Multicentre pilot study of a standard approach to document clinical pharmacy activity. Aust. J. Hosp. Pharm. **2000**, *30*, 150–156.

Clinical Pharmacy Scientist

Patricia Dowley Kroboth
Samuel M. Poloyac
Gary R. Matzke
University of Pittsburgh, Pittsburgh, Pennsylvania, U.S.A.

INTRODUCTION

A clinical pharmaceutical scientist is an independent investigator with education and training in pharmacotherapeutics who utilizes contemporary research approaches to generate new knowledge relevant to drug behavior in humans, to therapeutic interventions, and/or to patient outcomes.

The spectrum of research conducted by the community of clinical pharmaceutical scientists (CPSs) is broad (Fig. 1), with human research being common to all, as indicated by the shaded spheres of clinical research and outcomes research. Some clinical scientists may spend their entire careers focused on human interventional or observational trials. However, CPSs may also extend the programmatic scope of their work into adjacent research spheres, including preclinical research. Fig. 1 also demonstrates that, although individuals who spend their entire careers exclusively in the preclinical sphere are indeed pharmaceutical scientists, they do not fit the definition of a clinical pharmaceutical scientist. It is the nature and the scope of research that defines the clinical pharmaceutical scientist.

ORIGINS AND EVOLUTION OF THE CONCEPT

The term *clinical pharmaceutical scientist* was originally developed within the profession of pharmacy and was applied to pharmacy practitioners who became scientists. The concept of a CPS originated with the Millis Commission in 1975, which described the need for "people who are equally skilled and trained in a science and in pharmacy practice." This commission proposed the vision of "training skilled pharmacy practitioners in research to increase the number and variety of clinical pharmacists."[1] Since the original definition, the concept has evolved, and several subsequent definitions have been published (Table 1). These reflect not only the evolution of the CPS, but also how the profession of pharmacy has changed.

Since the first definition of the CPS, there have been substantial changes in the arenas of both clinical care and research. The way that pharmacists interact with patients has changed—the clinical context has matured from pharmacy practice to clinical pharmacy, and subsequently to pharmaceutical care. Managed care has emerged, and with it has come a greater focus on patient outcomes and on quality and cost of care. Pharmacokinetics and pharmacodynamics, which once represented cutting edge research, have become basic skills critical to the drug development process. An incredible array of technological advances has increased the spectrum of research possibilities. Finally, the mapping of the human genome has opened the vista of genetic research to better understand a patient's predisposition to both the beneficial and adverse effects of specific pharmaceutical interventions.

Together, these changes have expanded the spectrum of research opportunities for clinical pharmaceutical scientists and have fostered the continued evolution of the definition. The term *translational research* has evolved to define scientific endeavors that provide a critical link between research theory and human application. The result is that today, there is "variety" in the types of clinical pharmaceutical scientists to be found, which is in fulfillment of the Millis Commission imperative.

EVOLUTION OF TRAINING

Historically, some CPSs began their careers as clinical pharmacists who had earned a postbaccalaureate degree (PharmD or MS); in their pursuit of answers to therapeutic questions, they turned to scientific endeavors and conducted clinical research, often in the areas of pharmacokinetics and pharmacodynamics. These experiences sometimes resulted in individuals developing laboratory-

Encyclopedia of Clinical Pharmacy
DOI: 10.1081/E-ECP 120006354

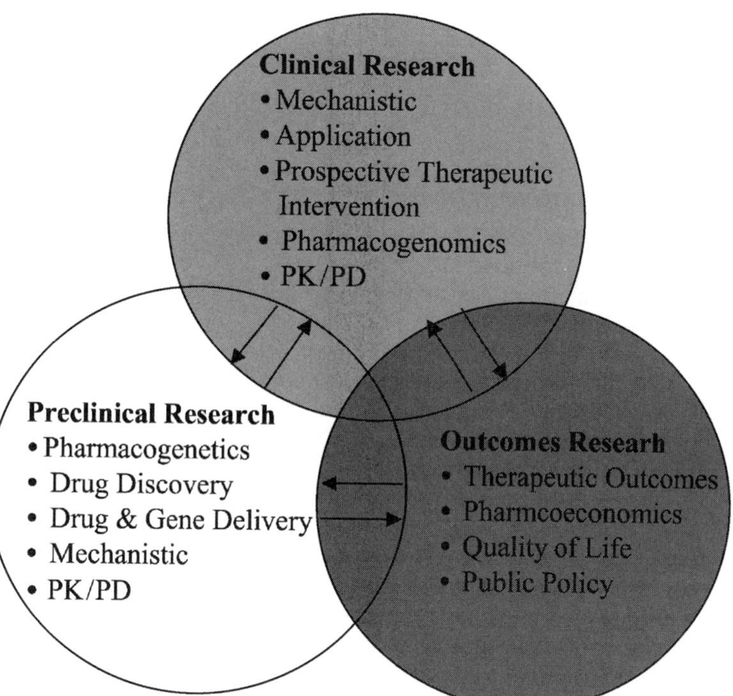

Fig. 1 This figure represents the continuum of research endeavors within the pharmaceutical sciences. The shaded areas represent the research areas encompassed by the definition of a CPS. The arrows indicate the potential movement from one sphere of research to another. The overlap with genetics, discovery, drug delivery, and animal mechanistic studies represent areas of translational research that provide the critical link between research theory and human application. Abbreviation: PK/PD, pharmacokinetics/pharmacodynamics.

based research as a means of gaining mechanistic or theoretical insight. Other clinical pharmaceutical scientists began their careers in the research laboratory and extended their research into normal volunteer and patient studies; these were often individuals with PhD degrees. Both career paths resulted in the generation of a CPS, and both were essentially ''bootstrap'' methods of becoming a clinical scientist.

The initial formal CPS training programs emerged in the 1980s. These programs required that the trainees conduct related patient-oriented and laboratory research. Historically, most of these programs were fellowships, although a few clinically oriented MS and PhD programs also emerged. Although most educational and training programs were in schools of pharmacy, some excellent fellowships are found in research institutes and health care organizations.

CURRENT EDUCATIONAL BACKGROUND AND TRAINING OPTIONS

Like the definition of the CPS, the background education and training options for becoming a CPS have also evolved. Common to most current CPS programs is the requirement for training in clinical pharmacotherapeutics. Stated in this way, clinically trained health care professionals other than pharmacists have the potential to become CPSs, given that they have an interest in developing research careers within the clinical pharmaceutical sciences. It is the nature and scope of research and not professional background that serves as the foundation for the definition of the CPS.

As stated in the definition, a CPS *is an independent investigator with education and training in pharmacotherapeutics who utilizes contemporary research approaches to generate new knowledge relevant to drug behavior in humans, to therapeutic interventions, and/or to patient outcomes.*

CPS training options that have been developed since the mid-1970s are summarized in Table 2. Each CPS training option has unique advantages and disadvantages, some of which have been debated previously.[3,8,9] Despite the differing opinions on the best method of training, the goal of these training programs is to develop individuals with the skills and confidence to conduct clinical research within their chosen career path.

Table 1 Published definitions of clinical pharmaceutical scientists

Year	Definition	Author
1975	People who are equally skilled and trained in a science and in pharmacy practice. The goal was to train skilled pharmacy practitioners in research to increase the number and variety of clinical scientists.	Millis Commission[1]
1984	Clinically experienced pharmacist capable of initiating and completing drug related research.	Smith et al.[2]
1987	1) An independent investigator conducting research relevant to the behavior of drugs in man. 2) Person has formal training as a clinical pharmacist.	Evans[3]
1987	A balanced clinical scientist who responds to clinical needs of patients and initiates research. Person has a balance of clinical and scientific skill.	Smith[4]
1987	Pharmacist with clinical skills, coupled with a sufficient background in pharmaceutics to perform clinical phamacokinetics and pharmacodynamics research.	Juhl and Kroboth[5]
1987	An individual with clinical service responsibilities who conducts clinical trials on human subjects and is the PI on grants.	Schwartz[6]
1991	A pharmacy-trained specialist who independently derives new knowledge through observation, study, and experimentation focused on drug therapy outcomes in patients, and the factors and mechanisms determining those outcomes.	Blouin et al.[7]
2001	An independent investigator with education and training in pharmacotherapeutics who uses contemporary research approaches to generate new knowledge relevant to drug behavior in humans, therapeutic interventions, and/or patient outcomes.	Kroboth et al.

Fellowship programs have provided advanced practice experience for clinicians for decades and have prepared them for board certification in areas of specialty in pharmacy and medicine.[10] Those individuals who desire a research component to their experience have either added an additional 1 to 2 years of training or sought mentors who provided a large research component to their core experiences. In the early 1980s and the 1990s, a small number of institutions developed postdoctoral graduate programs in the clinical pharmaceutical sciences.

Table 2 Clinical pharmaceutical scientist training options

Training option	Typical duration (yr)	Goal of training
Certificate in clinical pharmaceutical research	1	Develop individuals capable of conducting human interventional or observational research.
Master's in clinical pharmaceutical research	2–3	Develop individuals capable of designing and conducting human interventional or observational research.
Fellowship training	2–3	Develop independent scientists capable of designing and conducting clinical pharmaceutical research within academia, industry, or government.
PhD in clinical pharmaceutical sciences	4–5	Develop independent scientists capable of designing and conducting clinical pharmaceutical research within academia, industry, or government.

Some of these programs conferred an MS degree (e.g., University of Iowa, University of Minnesota), whereas others conferred a PhD (University of Pittsburgh, Virginia Commonwealth University, University of Kentucky). The core content of fellowship, MS, and PhD programs is similar, with the primary difference being the greater breadth and depth of experience associated with the PhD programs, which is paralleled by the greater duration of training.

Certificate programs in clinical pharmaceutical research have only emerged more recently. These are usually offered as a 1-year or less training option. Many of these programs provide graduate academic courses in which the individual can develop the skills to conduct clinically oriented research. Typically, these programs are very focused on human clinical interventional or observational studies. Independent research skills are limited due to the short training period.

The purpose of an educational program is to provide the optimal experience that will develop the skills for that individual to succeed within this chosen career. However, success is predicated not only on training, but also on the innate motivation and talents of the individuals, as well as their commitment to lifelong learning and perseverance in research after the formal training is complete. Each of the training options identified has resulted in successful CPSs. The decision as to the ''best'' training option is highly individualized and dependent on a multitude of factors, including the given student's career goals and aspirations. It is important for interested students to evaluate each training option extensively, realizing that the degree of development and the ability to conduct independent research increases directly with the duration and intensity of the various training options.

SKILL SETS

The world in which science is conducted has changed, and so have the skills necessary for scientists to function effectively. In a report published in 1995, essential skills and characteristics of scientists were addressed by The Committee on Science, Engineering, and Public Policy (COSEPUP), which was a joint committee of the National Academy of Science, National Academy of Engineering, and the Institute of Medicine.[11] The COSEPUP report states that ''a world of work that has become more interdisciplinary, collaborative, and global requires young people who are adaptable, flexible, as well as technically proficient.''[11] Subsequently, the 1996 report of the Research and Graduate Affairs Committee of the American Association of Colleges of Pharmacy emphasized the demand for ''scientists (who) possess excellent verbal and written communication skills, team-building aptitudes, critical thinking, problem-solving skills, leadership ability, and scientific integrity.''[9]

The primary goal of CPS training programs is to develop critical thinking, clinical acumen, and technical skills in a specialty research field, so as to develop the individual's ability to contribute to the knowledge that serves as the basis of clinical pharmaceutical science. It has been stated that ''for the most part, graduate education has produced technical proficiency and mastery of a specific discipline.''[9]

The common denominator for all training options, as well as for each research sphere (clinical, outcomes, and preclinical), is the skill set that one must acquire to successfully establish a clinical or translational research program. These skill sets can be grouped into six major categories. These categories include, but are not limited to, literature tracking and evaluation, critical scientific thinking and creativity, behavioral development, communication skills, technical proficiency, and research ethics and integrity. Mechanisms of attaining these skills vary with the different training options; however, the majority of these skills are attained via combination of graduate-level courses and mentored research experiences. Note that the role of advising and mentoring young scientists is viewed as sufficiently important that the National Academy of Sciences, National Academy of Engineering, and Institute of Medicine convened a committee that published a document regarding advising in 1997.[12]

Literature Tracking and Evaluation

The core of scientific research is based on the ability to review large volumes of published literature. To effectively pursue these research endeavors, the CPS must develop the ability to find, interpret, and critique the scientific literature. The trainee must become adept at using the existing body of information as the foundation for well-planned research. Once the research direction has been set, continual tracking of the literature is critical to maintaining currency and a successful research program.

Critical Scientific Thinking and Creativity

The development of critical and independent scientific thinking is paramount to the success of building a research program within each CPS research sphere. Using published literature as the foundation, scientists must be able to identify the next frontier for scientific exploration. They must also integrate the results of their own experiments with relevant existing literature in an iterative process of literature tracking, assimilation, and creativity to form the basis of new research ideas, develop new hypotheses, and design methods to test their validity. It is this creative process that requires critical scientific thinking that leads to new scientific discoveries. Implicit in hypothesis generation and study design is the ability to use accepted methods or develop new technology to quantify study endpoints.

Like most skills, this skill is best honed by practice. Through exposure to the ongoing work of several scientists and peers, the development of these skills and characteristics can be enhanced.

Behavioral Development

As discussed previously, the CPS is a clinically trained individual capable of conducting independent research. The combined clinical and research educational background ideally facilitates interactions of the CPS with both researcher and clinician colleagues, thereby developing a collaborative and interdisciplinary team approach to develop and test research hypotheses. For the CPS to develop both independent and collaborative research, it is critical to develop decision-making, problem-solving, team building, and leadership skills. These skills are not only important for collaborative development, but also for the development of leadership and managerial skills that will be necessary for the CPS to effectively supervise research assistants, graduate students, and postdoctoral students, and to lead a team of investigators. For CPSs or any other scientist to develop these traits, trainees should ideally be exposed to diverse research environments where scientists exhibit these traits.[9]

Communication Skills

CPSs must develop the ability to effectively communicate their ideas to colleagues, collaborators, and students, as well as granting agencies, regulatory agencies, and peer-reviewed publication. Frequent writing experiences enhance written skills. Venues including the preparation of Institutional Review Board and Institutional Animal Care and Use Committee proposals, grant applications, and manuscripts are essential components of the training experience. Peer review of manuscripts for journals and report preparation are also vehicles for enhancing written communications skills. Frequent written communication is essential. All written materials should be viewed as vehicles for enhancing communications skills, providing that a mentor gives feedback.

Frequent oral presentations to peer and mixed audiences develops verbal communication skills. Large- and small-group teaching and research presentation skills each require different talents that should be developed during the training period. Furthermore, the ability to verbally defend one's research ideas and results must be developed irrespective of the clinical pharmaceutical training path that has been chosen. Mastery of the ability to verbally defend one's research occurs only after the trainee has fully developed multiple other skill sets including critical thinking, hypothesis derivation, literature mastery, and technical proficiency.

Technical Proficiency

Concordant with hypothesis generation and study design development are the processes associated with method establishment and/or development, data generation, and data analysis, all collectively referred to as technical proficiency. Establishing a method for generating data using previously established methods allows for general technical skill development, as do data analysis and interpretation. These skills are generally obtained using an apprenticeship technique, which involves learning from someone already working in the laboratory. However, it is important for the individual to develop technical mastery of the methods used within the mentor's laboratory, as well as be able to identify, establish, and validate valuable methods described in the literature. This develops confidence to develop de novo techniques, which can move the entire discipline forward. Similarly, to complete the process, the appropriate statistical comparisons must be planned and applied to interpret the results of the tested hypotheses.

Research Ethics and Integrity

Ethics and integrity associated with the conduct of research is important in the development of every scientist. However, the direct human impact of the research adds a layer of complexity to the training of a CPS. Federal regulations mandate that study protocols be carefully evaluated for subject/patient safety by the investigator and the local Institutional Review Board. Regulations also mandate that all scientists funded by the National Institutes of Health (NIH) provide certification of training in the protection of human subjects. This training can be obtained either locally or through the NIH Office of Human Subjects Research, which maintains a web site for computer-based training on the

''Protection of Human Research Subjects.'' The URL for this site is http://ohsr.od.nih.gov/cbt/.

Concerns regarding confidentiality of patient information have heightened, particularly with the advent of human genetic research. There are also ethical issues that relate to accuracy of data collected and its security in a database, which relates to either technical proficiency in database management or the knowledge that a database manager needs to be part of the investigative team.

CAREER OPTIONS

The demand for researchers who focus on the clinical pharmaceutical sciences has exceeded the supply. Individuals graduating from CPS programs have secured successful careers within academic, industrial, and governmental institutions (Fig. 2). Academic institutions are major employers of CPSs within the clinical or basic science divisions in the various health-related schools. The CPSs' unique blend of clinical and research skills allow these individuals to develop interdisciplinary collaborations and thus are aggressively sought by research intensive institutions and those that are expanding their research focus.

Similarly, these individuals are capable of functioning in the preclinical or clinical areas of drug development within the pharmaceutical industry. In particular, the CPS has unique talents to contribute to the decision-making process and study design at the preclinical/clinical interface, which is a critical juncture of the drug development process. In addition to the traditional pharmaceutical industry, contract research organizations and site management organizations have become essential contributors to the drug, biological, and device development process. Many CPSs who have experience within the pharmaceutical industry have guided the growth and development of this industry. It is not surprising that the design, monitoring, analysis, and evaluation functions provided by these organizations draw heavily on the strengths of the CPS.

CPSs are prime candidates for positions in governmental and regulatory agencies. The CPSs' critical thinking and analysis skills, plus their practical experience in conducting clinical investigations, are highly desirable attributes for individuals entrusted with the evaluation of

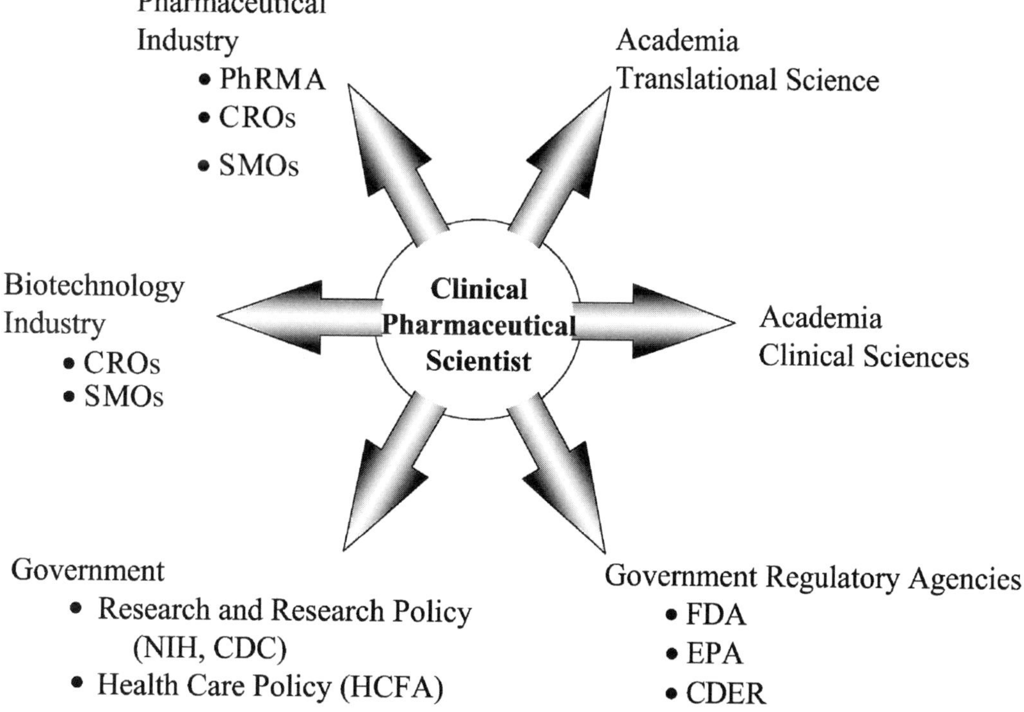

Fig. 2 This figure represents the career options for individuals interested in the clinical pharmaceutical sciences. Abbreviations: PhaRMA, Pharmaceutical Research Manufacturers; CROs, Contract Research Organizations; SMOs, Site Management Organizations; NIH, National Institutes of Health; CDC, Center for Disease Control and Prevention; HCFA, Health Care Financing Administration; FDA, Food and Drug Administration; EPA, Environmental Protection Agency; CDER, Center for Drug Evaluation Research; CBER, Center for Biological Evaluation Research.

the efficacy and safety of new drugs, biologicals, and devices in agencies such as the Center for Drug Evaluation Research, Center for Biological Evaluation Research, Environmental Protection Agency, and so on. Governmental health policy decisions are often based on critical review of scientific and economic evidence, along with a projection of their impact on the health care system. The CPS is similarly well suited to engage in this critical policy decision-making process at the national or local level.

REFERENCES

1. Pharmacists for the Future. In *The Report of the Study Commission on Pharmacy*; Health Administration Press: Ann Arbor, Michigan, 1975; 123–125.
2. Smith, R.V.; Cohen, J.L.; Kenyon, J.L.; Powell, J.R.; Rutledge, C.O.; Svarstad, B.L. Am. J. Pharm. Educ. **1984**, *48*, 439.
3. Evans, W.E. Training of Clinical Pharmaceutical Scientists. In *American College of Clinical Pharmacy Report*; Burckart, G.J., Ed.; ACCP: 1987; Vol. 6 (7), S22–S24.
4. Smith, R.V. Development of clinical scientists. Drug Intell. Clin. Pharm. **1987**, *21*, 101–103.
5. Juhl, R.P.; Kroboth, P.D. Conceptual issues in designing a clinical scientist program. Drug Intell. Clin. Pharm. **1987**, *21*, 103–106.
6. Schwartz, M.A. Are academic pharmacists meeting the clinical scientist role? Drug Intell. Clin. Pharm. **1987**, *21*, 114–117.
7. Blouin, R.A.; Cloyd, J.C.; Ludden, T.M.; Kroboth, P.D. Central issues relevant to clinical pharmaceutical scientist training programs. Pharmacotherapy **1991**, *11*, 257–263.
8. Juhl, R.P. Training of Clinical Pharmaceutical Scientists. In *American College of Clinical Pharmacy Report*; Burckart, G.J., Ed.; ACCP: 1987; Vol. 6 (7), S24.
9. Borchardt, R.T. Chair report of the Research and Graduate Affairs Committee. Am. J. Pharm. Educ. **1996**, *60*, 18S–22S.
10. *Directory of Residencies and Fellowships*; American College of Clinical Pharmacy: Kansas City, Missouri, 2001; 1–274.
11. *Reshaping the Graduate Education of Scientists and Engineers*; Committee on Science, Engineering and Public Policy; National Academy Press: Washington, DC, 1997.
12. *Adviser, Teacher, Role Model, Friend: On Being a Mentor to Students in Science and Engineering*; National Academy of Science, National Academy of Engineering, and Institute of Medicine; National Academy Press: Washington, DC, 1997; Available online at http://www.nap.edu/readingroom/books/mentor/.

Cochrane Library, The

Elaine Chiquette
San Antonio Cochrane Center, San Antonio, Texas, U.S.A.

INTRODUCTION

The Cochrane Library is a relatively new and growing electronic library that provides more than 850 summaries of published literature about pharmaceutical and other interventions to improve health. The Library adds new titles four times a year to its cumulative online and CD versions (the latter, available by subscription, offers more databases). The Library's 2000 Issue 3 contains evidence on dozens of clinical dilemmas, such as antibiotic treatment for traveler's diarrhea, antileukotriene agents compared to inhaled corticosteroids in the management of recurrent and/or chronic asthma, opioid antagonists for alcohol dependence, and bromocriptine versus levodopa in early Parkinson's disease. The Cochrane Library also updates earlier reviews when important new evidence becomes available. Among the newest updates are tacrine for Alzheimer's disease, tricyclic and related drugs for nocturnal enuresis in children, and nicotine replacement therapy for smoking cessation.

BRIEF HISTORY

The Library is the product of a grassroots network, the Cochrane Collaboration, which began in 1993. This international, nonprofit organization represents a worldwide network of more than 4000 healthcare professionals, researchers, and consumers working together toward a similar goal: to prepare, maintain, and disseminate systematic reviews of the effects of healthcare.[1]

One way to begin understanding the Collaboration is to read their ten guiding principles: 1) Maintain a collaborative spirit. 2) Build on the enthusiasm of individuals by encouraging people of different expertise, backgrounds, and culture to participate. 3) Avoid duplication by encouraging international collaboration and maximizing efforts. 4) Minimize bias by using rigorous scientific methods. 5) Keep up-to-date by assuring that Cochrane Reviews are updated as new evidence becomes available. 6) Strive for relevance by selecting outcomes that are clinically useful. 7) Promote worldwide access. 8) Ensure quality by developing systems for quality control and quality improvement. 9) Assure continuity of the Collaboration infrastructure. 10) Enable wide participation by reducing barriers to contributions.

The Cochrane Collaboration is named after Archie Cochrane, a British epidemiologist. He emphasized that the effectiveness of healthcare interventions should be based on evidence from randomized controlled trials. He argued that evidence-based healthcare could encourage the wise use of resources. Cochrane also recognized that people who want to make informed decisions about healthcare did not have access to reliable reviews of the available evidence when in 1979, he wrote: "It is surely a great criticism of our profession that we have not organized a critical summary, by specialty or subspecialty, adapted periodically, of all relevant randomized controlled trials."[2]

Even today, healthcare providers, consumers, researchers, and policymakers are overwhelmed by the vast amount of published research. Too often, the results of randomized controlled trials are ignored or lost in the information overload and helpful interventions are not identified promptly, while useless healthcare practices are continued. Review articles are needed to summarize all the relevant findings on a given topic or in a given field.[3] Systematic reviews, as suggested by Archie Cochrane, can help direct current therapeutic decisions and plan future research.

COLLABORATION STRUCTURE

Steering Group

A steering group of 14 elected members—clinicians, consumer advocates, researchers, and administrators representing all Cochrane entities—provide guidance to the Collaboration.

Cochrane Centers

The Cochrane Collaboration's two main entities are Cochrane Centers and Collaborative Review Groups (Fig. 1). Spanning the globe are 14 Cochrane Centers: Australasian,

Encyclopedia of Clinical Pharmacy
DOI: 10.1081/E-ECP 120006288

Brazilian, Canadian, Chinese, Dutch, French, German, Italian, Nordic, North American, South African, South American, Spanish, and United Kingdom. Each center has general responsibilities, such as helping to maintain a directory of contributors to the Collaboration, offering training in the process of producing a Cochrane review, and coordinating handsearches for healthcare journals. The centers are not responsible for preparing and/or maintaining systematic reviews. This is the role of the Collaborative Review Groups.

Cochrane Collaborative Review Groups

A review group is formed by researchers, healthcare professionals, consumers, and others who share a common interest in a particular health problem. As of November 2000, about 50 review groups cover the major areas of healthcare (Table 1). The groups' core function is to prepare systematic reviews that are evidence-based, internationally developed, quality controlled, and clinically useful. Creating a Cochrane review involves the systematic assembly, critical appraisal, and synthesis of all relevant studies that address a specific clinical question. Reviewers use strategies that limit bias and random error.[4] These strategies include a comprehensive search for potentially relevant articles and selection of relevant articles, using explicit, reproducible criteria. Reviewers critically appraise research designs and study characteristics during synthesis and interpretation of results. When appropriate, they integrate the results using meta-analysis. The unique value of Cochrane reviews is the commitment to regular updating. Reviews published in journals are often out-of-date by the time they are published. By updating reviews as new evidence becomes available, Cochrane Colla-

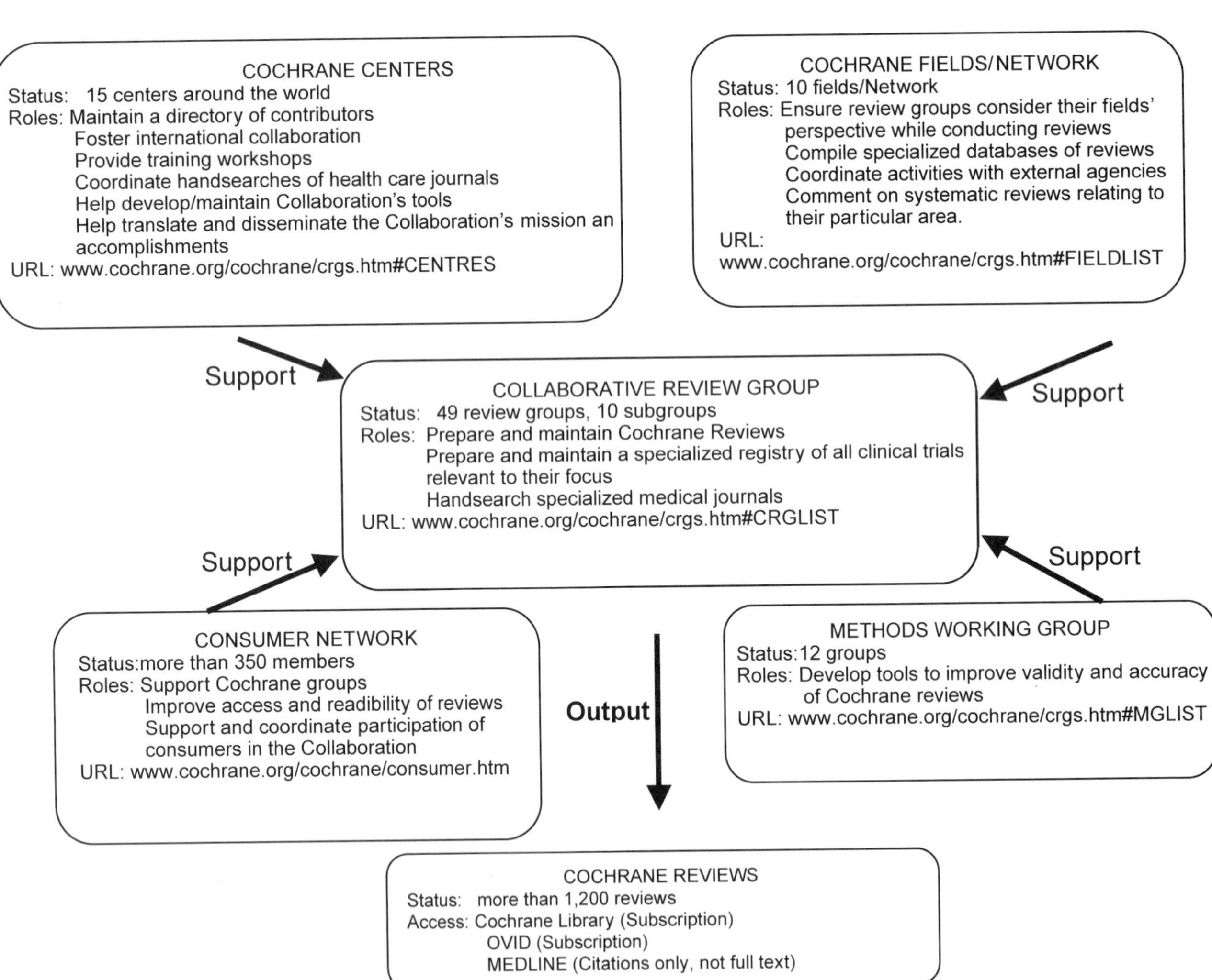

Fig. 1 The Cochrane collaboration structure.

Table 1 Cochrane collaborative group

Focus (No. of completed reviews)	URL address (assessed November, 2000)
Acute respiratory infection group (26)	http://nceph.anu.edu.au/user/rnd868/arigroup.html
Airways group (53)	http://www.cochrane-airways.ac.uk
Anesthesia group (2)	http://www.cochrane-anaesthesia.suite.dk
Back group (15)	http://www.iwh.on.ca/Pages/Cochrane/index.htm
Breast cancer group (4)	Not available
Colorectal cancer group (2)	http://www.cccg.dk
Consumers and communication group (2)	Not available
Cystic fibrosis and genetic disorders group (21)	http://web.bham.ac.uk/walterss/CFcochrane1.htm
Dementia and cognitive Impairment group (20)	http://www.jr2.ox.ac.uk/cdcig
Depression anxiety and neurosis group (10)	www.ccdan.auckland.ac.nz
Developmental, psychosocial and learning problems (5)	Not available
Drugs and alcohol group (5)	Not available
Ear, nose and throat disorders group (3)	http://www.entgroup.demon.co.uk/
Effective practice and organization of care group (17)	http://www.abdn.ac.uk/public_health/hsru/epoc/
Epilepsy group (7)	http://www.liv.ac.uk/epilepsy
Eyes and vision group (8)	http://www.archie.ucl.ac.uk
Fertility regulation group (6)	Not available
Gynaecological cancers group (16)	http://www.soton.ac.uk/~jps/gyn1.htm
Haematological malignancies group	Not available
Heart group (3)	http://www.epi.bris.ac.uk/cochrane/heart.htm
Hepato-biliary group (4)	http://inet.uni2.dk/~ctucph/chbg/index.htm
HIV/AIDS group (4)	http://hivinsite.ucsf.edu/cochrane/
Hypertension group (3)	Not available
Incontinence group (14)	http://www.otago.ac.nz/cure/
Infectious diseases group (38)	http://www.liv.ac.uk/lstm/nwc-id1.html
Inflammatory bowel disease group (7)	http://hiru.mcmaster.ca/cochrane/centres/canadian/IBD/IBD.htm
Injuries group (18)	http://www.cochrane-injuries.ich.ucl.ac.uk
Lung cancer group (2)	http://www.cochrane.es/English/LCG/
Menstrual disorders and subfertility group (44)	Not available
Metabolic and endocrine disorders group (5)	http://www.uni-duesseldorf.de/WWW/MedFak/MDN/Cochrane/ccset.htm
Movement disorders group (11)	Not available
Multiple sclerosis group	Not available
Musculoskeletal group (30)	http://www.arthritis.ca/cmsg/
Musculoskeletal injuries group (29)	Not available
Gout subgroup	
Lupus erythematous subgroup	
Osteoarthritis subgroup	
Osteoporosis subgroup	
Pediatric rheumatology subgroup	
Rheumatoid arthritis subgroup	
Scleroderma subgroup	
Soft tissue rheumatism subgroup	
Spondylarthropathy subgroup	
Vasculitis subgroup	
Neonatal group (91)	http://hiru.mcmaster.ca/cochrane/centres/canadian/neonatal/
Neuromuscular disease group (3)	Not available
Oral health group (3)	http://www.cochrane-oral.man.ac.uk
Pain, palliative care and supportive care group (8)	http://www.jr2.ox.ac.uk/Cochrane/
Peripheral vascular diseases group (14)	http://www.med.ed.ac.uk/pvd/
Pregnancy and childbirth group (161)	Not available

(Continued)

Table 1 Cochrane collaborative group (*Continued*)

Focus (No. of completed reviews)	URL address (assessed November, 2000)
Prostatic and urologic cancers group (7)	Not available
Renal group (4)	Not available
Schizophrenia group (52)	http://cebmh.warne.ox.ac.uk/csg/
Skin group (6)	http://www.nottingham.ac.uk/~muzd/index.htm
STD group	Not available
Stroke group (44)	http://www.dcn.ed.ac.uk/csrg/
Tobacco addiction group (20)	http://www.dphpc.ox.ac.uk/cochrane_tobacco/index.html
Upper gastrointestinal and pancreatic diseases group (7)	Not available
Wounds group (8)	http://www.york.ac.uk/depts/hstd/centres/evidence/ev-intro.htm#cochrane-wounds-group

borative Review Groups seek to provide the current best evidence for healthcare decision makers.

To create a comprehensive review on a given topic, Cochrane reviewers need access to all relevant randomized controlled trials. To assist the reviewers in this process, each Collaborative Review Group maintains a specialized registry of all (English and non-English; published and unpublished) randomized controlled trials pertinent to its particular focus. Trials are identified several ways: 1) electronic and manual searching of bibliographic databases, 2) contacting the pharmaceutical industry for unpublished trials, 3) handsearching hundreds of medical journals.

Three other Cochrane entities have much broader interests and focus on other dimensions other than specific healthcare problems. These are fields, methods groups, and networks.

Cochrane Fields

Fields serve to ensure that Collaborative Review Groups consider healthcare issues other than interventions, e.g., healthcare settings, types of consumers, and types of providers. For example, the field devoted to the healthcare of elderly people does the following: 1) assist in the handsearching activity of specialist journals, 2) ensure that Collaborative Review Groups address their issues and concerns, 3) compile a specialized database of reviews relevant to elder healthcare, and 4) establish internal and external partnerships.[5] A Cochrane Pharmaceuticals Field is being considered.

Cochrane Methods Groups

The Collaborative Review Groups are further assisted by Cochrane Methods Groups, which develop tools and assess new methodologies to improve the validity and accuracy of systematic reviews. For example, the informatics methods group played an important role in the development of the review manager software (REV-MAN).[6] The Cochrane review manager software assists Cochrane reviewers in conducting reviews (submitting review protocol, entering data for analysis, and writing results) in the structured format for publication in the Cochrane Library.

Consumer Network

Consumers, users of the healthcare system, participate throughout most entities of the collaboration. Consumers' input and feedback helps identify clinical questions that

Table 2 How to subscribe to the Cochrane Library

Update Software Ltd. Summertown Pavilion, Middle Way, Oxford OX2 7LG, U.K. Tel.: +44-1865-513902 Fax: +44-1865-516918 E-mail: info@update.co.uk URL address for information on the Cochrane Library: www.cochrane.org/cochrane/cdsr.htm	Update Software Inc. 936 La Rueda Vista, California 92084, U.S.A. Tel.: +1-760-727-6792 Fax: +1-760-734-4351 E-mail: info@updateusa.com

Table 3 Information on Cochrane Centers

Australasian Cochrane Centre
Monash Institute of Public Health and Health Services Research, Monash Medical Centre, Locked Bag 29,
Clayton Vic 3168, Australia
Tel.: +61-3-9594-7350
Fax: +61-3-9594-7554
E-mail: cochrane@med.monash.edu.au

Canadian Cochrane Centre
Health Information Research Unit,
McMaster University Medical Centre,
1200 Main Street West, Hamilton Ontario
L8N 3Z5, Canada
Tel.: +1-905-525-9140 ext. 22738
Fax: +1-905-546-0401
E-mail: cochrane@fhs.mcmaster.ca

Centro Cochrane do Brasil
Rua Pedro de Toledo 598, Vila Clementino
São Paulo, SP Brasil CEP
04039-001, Brazil
Tel.: +55-11-575-2970
Fax: +55-11-570-0469
E-mail: cochrane.dmed@epm.br

Centre Cochrane Français
Centre Léon Bérard, 28 Rue Laënnec,
69373 Lyon Cedex 08, France
Tel.: +33-478-78-28-34
Fax: +33-478-78-28-38
E-mail: ccf@upcl.univ-lyon1.fr

Centro Cochrane Iberoamericano
(Formerly called Centro Cochrane Español)
Hospital de la Santa Creu i Sant Pau,
Casa de Convalescència,
Sant Antoni M. Claret 171, 08041 Barcelona, Spain
Tel.: +34-93-291-95-27
Fax: +34-93-291-95-25
E-mail: Cochrane@cochrane.es

Centro Cochrane Italiano
Mario Negri Institute, Via Eritrea 62,
20157 Milano, Italy
Tel.: +39-02-39014327
Fax: +39-02-33200231
E-mail: cochrane@irfmn.mnegri.it

Deutsches Cochrane Zentrum
Abteilung für Medizinische Informatik,
Institut für Medizinische Biometrie und
Medizinische Informatik, Stefan Meier Str 26,
D-79104 Freiburg i. Br, Germany
Tel.: +49-761-203-6715
Fax: +49-761-203-6712
E-mail: mail@cochrane.de

Dutch Cochrane Centre
Academic Medical Centre, Meibergdreef 15, J2-221,
Postbus 22700, 1100 DE Amsterdam, The Netherlands
Tel.: +31-20-566-5602
Fax: +31-20-691-2683
E-mail: cochrane@amc.uva.nl

New England Cochrane Center
Division of Clinical Care Research, New England
Medical Center, 750 Washington Street, Box 63,
Boston, Massachusetts 02111, U.S.A.
Tel.: +1-617-636-5133
Fax: +1-617-636-8023
E-mail: cochrane@es.nemc.org
and
Providence Office, Brown University, Box G-S2,
Providence, Rhode Island 02912, U.S.A.
Tel.: +1-401-863-9950
Fax: +1-401-863-9944
E-mail: cochrane@brown.edu

Nordic Cochrane Centre
Rigshospitalet, Dept.7112, Blegdamsvej 9,
2100 Copenhagen O, Denmark
Tel.: +45-35-45-5571
Fax: +45-35-45-7007
E-mail: general@cochrane.dk

San Francisco Cochrane Center
University of California, San Francisco,
Suite 420, Box 0613, 3333 California Street,
San Francisco, California 94118, U.S.A.
Tel.: +1-415-502-8204
Fax: +1-415-502-0792
E-mail: sfcc@sirius.com

South African Cochrane Centre
Medical Research Council, Francie van Zijl Drive,
Parowvalley, PO Box 19070, Tygerberg,
7505 Cape Town, South Africa
Tel.: +27-21-938-0438
Fax: +27-21-938-0836
E-mail: cochrane@eagle.mrc.ac.za

(*Continued*)

Table 3 Information on Cochrane Centers (*Continued*)

Chinese Cochrane Centre The First University Hospital, West China University of Medical Sciences, Chengdu, Sichuan 610041, P.R. China Tel.: +86-28-5422078/5422079 Fax: +86-28-5582944 E-mail: cochrane@public.sc.cninfo.net	**U.K. Cochrane Centre** Summertown Pavilion, Middle Way, Oxford OX2 7LG, U.K. Tel.: +44-1865-516300 Fax: +44-1865-516311 E-mail: general@cochrane.co.uk

matter. Consumers also help disseminate and translate the results of Cochrane reviews to a broader audience.

THE COCHRANE LIBRARY

Cochrane reviews are the key component of the Cochrane Library, but not its only jewel. In addition to the Cochrane Database of Systematic Reviews (CDSR), there are three other databases: Cochrane Controlled Trials Register, Database of Abstracts of Reviews of Effectiveness, and Cochrane Review Methodology Register.

Cochrane Database of Systematic Reviews

Once completed, Cochrane reviews undergo internal and external peer review before electronic publication. More than 850 Cochrane reviews are currently available (2000, Issue 3) with more than 750 reviews in progress. As new reviews are added with each issue of the Cochrane Library, eventually all areas of healthcare will be covered.

The Cochrane Controlled Trials Register

The Cochrane Controlled Trials Register contains bibliographic details of more than 270,000 controlled trials identified by contributors to the Cochrane Collaboration. The register aims to be the most comprehensive source of trials to assist the reviewers in conducting systematic reviews. To accomplish this goal, contributors around the world systematically handsearch healthcare journals to identify randomized controlled trials (published and unpublished). The handsearching efforts are done in collaboration with the National Library of Medicine (MEDLINE) and Reed Elsevier (EMBASE). More than 50,000 trials have been identified as randomized controlled trials and sent for proper tagging in MEDLINE and EMBASE. Each Collaborative Review Group's specialized registry is included and respectively tagged in the Cochrane Controlled Trials Register.

The Database of Abstracts of Reviews of Effectiveness (DARE)

This database, produced by the National Health Services Center for Review and Dissemination at the University of York, contains more than 2500 structured abstracts of good-quality published reviews about the effectiveness of health interventions. The database also can be accessed free on the Internet at www.york.ac.uk/inst/crd.

The Cochrane Review Methodology Register

This bibliographic database of more than 1300 references addresses methodological aspects relevant to conducting systematic reviews. It assists novice reviewers in finding good-quality articles summarizing important methodological challenges encountered in conducting systematic reviews.

The Cochrane Library represents one of the most comprehensive sources of evidence about healthcare. The Library can be purchased via subscription by contacting Update Software (Table 2). OVID Technologies includes the Cochrane Database of Systematic Reviews as one of their subscription databases. The National Library of Medicine's MEDLINE also indexes Cochrane Reviews. This allows MEDLINE users to identify Cochrane Reviews relevant to their search strategies.[7]

FUNDING SOURCES

Most Collaboration members contribute time and effort without monetary compensation. More than 4000 volunteers help prepare, maintain, and disseminate Cochrane Reviews. Over the past seven years, a long list of government agencies, foundations, universities, and others have provided financial support, though the level of support varies considerably. The Collaboration is exploring ways to secure continuing financial support for its infrastructure to achieve sustainability.

PARTICIPATION

Expertise or interest in a particular healthcare field is the qualification requirement for a Cochrane reviewer or an editor for a Collaborative Review Group. Expertise in statistics or trial methodology qualifies membership in a Methods Working Group. Most review groups and Centers welcome volunteer handsearchers and will provide the necessary training and support. The Italian Cochrane Center coordinates the efforts of people willing to translate reports of non-English medical trials for those preparing Cochrane reviews. Membership in a Cochrane entity is not based on formal qualifications. There are no membership fees. The key requirements are a willingness to volunteer and a sharing of the Collaboration's goals and collaborative spirit. Additional information on participation can be obtained by contacting the nearest Cochrane Center (Table 3).

NOTE ADDED IN PROOF

This article was written and submitted in early 2001. The author expects to submit a fully updated version of this article, available online, in the first quarter of 2003.

REFERENCES

1. The Cochrane Collaboration. www.cochrane.org (assessed November 2000).
2. Cochrane AL. *Effectiveness and Efficiency*: Random Reflections on Health Services, London: Nuffield Provincial Hospitals Trust, 1972 (reprinted in 1989 in association with the BMJ).
3. Mulrow, CD. Rationales for systematic reviews. BMJ **1994**, *309*, 597–599.
4. The Cochrane Collaboration. *The Cochrane Collaboration Handbook*, Oxford, England: The Cochrane Collaboration, 1999.
5. Dickinson, E.; Rochon, P. Cochrane Collaboration in health care of elderly people. Age and Ageing. **1995**, *24*, 265–266.
6. The Cochrane Collaboration. Review Manager Software.
7. Clarke, M.; Oxman, A. Cochrane reviews will be in Medline. BMJ **1999**, *319*, 1435.

PHARMACY PRACTICE ISSUES

Collaborative Drug Therapy Management by Pharmacists (ACCP)

American College of Clinical Pharmacy
Kansas City, Missouri, U.S.A.

INTRODUCTION

The traditional system of providing drug therapy to patients, in which only certain health care professionals are authorized to initiate drug therapy, is under attack at many levels. The processes of drug prescribing, dispensing, administration, monitoring, and dosage adjustment, as practiced in this traditional system, occur in a disjointed fashion that frequently results in avoidable drug-related problems that contribute significantly to poor patient outcomes and increased medical costs.[1]

Collaborative drug therapy management, characterized by an interdisciplinary approach to patient care, is emerging as a solution that can maximize the patient's health-related quality of life, reduce the frequency of avoidable drug-related problems, and improve societal benefits from pharmaceuticals. In this approach to care, drug therapy decision making and management are coordinated collaboratively by pharmacists, physicians, other health care professionals, and the patient.

Many pharmacists with sufficient clinical training have or are willing to assume this level of responsibility for the patients they serve. When participating in collaborative drug therapy management, pharmacists share the responsibility for patient outcomes, not just by providing basic dispensing functions and drug information services, but by solving patient- and medication-related problems and by making decisions regarding drug prescribing, monitoring, and drug regimen adjustments.

This statement represents the position of the American College of Clinical Pharmacy (ACCP) on the role of pharmacists in collaborative drug therapy management. Furthermore, a model for collaborative management of drug therapy is described and endorsed as a way to enhance the quality of patient care within health care systems.

ACCP POSITION STATEMENT

The ACCP advocates the role of qualified pharmacists as capable collaborative drug therapy managers. Furthermore, ACCP supports the pharmacists' role in collaborative drug therapy management to improve patient outcomes and increase efficiencies in the health care system. To participate in collaborative drug therapy management, pharmacists must have access to patients and patient health information, conduct patient assessments, document activities, and undergo quality assurance programs on these activities. Scope of practice statements, identifying pharmacists' professional authority and responsibility, will be based on the pharmacist's credentials and the nature of the collaborative arrangement within the health care environment or system.

HISTORY OF PHARMACIST PRESCRIBING IN THE UNITED STATES

Regulation of pharmacist prescribing in the modern health care system of the United States can be traced to passage of the Federal Food, Drug, and Cosmetic (FDC) Act of 1938. This act was introduced to address concerns surrounding the availability of a growing therapeutic armamentarium of antimicrobial agents, led by introduction of the sulfonamides in 1935. Following a disaster in which 107 people died from consuming a toxic base used to compound a sulfanilamide elixir, Congress passed the FDC Act of 1938. The Food and Drug Administration (FDA) then issued regulations to enforce this legislation. The 1938 act deemed as misbranded any drug that failed to carry adequate directions for use or failed to warn patients about potential lack of safety. Any drug could be exempt from the requirement of adequate directions for

Encyclopedia of Clinical Pharmacy
DOI: 10.1081/E-ECP 120006410

use if, because of its potential for toxicity or misuse, it was to be used under the supervision of a physician. Regulations mandated these exempted agents carry the wording, "Caution: to be used only by or on the prescription of a physician, dentist, or veterinarian." Another provision was the working, "Warning—may be habit forming," required on certain narcotic and hypnotic drugs. These regulations became the forerunner to our present-day system for designating prescription drugs and controlled substances. Until this time, pharmacists had been able to prescribe medications legally.

The activity of pharmacists refilling, and thereby continuing, a patient's medications without authorization from the patient's physician was a secondary issue in the 1938 FDC debates. Although not defined as unlawful in 1938, the practice of pharmacists providing refills of medications directly to patients was not favored by the FDA. No definition had differentiated a prescription drug from a nonlegend, over-the-counter, drug. The two classes of drugs were not legally differentiated until passage of the Durham–Humphrey Amendment in 1951. At that time, it became illegal for pharmacists to refill legend drugs without authorization from the patient's physician.[2,3] Thus, the practice of physician prescribing and pharmacist dispensing became law. Many regulations endorsed by today's state boards of pharmacy are resultant attempts to define these distinctions clearly.

During this same period, the preparation of medications was increasingly assumed by pharmaceutical manufacturing companies, thereby lessening the role of individual pharmacists in production manufacturing. Thus, pharmacists were no longer taking an active role in initiating or continuing prescription drug therapy, and were also spending less time in the final preparation of the pharmaceutical product.

In the 1960s and 1970s, pharmacists began to assume roles as direct patient care providers in rural settings within the Indian Health Service. The activity of pharmacist prescribing was first documented in this setting. As early as 1977, Brands described pharmacist practitioners in the Indian Health Service who were trained to diagnose and treat acute, self-limiting diseases and chronic diseases in ambulatory patients.[4] A 1-year review of patients cared for by this arrangement found that 70% of the patients in this group were cared for solely by pharmacists. Quality of care was satisfactory and patient acceptance was excellent. In a similar fashion, Erickson described a program in the same Indian Health Service setting that demonstrated pharmacists were able to provide patient monitoring between physician visits and were also able to extend the interval between physician visits.[5]

In 1972, individual states began exploring the issue of pharmacist prescribing, heralded by the Health Manpower Experimental Act of 1972, a unique experiment in California. Health Manpower Pilot Projects were created with the purpose of training students of the allied health professions in areas that were then beyond their legal scope of practice. To include prescribing by pharmacists, nurses, and physician assistants in these pilot projects, the California Assembly Bill 717 was introduced in 1977, with a provision for sunsetting in 1983. The bill authorized prescriptive authority only to those directly involved with the pilot projects. The project was so successful in saving health care dollars[6] that the California Pharmacists Association, with assistance from the California Society of Hospital Pharmacists, introduced legislation in 1981 to enable prescribing by all pharmacists in the state. This legislation allowed registered pharmacists functioning in licensed acute and intermediate health care facilities to adjust the dosage of a patient's drug regimen pursuant to a prescriber's authorization, order laboratory tests, perform physical assessments, and administer medications. This law has been expanded twice since then and now enables pharmacists to initiate drug therapy (1983) and expands the types of practice sites to include clinics and systems licensed as health care service plans (e.g., managed care organizations; 1994). The specific duties outlined by each protocol are site- and practice-specific. Traditionally, they have ranged from pharmacist-managed nutritional support prescribing in the inpatient setting to antihypertensive medication management in the outpatient setting.[7–10]

Eventually, pharmacists have gained recognition as drug therapy experts at the national level. In 1974, the Department of Health, Education, and Welfare enacted a drug regimen review regulation for nursing homes in an attempt to improve the quality of drug prescribing in that health care setting. In 1984, Thompson and associates published the results of a study of clinical pharmacists who prescribed under physician protocol in a skilled nursing facility.[11] The findings of this controlled study indicated that patients in the prescribing clinical pharmacists' group had significantly fewer deaths, more patients discharged to lower levels of care, and fewer drugs per patient than the patients in the traditional care group. The estimated health care savings due to clinical pharmacists prescribing in a skilled nursing facility were $70,000 annually (in 1984 dollars) for every 100 beds.

Legislation enabling pharmacists to prescribe under protocol was first passed in the state of Washington in 1979. Since then, it has been amended several times to clarify or expand the types and numbers of protocols. Currently, the Washington State Board of Pharmacy has

Table 1 Attributes of state and federal regulations governing pharmacist prescribing

State	Arkansas	California	Florida	Indiana	Kentucky	Michigan	Mississippi
Year	1997	1981	1986	1996	1996	1991, under state public health code	1987
Types of Collaborative Practice Agreements	Protocol for each specific patient	Policies, procedures, protocols	Formulary only; legislation to establish protocols introduced in 1997	Policies, procedures, protocols	Collaborative care agreements	Responsibility delegated by MD or DO	Guidelines, protocols
Level of Review or Approval Required	Physician	Facility	None	Hospital and admitting practitioner	Yet to be determined by Board of Pharmacy	None	Board of Pharmacy
Medications Included	All	All	Specified formulary only; no narcotics or injectables	All, except narcotics	All; narcotics not specified	All, except C-II drugs and anabolic steroids	All
Environments	All settings	Licensed health care facilities, licensed clinics, providers who contract with licensed health care service plans	Pharmacies	Acute care settings, private mental health institutions	All settings	All settings	Institutional settings; in outpatient settings, specific signed protocols required for each patient
Educational Requirements/ Demonstrated Competencies	Those completing diabetes mellitus training eligible for reimbursement from insurance companies	Clinical residency or clinical experience as specified by the facility	No additional	No additional	No additional	None specified	Study course (of at least 20 CEUs) approved by Board of Pharmacy
Other Aspects Addressed	Completion of course approved by Board of Pharmacy enables pharmacist to administer certain medications, including immunizations and vaccinations, to patients age 18 yr or older	Administering injections; patient assessment; laboratory tests; initiating and adjusting drug regimens	No pregnant or nursing women; only drug supplies for less than 34 d; no refills	Changing duration of therapy, drug strengths, dosage forms, frequencies or routes of administration; stopping and adding drugs	Physical assessment; ordering clinical tests; initiating, continuing, or stopping drug therapy; drug modification and monitoring; therapeutic interchange	Pharmacist must record the name of the delegating MD or DO on the prescription	Initiating and modifying drug therapy

Table 1 Attributes of state and federal regulations governing pharmacist prescribing (*Continued*)

Nevada	New Mexico	North Dakota	Oregon	South Dakota	Texas	Washington	Federal government
1990	1993	1995	1980	1993	1995	1979	1995
Protocols	Protocols	Collaborative agreement with licensed physician	Protocols or on a case-by-case basis	Protocols	Written protocols with specific physicians	Protocols	Protocols within scope of practice
Available for inspection by Board of Pharmacy	Board of Pharmacy approves practitioner license	Board of Pharmacy and Board of Medical Examiners	None	Practitioner or the legal authority of the licensed health facility	Must be available for inspection by Board of Pharmacy	Board of Pharmacy	Appropriate facility-based authorizing body or chief of staff
All, except narcotics	All	All, except narcotics	All	All, except narcotics	All	All	All, except narcotics
Licensed medical facilities (hospitals, hospices, managed care settings, home health care, skilled nursing facilities)	All settings	Institutional settings (hospitals, skilled nursing facilities, swing bed facilities)	All settings	All settings	All settings	All settings	All settings
No additional	Additional training equivalent to that of a physician assistant (60 hr of physical assessment; 9 mo of clinical experience or MD preceptorship)	No additional	No additional	No additional	Specific clinical continuing education	No additional	MS degree, PharmD degree, accredited residency, specialty board certification, or 2 yr of clinical experience
Initiating, modifying, and monitoring drug therapy	Monitoring drug therapy; ordering laboratory tests; patient assessment; prescribing and modifying drug therapy	Pharmacist must notify physician when they initiate or modify drug therapy	Further rulings expected in 1997	Administering, initiating, and modifying drug therapy; research investigators	Written protocol defined as a physician's order, standing order, standing delegation order, or other protocol	Initiating and modifying drug therapy; protocols must be renewed every 2 yr	No protocol or cosignature required within scope of practice; policies required to assure practice is within identified scope of practice

over 70 protocols on file, conducted by over 425 pharmacists practicing in 60 locations throughout the state. Although the protocols were initially used in institutions, most are now used in managed care and community settings. In clinic settings, these protocols have been found to create efficiencies in prescribing antimicrobial and anticoagulation regimens.[12,13] In the community pharmacy setting, protocols are used for prescribing refills and for monitoring drug therapy of chronic disease states.

The third state to provide prescriptive authority to pharmacists was Florida. Taking a different approach, the Florida legislature created a third class of drugs in 1986. In contrast to the California and Washington provisions for prescribing under protocol, Florida pharmacists enjoy independent prescribing from within a limited formulary. Certain drugs within the following categories are included in this formulary: oral, urinary, and otic analgesics; hemorrhoid medications; antinausea preparations; antihistamines and decongestants; anthelmintics; topical antifungals and antimicrobials; topical antiinflammatory preparations; otic antifungals and antimicrobials; keratolytics; vitamins with fluoride; lindane shampoos; antidiarrheals; smoking cessation products; and ophthalmics. The formulary is subject to specific conditions spelled out in the state's pharmacy practice act. The legislation has been amended frequently.[14]

In 1995, the Veterans Health Administration (VHA) updated the granting of prescribing authority for practitioners in the Veterans Affairs (VA) system. "General guidelines for establishing medication prescribing authority for clinical nurse specialists, nurse practitioners, clinical pharmacy specialists, and physician assistants," VHA Directive 10-95-019, reviews and clarifies the prescribing role of these practitioners within the VA health care system. Clinical pharmacy specialists are defined as those with Master of Science or Doctor of Pharmacy degrees, pharmacists who have completed an accredited residency, specialty board-certified pharmacists, or pharmacists with equivalent experience. The scope of practice for each type of practitioner is determined by the practice site. The scope of practice statement identifies each individual's prescriptive authority and describes routine and nonroutine professional duties and general areas of responsibility. Prescriptions written by authorized practitioners within their approved scope of practice do not require a physician cosignature. Because states cannot regulate the activities of the federal government or its employees when acting within the scope of their employment, state laws and regulations related to medication orders and prescriptions do not affect scope of practice statements in the VA system.

With early models in place and numerous studies documenting success, momentum has mounted to support the pharmacist's role in collaborative drug therapy management. States are continuing to enact or pursue legislation to enable pharmacists to prescribe as part of collaborative drug therapy management agreements. Currently, 14 states and the federal government have enacted legislation allowing some form of collaborative prescribing for pharmacists. Table 1 provides some specific attributes of these laws.

IMPACT OF PHARMACISTS PERFORMING COLLABORATIVE DRUG THERAPY MANAGEMENT

Since the late 1970s, many studies have been published that document the success of pharmacists' management of specific types of patients, drugs, disease states, and specific patient problems and issues. Outcomes measured have included increased patient safety and satisfaction, reduced health care costs, and improved efficiencies.[15–22]

Recently, a summary and critique of 104 studies that assessed the economic outcomes of clinical pharmacy services from 1988–1995 was published.[23] The clinical pharmacy services evaluated could be classified into four main categories—disease state management (4%), general pharmacotherapeutic monitoring (36%), pharmacokinetic monitoring (13%), and targeted drug programs (47%). The services were provided in a variety of health care settings, including university, community, and government hospitals; health maintenance organizations; and community pharmacies.

Outcomes, or consequences, of the services described were considered in all 104 papers. Nineteen (18%) of the papers were found to be full economic analyses because they considered two or more alternatives to care and measured both input costs and outcomes. The most common outcomes measured were drug costs avoided, length of hospital stay, use of nonpharmaceutical resources, rates of adverse drug reactions, frequency of pharmacist-driven therapeutic interventions, and qualitative changes in prescribing patterns. In 93 (89%) of the papers, beneficial financial impacts of clinical pharmacy services were described.

In seven papers, the study design was sufficiently rigorous to allow the results to be expressed as a benefit-to-cost ratio. The calculated benefit-to-cost ratios for these seven studies ranged from 1.08:1 to 75.84:1 (mean 16.7:1). In other words, for every dollar invested in clinical pharmacy services, on average, $16.70 of benefit was realized. Overall, the body of literature contains a wealth of information pertinent to the value of the clinical practice of pharmacy.

EVOLVING VIEW OF HEALTH CARE

In November 1995, the Pew Health Professions Commission released its third report describing the future of the health professions in the United States.[24] The changes foreseen by the Pew Commission come from the backdrop of failed government-driven health care reform and the emergence of market-driven health care reform. Table 2 illustrates the shifting paradigm in health care as outlined by the Pew Commission.

The driving force behind health care reform in the United States is the trillion-dollar health care market and the rate of growth of this market. The rate of growth of health care resource utilization competes for other needed programs in both the private and public sectors. These expenditures are brought to the forefront by the fact that, compared with all other industrialized countries, the United States spends more of its gross national product on health care (nearly $3000/person versus $2000/person or less in all other countries), yet realizes no proportional improvement in quality of life.[25] In a market-driven health care economy, three principal values exist: 1) holding or lowering costs; 2) increasing patient satisfaction; and 3) improving the quality of patient outcomes.

The shift to create this new system will be accomplished by more integration and collaboration, as opposed to fragmentation. The steps in this change are occurring at an increasingly rapid pace. This is evidenced by the current movement of health care into a managed care environment. What these changes mean for health care systems and for pharmacists, in particular, are not absolutely clear, but the implications are that the next generation of health professionals will be practicing in an environment that is more intensively managed. In addition, exploration into changing the roles of health professionals to provide a more diverse skill mix within the health care team and more efficient delivery of integrated health care appears to be essential.

Table 2 The shifting paradigm in health care

1945–Present	Future
Specialization	Primary care
Cost unaware	Cost accountable
Technology driven	Humanely balanced
Institution based	Community focused
Professionally driven	Managerially driven
Individual care	Population health
Acute	Chronic
Treatment	Management and prevention
Individual providers	Team providers
Competitive	Collaborative

The Pew Commission has suggested that to meet these challenges, health professionals will have to redesign the way their work is organized, reregulate the ways in which they are permitted to practice, right-size the health professional workforce, and restructure health professional education.

This reregulation of health professions has direct bearing on the need for collaborative drug therapy management and prescriptive authority for pharmacists. As discussed earlier, our present prescriptive authority regulations evolved to protect consumers from misbranded and dangerous medications. However, at this juncture, the current process of drug prescribing, dispensing, administration, and consumption may, in fact, actually provide barriers to effective and efficient health care delivery. Current practice acts do not recognize overlapping or innovative scopes of practice based on demonstrated competency.[24] In addition, the current health care system is not oriented toward managing and monitoring chronic medication therapy. Rather, the focus has traditionally been toward managing acute medical events.[26]

Although pharmacists have traditionally assessed patients and assisted in drug therapy decision making, they have been given little autonomy to manage common and chronic disease states without the direct concurrence of a physician. Without authority to initiate and change medication regimens, many pharmacists must still contact a licensed prescriber as a step in solving drug-related problems they have identified. Scope of practice statements defining professional duties and general areas of responsibility are a logical way to improve access and continuity of patient care. Once considered only a hindrance to practicing disease and drug management, the inability of pharmacists to prescribe medications may well be considered both time and cost impediments to the delivery of quality and cost-efficient patient care in evolving health care delivery systems.

Pharmacy has embraced the philosophy that the provision of pharmaceutical care represents the principal mission of the profession.[27] Core activities of pharmacists who provide pharmaceutical care include the following: 1) participating in drug therapy decisions; 2) selecting drug products; 3) determining doses and dosage schedules; 4) preparing and providing drug products; 5) providing drug information and education; and 6) monitoring and assessing outcomes of drug therapy.

These types of activities can help solve significant problems in our health care system. Some examples of tasks associated with the provision of pharmaceutical care are listed in Table 3.[28] Many of these examples are necessary to help patients to use their medications

Table 3 Tasks associated with provision of pharmaceutical care

Interview patients to obtain information pertinent to product selection, dosage determination, and usage of current and past prescription and over-the-counter products.
Initiate requests for, or perform, and interpret results from appropriate laboratory and other diagnostic studies needed to select, initiate, monitor, and modify drug therapy.
Renew or rewrite prescriptions for continuation of drug therapy in accordance with established therapeutic endpoints or patient appointment status.
Measure vital signs and perform physical examinations of relevant organ systems and other patient assessments for the purpose of initiating, monitoring, and adjusting drug therapy.
Evaluate the patient's responses to therapy.
Provide oral and written recommendations for corrective actions for drug-related problems.
Document all patient care activities through orders and notes in the patient's medical record.
Select, initiate, monitor, continue, modify, and administer medication therapy to prevent disease or adverse reactions; resolve drug-related problems; or improve cost effectiveness.
Implement treatment guidelines, protocols, formulary changes, or critical pathways for therapy, as approved by an authorized health system provider or committee.
Provide patient education, identify expected outcomes of therapy, select monitoring parameters, and develop follow-up plans for drug therapy.
Provide direct patient care for appropriate disease management, either under protocol, policy, or guidelines.
Provide highly specialized inservice education and training to other health care professionals.
Develop medication use evaluation criteria and other quality improvement measures to assess the use of drug therapy by other providers.
Design, conduct, and coordinate clinical research projects under FDA guidelines and procedures of the institutional review board.

(Adapted from Ref. [28].)

optimally, but are prohibited by some state pharmacy statutes and regulations.

EVOLVING VIEW OF PRESCRIBING

Defining Prescribing

Today, prescribing is no longer the act of writing medication instructions. Prescribing encompasses multiple complex tasks, and as a term, it inadequately describes the numerous activities needed to provide drug therapy that achieves the defined outcomes that improve a patient's quality of life. The process of prescribing is more appropriately described by a broad set of activities that include selecting, initiating, monitoring, continuing, modifying, and administering drug therapy. Table 4 provides definitions of these prescribing activities. To select, initiate, and monitor drug therapy, the practitioner must be able to order and interpret laboratory tests, and perform patient assessments related to drug therapy management. This set of prescribing activities suggests that the focus of a practitioner's responsibility is on drug therapy management to improve patient outcomes.

Table 4 Definitions of prescribing activities

Select	When pharmacotherapy is necessary, and after review of an individual patient's history, medical status, presenting symptoms, and current drug regimen, the clinician chooses the best drug regimen among available therapeutic options.
Initiate	After selecting the best drug therapy for an individual patient, the clinician also determines the most appropriate initial dose and dosage schedule and writes an order or prescription.
Monitor	Once drug therapy is initiated, the clinician evaluates response, adverse effects, therapeutic outcomes, and adherence to determine if the drug, dose, or dosage schedule can be continued or needs to be modified.
Continue	After monitoring the current drug therapy of a patient, the clinician decides to renew or continue the same drug, dose, and dosage schedule.
Modify	After monitoring a patient's drug therapy, the clinician decides to make an adjustment in dose and/or dosage schedule, or may add, discontinue, or change drug therapy.
Administer	Regardless of who initiates a patient's drug therapy, the clinician gives the drug directly to the patient, including all routes of administration.

Defining Collaborative Relationships

Some individuals have advocated that pharmacists be granted independent prescriptive authority—that is, authority to prescribe medications independent of a defined collaborative relationship with an individual physician or medical group. Indeed, the system operative in Florida represents a form of independent prescriptive authority for pharmacists, albeit limited to a select formulary of drugs. Others have argued that pharmacists should function in a dependent role where prescriptive authority is delegated by a physician or other independent prescriber to another health care professional whom that prescriber believes possesses the professional skills and judgment necessary to perform these delegated duties.

However, the terms *dependent* and *independent prescribing authority* do not adequately reflect the collaborative relationship needed for pharmacists to contribute fully to the drug use process. A collaborative practice maximizes physician training and expertise in diagnosis, and pharmacist training and expertise in drug therapy and disease management. In most successful examples, the pharmacist and the physician have entered into a collaborative practice agreement or protocol under which the physician diagnoses and may make an initial treatment decision, and then authorizes the pharmacist to select, monitor, modify, and discontinue medications as necessary to achieve favorable patient outcomes. The physician and pharmacist then share the risk and responsibility for patient outcomes.[29]

Two additional factors support collaborative, rather than independent, management of patients by pharmacists. First, pharmacists have limited training in diagnosis. While physical diagnosis is a systematic process of organ system review, the pharmacist's assessment of physical findings is often targeted to a specific organ system or disease state. Except for acute self-limiting diseases or conditions identified during drug therapy monitoring, such as adverse drug reactions or inadequate responses, pharmacists are not trained to be diagnosticians. Second, a collaborative environment is the nature of current and future health care delivery systems. In fact, the future holds a marked increase in the extent of collaborative and managed health care delivery for all providers. All health care providers will be interdependent and will function in a collaborative fashion. The debate regarding dependent versus independent practice should be put to rest; instead, pharmacists should strive for collaboration with shared responsibilities and risks.

Prescriptive authority is not necessary to perform many duties involved in selecting, initiating, monitoring, continuing, modifying, and administering drug therapy. Nor is the ability to initiate drug therapy a prerequisite condition for pharmacists to establish a therapeutic relationship with a patient, solve drug-related problems, assume responsibility for therapeutic outcomes, or improve a patient's quality of life. However, when legally available, initiating drug therapy changes through collaborative drug therapy management agreements makes provision of care easier, more efficient, and convenient. Given the complexity of drug therapy decision making, evolving health care systems, and historic development of prescriptive authority, it may benefit society to review the scopes of practice of all health professionals, including the efficiencies gained by a collaborative health care team.

This discussion has focused on collaboration between pharmacists and physicians. However, optimal patient care and efficiency are most likely to result when effective collaboration exists among all the health professions. For example, there is no reason why nurse practitioners and pharmacists, or physician assistants and pharmacists, cannot collaboratively provide care for many patients with acute and chronic illnesses.

REQUIREMENTS FOR COLLABORATIVE DRUG THERAPY MANAGEMENT

For pharmacists to participate effectively in collaborative drug therapy management in a timely and cost-efficient manner, several conditions must exist: 1) a collaborative practice environment; 2) access to patients; 3) access to medical records; 4) knowledge, skills, and ability; 5) documentation of activities; and 6) compensation for their activities.

Collaborative Practice Environment

The pharmacist wanting to participate in collaborative drug therapy management first needs to identify a physician or practitioner group who wants to collaborate with the pharmacist. The physician or health system will identify patient populations, disease states, specific drugs, and certain drug-related issues in which other health professionals want to practice collaboratively with pharmacists. A description of routine and nonroutine professional duties and general areas of responsibility become the approved scope of practice for that pharmacist. The physician or health system needs to be willing to share responsibility for the pharmacist's actions. The environment may be an acute care hospital, a transitional care facility, a nursing home, a clinic, or a community pharmacy, as long as the remaining conditions are also met.

Table 5 Areas and content of core pharmacy curriculum adopted in 1997 by the American Council on Pharmaceutical Education

Biomedical Sciences	Anatomy, physiology, pathophysiology, microbiology, immunology, biochemistry, molecular biology, biostatistics
Pharmaceutical Sciences	Medicinal chemistry, pharmacognosy, pharmacology, toxicology, pharmaceutics, biopharmaceutics, pharmacokinetics
Behavioral, social, and administrative pharmacy sciences	Health care economics, pharmacoeconomics, practice management, communications, pharmacy history, ethics, social and behavioral applications and laws of practice
Pharmacy practice	Dispensing, drug administration, epidemiology, pediatrics, geriatrics, gerontology, nutrition, health promotion and disease prevention, physical assessment, emergency first care, clinical laboratory medicine, clinical pharmacokinetics, patient evaluation and ordering medications, pharmacotherapeutics, disease state management, outcomes documentation, self care and nonprescription drugs, drug information and literature evaluation
Professional experience	Introductory and advanced practice experiences throughout the curriculum as a continuum, in a variety of practice settings

(Adapted from Ref. [30].)

Access to Patients

Direct communication with patients is imperative for pharmacists to function successfully as collaborative drug therapy managers. In fact, it is best to establish an agreement with the patient describing the ideal conditions under which care should be rendered. Within this relationship, the patient grants the pharmacist responsibility, and the pharmacist in turn promises competency to perform the service, along with a willingness to assume responsibility, to the patient. This agreement codifies the direct relationships between patients and pharmacists, and heightens awareness of both groups to the responsibility assumed by the pharmacist in caring for the patient. The goal should be the establishment of a permanent and ongoing relationship that takes place over time. These relationships should complement, but not replace, those of patients and physicians.

Access to Medical Records

Access to a patient's medical records is essential to the provision of collaborative drug therapy management. In fact, it is only under these conditions, wherein the pharmacist has adequate knowledge of the patient and the patient's history, disease states, drug therapy, and laboratory and procedure results, that quality care can be rendered. Much work is being done in this area, via computerization of medical records and network facilitation of electronic data, to ensure this key element is in place to facilitate patient care by health care providers.

Knowledge, Skills, and Ability

In many ways, the pharmacist is uniquely trained for the task of collaborative drug therapy management. Contemporary pharmacy education has provided pharmacists with more extensive and indepth training in pharmacology and drug therapy management than any other health professional. Other health professionals who have prescriptive authority, such as nurse practitioners and physician assistants, have far less education in drug therapy management. Areas and examples of core curricula required under the 1997 American Council on Pharmaceutical Education requirements for Doctor of Pharmacy programs are listed in Table 5.[30]

Documentation of Activities

When pharmacists participate in any aspect of collaborative drug therapy management, they must document their activities in the patient's medical record. This information should, in turn, be available to other care providers within the health care system. Within the collaborative drug therapy management agreement, the frequency of communication with the collaborative team should also be established.

Compensation

In a vertically integrated managed health care system, the historical fee-for-service system of compensation is not operative. Therefore, pharmacists, either as primary care providers or as disease management specialists within a provider group, should expect to join with other health professionals on a collaborative team. Within a managed care contract, the pharmacist, along with other team members, assumes risk and responsibility for providing health care to patients in that system. Compensation from managed care payers will be on a contractual basis for team services. Demonstration of improved

outcomes will be integral to continuing contracts.[31] Specific duties and privileges will be defined by the scope of practice within the specific health care system, partly based on the mix of health care providers present and the type of patients for whom the system provides care. Collaborative drug therapy management will not lead to a fee-for-service form of compensation for clinical pharmacy services within a managed care environment. It is possible that it may do so in other types of health care systems.

COMPETENCIES, SETTING, CREDENTIALING, AND QUALITY ASSESSMENT

Competence assessment is essential when pharmacists assume collaborative drug therapy management activities, especially when such activities are new. Many methods exist to certify competence, such as granting clinical privileges or determining scope of practice in a health system via committee,[32] completing certificate programs for specific disease states, demonstrating knowledge and patient care skills, or earning national certification in a specialty via competency-based processes. The nature of the collaborative relationship will determine the appropriate mechanism for assessing competence. In addition, competencies may vary based on which prescribing activities are needed or how the scope of practice for each pharmacist is written. For example, initiating and modifying drug therapy may require competencies different than those necessary for administering, continuing, or monitoring drug therapy.

Pharmacists, by nature of their education and licensure, should be able to perform many of these functions without any additional demonstration of competence. The entire spectrum of prescribing activities is appropriate for any qualified licensed pharmacist in any practice setting as long as a collaborative relationship with other health care providers is established, access to relevant patient information exists, and ongoing competence and quality are assessed.

Pharmacists engaged in collaborative drug therapy management activities should be held accountable to the same quality assurance monitors and measures as other health professionals in their setting. Thus, supervision and quality assessment of activities are setting specific and will differ greatly among settings and health systems. Mechanisms to measure and ensure quality should be developed and put into place at the time the collaborative arrangement is established. These mechanisms should follow the same outline as those developed and used for other health professionals.

CONCLUSION

The practice of pharmacy and the provision of health care in the United States have changed dramatically over the past 60 years. Reports in the literature documenting pharmacists functioning in primary care roles and as prescribers of medications appeared as early as the 1970s. Reports of these early efforts, now renamed as efforts in collaborative drug therapy management, have demonstrated increased efficiencies in the health care system, while maintaining quality of care and patient satisfaction. At least 14 states and the federal government have authorized some form of pharmacist involvement in collaborative drug therapy management, and many other states are seeking to institute enabling legislation and regulations. Opportunities for pharmacists to increase efficiencies, decrease drug-related morbidity, and improve patient outcomes are abundant.

Not only has the role of the pharmacist evolved, but market-driven forces have caused the entire health care system in the United States to become more collaborative in nature. Pharmacists now have an opportunity to participate in collaborative drug therapy management and contribute to the quality of patient care in concert with other health care professionals.

To function successfully in a collaborative environment, the pharmacist must practice in a setting where teamwork is fostered, be able to establish a convenantal relationship with the patient, and have access to the patient's medical records. Because collaborative drug therapy management involves multiple complex tasks, the process may be more easily defined by describing the activities involved in the process—selecting, initiating, monitoring, continuing, modifying, and administering drug therapy. Ideally, these responsibilities should also include ordering, performing, and interpreting medication-related laboratory tests and procedures, along with performing patient assessment tasks related to drug therapy. By virtue of their extensive training in all relevant aspects of drug therapy management, pharmacists are well qualified and well equipped to provide collaborative drug therapy management services to patients.

Collaborative drug therapy management is most successful when the nature of the collaborative arrangement, the competencies and credentialing required, and the quality assurance checks that will be used to assess performance are defined at the outset in each specific setting.

In this era of rapid evolution in health care, the provision of collaborative drug therapy management by pharmacists can contribute to the efficacious, efficient, and cost-effective use of health care resources to improve patient outcomes in the United States.

ACKNOWLEDGMENTS

The authors want to acknowledge the contribution of Jeffrey C. Fay, PharmD, in preparing Table 1.

From *Pharmacotherapy* 1997, 17(5):1050–1061, with permission of the American College of Clinical Pharmacy.

REFERENCES

1. Webb, E.C. Prescribing medications: Changing the paradigm for a changing health care system. Am. J. Health-Syst. Pharm. **1995**, *52*, 1693–1695.
2. Marks, H.M. Revisiting "the origins of compulsory drug prescriptions". Am. J. Public Health **1995**, *85*, 109–115.
3. Swann, J.P. FDA and the practice of pharmacy: Prescription drug regulation before the Durham–Humphrey amendment of 1951. Pharm. Hist. **1994**, *36*, 55–70.
4. Brands, A.J. Treating ambulatory patients. U.S. Pharm. **1977**, *2*, 70–74.
5. Erickson, S.H. Primary care by a pharmacist in an outpatient clinic. Am. J. Hosp. Pharm. **1977**, *34*, 1086.
6. Health Manpower Pilot Projects. Final Report to the Legislature, State of California and to the Healing Arts Licensing Boards. In *Prescribing and Dispensing Pilot Projects*; Office of Statewide Health Planning and Development, Division of Health Professions Development, November, 1982.
7. An interview with Gordon Duffy, assemblyman, 32nd district. CSHP Voice **1978**, *5*, 11.
8. California Assembly Bill 1868, 1981.
9. California Senate Bill 502, 1983.
10. California Assembly Bill 1759, 1994.
11. Thompson, J.F.; McGhan, W.F.; Ruffalo, R.L.; Cohen, D.A.; Adamcik, B.; Segal, J.L. Clinical pharmacists prescribing drug therapy in a geriatric setting: Outcome of a trial. J. Am. Geriatr. Soc. **1984**, *32*, 154–159.
12. Christensen, D.; Fuller, T.; Williams, D. Prescriptive Authority Protocols. In *1993 Washington State Survey*; Washington State Board of Pharmacy, 1993.
13. Christensen, D. In *Prescriptive Authority for Pharmacists: Current Status and Future Opportunities*, Presented at the American Society of Hospital Pharmacists Annual Meeting, Reno, Nevada, June 5–9, 1994.
14. Flanagan, M.E. Update on state prescribing authority. Am. Pharm. **1995**, *NS35*, 13–18.
15. Borgsdorf, L.R.; Miano, J.S.; Knapp, K.K. Pharmacist-managed medication review in a managed care system. Am. J. Hosp. Pharm. **1994**, *51*, 772–777.
16. Bjornson, D.C.; Hiner, W.O.; Potyk, R.P., et al. Effect of pharmacists on health care outcomes in hospitalized patients. Am. J. Hosp. Pharm. **1993**, *50*, 1875–1884.
17. Chenella, F.C.; Klotz, T.A.; Gill, M.A., et al. Comparison of physician and pharmacist management of anticoagulant therapy of inpatients. Am. J. Hosp. Pharm. **1983**, *40*, 1642–1645.
18. Conte, R.R. Training and activities of pharmacist prescribers in a California pilot project. Am. J. Hosp. Pharm. **1986**, *43*, 375–380.
19. Ellis, R.F.; Stephens, M.A.; Sharp, G.B. Evaluation of a pharmacy-managed warfarin-monitoring service to coordinate inpatient and outpatient therapy. Am. J. Hosp. Pharm. **1992**, *49*, 387–394.
20. Hawkins, D.W.; Fiedler, F.P.; Douglas, H.L.; Eschbach, R.C. Evaluation of a clinical pharmacist in caring for hypertensive and diabetic patients. Am. J. Hosp. Pharm. **1979**, *36*, 1321–1325.
21. Menard, P.J.; Kirshner, B.S.; Kloth, D.D.; Pyka, R.S.; Hill, L.R.; Venkataraman, K. Management of the hypertensive patient by the pharmacist prescriber. Hosp. Pharm. **1986**, *21*, 20–35.
22. Stimmel, G.L.; McGhan, W.F.; Wincor, M.Z.; Deandrea, D.M. Comparison of pharmacist and physician prescribing for psychiatric inpatients. Am. J. Hosp. Pharm. **1982**, *39*, 1483–1486.
23. Schumock, G.T.; Meek, P.D.; Ploetz, P.A.; Vermeulen, L.C.; Publications Committee of the American College of Clinical Pharmacy. Economic evaluations of clinical pharmacy services—1988–1995. Pharmacotherapy **1996**, *16* (6), 1188–1208.
24. Pew Health Professions Commission. *Critical Challenges: Revitalizing the Health Professions for the Twenty-First Century*; UCSF Center for the Health Professions: San Francisco, 1995.
25. Schieber, G.S.; Poullier, J.P.; Greenwald, L.M. Health system performance in OECD countries, 1980–1992. Health Aff. **1994**, *13*, 100–112.
26. Holdford, D.A. Barriers to disease management. Am. J. Health-Syst. Pharm. **1996**, *53*, 2093–2096.
27. Trinca, C.E. In *The Pharmacist's Progress Toward Implementing Pharmaceutical Care*, Presented at the Third Strategic Planning Conference for Pharmacy Practice: Pharmacy in the 21st Century, Lansdowne, Virginia, October 7–10, 1994.
28. Carmichael, J. Do pharmacists need prescribing privileges to implement pharmaceutical care? Am. J. Health-Syst. Pharm. **1995**, *52*, 699–701.
29. Galt, K.A. The key to pharmacist prescribing: Collaboration. Am. J. Health-Syst. Pharm. **1995**, *52*, 1696–1699.
30. American Council on Pharmaceutical Education. *Accreditation Standards and Guidelines for the Professional Program in Pharmacy Leading to the Doctor of Pharmacy Degree. Adopted June 14, 1997*; American Council on Pharmaceutical Education, Inc.: Chicago, 1997.
31. Oddis, J.A. Pharmacy in integrated health care systems. Am. J. Health-Syst. Pharm. **1996**, *53* (Suppl. 1), S1–S49.
32. American College of Clinical Pharmacy. Establishing and evaluating clinical pharmacy services in primary care. Pharmacotherapy **1994**, *14* (6), 743–758.

Collaborative Practice Agreements (Collaborative Drug Therapy Management)

Jannet M. Carmichael
VA Sierra Nevada Health Care System, Reno, Nevada, U.S.A.

INTRODUCTION

This article examines how the practice of pharmacy can be improved by the legal and institutional recognition of Collaborative Drug Therapy Management (CDTM). Further, the development of Collaborative Practice Agreements or defining a specific Scope of Practice that allows the pharmacist and other health professionals to focus more on integration and collaboration is discussed.

DEFINITIONS

Pharmaceutical care: The responsible provision of drug therapy for the purpose of achieving a definite outcome that improves the patient's quality of life.[1]

Collaborative Drug Therapy Management (CDTM): The provision of pharmaceutical care in a collaborative and supportive practice environment that allows the qualified pharmacist legal, regulatory, and ethical responsibility to solve drug related problems when discovered.

Scope of practice: The boundaries within which a health professional may practice. For pharmacists, the scope of practice is generally approved by the board, agency, or committee that regulates the profession in a given state or organization.

Credentialing: The process by which an organization or institution obtains, verifies, and assesses a pharmacist's qualifications to provide patient care services.

Privileging: The process by which a healthcare organization, having reviewed an individual healthcare provider's credentials and performance and having found them satisfactory, authorizes that individual to perform a specific scope of patient care service within that organization.

Prescribing activities:[2]

- *Select*: When pharmacotherapy is necessary, and after review of an individual patient's history, medical status, presenting symptoms, and current drug regimen, the clinician chooses the best drug regimen among available therapeutic options.
- *Initiate*: After selecting the best drug therapy for an individual patient, the clinician determines the most appropriate initial dose and dosage schedule and writes an order or prescription.
- *Monitor*: Once drug therapy is initiated, the clinician evaluates response, adverse effects, therapeutic outcomes, and adherence to determine if the drug, dose, or dosage schedule can be continued or needs to be modified.
- *Continue*: After monitoring the current drug therapy for a patient, the clinician renews or continues the same drug, dose, and dosage schedule.
- *Modify*: After monitoring a patient's drug therapy, the clinician makes an adjustment in dose and/or dosage schedule, or adds, discontinues, or changes drug therapy.
- *Administer*: Regardless of who initiates a patient's drug therapy, the clinician gives the drug directly to the patient, including all routes of administration.

CURRENT PHARMACY PRACTICE ENVIRONMENT

As the volume and potency of prescription medications have increased, the nation's attention has become focused on the staggering human and economic cost of medication errors. Drug-related morbidity and mortality in the United States has a vast economic impact on the healthcare system. The nation was stunned in 1999 when the Institute of Medicine (IOM) issued a report that concluded that medical errors account for between 44,000 and 98,000 deaths annually.[3] The IOM report noted that "[b]ecause

Encyclopedia of Clinical Pharmacy
DOI: 10.1081/E-ECP 120006316

of the immense variety and complexity of medication now available, it is impossible for nurses or doctors to keep up with all of the information required for safe medication use. The pharmacist has become an essential resource in modern hospital practice.''[4]

Critical to understanding how to reduce the rate of medication errors is understanding what, how, and when drug-related problems arise.[5] Approximately 39% of medication errors occur in the prescribing phase.[6] Another 50% occur during order transcription and drug administration, and 11% occur during dispensing. The single proximal cause of medication errors (22%) is a lack of complete knowledge about drugs during the prescribing, order transcription, and drug administration stages. The direct presence and involvement of pharmacists during these stages can reduce medication errors as much as 66%.[7]

The traditional system of providing patient care—wherein physicians initiate drug therapy, pharmacists dispense medications, and nurses administer medications—is often run in a disjointed fashion. This results in potentially avoidable adverse drug events that contribute to poor patient outcomes and increased medical costs.[8] Efforts aimed at modifying the current processes of care to enhance efficiency of workflow, improve patient outcomes, and reduce medication errors are needed.

Pharmaceutical care has become the philosophy of pharmacy practice in recent years, as well as a mission or purpose for the profession. However, another term that has come into use to simultaneously describe pharmaceutical care and the system for the medication use process—CDTM. As the healthcare system grows in complexity, pharmaceutical care becomes increasingly important. It becomes more and more necessary for pharmacists, physicians, and other health professionals to work together in collaboration to assure safe use of medication. CDTM is an innovative approach in which pharmaceutical care services can be provided by a pharmacist in a supportive healthcare environment after a process of credentialing, privileging, and approval of a Collaborative Practice Agreement.

In a system utilizing CDTM, the American Society of Hospital Pharmacists (ASHP) supports the activities of a pharmacist that may include, but are not limited to: 1) initiating, modifying, and monitoring a patient's drug therapy; 2) ordering and performing laboratory and related tests; 3) assessing patient response to therapy; 4) counseling and educating a patient on medications; and 5) administering medications.[9] The American College of Clinical Pharmacists (ACCP) advocates the role of qualified pharmacists as capable collaborative drug therapy managers. Further, ACCP supports the pharmacists' role in CDTM to improve patient outcomes and increase efficiencies in the healthcare system. ACCP's core requirements for CDTM are:[2]

- A collaborative practice environment.
- Access to patients.
- Access to medical records.
- Knowledge, skills, and ability.[10]
- Documentation of activities.
- Compensation for these activities.[11]

CDTM is an interdisciplinary approach wherein pharmacists are integrated into the medical team to solve patient and medication-related problems and to share the responsibility for outcomes. A collaborative practice maximizes physician training and expertise in diagnosis, as well as the pharmacist training and expertise in drug therapy management. As of June 2002, in 38 states and in the federal government,[12] pharmacists share prescriptive authority with other healthcare professionals.[13] Pharmacists will find themselves in new legal and ethical positions as a result of these expanded prescribing roles.[14]

In most successful examples, pharmacists and physicians enter into a collaborative practice agreement. There is a specific Scope of Practice for the pharmacist in which, physicians diagnose and make initial treatment decisions, then authorize the pharmacist to continue, select, initiate, monitor, modify, or discontinue medications as necessary to achieve established therapy goals and favorable patient outcomes. While such pharmaceutical care goals can be provided through conventional pharmacy approaches, when legally feasible, CDTM agreements make provision of care simpler, more efficient, and more convenient than through traditional means. In addition, the literature is now rich with data proving that every dollar invested in clinical pharmacy services returns financial rewards and reduces patient mortality.[15]

Some examples of tasks associated with the provision of pharmaceutical care through CDTM have been published.[16] Many of these tasks are necessary to help patients use their medication optimally, but may be prohibited for pharmacists to perform independently by some state pharmacy statutes and regulations. CDTM may also be prohibited in some current practice sites of traditional pharmacy because they lack some core requirements from the list given above.[17] Note that we are discussing a fundamental change in the medication use *system*.

Next, we discuss the writing of a Collaborative Practice Agreement. This discussion assumes that core requirements for CDTM have been met or exceeded. Legal requirement may differ from state to state and from one practice environment to another. Credentialing re-

quirements might also be very different depending on the environment of care and the tasks the pharmacist wishes to perform. To confirm that all core requirements are met, most health systems require "privileging" or approval by a health system committee. During this process a healthcare organization, having reviewed an individual healthcare provider's credentials[10] and performance and found them satisfactory, authorizes that individual to perform a specific scope of patient care service within the organization. Credentialing and privileging are processes fundamental to CDTM.

Even though many pharmacists possess all of the core requirements, the widespread implementation of CDTM has not yet occurred. Although progress is being made in many environments, integrated health systems are among the fastest moving. Perhaps this is because these systems are rapidly incorporating the factors that support pharmaceutical care. Five enabling factors have been identified as features of a system or infrastructure that would likely facilitate CDTM or pharmaceutical care:

- Presence of an integrated electronic medical records system.
- Availability of an automated dispensing system for ambulatory care prescriptions.
- Pharmacist participation on multidisciplinary care teams.
- Support from medical staff.
- Support from senior management.

These trends in ambulatory care pharmacy have been studied by two surveys of the ASHP Managed Care and Ambulatory Care Pharmacy in Integrated Health Systems.[18,19]

COLLABORATIVE PRACTICE AGREEMENTS

The document describing the specific routine and nonroutine professional duties to be performed (the boundaries of practice) and the general areas of responsibility for each pharmacist practitioner is called a Scope of Practice. When this scope of practice has concurrence by the physician(s) or other collaborating practitioners in the patient care/program area in which the pharmacist functions, it is called a Collaborative Practice Agreement (CPA). The health system will generally have written policies to address all aspects of scope of practice issues for pharmacists including medication prescribing authority, quality assurance, and peer review. In addition, a process of credentialing and privileging will be outlined in policy, and CPAs will be reviewed at predetermined frequencies by appropriate boards, individuals, and committees.[20,21] An example of a CPA for primary care pharmacists is shown in Fig. 1.

TYPES OF COLLABORATIVE PRACTICE AGREEMENTS

CPAs can be written as process specific, or disease-state specific, or both. Process documents describe the routine duties of the pharmacist in global terms; e.g., write prescriptions, order laboratory tests needed to monitor medication, order certain radiological tests, take medication histories, record information in the medical record, order consults, etc. Disease specific CPAs give examples of the specific patient populations the pharmacist will see, and may include protocols for patient management. These CPAs may describe comprehensive, interim-care, and unscheduled or acute-care practice models.

Provision of *longitudinal comprehensive pharmaceutical care* using CPAs involves evaluation of patients' drug-related problems during an ongoing relationship with other care providers. One example of a pharmacist's role in CDTM programs focuses on identification of drug-related problems in the comprehensive review of a patient's medical record and by interviews with the patient. This review can be conducted when the patient is admitted to a general inpatient facility or is in an outpatient, primary-care clinic. In this model, the pharmacist uses a problem-based approach to determine the presence of medication-therapy problems and composes a problem list considering and incorporating disease state, adverse effects, cost, and compliance issues. The pharmacist then establishes treatment goals, monitoring parameters, and plans patient follow-up. Documentation in the medical chart of these actions is a critical component of any CDTM model. Interventions using this comprehensive model have been described.[22]

Interim care is defined as frequent care for specified patient populations and close patient monitoring between visits to the primary provider. Interim care models follow a similar process for delivering care as do comprehensive pharmaceutical care models, but on a disease-state specific basis. Many examples of this exist in the CDTM model. Examples of interim-care CDTM models include: anticoagulation/heparin clinics and clinics to treat asthma, seizures, pain, hypertension, diabetes, HIV, dyslipidemia, congestive heart failure, and other chronic-disease conditions.

Pharmacists working within *unscheduled acute care* or *urgent care models* handle patient issues that require immediate attention between scheduled visits. Some

SCOPE OF PRACTICE

Clinical Pharmacy Specialist

PHARMACIST: ________________________________

I am educationally competent and physically capable of performing the activities that I have requested. Where indicated below, it is fully understood that my scope is defined by approved protocols/procedures.

Signature of Applicant Date

RECOMMEND APPROVAL/DISAPPROVAL

Chief, Pharmacy Service Date

Reviewed by Credentialing Subcommittee on ____________________ APPROVED/DISAPPROVED

RECOMMEND APPROVAL/DISAPPROVAL

Chief of Staff Date

RECOMMEND APPROVAL/DISAPPROVAL

Medical Center Director Date

Granted on: ______________

To be reviewed on or before: ____________________

References:
VHA Directive 10-95-019 - General Guideline for Furnishing Medication Prescribing Authority for Clinical Nurse Specialists, Nurse Practitioners, Clinical Pharmacy Specialists and Physician Assistants, dated March 3, 1995
VHA Directive 96-034 – Scope of Practice for Clinical Pharmacy Specialists, dated May 7, 1996

Fig. 1 Example of a CPA for primary care pharmacists.

C

1. PURPOSE:

To identify scope of practice privileges for the clinical pharmacy specialist (CPS) at the VA Medical Center, and to define criteria for the qualifications for these privileges. The CPS will be qualified and authorized to perform specific clinical duties to assure high quality health care and appropriate pharmaceutical care is provided to the veterans.

2. POLICY:

Scope of practice guidelines for CPS shall be delineated in writing and will follow established protocols.

3. QUALIFICATIONS:

The CPS is trained in clinical pharmacy, clinical pharmacokinetics and clinical pharmacology. He/she is a Masters or Pharm. D. graduate, has completed an accredited pharmacy residency, is a specialty board certified pharmacist, or has equivalent education, training and experience functioning as a clinical pharmacist.

4. KEY FUNCTIONS:

- Conduct comprehensive appraisals of patients' health status by taking health histories, drug histories and performing physical examinations necessary to assess drug therapy
- Document relevant findings of a patients' health status in the patients' medical record
- Evaluate drug therapy through direct patient care involvement, with clinical assessment, subjective and objective findings relating to patient's responses to drug therapy and communicating and documenting those findings and recommendations to appropriate individuals and in appropriate records (i.e., patient's medical record)
- Develop, document and execute therapeutic plans utilizing the most effective, least toxic, and most economical medication treatments as per national or VA guidelines or VISN protocol or established local protocol
- Provide ongoing primary care for chronic stable or minor acute health problems as delineated in protocols/procedures
- Provide patient and health care professional education and medication information
- Evaluate and document patients' and caregivers ability to understand medication instructions and provide oral and written counseling on their medications
- Refer patients by consult to specialty clinics, order appropriate laboratory tests and other diagnostic studies necessary to monitor and support the patient's drug therapy
- Perform venipuncture or finger sticks for the purpose of withdrawing blood for clinical laboratory test
- Prescribe medications, including initiation, continuation, discontinuation, and altering therapy, based upon established formulary or protocols
- Conduct and coordinate research drug investigations and research under FDA guidelines and regulations and approval by appropriate local officials

Fig. 1 Example of a CPA for primary care pharmacists (*Continued*).

- Analyze laboratory and diagnostic test data so as to modify drug therapy and dosing as necessary.
- Perform physical measurement necessary to assure the patients responses to drug therapy
- Implement protocols approved by the Pharmacy and Therapeutics Committee or other Medical Center Committees regarding drug therapy
- Assist in the management of medical emergencies, adverse drug reactions, and acute and chronic disease states
- Administer medication according to pre-established protocol when requested by physicians
- Identify and take specific corrective action for drug-induced problems
- Serve as clinical managers of drug and drug-related programs in clinics and wards in conjunction with the attending physician

5. FURNISHING MEDICATIONS AND SUPPLIES:

A. The ability to prescribe non-controlled medications has been outlined in the General Guideline for Furnishing Medication Prescribing Authority for Clinical Nurse Specialists, Nurse Practitioners, Clinical Pharmacy Specialists and Physician Assistants in VHA Directive 10-95-019. CPS will initiate, continue, modify and monitor medication therapy as outlined in approved treatment protocols listed below or in policies and procedures of the Medical Center. The CPS will prescribe all drugs except narcotics.

B. Equipment and non-medication supplies issued by Pharmacy and Prosthetics/Materials Management Services may be ordered without co-signature of a physician.

6. SUPERVISION:

A collegial relationship with mutual consultation and referral exists with the physicians and the CPS. Consultation with the physician or referring practitioner is outlined and co-signature is required for practice outside approved procedures/protocols. The CPS will provide patient care as a Non-Physician Clinician (NPC). A physician is available at all times by telephone or in person for consultation. Periodic chart and peer reviews, and annual evaluations provide ongoing medication use evaluation. The CPS prescribing practices are included in the medication use evaluation process.

7. CLINICAL CONDITIONS:

The following is a list of clinical conditions that the CPS at the VA Medical Center would commonly be referred for evaluation and management. The most recent version of the protocols listed will be the working copy of the CPS protocols.

VISN 21 Protocol

	Requested		**Approved**		**Approved with Supervision**	
	No	Yes	No	Yes	No	Yes
Anticoagulation Clinic	☐	☐	☐	☐	☐	☐
Medication Renewal Clinic	☐	☐	☐	☐	☐	☐

Fig. 1 Example of a CPA for primary care pharmacists (*Continued*).

VHA Treatment Guidelines Pharmacologic Management of:
http://www.dppm.med.va.gov/PBM/menu.htm

	Requested		Approved		Approved with Supervision	
	Yes	No	Yes	No	Yes	No
Chronic Heart Failure	☐	☐	☐	☐	☐	☐
Chronic Obstructive Pulmonary Disease	☐	☐	☐	☐	☐	☐
Depression	☐	☐	☐	☐	☐	☐
Type 2 Diabetes Mellitus	☐	☐	☐	☐	☐	☐
Gastroesophageal Reflux Disease	☐	☐	☐	☐	☐	☐
H. pylori in PUD and Dyspepsia	☐	☐	☐	☐	☐	☐
Hyperlipidemia	☐	☐	☐	☐	☐	☐
Hypertension	☐	☐	☐	☐	☐	☐

VHA Clinical Practice Guidelines for the Management of:
http://www.va.gov/HEALTH/clinical.htm

	Requested		Approved		Approved with Supervision	
	Yes	No	Yes	No	Yes	No
COPD/Asthma	☐	☐	☐	☐	☐	☐
Major Depressive Disorder	☐	☐	☐	☐	☐	☐
Diabetes Mellitus	☐	☐	☐	☐	☐	☐

Other National Guidelines

	Requested		Approved		Approved with Supervision	
	Yes	No	Yes	No	Yes	No
HIV/AIDS Antiretroviral Therapy	☐	☐	☐	☐	☐	☐

Other Clinical Conditions (specify):

	Requested		Approved		Approved with Supervision	
	Yes	No	Yes	No	Yes	No
______________________	☐	☐	☐	☐	☐	☐
______________________	☐	☐	☐	☐	☐	☐

Fig. 1 Example of a CPA for primary care pharmacists (*Continued*).

patients may be new with no previously scheduled appointment. With this approach, pharmacists solve specific, urgent-care needs. Comprehensive evaluations, drug-therapy reviews, and extensive suggestions for treatment-plan modifications are not routine activities associated with this model, as pharmacists may not have an opportunity to perform follow-up with these patients. Examples of acute-care CDTM models include:

1. Refill/triage clinics.
2. Code teams.
3. Operating room protocols.
4. Polypharmacy consults.
5. Renal function drug dosing program.
6. Parenteral to oral route switch programs.

CONCLUSION

Pharmacists are assuming new roles in the healthcare system. Pharmacists with roles as direct patient care providers with expertise in identifying, resolving, and preventing drug-related problems are providing pharmaceutical care through CDTM. Essential components for the provision of CDTM have been identified. As examples and models of pharmacists' contributions in CDTM programs continue to grow, new proactive models will allow pharmacists to become key members of a system to decrease medication errors and their related costs.

REFERENCES

1. Hepler, C.D. The third wave in pharmaceutical education: The clinical movement. Am. J. Pharm. Educ. **1987**, *51*, 369–385.
2. Carmichael, J.M.; O'Connell, M.B.; Devine, B.; Kelly, W.; Ereshefsky, L.; Linn, W.D.; Stimmel, G.L. Collaborative drug therapy management by pharmacists. Pharmacotherapy **1997**, *17* (5), 1050–1061.
3. Institute of Medicine Division of Health Care Services Committee on Quality of Health Care in America. *To Err is Human: Building a Safer Health System*; National Academy Press: Washington, DC, 1999. http://www.nap.edu.
4. *ASHP Issue Paper. Issue: Patient Safety and Medical Errors*; Bethesda, Maryland, 2000. http://www.ashp.org/public/proad/psme.pdf.
5. Strand, L.M.; Morley, P.C.; Cipolle, R.J.; Ramsey, R.; Lamsam, G.C. Drug related problems: Their structure and function. DICP, Ann. Pharmacother. **1990**, *24*, 1093–1097.
6. Leape, L.L.; Bates, D.W.; Cullen, D.J.; Cooper, J.; Demonaco, H.J.; Gallivan, T.; Hallisey, R.; Ives, J.; Laird, N.; Laffel, G.; Nemeskal, R.; Petersen, L.A.; Porter, K.; Servi, D.; Shea, B.F.; Small, S.D.; Sweitzer, B.J.; Thompson, B.T.; Vander Vliet, M. Systems analysis of adverse drug events. JAMA, J. Am. Med. Assoc. **1995**, *274*, 35–43.
7. Leape, L.L.; Cullen, D.J.; Clapp, M.D.; Burdick, E.; Demonaco, H.J.; Erickson, J.I.; Bates, D.W. Pharmacist participation on physician rounds and adverse drug events in the intensive care unit. JAMA, J. Am. Med. Assoc. **1999**, *282*, 267–270.
8. Webb, E.C. Prescribing medications: Changing the paradigm of the changing health care system. Am. J. Health-Syst. Pharm. **1995**, *52*, 1693–1695.
9. *ASHP Issue Paper on Collaborative Drug Therapy Management*; http://www.ashp.org/public/proad/cdtm.pdf.
10. The Council on Credentialing in Pharmacy, X. *Credentialing in Pharmacy, Washington, DC*; 2000; 1–16, http://www.pharmacycredentialing.org/ccp/whitepaper.htm.
11. ECP article on Compensation for Clinical Pharmacy Services.
12. Billups, S.J.; Okano, G.; Malone, D., et al. Assessing the structure and process for providing pharmaceutical care in Veterans Affairs medical centers. Am. J. Health-Syst. Pharm. **2000**, *57*, 29.
13. ECP article on Pharmacist Prescribing.
14. Boatwright, D.E. Legal aspects of expanded prescribing authority for pharmacists. Am. J. Health-Syst. Pharm. **1998**, *55*, 585–594.
15. Proceedings 1st International Congress of Clinical Pharmacy. In *Documenting the Value of Clinical Pharmacy Services*; Pharmacotherapy, 2000; 20, 235S–344S.
16. Carmichael, J.M. Do pharmacist need prescribing privileges to implement pharmaceutical care? Am. J. Health-Syst. Pharm. **1995**, *52*, 699–701.
17. *NACDS Report on Pharmacy Activity Cost and Productivity Study*; NACDS; 1999; http://www.nacds.org/news/andersen.html.
18. Knapp, K.K.; Blalock, S.J.; O'Malley, C.H. ASHP survey of ambulatory care responsibilities of pharmacists in integrated health systems—1999. Am. J. Health-Syst. Pharm. **1999**, *56*, 2431–2443.
19. Reeder, C.E.; Kozma, C.M.; O'Malley, C. ASHP survey of ambulatory care responsibilities of pharmacists in integrated health systems—1997. Am. J. Health-Syst. Pharm. **1998**, *55*, 35–43.
20. *General Guidelines for Establishing Medication Prescribing Authority for Clinical Nurse Specialist, Nurse Practitioners, Clinical Pharmacist*; Department of Veterans Affairs: Washington, DC, 1995; http://vaww.va.gov/publ/direc/health/direct/195019.htm.
21. *Scope of Practice for Clinical Pharmacy Specialists*; Department of Veterans Affairs: Washington, DC, 1996; http://vaww.va.gov/publ/direc/health/direct/196034.doc.
22. Ellis, S.L.; Billups, S.J.; Malone, D.C.; Carter, B.L.; Covey, D.; Mason, B.; Jue, S.; Carmichael, J.M.; Guthrie, K.; Sintek, C.D.; Dombrowski, R.; Geraets, D.R.; Amato, M. Types of interventions made by clinical pharmacists in the IMPROVE study. Pharmacotherapy **2000**, *20* (4), 429–435.

College of Psychiatric and Neurologic Pharmacists

Alex A. Cardoni
The Institute of Living, Hartford, Connecticut, U.S.A.

INTRODUCTION

The College of Psychiatric and Neurologic Pharmacists (CPNP) was founded on March 24, 1998 when the network of pharmacists formerly known as the Conference of Psychiatric and Neurologic Pharmacists became an official professional society.

The formation of CPNP was the culmination of efforts of many pharmacists practicing in the psychiatry and neurology specialties over the past 30 years. But serious discussion for a formal professional society began at a strategic planning meeting in Austin, Texas, in October 1994. Forty neuropsychiatric pharmacists held a post-CE-program planning conference that would eventually give birth to CPNP. In the summer of 1997, a smaller group of these pharmacists drafted a constitution and bylaws that were eventually approved by the founding membership. A call was issued for founding members in the fall of 1997, which generated 60 additional members. Eventually, 116 founding members joined ranks to create the College of Psychiatric and Neurologic Pharmacists. These members then ratified the constitution and bylaws and nominated candidates for offices. The first officers were sworn in at the First Annual Meeting of the College in Orlando, Florida, April 23–26, 1998. They were President: Gary M. Levin (Albany, New York), president-elect: Alex A. Cardoni (Storrs, Connecticut), Treasurer: James E. Wilson (Omaha, Nebraska), Secretary: Cherry W. Jackson (Charleston, South Carolina), Director-at-Large: Lawrence J. Cohen (Oklahoma City, Oklahoma), and Director-at-Large: Sally K. Guthrie (Ann Arbor, Michigan).

ORGANIZATIONAL STRUCTURE AND GOVERNANCE

The CPNP constitution and bylaws provide for the structure and governance of the organization. Membership consists of Active Members, Founding Members, and Corporate Members.

Active Members are dues-paying members who are interested in advancing the specialty of psychiatric or neurologic pharmacy.

Founding Members include original members instrumental in creating CPNP who have paid Founding Member dues.

Corporate Members consist of corporations and members of corporations who are interested in supporting the goals and objectives of CPNP.

The officers of CPNP are president, immediate past-president, president-elect, secretary, and treasurer. The president-elect is elected by the general membership for a one-year term and ascends successively to the office of president. The president ascends to the office of immediate past-president. The secretary and treasurer are elected on alternate years for a two-year term of office.

The president chairs the Board of Directors and presides at all meetings of the Board as well as general membership meetings. The president appoints all chairs and members of committees and is an ex-officio member of each committee.

The immediate past-president presides at meetings in the absence of the president and the president-elect.

The president-elect executes the duties of the president in the president's absence and is vice-chair of the Board of Directors.

The secretary records minutes of all meetings, maintains the membership roll, receives and prepares all correspondence, and approves all membership applications.

The treasurer serves as the custodian of all funds, receives and keeps account of all monies received as dues or from other sources, and disburses monies at the discretion of the Board of Directors.

The Board of Directors of CPNP is composed of the elected officers and two Directors-at-Large elected by the membership. The Board represents the organization as the official voice of all members. The Board has charge of property and authority to control and manage the affairs and funds of the organization, and also to supervise all publications and to select editors for publications. The Board makes ultimate decisions regarding actions of

Encyclopedia of Clinical Pharmacy
DOI: 10.1081/E-ECP 120006203

committees and officers on professional and administrative matters.

CPNP committees function in an advisory capacity to the Board of Directors, developing and implementing programs and policies authorized by the Board in the major areas of interest to which it is assigned.

The following are standing committees of CPNP: Communication and Information—responsible for publishing the CPNP Newsletter and maintaining the web site cpnp.org; Community Resource—acts as a liaison to extramural groups and organizations on issues relating to psychiatry and neurology; Program—responsible for planning the annual meeting as well as other continuing education programs; and Membership—responsible for the recruitment and retention of members of CPNP.

CPNP officers since its founding are summarized: 1998—President: Gary Levin; president-elect: Alex Cardoni; Secretary: Cherry Jackson; and Treasurer: James Wilson. 1999—President: Alex Cardoni; president-elect: Roger Sommi; Immediate Past-President: Gary Levin; Secretary: Cherry Jackson; and Treasurer: James Wilson; and 2000—President: Roger Sommi; president-elect: Cherry Jackson; Immediate Past-President: Alex Cardoni; Secretary: Judith Curtis; and Treasurer: James Wilson. The following members were elected as Directors-at-Large: 1998–2000—Lawrence Cohen; 1998–2001—Sally Guthrie; and 2000–2002—Charles Caley.

MEMBERSHIP

As of fall 2000, the College of Psychiatric and Neurologic Pharmacists has a membership of approximately 300 individuals. In a survey of CPNP membership conducted in 1999, the following profile emerged:

- 60% between ages of 31–50.
- 52% female.
- 45% based in hospitals.
- 57% in practice less than 10 years.
- 50% with psych/neuro specialty residency training.
- 20% with fellowship training.
- 70% working in psychiatric pharmacy; 10% in neurologic pharmacy.
- 55% board certified in psychiatric pharmacy.

MISSION

The mission of the College of Psychiatric and Neurologic Pharmacists is to advance neuropsychiatric pharmacy practice, education, and research and to optimize the health of individuals affected by psychiatric and neurologic disorders.

Objectives include the following:

- Facilitate dissemination of information regarding psychotherapeutic pharmacotherapy, patient care, and community support.
- Endorse the Psychiatric Pharmacy Certification Exam process and support programs for the preparation of candidates for the exam.
- Facilitate programming in the areas of psychiatric and neurologic pharmacy at national meetings and with organizations that support our interests.
- Improve patient care.
- Promote research in patient care.

CURRENT MAJOR INITIATIVES

The Board of Directors has identified several priority initiatives in 2000:

- Further develop and refine ''cpnp.org'' web site.
- Retain the services of a management consultant to assume responsibility for routine administrative operations of the organization.
- Implement and facilitate a board recertification process for board certified psychiatric pharmacists by working with existing professional organizations.
- Initiate strategic planning for the organization that will set goals and objectives and a budget for the next several years.
- Stimulate development of pharmaceutical care psychiatric and neurologic clinical services by establishing a competitive ''visiting expert'' grant program.
- Stimulate development and growth of regional affiliates to facilitate attainment of CPNP goals and objectives.

MAJOR MEETINGS

CPNP holds its annual meeting in the spring, usually late March or early April. The site of the meeting changes to include East Coast, mid-America, and West Coast. The meeting is 2–2 1/2 days in length and includes sessions which focus on contemporary clinical and research topics in psychiatric and neurologic pharmacy. A poster session allows for presentation of research and clinical practice activities of members.

CPNP members also traditionally participate in the annual NCDEU meeting sponsored by the National Institutes of Mental Health. This meeting is held in the spring as well, usually in late May or early June. The location has been (with few exceptions) in Boca Raton, Florida. Cutting-edge programming characterizes this meeting, with presentations from leading NIH researchers as well as others in the United States and abroad. A poster session also facilitates reporting of research findings.

Regional programming for CPNP members and other psychiatric and neurologic pharmacists is provided by local "chapters" of CPNP throughout the country. Annual pharmacotherapy updates are held in the Northeast region (fall), Georgia-Southeast (winter), Midwest region (fall), Texas (fall), Arizona-Southwest (winter), and in Montana-Northwest (spring). These sessions bring high-quality programming to members and nonmembers who may not be able to attend the annual meeting.

Commission to Implement Change in Pharmacy Education, AACP

Harold H. Wolf
University of Utah, Salt Lake City, Utah, U.S.A.

INTRODUCTION

In July, 1989, Dr. William A. Miller, President of the American Association of Colleges of Pharmacy (AACP), appointed the Commission to Implement Change in Pharmacy Education to consider, debate, and offer recommendations on a series of issues pivotal to the further development of pharmacy education in the United States.

DISCUSSION

Members of the Commission included C. Douglas Hepler, Ph.D.; Mary Anne Koda-Kimble, Pharm.D.; David A. Knapp, Ph.D.; Kenneth W. Miller, Ph.D.; Milap C. Nahata, Pharm.D.; Charles O. Rutledge, Ph.D.; William E. Smith, Ph.D.; John H. Vandel; Victor A. Yanchick, Ph.D.; Charles A. Walton, Ph.D.; and Harold H. Wolf, Ph.D. (Chairman). Richard P. Penna, Pharm.D. artfully crafted the discussions of the Commission into documents.

Key questions to be addressed by the Commission included: What is the mission of pharmacy practice that should serve as the basis for pharmaceutical education? What types of pharmacy manpower are needed to fulfill this mission? What should be the curricular emphasis in an entry-level pharmacy practice degree? What should be the length of the curriculum and title of the degree? When should differentiation occur in the continuum of pharmaceutical education and training, and what are the roles of postgraduate educational experiences? What are the needs of the pharmacy enterprise for pharmaceutical scientists and clinical scientists, and what are appropriate models for such training? What changes in accreditation standards should be made by the American Council on Pharmaceutical Education to facilitate the broad implementation of such reforms?

The Commission met 13 times over a 3-year period and produced a series of position papers containing policy statements for AACP as well as recommendations for actions directly impacting pharmacy education that were to be implemented at individual schools or colleges.

Initially, the Commission examined the purpose of pharmaceutical education in contemporary society. To do this, it developed working statements for the mission of the profession of pharmacy and for pharmacy practice.[1] In crafting the former, the Commission offered the view that ''The pharmacy profession is a major part of a system that discovers, develops, produces, and distributes drug entities and drug products. It creates, and disseminates knowledge related to drug entities, drug products, and drug distribution systems. The major outputs of the profession are pharmaceutical care, knowledge, drug entities and drug products. The primary personnel in the profession that produce these outputs are practitioners, educators, researchers, and those involved with the manufacture and distribution of drug products (p. 375).''

In designing a working mission statement for pharmacy practice, the Commission expressed its belief that ''. . . the mission of such practice is to render pharmaceutical care with the view that such care focuses pharmacists' attitudes, behaviors, commitments, concerns, ethics, functions, knowledge, responsibilities, and skills on the provision of rational drug therapy. The goal of such therapy being outcomes which improve the quality of patients' lives (p. 376).''

While the Commission clearly recognized that the mission of pharmaceutical education certainly derives from the mission of the profession and thus must be consistent with the mission of pharmacy practice, it also recognized that the enterprise of pharmacy education is responsible for fulfilling that portion of the profession's mission that relates to research and education. Thus, the Commission described the mission of pharmaceutical education, in part, as follows:

> Pharmaceutical education is responsible for preparing students to enter into the practice of pharmacy and to

Encyclopedia of Clinical Pharmacy
DOI: 10.1081/E-ECP 120006188

function as professionals and informed citizens in a changing health care system. It does this by maintaining a dynamic, challenging, and comprehensive curriculum. It is also responsible for generating and disseminating new knowledge about drugs and about pharmaceutical care systems.

Pharmaceutical education inculcates students with the values necessary to serve society as caring, ethical, learning professionals and enlightened citizens. It provides students with scientific fundamentals and fosters attitudes necessary to adapt their careers to changes in health care over a lifetime. It also encourages students prior to and after graduation to take active roles in shaping policies, practices, and future directions of the profession.

Pharmaceutical education promotes advances in pharmaceutical care by fostering postgraduate residencies and fellowships in the clinical sciences and differentiated areas of pharmacy practice. It provides structured postgraduate education and training through which practitioners maintain their competence and acquire new competencies to serve the changing needs of society.

Pharmaceutical education is responsible to the profession and to society for generating new knowledge about drugs, drug products, drug therapy, and drug use through the conduct of basic and applied research. It promotes the pharmaceutical sciences by fostering graduate education and research within its schools and colleges. Pharmaceutical education is responsible for both professional education and graduate education for research. The latter focuses on preparing students to discover new knowledge, primarily by use of the scientific method. The goal is to prepare scholars to perform independent, creative research that addresses important questions related to the discovery and use of drugs.

Pharmaceutical education continually evaluates its mission, objectives, goals, and outcomes and determines and implements necessary changes in the nature and scope of education and research performed within the purview of pharmaceutical education (p. 376).

Perhaps the most substantive portion of the Commission's work resides in its second position paper (Background Paper II) which dealt with issues of entry level, curricular outcomes, curricular content, and educational process.[2] In considering what is "entry level," the Commission embraced the view that while a system of pharmaceutical care requires the participation of both generalists and specialists, students prepared at the entry level are general practitioners who coordinate and render pharmaceutical care. And they must be able to do this at a level commensurate with the evolving mission of pharmacy practice (see above).

Based on this assumption, the Commission outlined the major practice functions that comprise pharmaceutical care as rendered at the entry level and offered recommendations for the educational outcomes and competencies that are necessary to perform pharmaceutical care functions.

The practice functions identified were:

- Participate in the drug use, decision-making process.
- Select the appropriate dosage form, formulation, administration, and delivery system of specific drug entities.
- Select the drug product source of supply.
- Determine the dose and dosage schedule.
- Prepare medication for patient use.
- Provide drug products to patients.
- Counsel patients.
- Monitor patients to maximize compliance.
- Monitor patients' progress with regard to therapeutic objectives.
- Monitor patients to prevent adverse drug reactions and drug interactions.

Several general outcomes and competencies were described that underlie the education of a professional person and citizen. These included:

- Thinking abilities involving scientific comprehension and critical thinking.
- Communication abilities involving communication competence and aesthetic sensitivity.
- Facility with values and ethical principles involving professional ethics.
- Personal awareness and social responsibility involving contextual competence and professional identity.
- Self-learning abilities and habits involving adaptive competence, scholarly concern for improvement, and motivation for continued learning.
- Social interaction and citizenship including effective, interpersonal, and intergroup behaviors as well as leadership competence.

Additional professional outcomes and competencies were identified as essential to perform the functions that support practice. These included the broad skills necessary to solve problems and make decisions; manage; learn; communicate, teach, and collaborate; and participate in policy formulation and professional governance.

The Commission than proceeded to describe a core curriculum to serve as a guide for pharmacy faculty at individual schools and colleges to design the content of a specific curriculum felt likely to engender the competencies and outcomes necessary to render pharmaceutical care.

A major portion of the Commission's efforts in the area of curriculum was devoted to recommendations relating to the educational process. Particular suggestions were offered to assist faculty in teaching problem solving, fundamental information, communication skills, and practice skills.

Certainly the most controversial issues examined by the Commission were those contained in its position paper dealing with the standards of educational quality necessary for the entry-level curriculum, the length of that curriculum, and the title of the ensuing degree.[3] The Commission concluded that AACP must advocate the outcomes, competencies, content, and processes contained in Background Paper II before the American Council on Pharmaceutical Education (ACPE) for incorporation into the revised entry-level program accreditation standards that the Council was then developing. It emphasized that this should be done for all programs, including existing Pharm.D. offerings. The Commission went on to say that at least one additional year of professional education (beyond the 5-year entry-level programs commonly in place) was needed to accomplish the educational objectives previously described, and thus, proposed that AACP endorse an entry-level program that is at the doctoral level, is at least four professional, academic years in length, and follows preprofessional instruction of sufficient quality and length (2-year minimum) to prepare applicants for doctoral level education. Finally, the Commission proposed that AACP support the doctor of pharmacy (Pharm.D.) degree as the sole degree for entry into pharmacy practice and offered several recommendations to overcome the barriers that could impede the process of implementing such needed changes in pharmaceutical education.

The balance of the Commission's efforts related to discussions surrounding faculty scholarship, graduate education, fellowships, and postgraduate professional education.[4] Particular attention was focused on issues of fostering scholarship, assessing graduate programs, preparing clinical scholars, developing mid-career residencies, and considering the role of distance learning in sustaining up-to-date competence. Numerous recommendations were advanced, with the intent that these profound responsibilities of the enterprise of pharmaceutical education receive the attention necessary to catalyze required change.

The Commission was reappointed in 1995 by AACP President Mary Anne Koda-Kimble to analyze and assess how a range of rapid and extensive changes in health care delivery, education, and research might alter its original observations and recommendations. The Commission met twice, reaffirmed the contemporary value of its original views, and, in 1996, encouraged all schools and colleges of pharmacy to accelerate their plans for curricular reform based on recommendations made in its previous reports.[5]

While history will judge the overall impact of the Commission's work, there is little doubt that the effort was a catalyst for major change in pharmaceutical education.[6] Subsequent to the release of the Commission's reports and adoption of most of its recommendations by AACP in 1992, major, nationwide energies were, and continue to be, directed toward changes both in curricular structure based on educational outcomes as well as in the process of teaching within the curriculum. Moreover, there occurred a substantial increase in the number of schools offering the Pharm.D. as the sole professional degree. At the time the Commission's recommendations were adopted, 19% of pharmacy schools offered the Pharm.D as the sole professional degree. As of fall 1995, 37% of schools were admitting students into all Pharm.D. programs. In fall 1996, that percentage increased to over 50%. Finally, as a result of the Accreditation Standards and Guidelines for the Professional Program in Pharmacy Leading to the Doctor of Pharmacy Degree adopted June 14, 1997 by the American Council on Pharmaceutical Education,[7] all schools of pharmacy will only admit students into a Pharm.D. program by 2002. Thus, an issue that had been hotly debated within and outside of AACP for some 42 years is finally resolved.

REFERENCES

1. Wolf, H.H.; Walton, C.A.; Hepler, C.D.; Koda-Kimble, M.A.; Knapp, D.A.; Miller, K.W.; Nahata, M.C.; Rutledge, C.D.; Smith, W.E.; Vandel, J.H. Background paper I: What is the mission of pharmaceutical education? Am. J. Pharm. Educ. **1993**, *57* (4), 374–376.
2. Wolf, H.H.; Walton, C.A.; Hepler, C.D.; Koda-Kimble, M.A.; Knapp, D.A.; Miller, K.W.; Nahata, M.C.; Rutledge, C.D.; Smith, W.E.; Vandel, J.H. Background paper II: Entry level, curricular outcomes, curricular content and educational process. Am. J. Pharm. Educ. **1993**, *57* (4), 377–385.
3. Wolf, H.H.; Walton, C.A.; Hepler, C.D.; Koda-Kimble, M.A.; Knapp, D.A.; Miller, K.W.; Nahata, M.C.; Rutledge,

C.D.; Smith, W.E.; Vandel, J.H. Entry-level education in pharmacy: Commitment to change. Am. J. Pharm. Educ. **1993**, *57* (4), 366–374.

4. Wolf, H.H.; Hepler, C.D.; Koda-Kimble, M.A.; Knapp, D.A.; Miller, K.W.; Nahata, M.C.; Rutledge, C.D.; Smith, W.E.; Vandel, J.H.; Yanchick, V.A. The responsibility of pharmaceutical education for scholarship, graduate education, fellowships, and postgraduate professional education and training. Am. J. Pharm. Educ. **1993**, *57* (4), 386–399.
5. Wolf, H.H.; Hepler, C.D.; Koda-Kimble, M.A.; Knapp, D.A.; Miller, K.W.; Nahata, M.C.; Rutledge, C.D.; Smith, W.E.; Vandel, J.H.; Yanchick, V.A. Maintaining our commitment to change. Am. J. Pharm. Educ. **1996**, *60* (4), 378–384.
6. Buerki, R.A. In search of excellence: The first century of the American Association of Colleges of Pharmacy. Am. J. Pharm. Educ. **1999**, *63*, 181–184, (Fall Suppl.).
7. *Accreditation Standards and Guidelines for the Professional Program in Pharmacy Leading to the Doctor of Pharmacy Degree Adopted June 14, 1997*; American Council On Pharmaceutical Education: Chicago, Illinois, 1997; 1–51.

Computer Software for Clinical Pharmacy Services

Bill G. Felkey
Brent I. Fox
Auburn University, Auburn, Alabama, U.S.A.

INTRODUCTION

The decision-making process for purchasing computer software for use by pharmacists used to be a lot simpler. A small task force was formed with the goal of formulating a recommendation of the most functional software solution for any individual pharmacy department, outpatient pharmacy, or community pharmacy. A few site visits would take place, perhaps, and a budget would be drawn up for acquisition cost, training requirements, and ongoing/monthly update costs. The decision to purchase would be made, and then development of the necessary interfaces would begin (usually as an afterthought). The development of these interfaces would often consist of an employee looking at the screen of one system while typing those data into a second system.

The pendulum has now swung to another extreme where integration is a primary consideration for purchase decisions. Mainframe computers, which were nearly pronounced dead, are now being looked at favorably as the repository for a data warehouse that contains all necessary clinical information. The data repository concept is providing the thrust to move organizations into a future driven by an information-rich environment. Unfortunately, health system pharmacy departments are often mandated departmental solutions that are functionally inferior to their stand-alone systems, but are sold as globally integrated within the healthcare institution. Some of these systems are so terrible that pharmacists operate manually all day and then hire clerks to input transactions that occurred during the day in an after-hours session.

SYSTEM OVERVIEW

The decision to purchase pharmacy departmental software includes additional vendor-related factors that must be considered. There is a growing trend among many of the larger vendors to minimize the importance of functionality based on the opinion that many of the software systems are beginning to look and act alike. If this is indeed true, it is potentially more important to focus on the ability of the different vendors to implement their applications. The common conception that larger vendors provide better products and services must also be reconsidered. With a larger customer base comes less customization and specialization for the individual customer. While larger companies do have increased financial stability, the ability of a smaller company to tailor a product to a customer's specific usage cannot be overstated. Finally, we have seen from direct experience that many of the vendors in today's market have more than enough clients to get them through the coming year. Decision makers within health systems must recognize this fact and plan accordingly. Agree upon a timetable suitable for your system yet feasible for your vendor. And most importantly, recognize that the vendor's responsibility is to design and install the system, as well as to train your personnel. It is your responsibility to implement the system and put it to use as best applies to your health system.[1]

Today, the purchase of information technology solutions for pharmacy confronts decision makers with a vast array of decisional factors to consider. Some of these include:

- Identification of the core competency of vendors being considered.
- Comparison of the largest installation and installed user base of vendors.
- Comparison of budgetary constraints and prices of various solutions.
- Careful consideration of operating system platforms that are supported.
- Level of Transmission Control Protocol/Internet Protocol (TCP/IP) and Health Level 7 (HL7) support.
- System interfaces that include admission discharge transfer (ADT), pharmacy charges, automation, labs, etc.

Encyclopedia of Clinical Pharmacy
DOI: 10.1081/E-ECP 120006318

- Identification of value-added features and enhancements that discriminate one choice from other solutions.
- Vendor ability to personalize the solution for a health system's specific needs.
- Determination of quality assurance and safety features.
- Satisfaction that security features are adequate and responsive to the organizational structure.
- Identification of the frequency of significant system upgrades.
- Determination of the assistance available for data migration and the system rollout planning.
- Strategic and tactical consideration for how the pharmacy software solution matches up with the clinical information systems, Internet solutions, health resource planning, access management, decision support, home care, managed care, and infrastructure applications.

These considerations are primarily enterprise-wide in scope. With integration being of primary importance to healthcare in general, it is still necessary to obtain best-of-class software solutions for specific pharmacy activities. Recommendations that will help select appropriate applications and technologies that are backed by reliable implementation, support, and services will be the focus of the remainder of this chapter. Refer to Tables 1 and 2 (used with permission from *ComputerTalk*) as you read this chapter. These figures contain an extensive evaluation of the current information systems market.

THE HARDWARE ARRAY

Decision-makers now have an interesting array of hardware options with which to manage the data processing for pharmacy operations. Client/server architecture is still the dominant means of configuration, but Application Service Providers (ASPs) are expected to be utilized more as both information technology (IT) professionals and corporate executives begin to understand what these services offer. As pharmacists have moved from the central pharmacy into satellite pharmacies, home care, and other modalities that demand a mobile solution, other hardware must be considered for a total solution. Initially, pharmacists employed the use of notebook computers that operated offline for data storage and retrieval. These devices were limited because they did not provide real-time access to the health system's information system. Then, progressive hospitals began to develop the necessary infrastructure and to connect these devices wirelessly. More recently, however, enough resources have been developed in the PDA (personal digital assistant) market to have this platform considered fully in any purchase decision. PalmPilots and other compatible devices now comprise over 80 percent of the PDA market. Wireless versions using a variety of bandwidths and frequencies are making it possible to provide connectivity to devices that easily fit within the shirt pocket, lab coat pocket, and purse. As memory increases in these PDA devices, it is now possible to have complete medication references as well as network access, a bar-code scanner, a pager, a digital voice recorder, and a cellular telephone combined into a single device.

One must also consider the functional specifications of workstation hardware. While we're waiting for the perfectly integrated system to be realized, it will be necessary to consider terminal emulation as a first step in communicating with diverse systems. Previously, "dumb" terminals, which consisted of a keyboard and monitor connected to a server, dominated the workstations used in most pharmacies. Now a combination of terminal emulation running on microcomputers and Internet thin client workstations are beginning to proliferate. With the use of terminal emulation it is possible to have three or more patient sessions running on a taskbar of a Microsoft Windows workstations and, through multitasking, access Internet-based and information applications simultaneously. Due to the volume of data that must be transmitted, some functions still tend to run better offline during peak network congestion times in an organization. As bandwidth increases, this problem may diminish.

The importance of integration between systems is quite high. Historically, pharmacy departments purchased point-to-point interfaces at a cost of $1500 to $15,000 each. Interfaces are now more popular and often are more cost-effective for an organization. So important is the functioning of these interfaces that many pharmacists wear pagers that alert them when an interface has gone down. This occurs automatically as pages are sent from the information technology systems as problems are identified. Another interface consideration is the ability to remotely address problems such as these. Programs that allow system access and control from any computer in the world are becoming normal. Of course, firewall protection from unwanted intruders is necessary in all aspects of data management, including remote access.

Telecommuting presents unique opportunities and challenges for the health system and the pharmacy department, especially. As the pharmacist shortage continues, with one pharmacy chain reportedly building enough stores to hire every pharmacy school graduate for the next ten years, new approaches to practice must be explored. Telecommuting will be one response to this shortage, bringing the work to the worker. When work can be brought to the worker instead of bringing workers to the

Table 1 Entry-level prices for products and services

See page		Retail Pharmacy Mgmt.	Chain Central Mgmt.	Automated Dispensing/ Robotic Systems	HME/ Diabetes Type II Billing	E-Prescriptions	POS	Pharmaceutical Care	Self-Care/ Nutritional
65	Allwin Data	-	-	-	c/p[1]	-	-	-	-
65	AppDC	-	-	-	-	-	-	-	-
7	ateb, Inc.	-	-	-	-	c/p	-	-	-
39	Cam Commerce Solutions	-	-	-	-	-	$8,157[3]	-	-
8	CarePoint, Inc.	$9,500[S]	c/p	c/p	inc	c/p	$3,500[S]	inc.	-
9	ComCoTec	c/p	c/p	c/p	inc.	inc.	c/p	inc.	c/p
10	DAA Enterprises	$4,995	-	-	-	-	-	inc.	-
53	Delphi	$9,000[S]	-	-	-	-	-	$2,000[S]	-
65	Digital Simplistics	$6,950	-	-	-	-	-	-	-
40	D.P. Hamacher	-	-	-	-	-	-	-	-
58	Etreby Computer	$3,995[S]	-	-	inc.	inc.	$3,000[S]	$4,995[S]	-
60	First DataBank	-	-	-	-	-	-	-	-
41	Freedom Data Systems	$8,000	$14,000[7]	-	-	-	$13,500	-	-
65	GVP Medicare Billing	-	-	-	$149	-	-	-	-
13	HBS, Inc.	$7,995	$140,000[S]	c/p	-	inc.	$11,000[8]	-	$50/mo.
14	HCC	$2,995[S]	$7,500[S]	$1,500[S]	inc.	u/d	$3,995[S]	$2,995[S]	$50/mo.
15	Innovation Associates	-	-	$40,000	-	-	-	-	-
19	Interactive Systems	$9,995	c/p	c/p	[10]	c/p	$5,700[10]	[10]	-
66	The JAG Group	-	-	-	-	-	-	-	c/p
22	jASCorp	$5,995	c/p	-	-	-	c/p	$4,995	c/p
66	Liberty Computer	c/p	c/p	c/p	-	-	c/p	-	-
23	McKessonHBOC	$11,995[11]	$29,995[11]	c/p	-	inc.	$13,995[11]	-	-
61	Micromedex	-	-	-	-	-	-	-	-
66	OmniSYS	$995[S]	-	-	[12]	-	-	-	-
67	OPUS Core Corporation	$3,000[S]	$6,500[S]	$3,500[S]	$1.25/cl.	$.25/trx.	$5,500[S]	-	$40/mo.
25	pc I	$4,000[S]	-	i/f inc.	inc.	inc.	$13,500[H/S]	$5,000[S]	-
26	PDX-NHIN	$10,500[S]	$50,000[S]	i/f inc.	-	inc.	c/p	$5,995[S]	-
62	Pharmex	-	-	-	-	-	-	-	-
28	QS/1 Data Systems	$8,000[S]	$12,000[S]	c/p	$2,500[S]	inc.[14]	$6,000[S]	inc.[14]	inc.[14]
55	RNA	$9,995[S]	-	$4,500[S]	inc.	$9,000[S15]	inc.[IF]	inc.	-
56	Rescot Systems Group	-	inc.	c/p	c/p[16]	-	-	inc.	-
30	Retail Mgmt. Products	-	-	-	-	-	-	-	-
46	RMS	-	$10,000[18]	-	-	-	$6,500	-	-
67	Rx-Net-Inc.	-	-	-	-	-	-	-	-
36	Rx30	c/p	c/p	c/p	c/p	c/p	c/p	c/p	c/p
33	SRS	$8,200[20]	$12,950[20]	-	inc.	-	-	-	-
67	ScripMaster	$9,000	c/p	-	-	-	$20,000	-	$400/yr.
31	ScriptPro	-	-	c/p	-	-	-	-	-
32	Smart Solutions	-	-	-	-	-	-	-	-
68	SymRx	-	-	-	-	$60-75/mo.	-	inc.	inc.
34	TechRx	c/p	c/p	c/p	-	c/p	c/p	-	-
68	TMT	-	-	-	-	-	-	-	-
64	Two Point Conversions	-	-	-	-	-	-	-	-
68	Voice-Tech	-	-	-	-	-	-	-	-

Table 1 Entry-level prices for products and services (*Continued*)

Central Processing/Fill	Compounding Software	IVR	Workflow Mgmt.	Assisted Living	Long Term Care	Outpatient Hospital	HMO Pharmacy	Mail Order	Other
-	-	-	-	-	-	-	-	-	-
-	$1,975[2]	-	-	-	-	-	-	-	-
c/p	-	c/p	c/p	-	-	-	-	-	-
-	-	-	-	-	-	-	-	-	-
c/p	inc.	c/p	c/p	inc.	$3,000[S]	$9,500[S]	-	-	-
c/p	inc.	c/p	inc.	c/p	c/p	c/p	c/p	c/p	4
-	-	-	-	-	$500[S]	-	-	$1,295[S]	-
-	inc.	-	-	-	$10,000[S5]	-	-	-	-
-	-	-	-	-	inc.	-	-	-	-
-	-	-	-	-	-	-	-	-	c/p
-	-	-	-	inc.	$5,995[S5]	$3,995[S]	$6,995[S]	$6,995[S]	-
-	-	-	-	-	-	-	-	-	c/p[6]
-	-	-	-	-	-	-	-	-	-
-	-	-	-	-	-	-	-	-	-
$300,000[S]	-	$2,995[S]	inc.	inc.	$7,500[S]	inc.	-	$75K-275k	-
inc.[9]	inc.	$6,995[H/S]	-	-	inc.	inc.	inc.	inc.	9
$40,000	-	-	$10,000[S]	-	-	$40,000	$40,000	$40,000	-
c/p	-	$5,995[10]	c/p	-	$5,000[10]	c/p	c/p	-	-
-	-	-	-	-	-	-	-	-	-
c/p	inc.	c/p	c/p	c/p	-	c/p	-	-	-
-	c/p	-	-	-	-	c/p	-	c/p	-
-	inc.	$2,400[11]	c/p	-	$5,995[11]	inc.	-	inc.	-
-	-	-	-	-	-	-	-	-	c/p[6]
-	inc.	-	-	-	-	-	-	-	-
-	$500[S]	$2,000[S]	-	$1,500[S]	$2,500[S]	$5,500[S]	-	$3,500[S]	-
-	inc.	i/f inc.	-	inc.	$500[S]	inc.	inc.	inc.	-
c/p	-	$150[S]	c/p	-	$1,800[S]	-	-	-	-
-	-	-	-	-	-	-	-	-	c/p[13]
$12,000[14]	inc.	$1,500[S]	inc.	$3,000[S]	$12,000[S]	inc.[14]	inc.[14]	inc.[14]	-
$7,500[S]	inc.	-	inc.	inc.	inc.	inc.	inc.	inc.	-
inc.	$260[16]	-	c/p	inc.	$885[16]	-	inc.	inc.	-
-	-	-	-	-	-	-	-	-	17
-	-	-	-	-	-	-	-	-	-
-	-	-	-	-	-	-	-	-	c/p[19]
c/p	c/p	c/p	-	c/p	c/p	c/p	c/p	c/p	-
$17,900[20]	-	-	-	$1,200[S]	-	Inc.	-	-	20
-	-	c/p	-	-	-	-	-	c/p	-
c/p	-	-	c/p	-	-	c/p	c/p	c/p	-
-	-	$3,500	$1,500	-	-	-	-	-	-
-	-	-	-	-	-	-	-	-	-
c/p	-	c/p	c/p	c/p	c/p	c/p	c/p	c/p	-
-	-	$4,700	-	-	-	-	-	-	-
-	-	-	-	-	-	-	-	-	c/p[21]
-	-	$4,400[22]	-	-	-	-	-	-	-

Footnotes

1 Price based on transactions.
2 Includes software and support.
3 Price includes training.
4 System includes automatic online split billing.
5 Price includes assisted living.
6 Clinical drug information.
7 POS system for chain central management.
8 $11-12,000 —1 lane; $14-15,000 — 2 lane system.
9 Central fill is under development; "other" includes Dr. Fax, for $1,295.
10 HME is through OmniSYS CareCLAIM; POS price is per register, plus server. Pharm. Care through The jASCorp. IVR price is for four-line system. LTC is additional to retail system.
11 Discounts available to McKessonHBOC customers.
12 No charge for software, but there is a transaction charge.
13 Software-driven warning labels.
14 Included in retail and LTC systems. Central-processing/fill price reflects a host/remote only.
15 Price includes setup as Internet host.
16 Price represents monthly charge that includes license and support fees for software. Company charges on per-claim basis for HME and third-party claims submissions. All categories represent long-term-care pharmacy-management applications.
17 Barcode scanning and verification systems, $399 to $1,495.
18 POS-host system.
19 Pricing service.
20 Price includes conversion, training, scanner, and laser report program. Chain-central management and central-processing/fill systems are scheduled for release 11/2001. "Other" price is $1,400 for major-medical accounts receivable.
21 Data conversion and database-management services.
22 $4,400 is for retail system; $7,500 for hospital/mailorder.

LEGEND

S = software only
H/S = hardware and software
i/f = interface
c/p = call for pricing
inc. = included in price of pharmacy-management system unless otherwise noted.
u/d = under development

Table 2 Installed base and functionality offered

See page		Installed Base	Segments Served					Functionality		
			Independents	Chains	Outpatient Hospitals	Mail Order	HMO Pharmacy	Retail Pharmacy Mgmt.	Chain Central Mgmt.	Auto. Dispensing/ Robotic Systems
65	Allwin Data	8,500[1]	•	•	•	•	•	-	-	-
65	AppDC	68	•	•	•	•	•	-	-	-
7	ateb, Inc.	10,000+[2]	•	•	•	•	•	-	-	-
39	Cam Commerce Solutions	200	•	•	•	-	-	-	•[3]	-
8	CarePoint, Inc.	110	•	•	•	•	•	•	•	i/f
9	ComCoTec	1,700	•	•	•	•	•	•	•	i/f
10	DAA Enterprises	475	•	-	•	•	-	•	-	-
53	Delphi	38[4]	•	•	•	-	-	•	-	i/f
65	Digital Simplistics	250	•	-	•	-	-	•	-	i/f
40	D.P. Hamacher	n/a[5]	•	•	-	-	-	-	-	-
5	Eatonform	300+[6]	•	-	-	-	-	-	-	-
58	Etreby Computer	1,336	•	-	•	•	•	•	-	-
60	First DataBank	n/a[7]	•	•	•	•	•	-	-	-
41	Freedom Data Systems	363	•	•	•	-	-	•	•[8]	i/f
65	GVP Medicare Billing	n/s	•	•	•	-	-	-	-	-
13	HBS, Inc.	1,000+[9]	•	•	•	•	•	•	•	i/f
14	HCC	3,000+	•	•	•	•	•	•	•	i/f
15	Innovation Associates	140+[10]	•	•	-	-	-	-	-	•
19	Interactive Systems	687[11]	•	-	•	-	•	•	•	i/f
66	The JAG Group	1,800	•	•	•	•	•	-	-	-
22	jASCorp	352	•	•	•	-	-	•	•	-
66	Liberty Computer	n/s	•	•	•	•	-	•	•	i/f
23	McKessonHBOC	2,900	•	•	•	•	-	•	•	i/f
61	Micromedex	n/a[12]	•	•	•	•	•	-	-	-
66	OmniSYS, Inc.	320[13]	•	•	-	-	-	•	-	i/f
67	OPUS Core Corporation	n/s	•	•	•	•	•	•	•	i/f
25	pc I	378[14]	•	•	•	•	-	•	-	i/f
26	PDX-NHIN	9,000	•	•	-	u/d	-	•	•	i/f
62	Pharmex	n/a[15]	•	•	•	•	•	-	-	-
28	QS/1 Data Systems	8,182	•	•	•	•	•	•	•	i/f
55	RNA	495	•	•	•	•	•	•	-	i/f
56	Rescot Systems Group	315[16]	•	•	•	-	-	-	•	i/f
30	Retail Mgmt. Products	3,000[17]	•	•	•	•	•	-	-	-
46	RMS	n/s[18]	•	•	•	-	-	-	•	-
67	Rx-Net-Inc.	n/a[19]	•	•	•	-	-	-	-	-
36	Rx30	1,850	•	•	•	•	•	•	•	i/f
33	SRS	140[20]	•	•	•	-	-	•	•	-
67	ScripMaster	350	•	•	-	•	-	•	•	i/f
31	ScriptPro	2,000[21]	•	•	•	•	•	-	-	•
32	Smart Solutions	4,000	•	•	•	•	•	-	-	-
68	SymRx	1,800[22]	•	•	-	-	-	-	-	-
34	TechRx	n/s[23]	•	•	•	•	•	•	•	i/f
68	TMT	430	•	•	•	•	•	-	-	-
64	Two Point Conversions	n/a[24]	•	•	•	-	-	-	-	-
68	Voice-Tech	n/s	•	•	•	•	-	-	-	-

Table 2 Installed base and functionality offered (*Continued*)

Functionality

HME/Diabetes Type II Billing	E-Prescriptions	POS	Pharmaceutical Care	Self-Care/Nutritional	Central Processing/Fill	Compounding	IVR	Workflow Mgmt.	Assisted Living	Long Term Care
•	-	-	-	-	-	-	-	-	-	-
-	-	-	-	-	-	•	-	-	-	-
-	•	-	-	-	•	-	•	•	-	-
-	-	•	-	-	-	-	-	-	-	-
•	•	i/f	•	-	•	•	i/f	•	•	•
•	•	i/f	•	•	•	•	i/f	•	•	•
-	-	i/f	•	-	-	-	-	-	-	•
-	-	-	•	-	-	•	-	-	•	•
•	-	i/f	-	-	-	-	i/f	-	-	•
-	-	-	-	-	-	-	-	-	-	-
-	-	-	-	-	-	-	-	-	•	-
•	•	i/f	•	-	-	-	-	-	•	•
-	-	-	-	-	-	-	-	-	-	-
-	-	•	-	-	-	-	-	-	•	•
•	-	-	-	-	-	-	-	-	-	-
•	•	i/f	-	•	•	•	i/f	•	•	•
•	u/d	•	•	•	u/d	•	i/f	-	•	•
-	-	-	-	-	•	-	-	•	-	-
•	•	•	•	-	•	-	i/f	•	-	•
-	-	-	-	•	-	-	-	-	-	-
-	-	i/f	•	•	-	•	i/f	•	•	-
-	-	•	-	-	-	•	-	-	-	-
-	•	•	-	-	-	•	i/f	i/f	-	•
-	-	-	-	-	-	-	-	-	-	-
•	-	i/f	-	•	-	-	•	-	-	-
•	•	•	•	•	-	-	i/f	•	•	•
•	•	i/f	•	-	-	•	i/f	-	•	•
•	•	i/f	•	-	•	•	i/f	•	-	•
-	-	-	-	-	-	-	-	-	-	-
•	•	•	•	•	•	•	i/f	•	•	•
•	•	i/f	•	-	•	•	-	•	•	•
-	-	-	-	-	•	•	-	•	•	•
-	-	-	-	-	-	-	-	•	-	-
-	-	•	-	-	-	-	-	-	-	-
-	-	-	-	-	-	-	-	-	-	-
•	•	i/f	•	•	•	•	i/f	-	•	•
-	-	-	-	-	-	-	-	-	•	-
•	-	•	-	•	-	-	i/f	•	-	•
-	-	-	-	-	•	-	-	•	-	-
-	•	-	-	-	-	-	•	•	-	-
-	•	-	•	•	-	-	-	-	-	-
-	•	•	-	-	•	-	i/f	•	•	•
-	-	-	-	-	-	-	•	-	-	-
-	-	-	-	-	-	-	-	-	-	-
-	-	-	-	-	-	-	•	-	-	-

Footnotes

1 Internet-based Medicare billing and CMN documentation-management system. Markets served also include infusion pharmacy.
2 IVR system designed to handle electronic prescriptions and interface with central-processing and central-fill systems.
3 Chain central management for POS system.
4 Installations represent primarily large, long-term-care providers. Also serve correctional institutions.
5 Database for front-store systems.
6 Medication-compliance system.
7 Clinical drug data.
8 Chain central management for POS system.
9 Also systems to LTC market and offers ASP service.
10 Also have installations at military locations.
11 Systems also installed in clinics, correctional institutions, and long-term-care facilities.
12 Clinical drug data.
13 Also serve HME pharmacy market, with billing service that has 8,422 clients.
14 Also have systems installed in veterinary practices.
15 Products use by Dept. of Defense/VA as well. Company sells software-driven warning labels.
16 Company is dedicated soley to the long-term-care market. Installation figure represent 6,684 users serving in excess of 1 million LTC beds. All functionality offered is for long-term-care pharmacy. Chain installations represent long-term-care chain providers.
17 Represents installations of barcode scanning for data entry and verification.
18 Company specializes in POS systems. Chain central management is for POS system.
19 Customized pricing service for retail pharmacy.
20 Chain central management system will be released Nov. 2001.
21 Installed base also includes military installations.
22 Web design and management for retail pharmacy. Installations represent CornerDrugstore.com sites.
23 A core product is an ASP service.
24 Data conversion and database-management services.

LEGEND

i/f = interface
n/a = not applicable
n/s = not specified
u/d = under development

work, new and unique hardware challenges will be presented. Bandwidth to the home represents one of the greatest challenges for telecommuting. Services such as DSL and cable modems offer potential solutions to this bandwidth problem. Productivity gains as high as 30 percent are reported as an incentive for investigation of this area.

POINT-OF-CARE SOFTWARE

If one asks the question, "What is my computer supposed to be doing when I'm providing pharmaceutical care?" the answer will not only describe the appropriate hardware or device that matches the needs of the professional providing the care, but should also describe the optimal software that will support the provision of pharmaceutical care. We define the point-of-care as the place where a pharmacist provides pharmaceutical care to a patient or assists a colleague (pharmacist, physician, or nurse) in the provision of care. Many kinds of software available on the market today focus solely on transaction processing, with minimal decision support available through prospective drug utilization review (DUR) modules.

Because the clinical environment demands real-time or near-real-time decisions, a different kind of computer support is required. Pharmacy is like other healthcare disciplines in that we face the problem of having large volumes of information but a lack of information services that are able to translate this information into better outcomes for patients.[2] A clinical practitioner requiring decision support wants this support to be presented in a succinct manner that facilitates a timely response to the problems routinely encountered in his or her practice. Specific characteristics of successful decision-support systems include the provision of patient-specific recommendations, delivery of measurable time savings, and seamless integration into the daily work activities of the clinical setting.[3] Documentation should occur as a byproduct of the interactions between clinical practitioners and their patients or clients. Access to patient records should not only be provided instantaneously through electronic means, but the ability to customize the information provided into a format desired by the individual practitioner should be allowed. When pharmacokinetic calculations are required, known demographic values such as body weight or serum creatinine levels should be prepopulated into calculation variables.

Clinicians will often desire to examine historical data or use relevant references, or primary or secondary literature sources. The software design should include these aspects at a minimum. When prospective drug utilization review flags are presented, false positive warnings should be minimal to prevent practitioners from getting in the habit of simply ignoring them. System oversight for monitoring the potential of medication error should occur throughout the process, beginning with point-of-prescribing through point-of-administration. Bar codes and other technologies will be needed to facilitate this process.

DOCUMENTATION

It has been said, "If you didn't document it, you didn't do it." In the litigious environment in which we live, documentation is paramount to professional survival. Without documentation, reimbursement can be challenged. Personnel reductions are almost assured without documentation to demonstrate the impact of clinical services. Without documentation, unnecessary redundancies and events will be exacerbated.

Ideally, documentation should occur as a natural byproduct of rendering care to patients. In these times where the integration of care (care management) is being sought, the ability of clinical software to access and populate a clinical data repository is a key evaluation criterion. Increasingly, integration with clinical practice protocols is facilitating more effective and more comprehensive delivery of care. A major question is, "Will the patient and the profession of pharmacy be best served by accessing, on a read/write basis, an electronic medical record that is seen by all other disciplines, or should pharmacists to continue to have a pharmacy-specific software solution?"

It may be mission critical to the profession for pharmacists to gain or maintain read/write privileges where all pharmaceutical care contributions can be viewed by all caregivers. Additionally, pharmacists will need to be able to access diagnosis, laboratory, and other charted information such as demographics on a common medical record. Thus, at a minimum, it will be necessary for all pharmacy software to be able to be integrated into the electronic medical records as they emerge.

Orthopedics has recognized the importance of measuring outcomes in terms of quality-adjusted life-years instead of length of implant survival.[2] Similarly, pharmacy must implement software documentation solutions that facilitate outcomes monitoring beyond cost savings. Software is needed with the ability to calculate, in a cost-benefit analysis, the clinical impact of pharmacist interventions as they affect therapeutic, financial, and humanistic outcomes. The current array of products could be better integrated into documentation software to facilitate tabulation of these data. With the power of the Internet to manipulate data in a dynamic database, it would even be possible for hospitals to compare their outcomes on a local, regional, or national basis. Furthermore, the database could

be utilized to gain new insights into additional interventions that could be implemented by clinical personnel.

SPECIFIC CLINICAL SOFTWARE ATTRIBUTES

Distribution software has had many years to evolve and improve. Software that supports the provision of pharmaceutical care is still maturing. The provision of pharmaceutical care is a process. Whether this provision occurs in a community, health system, long-term care facility, or other pharmacy practice setting, there is a process that underlies each practice. The practice of pharmaceutical care begins with the appraisal of the patient. Based on the findings of that appraisal, the pharmacist will perform one or several interventions. Having documented an intervention, the pharmacist will then need to evaluate the outcomes of these interventions. Once the desired outcomes have been achieved and documented, a suitable follow-up and monitoring schedule should be established.

Software that supports this process to the highest degree should be sought, identified, evaluated, purchased, and through user feedback, enhanced continually. The best-of-class clinical software available helps pharmacists assure that an efficient, comprehensive, and cost-effective rendering of pharmaceutical care is provided. Excellent pharmaceutical care software will be evidence-based in all aspects of support, including practice protocols and decision support tools. Due to the importance of financial outcomes, multiple aspects of care provision would be covered by these applications, including medications, nondrug therapies, steps for prevention, lifestyle issues, and alternative medicine. The software would also include suggested outcomes to be measured and appropriate scheduling considerations.

Ideally, clinical software should help prompt practitioners through the provision of the care process. Automatic to-do lists will assure consistency in care provision. Integration with all other aspects of the management and distribution side of pharmacy practice must be assured so that the coordination of care within a pharmacy operation is assured. Prospective drug utilization review assets and succinct decision-support resources should be internalized within the software to minimize the necessity of using multiple applications for every problem-solving exercise. Again, measurement of the potential impact for pharmaceutical care interventions should be done as a by-product of rendering care. This means that as a pharmacist uses clinical software the application deductively calculates potential calamities that were averted and dollar savings that were attained.

Thus far we have only focused on patient care clinical software. An increasingly important alternative focus would include population-based patient management. Some health systems call this care management, and it is largely a nursing-involved activity. Because pharmacists use automation to a greater degree than other healthcare disciplines, it is possible to generate a clinical data repository revolving around drug-related problems in a timelier manner than is possible with a total system approach. Data mining of this repository can yield significant management information resources on both a strategic and tactical level. Selection of clinical software should never ignore this population-based aspect of data analysis. Although pharmacist involvement in this area is currently not as widespread as it could be, there are many opportunities available for expansion.

ASSOCIATED PERFORMANCE-ENHANCEMENT TOOLS

There are several complementary computer applications that deserve mention. Part of the difficulty associated with the increased documentation necessary from pharmacy can be alleviated, in part, through the use of continuous speech recognition applications. This chapter was authored using a speech recognition program, which allowed the authors to speak at 160 words per minute with about 99 percent accuracy. It recognizes medical vocabulary and will allow the user, after only a five-minute training session, to speak directly into any Windows-based program.

Other applications that should be considered for integration involve access to tertiary literature sources directly from the clinical application. The purpose of information is to reduce uncertainty. The ability for the pharmacist to access and validate decisions based on evidence found in the literature is an important skill and a necessary requirement for clinical software support. This kind of support can be provided in palm-top form, from network resources, and as intranet applications.

A movement is underway in medicine and nursing that will allow a concept known as just-in-time continuing education to become mainstream. When pharmacists need to access reference information while solving a clinical problem, it is appropriate that this activity be accredited as continuing education. When pharmacists accumulate one hour of this continuing education activity, a posttest would be offered at a convenient time. A score of 75 percent would result in 0.1 CEUs being awarded. Currently, WebMD is providing credit to physicians who use its Web site to keep up with current issues in medicine. Pharmacists will be the focus of a similar effort if funding permits.

SECURITY, PRIVACY, AND CONFIDENTIALITY

Even though patients distrust the implications surrounding their medical records being available electronically, the benefits for this movement should outweigh the potential risks. There are currently 14 layers of technology to help ensure privacy, security, and confidentiality of the medical records of patients. Encryption, user tracking, biometric authentication, and other measures make it possible, beyond a reasonable certainty, to give patients the assurance necessary for them to sign an informed consent document allowing their records to be online. The basis for moving forward in this effort will be founded on the trust relationship between individual practitioners and individual patients.

Research based on widespread use of electronic medical records in three British hospitals has demonstrated that these records can be practical while ensuring patient privacy. The first step taken was to develop an access control list to identify which individual caregivers are responsible for a patient, and therefore, can access his/her records. The system also documents all occasions when a record is accessed, whether or not information in it was modified. As not all caregivers will be acknowledged as providing care to a particular patient, certain users are given override privileges that allow them to access records when the system is not aware that they are, indeed, providing care to this patient. The user is warned that his actions are being documented when this override procedure is used. The ability to collate enterprise-wide clinical data has posed a problem in this system. When caregivers want to gather data on a specific condition, they are only able to gather information on those patients under their care. This is an acknowledged limitation of this current system. It is important to note that in the five years this system has been in use, no patients have requested a report of all accesses of their record.[4] Again, trust in caregivers translates into trust for the electronic medical record.

RECOMMENDATIONS

There is an old joke about a man, who, upon approaching the gates of heaven, sees that the people in heaven are singing and playing harps, and the people in hell are playing golf and tennis, and watching sports events. Because he doesn't sing very well or play a harp, he decides to opt for hell. Upon arriving, he finds himself enveloped by fiery brimstone and demons that are poking him with their pitchforks. He sees the devil walking by and asks where all the golfing and tennis went? The devil replies, ''Oh, you must have seen our demo.'' One of the first recommendations we make is, during an evaluation of a piece of software, stop allowing the salesperson to show you the ''power path'' way their software can solve all of your problems. Each salesperson knows the best set of circumstances to show off all of the unique features available from an application.

We recommend, instead, that you build a matrix whereby you will place competing systems in rows on the matrix and place all of the features and benefits offered by each system in the columns of the matrix. Next, devise a rating system where 3 would equal excellent, 2 would equal moderately available, 1 would equal minimally available, and 0 would equal missing. A simple priority system can help weight each feature by a priority to the pharmacy operation. A calculation using feature score and feature weight would help create a selection of the most powerful application, with the greatest score identifying the most suitable system from those compared. Subjective assessments of user-friendliness, screen designs, and number of keystrokes necessary to perform the most common tasks can be similarly evaluated.

In clinical applications, the best recommendation for testing available applications would be to use a case-based methodology. We recommend that five or six complicated cases, which would represent a cross section of the patient population served by the pharmacy, be used to test the application. In this way, the clinician will see how the application performs throughout an entire care process and avoid the power path demonstration. In this information age, selecting clinical software is an extremely important task. The explosion of capabilities offered by the Internet can make the selection process both exciting and confusing. A careful analysis of options will usually be rewarded by better results, but wary buyers need to prepare themselves to revisit the marketplace more frequently than they might have in the past to identify innovative alternatives.

REFERENCES

1. Capron, B.; Kuiper, D. The vendor relationship. Manuf. Syst. **1998**, (Suppl. A), 14–16.
2. Hurwitz, S.R.; Slawson, D.; Shaughnessy, A. Orthopaedic information mastery: Applying evidence-based information tools to improve patient outcomes while saving orthopaedists' time. J. Bone Jt. Surg. Am. **2000**, *82-A* (6), 888–894.
3. Payne, T.H. Computer decision support systems. Chest **2000**, *118-S* (2), 47S–52S.
4. Denly, I.; Smith, S.W. Privacy in clinical information systems in secondary care. BMJ **1999**, *318* (7194), 1328–1330.

Credentialing in Pharmacy

Richard J. Bertin
Council on Credentialing in Pharmacy, Washington, D.C., U.S.A.

INTRODUCTION

Pharmacist credentialing has become a topic of important discussions in the profession of pharmacy in recent years. These discussions, inherently complex, have sometimes been further complicated by the lack of a common lexicon. The situation is understandable. Many different words are used to describe the process by which pharmacists are educated, trained, licensed, and otherwise recognized for their competence and achievements. Many different organizations—public and private—are involved in assessing pharmacists' knowledge and skills, granting credentials, and accrediting programs and institutions.

Purpose of This Paper

The purpose of this paper is to create a common frame of reference and understanding for discussions concerning pharmacist credentialing. It begins with definitions of several terms that are essential to any discussion of credentialing. This is followed by a short section highlighting the importance of credentialing to pharmacists. The next three sections, which form the body of the paper, discuss in detail the three types of credentials that pharmacists may earn:

- Credentials needed to prepare for practice (i.e., academic degrees).
- Credentials needed to enter practice (i.e., licensure) and to update professional knowledge and skills (i.e., relicensure) under state law.
- Credentials that pharmacists voluntarily earn to document their specialized or advanced knowledge and skills (i.e., postgraduate degrees, certificates, certification).

Each of these sections contains, as applicable, information about the credential awarded, the training site, whether the credential is voluntary or mandatory, the credentialing body, and the agency that accredits the program. Particular attention is given to pharmacist certification programs, an area that has engendered much of the current interest in pharmacist credentialing.

The paper also includes a brief section on credentialing of pharmacy supportive personnel. It concludes with two appendices. Appendix A contains a comprehensive glossary of key terms relating to pharmacist credentialing. Appendix B is an alphabetical list of organizations involved in pharmacist credentialing and program accreditation. The list contains names, addresses, and uniform resource locators (URLs).

Council on Credentialing in Pharmacy

"Credentialing in Pharmacy" has been created by the Council on Credentialing in Pharmacy (CCP), a coalition of 11 national pharmacy organizations founded in 1999 to provide leadership, standards, public information, and coordination for professional voluntary credentialing programs in pharmacy. Founding members of the CCP include the following organizations:

- Academy of Managed Care Pharmacy.
- American Association of Colleges of Pharmacy.
- American College of Apothecaries.
- American College of Clinical Pharmacy.
- American Council on Pharmaceutical Education.
- American Pharmaceutical Association.
- American Society of Consultant Pharmacists.
- American Society of Health-System Pharmacists.
- Board of Pharmaceutical Specialties.
- Commission for Certification in Geriatric Pharmacy.
- Pharmacy Technician Certification Board.

SIX ESSENTIAL DEFINITIONS

Discussions of credentialing are often complicated by a lack of common understanding of key terms and the contexts in which they are used. To clarify these misunder-

Encyclopedia of Clinical Pharmacy
DOI: 10.1081/E-ECP 120006319

standings, one must first distinguish between processes (e.g., credentialing) and titles (a credential). Distinctions must also be made between processes that focus on individuals (e.g., credentialing and certification) and those that focus on organizations (accreditation). Finally, it is essential to understand that for practicing pharmacists, some credentials are required (e.g., an academic degree or a state license), while others are earned voluntarily (e.g., certification).

Beyond these distinctions, it is also necessary to understand the definitions of the words that commonly come up in discussions of credentialing and to be able to distinguish the sometimes subtle differences among them. A comprehensive glossary of such words and their definitions appears in Appendix A. The following definitions are provided here, because an understanding of these terms is a prerequisite to any meaningful discussion of credentialing in pharmacy.

- A *credential* is documented evidence of a pharmacist's qualifications. Pharmacist credentials include diplomas, licenses, certificates, and certifications. These credentials are reflected in a variety of abbreviations that pharmacists place after their names (e.g., Pharm.D. for ''doctor of pharmacy,'' an earned academic degree; R.Ph. for ''registered pharmacist,'' which indicates state licensure; and acronyms such as BCNSP for ''Board-Certified Nutrition Support Pharmacist,'' which indicates that an individual has demonstrated advanced knowledge or skill in a specialized area of pharmacy).
- *Credentialing* is the process by which an organization or institution obtains, verifies, and assesses a pharmacist's qualifications to provide patient care services.
- *Accreditation* is the process by which a private association, organization, or government agency, after initial and periodic evaluations, grants recognition to an organization that has met certain established criteria.
- A *certificate* is a document issued to a pharmacist upon successful completion of the predetermined level of performance of a certificate training program or of a pharmacy residency or fellowship.
- A *statement of continuing education credit* is a document issued to a pharmacist upon participation in an accredited continuing education program.
- *Certification* is a voluntary process by which a nongovernmental agency or an association grants recognition to a pharmacist who has met certain predetermined qualifications specified by that organization. This formal recognition is granted to designate to the public that this pharmacist has attained the requisite level of knowledge, skill, or experience in a well-defined, often specialized, area of the total discipline. Certification usually requires initial assessment and periodic reassessments of the individual's qualifications.

IMPORTANCE OF CREDENTIALS IN PHARMACY

''Credential'' and ''credentialing,'' like the words ''creed'' and ''credence,'' derive from the Latin verb *credere*, which means ''to trust,'' ''to entrust,'' or ''to believe.'' A pharmacist's credentials are indicators that he or she holds the qualifications needed to practice the profession of pharmacy and is therefore worthy of the trust of patients, of other health care professionals, and of society as a whole.

In the profession of pharmacy, the interest in credentials has been catalyzed in recent years by several factors. First among them is the pace of change and the increasing complexity of health care. A second factor is the pharmacist's expanding clinical role. Interest in credentialing has likewise been stimulated by the growing trend toward specialization in pharmacy practice and by the need to document the pharmacist's ability to provide specialty care.

Another contributing factor has been the need to help ensure lifelong competence in a rapidly changing, technologically complex field. The need to provide a means of standardization of practice has also had a role. Such a motivation was key, for example, to the development of the Federal Credentialing Program, which is creating a national database of health professionals that will include pharmacists.

Finally, economic realities enter the picture. Pharmacists who are providing cognitive services or specialized care need to be reimbursed for the services they provide. Payers rightfully demand validation that pharmacists are qualified to provide such services. Credentials, and in many cases, more specifically, certification, can help provide the documentation that Medicare and Medicaid, managed care organizations, and other third-party payers require of pharmacists today and in the future.

OVERVIEW OF CREDENTIALING IN PHARMACY

Pharmacist credentials may be divided into three fundamental types:

- The first type—college and university degrees—is awarded to mark the successful completion of a pharmacist's academic training and education.

- The second type—licensure and relicensure—is an indication that the pharmacist has met minimum requirements set by the state in which he or she intends to practice.
- The third type of credential—which may include advanced degrees and certificates—is awarded to pharmacy practitioners who have completed programs of various types that are intended to develop and enhance their knowledge and skills, or who have successfully documented an advanced level of knowledge and skill through an assessment process.

These three paths to pharmacist credentialing are illustrated in Fig. 1. The sections that follow provide information on each of the credentials offered in pharmacy, the credentialing or accreditation body involved, whether the credential is mandatory or voluntary, and other related information.

Preparing for the Pharmacy Profession

- *Credential earned*: Bachelor of Science degree in Pharmacy; Doctor of Pharmacy degree.
- *Credential awarded by*: School or college of pharmacy.
- *Accreditation body for professional programs in pharmacy*: American Council on Pharmaceutical Education (ACPE). The U.S. Department of Education has recognized the ACPE accreditation of the professional degree program in pharmacy.

Until July 1, 2000, an individual who wished to become a pharmacist could enroll in a program of study that would lead to one of two degrees: a bachelor of science degree in pharmacy (B.S.Pharm. or Pharm.B.S.) or a doctor of pharmacy (Pharm.D.) degree.

As of 1998, two-thirds of all students studying in professional programs in pharmacy were enrolled in Pharm.D. programs. The Pharm.D. degree became the sole degree accredited by ACPE for pharmacists' entry into practice in the United States, as of July 1, 2000, with the institution of new ACPE professional program accreditation standards. Pharm.D. programs typically take six years to complete and generally involve two years of preprofessional coursework and four years of professional education. A few programs

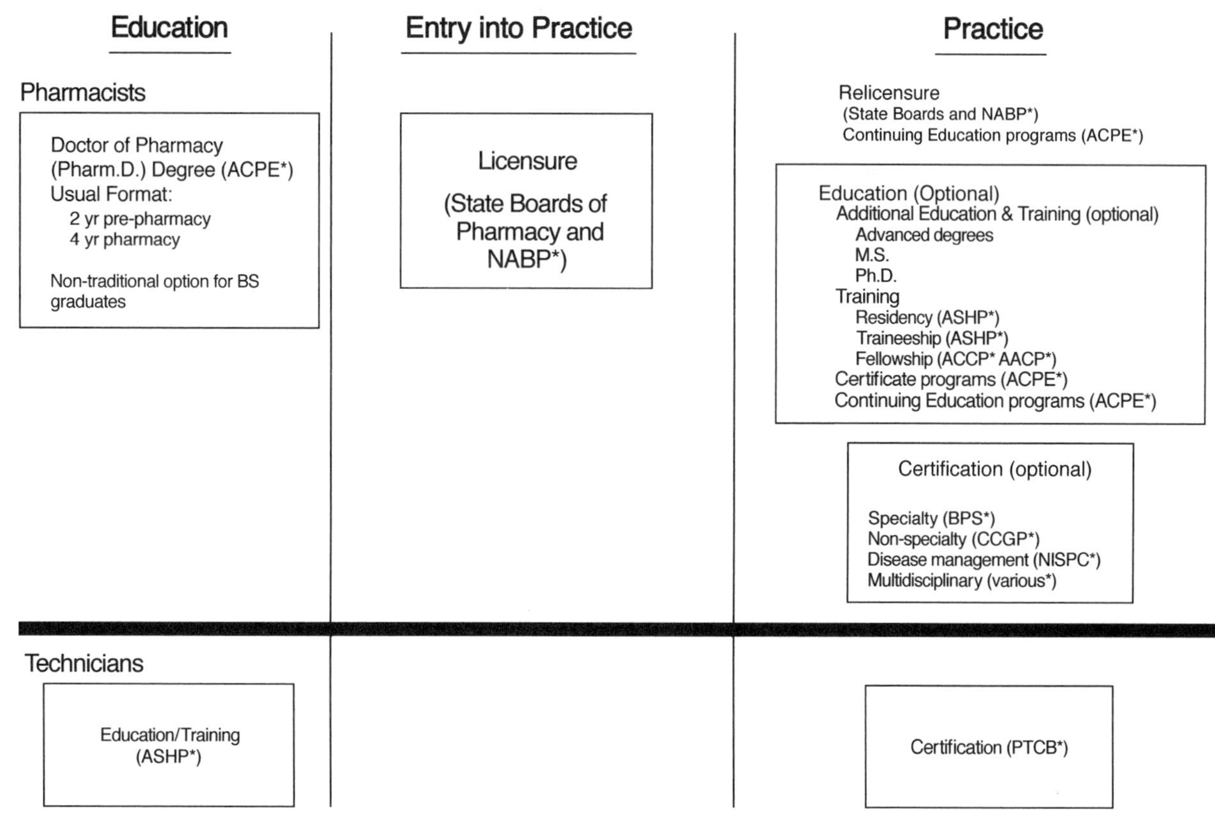

Fig. 1 U.S. pharmacy credentials and oversight bodies. (*Oversight bodies are described in text.)

offer the professional education over three years of full-time education.

B.S. level pharmacists who have been in the workforce may also return to a college or school of pharmacy to earn the Pharm.D. degree. These programs, which are tailored to the individual's background and experience, may follow ''nontraditional'' pathways; however, they must produce the same educational outcomes as does the entry-level Pharm.D. degree.

State boards of pharmacy require a Pharm.D. or B.S. degree from a program approved by the boards (almost always an ACPE-accredited program) for a candidate to be eligible to take the state licensing examination. A listing of accredited professional programs offered by colleges and schools of pharmacy is published annually by the ACPE, and is available on the ACPE web site (www.acpe-accredit.org).

Entering Practice and Updating Professional Knowledge and Skills

- *Credentials earned*: Licensure as registered pharmacist (R.Ph.); relicensure.
- *Credential awarded by*: State board of pharmacy.
- *Licensure process overseen by*: State regulatory authorities.

Before a graduate of a school or college of pharmacy can practice pharmacy in the United States, he or she must become licensed. The licensure process is regulated at the state level by the boards of pharmacy.

Candidates for licensure in all states but California must pass the North American Pharmacist Licensure Examination™ (NAPLEX®), a computer-adaptive, competency-based examination that assesses the candidate's ability to apply knowledge gained in pharmacy school to real-life practice situations. California administers a unique examination process. Most states also require candidates to take a state-specific pharmacy law examination. Currently, 36 states use the Multistate Pharmacy Jurisprudence Examination™ (MPJE™), a computer-adaptive assessment that tailors each examination to address the pharmacy law and regulations of the state in which the candidate is seeking licensure.

Both the NAPLEX and the MPJE are developed by the National Association of Boards of Pharmacy (NABP) for use by the boards of pharmacy as part of their assessment of competence to practice pharmacy. Development of these examinations is directly related to NABP's mission, which is to assist its member boards and jurisdictions in developing, implementing, and enforcing uniform standards for the purpose of protecting the public health. The NAPLEX and MPJE examinations are administered by appointment, daily, throughout the year, at a system of test centers located in all 50 states.

In addition to the NAPLEX and MPJE, some states require a laboratory examination or an oral examination before licensure is conferred. All state boards also require that candidates complete an internship before being licensed. The internship may be completed during the candidate's academic training or after graduation, depending upon state requirements.

State licensure is an indication that the individual has attained the basic degree of competence necessary to ensure the public health and welfare will be reasonably well protected. The names of individuals who have received a license may use the abbreviation ''R.Ph.'' (for ''registered pharmacist'') after their names.

Nearly all state boards of pharmacy also require that registered pharmacists complete a certain number of continuing education units (CEUs) before they can renew their licenses. The CEUs must be earned through participation in a continuing education (CE) program whose provider has been approved by the American Council on Pharmaceutical Education (ACPE). The symbol used by the American Council on Pharmaceutical Education to designate that the continuing education provider is approved is ACPE.

Note that ACPE approves providers of continuing education, not individual CE programs. CEUs may be secured by attending educational seminars, teleconferences, and meetings; reading journal articles; or completing traditional home study courses or computer-based education programs. Receipt of a satisfactory score on an assessment that is created by and submitted to the CE provider is sometimes required as a documentation of completion of a CE program. ACPE publishes an annual directory of approved providers of continuing pharmaceutical education, which is available on the ACPE web site (www.acpe-accredit.org).

Licensure and relicensure are mandatory for pharmacists who wish to continue to practice their profession.

In their regulatory role, state boards of pharmacy are ultimately responsible to the state legislature.

Developing and Enhancing Knowledge and Skills

Pharmacy practitioners who wish to broaden and deepen their knowledge and skills may participate in a variety of postgraduate education and training opportunities. They include the following:

Academic postgraduate education and training

Pharmacists who wish to pursue a certain field of study in depth may enroll in postgraduate master's or doctor of philosophy (Ph.D.) programs. Common fields of study for master's candidates include business administration, clinical pharmacy, and public health. Common fields for Ph.D. studies include pharmacology, pharmaceutics, pharmacy practice, and social and administrative sciences.

Residencies

- *Credential earned*: Residency certificate.
- *Credential awarded by*: Residency training program.
- *Program accreditation*: The American Society of Health-System Pharmacists (ASHP). (independently or in collaboration with other pharmacy organizations).

ASHP is the chief accreditation body for pharmacy practice and specialty residency programs in pharmacy. A total of 505 programs nationwide now hold ASHP accreditation. ASHP also partners with other organizations, including the Academy of Managed Care Pharmacy, the American College of Clinical Pharmacy, the American Pharmaceutical Association, and the American Society of Consultant Pharmacists, in accrediting residency programs.

The majority of pharmacists who pursue residency training do so in the area of pharmacy practice. These residencies sometimes focus on a particular practice setting, such as ambulatory care. Pharmacists may also pursue specialty training in a certain topic (e.g., pharmacokinetics), in the care of a specific patient population (e.g., pediatrics), or in a specific disease area (e.g., oncology).

Residency programs last one to two years. The typical training site is a practice setting such as an academic health center, a community pharmacy, a managed care organization, a skilled nursing facility, or a home health care agency.

The Health Care Financing Administration (HCFA), an agency of the federal government, recognizes residency accreditation bodies within the health professions.

Fellowships[a]

- *Credential earned*: Fellowship certificate.
- *Credential awarded by*: Fellowship training program.
- *Program accreditation*: No official accreditation body.

A fellowship is an individualized postgraduate program that prepares the participant to become an independent researcher. Fellowship programs, like residencies, usually last one to two years. The programs are developed by colleges of pharmacy, academic health centers, colleges and universities, and pharmaceutical manufacturers.

There is no official accreditation body for fellowship programs; however, the American Association of Colleges of Pharmacy and American College of Clinical Pharmacy have issued guidelines that are followed by many fellowship program directors.

Certificate Training Programs

- *Credential earned*: Certificate of Completion.
- *Credential awarded by*: Educational institutions and companies, pharmacy organizations, and others.
- *Provider accreditation*: American Council on Pharmaceutical Education.

A certificate training program is a structured and systematic postgraduate continuing education experience for pharmacists that is generally smaller in magnitude and shorter in duration than degree programs. Certificate programs are designed to instill, expand, or enhance practice competencies through the systematic acquisition of specified knowledge, skills, attitudes, and behaviors. The focus of certificate programs is relatively narrow; for example, the American Pharmaceutical Association offers programs in such areas as asthma, diabetes, immunization delivery, and management of dyslipidemias.

Certificate training programs are offered by national and state pharmacy organizations and by schools and colleges of pharmacy and other educational groups. The programs are often held in conjunction with a major educational meeting of an organization. The American Council on Pharmaceutical Education (ACPE) approves providers of such programs. The symbol used by the ACPE to designate that a certificate training program is provided by an accredited provider is 
.

---

[a]Several pharmacy organizations, including the American College of Clinical Pharmacy, the American Society of Health-System Pharmacists, and the American Pharmaceutical Association, award the honorary title of "Fellow" to selected members as a means of publicly recognizing their contributions to the profession. A Fellow of ASHP, for example, may write "FASHP" for "Fellow of the American Society of Health-System Pharmacists," after his or her name. The two uses of the word "fellow"—one denoting an individual participating in a postgraduate training program and the other denoting receipt of an honorary title—should be clearly distinguished.

Traineeships

Traineeships, in contrast to certificate training programs, are defined as intensive, individualized, structured postgraduate programs intended to provide the participant with the knowledge and skills needed to provide a high level of care to patients with various chronic diseases and conditions. Traineeships are generally of longer duration (about five days) and involve smaller groups of trainees than certificate training programs do. Some are offered on a competitive basis, with a corporate sponsor or other organization underwriting participants' costs. Pharmacy organizations currently offering traineeships include the American College of Apothecaries, the American Society of Consultant Pharmacists, and the American Society of Health-System Pharmacists' Research and Education Foundation.

Certification

Certification is a credential granted to pharmacists and other health professionals who have demonstrated a level of competence in a specific and relatively narrow area of practice that exceeds the minimum requirements for licensure. Certification is granted on the basis of successful completion of rigorously developed eligibility criteria that include a written examination and, in some cases, an experiential component. The certification process is undertaken and overseen by a nongovernmental body.

The development of a certification program includes the following steps:

- *Role delineation.* The first step is to define the area in which certification is to be offered. This is done through a process called role delineation or "task analysis." An expert panel of individuals in the proposed subject area develops a survey instrument to assess how practitioners working in the area rate the importance, frequency, and criticality of specific activities in that practice. The instrument is then sent to a sample of pharmacists who are practicing in that field.
- *Development of content outline.* On the basis of responses to the survey, a content outline for the certification program is developed.
- *Preparation of examination.* The written examination component of the certification program is developed on the basis of the content outline.
- *Other activities.* Appropriate measures are taken to ensure that security and confidentially of the testing process are maintained, that the examination and eligibility criteria are appropriate, and that the knowledge and skills of those who are certified do, in fact, reflect competence.

A professional testing company typically assists in the development of the role delineation and the examination to ensure that the examination meets professional standards of psychometric soundness and legal defensibility.

Certifying agencies for pharmacists only

Three groups—the Board of Pharmaceutical Specialties, the Commission for Certification in Geriatric Pharmacy, and the National Institute for Standards in Pharmacist Credentialing—offer certification to pharmacists.

Board of Pharmaceutical Specialties (BPS). Established in 1976 by the American Pharmaceutical Association, BPS is the only agency that offers certification at the specialty level in pharmacy. It certifies pharmacists in five specialties: nuclear pharmacy, nutrition support pharmacy, oncology pharmacy, pharmacotherapy, and psychiatric pharmacy. As of June 2002, nearly 3500 pharmacists held BPS certification, distributed across the five specialties as follows:

Nuclear Pharmacy—471
Nutrition Support Pharmacy—425
Oncology Pharmacy—288
Pharmacotherapy—1843
Psychiatric Pharmacy—387

Pharmacists who wish to retain BPS certification must be recertified every seven years.

The recognition of each specialty is the result of a collaborative process between the Board and one or more pharmacy organizations, which develop a petition to support and justify recognition of the specialty. This petition must meet written criteria established by the BPS.

The BPS is directed by a nine-member board that includes six pharmacists, two health professionals who are not pharmacists, and one public/consumer member. A specialty council of six specialist members and three pharmacists not in the specialty direct the certification process for each specialty.

BPS examinations are administered with the assistance of an educational testing firm, resulting in a process that is psychometrically sound and legally defensible. Each of the five specialties has its own eligibility criteria, examination specifications, and recertification processes. All

five examinations are given on a single day once a year in approximately 25 sites in the United States and elsewhere.

In 1997, BPS introduced a method designed to recognize focused areas within pharmacy specialties. A designation of ''Added Qualifications'' denotes that an individual has demonstrated an enhanced level of training and experience in one segment of a BPS-recognized specialty. Added qualifications are conferred on the basis of a portfolio review to qualified individuals who already hold BPS certification. The first added qualification to receive BPS approval was infectious diseases, within the pharmacotherapy specialty.

Commission for Certification in Geriatric Pharmacy (CCGP). In 1997, the American Society of Consultant Pharmacists (ASCP) Board of Directors voted to create the CCGP to oversee a certification program in geriatric pharmacy practice. CCGP is a nonprofit corporation that is autonomous from ASCP. It has its own governing Board of Commissioners. The CCGP Board of Commissioners includes five pharmacist members, one physician member, one payer/employer member, one public/consumer member, and one liaison member from the ASCP Board of Directors.

Pharmacists who meet CCGP's requirements are entitled to use the designation Certified Geriatric Pharmacist, or CGP. As of June 2002, approximately 800 pharmacists have earned the CGP credential. Pharmacists who wish to retain their CGP credential must recertify every five years by successfully completing a written examination.

CCGP contracts with a professional testing firm to assist in conducting the role delineation or task analysis and in developing and administering the examination. The resulting process is psychometrically sound and legally defensible; it also meets nationally recognized standards. The CGP certification exams are administered twice a year at multiple locations in the United States, Canada, and Australia. CCGP publishes a candidate handbook that includes the content outline for the examination, eligibility criteria for taking the examination, and the policies and procedures of the certification program.

National Institute for Standards in Pharmacist Credentialing (NISPC). The NISPC was founded in 1998 by the American Pharmaceutical Association, the National Association of Boards of Pharmacy (NABP), the National Association of Chain Drug Stores, and the National Community Pharmacists Association. The purpose of NISPC is to ''promote the value and encourage the adoption of National Association of Boards of Pharmacy disease-specific examinations as the consistent and objective means of documenting the ability of pharmacists to provide disease state management services.''

NISPC offers certification in the management of diabetes, asthma, dyslipidemia, and anticoagulation therapy. At the time of its founding, the organization's immediate objective was to design a process that would document the competence of pharmacists providing care for patients with these disease states. The NISPC credential was first recognized in the state of Mississippi, where it was used to enable pharmacists to qualify for Medicaid reimbursement as part of a pilot project in that state. The NABP developed the competency assessment examinations and oversees their administration. As of June 2002, 1340 pharmacists hold NISPC certification: 771 in diabetes, 314 in asthma, 120 in dyslipidemia, and 135 in anticoagulation therapy.

The NISPC tests are administered nationally as computerized examinations and are available throughout the year.

Multidisciplinary certification programs

Some certification programs are available to professionals from many health disciplines, including pharmacists. Areas in which such certification programs are available include diabetes education, anticoagulation therapy, pain management, and asthma education. Some of these programs are still in the early stages of development. Several of these providers are listed in Appendix B; however, the information is not intended to be exhaustive.

PHARMACY SUPPORTIVE PERSONNEL

A pharmacy technician is an individual who assists in pharmacy activities that do not require the professional judgment of a pharmacist. For example, pharmacy technicians may accept orders from patients, prepare labels, enter drug information into the pharmacy's computer system, and retrieve medications from inventory. As pharmacists assume an increasing number of clinical roles, pharmacy technicians are taking more and more responsibility for distributive functions in pharmacies in all settings.

The exact functions and responsibilities of pharmacy technicians are defined by state laws and regulations and are also determined by the willingness of pharmacists to delegate the nonjudgmental activities of their practice. Pharmacy technicians always work under the supervision of a licensed pharmacist.

The education and training, certification, and continuing education of pharmacy technicians are similar in some ways to those of pharmacists.

Education and Training

Most pharmacy technicians today have been trained on the job, either formally or informally. As the responsibilities of pharmacy technicians grow, however, more and more individuals are enrolling in formal training programs. These programs are generally affiliated with a community college, a four-year college, a hospital, or another health care organization. Graduates of these programs may be awarded an associate's degree or a certificate of completion.

ASHP is the accreditation body for pharmacy technician training programs. Sixty programs were accredited as of 1999.

Regulation

State boards of pharmacy oversee the registration of pharmacy technicians. Practices differ substantially from state to state.

Certification

The Pharmacy Technician Certification Board (PTCB) was established in 1995 as a national voluntary certification program for pharmacy technicians. Its founders were the American Pharmaceutical Association, the American Society of Health-System Pharmacists, the Illinois Council of Health-System Pharmacists, and the Michigan Pharmacists Association.

In collaboration with testing experts, the PTCB developed a national examination, the Pharmacy Technician Certification Examination (PTCE). The examination is designed to assess the candidate's knowledge and skill base for activities that are most commonly performed by a pharmacy technician, as determined by a national task analysis.

The Board administers the PTCE three times a year at more than 120 sites across the nation. A technician who passes the PTCE is designated as a Certified Pharmacy Technician (CPhT). As of June 2002, more than 100,000 pharmacy technicians have earned PTCB certification.

Pharmacy technicians must renew their certification every two years. To qualify for recertification, they must participate in at least 20 hr of pharmacy-related continuing education that includes an hour of pharmacy law.

APPENDIX A

Glossary

These definitions have been developed by a variety of organizations involved in credentialing and are generally accepted by those in the pharmacist credentialing arena.

Accreditation: The process whereby an association or agency grants public recognition to an organization that meets certain established qualifications or standards, as determined through initial and periodic evaluations.

Certificate Training Program: A structured, systematic postgraduate education and continuing education experience for pharmacists that is generally smaller in magnitude and shorter in duration than a degree program. Certificate programs are designed to instill, expand, or enhance practice competencies through the systematic acquisition of specific knowledge, skills, attitudes, and performance behaviors.

Certified: Adjective that is used to describe an individual who holds certification and that is incorporated into the name of the credential awarded that individual. For example, someone who has earned BPS certification in oncology is a ''Board-Certified Oncology Pharmacist.''

Certificate: A certificate is a document issued to a pharmacist upon successful completion of the predetermined level of performance of a certificate training program or of a pharmacy residency or fellowship. See also ''statement of continuing education credit.''

Certification: The voluntary process by which a nongovernmental agency or association formally grants recognition to a pharmacist who has met certain predetermined qualifications specified by that organization. This recognition designates to the public that the holder has attained the requisite level of knowledge, skill, or experience in a well-defined, often specialized, area of the total discipline. Certification entails assessment, including testing, an evaluation of the candidate's education and experience, or both. Periodic recertification is usually required to retain the credential.

Clinical privileges: Authorization to provide a specific range of patient care services. *See* also Privileging.

Competence: The ability to perform one's duties accurately, make correct judgments, and interact appropriately with patients and with colleagues. Professional competence is characterized by good problem-solving and decision-making abilities, a strong knowledge base, and the ability to apply knowledge and experience to diverse patient-care situations.

Competency: A distinct skill, ability, or attitude that is essential to the practice of a profession. Individual competencies for pharmacists include, for example, mastery of aseptic technique and achievement of a thought process that enables one to identify therapeutic duplications. Pharmacists must master a variety of competencies in order to gain competence in a profession.

Continuing education: Organized learning experiences and activities in which pharmacists engage after they have completed their entry-level academic education and training. These experiences are designed to promote the continuous development of the skills, attitudes, and knowledge needed to maintain proficiency, provide quality service or products, respond to patient needs, and keep abreast of change.

Credential: Documented evidence of professional qualifications. For pharmacists, academic degrees, state licensure, and Board certification are all examples of credentials.

Credentialing: 1) The process by which an organization or institution obtains, verifies, and assesses a pharmacist's qualifications to provide patient care services; 2) The process of granting a credential (a designation that indicates qualifications in a subject or an area.)

Fellowship: A directed, highly individualized postgraduate program designed to prepare a pharmacist to become an independent researcher.

License: A credential issued by a state or federal body that indicates that the holder is in compliance with minimum mandatory governmental requirements necessary to practice in a particular profession or occupation.

Licensure: The process of granting a license.

Pharmacy technician: An individual who, under the supervision of a licensed pharmacist, assists in pharmacy activities not requiring the professional judgment of the pharmacist.

Privileging: The process by which a health care organization, having reviewed an individual health care provider's credentials and performance and found them satisfactory, authorizes that individual to perform a specific scope of patient care services within that organization.

Residency: An organized, directed, postgraduate training program in a defined area of pharmacy practice.

Registered: Adjective used to describe a pharmacist who has met state requirements for licensure and whose name has been entered on a state registry of practitioners who are licensed to practice in that jurisdiction.

Scope of practice: The boundaries within which a health professional may practice. For pharmacists, the scope of practice is generally established by the board or agency that regulates the profession in a given state or organization.

Statement of continuing education credit: A document issued to a pharmacist upon completion of a continuing education program provided by an organization approved by the American Council on Pharmaceutical Education.

Traineeship: A short, intensive, clinical, and didactic postgraduate educational program intended to provide the pharmacist with knowledge and skills needed to provide a high level of care to patients with specific diseases or conditions.

APPENDIX B

Referenced Pharmacy Organizations and Certification Bodies

Pharmacy organizations

Academy of Managed Care Pharmacy (AMCP)
100 North Pitt Street, Suite 400; Alexandria, Virginia 22314; (800) 827-2627
www.amcp.org

American Association of Colleges of Pharmacy (AACP)
1426 Prince Street; Alexandria, Virginia 22314-2841; (703) 836-8982
www.aacp.org

American College of Apothecaries (ACA)
P.O. Box 341266; Memphis, Tennessee 38184; (901) 383-8119
www.acaresourcecenter.org

American College of Clinical Pharmacy (ACCP)
3101 Broadway, Suite 380; Kansas City, Missouri 64111; (816) 531-2177
www.accp.com

American Council on Pharmaceutical Education (ACPE)
20 North Clark Street, Suite 2500; Chicago, Illinois 60610; (312) 664-3575
www.acpe-accredit.org

American Pharmaceutical Association (APhA)
2215 Constitution Avenue, NW; Washington, D.C. 20037-2985; (202) 628-4410
www.aphanet.org

American Society of Consultant Pharmacists (ASCP)
1321 Duke Street; Alexandria, Virginia 22314-3563; (703) 739-1300
www.ascp.com

American Society of Health-System Pharmacists (ASHP)
7272 Wisconsin Avenue; Bethesda, Maryland 20814; (301) 657-3000
www.ashp.org

National Association of Boards of Pharmacy (NABP)
700 Busse Highway; Park Ridge, Illinois 60068; (847) 698-6227
www.nabp.net

National Association of Chain Drug Stores (NACDS)
413 N. Lee Street, P.O. Box 1417-D49; Alexandria, Virginia 22313-1480; (703) 549-3001
www.nacds.org

National Community Pharmacists Association (NCPA)
205 Daingerfield Road; Alexandria, Virginia 22314; (703) 683-8200
www.ncpanet.org

Certification bodies for pharmacists or pharmacy technicians (may be multidisciplinary)

Anticoagulation Forum
88 East Newton Street, E-113; Boston, Massachusetts 02118-2395; (617) 638-7265
www.acforum.org

Board of Pharmaceutical Specialties (BPS)
2215 Constitution Avenue, NW; Washington, D.C. 20037-2985; (202) 429-7591
www.bpsweb.org

Council on Certification in Geriatric Pharmacy (CCGP)
1321 Duke Street; Alexandria, Virginia 22314-3563; (703) 535-3038
www.ccgp.org

National Asthma Educator Certification Board
American Lung Association
1740 Broadway; New York, New York 10019-4374; (212) 315-8865
www.lungusa.org

National Certification Board for Diabetes Educators (NCBDE)
330 East Algonquin Road, Suite 4; Arlington Heights, Illinois 60005; (847) 228-9795
www.nbcde.org

National Institute for Standards in Pharmacist Credentialing (NISPC)
P.O. Box 1910; Alexandria, Virginia 22313-1910; (703) 299-8790
www.nispcnet.org

Pharmacy Technician Certification Board (PTCB)
2215 Constitution Avenue, NW; Washington, D.C. 20037-2985; (202) 429-7576
www.ptcb.org

Critical Care Pharmacy Practice

Judith Jacobi
Methodist Hospital/Clarian Health, Indianapolis, Indiana, U.S.A.

INTRODUCTION

Pharmacists have been practicing in critical care since the 1970s when the new breed of clinical pharmacists were expanding the horizons of care to include more than dispensing services. These early practitioners recognized, as we do today, that critically ill patients are treated with a large number of different drugs that have significant alterations in their pharmacokinetics and pharmacodynamics, and great potential for drug misadventure.

Today, a growing number of pharmacists practice exclusively in critical care settings. They are recognized members of the critical care team, along with the intensivist, nurses, dietitians, respiratory therapists, and others. The exact number of critical care pharmacists is not known, but more than 450 pharmacists are members of the Society of Critical Care Medicine (SCCM). This number may only represent a fraction of the total number because many other pharmacists work in critical care satellites and provide a broad range of pharmaceutical care services.

HISTORY OF CRITICAL CARE PHARMACY

The 1985 textbook, *Practice of Critical Care Pharmacy* was the gateway for many young practitioners. This book contained chapters written by many of the groundbreaking practitioners in critical care.[1] These chapters detailed the roles these and other influential critical care practitioners had developed and served as a primer for practice in various intensive care units (ICU). These clinicians deserve recognition for this and other contributions to the development of critical care pharmacy practice in a variety of practice settings: Dave Angaran, MS (cardiovascular); Deborah Armstrong, PharmD (pediatric); Joyce Comer, PharmD (medical); Gary Cupit,PharmD (pediatric and neonatal); Joseph Dasta, MS (surgical); Robert Elenbaas, PharmD (emergency medicine); Thomas Majerus, PharmD (trauma); and Christine Quandt, PharmD (neurosurgical). These same practitioners fostered the expansion of the specialty through pharmacy residency and fellowship training programs.

The need for ongoing education in a rapidly evolving specialty led these practitioners to the SCCM, an organization whose membership includes critical care practitioners from every discipline. As the number of pharmacist members grew, a section was formed within SCCM in 1989, and its membership has expanded progressively. Although the pharmacy organizations, the American College of Clinical Pharmacy (ACCP) and American Society of Health-Systems Pharmacists (ASHP) have sections or interest groups devoted to critical care, the SCCM has been a focal point for the organizational activity of many of these pharmacists. Many critical care pharmacists have made significant contributions to these national organizations through service on the governing boards and committees.

Bart Chernow, MD, a member of SCCM, was very influential in the initiation of the Clinical Pharmacy and Pharmacology Section of SCCM. His textbook on critical care pharmacology included a chapter entitled, "The Role of the Pharmacist in Caring for the Critically Ill Patient."[2] This was a vehicle to introduce the benefits of the pharmacist as a member of the critical care team to critical care physicians. Critical care physicians have been strong advocates for the role of the critical care pharmacist, were essential champions for the early practitioners, and continue to be today. The SCCM recognizes the critical care pharmacist as an essential member of the multidisciplinary ICU team.[3] These guidelines recommend that pharmacists provide pharmaceutical care through monitoring of dosing, adverse drug reactions, drug interactions, and education to create cost-optimized regimens. In addition, the guidelines also recognize the important role of critical care pharmacists in the care of the most complex patients treated in tertiary care centers. The SCCM also includes a permanent position for a critical care pharmacist on its governing council.

Additional recognition of the importance of the critical care pharmacist has been the selection of numerous pharmacists as fellows of the American College of Critical Care Medicine (ACCM). This designation is awarded to critical care practitioners who have demonstrated a high level of practice, research, education, and service to SCCM and their local organizations.

Encyclopedia of Clinical Pharmacy
DOI: 10.1081/E-ECP 120006226

Pharmacists have contributed to our understanding of the needs of the critically ill patient through descriptive reports characterizing patterns of drug utilization. In surgical and trauma patients, an average total of 7.6 and 9.1 medications per patient were reported, respectively.[4,5] These data reaffirm reports that a growing number of medications are used in ICU patients, from an average of 4.2 to 7 ± 4.6 drugs per patient in general ICUs.[6,7] In 1988, the pharmacoeconomic impact of the large number of medications used in surgical ICU patients was significant, averaging 13.6% of total hospital charges.[8]

Hazards of these complex drug therapy regimens are becoming well documented. Concern with medication errors is a frequent topic in the medical and lay literature.[9] A general adult ICU reported that a medication error was made in more than 50% of the patients, but the overall error rate was low, detected in 2.2% of the doses dispensed or administered.[10] However, this unit reported a generous 1:1 nurse-to-patient ratio and relied on a retrospective medical record review to detect errors. Current staffing patterns with 1:2 or 1:3 nurse-to-patient ratios are becoming more common and may adversely affect error frequency. Higher rates have been detected in other studies, due to differences in the definition of an error and the method of detection. New therapies and technologies may increase the risk of critical errors.[11] The ability of the pharmacist to reduce medication errors and adverse drug events has been reported.[12] Pharmacists participating on physician rounds on a part-time basis were shown to decrease all adverse drug events and, in particular, to decrease preventable adverse drug events by 66%.[12] The pharmacist primarily detected prescribing errors such as incomplete orders, incorrect doses or frequency, and therapy duplication; however, the pharmacist also avoided inappropriate drug selection. Alternative therapies were recommended that were safer (avoided a drug interaction or allergy) or less expensive. Participation of a pharmacist in the medication ordering process in a teaching hospital appears to be essential to avoid prescribing errors.

Pharmacists have used avoidance of medication errors to justify expanding services. A pediatric critical care satellite was opened to reduce the rate of errors from a total of 17.4% in an intensive care nursery and 38% in a pediatric ICU.[13] A large number of the errors (86.5%) occurred with medications possessing a high potential for serious adverse consequences.

Avoidance of medication administration errors is another potential contribution of critical care pharmacists. A multicenter analysis of medication errors from five ICUs revealed that medication errors occurred most commonly with vasoactive agents and sedative-analgesics.[14] Incorrect infusion rate was the most common error. The overall error rate was 3.3% (187 of 5744 observations), which is lower than previously reported. The majority of the errors caused no harm, but some required additional patient monitoring or intervention. Other errors were related to omission of doses or administration at an incorrect time. Critical care pharmacists observed these medication errors during their routine daily activities. The presence of these trained specialists may have influenced the number and type of errors. The process in this report differs from prior observational studies where a trained observer recorded medication administration practices and recorded errors. However, calculation of infusion rates was identified as an area for potential quality improvement.

CURRENT PRACTICE

Pharmacists have developed specialty practices in every facet of critical care and emergency medicine. These include burn, cardiac care, cardiovascular surgery, medical, neurological, neurosurgical, renal, respiratory, surgical, trauma, and pediatric and neonatal critical care. Pediatric specialty units in the same areas may also have pharmacists dedicated to those units. Descriptive reports of their contributions to these units have been published.[15–23]

Although many of these practitioners work for larger hospitals or have academic appointments, there is a growing number of pharmacists focusing on critical care patients in smaller hospitals. In 1989, a survey of hospitals with >100 beds demonstrated that 34% of medium-size hospitals had pharmacy satellites, as did 61% of large hospitals.[24] The majority (76%) of pharmacists who staffed these satellites held BS degrees. However, one-third of the pharmacists spent more than 50% of their day providing clinical services. Hopefully, this level of pharmacist involvement has been maintained or has grown since the early 1990s. Although not a complete measure, the number of critical care pharmacists who are members of SCCM has grown progressively from approximately 80 in the early 1990s to the current number of more than 450 pharmacist members. It is hoped that this reflects a larger number of specialty practitioners in critical care.

The majority (63.2%), of critically ill patients are cared for in hospitals without full-time or consultative intensivists.[25] Patients are more likely to be cared for by their primary physician or a single or multiple consultants (pulmonary or cardiology), although hospitalists are becoming more available to care for ICU patients (internal medicine specialists who care for hospitalized patients full-time). An intensivist is a physician who is board

certified in critical care but may have received their initial medical training in medicine, surgery, anesthesia, or pediatrics. Board certification is achieved through the subspecialty organizations after completion of a written examination. The SCCM model of critical care is that of an intensivist-led multidisciplinary team. An organized critical care team has been a predictor of improved patient care.[26] Despite the apparent benefit of the intensivist-led team, workforce projections predict a significant manpower shortfall by 2020, due to a relatively constant supply but an increasing demand for critical care services.[25] Critical care pharmacists may have increasing opportunities to affect the care of these patients as a result of this imbalance.

RESEARCH IN CRITICAL CARE

There are many challenges to performing research in a critically ill patient. Informed consent may be difficult to obtain if there is a narrow window for therapy initiation. The ICU population may have numerous other injuries or organ dysfunction that would disqualify the patient. Finally, an adequate number of patients may be difficult to recruit. As a result, animal research models have been used by critical care pharmacists to develop the framework for clinical trials and control the large number of patient variables present in a critically ill patient.

Numerous avenues exist for critical care research. Critical care pharmacists have contributed to the understanding of pharmacotherapy of multiple disease states and organ systems. Evaluation of pharmacokinetics and pharmacodynamics in the critically ill patient has facilitated the design of therapeutic regimens in these complicated patients. For example, the variations in hepatic metabolic rate following head trauma or hemorrhagic shock have been characterized.[27,28] However, much additional research is needed to further characterize the impact of changes in organ function on pharmacokinetics in this complex and heterogeneous population.

Clinical case reports of unusual treatments or response to therapy have presented a plethora of questions that remain to be answered. Case series and descriptions of experience with treatment protocols or the impact of pharmacist interventions are useful contributions to the literature. Evaluation of economics and outcomes has been an important area of research in critical care. The critically ill patient patient typically receives a large number of different and often expensive medications, and is monitored with expensive devices. Pharmacists have characterized various aspects of the cost of care, although comprehensive pharmacoeconomic outcome research is difficult to accomplish in complex patients where numerous factors can influence outcome, but is essential to the integration of novel therapies and improved utilization of existing agents.

Critical care pharmacists have also been active in the area of collaborative disease-state management and quality improvement projects, and have documented the impact of these on patient outcome. Pharmacists often take a leadership role in these efforts. Examples of the impact of these programs include reductions in the use of laboratory tests,[29,30] cost saving through improved antibiotic utilization,[18] improved utilization of sedative and neuromuscular blocking agents,[31–33] improved monitoring of sedation, and avoidance of adverse drug events.[12]

Training in critical care research is available through fellowships cosponsored by the pharmaceutical industry through organizations such as SCCM and ACCP, as well as through a number of clinical training centers.

KNOWLEDGE BASE

Although critical care is considered a specialty area, there is a challenge to define the body of knowledge encompassed by this field. Critical care patients as a whole are a very heterogeneous group. As a result, many institutions provide care for a more homogenous group of patients in geographically distinct units. Whether these patients are in

Table 1 Important components of critical care pharmacist knowledge base for adult and pediatric patients

Pharmacokinetic alterations in the critically ill
Analgesia, sedation, neuromuscular blockade
Cardiovascular pathology and therapeutics
Endocrine pathology and therapeutics
Gastrointestinal pathology and therapeutics
Hemodynamic monitoring/manipulation
Hepatic pathology and therapeutics
Hematologic pathology and therapeutics
Infection control/antimicrobial therapy
Inflammatory injury/multiple organ system dysfunction
Neurological pathology and therapeutics
Nutritional support
Patient/ventilator interface
Psychiatric therapeutics
Renal pathology and therapeutics
Respiratory pathology and therapeutics
Resuscitation therapeutics
Shock and related problems
Thrombosis/hemostasis and therapeutics
Toxicologic therapeutics

a specialty unit or a general ICU, the critical care practitioner must focus on a variety of complex patient problems and therapeutic areas (Table 1).

Understanding the potential pharmacokinetic changes experienced by critically ill patients is essential for the optimal dosing and monitoring of drug therapy in the critically ill patient.[34] Altered organ blood flow, dysfunction of drug-eliminating organs, and changes in fluid compartment volumes often dictate the need for individualized approaches to drug dosing.[35–38] Pharmacists are ideally trained to provide comprehensive therapeutic drug monitoring and optimize expenditures for serum drug concentrations.[29]

Universal concern with the comfort of the patient requires knowledge of analgesics and sedatives. Guidelines have been published to guide the optimal use of these agents.[39] These therapies must be prescribed in a manner that does not adversely affect the respiratory status of the patient and, in many cases, is used to facilitate adequate ventilation of patients with acute lung injury. An understanding of lung injury and mechanical ventilation is essential. Use of pharmacologic agents to induce paralysis may be an essential part of this care for some patients.[40] Prevention of common adverse events such as stress-related gastrointestinal bleeding and deep venous thrombosis requires knowledge of the gastrointestinal and coagulation systems and therapeutics. Nutrition support consultation is also provided by critical care pharmacists in many settings. Critically ill patients may be highly catabolic and optimal provision of macro- and micronutrients may enhance their recovery. Proper application of immune-enhancing nutrients has added complexity to the nutritional support of critically ill patients.

Critically ill patients are at high risk for a variety of nosocomial infections. Knowledge of infection control techniques, as well as proper use of prophylactic and empiric antibiotics, is an important component of critical care pharmacy practice. Inappropriate antibiotic use can lead to antimicrobial resistance and outbreaks of nosocomial infection that are difficult to treat with conventional therapies. Critical care pharmacists work with infection control staff and infectious disease pharmacists to optimize the use of antimicrobial therapies.

The hemodynamic stability of patients is another universal concern for critical care pharmacists. Patients with cardiac diseases or postcardiac surgery are obvious candidates for inotropic and vasoactive therapy (vasopressors and vasodilators). However, patients of all ages with severe injury, sepsis, or the systemic inflammatory response syndrome require vigorous resuscitation with fluids and vasoactive agents. Guidelines have been written to guide the management of these challenging patients.[41] Understanding the cardiovascular system and therapeutics is another fundamental portion of a critical care pharmacists' knowledge base. In addition, expertise in the resuscitation of patients experiencing an acute cardiac or respiratory event is an important skill and knowledge set. Pharmacist knowledge of the current guidelines and participation in cardiopulmonary resuscitation teams is a clinical pharmacy service shown to be associated with reduced hospital mortality.[42,43]

Disease-specific therapies and other fundamental components of the knowledge-base required to practice as a critical care specialist is outlined in the ASHP requirements for critical care residency training.[44] In addition to direct patient care, the critical care resident should receive training or experience in drug information and drug policy development, practice management, and participate in the management of drug distribution systems in the critical care setting.

There are currently 21 critical care residencies listed in the ACCP *2001 Guide to Residencies and Fellowships*. Eight of these residencies are accredited by the ASHP. In addition, 11 critical care fellowships are available.[45]

IMPACT OF CRITICAL CARE PHARMACISTS

Critical care pharmacists have demonstrated numerous contributions to the cost-effective care of ICU patients. Pharmacist initiated interventions in a 1200-bed teaching hospital were demonstrated to lower the drug costs by 41% compared with a control group (mean $73.75 vs. $43.40; $p<0.001$).[46] Approximately, fifty percent of the 259 patients were in critical care units. The majority (79%) of the interventions over the 30-day trial period were aimed at improving the quality of care, whereas the remaining 21% were to provide equivalent care at a lower cost. There was no impact on length of stay or readmission rates, but there was a trend toward lower hospital mortality in the intervention group.

Similarly, a multidisciplinary performance improvement team that included a pharmacist established a series of patient care protocols and tested the impact on the costs of care and outcome.[30] Protocols were developed that eliminated many standing orders for laboratory tests, electrocardiograms, and chest x-ray films. Other protocols were directed toward the use of sedatives, analgesics, neuromuscular blocking agents, and ventilator weaning. The outcome and costs from a baseline evaluation of 72 patients were compared with 85 patients in the follow-up phase. Application of the guidelines reduced costs for laboratory tests by 65%, and the number of chest x-rays were reduced by 56%. The cost of neuromuscular block-

ers was reduced by 75%. In addition, the length of ICU stay and duration of mechanical ventilation were reduced. There was no change in mortality. If these results can be extrapolated to a larger population and maintained, significant economic and outcome benefits could be realized. Another publication from this same group focused on the impact of guidelines for the use of analgesics, sedatives, and neuromuscular blocking agents in the same population.[47] The pharmacist was involved in protocol development, education and implementation, and ongoing intervention when practice did not meet guidelines. A reduction of direct drug costs, ventilator time, and length of stay were accomplished by using the protocol. Pharmacists have demonstrated similar results using sedation protocols elsewhere.[48]

Others have shown a positive impact of critical care pharmacists. A clinical pharmacist working part-time (daily rounds, average 10 hours per week) in a medical ICU over 8 weeks demonstrated a net benefit of $101 per day, considering cost avoidance, cost savings, increased drug costs, and the pharmacist's salary.[49] Although the medical ICU was unable to increase staffing levels to maintain this service, they redistributed pharmacist and technician workloads to perpetuate the clinical activities. Similarly, a clinical pharmacist with 50% teaching responsibility was assigned to participate in daily work rounds with the (medical-surgical patients) critical care team for 13 weeks.[50] A cost saving using drug costs only (no personnel costs) was $69.11 per patient day. Inclusion of personnel costs, for an average of 3 hours per day, did not negate the cost benefit. High-cost drugs were targeted, so it may be difficult to maintain this level of cost savings once the prescribing habits are modified or as protocols are implemented.

Table 2 Selected critical care pharmacist activities

Fundamental
Dedicated ICU pharmacist providing pharmaceutical care
Order evaluation and intervention
Adverse drug event and medication error management and prevention
Documentation of impact
Medication use policy implementation and support
Desirable
Rounding with the critical care team
Medication history review
Resuscitation response
Education of pharmacy students and residents
Implements and evaluates drug therapy protocols or pathways
Participates in clinical research
Optimal
Formal and informal education of the critical care team
Advanced cardiac life support education
Residency or fellowship development
Conducts research and presents and publishes findings

(From Ref. [51].)

ONGOING CHALLENGES

A focal point for critical care pharmacists and hospitals with critical care units is a position paper on critical care pharmacy services jointly developed and published by ACCP and SCCM.[51] This paper identifies and describes the fundamental, desirable, and optimal activities that define the scope of practice of the critical care pharmacist (Table 2). Fundamental activities are deemed vital to the safe provision of pharmaceutical care to the critically ill patient. The fundamental responsibilities of critical care pharmacists include a full-time commitment to critical care patients, evaluation of all drug therapy, identification of adverse events, individualized drug dosing, provision of drug information, documentation of activities, and participation in quality improvement activity. At a higher level of practice, the desirable activities additionally include critical care-specific pharmacotherapeutic services. Additional desirable activities may include rounding with a critical care team, review of medication histories, participation in resuscitation events, student and resident education, protocol development, participation in research, and outcome analysis. At the highest level, the optimal activities of a critical care pharmacy specialist include provision of education to families, pharmacists, and physicians, development of research protocols, new pharmacy services, and publication of the results of these programs. A single pharmacist cannot provide all these services, but rather should function within a team to meet these goals.

A similar model is used to present the recommended levels of service and personnel from a pharmacy department and hospital perspective.[51] Fundamental service includes the use of patient profiles, provision of ''ready to administer'' medications and parenterals, and adequate quality improvement programs. Desirable pharmacy services include computerized information management systems and an ICU satellite. Optimal pharmacy department services include a 24-hour satellite, physician order entry, and continuous availability of pharmaceutical care services. This document challenges practitioners and institutions to measure their progress and strive for the delivery of optimal pharmaceutical care services to critically ill patients.

REFERENCES

1. *Practice of Critical Care Pharmacy*; Majerus, T.C., Dasta, J.F., Eds.; Aspen Publications: Rockville, Maryland, 1985; 315 pp.
2. Dasta, J.F.; Jacobi, J.; Armstrong, D.K. Role of the Pharmacist in Caring for the Critically Ill Patient. In *The Pharmacologic Approach to the Critically Ill Patient*, 3rd Ed.; Chernow, B., Ed.; Williams & Wilkins: Baltimore, Maryland, 1994; 156–166.
3. American College of Critical Care Medicine and the Society of Critical Care Medicine Critical care services and personnel: Recommendations based on a system of categorization into two levels of care. Crit. Care Med. **1999**, *27* (3), 422–426.
4. Dasta, J.F. Drug use in a surgical intensive care unit. Drug Intell. Clin. Pharm. **1986**, *20* (10), 752–756.
5. Boucher, B.A.; Kuhl, D.A.; Coffey, B.C.; Fabian, T.C. Drug use in a trauma intensive-care unit. Am. J. Hosp. Pharm. **1990**, *47* (4), 805–810.
6. Farina, M.L.; Levati, A.; Tognoni, G. A multicenter study of ICU drug utilization. Intensive Care Med. **1981**, *7*, 125–131.
7. Buchanan, N.; Cane, R.D. Drug utilization in a general intensive care unit. Intensive Care Med. **1978**, *4*, 75–77.
8. Dasta, J.F.; Armstrong, D.K. Pharmacoeconomic impact of critically ill surgical patients. Drug Intell. Clin. Pharm. **1988**, *22* (12), 994–998.
9. *To Err is Human, Building a Safer Health System*; Kohn, L.T., Corrigan, J.M., Donaldson, M.S., Eds.; National Academy Press: Washington, DC, 2000.
10. Girotti, M.J.; Garrick, C.; Tierney, M.G.; Chesnick, K.; Brown, S.J. Medication administration errors in an adult intensive care unit. Heart Lung **1987**, *16* (4), 449–453.
11. Wright, D.; Mackenzie, S.J.; Buchan, I.; Cairns, C.S.; Price, L.E. Critical incidents in the intensive therapy unit. Lancet **1991**, *338*, 676–678.
12. Leape, L.L.; Cullen, D.J.; Clapp, M.D.; Burdick, E.; Demonaco, H.J.; Erickson, J.I.; Bates, D.W. Pharmacist participation on physician rounds and adverse drug events in the intensive care unit. JAMA, J. Am. Med. Assoc. **1999**, *282* (3), 267–270.
13. Tisdale, J.E. Justifying a pediatric critical-care satellite pharmacy by medication-error reporting. Am. J. Hosp. Pharm. **1986**, *43* (2), 368–371.
14. Calabrese, A.D.; Erstad, B.L.; Brandl, K.; Barletta, J.F.; Kane, S.L.; Sherman, D.S. Medication administration errors in adult patients in the ICU. Intensive Care Med. **2001**, *27* (8), 1592–1598.
15. Barber, N.; Jacklin, A. CCU drug costs—the pharmacist's role. Intensive Care World **1987**, *4* (3), 80–82.
16. White, C.M.; Chow, M.S. Cost impact and clinical benefits of focused rounding in the cardiovascular intensive care unit. Hosp. Pharm. **1998**, *33* (4), 419–423.
17. Herfindal, E.T.; Bernstein, L.R.; Kishi, D.T. Impact of clinical pharmacy services on prescribing on a cardiothoracic/vascular surgical unit. Drug Intell. Clin. Pharm. **1985**, *19* (6), 440–444.
18. Peterson, C.D.; Lake, K.D. Reducing prophylactic antibiotic costs in cardiovascular surgery: The role of the clinical pharmacist. Drug Intell. Clin. Pharm. **1985**, *19* (2), 134–137.
19. Levy, D.B. Documentation of clinical and cost-saving pharmacy interventions in the emergency room. Hosp. Pharm. **1993**, *28* (7), 624–653.
20. Katona, B.G.; Ayd, P.R.; Walters, J.K.; Caspi, M.; Finkelstein, B.W. Effect of a pharmacist's and a nurse's interventions on cost of drug therapy in a medical intensive-care unit. Am. J. Hosp. Pharm. **1989**, *46* (6), 1179–1182.
21. Ellinoy, B.R.; Clarke, J.E.; Wagers, P.W.; Swinney, R.S. Comprehensive pharmaceutical services in a medical intensive-care unit. Am. J. Hosp. Pharm. **1984**, *41* (11), 2335–2342.
22. Miyagawa, C.I.; Rivera, J.O. Effect of pharmacist interventions on drug therapy costs in a surgical intensive-care unit. Am. J. Hosp. Pharm. **1986**, *43* (12), 3008–3013.
23. Montazeri, M.; Cook, D.J. Impact of a clinical pharmacist in a multidisciplinary intensive care unit. Crit. Care Med. **1994**, *22* (6), 1044–1048.
24. Dasta, J.F.; Segal, R.; Cunningham, A. National survey of critical-care pharmaceutical services. Am. J. Hosp. Pharm. **1989**, *46* (11), 2308–2312.
25. Angus, D.C.; Kelley, M.A.; Schmitz, R.J.; White, A.; Popovich, J. Current and projected workforce requirements for care of the critically ill and patients with pulmonary disease, can we meet the requirements of an aging population? JAMA, J. Am. Med. Assoc. **2000**, *284* (21), 2762–2770.
26. Hanson, C.W.; Deutschman, C.S.; Anderson, H.L. Effects of an organized critical care service on outcomes and resource utilization: A cohort study. Crit. Care Med. **1999**, *27*, 270–274.
27. Boucher, B.A.; Kuhl, D.A.; Fabian, T.C.; Robertson, J.T. Effect of neurotrauma on hepatic drug clearance. Clin. Pharmacol. Ther. **1991**, *50* (11), 487–497.
28. DiPiro, J.T.; Hooker, K.D.; Sherman, J.C.; Gaines, M.G.; Wynn, J.J. Effect of experimental hemorrhagic shock on hepatic drug elimination. Crit. Care Med. **1992**, *20* (6), 810–815.
29. Crisp, C.B.; Lane, J.R.; Murray, W. Audit of serum drug concentration analysis for patients in the surgical intensive care unit. Crit. Care Med. **1990**, *18* (7), 734–737.
30. Marx, W.H.; DeMaintenon, N.L.; Mooney, K.F.; Mascia, M.L.; Medicus, J.; Franklin, P.D.; Sivak, E.; Rotello, L. Cost reduction and outcome improvement in the intensive care unit. J. Trauma **1999**, *46* (4), 625–630.
31. Bair, N.; Bobek, M.B.; Hoffman-Hogg, L.; Mion, L.C.;

Slomka, J.; Arroliga, A.C. Introduction of sedative, analgesic, and neuromuscular blocking agent guidelines in a medical intensive care unit: Physician and nurse adherence. Crit. Care Med. **2000**, *28* (3), 707–713.

32. Brook, A.D.; Ahrens, T.S.; Schaiff, R.; Prentice, D.; Sherman, G.; Shannon, W.; Kollef, M.H. Effect of a nursing-implemented sedation protocol on the duration of mechanical ventilation. Crit. Care Med. **1999**, *27* (12), 2609–2615.
33. Riker, R.R.; Fraser, G.L.; Simmons, L.E.; Wilkins, M.L. Validating the sedation-agitation scale with the bispectral index and visual analog scale in adult ICU patients after cardiac surgery. Intensive Care Med. **2001**, *27*, 853–858.
34. Wolfe, T.; Dasta, J.F. Pharmacokinetic issues in the critically ill patient. Curr. Opin. Crit. Care **1995**, *1*, 272–278.
35. Mann, H.J.; Fuhs, D.W.; Cerra, F.B. Pharmacokinetics and pharmacodynamics in critically ill patients. World J. Surg. **1987**, *11* (2), 210–217.
36. McKindley, D.S.; Hanes, S.; Boucher, B.A. Hepatic drug metabolism in critical illness. Pharmacotherapy **1998**, *18* (4), 759–778.
37. Boucher, B.A.; Hanes, S. Pharmacokinetic alterations after severe head injury, clinical relevance. Clin. Pharmacokinet. **1998**, *35* (3), 209–221.
38. Robert, S.; Zarowitz, B.J. Is there a reliable index of glomerular filtration rate in critically ill patients? DICP, Ann. Pharmacother. **1991**, *25* (2), 169–178.
39. Task Force of the American College of Critical Care Medicine, Society of Critical Care Medicine in collaboration with the American Society of Health-System Pharmacists Clinical practice guidelines for the sustained use of sedatives and analgesics in the critically ill adult. Am. J. Health-Syst. Pharm. **2002**, *60*.
40. Task Force of the American College of Critical Care Medicine, Society of Critical Care Medicine in collaboration with the American Society of Health-System Pharmacists Clinical practice guidelines for the sustained use of neuromuscular blocking agents in the critically ill adult. Am. J. Health-Syst. Pharm. **2002**, *60*.
41. Task Force of the American College of Critical Care Medicine, Society of Critical Care Medicine Practice parameters for hemodynamic support of sepsis in adult patients with sepsis. Crit. Care Med. **1999**, *27* (3), 639–660.
42. The American Heart Association in Collaboration with the International Liaison Committee on Resuscitation Guidelines 2000 for cardiopulmonary resuscitation and emergency cardiac care, an international consensus on science. Circulation **2000**, *102* (8), I1–I384.
43. Bond, C.A.; Raehl, C.L.; Franke, T. Clinical pharmacy services, pharmacist staffing, and drug costs in United States hospitals. Pharmacotherapy **1999**, *19* (12), 1354–1362.
44. ASHP Supplemental Standard and Learning Objectives for Residency Training in Critical Care Pharmacy Practice. www.ashp.com/public/rtp/SUPPSTND.html (accessed Nov. 2000).
45. *2001 Directory of Residencies and Fellowships*; American College of Clinical Pharmacy: Kansas City, Missouri, 2001; 69–85.
46. McMullin, S.T.; Hennenfent, J.A.; Ritchie, D.J.; Huey, W.Y.; Lonergan, T.P.; Schaiff, R.A.; Tonn, M.E.; Bailey, T.C. A prospective, randomized trial to assess the cost impact of pharmacist-initiated interventions. Arch. Intern. Med. **1999**, *159* (19), 2306–2309.
47. Mascia, M.F.; Koch, M.; Medicus, J.J. Pharmacoeconomic impact of rational use guidelines on the provision of analgesia, sedation, and neuromuscular blockade in critical care. Crit. Care Med. **2000**, *28* (7), 2300–2306.
48. Devlin, J.W.; Holbrook, A.M.; Fuller, H.D. The effect of ICU sedation guidelines and pharmacist interventions on clinical outcomes and drug cost. Ann. Pharmacother. **1997**, *31* (6), 689–695.
49. Baldinger, S.L.; Chow, M.S.S.; Gannon, R.H.; Kelly, E.T. Cost savings from having a clinical pharmacist work part-time in a medical intensive care unit. Am. J. Health-Syst. Pharm. **1997**, *54*, 2811–2814.
50. Chuang, L.C.; Sutton, J.D.; Henderson, G.T. Impact of a clinical pharmacist on cost saving and cost avoidance in drug therapy in an intensive care unit. Hosp. Pharm. **1994**, *29* (3), 215–218.
51. Task Force of the Clinical Pharmacy and Pharmacology Section of the Society of Critical Care Medicine and the Critical Care Practice and Research Network of the American College of Clinical Pharmacy Position paper on critical care pharmacy services. Pharmacotherapy **2000**, *20* (11), 1400–1406.

Critical Care Pharmacy Services (ACCP)

American College of Clinical Pharmacy
Kansas City, Missouri, U.S.A.

INTRODUCTION

The object was to identify and describe the scope of practice that characterizes the critical care pharmacist and critical care pharmacy services. Specifically the goals were to define the level of clinical practice and specialized skills characterizing the critical care pharmacist as clinician, educator, researcher, and manager; and to recommend fundamental, desirable, and optimal pharmacy services and personnel requirements for the provision of pharmaceutical care to critically ill patients. Hospitals having comprehensive resources as well as those with more limited resources were considered.

Consensus of critical care pharmacists from institutions of various sizes providing critical care services within several types of pharmacy practice models was obtained, including community-based and academic practice settings. Existing guidelines and literature describing pharmacy practice and drug use processes were reviewed and adapted for the critical care setting.

By combining the strengths and expertise of critical care pharmacy specialist with existing supporting literature, these recommendations define the level of clinical practice and specialized skills that characterize the critical care pharmacist as clinician, educator, researcher, and administrator. Recommendations include fundamental, desirable, and optimal pharmacy services as well as personnel requirements or the provision of pharmaceutical care to critically ill patients.

HISTORICAL BACKGROUND

The discipline of critical care pharmacy practice evolved since the mid-1970s to become an essential component of the multidisciplinary team in the intensive care unit (ICU).[1–3] In the early 1970s, there were a few practitioners in critical care who were members of surgical or trauma services and cardiac arrest teams. During the next decade, pharmacy services expanded to various ICU settings (both adult and pediatric), the operating room, and the emergency department. In these settings, pharmacists established clinical practices consisting of therapeutic drug monitoring, nutrition support, and participation in patient care rounds. Pharmacists also developed efficient and safe drug delivery systems with the evolution of critical care pharmacy satellites and other innovative programs.

In the 1980s, critical care pharmacists designed specialized training programs and increased participation in critical care organizations. The number of critical care residencies and fellowships doubled between the early 1980s and the late 1990s. Standards for critical care residency were developed,[4] and directories of residencies and fellowships were published.[5,6] Several professional pharmacy organizations formed specialty groups consisting of critical care pharmacists. These include the American College of Clinical Pharmacy (ACCP), American Society of Health-System Pharmacists, and the Operating Room Satellite Pharmacy Association. In 1989, the Clinical Pharmacy and Pharmacology Section was formed within the Society of Critical Care Medicine, the largest international, multidisciplinary, multispecialty critical care organization. This recognition acknowledged that pharmacists are necessary and valuable members of the physician-led multidisciplinary team.

The Society of Critical Care Medicine Guidelines for Critical Care Services and Personnel deem that pharmacists are essential for the delivery of quality care to critically ill patients. These guidelines recommend that a pharmacist monitor drug regimens for dosing, adverse reactions, drug–drug interactions, and cost optimization for all hospitals providing critical care services.[1] The guidelines also advocate that a specialized, decentralized pharmacist provide expertise in nutrition support, cardiorespiratory resuscitation, and clinical research in academic medical centers providing comprehensive critical care.[1]

Since the early 1990s, clinical pharmacy became increasingly specialized and developed specialty board certification.[7] The growth of critical care pharmacy practice paralleled this development. Pharmacists assumed in-

From *Pharmacotherapy* 2000, 20(11):1400–1406, with permission of the American College of Clinical Pharmacy.

Encyclopedia of Clinical Pharmacy
DOI: 10.1081/E-ECP 120006407

creased responsibility for monitoring patient outcomes as well as supervising drug distribution services.[3]

Pharmacists have demonstrated a role in the management of drug costs and reductions in morbidity and mortality.[2,3,6–19] Clinical pharmacy services such as clinical research, provision of drug information, drug admission histories, and participation on a cardiopulmonary resuscitation (CPR) team have been associated with reduced mortality.[11] Prospective, controlled trials demonstrated that when pharmacists assume responsibility for pharmacotherapy as part of a multidisciplinary health care team, significant reductions in adverse drug events (ADEs) and length of stay are realized.[12–16] Many of these findings have been documented in specialized critical care populations.[14–20] The ACCP estimates that a benefit of $16.70 is realized for every $1.00 invested in clinical pharmacy programs.[17] A landmark study involving critical care pharmacists confirmed that pharmacist rounding in the ICU with the multidisciplinary team reduces preventable ADEs and associated costs caused primarily by prescribing errors.[16] Pharmacist intervention during prescribing decreased the rate of preventable ADEs by 66% from 10.4 to 3.5/1000 patient-days ($p < 0.001$). Pharmacist involvement was categorized as drug order clarification (45%), provision of drug information (25%), and recommendations for alternative therapy (12%). Based on an estimated cost of $4685/preventable ADE, the annualized financial impact in the unit studied would be $270,000 (1995 dollars).

Despite the growing evidence supporting the critical care pharmacist's contribution to patient care, many ICUs have not taken full advantage of this vital resource.[18] A description of pharmacy services and pharmacist activities in a critical care setting will assist practitioners and administrators in establishing or advancing these specialized pharmacy services. This article may be used to educate other health care providers, administrators, and developers of health care policy on the role of pharmacists and pharmacy services in the care of the critically ill. Furthermore, the application of the elements in this article will allow researchers to further document the effect of critical care pharmacy services on improving patient outcomes.

PURPOSE

This article identifies and describes the scope of pharmacy practice of the critical care pharmacist and critical care pharmacy services. Specifically, the aims of the Task Force on Critical Care Pharmacy Services were:

1. To define the level of clinical practice and specialized skills characterizing the critical care pharmacist as clinician, educator, researcher, and manager.
2. To recommend levels of service and personnel requirements for the provision of pharmaceutical care to critically ill patients. The levels will be defined as fundamental, desirable, or optimal.

METHODS

The Task Force on Critical Care Pharmacy Services consisted of members from the Clinical Pharmacy and Pharmacology Section of the Society of Critical Care Medicine and the Critical Care Practice and Research Network of the ACCP. Members of the task force were from institutions of various sizes and they provide critical care services within a variety of pharmacy practice models. Practitioners from both community-based and academic practice settings were included.

The formulation of these recommendations, including discussion and development of consensus, took place between October 1997 and September 1999. Task force members were charged with developing graded parameters within six domains: clinical activities, drug distribution, education, research, documentation, and administration. This article was organized into pharmacist activities and pharmacy services. Drafts were reviewed and evaluated by all members of the task force, and a consensus was reached. When differences in opinion were expressed, they were resolved using a modified Delphi method.[21] The document was reviewed externally by three established leaders in critical care pharmacy and by 18 pharmacy and hospital administrators for appropriateness of categorization of pharmacy activities and services. The article was further reviewed by select members and the governance of both the Clinical Pharmacy and Pharmacology Section of the Society of Critical Care Medicine and the Critical Care Practice and Research Network of the ACCP. Before organizational endorsement, the article underwent internal review by both the Council of the Society of Critical Care Medicine and the Board of Regents of the ACCP.

Existing guidelines and literature for pharmacy practice and drug use processes were reviewed and adapted for the critical care setting.[7,22–24] The needs of hospitals with comprehensive resources as well as those with more limited resources were considered. The task force created three gradations of pharmacist responsibilities and departmental services as fundamental, desirable, and optimal. Classification of the elements into each category was the result of the consensus process. For the purposes of this article, the following definitions were used. Fundamental activities are vital to the safe provision of pharmaceutical care to the critically ill patient. Desirable activities include fundamental activities and critical care-specific pharmacotherapeutic services. Optimal activities encompass the range of fundamental to desirable services and, additionally, reflect an integrated, specialized, and dedicated model of critical care that aims to optimize pharmacotherapeutic outcomes through the highest level of teaching, research, and pharmacotherapy practice. Fundamental services should not be interpreted as an acceptable minimum level of service. Each institution and practitioner continually should strive for the highest level of service possible.

A single pharmacist cannot perform all the fundamental activities on all patients every day. Rather, these critical care pharmacy activities will require varying levels of involvement from multiple pharmacists and trained technicians acting as a team, along with support from pharmacy and hospital administrators, and other personnel. The exact allocation of labor and the pharmacist-to-patient ratio will vary by institution and depend on the level of care, the acuity of patients, and the degree of specialization of the institution.

"The pharmacist," as used herein, refers to the team of licensed pharmacy practitioners with specialized training or practice experience focusing on the unique characteristics and needs of critically ill patients. Although various practice models exist, the pharmacist practices within the framework of a multidisciplinary team. In collaboration with other members of the patient care team, pharmacists share the responsibility for patient care outcomes, not just by providing basic dispensing functions and drug information services, but by solving patient- and drug-related problems and by making decisions regarding drug prescribing, monitoring, and drug regimen adjustments.[25] The pharmacist's practice may integrate varying elements of patient care, teaching, and research activities, depending on the nature of the institution and the pharmacist's training.

The task force recognizes the varied educational backgrounds of practicing critical care pharmacists. Having the qualifications and competence necessary to provide pharmaceutical care in the ICU is essential and may be achieved by a variety of means including advanced degrees, residencies, fellowships or other specialized practice experiences.

The term pharmacy and hospital services refers to departmental and institutional/organizational components of the infrastructure that support the pharmacist's activities. They consist of systems, operations, and personnel who facilitate and support the provision of patient care, teaching, and research to optimize safe and effective pharmaceutical care of the critically ill.

This article is not intended to be a standard of practice; however, we envision that it will serve as a guideline for hospitals of varying resources to optimize the delivery of pharmaceutical care to the critically ill. It is expected that these recommendations will continue to be reviewed at intervals of approximately 5 years as critical care pharmacy services, clinical pharmacy, and critical care medicine evolve.

CRITICAL CARE PHARMACIST ACTIVITIES

Fundamental Activities

1. The pharmacist's time is dedicated to critical care patients, with few commitments outside the ICU area.
2. The pharmacist prospectively evaluates all drug therapy for appropriate indications, dosage, drug interactions, and drug allergies; monitors the patient's pharmacotherapeutic regimen for effectiveness and ADEs; and intervenes as needed.
3. In conjunction with the clinical dietitian, the pharmacist evaluates all orders for parenteral nutrition and recommends modifications as indicated to optimize the nutritional regimen.
4. The pharmacist identifies ADEs and assists in their management and prevention, and develops process improvements to reduce drug errors and preventable ADEs.
5. The pharmacist uses the medical record as one means to communicate with other health care professionals and to document specific pharmacotherapeutic recommendations.
6. The pharmacist provides pharmacokinetic monitoring when a targeted drug is prescribed.
7. The pharmacist provides drug information and intravenous compatibility information to the ICU team and uses the regional poison information center when indicated.
8. The pharmacist maintains current tertiary drug references.
9. The pharmacist provides drug therapy-related education to ICU team members.
10. The pharmacist participates in reporting ADEs to institutional committees and to the Food and Drug Administration's MedWatch program.
11. The pharmacist documents clinical activities that include, but are not limited to, disease state management, general pharmacotherapeutic monitoring, pharmacokinetic monitoring, ADEs, education, and other patient care activities.
12. The pharmacist acts as a liaison between pharmacy, nursing, and the medical staff to educate health professionals regarding current drug-

related procedures, policies, guidelines, and pathways.

13. The pharmacist contributes to the hospital newsletters and drug monographs on issues related to drug use in the ICU.
14. The pharmacist implements and maintains departmental policies and procedures related to safe and effective use of drugs in the ICU.
15. The pharmacist collaborates with nursing, medical staff, and hospital administration to prepare the ICU for the Joint Commission on the Accreditation of Healthcare Organizations (JCAHO) survey and responds to any deficiencies identified.
16. The pharmacist provides consultation to hospital committees, such as Pharmacy and Therapeutics, when critical care pharmacotherapy issues are discussed.
17. The pharmacist identifies how drug costs may be minimized through appropriate use of drugs in the ICU and through implementation of cost-containment measures.
18. The pharmacist participates in quality assurance programs to enhance pharmaceutical care.

Desirable Activities

1. The pharmacist regularly makes rounds as a member of the multidisciplinary critical care team (if available) to provide pharmacotherapeutic management for all ICU patients.
2. The pharmacist maintains knowledge of current primary references pertinent to critical care pharmacotherapy.
3. The pharmacist reviews a patient's drug history to determine which maintenance drugs should be continued during the acute illness.
   a. The pharmacist clarifies previously effective dosages and dosage regimens.
   b. For all suspected drug-related ICU admissions, the pharmacist assesses the patient drug history for causality and documents in the medical record any findings that will impact patient management.
4. In collaboration with the clinical dietitian, the pharmacist provides formal nutrition consultation on request and responds within 24 hours.
5. The advanced cardiac life support-certified (or pediatric advanced life support-certified) pharmacist responds to all resuscitation events in the hospital 7 days/week, 24 hours/day.
6. The pharmacist provides didactic lectures to health professional students in critical care pharmacology and therapeutics, where applicable.
7. The pharmacist participates in training pharmacy students, residents, and fellows through experiential critical care rotations, where applicable.
8. The pharmacist coordinates the development and implementation of drug therapy protocols and/or critical care pathways to maximize benefits of drug therapy.
9. The pharmacist uses a documentation program that attaches both a clinical significance and an economic value to clinical interventions.
10. The pharmacist is actively involved in critical care pharmacotherapy research by assisting in the screening and enrollment of patients and by serving as a study coordinator or contact person, where applicable.
11. The pharmacist participates in research design and data analysis, where applicable.
12. The pharmacist contributes to the pharmacy and medical literature, e.g., case reports, letters to the editor, and therapeutic, pharmacokinetic, and pharmacoeconomic reports.
13. The pharmacist is involved in nonpatient care activities including multidisciplinary committees and educational in-services.

Optimal Activities

1. The pharmacist assists physicians in discussions with patients and/or family members to help make informed decisions regarding treatment options.
2. The pharmacist provides formal accredited educational sessions, such as medical grand rounds or intensive care rounds, for medical staff, students, and residents.
3. The pharmacist participates in teaching advanced cardiac life support.
4. The pharmacist develops residencies and/or fellowships in critical care pharmacy practice.
5. The pharmacist develops and implements pharmacist and pharmacy technician training programs for personnel working in the ICU.
6. The pharmacist identifies and educates lay groups and medical personnel in the community about the role of pharmacists as part of the multidisciplinary health care team in the ICU.
7. The pharmacist independently investigates or collaborates with other critical care practitioners to evaluate the impact of guidelines and/or protocols used in the ICU for drug administration and management of common disease states.
8. The pharmacist uses pharmacoeconomic analyses to prospectively evaluate existing or new phar-

macy services and the place of new drugs in critical care pharmacotherapy.

9. The pharmacist is proactive in designing, prioritizing, and promoting new pharmacy programs and services.
10. The pharmacist secures funds for conducting research.
11. The pharmacist reports results of clinical research and pharmacoeconomic analyses to the pharmacy and medical community at regional and national meetings.
12. The pharmacist publishes in peer-reviewed pharmacy and medical literature as a result of any of the following activities:
    a. Clinical research or other original research that qualitatively and quantitatively evaluates drug therapy and the provision of pharmacy services.
    b. Investigator-initiated grants and contracts.
    c. Pharmacoeconomic and outcomes research.

PHARMACY AND HOSPITAL SERVICES

Fundamental Services

1. Drug use systems can do the following:
    a. Create and maintain patient drug profiles.
    b. Interface with patient laboratory data.
    c. Alert users to drug allergies.
    d. Alert users to maximum dosage limits.
    e. Alert users to drug–drug and drug–food/nutrient interactions.
2. If manual drug administration records are the only available drug administration document, quality assurance[1] systems are in place to verify the accuracy of this process.
3. A ''ready to administer'' (unit-dose) drug distribution system is available in the ICU with no more than a 24-hour supply for each patient.
4. Large- and small-volume parenteral products are prepared in the pharmacy and delivered at regularly scheduled times to the patient care area 7 days/week.
5. Pharmacy space and facilities in the ICU are assessed routinely to determine whether efficiency can be improved, where applicable.
6. Procurement, storage, inventory, and distribution of investigational drugs, where applicable, are under the supervision of a pharmacist.
7. The pharmacy department is represented on the Institutional Review Board and/or Scientific Review Board, as applicable.

Desirable Services

1. The hospital information management system is computerized, can comply with the requirements listed for drug use processes (see Fundamental Services, Item 1), and can do the following:
    a. Alert users to disease state–drug interactions.
    b. Provide intravenous admixture information (e.g., compatibility, stability, preparation).
    c. Provide online drug and poison information.
    d. Document clinical pharmacy patient care interventions.
2. Computerized drug administration records are generated. Manual records are used only in emergencies.
3. An ICU satellite pharmacy with unit-dose drug distribution and intravenous admixture capabilities is open a minimum of 40 hours/week.

Optimal Services

1. The computerized hospital information management system serving the ICU has the following additional capabilities:
    a. Direct physician drug order entry at patient bedside.
    b. Interface with bedside clinical information system.
2. An ICU satellite pharmacy with unit-dose drug distribution and intravenous admixture capabilities is open 24 hours/day, 7 days/week.
3. Pharmacotherapeutic, pharmacokinetic, and nutrition consultation are available 24 hours/day, 7 days/week.

ACKNOWLEDGMENTS

The task force acknowledges the following individuals for their review of this manuscript: Bradley A. Boucher, Pharm.D., FCCP, FCCM, BCPS, University of Tennessee, Memphis, TN; Joseph F. Dasta, M.S., FCCP, FCCM, Ohio State University, Columbus, OH; and Barbara J. Zarowitz, Pharm.D., FCCP, FCCM, BCPS, Henry Ford Health System, Bingham Farms, MI.

This position paper was developed by a task force of the Clinical Pharmacy and Pharmacology Section of the Society of Critical Care Medicine and the Critical Care Practice and Research Network of the American College of Clinical Pharmacy. It was approved by the Council of the Society of Critical Care Medicine on February 10, 2000, and the Board of Regents of the American College of Clinical Pharmacy on October 5, 1999.

Task force members were Maria I. Rudis, Pharm.D., University of Southern California, Los Angeles, CA (Chair); Henry Cohen, Pharm.D., Long Island University, New York, NY; Bradley E. Cooper, Pharm.D., Hamot Medical Center, Erie, PA; Luis S. Gonzalez, III, Pharm.D., Conemaugh Medical Center, Erie, PA; Erkan Hassan, Pharm.D., FCCM, University of Maryland, Baltimore, MD; Christian Klem, Pharm.D., Tampa General Healthcare, Tampa, FL; Vanessa L. Kluth-Land, Pharm.D., SmithKline Beecham Pharmaceuticals, Houston, TX; Katherine M. Kramer, Pharm.D., University of New Mexico, Las Cruces, NM; Angela M. Swerlein, Pharm.D., Grant/Riverside Methodist Hospitals, Columbus, OH; Julie Ann Whippel, Pharm.D., Waukesha Memorial Hospital, Waukesha WI. At the time of manuscript preparation, Dr. Kluth-Land was at Hermann Hospital, Houston, TX.

REFERENCES

1. American College of Critical Care Medicine and the Society of Critical Care Medicine. Critical care services and personnel: Recommendations based on a system of categorization into two levels of care. Crit. Care Med. **1999**, *27*, 422–426.
2. Dasta, J.F.; Jacobi, J. The critical care pharmacist: What you get is more than what you see. Crit. Care Med. **1994**, *22*, 906–909.
3. Dasta, J.F.; Jacobi, J.; Armstrong, D.K. Role of the Pharmacist in Caring for the Critically Ill Patient. In *The Pharmacologic Approach to the Critically Ill Patient,* 3rd Ed.; Chernow, B., Ed.; Williams & Wilkins: Baltimore, 1994; 156–166.
4. American Society of Hospital Pharmacists. Supplemental standard and learning objectives for residency training in critical care pharmacy practice. Am. J. Hosp. Pharm. **1990**, *47*, 609–612.
5. Society of Critical Care Medicine. Directory of Critical Care Residencies and Fellowships. In *Clinical Pharmacy and Pharmacology Section*; Society of Critical Care Medicine: Anaheim, California, 2000.
6. American College of Clinical Pharmacy. *Directory of Residencies and Fellowships*; American College of Clinical Pharmacy: Kansas City, Missouri, 2000.
7. American College of Clinical Pharmacy Practice Affairs Committee. Practice guidelines for pharmacotherapy specialists. Pharmacotherapy **2000**, *20*, 487–490.
8. Hatoum, H.T.; Hutchinson, R.A.; Witte, K.W., et al. Evaluation of the contribution of clinical pharmacists: Inpatient care and cost reduction. Drug Intell. Clin. Pharm. **1988**, *22*, 252–259.
9. Bond, C.A.; Raehl, C.L.; Pittele, M.E., et al. Health care professional staffing, hospital characteristics, and hospital mortality rates. Pharmacotherapy **1999**, *19*, 130–138.
10. Chuang, L.C.; Suttan, J.D.; Henderson, J.P. Impact of the clinical pharmacist on cost saving and cost avoidance in drug therapy in an intensive care unit. Hosp. Pharm. **1994**, *29*, 215–221.
11. Bond, C.A.; Raehl, C.L.; Franke, T. Clinical pharmacy services and hospital mortality rates. Pharmacotherapy **1999**, *19*, 556–564.
12. Bjornson, D.C.; Hiner, W.O.; Potyk, R.P., et al. Effect of pharmacists on health care outcomes in hospitalized patients. Am. J. Hosp. Pharm. **1993**, *50*, 1875–1884.
13. Boyko, W.L.; Yurkowski, P.J.; Ivey, M.F., et al. Pharmacist influence on economic and morbidity outcomes in a tertiary care teaching hospital. Am. J. Health-Syst. Pharm. **1997**, *54*, 1591–1595.
14. Kelly, W.N.; Meyer, J.D.; Flatley, C.J. Cost analysis of a satellite pharmacy. Am. J. Hosp. Pharm. **1986**, *43*, 1927–1930.
15. Smythe, M.A.; Shah, P.P.; Spiteri, T.L., et al. Pharmaceutical care in medical progressive care patients. Ann. Pharmacother. **1998**, *32*, 294–299.
16. Leape, L.L.; Cullen, D.J.; Clapp, M.D., et al. Pharmacist participation on physician rounds and adverse drug events in the intensive care unit. JAMA **1999**, *282*, 267–270.
17. Schumock, G.T.; Meek, P.D.; Ploetz, P.A., et al. Economic evaluations of clinical pharmacy services—1988–1995. Pharmacotherapy **1996**, *16*, 1188–1208.
18. Matuszewski, K.A.; Vlasses, P.H. Survey results from academic health center intensive care units: Considerations for departments of pharmacy. Clin. Ther. **1995**, *17*, 517–525.
19. Montazeri, M.; Cook, D.J. Impact of a clinical pharmacist in a multidisciplinary intensive care unit. Crit. Care Med. **1994**, *22*, 1044–1048.
20. Miyagawa, C.I.; Rivera, J.O. Effect of pharmacist interventions on drug therapy costs in a surgical intensive care unit. Am. J. Hosp. Pharm. **1986**, *43*, 3008–3113.
21. Dalkey, N.C. *The Delphi Method: An Experimental Study of Group Opinion*; Rand Corporation: Santa Monica, California, 1969.
22. Joint Commission on the Accreditation of Healthcare Organizations. *Standards for Hospitals*; JCAHO: Chicago, Illinois, 1998.
23. American Pharmaceutical Association. *Pharmacy Practice Activity Classification*; APhA: Washington, DC, 1998.
24. Anonymous. Over-reliance on pharmacy computer systems may place patients at great risk. ISMP Med. Saf. Alert. **1999**, *4*, 1, Available from: http://www.ismp.org/MSArticles/Computer.html.
25. American College of Clinical Pharmacy. Collaborative drug therapy management by pharmacists. Pharmacotherapy **1997**, *17*, 1050–1061.

PHARMACY PRACTICE ISSUES

Cytochrome P450

David I. Min
University of Iowa, Iowa City, Iowa, U.S.A.

INTRODUCTION

The cytochrome P450 (CYP) is a major hemo (iron-containing) protein family that catalyzes drug and xenobiotic metabolism. It is present in the microsomes (tiny membrane vesicles of endoplasmic reticulum) of many different cells in the body, but it is at highest concentration in liver.[1] There are two types of microsomal enzymes in the body: those catalyzing mainly oxidations (termed the phase I enzymes) and those catalyzing conjugations (termed the phase II enzymes).[2] The CYP is the most important enzyme system catalyzing phase I metabolism reactions such as oxidation, reduction, and hydrolysis. It generally serves as a detoxification mechanism for lipophilic drugs and xenobiotics by converting them to more water-soluble compounds.[3] However, this enzyme system occasionally transforms nontoxic chemicals or drugs into toxic reactive intermediates, or procarcinogens into carcinogens. In addition, it converts hormones and steroids into more active forms.

ORIGIN AND NOMENCLATURE

In the late 1950s, it was discovered that when rat liver microsomes were treated in a certain condition, a strong absorption band occurred at approximately 450 nm wavelength in the spectrophotometer, which was very unusual for the pigments.[3] The red pigment responsible for this phenomenon was called P (for *pigment*) 450. It was later named ''cytochrome P450'' because it was believed to be similar to mitochondrial cytochromes.[4]

At first, it was believed that the P450 was a single protein, but soon it became apparent that it was not a single protein but comprised a number of different proteins. Each human CYP protein identified appears to be the expression of an unique gene. There are more than 1000 unique genes for CYP identified among prokaryotes and eukaryotes to date.[5] There are significantly common amino acid sequences among all CYPs in a few regions of the proteins, suggesting a common ancestry. For this reason, the CYP is referred to as a supergene family.[6,7]

Accordingly, a recommended nomenclature system has been devised based on the evolutionary relations of these enzymes.[7] According to this system, the deduced amino acid sequences from the genes are compared and divided into families, which comprise those CYPs that share at least 40% identity and designated by Arabic number after CYP (i.e., CYP1, 2, 3, etc.). These families are divided further into subfamilies, which comprise those forms that are at least 55% related by their deduced amino acid sequences, and are designated by a capital letter after the Arabic number (i.e., CYP1A, CYP2D, etc.). Each individual enzyme is designated by Arabic number after subfamily (i.e., CYP1A2, CYP3A4).

MAJOR ISOENZYMES OF CYP IN HUMAN DRUG METABOLISM

There are numerous CYPs identified in humans, animals, and plants, but currently three P450 gene families, including CYP1, CYP2, and CYP3, are responsible for most of human drug metabolism.[8] These three CYP gene families, their subfamilies, and their major substrates are illustrated in Table 1. In drug metabolism, CYP1A2, CYP2C9, CYP2C19, CYP2D6, and CYP3A4 are important. Especially, approximately 75% of all therapeutic agents are metabolized by CYP2D6 and CYP3A4 together.[9]

INDIVIDUAL DIFFERENCES OF DRUG RESPONSE AND GENE POLYMORPHISM

It is well known that each patient may respond variably even when each patient receives the same dose of the

Encyclopedia of Clinical Pharmacy
DOI: 10.1081/E-ECP 120006216

Table 1 Major CYP enzymes, their substrates, inducers, inhibitors, and phenotype markers

Enzyme	Representative substrates	Known inducer	Inhibitor	Polymorphism	Noninvasive test marker
CYP1A2	Caffeine Theophylline	Tobacco Charcoal-broiled meat	Fluvoxamine	(Yes)[a]	Caffeine
CYP2C9	Warfarin Tobutamide	Barbiturates Rifampin	Sulfaphenazole	(Yes)[a]	Tobutamide
CYP2C19	Mephenytoin Omeprazole	Barbiturates Rifampin		Yes	(S)-Mephenytoin
CYP2D6	Metopronolol Imipramine Encainide	Barbiturates Rifampin in EMs only	Quinidine	Yes	Dextromethorphan Debrosoquine
CYP2E1	Ethanol Acetaminophen	Ethanol Isoniazid	Disulfiram	(Yes)[a]	Chlorzoxazone
CYP3A3/4	Cyclosporine Verapamil Nifedipine blockers	Phenytoin Rifampin Barbiturates	Erythromycin Verapamil Ketoconazole Grapefruit juice	(Yes)[a]	Erythromcyin Midazolam
CYP3A5	Cyclosporine	Barbiturates Rifampin	Ketoconazole Erythromycin	Yes	Erythromycin

[a]There are no conclusive evidences between genotype and phenotype, although some phenotype or genotype differences are detected in the population.

same medication. Many factors involve this inter- and intrapatient variability of drug response. The recommended dose of each therapeutic agent is determined based on clinical study results from a small number of patients who meet a narrowly defined criteria. However, the optimal daily dose in clinical practice can vary widely among patients because of factors such as age, size of the patient, gender, ethnicity, concurrent drug therapy, food intake, and the patient's own disease conditions (i.e., renal function or liver function).[10] It should be noticed that each individual patient has his or her own optimal dose of drug therapy in a specific clinical condition. Because the CYP is an important enzyme system for various drug classes, large differences in the activities of the CYP among individuals explain some of this wide variability in the dosing requirements of various drugs.[11]

Most drugs are metabolized by multiple metabolic pathways, which is necessary because this may protect the body from the toxic effects of drugs in case one metabolic pathway is shut down. However, in certain cases, single enzyme activity largely determines drug response. For example, the therapeutic effect of the drug may be correlated with the blood levels of the parent drug, and this may depend largely on the rate of metabolism of the drug catalyzed by a single CYP. Although it is not always true, there are now at least several important examples in which the relation between drug dose, blood levels, and therapeutic response in an individual patient is largely determined by the catalytic activity of single enzyme, CYP2D6. For example, patients with the poor metabolizer phenotype for CYP2D6 demonstrate significantly high area under the plasma concentration time curve for metoprolol, compared with extensive metabolizers of CYP2D6.[12] As a result, poor metabolizers generally attain therapeutic effects from these drugs at significantly reduced daily doses. If the same dose as a rapid metabolizer is given, the patients will develop severe toxicity.

The activity of CYP3A varies at least 10-fold among patients, and the activity level in a given patient appears to be related to the dosing requirements of a certain substrate metabolized by CYP3A.[13,14] It has been shown that the liver activity of CYP3A largely predicts blood levels of cyclosporine in patients receiving the drug for treatment of psoriasis;[13] that is, patients with higher CYP3A activity have lower blood levels of cyclosporine at any given daily dose of the drug.[14,15]

There are significant differences in CYP enzyme activities in the general population, which are mainly determined by genetics, although some environmental factors such as enzyme inducers or inhibitors are involved. One of the best known CYP enzyme inducer is cigarette smoking, which generally induces CYP1A2 and other CYP enzymes.[16,17] In a certain population, there is genetically a lack of CYP genes (i.e., CYP2C19, CYP2D6), and these populations may develop significant toxicity if the standard dose of a drug with a narrow therapeutic index is given.

ROLE OF CYP IN SYSTEMIC AVAILABILITY OF DRUG

The CYP is present not only in the liver, but also in the intestine. It appears that the CYP is located mainly at the apex of the mature enterocytes, lying in a band just below the microvillous border.[18] In humans, the major enterocyte CYP appears to be the CYP3A4, which accounts for more than 70% of CYP activity in the intestine. Interestingly, CYP3A4 is located along with P-glycoprotein, a cell membrane efflux pump.[19] This may indicate that CYP3A4 along with P-glycoprotein are intended to prevent the environmental toxins, or xenobiotics such as drugs, from entering the body. The intestinal metabolism of many lipophilic drugs metabolized by CYP3A4 is estimated to be as much as one-half of the administered dose.[20] Previously, many CYP inhibitors were thought to act only on liver CYP enzymes, but it was found that they affect on both liver and intestinal CYP.[19,20] For example, ketoconazole, which is a potent inhibitor of CYP3A4, increases area under the curve of cyclosporine not only by inhibiting hepatic CYP3A4, resulting in reducing metabolism of cyclosporine, but also inhibiting intestinal CYP3A4, subsequently increasing bioavailability of cyclosporine.[21] Some food components such as grapefruit juice inhibit CYP3A in the intestine and, when oral felodipine is given with grapefruit juice, its AUC and Cmax are increased by 250% and 150%,[22] respectively. However, when intravenous felodipine is given with grapefruit juice, there is no significant difference.[22]

ROLE OF CYP IN DRUG–DRUG INTERACTION

Drug interactions constitute a major problem in chronic multiple drug therapy.[23] Although interactions affecting the pharmacodynamics of a drug can be reasonably predicted (i.e., additive effect or synergistic effect), those affecting its pharmacokinetics are difficult to predict. These might result from various contributions involving absorption, transportation, distribution, metabolism, and excretion.[23] Among these, metabolism in liver as well as intestine appears to represent the major source of drug–drug interactions. Because CYP enzymes are known to be induced or inhibited by, and involved in the oxidation of, a number of currently used drugs, they are likely to be responsible for numerous drug interactions in humans.[2] Because CYP3A4 metabolizes more than 50% of all therapeutic agents, its inhibitors and inducers have significant impact. Clinically important CYP3A4 inhibitors include ketoconazole, itraconazole, erythromycin, clarithromycin, nefazodone, ritonavir, and grapefruit juice.[23,24]

Torsades de pointes, a life-threatening ventricular arrhythmia associated with QT prolongation, can occur when these inhibitors are coadministered with terfenadine, astemizole, or cisapride because they inhibit these agents from converting the parent compound into nontoxic, pharmacologically active metabolites.[25–27] As a result, the proarrhythmic parent compound accumulates in the body, which causes toxic effects. Because of this serious drug interactions, these drugs have to be withdrawn from the market. Cyclosporine, an important immunosuppressant, has clinically been shown to be involved in multiple drug interactions.[23] Because cyclosporine is extensively metabolized in human liver and enterocytes by CYP3A4, any inducer of CYP3A4 (e.g., ripampin) should cause a decrease in cyclosporine levels, whereas any substrate or inhibitor (e.g., ketoconazole) of this CYP should elicit the opposite effect. Indeed, this has been clearly demonstrated clinically[28] as well as in an experimental model.[29]

Some drugs with multiple metabolic pathways are affected by many different inhibitors. For example, codeine is metabolized by CYP2D6 and CYP3A4, which act on different sites of action.[30] The O-demethylation of codeine is catalyzed by CYP2D6 and N-demethylation is catalyzed by CYP3A4.[31] The substrates of CYP2D6, such as thioridazine, amitriptyline, and metoprolol inhibit the O-demethylation of codeine preferentially,[30] whereas substrates of CYP3A4, such as ketoconazole, are strong inhibitors of the N-demethylation of codeine.[31]

Not all predicted drug interactions are expected to be clinically significant. For example, nifedipine and cyclosporine are both CYP3A4 substrates, but there is no clinically important drug interaction noticed.[23] To predict the drug interaction, several parameters are expected to be important. These include notably:[32] 1) relative affinity of CYP enzyme on both drugs (km); 2) dose and local concentration of each drug either in enterocytes or hepatocytes; 3) duration of concurrent therapy; and 4) CYP enzyme in the liver or intestine of the patient. The level of CYP3A is highly variable in each individual. It is possible that, in one patient with a low CYP3A level, all the cytochrome would

be saturated by the coadministered drugs, but not in another with higher CYP3A level; the consequence is that the interaction should occur in the former but not the latter.

NONINVASIVE MEASUREMENT OF CYP ACTIVITY AND PREDICTION OF DRUG RESPONSE

It would be great to measure the activity of individual CYP enzymes and predict drug response or drug interaction in individual patients because of the CYP enzymes involved in the metabolism of various therapeutic agents.[33] A reliable in vivo probe for phenotyping CYP3A4 would make it possible to identify individuals at greatest risk of toxicity due to high blood levels and inefficacy due to subtherapeutic blood levels, and to detect potentially dangerous drug–drug interactions.[34]

CYP3A is the predominant drug-metabolizing enzyme in humans. Thus, there have been considerable efforts to develop a simple, safe, and reliable phenotyping procedure for CYP3A activity; however, these efforts are largely suboptimal.[33] Among candidate probes, two procedures have shown clinical utility, intravenous midazolam clearance[35] and the erythromycin breath test (ERBT).[36,37] There is strong evidence that the clearance of midazolam provides an estimate of liver CYP3A activity.[38] However, it has a potent sedative effect, and the multiple blood sampling required for pharmacokinetic evaluation makes it inappropriate for widespread use in an outpatient setting.

The ERBT has been most widely studied[36,37] and has a significant correlation with the pharmacokinetics of the CYP3A substrate, cyclosporine.[14] The ERBT correlates with trough blood concentration of cyclosporine in patients with psoriasis[14] and with the oral clearance of cyclosporine in transplant recipients.[15] However, as a probe for CYP3A activity, the ERBT has significant limitations. Not only does it require intravenous access, which may exclude the fraction of gastrointestinal metabolism by CYP3A, but it also requires the administration of a radioactive substance, a potential safety concern.[33] In addition, ERBT fails to show a significant correlation with other known CYP3A4 substrates, such as alfentanil[39] or dapsone clearances.[40] Other markers such as dapsone, or the 6β-hydroxy cortisol urine test, have been tried with variable results.[40] Dextromethorphan shows some promise as a probe simultaneously measuring both CYP2D6 and CYP3A4 acitivity, but its urine metabolic ratios failed to predict CYP3A4 activity.[41] In the near future, genotyping information regarding individual CYP will be readily available, which may better explain individual variability of drug pharmacokinetics and pharmacodynamics.

REFERENCES

1. Guengerich, F.P. Reactions and significance of cytochrome P-450 enzymes. J. Biol. Chem. **1991**, *266*, 10019–10022.
2. Watkins, P.B. Drug metabolism by cytochromes P450 in the liver and small bowel. Gastroenterol. Clin. North Am. **1992**, *21* (3), 511–526.
3. Guengerich, F.P. Enzymatic oxidation of xenobiotic chemicals. Biochem. Mol. Biol. **1990**, *25*, 97–153.
4. Eatabrook, R.W. Cytochrome P450: from a Single Protein to a Family of Proteins—with Some Personal Reflections. In *Cytochromes P450: Metabolic and Toxicological Aspects,* 1st Ed.; Ioannides, Costas, Ed.; CRC Press, Inc.: Boca Raton, Florida, 1996; 3–28.
5. Harding, B.W.; Wong, S.H.; Nelson, D.H. Carbon monoxide-combining substances in rat adrenal. Biochem. Biophys. Acta **1964**, *92*, 415–421.
6. Nebert, D.W.; Carvan, M.J. Ecogenetics: From ecology to health. Toxicol. Ind. Health **1997**, *13*, 163–192.
7. Nebert, D.W.; Nelson, D.R.; Coon, M.J.; Estabrook, R.W.; Feyerseisen, R.; Fujii-Kuriyama, Y.; Gonzalez, F.J.; Guengerich, F.P.; Gunsalus, I.C.; Johnson, E.F.; Loper, J.C.; Sata, R.; Waterman, M.R.; Waxman, D.J. The P450 superfamily: Update on new sequences, gene mapping, and recommended nomenclature. DNA Cell Biol. **1991**, *10*, 1–14.
8. Wrighton, S.A.; Stevens, J. The human hepatic cytochromes P450 involved in drug metabolism. Crit. Rev. Toxicol. **1992**, *22* (1), 1–21.
9. Benet, L.Z.; Kroetz, D.L.; Sheiner, L.B. The Dynamic of Drug Absorption, Distribution, and Elimination. In *Goodman and Gilman's Pharmacological Basis of Therapeutics,* 9th Ed.; Hardman, J.G., Limbird, L.E., Molinoff, P.B., Ruddon, R.W., Eds.; McGraw-Hill, 1996; 3–27.
10. Lindholm, A. Factors influencing the pharmacokinetics of cyclosporine in man. Ther. Drug Monit. **1991**, *13*, 465–477.
11. Skoda, R.C.; Gonzalez, F.J.; Demierre, A.; Meyer, U.A. Two mutant alleles of the human cytochrome P-450dbl gene (P450C2D1) associated with genetically deficient metabolism of debrisoquine and other drugs. Proc. Natl. Acad. Sci. **1988**, *85*, 5240–5243.
12. Lennard, M.S.; Silas, J.H.; Freestone, S.; Trevethick, J. Defective metabolism of metoprolol in poor hydroxylators of debrisoquine. Br. J. Clin. Pharmacol. **1982**, *14*, 301–303.
13. Kahan, B.D. Cyclosporine. N. Engl. J. Med. **1989**, *321*, 1725–1738.
14. Watkins, P.B.; Hamilton, T.A.; Annesley, T.M.; Ellis, C.N.; Kolars, J.C.; Voorhees, J.J. The erythromycin breath

test as a predictor of cyclosporine blood levels. Clin. Pharmacol. Ther. **1990**, *48*, 120–129.
15. Turgeon, D.K.; Normolle, D.P.; Leichtman, A.B.; Annesley, T.M.; Smith, D.E.; Watkins, P.B. Erythromycin breath test predicts oral clearance of cyclosporine in kidney transplant recipients. Clin. Pharmacol. Ther. **1992**, *52*, 471–478.
16. Nakachi, K.; Imai, K.; Hayashi, S.-I.; Watanabe, J.; Kawajiri, K. Genetic susceptibility to squamous cell carcinoma of the lung in relation to cigarette dose. Cancer Res. **1991**, *51*, 5177–5180.
17. Nebert, D.W. Role of genetics and drug metabolism in human cancer risk. Mutat. Res. **1991**, *247*, 267–281.
18. Kolars, J.C.; Schmiedlin-Ren, P.; Dobbins, W.O.; Schuetz, J.; Wrighton, S.A.; Watkins, P.B. Heterogeneity of cytochrome P450IIIA expression in rat gut epithelia. Gastroenterology **1991**, *102*, 1186–1198.
19. Wacher, V.J.; Wu, C.-Y.; Benet, L.Z. Overlapping substrate specificities and tissue distribution of cytochrome P450 3A and P-glycoprotein: Implications for drug delivery and activity in cancer chemotherapy. Molecular Carcinogenesis. **1995**, *13*, 129–134.
20. Wu, C.-Y.; Benet, L.Z.; Hebert, M.F.; Gupta, S.K.; Rowland, M.; Gomez, D.Y.; Wacher, V.J. Differentiation of absorption and first-pass gut and hepatic metabolism in humans: Studies with cyclosporine. Clin. Pharmacol. Ther. **1995**, *58*, 492–497.
21. Gomez, D.Y.; Wacher, V.J.; Tomlanovich, S.J.; Hebert, M.F.; Benet, L.Z. The effects of ketoconazole on the intestinal metabolism and bioavailability of cyclosporine. Clin. Pharmacol. Ther. **1995**, *58*, 15–19.
22. Baily, D.G.; Arnold, M.O.; Munoz, C.; Spence, J.D. Grapefruit juice-felodipine interaction: Mechanism, predictability, and effect of naringin. Clin. Pharmacol. Ther. **1993**, *53*, 637–642.
23. Yee, G.C.; McGuire, T.R. Pharmacokinetc drug interactions with cyclosporine. Clin. Pharmacokinet. **1990**, *19*, 319–332 and 400–415.
24. Ku, Y.; Min, D.I.; Flanigan, M. Effect of grapefruit juice on microemulsion cyclosporine and its metabolites, M1 and M17 pharmacokinetics in healthy volunteers. J. Clin. Pharmacol. **1998**, *38*, 959–965.
25. Honig, P.K.; Worthan, D.C.; Zamani, K.; Conner, D.P.; Mullin, J.C.; Cantilena, L.R. Terfenadine–ketoconazole interaction: Pharmacokinetic and electrocardiographic consequences. JAMA, J. Am. Med. Assoc. **1993**, *269*, 1513–1518.
26. Zechnich, A.D.; Hedges, J.R.; Eiselt-Proteau, D.; Haxby, D. Possible interactions with terfenadine or astemizole. West. J. Med. **1994**, *160* (4), 321–325.
27. Chan-Tompkins, N.H.; Babinchak, T.J. Cardiac arrhythmias associated with coadministration of azole compounds and cisapride. Clin. Infect. Dis. **1997**, *24* (6), 1285–1313.
28. Jensen, C.W.B.; Flechner, S.M.; Van Buren, C.T.; Frazier, O.H.; Cooley, D.A.; Lorber, M.I.; Kahan, B.D. Exacerbation of cyclosporine toxicity by concomitant administration of erythromycin. Transplantation **1987**, *43* (2), 263–269.
29. Wang, R.W.; Newton, D.J.; Liu, N.; Atkins, W.M.; Lu, A.Y.H. Human cytochrome P-450 3A4: in vitro drug–drug interaction patterns are substrate-dependent. Drug Metab. Dispos. **2000**, *28* (3), 360–366.
30. Dayer, P.; Desmeules, J.; Leemann, T.; Striberni, R. Bioactivation of the narcotic drug codeine in human liver is mediated by the polymorphic monooxygenase catalyzing debrisoquine 4-hydroxylation of debrisoquine. Biochem. Biophys. Res. Commun. **1988**, *152*, 411–416, 30.
31. Pellinen, P.; Honkakoski, P.; Stenback, F.; Niemitz, M.; Alhava, E.; Pelkonen, O.; Lang, M.; Pasanen, M. Codeine N-demethylation and the metabolism-related hepatotoxicity can be prevented by cytochrome P450 3A inhibitors. Eur. J. Pharmacol., Environ. Toxicol. Pharmacol. Sect. **1994**, *270*, 35–43.
32. De Waziers, I.; Cugnenc, P.H.; Yang, C.S.; Leroux, J.P.; Beaune, P.H. Cytochrome P450 isoenzymes epoxide hydrolase and glutathione transferases in rat and human heaptic and extrahepatic tissues. J. Pharmacol. Exp. Ther. **1990**, *253*, 387–394.
33. Watkins, P.B. Noninvasive tests of CYP3A enzymes. Pharmacogenetics **1994**, *4*, 171–184.
34. Lown, K.; Kolars, J.; Turgeon, K.; Merion, R.; Writhton, S.A.; Watkins, P.B. The erythromycin breath test selectively measures P450IIIA in patients with severe liver disease. Clin. Pharmacol. Ther. **1992**, *51*, 229–238.
35. Thummel, K.E.; Shen, D.D.; Podoll, T.D.; Kunze, K.L.; Trager, W.F.; Hartwell, P.S.; Raisys, V.A.; Marsh, C.L.; McVicar, J.P.; Barr, D.M. Use of midazolam as a human cytochrome P450 3A probe: I. In vitro–in vivo correlations in liver transplant patients. J. Pharmacol. Ther. **1994**, *271* (1), 549–556.
36. Lown, K.; Thummel, K.E.; Benedict, P.E.; Shen, D.D.; Turgeon, D.K.; Berent, S.; Watkins, P.B. The erythromycin breath test predicts the clearance of midazolam. Clin. Pharmacol. Ther. **1995**, *57*, 16–24.
37. Watkins, P.B. Erythromycin breath test and clinical transplantation. Ther. Drug Monit. **1996**, 18, 368–371.
38. Kronbach, T.; Mathys, D.; Umeno, M.; Gonzalez, F.J.; Meyer, U.A. Oxidation of midazolam and triazolam by human liver cytochrome P450IIIA4. Mol. Pharmacol. **1989**, *36*, 89–96.
39. Krivoruk, Y.; Kinirons, M.T.; Wood, A.J.; Wood, M. Metabolism of cytochrome P4503A substrates in vivo administered by the same route: Lack of correlation between alfentanil clearance and erythromycin breath test. Clin. Pharmacol. Ther. **1994**, *56*, 608–614.
40. Kinirons, M.T.; O'Shea, D.; Downing, T.E.; Fitzwilliam, A.T.; Joellenbeck, L.; Groopman, J.D.; Wilkinson, G.R.; Wood, A.J. Absence of correlations among three putative in vivo probes of human cytochrome P4503A activity in young healthy men. Clin. Pharmacol. Ther. **1993**, *54*, 621–629.
41. Min, D.I.; Ku, Y.; Vichiendilokkul, A.; Fleckenstein, L. A urine metabolic ratio of dextromethorphan and 3-methoxymorphinan as a probe for CYP3A4/5 activity and prediction of cyclosporine clearance in healthy volunteers. Pharmacotherapy **1999**, *19*, 753–759.

Department of Health and Human Services

Lauren Schlesselman
Niantic, Connecticut, U.S.A.

INTRODUCTION

The Department of Health and Human Services (DHHS) is the principal U.S. government agency assigned to protect the health of all Americans and for providing essential human services for those unable to help themselves. The goals of the DHHS, according to the strategic plan for fiscal years 2001–2006,[1] include reducing the major threats to the health and productivity of all Americans, improving economic and social well-being, improving access to health services, improving the public health systems, and strengthening the nation's health scence research productivity. The DHHS accomplishes this mission through more than 300 programs under the leadership of the Office of the Secretary. DHHS programs are administered through 11 operating divisions utilizing nearly 62,000 employees and a budget approaching $400 billion.[1]

BACKGROUND

Many of DHHS' divisions are managed by the Public Health Service's (PHS) commissioned officers with the assistance of civil service employees. The commissioned officers corps consists of pharmacists, physicians, dentists, nurses, and other health care professionals. These officers and employees engage in clinical care, medical research, and disease surveillance through the DHHS divisions. Before the formation of the commissioned corps, pharmacists were already serving the American population through the Marine Hospital Service, which cared for merchant seaman in large seaport cities.[2] The services of the PHS expanded beyond seaports when Congress discovered the poor health care and living conditions of Native Americans under the authority of the Department of Interior's Bureau of Indian Affairs. In 1954, Congress transferred the care of all Native Americans from the Department of the Interior to the Department of Health, Education, and Welfare. With the creation of the Department of Education through the signing of the Department of Education Organization Act in 1979, the current DHHS officially succeeded the Department of Health, Education, and Welfare.

Throughout history, the divisions of DHHS have influenced many aspects of pharmacy practice and continue to have an impact on it today. Pharmacy practice is influenced by legislation administered through the DHHS divisions, through the funding of grants to support health care for the underprivileged and through research and the funding of research to monitor and improve health services. Each division plays a unique role in providing and improving health care.

AGENCIES OF THE DHHS

National Institutes of Health (NIH)

Although now considered the premier medical research organization, NIH's roots began in 1887 as a one-room laboratory, known as the Hygienic Laboratory, on Staten Island, New York. The laboratory was opened under the direction of Surgeon General John Hamilton to study major epidemics of the nineteenth century, including cholera, yellow fever, Rocky Mountain spotted fever, and hookworm.[2] The importance of the Hygienic Laboratory's work prompted legislation to move it to Washington, DC. Finally, in 1930, the Ransdell Act created the NIH to replace the Hygienic Laboratory. NIH researchers continue to investigate the causes of and cures for the nation's most devastating diseases. Currently, the 17 separate health institutes of the NIH are focusing large amounts of its nearly $18 billion budget on cancer, Alzheimer's disease, diabetes, arthritis, and acquired immunodeficiency syndrome (AIDS).[3] Along with performing research, NIH also supports nearly 40,000 research programs nationwide.

Food and Drug Administration (FDA)

The FDA is responsible for assuring the safety of foods and cosmetics, along with the safety and efficacy of phar-

Encyclopedia of Clinical Pharmacy
DOI: 10.1081/E-ECP 120006180

maceuticals, biological products, and medical devices.[4] This authority to monitor medications and foods was first granted by Congress with the Food and Drug Act in 1906.[5,6] Assuring compliance with this important act remains a key function of the FDA.

Since 1906, various amendments to the Food and Drug Act have greatly influenced the practice of Pharmacy. The Sherley Amendment of 1912 was the first legislation to regulate the labeling of medications. The amendment mandated a guarantee against adulteration and misbranding from manufacturers.

The Delaney Clause, named for Congressman James Delaney, remains an important part of the 1958 Food Additives Amendment and the 1960 Color Additive Amendments to the Food, Drug, and Cosmetic Act. The clause states that "no additive shall be deemed to be safe if it is found to induce cancer when ingested by man or animal"[5] at any dose. The clause also recognizes and accepts that evidence of carcinogenicity in animals is sufficient to correlate to a risk in man. Examples of the FDA invoking the Delaney Clause include the removal of cyclamates, aminotriazole, and DDT from human food. Modernization of the act, to allow for a negligible risk standard, rather than the current zero risk, is currently being pursued.

The Kefauver–Harris Amendments of 1962 gave the FDA control over prescription drug advertising. According to the amendment, all advertisements and printed matter issued by a manufacturer must include the medication name, strength, side effects, contraindications, and information on effectiveness. Another major change to the act was the requirement that all medications must be shown to be effective, as well as safe. After this amendment, all new drug applications submitted to the FDA must contain research proving the effectiveness of the product. At that time, control over investigational medications and the inspection of factories was also transferred to the FDA.

Another amendment to the Food, Drug, and Cosmetic Act is the Nutrition Labeling Health and Education Act (NLHEA) of 1990. NLHEA is intended to provide consumers with information to help maintain healthy dietary practices and to protect consumers from unfounded health claims. NLHEA provides information to consumers by requiring nutrition labeling on all foods and dietary supplements. These nutrition labels must include the serving size, and number of servings per package, along with the amount of calories, fat, saturated fat, cholesterol, sodium, carbohydrates, sugars, dietary fiber, and total protein per serving. NLHEA also ensures the validity of nutrition claims by reviewing research submitted by manufacturers to ensure the claim meets with significant scientific agreement.

Centers for Disease Control and Prevention (CDC)

The roots of the CDC, the agency responsible for protecting health, are traced back to World War II. At that time the Malaria Control in War Areas (MCWA) attempted to control the spread of malaria among servicemen, along with preventing the introduction of the disease into the civilian population.[2] After the war, the importance of continued monitoring of infectious diseases prompted the conversion of MCWA to the Communicable Disease Center in 1946, the predecessor of the modern CDC. Today, the CDC monitors disease trends, investigates outbreaks and health risks, fosters healthy environments, and implements illness prevention measures and standards. The research performed by the CDC is primarily field research, as compared with the laboratory research that is performed by the NIH. More than 7500 employees and $3 billion per year are necessary to accomplish these goals.

Indian Health Services (IHS)

Although federally funded health services for Native Americans began in the early 19th century, the Transfer Act of 1954 propelled Native American health toward its modern form. This law transferred responsibility for the health care of Native American and Alaska Natives to the PHS. Soon after the transfer the PHS was directed by Congress to conduct health surveys of Native American populations. The first study was the Trachoma Study. This study found a widespread trachoma epidemic, along with increased incidence of other infectious diseases, including tuberculosis, among this population. These results prompted moves to improve sanitary living conditions and expand the provision of health care available to Native Americans.

Another PHS survey, the Meriam Report, also pushed for advances in Native American health care. Among the Meriam Report findings were that 1 out of 10 Native Americans had tuberculosis and over one-third of all Native American deaths were children under 3 years of age. These findings prompted moves for stronger health program supervision with more qualified staff and the establishment of health clinics on Native American reservations.

The findings of all the PHS surveys led to the formation of the current Indian Health Services as the fed-

eral agency responsible for providing health services to Native American and Alaska Natives. These services are currently provided to nearly 1.5 million persons in more than 550 federally recognized tribes in 35 states[7] with the goal to assure that comprehensive, yet culturally acceptable, personal and health services are available and accessible. The IHS currently maintains 36 hospitals, 58 health centers, 4 school health centers, and 44 health stations. With the health care provided by the IHS, the Native American life expectancy has increased 12 years since 1973,[7] with decreased infant and maternal pneumonia and influenza, tuberculosis, and gastrointestinal mortality. Despite these advances, IHS continues to work to reduce deaths due to alcoholism, accidents, diabetes mellitus, homicide, and suicide. The rates of death due to these causes remain significantly higher in the Native American population than the rest of the U.S. population.

Health Resources and Services Administration (HRSA)

HRSA provides the leadership necessary to achieve integration of service delivery to meet the health needs of Americans. This is done through the provision of personnel, educational, physical, and financial resources. Part of HRSA's $4.8 billion budget funds more than 3000 health clinics to provide medical care to more than 9 million individuals in underserved communities each year. HRSA also administers the Migrant Health Program, which provides grants to communities to support culturally based medical services to migrant and seasonal farmworkers and their families.

Although HRSA administers many diverse programs, one of the major programs is the Ryan White Comprehensive AIDS Resources Emergency (CARE) Act, Public Law 101-381.[8] The Ryan White CARE Act is named in memory of an Indiana teenager who increased awareness about the needs of people with AIDS while suffering from the disease himself. This act helps states, communities, and families to ease the burden of the AIDS epidemic. HRSA estimates 500,000 individuals with HIV and AIDS receive assistance through this act each year.[9]

The Ryan White CARE Act is divided into multiple parts with each part providing support to different segment of the AIDS community. The first part of the CARE Act, Title 1, provides grants to cities and large numbers of low-income, underinsured, or uninsured individuals with HIV and AIDS. These grants are intended to provide outpatient health care, prescription medications, home health services, hospice care, counseling services, and housing and transportation assistance. Title 2 of the CARE Act provides grants to states, Washington, DC, Puerto Rico, and other United States territories to provide health care to individuals living with HIV and AIDS. Title 2 is aimed at prolonging life and preventing hospitalization, particularly through assistance with obtaining medications through the AIDS Drug Assistance Program. With more than $150 million in funding from the CARE Act, the AIDS Drug Assistance Program allows states to establish programs to purchase and distribute antiretroviral therapy for low-income individuals. The third section of the act, Title 3, provides funds to public and nonprofit organizations to support early intervention services for low-income, medically underserved people at risk for HIV. These services are designed to slow the spread of HIV through education, counseling, testing, and early treatment. Title 4 provides grants to establish services for children, women, and families. In 1996, Part F was added to the CARE Act to combine other existing AIDS programs under the HRSA umbrella. Included in Part F are AIDS Education and Training Centers that train health care providers about the necessity of early intervention and appropriate treatment, Dental Reimbursement Programs that provide grants to dental schools to assist in covering costs incurred in providing treatment to HIV patients, and the Special Projects of National Significance Program that provides grants to develop models for providing care to persons with HIV in special populations.

Substance Abuse and Mental Health Services Administration (SAMHSA)

Although the Narcotics Division of the PHS (later renamed the Mental Hygiene Division) was created in 1929 to treat and study addiction, the National Mental Health Act of 1946 was the first legislation to authorize research and aid for mental health services. Starting in 1973, this act was administered by the Alcohol, Drug Abuse and Mental Health Administration (ADAMHA) through the National Institute of Mental Health (NIMH), the National Institute of Alcohol and Abuse and Alcoholism, and the National Institute on Drug Abuse. The current SAMHSA did not replace ADAMHA until 1992. SAMHSA continues ADAMHA's work to improve the quality and availability of substance abuse prevention, addiction treatment, and mental health services. The goal of SAMHSA is to reduce illness, disability, and death, along with the cost to society, which result from substance abuse and mental illness. SAMHSA is able to provide federal grants to states to support programs intended to eliminate the stigma associated with substance abuse and mental illness, to disseminate information to improve available services, and to develop standards for the treatment of addicted and mentally ill persons.

Agency for Toxic Substances and Disease Registry (ATSDR)

ATSDR, one of DHHS' newest agencies, works to prevent exposure to hazardous substances from waste sites. The agency develops toxicological profiles of hazardous chemicals found at waste sites on the U.S. Environmental Protection Agency's National Priorities List. Its 400 employees also provide health education training in communities near these waste sites.

Agency for Healthcare Research and Quality (AHRQ)

Since its establishment in 1989, AHRQ has sponsored and conducted research to improve the quality of health care, reduce its cost, and increase access. It also supports research to address patient safety issues and medication errors. AHRQ's goal is to provide information that allows people to make better decisions about healthcare.

HUMAN SERVICES OPERATING DIVISIONS

Centers for Medicare and Medicaid Services (CMS)

The primary responsibilities of CMS, formerly known as Health Care Financing Administration (HCFA), include administration of Medicare and Medicaid programs. Since 1965, Medicaid has provided health coverage for low-income persons, while Medicare has provided for the elderly and disabled. Medicaid currently provides coverage for more than 34 million people, including nearly 18 million children. Medicare currently provides coverage for more than 39 million elderly and disabled Americans.[10] CMS requires a budget of $325.4 billion to provide these and other services. Among CMS other responsibilities is administration of the Children's Health Insurance Program. The Children's Health Insurance Program provides reduced or no-cost health coverage for more than 2 million children under the age of 19 whose families earn too much to be eligible for Medicaid but do not earn enough to afford private insurance.

Administration for Children and Families (ACF)

The ACF, established in 1991, maintains more than 60 programs that promote the economic and social well-being of children, families, and communities. Many of ACF's programs are aimed at helping children, with the most widely recognized being the Head Start Program. Head Start works with children from birth to 5 years of age, pregnant women, and their families to increase the school readiness of children from low-income families. ACF also funds programs to prevent child abuse and domestic violence. ACF continues to administer a national enforcement system that works to collect child support payments from noncustodial parents.

Administration on Aging (AoA)

Establishment of the AoA was mandated as part of the Older Americans Act of 1965. The Older Americans Act was passed as a means to organize, coordinate, and provide community-based services and opportunities for older Americans and their families. Although AoA programs are available to all Americans 60 years of age or older, priority is given to those with the greatest need. AoA's work is intended to protect the rights of vulnerable and at-risk persons, educate the community about the danger of elder abuse, and provide employment opportunities for older Americans.

Office of the Secretary of Health and Human Services (OS)

The Office of the Secretary provides leadership for the entire DHHS. It is responsible for advising the President on issues relating to health and welfare. The most recent expansion in the OS is the formation of the Office of Public Health Preparedness (OPHP) in late 2001. This office was created in response to the terrorist attacks on September 11, 2001. The OPHP directs the DHHS' activities aimed at protecting the population from acts of bioterrorism and other public health emergencies. Working with the Office of Homeland Security, OPHP's efforts are aimed at coordinating the preparation for and recovery from such events.

Program Support Center (PSC)

The final DHHS agency, PSC, provides administrative support for the DHHS. PSC is a self-supporting division that operates as a business-like enterprise. Their mission

is to provide ''qualitative and responsive 'support services' on a cost-effective, competitive, 'service-for-fee' basis''[11] to DHHS agencies and other federal agencies. Services available through PSC include personnel, grants, information technology, and administrative services.

Through the many divisions of the DHHS, the pharmacists of the PHS have worked to improve and protect the health of Americans, particularly those who are unable to care for themselves. Not only do pharmacists play an important role in the DHHS, but the DHHS influences pharmacist on a daily basis. Pharmacy practice is continually influenced by legislation administered through the DHHS divisions, through the funding of grants to support health care for the underprivileged and through research and the funding of research to monitor and improve health services.

The DHHS influence is apparent on every medication bottle delivered from the manufacturer, in the patient counseling techniques utilized, and treatment guidelines and research protocols administered throughout the country.

REFERENCES

1. U.S. Department of Health and Human Services. www.hhs.gov. September 2000.
2. Mullan, F. *Plagues and Politics: The Story of the United States Public Health Service*; Basic Books: New York, 1989.
3. U.S. Department of Health and Human Services National Institute of Health. www.nih.gov. September 2000.
4. FDA: U.S. Food and Drug Administration. www.fda.gov. September 2000.
5. Federal Food, Drug, and Cosmetic Act, 21CFR§201(1906).
6. Gennaro, A.R. In *Remington's Pharmaceutical Sciences*; Osol, A., Ed.; Mack Publishing: Easton, Pennsylvania, 1990.
7. Indian Health Services. www.ihs.gov. September 2000.
8. Ryan White CARE Act, Pub. L No. 101-381.
9. HRSA: Department of Health and Human Services: U.S. Public Health Service. www.hrsa.gov. September 2000.
10. HCFA: Health Care Financing Administration. www.hcfa.gov. September 2000.
11. Program Support Center. http://www.psc.gov/concept.html (accessed September 2000).

PROFESSIONAL DEVELOPMENT

Diabetes Care, Pharmacy Practice in

Tommy Johnson
University of Georgia College of Pharmacy, Athens, Georgia, U.S.A.

INTRODUCTION

When people talk about pharmacy practice in diabetes care, the first thought that comes to most peoples' minds is the community pharmacist dispensing a prescription for a blood glucose lowering medication. However, pharmacists are involved at a much deeper level in the care of patients with diabetes. In this article, examples of different ways pharmacists are involved in the care of patients with diabetes are provided. Resources to learn more about diabetes, as well as tools that will assist you in providing care, are also indicated.

COMMUNITY PHARMACY

Diabetes services in community pharmacies range from basic to complex. Basic services include the following:

1. The dispensing of medications and counseling about their proper use, storage, side effects, and potential drug interactions is the minimal involvement that pharmacists in this setting should have.
2. Educating patients about the proper use of ketone strips, lancing devices, and the proper selection of over-the-counter (OTC) products is included as a basic service.
3. Pharmacists may want to devote a section of the pharmacy, or at least shelf space, to specific products for patients with diabetes, or to enroll in one of the franchises that sells these products to pharmacies to help them with marketing and signage.
4. Patient education flyers, pamphlets, and video tapes may be ways that pharmacists try to improve the knowledge of their customers with diabetes.

More in-depth services include the following:

1. Educating patients about the proper use of blood glucose monitoring products and discussing before and after meal blood glucose target ranges and helping patients to determine causes of above and below target range readings requires more time.
2. Holding blood sugar screening programs in a store to increase diabetes awareness and potentially identify undiagnosed patients. This can be performed by the pharmacist or in conjunction with a local diabetes education program.
3. Performing talks for civic and diabetes support groups is a way to let people know that there is a pharmacist that is knowledgeable about diabetes in their community.
4. Having diabetes days in the pharmacy where different local diabetes educators and health professionals discuss or perform services in the store.
5. Fitting and selling orthotic shoes.
6. Many pharmacists, especially those who own their own stores, have turned to compounding as a means for financial and professional satisfaction. Compounding topical products for peripheral neuropathy, wounds, periodontal, and retinopathy are a few examples.

Complex services include the following:

1. Develop and run a diabetes education/management program through a pharmacy. This can be performed on a basic level where patient assessment to identify areas that the patient needs education in and performance of certain educational components occurs in the pharmacy. Components that the pharmacist may perform for this level of service are medication review, discussion of the differences in Type 1 and Type 2 diabetes, reasons that the person or family member developed diabetes, signs/symptoms, causes and treatments of hyperglycemia and hypoglycemia, basic foot screen, and reminder and education about tests that should be performed and their frequency. Areas such as nutrition counseling, foot care, and medication adjustments are referred to other providers.
2. Pharmacists that are more comfortable with their diabetes knowledge and counseling skills may per-

Encyclopedia of Clinical Pharmacy
DOI: 10.1081/E-ECP 120006227

form basic nutrition assessments, or educate patients on carbohydrate counting and the exchange systems of meal planning. They may discuss medication adjustments, specifically increases or decreases in insulin dosage, based on blood glucose readings and the carbohydrate content of the next meal. In-depth discussions about the cause, prevention, and treatment of complications of diabetes may be part of the education provided to patients. Pharmacists may provide these services by themselves or hire a nurse or dietitian to work with them through the pharmacy. The creation of educational rooms where individual and group sessions can occur are often created to give the pharmacist, educators, and patients privacy. By having a nurse and dietitians on staff as part- or full-time employees, pharmacists can apply to become American Diabetes Association Recognized Outpatient Education Providers and to be subsequently reimbursed by medicare for their educational services.

HOSPITAL PHARMACY

Pharmacists can provide diabetes care in the hospital setting in several ways. One way is to perform in-services to the nursing and hospital staff on medication used in treating diabetes and comorbidities. Which blood pressure medication should be used in patients with microalbuminuria, and why? Which medications when used in patients with diabetes can cause an increase or decrease in blood sugar levels? What contraindications should they look out for in patients in the hospital with diabetes? Another way is to actively participate in patient education of inpatients or outpatients.

In most instances, pharmacists are relegated to the medication or blood glucose monitor counseling aspect only. In some hospital programs, pharmacists are the diabetes coordinators and perform all areas of administration and patient education. Preparing IVs for patients with diabetes undergoing surgery, those admitted with diabetic ketoacidosis (DKA), or newly diagnosed patients are common areas of pharmacist involvement in hospitals.

CLINICS

Pharmacists may be involved in a variety of clinics. Armed forces clinics or specialty clinics that are part of hospitals are the largest types of freestanding clinic for patients with diabetes. Pharmacists are involved with the dispensing, medication counseling, and to some degree, counseling about some aspect of diabetes. Clinics for indigent patients are becoming more common with the number of working poor increasing. Pharmacists may be involved with collaborative practice arrangements with physicians where medication changes are made based on the pharmacist assessment in some cases. This type of setting tends to give the pharmacist flexibility to perform diabetes education and management services.

PHYSICIANS' OFFICES

Physicians are being overburdened by patient visits and the necessity to follow the Health Plan Employer Data and Information Set (HEDIS) and other practice guidelines. Pharmacists can perform chart reviews to see if patients with diabetes have received regularly scheduled test for A1C, urinary microalbumin, lipid measurements, referral for dilated eye exams, foot assessment and foot care, and blood pressure measurements.

Pharmacists can assist the physician by assessing clinical outcomes of diabetes, hypertension, thyroid disorders, and lipids, and making recommendations to the physician about the potential need for adjustments in medications. Pharmacists can also educate individuals and groups of the physician's patients on diabetes within the office setting.

PRIVATE PRACTICE

Some pharmacists are confident in their counseling and business skills to where they develop their own private practice. However, in the United States, this is not common for pharmacists to do and heavily relies on individual state's reimbursement for diabetes education and management services. Services are provided in clinic-type settings, in other pharmacist's practices, and even over the Internet and phone. This area will expand when reimbursement improves for the provision of these services. Examples of a few of these services are The Diabetes Center in Connecticut, and Diabetes In Control, which is an Internet business.

NURSING HOMES AND ASSISTED LIVING FACILITIES

Type 2 diabetes is common in the elderly. Most nursing home and assisted living facilities are staffed by Certified Nursing Assistants. These staff need to be trained in the proper care of patients with diabetes. Training on the identification of the signs/symptoms, causes, and treatments of high and low blood glucose is a basic training skill that all staff should know, but few do. A written protocol should be available and accessible to staff. Inservices, including medications, blood sugar reading assessments, and foot and skin care, should be covered quarterly with staff and even more frequently in some homes due to the high turnover rate of staff. It should be included in all new hire training.

PHARMACEUTICAL INDUSTRY AND DIABETES PRODUCT SALES

Pharmacists that work for pharmaceutical companies may be involved in diabetes care either as salesmen, clinical education consultants, or researchers. The number of products used in the treatment of diabetes is expanding as we learn more about the underlying causes of the disease. Since the late 1990s, more than five new oral agents and three new insulins have come into the marketplace. Pharmacists have played an integral part in educating physicians, other pharmacists, and other health care personnel on actions and uses of these new products. Clinical education consultants or medical liaisons for pharmaceutical companies take this education a step further by providing continuing education and clinical assistance to the physicians in the treatment of their patients with diabetes. It is evident with the development of the alpha glucosidase inhibitors, meglitinides, thiazolidinediones, and new insulin formulations such as lispro and glargine that researchers have been trying to develop products that improve the outcomes of these patients. The number of products used for patients to check their blood sugar has mushroomed since the early 1990s. Blood glucose monitor technology has allowed patients to perform these tests with minimal invasion. The development of truly noninvasive blood glucose monitoring; testing devices for home use for blood pressure, cholesterol, and A1c; and other diabetes-related devices will require sales personnel with more technical and medical knowledge that those used in the past.

MANAGED CARE AND PRESCRIPTION DRUG BENEFIT ADMINISTRATORS

With passage and implementation of national medicare prescription drug coverage, pharmacists will need to take a more active role in the development of reasonable, effective formularies of medications used to treat diabetes and the supplies necessary for patients to achieve optimal outcomes. Pharmacists working for managed care organizations may be in decision-making positions that determine the frequency and type of diabetes education that particular insurance companies will provide to their cardholders. Pharmacists that have been involved in diabetes care know of the importance of individual assessment and periodic follow-up to assess maintenance of optimal therapeutic and personal outcomes. Pharmacists without this background may only look at products and educational services as a current cost without taking long-term benefits into consideration.

Comprehensive diabetes management programs that have showed positive clinical and financial outcomes extend past the examples of the Asheville Pharmacy Project, the Mississippi Medicaid Project, and the South Carolina Pharmacists Diabetes Management Programs. The degree of reimbursement for diabetes education services often differs by state.

CONCLUSION

Pharmacists can be involved in a variety of areas of diabetes care. These areas can range from direct, with personal intervention and counseling, to indirect by deciding what services and products a patient may obtain. With any pharmacist practice, the environment, financial constraints, time limitations, desire, and competence of the pharmacist each play a role as to the involvement a pharmacist has with a person with diabetes. With the number of cases of diabetes expected to increase, pharmacists can and should play a more prominent role in assisting patients with diabetes.

Resources for information about diabetes products and management are abundant. Below are some of the many informative web sites available to patients and pharmacists that will enable them to increase their knowledge about diabetes.

DOCUMENTATION FORMS

Forms to document patient assessments and educational session content are abundant. Individual practices can modify these forms to meet their specific locations needs. Examples of these forms are often included in certificate programs such as those offered by the National Community Pharmacist Association, American Pharmaceutical Association, American Association of Diabetes Educators, and state pharmacy organizations. These forms are also found on the different web sites, such as www.bd.com, www.novo.dk, and www.humulinpen.com.

SOME DIABETES-RELATED WEB SITES

www.aadenet.org
www.pharminfo.com/disease/immun/#iddm
www.ezdiabetes.com/
www.afpafitness.com/FACTINDX.HTM
http://medicine.ucsf.edu/resources/guidelines/guidedm.html
www.pfizer.com/main.html
www.cdc.gov/diabetes/index.htm
www.diabetes.org/
www.avandia.com
www.actos.com/
www.novo.dk/health/dwk/info/ydww/index.asp
http://diabetes.lilly.com
www.eatright.org/
www.diabetesmonitor.com/tx_tin2/sld001.htm
www.joslin.harvard.edu/education/library/oha.html
www.lifescan.com
www.intelihealth.com/IH/ihtlH?t=21054
www.niddk.nih.gov/health/diabetes/diabetes.htm
www.cdc.gov/nccdphp/cdnr.htm
www.aace.com/indexnojava.htm
www.bms.com/products/index.html
www.aventispharma-us.com
www.aafp.org/acf/1999/resource.html
www.mendosa.com/insulin.htm
www.guidelines.gov/index.asp
www.diabetesincontrol.com
www.kunkelrx.com
www.edu-center.com

BIBLIOGRAPHY

Coast-Senior, E.A.; Kroner B.A.; Kelley C.L.; Trilli L.E. Management of patients with Type 2 diabetes by pharmacists in primary care clinics. Annals of Pharmacotherapy **1998 Jun**, *32* (6), 636–641.

McDermott, J.H.; Christensen D.B. Provision of pharmaceutical care services in North Carolina: A 1999 survey. Journal of the American Pharmaceutical Association **2002 Jan–Feb**, *42* (1), 26–35.

Monroe, W.P.; Kunz, K.; Dalmady–Israel, C.; Potter, L.; Schonfield, W. Economic evaluation of pharmacist involvement in disease management in a community setting. Clinical Ther. **1997**, *19*, 113–123.

Schapansky, L.M.; Johnson J.A. Pharmacists' attitudes toward diabetes. Journal of the American Pharmaceutical Association **2000 May–Jun**, *40* (3), 371–377.

Setter, S.M.; Corbett C.F.; Cook D. Johnson SB exploring the clinical pharmacist's role in improving home care for patients with diabetes. Home Care Provid. **2000 Oct**, *5* (5), 185–192.

Swain, J.H.; Macklin R. Individualized diabetes care in a rural community pharmacy. Journal of the American Pharmaceutical Association **2001 May–Jun**, *41* (3), 458–461.

Dietary Supplement Health and Education Act

Gayle Nicholas Scott
Medical Communications and Consulting, Chesapeake, Virginia, U.S.A.

INTRODUCTION

President Clinton made history when he signed the Dietary Supplement Health and Education Act (DSHEA) into law in 1994. The DSHEA amended the Federal Food, Drug, and Cosmetic Act to create a new regulatory category of products: dietary supplements. The DSHEA exempts dietary supplements from laws regulating drugs, as long as the manufacturer does not claim that the supplement can diagnose, mitigate, treat, cure, or prevent disease. This monograph describes the history and provisions of the DSHEA, and its importance to pharmacists. This monograph will also compare the regulation of drugs and dietary supplements, address the minimal FDA scrutiny and inadequate safeguards required by the DSHEA, and the implications for pharmacists.[1,18,19]

HISTORY OF THE DSHEA

The Federal Food, Drug, and Cosmetic Act was passed in 1938 after the use of diethylene glycol as an ingredient in a sulfanilamide elixir resulted in the deaths of almost 100 people. This law empowered the FDA to require New Drug Applications (NDA) to have evidence of safety and efficacy from the manufacturer before a product could be marketed. Although the law intended more FDA regulation for dietary supplements than foods, the regulations promulgated by the FDA left unanswered questions about how products for ''special dietary uses'' should be classified and regulated. During the following decades, policy on dietary supplements was created largely through FDA litigation. In 1962, following public alarm at thalidomide-induced birth defects in Europe, Congress passed the Kefauver–Harris Act, which required drug manufacturers to provide scientific proof that a drug was safe and effective before marketing, and tightened control over products classified and sold as drugs.[4,5]

Following the passage of the Nutrition Labeling and Education Act in 1990 that granted additional labeling authority to the FDA, dietary supplements would be subject to stricter criteria for health claims under proposed new regulations. Congress also began deliberation on bills that would increase the FDA's enforcement powers and amend the Federal Trade Commission Act to prohibit advertising nutritional or therapeutic claims that were not on supplement labels. Fearing the impact of these pending regulations and laws, the health food industry mounted a massive grass-roots effort to limit FDA jurisdiction of dietary supplements. Additionally, heightened concerns about the escalating costs of traditional medicine and a cultural climate promoting self-care and healthy lifestyle fueled demand for greater access to self-treatment. Congress passed the Dietary Supplement Health and Education Act of 1994 citing improvement in the health status of U.S. citizens as a top government priority.[1,3,6,7]

ANATOMY OF THE DSHEA: INTRODUCTORY SECTIONS (SECTIONS 1–4)

The DSHEA contains 13 sections that define dietary supplements, set forth regulatory requirements, and provide for the administration of the DSHEA. The provisions of the DSHEA for regulation of dietary supplements are vastly different from the regulation of drugs (Table 1). The first and second sections include an overview and rationale for the DSHEA. Section 3 defines dietary supplement as a product intended to supplement the diet that contains one or more of the following ingredients: vitamins, minerals, herbs or other botanicals, amino acids, dietary substances intended to supplement the diet by increasing dietary intake, and any concentrate, metabolite, constituent, extract, or combination of any of these ingredients. Interestingly, the DSHEA does not require that the substance must provide dietary or nutritional benefit, despite the fact that it is intended to supplement the diet. The DSHEA specifies that a dietary supplement is for oral use in tablet, capsule, powder, softgel, gelcap, or liquid form. The DSHEA explicitly excludes tobacco, meal replacement products, and substances that have previously been approved as a new drug, antibiotic, or biologic. The DSHEA also specifically excludes dietary supplements from the definition of food additive, reversing prior re-

Encyclopedia of Clinical Pharmacy
DOI: 10.1081/E-ECP 120006399

Table 1 Comparison of dietary supplement versus drug regulation

Distinguishing feature	Dietary supplement	Drug
Description	Vitamins, herbs or other botanicals, minerals, amino acids, substances intended to supplement the diet	Substances approved by the FDA as prescription or nonprescription drugs
Route of administration	Oral	Oral, parenteral, topical
Safety standard	Reasonable expectation of safety	Reasonable certainty of safety
Safety data requirement	None required for any product sold before 1994 (most products) New dietary ingredients require safety data to be submitted to FDA prior to marketing	Submitted to FDA prior to marketing
Adulteration	Burden of proof on FDA	Burden of proof on manufacturer
Nutritional labeling	Required	Not required
Labeling	No FDA review required as long as labeling is not attached to product	FDA review required on all labeling prior to distribution
Indication	To treat a nutrient deficiency, to affect the structure or function of the body, to maintain well-being	To prevent, treat, cure, or diagnose disease or other pathologic conditions
Ingredient listing	No requirement that all ingredients, active and inert, be listed on product label. In multiingredient proprietary mixtures, quantities of individual ingredients are not required	All ingredients of active constituents must be listed with quantity. Inert ingredients must also be labeled
Good manufacturing practices	Standards set by industry groups or individual manufacturer (not yet established by FDA)	Set by FDA

(From Refs. [1,4,8,9].)

gulations and legal decisions by which the FDA had prohibited dietary supplements as unapproved food additives.[1,8,9]

Section 4 of the DSHEA establishes adulteration provisions for dietary supplements. The DSHEA sets considerably less stringent safety standards for dietary supplements than those required for drugs or food additives. The FDA safety standard for drugs and food additives is a ''reasonable certainty'' that a substance is not harmful. In contrast, the DSHEA requires a ''reasonable expectation'' of safety for dietary supplements. Manufacturers are not required to submit safety data *for most products* to the FDA prior to marketing dietary supplements; the product is presumed safe. The burden of proof to show that a dietary supplement is adulterated or unsafe is the responsibility of the FDA. Additionally, the DSHEA defines a dietary supplement as adulterated if an ingredient presents ''a significant or unreasonable risk of illness or injury'' when used as directed on the label. The adulteration definition for dietary supplements focuses on the *toxicity for a labeled use*, unlike standards for drugs and food additives which focus on the toxicity of product itself, regardless of labeled use. For example, a dietary supplement that is used as a substance of abuse cannot be removed from the market unless the FDA can prove that it is unsafe for its labeled use.[1,4,10]

ANATOMY OF THE DSHEA: MARKETING AND LABELING OF DIETARY SUPPLEMENTS (SECTIONS 5–7)

Section 5 of the DSHEA addresses dietary supplement claims and marketing. Unlike drugs for which any advertising, informational, or promotional material is considered labeling by law and is subject to FDA review before distribution, dietary supplement literature is not deemed labeling. ''A publication, including an article, a chapter in a book, or an official abstract of a peer-reviewed scientific publication that appears in an article and was prepared by the author or editors of the publication, which is reprinted in its entirety'' is not considered labeling under the provisions of the DSHEA. The DSHEA requires that the information presented must not be false or misleading, cannot promote a specific supplement brand, must be displayed with other similar materials to present a balanced view, must be displayed separate from supplements, and must not have other information attached, such as product promotional information. The DSHEA relies on good faith marketing by the manufacturer to adhere to these requirements.[1,10,18,19]

Section 6 of the DSHEA amends the Nutrition Labeling and Education Act to allow four types of label claims on dietary supplements without obtaining premarketing approval by the FDA. A product may claim a benefit related to a classical nutrient deficiency, as long as the U.S. disease prevalence is disclosed. The label may also describe the role of a nutrient or dietary ingredient that is intended to affect the structure or function of the human body (so-called structure and function claim), or characterize the documented mechanism by which a nutrient or dietary ingredient acts to maintain such structure or function. The label may also include a statement about general well-being from consumption of a nutrient or dietary ingredient. If any of these claims is made, the product must also include the following statement: ''This statement has not been evaluated by the Food and Drug Administration. This product is not intended to diagnose, treat, cure, or prevent any disease.'' It is the manufacturer's responsibility to substantiate these claims and submit the information to the FDA within 30 days after marketing, but, unlike requirements for drugs, FDA approval of claims is not required before marketing. The definition of what information is required to substantiate a claim is not addressed by the DSHEA.[1,11]

In subsequent rulemaking, the FDA clarified structure and function claims, which are widely used by supplement manufacturers. FDA rules prohibit specific disease claims, such as prevents osteoporosis and implied disease claims, such as prevents bone fragility in postmenopausal women, without prior FDA review. Express and implied disease claims are allowed through the name of the product, for example, ''Carpaltum'' or ''CircuCure.'' The use of pictures, vignettes, or symbols, such as electrocardiogram tracings, is also permitted. Additionally, health maintenance claims, such as maintains a healthy circulatory system, and nondisease claims, such as for muscle enhancement or helps you relax, are allowed. The FDA also clarified structure and function claims to include for common, minor symptoms associated with life stages. For example, common symptoms of PMS or hot flashes are permissible structure and function claims.[1,11,18,19]

Section 7 of the DSHEA addresses dietary supplement ingredient labeling and nutrition information labeling. The label must identify the product as a dietary supplement. To avoid misbranding, supplement labels must include the name and quantity of each active ingredient. If the product is a proprietary blend, the total quantity of all ingredients in the blend (without listing quantities of individual ingredients) may be used. If the product contains botanical ingredients, the label must state the part of the plant used in the supplement. Listing of inert ingredients is not required. Supplements that claim to conform to the standards of an official compendium, such as the U.S. Pharmacopeia (USP) or National Formulary (NF), must meet the specifications of the compendium to avoid misbranding.[1,10,11]

Dietary supplement labels must also include nutrition labeling. Ingredients for which the FDA has established Reference Daily Intake (RDI) or Daily Reference Value (RDV) are listed first, followed by ingredients with no daily intake recommendations. If an ingredient is listed in the nutrition labeling, it does not have to be included again in the list of ingredients. Dietary ingredients that are not present in significant amounts do not need to be listed. Significant amounts are not defined by the DSHEA. The label must state a suggested quantity (dose) per serving.[1,10,11]

ANATOMY OF THE DSHEA: ADMINISTRATION (SECTIONS 8–13)

Section 8 of the DSHEA is a grandfathering clause. Substances in use prior to October 15, 1994 are not subject to the standard of reasonably expected to be safe. Substances marketed after this date are considered new dietary ingredients. Unless the dietary supplement was ''present in the food supply as an article used for food in a form in which the food has not been chemically altered,'' the manufacturer must notify the FDA at least 75 days before marketing the product. The manufacturer must supply the FDA with information based on history of use or other evidence of safety to support that the product will reasonably be expected to be safe for the stated use. There are no guidelines in the DSHEA for what constitutes history of use or other evidence of safety.[1,8,9]

Section 9 gives the FDA the authority to establish good manufacturing practice (GMP) regulations to control the preparation, packing, and storage of dietary supplements. The DSHEA specifies that the GMP regulations for dietary supplements should be modeled after current GMP regulations for the food industry. To date, the FDA has not established GMP regulations for dietary supplements.[1,4,10]

The remaining four sections of the DSHEA are administrative provisions to implement and support the DSHEA. Sections 10 and 11 override prior legislative and regulatory actions that conflict with the DSHEA. Section 12 set up a Commission on Dietary Supplement Labels, which was composed of nutritionists, industry representatives, a pharmacognosist, and attorneys to make re-

Table 2 Recommended dietary supplement references for pharmacists

Pharmacist's Letter/Prescriber's Letter Natural Medicines Comprehensive Database, 4th ed., Jeff M. Jellin, Philip Gregory, Forrest Batz, and Kathy Hitchens, ed. Stockton, CA: Therapeutic Research Faculty, 2002. Also online at www.NaturalDatabase.com (updated daily).
Tyler's Herbs of Choice: The Therapeutic Use of Phytomedicinals. James E. Robbers and Varro E. Tyler. Binghampton, NY: Hawthorn Herbal Press, 1999.
The Review of Natural Products. Ara DerMarderosian, ed. St. Louis, MO: Facts and Comparisons, Inc. (published monthly).
The Cochrane Library, 2002. Oxford: Update Software. Online at www.update-software.com (updated quarterly).
Herbal Medicine: Expanded Commission E Monographs. Mark Blumenthal, ed. Newton, MA: Integrative Medicine Communications, 2000.

commendations for dietary supplement label claims. The Commission submitted its findings to the president and Congress in 1997.[12]

The last section of DSHEA, Section 13, establishes an Office of Dietary Supplements (ODS) within the National Institutes of Health (NIH). The purpose of ODS is to conduct and coordinate scientific study within NIH relating to supplements in maintaining health and preventing disease and to collect and compile scientific research, including data from foreign sources and the NIH Office of Alternative Medicine. The OCS is also responsible for serving as the principal advisor to other government agencies on issues relating to dietary supplements, compiling a database on scientific research on dietary supplements and individual nutrients, and coordinating NIH funding relating to dietary supplements.[1,13]

The FDA Center for Food Safety and Applied Nutrition published a 10-year plan for fully implementing the DSHEA. The goal of the plan is, ''By the year 2010, have a science-based regulatory program that fully implements the Dietary Supplement Health and Education Act of 1994, thereby providing consumers with a high level of confidence in the safety, composition, and labeling of dietary supplement products.'' In the plan, the FDA details strategy to improve safety and labeling; clarify structure and function claims, and differences among dietary supplements and foods and drugs; improve enforcement of the DSHEA provisions; enhance science and research capabilities; and improve communication with the public. Pharmacists should be familiar with the DSHEA and FDA rules concerning dietary supplement products to be effective conveyors of consumer information.[20]

IMPLICATIONS FOR PHARMACISTS

Dietary supplement sales have grown from $8.8 billion since the passage of the DSHEA in 1994 to a projected $15.7 billion in 2000. Nearly half of Americans surveyed report using vitamins, herbal products, or other supplements. The DSHEA exempts dietary supplements, most of which are nonpatentable, from the multimillion dollar FDA drug approval process and simultaneously shifts the burden of proof of safety from the manufacturer to the FDA. The DSHEA allows marketing of substances with safety standards that predate the Food, Drug, and Cosmetic Act of 1938.[2–4,11]

Pharmacists should be aware of the differences in safety standards and regulatory control between drugs and dietary supplements (Table 1). When counseling people about dietary products, pharmacists must be aware that the DSHEA allows the promotion of substances that may have variable potency, unidentified components, unproven efficacy, and unknown adverse effects. The DSHEA does not require warnings about drug interactions or medical conditions under which a dietary supplement should not be used. In view of the liberal labeling provisions of the DSHEA, pharmacists cannot trust dietary supplement company literature and should consult reliable information sources (Table 2).[1,17]

CONCLUSION

Although the passage of DSHEA was hailed as a victory for consumer access to dietary supplements and a defeat of government overregulation, the DSHEA has been widely criticized by medical, legal, and public groups as being deficient in safety provisions and requirements for scientifically proven claims. Citing reports of serious toxicity caused by substances regulated as dietary supplements, critics point out that Congress passed the Food, Drug, and Cosmetic Act of 1938 as a consequence of poisoning by sulfanilamide elixir and the Kefauver–Harris Amendments in 1962 in reaction to the thalidomide tragedy in Europe. Barring public outcry for congressional action over a disastrous toxic effect, the slow process of FDA rule-making and litigation between the FDA and the dietary supplement industry will define the broad-based language of the DSHEA. By counseling consumers about possible lax manufacturing standards and potential drug interactions and adverse effects of dietary supplements, pharmacists can circumvent some of the inadequate safeguards of the DSHEA.[2,4,8,10,14–17]

REFERENCES

1. *Dietary Supplement Health and Education Act of 1994*; http://thomas.loc.gov/cgi-bin/query/D?c103:6:./temp/~c103705qih:e13550. (accessed September, 2000).
2. The Washington Post. *Health Concerns Grow Over Herbal Aids*; March 19, 2000. http://washingtonpost.com/wp-dyn/articles/A32685-2000Mar17.html. (accessed September, 2000).
3. Balluz, L.S.; Kieszak, S.M.; Philen, R.M.; Mulinare, J. Vitamin and mineral supplement use in the United States. Results from the third National Health and Nutrition Examination Survey. Arch. Fam. Med. **2000**, *9*, 258–262.
4. Kaczka, K.A. From herbal Prozac to Mark McGwire's tonic: How the Dietary Supplement Health and Education Act changed the regulatory landscape for health products. J. Contemp. Health Law Policy **2000**, *16*, 463–499.
5. Simmons, C.; Simmons, M. Drugs and dietary supplements: Ramifications of the Food Drug and Cosmetic Act and the Dietary Supplement Health and Education Act. W. Va. J. Law Tech. **1998**, *2*, (February 14, 1998) http://www.wvjolt.wvu.edu/v2i1/simmons.htm. (accessed September, 2000).
6. *Food, Drug, Cosmetic, and Device Enforcement Amendments of 1991 (H.R. 2597, 102nd Congress)*; http://thomas.loc.gov/cgi-bin/query/z?c102:H.R.2597. (accessed September, 2000).
7. *Nutrition Coordinating Act of 1991 (H.R. 1662, 102nd Congress)*; http://thomas.loc.gov/cgi-bin/query/z?c102:H.R.1662. (accessed September, 2000).
8. Burdock, G.A. Dietary supplements and lessons to be learned from GRAS. Regul. Toxicol. Pharmacol. **2000**, *31*, 68–76.
9. Young, A.L.; Bass, I.S. The Dietary Supplement Health and Education Act. Food Drug Law J. **1995**, *50*, 285–292.
10. Anon. *Dietary Supplement Health and Education Act of 1994*; U.S. Food and Drug Administration, December 1, 1995; http://vm/cfsan.fda.gov/~dms/dietsupp.html. (accessed September, 2000).
11. Commission on Dietary Supplement Labels. *Major Issues and Recommendations Related to Labeling of Dietary Supplements*; http://www.health.gov/dietsupp/ch3.htm (accessed September, 2000).
12. Commission on Dietary Supplement Labels. http://web.health.gov/dietsupp (accessed September, 2000).
13. Office of Dietary Supplements. http://odp.od.nih.gov/ods/default.html (accessed October, 2000).
14. Kessler, D.A. Cancer and Herbs. NEJM **2000**, *342*, 1742–1743.
15. Quackwatch. *''How the Dietary Supplement Health and Education Act of 1994 Weakened the FDA'' by Stephen Barrett, MD*; http://www.quackwatch.com/02ConsumerProtection/dshea.html (accessed September, 2000).
16. Anon. Herbal roulette. Consumer Reports **1995**, 698–705, November.
17. Hasegawa, G.R. Uncertain quality of dietary supplements: History repeated. Am. J. Health-Syst. Pharm. **2000**, *57*, 951.
18. Regulations on statements made for dietary supplements concerning the effect of the product on the structure or function of the body; final rule. Federal Register **2000**, *65*, 999–1050, http://vm.cfsan.fda.gov/~lrd/fr000106.html (accessed October, 2000).
19. FDA Finalizes Claims for Claims on Dietary Supplements. In *FDA Talk Paper T00-1*; Jan 5, 2000. http://vm.cfsan.fda.gov/~lrd/tpdsclm.html (accessed October, 2000).
20. U.S. Food and Drug Administration; Center for Food Safety and Applied Nutrition. *FDA Dietary Supplement Strategy (Ten Year Plan)*; http://vm.cfsan.fda.gov/~dms/ds-strat.html (accessed October, 2000).

Directions for Clinical Practice in Pharmacy (Hilton Head Conference)

Mae Kwong
American Society of Health-System Pharmacists, Bethesda, Maryland, U.S.A.

INTRODUCTION

The ASHP Research and Education Foundation and the American Society of Hospital Pharmacists (ASHP) conducted an invitational consensus conference entitled ''Directions for Clinical Practice in Pharmacy'' on February 10–13, 1985. The conference was held in Hilton Head Island, South Carolina, and has come to be known as the Hilton Head Conference.[1] The conference included approximately 150 pharmacy practitioners and educators; in addition, others from medicine, nursing, and hospital administration were invited as observers.

SYNOPSIS

The goals for the development of consensus statements were to determine the status of the clinical pharmacy movement and to help the pharmacy profession to continue advancing clinical practice. The principal objectives of the conference were: 1) to examine to what extent the profession had established goals with respect to clinical practice, 2) to assess the current status of the clinical practice of pharmacy and pharmacy education, and 3) to identify practical ways by which clinical pharmacy could be advanced.

The keynote address, presented by Paul F. Parker, Sc.D., a consultant and retired director of pharmacy at the University of Kentucky, was entitled ''Clinical Pharmacy's First 20 Years.'' Parker described clinical pharmacy as the most important practice, education, and professional philosophy in the history of pharmacy. He noted that clinical pharmacy will advance only by meeting the goals of quality care.

Accomplishing the conference goals required an exploration of four key topics: 1) pharmacy as a clinical profession, 2) barriers to clinical practice, 3) the symbiosis of clinical practice and education, and 4) building pharmacy's image. Plenary presentations were given on each topic, and these were followed by workshop discussions.

Charles D. Hepler, Ph.D., presented ''Pharmacy as a Clinical Profession.'' His assessment of clinical pharmacy focused on the role of clinical pharmacy, how professional services are provided, the need to obtain professional authority, and the patient-oriented focus of clinical pharmacy. The workshops were charged with two tasks: 1) to determine whether there is a need for the term clinical pharmacy and, if so, to conceptually distinguish it from pharmacy and 2) to consider what steps are needed to establish pharmacy as a clinical profession. A total of 18 consensus statements resulted. The statements with the highest consensus among participants emphasized that the profession of pharmacy has a fundamental purpose to serve society for safe, appropriate, and rational use of drugs; to provide leadership to other healthcare professionals and the public to ensure responsible drug use; to provide authoritative, usable drug information; and to work collaboratively with other healthcare professions on health promotion and disease prevention through the optimal use of drugs. According to another statement, pharmacists are the professionals ultimately responsible for drug distribution and control, and the use of technicians, automation, and technology should be maximized to free time for pharmacists to perform clinical services. Ultimately, the purposes and goals of clinical pharmacy are the same as those of pharmacy, but clinical pharmacy stresses patient-oriented services and the association with patient outcomes.

To address the barriers to clinical practice, a panel discussed the ''Realities of Contemporary Practice.'' The panelists were Chip Day, Robert P. Fudge, Teresa Volpone McMahon, Pharm.D., and Steven L. Smith, Pharm.D., and the session moderator was Dennis K. Helling, Pharm.D. Each panelist described routine activities in his or her practice and identified several barriers to clinical pharmacy, notably a lack of time. Ultimately, the panelists

Encyclopedia of Clinical Pharmacy
DOI: 10.1081/E-ECP 120006191

agreed that practicing clinical pharmacy had become easier within the past few years, mostly because of increased recognition by other healthcare professionals of the pharmacist's role in patient care. The workshop groups produced 37 consensus statements on barriers to clinical practice. According to the statements receiving the highest consensus, pharmacy directors are unable to provide effective leadership to their staff, a widely agreed-upon philosophy of pharmacy practice is lacking, there is no concurrence on what the standard of practice in pharmacy should be, consumer demand for clinical pharmacy services is weak because the public has a poor understanding of the services pharmacists can offer, and the value of clinical pharmacy services has not been adequately demonstrated.

To discuss the symbiosis of clinical practice and education, Charles A. Walton, Ph.D., presented the educator's perspective and Marianne F. Ivey presented the practitioner's perspective. The presenters believed that both pharmacy practitioners and educators should share in advancing the profession through the establishment and provision of clinical pharmacy services, through education and training of pharmacy students and pharmacists, and through clinical research. The objectives for the workshop groups were: 1) to identify steps for making more effective use of clinical pharmacy faculty in improving the level and quality of clinical pharmacy services and 2) to use pharmacy staff more effectively in clinical education. A total of 33 consensus statements were developed for objective 1 and 18 for objective 2. With respect to using clinical pharmacy faculty, it was agreed that there is a need to clearly define a shared philosophy between clinical faculty members and pharmaceutical services staff, the clinical service responsibilities of clinical faculty, and the clinical education missions of both the college and the pharmacy department. In addition, orienting deans and other academics to the roles of clinical faculty would provide a basis for balancing teaching, research, and service responsibilities and would help acknowledge the scholarly activity and clinical research that occur in clinical practice.

With respect to using pharmacy staff more effectively in clinical education, the major statements identified that staff should be recognized for their teaching activities; that staff involved in clinical instruction should participate in the evaluation of students; that hospital administrators, pharmacy directors, and staff should recognize their respective roles in pharmacy education and have a thorough understanding of the clinical faculty's responsibilities; and that educational programs should be developed to train pharmacists to manage clinical services.

William A. Miller, Pharm.D., presented the final plenary session on building pharmacy's image. According to Miller, building pharmacy's image as a clinical profession would occur simply by providing clinical services. Pharmacy would be advanced as a clinical profession by establishing goals for pharmaceutical services; creating standards for pharmacy practice; planning, implementing, and managing pharmaceutical service, education, and research programs; providing financial management; and assessing the quality of pharmaceutical services and drug use within the institution. The workshop groups sought to characterize the type of relationship pharmacy should establish with medicine, nursing, hospital administration, and the public. Eight consensus statements were written. The major consensus statement was that pharmacy should establish a public image of advocacy in all matters related to the use of drugs. Other statements expressed that pharmacist input should be a required component of the drug-use process, that pharmacy should be viewed as a clinical service, and that pharmacy is a colleague with nursing and medicine in patient care.

DISCUSSION

The Hilton Head Conference affirmed that pharmacy is a clinical profession committed to clinical practice and the patient. Pharmacy is fundamentally a healthcare profession with a responsibility for safe and effective drug use in society.

The conference provided a forum for pharmacists to discuss the past, present, and future of clinical pharmacy. Even though the conference occurred in 1985, many of the conclusions reached still apply to practice today. For instance, some of the barriers identified with respect to leadership and substantiation of the value of clinical pharmacy services still exist. Also, there continues to be a need to educate the public and gain the support of other healthcare professionals for clinical pharmacy practice.

REFERENCE

1. Proceedings from the conference were published in the American Journal of Hospital Pharmacy **1985**, *42*, 1287–1342.

Disease Management

Leigh Ann Ramsey
University of Mississippi, Jackson, Mississippi, U.S.A.

Brendan S. Ross
Department of Veterans Affairs Medical Center, Jackson, Mississippi, U.S.A.

INTRODUCTION

Pharmacy practice has evolved from a focus on the responsible dispensing of medications to a patient-oriented profession concerned with the optimum use of pharmaceutical products in the management of disease states. This new practice model, which is known as *pharmaceutical care*, emphasizes the role of pharmacists in meeting the health needs of patients through medication-related care.[1] Pharmaceutical care necessitates an ongoing collaboration with physicians and is often referred to as *collaborative drug therapy management*.[2] Pharmaceutical care is a form of *disease management*, a phrase which broadly encompasses coordinated healthcare by providers from complementary professions whose shared goal is the improvement of patient well-being.[3] Federal law requires that pharmacists offer medication counseling to patients receiving Medicaid benefits in the belief that such education will lead to more effective drug therapy.[4] In most states, this requirement is interpreted as a mandate compelling pharmacists to counsel all patients. Disease management extends the traditional duties of pharmacists from dispensing medications and counseling patients to include a more significant role in securing the success of drug therapy.

Managed care often forces health plan administrators to limit costs by limiting enrollee access to physicians, which creates a need for nonphysician involvement in patient care.[5] Chronic diseases are often accompanied by complex drug regimens that may lead to patient confusion and poor outcomes that further increase healthcare costs.[6] The knowledge and training of pharmacists in drug and disease interactions uniquely qualify them to assist in medication management. The distribution of pharmacists in the community enhances patient contact, and ideally situates pharmacists to assess therapeutic responses to prescribed therapies. Through education and accessibility, pharmacists can improve medication compliance and diminish the risk of adverse drug effects; and by monitoring and modifying drug therapy, pharmacists can assure that patients increase their chances for achieving favorable outcomes.[7] The potential to decrease healthcare costs provides a pharmacoeconomic incentive for involving pharmacists in disease management.[8]

SCOPE OF PRACTICE

Pharmacy involvement in disease management may entail educating patients on the desirable and undesirable effects of pharmaceutical products and on proper drug administration, therapeutic drug monitoring through laboratory testing and interpretation, or initiating and modifying medication regimens based upon ongoing assessments of physiologic response.[9] The disease states amenable to pharmaceutical care include asthma, diabetes mellitus, cardiovascular risk reduction, chronic pain management, mental health disorders, epilepsy, women's health concerns, infectious diseases, and anticoagulation therapy.[10] However, pharmacy involvement in any field of clinical care is only limited by the needs of patients and providers and the willingness and competence of pharmacists to participate. Although practiced in acute care hospitals, critical analyses of the potential benefits of pharmaceutical care have focused on drug therapy management of chronic diseases by community pharmacists in ambulatory settings. Pharmacy participation in the direct care of patients can be demonstrated in a concise review of several current programs.

The University of Mississippi School of Pharmacy has been an innovator in disease management for over a decade. One of its more successful Pharmaceutical Care Clinics addresses the community need to improve asthma management so that the overutilization of emergency care is curtailed. The backbone of the clinic is a protocol dictating diagnostic and therapeutic algorithms that were

Encyclopedia of Clinical Pharmacy
DOI: 10.1081/E-ECP 120006243

adapted to local conditions from established national practice guidelines.[11] The care is fleshed out by educating patients as to the pathogenesis of asthma, the signs and symptoms of airway decompensation, and the pharmacology underlying medication options. Both short-term goals, lifestyle modifications such as smoking cessation and allergen avoidance, and long-term goals, such as decreased rates of school or work absenteeism, are set and reviewed. In concord with the physician-supervised protocol, an individualized asthma action plan is developed for each referred patient. Pharmacists train patients to use peak flow meters and to monitor and self-adjust drug therapy. Pharmaceutical care is intended to supplement regularly scheduled physician appointments, to identify and respond to intervening pathophysiology, and to mitigate the need for urgent medical attention.

Outcome analysis reveals that the Asthma Care Clinic at the University of Mississippi is achieving its stated goals.[12] Utilizing enrolled patients as historical controls, this disease-management intervention resulted in fewer emergency room visits or hospitalizations for asthma decompensation. An annualized cost saving of approximately 60 percent for these hospital services has been realized. Cost savings are sustained even though additional clinical funds are expended on pharmaceutical care. As a result of these salutary findings, all patients presenting for the emergency treatment of asthma-related bronchospasm at the University Medical Center are subsequently considered for disease-management assessment.

Similarly encouraging results of collaborative drug therapy are reported in the medical literature for a number of economically burdensome chronic diseases. Project ImPACT (Improve Persistence and Compliance with Therapy): Hyperlipidemia assessed the contributions of community pharmacists to the care of patients with lipid disorders requiring pharmacologic intervention.[13] During this three-year project, the observed rate for compliance with lipid-lowering medication therapy improved to approximately 90 percent. The impact of these Virginia pharmacists was significant, as nearly two-thirds of participants achieved and maintained nationally recognized treatment goals. The City of Asheville, North Carolina, and the largest private employer in western North Carolina, the Mission St. Joseph Health System, contracted with trained community pharmacists to manage the drug therapy of their employees with diabetes.[14] Patient interaction with providers increased with the advent of pharmaceutical care, while metabolic indices of disease control improved. Moreover, payer expenditures for the total cost of ambulatory and inpatient diabetes care decreased. When absentee rates were compared to prior years, participants worked an average of 6.5 days more per year during the project. Pharmacy disease management favorably impacted both direct and indirect medical costs. The majority of employees were highly satisfied with their care, as they reported improvements in functional status and quality of life.

A critical review of pharmacy disease management programs is hindered by the lack of statistical design rigor and robust cost analyses found in many published reports. The heterogeneity of studies with regard to clearly defined and widely accepted outcome measures also hampers systematic assessment. The authors of a review of 55 comparative studies representing 50 programs in which pharmacists provided support for ambulatory care providers in outpatient clinics and community pharmacies found that prescription monitoring led to a general trend toward cost savings, enhanced timeliness of care, and improved clinical outcomes.[15] However, this review noted no consistent improvement in disease knowledge or patient satisfaction and little improvement in quality of life among pharmaceutical care enrollees. The National Institutes of Health (NIH) through the Agency for Healthcare Research and Quality (AHRQ) is funding studies to address these deficiencies in the evidence base.[16] This federal interest in data documenting the costs and benefits associated with disease management bodes well for the acceptance of pharmaceutical care into the medical mainstream.

CREDENTIALING AND CERTIFICATION

As pharmacy embraces disease management, the profession must reassure skeptics that pharmacists possess the necessary knowledge and skills to provide these services. For pharmacists desiring to broaden their scope of practice, training beyond that required to obtain a pharmacy degree or license may be necessary. Although credentialing is a controversial topic, it is increasingly evident that a nationally recognized process is necessary to bolster the professional stature of pharmacists and to identify clinical specialists who are capable of providing reimbursable pharmaceutical care.

Credentialing is defined in a medical context as the process by which an organization or institution obtains, verifies, and assesses an applicant's qualifications to provide a particular patient-care service. The Council on Credentialing in Pharmacy (CCP), a coalition of 11 national organizations founded in 1999 as a coordinating body for credentialing programs, delineates three avenues for credentialing in pharmacy: 1) credentials required to enter the profession—academic degrees, 2) credentials required to enter practice—licenses, and 3) optional cre-

dentials documenting specialized knowledge and skills—advanced academic degrees or certificates.[17] Certification involves granting a credential to a pharmacist who has demonstrated a level of competence in a specific and relatively narrow area of practice. Postlicensure certification usually requires an initial assessment and periodic reassessments of a grantee's qualifications. There are three agencies that offer certification to pharmacists: the Board of Pharmaceutical Specialties (BPS), the Commission for Certification in Geriatric Pharmacy (CCGP), and the National Institute for Standards in Pharmacist Credentialing (NISPC). BPS was established by the American Pharmaceutical Association (APhA) in 1976 and certifies pharmacists in five practice concentrations: nuclear pharmacy, nutrition support pharmacy, oncology pharmacy, psychiatric pharmacy, and pharmacotherapy. An ''Added Qualifications'' in either infectious diseases or cardiovascular pharmacy is available for pharmacists certified in pharmacotherapy. The CCGP was established by the American Society of Consultant Pharmacists (ASCP) in 1997 and supervises the certification program in geriatric pharmacy practice. NISPC was founded in 1998 to oversee pharmacist credentialing in disease management.

NISPC is composed of four member organizations: the APhA, the National Association of Boards of Pharmacy (NABP), the National Association of Chain Drug Stores (NACDS), and the National Community Pharmacy Association (NCPA).[18] NISPC was charged with coordinating the development of a nationally recognized testing program to credential pharmacists in disease-specific pharmaceutical care. NISPC utilized NCPA's National Institute for Pharmacist Care Outcomes (NIPCO) model as a resource for constructing examinations to test disease-management competencies. An expert panel drawn from community practitioners, academicians, pharmacy benefits managers, and state board of pharmacy members develops the standards and objectives for each disease-management examination. Panel members ensure that the content of the examinations reflects the knowledge base expected of pharmacists providing care at an advanced practice level.

The first pharmacy disease-management examinations were offered in 1998 as pencil-and-paper tests in the states of Arkansas, North Dakota, and Mississippi.[19] Certification was offered in four disease states: asthma, dyslipidemia, diabetes, and anticoagulation therapy. Since that time, the examinations have been adapted for computer administration at multiple test sites any time of the year. However, non-electronic testing is offered annually at the APhA national meeting. Due to the specialized funds of knowledge required for successful certification, pharmacists are strongly encouraged to have at least 2 years of experience in the field being tested prior to applying for examination. States may require that prerequisites, such as a training program, be completed before permission for testing is granted. The NISPC-sanctioned examinations are graded as either pass or fail; a score of 75% or greater yields a passing grade. Pharmacists receiving a passing score are eligible for this recognition to be listed on the NABP's Pharmacist and Pharmacy Achievement and Discipline (PPAD) Web site database (Table 1).

Since 1998, NISPC has awarded credentials to over 1,200 pharmacists in the United States. In 2001, NISPC adopted the designation of Certified Disease Manager (CDM) for those pharmacists successfully completing one of the disease-management exams. NISPC hopes the CDM credential will gain national recognition by both patients and payers. NISPC certification must be renewed every 3 years. For pharmacists awarded a CDM credential in 2000 or later, 30 hours of American Council on Pharmaceutical Education (ACPE)–approved continuing education in the credentialed disease state must be documented within the 3-year recertification period. Ten of the required 30 hours must be obtained during the third year.

The availability of a cadre of pharmacists certified in disease management does not assure reimbursement for pharmaceutical care. NISPC formed a Standards Board and a Payer Advisory Panel to ensure public trust in the care provided through collaborative drug therapy by credentialed pharmacists. The Standards Board has identified a need to improve the communication skills of pharmacists so that their collaborative work is enhanced, to train pharmacists regarding the benefits of nonpharmacologic therapies, and to adopt a regular review and modification process for disease-management competencies. The Payer Advisory Panel was tasked with advising NISPC on the needs of the payer community as pharmaceutical care penetrates the marketplace. The Panel has stressed the need to clearly define the package of clinical services credentialed pharmacists provide, to involve other allied health professionals in collaborative management, to develop standard outcome measures, and to establish an accessible databank for credentialed pharmacists. The work of the Standards Board and Payer Advisory Panel should significantly contribute to the stature of pharmaceutical care.

REIMBURSEMENT

In 1998, Mississippi became the first state to secure government reimbursement for pharmaceutical care.[20]

Table 1 Disease management resources

Professional Organizations		E-mail	Phone
Academy of Managed Care Pharmacy	AMCP	www.amcp.org	800-827-2627
American Association of Colleges of Pharmacy	AACP	www.aacp.org	703-739-2330
American College of Apothecaries	ACA	www.acaresourcecenter.org	901-383-8119
American College of Clinical Pharmacy	ACCP	www.accp.com	816-531-2177
American Council on Pharmaceutical Education	ACPE	www.acpe-accredit.org	312-664-3575
American Pharmaceutical Association	APhA	www.aphanet.org	202-628-4410
American Society of Consultant Pharmacists	ASCP	www.ascp.com	703-739-1300
			800-355-2727
American Society of Health-System Pharmacists	ASHP	www.ashp.org	301-657-3000
National Association of Boards of Pharmacy	NABP	www.nabp.net	847-698-6227
National Association of Chain Drug Stores	NACDS	www.nacds.org	703-549-3001
National Community Pharmacists Association	NCPA	www.ncpanet.org	703-683-8200
			800-544-7447
Certifying Bodies			
Anticoagulation Forum	ACF	www.acforum.org	617-638-7265
Board of Pharmaceutical Specialties	BPS	www.bpsweb.org	202-429-7591
Commission for Certification in Geriatric Pharmacy	CCGP	www.ccgp.org	703-535-3038
Association of Asthma Educators	AAE	www.asthmaeducators.org	888-988-7747
National Certification Board for Diabetes Educators	NCBDE	www.ncbde.org	847-228-9795
National Institute for Standards in Pharmacist Credentialing	NISPC	www.nispcnet.org	703-299-8790
Industry Liaisons			
Disease Management Association of America	DMAA	www.dmaa.org	202-861-1490
Disease Management Purchasing Consortium and Advisory Council	DMC[2]	www.dismgmt.com	781-237-7208

(Adapted from Ref. [17].)

The Health Care Financing Administration (HCFA) approved payment through the Mississippi Division of Medicaid to pharmacists for disease-management services provided to patients enrolled in the Medicaid program. The components of a reimbursable service are patient evaluation, patient or caregiver education, drug therapy review and compliance assessment, and disease management under protocol according to clinical practice guidelines. NISPC credentialing is currently required for pharmacists to apply for a Mississippi Medicaid provider number, which in turn is necessary to bill for pharmaceutical care under the Other Licensed Practitioner designation.

In addition to obtaining a Medicaid provider number, a pharmacist must produce two other documents prior to providing pharmaceutical care in Mississippi: a written evaluation and treatment protocol and a referral from a physician. The protocol must define the collaborative agreement between the pharmacist and the referring physician and be on file with the State Board of Pharmacy. The nature of protocol requirements differs among practice sites. Within an institution, one protocol agreement may be submitted for all the physicians practicing at that site for use by all their referred patients. In the community setting, a separate protocol from each referring physician must be completed for each patient. Pharmacists are paid a flat fee for each 15- to 30-minute patient encounter. Currently, pharmacists can be reimbursed for up to 12 visits per year per patient for all disease states managed. These pharmaceutical care visits are in addition to the annual allotment of reimbursed physician visits provided by Mississippi Medicaid. No restrictions exist as to the number of patients a pharmacist can manage.

Mississippi is not alone in explicitly recognizing the important contributions of pharmacists to disease management with state funding. In 2000, the Iowa Division of Medicaid initiated a reimbursement program for pharmaceutical case management.[21] During this 2-year pilot project, patients who are candidates for pharmaceutical care are identified and participating pharmacists are notified of their eligibility. The pharmacist performs an initial disease assessment and develops an individualized therapeutic plan for each patient, which is subsequently reviewed by a physician collaborator. Pharmacists must meet criteria outlined by the project's advisory commit-

tee, and complete a training program approved by the Iowa Department of Human Services. New Mexico also has a demonstration project that allows for pharmaceutical care under physician-supervised protocol.[22] Assessments of the medical and cost outcomes of these projects will determine the future of these initiatives.

State support for pharmaceutical care is bolstered by managed care imperatives. States are increasingly willing to fund programs that maintain the health of their insured populations, if this care can be proven to be efficacious and to control medical costs. States are confronting the same financial challenges faced by private health plan leaders such as Humana and Kaiser Permanente, who were early adopters of disease management. Health Maintenance Organizations (HMOs) employ pharmaceutical care to improve the health of their enrollees and thus limit the need for costly medical interventions.[23] HMOs have invested considerable time and effort into developing multidisciplinary treatment pathways and algorithms for many pharmacy-intensive disorders and diseases.[24] By standardizing processes of care, providers become accountable for fully implementing therapies proven to favorably impact patient outcomes. The dedication of organized pharmacy to disease management is likely to lead states to commit greater resources to pharmaceutical care.

As Medicare becomes structured to support health maintenance, federal interest in disease management is coming full circle. The Indian Health Service was a pioneer in pharmaceutical care, and the U.S. Armed Forces and Veterans Affairs healthcare sectors are now leaders in collaborative drug therapy.[25] In 1999, HCFA recognized the provider status of nonfederally employed pharmacists to participate in diabetes management.[26] Congress and the Medicare Trust administrators are actively weighing the benefits of extending coverage to include disease management by pharmacists. Pharmacists are likely to be accorded enhanced provider status given the developing affirmative body of research and patient willingness to embrace pharmaceutical care. Indeed, a recent patient survey indicated that a majority would pay for disease management by pharmacists, if accredited services were widely available.[27]

Although reimbursement is often cited by pharmacists as the paramount barrier to the widespread dissemination of disease management, other troublesome yet surmountable obstacles exist.[28] Even when reimbursement is assured, the requirements accompanying billing can be time-consuming and costly. As in other medical fields, the paperwork required to document encounters and apply for pharmacy service reimbursement from various payers in different practice settings is not uniform. The information systems in place in most community pharmacies are often inadequate to respond to the additional demands of disease management.[29] These administrative concerns compound the stresses placed on pharmacists by the high volume of medication dispensing and the need for technician supervision characteristic of retail pharmacy practice. These issues will need urgent attention so that the willingness of the public and payers to support pharmaceutical care is not hindered.

REGULATORY AND ETHICAL ISSUES

Laws pertaining to disease management differ from state to state. Most states provide the Board of Pharmacy with statutory authority to regulate pharmaceutical care. Thirty-three states currently allow pharmacists to initiate or modify drug therapy pursuant to a collaborative practice agreement or protocol; other states are in the process of amending their practice acts to incorporate pharmaceutical care services (Table 2). Pharmacists must adhere to the restrictions imposed by state practice agreements or they assume a greater risk of liability. Exposure to administrative or criminal penalties can be diminished if pharmacists fully acquaint themselves with the boundaries limiting pharmaceutical care in their state.

One legislative initiative is worthy of note. North Carolina allows a pharmacist who provides disease management to be designated as a Clinical Pharmacy Practitioner.[30] Pathways to attain this professional recognition are available to either bachelor or doctorate degree–holding pharmacists. Applicants for this designation must submit a collaborative practice agreement that delineates the dimensions of their pharmaceutical care proposal for formal review. Approval is granted after appraisal by both the Board of Pharmacy and State Medical Board. Many in organized medicine have joined pharmacy in endorsing a grant of proscribed prescriptive authority to pharmacists through such novel state provisions. However, the American Medical Association (AMA) opposes nonphysician groups that seek independent prescribing rights as they believe this will further fragment healthcare.[31] A more recent Position Paper outlining the stance of the American College of Physicians-American Society of Internal Medicine (ACP-ASIM) with regard to the increasing scope of pharmacy practice endorses further research on pharmaceutical care programs, yet opposes independent pharmacist prescriptive privileges and the initiation of drug therapy.[32]

Some within the pharmacy profession also question whether direct patient care is a proper role for pharmacists.[33] They argue that disease management requires

Table 2 Disease management by state

State	Practice pursuant to collaborative agreement or protocol	Practice setting	Comments
Alaska	Yes	All	Drug administration allowed
Arizona	Yes	Institutional settings, community health center	Drug administration allowed
Arkansas	Yes	All	Administer drugs by injection allowed
California	Yes	Licensed healthcare facilities, clinics, home care settings	Administer drugs by injection allowed
Florida	Yes	All	Prescriptive authority restricted to formulary drugs Protocol-based dependent prescribing pending
Georgia	Yes	All	Restricted to modifying drug regimen
Hawaii	Yes	Licensed acute care hospitals	Restricted to modifying drug regimen Administer drugs by injection allowed
Idaho	Yes	All	Drug administration allowed
Indiana	Yes	Acute care settings, private mental health institutions	Drug administration not prohibited
Kansas	Yes	All	By delegation of physician
Kentucky	Yes	All	Drug administration allowed
Louisiana	Yes	All	Drug administration allowed
Michigan	Yes	All	Drug administration allowed
Minnesota	Yes	All	Restricted to modifying drug regimen Administer drugs by injection allowed in first doses and in medical emergencies
Mississippi	Yes	All	Administer drugs by injection allowed
Montana	Yes	All	Drug administration allowed
Nebraska	Yes	All	Restricted to monitoring drug therapy Administer drugs by injection allowed
Nevada	Yes	Licensed medical facilities	Immunizations by protocol
New Mexico	Yes	All	Limited to Pharmacist Clinician, eligible to register with Drug Enforcement Administration
North Carolina	Yes	All	Recognizes Clinical Pharmacy Practitioner
North Dakota	Yes	Institutional settings, clinics	Drug administration allowed
Ohio	Yes	All	Restricted to modifying drug regimen Drug administration allowed
Oregon	Yes	All	Restricted to modifying drug regimen Drug administration allowed
Rhode Island	Yes	Hospital (including outpatient clinics), nursing homes	
South Carolina	Yes	All	Drug administration allowed
South Dakota	Yes	All	Administer drugs by injection allowed
Texas	Yes	All	Administer drugs by injection allowed
Utah	Yes	All	Physician Licensing Board approval necessary for outpatient services
Vermont	Yes	Institutional settings	Restricted to modifying drug regimen Drug administration allowed
Virginia	Yes	All	Restricted to modifying drug regimen Drug administration allowed
Washington	Yes	All	Administer drugs by injection allowed
Wisconsin	Yes	All	Guideline Drug administration allowed under protocol with prescriber
Wyoming	Yes	All	Administer drugs by injection allowed

(Adapted from Refs.[2],[42–45].)

proficiency in differential diagnosis, analytical thinking, and patient interaction skills that are not within the purview of pharmacists. They believe that pharmacy should retain its traditional focus on quality assurance in medication delivery and cede responsibility for patient care to physicians rather than join the ranks of other mid-level practitioners. These concerns are being addressed in undergraduate and postgraduate pharmacy education. Pharmacy schools are moving away from passive teaching models to active curricula founded on problem-based learning and from didactic lectures to clinical pharmacy preceptorships in practice environments. Pharmacy leaders are also addressing the lack of readily available advanced clinical training for community pharmacists.

Pharmaceutical care is not without its extramural critics as well. Consumer advocates question whether disease management is a sound public health policy.[34] According to these critics, such programs concentrate healthcare expenditures on high-risk patients to control short-term costs and thus redirect scarce resources needed for health promotion and disease prevention. They believe that disease management has been promoted by the pharmaceutical industry as a way to augment drug sales. Drug manufacturers are accused of organizing disease-management programs to gain access to restricted formularies and to ensure control over medication demands rather than to improve patient care.[35] The research community also raises ethical concerns, as it objects to the lack of public reporting of the outcomes from commercial pharmaceutical care programs.[36] It cautions that while exclusive access to proprietary information may be necessary to preserve a company's competitive advantage, it may hinder medical progress. Pharmacists must be wary of uncritically adopting pharmaceutical care protocols developed by the for-profit sector and be vigilant to unethical inducements to prescribe unnecessary or inappropriate medication therapies.

Public regard for the honesty and ethical character of pharmacists is greater than for either physicians or the clergy.[37] Pharmacy must guard against a decline in consumer confidence as it expands its spectrum of services. Professional codes are being challenged by the new relationships developing between pharmacists and those they serve. As pharmacists become more involved in direct patient care, their ethical obligations extend beyond professional dictates to maintain knowledge and skills to uphold the welfare of patients. Of significant patient concern are the related issues of privacy and confidentiality. Therapeutic relationships are built on a foundation of trust. Clinicians are entrusted with sensitive, personal information by patients and they are expected to hold these private communications in strict confidence by the canons of medical ethics.[38] Additionally, respect for the dignity of patients entails promoting their autonomy: the right of patients to inform and to govern their own healthcare decisions. However, pharmacists practicing in collaborative arrangements have ethical duties to each party to the agreement, to both patients and physicians. Conflicts can arise as patients may provide information to a pharmacist that they are unwilling to share with their physician.[39] A pharmacist may be faced with the moral dilemma of disclosing information provided in confidence or withholding data pertinent to medical decision making. In a purely consultant relationship, the primary duty of the consultant is owed to the party requesting the consultation and professional standards hold that full disclosure is warranted. Disease management is an effort by both pharmacists and physicians on the behalf of patients and thus the desires of patients for confidential interactions may be ethically problematic. The importance of these ethical principles is reflected in recent federal legislation; the Health Insurance Portability and Accountability Act requires the establishment of health privacy regulations to protect the confidential information yielded by patients.[40] Pharmaceutical care is more covenant than contract; when the inevitable conflicts arise, pharmacists must recognize that resolution may require choices based upon individual patient values rather than on a reflexive recourse to an objective standard. As disease management takes hold, comprehensive pharmacy education will need to encompass legal and ethical training so that pharmacists retain the good will of the public they currently enjoy.

SUMMARY AND CONCLUSION

National surveys estimate that one-third of adults in the United States suffers from a chronic disease, yet most fail to achieve treatment goals promulgated by consensus care guidelines; fewer than one half of hypertensives have well-controlled blood pressure and less than one-quarter of patients with coronary artery disease have lipid levels within optimal limits.[41] The current healthcare model is geared to acute disorders, rather than tooled for the systematic care of chronic diseases. Pharmacy disease management is the multidisciplinary process of selecting appropriate drug therapy and continually monitoring patient outcomes to that therapy. It is a response to the demands of health-conscious consumers and cost-conscious payers. The value of pharmaceutical care can be promoted by ensuring the knowledge and judgment of practitioners through training and credentialing and by measuring their impact on health outcomes through rigorously designed clinical trials. Healthcare is in transition, and new models of delivery will likely be accompanied by a broader pharmacoeconomic perspective, one that does not isolate drug costs, but views the cost of pharmaceutical care in the overall context of medical care. Pharmacy is favorably situated to contribute to disease management and to profit from this fundamental shift in the healthcare paradigm.

REFERENCES

1. Hepler, C.D.; Strand, L.M. Opportunities and responsibilities in pharmaceutical care. Am. J. Hosp. Pharm. **1990**, *47*, 533–543.
2. American College of Clinical Pharmacy. Collaborative drug therapy management by pharmacists. Pharmacotherapy **1997**, *17* (5), 1050–1061.
3. Armstrong, E.P.; Langley, P.C. Disease management programs. Am. J. Health-Syst. Pharm. **1996**, *53*, 53–58.
4. Omnibus Reconciliation Act of 1990, Section 4401. Fed. Regist. **1992**, *57* (212), 49397–49412.
5. Bodenheimer, T.; Sullivan, K. How large employers are shaping the health care marketplace. N. Engl. J. Med. **1998**, *338*, 1003–1007.
6. Johnson, J.A.; Bootman, J.L. Drug-related morbidity and mortality: A cost-of-illness model. Arch. Intern. Med. **1995**, *155*, 1949–1956.
7. Nichols-English, G.; Poirier, S. Optimizing adherence to pharmaceutical care plans. J. Am. Pharm. Assoc. **2000**, *40* (4), 475–485.
8. Fischer, L.R.; Scott, L.M.; Boonstra, D.M.; DeFor, T.A.; Cooper, S.; Eelkema, M.A.; Hase, K.A.; Wei, F. Pharmaceutical care for patients with chronic conditions. J. Am. Pharm. Assoc. **2000**, *40* (2), 174–180.
9. Galt, K.A. The key to pharmacist prescribing: Collaboration. Am. J. Health-Syst. Pharm. **1995**, *52*, 1696–1699.
10. de Gier, J.J. Clinical pharmacy in primary care and community pharmacy. Pharmacotherapy **2000**, *20* (10 Pt. 2), 278S–281S.
11. *National Asthma Education and Prevention Program Expert Panel Report 2: Guidelines for the Diagnosis and Management of Asthma*; National Heart, Lung, and Blood Institute: Bethesda, Maryland, 1997, NIH Publication No. 97-4051.
12. Ramsey, L.A. *University of Mississippi Pharmaceutical Care Clinics: Asthma Outcomes*; Private communication.
13. Bluml, B.M.; McKenney, J.M.; Cziraky, M.J. Pharmaceutical care services and results in Project ImPACT: Hyperlipidemia. J. Am. Pharm. Assoc. **2000**, *40* (2), 157–165.
14. Schwed, D.; Miall, J.; Harteker, L.R. Pharmacy risk-sharing contracts: Do's and don'ts. J. Am. Pharm. Assoc. **1999**, *39* (6), 775–784.
15. Tully, M.P.; Seston, E.M. Impact of pharmacists providing a prescription review and monitoring service in ambulatory care or community practice. Ann. Pharmacother. **2000**, *34*, 1320–1331.
16. *Outcomes of Pharmaceutical Therapy (OPT) Program Update*; Agency for Health Care Policy and Research: Rockville, Maryland, 1995, AHCPR Publication No. 950-0007.
17. The Council on Credentialing in Pharmacy. Credentialing in pharmacy. Am. J. Health-Syst. Pharm. **2001**, *58* (1), 69–76.
18. Credentialing in pharmacy: No simple matter. Am. J. Health-Syst. Pharm. **2000**, *33* (2), 84–85.
19. National Association of Boards of Pharmacy. *Disease State Management Examinations Registration Bulletin*; National Association of Boards of Pharmacy: Park Ridge, Illinois, 2000.
20. Posey, L.M. Mississippi DSM program is model for nation. Pharm. Today **2000**, *6* (6), 1 and 20.
21. Clarke, C. Iowa Medicaid pharmaceutical case management evaluation plan. J. Iowa Pharm. Assoc. **2000**, 15–17, Sept./Oct.
22. *Professional Practice Act Revisions: New Mexico Senate Bill 353*; www.ashp.org/public/proad/state/may_2001-2.html (accessed March 2002).
23. Anderson, G.F.; Zhang, N.; Worzala, C. Hospital expenditures and utilization: The impact of HMOs. Am. J. Managed Care **1999**, *5*, 853–864.
24. Pearson, S.D.; Goulart-Fisher, D.; Lee, T.H. Critical pathways as a strategy for improving care: Problems and potential. Ann. Intern. Med. **1995**, *123*, 941–948.
25. Ellis, S.L.; Carter, B.L.; Malone, D.C.; Billups, S.J.; Okano, G.J.; Valuck, R.J.; Barnette, D.J.; Sintek, C.D.; Covey, D.; Mason, B.; Jue, S.; Carmichael, J.; Guthrie, K.; Dombrowski, R.; Geraets, D.R.; Amato, M. Clinical and economic impact of ambulatory care clinical pharmacists in management of dyslipidemia in older adults: The IMPROVE study. Impact of Managed Pharmaceutical Care on Resource Utilization and Outcomes in Veterans Affairs Medical Centers. Pharmacotherapy **2000**, *20* (12), 1508–1516.
26. *Part B Answer Book: Medicare Billing rules from A to Z*; United Communications Group: Rockville, Maryland, 2000.
27. Larson, R.A. Patients' willingness to pay for pharmaceutical care. J. Am. Pharm. Assoc. **2000**, *40* (5), 618–624.
28. Norwood, G.J.; Sleath, B.L.; Caiola, S.M.; Lien, T. Costs of implementing pharmaceutical care in community pharmacies. J. Am. Pharm. Assoc. **1998**, *38*, 755–761.
29. Shane, R. The health care puzzle: Where does pharmacy fit? Am. J. Health-Syst. Pharm. **1997**, *54*, 2503–2505.
30. Clinical Pharmacy Practitioner. North Carolina Board of Pharmacy, 21 NCAC 46.3101. www.ncbop.org/cpp.htm (accessed February 2001).
31. American Medical Association Council on Medical Service. *AMA House of Delegates. Interim Meeting, San Diego, California*; 1999.
32. Keely, K.L. American College of Physicians-American Society of Internal Medicine Position paper: Pharmacist scope of practice. Ann. Intern. Med. **2002**, *136*, 79–85.
33. Galt, K.A.; Narducci, W.A. Integrated pharmaceutical care

services: The product is part of the care. Pharmacotherapy **1997**, *17* (4), 841–844.

34. Kassirer, J.P. Managing care. N. Engl. J. Med. **1998**, *339*, 397–398.
35. Bodenheimer, T. Disease management in the American market. BMJ **2000**, *320*, 563–566.
36. Bodenheimer, T. Disease management—promises and pitfalls. N. Engl. J. Med. **1999**, *340* (15), 1202–1205.
37. Metge, C.J.; Hendricksen, C.; Maine, L. Consumer attitudes, behaviors, and perceptions about pharmacies, pharmacists, and pharmaceutical care. J. Am. Pharm. Assoc. **1998**, *38*, 37–47.
38. Beauchamp, T.L.; Childress, J.F. *Principles of Biomedical Ethics,* 4th Ed.; Oxford University Press: New York, New York, 1994.
39. Miyahara, R.K. Confidentiality of a patient's disclosure of symptoms. Am. J. Hosp. Pharm. **1993**, *50*, 953–957.
40. Standards of privacy of individually identifiable health information: Proposed rule. 64. Fed. Regist. **1999**, 55918–60065.
41. American Heart Association. *2000 Heart and Stroke Statistical Update*; American Heart Association: Dallas, Texas, 1999.
42. National Association of Boards of Pharmacy. *Survey of Pharmacy Law—2001–2002*; National Association of Boards of Pharmacy: Park Ridge, Illinois, 2000; 82–83.
43. American Society of Health-System Pharmacists Government Affairs Division. *Collaborative Drug Therapy Management*; www.ashp.org/public/proad/state (accessed March 2002).
44. American Society of Health-System Pharmacists. *Collaborative Drug Therapy Management Supplement to State Laws and Regulations Packet*; American Society of Health-System Pharmacists: Bethesda Maryland, 2000.
45. Koch, K. Trends in collaborative drug therapy management. Drug Benefit Trends **2000**, *12*, 45–54.

PROFESSIONAL DEVELOPMENT

Doctor of Pharmacy

George E. Francisco
University of Georgia College of Pharmacy, Athens, Georgia, U.S.A.

INTRODUCTION

The doctor of pharmacy (Pharm.D.) degree is the professional degree awarded to graduates of a U.S. school or college of pharmacy who have completed a minimum of 6 years of academic course work. The degree is required to take a board examination to become licensed as a pharmacist.

HISTORY OF THE Pharm.D. DEGREE

The awarding of the Pharm.D. degree in the United States dates back to the turn of the twentieth century. Prior to the 1900s, there were no standardized education requirements to become a pharmacist in the United States. One needed only the requisite knowledge to pass a board exam, and this knowledge could be obtained through apprentice programs, correspondence courses, ''cram schools,'' home study programs, or other means. Each course of study offered its own credentials for successful completion. In 1892, the University of Wisconsin introduced the first bachelor's course in pharmacy, and in 1904, New York became the first state to require a school diploma to practice pharmacy.[1] By 1900, there were about 55 pharmacy schools in the United States, but less than 10% of pharmacists had attended a school or college of pharmacy. Common pharmacy ''degrees'' awarded in the United States at this time included the Ph.G. (Graduate in Pharmacy); Ph.C. (Pharmaceutical Chemist); Ph.D., Phm.D., Pharm.D., or P.D. (Doctor of Pharmacy); Pharm.M. or Ph.M. (Master of Pharmacy); and B.S. (Bachelor or Science in Pharmacy).[2]

In the 1920s, pharmacy education became more standardized. In 1927, the American Association of Colleges of Pharmacy (AACP) adopted ''Basic Material for a Pharmaceutical Curriculum,'' and in 1928, its member schools approved a resolution requiring at least 4 full college years of at least 30 weeks each for graduation from a pharmacy program.[2] In 1932, the American Council on Pharmaceutical Education (ACPE) was formed, and pharmacy schools now had a formal mechanism by which to be evaluated for purposes of accreditation.

During the 1920s and 1930s, the Pharm.D. degree underwent several transformations. In 1925, institutions were allowed to offer the Pharm.D. degree after candidates completed not less than 3 years of graduate work. In 1932, the degree was officially defined and required not less than 3 years of graduate work for a total of 7 years of undergraduate and graduate work. In 1934, members schools of AACP restricted schools from offering pharmacy degrees other than a B.S. or B.S. Pharmacy. It was decided at that time that the Pharm.D. degree could not be awarded after 1938.[1,2]

In the 1940s, there was talk among pharmacy educators to increase the length of the baccalaureate degree from 4 to 5 years and perhaps to resurrect the Pharm.D. degree. World War II had begun, and just as in World War I, pharmacy was not viewed by the U.S. armed forces as an academic profession. Pharmacy students were refused deferments, and pharmacy graduates were refused commissioned officer status.[2,3] Concurrently, sulfa drugs and other antibiotics, as well as hormones, had been developed into dosage forms, and there was an emerging knowledge explosion in the pharmaceutical sciences. These events prompted educators to define a 6-year doctoral program comprised of an expanded prepharmacy curriculum of general courses and a highly science-based pharmacy curriculum. This, in turn, led to ''The Pharmacy Survey of 1946–48,'' in which the ACPE surveyed pharmacy faculty and practitioners to determine future directions for pharmacy education.[2]

As the 1940s came to a close, both AACP and ACPE granted approval to reinstate the Pharm.D. degree for a 6-year course of study, and in 1950, the University of Southern California became the first pharmacy school to adopt the 6-year Pharm.D. as its only professional degree leading to licensure as a pharmacist. Discussion among pharmacy educators continued regarding the length of the baccalaureate degree. In the early 1960s, AACP recom-

Encyclopedia of Clinical Pharmacy
DOI: 10.1081/E-ECP 120006355

mended that the baccalaureate curriculum in pharmacy be extended to 5 years, and in 1965, the 5 year degree became the minimum standard.[2] Pharmacy schools were now able to offer a baccalaureate or a Pharm.D. degree as their entry-level degree into the profession. They could also offer the PharmD as a postbaccalaureate degree, a 1- to 3-year program after completion of the B.S. degree. A third option was the ''track in'' PharmD, whereby all students began as baccalaureate students, and a few students were allowed to enter the Pharm.D. curriculum after completing a certain number of their baccalaureate courses. For the next 25 years, debates raged over the length of study and titles of pharmacy degrees.

The 1960s through the 1980s witnessed the slow transformation of pharmacists' roles from solely dispensing to more patient and information centered. During the early 1960s, a few pharmacists—including Donald Brodie of the University of California–San Francisco, Donald Francke of the University of Michigan, and Paul Parker of the University of Kentucky—envisioned new roles for pharmacists. They saw the pharmacist working side by side with physicians to provide information about the potent new drugs that were being manufactured. Thus began the clinical pharmacy movement. By 1967, the first journal devoted to clinical pharmacy, *Drug Intelligence and Clinical Pharmacy*, was published. In 1972, two therapeutics textbooks that were centered around clinical pharmacy were published. By the mid-1970s, the U.S. government began to recognize the clinical contributions of pharmacists and passed legislation that required monthly reviews of drug regimens for patients residing in skilled nursing facilities.[3]

By the early 1970s, debates over pharmacy manpower, pharmacists' roles, and pharmacy education were heightened. In 1972, AACP President Arthur Schwarting recommended the formation of a ''Commission on Pharmacy'' to study the scope of pharmacy services in health care and to project the educational requirements needed to train pharmacists to provide these services. The commission was chaired by John S. Millis, President of the National Fund for Medical Education. The Millis Commission's report, ''Pharmacists for the Future,'' was published in 1975.[2,3] The report contained 14 recommendations for pharmacy practice and education. Among these were continued movement of pharmacy as a knowledge-based clinical profession, increased development of clinical practice sites for pharmacy school faculty, and development of a national board licensing exam for pharmacists.

As a result of the Millis Commission's report, many pharmacy educators believed the time was right to adopt the Pharm.D. degree as the sole degree leading to licensure. In July 1978, the AACP's House of Delegates voted on the entry-level degree issue. By an almost two-to-one majority, the delegates voted to retain both the baccalaureate and Pharm.D. dual-degree structure.

In an article published in 1987 titled, ''The Third Wave in Pharmaceutical Education: The Clinical Movement,'' pharmacy educator Dr. Charles D. Helper described his vision of pharmacy education and practice.[4] He described clinical pharmacy as adding new knowledge for the patient's welfare. This knowledge had limitations because it was provided to other health care providers for the patient's benefit, instead of being provided directly to the patient. Hepler wrote:

> As pharmacy further clarifies its clinical role, it should underscore its acceptance of as much responsibility for drug use control as its social authority (under law) will support. This ideal can be called pharmaceutical care: a covenantal relationship between a patient and a pharmacist in which the pharmacist performs drug use control functions (with appropriate knowledge and skill) governed by awareness of and commitment to the patient's interest. The term is intended to invoke analogies with the ideals of medical care and nursing care.

It was this concept of pharmaceutical care that rekindled the discussion of pharmacy curricula, degrees, and practice.

By 1989, over 50% of all U.S. pharmacy schools still offered only the baccalaureate degree as their entry-level degree, 14% offered only the Pharm.D. degree, and 30% offered both degrees.[2] AACP President William Miller appointed a task force (which was termed the Commission to Implement Change in Pharmaceutical Education) to develop recommendations to guide pharmaceutical education to meet the changing demands of the profession, the health care system, and society. During the next 2 years, the task force addressed the educational standards necessary for the entry-level curriculum for pharmacy students, the length of the curriculum, and the title of the degree granted for completing the curriculum. The commission's recommendations were published as a two-part series in 1991.[5,6] The recommendations included ''an entry-level educational program for pharmacy practice that is at the doctoral level, is a least four professional, academic years in length, and follows pre-professional instruction of sufficient quality and length (two-year minimum) to prepare applicants for doctoral level education.'' Furthermore, the task force recommended the Pharm.D. degree as the sole degree for entry into pharmacy practice. In addition, schools and colleges that currently offered Pharm.D. programs were urged to examine, analyze, and revise their curricula to ensure that they were based on and reflected the philosophy of pharmaceutical care.

In late 1989, the ACPE, in its regular periodic review of accreditation standards and guidelines, issued a "declaration of intent."[3] In this declaration, ACPE stated that its intent was to accredit only Pharm.D. degree programs as the entry-level degree into the profession and suggested the year 2000 as a probable target date. This declaration fueled much discourse among pharmacy educators, practitioners, and organizations. Educators were skeptical of obtaining adequate resources to add another year to their curricula. Practitioners were fearful that baccalaureate practitioners would be disenfranchised if pharmacy schools and colleges produced only doctoral-level graduates. Various pharmacy organizations were wary of the economic and political ramifications of such a decision.

During the next 3 years, a number of meetings were held and articles written regarding the future of pharmacy education, particularly with respect to the Pharm.D. degree. A joint statement by the American Pharmaceutical Association, the American Society of Hospital Pharmacists (now the American Society of Health-System Pharmacists), and NARD (now the National Community Pharmacists Association) supported a new Pharm.D. degree as the entry-level degree for practice in the profession of pharmacy and outlined methods of degree equivalence for current practitioners. In defining the new degree, the joint statement stressed, "It is the responsibility of pharmaceutical education to provide a graduate prepared for immediate licensure and commencement of a career in any area of pharmacy practice." Therefore, the joint statement urged that the degree requirements prepare pharmacists for entry into the practice rather than for specialty practice.[7]

Debates about the PharmD as the entry-level degree had continued for more than half a century, but the issue was finally resolved in July 1992, in Washington, D.C., at the annual meeting of the AACP. With every school and college of pharmacy casting one administrative and one faculty vote each, the delegates voted overwhelmingly to endorse the Pharm.D. degree as the sole degree leading into the practice of pharmacy. It was then up to the ACPE to finalize the accreditation standards and up to the individual colleges and schools of pharmacy to revise their respective curricula.

THE Pharm.D. CURRICULUM

As defined in the ACPE's standards for accreditation, the purpose of the Pharm.D. curriculum is to prepare students to become generalist practitioners of pharmacy.[8] Specifically stated in the accreditation standards,

> The goals and objectives of the curriculum in pharmacy should embrace the scope of contemporary practice responsibilities as well as emerging roles that ensure the rational use of drugs in the individualized care of patients as well as in patient populations. The organized program of study should provide students with a core of knowledge, skills, abilities, attitudes, and values that are necessary to the provision of pharmaceutical care and should provide opportunity for selection by students of courses and professional experiences in keeping with particular interests and goals. The need for life-long learning should be reflected as an integral theme of the curriculum.

The Pharm.D. degree requirements include postsecondary preprofessional courses and requirements, as well as a minimum of 4 academic years to achieve professional competencies. Most pharmacy schools and colleges require a minimum of 6 academic years to complete all the degree requirements. The preprofessional requirements include basic sciences (e.g., general chemistry, organic chemistry, biological sciences, mathematics, computer technology, physical sciences). In addition, the student should have adequate preparation in general education requirements such as humanities, behavioral sciences, social sciences, and communication skills.

The professional courses in the Pharm.D. curriculum consist of didactic material, laboratory courses, and practical experiences in patient care environments. The overall curriculum is structured to provide instruction in the following core areas: biomedical sciences (including anatomy, physiology, pathophysiology, microbiology, immunology, biochemistry, molecular biology, and biostatistics); pharmaceutical sciences (including medicinal chemistry, pharmacognosy, pharmacology, toxicology, and pharmaceutics); behavioral, social, and administrative pharmacy sciences (including health care economics, pharmacoeconomics, practice management, communications applicable to pharmacy, the history of pharmacy, ethical foundations to practice, and social and behavioral applications and laws pertaining to practice); pharmacy practice (including prescription processing, compounding and preparation of dosage forms, drug distribution, drug administration, epidemiology, pediatrics, geriatrics, gerontology, nutrition, health promotion and disease preention, physical assessment, emergency first care, clinical laboratory medicine, clinical pharmacokinetics, patient evaluation and ordering medications, pharmacotherapeutics, disease state man-

Table 1 American council on pharmaceutical education: standard for professional competencies and outcome expectations

Professional competencies that should be achieved through the College or School of Pharmacy's curriculum in pharmacy are an ability to

1. Evaluate drug orders or prescriptions accurately and safely, compound drugs in appropriate dosage forms, and package and dispense dosage forms.
2. Manage systems for storage, preparation, and dispensing of medicines, and supervise technical personnel who may be involved in such processes.
3. Manage and administer a pharmacy and pharmacy practice.
4. Apply computer skills and technological advancements to practice.
5. Communicate with health care professionals and patients regarding rational drug therapy, wellness, and health promotion.
6. Design, implement, monitor, evaluate, and modify or recommend modifications in drug therapy to ensure effective, safe, and economical patient care.
7. Identify, assess, and solve medication-related problems, and provide clinical judgment as to the continuing effectiveness of individualized therapeutic plans and intended therapeutic outcomes.
8. Evaluate patients and order medications and/or laboratory tests in accordance with established standards of practice.
9. Evaluate patient problems and triage patients to other healthcare professionals as appropriate.
10. Administer medications.
11. Monitor and counsel patients regarding the purposes, uses, and effects of their medications and related therapy.
12. Understand relevant diet, nutrition, and nondrug therapies.
13. Recommend, counsel, and monitor patient use of nonprescription drugs.
14. Provide emergency first care.
15. Retrieve, evaluate, and manage professional information and literature.
16. Use clinical data to optimize therapeutic drug regimens.
17. Collaborate with other healthcare professionals.
18. Evaluate and document interventions and pharmaceutical care outcomes.

(From Ref. [8].)

agement, outcomes documentation, self-care/nonprescription drugs, and drug information and literature evaluation); and professional experience (including introductory and advanced practice experiences acquired throughout the curriculum). Specific professional competencies common to all pharmacy curricula are listed in Table 1.

As depicted in Fig. 1, the pharmacy curriculum is grounded in the philosophy of providing pharmaceutical care to patients. The process begins by identifying therapeutics goals and outcomes for a patient's medical problem. Although this is usually done by physicians and other diagnosticians, pharmacists may provide information to assist with identifying these goals. In addition, pharmacists must make their own therapeutic decisions with regard to recommending nonprescription drug therapy. Once the role of medication has been determined, the correct drug, dosage form, dose, route of administration, and dosing schedule must be determined. Pharmacists must determine that these parameters are consistent with the patient's medication condition(s) and individual characteristics. The medication order is then filled and dispensed, with appropriate information regarding medication administration, storage, and side effects being given to the patient, their caregiver, or another healthcare professional. The effects of the drug must be monitored to determine whether the medication is working and whether it is producing any undesirable effects. Based on all this information, the patient's therapeutic goals may need to be readjusted.

A typical Pharm.D. curriculum contains didactic and laboratory courses as well as practice experiences. The didactic courses can be taught via traditional classroom lectures, through technology-based applications (e.g., computer applications, Web-based instruction, or other means of distance learning), or as independent study courses. They may represent discrete academic disciplines within pharmacy education (e.g., pharmacology, medicinal chemistry, pharmaceutics, pharmacy administration, pharmacy practice), or the curricular material may be integrated across disciplines. Courses are often designed to transform the student from a dependent to an independent learner so that the graduate is prepared for continuous lifelong learning throughout their career.

The focus of laboratory courses is to develop students' skills in various areas of pharmacy practice. In labs, students practice dosage form preparation and administration, product selection, medication dispensing, patient counseling, physical assessment, and other components of delivering pharmaceutical care. Laboratory experiences may be components of didactic courses or may exist as stand-alone courses.

The experiential portion of the curriculum combines the student's knowledge and skills to enable the student to provide actual care to patients and their caregivers, as well as to interact with other healthcare providers. Experiential training is provided throughout all 4 years of the

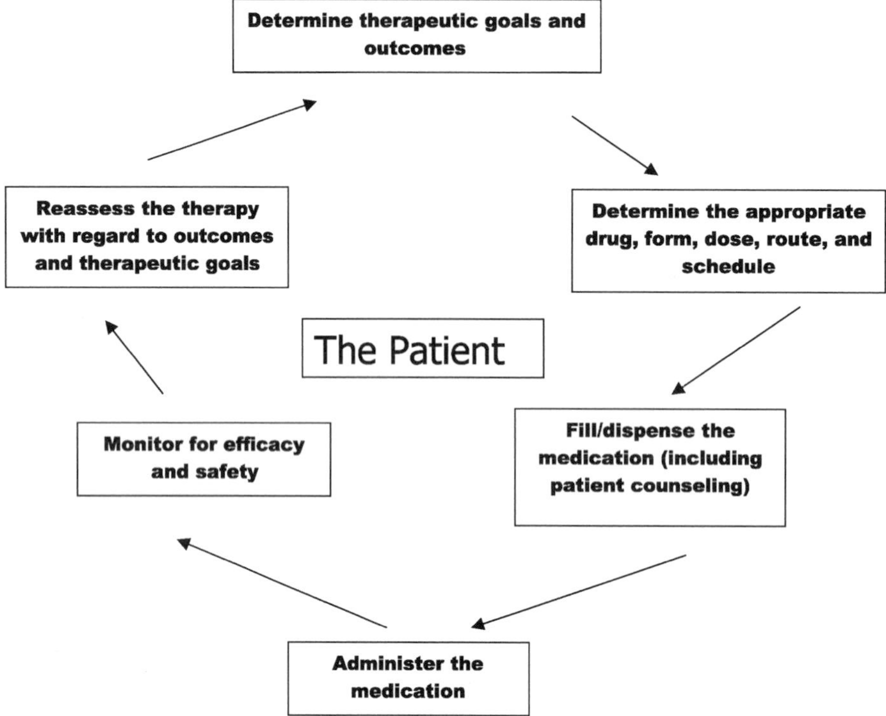

Fig. 1 The process of providing pharmaceutical care.

professional curriculum. Experiences that are taught earlier in the curriculum are usually brief in nature and are designed to introduce students to the healthcare system in general, to the pharmacy profession, or to various aspects of patient care through a combination of observation and participation. The later experiences consist of rotations of several weeks in length through various aspects of pharmacy practice in which students spend all their time learning to manage the practice setting, interacting with other healthcare professionals, and providing direct care to patients. Each rotation may typically last several weeks and include experiences in outpatient and inpatient settings, managed care organizations, pharmacy associations, or the pharmaceutical industry. These advanced rotations are considered capstone courses in which students, through direct practice experience, develop their competence and confidence to practice pharmacy.

CAREER OPPORTUNITIES

There are literally hundreds of different career opportunities for pharmacy graduates. The majority of graduates are employed in traditional community and hospital pharmacies. Many pharmacists, however, seek employment in other patient care areas, including nutritional support, ambulatory care, primary care, pharmacokinetics, pediatrics, and a variety of medical subspecialty areas (e.g., oncology, nephrology, pulmonology, hematology, critical care, infectious disease, emergency medicine, gastroenterology, psychiatry, cardiology). Pharmacists may work in areas of less direct patient care, such as nuclear medicine, drug information, or medical writing. They may be employed in all aspects of managed care, including pharmacy benefits management or formulary development and control, as well as in patient care areas. Pharmaceutical companies employ pharmacists to manage clinical trials, serve as medical liaisons to physicians and other healthcare providers, or work in pharmaceutical sales. There are also careers in academic fields to teach pharmacy students and students in other health disciplines and to conduct research or serve as role models in providing healthcare to patients.

Some career opportunities require additional training or credentialing through residencies, fellowships, certificate programs, and credentialing examinations. Residencies provide intense 1- to 2-year learning opportunities for

continued development of patient care and managerial skills. Fellowships develop pharmacists' research skills to prepare them for careers in the pharmaceutical industry or academia. Credentialing and certification provide specialized training in distinct areas such as nuclear pharmacy and disease management, and may be beneficial to the pharmacist to receive reimbursement for providing these specialized services.

REFERENCES

1. Kremers, E. *Kremers and Urdang's History of Pharmacy*; Lippincott: Philadelphia, 1976; 221–280.
2. Buerki, R. In search of excellence: The first century of the American Association of Colleges of Pharmacy. Am. J. Pharm. Educ. **1999**, *63* (Suppl.), 1–210.
3. Posey, L.M. Pharmaceutical Care: The Reprofessionalization of Pharmacy. In *Pharmacy Cadence 1992*; Posey, L.M., Ed.; PAS Pharmacy/Association Services: Athens, Georgia, 1992; 15–22.
4. Hepler, C.D. The third wave in pharmaceutical education: The clinical movement. Am. J. Pharm. Educ. **1987**, *51*, 369–385.
5. Wolf, H.H.; Walton, C.A.; Hepler, C.D.; Koda-Kimble, M.A.; Knapp, D.A.; Miller, K.W.; Nahata, M.C.; Rutledge, C.O.; Smith, W.E.; Vandel, J.H. Commission to implement change in pharmaceutical education: A position paper. AACP News **Nov. 1991**, 1–13.
6. Wolf, H.H.; Walton, C.A.; Hepler, C.D.; Koda-Kimble, M.A.; Knapp, D.A.; Miller, K.W.; Nahata, M.C.; Rutledge, C.O.; Smith, W.E.; Vandel, J.H. Commission to implement change in pharmaceutical education: Background paper II. **Mar. 1991**, 1–10.
7. Joint statement on the entry-level Doctor of pharmacy degree. Am. J. Hosp. Pharm. **1992**, *49,* 244–251.
8. *Accreditation Standards and Guidelines for the Professional Program in Pharmacy Leading to the Doctor of Pharmacy Degree*; American Council on Pharmaeutical Education; ACPE: Chicago, 1997; 12–20.

PROFESSIONAL ORGANIZATIONS

Drug Enforcement Agency

Claire E. Gilmore
Phillips Group Oncology Communications, Philadelphia, Pennsylvania, U.S.A.

INTRODUCTION

The Drug Enforcement Administration (DEA), an organization of the U.S. Department of Justice, is a federal agency whose mission is to enforce U.S. controlled substances laws and regulations and bring violators of these laws to the criminal justice system. The DEA maintains 78 offices in 56 countries throughout the world. Table 1 lists contact information for the DEA.

The DEA has several missions. First, the DEA enforces U.S. controlled substances laws as they pertain to the manufacture, distribution, and dispensing of controlled substances. Second, the DEA investigates and prepares for prosecution organizations and principal members of organizations involved in the growing, manufacture, or distribution of controlled substances for illicit traffic. Third, the DEA liaises with the United Nations, Interpol, and other organizations to reduce the availability of illicit controlled substances, both domestically and internationally.

HISTORY

In 1973, under the administration of President Richard Nixon, several federal drug agencies of various departments of the U.S. government united to form the DEA. The DEA predecessor agencies included the Bureau of Narcotics and Dangerous Drugs, the Office of Drug Abuse Law Enforcement, the Office of National Narcotics Intelligence, the Narcotics Advance Research Management Team, and the Drug Investigations division of U.S. Customs.

DEA PROGRAMS

The DEA operates many programs in an effort to fight the battle of illegal drug use (Table 2). Currently, two major drug threats in the United States are heroin and methamphetamine. More recently, the DEA successfully concluded several operations to address the drug problem in the United States. Operation Tar Pit, for example, successfully targeted a Mexico-based black tar heroin

Table 1 DEA contact information

Type of assistance needed	Contact information
DEA web site	www.usdoj.gov/dea
General comments or questions	DEA Information Services Section; 700 Army Navy Drive; Arlington, Virginia 22202
Physician registration or the Controlled Substances Act	DEA Office of Diversion Control; 600 Army Navy Drive; Arlington, Virginia 22202

Key: DEA, Drug Enforcement Administration.

Table 2 Select DEA programs

Program	Description
Asset Forfeiture	Executes forfeiture of profits and proceeds of designated crimes or property used in drug trafficking crimes
Diversion Control	Prohibits diversion of licit controlled substances and diversion of controlled chemicals into illegal trade
Intelligence	Collects, analyzes, and disseminates drug-related intelligence information in coordination with other law enforcement organizations
Laboratories	Provide forensic drug analysis to law enforcement agencies
Organized Crime Drug Enforcement Task Forces	Fight organized crime and drug traffickers
Marijuana Eradication Program	Funds cannabis eradication program in the United States

Encyclopedia of Clinical Pharmacy
DOI: 10.1081/E-ECP 120006181

trafficking organization. Another operation, Operation Green Air, successfully halted marijuana trafficking activities of an organization that exclusively used a commercial shipment company, FedEx, to transport the drug. Two additional problems for which the DEA is responsible are the diversion of controlled pharmaceuticals and the diversion of controlled chemicals.

INTERFACE WITH CLINICAL PHARMACY PRACTICE

Although intended for legitimate medical use, narcotics, stimulants, and depressants are frequently abused; therefore, controls have been established by the DEA to prevent their illegal distribution. Registration with the DEA is required of all health professionals entitled to dispense, administer, or prescribe controlled substances and of all pharmacies dispensing controlled substances. Strict regulatory standards relating to drug security and records accountability are required of these groups. Clinical pharmacists should be aware of the potential for drug diversion and be alert to the various ways individuals divert controlled substances. Examples of drug diversion schemes include physicians who sell prescriptions to drug dealers or abusers, pharmacists or nurses who falsify records to steal drugs to sell, employees who steals narcotics from inventory, prescription forgers, patients who obtain controlled substances from multiple physicians, and individuals who falsify narcotic orders to hide illicit sales. Research studies involving controlled substances or investigational controlled substances are subject to strict accountability per DEA regulations.

BIBLIOGRAPHY

Briefs and Backgrounds: Inside the DEA. Drug Enforcement Agency Website. www.dea.gov.

Drug History

Christi Cawood Marsh
University Health Care System, Augusta, Georgia, U.S.A.

INTRODUCTION

Performing a patient drug history involves gathering detailed patient medication information and is an important component of the patient's medical history. The patient drug history provides a thorough understanding of the patient's medication experience, with an emphasis on patient's current medications. The goals of the history are to obtain information on: 1) prescription and nonprescription medications (including dose, frequency, route, indication, and length of therapy); 2) perceived benefit or adverse effects of the therapy; and 3) medication allergies or intolerance. The pharmacist can identify potential medication problems by conducting the drug history. The need for intervention can then be discussed with the health care providers and physicians involved with that patient's care.[1]

Patient Drug History Versus Patient Medical History

The patient drug history focuses on medication therapy and is part of the patient medical history. The medical history encompasses information regarding: 1) past medical and surgical history; 2) acute and chronic medical problems; 3) social and family history; 4) other relevant health information; and 5) the drug history.[1] Depending on the setting, the pharmacist may incorporate elements of the medical history with the drug history. These elements can be used to develop a more thorough assessment and pharmaceutical care plan.

The practice setting and type of patient information available to the practitioner will largely determine the extent and scope of the patient history. Better patient outcomes are ensured by having adequate knowledge about patient's medical problems.[2]

Rationale for a Pharmacist-Conducted Patient Drug History

Traditionally, the physician completes the patient drug history. The physician takes the drug history as part of the medical history in the office or clinic, or upon admission in an acute care setting. Doctors often rely on hospital records, referral letters, or office records as the primary source of information regarding patient drug treatment. Many studies agree that physician histories are lacking, and show that pharmacists' drug histories are more accurate and complete.[3–5] One study documented that 11% of pharmacist-conducted histories contained important clinical information overlooked by the physician.[4] Pharmacists have been involved in obtaining patient drug histories in retail, ambulatory clinic, acute care, and long-term care settings.[3] Pharmacists improve the care provided to patients by becoming involved with the patient history process in their practice setting.

Patient Data Records

The records kept on the patient drug history need to be tailored to the setting. Some patient histories may be a permanent part of a medical record for one individual admission to the hospital, whereas other histories may be part of a continuum of the patient care (e.g., anticoagulation clinic). Records may be stored in a computer database and updated as patients return for follow-up in that care setting (e.g., retail/clinic pharmacy setting). Various forms have been used for record keeping both in written and in computer software formats.

TYPES OF DATA

Subjective Data

There are two ways to classify the types of data collected—subjective and objective. Subjective data refer to all information provided by the patient that cannot be confirmed independently.[1] The weakness of subjective data is that it cannot be confirmed, observed, or measured by the interviewer. However, it can be validated by other means. For example, patient compliance with a medication regimen can be supported by talking with a family caregiver; however, this is also subjective.

Encyclopedia of Clinical Pharmacy
DOI: 10.1081/E-ECP 120006322

Objective Data

These data are measurable or can be observed. Laboratory values and vital signs such as blood pressure are examples of objective data.[2] Objective data are not influenced by opinion or perception of the patient. Objective data are not infallible and can be limited. For example, if a patient has their blood level checked for drug therapy management, a laboratory error can occur. Objective data such as pharmacy refill records can be used to verify subjective patient information.

SOURCES OF PATIENT DATA

Patients

The patient is the most important source of information regarding their medication therapy. Although the data from the patient is subjective, the interview process can provide clarification on medications taken, knowledge of therapy, and barriers to education or compliance.

Medical Records

The medical record is another source of medication and health-related information. Access to this record may be limited in certain practice settings; however, it can be a valuable tool to review prior to conducting your patient drug history interview. Some practitioners use medical release forms to obtain medical record information such as laboratory data from other institutions required for drug therapy monitoring.[2]

Pharmacy Dispensing Records

Pharmacy refill records can be a valuable source for assessing what the patient is prescribed and how often the patient refills the prescriptions. Clarification of medication usage should be verified by refill records in your practice setting or by contacting pharmacies that the patient uses. Inpatient pharmacists can provide valuable patient information to the outpatient or retail pharmacists upon hospital discharge. This can prevent duplication and medication errors.[5]

Healthcare Providers

Other sources of information regarding medications/therapy can be obtained from home healthcare providers, long-term care facilities, and physician's offices.[5]

Caregiver/Family Member

Many patients rely on a caregiver or family member to assist them with their medications. These individuals can be a valuable source for patient drug history data.[5]

INTERVIEWING THE PATIENT

Setting

The location of the interview should be in a quiet environment free of distractions and allowing patient privacy. Avoid barriers between you and the patient. Respect patient privacy, and discuss the patient's health issues only with those directly involved with the patient's care.[2,5]

Communication

Introduce yourself initially, and describe your intentions and role in the patient's care. Always maintain good eye contact and avoid negative body language. For example, crossed arms or negative facial expressions will not make the patient feel at ease. It is important to record the history data; however, do not let your record taking distract from listening to the patient. Maintaining the continuity of the interview and listening are key to developing the patient's trust.[2]

The history is often affected simply by the way in which we ask the patient about their health problems and medications. Using open-ended questions (i.e., cannot be answered as ''yes'' or ''no'') versus closed-ended questions will require the patient to explain and inform you about their therapy. Open-ended questioning helps the practitioner quickly assess the depth of the patient's knowledge about their therapy and health.[1]

The basic format of the history interview will apply to all settings, including acute care, long-term care, ambulatory care, and retail, and can be adjusted to the specific needs of that setting. Utilization of patient data collection forms may be useful for documentation purposes and for guiding the flow and consistency of the interview. There are many sources for the format of data collection forms, which are discussed in a later section.[2,5]

Considerations in Special Patient Populations

''As people develop, have families, and age they provide you with special opportunities and require certain adaptations in interviewing style.''[6] In general, open the interview with the focus being on the patient (ask about school, friends, hobbies, work, family, etc.) to show interest in them personally. Once interest in ''them'' is established, the patient will usually open up to questioning.[6]

Infants and children younger than 5 years of age: Interviewing the parent will be required, but with the infant or child present. It is always best to refer to the infant/child by name and to the parent by ''Mr.'' or ''Mrs.'' to show both interest and respect.

The information obtained from the parent is third party, but is fairly accurate. Of note, however, the parent may have preconceived ideas about the nature of the child's problem. Practitioners must remember to be supportive rather than judgmental when interviewing the parent of the child. Avoid questions such as ''why did you give the child that medicine?'' This would imply judgment and that the parent did not have the child's best interest in mind.[6]

Children older than 5 years of age: Avoid talking ''down'' to children, but rather speak to them normally. The child can be interviewed about their health and medications, both with and without the parent present. First, ask the basic past medical history questions of the parent, then ask to speak to the child alone. Often, the child can tell you in more detail the severity of a problem or perception of medication treatment benefit than the parent.[6]

Adolescents: This population can be difficult to question at times. It is best to be straight forward, and ''real'' with this age group.

Aging patients: These patients can be visually or hearing impaired, have poor memory, or be slow to answer questions. Be sure to speak slowly and in a lower voice, and give extra time for a response to your questions. Most elderly patients may not be at ease with their medical problems, so be sensitive to them and really ''listen.'' Interview the patient in a comfortable setting free of noise and barriers. If the patient is cognitively impaired, you may have the caregiver and the patient present together. Remember to include the patient in the discussion by acknowledging them and establishing a relationship with them, even if the care provider has to answer questions regarding medication administration, etc.[6]

Language barriers: An interpreter may be needed in special situations. Many pharmaceutical companies provide medication literature in other languages that may be beneficial to have when counseling patients that do not speak the same language.

COMPONENTS OF A PATIENT DRUG HISTORY

Demographic and Patient Financial/ Insurance Information

This section of the history should include the patient's age, date and place of birth, any nicknames, names of both parents, work contact information, gender, ethnicity, address, phone, emergency contact information, names of the pharmacy the patient uses, and insurance information.[5]

Most patients are used to providing this type of information for their doctor's office visits, but may question the pharmacist's need to inquire. The pharmacist should explain that updated information will assist in providing better care for the patient. For example, when a patient's insurance does not cover the medication the patient was prescribed upon hospital discharge, the cost may prevent the patient from taking the medication. Obtaining insurance information prior to patient discharge as part of the history can prevent this type of problem.

Medication Allergies and Intolerances

A medication allergy is a hypersensitivy reaction to the allergen (drug) that provokes characteristic symptoms (rash, urticaria, bronchospasm, or dermatitis) upon subsequent exposure. A medication allergy may be delayed or not seen with initial administration, but after repeated exposure and antibody development the reaction occurs. A drug intolerance is different in that the reaction is not due to an antibody/hypersensitivity response. Intolerance is the inability of the patient to tolerate the particular medication due to a side effect of the medication.[6] Examples of drug intolerance are nausea from codeine or constipation related to an antihypertensive medication. Ask the patient to describe any drug allergies or intolerances using open-ended questions when possible so that they can describe the reaction rather than simply answering with ''yes'' or ''no'' to a question. Patients can often confuse medication intolerance with an allergy. The pharmacist can be valuable in clarifying this for the patient record. The information should be as specific as possible, including the description, treatment, and date of

the intolerance/allergy. For infants, children, and adolescent, patients give primary attention to any allergies prevalent during infancy or childhood.[6]

Immunizations

Immunization status is an important part of the medical history. Recording dates of childhood immunizations is pertinent so that ongoing boosters can be scheduled throughout childhood and adolescence.[6]

Adult immunizations are important to document as well and include vaccines such as pneumococcusl (for elderly and those at risk for pneumonia), influenza, hepatitis B, and tetanus. Although not an immunization, skin testing for tuberculosis might also be included under this section in high-risk patients (elderly, health care worker, or immunocompromised patient).

Medications

This list should include all prescription and nonprescription medications (including nutritional supplements, vitamins, and herbal remedies) the patient is taking. Information regarding the dosage strength, frequency, length of therapy, indication for use, and adherence must be obtained. Perceived benefit from the medication or any adverse experiences due to the medication should also be noted. Remember to inquire using open-ended questioning with patients using words such as ''how,'' ''what,'' and ''when.''[1,2,5]

Examples of open-ended questioning are ''What are you taking this medication for?'', ''How do you take your medication?'', and ''What do you do when you miss a dose of your medication?'' Closed-ended questions will not really tell the interviewer how much the patient understands about the dosing and purpose of medications without further questioning. Avoid questions such as ''Do you take all of your medicine once a day?'', ''Do you miss any doses?'', and ''Did your doctor tell you what this is for?''. All of these questions could be answered with either ''yes'' or ''no,'' and additional questioning would then be required for clarification. The open-ended style is efficient in that one type of question tells the interviewer all the patient's strong knowledge points and also pinpoints weak areas.[2]

Additional Home Monitoring and Compliance Aids

Establish records on patient use of any monitoring devices (i.e., blood glucose monitor) or compliance aids. This information helps understand the need for additional compliance aids or education on monitoring devices to improve therapy outcomes.[5]

Barriers to Compliance

Barriers to compliance must be identified during the history. Emotions, cognitive function, and physical ability can affect patient adherence to therapy. If a patient suffers from depression (emotional barrier), schizophrenia or dementia (cognitive barrier), or severe arthritis of the hands (physical barrier), compliance can diminish. Special attention should be given to these three areas, and barriers should be indicated on the history record. This process directs the implementation of specific aids to improve compliance.[5]

ADDITIONAL INFORMATION FOR PATIENT HISTORIES

The following sections are typically part of the broader medical history. These sections may be included to provide a more thorough assessment of the patient's therapy and health needs.[1–3,5,6]

Social History

The focus of the history is the patient's occupation, lifestyle, family relationships, and support system. Points of inquiry include job, marital status, diet, social drug use (i.e., alcohol, tobacco, illicit drug use), and religious beliefs related to health care. Asking patients about their use of alcohol and illicit drugs can be difficult for practitioners. It is not our role to pass judgment on the use of these agents, rather it is our job to gather the information to properly assess the patient and their health. Explaining to the patient that health outcomes are often affected by lifestyle choices and the family support for the individual may help with this part of the interview. For example, a visually impaired patient would need assistance with drawing up insulin. The support systems in place to assist the patient with the insulin preparation and administration need to be identified.

Acute and Chronic Medical Problems

Knowledge of the patient's health status will help the practitioner understand the purpose of the prescribed therapy, select optimal therapies for the patient, and help prevent adverse drug–disease state interactions. For example, a pharmacist would want to avoid recommend-

Name: ________________ **MR#**_____________ **Date:** ________________
Age: _________ **Gender:**_____ **Ethnicity:** _____ **R.Ph. :** ________________

INSURANCE INFORMATION: ______________________________________

PMH (acute/chronic medical problems):

____

____

____

PSH (past surgical history): ____

SH (social history):
___ EtOH ___Tobacco use (amounts)____________________
____ illicit drugs (list type if yes)__________________________
____ job status (list type of work)_____________
____ marital status ____children (number)___________

____

____

Allergies/Immunizations:

Medication	Type of Reaction	Allergy or Intolerance

Immunizations:
___**Pneumococcal vaccine: (dates)**_____________
___**Hepatitis B: (dates)**_____________________
___**Influenza: (dates)**______________________
___**Tetanus: (dates)**_______________________
___**MMR: (dates)**_____________
Other: ___________________________________

Prescription Medications	Dose/Frequency	Therapy Dates

Nonprescription Medications	Dose/Frequency	Therapy Dates

Herbal/Nutritional Supplements	Dose/Frequency	Therapy Dates

COMMENTS/ NOTES:

Fig. 1 Patient history form.

ing a nonsteroidal antiinflammatory drug for a patient with a recent gastrointestinal bleed.

Information gathered should include all medical problems the patient receives treatment for such as hypertension and diabetes. Problems the patient has been treated for in the past and past surgical procedures should also be noted. In children, it is important to include childhood illnesses (i.e., mumps, chicken pox) and exposure to these as well.

The medications prescribed will help prompt the interviewer to ask about specific medical problems such as antihypertensives being prescribed for the patient who has hypertension. Open-ended questions help the interviewer to become the listener and the patient to become the information provider.

USE OF DATA COLLECTION FORMS

Specific forms for patient drug histories are not required, but may benefit the history-taking process. The advantages of a patient drug history data collection form are: 1) it establishes a record (written or computerized) for the pharmacist's future use; 2) it provides a format for prompting questions during the interview; 3) its consistent format fosters organized flow of questioning; and 4) it prevents duplication of questioning in the future. The data recording process should never detract from the interaction with the patient.

The format can vary, but most forms will contain lines, tables, or checklists for the patient history components discussed in this entry: demographics, social information, allergy information, medical problems and procedures, and patient prescription and nonprescription medications.

Many sources have good examples of patient data collection forms.[2,5] An example of a patient history form is given in Fig. 1. The format of the form will require modification for the specific care setting and goals of the individual practitioner.

Some drug therapy management clinics use computer databases that store the patient history information and can print out the profiles when needed.[8–10] A few examples of data management software programs include CoumaCare® Patient management system, Anticoagulation Management Program (AMP) Anticoagulation, and Information Manager (AIM).[7–9] Pharmacist-managed anticoagulation and lipid clinics often use software programs to store and update patient history information.

MORE INFORMATION

Some key web sites to visit to obtain additional information on patient history taking as related to pharmaceutical care practice would be www.ashp.org (under practice standards or primary/ambulatory care), www.aphanet.org (under pharmaceutical care), www.auburn.edu (under case presentation guidelines), and www.altimed.com (under focus on patient communication). Medical-affiliated web sites that have some information on patient histories are www.ama-assn.org, www.acponline.org, and www.med.stanford.edu (under shs/smg/tools for pt. History forms). A web site for interpreter-guided interviews is www.hslib.washington.edu (under/clinical/ethnomed/intrprt). For more in-depth discussion, three referenced publications provide an excellent review on conducting patient medical and drug histories.[1,2,5]

REFERENCES

1. Young, L.Y.; Koda-Kimble, M.A. Assessment of Therapy and Pharmaceutical Care. In *Applied Therapeutics: The Clinical Use of Drugs,* 6th Ed.; Young, L.Y., Koda-Kimble, M.A., Eds.; Applied Therapeutics: Vancouver, Washington, 1995; 1–4.
2. Rovers, J.P.; Currie, J.D.; Hagel, H.P., McDonough, R.P.; Sobotka, J.L. Patient Data Collection. In *A Practical Guide to Pharmaceutical Care*; American Pharmaceutical Association: Washington, DC, 1998; 26–55.
3. Titcomb, L.C. The pharmacist role in drug history taking. Br. J. Pharm. Prac. **1989**, *11*, 186–195, (Jun).
4. Claoue, C.; Elkington, A.R. Informing the hospital of patients' drug regimens. Br. Med. J. **1986**, *292*, 101.
5. Munroe, W.P.; Briggs, G.C.; Dalmady-Israel, C. Establishing a Relationship with Patients and Identifying Needed Information. In *Ambulatory Clinical Skills Program: Core Module*; American Society of Health-System Pharmacists, Inc.: Gaithersburg, Maryland, 1998; 1–22.
6. Bates, B. Interviewing and Health History. In *A Guide to Physical Examination and History Taking,* 5th Ed.; J.B. Lippincott Company: Philadelphia, 1991; 1–26.
7. CoumaCare Patient Management System. DuPont Pharmaceuticals, Chestnut Run Plaza, PO Box 80723, Wilmington, Delaware 19880-0723. 1-800-474-2762.
8. Anticoagulation Information Manager. Wellersoft, 5416 Parkgrove, Ann Arbor, Michigan 48103. (734) 213-5360.
9. Anticoagulation Management Program. Telehealth Systems, Inc., 520 N. State Road, 135 Suite M78, Greenwood, Indiana 46142. (317) 535-6161.

PROFESSIONAL DEVELOPMENT

Drug Information Pharmacy Practice

Patrick M. Malone
Creighton University, Omaha, Nebraska, U.S.A.

INTRODUCTION

Drug information is an area of pharmacy practice that deals with obtaining, managing, and evaluating information to prepare and disseminate it in a suitable format, wherever and whenever it is needed or in anticipation of need. This area of practice is one of the oldest in clinical pharmacy, with the decision to establish a drug information center being made in 1959 at Ohio State University[1] and 1960 at the University of Kentucky,[2] with the latter one being the first to open in 1962.[3] The drug information specialist has not specifically been considered a documentation specialist, as is a librarian, but was originally defined as a subject-oriented specialist in the area of drug knowledge.[4] Usually, this definition is broadened to include both pharmaceutical and therapeutic knowledge,[1] which has lead to controversy over the true name for the area of practice.

Usually, the term *drug information* is coupled with terms such as *specialist*, *center*, or *service*. Some people have substituted the words *medication* or *biomedical* for *drug*, due to the negative connotation that the latter term has in society. Others have substituted the word *informatics* for *information*, to better acknowledge the increased role of computers in information management. Unfortunately, a better term or phrase that indicates that people working in this area deal with information relating both to drugs (e.g., therapeutics, adverse drug reactions) and to pharmacy (e.g., how to perform various pharmacy tasks) has not been identified.

Individuals working in the area of drug information possess some of the skills of a documentation specialist because these skills are necessary to manage information. However, unlike pure documentation specialists, the drug information practitioner has the ability to adequately understand the initial problem and, after locating the information using the skills of documentation specialists, can evaluate the information and use it to formulate a solution to a particular pharmacy or medication-related situation. Because of this ability, drug information practitioners have occasionally been referred to as the "ultimate generalist" in pharmacy. They do not necessarily know the in depth information that a specialist in a particular clinical or practice area would have about that particular specialty, but the drug information practitioner has the in depth knowledge of how to obtain the necessary information and use it to address specific problems or concerns in most areas.

The need for drug information practitioners is likely to increase due to the rapid increases and improvements in information and the technology to manage it, particularly due to Internet technology and information sources. Skills expected of drug information practitioners in the 1980s are now expected of many pharmacists; drug information practitioners now are expected to have even greater skills, and the ability to handle larger and more complex information management situations.

CLINICAL PHARMACY OPPORTUNITIES

It is difficult to describe clinical pharmacy opportunities in relation to drug information because drug information skills are at the core of clinical pharmacy practice. It is impossible for anyone to know everything they need to know about pharmacy practice, and new information is becoming available so rapidly, particularly with the advent of the Internet, that even adequately keeping up with methods of obtaining and managing information is difficult. Given that, it can be stated that in any pharmacy environment there are opportunities for individuals who specialize in information management, whether that be in community, institutional, academic, industry, managed care (including health maintenance organizations and pharmacy benefit managers), insurance companies, associations, government, or other environments.

MODEL CLINICAL PRACTICES

Community

Few community pharmacy-based drug information practices have ever existed. Occasionally, however, there have

Encyclopedia of Clinical Pharmacy
DOI: 10.1081/E-ECP 120006307

been such practices to provide information to patients. These were sometimes established using grant funding, but failed financially afterward. There have been some practices for a fee (e.g., 900 phone numbers), with good results. Now, more drug information is being provided via the Internet. This includes services such as those run by Internet pharmacies (e.g., http://www.rx.com), pharmacy organizations (e.g., http://www.pharmacyandyou.org), and even individual pharmacists (e.g., http://www.medconsultant. com/index.shtml).

Functions of those services tend to center around providing prepared drug information documents (e.g., patient information sheets) and answering specific medication-related questions. These are seldom money-making operations, but are often provided as a service to attract customers to a pharmacy or as a public service.

Institutional and Academic Practice

These two environments are grouped because they are similar in nature and are often combined. Typically, practitioners here are located in a dedicated drug information center that resides in a hospital pharmacy or medical library. Typically, such services were begun to provide literature searches and answers for specific questions and to perform formulary management.[5] Given the greater concern for the cost of services, the former service is now sometimes deemphasized. Instead, services that will decrease hospital costs (including liability), increase income, or provide functions that are required by legal or regulatory bodies are often performed. Overall, it has been shown that having a drug information service may save 2.9 to 13.2 times its cost.[6]

The following functions are performed by drug information practitioners in the institutional and academic environments:[7]

- Answer questions and perform literature searches.
- Drug formulary management (e.g., evaluating drugs for addition or deletion from the formulary, preparing use guidelines and policies and procedures, pharmacoeconomic analysis), including publication of a drug formulary book, whether in hard copy or electronic format.
- Quality assurance activities (e.g., departmental quality assurance, drug usage evaluation, medication usage evaluation). This includes setting up, managing, and evaluating the data from such activities.
- Development and/or modification of evidence-based clinical guidelines. This includes the concepts of disease state management and outcomes management.
- Development and/or modification of policies and procedures.
- Adverse drug reaction/medication error tracking and reporting.
- Investigational drug information (e.g., Institutional Review Board activities, central depository of study protocols, providing patients and practitioners with information about investigational drugs, managing medication studies).
- Poison information—occasionally, drug information services are run in conjunction with poison information services.
- Management of department information equipment, software, and procedures.
- Provision of educational programs and materials, which may include newsletters and web sites.
- Potential contract services with industry, managed care, or insurance companies and other groups to provide specific information services (see the information listed under those environments for further information on necessary services).[8]
- A major function for academic, and occasionally institutional, drug information centers is education. This can include pharmacy students, residents (general or drug information specialty residents), and fellows.

Drug information practitioners in institutional and academic environments may work within a single institution or may be involved in a hospital system that requires services to multiple institutions, perhaps over a wide geographic region.

Industry

Within the pharmaceutical industry there is a major need for drug information specialists for a variety of functions:

- Answering information requests from health care professionals, employees, and occasionally patients.
- Preparation and management of information databases for employees.
- Preparation of materials to be distributed directly to health care professionals, employees, and patients.
- Setting up and managing clinical drug research and the information derived from it.
- Preparing Food and Drug Administration (FDA)-required information, such as New Drug Applications.
- Collecting, collating, and using adverse drug reaction information.
- Provision of training to pharmacy students and residents.

It is important to note that in the industrial environment, physicians often manage the drug information services or other areas that use drug information practitioners, while they are staffed by some combination of physicians, pharmacists, nurses, or others.

Managed Care and Insurance Companies[9]

Drug information services in a managed care or insurance company environment often deal with issues concerning providing the lowest cost therapy for patients (i.e., keeping reimbursement cost low). At one time, this might have amounted to individuals (including nonpharmacists) simply reviewing and comparing the costs of drugs within a therapeutic class. The assumption was that they were all interchangeable. Fortunately, it has become more widely recognized that many factors are involved in providing the best and least expensive therapy to patients. This includes the efficacy of the drug, adverse effect frequency and severity, cost of monitoring, the need for additional care, the length of therapy, and a variety of other therapeutic, ethical, legal, and patient issues. A full pharmacoeconomic analysis is necessary to ensure that all aspects are evaluated. It is not unusual that a drug product that initially looks to be the least expensive may actually be the most expensive due to a variety of reasons, such as a need for increased monitoring, lower efficacy, more severe adverse effects, etc. Drug information centers may evaluate whether reimbursement is available for drugs or the disease state, what copays might be required, restrictions or authorizations that are needed before medication use, and other information.

Also, drug information centers in these environments may spend a lot of time preparing information on the best way to treat disease states to produce optimal outcomes for the lowest cost (i.e., disease state management, outcomes management). This information may be used either actively or passively to educate health care practitioners, particularly physicians and pharmacists.

Drug information practitioners in these areas may also perform other drug information activities, such as answering questions, quality assurance, and electronic information interchange on a national level. Other activities listed under institutional practice may also be carried out.

Associations

Various professional associations have drug information needs. This may have to do with association publications; researching items of interest to the association; providing information for association members or other interested people; and preparation of statements, guidelines (including evidence-based, clinical guidelines), and other official documents.

Government

Government organizations at the national or state levels have a need for drug information specialists. For example, FDA (http://www.fda.gov) can use the services of drug information practitioners in the collection, organization, management, and distribution of information on drugs (both investigational and marketed).

At a state level, drug information practitioners may be involved with drug utilization review, whereby data on drug usage patterns is collected and analyzed to determine ways by which drug therapy may be improved.

Other activities similar to those listed for managed care organizations can also be performed by government drug information practitioners. Also, some state and foreign governments have drug formularies, which drug information practitioners would be involved in managing.

MATERIALS USEFUL FOR DRUG INFORMATION PRACTITIONERS

There are a variety of resources that may be of value to drug information practitioners, both in learning how to perform the various necessary skills and in carrying out the responsibilities.

There is currently one general reference to guide drug information practitioners and those who would like to learn the skills. Other general references are currently out of print, although some may still be obtainable. They are as follows:

- Malone PM, Mosdell KW, Kier KL, Stanovich JE. *Drug Information—A Guide for Pharmacists*, 2nd ed. New York: McGraw-Hill, 2001. This is the most complete and up-to-date reference, covering all aspects of drug information practice, including formulary management and quality assurance. It includes extensive lists of references and Internet sites that are of use to individuals who are trying to obtain drug information. Also, this reference covers the evaluation of all types of literature, rather than just clinical studies, which many other drug information books are limited to covering.

The June and August 1998 issues of *Journal of Pharmacy Practice* are also devoted to the practice of drug information and contain a great deal of useful

information, similar to the contents of the previously mentioned books.

There are also references that cover specific aspects of drug information practice:

- Galt KA. *Analyzing and Recording a Drug Information Request.* Bethesda, MD: American Society of Hospital Pharmacists, Inc., 1994. This first module in a series of three deals with the skills needed to initially take a drug information request, mostly from a general practitioner's point of view.
- Smith GH, Norton LL, Ferrill MJ. *Evaluating Drug Literature.* Bethesda, MD: American Society of Health-System Pharmacists, Inc., 1995. This second module deals specifically with the skills necessary to evaluate drug literature.
- Galt KA, Calis KA, Turcasso NM. *Preparing a Drug Information Response.* Bethesda, MD: American Society of Health-System Pharmacists, Inc., 1995. This third module takes the information obtained and evaluated using methods in the first two books and describes methods to effectively distribute it. Again, it addresses the subject from the point of view of the average pharmacy practitioner.
- Ascione FJ. *Principles of Scientific Literature Evaluation: Critiquing Clinical Drug Trials.* Washington, DC: American Pharmaceutical Association, 2001. While previous editions of this book were somewhat broader in scope, the current edition specifically covers the evaluation and interpretation of scientific papers describing clinical trials.
- Slaughter RL, Edwards DJ. *Evaluating Drug Literature—A Statistical Approach.* New York: McGraw-Hill, 2001. This book covers a wider area of drug literature evaluation than the previous reference, including some information on other topics, such as performing a literature search.
- Snow B. *Drug Information—A Guide to Current Resources.* Lanham, MD: Scarecrow Press, Inc., 1999. This is an extremely comprehensive book that lists and describes multiple sources of drug information. The focus is very limited, but no other book covers this subject as completely.

There are many resources available to the drug information practitioner. They will not be presented here due to their vast number, but they are described in some of the previous references. It should be noted that many of those resources are now available electronically (e.g., evidence-based clinical practice guidelines are available at http://www.guideline.gov), allowing wider access and easier use. Also, references dealing with the electronic management of information are particularly helpful to drug information practitioners.

PROFESSIONAL NETWORK OPPORTUNITIES

There are a variety of professional networking opportunities available through professional associations for drug information practitioners. They are presented here in alphabetical order:

- *American Medical Informatics Association (AMIA)*—This group is concerned with health information technology. It consists of members in a wide variety of professional areas. Some professions, such as dentists and nurses, have specific working groups in the association. Some drug information pharmacists are members, but there is not yet a working group for those individuals. Further information about this organization can be found at http://www.amia.org.
- *American Society of Health-System Pharmacists (ASHP)—Clinical Practice Section—Drug Information/Pharmacoeconomics Network*—Members of ASHP can also become members of the Clinical Practice Section, which has many practitioner networks. One of these is for drug information and pharmacoeconomics. This network sponsors continuing education programs at the ASHP annual and midyear clinical meetings. Members are also provided a time and place to gather at the midyear clinical meeting, and sometimes the annual meeting, to discuss topics in their area of interest. In addition, an e-mail listserve is available for communications among members of this network and information is available for members at http://www.ashp.org/clinical/index.html. This group would be of most interest to institutional and academic drug information practitioners. A variety of guidelines, as well as position statements of interest to drug information practitioners on such subjects as formulary management and medication use evaluation, are available at http://www.ashp.org/bestpractices/index.html.
- *Consortium for the Advancement of Information, Policy and Research (CAMIPR)*—This is the newest of the drug information associations, formed in 1994 to better serve the needs of institutional and academic drug information pharmacists.[10] This group generally meets in conjunction with the ASHP midyear clinical meeting. There is no cost involved in joining the organization; it is only necessary to join their listserve. Information on joining and compilations of previous

listserve discussions is available at http://druginfo.creighton.edu/camipr.

- *Drug Information Association* (*DIA*)—The DIA is a group devoted entirely to drug information specialists, including physicians, pharmacists, and others. It is mostly involved with drug information practitioners in industry practice. Information on the organization and its services can be found at http://www.diahome.org.

Other smaller drug information groups also exist.[11]

REFERENCES

1. Anderson, R.D.; Latiolais, C.J. The Drug Information Center at the Ohio State University Hospitals. Am. J. Hosp. Pharm. **1965**, *22*, 52–57.
2. Parker, P.F. The University of Kentucky Drug Information Center. Am. J. Hosp. Pharm. **1965**, *22*, 42–47.
3. Amerson, A.B.; Wallingford, D.M. Twenty years' experience with drug information centers. Am. J. Hosp. Pharm. **1983**, *40*, 1172–1178.
4. Francke, D.E. The role of the pharmacist as a drug information specialist. Am. J. Hosp. Pharm. **1966**, *23*, 49.
5. Wittrup, R.D. The responsibility of the hospital for drug information services. Am. J. Hosp. Pharm. **1965**, *22*, 58–61.
6. Kinky, D.E.; Erush, S.C.; Laskin, M.S.; Gibson, G.A. Economic impact of a drug information service. Ann. Pharmacother. **1999**, *33*, 11–16.
7. Rosenberg, J.M.; Fuentes, R.J.; Starr, C.H.; Kirschenbaum, H.L.; McGuire, H. Pharmacist-operated drug information centers in the United States. **1995**, *52*, 991–996.
8. Forrester, L.P.; Scoggin, J.A.; Velle, R.D. Pharmacy management company-negotiated contract for drug information services. Am. J. Health-Syst. Pharm. **1995**, *52*, 1074–1077.
9. Redman, R.L.; Mays, D.A. Drug information services in the managed care setting. Drug Benefit Trends **1997**, *9* (Aug), 28–30, 36–40.
10. Vanscoy, G.J.; Gajewski, L.K.; Tyler, L.S.; Gora-Harper, M.L.; Grant, K.L.; May, J.R. The future of medication information practice: a consensus. Ann. Pharmacother. **1996**, *30*, 876–881.
11. O'Brien, E. Network helps pharmacists access drug information. Hosp. Pharm. Rep. **1997**, *11* (8), 55.

Drug Samples

Nanette C. Bultemeier
Dean G. Haxby
Oregon State University, Portland, Oregon, U.S.A.

INTRODUCTION

Drug sampling is an important, although controversial, marketing technique used by pharmaceutical companies. Over half of the $13.9 billion spent in 1999 on the marketing of prescription drugs by pharmaceutical companies was for "free" medication samples.[1]

Drug sample availability is intended to increase market share of products by directly influencing providers' prescribing habits. Sampling also opens doors for pharmaceutical representatives to gain access to busy prescribers. Sampling gives representatives a reason to visit prescribers' offices and provides prescribers an incentive to permit representatives to visit. The influence of pharmaceutical representatives on physician behavior is well established.[2]

Inventories of drug samples valued at tens of thousands of dollars fill closets and even entire rooms of many outpatient clinics, physician offices, and emergency rooms.[3–6] Reports suggest that prescribers dispense samples at 10–20% of patient encounters, although there is wide variation between and among practices.[3,5] The most commonly dispensed samples in primary care settings include pulmonary medications, anti-infective agents, analgesics and anti-inflammatory drugs, allergy medications, cardiovascular agents, gastrointestinal agents, and oral contraceptives.[3–5]

REGULATORY ISSUES

Before the implementation of the Prescription Drug Marketing Act (PDMA) of 1987, record keeping of sample distribution was not required.[7,8] Abuses in the system of distributing samples that "resulted in the sale to consumers of misbranded, expired, and adulterated pharmaceuticals" led to the passage of the PDMA.[8] The PDMA prohibits the selling of samples. The Act requires signed requests for samples by practitioners licensed to prescribe such drugs. The PDMA includes provisions to the manufacturer on record keeping and reporting to the Food and Drug Administration. It also requires storage conditions that maintain the stability, integrity, and effectiveness of sample products and keep products free of contamination and deterioration.[8] Because state laws and regulations on samples vary, Boards of Pharmacy should be contacted for state-specific information.

Clinics and emergency rooms of hospitals and health systems must follow standards set by the Joint Commission on the Accreditation of Healthcare Organizations (JCAHO) for sampling and medication use. JCAHO requires policy and procedures related to the control of drug samples.[9] In addition, all other JCAHO standards that are applicable to medication use, including uniformity in processes, apply to drug samples to the same extent as they apply to regular prescription medications dispensed by the hospital pharmacy.[9] Specific methods or processes for controlling samples are not dictated by JCAHO, although certain features of the sampling system may be inspected[9] (Table 1).

BENEFITS

The availability of drug samples can benefit prescribers and patients. Prescribers often report using drug samples to avoid medication costs to the patient.[5,10,11] Samples may be used before a full prescription is purchased so that safety, tolerability, and effectiveness can be evaluated, or doses titrated.[11,12] Samples may be used to partially or fully offset the cost of drugs for indigent patients or to avoid formulary restrictions or prior authorization requests.[5,10,11] Prescribers may use samples to initiate therapy immediately in the office, allowing patients to avoid a trip to the pharmacy. This may be important for urgent and painful conditions, and for after-hours care.[11] Samples can be helpful for demonstrating appropriate use of inhalers, topical products, and other medications.[13] Prescribers frequently report that sampling is beneficial, because it allows them to gain experience with new

Encyclopedia of Clinical Pharmacy
DOI: 10.1081/E-ECP 120006394

Table 1 Criteria JCAHO surveyors generally look for in evaluating compliance

- There is a system, defined by policy and procedure, for the control, accountability, and security of all drug samples throughout the organization.
- The drug samples are properly stored.
- Drug sample storage areas are routinely inspected.
- Drug samples are secure.
- Drug samples are labeled and dispensed according to the same standardized method that the hospital uses for nonsample prescription medications.
- Documentation requirements for sample drugs should be the same as other nonsample medications ordered and dispensed by the clinic or hospital.
- There must be an effective recall mechanism for drug samples.

(From Ref. [8].)

drugs.[11] Other factors for dispensing samples include improving patient satisfaction and adherence.[10,14]

CONCERNS

Despite the many apparent benefits of sampling, the practice has been heavily criticized. Concerns surrounding sampling include patient safety, product integrity, security and control of samples, ethical issues, influence on prescribing habits, and costs to the healthcare system.[3,5,9,10,15–24]

Important safety controls are lost when samples are used. Sampling bypasses the safeguards of pharmacist medication regimen review and counseling. Drug interactions, which are screened when prescriptions are dispensed at pharmacies, may go undetected when samples are used.[15,16] Labeling samples with patients' names and instructions for use is often inconsistent, if it is done at all.[3,5,10,16] One study found that instructions to the patient accompanied less than 50% of patient encounters involving sample dispensing and were predominantly verbal in nature.[3] Patient information sheets are infrequently provided by the manufacturer in sample packaging, and most providers do not have systems to generate the extensive printed information that is provided to patients at pharmacies.[17] There is also concern about increasing prescribing errors, because formulary and nonformulary drugs with which prescribers and staff may not be familiar are often delivered by representatives.[16]

The potential for inadequate sample inventory management in prescribers' offices and inappropriate storage of products by pharmaceutical representatives raises concern about product potency and stability. Expiration dates and product recalls may go unnoticed.[15,16] Products might be stored in garages and automobile trunks of pharmaceutical representatives, potentially exposing samples to extremes in temperatures or humidity.

Significant deficiencies in the security and control of samples have been well documented.[5,18–21] In fact, it has been estimated that just over half of samples actually reach patients.[5] Samples may be used by prescribers and staff, or they may be diverted. Personal use of drug samples by physicians and other healthcare providers raises ethical concerns and is not without risk.[18,19] Limaye and Paauw described three medical residents who self-prescribed antimicrobials and were subsequently diagnosed with *Clostridium difficile* infection.[19] Tong and Lien reported self-medication with samples and distribution of samples to nonphysicians by almost 60% of pharmaceutical representatives surveyed at a Canadian family practice office.[20] A contributing factor to some of these issues is that institutional or facility sample policy and procedures are often absent, or compliance is poor. One institution found only 10% compliance when the inventory of samples was compared with the required written documentation. Even after an educational program in which the policy was explained to the house staff, a second audit found only 26% compliance.[21] Poor compliance with policy and procedure may jeopardize patient safety, as well as put the institution at risk for JCAHO recommendations or Board of Pharmacy penalties.

Another concern is that drug choices may be dictated more by what is available in the sample closet than by evidence-based recommendations or by known cost-effectiveness. The influence of sampling has potential implications for patient care and healthcare costs. While studies have shown that sampling may increase subsequent prescription of the sampled drugs, research on the quality of prescribing related to sampling is sparse.[5,9,10] A survey by Chew et al. of physicians' self-reported prescribing patterns for three clinical scenarios found that the availability of drug samples led physicians to dispense and subsequently prescribe drugs that differ from their preferred drug choice.[9] In addition, the study found that when drug samples were made available, 27% of physicians indicated that they would dispense a drug sample not recommended as a first-line agent by the Joint National Committee on Hypertension.[9]

Because sampling is labor intensive and is subject to industry and institutional regulations, it is a very costly practice. In addition to the wholesale value of samples, there are other costs, such as packaging, distribution via representatives, prescriber and staff time interacting with

representatives and handling samples, and institutional administration of sample programs. The bulky cardboard packaging that is characteristic of drug samples not only creates an inordinate amount of waste, it also takes up valuable office space.[22,23] Health systems likely incur the financial burden of giving out ''free'' samples when less expensive medications are available, although there is little evidence supporting this. A sample is only free in the sense that neither the prescriber nor the patient paid cash for it when it was received.

Despite the many shortcomings of sampling, it is often continued, because it enables prescribers to provide medications to indigent patients. However, sampling is an inefficient method for helping patients in need of medications.[24] Supplies may be inconsistent, and multiple packages of samples are usually required on a frequent basis to maintain patients' needs. Patients often do not get as much drug as needed, or they are switched from brand to brand based on the samples that are available, creating confusion for patients and prescribers.

IMPROVING THE PROCESS

Efforts to address concerns about sampling range from development of policies and guidelines for sample use to restrictions and banning of samples. Pharmacists are involved in, and many times spearheading, sample practice changes.[14,16,21,25–27]

In facilities where samples are used, important first steps to bringing sample practices into compliance include consulting state and federal regulations and JCAHO standards (Table 1). Guidelines and recommendations from organizations such as the Society of Teachers of Family Medicine and the Institute for Safe Medication Practices (ISMP) can be consulted.[16,28]

Other innovative approaches to improve sampling processes have been developed. Multiple-part carbonless adhesive forms with space for patient name, date, medication name and strength, quantity, directions for use, lot number, expiration date, and physician signature have been created for signing out samples. One copy goes to the patient, one goes to the chart, and one is kept for record maintenance. This type of system helps ensure written directions, provides a double check on expiration date, and becomes a log with lot number in the event of a recall. Although it is time-intensive, maintaining a perpetual inventory or auditing closets and sign-in/sign-out logs on a regular basis is a way to determine if samples are being stolen. Posting information about unaccounted for samples may heighten awareness of security and compliance issues.

Some pharmacists dispense samples in the clinic when requested by prescribers. A full range of services, including assisting with product selection, labeling the product, and counseling the patient, may be provided. Sometimes pharmacists supervise nursing staff that order, stock, label, and discard samples.[25] Educational efforts to promote sample compliance and appropriate prescribing are often done or coordinated by pharmacists. Providing reviews and summaries of treatment guidelines and evidenced-based pharmacotherapy that include formulary and cost information to small groups, via newsletters and in postings in sample closets, are common academic detailing activities used to counter sampling practices.

To further encourage prescribing of cost-effective drugs, some clinics request generic samples or prepackaged first-line medications. Generic samples are available from some manufacturers, and at least one pharmacy benefit management company is planning to provide samples of generic drugs in its efforts to encourage the use of generic drugs. Unfortunately, there has been limited success in getting prescribers to use generic samples or prepackaged first-line medications.[21]

Technology is being used to improve sample control and patient safety. Some facilities have developed databases to generate labels and to log lot numbers of samples dispensed. Sophisticated systems that include computer-controlled dispensing cabinets are marketed by companies like www.drugsampling.com. These systems include fingerprint-recognition technology to open the cabinets and touch screen monitors that can be used to generate medication labels, patient education, and required dispensing documentation. The systems are paid for by renting space in the cabinet to drug companies. The value for the manufacturer, according to a company that markets this type of system, is to maintain sampling privileges in clinics where sampling is at risk of being banned because of poor control and to provide sample usage information to the manufacturer.

Restricting samples to drugs available on or preferred by the organization's formulary is an approach used by some organizations to discourage prescribing of nonformulary drugs. Some clinics appoint a committee, which usually includes a pharmacist, to develop a formulary of requested samples based on safety, efficacy, and cost.[16,25] General guidelines that one clinic used in selecting formulary samples included: 1) stocking one or two of the least expensive drugs in each therapeutic class; 2) delaying the addition of new agents until adverse reactions and drug interactions are clinically demonstrated; 3) refraining from adding ''me too'' drugs unless they have clear advantages; and 4) accepting drugs that have generic equivalents offering long-term savings to patients.[25] Some clinics simply restrict samples to those that address the most common needs of patients treated in the practice.[25] Placing restrictions on the types and quantities of samples requested promotes efficient use of storage space by reducing the number of unused and expired products.[16,25]

Despite implementing many of the controls mentioned above, more and more facilities are banning samples.[14,21] Most cite concerns about complying with regulations and the promotion of poor prescribing habits that lead to increased costs. Vigilance in enforcing policies prohibiting samples is necessary, because it is likely that samples will find their way into clinics.[16] Interestingly, some institutions provide exemptions from the sample bans when a pharmacist in the clinic is responsible for ensuring compliance.[14]

ALTERNATIVES TO SAMPLES

Coupons or voucher systems have been proposed as an alternative to samples.[14,16] Voucher systems rely on prescribers to issue coupons to patients who then present the coupons along with their prescriptions to the pharmacy of their choice. With the information provided on the voucher, prescriptions are paid for through on-line pharmacy claims. Medications are dispensed to patients fully labeled and with counseling.

Medications for indigent patients may be obtained through pharmaceutical company medication assistance programs.[29] These programs provide brand name medications to patients based on financial need. Each company determines the eligibility criteria for its program. The application processes and the amount of medication supplied vary. While medication assistance programs help thousands of patients obtain medications, it requires cumbersome paperwork and frequent reapplication, and there can be considerable delays in getting medication to patients.[30] Like sampling, medication assistance programs are limited in their ability to help indigent patients.

CONCLUSION

Drug sampling is a controversial marketing technique used to promote pharmaceuticals. Pharmacists should be encouraged to get involved in efforts to promote safe and appropriate use of samples and ensure control and security. Clinics and institutions should have and enforce policies and procedures for managing samples. Controlling which samples will be requested and ensuring appropriate labeling, documentation screening for drug interactions, and patient education will help improve the use of drug samples.

REFERENCES

1. Pharmaceutical Research and Manufacturers of America. *Backgrounders and Facts: Marketing and Promotion of Pharmaceuticals*; Available at: http://www.phrma.org/publications/documents/backgrounders//2000-10-23.184.phtml (accessed January 8, 2001).
2. Wazana, A. Physicians and the pharmaceutical industry: Is a gift ever just a gift? JAMA, J. Am. Med. Assoc. **2000**, *283* (3), 373–380.
3. Backer, E.L.; Lebsack, J.A.; Van Tonder, R.J.; Crabtree, B.F. The value of pharmaceutical representative visits and medication samples in community-based family practices. J. Fam. Pract. **2000**, *49* (9), 811–816.
4. Haxby, D.G.; Rodriguez, G.S.; Zechnich, A.D.; Schuff, R.A.; Tanigawa, J.S. Manufacturers' distribution of drug samples to a family medicine clinic. Am. J. Health-Syst. Pharm. **1995**, *52*, 496–499.
5. Morelli, D.; Koenigsberg, M.R. Sample medication dispensing in a residency practice. J. Fam. Pract. **1992**, *34* (1), 42–48.
6. Wolf, B.L. Drug samples: Benefit or bait? [letter]. JAMA, J. Am. Med. Assoc. **1998**, *279* (21), 1698–1699.
7. *Prescription Drug Marketing Act of 1987, 102 Stat. 95*; 1987.
8. Joint Committee on Accreditation of Healthcare Organizations. *Pharmacy FAQs*; Available at: http://www.jcaho.org/standards_frm.html (accessed on January 8, 2001).
9. Chew, L.D.; O'Young, T.S.; Hazlet, T.K.; Bradley, K.A.; Maynard, C.; Lessler, D.S. A physician survey of the effect of drug sample availability on physicians' behavior. J. Gen. Intern. Med. **2000**, *15* (7), 478–483.
10. Shaughnessy, A.F.; Bucci, K.K. Drug samples and family practice residents. Ann. Pharmacother. **1997**, *31*, 1296–1300.
11. U.S. Senate Committee on Labor and Human Resources. *Hearings on Advertising, Marketing, and Promotional Practices of the Pharmaceutical Industry. 101st Cong., 2nd Sess., December 11 and 12*; 1990.
12. Weary, P.E. Free drug samples. Use and abuse. Arch. Dermatol. **1988**, *124*, 135–137.
13. Bastiaens, L.; Chowdhury, S.; Gitelman, L. Medication samples and drug compliance. Psychiatr. Serv. **2000**, *51* (6), 819.
14. Page, L. More clinics ban drug samples, citing cost, safety concerns. Am. Med. News **2000**, *43* (39), 1–2.
15. Institute for Safe Medication Practices. Safety briefs. ISMP Medication Safety Alert! **1996**, *1* (5), 1.
16. Institute for Safe Medication Practices. Sample medica-

tions: Safe management is a difficult but necessary process. ISMP Medication Safety Alert! **1999**, *4* (14), 1.
17. Dill, J.L.; Generali, J.A. Medication sample labeling practices. Am. J. Health-Syst. Pharm. **2000**, *57*, 2087–2090.
18. Westfall, J.M.; McCabe, J.; Nicholas, R.A. Personal use of drug samples by physicians and office staff. JAMA, J. Am. Med. Assoc. **1997**, *278* (2), 141–143.
19. Limaye, A.P.; Paauw, D.S. Personal use of drug samples by physicians and office staff [letter]. JAMA, J. Am. Med. Assoc. **1997**, *278* (19), 1568–1569.
20. Tong, K.L.; Lien, C.Y. Do pharmaceutical representatives misuse their drug samples? Can. Fam. Physician **1995**, *41*, 1363–1366.
21. O'Young, T.; Hazlet, T.K. Removal of drug samples from two teaching institutions. Am. J. Health-Syst. Pharm. **2000**, *57*, 1179–1180.
22. Donohoe, M.T.; Matthews, H. Wasted paper in pharmaceutical samples. N. Engl. J. Med. **1999**, *340* (20), 1600.
23. Pai, M.P.; Graci, D.M.; Bertino, J.S., Jr. Waste generation of drug product samples verses prescriptions obtained through pharmacy dispensing. Pharmacotherapy **2000**, *20* (5), 593–595.
24. Storrs, F.J. Drug samples. A conflict of interest? Arch. Dermatol. **1988**, *124*, 1283–1285.
25. Sigmon, J.L.; Anastasio, G.D. Drug sample closet. J. Fam. Pract. **1992**, *34*, 262–263.
26. McSherry, T.J.; Tonnies, F.E. Dispensing medication samples by the pharmacist in an institutional setting. J. Am. Coll. Health **1992**, *40*, 136–235.
27. Erickson, S.H.; Cullison, S. Closing the sample closet. Fam. Pract. Manag. **1995**, 43–47.
28. STFM Group on the Pharmaceutical Industry in Family Medicine. *Guidelines for Residency Program Relationships with Pharmaceutical and Other Proprietary Companies*; Society of Teachers of Family Medicine: Kansas City, Missouri, 1994.
29. Pharmaceutical Research and Manufacturers of America. *Directory of Prescription Drug Patient Assistance Programs*; Available at: http://www.phrma.org/searchcures/dpdpap/ (accessed January 8, 2001).
30. Prutting, S.M.; Cerveny, J.D.; MacFarlane, L.L.; Wiley, M.K. An interdisciplinary effort to help patients with limited prescription drug benefit afford their medication. South. Med. J. **1998**, *91* (9), 815–820.

Economic Evaluations of Clinical Pharmacy Services (ACCP)

American College of Clinical Pharmacy
Kansas City, Missouri, U.S.A.

INTRODUCTION

The objectives of this effort were to summarize and critique original economic assessments of clinical pharmacy services published from 1988–1995, and to make recommendations for future work in this area. A literature search was conducted to identify articles that were then blinded and randomly assigned to reviewers to confirm inclusion, abstract information, and assess the quality of study design. The 104 articles fell into four main categories based on type of service described: disease state management (4%), general pharmacotherapeutic monitoring (36%), pharmacokinetic monitoring services (13%), and targeted drug programs (47%). Articles were categorized by type of evaluation; 35% were considered outcome analyses, 32% outcome descriptions, and 18% full economic analyses. A majority (89%) of the studies reviewed described positive financial benefits from the clinical services evaluated; however, many (68%) did not include the input costs of providing the clinical service as part of the evaluation. Studies that were well conducted were most likely to demonstrate positive results. Commonly, results were expressed as net savings or costs avoided for a given time period or per patient. Seven studies expressed results as a benefit : cost ratio (these ranged from 1.08 : 1 to 75.84 : 1, mean 16.70 : 1). Overall, this body of literature contains a wealth of information pertinent to the value of the clinical practice of pharmacy. Future economic evaluations of clinical pharmacy services should incorporate sound study design and evaluate practice in alternative settings.

In 1989, the American College of Clinical Pharmacy (ACCP) published a position statement entitled ''Prospectus on the Economic Value of Clinical Pharmacy Services.''[1] The document summarized literature published prior to 1988 that supported the economic value of clinical pharmacy services and as such provided a resource to the profession in efforts to advance the clinical practice of pharmacy. A similar review was published in 1986.[2] These papers have proved to be valuable indexes of the literature and have been referred to by many in the profession on points pertinent to the economic value of clinical pharmacy.

In the time that has passed since the original ACCP prospectus, the literature has continued to grow in both depth and breadth of evidence supportive of the financial justification of clinical pharmacy services. New service models and philosophies of practice have developed in the past 6 years, the most notable being that of pharmaceutical care.[3] In addition, our ability to evaluate scientifically and measure the impact of clinical services on costs and outcomes has matured with the increased understanding and use of analytical techniques in health economics and pharmacoeconomics.[4,5] The effect of these advances on the quality and quantity of literature is unknown. The ACCP Board of Regents thus asked the ACCP Publications Committee to update this prospectus.

The committee reviewed, summarized, and critiqued the literature published between January 1988 and December 1995 that included original economic assessment of clinical pharmacy services or programs, thereby serving to update the original position statement of ACCP. Further intentions were to provide a barometer of the degree to which accepted techniques of economic analysis have been incorporated into this literature, and to make recommendations for future work in this area.

METHODS

A search of two major data bases (MEDLINE, International Pharmaceutical Abstracts) was conducted to identify articles published between January 1988 and December 1995. The beginning date of January 1988 was selected because the original ACCP prospectus was inclusive through December 1987. Both MeSH and free text search terms were used to identify English language articles assessing the value of clinical pharmacy services. Search terms were clinical pharmacy services, pharmacy

Encyclopedia of Clinical Pharmacy
DOI: 10.1081/E-ECP 120006412

services, program, economic evaluation, cost justification, cost, cost effectiveness, cost–benefit, cost analysis, cost–consequence analysis, and cost–utility analysis. Review articles, editorials, and other unoriginal reports were excluded from the search. All citations identified were screened for inclusion by review of titles and abstracts. Those articles for which abstracts were not available from the computerized databases were collected manually and screened for inclusion.

Inclusion criteria were English language, original evaluation, publication between January 1988 and December 1995 inclusive, assessment of a clinical pharmacy service (defined as patient-level interaction, and not including policy-type interventions unless accompanied by a patient-level interaction), and some economic assessment. Exclusion criteria were reviews, editorials, and letters, and studies published in abstract form only. All papers suspected of meeting the inclusion criteria were submitted to full review. In addition, the authors examined personal files, and a secondary search of the titles of articles cited in papers meeting the inclusion criteria was conducted. Papers identified through this search were again collected and screened for inclusion, and added to the set of papers subjected to full review.

In the full review process, a modified block randomization scheme was used to confirm inclusion and to abstract information and assess the quality of each article. Each paper was randomly assigned to two of four reviewers. Reviewers were blinded to original authors' names, affiliations, and journal of publication. Reviews were recorded on a standard case report form and entered into a database for analysis. Discrepancies between reviewers were arbitrated by group consensus. Reviewers first made a final check of inclusion and exclusion criteria to exclude further any nonapplicable articles. Reviewers recorded the study setting, objectives, methods, results, and any additional comments.

Each article was assessed for the type of evaluation and categorized (Table 1). Two factors were considered in determining the type of evaluation: the presence of two or more alternatives, and the consideration of both input (costs) and outcomes. Evaluations that included two or more alternatives (i.e., concurrent control group, historical control, preintervention and postintervention design) were considered true analyses, whereas those that did not include a comparison were labeled descriptions. A description of the type of analysis was assigned to the evaluation and included the options of cost or outcome description, cost or outcome analysis, cost and outcome description, and true clinical economic evaluation. Those articles considered true clinical economic evaluations were subcategorized by type, options including cost-minimization analysis, cost–benefit analysis, cost-effectiveness analysis, and cost–utility analysis.[6]

Descriptive statistics were used to profile and characterize the articles within each data field abstracted by the reviewers, including the type of clinical service performed, the site of the study or evaluation, and the type of analysis performed.

RESULTS

The results of the search and screen process used are illustrated in Fig. 1. A total of 575 articles were found through the original search. A preliminary review of the abstracts of these articles identified 444 that did not involve the justification of clinical pharmacy services, and these were deleted from the set. Seven articles were added from the files of the authors, and 46 were identified through the secondary search of the articles found. Thus, 184 articles were subjected to full review. During full review, an additional 80 articles were found that did not meet the inclusion criteria: 44 did not review a clinical

Table 1 Criteria for assessing type of analysis

		Were both cost and outcomes considered?	
		No	**Yes**
Were two or more alternatives considered?	No	Cost description or outcome description	Cost and outcome description
	Yes	Cost analysis or outcome analysis	True clinical economic analysis Subcategories Cost-minimization analysis Cost–benefit analysis Cost–effectiveness analysis Cost–utility analysis

(Adapted from Ref. [6].)

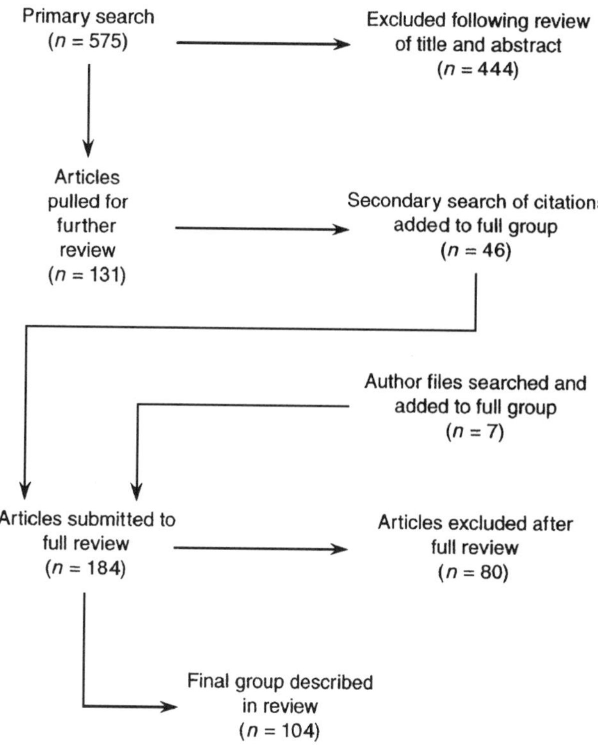

Fig. 1 Literature search method and results.

pharmacy service, 20 did not describe original work, and 16 failed on both points. An analysis of the final set of 104 articles is shown in Appendix 1.[7–110]

Articles are sorted in Appendix 1 by the type of clinical pharmacy service described in the evaluation. Four major categories were used in grouping articles by type of clinical pharmacy service: 1) disease state management, defined as clinical pharmacy services primarily directed at patients with a specific disease state or diagnosis; for example, a renal dosing program; 2) general pharmacotherapeutic monitoring, defined as clinical pharmacy services that encompass a broad range of activities based primarily on the needs of a geographically assigned group of patients; services provided may include patient drug regimen review, adverse drug reaction monitoring, drug interaction assessment, formulary compliance, or rounding with physicians; 3) pharmacokinetic monitoring services, defined as clinical pharmacy services that primarily involve evaluation of anticipated or actual serum drug concentrations and provision of subsequent dosing recommendations; and 4) targeted drug programs, defined as clinical pharmacy services that are primarily focused on a single drug or class of drugs and include predefined guidelines for provision of alternative therapy or dosing recommendations; for example, recommended switch from intravenous to oral administration of histamine$_2$-receptor antagonists (H_2RAs). Because of the number of articles describing targeted drug programs, those articles are further subcategorized in Appendix 1 based on the class of drug involved.

Provided in Appendix 1 are the following data for each article: 1) reference number; 2) the setting in which the evaluation was conducted; 3) a summary of the primary intent or objective; 4) a description of the analytical method of the evaluation; 5) number and type of alternatives included in the evaluation; 6) input cost components included in the evaluation; 7) outcomes evaluated; 8) a summary of the main results of the evaluation; and 9) miscellaneous comments about the evaluation made by the reviewer.

Articles from pharmacy-based journals dominated the set of articles. The most common journal source was the *American Journal of Health-System Pharmacy* ($n = 32$, 30%). *DICP/Annals of Pharmacotherapy, Hospital Pharmacy*, and *Hospital Formulary* were also common ($n = 19$, $n = 15$, and $n = 7$, respectively). Several foreign journals also provided articles.

The most common type of pharmacy service was targeted drug programs ($n = 49$, 47%). The specific drug classes described in targeted drug programs were most likely to be antimicrobials ($n = 27$) or H_2RAs ($n = 17$). Articles classified as general pharmacotherapeutic monitoring made up 36% ($n = 38$), pharmacokinetic monitoring services 13% ($n = 13$), and disease state management 4% ($n = 4$).

Table 2 summarizes the settings of the studies included in this evaluation. The settings of most studies were university or community hospitals ($n = 33$ and $n = 25$, respectively). University-affiliated community hospitals and government hospitals were also common ($n = 12$ and $n = 10$, respectively). Less common settings were ambu-

Table 2 Settings of cost-justification studies

Setting	Number of studies
University hospital	33
Community hospital	25
University-affiliated teaching community hospital	12
Government hospital	10
University-affiliated ambulatory clinic	8
Government-affiliated ambulatory clinic	5
Health maintenance organization clinic	4
Multicenter, multisite	3
Community pharmacy	2
University-affiliated government hospital	2

Table 3 Analytic methods of cost-justification studies[a]

Method	Number of studies
Outcome analysis	37
Outcome description	33
Economic analysis	19
Cost and outcome description	13
Cost analysis	1
Cost description	1

[a]Refer to Table 1 for classification analysis.

latory clinics of various affiliations, health maintenance organizations, and community pharmacies.

Table 3 summarizes the analytic methods used in the included articles. Although 19 (18%) articles were considered full economic analyses (by definition, considering two or more alternatives and measurement of both input costs and outcomes), most were less rigorous. The most common types of studies were outcome analyses ($n = 37$, 35%), which considered two or more alternatives but excluded consideration of the costs of providing the service, and outcome descriptions ($n = 33$, 32%), which failed to consider two or more alternatives and did not consider the cost of providing the service.

The study design of the included articles was further analyzed by individually considering the use of a comparison group (alternative) and by the types of input costs and outcomes measured. Sixty-one (59%) studies included a comparison group, whereas 43 (41%) did not and were therefore considered to be descriptive. The study designs used in papers that had a comparison group were a concurrent control group ($n = 21$), a historical control group ($n = 10$), and preintervention and postintervention groups ($n = 30$). Precontrols and postcontrols were differentiated from historical control designs in the temporal relationship to the intervention. If a study compared measurements taken immediately prior to an intervention and immediately after, it was coded as a pre/post design. If a longer period of time elapsed between comparison groups (e.g., comparing data from the study period to the same month 1 year earlier), it was defined as a historical control.

Seventy-one studies (68%) did not evaluate the cost of providing the clinical service as part of the economic evaluation of that service. Most commonly, costs were considered as an outcome or consequence of the service (i.e., as in drug costs avoided) rather than as an input (i.e., as in the investment required to establish and maintain the program under study). Of the 33 (32%) studies that did consider some input costs, the most common cost assessed was personnel ($n = 25$). In these cases, the costs of the program under study were quantified in terms of salary and/or benefits associated with providing the program or service. Some studies used charges (i.e., hospital room, emergency room) rather than true costs.

Outcomes or consequences of the services described were considered in all the articles. The most common ($n = 80$, 77%) outcome measured was drug costs avoided (i.e., the impact of the program on reducing use or cost of a particular drug). Other nonfinancial outcomes were also measured, including length of hospital stay ($n = 14$, 13%), use of nonpharmaceutical resources, rates of adverse drug reactions, frequency of pharmacist-driven therapeutic interventions, and qualitative changes in prescribing patterns. True clinical patient outcomes were considered in few studies.

Ninety-three (89%) of the articles described beneficial financial impact of the clinical pharmacy service described. Many provided either gross cost savings or, in those that did consider input costs, net savings. Of the 33 studies that considered input costs, 31 (94%) demonstrated positive findings. Results of these were presented a number of different ways (Table 4).

Commonly these articles expressed net savings on an annual basis or for the time period of the study. For example, a study in 1992 described annual net cost savings of $221,056 for clinical pharmacy services provided in an ambulatory care clinic.[25] It did not, however, include a control group. In other cases, savings were expressed per patient admission or per patient-day. In 1993, a well-conducted and controlled evaluation described an average net savings of $377 per patient admission as a result of clinical pharmacists assigned to selected inpatient medical services.[14]

In seven articles, results were expressed as benefit:cost ratios. They differed in type of clinical pharmacy service, site of provision of service, and resources invested in the service (Table 5). Nevertheless, the results were impress-

Table 4 Studies that considered input costs of providing service

Method of expressing results	References[a]
Net savings annualized or for time period of study	[8,9,11,18,20,24,25,31,36, 45,51,53,55,68,79,82,91, 94,98,104,110]
Net savings/patient-day or patient admission	[13–15,20,38,52,60,71]
Benefit : cost ratio	[11,14,15,41,51,60,98],
Other	[10,29]

[a]References may be listed more than once if results were expressed in different formats.

Table 5 Studies allowing calculation of benefit:cost ratio

Setting	Clinical service	Objective	Benefit:cost ratio
University hospital[11]	Pharmacotherapeutic monitoring	To examine cost benefit of clinical pharmacy intervention and documentation system	1.98:1
Government hospital[14]	Pharmacotherapeutic monitoring	To study effect of clinical RPh on health care outcomes	6.03:1
HMO clinic[15]	Pharmacotherapeutic monitoring	To measure impact of pharmaceutical services on overall health care costs and to estimate RPh productivity	3.2:1
University hospital[41]	Pharmacotherapeutic monitoring	To evaluate impact of clinical pharmacy service on hospital costs using cost–benefit analysis	1.08:1 and 1.59:1
University-affiliated[51] community hospital	Pharmacokinetic monitoring	To determine cost benefit of pharmacokinetic services for patients receiving aminoglycosides	75.84:1 and 52.25:1
University hospital[60]	Pharmacokinetic monitoring	To evaluate impact of computer-assisted aminoglycoside dosing	4.09:1
HMO clinic[98]	Target drug program	To evaluate impact of clinical RPh intervention program on cost of H_2RA therapy	4.3:1

HMO, health maintenance organization; H_2RA, histamine$_2$-receptor antagonist.

ively positive, with calculated benefits to cost ranging from 1.08 : 1 to 75.84 : 1 (mean 16.70 : 1).

DISCUSSION

Assessment of the Literature

The conclusions drawn from our review and evaluation of literature assessing the economic value of clinical pharmacy services published from 1988–1995 are multifocal. The total number of articles published on this topic has grown, as demonstrated by the number in this review (104, average 13/yr) versus the original prospectus (58, average 4/yr), which included articles published from 1974–1987. Although the number of published articles on this topic appears sufficient, an opportunity does exist for improvement in the quality of study design.

A large percentage (41%) of the articles we reviewed did not include a comparison group. They did not incorporate a study design that would allow one to control variance, which therefore makes it difficult for the reader to confirm the validity or extrapolate the results to other practice settings. This is not to say that these articles are without value, however. Many are excellent descriptive reports that provide insight and experience from which others may learn.

Sixty-eight percent of studies did not consider the costs associated with providing clinical pharmacy services as a factor in the economic evaluation or justification of that service, thus making it difficult to demonstrate true economic justification of the service. For those studies that did consider some input costs, personnel costs were often singularly included, with nonlabor costs (i.e., overhead) being omitted. Furthermore, when charges were used, they were often misinterpreted as costs.

The outcomes measured tended to focus on financial consequences and not to include clinical or patient consequences. Without consideration of clinical outcomes, or without being able to make an assumption that clinical outcomes are unchanged, the true economic impact of the services studied could not be proved.

Despite the limitations of many of the articles as true economic evaluations, this literature contains a wealth of information pertinent to the clinical practice of pharmacy that serves to document innovative and successful experiences and programs. Of importance, we did find that when studies were well conducted (considered true economic evaluations), the results were likely to be favorable; that is, the studies were able to demonstrate net savings or positive benefit : cost ratios. Because of lack of standardization in reporting of results and variability in study design, it is difficult to make a general statement as to the degree of benefit derived from clinical pharmacy services. However, we were able to abstract calculated benefit : cost ratios from the seven applicable studies and describe a range of value from 1.08:1 to 75.84:1 (mean 16.70:1). In other words, for every dollar invested in clinical services, on average $16.70 was saved.

These seven studies were conducted in a variety of practice environments—university hospitals (3), university-affiliated community hospital (1), governmental hospital (1), and health maintenance organization clinics (2). They evaluated a spectrum of pharmacist-delivered services including pharmacotherapeutic monitoring (4), pharmacokinetic monitoring (2), and targeted drug programs (1). Both of these considerations speak to what we believe to be the broad applicability of the studies' results.

Limitations

We undertook this review and evaluation with the intent of providing the reader a resource to access original literature published assessing the economic value of clinical pharmacy services, and to evaluate the quality of that literature. The articles included in this review represent only those published in standard literature. We did not consider unpublished studies and therefore our results may be subject to inherent publication bias (so-called ''file drawer'' effect). We included only articles that contained some consideration of the financial impact of clinical pharmacy services. Certainly, many useful articles describe and evaluate clinical pharmacy services, but focus on nonfinancial outcomes and impact, and are worthy of review. Finally, our review of the literature, although intended to be systematic and thorough, may not have captured all the published literature on this topic.

Recommendations

Having reviewed and evaluated the published literature on the economic value of clinical pharmacy services, we make the following recommendations to clinicians, investigators, authors, reviewers, and journal editors:

1. Future economic evaluations should incorporate sound methodology and study designs. Study designs should control for variance by using a comparison group such as a historical control, concurrent control, or pre- and postintervention measurement.
2. Consideration should be given to the input costs, that is, the costs of providing the service, as part of the economic evaluation. These costs should include direct and indirect costs if possible. Where charges are used, they should be appropriately labeled and interpreted as such.
3. Outcome measurements should include more than just drug costs avoided. Nonfinancial outcomes such as clinical patient outcomes are important and should be part of the evaluation of any service that affects patient care. Using a disease state management approach rather than the targeted drug approach to cost justification may help to identify important outcome measurements that should be considered.
4. The concept of opportunity costs (i.e., money spent on one resource that cannot be spent for other purposes) should be explored. The value of any given service should be weighed against the possible services that might be provided. The concept of opportunity costs becomes even more important as health care downsizing and restructuring occur.
5. Clinical pharmacy services provided in settings outside the traditional hospital should be included in future economic evaluations.

CONCLUSION

It is hoped that the data summarized in this article will assist individual pharmacists, departmental managers, and health system administrators to document and recognize the cost effectiveness of pharmacists' clinical services. Pharmacy practitioners should take pride in both the quantity and strength of this literature, and feel empowered to use it to justify further expansion or refinement of their caregiving responsibilities. Attention to our recommendations regarding the design and performance of future economic evaluations of clinical pharmacy services will further add to the strength of this literature and the conclusions that may be drawn from it.

ACKNOWLEDGMENTS

Members of the 1995 and 1996 Publications Committee of the American College of Clinical Pharmacy were Brian Alldredge, Guy Amsden, Douglas Anderson, Edward Bednarczyk (Chair, 1995), S. Diane Goodwin (Chair, 1996), Linda Jaber, David Knoppert, Bruce Mueller, Michael Otto, Therese Poirier, Jay Rho, Richard Scheife, Glen Schumock, Maureen Smythe, Wilkinson Thomas, Dennis Thompson, Donald Uden, and Eva Vasquez.

Endorsed by the ACCP Board of Regents on August 2, 1996.

From Schumock GT, Meek PD, Ploetz, PA, Vermeulen LC. Economic evaluations of clinical pharmacy services: 1988–1995. *Pharmacotherapy* 1996, 16(6): 1188–1208, with permission of the American College of Clinical Pharmacy.

Appendix 1 Evaluations of economic value of clinical pharmacy services—1988–1995

Setting	Objective (as stated by authors)	Analytic method	Comparison group	Input costs	Outcomes included	Results measured	Comments
Disease state management							
CH[7]	To evaluate impact of benzodiazepine guidelines on cost and quality of care of patients hospitalized for alcohol withdrawal	OA	Control group	None	DCA, LOS	Mean drug cost decreased from $1008/day to $59/day/patient; mean ICU LOS decreased from 4.1 to 1.1 days	Input costs not considered
UH[8]	To evaluate impact of clinical RPh on cost savings and patient outcome in asthma clinic	CBA	Historical control	Cost of clinic visit offset other savings	Cost of emergency room visits for asthma exacerbation	Cost savings $30,693 and $68,393 between study period and each of two control periods; savings derived from reduction in ER visits	Drug costs not considered; economic value of clinical outcomes (beyond ER visits) not assessed; no ratio calculated
CH[9]	To evaluate impact of renal function monitoring program, focusing on appropriate dosages of renally eliminated agents	COD	None	Personnel costs	DCA	Cost savings $5040 noted, with program cost $2700 for labor	No control group; clinical outcomes not considered; measured only what the cost of therapy would have been without intervention
UACH[10]	To conduct time and motion analysis of PCA vs. i.m. analgesia and evaluate impact on cost and quality of pain control	CBA	Historical control	Costs of drug, RPh, and nursing labor	LOS, cost of ADRs, quality of analgesia	Quality of analgesia increased with PCA, but so did cost and time required	Evaluated both RPh and nursing time; did not provide ratio
General pharmacotherapeutic monitoring							
UH[11]	To examine cost benefit of clinical pharmacy intervention and documentation system	COD	None	Personnel costs	DCA, type of intervention	Cost savings of $1.98/$1 invested, with total annual savings $7100	Missing relevant costs and outcomes

(*Continued*)

Appendix 1 Evaluations of economic value of clinical pharmacy services—1988–1995 (*Continued*)

Setting	Objective (as stated by authors)	Analytic method	Comparison group	Input costs	Outcomes included	Results measured	Comments
CH[12]	To assess the quality and cost avoidance of RPh interventions using physician assessors	OD	None	None	DCA, LOS	Positive impact on patient care, estimated reduced LOS by 3.7 days	Physician reviewers estimated reduction in LOS resulting from interventions
UH[13]	To cost justify clinical pharmacy service on general surgery team	COD	None	Personnel costs	DCA, type of intervention, clinical impact of intervention	Positive impact on outcomes; net cost avoidance of $441.46/patient	Small sample
GH[14] (Army)	To study effect of clinical RPh on health care outcomes	CBA	Control group	Personnel costs	LOS, drug costs/ admission	Average net savings $377/patient admission; cost : benefit ratio 6.03 : 1	Control group included
HMOC[15]	To measure impact of pharmaceutical services on overall health care costs, and to estimate RPh productivity	COD	None	Personnel costs, direct costs, overhead	Percentage of problematic drugs, use of service, DCA	Average total cost savings $644/patient; cost : benefit ratio 3.2 : 1	
GH[16] (VA)	To evaluate clinical RPh recommendations on number and costs of drugs	OD	Control group	None	DCA	Decreased average monthly drug cost/patient	Input costs not considered
UACH[17]	To describe program and determine cost savings from clinical pharmacy services provided in rehabilitation clinic	OD	None	None	DCA	Reduced hospital drug costs by $2700 during 6-mo study	Input costs not considered
CH[18]	To evaluate clinical pharmacy services and determine cost savings and justification for additional pharmacy staff	COD	None	Personnel costs	DCA	Annual net savings $25,862	
CH[19]	To evaluate impact of a clinical coordinator on costs avoided by the institution from clinical clinical intervention program	OA	Pre/post	None	DCA, NOI	Average monthly net savings $3739 and $4644 before and after clinical coordinator	

(*Continued*)

E

Appendix 1 Evaluations of economic value of clinical pharmacy services—1988–1995 (*Continued*)

Setting	Objective (as stated by authors)	Analytic method	Comparison group	Input costs	Outcomes included	Results measured	Comments
UH[20]	To describe interventions made by clinical RPh and evaluate cost savings and cost avoidance impact	COD	None	Personnel costs	DCA, NOI	Cost savings of $69.11/patient-day; annual net savings $300,079	
UH[21]	To compare cost and quality of decentralized vs. centralized pharmaceutical services	OA	Pre/post	None	LOS, total cost/admission	Decreased average total cost/admission by $1293; decreased average pharmacy cost/admission by $155 for decentralized	
CP[22]	To examine value of clinical pharmacy intervention program in a community pharmacy setting and determine economic value	OD	None	None	DCA, NOI	Cost avoided of $3.47/prescription processed	
UACH[23]	To describe program to develop clinical pharmacy staff and determine cost avoidance to hospital resulting from the service	OD	None	None	DCA	Average estimated cost avoidance $9306/mo over 5 yrs	Input costs not considered
UH[24]	To evaluate and document impact of clinical RPh on costs avoided at tertiary care teaching hospital	COD	None	Personnel costs	DCA	Net annualized cost avoidance $897,350	
UAAC[25]	To evaluate impact of clinical RPh on cost and quality of patient care in ambulatory care clinics	COD	None	Personnel costs	DCA	Net annualized cost avoidance $221,056	Emphasized need for documenting interventions
UH[26]	To evaluate impact of clinical RPh on medical team	OD	None	None	Interventions documented	27% of interventions prevented serious effects	Input costs not considered

(*Continued*)

Appendix 1 Evaluations of economic value of clinical pharmacy services—1988–1995 (*Continued*)

Setting	Objective (as stated by authors)	Analytic method	Comparison group	Input costs	Outcomes included	Results measured	Comments
MC, CH, MHF, SNF[27]	To evaluate impact of reactive clinical pharmacy interventions on cost and quality of patient care	OD	None	None	Cost impact of interventions documented	2.9% of pharmacy interventions prevented potential medical harm; limited cost impact	Input costs not considered; physicians assessed RPh service, introducing potential bias
GH[28] (VA)	To evaluate daily data collection of decentralized clinical pharmacy services	OD	None	None	DCA	Total savings $126,504 due to 2506 interventions provided	Input costs not considered; clinical outcomes not considered; no comparative group used to assess cost and outcome difference
GAAC[29]	To evaluate impact of clinical RPh's interventions on physician prescribing and costs in an ambulatory clinic	CBA	Control group	Personnel costs	Cost avoidance due to reduced number of prescriptions	Cost avoidance $4.63 for intervention group vs. $1.10 in control group; savings in prescription filling labor noted; labor costs associated with program offset by DCA	Clinical outcomes not considered; no ratio presented
UAAC[30]	To evaluate impact of ambulatory clinical pharmacy program and to justify personnel for the program	OD	None	None	Cost avoidance in drug and laboratory use	$19,000 in cost reduction for interventions, 184 patients; documented clinical outcomes after interventions	Discussed cost of personnel required for program, but did not factor cost into analysis; no comparison group for analysis
GAAC[31] (VA)	To evaluate impact of clinical RPh on cost and quality of patient care	CBA	Pre/post	Costs associated with program and dispensing prescriptions generated in the clinic	DCA	Total cost decrease of $22,241 during study period	Charts assessed for quality based on the rate of suggestion implementation, but actual patient outcomes not assessed
UACH[32]	To evaluate cost impact of clinical RPh in intensive care unit	COD	None	Personnel costs	DCA	Cost savings $10,010 (Canadian) documented over 3-mo study period; cost:benefit ratio 4 : 1	No control group; measured only what the cost of therapy would have been without intervention

(*Continued*)

Appendix 1 Evaluations of economic value of clinical pharmacy services—1988–1995 (*Continued*)

Setting	Objective (as stated by authors)	Analytic method	Comparison group	Input costs	Outcomes included	Results measured	Comments
UH[33]	To evaluate impact of pharmacy faculty providing clinical pharmacy interventions on drug costs and pharmacy department revenue	OD	None	None	DCA and service revenue generated	Impact of 278 interventions evaluated, demonstrating drug cost avoidance $1661, generation of $6000 in revenue from pharmacokinetic consultations	No control group; measured only what the cost of therapy would have been without intervention
GAAC[34] (VA)	To evaluate impact of clinical RPh on drug prescribing and cost savings	CBA	Control group	Personnel costs	DCA	Decreased total number of prescriptions and associated ADRs; total cost of prescriptions filled in study period $3872 less than during control period; total cost to administer program $2250	No ratio presented; mentioned but did not quantify value of prevented ADRs
CH[35]	To evaluate impact of documentation system for clinical pharmacy services	OD	None	None	DCA	Cost avoidance ranged $2341–$7762/quarter during study	Input costs not considered; no control group; clinical outcomes not considered
CH[36]	To evaluate cost impact of implementing clinical pharmacy services in intensive care unit	COD	None	Personnel costs	DCA	During 32 days, cost avoidance $1651, labor cost associated with program was $2599	No control group; clinical outcomes not considered; small sample size (number of pilot days assessed, and short period of time/day)
MC, UH[37]	To evaluate acceptance and cost savings resulting from 2-yr postbaccalaureate PharmD student interventions	OD	None	None	NOI, DCA, laboratory cost avoidance	Estimated annual drug savings $3891	Input costs not considered

(*Continued*)

Appendix 1 Evaluations of economic value of clinical pharmacy services—1988–1995 (*Continued*)

Setting	Objective (as stated by authors)	Analytic method	Comparison group	Input costs	Outcomes included	Results measured	Comments
CH[38]	To determine cost savings of clinical pharmacy service in a community hospital	CD	None	Personnel costs	DCA	Savings of $1.49/patient/day for clinical pharmacy services	Brief description of daily documentation activity to demonstrate cost savings
CH[39]	To describe impact of general clinical pharmacy interventions on hospital costs	OD	None	None	Physician acceptance, NOI, DCA	Total savings $15,525.81	Input costs not considered
CH[40]	To evaluate impact of comprehensive clinical pharmacy services on hospital costs	OA	Pre/post	None	DCA	Net cost savings $34.10/RPh-day	Input costs not considered; clinical outcomes not considered
UH[41]	To evaluate impact of clinical pharmacy service on hospital costs using cost–benefit analysis	CBA	Historical control	Cost of providing service	DCA	Cost:benefit ratios 1.08 and 1.59 for 2 ward-based groups	Clinical outcomes not considered
CH[42]	To determine impact of clinical interventions on cost and quality of patient care	OD	None	None	Number of inappropriate laboratory tests, DCA	Annual drug cost avoidance of $26,580	
UH[43]	To evaluate impact of PharmD student interventions	OD	None	None	NOI, physician acceptance	Decreased drug costs by 50.7%	
UH[44]	To document interventions of clinical RPh in emergency department	OA	Pre/post	None	DCA	Description of clinical and cost-saving interventions	Input costs not considered; clinical outcomes not considered
UAAC[45]	To evaluate impact of clinical pharmacy interventions on cost and quality of patient care	COD	None	Personnel costs	Physician acceptance, DCA, various quality indicators	Annual extrapolated cost savings $19,076	Documented cost and quality using daily patient data collection forms
UAAC[46]	To determine impact of clinical RPh on cost savings to the hospital and quality of patient care	OA	Control group	None	NOI, DCA	RPhs saved $176,724 annually	Extrapolated savings from 2-wk pilot

(*Continued*)

Appendix 1 Evaluations of economic value of clinical pharmacy services—1988–1995 (*Continued*)

Setting	Objective (as stated by authors)	Analytic method	Comparison group	Input costs	Outcomes included	Results measured	Comments
CP[47]	To evaluate cost savings to pharmacy from interventions of community RPh	OD	None	None	Assessment of value of RPh interventions, cost of medical care avoided	Value of avoided care was $122.98/ intervention; $2.32 savings/prescription screened	
UAAC[48]	To evaluate impact of clinical RPh on cost and quality of patient care	OD	None	None	Physician acceptance, patient outcome indicators, DCA	205 interventions made during 6-mo study; 80.9% made to increase quality; 18.1% to increase quality and decrease cost	
Pharmacokinetic monitoring service							
CH[49]	To determine effect of TDM program on inappropriate sampling times	OD	None	None	Unnecessary samples, patient charges	Charge avoidance $500,000 annually	Input costs not considered; charges vs. costs
UH[50]	To evaluate impact of educational efforts on use of SDCs	OA	Pre/post	None	DCA, number of drug assays	Increased number of drug levels ordered; decrease of $599 in hospital costs	Increased rational ordering of serum drug concentrations
UACH[51]	To determine cost benefit of pharmacokinetic services for patients receiving aminoglycosides	CBA	Control group	Variable costs, personnel costs, fixed costs	LOS, clinical response	Decreased LOS; decreased duration of febrile period; benefit:cost ratio 75.84:1 and 52.25:1	
CH[52]	To determine physician acceptance and impact of clinical pharmacokinetic recommendations on cost and quality of patient care	CBA	Control group	Variable costs, personnel costs, fixed costs	Acceptance by physicians, LOS, DCA, clinical response	Decreased LOS; decreased febrile period; decreased direct costs; cost of service $85/patient	
CH[53]	To evaluate impact of clinical pharmacokinetic service on cost and quality of patient care	CBA	Control group	Variable costs, fixed costs	LOS, clinical response, patient charges	Decreased length of treatment; decreased LOS; annual cost savings $113,934	Used charges rather than costs
CH[54]	To evaluate costs associated with clinical pharmacokinetic dosing service	OA	Pre/post	None	LOS, DCA	Cost reduction $107,000 associated with decrease in LOS; reduction of $14,000 in drug costs associated with program	Mentioned but did not value cost of system

(*Continued*)

Appendix 1 Evaluations of economic value of clinical pharmacy services—1988–1995 (*Continued*)

Setting	Objective (as stated by authors)	Analytic method	Comparison group	Input costs	Outcomes included	Results measured	Comments
UH[55]	To evaluate impact of clinical RPh on appropriate serum drug concentration ordering	CBA	Historical control	Personnel costs	Cost of laboratory testing avoided	Increased appropriateness of serum drug concentration determination; cost of $1000 with savings of $3000	Clinical outcomes not considered; no ratio presented
UH[56]	To evaluate impact of pediatric pharmacokinetic service using guidelines as basis for appropriate monitoring	CA	Control group	None	Costs avoided through decrease in inappropriate monitoring	Annual cost avoidance $12,325 based on fewer inappropriate laboratory assays	Input costs not considered
CH[57]	To evaluate effectiveness of serum digoxin concentration monitoring, and determine cost impact of service	OD	None	None	NOI, timing of digoxin serum concentrations, laboratory costs avoided	Decreased number of digoxin serum drug concentrations ordered	Input costs not considered
UH[58]	To analyze need for therapeutic drug monitoring program for phenytoin	OA	Control group	None	Number and cost of drug assays, LOS and readmission rate	Overall cost savings after 1 yr of program $100.00	Charges vs. costs
UH[59]	To evaluate impact of therapeutic drug monitoring program for theophylline	OA	Control group	None	Number and cost of drug assays, LOS	Equal cost of RPh monitoring and savings after 1 yr	Charges vs. costs
UH[60]	To evaluate impact of computer-assisted aminoglycoside dosing	CBA	Control group	Service cost	LOS, room charge, DCA	$1311 savings/ patient in study group; CBA ratio of 4.09 : 1 in favor of study group	Used charges rather than costs
CH[61]	To compare RPh vs. physician dosing of aminophylline	OA	Control group	None	LOS, room charges, cost of concomitant drugs	Decreased LOS of 1.96 days; $490 savings/ patient in study groups	Used charges rather than costs
Target drug programs: Antiemetic agents							
UH[62]	To evaluate impact of prescribing guidelines for use of ondansetron on drug costs	OA	Pre/Post	None	DCA	15% reduction in amount of ondansetron dispensed from period before guideline implementation	Input costs not considered; clinical outcomes not considered

(*Continued*)

E

Appendix 1 Evaluations of economic value of clinical pharmacy services—1988–1995 (*Continued*)

Setting	Objective (as stated by authors)	Analytic method	Comparison group	Input costs	Outcomes included	Results measured	Comments
Target drug programs: Antihypertensives							
HMOC[63]	To evaluate impact of clinical RPh consultation on cost of antihypertensive therapy in HMO family practice clinic	OA	Control group	None	Average daily drug costs	Decreased drug costs of $20.61/ patient-year	Input costs not considered
Target drug programs: Antimicrobials							
UAAC[64]	To assess impact of fluconazole guidelines and concurrent RPh intervention	OA	Historical control	None	Appropriate use, ADRs, DCA	Annual cost avoidance $65,520	Input costs not considered
UACH[65]	To describe experience with program for modifying dosing regimens of mezlocillin	OD	None	None	DCA	Annual cost savings $33,000 or $49.47/patient	Input costs not considered
UH[66]	To document cost containment of RPh antibiotic streamlining program	OD	None	None	DCA	Annual cost savings $47,700	Input costs not considered
UH[67]	To evaluate educational and intervention program promoting use of metronidazole for antibiotic-associated colitis	OD	Historical control	None	DCA	Estimated annual savings $38,829 based on decreased drug costs	Input costs not considered; clinical outcomes not considered
CH[68]	To evaluate impact of therapeutic intervention to alter metronidazole dosing	COD	Pre/post	Personnel costs	DCA	Annual savings $28,000	Input costs not considered
GH[69] (VA)	To describe antibiotic monitoring program and determine costs avoided to hospital from rational antibiotic use	OD	None	None	DCA, appropriateness	Total cost avoidance $42,512 during study period	Input costs not considered
UH[70]	To evaluate impact of target drug monitoring program for clindamycin on hospital costs	OA	Historical control	None	DCA	Cost avoidance $16,000 annually	Input costs not considered

(*Continued*)

Appendix 1 Evaluations of economic value of clinical pharmacy services—1988–1995 (*Continued*)

Setting	Objective (as stated by authors)	Analytic method	Comparison group	Input costs	Outcomes included	Results measured	Comments
GH[71] (VA)	To evaluate impact of clinical RPh monitoring on i.v. ceftriaxone use (conversion to oral cefpodoxime)	CBA	Control group	Cost of treatment	Cost of treatment outcome	Cost savings $46.05/patient achieved, 1-day decrease in LOS	Input costs not considered; small sample
UH[72]	To evaluate antimicrobial management program and evaluate impact on cost and quality of patient care	OA	Historical control	None	DCA	Gross savings in antibiotic acquisition cost $483,032/yr	Cost associated with service considered, but not quantified
GH[73]	To evaluate cost impact of two DUE activities performed by undergraduate pharmacy students	OD	Historical control	None	DCA	Cefazolin dosing modification (q6h to q8h) resulted in savings of $18,000; substitution of metronidazole for clindamycin saved $21,000	Input costs not considered; clinical outcomes not considered
UH[74]	To evaluate cost impact of pharmacy-based antibiotic optimization program	OA	Pre/post	None	DCA	Savings of $12,640 realized after program implementation	Input costs not considered; clinical outcomes not considered
GH[75] (State)	To evaluate impact of RPh participating in patient care rounds on costs associated with antimicrobial drug use	OA	Pre/post	None	DCA	Cost reduction of $29,800 greater in study period vs. prestudy period	Input costs not considered
UACH[76]	To evaluate impact of clinical RPh-based antibiotic management program	OA	Control group	None	Drug and ancillary cost avoidance	Estimated cost savings $40,000 associated with drug cost avoidance and appropriate use of laboratory data	Input costs not considered; clinical outcomes not considered
UACH[77]	To evaluate impact of renal function monitoring program, focusing on appropriate dosages of imipenem	OD	None	None	DCA	Potential to save $11,500 annually by adjusting imipenem dosages on basis of renal function	Input costs not considered; no control group; clinical outcomes not considered

(*Continued*)

E

Appendix 1 Evaluations of economic value of clinical pharmacy services—1988–1995 (*Continued*)

Setting	Objective (as stated by authors)	Analytical method	Comparison group	Input costs	Outcomes included	Results measured	Comments
UACH[78]	To evaluate cost impact of computerized antibiotic monitoring program	OA	Historical control	None	DCA	Predicted cost avoidance approximately $80,000 in control vs. study periods, but actual cost reduction attributed to program >$200,000	Cost associated with providing program mentioned but not quantified
UH[79]	To evaluate impact on hospital costs of antibiotic program using education and antimicrobial restriction	CBA	Pre/post	Costs of drug, labor, and program monitoring and implementation	LOS, infection frequency	Cost savings $14,250 annually with quality of care remaining constant	No ratio presented
MC, UH[80]	To conduct retrospective DUE to determine potential cost savings of ceftazidime dosage adjustment	OD	None	None	DCA	Ceftazidime dosing in elderly found to be in excess of labeled dosing because renal function not considered	Input costs not considered; clinical outcomes not considered
UH[81]	To evaluate impact of clinical RPh's intervention on antibiotic costs	OA	Pre/post	None	LOS, DCA	Audit results 3 mo before and after intervention revealed $3498.40 reduction in drug costs	
UH[82]	To determine impact of antibiotic monitoring program	CBA	Pre/post	Cost of printing intervention form	DCA	Net savings $17,000 annually	Clinical outcomes not considered; personnel costs not considered; no ratio presented
UAGH[83]	To evaluate impact of compliance with guidelines for third-generation cephalosporins	OA	Pre/post	None	Clinical and microbiologic indicators; DCA	Documented reduction of $27,000 over 6 mo in pharmacy expenditure for antibiotics	Input costs not considered
UACH[84]	To evaluate impact of antimicrobial intervention program	OD	None	None	Clinical and microbiologic indicators, laboratory costs, DCA	Savings $38,920 over 7 mo; projected annual savings $107,000	Input costs not considered; assumed quality and clinical outcome to be equal

(*Continued*)

Appendix 1 Evaluations of economic value of clinical pharmacy services—1988–1995 (*Continued*)

Setting	Objective (as stated by authors)	Analytical method	Comparison group	Input costs	Outcomes included	Results measured	Comments
GH[85] (VA)	To evaluate impact of antibiotic policy on hospital costs and quality of patient care	OA	Pre/post	None	DCA, duration of antibiotics, LOS, mortality	Decreased monthly antibiotic costs by $7600; average savings $91,200 annually; fewer deaths; decreased LOS	
CH[86]	To describe cost savings to hospital resulting from clinical RPh and nursing antibiotic prescribing interventions	OD	None	None	DCA, NOI	Cost avoidance $23,993 during study period	Input costs not considered
UH[87]	To describe and evaluate dosing intervention program for imipenem	OA	Pre/post	None	ADRs, DCA	Decreased number of seizure episodes; cost savings due to dosage change	Retrospective chart review
GH[88] (VA)	To evaluate impact of concurrent antibiotic use program	OA	Pre/post	None	Length of antibiotic therapy, mortality, DCA, pharmacy cost, nursing cost	Decreased number of antibiotic doses/patient by 24%; 32% reduction in drug costs	Input costs not considered
UH[89]	To conduct DUE of prophylactic antibiotic therapy and determine cost savings to hospital	OA	Pre/post	None	DCA, number of inappropriate orders	Projected annual cost savings $25,000	Input costs not considered
UACH[90]	To evaluate impact of antibiotic therapeutic interchange program	OA	Pre/post	None	Efficacy indicators, ADRs, DCA	Decreased cost of daily antibiotic therapy in study group	Input costs not considered
Target drug programs: Acid-reduction therapy							
CH[91]	To document inappropriate use of i.v. H_2RAs and calculate cost avoided with oral conversion	COD	None	Personnel costs, direct costs	DCA	Cost avoidance range $606–8668 annually	No control group
CH[92]	To describe and evaluate the development of renal dosing intervention strategy for intermittent i.v. H_2RAs	OA	Pre/post	None	DCA	Decreased hospital cost/patient treatment day by 33% equal to $8053/yr	

(*Continued*)

Appendix 1 Evaluations of economic value of clinical pharmacy services—1988–1995 (*Continued*)

Setting	Objective (as stated by authors)	Analytical method	Comparison group	Input costs	Outcomes included	Results measured	Comments
CH[93]	To evaluate cost savings to hospital resulting from clinical RPh recommendations for dosing i.v. H_2RAs	OA	Pre/post	None	DCA	Treatment cost decreased by $1.27/day; annual savings $838	Input costs not considered; clinical outcomes not considered
GH[94] (VA)	To evaluate impact of educational intervention with guideline implementation	CBA	Pre/post	Personnel costs	DCA	Annual cost avoidance of $25,000 associated with decreased use of acid-reducing therapy; estimated cost of program $3000	Clinical outcomes not considered; no ratio presented
GAAC[95] (State)	To evaluate impact of concurrent DUE program on costs associated with acid-reducing therapy	OA	Pre/post	None	DCA; clinical outcomes including antacid use and ordering of gastro-intestinal tests	Cost avoidance of $327,273 attributed to program, with no significant increase in antacid use of number of upper gastrointestinal studies	Input costs not considered
UH[96]	To evaluate cost impact of program authorizing clinical RPh conversion of drugs from parenteral to oral route	OA	Control group	None	DCA	Cost avoidance $53,950 with decrease in length of parenteral therapy	Clinical outcomes not considered; mentioned but did not quantify labor cost associated with program; mentioned but did not calculate ratio
UAAC[97]	To evaluate impact of guideline-based intervention program on cost of H_2RA therapy	OD	None	None	DCA	Total cost avoidance $47,672 during first 6 mo	Input costs not considered; no control group; clinical outcomes not considered
HMOC[98]	To evaluate impact of clinical RPh intervention program on cost of H_2RA therapy	CBA	Pre/post	Personnel costs	DCA	Annual savings $14,600, with labor costs of $3400; calculated cost : benefit ratio 4.3 : 1	Clinical outcomes not considered; useful model for justification of program provided outcomes considered

(*Continued*)

Appendix 1 Evaluations of economic value of clinical pharmacy services—1988–1995 (*Continued*)

Setting	Objective (as stated by authors)	Analytical method	Comparison group	Input costs	Outcomes included	Results measured	Comments
CH[99]	To evaluate cost impact of therapeutic interchange program for H_2RA therapy	OD	None	None	Drug and ancillary cost avoidance	Estimated cost avoidance $37,565/yr	Input costs not considered; no control group; clinical outcomes not considered; included sunk costs (nursing costs associated with additional doses of drug) as costs avoided
CH[100]	To evaluate impact of therapeutic interchange program for H_2RA therapy	OD	None	None	DCA	Total $145,557 in cost avoidance in first yr of program	Input costs not considered; no control group; clinical outcomes not considered
HMOC[101]	To evaluate cost impact of educational interventions in improving use of H_2RA therapy	OA	Pre/post	None	DCA	Study group had fewer prescriptions, less expensive prescriptions, and more appropriate prescriptions after educational interventions than control group	Input costs not considered; clinical outcomes not considered; small sample (number of prescribers involved in intervention)
UACH[102]	To describe impact of therapeutic interchange program for H_2RAs on cost and quality of patient care	OD	None	None	DCA, ADRs, assessment of treatment failure	Estimated annual cost savings $16,000; reduced parenteral H_2RA use	Retrospective analysis; no evidence of increased treatment failure or adverse patient outcome
UH[103]	To evaluate impact of ranitidine i.v. to oral conversion project on cost savings to hospital	OD	None	None	DCA	Decreased number of days of i.v. acid-reducing agents; annual savings $23,425	
UH[104]	To evaluate impact of clinical RPh monitoring and intervention program on i.v. H_2RA therapy	CBA	Control group	Personnel costs	Number of i.v. doses and days of i.v. drug, DCA	Lower mean number of inappropriate doses in study group; projected net annual savings $15,766.37	No ratio presented

(*Continued*)

E

Appendix 1 Evaluations of economic value of clinical pharmacy services—1988–1995 (*Continued*)

Setting	Objective (as stated by authors)	Analytical method	Comparison group	Input costs	Outcomes included	Results measured	Comments
UH[105]	To conduct prospective cost analysis of educational efforts to change inappropriate prescribing of H_2RAs	OA	Pre/post	None	Physician prescribing pattern, DCA, number of drug interactions	Savings of $250,000 estimated for 1st yr of program	Input costs not considered
UAGH[106]	To evaluate impact of i.v. to oral switch program for ranitidine	OA	Pre/post	None	DCA, pharmacy preparation costs	Cost avoidance $4214	Input costs not considered
UH[107]	To evaluate impact of H_2RA program on cost and quality of patient care	OA	Pre/post	None	Patient outcome, ADRs, drug interactions, DCA	Decreased cost but preserved quality	Input costs not considered
Target drug programs: NSAIDs							
GAAC[108] (VA)	To evaluate impact of clinical RPh activities in an ambulatory clinic	OA	Control group	None	DCA	Greater reduction in NSAID use in clinic staffed by RPh, resulted in cost savings of $38,776 more than control group	Input costs not considered; clinical outcomes not considered; data collected in 1985–1986, report not published until 1991
CH[109]	To describe target DUE program and determine impact on drug and labor costs	OA	Pre/post	None	DCA, NOI	Net annual savings $18,756	Considered personnel costs
UAAC[110]	To evaluate effect of pharmacist-managed anticoagulation clinical on therapeutic outcomes and costs	CMA	Control group	Charge for service	Hemorrhagic events, thromboembolic events, frequency and charge for clinic visits, ER visits, hospital admissions	Improved clinical outcomes, charge avoidance $4073/person-year	Included clinical outcomes, used charges rather than costs

CA, cost analysis; CBA, cost–benefit analysis; CD, cost description; COD, cost/outcome description; CMA, cost-minimization analysis; OA, outcome analysis; OD, outcome description; CH, community hospital; CP, community pharmacy; ER, emergency room; GAAC, government-affiliated ambulatory clinic; GH, government hospital; HMOC, health maintenance organization clinic; MC, multicenter; MHF, mental health facility; SNF, skilled nursing facility; UAAC, university-affiliated ambulatory clinic; UACH, university-affiliated community hospital; UAGH, university-affiliated government hospital; UH, university hospital; DCA, drug costs avoided; DUE, drug use evaluation; NOI, number of interventions or recommendations; ADRs, adverse drug reactions; H_2RA, histamine$_2$-receptor antagonist; ICU, intensive care unit; LOS, length of hospital stay; NSAIDs, nonsteroidal anti-inflammatory drugs; RPh, pharmacist; SDC, serum drug concentration; TDM, therapeutic drug monitoring.

REFERENCES

1. Willett, M.S.; Bertch, K.E.; Rich, D.S.; Ereshefsky, L. Prospectus on the economic value of clinical pharmacy services. Pharmacotherapy **1989**, *9* (1), 45–56.
2. Hatoum, H.T.; Catizone, C.; Hutchinson, R.A.; Purohit, A. An eleven-year review of the pharmacy literature: Documentation of the value and acceptance of clinical pharmacy. Drug Intell. Clin. Pharm. **1986**, *20*, 33–41.
3. Penna, R.P. Pharmaceutical care: Pharmacy's mission for the 1990s. Am. J. Hosp. Pharm. **1990**, *47*, 543–549.
4. Elixhauser, A.; Luce, B.R.; Taylor, W.R.; Reblando, J. Health care CBA/CEA: An update on the growth and composition of the literature. Med. Care **1993**, *31* (7), JS1–JS11.
5. Bradley, C.A.; Iskedjian, M.; Lanctot, K.L., et al. Quality assessment of economic evaluations in selected pharmacy, medical, and health economic journals. Ann. Pharmacother. **1995**, *29*, 681–689.
6. Drummond, M.F.; Stoddard, G.; Torrance, G.W. *Methods for Economic Evaluation of Health Care Programmes*; Oxford University Press: Oxford, 1992; 8.
7. Hoey, L.L.; Nahum, A.; Vance-Bryan, K. A prospective evaluation of benzodiazepine guidelines in the management of patients hospitalized for alcohol withdrawal. Pharmacotherapy **1994**, *14*, 579–585.
8. Pauley, T.R.; Magee, M.J.; Cury, J.D. Pharmacist-managed, physician-directed asthma management program reduces emergency department visits. Ann. Pharmacother. **1995**, *29*, 5–9.
9. Peterson, J.P.; Colucci, V.J.; Schiff, S.E. Using serum creatinine concentrations to screen for inappropriate dosing of renally eliminated drugs. Am. J. Hosp. Pharm. **1991**, *48*, 1962–1964.
10. Smythe, M.; Loughlin, K.; Schad, R.F.; Lucarroti, R.L. Patient-controlled analgesia versus intramuscular analgesic therapy. Am. J. Hosp. Pharm. **1994**, *51*, 1433–1440.
11. Baciewicz, A.M.; Cowan, R.I.; Michaels, P.E.; Kyllonen, K.S. Quality and productivity assessment of clinical pharmacy interventions. Hosp. Formul. **1994**, *29*, 773, 777–779.
12. Bayliff, C.D.; Einarson, T.R. Physician assessment of pharmacists' interventions—method of estimating cost avoidance and determining quality assurance. Can. J. Hosp. Pharm. **1990**, *43*, 167–171, 195.
13. Bertch, K.E.; Hatoum, H.T.; Willett, M.S.; Witte, K.W. Cost justification of clinical pharmacy services on a general surgery team: Focus on diagnosis-related group cases. Drug Intell. Clin. Pharm. **1988**, *22*, 906–911.
14. Bjornson, D.C.; Hiner, W.O., Jr.; Potyk, R.P., et al. Effect of pharmacists on health care outcomes in hospitalized patients. Am. J. Hosp. Pharm. **1993**, *50*, 1875–1884.
15. Borgsdorf, L.R.; Miano, J.S.; Knapp, K.K. Pharmacist-managed medication review in a managed care system. Am. J. Hosp. Pharm. **1994**, *51*, 772–777.
16. Britton, M.L.; Lurvey, P.L. Impact of medication profile review on prescribing in a general medicine clinic. Am. J. Hosp. Pharm. **1991**, *48*, 265–270.
17. Brown, W.J. Pharmacist participation on a multidisciplinary rehabilitation team. Am. J. Hosp. Pharm. **1994**, *51*, 91–92.
18. Catania, H.F.; Catania, P.N. Using clinical interventions to cost-justify additional pharmacy staff. Hosp. Pharm. **1988**, *23*, 544, 546–548.
19. Catania, H.F.; Yee, W.P.; Catania, P.N. Four years' experience with a clinical intervention program: Cost avoidance and impact of a clinical coordinator. Am. J. Hosp. Pharm. **1990**, *47*, 2701–2705.
20. Chuang, L.C.; Sutton, J.D.; Henderson, G.T. Impact of a clinical pharmacist on cost saving and cost avoidance in drug therapy in an intensive care unit. Hosp. Pharm. **1994**, *29*, 215–218, 221.
21. Clapham, C.E.; Hepler, C.D.; Reinders, T.P.; Lehman, M.E.; Pesko, L. Economic consequences of two drug-use control systems in a teaching hospital. Am. J. Hosp. Pharm. **1988**, *45*, 2329–2340.
22. Dobie, R.L.; Rascati, K.L. Documenting the value of pharmacist interventions. Am. Pharm. **1994**, *NS34*, 50–54.
23. Garrelts, J.C.; Smith, D.F. Clinical services provided by staff pharmacists in a community hospital. Am. J. Hosp. Pharm. **1990**, *47*, 2011–2015.
24. Hatoum, H.T.; Hutchinson, R.A.; Witte, K.W.; Newby, G.P. Evaluation of the contribution of clinical pharmacists: Inpatient care and cost reduction. Drug Intell. Clin. Pharm. **1988**, *22*, 252–259.
25. Hatoum, H.T.; Witte, K.W.; Hutchinson, R.A. Patient care contributions of clinical pharmacists in four ambulatory care clinics. Hosp. Pharm. **1992**, *27*, 203–206, 208–209.
26. Haslett, T.M.; Kay, B.G.; Weissfellner, H. Documenting concurrent clinical pharmacy interventions. Hosp. Pharm. **1990**, *25*, 351–355, 359.
27. Hawkey, C.J.; Hodgson, S.; Norman, A.; Daneshmend, T.K.; Garner, S.T. Effect of reactive pharmacy intervention on quality of hospital prescribing. BMJ **1990**, *300*, 986–990.
28. Hoolihan, R.J.; Skoogman, L.H. Documentation of pharmacist interventions in a decentralized unit dose system. Hosp. Pharm. **1991**, *26*, 875–879.
29. Kao, R.A.; Rodriquez, L.R.; Weber, C. Clinical role of pharmacists in hospital ambulatory care services. Top. Hosp. Pharm. Manage. **1988**, *8*, 1–8.
30. Lepinski, P.W.; Woller, T.W.; Abramowitz, P.W. Implementation, justification, and expansion of ambulatory clinical pharmacy services. Top. Hosp. Pharm. Manage. **1992**, *11*, 86–92.
31. Lobeck, F.; Traxler, W.T.; Bobinet, D.D. The cost-effectiveness of a clinical pharmacy service in an outpatient mental health clinic. Hosp. Commun. Psychiatry **1989**, *40*, 643–645.

32. Montazeri, M.; Cook, D.J. Impact of a clinical pharmacist in a multidisciplinary intensive care unit. Crit. Care Med. **1994**, *22*, 1044–1048.
33. Mueller, B.A.; Abel, S.R. Impact of college of pharmacy-based educational services within the hospital. Drug Intell. Clin. Pharm. **1990**, *24*, 422–425.
34. Phillips, S.L.; Carr-Lopez, S.M. Impact of a pharmacist on medication discontinuation in a hospital-based geriatric clinic. Am. J. Hosp. Pharm. **1990**, *47*, 1075–1079.
35. Poirier, T.I.; Ceh, P. Documenting the outcomes of pharmacists' clinical interventions. P&T **1994**, *19*, 132–134, 137–138.
36. Rosenbaum, C.L.; Fant, W.K.; Miyagawa, C.I.; Armitstead, J.A. Inability to justify a part-time clinical pharmacist in a community hospital intensive care unit. Am. J. Hosp. Pharm. **1991**, *48*, 2154–2157.
37. Slaughter, R.L.; Erickson, S.R.; Thomson, P.A. Clinical interventions provided by doctor of pharmacy students. Ann. Pharmacother. **1994**, *28*, 665–670.
38. Taylor, J.T.; Kathman, M.S. Documentation of cost savings from decentralized clinical pharmacy services at a community hospital. Am. J. Hosp. Pharm. **1991**, *48*, 1467–1470.
39. Torchinsky, A.; Landry, D. An analysis of pharmacist interventions. Can. J. Hosp. Pharm. **1991**, *44*, 245–248, 270.
40. Torok, N.; Brown, G. The economic impact of clinical pharmacists' unsolicited recommendations. Hosp. Pharm. **1992**, *27*, 1052–1053, 1056–1058, 1060.
41. Warrian, K.; Irvine, M.J. Cost-benefit of a clinical services integrated with a decentralized unit dose system. Can. J. Hosp. Pharm. **1988**, *41*, 109–112.
42. Bearce, W.C. Documentation of clinical interventions: Quality of care issues and economic considerations in critical care pharmacy. Hosp. Pharm. **1988**, *23*, 883–890.
43. Briceland, L.L.; Kane, M.P.; Hamilton, R.A. Evaluation of patient-care interventions by Pharm.D. clerkship students. Am. J. Hosp. Pharm. **1992**, *49*, 1130–1132.
44. Levy, D.B. Documentation of clinical and cost saving pharmacy interventions in the emergency room. Hosp. Pharm. **1993**, *28*, 624–627, 630–634, 653.
45. Lobas, N.H.; Lepinski, P.W.; Abramowitz, P.W. Effects of pharmaceutical care on medication cost and quality of patient care in an ambulatory care clinic. Am. J. Hosp. Pharm. **1992**, *49*, 1681–1688.
46. Mason, J.D.; Colley, C.A. Effectiveness of an ambulatory care clinical pharmacist: A controlled trial. Ann. Pharmacother. **1993**, *27*, 555–559.
47. Rupp, M.T. Value of community pharmacists interventions to correct prescribing errors. Ann. Pharmacother. **1992**, *26*, 1580–1584.
48. Tang, I.; Vrahnos, D.; Hatoum, H.T.; Lau, A. Effectiveness of a clinical pharmacist's interventions in a hemodialysis unit. Clin. Ther. **1993**, *15*, 459–464.
49. Ambrose, P.J.; Nitake, M.; Kildoo, C.W. Impact of pharmacist scheduling of blood-sampling times for therapeutic drug monitoring. Am. J. Hosp. Pharm. **1988**, *45*, 380–382.
50. D'Angio, R.G.; Stevenson, J.G.; Lively, B.T.; Morgan, J.E. Therapeutic drug monitoring: Improved performance through educational intervention. Ther. Drug Monit. **1990**, *12*, 173–181.
51. Destache, C.J.; Meyer, S.K.; Bittner, M.J.; Hermann, K.G. Impact of a clinical pharmacokinetic service on patients treated with aminoglycosides: Cost-benefit analysis. Ther. Drug Monit. **1990**, *12*, 419–426.
52. Destache, C.J.; Meyer, S.K.; Rowley, K.M. Does accepting pharmacokinetic recommendations impact hospitalization? Cost-benefit analysis. Ther. Drug Monit. **1990**, *12*, 427–433.
53. Destache, C.J.; Meyer, S.K.; Padomek, M.T.; Ortmeier, B.G. Impact of a clinical pharmacokinetic service on patients treated with aminoglycosides for gram-negative infections. Drug Intell. Clin. Pharm. **1989**, *23*, 33–38.
54. Jorgenson, J.A.; Rewers, R.F. Justification and evaluation of an aminoglycoside pharmacokinetic dosing service. Hosp. Pharm. **1991**, *26*, 605, 609–611, 615.
55. Klamerus, K.J.; Munger, M.A. Effect of clinical pharmacy services on appropriateness of serum digoxin concentration monitoring. Am. J. Hosp. Pharm. **1988**, *45*, 1887–1893.
56. Kraus, D.M.; Calligaro, I.L.; Hatoum, H.T. Multilevel model to assess appropriateness of pediatric serum drug concentrations. Am. J. Dis. Child. **1991**, *145*, 1171–1175.
57. Wade, W.E.; Griffin, C.R.; Bennett, T.A.; Robertson, L.M. Educational effort and CQI program improves ordering of serum digoxin levels. Hosp. Formul. **1994**, *29*, 657–659.
58. Wing, D.S.; Duff, H.J. Impact of a therapeutic drug monitoring program for phenytoin. Ther. Drug Monit. **1989**, *11*, 32–37.
59. Wing, D.S.; Duff, H.J. Evaluation of therapeutic drug monitoring program for theophylline in a teaching hospital. Drug Intell. Clin. Pharm. **1987**, *21*, 702–706.
60. Burton, M.R.; Ash, C.L.; Hill, D.P.; Handy, T.; Shepherd, M.D.; Vasko, M.R. A controlled trial of the cost-benefit of computerized Bayesian aminoglycoside administration. Clin. Pharmacol. Ther. **1991**, *49*, 685–694.
61. Cook, R.L.; Bush, N.D. Cost-saving impact of pharmacist-calculated IV aminophylline dosing schedules. Hosp. Formul. **1989**, *24*, 716–721.
62. Lesar, T.; Belemjian, M.; Harrison, C.; Dollard, P.; Snow, K. Program for controlling the use of ondansetron injection. Am. J. Hosp. Pharm. **1994**, *51*, 3054–3056.
63. Forstrom, M.J.; Ried, L.D.; Stergachis, A.S.; Corliss, D.A. Effect of a clinical pharmacist program on the cost of hypertension treatment in an HMO family practice clinic. Drug Intell. Clin. Pharm. **1990**, *24*, 304–309.
64. Anassi, E.O.; Egbunike, I.G.; Akpaffiong, M.J.; Ike, E.N.; Cate, T.R. Developing and implementing guidelines to promote appropriate use of fluconazole therapy

in an AIDS clinic. Hosp. Pharm. **1994**, *29*, 576–578, 581–582, 585–586.

65. Briceland, L.L.; Nightingale, C.H.; Quintiliani, R.; Cooper, B.W. Multidisciplinary cost-containment program promoting less frequent administration of injectable mezlocillin. Am. J. Hosp. Pharm. **1988**, *45*, 1082–1085.
66. Briceland, L.L.; Lesar, T.S.; Lomaestro, B.M.; Lombardi, T.P.; Gailey, R.A.; Kowalsky, S.F. Streamlining antimicrobial therapy through pharmacists' review of order sheets. Am. J. Hosp. Pharm. **1989**, *46*, 1376–1380.
67. Briceland, L.L.; Quintiliani, R.; Nightingale, C.H. Multidisciplinary cost-containment program promoting oral metronidazole for treatment of antibiotic-associated colitis. Am. J. Hosp. Pharm. **1988**, *45*, 122–125.
68. Bunz, D.; Gupta, S.; Jewesson, P. Metronidazole cost containment: Two-stage intervention. Hosp. Formul. **1990**, *25*, 1167–1169.
69. Capers, C.C.; Bess, D.T.; Branam, A.C., et al. Antibiotic surveillance: The results of a clinical pharmacy intervention program. Hosp. Pharm. **1993**, *28*, 206–210, 212.
70. Gin, A.S.; Lipinski, L.A.; Honcharik, N. Impact of a target drug monitoring program on the usage of clindamycin. Can. J. Hosp. Pharm. **1994**, *47*, 53–58.
71. Hendrickson, J.R.; North, D.S. Pharmacoeconomic benefit of antibiotic step-down therapy: Converting patients from intravenous ceftriaxone to oral cefpodoxime proxetil. Ann. Pharmacother. **1995**, *29*, 561–565.
72. Hirschman, S.Z.; Meyers, B.R.; Bradbury, K.; Mehl, B.; Kimelblatt, B. Use of antimicrobial agents in a university teaching hospital: Evolution of a comprehensive control program. Arch. Intern. Med. **1988**, *148*, 2001–2007.
73. Julius, H.E. Cost-effective management of DUE policy through a pharmacy clerkship. Hosp. Formul. **1993**, *28*, 789–791.
74. Jung, B.; Andrews, J.D. Effectiveness of an antibiotic cost containment measure. Can. J. Hosp. Pharm. **1990**, *43*, 116–122.
75. Karki, S.D.; Holden, J.M.; Mariano, E. A team approach to reduce antibiotic costs. Drug Intell. Clin. Pharm. **1990**, *24*, 202–205.
76. Pastel, D.A.; Chang, S.; Nessim, S.; Shane, R.; Morgan, M.A. Department of pharmacy-initiated program for streamlining empirical antibiotic therapy. Hosp. Pharm. **1992**, *27*, 596–603, 614.
77. Ritchie, D.J.; Reichley, R.M.; Canaday, K.L.; Bailey, T.C. Evaluation and financial impact of imipenem/cilastatin dosing in elderly patients based on renal function and body weight. J. Pharm. Technol. **1993**, *9*, 160–163.
78. Schentag, J.J.; Ballow, C.H.; Fritz, A.L., et al. Changes in antimicrobial agent usage resulting from interactions among clinical pharmacy, the infectious disease division, and the microbiology laboratory. Diagn. Microbiol. Infect. Dis. **1993**, *16*, 255–264.
79. Smith, K.S.; Quercia, R.A.; Chow, M.S.; Nightingale, C.H.; Millerick, J.D. Multidisciplinary program for promoting single prophylactic doses of cefazolin in obstetrical and gynecological surgical procedures. Am. J. Hosp. Pharm. **1988**, *45*, 1338–1342.
80. Vlasses, P.H.; Bastion, W.A.; Behal, R.; Sirgo, M.A. Ceftazidime dosing in the elderly-economic implications. Ann. Pharmacother. **1993**, *27*, 967–971.
81. Ware, G.J.; Ford, D.J. Cost benefit of pharmacy audit and nonrestrictive antibiotic policy. N. Z. Med. J. **1993**, *106*, 160.
82. Avorn, J.; Soumerai, S.B. Reduction of incorrect antibiotic dosing through a structured educational order form. Arch. Intern. Med. **1988**, *148*, 1720–1724.
83. Bamberger, D.M.; Dahl, S.L. Impact of voluntary vs. enforced compliance of third-generation cephalosporin use in a teaching hospital. Arch. Intern. Med. **1992**, *152*, 554–557.
84. Briceland, L.L.; Nightingale, C.H.; Quintiliani, R.; Cooper, B.W.; Smith, K.S. Documentation of clinical interventions: Quality of care issues and economic considerations in critical care pharmacy. Arch. Intern. Med. **1988**, *148*, 2019–2022.
85. Coleman, R.W.; Rodondi, L.C.; Kaubisch, S.; Granzella, N.B.; O'Hanley, P.D. Cost-effctiveness of prospective and continuous parenteral antibiotic control: Experience at the Palo Alto Veterans Affairs Medical Center from 1987 to 1989. Am. J. Med. **1991**, *90*, 439–444.
86. Katona, B.G.; Ayd, P.R.; Walters, J.K.; Caspi, M.; Finkelstein, P.W. Effect of a pharmacist's and a nurses interventions on cost of drug therapy in a medical intensive-care unit. Am. J. Hosp. Pharm. **1989**, *46*, 1179–1182.
87. Newcomb, H.W.; Hill, E.M.; McCarthy, I.D.; Mostow, S.R. Imipenem-cilastin dosing intervention program by pharmacists. Am. J. Hosp. Pharm. **1992**, *49*, 1133–1135.
88. O'Hanley, P.; Rodondi, L.; Coleman, R. Efficacy and cost-effectiveness of antibiotic monitoring at a Veterans Administration hospital. Chemotherapy **1991**, *37*, 22–25.
89. Zhanel, G.G.; Gin, A.S.; Przybylo, A.; Louie, T.J.; Otten, N.H. Effect of interventions on prescribing of antimicrobials for prophylaxis in obstetric and gynecologic surgery. Am. J. Hosp. Pharm. **1989**, *46*, 2493–2496.
90. Smith, K.S.; Briceland, L.L.; Nightingale, C.H.; Quintiliani, R. Formulary conversion of cefoxitin usage to cefotetan: Experience at a large teaching hospital. Drug Intell. Clin. Pharm. **1989**, *23*, 1024–1029.
91. Algozzine, G.J.; Sprenger, R.L.; Caselnova, D.A., III; Proper, P. Pharmacy intern intervention to reduce costs associated with histamine H_2-antagonist therapy. Am. J. Hosp. Pharm. **1989**, *46*, 1183–1184.
92. Connelly, J.F. Adjusting dosage intervals of intermittent intravenous ranitidine according to creatinine clearance: Cost-minimization analysis. Hosp. Pharm. **1994**, *29*, 992, 998, 1001.
93. Holt, R.T.; Graves, L.J.; Scheil, E. Reducing costs by adjusting dosage intervals for intravenous ranitidine. Am. J. Hosp. Pharm. **1990**, *47*, 2068–2069.
94. Kane, M.P.; Briceland, L.L.; Garris, R.E.; Favreau, B.N. Drug use review program for concurrent histamine H_2

receptor antagonist-sucralfate therapy. Am. J. Hosp. Pharm. **1990**, *47*, 2007–2010.
95. Keith, M.R.; Cason, D.M.; Helling, D.K. Antiulcer prescribing program in a state correctional system. Ann. Pharmacother. **1994**, *28*, 792–796.
96. Kirking, D.M.; Svinte, M.K.; Berardi, R.R.; Cornish, L.A.; Ryan, M.L. Evaluation of direct pharmacist intervention on conversion from parenteral to oral histamine H_2-receptor antagonist therapy. Ann. Pharmacother. **1991**, *25*, 80–84.
97. Malcolm, K.E.; Griffin, C.R.; Bennett, T.A.; Robertson, L.M. Pharmacist adjustment of H_2-receptor antagonist dosage to meet medical staff-approved criteria. Am. J. Hosp. Pharm. **1994**, *51*, 2152–2154.
98. Mead, R.A.; McGhan, W.F. Use of histamine$_2$-receptor blocking agents and sucralfate in a health maintenance organization following continued clinical pharmacist intervention. Drug Intell. Clin. Pharm. **1988**, *22*, 466–469.
99. Oh, T.; Franko, T.G. Implementing therapeutic interchange of intravenous famotidine for cimetidine and ranitidine. Am. J. Hosp. Pharm. **1990**, *47*, 1547–1551.
100. Quercia, R.A.; Chow, M.S.; Jay, G.T.; Quintiliani, R. Strategy for developing a safe and cost-effective H_2-receptor antagonist program. Hosp. Formul. **1991**, *26* (Suppl. D), 20–24.
101. Raisch, D.W.; Bootman, J.L.; Larson, L.N.; McGhan, W.F. Improving antiulcer agent prescribing in a health maintenance organization. Am. J. Hosp. Pharm. **1990**, *47*, 1766–1773.
102. Wetmore, R.W.; Jennings, R.H. Retrospective analysis of formulary restriction demonstrates significant cost savings. Hosp. Formul. **1991**, *26*, 30–32.
103. Baciewicz, A.M. Conversion of intravenous ranitidine to oral therapy. Ann. Pharmacother. **1991**, *25*, 251–252.
104. Dannenhoffer, M.A.; Slaughter, R.L.; Hunt, S.N. Use of concurrent monitoring and a preprinted note to modify prescribing of i.v. cimetidine and ranitidine therapy to oral therapy. Am. J. Hosp. Pharm. **1989**, *46*, 1570–1575.
105. Fudge, K.A.; Moore, K.A.; Schneider, D.N.; Sherrin, T.P.; Wellman, G.S. Change in prescribing patterns of intravenous histamine$_2$-receptor antagonists results in significant cost savings without adversely affecting patient care. Ann. Pharmacother. **1993**, *27*, 232–237.
106. Santora, J.; Kitrenos, J.G.; Green, E.R. Pharmacist intervention program focused in i.v. ranitidine therapy. Am. J. Hosp. Pharm. **1990**, *47*, 1346–1349.
107. Foulke, G.E.; Siepler, J. Antiulcer therapy: An exercise in formulary management. J. Clin. Gastroenterol. **1990**, *12*, S64–S68.
108. Jones, R.A.; Lopez, L.M.; Beall, D.G. Cost-effective implementation of clinical pharmacy services in an ambulatory care clinic. Hosp. Pharm. **1991**, *26*, 778–782.
109. Chrymko, M.M.; Meyer, J.D.; Kelly, W.N. Target drug monitoring: Cost-effective service provided by staff pharmacists. Hosp. Pharm. **1994**, *29*, 347, 350–352.
110. Wilt, V.M.; Gums, J.G.; Ahmed, O.I.; Moore, L.M. Outcome analysis of a pharmacist-managed anticoagulation service. Pharmacotherapy **1996**, *15* (6), 732–739.

Electronic Prescribing

Woodie M. Zachry III
Edward P. Armstrong
University of Arizona, Tucson, Arizona, U.S.A.

INTRODUCTION

More than 17,000 brand and generic names for medications are currently approved for prescribing in North America.[1] Of those 17,000 chemical entities, a surprising amount have similar dosages. Furthermore, many names of the medications prescribed today are spelled or pronounced in similar ways. This can lead to a substantial number of errors due to the misinterpretation and/or misuse of abbreviations, chemical names, and dosages.[2] A study by Lesar et al. evaluated 696 clinically important errors in a 631-bed tertiary hospital and found that errors of nomenclature (incorrect drug name, dosage form or abbreviation) accounted for 13.4% of all medication errors. The authors further found that one in six errors involved the miscalculation of dosages, incorrect placement of a decimal, incorrect unit of measure, or an incorrect administration rate.[3] Although poor transcription of a medication order is an obvious contributing factor for these types of errors, other factors at the point of prescribing also play a role. Lesar et al. found that the most common types of errors made were due to the inappropriate application of drug therapy knowledge (30%) and the inappropriate use of knowledge regarding patient factors related to drug therapy (29.2%).[3]

Physician order entry has been recommended as one possible solution to help to prevent these types of medication errors.[1] Initially, the goal of prescribing automation was to decrease the potential for error due to the misinterpretation of handwritten orders. However, the capabilities of computers used to aid in medication order entry now exceed common word-processing duties. Newer systems have allowed clinicians to link patient data to the prescribing process. Clinicians can use these data to ensure that the drug dose, timing, and dosage form are correct, while checking for drug interactions, duplicate therapy, allergies, or disease-state contraindications. A study by Bates et al. found a greater than 50% reduction (10.7–4.86 events per 1000 patient days) in nonintercepted serious medication errors after a hospital-implemented direct physician order entry.[4] Another study found a significant reduction in errors due to allergies (76%) and excessive drug dosages (78.5%) after implementation of a computerized antiinfective management program.[5] Due to these and other study results, the National Patient Safety Partnership has recommended implementation of direct order entry strategies.[1]

DESCRIPTION OF ELECTRONIC PRESCRIBING

Direct order entry, or electronic prescribing, is not limited to the inpatient setting. Electronic prescribing encompasses all computer-driven automated processes used to write a prescription for a patient. Within the past few years, technological advances have allowed electronic prescribing to be performed in an ambulatory setting. This process is executed in many ways. Early versions of electronic prescribing devices consisted of a stand-alone computer terminal located at fixed points in physicians' offices.[6,7] These fixed terminals have expanded to use Internet web-based interfaces to access patient level information from a health plan, write prescriptions, and send prescriptions to a pharmacy to be filled.[8]

More recent advances in technology made possible by the personal digital assistant (PDA) have allowed physicians to electronically prescribe at the point of care. PDAs are handheld computers that typically run using a Windows- or other proprietary-based platforms. These PDAs use a touch-sensitive screen to maneuver through a menu-driven prescribing process that can execute a prescription in as little as three stylus taps.[9] The PDAs or other proprietary devices then upload the prescription via a network connection or modem to be printed, faxed, or electronically transmitted to a pharmacy.

INFORMATION AVAILABLE TO CLINICIANS

Electronic prescribing devices provide several sources of information to prescribers at the point of care provided to

Encyclopedia of Clinical Pharmacy
DOI: 10.1081/E-ECP 120006404

patients. Depending on the level of programming sophistication, and the database links built into the prescribing device, the clinician can access patient-specific formulary lists, manufacturer recalled medications, and a host of clinical references while choosing a therapy. The devices can also be used to review any managed care disease treatment protocols at the point of prescribing. It is also possible for the prescriber to perform drug utilization review (DUR) analyses to detect any possible drug–drug interactions, therapeutic duplications, drug–disease contraindications, drug allergies, past adverse reactions, and inappropriate dosing levels. These therapy edits are either provided real-time or as possible problems detected upon transmittal to the electronic prescribing vendor's server. Finally, electronic prescribing devices allow the user to provide informational leaflets to patients about their specific therapy.

PRESCRIPTION DESTINATION

Once the prescription has been entered, most electronic prescribing systems allow prescribers to transmit prescriptions directly to retail or mail order pharmacies electronically or by facsimile. However, some systems use an intermediary server to process prescriptions before sending them to a pharmacy. The limiting factor for electronic disposition of prescriptions is the ability to receive the data. Currently, a large percentage of pharmacies are not web enabled, and an even larger number of pharmacies do not operate on an electronic data interface that can speak to a prescriber's electronic prescribing devise. The solution rapidly being accepted to reconcile these inequities is a standard data transfer protocol called SCRIPT created by the National Council for Prescription Drug Programs. This standard (approved by the American National Standards Institute) has been accepted by most electronic prescribing device companies, and is rapidly being adopted by large chain drug stores.[10,11]

Who ultimately pays for the electronic prescribing capability is dependent on the electronic prescribing vendor. Some companies charge prescribers a basic monthly fee that ranges from \$20–\$250 per prescriber per month, depending on the level of information provided at the point of prescribing. This fee typically includes hardware, software, network connectivity devices, upgrades, and a local server.[9,10] Other companies provide hardware and software free of charge to prescribers and charge a second party for the use of the system. This second party is typically a pharmacy benefit manager or pharmacy, and the fees range from \$.10–\$.20 per prescription.[10]

ADVANTAGES OF ELECTRONIC PRESCRIBING

Electronic prescribing technology promises to bring many benefits to the current system of prescribing. The technology promises to bring greater efficiency to the prescribing process and reduces the likelihood of medication nomenclature errors. The following points highlight the potential benefits of adopting an automated prescribing system:

- Current, unbiased drug information and references can be provided real-time to clinicians, including educational updates for existing or new chemical entities and manufacture recalls of medications. This information could include recommended dosing, available routes of administration, and patient educational materials.[1,12]
- Patient-specific insurance information can be provided to prescribers at the point of care, including formulary lists and disease protocol information.[13]
- Patient-specific medical histories can be provided to prescribers at the point of care, including last filled medications, past adverse events, drug allergies, and medical conditions.[13]
- Pharmacies and physicians will need to spend less time contacting each other and insurance companies to overcome formulary restrictions and problems found upon drug utilization review, and to clarify illegible handwriting.[1,14]
- Physicians and pharmacists can expedite refill requests electronically rather than through person-to-person communication.[1]
- Computers can expedite data exchange between health care professionals who represent other parts of the patient's health care management team. The sharing of patient data could lead to less preventable adverse drug reactions and therapeutic duplications. The provision of diagnosis data along with prescription information also allows other heath care providers to check for therapy–diagnosis mismatches.[1,14]
- Computers can inform prescribers about lower-cost alternatives and generic availability at the time of prescribing.[1]

DISADVANTAGES OF ELECTRONIC PRESCRIBING

Conversely, electronic prescribing has a few potential disadvantages. Most of these disadvantages stem from the potential of the technology to be used for other purposes apart from which it was intended. The following is a summary of potential misuses of the new technology:

- The potential exists for a patient's confidentiality to be violated. Some of the companies offering electronic prescription solutions download patient information to a vendor-based server for DUR checks. The security of this information and what it is used for beyond the prescribing process creates the potential to impinge upon the privacy of the patient's medical information.
- The receipt of a prescription can be subject to several market barriers. First, the pharmacy must have the electronic capability to receive the data. Second, the pharmacy must accept the patient's prescription drug plan and be willing to operate under the financial constraints imposed by the electronic prescribing provider. Finally, the potential exists for pharmacy benefit managers (PBMs) to use electronic prescribing technology to route prescriptions to preferred pharmacies such as mail order companies.
- Physicians will be prompted to adhere to formulary restrictions and PBM-driven disease protocols more frequently. As a result, evidence-based prescribing may become more dependent on the use of appropriate clinical knowledge by PBMs rather than health care providers.
- This technology can provide a false sense of security concerning the clinical judgment of the software programming. The programming is limited to the data it receives and the problems it is designed to detect. The innate ability of clinicians to question and rationalize is integral to the process of appropriate prescribing. However, electronic prescribing technology will make it easier to overlook the clinician's importance to the process.
- Theoretically, it is possible that electronic prescribing devices will allow unimpeded access to physicians by whoever is willing to pay for that access. Physician detailing may become more prevalent through these devices and could possibly be confused with unbiased medication information.

IMPACT ON PRACTICE OF PHARMACY

The advent of electronic prescribing will decrease pharmacists' roles in many areas. In dispensing roles, pharmacists will have less responsibility for order entry, PBM formulary management, and disease protocol adherence. Furthermore, a large number of DUR functions will be taken care of before the patient's order is received in the community or hospital pharmacy. However, the dispensing pharmacy may still function as a redundancy check on these issues, continuing to act as a patient advocate to manage the appropriateness of patients' drug therapy. The pharmacist will still operate as an integral check and balance concerning overlooked problems and missed patient information pertinent to a patient's effective drug treatment.

The functions performed by the electronic prescribing technology will most likely lessen the technical burden of the pharmacist, while augmenting the need for nontechnical clinical judgment. This augmentation of clinical judgment should manifest primarily in the review of a patient's situation and pharmacotherapy plan to identify barriers to the desired patient outcomes.[15] Although the more obvious problems will have a higher likelihood of being addressed at the point of prescribing, the pharmacist will still be needed to identify missed pharmaceutical errors related to dosage route, timing, duration, frequency, interaction, contraindication, and allergies. The main emphases of the pharmacist will likely shift to identifying and treating mismatched medications and indications, drug overuse and abuse, drug-induced problems, improper drug use, and potential medication errors.

With a decreased need for pharmacists to identify obvious problems associated with pharmaceutical therapy, the pharmacist should be free to concentrate on patient-centered therapy issues. Pharmacists can spend more time with patients identifying barriers that might prevent a patient reaching an optimal outcome. Pharmacists can then address these issues with education and proactive adjustments in the patient's therapy. The pharmacist can concentrate more time on educating patients to better monitor their therapy to increase the likelihood of maximal therapeutic benefit without troublesome misadventures. Furthermore, the pharmacist could concentrate on therapeutic outreach programs such as ''brown bag'' clinics, diabetic care clinics, and asthma screening.

In a hospital setting, pharmacists can shift their focus away from dispensing roles, and take a more proactive role at the point of care. Lieder reported that the implementation of physician electronic prescribing at Vanderbilt University Medical Center (VUMC) allowed pharmacists to have a greater role in the prescribing process. Pharmacists reported that clinical evaluations were easier with electronic records available at the touch of a key. Pharmacists felt free to pursue other areas of need such as cost-saving issues (e.g., intravenous to oral conversions of medications). The technology seemed to promote the presence of pharmacists on the floors to provide drug information to other health care professionals. The VUMC pharmacy actually maintained the electronic prescribing

system and provided educational enhancements directed at physicians as the need for interventions in therapeutic areas arose. Furthermore, the pharmacy planned to expand its services to include an inpatient anticoagulant management program.[16]

CONCLUSION

The future appears very bright for electronic prescribing. Certainly, the upfront costs for implementing programs, and the refinement of hardware and software specifics are important issues to resolve. However, the benefits of improved care, streamlined workflow, and more efficient use of clinicians' time are important enhancements that have continued to encourage expansion of these technologies. As wider audiences use these applications, continued research is needed to assess the use and refinements necessary to optimally apply these important systems.

REFERENCES

1. Institute for Safe Medication Practices. *A Call to Eliminate Handwritten Prescription Within 3 Years*; Institute for Safe Medication Practices: 2000; 1–12.
2. Institute of Medicine. *To Err Is Human; Building a Safer Health System*; National Academy Press: Washington, DC, 2000.
3. Lesar, T.S.; Briceland, L.; Stein, D.S. Factors related to errors in medication prescribing. JAMA, J. Am. Med. Assoc. **1997**, *277* (4), 312–317.
4. Bates, D.W.; Leape, L.L.; Cullen, D.J.; Laird, N.; Petersen, L.A.; Teich, J.M.; Burdick, E.; Hickey, M.; Kleefield, S.; Shea, B.; Vliet, M.V.; Seger, D. Effect of computerized physician order entry and a team intervention on prevention of serious medication errors. JAMA, J. Am. Med. Assoc. **1998**, *280* (15), 1311–1316.
5. Evans, R.S.; Pestotnik, S.L.; Classen, D.C.; Clemmer, T.P.; Weaver, L.K.; Orme, J.F.; Lloyd, J.F.; Burke, J.P. A computer-assisted management program for antibiotics and other antiinfective agents. N. Eng. J. Med. **2001**, *338* (4), 232–238.
6. Rivkin, S. Opportunities and challenges of electronic physician prescribing technology. Med. Interface **1997**, *83*, 77–83.
7. Sardinha, C. Electronic prescribing: The next revolution in pharmacy? J. Managed Care Pharm. **1998**, *4* (1), 35–39.
8. Pankaskie, M.; Sullivan, J. New players, new services: E-scripts revisited. J. Am. Pharm. Assoc. **2000**, *40* (4), 566.
9. Martin, R.D. Digital prescription pads; bad penmanship? Essent. Inf. **2000**, *2* (1), 3–4.
10. Ukens, C. Are you ready? Drug Top. **2001**, *39*, 34–36.
11. Staniec, D.J.; Goodspeed, D.; Stember, L.A.; Schlesinger, M.; Schafermeyer, K., et al. The National Council for Prescription Drug Programs: Setting standards for electronic transmission of pharmacy data. Drug Benefit Trends **1997**, *1*, 29–35.
12. Venot, A. Electronic prescribing for the elderly; will it improve medication usage. Drugs Aging **2001**, *15* (2), 77–80.
13. Armstrong, E.P. Electronic prescribing and monitoring are needed to improve drug use. Arch. Int. Med. **2000**, *160* (18), 2713–2714.
14. Komshian, S. Electronic prescribing; system helps physicians avoid errors and offer better service. Phys. Comput. **2000**, 12–15.
15. Canaday, B.R.; Yarborough, P.C. Documenting pharmaceutical care: Creating a standard. Ann. Pharmacother. **1994**, *28*, 1292–1296.
16. Lieder, T.R. Computerized prescriber order entry changes pharmacists' roles. Am. J. Health-Syst. Pharm. **2001**, *58* (10), 846–851.

Ethical Issues in Clinical Pharmacy

Teresa Requena Caturla
Hospital "Principe de Asturias", Madrid, Spain

INTRODUCTION

The *Encyclopedia of Bioethics* defines "bioethics" as: "The systematic study of the moral-dimensions—including moral vision, decisions, conduct and policies—of the life sciences and healthcare, employing a variety of ethical methodologies in an interdisciplinary setting".[1] "Clinical ethics" is considered to be a subspecialty of bioethics and refers to the daily decision making of those who care for the patient.

CLINICAL ETHICS—THE PROFESSIONAL RELATIONSHIP BETWEEN THE HEALTH PROFESSIONAL AND PATIENT

As emphasized by Diego Gracia,[2] the professional relationship between the health professional and patient is a social one, although it seems that no one else is involved. When speaking of "third parties," one delimits within a generic concept of society, a more precise one. In the professional relationship between the health professional and patient, there are "three parties." The relationship is not lineal but rather triangular with three vertices: the *patient, the health professional* (physician, pharmacist, or nurse), and the *society* (social structures: health institutions, health insurance, etc.).

One might think that the health professional and the patient make, in accordance with the principles of nonmaleficence and autonomy, the decisions they find to be pertinent. The third parties put them into practice, as if these were means or instruments to reach an "end": the health professional–patient decision. But the "third parties" are structures with their own entity. So much so that they are guided by a third principle distinct from that of the health professionals nonmaleficence principle and that of the patient's autonomy. The principle of the third parties or that of the society is that of "justice." The principle of justice has embodied itself in a political tradition.

Changes in the healthcare model can generate ethical conflicts. If healthcare is made universal, it covers the entire population. Due to economic crises and scarce resources, it is not possible to meet all needs, just the basic ones or those that can be legally claimed. In any case, the system should guarantee equal access to and fair distribution of limited health resources.

ETHICAL THEORIES AND PRINCIPLES AS A FRAMEWORK FOR DECISION MAKING

Despite the fact that the new codes of pharmaceutical ethics[3,4] include the basic principles upon which bioethics is based (i.e., beneficence, autonomy, and justice), they are not complete enough to serve as a framework for making decisions in concrete situations where the basic principles come into conflict. In this case, an ethical foundation and a method are necessary.

The primary foundations are summed up in three theories: the theory of virtue, the deontological theory, and the consequentialist theory. (The reader is referred to other sources for more information.[1])

Bioethics, basing itself on the moral canon of the human being and on the necessity, as a rational being, to morally justify one's own acts, adopts the four ethical principles: *autonomy* and *beneficence* which pertain to the private sphere of the individual and *nonmaleficence* and *justice* which pertain to the public sphere.[5]

Decision-Making Procedures in Clinical Ethics

For several years, decision trees have been used in clinical ethics, although generally in a simplified form without carriyng out a detailed calculation of probabilities. One of the first to use this procedure was Baruch Brody, but the model was more widely accepted due to its simplicity was that of David C. Thomasma. Albert Jonsen developed a procedure based on the language of "cases" and "maxims." Sir David Ross, a great English ethicist at the beginning of the twentieth century, established the principalist method of the analysis of concrete cases. In this method, he establishes two moments in the moral judgment. First, that of the *prima facie* obligations and then

Encyclopedia of Clinical Pharmacy
DOI: 10.1081/E-ECP 120006385

that of *actual* obligation—that which is a true duty in a concrete circumstance. In other words, the *prima facie* obligations are objectives that can be canceled by other *prima facie* obligations of greater urgency. According to D. Gracia, their present application consists of:[6]

- The "a priori" moment: The *prima facie* principles of autonomy, beneficence, nonmaleficence, and justice.
- The "a posteriori" moment: Real and effective principles where the *prima facie* principles that are in conflict are arranged in order of importance, taking into account the concrete situation and the foreseen consequences. The hierarchy can vary according to each person's perception of a concrete situation. For this reason, it is best to keep in mind the greatest number of possible viewpoints in an attempt to enrich the analysis as much as possible before making a decision.

Such is the primary objective of the so-called "Institutional Committees of Ethics."

Professor Diego Gracia uses a procedure based on the analysis of the principles and consequences, like that suggested by David Ross, and applies it to clinical ethics.

Decision procedure in clinical ethics[6]

1. Analysis of clinical history by problems (biological, social).
2. Analysis of the clinical biological data and discussion of findings.
3. Identification of possible ethical problems—differentiate, count, and define all the ethical problems found in the clinical history.
4. Selection of the problem that causes a fundamental conflict of values.
5. Study of the possible courses of action.
6. Selection of the optimum possibility, that which saves the most values in conflict.
7. Decision on the course of action to be taken.
8. Analysis of the strong arguments against the decision, as well as the reasons for the decision (ability to defend it publicly).

ETHICAL PROBLEMS IN THE PHARMACIST'S CLINICAL PRACTICE

Relationship Between Physician, Pharmacist, and Patient

The pharmacist, as a health professional, can become immersed in various ethical problems. These are not unique to the pharmacist; many health professionals must deal with these same problems.[7] Such conflicts develop within the framework of the relationship between health professional and patient discussed earlier. For teaching purposes and because therapy with medication is used on almost all patients, this relationship triangle could be modified. It could be given a new dimension by converting it into a tetrahedron with the relationship physician–pharmacist–patient at the base and the society at the upper vertex (Fig. 1). Neither nursing nor the family is being excluded, as they are included with the physician and patient, respectively.

Professionals within the clinical relationship should work within a legal framework that defines the domain of each and respects the following patient rights:

- Confidentiality is the obligation of all health professionals to not reveal to others, without permission of the patient, information relative to the sick person or the illness, which goes along with the right to confidentiality of the patient. But this is a *prima facie* obligation, not an absolute one. Thus, when another person is in danger or the law calls for it, an exception should be made.
- Privacy, a patient right, dictates that no nonauthorized persons have access to their room, clinical history, or databases where pertinent information can be found.
- Revealing clinical, diagnostic, therapeutic, and prognostic information to the patient, as long as legislation does not say anything to the contrary, is in the domain of the treating physician. This fact does not mean that the pharmacist cannot give the patient information on the prescribed medication. But, for the benefit of the patient, it is best that this be done within the framework agreed upon for the collaboration between physician and pharmacist.

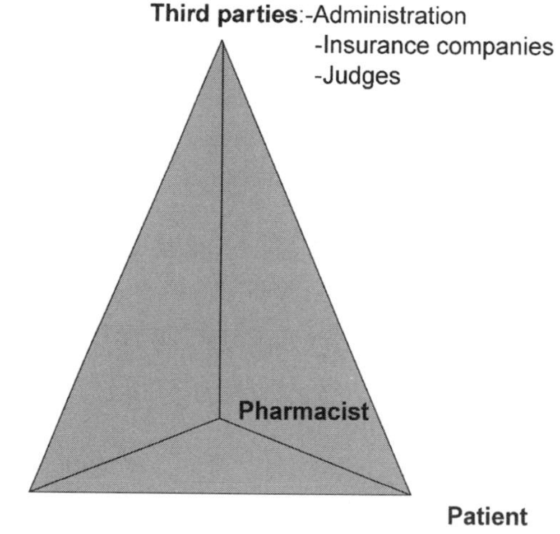

Fig. 1 Relationship between physician, pharmacist, and patient.

Definition of the Ethical Problem in Pharmacotherapy

T. L. Beauchamp and J. F. Childress define an ethical problem as a conflict between two moral obligations or norms. In general, there are two types of ethical problems:[8]

- Those originating from doubts about the morality of the act in itself in the face of strongly opposed arguments.
- Those originating from doubts about the decision whether to do one thing or another, both being mutually exclusive and implying a moral obligation.

The more specific problems in the pharmacist's clinical practice within this relationship are derivatives of the therapy with medication, nutrition, hydration, and placebo treatments.

We can define the ethical problem in pharmacotherapy as the conflict between moral obligations or norms that can put in danger the pharmacological treatment that is best for the patient.

Classification of Ethical Problems in Pharmacotherapy

The ethical problems in pharmacotherapy can be classified in the following manner.

Pharmacotherapeutic decisions

These are problems brought about by interprofessional differences (physician–pharmacist–nurse) in the making of pharmacotherapeutic decisions:

- In the evaluation of the benefits and risks of the necessary pharmacotherapy or that prescribed by a physician for a patient.
- In the inclusion of patient preferences in the pharmacotherapeutic decisions.

The analysis of these problems identifies a conflict of values or norms. On the one hand, in the first case it is the moral obligation of the pharmacist to promote the optimum treatment for the patient. In the second case, it is the obligation of the pharmacist to respect the autonomy and dignity of the patient.

The most adequate therapeutic decision is the selection of the therapeutic option that is most valid, taking into account the patient's circumstances in view of a highly probable diagnosis and prognosis, which is furthermore then accepted by the patient.

For this reason, as a precursor to the problem, it is assumed that the pharmacist will maintain professional competence, and that the pharmacist knows the clinical history of the patient as well as the circumstances of the case and preferably of the patient.

A conflict is generated when once the discrepancies have been discussed with the physician, it is socially expected that the pharmacist follow the medical order and dispense the medication prescribed.

This type of conflict can come about in the following circumstances:

- Omission of a validated and clearly suitable therapy.
- Prescription of nonvalidated therapies, which are considered to be neither suitable nor nonsuitable.
- Therapies that are clearly nonsuitable.
- The imposition of therapies on the patient on the part of the health professionals.
- Patient demand for a therapy not recommended by the physician.

Practical examples from scientific literature include obligatory sedation,[9] toxic analgesia,[10] the withdrawal of treatments (antibiotics, nutrition, hydration),[11,12] and the use of a placebo.[13]

Unavailability of medication

This is an ethical problem brought about due to lack of access to or unavailability of medication which is clearly suitable, with no equally efficient alternatives for a specific patient, orphan drugs, etc.

The present availability of scientific literature to all professionals in industrialized countries can lead to the knowledge of the existence of medications that are not commercially available in our countries. The professional could feel that it is more appropriate for the patient, but the administration does not approve its importation.

Another case would be when there is a lack of medicines in a given moment. This rationing would then imply the selection of a population to be treated, and it would be required that clear and fair criteria be used, such as the objective criteria of greatest benefit or due to prognostic factors or even by drawing lots.

The analysis of this problem introduces, on the one hand, the obligation of the pharmacist to promote the optimum treatment for the patient, and on the other hand, the obligation of the administration to establish explicit criteria for access to or availability of medicines being researched for severe illnesses or those which are life-threatening without satisfactory alternative treatments[8] (such as policies on orphan medicines, magistral formulation of nonregistered active ingredients, etc.).

A conflict can arise between the standard of evidence considered necessary by the administration, the randomized and controlled clinical study (RCT), and the desire of the patient to participate in an open trial, compassionated use (CU). This would mean a conflict between the principle of autonomy (patient) and that of beneficence (administration).

In favor of the open trials CU, it is argued that a minimum is being required (the RCT), which the patient does not want, and thus falls into a social paternalism. Furthermore, it is argued that the investigation of the clinical practice is possible, carrying out studies of results, without having to do studies with a control-arm or placebo.

In favor of RCTs, it is argued that since a vulnerable population is being dealt with, there could be a commercial exploitation upon introducing a medication in a pathology that does not have therapeutic alternatives, without having obtained a minimum standard of scientific evidence. If all of the patients with this pathology are offered this medication, no comparison can be made between this alternative and a placebo. Thus, there will be no certainty of its efficacy, and no other posterior therapy can be compared with a placebo.

Discrimination

This ethical problem is brought about due to a possible discrimination either in the use of or the cost for the patient of the pharmacotherapy.

Negative Discrimination in the Use of the Pharmacotherapy. This refers to the nonutilization of suitable therapies for elderly patients or women without situations of comorbidity which justify it.[14,15] The Committee of Ethical and Judicial Affairs of the American Medical Association has written reports about age-base rationing, gender, and black–white disparities in clinical decision making.[16]

In reality, negative descrimination does not produce any ethical conflict. It is not ethical in itself, as it does not respect the principles of nonmaleficence and justice.

Positive Discrimination in the Use of or in the Cost of the Pharmacotherapy. An example is the use of epoetin in patients who do not accept blood transfusions for religious and other reasons. The conflictive principles in this case could be beneficence and justice. Its use could be justified if justice is understood as equity, using the following argument: Blood transfusion is clearly against the beliefs of this group. These beliefs have been repeatedly infringed upon. According to the principle of equity, more should be given to the most needy, always applying explicit and transparent criteria.

As far as cost is concerned, positive discrimination occurs when the administration decides in favor of public financing of complete therapies for certain pathologies.[17]

Rationing

These ethical problems are brought about by the denial or restriction of medicines due to cost.

Rationing according to cost is the systematic and deliberate denial of some resources, although they could be very beneficial, because they are considered very expensive. Those cases for which there are less expensive alternative therapies, which are equally efficient and safe, are excluded. This would clearly be the most just (principles of rationality and distributive justice) and suitable therapy.

- Rationing of a clearly suitable therapy that does not have an alternative that is equally efficient and safe. The principles in conflict here would be those of nonmaleficence and justice. The rationing should be equitable and not infringe upon the ''decent minimum.'' This is ethically acceptable when the rationing criteria are explicit and known to those potentially affected. This is understood within a framework of scarce resources in which all of the measures have been adopted for the rationalization of these.
- Rationing of therapies that are thought to be neither suitable nor nonsuitable (there is no proof for or against) which are restricted or denied due to their elevated cost. The conflict in this situation comes about between the principle of beneficence (if the physician orders the treatment) or the principle of autonomy (the patient wants the therapy) and that of justice. No conflict exists if the patient finances his/her own treatment, but it does exist if it is financed by the public health service. Generally, the principle of justice prevails over the other two, and all exceptions should be justifiable. For decisions for rationing to be just (distributive justice), they need to be adopted by the Health Administration.

REFERENCES

1. Reich, W.T. *Encyclopedia of Bioethics (CD-Rom Revised Edition)*; MacMillan Library Reference: New York, 1995.
2. Gracia, D. La relación clínica. Rev. Clin. Esp. **1992**, *191* (2), 61–63.
3. American Pharmaceutical Association. Code of ethics for pharmacists. Am. J. Health-Syst. Pharm. **1995**, *52*, 2131.

4. FIP Code of ethics 1997. www.fip.nl/publication/publication1.htm. (accessed Oct. 2000).
5. Gracia, D. *Fundamentos de Biotica,* 1st Ed.; Eudema: Madrid, 1989.
6. Beauchamp, T.L.; Childress, J.F. *Principles of Biomedical Ethics,* 4th Ed.; Oxford University Press: New York, 1994.
7. Gracia, D. *Procedimientos en Ètica Clínica,* 1st Ed.; Eudema: Madrid, 1991.
8. Buerki, R.A.; Vottero, L.D. *Ethical Responsability in Pharmacy Practice*; American Institute of the History of Pharmacy: Wisconsin, Madison, 1996.
9. Manolakis, M.L.; Uretsky, S.D.; Veatch, R.M. Sedation of an unruly patient. Am. J. Hosp. Pharm. **1994**, *51*, 205–209.
10. van der Heide, A.; van der Maas, P.J.; van der Wal, G.; Kollee, L.A. Using potencially life-shortening drugs in neonates and infants. Crit. Care Med. **2000**, *28* (7), 2595–2599.
11. Winker, M.A.; Flanagin, A. Caring for patients at the end of life. JAMA **1999**, *282* (20), 1965.
12. Council on Ethical and Judicial Affairs American Medical Association. Medical futility in end-of-life care. JAMA **1999**, *281*, 937–941.
13. Lachaux, B.; Placebo, L.P. *Un Medicamento que Busca la Verdad,* 1st Ed.; McGraw Hill: Madrid, 1989.
14. Pettersen, K.I. Age-related discrimination in the use of fibrinolitic therapy in acute myocardial infarction in Norway. Age and Aging **1995**, *24*, 198–203.
15. Miller, M.; Byington, R.; Hunninghake, D.; Pitt, B.; Fuberg, C.D. Sex bias and underutilization of lipid-lowering therapy in patients with coronary artery disease at academical medical centers in the United States and Canada. For the Prospective Randomized Evaluations of the Vascular Effects of Norvasc Trial (PREVENT) Investigators. Arch. Intern. Med. **2000**, *160* (3), 343–347.
16. www.ama-assn.org/ama/pub/category/2513.html (accessed Oct. 2000).
17. Rothman, D.J. The rising cost of pharmaceuticals: An ethicists perspective. Am. J. Hosp. Pharm. **1993**, *50* (Suppl. 4), 10–12.

Ethical Issues Related to Clinical Pharmacy Research (ACCP)

American College of Clinical Pharmacy
Kansas City, Missouri, U.S.A.

INTRODUCTION

Bioethics is a relatively new field of study concerning the investigation of ethical issues in medicine, health care, and the life sciences. From the standpoint of bioethics, clinical pharmacy research presents no novel ethical questions; however, the type and scope of issues involved differ from those faced by other practitioners. It is important for pharmacists to be aware of the ethical issues, give thoughtful consideration to them, and be sensitive to how they may affect their involvement in research. The current Code of Ethics for the practice of pharmacy virtually neglects issues encountered by pharmacists as they conduct clinical research.[1]

Pharmacists are expanding their responsibilities as health care practitioners by initiating and participating in clinical research.[2] These activities range from custodian of nonclinical and clinical trial information to principal investigator engaged in original research. For a discipline to survive as an entity, it must expand its body of knowledge continuously, rather than relying on other disciplines to create its knowledge base, including generating data that propose of confirm theories, principles, or relationships.

Because of the nature of ethics, this article presents more questions than it provides answers; it is difficult to predefine the right answers to ethical questions. Most experienced investigators will recognize the circumstances described and will have developed their own solutions. The article however, should prove useful to new investigators or trainees, perhaps as a mechanism to introduce discussion with mentors. It identifies ethical issues and questions in clinical pharmacy research regarding protection of human subjects, informed consent, conflicts of interest, clinical trial design, investigator independence, and scientific integrity.

HISTORICAL PERSPECTIVE

The Nuremberg Code[3,4] and the Declaration of Helsinki[5] are accepted international documents guiding the conduct of human clinical research (Appendices 1 and 2). The Nuremberg Code, established in 1948 after the war crimes trials of 1946, was the first internationally recognized code for human research. During the early 1950s, ethics committees for clinical research appeared in the United States. Until then, physician investigators and research institutes autonomously determined when investigations became dangerous and to what extent research subjects should be informed. Later the Department of Health, Education, and Welfare [the present Department of Health and Human Services (DHHS)], in response to reported abuses of the rights of individuals participating in certain federally supported research endeavors, mandated that all protocols be screened by institutional committees responsible for the protection of human subjects. The federal government committed itself when Congress established the National Commission for the Protection of Human Subjects of Biomedical and Behavioral Research in 1974. The commission issued the Belmont report in 1978;[6,7] with that, institutional review boards (IRBs) were born and principles of protecting the rights of human subjects participating in research began to evolve.

The Belmont report describes the basic ethical principles that underlie research involving human subjects: respect for persons, beneficence, and justice. The report discusses application of informed consent, assessment of risks and benefits, and selection of subjects. Its regulations require that IRBs have not fewer than five members who have the capability to judge research proposals in terms of community attitudes. Therefore, IRBs must include people whose primary concerns lie in the areas of legal, professional, and community acceptance rather than in the overall scientific design.

During the early 1980s, the DHHS developed and published rules and regulations for the protection of

Encyclopedia of Clinical Pharmacy
DOI: 10.1081/E-ECP 120006406

human research subjects' participation in federally funded research known as the Code of Federal Regulations. The Food and Drug Administration published similar regulations governing human research and investigations that are intended to support marketing permits for drugs, food additives, medical devices, biologic products, and electronic devices. These sets of regulations serve as the cornerstone for the oversight of safe human experimentation and guide all who participate in clinical research, (e.g., IRBs, investigators, research sponsors, research subjects). Generally, state agencies adopt the federal standards, and local research institutions interpret and apply them to all research activities involving humans.

PROTECTION OF HUMAN SUBJECTS

The IRB is charged, by federal, state, and local institutions with ensuring that principal investigators adequately protect the health and well-being of individuals whose participation may cause them to be at increased risk to hazards, defined broadly as physical, psychologic, sociologic, and legal. Thus, it is impossible to conduct clinical research in humans that would not affect one or more of these areas.

If local institutions receive any federal research money, all human research must be approved by the IRB. This is not the basis for IRB review but provides the incentive for local institutions to conduct studies that are ethical. The committee becomes involved in matters such as confidentiality, anonymity, and moral issues related to experimental activities. Approval from an IRB, however, does not relieve the principal investigator from the basic responsibility of safeguarding the health and welfare of participating individuals. This is a moral and professional responsibility that cannot be delegated.

INFORMED CONSENT

Informed consent comprises two distinct concepts. *Informed* means that the researcher provides something (information, assistance with a decision) to the subject. *Consent* means that there is something (permission) that the researcher requests from the subject. Consent must be given freely.

The informed consent process answers the moral question, when is it permissible to include competent people as research subjects? The answer is, if, and only if, they have given their free and informed consent. Inherent in this statement is the idea that investigators should ask for or request consent, not simply to get or obtain it. The mere existence of a signed form does not guarantee that the informed consent process worked for the benefit of the subject, but it can facilitate the process. The rights of the subject must be protected, and informed consent must be requested and obtained.

A risk–benefit assessment must be performed before a proposed investigation is submitted to the IRB. Because the true risk-to-benefit ratio generally is unknown, clear evidence for a favorable outcome must exist. Benefits may be gained by individual participants or by society as a consequence of the proposed activity.

Potential participants must agree in writing to the conditions of the study after receiving a complete and understandable explanation of the conditions of participation, the purpose of the activity, and the possible hazards involved. They must have the right to ask questions and to withdraw their consent at any stage of the activity.

Ethical Questions Concerning Informed Consent

Informed consent assumes that accurate information is being given, that the subject comprehends the information, and that the subject volunteers to participate. Do investigators emphasize each aspect appropriately? For example, how does the investigator ensure that the subject comprehends? Examples of methods used are having the subject repeat back in her or his own words the information immediately, and repeat it at some future time while involved in the research; and using a witness to sign the consent form.

While preoccupied with the informed consent form, investigators may neglect using required and appropriate language. How do pharmacists ensure that eighth-grade language is used on the consent form, and is it ever verified? If non-English-speaking people are being requested to participate in research studies, the informed consent form should also be written and presented in a language they understand. Computer programs and English teachers may be used to facilitate this process. Thus, two informed consent forms would be prepared, one in English and one in the appropriate non-English language. Investigators should ensure that these issues are not neglected.

How much information is necessary for potential subjects to be informed? Should investigators tell the subjects how much money they are compensated per subject recruited? It is probably unnecessary for subjects to understand how clinical research is funded (e.g., overhead fees, fees for certain services), unless this information would influence any reasonable person to participate (or not to participate). The pharmacy profes-

sion and society as a whole determine what reasonable people usually do, and this is susceptible to change over time. Research subjects have a right to know what is known, including the views of the investigators specifically and the pharmacy and medical professions in general. At minimum, investigators should give subjects the information that the average reasonable person would want to know.

Finally, how informed could a subject be about a new chemical entity when the aim of the study is to gain information for the first time in humans? To balance the apparent lack of information, the investigator is responsible for carefully monitoring the subject during all stages of the investigation.

Payment to Study Volunteers

The informed consent process raises ethical questions. The informed consent form may state that volunteers will be paid for their services, yet when does the payment become simply an inducement? Payment to volunteers for participation in drug trials is common and usually can be subdivided into two types, reimbursement for expenses incurred incidentally and wage payments. Reimbursement might cover expenses such as transportation costs, costs incurred by participation (e.g., extra blood sampling, new drugs or devices being used), and lost work time. Wage payments involve remuneration for services provided in serving as a research subject. These payments could be based on a number of factors, such as time commitment required, nature, and number of procedures performed, or to facilitate recruitment in a timely fashion. Payment should not constitute an inducement.

When ill persons are offered money over and above expenses to enter a clinical trial of a new drug therapy, the possibility of coercion exists. The reasoning is that if the patient is poor, they might not be able to afford the therapy without entering the trial. Contrast this experience with renally impaired volunteers recruited for a pharmacokinetic study of an antibiotic. Renal failure is not the target of therapy. The subjects receive no therapeutic benefit from participating and are paid as volunteers.

The IRBs should review research funding for appropriateness and possible coercion, specifically as it applies to subject recruitment. If the amount of payment is so high as to induce any reasonable person to participate, regardless of the risk, it is obviously too high. It becomes difficult, however, to determine when coercion is present because the majority of cases are not this obvious. Investigators should be able to justify any payment to research subjects. Several factors should be considered in justifying payment, such as the intensity of the protocol, whether it is funded and by whom, and the degree of benefit to subjects other than monetary.

Influence of Drug Therapy

Little information exists about how a patient's drug therapy influences the informed consent process. For example, can a patient who has had several doses of intravenous morphine give consent to participate in an acute myocardial infarction protocol? Sedated patients may not understand adequately what they are being told; therefore, they cannot make up their minds freely. As another example, how informed can patients be who are experiencing blurred vision from atropine? Does drug exposure influence continued participation or future consent? If there are any doubts, a family member, guardian, or patient advocate should be involved in the informed consent process.

Adverse Effects

In the context of a clinical trial, informing the patient of possible side effects could influence the outcome of the study. However, subjects have the right to know what may be expected to occur during participation. They must be informed of all possible adverse effects consistent with the information in the package insert (if available) and the information known from other studies.

ETHICAL QUESTIONS CONCERNING MORAL PRINCIPLES

Pharmacists, like physicians, have to be aware of the sovereignty of the patient. Although the protection of human subjects is critical, there is little opposition to the protection of human rights. However, opposition to other critical issues does exist to various degrees.

Questions of Fairness

When should we encourage repeated volunteering? Could studying the same pool of patients have a negative impact on the care of others? In other words, volunteering over and over again may; 1) deny the benefit of that research to others; 2) make research subjects bear too great a burden themselves; and 3) result in data that cannot be generalized to the rest of the population. Careful examination of the purposes of each investigation must be made to ensure that repeated volunteering is beneficial to subjects or to the experimental purpose. Thus, mere expediency of enrolling subjects does not justify studying the same individuals routinely. In some instances, such as pharmacokinetic studies, repeated use of the same subjects may be acceptable.

Therapeutic research is intended to benefit those who are the subjects of that research. What are the proper criteria for inclusion and exclusion that would ensure that everyone has a fair chance of benefiting from participating within the scope of the hypothesis being tested? The principle of justice or fairness dictates that subjects be selected equitably, in other words, giving everyone an opportunity, and not concentrating on individuals with or without certain diseases, those located in close proximity to the service institution, or those of a particular gender. For example, patients with liver dysfunction commonly are excluded from research protocols, but in fact are frequently the ones who receive the study drugs. Consider also, investigations using predominantly individuals of one race or ethnic minority simply because of their availability. Should we encourage the investigation of drug disposition in these patients, especially as they relate to the problem being studied?

Thus, selection of subjects has the potential to be an ethical dilemma. The Belmont Commission's interpretation of the requirement of justice[6–8] is seen in the following statement:

> The selection of research subjects needs to be scrutinized in order to determine whether some classes (e.g., welfare patients, particular racial and ethnic minorities, or persons confined to institutions) are being systematically selected simply because of their easy availability, their compromised position, or their manipulability, rather than for reasons directly related to the problem being studied. Finally, whenever research supported by public funds leads to the development of therapeutic devices and procedures, justice demands both that these not provide advantages only to those who can afford them and that such research should not unduly involve persons from groups unlikely to be among the beneficiaries of subsequent applications of the research.

If there are known populations of people in whom drug disposition and effect differ, should we neglect enrolling them in clinical trials? Certainly it is expedient to develop protocols that control for factors that may be a source of variability. However, in doing this, investigators must not systematically neglect important segments of the population.

Research involving healthy volunteers rarely benefits the subjects directly, yet may be harmful to them. Should pharmacists then encourage the development and use of new technologies or methods (e.g., noninvasive) to reduce risks while maintaining the scientific integrity of projects?[9] A simple venipuncture exposes both the subject and the investigator to a degree of risk above that which occurs in daily life.[6–9] If the study drug possesses a saliva : plasma concentration of approximately 1, are we justified in obtaining plasma samples? If the study intent is to screen for substances present, investigators should use noninvasive methods when possible rather than those requiring venipuncture.

Conflicts of Interest

Conflicts of interest issues are morally relevant because they represent temptations to do wrong. Million-dollar budgets have ways of creating ethical dilemmas for investigators. A prevalent problem is the influence of commercial interests on independent drug research. Medicine has emphasized disclosure to minimize this problem, but disclosure does not guarantee elimination of ethical dilemmas.

The American College of Clinical Pharmacy offers recommendations to minimize conflicts of interest in the accompanying position statement ''Pharmacists and the pharmaceutical industry: guidelines for ethical interactions.'' The statement addresses questions such as, when is it permissible to accept an honorarium from a sponsor for providing a research talk, contributing to a symposium, or arranging a research-oriented training session? It also discusses the type of research that is appropriate to be funded. For example, it is unethical to perform a phase IV study for the sole purpose of familiarizing practitioners with a drug so that they will prescribe or recommend it frequently in the future. Ultimately, the pharmacist has the responsibility to maintain objectivity through the unprejudiced and unbiased performance of research activities regardless of the potential for personal financial gain.

Another example of a potential conflict of interest is the use of finder's fees to help to identify research subjects. A finder's fee is a fee paid to individuals, usually nurses, physicians, and pharmacists, who assist in locating potential research subjects. It may not be wrong to offer such a fee, but it is probably wrong for investigators to demand it. It would be unethical to deny a patient the opportunity to benefit from a study simply because the investigator would not receive the money. In lieu of paying finder's fees directly, some institutions provide credit to a bookstore account or payment to a special account whose funds can be used only for educational purposes.

Individual Versus Social Interest

When is it permissible to deny some benefits, or put some subjects at risk, for the sake of research and the benefits it promises? For example, when is it permissible to perform cost-containment research, and what type of peer review and informed consent is necessary?[10] This is particularly relevant for pharmacists because many are involved in this type of data collection and analysis. It is possible that some subjects may receive a lower standard of care than that to which they are accustomed. Thus, experimental strategies that reduce services may expose subjects to the possibility of harm without benefit.

Political or public policy agendas may exist that do not necessarily reflect the best interests of research subjects. If so, pharmacists must maintain the highest standards of integrity. This may require them to become more involved in establishing research priorities at federal, state, and local levels. We should address under what circumstances it is appropriate to encourage studies that are risky, potentially unfundable, or would require extensive time or commitment (which usually means a long delay before publishable results are generated). The probability of funding should not determine the direction of research.

Under what conditions is it permissible to delay the publication of promising results until more substantial evidence is available? The reverse question is an ethical dilemma as well. That is, under what conditions is it permissible to publish promising results even though, according to accepted standards, more evidence is needed to validate the results? The increasing newsworthiness of medical research has given this issue much attention, and conflicts directly with the established, albeit time-consuming, publication process: manuscript preparation, peer review, and revision.

Some have criticized the Ingelfinger rule.[11] Over a decade ago, the editor of the *New England Journal of Medicine*, Franz Ingelfinger, ruled that no medical research would be published if it had been published previously, whether in the scientific or lay press. (The rule permits previous publication of abstracts or presentations at meetings.) Most major scientific journals have similar policies.

Vocal patient groups, the lay press, and the public want medical news as fast as possible; results of new research are seen or heard daily in the news. The National Institutes of Health has begun releasing some research results (e.g., Clinical Alert) directly to health care providers and the public before the results are published. They deem the results too urgent for the public's health to be delayed by the publication process.[11] What are the ethical issues of such early release of research results, and who is the appropriate authority to decide what is urgent? How complete should the prepublication release of medical research be? What is the track record of prepublication releases? Is it unethical that some journals take months to print research results because of their peer review process? Policies should be developed that define appropriate mechanisms for early release of research findings, and their effectiveness and impact should be evaluated.

CLINICAL TRIAL DESIGN

The design of randomized clinical trials introduces ethical issues.[12] Usually, study designs prevent the treatment from being modified because of the need to collect sufficient data to allow valid statistical inference. Ethically, clinicians are required to provide their patients with the best available treatment; however, the justification for a randomized clinical trial is simply that the best treatment is not yet known.

What is the proper role for placebo controls? It has been suggested that, ''apart from needing to be both valid and valuable,''[13] they must satisfy two premises: there exists (or there is the likelihood to exist) a controversy among expert clinicians concerning the relative therapeutic merits of each treatment, including the placebo; and the design of the study must warrant confidence that the results will show which of the regimens is superior and therefore will influence clinical practice.[13]

Placebo controls can be justified if the trial is conducted in an area that falls within one of four broad categories:

1. Conditions for which no standard therapy exists at all.
2. Conditions for which standard therapy has been shown to be no better than placebo.
3. Conditions for which standard therapy has been called into question by new evidence, creating doubt concerning its presumed net therapeutic advantage.
4. Conditions for which validated optimum treatment is not made freely available to patients because of cost constraints or other considerations (e.g., physical location of treatment centers).

These categories should be used as initial guidelines. The federal Food and Drug Administration desires that studies of a new chemical entity be compared with placebo in small groups of patients during phase II testing.

At what point should a clinical trial be stopped prematurely because enough evidence has been gathered to show that some treatment is efficacious? Investigators, with the help of statisticians, should develop guidelines that answer this question before the protocol is submitted to the IRB (or at least prior to data collection). These guidelines should be communicated to all persons involved in the research effort directly (investigators and research subjects). A safety committee should be responsible for monitoring data collected during a trial and stopping the trial if a predefined boundary is crossed, whether for early evidence of benefit or unacceptable toxicity.

INVESTIGATOR INDEPENDENCE

Investigator independence is another important issue. Almost all industry-funded research is reviewed by the sponsor prior to publication, and if the results are not favorable, pressure not to publish may be considerable. Some protocols forbid the investigator to publish results without permission; they cite the availability of confidential commercial information as the reason. The implied threat is that if the results are published, the investigator will not receive funding in the future. Pharmacists should be independent investigators with the right and authority to publish research findings. The intellectual property is owned by the investigators and their institution, not the funding agency.

SCIENTIFIC INTEGRITY

Integrity is a complex concept with associations to conventional standards of morality and personal beliefs about truth telling, honesty, and fairness. Unintentional investigator bias is a scientific error. Intentional investigator bias is a form of fraud. Fraud is the deliberate reporting of what one believes to be false with the intention of deceiving others.[14] Within a research program or institution, mechanisms should exist that check for data trimming, selective reporting, quality control, and originality. Sloppy research is unethical; examples are inconsistencies in record keeping involving research subject files, sample preparation and other analytical procedures, raw data files, and statistical analysis files. Plagiarism is another serious offense that compromises scientific integrity and is not acceptable.

Negative data should be published if they are scientifically sound, particularly when they fill gaps in current knowledge. They also may decrease redundancy in future investigations. Investigators should publish complete information when possible; fragmenting data sets is discouraged.

CONCLUSION

The research process introduces many ethical questions particularly relevant to clinical pharmacy investigators. Most important, investigators must be aware of their moral responsibility to safeguard the health and welfare of individuals who participate in research. The informed consent process is used to ensure that study subjects understand the conditions of their participation, the purpose of the study, and the possible hazards involved; and to ensure that consent is given freely. Investigators and IRBs must be certain that payments to study volunteers are not excessive or coercive. Finally, clinical pharmacist investigators must avoid or minimize potential conflicts of interest by establishing themselves as independent investigators performing studies with utmost scientific integrity.

ACKNOWLEDGMENTS

The subcommittee appreciates the helpful comments and critical review of the following people: David Brushwood, J.D., Peter Iafrate, Pharm.D., Bruce Russell, Ph.D., Craig Svennson, Pharm.D., Ph.D., and Russel Thomas, Pharm.D.

This document was written by the following subcommittee of the 1991–1992 ACCP Research Affairs Committee: David R. Rutledge, Pharm.D., Chair; Clinton Stewart, Pharm.D., Vice Chair; and Daniel Wermeling, Pharm.D. Other members of the committee were Kathryn Blake, Pharm.D.; Jacquelien Danyluk, Pharm.D.; Jimmi Hatton, Pharm.D.; Dennis Helling, Pharm.D., FCCP; K. Dale Hooker, Pharm.D.; David Knoppert, M.Sc. Pharm.; Patrick McCollam, Pharm.D.; Christopher Paap, Pharm.D.; Michael J. Rybak, Pharm.D., FCCP; Kathleen Stringer, Pharm.D.; and Joseph DiPiro, Pharm.D., FCCP, Board Liaison. Staff editor: Toni Sumpter, Pharm.D. Approved by the Board of Regents on May 21, 1993.

From *Pharmacotherapy* 1993, 13(5):523–530, with permission of the American College of Clinical Pharmacy.

APPENDIX 1—THE NUREMBERG CODE[3,4]

1. The voluntary consent of the human subject is absolutely essential.
2. The experiment should be such as to yield fruitful results for the good of society, unprocurable by other methods or means of study, and not random and unnecessary in nature.
3. The experiment should be designed and based on the results of animal experimentation and a knowledge of the natural history of the disease or other problem under study that the anticipated results will justify the performance of the experiment.
4. The experiment should be so conducted as to avoid all unnecessary physical and mental suffering and injury.
5. No experiment should be conducted where there is a priori reason to believe that death or disabling injury will occur except, perhaps, in those experiments where the experimental physicians also serve as subjects.
6. The degree of risk to be taken should never exceed that determined by the humanitarian importance of the problem to be solved by the experiment.
7. Proper preparations should be made and adequate facilities provided to protect the experimental subject against even remote possibilities of injury, disability, or death.
8. The experiment should be conducted only by scientifically qualified persons. The highest degree of skill and care should be required through all stages of the experiment of those who conduct or engage in the experiment.
9. During the course of the experiment the human subject should be at liberty to bring the experiment to an end if he has reached the physical or mental state where continuation of the experiment seems to him to be impossible.
10. During the course of the experiment the scientist in charge must be prepared to terminate the experiment at any stage, if he has probable cause to believe, in the exercise of the good faith, superior skill, and careful judgment required of him that a continuation of the experiment is likely to result in injury, disability, or death to the experimental subject.

APPENDIX 2—MAJOR COMPONENTS OF THE WORLD MEDICAL ASSOCIATION DECLARATION OF HELSINKI[5]

Basic Principles

1. Biomedical research involving human subjects must conform to generally accepted scientific principles and should be based on adequately performed laboratory and animal experimentation and on a thorough knowledge of the scientific literature.
2. The design and performance of each experimental procedure involving human subjects should be clearly formulated in an experimental protocol which would be transmitted to a specially appointed independent committee for consideration, comment, and guidance.
3. Biomedical research involving human subjects should be conducted only by scientifically qualified persons and under the supervision of a clinically competent medical person. The responsibility for the human subject must always rest with a medically qualified person and never rest on the subject of the research, even though the subject has given his or her consent.
4. Biomedical research involving human subjects cannot be legitimately carried out unless the importance of the objective is in proportion to the inherent risk to the subject.
5. Every biomedical research project involving human subjects should be preceded by careful assessment of predictable risks in comparison with foreseeable benefits to the subject or to others. Concern for the interest of the subject must always prevail over the interest of science and society.
6. The right of the research subject to safeguard his or her integrity must always be respected. Every precaution should be taken to respect the privacy of the subject and to minimize the impact of the study on the subject's physical and mental integrity and on the personality of the subject.
7. Doctors should abstain from engaging in research projects involving human subjects unless they are satisfied that the hazards involved are believed to be predictable. Doctors should cease any investigation if the hazards are found to outweigh the potential benefits.
8. In publication of the results of his or her research, the doctor is obliged to preserve the accuracy of the results. Reports of experimentation

not in accordance with the principles laid down in the Declaration should not be accepted for publication.

9. In any research on human beings, each potential subject must be adequately informed of the aims, methods, anticipated benefits, and potential hazards of the study and the discomfort it may entail. He or she should be informed that he or she is at liberty to abstain from participation in the study and that he or she is free to withdraw his or her consent to participation at any time. The doctor should then obtain the subject's freely given informed consent, preferably in writing.
10. When obtaining informed consent for the research project the doctor should be particularly cautious if the subject is in a dependent relationship to him or her or may consent under duress. In that case the informed consent should be obtained by a doctor who is not engaged in the investigation and who is completely independent of this official relationship.
11. In case of legal incompetence, informed consent should be obtained from the legal guardian in accordance with national legislation. Where physical or mental incapacity makes it impossible to obtain informed consent, or when the subject is a minor, permission from the responsible relative replaces that of the subject in accordance with national legislation.
12. The research protocol should always contain a statement of the ethical considerations involved and should indicate that the principles enunciated in the present Declaration are complied with.

Medical Research Combined with Professional Care (Clinical Research)

1. In the treatment of the sick person, the doctor must be free to use a new diagnostic and therapeutic measure, if in his or her judgment it offers hope of saving life, reestablishing health, or alleviating suffering.
2. The potential benefits, hazards, and discomforts of a new method should be weighed against the advantages of the best current diagnostic and therapeutic methods.
3. In any medical study, every patient—including those of a control group, if any—should be assured of the best proven diagnostic and therapeutic method.
4. The refusal of the patient to participate in a study must never interfere with the doctor–patient relationship.
5. If the doctor considers it essential not to obtain informed consent, the specific reasons for this proposal should be stated in the experimental protocol for transmission to the independent committee.
6. The doctor can combine medical research with professional care, the objective being the acquisition of new medical knowledge, only to the extent that medical research is justified by its potential diagnostic or therapeutic value for the patient.

Nontherapeutic Biomedical Research Involving Human Subjects (Nonclinical Biomedical Research)

1. In the purely scientific application of medical research carried out on a human being, it is the duty of the doctor to remain the protector of the life and health of that person on whom biomedical research is being carried out.
2. The subjects should be volunteers—either healthy persons or patients for whom the experimental design is not related to the patient's illness.
3. The investigator or the investigating team should discontinue the research if in his or her or their judgment it may, if continued, be harmful to the individual.
4. In research on man, the interest of science and society should never take precedence over considerations related to the well-being of the subject.

REFERENCES

1. American Pharmaceutical Association. *Code of Ethics*; Washington, DC, 1981.
2. Cloyd, J.C.; Oeser, D.E. Clinical pharmacists in drug research and development: A historical perspective. Drug Intell. Clin. Pharm. **1987**, *21*, 93–97.
3. Anonymous. *Trials of War Criminals Before the Nuremberg Military Tribunals Under Control Council Law No. 10, Vol. 2*; US Government Printing Office: Washington, DC, 1949; 181–182.
4. Anonymous. The Nuremberg Code, Appendix 3. In *Ethics*

and Regulation of Clinical Research, 2nd Ed.; Levine, R.J., Ed.; Urban & Schwarzenberg: Baltimore, 1986; 425–426.

5. Anonymous. World Medical Association Declaration of Helsinki: Recommendations Guiding Medical Doctors in Biomedical Research Involving Human Subjects, Appendix 4. In *Ethics and Regulation of Clinical Research*, 2nd Ed.; Levine, R.J., Ed.; Urban & Schwarzenberg: Baltimore, 1986; 427–429.
6. Department of Health and Human Services. *Code of Federal Regulations, 45 C.F.R. 46. Protection of Human Subjects*; Washington, DC, 1978.
7. National Commission for the Protection of Human Subjects of Biomedical and Behavioral Research. *The Belmont Report: Ethical Principles and Guidelines for the Protection of Human Subjects of Research. DHEW Publication 05-78-0012*; US Government Printing Office: Washington, DC, 1978.
8. Levine, R.J. Basic Concepts and Definitions. In *Ethics and Regulation of Clinical Research*, 2nd Ed.; Levine, R.J., Ed.; Urban & Schwarzenberg: Baltimore, 1986; 1–18.
9. Svensson, C.K. Ethical considerations in the conduct of clinical pharmacokinetic studies. Clin. Pharmacokinet. **1987**, *4*, 217–222.
10. Brett, A.; Grodin, M. Ethical aspects of human experimentation in health services research. JAMA **1991**, *265*, 1854–1857.
11. Fletcher, S.W.; Fletcher, R.H. Early release of research results. Ann. Intern. Med. **1991**, *114*, 698–700.
12. Hellman, S.; Hellman, D.S. Sounding board: Of mice but not men. Problems of the randomized clinical trial. N. Engl. J. Med. **1991**, *324*, 1585–1589.
13. Freddman, B. Placebo-controlled trials and the logic of clinical purpose. IRB: A Review of Human Subjects Research **1985**, *7* (2), 1–4.
14. Levine, R.J. Ethical Norms and Procedures. In *Ethics and Regulation of Clinical Research*, 2nd Ed.; Levine, R.J., Ed.; Urban & Schwarzenberg: Baltimore, 1986; 19–35.

European Society of Clinical Pharmacy

Annemieke Floor-Schreudering
European Society of Clinical Pharmacy,
Leiden, The Netherlands

Yechiel Hekster
University Medical Centre, Nijmegen, The Netherlands

INTRODUCTION

In the 20th century, a conviction developed within the pharmacy profession that the professional knowledge of pharmacists was not used to its full potential. Activities to assure the safe and appropriate use of drugs became a new target, leading to activities in the direction of more patient-related aspects of drug therapy. This perception was present at about the same time on both sides of the Atlantic. It was logically named "Clinical Pharmacy," meaning a pharmacy activity directed to and in contact with the patient. The leaders of this new approach wanted to reinforce their message by founding professional organizations preoccupied with the teaching and practical development of Clinical Pharmacy. In 1979, the birth of the American College of Clinical Pharmacy (ACCP) and the European Society of Clinical Pharmacy (ESCP) took place simultaneously.[1]

WHAT IS ESCP?

The European Society of Clinical Pharmacy (ESCP) is an international society founded by clinical practitioners, researchers, and educators from various countries in Europe, which constantly looks for new areas of professional practice. Since the formation of the Society, there has been a gradual and sustained growth of clinical pharmacy in many European countries.

Overall Aim

The overall aim of the Society is to develop and promote the rational and appropriate use of medicines (medicinal products and devices) by the individual and by society.

Goal

The goal of ESCP is to encourage the development and education of clinical pharmacists in Europe.

The Society tries to achieve this goal by:

1. *Membership activities*:
   - Providing a forum for the communication of new knowledge and developments in clinical pharmacy.
   - Developing links with national and international organizations of pharmacists, teachers, and students interested in the development of clinical pharmacy.
2. *External relations*:
   - Promoting the value of clinical pharmacy services among other health care professionals, among scientific societies that share the same interest, organizations such as WHO (World Health Organization) and EMEA (European Agency for the Evaluation of Medicinal Products), and generally within the health service.
3. *Educational activity*:
   - Enforcing the formation of activities in the field of clinical pharmacy and pharmacotherapy through conventions and specific courses.
   - Promoting the inclusion of clinical pharmacy teaching at pre- and postgraduate levels.
4. *Training*:
   - Providing accrediting centers, where clinical pharmacy activities are carried out and which are prepared to host visiting pharmacists or pharmacy students in each European country.

Encyclopedia of Clinical Pharmacy
DOI: 10.1081/E-ECP 120006205

5. *Research*:
   - Promoting multicenter research in all areas of clinical pharmacy.
   - Promoting the participation of pharmacists in clinical trials and pharmacoeconomic studies.
6. *Publications*:
   - Producing a number of publications on clinical pharmacy.
   - Promoting a more widespread use of existing clinical pharmacy publications.

CLINICAL PHARMACY

Clinical pharmacy is a health specialty, which describes the activities and services of the clinical pharmacist to develop and promote the rational and appropriate use of medicinal products and devices.

Clinical pharmacy includes all the services performed by pharmacists practicing in hospitals, community pharmacies, nursing homes, home-based care services, clinics, and any other setting where medicines are prescribed and used.[2]

Activities of the clinical pharmacist are consulting, selecting drugs, providing drug information, formulating and preparing medicinal products and devices, conducting drug use studies/pharmacoepidemiology/outcome research/pharmacovigilance and vigilance in medical devies, studying pharmacokinetics/therapeutic drug monitoring, conducting clinical trials, being aware of the pharmacoeconomy, dispensing and administrating medicinal products and devices, and providing pre- and post-graduated teaching and training activities to provide training and education programs for pharmacists and other health care practitioners.[1,3]

ACTIVITIES OF ESCP

To obtain the goals and objectives, ESCP organizes different types of activities.

Conferences and Symposia

Every year in autumn, the Society's European Symposium on Clinical Pharmacy is held. ESCP also organizes Spring Conferences, focused on specific themes to provide professional education. During these conferences, workshops play an important role.

Education and Research

On the day prior to the Annual Symposium, ESCP organizes a one-day full immersion course, ''Masterclass in Search of Excellence,'' on specific topics of interest. ESCP and EPSA (European Pharmaceutical Students' Organization) jointly organize a Students' Symposium, which aims to bring the clinical pharmacists and pharmacy students together to learn from each other's perspectives and experiences.

Different educational and research programs have been developed and are planned for the coming years (see ESCP Calendar of Events mentioned below and at www.escp.nl).

Several collaborative studies, particularly in the field of drug utilization review and drug evaluation, among member countries have been or are still in progress.

ESCP offers awards to individual researchers in clinical pharmacy fields in collaboration with sponsors.

A number of accredited centers have been established to enable European clinical pharmacists to gain experience in a range of clinical pharmacy specialties.

ESCP has produced a database of clinical pharmacy courses in Europe. Moreover, a long distance Pharmacotherapy Self-Assessment Program (PSAP) published by ACCP is available at ESCP.

Publications

The editing and issuing of publications and journals is an important task undertaken by ESCP and comprises the publication of the *Proceedings of the Annual Symposium in Pharmacy World and Science* (PWS). The Society has adopted a scientific journal *Pharmacy World and Science*, where research papers are published and are retrievable.

ESCP Newsletter is a bimonthly publication, serving as a link between the Society and their members, with news about the activities of ESCP and of the members.

In addition, ESCP selects existing clinical pharmacy publications for promotion among ESCP members.

Related Organizations

To promote the value of clinical pharmacy services among other health care professionals and scientific societies, ESCP has established a relationship with societies that share the same interests: American College of Clinical Pharmacy (ACCP), American Society for Health Care Systems (ASHP), European Association of Hospital Pharmacists (EAHP), European Pharmaceutical Students'

Association (EPSA), Royal Dutch Association for the Advancement of Pharmacy (KNMP), and the United Kingdom Clinical Pharmacy Association (UKCPA). ESCP has been recognized by the Efficacy Working Party of the European Agency for the Evaluation of Medicinal Products (EMEA) as contributor in the consulting process. Within the European Forum of Pharmaceutical Associations and the World Health Organization Regional Office for Europe (EuroPharm Forum), ESCP is recognized as an observer organization.

ORGANIZATION OF ESCP

The European Society of Clinical Pharmacy International Office is located at Theda Mansholtstraat 5b, 2331 JE Leiden, The Netherlands (Phone: +31 (0)71 5722430; Fax: +31 (0)71 5722431; E-mail: office@escp.nl; Internet: www.escp.nl).

General Committee

The Society is conducted by a General Committee consisting of 12 members. They represent individual countries or, where appropriate, groups of countries. General Committee members are elected by the membership. The General Committee meets twice a year, before the Annual Symposium and Spring Conference. (See Table 1 for more information about the General Committee.)

Executive Committee

The General Committee elects the Executive Committee, which implements the resolutions passed by the General Committee and by the General Assembly. The Executive Committee, composed of the President, Past-President, Vice-President, Treasurer, and Chair of the Research and Education Committee is responsible for the day-to-day coordination of ESCP activities.

Table 1 General committee members 2001–2002

Professor M. Alos Almiñana
Hospital General Castellon, Avda. Benicassim s/n, 12004 Castellon, Spain

Ms. C.M. Clark
Brandlesholme, 9 Salthouse Close, BL8 1HD Bury, United Kingdom

Ms. F. Falcao
Hospital de Sao Francisco Xavier, Sevicos Farmaceuticos, Estrada do Forte do Alto Duque, 1495 Lisbon, Portugal

Dr. J. Grassin
Trousseau Hospital, Pharmacy Logipole, Route de Loches, 37170 Chambray les Tours, France

Dr. E. Grimm Bättig
Sonnhaldenweg 28, 4450 Sissach, Switzerland

Professor Dr. Y.A. Hekster
University Medical Centre, Clinical Pharmacy Department, KF 533, P.O. Box 9101, 6500 HB Nujmegen, The Netherlands

Mr. Y. Huon
University Hospital Sart Tilman, Pharmacy Department B 35, 4000 Liege, Belgium

Ms. H. Kreckel
University Hospital Justus-Liebig, Pharmacy Department, Schubertstrasse 89-99, 35392 Giessen, Germany

Mr. K. Linnet
Reykjavik Hospital, Pharmacy Department, Fossvogi, 108 Reykjavik, Iceland

Ms. H. Stenberg-Nilson
Rikshospitalet, Pharmacy, Relis Sor, Holbergs Terrasse, 0027 Oslo, Norway

Dr. F. Venturini
Pharmacy Interna, Policlinico GB Rossi, Piazzale L.A. Scuro, 10, 37134 Verona, Italy

Dr. J. Vlcek
Charles University, Faculty of Pharmacy, Heyrovskeho 1203, 50005 Hradec Kralove, Czech Republic

Research and Education Committee

The Research and Education Committee is in charge of the coordination of educational activities, stimulates and initiates research project, and takes care of the scientific level of these activities.

Special Interest Groups

The Special Interest Groups (SIGs) of ESCP are intended to help ESCP meet the evolving needs of its members and fulfill a growing need for providing targeted services to ESCP members with similar interests.

The goal is to provide a focal point to gather ESCP members with common interests and needs in practice, research, and education, to create a network for:

- Professional interaction.
- Problem solving and discussion of professional issues.
- Continuing education.
- Research.
- Publications.

The following SIGs are currently active: Cancer Care, Drug Information, Education and Training, Geriatrics, Infectious Diseases, Integrated Primary Care, Nutritional Support, Pediatrics, Pharmacoeconomics, Pharmacoepidemiology, and Pharmacokinetics.

International Office

The Society has an International Office which coordinates the total operations of the Society, administers the activities of the Society, and implements new policies and strategies. The staff of the International Office consists of a director, who is a pharmacist, and two secretaries. The director of the ESCP International Office is Annemieke Floor-Schreudering.

Members

ESCP has about 850 members from 48 countries. Members practice in hospitals, clinics, universities, community pharmacies, governmental settings, drug information centers, pharmaceutical industry, and any other places where clinical pharmacists are employed. The Society has four different classes of members: *ordinary members* are individuals who are actively involved in pursuing the objectives of the Society; *honorary members* are those who have distinguished themselves in a particularly honorable way toward the Society; *patrons and sponsors* are individuals or corporate bodies, who have expressed their willingness to support the Society financially; and *student members* are individual students or educational institutions.

During its Annual Symposium, ESCP holds a General Assembly for all members and patrons of the Society.

CALENDAR OF EVENTS

October 2002	Florence, Italy	31st European Symposium on Clinical Pharmacy
May 2003	Portugal	4th Spring Conference on Clinical Pharmacy

REFERENCES

1. Zelger, G.L.; Scroccaro, G.; Hekster, Y.A.; Floor-Schreudering, A. Introduction to the proceedings. Pharmaceutical care, hospital pharmacy, clinical pharmacy—what is the difference? Pharm. World Sci. **1999**, *21* (3), 1A, A2–A3.
2. Scroccaro, G.; Alós Almiñana, M.; Floor-Schreudering, A.; Hekster, Y.A.; Huon, Y. The need for clinical pharmacy. Pharm. World Sci. **2000**, *22* (1), 27–29.
3. ESCP website. www.escp.nl.

Evidence Based Practice

Christine M. Bond
Margaret C. Watson
University of Aberdeen, Aberdeen, U.K.

INTRODUCTION

In 1992, a group led by Gordon Guyatt at McMaster University in Canada[1] first articulated the term ''evidence based medicine.'' Evidence-based medicine (EBM) was defined more recently as ''the integration of best research evidence with clinical expertise and patient values.''[2] Despite its recent recognition, EBM has probably always been practiced by health professionals, but what has changed is that the quality of evidence and the clinical benefit of applying it, are now looked at critically and systematically.

Historically, personal experience, the advice of a professional colleague or data presented in an article in a health journal might have been considered sufficient evidence on which to base a clinical decision. Nowadays, the importance of using ''best evidence'' to underpin practice is recognized, thereby increasing the likelihood that an effect can be predicted with confidence. The growth in EBM has been accompanied by a greater understanding of the different levels of evidence.

The demand for healthcare increases relentlessly, therefore, it is essential that decision makers operate at both patient and population levels within an evidence-based framework. Evidence is needed for diagnostic tools, management options (including drug treatments), the introduction of healthcare models, and patients' values regarding their health service. Scarce resources should not be spent on treatments which provide little benefit or which may even do harm. The relative effectiveness of treatments needs to be assessed where there is competition for limited resources. Valid and reliable information on the clinical and cost-effectiveness of different options is therefore needed.

Another reason for the need for EBM is the accelerating pace with which new procedures and treatments are introduced, with the result that knowledge gained during training quickly becomes redundant. It is essential, therefore, to have up-to-date information about best clinical practice.

This article describes how to find and understand the evidence, and how to apply it in the healthcare setting.

FINDING THE EVIDENCE

The first stage in practicing EBM is to define the precise question to which an evidence-based answer is required. A carefully focused question will inform the search for relevant evidence, and should (hopefully) avoid excessive retrieval of irrelevant publications and other information sources. For example, a clinician who wishes to know whether it is best to use oral or topical antifungals for the treatment of vaginal candidiasis could articulate the question as ''What is the relative effectiveness of oral versus intra-vaginal antifungals for the treatment of uncomplicated vulvovaginal candidiasis?''

There is a hierarchy[3] of trial evidence:

- Ia Evidence obtained from meta-analysis of randomized controlled trials.
- Ib Evidence obtained from at least one randomized controlled trial.
- IIa Evidence obtained from at least one well-designed controlled study without randomization.
- IIb Evidence obtained from at least one other type of well-designed quasi-experimental study.
- III Evidence obtained from well-designed nonexperimental descriptive studies, such as comparative studies, correlation studies, and case studies.
- IV Evidence obtained from expert committee reports or opinions and/or clinical experiences of respected authorities.

The above ranking depends not only on the type, but also the quality of the studies. Therefore, a badly conducted randomized controlled trial could be less robust than a well-conducted controlled clinical trial.

Encyclopedia of Clinical Pharmacy
DOI: 10.1081/E-ECP 120006244

E

Table 1 Quality "questions" for assessing RCTs

- Were subjects randomly assigned to treatment?
- Was randomization done blindly?
- Were all subjects analyzed?
- Was analysis according to unit of randomization?
- Were researchers blind to group allocation?
- Apart from the intervention, were the two groups treated equally?
- Were the groups similar at baseline?

It is important to ensure that all the relevant information is identified and critically appraised. This is easier said than done! Evidence that is unpublished or that is not in the public domain is difficult to identify and retrieve. Pharmaceutical companies might not publish unfavorable results of drug trials, therefore, the clinician or reviewer is reliant upon the cooperation of the company to provide all relevant trial data for its specific drug. Trials reported in the English language[4] and those with positive outcomes are more likely to be published. Problems can also arise if trial results have been accepted by a medical journal that has a long time lag before publication. It may be months or years before the results are published. The sources and avoidance of bias are discussed elsewhere. It is important to attempt to minimize the effects of bias when reviewing evidence. It is also useful to contact experts on the subject of interest, as they will be able to advise on sources of relevant data and contact details of researchers conducting trials in the area. Other useful methods of identifying potentially relevant information include placing notices about the literature review in professional journals and on web site noticeboards and searching conference abstracts and lists of grant awards.

Once the literature search is complete, the identified trials need to be retrieved and reviewed critically to decide whether they satisfy specific standards for inclusion in the review. A list of some important quality criteria for randomized controls is shown in Table 1. It is essential that studies that do not meet the necessary quality standards be excluded from the final analysis.

UNDERSTANDING THE EVIDENCE

The results of trials can be used for different purposes. They could be combined and reviewed descriptively, or, if

Table 2 Relative risk

	Outcome		
	Yes	No	Total
Drug A	a	b	a + b
Drug B	c	d	c + d

Where,

a = the number of subjects receiving Drug A with the outcome

b = the number of subjects receiving Drug A without the outcome

c = the number of subjects receiving Drug B with the outcome

d = the number of subjects receiving Drug B without the outcome

If the outcome was cure then the relative risk of cure would be calculated as follows:

The risk of cure with Drug A = a / a + b,

divided by the risk of cure with Drug B = c / c + d

with Drug

B = c / c + d.

	Outcome		
	Yes	No	Total
Durg A	10	20	30
Durg B	5	50	55

Relative risk = (a / a + b) ÷ (c / c + d) = (10 / 30) ÷ (5 / 55) = 3.7

This means that cure is 3.7 times more likely with Drug A than Drug B.

Table 3 Number needed to treat

Absolute risk reduction (ARR) = (a/(a + b)) − (c/(c + d))
For the above example, using the hypothetical values in Table 2, ARR = (10/(10 + 20)) − (5/(5 + 50)) = 0.24
Therefore, the NNT = 1/0.24 = 4
This means that for every four people treated with Drug A, one additional cure is likely to occur.

the trials are similar enough, their data can be combined in the form of a meta-analysis. This technique allows reporting the results to a greater level of statistical confidence because of the increased numbers of subjects included in the analysis.

An alternative statistic which is sometimes quoted is the odds ratio (OR). This is the odds of an event occurring in a patient in one treatment group relative to the odds of the same event occurring in a patient in an alternative treatment group.

The results of randomized controlled trials comparing two drugs can be used to generate a statistic called the relative risk (RR) (Table 2). This is a ratio of the risk of an outcome with one treatment and the risk of the same outcome with the other treatment.

While the relative risk is a standard statistic that can be used to compare treatments, it can be difficult to understand and to relate to practice. For example, although the relative risk of 3.7 that was calculated above indicates that Drug A is associated with nearly four times the risk of cure compared with Drug B, this gives no indication of the practical implications. For this reason, effects are often quoted as the ''Number Needed to Treat'' (NNT). The NNT is calculated as the reciprocal of the absolute risk reduction (ARR). In the example in Table 3, the NNT refers to the number of patients who need to receive Drug A before an additional cure is likely to occur.

APPLYING THE EVIDENCE

Having identified the evidence from the available information and interpreted it in the context of the original question, the next step is to apply it to practice. This is a complex and challenging task. The evidence may suggest benefits from discontinuing existing treatments or changing to alternative therapy, e.g., using a beta-blocker in hypertensive patients following a myocardial infarction.[5] Alternatively, the evidence may recommend against adopting a new ''miracle'' drug such as the anticholinesterase inhibitors for Alzheimer's disease.[6]

Currently, much clinical practice is based on established practice and personal experience. Producing changes in practice will involve the dissemination of information to individual clinicians and persuading them that, sometimes against their better judgment, there is a benefit in adopting a new approach. Evans and Haines[7] cite 12 initiatives to introduce evidence-based practice, and they are refreshingly honest in identifying the barriers that are encountered. These included the time required to support change; the resources needed from existing budgets; a failure to always demonstrate quantifiable gains in the real world; a failure to give ownership to all parties; and, probably the most difficult and complex of all, changing professional behavior. This last area is a research topic in its own right and is discussed later in this article.

Patient resistance to change, as well as professional resistance, also needs to be addressed. For example, new evidence may require changes to be made to a patient's current long-term medication. Patients previously satisfied with their treatment may be reluctant to try a new drug, despite evidence of greater benefit. A concordant and patient-centered approach is being promoted.[8] The clinician has a responsibility to involve their patients in treatment decisions and to ensure that they understand and agree with any changes that are made, as well as address any concerns that they may have. In the interests of maximizing patient outcomes and cost-effective use of medicines, it is paramount that patients understand and agree with new or existing treatments. Within this framework, management decisions may not be in line with current best evidence, giving rise to a debate about the legal implications and professional ethical issues of this scenario.

It is important to remember that EBM applies to a range of providers at a variety of levels. Thus, it should be used to support decision making by all healthcare providers, not just medical clinicians. It is for this reason that the term Evidence-Based Practice (EBP) is increasingly used. Pharmacy, nursing, physiotherapy, and all other professions allied to medicine should, where possible, be providing evidence-based treatment at an individual and service level. For example, evidence can support decisions about whether to treat stroke patients in a dedicated stroke unit or as part of a general ward.[9]

CRITICISMS OF EVIDENCE-BASED MEDICINE

There are two levels of criticism applied to evidence-based medicine. The first relates to the widespread dependence on the randomized controlled trial, and the second relates to the patient–population dichotomy.

Concern has been expressed that gold standard evidence, i.e., the RCT, may not be as robust as it first appears. Critics of this study design argue that the patient populations are highly selected. Randomized controlled trials often exclude patients above a certain age or those who are taking other concomitant medications or who have significant comorbidities. Additionally, participants in RCTs often have intensive support from medical, nursing, and research staff, contrary to the normal situation. The reasons for these exclusions and enhanced care are self-evident, but they may mean that the results are not generalizable to the wider patient population. A comparison of randomized and nonrandomized studies has also identified that subjects excluded from RCTs tend to have worse prognosis than those who are included.[10] Furthermore, subjects entered into RCTs for evaluation of treatment for existing conditions may be less affluent, less educated, and less healthy then those who are not. The opposite is true for trials of preventive interventions.[10]

Secondly, clinicians have argued that evidence-based guidelines do not accommodate individual patients and their specific circumstances or needs. It may be necessary to remind clinicians that guidelines ''are not tramlines''—they apply to a specific population, and their recommendations should be tailored to the needs of their individual patients. This is discussed later in this article.

CHALLENGES OF BASING DECISIONS ON EVIDENCE AT POLICY AND INDIVIDUAL PATIENT LEVELS

With increasing healthcare costs, particularly in the field of drug treatments, decisions regarding the uptake of new drugs may be made at organizational rather than individual clinician or patient level. In the United Kingdom, this is particularly true in areas where NHS budgets constrain both the choice of treatment and patient selection. EBM can be used to inform these policy decisions, as it can assess both the cost-effectiveness and clinical effectiveness of treatments. The final decision can take into account the wider ramifications of alternative treatments, such as the possible need for residential or surgical care or the impact on lay carers. A decision may be made at a population level that a new drug should not be introduced because of the adverse overall health economic balance, whereas at an individual level, it could be worth trying.

An example of this patient versus the population dilemma is illustrated by the use of the expensive interferon-beta-1b to treat secondary progressive multiple sclerosis (MS). The evidence tells us that treatment with interferon-beta-1b will delay time to wheelchair dependence and prevent relapses in some subjects. However, the NNT is 18 and at a population level, the economics mitigate against making this a recommended treatment.[11] Conversely, despite their cost, there has been considerable use of statins as lipid-lowering agents to reduce cholesterol levels in targeted patients.[12] This is because the evidence shows long-term reduction in further coronary events, and the exact health gain can be calculated and is deemed worthwhile.[13] This intervention is both clinically and cost effective.

Ultimately, it is the clinician who has to weigh the costs and benefits for each individual patient, taking into account the evidence but also considering patient factors. This has been summarized as ''conscientious, explicit and judicious use of current best evidence in making decisions about the care of individual patients.''[14]

WHAT TO DO WHEN THERE IS NO EVIDENCE OR EVIDENCE IS INCOMPLETE

The EBM movement is a relatively recent endeavor. With such a wide range of treatments available and numerous conditions, it is inevitable that there will not always be evidence to inform decision making. This may be due to a lack of collation of the available research evidence or a lack of research per se. In these instances, there are several options depending on the immediacy of the decision.

If a decision needs to made quickly, advice should be sought from the most experienced practitioner on the subject. This advice should be interpreted with caution and considered in light of whatever published literature exists. This should be judged on the basis of the ranked levels of evidence included earlier in this article. New drugs may be tried in the context of local clinical trials. If this is the case, these trials should be expertly designed and conducted in collaboration with other colleagues. This means that while a treatment may not ultimately be the best, it will have been used in a controlled way such that it has contributed to generating future evidence.

CLINICAL EFFECTIVENESS AND CLINICAL GOVERNANCE

There is a growing emphasis on the accountability of individual clinicians and organizations that provide healthcare. EBM contributes to the definition of criteria used for clinical performance indicators. This forms the basis of assessing the clinical effectiveness of services. Increasingly, clinicians and their corporate managers are held responsible for the delivery of quality care; this is known as clinical governance. Despite the caveats for EBM summarized above, the knowledge and understanding it has promoted now underpin the healthcare infrastructures that exist today.

THE PHARMACIST'S ROLE

Pharmacists can contribute to the delivery of evidence-based care.[15] At a population level, pharmacists' clinical knowledge and analytical strengths can be used to facilitate the production of systematic reviews, the interpretation and analysis of findings, and the development of guidelines. At a patient level, pharmacists are consulted in both primary and secondary care, and may be a useful vehicle for transfer of evidence-based information to the clinician, being able to give a more objective decision than the doctor faced with a patient with alternative expectations.[16] Pharmacists can influence the choice of prescribed drugs mediated either through the GP to the patient, or face to face with the patient.[17]

In many countries, a wider range of drugs is available for purchase from pharmacies without the need for a prescription. This has enabled pharmacists to provide treatment and advice for a greater range of minor illnesses. Although there have been concerns that pharmacists and their staff may give inappropriate advice,[18–21] the use of evidence-based guidelines to support their treatment of minor illness is currently being explored.[22]

RESOURCES FOR EVIDENCE-BASED MEDICINE

Electronic databases of peer-reviewed healthcare journals (primary references) include MEDLINE and EMBASE. The Cochrane Collaboration library contains a database of systematic reviews as well as a database of RCTs and controlled clinical trials. Medical librarians will be able to advise and perhaps provide training on performing literature searching and retrieval. Hospital-based drug information centers will likely have access to a range of electronic databases. The Royal Pharmaceutical Society of Great Britain's information center has a number of databases that can be searched for information that is of particular relevance to drug therapy and pharmaceutical care. It is likely that most national pharmaceutical organizations have similar resources.

One of the greatest resources for EBM is the World Wide Web. There are numerous sites that provide information on EBM, including literature retrieval and review, EB guidelines, and so on (Table 4).

GETTING EVIDENCE INTO PRACTICE: DISSEMINATION AND IMPLEMENTATION

The mere dissemination of information (i.e., evidence) is unlikely to achieve behavioral change.[23] In order for evidence to influence practice, active dissemination and implementation strategies need to be employed. It is recognized that "individual beliefs, attitudes and knowledge influence professional behavior" and that "other factors including organisational, economic and community environments of the practitioner are also important."[24] It has been suggested that implementation strategies that

Table 4 Suggested EBM-related web sites

Adept Programme	www.shef.ac.uk/~scharr/ir/adept
Agency for Healthcare Research and Quality (USA)	www.ahcpr.gov/
Bandolier	www.jr2.ox.ac.uk/bandolier/
Critical Appraisal Skills Programme	www.phru.org.uk/~casp/index.htm
National Guideline Clearing (USA)	www.guideline.gov/index.asp
Netting the Evidence	www.shef.ac.uk/~scharr/ir/netting/
NHS Centre for Evidence-Based Medicine	www.minervation.com/cebm/
National Institute for Clinical Excellence (UK)	www.nice.org.uk
Primary Care Clinical Practice Guidelines	www.medicine.ucsf.edu/resources/guidelines/
Scottish Intercollegiate Guidelines Network (SIGN)	www.sign.ac.uk
The NHS Centre for Reviews and Dissemination	www.york.ac.uk/inst/crd/welcome.htm
TRIP Database	www.tripdatabase.com/index.cfm
UK Cochrane Centre	www.cochrane.org/
Virtual library	www.shef.ac.uk/~scharr/ir/core.html

address barriers to change may be more effective than those that do not.[24] A comprehensive review of implementation strategies is presented in the *Effective Health Care Bulletin: Getting Evidence Into Practice*.[24]

The use of guidelines as a method of summarizing evidence is discussed elsewhere in this encyclopedia. There has been considerable evaluation of the effectiveness of different guideline implementation strategies as methods of eliciting behavior change among healthcare professionals. Most implementation research has targeted physician behavior. However, as greater emphasis is placed on multidisciplinary healthcare teams, strategies need to be identified, tested, and adopted, which are effective in promoting evidence-based practice among all health professional groups.

Mass media is a method commonly used to disseminate information to large audiences. This strategy usually involves the dissemination of printed materials (e.g., guidelines, therapeutic bulletins) to specific health professionals (e.g., physicians, pharmacists). There is little evidence to support the use of this method, as it is largely ineffective in influencing behavior change.[25]

Educational outreach visits (also known as academic detailing) have been used by the pharmaceutical industry for decades to influence the prescribing behavior of physicians. Although there is little published empirical evidence of the effect of the pharmaceutical industry's promotional activities on prescribing patterns, the investment of 57% of their pharmaceutical promotion budget on pharmaceutical representatives and 11% on promotional literature, gives some indication of its importance.[26] There is considerable research evidence of the effectiveness of educational outreach as a behavior change strategy for healthcare professionals.[27] It is no surprise (considering their origin) that educational outreach visits have been shown to be effective in achieving change in prescribing behavior among physicians.[27]

The use of opinion leaders as an implementation strategy has been evaluated in a number of studies, the results of which are inconclusive.[28] This method relies on persuasion (i.e., the persuasive ability of the opinion leader) to influence the behavior of the target audience. Further evaluation of this strategy is required, including methods of describing characteristics of opinion leaders and how to identify individuals who satisfy these criteria.

SUMMARY

Evidence-based practice is increasingly recognized as the best way to maximize the chances of individual patients receiving the most appropriate treatment. It is also used to inform policy making about both medical treatments and new services, including models of healthcare.

While there are still some caveats, some of which have been highlighted in this article, EBP is the goal to which all healthcare professionals should aspire.

REFERENCES

1. Evidence Based Medicine Working Group 1992. Evidence based medicine. A new approach to teaching the practice of medicine. J. Am. Med. Assoc. **1992**, *268*, 2420–2425.
2. Sackett, D.; Strauss, S.; Richardson, W.; Rosenberg, W.; Haynes, R.B. *Evidence Based Medicine: How to Practise and Teach EBM*, 2nd Ed.; Churchill Livingstone: Edinburgh, 2000.
3. US Department of Health and Human Services. *Agency for Health Care Policy and Research. Acute Pain Management: Operative or Medical Procedures and Trauma*; AHCPR: Rockville, Maryland, 1993.
4. Egger, M.; Zellweger-Zähner, T.; Schneider, M.; Junker, C.; Lengeler, C.; Antes, G. Language bias in randomised controlled trials published in English and German. Lancet **1997**, *350*, 326–329.
5. Scottish Intercollegiate Guidelines Network. *Secondary Prevention of Coronary Heart Disease following Myocardial Infarction*; 2000, Edinburgh.
6. Coelho Filho, J.M.; Birks, J. Cochrane Collaboration Physostigmine for Alzheimer's Disease. In *The Cochrane Library, Issue 3*; 2002, Oxford: Update Software.
7. Evans, D.; Haines, A. *Implementing Evidence-Based Changes in Health Care*; Radcliffe Medical Press: Oxford, 2000.
8. Working Party: Royal Pharmaceutical Society of Great Britain. From Compliance to Concordance: Achieving shared goals in medicine taking. RPSGB and Merck Sharp & Dohme, 1997.
9. Stroke Unit Trialists' Collaboration. Organised Inpatient (Stroke Unit) Care for Stroke (Cochrane Review). In *The Cochrane Library, Issue 3*; 2002, Oxford: Update Software.
10. McKee, M.; Britton, A.; Black, N.; McPherson, K.; Sanderson, C.; Bain, C. Interpreting the evidence: Choosing between randomised and non-randomised studies. Br. Med. J. **1999**, *319*, 312–315.
11. Forbes, R.; Lees, A.; Waugh, N.; Swingler, R. Population based cost utility study of interferon beta-1b in secondary progressive multiple sclerosis. Br. Med. J. **1999**, *319* (7224), 1529–1533.
12. Scottish Intercollegiate Guidelines Network. *Lipids and the Primary Prevention of Coronary Heart Disease*; SIGN: Edinburgh, 1999.

13. Reckless, J. The 4S study and its pharmacoeconomic implications. PharmacoEconomics **1996**, *9* (2), 101–105.
14. Sackett, D.; Rosenberg, W.; Gray, J.; Haynes, R.; Richardson, W. Evidence based medicine: What it is and what it isn't. Br. Med. J. **1996**, *312*, 71–72.
15. Bond, C. *Evidence-Based Pharmacy,* 1st Ed.; Pharmaceutical Press: London, 2000.
16. Krska, J.; Cromarty, J.; Arris, F.; Jamieson, D.; Hansford, D.; Duffus, P.; Downie, G.; Seymour, G. Pharmacist led medication review in patients over 65: A randomised controlled trial in primary care. Age Ageing **2001**, *30*, 215–221.
17. Beney, J.; Bero, L.A.; Bond, C. Expanding the Roles of Outpatient Pharmacists: Effects on Health Services Utilisation, Costs, and Patient Outcomes (Cochrane Review). In *The Cochrane Library, Issue 3*; 2002, Oxford: Update Software.
18. Goodburn, E.; Mattosinho, S.; Mongi, P.; Waterston, A. Management of childhood diarrhoea by pharmacists and parents: Is Britain lagging behind the Third World? Br. Med. J. **1991**, *302*, 440–443.
19. Krska, J.; Greenwood, R.; Howitt, E. Audit of advice provided in response to symptoms. Pharm. J. **1994**, *252*, 93–96.
20. Anonymous. Counter advice. Consumers are still not receiving the right advice when buying medicines from pharmacists. Which **1999**, 22–25 April.
21. Mobey, N.; Wood, A.; Edwards, C.; Jepson, M.H. An assessment of the response to symptoms in community pharmacies. Pharm. J. **1986**, *237*, 807.
22. Watson, M.C.; Bond, C.; Grimshaw, J.M.; Mollison, J.; Ludbrook, A. Educational Strategies to Promote Evidence-Based Practice: A Cluster Randomised Controlled Trial (RCT). In *Health Services Research and Pharmacy Practice Conference Proceedings*; 2001.
23. Mittman, B.S.; Tonesk, X.; Jacobson, P.D. Making good the promise: Disseminating and implementing practice guidelines. Qual. Rev. Bull. **1992**, *18*, 413–422.
24. Anonymous. Getting evidence into practice. Eff. Health Care Bull. **1999**, *5* (1).
25. Freemantle, N.; Wolf, F.; Grimshaw, J.M.; Grilli, R.; Bero, L. Printed Educational Materials: Effects on Professional Practice and Health Care Outcomes (Cochrane Review). In *The Cochrane Library, Issue 2*; 2001.
26. Association of the British Pharmaceutical Industry. *PHARMA Facts & Figures*; ABPI: London, 1997.
27. Thompson O'Brien, M.; Oxman, A.; Davis, D.; Haynes, R.; Freemantle, N.; Harvey, E. Educational Outreach Visits: Effects on Professional Practice and Health Care Outcomes (Cochrane Review). In *The Cochrane Library, Issue 3*; 2002, Oxford: Update Software.
28. Thompson O'Brien, M.A.; Oxman, A.D.; Haynes, R.B.; Davis, D.A.; Freemantle, N.; Harvey, E.L. Local Opinion Leaders: Effects on Professional Practice and Health Care Outcomes (Cochrane Review). In *The Cochrane Library, Issue 3*; 2002, Oxford: Update Software.

Fellowships in Pharmacy

Joseph T. DiPiro
*University of Georgia College of Pharmacy,
Athens, Georgia, U.S.A.*

INTRODUCTION

The term *fellowship* is used to designate training programs or to indicate status within a profession or professional organization. In pharmacy, the accepted definition for a fellowship is "a directed, highly individualized, postgraduate program designed to prepare the participant to become an independent researcher." This definition was adopted by a coalition of seven national pharmacy organizations to distinguish fellowship from residency training.[1] This definition is contrasted with that for a *residency* which is "an organized, directed, postgraduate training program in a defined area of pharmacy practice." Training fellowships may occur at any stage of education and are commonly referred to as *predoctoral* (usually at the undergraduate level) or *postdoctoral* (*postgraduate*). This article includes discussion of fellowship as a postdoctoral research training program.

DEFINITIONS

A member of a professional organization may be designated as a fellow to recognize accomplishments, experience, or some other laudable standing in the profession. For example, a person may be a Fellow of the American College of Clinical Pharmacy (ACCP) or the American Society of Health-System Pharmacists (ASHP). This designation does not indicate completion of a training program nor proficiency in research.

Fellowships are offered by many institutions, including colleges and universities, government entities such as the National Institutes of Health and the Centers for Disease Control and Prevention, pharmaceutical manufacturers, healthcare systems, and professional organizations. Most pharmacy fellowship training programs are offered by colleges of pharmacy or academic medical centers.

Generally, fellowships are generally highly individualized programs to develop competency in research, including conceptualizing a research problem, planning and conducting research processes and experiments, analyzing data, and reporting of results. These programs are conducted under the close supervision of an experienced research mentor or preceptor. More so than most residencies, a fellowship is guided by one person or a small group of individuals. Fellowships are generally 12 or 24 months in duration and fellows often complete formal courses in selected topics such as research design, statistics, or research methods before or during a fellowship. Fellows should possess basic pharmacy practice skills relevant to the knowledge area of the fellowship. These skills are acquired through training in a Pharm.D. program, a residency, or practice experience. For most individuals, a residency should be completed before beginning a fellowship.

The goal of fellowship training is to produce an individual capable of conducting collaborative research or functioning as a principal investigator. A fellowship-trained individual will usually work for a college of pharmacy, academic medical center, pharmaceutical company, or contract research organization. Research-intensive positions often indicate a hiring preference for those with fellowship training.

GUIDELINES FOR CLINICAL FELLOWSHIP TRAINING PROGRAMS

In 1987 a document with specific guidelines for clinical fellowship training programs in pharmacy was approved by ACCP and the American Association of Colleges of Pharmacy.[2] The guidelines, which have been updated by ACCP, relate to the training program overall, preceptor qualifications, fellow qualifications, and the fellowship experience, as follows.

Encyclopedia of Clinical Pharmacy
DOI: 10.1081/E-ECP 120006357

Training Program Requirements

1. In general, a commitment of 80% of fellowship training time to research activities over a period of at least two years.
2. Administrative institutional support for the preceptor's research program and the fellowship training program.
3. Availability of graduate-level course work in the area of the fellowship.
4. Availability of personnel to teach laboratory-based and clinical research skills.
5. Ready access to a medical library and computer facilities.

Preceptor Qualifications

1. A clinical scientist with an established record of research accomplishments, which may be exemplified by:

   a. Fellowship training or equivalent experience.
   b. Principal or primary investigator on research grants.
   c. Published research papers in peer-reviewed pharmacy/medical literature where the preceptor is primary or senior author.

2. Active collaborative research relationships with other scientists.
3. Expertise in pharmacotherapeutics in the area of specialization.

Fellowship Applicant Requirements

1. Pharm.D. or equivalent experience.
2. Residency or equivalent experience.
3. High level of motivation for a research career.

Fellowship Experience

The initiation and completion of a research project, including:

1. Development of at least one scientific hypothesis and experimental methods to test hypothesis.
2. Preparation and submission of a grant proposal.
3. Submission of a protocol to the appropriate institutional review committee.
4. Research experiences including study conduct and data collection related to the field of specialization.
5. Experience in statistical analysis of data.
6. Preparation and submission of abstracts and manuscripts for publication in peer-reviewed journals.
7. Formal presentation of research at peer-reviewed scientific meetings.
8. Participation in journal clubs, research workshops, and seminar series.
9. Instruction in biomedical science ethics.

REVIEW OF FELLOWSHIPS

In an effort to improve fellowship training, ACCP instituted a program for peer review of research fellowships training programs to assure quality of these programs. This is a voluntary process conducted by an ACCP committee to determine whether a program meets the ACCP Guidelines for Research Fellowship Training Programs as detailed above. In this process, both the preceptor and the fellowship site are evaluated. A positive review indicates that the program meets the guidelines. At present, 15 fellowship programs have been recognized as meeting the guidelines.[3]

FELLOWSHIP RESOURCES

An excellent resource for information about pharmacy fellowships is the ACCP Directory of Residencies and Fellowships.[3] This source provides information on over 100 individual fellowship programs. Additional information on fellowships can be obtained from the Academy of Managed Care Pharmacy[4] and the American Pharmaceutical Association.[5] Currently, fellowships can be served in the following areas:

- Ambulatory care.
- Cardiology.
- Clinical pharmacology.
- Critical care.
- Drug development.
- Drug information.
- Family medicine.

- Geriatrics.
- Infectious diseases.
- Internal medicine.
- Managed care pharmacy.
- Nephrology.
- Neurology.
- Oncology.
- Outcomes research.
- Pediatrics.
- Pharmacoeconomics.
- Pharmacoepidemiology.
- Pharmacokinetics.
- Psychiatry.
- Pulmonary.
- Rheumatology.
- Translational research.
- Transplantation.

Funding for fellowships varies from year to year and has been available from pharmacy organizations including ACCP, ASHP, American Society of Consultant Pharmacists, and the American Foundation for Pharmaceutical Education.

F

REFERENCES

1. Anon. Definitions of residencies and fellowships. Am. J. Hosp. Pharm. **1987**, *44*, 1143–1144.
2. Anon. Guidelines for clinical fellowship training programs. Pharmacotherapy **1988**, *8*, 299.
3. *2001 Directory of Residencies and Fellowships*; American College of Clinical Pharmacy: Kansas City, 2001.
4. http://www.amcp.org/public/pubs/journal/vol5/current/reports.html (last accessed on June 19, 2001).
5. http://www.aphanet.org/ (last accessed June 19, 2001).

PROFESSIONAL RESOURCES

First DataBank, Inc.

Joan Kapusnik-Uner
First DataBank, Inc., San Bruno, California, U.S.A.

INTRODUCTION

Successful computerization and drug-related decision support is achieved to a significant degree within the profession of pharmacy. Pharmacy has been a leader among healthcare professions in embracing computerization. Drug databases and knowledge bases are now the backbone of pharmaceutical care or pharmaceutical decision support.

MISSION STATEMENT

Our mission is to be the best provider of point-of-care decision support knowledge bases that provide outstanding value to patients and to our customers. We are committed to exceeding our customers' expectations so that they regard us as the best company in our industry. We will achieve this position through the comprehensiveness and quality of our data, the responsiveness of our service, and our understanding of their business problems. While continuing to grow, we will provide an open, supportive, challenging, and team-oriented environment within which our staff members can achieve job satisfaction, professional and personal growth, and compensation based on company and individual performance. We will actively work to increase our impact on the quality of healthcare.

First DataBank is one of the world's leading suppliers of healthcare knowledge bases, supplying drug knowledge bases, as well as medical diagnostic and nutrition software to system vendors. First DataBank serves hospitals, hospital pharmacies and laboratories, retail pharmacies, physician clinics and group practices, insurers, managed care organizations, pharmacy benefits managers, claims processors, employers, utilization review organizations, government, pharmaceutical manufacturers, wholesalers, and all 50 state Medicaid programs.

First DataBank's professional staff is committed to deliver comprehensive and accurate information to physicians, pharmacists, nurses, dietitians, and other healthcare professionals to be useful in a variety of healthcare settings. Drug information to be used directly by consumers (i.e., patients and their families) is also another focus of our drug knowledge bases. These knowledge bases are updated continually and are available to accommodate any update schedule, providing the immediate access that businesses need to perform mission critical functions and to realize significant time and financial savings. Enhancements in the delivery of pharmaceutical care have increased the need for First DataBank to deliver clinically significant drug information in a timely manner. Therefore, updates to products containing clinical knowledge bases (i.e., drug–drug interactions or patient education materials) are made available on a *weekly* basis. Institutional drug buying practices, retail pharmacy services, and Pharmacy Benefit Manager functions have created the need for *daily* updates of drug pricing information.

First DataBank also has many international drug databases that were developed by a professional staff consisting of native language speakers who best understand how drugs are used in other countries. As the Internet and global travel make the world "a smaller place," these knowledge bases will become even more omnipresent and obligatory.

Experience is critical to develop and maintain a comprehensive database of drug information. First DataBank has been in this business for over 20 years, having spent virtually all of that time developing, maintaining, and enhancing the most comprehensive drug, medical, and nutrition knowledge bases in the world. Our knowledge bases have evolved along with new technologies in healthcare and continue to develop as the Internet and mobile devices become part of clinical practice. First DataBank offers customization for every client's file in terms of record format, formulary selection, media specifications, updates, and user-specific data elements. As

Encyclopedia of Clinical Pharmacy
DOI: 10.1081/E-ECP 120006382

a result, our customers save valuable programming and processing time.

DRUG KNOWLEDGE BASES AND THEIR CLINICAL MODULES

First DataBank creates and maintains some of the largest and most comprehensive drug and healthcare knowledge bases in the world, including NDDF Plus™, based on the industry standard National Drug Data File®. One of the industry's most trusted and widely used sources of up-to-date drug information, NDDF Plus combines descriptive and pricing data with a selection of advanced clinical support modules. NDDF Plus delivers information on every drug approved by the Food and Drug Administration (FDA). Clinical modules are available to support healthcare professionals in making critical decisions about dosing and orders, interactions, allergy alerts, disease contraindications, drug identification, and much more. Plus, several modules offer drug information specifically written for the consumer. NDDF Plus is used in a wide variety of applications, such as:

- Determining drug indications.
- Identifying potential contraindications.
- Helping prevent adverse drug events.
- Identifying drug–drug and drug–food interactions.
- Identifying potential drug interactions with alternative therapy agents.
- Offering printed patient education and counseling messages.
- Prioritizing medication warning labels for patients.
- Listing recommended doses for common drugs.
- Performing indication-specific dose range checking.
- Identifying undesired effects of drugs on lab tests.
- Supporting electronic medical records.
- Handling prescriber order entry.
- Analyzing drug pricing trends.
- Facilitating drug formulary management.
- Accelerating claims processing and adjudication.

First DataBank offers comprehensive international drug knowledge bases for several countries outside the United States, including Canada, Argentina, and Australia. First DataBank Europe, located in Exeter, England, develops drug knowledge base products for the United Kingdom.

Examples of clinical functionality in drug knowledge bases are described below for Patient Education, Drug Interactions, and Prescriber Order Entry modules. Many other pharmaceutical decision support modules are also available and include Drug/Disease Contraindications, Drug Indications, Pregnancy and Lactation Precautions, Geriatric/Pediatric Warnings, Minimum/Maximum Dose Checking, and Duplicate Therapy/Ingredient Checking. A few specific modules are highlighted.

Patient Education

Patient Education Monographs were written for consumers. They are both comprehensive and customizable, covering the most common prescription and OTC medications. The format of these patient education monographs is flexible and is available in English and Spanish. Other patient education materials are available including Prioritized Label Warnings that indicate which ancillary ''stickers'' should be placed on a medication being dispensed and Counseling Messages to be used as reminders for healthcare professionals.

Drug Interactions

First DataBank's drug interaction modules are meant to be able to detect all clinically significant drug–drug interactions for a given patient in either a prospective or retrospective manner. Drug–food interaction information is also available. Interactions are classified by severity, and documentation levels are also noted in coded fields for searching and filtering applications. Full text monographs describe the drug–drug interaction in detail and include reference citations in MEDLINE format. A ''consumerized'' version of the drug–drug interaction monograph has been created for systems that allow patients to monitor their medications.

Prescriber Order Entry

Prescriber Order Entry Module (POEM™) provides a database of the most common medication orders. These orders are specific to drug, route of administration, formulation, age, indication/use, and weight or body surface area, if applicable. This enables more accurate and efficient point-of-care computerized order entry applications to prevent errors at the prescribing stage of drug delivery.

INTEGRATED CONTENT SOFTWARE

Success in today's drug information marketplace requires products that can be developed quickly and economically, lowering the cost of entry into a given market. Toward that end, First DataBank offers a number of application development toolkits that minimize lead times and make more efficient use of scarce resources.

Drug Information Framework™

The Drug Information Framework™ enables developers to build healthcare solutions faster, using the time-tested NDDF Plus knowledge base and critical decision-support modules. The Framework gives developers a choice of technologies and access layers, so it can adapt to most platforms, operating systems, development tools, and relational databases. Application environments can include the Internet; client/server networks; stand-alone desktops; and handheld wireless devices.

Drug Information Framework components encapsulate drug information in intuitive objects, which shortens the typical programmer learning curve and development cycle. These components simplify system implementation, resulting in quicker, easier deployment of systems offering point-of-care, patient-specific drug information, as well as convenient access to full-text clinical monographs.

AHFS Framework™

The AHFS Framework™ enables developers to easily embed drug content into pharmacy and clinical information systems. It can be used to rapidly integrate two respected drug knowledge bases: the American Hospital Formulary Service (AHFS) Drug Information® monographs, and First DataBank's NDDF Plus. Combined, they allow healthcare professionals to have seamless access to comprehensive drug information, within their usual workflow systems.

Rx InHand™

With Rx Inhand™, developers have a powerful tool for creating stand-alone handheld applications. This drug navigation and drug utilization review (DUR) engine enables developers to easily create systems for use by physicians to write prescriptions and to screen them on a handheld device for possible medication errors, thus potentially minimizing adverse medical events.

RxWeb™

RxWeb™ provides instant access to drug information, navigation capabilities, and clinical-screening functionality over the Internet. Using the latest Web browser technology, RxWeb enables the developer to offer proven drug screening (via NDDF Plus) with little or no development time. With RxWeb, software developers can create Web-based applications that provide information on over 100,000 marketed drugs, as well as alternative therapies.

REFERENCE PRODUCTS

First DataBank has developed numerous drug reference products for healthcare applications, in both print and electronic forms. Most of these products can be deployed on an individual desktop, a local area network and, in some cases, over the Internet or intranet.

AHFS*first*™

AHFS*first*™ combines into one easy-to-use package two of the most widely used sources of unbiased drug information—NDDF Plus and the AHFS Drug Information® monographs from the American Society of Health-System Pharmacists®. It links over 100,000 drugs from NDDF Plus to the AHFS monographs, providing maximum coverage from two respected sources. This sophisticated reference makes it possible, in just seconds, to find information on drug interactions, contraindications, adverse reactions, and precautions. The AHFS*first* Web edition brings this capability to users of the Internet or intranets. AHFS Drug Information monographs are also available as a data-only product.

Evaluations of Drug Interactions™

First DataBank's Evaluations of Drug Interactions™ (EDI) provides the most comprehensive printed source of drug–drug interaction information available. Containing interactions on both prescription and over-the-counter drugs, this two-volume loose-leaf textbook of drug monographs is the only source endorsed by the American Pharmaceutical Association.

SPECIALTY SOFTWARE

In addition to drug information products, First DataBank has developed several interactive software products intended for direct use by healthcare specialists, including nutritionists and physicians.

Nutritionist Pro™

Nutritionist Pro™ software represents the next generation of nutrition-analysis tools from First DataBank.

With the most comprehensive food knowledge base and set of program features, Nutritionist Pro provides thorough analysis of diets, recipes, and menus. The intuitive user interface design and powerful functionality of Nutritionist Pro can help ease the workload and boost the productivity of nutrition professionals in virtually any healthcare delivery, food service, or educational setting.

ANTICIPATING FUTURE NEEDS

In healthcare, we are today at a crossroads of yet, another of many notable technical developments. Personal computers have become ubiquitous and easier to use for healthcare professionals and patients. The newly available mobile or handheld devices have become more practical for real-time computing. Through the Internet or handeld device, there is ready access to a patient's medical information. With these tools, the art of practicing medicine is truly about to change. An electronic information resource for the Internet and for handheld devices, as for other platforms, requires that the data meet specific standards of reliability. First DataBank information is "tried and true," a tested, authoritative source of such information.

First DataBank has anticipated this wave of technological change in medicine and has developed an array of software and middleware for ease and effectiveness of decision support implementation for these platforms. Time-to-market has become an absolutely critical factor in the success of healthcare IT applications. The Internet, cost containment, IT undercapitalization, and a host of other factors impacting the healthcare industry have created a demand for the rapid deployment of new applications.

First DataBank is not only meeting healthcare challenges but is also leading the industry into an era of greater patient safety and knowledge.

FIRST DATABANK AS A HEARST CORPORATION SUBSIDIARY

The visionary behind First DataBank is founder and President, Joseph L. Hirschmann, Pharm.D. Doctor Hirschmann started the company after a distinguished academic career at University of California, San Francisco. Another of his accomplishments that lives on today is the *Textbook of Therapeutics: Disease and Drug Management*, which was previously known as *Clinical Pharmacy and Therapeutics*.

First DataBank became part of the Hearst Corporation in 1980. The Hearst Corporation is one of the world's largest diversified communications companies, with interests in newspapers, magazine, book and business publishing, television and radio broadcasting, cable network programming, and new media activities. First DataBank remains under established independent leadership, while benefiting from the support and financial stability offered by The Hearst Corporation.

LOCATIONS

The First DataBank home office is located in San Bruno, California, just a few miles from the San Francisco airport. The company also has offices in St. Louis, Missouri; Exeter, England; and Indianapolis, Indiana.

Formulary Systems

J. Russell May
Medical College of Georgia Hospitals & Clinics, Augusta, Georgia, U.S.A.

INTRODUCTION

Formulary systems are an essential tool used in a variety of settings including hospitals, ambulatory clinics, health plans, pharmacy benefit management companies, and government agencies. This tool, if used correctly, promotes rational, clinically appropriate, safe, and cost-effective pharmaceutical care.

The term ''formulary'' has been used to describe a published list of medications used by an organization, from which prescribers can choose therapy for their patients. Historically, an ''open'' formulary implied that the list was fairly inclusive of any medications the prescribers wanted. A ''closed'' formulary was a finite list that reflected the clinical judgment of a group of physicians, pharmacists, and other health care professionals meeting regularly to choose the most appropriate drugs for the list. Most pharmacists have stopped using ''open'' and ''closed'' because few contemporary formularies are truly ''open.'' A formulary now typically refers to a book or on-line publication used by the organization that contains the approved drug list and other prescribing information deemed useful by its editors.

A formulary system goes much beyond a publication or list of drugs. A coalition of national organizations representing health care professionals, government and business leaders has offered this definition:

> Drug Formulary System—an ongoing process whereby a health care organization, through its physicians, pharmacists, and other health care professionals, establishes policies on the use of drug products and therapies that are the most medically appropriate and cost-effective to best serve the health interests of a given population.[1]

This review of formulary systems covers their history, structure, positive and negative outcomes, and possible future directions.

HISTORY

Formularies were first developed in hospitals during the 1950s. The pharmaceutical market was experiencing unprecedented growth. For example, 17 different companies were marketing 45 different oral penicillin preparations.[2] Institutional policies were developed that allowed pharmacists to dispense a generically equivalent drug for a brand name product prescribed by physicians.

The pharmaceutical industry and physicians, represented by the National Pharmaceutical Council and the American Medical Association (AMA) respectively, successfully worked to get state laws passed forbidding this substitution by pharmacists. While community pharmacists complied, hospital pharmacists resisted. In the late 1950s, the American Society of Hospital Pharmacists (ASHP) published a set of minimal standards for pharmacies in hospitals with guidelines for their interpretation. Among the standards developed was a call for the implementation of a formulary system. Interestingly in 1959, the successful launch of another ASHP publication, the American Hospital Formulary Service, a reference book reviewing the key characteristics of drugs, greatly advanced ASHP's financial status and added to the organization's sphere of influence.[3]

By the 1960s, many hospitals were successful in developing institutional procedures that gave prior consent for physician authorized pharmacists to select generic alternatives under what was called a formulary system.[4] The American Hospital Association (AHA) and ASHP issued joint statements on the legal basis of a hospital formulary system and the guiding principles for operating it. A few years later, the AMA and APhA participated with AHA and ASHP to revise the guidelines to the mutual satisfaction of all parties in a way that would not alienate the pharmaceutical industry.

In 1965, two significant actions occurred that promoted formulary systems. Medicare administrators borrowed freely from ASHP's publications to create standards for institutional health care resulting in a Medicare bill listing the use of a formulary system among the eligibility requirements of Medicare reimbursement. Also, the Joint Commission required an active pharmacy and therapeutics (P&T) committee for hospital accreditation.

Even with these supporting documents and accreditation standards, adoption of formulary systems was not as fast as many anticipated. In the 1970s, two surveys revealed surprising results. In the first, of the 172

Encyclopedia of Clinical Pharmacy
DOI: 10.1081/E-ECP 120006321

Medicare-approved hospitals responding, 31 did not have a formulary system in place even though Medicare required one.[5] The second, looking at academic medical centers with 500 beds or more, found that the majority of formularies analyzed were simple drug lists and were not used to guide prescribing decisions.[6]

As the value of formulary systems became apparent, their acceptance grew, at first just in the hospital setting but later expanding to ambulatory sites. In 1986, the Pharmaceutical Manufacturers Association officially accepted the concept of therapeutic interchange for hospital inpatients, but opposed its use in other settings. The AMA released a policy on drug formularies and therapeutic interchange in both inpatient and ambulatory care settings in 1994.[7] This brought the AMA's views on formularies into close alignment with ASHP. In the most recent survey of pharmacy practice in acute care settings, more than 90% of health-systems had P&T committees responsible for formulary system development and management.[8] Pharmacy directors reported using pharmacoeconomic and therapeutic information in their system's formulary development process. Today, drug formulary systems are considered an essential tool used routinely by health plans, pharmacy benefit management companies, self-insured employers, and government agencies.

THE STRUCTURE OF FORMULARY SYSTEMS

The development of a formulary system within an organization rests with a multidisciplinary committee. In the hospital and health system setting, this is typically called the P&T committee. Virtually all hospitals and health-systems have a P&T committee.[8] P&T committees usually meet six to eight times annually. An ASHP Position Statement on formulary management declares that decisions should be based on clinical, quality of life, and pharmacoeconomic factors that result in optimal patient care.[9] It advises against decisions solely based on economic factors. The Position Statement also recommends that decisions must include active and direct involvement of physicians, pharmacists, and other appropriate health care providers. This may include dieticians, nurses, administrators and quality management coordinators.

Formulary system management falls into three general categories: drug selection for formulary inclusion, formulary maintenance, and medication use evaluation.

Drug Selection

Drug evaluation for inclusion on a formulary should involve a careful assessment of scientific evidence, in particular, peer-reviewed medical literature, including randomized clinical trials, pharmacoeconomic studies, and outcomes research data. If a drug is a new pharmacologic class, unlike any other available drug, the review will focus on efficacy, safety, and the potential value to the organization's patient population. For drug's that are additions to an existing pharmacologic class, the evaluation takes on a more comparative nature. Reviewers look for studies that compare the new agent to the agent currently listed on the formulary. If these are limited or unavailable, the comparisons are difficult and more subjective. If two agents appear similar in all clinical respects, the decision may be a financial one. This process often results in class review as described later. Reviewers must remember that new agents coming on the market have been tested in a limited number of patients. Because a drug's full adverse effect profile may not be evident when first released, many committees choose to stay with the older drug already listed on the formulary until sufficient information is published.

Two key components of formulary drug selection are generic substitution and therapeutic interchange. Generic substitution is the substitution of one drug product for another when the products contain the same active ingredients and are chemically identical in strength, concentration, dosage form, and route of administration. The formulary will list the drug by its generic name, strength, and dosage form. The product dispensed will be the least expensive one. As the price of products change, the product dispensed may change as well. When this occurs, the pharmacist should inform the patient if the physical appearance (e.g., color, tablet size) of their medication has changed. The patient should be assured that the new medication is identical to the previous one.

Therapeutic interchange is more complex that generic substitution. The AMA defines therapeutic interchange as the authorized exchange of therapeutic alternates in accordance with previously established and approved written guidelines or protocols within the formulary system.[7] Therapeutic alternates are drugs with different chemical structures but which are of the same pharmacologic and/or therapeutic class. They can be expected to have similar therapeutic effects and adverse reaction profiles when administered to patients in therapeutically equivalent doses. The AMA does not support therapeutic substitution defined as dispensing a therapeutic alternate for the product prescribed without prior authorization of the prescriber. Therapeutic interchange in institutional health systems has been used successfully for years. Working out an acceptable procedure for therapeutic interchange by the P&T Committee may be easier in this setting. Therapeutic interchange in outpatient drug programs in less structured ambulatory and managed care settings may be more difficult and has been criticized.[10]

Formulary Maintenance

Maintaining a formulary is an ongoing process. Policies and procedures for requests to add and delete drugs from the formulary must be in place. This includes changing recommendations for therapeutic interchanges and components of drug use guidelines. As the medical evidence changes in the published literature, the formulary system must be able to quickly respond.

Therapeutic class reviews are an important part of formulary maintenance. The pharmacologic class of drugs selected for review should be prompted by criteria set by the P&T Committee.[11] These criteria may include the number of adverse drug reaction reports, new information in the medical literature, or drug class expenditures. Some groups may choose to review the class whenever a request is received to add a new drug from that class to the formulary. The goal is to always have the best agents within a class available based on the latest medical evidence. At the time of the review, new drug use or treatment guidelines may be considered.

The formulary system should include a mechanism for patients to receive a drug not listed on the formulary if it is truly needed. A review of these non-formulary drug requests may offer insight into areas where the formulary is not meeting the needs of the health system's patients. This is true if the review reveals that requests for a specific agent are justified and frequent. The review may show that education is needed for the prescriber to steer them toward a more rational formulary choice.

Medication Use Evaluation

Medication use evaluation (MUE) is a performance improvement method that is an important part of the formulary system. MUE focuses on evaluating and improving medication use processes with the goal of optimal patient outcomes.[12] It involves establishing criteria, guidelines, treatment protocols, and standards of care for specific drugs and drug classes and the medication use process (prescribing, preparing and dispensing, administering, and monitoring).

POSITIVE AND NEGATIVE OUTCOMES

The description of the formulary system leads one to believe that it would lead to positive outcomes. In the hospital setting, a significant association has been shown between decreased costs and a well-controlled formulary, therapeutic interchange, or both.[13] Hospitals that used either strategy spent 10.7% less for drugs than those that used neither. Hospitals using both spent 13.4% less than those that used neither. An estimated $100 million in pharmacy expenditures was saved by the Department of Veterans Affairs (VA) in two years by implementing a national formulary.[14] A committee of the Institute of Medicine found that for inpatient discharges for conditions likely to be affected by the VA formulary's limited drug list, no increases in hospitalizations were found.[15] The committee did recommend to increase physician representation on formulary committees and to abandon the requirement that a drug be marketed in the United States for a year before it could be admitted to the formulary. However, convincing research clearly documenting improved patient outcomes is scarce.

Managed care organizations have used formularies to rein in drug costs but a controversial study concluded that formularies produced an opposite effect.[16] Researchers found that restrictions on drug availability were linked to increases in other services shifting costs by increasing the use of either nonrestricted drugs or other health care services. This study included the use of the restriction method called prior authorization, a method used to discourage the routine use of an expensive drug by requiring an approval process before the agent could be prescribed. In general, the results showed that the more restrictive the formulary, the higher the drug costs and the higher the number of prescriptions, outpatient and emergency room visits, and hospitalizations per patient per year. The study design and conclusions have been highly criticized.[17]

A prior authorization technique involving non-steroidal anti-inflammatory drugs (NSAIDs) in Medicaid patients was shown to be highly effective.[18] NSAIDs not available generically were place on prior approval status. This lead to the increased use of generically available NSAIDs as first line therapy. For a two-year period, the result was a 53 percent decrease in expenditures ($12.8 million) with no concomitant increase in Medicaid expenditures for other medical care.

ETHICAL ISSUES

Few ethical questions have been raised in the hospital setting but in the outpatient setting, there may be concerns. Health plans may try to manage pharmacy costs by offering incentives to physicians for prescribing lower cost drugs. This may be depicted as unethical because the strategy appears to be purely cost driven and possibly

lowering the quality of care. However, such incentives may improve quality. For example rewarding physicians for following recent guidelines for treating hypertension (an inexpensive beta-blocker and a generic thiazide).[19] In presentation at the Joseph A. Oddis Colloquium on Ethics, it was suggested that the pharmacist play the role of a pharmacoethicist on P&T Committees.[20]

REFERENCES

1. http://www.ashp.org/public/news/breaking/DF_fix.pdf (accessed October 2000).
2. Higby, G.J. American pharmacy in the twentieth century. Am. J. Health-Syst. Pharm. **1997**, *54* (16), 1805–1815.
3. Talley, C.R.; Oddis, J.A. Influences and Achievements. Am. J. Health-Syst. Pharm. **1997**, *54* (16), 1815–1825.
4. Harris, R.R.; McConnell, W.E. The American Society of Hospital Pharmacists: A history. Am. J. Hosp. Pharm. **1993**, *50* (Suppl. 2), S1–43.
5. Rolands, T.F.; Williams, R.B. How drugs attain formulary listing. Hospitals **1975**, *49* (2), 87–89.
6. Rucker, D.T.; Visconti, J.A. Hospital formularies: Organizational aspects and supplementary components. Am. J. Hosp. Pharm. **1976**, *33* (9), 912–917.
7. American Medical Association. AMA policy on drug formularies and therapeutic interchange in inpatient and ambulatory patient care settings. Am. J. Hosp. Pharm. **1994**, *51* (14), 1808–1810.
8. Ringold, D.J.; Santell, J.P.; Schneider, P.J.; Arenberg, S. ASHP national survey of pharmacy practice in acute care settings: Prescribing and transcribing—1998. Am. J. Health-Syst. Pharm. **1999**, *56* (2), 142–157.
9. http://www.ashp.org/bestpractices/formulary/position/position.pdf (accessed December 2000).
10. Carroll, N.V. Formularies and therapeutic interchange: The health care setting makes a difference. Am. J. Health-Syst. Pharm. **1999**, *56* (5), 467–472.
11. http://www.ashp.org/bestpractices/formulary/guide/formulary.pdf (accessed December 2000).
12. http://www.ashp.org/bestpractices/formulary/guide/medication.pdf (accessed December 2000).
13. Hazlet, T.K.; Hu, T.W. Association between formulary strategies and hospital drug expenditures. Am. J. Hosp. Pharm. **1992**, *49* (9), 2207–2210.
14. Anon. National formulary a plus for VA. Am. J. Health-Syst. Pharm. **2000**, *57* (14), 1302,1309.
15. http://books.nap.edu/catalog/9879.html (accessed December 2000).
16. Horn, S.D. Unintended consequences of drug formularies. Am. J. Health-Syst. Pharm. **1996**, *53* (18), 2204–2206.
17. Curtiss, F.R. Drug formularies provide path to best care. Am. J. Health-Syst. Pharm. **1996**, *53* (18), 2201–2203.
18. Smalley, W.E.; Griffin, M.R.; Fought, R.L.; Sullivan, L.; Ray, W.A. Effect of a prior authorization requirement on the use of nonsteroidal antiinflammatory drugs by medicaid patients. N. Engl. J. Med. **1995**, *332* (24), 1612–1617.
19. Smartwood, D.E. Ethics and managed care formularies. Am. J. Health-Syst. Pharm. **2000**, *57* (9), 851.
20. Jonsen, A. Fortresses and formularies: Drugs, ethics, and managed care. Am. J. Health-Syst. Pharm. **2000**, *57* (9), 853–856.

Gene Therapy

Daren L. Knoell
The Ohio State University, Columbus, Ohio, U.S.A.

Jill M. Kolesar
University of Wisconsin, Madison, Wisconsin, U.S.A.

INTRODUCTION

Extraordinary in its scope and significance, the human genome project (HGP) has revealed the complete 3 billion base pair sequence that includes the estimated 35,000 genes of the human genetic blueprint.[1]

One important outgrowth of the HGP is the development of technologies for the transfer of therapeutic genes to humans. Undoubtedly, improved biomedical technology, coupled to a better understanding of the genetic basis for most human diseases, is resulting in the rapid identification of new disease targets and the development of innovative gene therapy strategies.[2]

The number of clinical trials involving human gene therapy has dramatically increased since the initiation of the first approved trial in the United States to treat adenosine deaminase (ADA) deficiency in 1990.[3] Since this time, more than 3500 patients have been enrolled in trials worldwide, with more than 2400 in the United States.[4] The pharmaceutical industry is actively supporting gene-based therapy by investing billions of dollars, and most major academic medical centers have developed gene therapy programs.[5] The majority of active trials involve gene therapy for malignancy (68%), AIDS (18%), and cystic fibrosis (8%).[4]

Valuable experience has been gained through recipients of gene therapy, documenting the technical feasibility of human gene therapy and demonstrating, in most trials, a relative lack of treatment-related adverse effects. In particular, patients receiving both ex vivo gene therapy, a procedure where cells are removed, transfected, and placed back into the host, and in vivo gene therapy, in which the gene vector is placed directly in the patient's body, have tolerated the administration procedures without acute adverse effects. Despite this, close attention has focused on the relative lack of proven efficacy from preliminary phase I and II trials. In general, clinical trials have demonstrated short-term expression of the gene product, overall low efficiency of gene expression in the tissue(s) of interest, and lack of clinical efficacy. For these reasons, the entire field of gene therapy has been critically evaluated at the National Institutes of Health (NIH). In particular, conclusions from one panel strongly encouraged a redirection back to basic scientific research, with a particular emphasis on improving vector design.[6]

ASPECTS OF GENE DELIVERY

Definition

Gene delivery is the introduction of genes or cells containing genes foreign to the human body for the purposes of prevention, treatment, diagnosis, or curing disease.

The introduction of exogenous deoxyribonucleic acid (DNA) into mammalian cells for therapeutic intention can be accomplished by several techniques that include physical, viral, and nonviral methods, each with advantages and disadvantages. The majority of clinical experience is derived from viral and nonviral vectors and is therefore discussed. In all cases, several fundamental attributes are required for a gene therapy vector to be suitable for human use. The vector should be safe to the recipient, capable of efficient gene delivery and expression in the targeted tissue, and capable of mass production for human use. Based on these major criteria, the "ideal" gene delivery system has yet to be identified. Of the more than 425 clinical trials conducted worldwide, the field remains dominated by retroviruses (37.6%), adenoviruses (20.2%), and plasmid-based, nonviral vectors such as cationic liposomes (17.6%).[4] Numerous other vectors and techniques are being used in phase I trials, but alone they do not comprise greater than 5%

Encyclopedia of Clinical Pharmacy
DOI: 10.1081/E-ECP 120006217

Table 1 Comparison of viral transfer techniques

	Retrovirus	Adenovirus	Naked DNA	Liposome
Genome transfer	RNA	DNA	DNA	Either
Virus titer	$10^6 - 10^9$	$10^{11} - 10^{12}$	NA[a]	NA
Purification	Difficult	Yes	Yes	Yes
Maximum size	8 kB	8 kB	50+ kB	50+ kB
In vivo	No	Yes	Yes	Yes
Integration	Yes	No	Low	Low
Efficiency	High	Very high	Moderate	Low
Safety issues	Mutagenesis	Immune reaction	?	?
Nondividing	No	Yes	?	?
Limitation	Cell division required	Transient expression	Low efficiency	Low efficiency

Key: DNA, deoxyribonucleic acid; RNA, ribonucleic acid.
[a]NA, not applicable.

of the market share and therefore are not reviewed. Each of the three major categories of vectors used in clinical trials has unique attributes and limitations that include, but are not limited to, DNA-carrying capacity, tropism for target cells, in vivo transfer efficiency, duration of gene expression, and potential to induce inflammation (Table 1).

VECTORS

The fundamental goal of gene therapy is to correct a singular genetic defect in the cells responsible for causing disease in the host. To accomplish this, the gene of interest must be isolated and packaged into a delivery vector and then introduced to the recipient. The therapeutic gene must enter into the cell intact and travel to the nucleus where it interacts with the host cell machinery, ultimately being turned into a therapeutic protein (Fig. 1). A major limitation of most gene therapy is poor transfer efficiency of the gene to the target cell population. To overcome this obstacle, scientists have turned to the most efficient, naturally occurring gene vectors known to human kind—viruses. The primary objective is to produce virus-based vectors that retain the essential "gene delivering" features, while also eliminating characteristics associated with infection and host toxicity. Due to the pathogenic nature of viruses, substantial effort has also been devoted to the development of synthetic vectors that chemically mimic the natural gene delivery features of viruses. The most common viral and nonviral vectors used in clinical trials share certain attributes but are quite distinct in many ways. As is discussed, these features have a substantial impact on therapeutic strategies and, in certain situations, limit the use of vectors in different disease states.

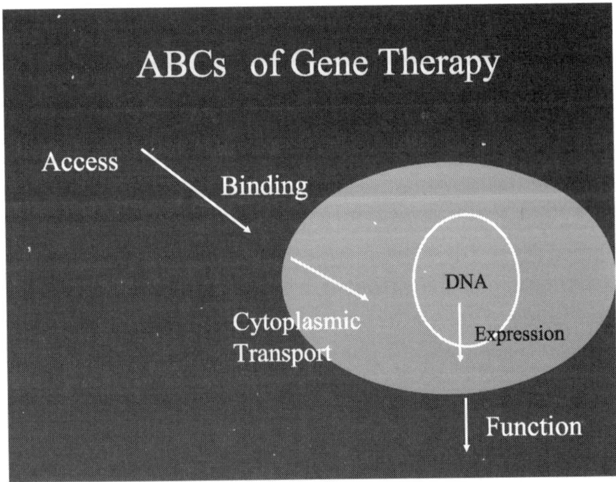

Fig. 1 Overview of gene therapy.

Retroviral Vectors

Retroviral vectors work by reverse transcribing their viral ribonucleic acid (RNA) genome, which includes the therapeutic gene insert, into a double-stranded DNA that becomes stably incorporated in the host cell genome (Fig. 2).[7] The virus components associated with replication are removed, thereby preventing infectious risk to the host and providing space for the inserted gene. The simplest type of retroviral construct is the single gene vector. In this system, the entire gene cassette of a functional gene is placed in the retroviral construct with gene expression controlled by the retrovirus gene promotor. The most widely used retroviral vectors in cli-

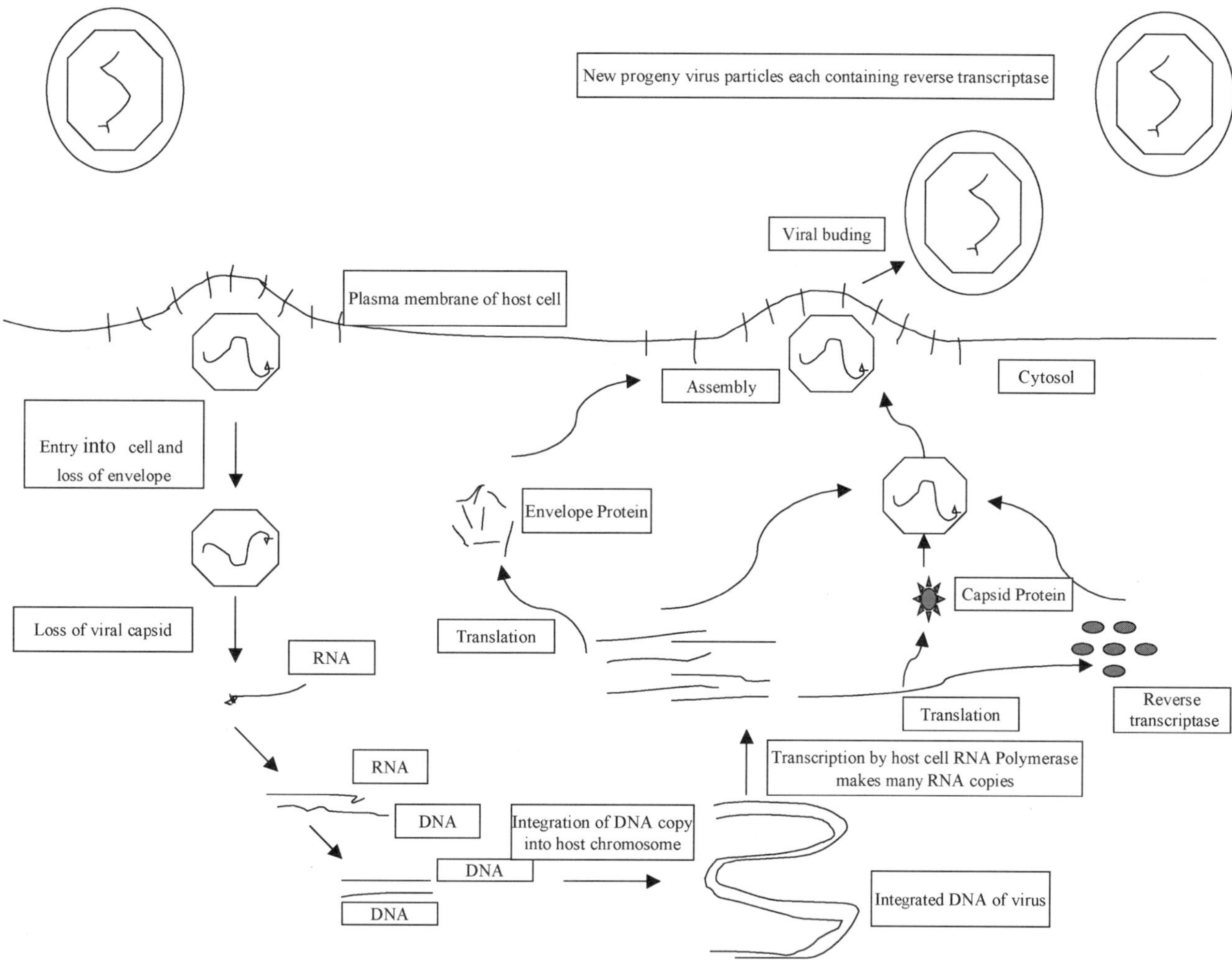

Fig. 2 Overview of retroviral vector administration.

nical trials are the double gene constructs. They possess the therapeutic gene, as well as a second marker gene, such as the neomycin phosphotransferase gene. A significant advantage of the double gene construct is that cells expressing the gene marker protein can be selected in culture and then readministered to the patient. Retroviruses integrate the gene insert into the host cell so they are particularly suited for chronic diseases that require long-term gene expression to correct the disease phenotype.[8] One of the biggest limitations of retroviruses is that they are relatively unstable following systemic administration.[7] For this reason, most human applications require removal of the target cells for ex vivo gene transduction. Retroviral vectors also require that cells undergo replication during transfection to stably integrate the gene of interest. Therefore, most clinical protocols involve induction of cell replication during ex vivo cell culture to enhance transfection efficiency. However, many cells exist in a differentiated state; that is, they do not replicate and may not readily be removed from the host, thereby preventing the use of retroviruses. An additional theoretical limitation of retroviral vectors involves insertional mutagenesis. Integration of the genetic material is random and may occur anywhere in the host genome. It is therefore theoretically possible that random integration could disrupt expression of other key proteins.

Adenoviral Vectors

The most extensively used adenoviruses are serotypes 2 (Ad2) and 5 (Ad5) because both are not associated with serious infectious disease in humans.[7] Similar to retroviral vectors, elements of adenovirus DNA genome are removed to prevent replication once inside the

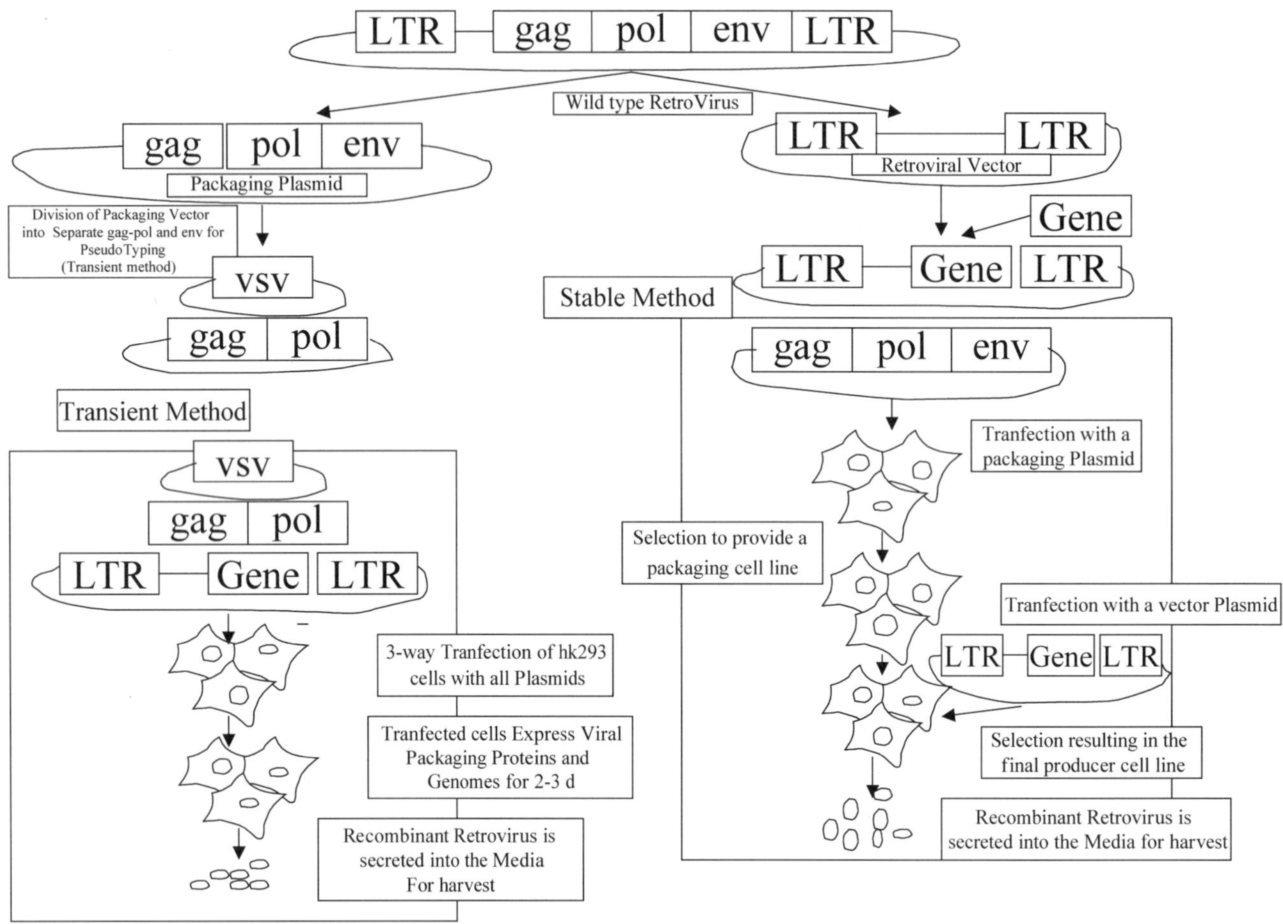

Fig. 3 Overview of gene therapy preparation.

mammalian cell. Removal of these elements also provides space for insertion of a therapeutic gene (Fig. 3). An additional reason for tailoring the adenoviral vector genome is to eliminate the expression of antigenic viral proteins that precipitate a host inflammatory response. A distinct advantage of adenoviral vectors is that they have broad cell tropism and can transfect nondividing cells. They can also be administered systemically via the intravenous, intramuscular, and intranasal routes. From a formulation standpoint, adenoviral vectors are superior because relatively high titers can be achieved (10^{10} colony-forming units/milliliter) to ensure convenient dosing in a minimal volume. Despite these attributes, the adenoviral vectors possess features that limit their utility. Unlike retroviral vectors, the gene cassette resides in the nucleus independent of the host cell genome. Because stable integration is not achieved, expression of the gene product is transient. This can be an advantage if temporary expression will correct the defect; however, in most strategies, persistent gene expression is required to correct the underlying disorder. Therefore, maintenance dosing of the vector is required to sustain therapeutic benefit to the patient. In this situation, the other major limitation of the adenoviral vectors, host toxicity, becomes a consideration.[9,10] The adaptive host response becomes important because memory is generated against the vector, thereby amplifying the immune response upon repeat administration and reducing the duration of gene expression. For these reasons, substantial effort is underway to eliminate adenoviral vector-induced inflammation by selectively removing key antigenic determinants associated with the host response. The fundamental challenge is to make a safe vector without removing the ability of the modified virus to efficiently deliver its genetic payload.

Plasmid-Based Vectors

Successful transfection depends on both the efficiency of DNA delivery to the cell (e.g., the fraction of DNA getting into the nucleus) and the efficiency of DNA expression (e.g., the amount of gene that is transcribed). In theory, the nonviral vector systems are attractive candidates due to their potential versatility. Despite their use in clinical trials, the lipid-based systems have several drawbacks. Many targeting strategies have been developed but few have worked in vivo. Once the nonviral plasmid expression cassette enters the nucleus, it exists as an episome, similar to the adenoviral vector; therefore, transient expression is achieved. In general, the lipid-based systems have a superior safety profile and gene therapy recipients tolerate high doses without any notable adverse events. The ''Achilles heel'' of the nonviral vectors have been inefficient introduction and expression of DNA into target cells when compared with viral vectors. Another major obstacle with these macromolecular aggregates that prevents efficient gene delivery in the host is opsonization.[11] They are recognized as large, hydrophobic macromolecules and are cleared within minutes from the circulation. Therefore, the fundamental challenge of the plasmid-based systems is to improve transfection efficiency in vivo. This will likely be achieved by incorporating more features of the most efficient gene delivery systems, viral vectors, while also maintaining a superior toxicity profile.

VECTOR PRODUCTION AND ADMINISTRATION

Large-scale production and purification of gene therapy vectors is critical in advancing the clinical utility of this new class of medicine. Under ideal circumstances, a highly purified vector stock should be manufactured with a relatively stable shelf-life, in a dosage form that is easy to dispense and ultimately administer to the patient. The ideal system does not currently exist for any of the vectors used in clinical trials.[12]

In general, vector production is analogous to generating a recombinant protein. The product is a macromolecule that must be derived from cultures of living prokaryotic or eukaryotic cells and purified on a large scale. The viral components required during vector production can be defined as cis and trans elements. Cis elements, for example, transcription initation promotors, must be carried by the virus itself. Trans elements are removed from the original viral genome to eliminate infectious risk but are required during production to formulate a functional viral vector. The trans elements are provided by a packaging mammalian cell line, thereby providing the necessary elements to build a functional vector during viral production but without producing an infectious viral particle. Distinct production, formulation, and patient administration skills are required for each unique product. From a manufacturing perspective, no standardized test currently exists that can be used to predict virulence or pathogenicity of each unique vector. Currently, each lot of vector must be individually evaluated by the manufacturer. All vectors are tested on three basic principles in preclinical development and large-scale production before use in clinical trials. Each vector must demonstrate evidence of safe vector system design, appropriate production of vector stocks under good manufacturing process guidelines, and documentation of purity under good laboratory practice guidelines. One review discussed this topic extensively.[13]

The final product must be free of adventitious agents that primarily include bacterial or viral pathogens and other biologic contaminates contributed during cell culture, such as DNA from prokaryotic or eukaryotic host cells and endotoxin. Purity testing for these factors must be performed throughout the production procedure and meet defined criteria before administration into humans.

The formulation and packaging of gene delivery vectors is labor intensive and places a potential burden on this rapidly advancing field. Vectors currently used in clinical trials are limited by short shelf-lives. The vector is then provided to the investigator(s) as a frozen, ready-to-use product typically in a glycerol–salt solution. The vector must be handled as a biohazard, with strict safety precautions enforced by all personnel prior to, during, and shortly after administering the vector to the patient. Significant effort is under way to develop convenient dosage forms for synthetic gene delivery vectors that will sustain potency on the shelf and allow convenient production, formulation, and patient administration. Therefore, this class of vector has a distinct advantage over modified viruses. Administration to the patient can be done either in vivo or ex vivo as already described. The major advantage of ex vivo gene therapy is that it ensures delivery of the gene to the intended cells. The major disadvantages include the amount of time, expertise, and specialized facilities required to accommodate this strategy. In contrast, in vivo gene therapy involves direct administration of the vector into the patient, which is much more convenient. However, this creates unique challenges because the product must be received fresh, handled as a biological hazard, and provided to the patient usually within hours of receiving the product. Currently, no guidelines have been published to address the safe preparation and handling of gene therapy products. Handling usually requires that therapies are prepared in biological cabinets under sterile conditions. Appropriate barriers, including gloves, gowns, and masks should be worn by those preparing and administering the dose. Gloves and other supplies used in preparation should be autoclaved or decontaminated by ultraviolet light prior to disposal in biohazardous waste containers. Patient secretions including blood, urine, feces, and respiratory secretions may be decontaminated with bleach prior to disposal.

PATIENT MONITORING

The pharmacokinetic and pharmacodynamic parameters of gene delivery vectors are largely uncharacterized in humans. However, essential concepts have been described regarding the many unique aspects inherent to in vivo distribution of macroparticulate DNA carrier systems. The distribution of most vectors is predictable and, in most cases, is limited by physical characteristics of a macromolecule. In general, all gene delivery systems are rapidly cleared from the systemic circulation, within minutes, after placement into the bloodstream. This often limits the capability of the vector to transfect cells in the targeted tissue. Fortunately, many of the vectors used in clinical trials can withstand physical manipulation, allowing site-specific administration in an attempt to enhance expression in defined tissues. Perhaps a greater challenge is to determine how long a particular gene will be expressed in a specific tissue once the vector has delivered the therapeutic gene to the targeted cells. Preliminary investigations have addressed this concern, but are limited to theoretic calculations from in vitro data.[14] Ultimately, this information is essential to develop a gene dosing regimen for a given patient, vector, and disease. Many trials involve treatment of chronic disorders, including AIDS, malignancy, and cystic fibrosis in which the gene is being delivered to differentiated cells with a limited life span. Therefore, it is presumed that many patients will require maintenance dosing of a vector to sustain expression of the therapeutic gene product over time.

Monitoring clinical efficacy of gene therapy has received little attention. For example, in published trials involving patients with ADA deficiency, the investigators routinely measured serum ADA protein concentrations to document sustained expression of the therapeutic gene.[15] In additional, the patients were extensively monitored for evidence of improved immune function and decreased number of infections. In the case of cystic fibrosis, the cystic fibrosis transmembrance conductance regulator (CFTR) protein is not released by transfected cells and remains associated with the cell membrane in patients. Knowles et al. had to physically remove nasal epithelial cells and then use advanced molecular techniques to document protein expression and function.[16] Gene therapy strategies undoubtedly create unique challenges for the clinician trying to determine when the next dose of a gene therapy vector should be administered. Measurement of sustained gene expression, or a lack thereof, will likely become common laboratory

Table 2 Monogenic diseases: phase I and II ongoing gene therapy clinical trials as of February 1, 2001

Indication	Gene	Number of open trials	Countries
Chronic granulomatous disease	P47 phox	2	U.S.A.
Cystic fibrosis	CTFR	10	France, U.K., U.S.A.
Fanconi's anemia	FACC	1	U.S.A.
Gaucher's disease	Glucocerebrosidase	1	U.S.A.
Hemophilia B	Factor IX	1	China
Hurler's syndrome	IDUA	1	U.K.
SCIDS	ADA	5	France, Italy, Japan, Netherlands, U.K.
SCIDS	MDR	1	Netherlands
Purine nucleoside phosphorylase deficiency	PNP	1	U.S.A.

Key: CTFR, cystic fibrosis transmembrane conductance regulator; FACC, factor C; IDUA, α-L-iduronidase; SCIDS, severe combined immunodeficiency; ADA, adenosine deaminase; MDR, multidrug resistance; PNP, purine nucleoside phosphorylase.

procedure for gene therapy recipients and require specialized molecular assay techniques.

GENE THERAPY CLINICAL TRIALS

Monogenic Disorders

Monogenic, or single gene disorders, are rare hereditary disorders usually identified in childhood. They represent the purest approach to gene therapy, where potentially, the correction of a single gene defect by gene therapy may lead to correction of the disease state. The major limitation of gene therapy for monogenic disorders is that the rarity of these conditions limits the number of patients able to participate in clinical trials. The majority of gene therapy trials for monogenic disorders has focused on severe combined immunodeficiency syndrome (SCIDS)[17,18] and cystic fibrosis (CF).[19,20] In addition, small trials are ongoing in Fanconi's anemia, hemophilia, and other diseases (Table 2).

SCIDS

Patients with SCIDS, a rare genetic disorder in which ADA is absent, have a greatly impaired immune system. The initial success in gene therapy came in 1989, with the report of the successful transfection of the normal ADA gene into T lymphocytes. In the two patients studied, both had normal immune function restored without adverse effects. Subsequent studies have demonstrated that both stem cells and CD34+ umbilical cord cells can be engineered to produce ADA and restore immune function. Although this disease is extremely rare, it represents the first successful clinical use of gene therapy.[17,18]

Cystic Fibrosis

CF should be the ideal candidate for gene therapy because it is a single gene defect and thus presents a clear target. The main clinical problem is in the lungs, and the likely target is the surface epithelium. Methods of topical delivery to the airway surface are already well developed. All the required components for gene therapy were in place, and CF gene therapy progressed rapidly from preclinical to clinical studies. The gene, although large, could easily be inserted into a virus or produced as a plasmid; cellular studies showed that CFTR gene transfer could produce functional chloride channels and subsequently showed that cystic fibrosis cell lines could be corrected. The next steps were the demonstration of relatively efficient gene transfer to the airway epithelium using reporter genes in rodents, followed by partial correction of the disordered airway electrophysiology in CF mice. Clinical trials soon followed and more than 150 volunteers with cystic fibrosis participated. The results have been both encouraging, as gene therapy appears to be possible, and frustrating, as it just do not work that well. There is good evidence of low levels of gene transfer and small changes in ion transport, but progress has been hampered by inefficient gene transfer, immunity to viral vectors, and a systemic inflammatory reaction provoked by plasmid DNA, resulting in no clinical benefit to date.[19,20]

Cancer

In contrast to monogenic disorders, cancer is generally caused by multiple genetic defects, providing no clear single target for gene therapy. However, because cancer is the second leading cause of death in the United States, gene therapy is under intensive investigation. Rather than correcting the multiple genetic defects found in tumors, cancer investigators have generally investigated approaches to conferring drug sensitivity, either by transvecting tumor cells with a gene encoding an enzyme such as herpesvirus thymidine kinase (HSV-TK)[21] that can metabolize a nontoxic drug to its toxic form (suicide genes) or with p53 (Table 3).[22]

The majority of gene therapy clinical trials are for cancer, with trials ongoing for almost all types of cancers. In addition, gene therapy for cancer is closest to the clinic, with both p53 and HSV-TK gene therapy in phase III clinical trials (Tables 4 and 5).

HSV-TK

The HSV-TK gene converts nontoxic nucleoside analogs such as ganciclovir into phosphorylated compounds that kill dividing cells. Therefore, cells genetically modified to express the HSV-TK gene can be killed by the administration of ganciclovir.[21]

This cytotoxic effect of transduced cells on nontransduced cells is termed the bystander effect.[23] Because only a small number of cells will be transduced with the cytotoxic gene, when these cells die, they release toxic products that in turn kill the surrounding (or bystander) cells. The TK-ganciclovir approach is currently used in several clinical trials for a variety of malignancies, including gliomas.[24]

Adenoviral (Ad)-mediated intrapleural HSV-TK-ganciclovir gene therapy has been tested primarily in phase I and II clinical trials in patients with mesothelioma,

Table 3 Oncology: phase I and II ongoing gene therapy clinical trials as of February 1, 2001

Indication	Gene	Number of trials	Country
Breast	c-erb-b2	1	U.K.
Cervical	HPV	1	U.K.
CML	HSV-TK	2	U.S.A.
Colon cancer	CC49 zeta TcR chimera	2	U.S.A.
Head and neck	INF	1	U.S.A.
Head and neck	IL-12	1	U.S.A.
Glioblastoma	HSV-TK	6	Finland, France, Spain, Switzerland, U.S.A.
Lymphoma	MDR1	1	U.K.
Lymphomas and leukemias	Specific idiotype	3	U.S.A., U.K.
Melanoma	IL-2	6	Germany, France, Italy, Netherlands, U.K., U.S.A.
Melanoma	IL-7, IL-12, Gm-CSF	3	Germany
Melanoma	IL-4	2	Italy
Melanoma	GM-CSF	1	Netherlands
Melanoma	IL-6	1	Poland
Melanoma	HLA-B7/beta 2 micro	2	U.S.A.
Melanoma	MART1 +gp100	2	U.S.A.
Mesothelioma	IL-2	1	Australia
Metastatic cancer	IL2	2	France, Switzerland
NSCLC	P53	1	U.S.A.
NSCLC	IL-2	1	U.S.A.
NSCLC	GM-CSF	1	U.S.A.
Ovarian	HLA-A2	1	Singapore
Ovarian	P53	2	U.S.A., U.K.
Ovarian, prostate, and breast	BRCA1	2	U.S.A.
Ovarian	Mov-gamma	1	U.S.A.
Pancreas	Cytochrome p450	1	Germany
Prostate	IL-2	1	U.S.A.
Prostate	PSA	3	U.S.A.
Prostate	P53	1	U.S.A.
Prostate	GM-CSF	2	U.S.A.
Prostate	HSV-TK	1	U.S.A.
Renal cell	IL-2+HLA B7	1	Germany
Renal cell	HLA B7/Beta 2 micro	2	U.S.A.
Superficial solid tumors	IL-2	1	Switzerland

glioblastomas, or ovarian cancer. The gene was administered intrapleurally in patients with mesothelioma or ovarian cancer and by direct injection during surgery in those with glioblastomas. In most phase I trials, the dose-limiting toxicity was not reached. Side effects have been minimal and included fever, anemia, transient liver enzyme elevations, and bullous skin eruptions, as well as a temporary systemic inflammatory response. Using RNA polymerase chain reaction (PCR), in situ hybridization, immunohistochemistry, and immunoblotting, HSV-TK gene transfer has been documented in approximately 50% of patients. Clinical activity has been minimal, although this may be related to the patient population studied, which is generally those with advanced refractory disease. Ongoing approaches are evaluating gene therapy in combination with chemotherapy.[24]

P53

P53 is the most frequently mutated gene in human cancer, with an up to 50% mutation frequency in solid tumors. Most commonly, these genetic changes are missense mutations in one allele, although deletions or chain termination mutations can occur.

Table 4 Oncology: phase III ongoing gene therapy clinical trials as of February 1, 2001

Indication	Gene	Number of trials	Country
Glioblastoma	HSV-TK	1	Multicountry
Head and neck	P53	1	U.S.A.
Melanoma	HLA-B7/Beta 2 microglobin	1	U.S.A.
Ovarian cancer	P53	2	U.K., U.S.A.

Key: HSV-TK, herpesvirus thymidine kinase.

Because normal or wild-type p53 is important in cell cycle regulation and apoptosis, restoration or modulation of p53 function is under intensive investigation for cancer therapy, with the hypothesis that restoration of p53 function may make cancer cells more susceptible to the effects of DNA damage inflicted by conventional chemotherapy or radiotherapy and able to undergo apoptosis.[22] Three main approaches are under evaluation. First, there is virus-mediated gene transfer in which a viral genome is engineered to contain foreign genes that are expressed in the host cell genome after infection. Second, there is the use of a cytolytic virus that can replicate only in cells that lack p53 function, and by targeting such cells could destroy tumors with mutant p53. Third, there is the discovery or design of small molecules that can interfere with the negative regulation of p53, pharmacologically activating the p53 response.

A single clinical trial using wild-type p53 gene transfer in nine patients with non-small cell lung cancer in whom conventional treatment had failed has been reported.[25] In this study, the LNSX retroviral vector was injected directly into the tumor either percutaneously with radiological guidance or via a bronchoscope. In situ hybridization and DNA PCR showed vector-p53 sequences in posttreatment biopsies, and apoptosis was more frequent in posttreatment than in pretreatment biopsies. No treatment-related toxicity was noted, and tumor regression occurred in three patients. Further extensive trials of adenovirus encoding wild-type p53 are currently underway.

The DNA tumor virus adenovirus produces a 55-kDa protein from the E1B region of its genome, which binds and inactivates p53. It was hypothesized that an adenovirus lacking E1B would not be able to replicate in normal cells but would in cancer cells lacking p53 function. For this reason, ONYX-015, an E1B gene-attenuated adenovirus was compared with normal adenovirus in human and colonic cancer cell lines with and without p53 function. As expected, the ONYX-015 virus replicated as efficiently as the normal virus in the cell line lacking wild-type p53, but not in the line with normal p53 function.[26] This vector is in early clinical trials.

Table 5 Infectious disease: phase I and II ongoing gene therapy clinical trials as of February 1, 2001

Indication	Gene	Number of studies	Country
EBV and CMV	CMV pp65	1	U.S.A.
HIV	HIV env/rev	3	U.S.A., Switzerland
HIV	CD-zeta TcR chimera	2	U.S.A.
HIV	Antisense to pol 1	2	U.S.A.
HIV	Rev + pol 1	2	U.S.A.

Key: EBV, Epstein-Barr virus; CMV, cytomegalovirus; HIV, human immunodeficiency virus.

Table 6 Cardiology: phase I and II ongoing gene therapy clinical trials as of February 1, 2001

Indication	Gene	Number of trials	Country
Coronary artery disease	VEGF	1	Finland
Coronary artery disease	FGF	1	U.S.A.
Peripheral artery disease	VEGF	7	Finland, U.S.A.

Key: VEGF, vascular endothelial growth factor; FGF, fibroblast growth factor.

Cardiovascular Disease

Angiogenesis, or growth of new blood vessels, appears essential in revascularization after myocardial infarction as well as in treating coronary artery disease and peripheral artery disease. Therefore, cardiovascular gene therapy has concentrated on vascular endothelial growth factor (VEGF) in these diseases[27] (Table 6).

Low-Density Lipoprotein (LDL) Receptor

Familial homozygous hypercholesterolemia is a rare hereditary monogenic disorder caused by mutations of the LDL receptor gene. Individuals have severe hypercholesterolemia associated with premature atherosclerosis. In a single study, patients were treated with gene therapy

Table 7 Other ongoing gene therapy clinical trials as of February 1, 2001

Indication	Gene	Studies	Countries
Amyotrophic lateral sclerosis	CNTF	1	Switzerland
Alzheimer's disease	Nerve growth factor	1	U.S.A.
Anemia of end-stage renal disease	EPO	1	U.S.A.
Cubital tunnel syndrome	HIGF-1	1	U.S.A.
Hip fracture	Parathyroid hormone	1	U.S.A.
Rheumatoid arthritis	HSV-TK	1	U.S.A.
Rheumatoid arthritis	IRAP	1	U.S.A.
Severe inflammatory disease of rectum	IL-4 and IL-10	1	Austria

Key: CNTF, ciliary neurotrophic factor; EPO, erythropoetin; HIGF, human insulin-like growth factor; HSV-TK, herpesvirus thylimidine kinase; IRAP, insulin responsive aminopeptidase.

with the LDL receptor. Expression of the receptor was documented, but LDL cholesterol levels remained substantially elevated 3 to 6 months after gene transfer, 611 ± 27 vs. 550 ± 51 mg/dL, before and after gene therapy, respectively.[28]

VEGF

Formation of new blood vessels by the angiodan VEGF is an experimental strategy for treating myocardial ischemia. The VEGF proteins function by interacting with specific receptors on endothelial cells, which initiates a cascade of events culminating in endothelial cell migration, proliferation, aggregation into tubelike structures, and networking of the arterial and venous systems.[27]

Gene transfer represents one approach to delivering an angiogen to the heart in which the carrier DNA (cDNA) coding for VEGF is delivered to the myocardium, with the myocardial cells used to secrete the VEGF. Studies in experimental animals have shown that replication-deficient, recombinant adenovirus (Ad) gene transfer vectors are advantageous for delivery of angiogens such as VEGF, in that Ad vectors provide a high transfection efficiency, remain highly localized, and express VEGF for a period of 1 to 2 weeks, which is sufficient to induce collateral vessels to relieve the ischemia but not long enough to evoke abnormal angiogenesis.[27]

In a phase I evaluation, VEGF121.10 was administered to 21 individuals by direct myocardial injection into an area of reversible ischemia either as an adjunct to conventional coronary artery bypass grafting or as sole therapy via a minithoracotomy. There were no adverse effects attributed to the gene transfer, and patients had decreased angina.[29]

Other trials of VEGF have been reported. A case report demonstrated improvement in blood supply to an ischemic limb after intra-arterial gene transfer of a plasmid encoding for VEGF.[30] The use of a plasmid-based gene delivery system, although inefficient, was reasonable in this situation because VEGF is a potent secreted product. A phase 1 trial of intramuscular delivery of a plasmid-encoding VEGF in the setting of severe peripheral vascular disease was reported.[31] Gene transfer was performed in 10 limbs in nine patients with nonhealing ischemic foot ulcers. Increased circulating VEGF levels were demonstrated after intramuscular gene delivery. Various measures, including ankle-brachial index and magnetic resonance angiography, showed qualitative evidence of improved distal flow in 8 limbs.

Multidrug Resistance (MDR)

In a therapeutic approach, stem cells may be isolated from patients and genetically modified to express the MDR gene.[32] These cells are then retuned to the patient prior to administration of chemotherapy, making the stem cells resistant to chemotherapy.

Other Diseases

Gene therapy is under evaluation for many diseases, ranging from rare inherited single gene defects to common disease such as HIV, deafness, autoimmune diseases, bone regeneration, and many others[4,33,34] (Tables 5 and 7).

ETHICAL ISSUES

The first death attributable to gene therapy occurred in September 1999, when an 18-year-old patient with ornithine transcarbamylase deficiency died, apparently as a direct result of the experimental gene therapy studies.[35,36] This prompted two senate hearings and resulted in recommendation for implementation of new policies by the Recombinant Advisory Council (RAC), Food and Drug Administration (FDA) and NIH, which require earlier review of researcher's plans for monitoring safety and

quarterly meetings.[37–39] A few of the safeguards implemented include thorough public evaluation of protocols before investigational new drug assignment for FDA and institutional review board (IRB) approval; the development of a single, uniform mechanism for reporting adverse events to the RAC, FDA, and other relevant agencies; establishment of a public database of all adverse events; and nonparticipation of investigators with financial interests in study outcomes in patient selection, the informed consent process, and direct management of clinical studies.

Further evaluation of this tragic event has identified that vector-associated toxicity was not the sole cause for this patient's death. The FDA determined that human subjects in this investigation were not adequately protected and that there was substantial financial conflict of interest. Subsequently, the NIH has discovered hundreds of unreported adverse events among volunteers enrolled in gene transfer experiments. These findings have catalyzed broad examination of the entire clinical research process, with the Secretary of Health and Human Services calling for broad reforms in informed consent, clinical monitoring, and conflict of interest.

CONCLUSION

Gene therapy is in its infancy. Early and ongoing success in SCID, combined with promising studies in cardiovascular and oncology therapies, supports optimism for these novel strategeis. However, several important issues remain, including the best vector for transfer and appropriate protection for both patients and health care providers. The Orkin–Motulsky report clearly stated that ''significant problems remain in all basic aspects of gene therapy. Major difficulties at the basic level include shortcomings in all current gene transfer vectors and an inadequate understanding of the biological interaction of these vectors with the host.''[6] As such, the report clearly identified key recommendations to ensure continued progress in this field. The recommendations were to 1) continue research at the basic level to improve vector design and studies that will further identify pathogenic mechanisms of disease, 2) improve trial design with quantitative and qualitative assessment of gene transfer and expression, 3) maintain adequate financial support for gene therapy studies and promote interdisciplinary collaborations at the basic and clinical levels, and 4) disseminate information to the public that clearly identifies limitations of the field as well as exciting new discoveries.

Many new gene delivery vectors and protocols are currently in developmental stages that aim to improve on the earlier prototypes. The relatively small number of vectors used in clinical trials underscores the complexity of DNA delivery and our lack of knowledge about how these macroparticulates are handled and expressed in the human body. It is hoped that the recent reprioritization of gene therapy studies will improve the design of vectors, enhance our understanding of the biological interactions between gene-carrying vectors and the body, eliminate adverse events, and improve information gained from future clinical trials. Assuming these events occur, experts still predict that gene therapy is still more than 5 to 10 years from routine use in patients.

REFERENCES

1. Emilien, G.; Ponchon, M.; Caldas, C.; Isacson, O.; Maloteaux, J.M. QJM **2000**, *93*, 391–423.
2. Anderson, W.F. Science **2000**, *288*, 627–629.
3. Blaese, R.M.; Culver, K.W.; Miller, A.D.; Carter, C.S.; Fleisher, T.; Clerici, M.; Shearer, G.; Chang, L.; Chiang, Y.; Tolstoshev, P., et al. Science **1995**, *270*, 475–480.
4. **2001**, http://www.wiley.co.uk/genetherapy/clinical/.
5. Bossart, J.; Pearson, B. Trends Biotechnol. **1995**, *13*, 290–294.
6. Verma, I.M.; Somia, N. Nature **1997**, *389*, 239–242.
7. Walther, W.; Stein, U. Drugs **2000**, *60*, 249–271.
8. Miller, D.G.; Adam, M.A.; Miller, A.D. Mol. Cell. Biol. **1990**, *10*, 4239–4242.
9. Zuckerman, J.B.; Robinson, C.B.; McCoy, K.S.; Shell, R.; Sferra, T.J.; Chirmule, N.; Magosin, S.A.; Propert, K.J.; Brown-Parr, E.C.; Hughes, J.V.; Tazelaar, J.; Baker, C.; Goldman, M.J.; Wilson, J.M. Hum. Gene Ther. **1999**, *10*, 2973–2985.
10. Yei, S.; Mittereder, N.; Tang, K.; O'Sullivan, C.; Trapnell, B.C. Gene Ther. **1994**, *1*, 192–200.
11. Pouton, C.W.; Seymour, L.W. Adv. Drug Deliv. Rev. **1998**, *34*, 3–19.
12. Lyddiatt, A.; O'Sullivan, D.A. Curr. Opin. Biotechnol. **1998**, *9*, 177–185.
13. Smith, K.T.; Shepherd, A.J.; Boyd, J.E.; Lees, G.M. Gene Ther. **1996**, *3*, 190–200.
14. Ledley, T.S.; Ledley, F.D. Hum. Gene Ther. **1994**, *5*, 679–691.
15. Bordignon, C.; Notarangelo, L.D.; Nobili, N.; Ferrari, G.; Casorati, G.; Panina, P.; Mazzolari, E.; Maggioni, D.; Rossi, C.; Servida, P., et al. Science **1995**, *270*, 470–475.
16. Knowles, M.R.; Hohneker, K.W.; Zhou, Z.; Olsen, J.C.; Noah, T.L.; Hu, P.C.; Leigh, M.W.; Engelhardt, J.F.; Edwards, L.J.; Jones, K.R., et al. N. Engl. J. Med. **1995**, *333*, 823–831.

17. Cavazzana-Calvo, M.; Hacein-Bey, S.; de Saint Basile, G.; Gross, F.; Yvon, E.; Nusbaum, P.; Selz, F.; Hue, C.; Certain, S.; Casanova, J.L.; Bousso, P.; Deist, F.L.; Fischer, A. Science **2000**, *288*, 669–672.
18. Dobson, R. BMJ **2000**, *320*, 1225.
19. Albelda, S.M.; Wiewrodt, R.; Zuckerman, J.B. Ann. Intern. Med. **2000**, *132*, 649–660.
20. Geddes, D.M.; Alton, E.W. Thorax **1999**, *54*, 1052–1054.
21. Carew, J.F.; Federoff, H.; Halterman, M.; Kraus, D.H.; Savage, H.; Sacks, P.G.; Schantz, S.P.; Shah, J.P.; Fong, Y. Am. J. Surg. **1998**, *176*, 404–408.
22. Steele, R.J.; Thompson, A.M.; Hall, P.A.; Lane, D.P. Br. J. Surg. **1998**, *85*, 1460–1467.
23. Aspinall, R.J.; Lemoine, N.R. Gut **2000**, *47*, 327–328.
24. Fueyo, J.; Gomez-Manzano, C.; Yung, W.K.; Kyritsis, A.P. Arch. Neurol. **1999**, *56*, 445–448.
25. Schuler, M.; Rochlitz, C.; Horowitz, J.A.; Schlegel, J.; Perruchoud, A.P.; Kommoss, F.; Bolliger, C.T.; Kauczor, H.U.; Dalquen, P.; Fritz, M.A.; Swanson, S.; Herrmann, R.; Huber, C. Hum. Gene Ther. **1998**, *9*, 2075–2082.
26. Roth, J.A.; Swisher, S.G.; Merritt, J.A.; Lawrence, D.D.; Kemp, B.L.; Carrasco, C.H.; El-Naggar, A.K.; Fossella, F.V.; Glisson, B.S.; Hong, W.K.; Khurl, F.R.; Kurie, J.M.; Nesbitt, J.C.; Pisters, K.; Putnam, J.B.; Schrump, D.S.; Shin, D.M.; Walsh, G.L. Semin. Oncol. **1998**, *25*, 33–37.
27. Simons, M.; Bonow, R.O.; Chronos, N.A.; Cohen, D.J.; Giordano, F.J.; Hammond, H.K.; Laham, R.J.; Li, W.; Pike, M.; Sellke, F.W.; Stegmann, T.J.; Udelson, J.E.; Rosengart, T.K. Circulation **2000**, *102*, E73–E86.
28. Grossman, M.; Raper, S.E.; Kozarsky, K.; Stein, E.A.; Engelhardt, J.F.; Muller, D.; Lupien, P.J.; Wilson, J.M. Nat. Genet. **1994**, *6*, 335–341.
29. Rosengart, T.K.; Lee, L.Y.; Patel, S.R.; Kligfield, P.D.; Okin, P.M.; Hackett, N.R.; Isom, O.W.; Crystal, R.G. Ann. Surg. **1999**, *230*, 466–470, discussion 470–472.
30. Losordo, D.W.; Vale, P.R.; Symes, J.F.; Dunnington, C.H.; Esakof, D.D.; Maysky, M.; Ashare, A.B.; Lathi, K.; Isner, J.M. Circulation **1998**, *98*, 2800–2804.
31. Isner, J.M.; Walsh, K.; Symes, J.; Pieczek, A.; Takeshita, S.; Lowry, J.; Rosenfield, K.; Weir, L.; Brogi, E.; Jurayj, D. Hum. Gene Ther. **1996**, *7*, 959–988.
32. Abonour, R.; Williams, D.A.; Einhorn, L.; Hall, K.M.; Chen, J.; Coffman, J.; Traycoff, C.M.; Bank, A.; Kato, I.; Ward, M.; Williams, S.D.; Hromas, R.; Robertson, M.J.; Smith, F.O.; Woo, D.; Mills, B.; Srour, E.F.; Cornetta, K. Nat. Med. **2000**, *6*, 652–658.
33. Rodan, G.A.; Martin, T.J. Science **2000**, *289*, 1508–1514.
34. Tsokos, G.C.; Nepom, G.T. J. Clin. Invest. **2000**, *106*, 181–183.
35. Leiden, J.M. Circ. Res. **2000**, *86*, 923–925.
36. Teramato, S.; Ishii, T.; Matsuse, T. Lancet **2000**, *355*, 1911–1912.
37. Friedmann, T. Science **2000**, *287*, 2163–2165.
38. Greenberg, D.S. Lancet **2000**, *355*, 1977.
39. Shalala, D.N. Engl. J. Med. **2000**, *343*, 808–810.

Generic Drugs and Generic Equivalency

Arthur H. Kibbe
Wilkes University, Wilkes-Barre, Pennsylvania, U.S.A.

INTRODUCTION

All drugs that are approved for sale generally carry at least two names. The drugs are given a proprietary or trade name given by the company that first develops them. These companies often are referred to as the innovator company. The drug is assigned a nonproprietary or generic name, which is agreed to by the WHO International Nonproprietary Nomenclature (INN) Committee and the U.S. Adopted Names Council (USAN). A new drug is usually first marketed with some patent protection and at a price that, at a minimum, recoups the cost of development over the remaining life of the patent or other exclusivity arrangement. Eventually, protection from competition is lost to other pharmaceutical companies, often companies or divisions of companies that specialize in marketing off-patent drugs. These companies or divisions are called generic companies. They can apply to the appropriate regulatory body such as the Food and Drug Administration (FDA) for permission to market the same active ingredient under its nonproprietary or generic name. The generic manufacturer is not required to do a complete clinical trial to prove effectiveness and safety because that has already been well established for the drug. However, it is required to show that the new drug product is equivalent to the original drug product. For the purposes of this article, we define the drug as the chemical that has the pharmacological effect and the drug product as a dosage form that contains the drug and other ingredients or excipients that allow the formulation of the dosage form. There is a large economic incentive for the development of generic drug products, especially for highly successful drug products. The pharmaceutical company that first brought the product to market maintains the price at the original level or higher to continue the cash flow into the company. This allows the other companies to develop a formulation of the drug and to win approval to market with the knowledge that, even at a fraction of the selling price of the innovator's product, the company can make a good profit. Some innovators defend their market share by arguing quality and reliability. The FDA must act as an impartial arbitrator of this debate. The debate is clearly about money, but is argued in a scientific forum. The key question is, "Are we sure that the two products, if used in the same way in the same patient, will yield the same result." If a drug product is subject to this debate, the innovator always says "no" and the second and subsequent manufacturers always say "yes." In the United States, the FDA sets the standards against which the question is resolved, and scientists take sides usually on the issue of "are the current FDA standards good enough." If the FDA gives an "A" rating to a drug product, it is in effect telling the prescriber that the drug product will yield the same therapeutic and side-effects profile as the innovator drug product. The Orange Book specifies the equivalence rating from the FDA. Almost all generic drug products currently marketed are rated A; the FDA has not approved a generic without an A rating in decades. Finally, the consumer pays the price, either in the unnecessarily high cost of drugs if unnecessary studies are performed and generic competition delayed or in risky drug substitution if the FDA is too relaxed in its standards. The tests required by the FDA have changed over the years. They have become more proscriptive and are based on sound statistical grounds. The FDA has also increased the level of oversight of the pharmaceutical companies that manufacture generic equivalents of innovator products. Thus, the regulatory process has become more stringent, and the level of assurance that the public has that a generic product is both safe and effective has gone up. The FDA has often stated that there are no known therapeutic failures from switching among products that have been ruled as equivalent by the FDA.

LEGISLATIVE AND REGULATORY HISTORY

In the early 1970s, most states had antisubstitution laws that required the dispensing of the innovator product when the prescriber wrote for a drug by trade name. Most physicians had learned only the trade name of the drug product, and these laws ensured that generic substitution would be at a minimum (1). The American Pharmaceutical Association (APhA) along with other groups pushed for the repeal of these laws and opened the way for the growth

Encyclopedia of Clinical Pharmacy
DOI: 10.1081/E-ECP 120006417

of the generic industry. The lack of bioequivalence data available at that time led to the formation of the Generic Drug Bureau within the Food and Drug Administration. As a result of the efforts of that group, the FDA produced a book, *Approved Drug Products With Therapeutic Equivalence Evaluations*, in the late 1960s. This became known as the Orange Book because of the cover color. The book has been published annually with monthly updates. The contents are now available on the FDA Website (2).

In 1984, the Drug Price Competition and Patent Term Restoration Act was passed. This act, also known as the Waxman–Hatch Bill of 1984, encouraged the development of new innovative drugs by established procedures, extended patent rights, and facilitated the FDA approval process for generic drugs (3). To address the first goal, the law created a mechanism to extend the period of patent protection for manufacturers of innovative new drugs generally ensuring at least 5 years of market exclusivity after approval. To address the second goal, the law established an Abbreviated New Drug Application (ANDA) for applications after 1962. Drugs chemically equivalent to those previously approved by a full application process need only be proven bioequivalent, not clinically equivalent. Depending on the drug, proof of bioequivalence can involve in vitro dissolution studies, in vivo single-dose bioavailability studies, in vivo multidose bioavailability studies, or a combination of these. However, in vitro dissolution studies alone are not adequate proof of bioequivalence for purposes of an ANDA.

SCIENTIFIC BASIS FOR GENERIC DRUG PRODUCT EQUIVALENCY: BIOAVAILABLITY–BIOEQUIVALENCY

The goal of the testing of generic products is not to establish the clinical usefulness of the drug but only to ensure that the generic product or new formulation has the same relative bioavailability as or is bioequivalent to the innovator product.

Bioavailablity has been defined as a measure of the rate and extent of absorption of a drug into the systemic circulation after administration of a dosage form. An intravenous i.v. dose is considered by definition to be 100% bioavailable. All other routes of administration will produce a total bioavailability less than or equal to that of the i.v. dose. Thus, only a drug that is completely absorbed into the systemic circulation can have the extent of bioavailability equal to the dose stated on the label. In addition to the extent of absorption, the rate of absorption plays a key role when evaluating the potential therapeutic impact of a particular dosage form. Knowledge of the time to onset of drug action, which is directly related to rate of absorption, is a significant concern, especially in acute clinical situations such as asthma attack, hyperglycemic shock, and pain.

The bioavailability of drugs from specific dosage forms is affected by the nature of the inactive ingredients or pharmaceutical excipients and the process used in its formulation. (For additional information, see Bioavailability of Drugs and Bioequivalency in this Encyclopedia.) When comparing similar dosage forms from different manufacturers or different lots from the same manufacturer, it is most useful to determine the relative bioavailability of the two products or lots. Some scientists have attempted to establish an in vitro test that could successfully predict in vivo bioavailability. However, to date, none has been developed.

Pharmacokinetics means the application of kinetics to drugs. It can be defined as the study of the time course and fate of drugs in the body. Teorell is often given credit for the origin of pharmacokinetics with his publications, *Kinetics of Distribution of Substances Administered to the Body* (4, 5). This science is the theoretical support for the use of bioequvalency testing to establish therapeutic equivalence among dosage forms of the same drug. The first approach to a pharmacokinetic understanding of drugs in the body, called compartment analysis, considered the body as a group of compartments through which the drug must pass. The compartment itself does not exist but represents the average of many processes that give rise to the observed phenomenon. The size of the imaginary compartment can be calculated and is useful in understanding the process of absorption, distribution, and elimination or metabolism of the drug. Regardless of the model used, a plot of the plasma concentration of the drug versus time yields a curve that can be described by a polyexponential equation. The area under that concentration–time curve (AUC) is directly related to the amount of drug absorbed. The time to reach peak concentration and the peak concentration itself are related to both the dose and the rate of absorption.

An important limitation of compartment analysis is that it cannot be applied universally to any drug. A simpler approach that is useful in the case of bioequavalency testing is the model independent method. It is based on statistical-moment theory. This approach uses the mean residence time (MRT) as a measure of a statistical half-life of the drug in the body. The MRT can be calculated by dividing the area under the first-moment curve (AUMC) by the area under the plasma curve (AUC) (6).

(See other articles in this Encyclopedia for more detailed discussion of these subjects.)

MEASUREMENT OF RELATIVE BIOAVAILABILITY OR BIOEQUIVALENCY

Drug products often undergo bioavailability testing in the early stages of development. Changes in formulation necessitated by results of clinical trials or stability testing or changes in the availability of excipients or changes in suppliers of excipients often require that the manufacturer perform a relative bioavalability or bioequivalency test to ensure that subsequent lots of a product will yield the same amount of active ingredient at the same rate as was possible in earlier formulations.

Bioequivalency studies are usually performed on young, healthy, male adult volunteers under controlled dietary conditions and fixed activity levels. This is because the goal of the study is not to establish the clinical usefulness of the drug but only to ensure that the two formulations have the same relative bioavailability or are bioequivalent.

Key Parameters

When assessing bioequivalence, the following three parameters that characterize the plasma or blood concentration–time profile of the administered drug are usually measured:

1. Peak height, $C_{\max}$, represents the highest concentration of the drug in the systemic circulation;
2. Time to peak, $t_{\max}$, represents the time for peak height to occur after the drug was administered;
3. Area under the curve, AUC, represents the total integrated area under the concentration–time curve.

The first two parameters are indicators of absorption rate, whereas the third is directly proportional to the extent of drug absorbed into the systemic circulation from the dosage form. Figure 1 is an example of a concentration–time curve for a single dose of drug to a subject.

Although it is theoretically possible to determine the rate and extent of absorption of a drug by measurement of the rate and extent of the appearance of the drug in the urine, this is not considered as reliable a method for evaluation of a drug product's bioequivalency as are blood level data. Thus, the studies commonly performed to demonstrate bioequvalence fall into two catagouries: single-dose and multidose or steady-state studies. There are advantages and disadvantages to each. Single-dose studies are less expensive and expose healthy volunteers to less drug during the course of the study. However, these studies require more sensitive analytical methods and have higher subject-to-subject variability. In both cases, a cross-over study design is used to control for sequence effects. The study is designed to control for or take into account as many variables as possible. The subjects are randomly assigned to groups. Blood samples are obtained from each subject before dosing and at fixed time intervals after dosing. Currently, the data are then analyized using appropriate statistical ANOVA. The results must meet FDA guidelines for mean and 90% confidence interval for each of the three key parameters. For oral solid dosage forms, the FDA requires that for a product to be considered bioequivalent, the ratio of the parameter for the two products, together with their 90% confidence interval, must fall between 0.8 and 1.25, using log-transformed data. This, in effect, means that drug products that differ by more than 10% in their rate and extent of absorption will not be approved as generic equivalents.

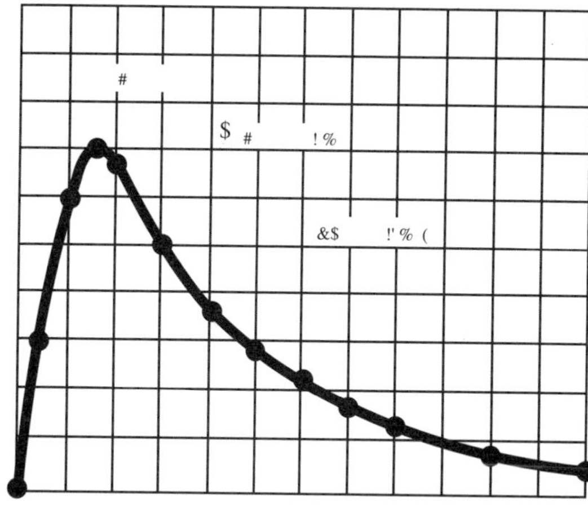

Fig. 1 Blood concentration curve.

CURRENT SCIENTIFIC ISSUES

Two issues have been raised recently with regard to the approval of generic drugs. The first has to do with the issue of "Narrow Therapeutic Index Drugs," and the second has to do with the use of individual bioequivalence in place of average bioequivalence. The former concern has been addressed in detail by Drs. Benet and Goyan (7). They concluded that narrow-therapeutic-range drugs were the least likely to have therapeutic failures among generic drugs, with proof of bioequivalency. The use of average bioequivalence data is under attack. This is because of the concern that there might be a significant subject–by-product interaction. Regulatory agencies now assume that this is not the case (8). The advantage of using individual bioequivalence studies is the reassurance that if subject/product interactions do occur, the study design would control for them, and a more statistically valid measure of the rate and extent of absorption of the drug from the two product would be determined. Some of the disadvantages associated with shifting from average to individual bioequivalence testing are cost, numbers of subjects needed, and diversity of the study population required. [See other articles that address the impact of the new metrics on the reliability and cost of the performance of bioequivalence testing (9–12).]

THE CHANGING POLITICAL ECONOMY OF GENERIC DRUGS

The modern generic drug industry in the United States really only dates from the passage of the Waxman–Hatch Act in 1984. Within 5 years of passage, generic drugs captured 40% of the market for prescriptions written inthe United States. Since that time, the generic drug market share has stablized between 40 and 50% of the prescriptions written. However, the dollars paid for generic drugs are only 10% of the total sales of drugs in the United States. That statistic alone tells us that the consumer receives enormous benefit from the substitution of therapeutically equivalent generic drugs when available.

A horrendous scandal hit the industry in the late 1980s, wherein firms representing 75% of the production of the generic industry pled guilty to one or more criminal charges involving filing false applications with the FDA, paying illegal gratuities to FDA personnel, and/or related crimes to gain an unfair competitive advantage in the emerging marketplace. Surprisingly, this scandal produced only a small delay in the market share march of generic drugs and only a temporary loss of consumer confidence in generic products.

The scandal was tied to a phenomenon that still dominates the business strategies of generic drug firms to this day: the need to obtain approval to manufacture and distribute before other firms enter the market. Because of the "commodity" nature of the business and the relative ease of entry into the industry, firms devote most of their resources and managerial talent to obtaining first or second approvals from the FDA for their products. Once a generic drug has four or more competitors, it is no longer profitable for additional generic companies to enter the market.

Generic drug manufacturers typically will continue to manufacture drugs that produce little or no profit because large purchasers that are their prime customers (chain drugs stores, buying groups for smaller community pharmacies, etc.) prefer to buy from companies that can supply most of the common generic drugs. For example, if a generic drug firm no longer produces amoxicillin because it can make more money by shifting its antibiotic production facilities to, for example, a cephalosporin drug for which it has less competition, a large chain may chose to buy its entire generic antibiotic line from another company that supplies both.

The profitable generic drug companies are profitable because they have found a strategy to maintain some control over the price of their products. In the early years (1984–1988), the best way to get "first approval" from the FDA apparently was to be first to file, to get assays or bioquivalence studies done on difficult to duplicate drugs, or to find some way to get an expedited approval from inside the agency. Unfortunately, this sometime involved payoffs to FDA review chemists (those FDA experts assigned the task of evaluating biostudy results, the crucial piece of a generic drug application, remained remarkably free of the scandal). More often, it involved submitting false information to the FDA (including, in a few cases, false biostudies). Many generic drug firms did not survive the scandal, and others survived only after the previous management and ownership were purged from the firms.

For a short period, it was believed that the profitable segment of the business involved not production but distribution. After all, if commodity prices approach marginal cost and the marginal cost of manufacturing drugs is minimal, but the price to the consumer remains significantly more than marginal, there must be middlemen somewhere making the money. Clearly, those middlemen were not in the retail pharmacy where profits continued to be squeezed. Distributors were thought to be the new profit centers. But a funny thing happened on the way to that particular bank... .

Consumers became outraged at the rapid increase in the price of pharmaceuticals as the innovator companies (and some generic firms) rushed to raise prices and as generic drug company after generic drug company was pushed out of the industry in the wake of investigations by a Congressional committee and a federal grand jury. Second, the Administration, in response to public concern about the cost of pharmaceuticals, pressured the pharmaceutical industry and forced lower prices and significant rebates to the federal and state government programs that paid for drugs. Wholesale distributors of all drugs subject to the federal rebates suffered.

Finally, the firms that thought they could profit most from the scandal entered the market. These were innovator firms, many of which had already played a significant role in the distribution of generics. Ultimately, the profit margins from generic drug sales were not sufficient to carry the overhead of the branded companies, and most left the market or returned to their distributor role. Even in the case of the firms manufacturing and marketing generic versions of their own branded products, giving them significant advantage over the remaining pure generic firms in developing and filing of the ANDAs with the FDA and the added advantage of relatively less scrutiny from the scandal-rocked agency, most had exited the marketed by the end of the decade.

Some innovator firms entered the generic drug market so that they could have a product line consistent with their new business strategy: disease state management. This strategy, a function of the rise of HMOs and the return of the concept of scarcity to prescription drug dispensing, was intended to involve the development of a continuum of drug therapies for the treatment of a specific illness (diabetes, depression, etc.), wherein the patient would be tried on the older, less-expensive drug first and, if it did not work, the next most cost-effective drug would be administered and so on until the least cost-effective drug would be the treatment of last resort. Unfortunately, the branded companies that selected this strategy found themselves competing with doctors, hospitals, and insurance companies for control of the treatment regime of individual patients, a losing proposition for the entity with the least amount of information about and access to the individual patient.

Another factor in reducing prices of all drugs that had some form of competition was the rise of the HMO and its pharmaceutical watchdog, the pharmacy benefit manager (PBM). These PBMs create a formulary of approved drugs (drugs for which they would reimburse partially or fully) based on bids from competing companies.

Much of the public's confusion regarding generic drugs arose from a practice of the PBMs to pressure doctors to substitute different chemical entities in the same therapeutic class for the prescribed medicine. Such a switch is called a theraputic substitution as opposed to the switching among manufacturers of therapeutically equivalent drugs (generics and the innovator drug or other FDA "AB"-rated substitutes). Therapeutic substitution involves a switch to a different drug, whereas generic substitution involves a switch to the same drug from a different manufacturer. If a patient is switched between FDA "AB"-rated drugs, the FDA offers the assurance that they can expect the same therapeutic and side–effect profile as the brand drug or another "AB"-rated generic drug. The FDA offers no such assurance if the switch occurs among different drugs, even if they are in the same therapeutic class. For example, aspirin and Tylenol may be equally effective in the treatment of headache, but the FDA makes no such certification, whereas it makes exactly that certification for Bayer aspirin and Safeway aspirin.

The dominance of the HMO (and related organizations) and their PBMs (and related organization types) served to accelerate the substitution of generic drugs at the turn of the 21st century. However, even that pressure could not slow the re-emergence of a high rate of price increase, greater than consumer or comparable wholesale prices as a whole, in prescription drugs. Innovator companies learned that establishing very high prices for "breakthorough" drugs could more than compensate for the loss of patent protection on a highly profitable drug.

Furthermore, the United States is the only developed country in the world that has chosen not to explicitly control the price of any drug product and has used its market power as a huge buyer relatively sparingly. Consequently, U.S. prices for drug products still under patent are usually substantially above those charged anywhere else in the world. Generic prices approach cost except for those few generics that have managed to eliminate or limit for a specific period competition from other generics.

Those generic drug firms that prospered in this restrictive price environment all had one or more niche drugs that were immune from corrosive price competition. Some companies mastered a manufacturing process that produced bioequivalent medicine that the innovator itself found difficult to master lot to lot. Others took advantage of certain exclusivity provisions in the law for those that challenged a product patent in court, ostensibly to cover the cost of litigation. In other cases, the settlement of those cases provided some form of licensing or distribution rights that permitted the sale of a generic product while the patent was still valid. Finally, a fortunate firm might find itself in possession of the exclusive right to purchase the raw material from the only source available to generic drug manufacturers.

All generic drug firms capable of generating the necessary cash to develop and market new drugs are moving toward that lucrative market. For the time being, the United States has chosen to use the market mechanism as its only important control on drug prices. Generics are the competition, and competition is our only real form of price control.

According to the Congresslonal Budget Office (CBO), consumers saved $8–10 billion in 1994 because of the use of generic drugs. In that same 1998 report, CBO cited the Waxman–Hatch Act, generic substitution laws passed by the states, and government health programs as seminal events leading to the acceptance of generic drugs and the resulting savings.

REFERENCES

1. Knoben, J.E.; Scott, G.R.; Tonelli, R.J. Overview of the FDA Publication Approved Drug Products with Therapeutic Equivalence Evaluations. Am. J. Hosp. Pharm. **1990**, *47* (12), 2696–2700.
2. Food and Drug Administration. *Approved Drug Products with Therapeutic Equivalence Evaluations*; http://www.fda.gov/cder/ob/ FDA: Rockville; Available from the Government Printing Office: Washington, DC, 2000.
3. Weaver, L.C. Drug Cost Containment—The Case for Generics: Situation in the U.S.A. J. Soc. Admi. Pharm. **1989**, *6* (1), 9–13.
4. Teorell, T. Kinetics of Distribution of Substances Administered to the Body. I. The Extravascular Modes of Administration. Archives Internationales de Pharmacodynamie et de Therapie. **1937**, *57*, 205–225.
5. Teorell, T. Kinetics of Distribution of Substances Administered to the Body. II. The Intravascular Mode of Administration. Archives Internationales de Pharmacodynamie et de Therapie. 1937, *57*, 226–240.
6. Yamaoka, K.; Nakagawa, T.; Uno, T. Statistical Moments in Pharmacokinetics. J. Pharmacokinet. Biopharm. **1978**, *6* (6), 547–558.
7. Benet, L.Z.; Goyan, J.E. Bioequivalence and Narrow Therapeutic Index Drugs. Pharmacotherapy **1995**, *15* (4), 433–440.
8. Patnaik, R.N.; Lesko, L.J.; Chen, M.L. Individual Bioequivalence. New Concepts in the Statistical Assessment of Bioequivalence Metrics. FDA Individual Bioequivalence Working Group. Clin. Pharmacokine. **1997**, *33* (1), 1–6.
9. Midha, K.K.; Rawson, M.J.; Hubbard, J.W. Individual and Average Bioequivalence of Highly Variable Drugs and Drug Products. J. Pharm. Sci. **1997**, *86* (11), 1193–1197.
10. Snikeris, F.; Tingey, H.B. A Two-Step Method for Assessing Bioequivalence. Drug Inf. J. **1994**, *28* (3), 709–722.
11. Holder, D.J.; Hsuan, F. A Moment-Based Criterion for Determining Individual Bioequivalence. Drug Inf. J. **1995**, *29* (3), 965–979.
12. Mohandoss, E.; Chow, S.C.; Ki, F.Y. Application of Williams' Design for Bioequivalence Trials. Drug Inf. J. **1995**, *29* (3), 1029–1038.

Government, Clinical Pharmacy Careers in

Stephen C. Piscitelli
Robert DeChristoforo
Virco Lab Inc., Rockville, Maryland, U.S.A.

INTRODUCTION

Approximately 7000 pharmacists serve the federal government in a variety of roles and organizations, including the Department of Veterans Affairs (VA), the Department of Defense (DOD), and the U.S. Public Health Service (PHS). Pharmacists in the uniformed services, Army, Navy, Air Force, and PHS, may be either commissioned officers or hired via the civil service system. Opportunities for clinical practice and research in the federal government represent a large, but relatively unknown option.

DEPARTMENT OF VETERANS AFFAIRS

The VA health care system now includes 4000 pharmacists, 173 medical centers, nearly 670 outpatient and community clinics, and 131 nursing home units. The VA is affiliated with more than 1000 schools across the United States, including pharmacy, medical, and dental schools. Each year, approximately 100,000 health professionals receive training at VA medical centers. The VA system has been a leader in opening new career pathways for pharmacists that reward the achievement of exceptional skills. For example, pharmacists can receive increases in pay by completion of advanced degrees or by passing the board certified pharmacotherapy specialist (BCPS) examination. There are a number of programs to provide additional training for VA pharmacists and transition them from distributive roles to clinical functions.

Veterans Affairs pharmacists serve in a number of clinical roles including, but not limited to, pharmacist-run ambulatory clinics, members of interdisciplinary care teams, patient education, pharmacokinetic evaluations, therapeutic consultation, and research.[1] These services are provided in various inpatient, long-term, and ambulatory patient care settings. Most clinical pharmacists will have advanced professional degrees (M.S. or Pharm.D.), postgraduate training, and/or sufficient professional experience. Clinical pharmacy specialists are advanced practitioners who provide clinical services for specialized services. These services include anticoagulation, psychiatry, geriatrics, diabetes, infectious diseases, and medication refill. They also may have prescribing authority within a defined scope of practice. There are 185 pharmacy residency programs at VA medical centers, many with a strong emphasis on ambulatory and primary care.

U.S. ARMED SERVICES

The mission of the medical departments in the Army, Navy, and Air Force is to provide effective health care to U.S. forces in times of conflict and to provide high-quality health care in peacetime.[2,3] There are currently approximately 1500 pharmacists working in these units, both as commissioned officers and civil service. Some pharmacists within the armed services are deployed with troops to provide pharmacy services during training missions or wars. Therefore, they must participate in training exercises and workshops designed to simulate these types of experiences. Other pharmacists work at military hospitals and outpatient clinics, providing more traditional clinical pharmacy services. Pharmacists participate in a variety of clinical roles, including patient rounds, drug information, and patient counseling. Some pharmacists undergo a credentialing process that gives them prescriptive authority and enables them to assume responsibility for the management of the patient within defined roles and limits. Armed services ambulatory care pharmacists play active roles in direct patient care within such therapeutic areas as diabetes, asthma, hyperlipidemia, and hypertension.

Pharmacists within the Army are also members of a bioterrorism readiness force that is prepared to respond to medical emergencies arising from the terrorist use of weapons of mass destruction. Many pharmacists within the armed services do not possess Pharm.D. or other advanced degrees, although there is a strong com-

DOI: 10.1081/E-ECP 120006176

mitment to support those individuals who pursue additional education. There are a number of residency and fellowship programs available to these pharmacists, and opportunities exist to attain a nontraditional PharmD degree.

U.S. PUBLIC HEALTH SERVICE

The PHS is organizationally part of the Department of Health and Human Services.[4] Pharmacists are probably most familiar with such agencies as the Centers for Disease Control and Prevention (CDC), the Food and Drug Administration (FDA), the Indian Health Service (IHS), and National Institutes of Health (NIH). In addition, the PHS has memorandums of agreement with the Federal Bureau of Prisons (BOP), Immigration and Naturalization Service, and U.S. Coast Guard (USCG), to provide primary health services. The Office of Emergency Preparedness and the National Disaster Medical System are also located within the Department of Health and Human Services.

Centers for Disease Control and Prevention

There are currently nine pharmacists who serve at the CDC coordinating the CDC Drug Service, which distributes 13 special immunobiological materials and drugs to physicians in the United States. Special biological and antiparasitic drugs that the CDC distributes include botulism and diptheria antitoxin, bithionol, ivermectin, pentostam, and other medications with restricted usage in the United States. These pharmacists also ensure procurement of drugs, maintenance of treatment investigational new drug applications (INDs), and timely reporting to the FDA. Other pharmacists who also possess a Master's degree in Public Health perform epidemiology and field work in foreign countries. The CDC has been charged to maintain a stockpile of pharmaceuticals that can be immediately deployed in response to chemical or biological terrorism events within the United States.

Food and Drug Administration

The FDA employs more than 250 pharmacists in all phases of the agency's regulation of drugs, biologics, medical devices, medical foods, and veterinary products. Pharmacists serve as reviewers for INDs, new drug applications (NDAs), and generic drug approvals, evaluating the safety, efficacy, packaging, and advertising of prescription and nonprescription drugs. They are also involved in adverse experience reporting and postmarketing surveillance, and function in many other positions ranging from field inspector to project managers, which are the liaison between the pharmaceutical industry and the FDA. Other pharmacists contribute to the FDA with respect to compendial standards, scientific investigations, manufacturing facility inspections, and the FDA's research laboratories. In addition, they also work with expert advisory committees and review panels. Most FDA pharmacists serve at the headquarters in Rockville, MD, but others are assigned to the many regional, district, and local offices throughout the United States that carry out inspection and enforcement activities. A PharmD degree is preferred but not generally required for many FDA positions.

Indian Health Service

The IHS employs more than 500 pharmacists who are part of a health care team that provides comprehensive care to 1.4 million Native Americans and Alaska Natives in hospitals and ambulatory clinics in 34 states. The IHS pioneered many of the clinical pharmacy services that are now considered standard practice. Pharmacists have direct access to the patient's medical record to ensure appropriateness of drug therapy, monitor for adverse effects, and conduct activities in health promotion and disease prevention. Indian Health Services pharmacists have long been involved in expanded roles such as primary care, and many have prescriptive authority under medical staff protocols. They are actively involved in drug selection, dosing, treatment, and evaluation of therapy. Patient consultation has been an integral part of the IHS pharmacy program for more than 30 years, and private consultation rooms are used to promote effective patient communication. The IHS offers three residency programs: American Society of Health-System Pharmacists (ASHP)-accredited programs in pharmacy practice and ambulatory care, and an American Pharmaceutical Association (APhA)-accredited residency in community pharmacy practice. The IHS also provides their pharmacists with the opportunity to pursue a PharmD degree through a relationship with Idaho State University.

National Institutes of Health

Opportunities for pharmacists exist in both the intramural and extramural programs. The extramural program accounts for nearly 90% of NIH funding and is

comprised of sites around the world, whereas the intramural program is located on the NIH campus in Bethesda, Maryland. The NIH Clinical Center is a 350-bed hospital devoted exclusively to patients of the intramural clinical research program. Its pharmacy is supported by 50 pharmacists in various roles, including nine clinical pharmacy specialists in the areas of oncology, infectious diseases, critical care, bone marrow and solid organ transplant, mental health, drug information, and ambulatory care. These pharmacists also serve as principal and associate investigators in various NIH studies. Clinical pharmacy specialists generally have a PharmD degree and postgraduate training in residency and/or fellowship programs. The staff also includes pharmacists with expertise in drug formulation, study design, analytical/quality control, and pharmacokinetics. The NIH also offers four ASHP-accredited residencies. There are also opportunities for radiopharmacists within the NIH Clinical Center's Nuclear Medicine and Positive Electron Tomography (PET) Departments.

The research program at NIH also uses pharmacists in many of its 14 institutes. Pharmacists in the National Cancer Institute's (NCI's) Pharmaceutical Management Branch are involved in anticancer drug development, protocol development, collection of clinical data, distribution of NCI investigational drugs and the Treatment Referral Center. In addition, the intramural program of the NCI has a pharmacokinetics laboratory where pharmacists perform basic and clinical research. The National Institute of Allergy and Infectious Diseases (NIAID) Division of AIDS pharmacists participate in protocol development and implementation, and act as consultants to more than 300 pharmacists involved in NIAID-sponsored AIDS clinical trials.

Federal Bureau of Prisons

The BOP employs more than 120 pharmacists who work in both hospital and ambulatory settings in 99 prisons in 38 states. Pharmacists fill medication orders directly from the inmate's medical record, thereby having access to full information on the patient. Pharmacists at the BOP are significantly involved in monitoring compliance, managing drug therapy, ordering and interpreting laboratory studies, and medication counseling for inmates in tuberculosis prophylaxis, mental health, HIV/AIDS, and other more traditional chronic disease clinics. Many pharmacists stationed in hospital settings have a presence on mental health and medical/surgery floors, round with physicians, and provide drug information services to the medical staff. Pharmacists at the BOP are also performing research in the area of patient counseling and compliance.

U.S. Coast Guard

Officers commissioned by the PHS deliver primary care services to USCG members and their families at 26 shore-based sites. Sixteen active-duty, PHS- commissioned corps pharmacists are detailed to the USCG. In the early 1990s, the USCG adopted the chart prescribing and prescription dispensing model developed by the IHS. The USCG pharmacy program is linked throughout the United States to the DOD Composite Health Care System for computerized dispensing functions.

COMMISSIONED OFFICER STUDENT TRAINING AND EXTERN PROGRAM (COSTEP)

The PHS offers students in medicine, nursing, pharmacy, and other allied health professions the chance to gain career experience at sites throughout the United States through a program called COSTEP. These salaried positions, available during vacation or elective time, provide students with valuable experience and insight into career opportunities within the PHS.

CONCLUSION

These programs represent the most common career paths for pharmacists in the U.S. government. However, there are additional federal agencies, such as the Centers for Medicare and Medicaid Services, where pharmacists serve in nontraditional roles. Although generally not considered by pharmacy practitioners and students, the federal government provides a number of innovative and unique practice areas for clinical pharmacists.

IMPORTANT GOVERNMENT WEBSITES

- Pharmacy programs within the PHS and related links http://www.hhs.gov/pharmacy/
- Links to numerous DHHS agencies http://www. hhs.gov/agencies

- VA
  http://www.va.gov
- U.S. Public Health Service Commissioned corps
  http://www.usphs.gov
- U.S. Army Pharmacy
  http://armypharmacy.org
- U.S. Air Force Pharmacy
  http://www.af-pharmacists. org/
- U.S. Navy Pharmacy
  http://navymedicine.med.navy. mil/navypharmacy
- DOD Pharmacoeconomic Center
  http://www.pec.ha. osd.mil/
- NIH Pharmacy Department
  http://www.cc.nih.gov/phar

REFERENCES

1. Ogden, J.E.; Muniz, A.; Patterson, A.A.; Ramirez, D.J.; Kizer, K.W. Pharmaceutical services in the department of Veterans Affairs. Am. J. Health-Syst. Pharm. **1997**, *54* (7), 761–765.
2. Williams, R.F.; Moran, E.L.; Bottaro, S.D., II; Dydek, G.J.; Caouette, M.L.; Thomas, J.D.; Echevarria, R. Pharmaceutical services in the United States army. Am. J. Health-Syst. Pharm. **1997**, *54* (7), 773–778.
3. Young, J.H. Pharmaceutical services in the United States air force. Am. J. Health-Syst. Pharm. **1997**, *54* (7), 783–786.
4. Paavola, F.G.; Dermanoski, K.R.; Pittman, R.E. Pharmaceutical services in the United States Public Health Service. Am. J. Health-Syst. Pharm. **1997**, *54* (7), 766–772.

Health Care Systems: Outside the United States

Albert I. Wertheimer
Temple University, Philadelphia, Pennsylvania, U.S.A.

Sheldon X. Kong
Merck & Co. Inc., Whitehouse Station, New Jersey, U.S.A.

INTRODUCTION

It is quite fascinating how the organization, structure, and financing of health care services can be so very diverse in different countries around the world. One might think that leaders and policymakers would be aware of each other's national health systems and, by emulating the best features, that they would tend to move toward harmonization and greater similarity.

Actually, this assumptions is false. National health care systems vary widely and are more related to variables in each country (1). In fact, the health system in a given country is a mirror of how that society functions at large. Health care delivery systems must be compatible with the: 1) *economic system*: socialist, capitalist, or mixed; 2) *political system*: major or minor role of degree of government centralization; 3) *wealth of the country*: use of primary care facilities, access to specialists and tertiary care facilities; 4) *traditions and conventions as seen in their history*—fundamental, visible things are difficult to change; 5) *geography*: whether the majority of the population is located in a few metropolitan areas, with the remainder scattered in rural areas, or whether the population is spread over hundreds of islands; 6) *infrastructure*: roads, communication systems, and air service; and 7) *extent of and belief in high technology* (2).

There are other factors as well: the system from a previous colonial power, extent of literacy and education, and relationships with outside countries, to name a few.

BACKGROUND

The remainder of this article examines the health care delivery systems in six very different countries. Even though Canada and the United States are similar countries with a shared border and language and with open communication, their health care delivery systems could not be any more different. Each side of the border is aware of what happens on the other side, however, a series of complex and powerful forces keep them moving in their own directions.

We look at six countries very briefly in this article to highlight the incredibly diverse approaches to health service organization and financing. In essence, most health systems fit into one of the following models:

1. State ownership and control—The best examples are the British National Health Service and the Swedish system in which clinics, hospitals, and most service providers are owned and operated by the government (3).
2. State health insurance program—Here, the government is the sole or major payer. However, some of the facilities and resources are in nongovernment hands. This is the case in much of Europe (4).
3. Mixed systems—This is seen in much of Asia and Central America and usually where there is a small wealthy class and a massive lower class. The lower class receives care from public facilities, and the small upper class uses private-sector, fee-for-service, and self-paid care.

Other scenarios fit into this category as well. The United States has several independent health care systems including the military, veterans, Medicaid (a federal program for the medically indigent), Medicare (a federal insurance program for those 65 years of age and older), private-sector for-profit, and not-for-profit clinics, hospital chains, managed-care organizations, religious, prison health, and university teaching facilities (5).

4. Exclusively private sector—This category is shrinking as nations realize that health maintenance and disease prevention/wellness are important to their national goals of strength and productivity. Switzerland would still fit into this category, where most health care resources are in private hands (6).

Encyclopedia of Clinical Pharmacy
DOI: 10.1081/E-ECP 120006421

SPECIMEN NATIONAL SYSTEMS

Canada

Organization

Canada uses a national health service, which provides medical services and hospital care to its entire population. The individual provincial governments operate health plans that conform to national legislation but can differ in various aspects. This "Medicare" program guarantees comprehensiveness, universal access, portability, and public administration (7).

Health Canada is the national, federal health agency; however, the operation of health service provision is delegated to the provincial governments, which control virtually 100% of Canada's hospitals. There is a gatekeeper primary health care system, with GPs (general practitioners) or primary care family doctors serving as the entry point. Access to specialists, diagnostic testing, hospitals, and others is through the GP. Individual citizens have the freedom to choose their own doctors, 95% of whom are self-employed in private practice. The provincial government pays these doctors on a fee-for-service basis.

The individual provincial governments offer different supplemental benefits not covered by the national Medicare program, such as drugs, dental care, and vision care to the poor, elderly, and other specific groups. Supplemental benefits for the typical, employed, and nonelderly person come from the purchase of supplemental health insurance from private sources (8).

Pharmaceuticals

Canada created the Patented Medicine Prices Review Board (PMPRB) in 1987 to guarantee that pharmaceutical products would not have excessive prices in Canada. The board reviews prescribed and over-the-counter (OTC) prices and publishes annual guidelines for manufacturers. Compliance with PMPRB guidelines is voluntary; however, since 1993, the board has the authority to reduce excessive prices and return the excess amount to the government, and to punish the manufacturer.

The PMPRB compares prices in Canada with those in seven industrialized nations (France, Germany, Italy, Sweden, Switzerland, the United Kingdom, and the United States) to ensure that Canadian prices are in line with those of comparable countries. There is some controversy that existing drug products are well-controlled regarding prices, but that such is not the case with newly introduced pharmaceuticals.

Further controls exist at the provincial level at which each province maintains a published formulary of drugs that are reimbursable along with the reimbursement level. Quebec, observers perceive, lists nearly all new drug products, whereas Ontario appears to be slow to list newly approved products. Each province has additional control mechanisms. Ontario requires the first generic drug to be at least 40% less costly than the branded originator product. Some components of the reference price system are seen in British Columbia and Newfoundland.

There is growing harmonization among the provinces; however, there is still no national, standardized, and interchangeable list of drugs for ambulatory care use. In hospitals, drugs that are administered are paid for by Medicare. Each province has interesting and different features in its drug benefit plan.

The Prince Edward Island plan pays for seniors; welfare recipients; nursing home patients; and those with rheumatic fever, diabetes, tuberculosis, multiple sclerosis, AIDS, and several other conditions. New Brunswick has an annual copayment cap for seniors and for organ transplant recipients and for selected other patient categories. A copayment is set at approximately $9 (Canadian) but is waived for some groups in Quebec, along with an annual copay ceiling of $750.

Other interesting features of the Canadian system include its 1998 mutual recognition agreement with the EU, prohibition of prescription drug advertising to consumers, a 20-year patent exclusivity period, and the establishment of the PMPRB to ensure fair pricing of medications (9, 10).

Republic of South Africa

Organization

The Republic of South Africa (RSA) has a most diverse health care environment, with world-class practice and facilities in wealthy urban areas and some of the most primitive care in poor remote villages, with a vast array between these extremes. Primary care is now the focus of the ANC government in an effort to correct years of neglect and undemocratic practices under the earlier apartheid-oriented regimes. Public health services are being brought to the Black townships as rapidly as resources permit (11).

However, there are virtually no funds for new drugs against HIV infection in patients, a problem most prevalent in the RSA. To maximize the value of its drugs budget, the RSA has enacted legislation to create an Essential Drugs List for the public sector, along with generic substitution authority, the removal of some pharmacists' unique

professional privileges, and legislation permitting the parallel importation of pharmaceutical products already registered in the RSA. Obviously, this conserves resources, stretching them for more patients, but this angers the RSA and multinational pharma firms.

South Africa is still the wealthiest country in Africa, with a (1997) GDP at approximately $130 billion. It must be noted, though, that aggregate numbers hide massive racial differences. It is improving, but the standard of living for Blacks is yet only slightly better than it is in neighboring countries, whereas whites enjoy a standard of living similar to that found in North America or Western Europe. An unemployment rate of over 30% (mostly among Blacks) exacerbates the fiscal situation (12).

Routine immunizations for children, conforming to the World Health Organization (WHO) recommended schedule is the governmental policy, but it is not yet accomplished in all regions. Infectious diseases including HIV remain a serious challenge. Planning and budgeting for resource allocation are difficult because accurate census figures do not exist. Total health expenditures appear to be in the area of $300 per person per year, and it is estimated that the private sector accounts for greater than 50% of total expenditures.

Public-sector expenditures emphasize primary care, lately, at the expense of tertiary care facilities. Private-sector spending is primarily through private "medical schemes." These are nonprofit organizations supported by employer associations and employees. There are slightly fewer than 200 of these schemes, providing insurance and care payment for nearly 3 million workers and their 5 million dependents (of a total estimated RSA population of 40 million). The largest area of medical scheme expenditure is for medicines, which causes the pressures on pharmaceutical pricing addressed below. After drugs, the next largest expenditures are for private hospitals, medical specialists, general practitioners, and dentists (13).

The RSA Department of Health (DOH) has totally restructured the previous apartheid system of racial and provincial health systems into a coordinated national health program operated through health regions and local health districts. Still, there are major differences in knowledge, education, expectations, and wealth within different subpopulations (14, 15).

Pharmaceuticals

Until recently, manufacturers were free to establish their desired price for a drug. Wholesalers and retailers added what they chose to reach the retail selling price for medications. In 1997, a proposed scheme of prices extending to the retailer was agreed on, but resistance was met from the Pharmaceutical Manufacturers Association(PMA). In the legislation, a pricing board composed of members selected by the Minister of Health would establish prices for each product and a maximum selling price. Public-sector primary care drugs are reimbursed 100% by the government. Hospital care outpatient drugs can have copayments. The Essential Drugs List would be the core of what is to be available at public facilities, but there appears to be a long way to go before most of these agents will be regularly available on a consistent basis at primary care centers or at public hospitals (13).

The parallel importation of RSA-registered drugs available at lower prices abroad is the basis for PMA litigation against the Drug Legislation of 1997. In addition to the price-setting committee, DOH efforts to encourage the use of generic drugs has proven to be a source of conflict. Other features of the new legislation bar dispensing samples or making bonus payments to dispensers of medicines; the creation of a Code of Ethics for pharmaceutical marketing; and a series of safety regulations, dealing primarily with limiting practice to fully qualified and licensed professionals.

There is a fast lane for new drug approvals if the product is already in at least one of the following jurisdictions: the United Kingdom, Canada, United States, Sweden, or Australia. Approxmately 85% (by value) of pharmaceuticals go through the nearly 3,000 community pharmacies. Yet, approxmately 80% of the population rely on the public sector for drugs, received through clinics, hospitals, primary care posts, or military facilities. Although there is a 20-year patent period of exclusivity/protection, the parallel imports option effectively defeats this protection.

It will be interesting to see how the access to drugs, price controls, and quality improvement forces will interact and what the actual situation will be in South Africa in the coming years, especially as the country complies with intellectual property and World Trade Organization policies and rules (16).

Japan

Organization

After North America and before Western Europe, Japan is the second largest pharmaceutical market in the world. Its population of 126 million spends $70 billion on pharmaceuticals each year. On average, each Japanese resident spends $2000 each year on health care with $550 of that on pharmaceuticals. Perhaps the primary single features of the Japanese market are the above-average proportion of elderly in the population and the higher than usual consumption of drugs. It has been estimated that by the year 2050, nearly 30% of the population will be older than 65 years of age. The high consumption rate is attributed to drugs being injected and/or sold by the physician, a practice used, in part, to increase the total price of an office visit (17).

The primary funding source for health services in Japan is the Social Insurance System (SIS), made up of employee programs that pay for nearly 55% of care. The Medical Service for the Aged program covers another 35% of care. Private expenditures and a very small portion for public health promotion and disease prevention make up the difference. The Ministry of Health and Welfare (MHW) maintains overall responsibility for health care services and functions via a number of bureaus. Numerous sources comment that regulations are difficult to understand and interpret, often overlapping, and that this serves as a barrier to foreign firms desiring to enter a market. Physicians, for example, are authorized to own and operate hospitals, effectively excluding corporate owners or physicians not licensed in Japan (18).

Universal health insurance was established in 1961. Nearly the entire population is covered through the employer plans or through programs for the unemployed, retired, or self-employed. Employees pay 10% of the cost of treatments, up to an annual ceiling, and also pay a portion of their premiums, with their employers.

Pharmaceuticals

The MHW sets prices for reimbursable drugs (those approved for the Social Insurance System). Physicians, clinics, and private hospitals are reimbursed at a price slightly higher than their actual acquisition cost. The government has scheduled annual reductions in the reimbursement prices to reduce this source of additional income to physicians. Patients make copayments of 20%, although for children and low-income elderly the copayment is waived, and recently a plan to eliminate copayments for persons 70 years of age and older was introduced.

The MHW reductions of 5–10% of the prices of existing drug products appear to have had the opposite of the intended impact. Doctors are prescribing more of the newest, high-priced pharmaceuticals that have not had their margins reduced yet, thereby earning a bigger amount from the wider difference between their actual cost and the listed reimbursement amount.

With regard to generic drugs, astute observers believe that the Japanese government wants its R&D-intensive firms to be successful. A regulation requires generics to be priced at not less than 40% of the innovator brand price. It is reasonable to assume that the margins (Yakkasa) for physicians are lower with generic drugs, and that these margins will continue into the future, as will the reference price scheme (19).

There is a Japanese pharmacopeia that sets official standards and diverse government agencies that perform tasks undertaken by an FDA. It is rumored that the Japanese will establish a Western-style FDA in the near future.

One of the most disliked regulations in the view of foreign and multinational pharmaceutical companies is the requirement for duplicative clinical trials with humans in Japan, because those carried out elsewhere are not recognized. Also of interest is the fact that Japan, like Korea and Taiwan, has no separation between prescriber and dispenser of drugs. Called "Bungyo," it is a major source of revenue for doctors and clinics. Fewer than 20% of prescriptions ever reach a pharmacy for dispensing (19).

Good post-marketing surveillance practices (GPMSP) rules have been in place since 1993. Postmarketing experience reports are to be sent to a government agency. Both GPMSP and periodic safety reporting requirements are in place that require a review of the product each year while it is in its re-examination period, immediately after marketing approval. Unlike in the United States, where a new drug application is approved for an indefinite period, in Japan, there is a periodic full reassessment. Such re-evaluations are conducted every 5 years once the initial re-examination period for a drug product has ended.

Drug products are distributed primarily via the 2000 wholesalers, and in addition, there exists a small second channel with drugs going directly to hospitals, GPs, and pharmacies. There are approximately 66,000 pharmacies, most of which are family-owned independents. There are chains as well. However, a growing market for OTCs is found in convenience stores.

Physicians administer and sell drugs to patients as a highly profitable sideline. The incentive is for the physician to use as much of the most costly drug products as possible. There is only a small OTC market, because physicians try to prescribe and dispense as much as is possible. Other than some concern about a drug lag, the pharmaceutical environment in Japan is robust. Periodically, there are calls to separate prescribing and dispensing; however, this is not likely in the near future given the powerful forces backing the status quo (20).

United Kingdom

Organization

With a population of more than 60 million and GDP per capita of more than US $22,000, the United Kingdom is one of the richest nations in the world. It is one of the G7 countries, a member of the European Union, and a member of the Organization for Economic Co-operation and Development (OECD).

In 1996, total health care expenditure in the United Kingdom was approximately 7.0% of the GDP. Public expenditure by the National Health Service (NHS) accounts for most of the health care costs. The NHS was set up after World War II, with the aim of unifying health care services by voluntary and local hospitals. The NHS offers free health services to all U.K. residents, funded through general taxation.

Two of the major characteristics of the U.K. health care system include health authorities responsible for hospital services and GP fundholders responsible for primary care. In 1996, 100 health authorities became operational in England, responsible for the provision of NHS hospital and community health services covering geographic boundaries with populations ranging from 125 thousand to over 1 million. There are four levels of hospital services. At the community level, community hospitals offer basic medical care for the treatment of acute cases and patients requiring convalescent and long-term/terminal care. General practitioners are the key staff here. At the district level, district general hospitals operate the key acute units, serving an average population of a quarter-million. At the regional level, major specialty services such as neurosurgery, open-heart surgery, and radiotherapy are provided. At the national level, highly specialized hospitals provide complex services for parts or for the entire country (21).

GPs are the gatekeepers and fundholders of the health care system. The principle of fundholding is that GPs manage their own budgets. They can obtain a defined range of services from hospitals and manage patients at the GP level whenever possible to reduce costs. In the late 1990s, GPs fundholders were organized into Primary Care Groups (PCGs). These networks of GPs cover wide geographic areas with an average population of 100,000. In 1999, there were 481 PCGs in England and Wales, and all have unified budgets (e.g., drugs, hospital care services). With a population of a small to medium-sized HMO in the United States, these PCGs have a very broad influence on patient health care and the selection of drugs through formularies.

Pharmaceuticals

The regulatory authority in the United Kingdom is the Medicines Control Agency (MCA) under the Department of Health. The agency's responsibilities include drug licensing, clinical trials licensing, pharmacovigilance and drug safety, communication and provision of information on medicines, inspection of facilities and enforcement of regulations, and the *British Pharmacopoeia*. The United Kingdom is a reference member state for the European Union mutual recognition procedure. The European Union's pharmaceutical registration system came into effect for all member countries in 1995. The aim of the EU system is to harmonize pharmaceutical regulations throughout the EU. The centralized registration procedure is handled by the European Medicines Evaluation Agency (EMEA). Authorization through the central registration procedure is immediately valid in all EU member countries. The decentralized procedure relies on the principle of mutual recognition. After registration has been obtained in a member country under the centralized procedure, application may be made for registration in one or more other member countries via the decentralized procedure (21).

The majority of pharmaceuticals are distributed through wholesalers to retail pharmacies, with large pharmacy chains now dominating the market. There are approximately 11,000 community pharmacies in the United Kingdom (21). In recent years, pharmacy services are increasingly available in supermarkets at the expense of local independent pharmacies.

Total expenditure on pharmaceuticals in the United Kingdom amounted to approximately 8650 million pounds in 1999, accounting for approximately 17% of the total health expenditure (21). The NHS covers prescription drugs. However, the government does not reimburse for over-the-counter (OTC) products. The Department of Health indirectly controls pharmaceutical prices. Because the price control scheme is related to profit control, rather than to the prices of individual products, pharmaceuticals are relatively free-priced in the United Kingdom. The government operates a negative list for products that are not reimbursable. The cost of most licensed prescription products is fully reimbursed. However, cost constraints and prescribing budgets mean that GPs will often prescribe a generic when one is available. As a result, new prescription drugs usually have a slower penetration rate in the United Kingdom than in the United States. The recently introduced National Institute for Clinical Excellence (NICE) will add more barriers to the introduction of new pharmaceutical products in the United Kingdom.

National Institute for Clinical Excellence

Funded by the government, the National Institute for Clinical Excellence (NICE) was set up as a Special Health Authority in the United Kingdom in 1999 and, as such, it is a part of the National Health Service (NHS). It was set up to "provide the NHS [patients, health professionals, and the public] with authoritative, robust and reliable guidance on current best practice." Its key functions are "to appraise the clinical benefits and the costs of those [health care] interventions and to make recommendations." Guidance is issued from each appraisal based on the clinical benefits, cost-effectiveness, and total economic impact on the

National Health Service. The government does not have to adhere to the recommendations by the NICE in its guidance and financial payment to health care providers. However, many believe that a negative recommendation from the NICE will have a detrimental impact on the pricing, reimbursement, and sales of the appraised product not only in the United Kingdom but also throughout Europe, Australia, and Canada.

The guidance covers both individual health technologies (including medicines, medical devices, diagnostic techniques, procedures, and health promotion) and the clinical management of specific conditions. The Institute may recommend a technology for general use, for specific indications, or for defined subgroups of patients. Based on the appraisal, a therapeutic intervention (e.g., drug) will be classified into one of three categories: category A, routine use in the NHS; category B, further trials needed; and category C, not recommended for routine use in the NHS.

The NICE has a board reflecting a range of expertise including the clinical professions, patients and user groups, NHS managers, and research bodies. The Board ensures that the NICE conducts its business on behalf of the NHS in the most effective manner. Details of the appraisal process and membership of the Appraisals Committee are available on the NICE Web site (www.nice.org.uk). Because the NICE was new at the time of completion of this article, its impact on the pharmaceutical industry is still not clear.

Germany

Organization

With a population of approximately 82 million in 1998 and a GDP per capita of more than $26,000, Germany is one of the world's largest economies and health care markets. The population enjoys a generally good standard of health with a high degree of public awareness about health-related issues. Life expectancy in Germany is among the highest in the world. In 1997, the life expectancy for males was 74 years and for females 80. Approximately 15.8% of the population were over 65 years in 1997, and it has been projected that by 2020, the number of German inhabitants aged over 60 years will be 28.2% (22).

In 1997, health expenditures in Germany totaled $298 billion, equal to 14.2% of the GDP. The health care system in Germany is decentralized, and health care expenditures are covered by a variety of sources/payers. The statutory insurance system (GKV) represents the biggest proportion of the total care coverage (for almost 50%). Employers, government budget, private households, private insurance, retirement insurance, and accident insurance cover the remaining 50% of the health care expenditures. The largest spending sector is hospital expenditure, representing 34.3% of the total GKV health care expenditures (22).

The federal government has little executive responsibility for the provision of health care in Germany. Its primary responsibility is to provide a regulatory framework within which the individual Länder have to operate. The health ministries of the individual Länder are responsible for implementing the federal legislation, enacting their own legislation, supervising subordinate authorities and the medical profession, hospital planning, and regional administration.

Hospitals in Germany can be classified into three major categories based on ownership: public, nonprofit, and private. In 1997, the public sector operated approximately 40% of general hospitals, and nonprofit organizations operated another 40%. However, the number of privately owned facilities has been increasing steadily over the past decade.

The number of practicing doctors has risen steadily for the past 10 years. More than 70% of the practicing doctors are specialists, with general medicine as the largest specialty. Fewer than 30% of doctors practice without any specialty.

Pharmaceuticals

Germany is a reference member of the EU pharmaceutical registration system. The European Medicines Evaluation Agency (EMEA) handles the centralized registration and the decentralized registration procedures in individual countries. After marketing authorization of a product with a new active substance has been granted in one country, the mutual recognition procedure is compulsory in other member countries. The mutual recognition procedure is also compulsory for line extensions and generic products. Marketing authorization approvals in Germany are valid for 5 years and renewable thereafter in 5 year periods.

Germany is the home of some major multinational pharmaceutical companies such as Aventis, BASF, Bayer, Boehringer Ingelheim, Merck KGaA, and Schering AG. VFA is the research-based manufacturers' association, whereas the Bundesverband de Pharmazeutischen Industrie (BPI) represents small and medium-sized companies. Because North America is the largest pharmaceutical market in the world, many of the VFA pharmaceutical companies locate their key operations in the United States. Exports to Western European countries represent a major source of income for many of the German pharmaceutical companies.

The pharmaceutical market in Germany is one of the largest in the world. Based on drug use per capita, Germany is second only to Japan in the consumption of pharmaceuticals. The principal distribution channels for pharmaceuticals in Germany are public retail pharmacies and hospital

pharmacies. In 1998, there were 47,322 pharmacists in Germany, equal to 0.6 pharmacists per thousand population (22). Public (retail) pharmacies employed 96% of all pharmacists in 1998 and they obtained their supplies primarily from wholesalers. Prescribed drugs, including both branded and generic products, can only be dispensed in a pharmacy with a doctor's prescription. The generics market in Germany is one of largest and fastest-growing in Western Europe, representing approximately one-third of the European generics markets. OTC products can be divided into three overlapping categories: prescription OTC medicines, nonprescription OTC medicines, and freely available OTC products that can be sold freely through all retail outlets such as health food stores, supermarkets, and other retail outlets.

Mexico

Organization

Mexico is a federal republic of 31 states and a federal district. The population was officially estimated to be 97.7 million in 1997. GDP per capita was estimated at approximately US $4400 in 1998. As a developing nation, communicable diseases are still one of the major causes of mortality, although chronic and degenerative diseases have become the leading cause of death during the past decade.

One of the major challenges for the government is to address the inadequacies of the Mexican health care system. Approximately 10 million people have virtually no access to regular basic health care services, and another 20 million people have less than adequate access. In 1996, the total health care expenditure in Mexico was equivalent to approximately 4.6% of GDP. Spending by the public sector accounted for approximately 60% in 1996 (23).

There are three sectors in the Mexican health care system: public, social security, and private. The public sector is primarily directed and operated by the Secretariat of Health. The public sector of health services is under the Secretariat of Health and is coordinated by over 200 health districts. The Federal District Department provides health care services to some 3.2 million people in Mexico City. The Mexican Social Security Institute (IMSS) Solidarity program covers another 10 million people in rural areas.

The social security system covers health services for government employees, managed by the Social Insurance Institute of State Employees (ISSSTE), and for private-sector workers, managed by the Mexican Social Security Institute (IMSS). The two agencies operate their own networks of hospitals and clinics and provide similar benefits. Some other smaller social security agencies exist, providing medical services for special groups such as the army, navy, and state oil company personnel.

The private (commercial) sector includes private hospitals, doctor's offices, and practitioners of traditional medicine. Charity organizations such as the Red Cross also play a role in the Mexican health care system.

Pharmaceuticals

The regulatory authority in Mexico is the Dirección General de Control de Insumos para la Salud (DIGECIS). The Health Secretariat issues pharmaceutical registration. Safety and efficacy must be proven by phase III clinical trials in Mexico to register drugs that are new to the Mexican market. All major pharmacopeia (*International Pharmacopoeia, US Pharmacopeia, British Pharmacopoeia, French Pharmacopoeia, Swiss Pharmacopoeia, European Pharmacopoeia*, and *Japanese Pharmacopoeia*) are acceptable in Mexico.

Most domestic producers in Mexico are wholly owned or licensed subsidiaries of multinational pharmaceutical firms. Exports have been growing fast, with other Latin American countries as the major destination markets. However, the United States is the major supplier of pharmaceutical imports in Mexico.

Pharmaceuticals in Mexico are subject to government price control. The private sector accounts for approximately 85% of the pharmaceutical market. Prescription drugs account for the majority of the pharmaceutical market, with antibiotics as one of the largest classes. Because the use of generics is still a relatively new phenomenon, most of the prescribed pharmaceuticals are branded products. OTC products represent approximately one-fifth of the total pharmaceuticals market.

SUMMARY

As presented, these six representative countries use vastly different organizations, financing mechanisms, goals, and provision structures. In fact, few systems around the world are identical because the systems represent the values and priorities and political as well as economic leanings and traditions of that country. If there were one perfect system, we would be seeing migration toward that model. However, because this is not the case, it is reasonable to assume that most of the various systems encountered around the world are at least satisfactory in their foundations and macrolevel characteristics, even if some of the operating details are not always popular (24).

The world is full of interesting additional approaches that a serious student of this subject might wish to explore further. Some of these include the "need clause" used in Norway, where, for example, their FDA had the authority to refuse to accept and review a new drug because Norway already had six benzodiazepines on the market. The FDA deemed that sufficient unless the sponsoring company knew of a new indication or other therapeutic breakthrough from its use. The Swedes bought all of the then-existing community pharmacies in the country in 1970 to rationalize distribution, and service level and to create a monopsonistic body for negotiating with manufacturers in price-setting. The French and others place new drugs into one of several reimbursement categories. Clearly, life-saving drugs are put in the 100% reimbursement (to the patient) category. Most others strive for the 70% reimbursement category; however, if the manufacturer cannot agree on a price satisfactory to the Social Security agency, the product will be placed in a lower reimbursement category, effectively hampering its market success. This is a powerful bargaining chip for the government to contain drug prices.

It will be interesting to watch the future in this area to see how medications previously requiring a doctor's prescription that move to OTC status are handled, and how nutraceuticals, herbals, homeopathic, and naturopathic drugs, without the benefit of rigorous, randomized clinical trial or outcome data are handled as well. Similarly, we can be certain that there will be excitement galore when the nations in Central America and the Middle East decide to control pharmaceuticals and to end the practice of lay-person purchases of virtually any product without the benefit of a physician's order. Separation of pharmacy and physician functions will occur in the Far East in the not too distant future, causing even more excitement or grief.

If logic dictates, we should expect to see in the future a trend to offer incentives for prescribers who use the most cost-beneficial products (bonuses) and disincentives for patients (reimbursement level co-payment differences) and physicians when less than optimal choices are made. Irrespective of whatever does actually occur, it will be most interesting to observe.

REFERENCES

1. Roemer, M.I. *National Health Systems of the World*; Oxford University Press: Oxford, UK, 1993; 2, 61.
2. Fry, J.; Farndale, W.A.J. *International Medical Care*; Washington Square East: Wallingford, PA, 1972; 367.
3. In *OECD Health Systems: Facts and Trends*; OECD: Paris, 1993; 100–161.
4. Hewitt, M. *International Health Statistics*; Office of Technology Assessment, U.S. Congress: Washington, D.C., 1993; 76–86.
5. Elling, R.H. *International Health Perspectives*; Springer: New York, 1977; 4, 17–25.
6. Joseph, S.C.; Koch-Weser, D.; Wallace, N. *Worldwide Overview of Health and Disease*; Springer: New York, 1977; 7–43.
7. Korman, R.A. *Academic Reference Manual: Canadian Health Care Information*; IMS Health: Mississauga, Ontario, 1999; 80–162.
8. *World Pharmaceutical Markets: Canada*; Espicom Business Intelligence: Chichester, UK, May, 1999; 9–48.
9. Alleyne, G.A.O. *Health Statistics from the Americas*; Pan American Health Organization: Washington, DC, 1998; 17–37.
10. Alleyne, G.A.O. *Health in the Americas*; Pan American Health Organization: Washington, DC, 1998; 1, 325–337.
11. Monekosso, G.L. *Eighth Report on the World Health Situation; African Region*; WHO: Brazzaville, 1994; 2, 2–37.
12. Nokagima, H. *Eighth Report on the World Health Situation*; 147–165 Global View: Geneva, 1993; 1, 37–59.
13. *World Pharmaceutical Markets: South Africa*; Espicom Business Intelligence: Chichester, UK, Feb 1999; 3–13.
14. *World Development Report 1993, World Bank*; Jamison, D.T., Ed.; 108–170 Oxford University Press: Oxford, 1993; 3–65.
15. *World Tables 1995*; World Bank: Washington, DC, 1995; 22–66.
16. Basch, P.F. *Textbook of International Health*; Oxford: New York, 1990; 144–326.
17. *Bartholomew's Mini World Factfile*; Harper Collins: London, 1995; 44, 102, 175.
18. *Eighth Report on the World Health Situation: Western Pacific Region*; World Health Organization: Manila, 1993; 7, 79–83.
19. *World Pharmaceutical Markets: Japan*; Espicom Business Intelligence: Chichester, UK, Jan 1999; 10–23.
20. SCRIP: London UK, 1998, 1999, and 2000.
21. *World Pharmaceutical Markets: United Kingdom*; Espicom Business Intelligence: Chichester, UK, Feb 1999; 17–70.
22. *World Pharmaceutical Markets: Germany*; Espicom Business Intelligence: Chichester, UK, Feb 1999.
23. *World Pharmaceutical Markets: Mexico*; Espicom Business Intelligence: Chichester, UK, Feb 1999.
24. Brudon, P. *The World Drug Situation*; WHO: Geneva, 1988; 7–108.

FURTHER READING

Alexander, T.J. *Internal Markets in the Making: Health Systems in Canada, Iceland, and the UK*; OECDF: Paris, 1995.

Saltman, R.B.; Figueras, J.; Sakellarides, C. *Critical Challenges for Health Care Reform in Europe*; Open University Press: Buckingham, UK, 1999.

Schneider, M.; Dennerien, R.K.; Kose, A.; Scholtes, L. *Health Care in the EC Member States*; Elsevier: Amsterdam, 1992.

Spivey, R.N.; Wertheimer, A.I.; Rucker, T.D. *International Pharmaceutical Services*; Haworth: Binghamton, NY, 1992.

The Use of Essential Drugs: 6th Report of the WHO Expert Committee; WHO: Geneva, 1995.

Van de Water, H.; Van Herten, L.M. *Health Policies on Target*; THO Prevention and Health: Leiden, 1998.

Health Care Systems: Within the United States

Henri R. Manasse, Jr.
American Society of Health-System Pharmacists, Bethesda, Maryland, U.S.A.

INTRODUCTION

A national health care system reflects the social, political, economic, and cultural character of a nation. A nation's historical roots and dominant values shape policies and directions for the organization, quality, financing, and access to health care services. These factors determine who gets what kind of care—at which locations, for what price, and paid by whom.

The distinctive historical antecedents of American cultural and social development have shaped the present health care system. These contexts have led to a health care system that is uniquely American in character and composition. Although the issues currently facing the American health care system bear some similarity to those in other developed, industrialized nations, many of the factors are unique to the United States.

Social values, that is, the collective societal beliefs about the nature of the human being and the structure of a society, play a strong role in the development of national policies. Political and economic decisions rest in large measure on the prevailing values held in a society. Hence, if a predominant social value rests on the notion that all societal members have a right to health care, political and economic policy developments will follow suit. One way of examining these contexts is to look at a spectrum of social values.

Donabedian (1) has proposed that such a spectrum might be considered from two polar positions: libertarianism versus egalitarianism. Dougherty (2) adds the dimensions of utilitarianism and contractarianism. The essence of these taxonomies of social values is that they characterize specific sets of beliefs and values held by a wide array of individuals.

Libertarian philosophical thought places major emphasis on personal achievement and freedom from political intervention. It holds that individuals should be free to exert their rational capacity to evaluate and determine what is good for them. They can then further act on these determinations for themselves from their own personal, fiscal, physical, and human resources. To this view, Dougherty (2) adds:

> Because they can think, persons can understand their circumstances and the alternatives available to them. Because they can choose, persons can act to affirm or change their circumstances. Because they can think and choose, persons are free to create their own life plans and the values of which they are made.

It follows then, that predominant libertarian values are deeply entrenched in the notion of the self-made person and that social rewards should only accrue if they are deserved and earned. The role of government is, therefore, limited to those functions that absolutely do not abridge the rights of the individual to exert his or her own will for what he or she believes to be best. Moreover, government's role would be limited to those functions and needs for which individuals could not provide (national defense, negotiation of treaties, etc.).

Egalitarian principles focus on the equal moral standing of all individuals regardless of achievement or station in life. This philosophy also centers on the right to equal opportunity and to the extent possible, to be free from need and want. Thus, egalitarianism (2) can be viewed as follows:

> Practically, this means an equal right to a reasonable share of those basic goods and services known to be necessary for a decent human life, including a right to a job, minimum income support, or provision in kind of the goods necessary for life, as well as a right to a range of social and health care services designed to prevent and minimize psychological and physical suffering, disabilities, and premature death.

Egalitarian values place specific demands on government and political policy to construct broad services and support systems so that all members of society are provided with equal opportunity designed to prevent and minimize psychological and physical suffering and disabilities, and to achieve one's life's aims. In this fashion, government would act on the entitlements due to all members of society. Such entitlements might be derived from legal or other forms of social consensus.

This range of social values from libertarianism to egalitarianism holds differing beliefs about equality,

Encyclopedia of Clinical Pharmacy
DOI: 10.1081/E-ECP 120006170

justice, opportunity, rights, and the functional responsibilities of government. When this spectrum of social values is held over health, health care, and the administration and financing of health care services, it is not surprising that a vastly different array of designs emerge. Because health and health care are so tightly wound into personal, cultural, and social beliefs, it is not surprising that such a vast array of health care systems and notions about health have emerged across the world.

America's historical foundations have leaned strongly to the libertarian philosophical viewpoint (3). The influence of the "Protestant ethic" from Europe, coupled with the opportunities that a fresh land provided to "become one's own person," are strongly borne out in American society. An unbridled, free-market economy and freedom from governmental intervention in the daily lives of the citizenry are strong values that have been integrated into the American lifestyle and American political economic thought. The notion of "pulling yourself up by your bootstraps" succinctly reflects these dominant social, political, and economic values. Such antecedents are reliable markers for characterizing America's health care system. Consequently, it is a fascinating mosaic of pluralistic approaches. It is a strongly market-driven, industrialized system, which, at the same time, may be described as one of the world's best and one of the world's most troubled systems.

The United States does not have a universal health insurance program characteristic of many developed nations. Nor does it have national health care services like that of the United Kingdom and other nations. Except for those persons in the United States who possess special legal entitlements, the American health care system is largely a private enterprise; in other words, an industrialized system. The providers, payers, and institutions of care represent a rich mixture of private agents, corporations, insurance systems, and governmental agencies. There is not a singular rationalizing source for setting broad-based national policy and direction for the health care system as a whole. Rather, the vast market place of ideas has a variety of options in order to implement any proposal for which someone will pay. Relman has termed this approach "the industrialization of health care" (4).

The role of the national and state governments in the health care system is limited to those entitlement programs that have been legislated into federal or state law or where there is a federal and state partnership. Federal involvement in the provision of health care services began with the U.S. Public Health Service (PHS), an agency of the U.S. government. The PHS was established in 1798 to provide essential health care services to merchant marine personnel and members of the U.S. armed forces. Subsequent federal involvement in the provision of and the payment for health care has incrementally increased to include care for individuals with special entitlements. The latter include veterans of the armed forces, the elderly, indigent people, Native Americans, persons with HIV/AIDS, certain disabled individuals, and qualifying persons with end-stage renal disease. For example, qualified veterans of the armed forces have access to a federal system of hospitals, clinics, and long-term care facilities under the Department of Veterans Affairs (a cabinet-level agency of the executive branch of the federal government). Since 1965, the federal government sponsors Medicare, a health insurance program for the elderly (65 years of age and over and later the disabled). The federal government also cost-shares with participating state governments to provide the Medicaid program (also enacted in 1965). The latter is an insurance program for health services directed toward qualifying indigent people. In Medicare and Medicaid, institutional and individual providers participate as contractors under a set of specific conditions for participation.

State and local (city and county) governments have limited roles in the provision of health care services. State, county, and city health departments are as differently organized and functioning, as there are states, counties, and cities in the United States. These agencies reflect and represent the special needs of the geographic areas and demographic compositions of their respective domains. Hence, the functioning and expanse of services offered by the New York City Department of Health is vastly different from a similar agency in rural Montana.

This unique approach to the application of a health care system must also be examined in light of the diversity of the demography and geography of the United States. Approximately 273 million people inhabit the United States across a geographic expanse of 3.5 million square miles of land. Ranging from the deserts of Nevada to the Rocky Mountains of Colorado and Wyoming to the tropics of Florida and the oceanic seaboards of the east, west, and southern coasts, American geography and topography is expansive (5). Hence, a substantial challenge to the delivery of health care services exists in this array of geographical areas.

The American population is equally diverse and expansive. There are almost 35 million people who are age 65 or older. African Americans constitute 12.8% of the population, Asian and Pacific Islanders 4%, American Indians 0.9%, and Caucasians make up 82% of the population (5). Because the United States is largely a nation of immigrants, there are literally hundreds of additional ethnic groups that are part of the American

social fabric. As of March 1997, 25.8 million individuals in the United States were foreign-born, which represents a 30% increase from 1990, when there were 19.8 million foreign-born individuals in the United States. Mexico was the place of origin for 7 million or 28% of the total foreign born population in 1997 (5). During 1996 and 1997, 1.3 million people moved to the United States from abroad, and 92% of those individuals moved to metropolitan areas. Additionally, during this time period, 3 million people left the central cities and 2.8 million moved to the suburbs (6). The health care system of the United States should then be viewed in the following context:

- A diverse spectrum of social values, which have historically pointed more toward libertarianism than egalitarianism
- Limited roles of the federal, state, and local governments in the provision of, and payment for, health care services
- A pluralistic, free-market approach to the provision of health care services
- A geographically diverse and substantive land mass
- A culturally diverse and numerically large population whose characteristics are changing toward more elderly and racial and ethnic heterogeneity

It is critical that the reader be sensitive to these contextual variables to understand the American health care system and how health care policy is shaped and implemented in the United States.

THE ORGANIZATION OF U.S. HEALTH CARE SERVICES

Health care services in the United States are provided by a broad array of facilities, which are financed from a variety of payment sources. As of 1998, there were 6021 hospitals (7), 1,012,582 hospital beds, 33,765,940 admissions, and 241,574,380 inpatient days. In 1998, the average length of stay in community hospitals was 6 days, whereas it was 7.7 days in 1975 (7).

It is also notable that the numbers of hospitals in urban and rural settings are shrinking. In 1993, there were 3012 urban hospitals and 2249 rural, whereas in 1998, there were 2816 urban and 2199 rural hospitals (13). The numbers of public acute care hospitals decreased from 1390 in 1993 to 1260 in 1997 (8). Closure of hospitals and simultaneous reductions in hospital beds has occurred in inner city areas where care is provided for large numbers of indigent patients. Such closures are related to the high costs of care, which are not concomitantly reimbursed by state and federal sources either because the individuals are not eligible or because payment rates do not cover the costs incurred. Small, isolated rural hospitals are facing similar economic and, hence, survival difficulties. The plight of rural hospitals is of special significance because their survival is often linked to the economic and social survival of a rural community.

While the world's population grows at an annual rate of 1.7%, the population over 65 increases by 2.5% per year. There are just fewer than 600 million people over the age of 60 in the world. Approximately 360 million of the world's over 60 population lives in the developing world, in which 7.5% of the population is elderly. In contrast, 18.3% of the population is elderly in the developed world. The most rapid changes are occurring in some developing countries where an increase of 200–400% in the elderly population is predicted over the next 30 years (9). Because of the growth of the elderly population, there has been an increase in the demand for geriatric and long-term care facilities. Over the next several decades, the elderly's health care consumption in the United States will be approximately $25,000 per person (in 1995 dollars) compared to $9200 in 1995 (10). In this respect, the United States is following the trends exhibited in most developed industrialized countries.

The increased utilization of health care services by the elderly is expected to put additional strains on an already besieged health care system. Increasing the life span, either through preventive measures or through other acts of distributive justice, solves some problems while creating others. This astounding paradox will assuredly complicate the political and social processes of decision making. Equally likely will be the burdens these phenomena add to an already overburdened national economy.

In the last several years, it is the substitutability that has been emphasized, as more and more procedures are performed in outpatient settings. Many services previously performed in the hospital now take place in physician offices. In 1996, there were 734,493,000 visits to the physician, with an average of 3.4 per person (11), and the most frequent principal reason for a visit was a general medical examination, with a total of 54.7 million in 1996. Also in 1996, there were 67.2 million visits to outpatient departments, and 40.3 million inpatient surgery procedures were performed (11). This analysis points to the increasing importance of the ambulatory care setting as a place for rendering care. The relevance of outpatient care will continue to grow as more medical procedures are performed outside hospitals and greater emphasis is placed on preventive care. Outpatient visits in community hospitals alone have advanced from 263,631,000 in 1986 to 301,329,000 in 1990 to 474,193,000 in 1998 (7).

The National Association of Home Care estimates that more than 20,000 providers deliver home care services to approximately 8 million individuals each year (12). According to the Health Care Financing Administration (HCFA), the average number of home health visits a year per Medicaid beneficiary was 80, compared to 27 visits in 1989. Additionally, the number of home health agencies participating in Medicare has increased from almost 5000 in 1988 to over 10,000 in 1997 (13). Care of patients in home settings is likely to expand as data further suggest reduced cost for such care without compromising quality. Technological and scientific developments related to providing sophisticated treatments in the home will also stimulate growth in this sector of health services.

TRENDS IN HEALTH INSURANCE COVERAGE

According to the President's Advisory Commission on Consumer Protection and Quality in the Health Care Industry, there are five trends that summarize the characteristics of health insurance plans of the late 1990s:

- Increased complexity and concentration of health plans
- Increased diversity of health insurance products
- Increased focus on network-based delivery
- Shifting financial structures and incentives between purchasers, health plans, and providers
- The development of clinical infrastructure for utilization management and quality improvement (14)

In response to rapidly increasing health care costs, private insurance companies and employers (who pay the premiums in whole or in part for their employees) have increased their part in implementing cost-containment strategies. A dramatic effort has been the application of business principles to purchasing and vendor selection and payment for and selection of health care providers and institutions of care.

Private employers, the federal government, and state and local governments invest significant financial resources in health care purchasing expenditures. In 1995, private employers contributed $183.8 billion to private health insurance premiums, whereas the federal government spent $11.3 billion on private health insurance premiums, and state and local government spent $47.1 billion (14). In 1995, more than 83% of the insured population was covered by private insurance, whereas about 31% was enrolled in a public program, such as Medicare or Medicaid.

Probably the most significant change in the American health care system in recent years is the development of managed care. In managed care settings, the covering company is responsible for providing services, whereas, at the same time, it is exposed to the financial risks of unanticipated services. Health Maintenance Organizations (HMOs) contract with hospitals and certain physician providers for services within a negotiated schedule of fees. HMOs and other such managed care organizations specify where and by whom care is to be given. The latter is a radical departure from the historically preeminent "freedom of choice" that patients and care providers enjoyed under the traditional indemnity and fee-for-service reimbursement programs. The traditional method of paying for medical services is fee-for-service when the provider charges a fee for each service provided, and the insurer pays all or part of that fee.

Managed care is an umbrella term for HMOs and all health plans that provide health care in return for preset monthly payments and coordinate care in a defined network of primary care physicians and hospitals. A network includes physicians, clinics, health centers, medical group practices, hospitals, and other providers that a health plan selects and contracts with to care for its members. An HMO is an organization that provides health care in return for preset monthly payments. Most HMOs provide care through a network of physicians, hospitals, and other medical professionals that their members must use in order to be covered for that care.

There are a number of different types of HMOs. A staff model HMO is an HMO in which the physicians and other medical professionals are salaried employees, and the clinics or health centers in which they practice are owned by the HMO. A group model HMO is made up of one or more physician group practices that are not owned by the HMO but operate as independent partnerships or professional corporations. The HMO pays the groups at a negotiated rate, and each group is responsible for paying its doctors and other staff as well as covering the cost of hospital care or care from outside specialists. An Independent Practice Association (IPA) generally includes large numbers of individual private practice physicians who are paid either a fee or a fixed amount per patient to take care of the IPA's members. A Preferred Provider Organization is a network of doctors and hospitals that provides care at a lower cost than through traditional insurance. PPO members have more health coverage when they use the PPO's network and pay higher out-of-pocket costs when they receive care outside the PPO network (15).

An integrated health system is a network that provides a coordinated continuum of services and is clinically and fiscally accountable for outcomes. There was a significant growth of integrated health systems during the late 1990s. In 1997, there were 228 integrated systems and, in 1998, there were 266, representing an increase of almost 17% (16). Simultaneously, there has been a disintegration of systems when mergers fail and disassemble. Iglehart comments on how managed care has changed the face of health care:

> Before the emergence of managed care, it was largely physicians, acting individually on behalf of their patients, who decided how most health care dollars were spent. They billed for their services, and third-party insurers usually reimbursed them without asking any questions, because the ultimate payers—employers—demanded no greater accounting. Now, many employers have changed from passive payers to aggressive purchasers and are exerting more influence on payment rates, on where patients are cared for, and on the content of care. Through selective contracting with physicians, stringent review of the use of services, practice protocols, and payment on a fixed, per capita basis, managed-care plans have pressured doctors to furnish fewer services and to improve the coordination and management of care, thereby altering the way in which many physicians treat patients. In striving to balance the conflicts that arise in caring for patients within these constraints, physicians have become "double agents." The ideological tie that long linked many physicians and private executives—a belief in capitalism and free enterprise—has been weakened by the aggressive intervention of business into the practice of medicine through managed care (17).

There has been a recent challenge to the core tenet of managed care that centralized decision making could deliver improved care at a reduced cost. In November 1999, a large health care company decided to allow physicians to choose what care patients need without the insurance company's intervention or approval. This action opens the door to further discussions about how managed care principles are utilized. Regardless of managed care's future course, cost containment measures will be necessary to prevent an explosion of health care costs. The demand for cost containment will need to be weighed against the imperative to insure that patients have access to care. Paul Ellwood, often referred to as the "father of the HMO," believes that there will be a new era in which patients, not employers and government purchasers, will have power (18). Regardless, the weight of political and consumer pressures, along with experience and economic efficiency, will determine the future of managed care.

HEALTH CARE FINANCING

The expenditures for health care in the United States have grown from $51 billion in 1967 (6.3% of GNP) to over $1 trillion in 1997 (14% of GDP).[a] In 1997, on a per capita basis, $4090 was spent on health care (19) and 0.64 per day/capita was spent on prescription drugs (20). This is substantially higher than that of other industrialized nations. When comparing health expenditures in the major industrialized countries comprising the Organization for Economic Cooperation and Development (OECD), for example, dramatic differences in per capita expenditures are noted (21). Such differences also exist in the percentage share of GDP spent on health care (21), and relative growth in health care expenditures over time varies greatly among these countries (21, 22).

The Health Care Finance Administration asserts that national health expenditures are projected to total $2.2 trillion and reach 16.2% of the GDP by 2008. The growth in health spending is projected to average 1.8 percentage points above the growth rate of the GDP for 1998–2008. This differential is higher than recent experience but remains below the historical average for 1960–1997, where growth in health spending exceeded growth in GDP by close to three percentage points. There are a number of factors that contribute to the projected acceleration, including:

- An increase in private health insurance underwriting cycle
- A slower growth in managed care enrollment
- A movement towards less restrictive forms of managed care
- A continued trend toward increased state and federal regulation of health plans

The growth of health care expenditures without a concomitant gain in health status of the population is receiving more and more attention on the governmental and corporate agenda. On the governmental level, an increasing proportion of federal and state budgets is being allocated to health care. In the private sector, corporations and individuals are bearing larger proportions of health care costs. Although no particular percentage of GDP has been determined to be an acceptable or unacceptable expenditure for health care services, the fact is that costs are increasing and the health care sector is gaining an increasing share of the economy. This follows several other interesting trends. During the period of 1961 to 1997,

[a] The GNP is the total annual flow of goods and services in a nation's economy. Most industrial countries now use GDP, which measures the value of all goods and services produced within a nation, regardless of the nationality of the procedure.

national health expenditures as a percentage of GNP rose from 5.4% to over 14%. In the same period, dramatic differences occurred in the source of revenues for health care expenditures. The pattern of spending these resources also changed significantly (13).

In 1960, 49% of health care revenues came from out-of-pocket payments from individuals. Out-of-pocket spending is defined as expenditures for coinsurance and deductibles required by insurers, as well as direct payments for services, which are covered by a third party. In 1990, individual consumers spent $144.4 billion directly for out-of-pocket payments for personal health services (23). This accounted for 38% of all personal health spending. In 1998, consumers spent $183.7 billion in out-of-pocket payments, which accounts for 33% of the $558.7 billion in personal health spending (23).

Consumers have spent and continue to spend less of their own personal money for health care services. This decrease in personal spending has been shifted largely to third parties, such as private health insurance, government programs, philanthropic organizations, and other sources. It is evident that the shift away from personal, out-of-pocket health spending has resulted in greater consumption of health care services. This transition reflects the general maxim in health care economics that the consumption of health care services is probably insatiable (24). Moreover, unlike other sectors of the economy and the laws of economics they obey, prices for health care services do not fall with increased consumption or purchasing.

According to Iglehart, the decline in personal spending is "attributed in large part to the growth in health maintenance organizations (HMOs), which traditionally offer broad benefits with only modest out-of-pocket payments. In the past few years, however, most HMO enrollees have had increased cost-sharing requirements, as employers and health plan managers have sought to constrain spending even further. Out-of-pocket payments are still considerably less in an HMO than with indemnity insurance (17)." However, "The overall declines in per capita out-of-pocket spending mask the financial difficulties of many poor people and families. A recent study estimated that Medicare beneficiaries over 65 years of age with incomes below the federal poverty level (in 1997 the level was $7755 for individuals and $9780 for couples) who were also eligible for Medicaid assistance still spent 35% of their incomes on out-of-pocket health care costs. Medicare beneficiaries with incomes below the federal poverty level who did not receive Medicaid assistance spent, on average, half their incomes on out-of-pocket health care costs (17)."

Historically, a lack of public insurance programs created obstacles to health care services. For those who could not afford to pay for private insurance, the costs associated with health care were larger than most could afford. After lengthy debate, the U.S. Congress passed legislation in 1965 that established Medicare and Medicaid. Medicare covers over 95% of the elderly in the United States as well as many individuals who are disabled. Coverage for the disabled began in 1973 and is divided in two parts: 1) hospital insurance and 2) supplementary medical insurance. The total disbursement for Medicare in 1997 was $213.575 billion, and there were 36,460,143 enrollees, of which 32,164,416 were elderly.

The total expenditure for the Medicaid program was $160 billion in 1996. Of the total amount spent in 1996, Medicaid payments for nursing facilities and home health care totaled $40.5 billion for more than 3.6 million recipients. The average cost per recipient in 1996 was $12,300, and almost 45% of the total cost of care for individuals using nursing homes and Medicaid was paid for home health care (13).

Since the enactment of Medicare and Medicaid, there have been various legislative and administrative changes. The Balanced Budget Act of 1997 enacted the most significant changes to Medicare and Medicaid since its inception, including a capped allocation of monetary resources to states and the addition of the Children's Health Insurance Program. The Children's Health Insurance Program set aside $24 billion over 5 years for states to provide health care to over 10 million children who are not eligible for Medicaid.

In 1960, public programs paid for one quarter (24.5%) of all health care spending; by 1988, this share had increased to 42.1%. Together Medicare and Medicaid financed $351 billion in health care services in 1996, which is more than one-third of the nation's total health care bill. Additionally, it represents three-quarters of all public spending on health care. There has been a significant increase in Medicare managed care enrollment—from 3.1 million at the end of 1995 to 6.3 million in 1999, leaving approximately 33 million beneficiaries in a traditional fee-for-service Medicare program.

An area of controversy is the limitation on coverage for prescription drugs. Spending on prescription drugs is the fastest-growing piece of personal health expenditures, amounting to $78.9 billion in 1997. Additionally, spending for prescription drugs has increased at double-digit rates: 10.6% in 1995, 13.2% in 1996, and 14.1% in 1997 (17). The reason for this rapid growth, according to Iglehart, includes: "Broader insurance coverage of prescription drugs, growth in the number of drugs dispensed, more approvals of expensive new drugs by the Food and Drug Administration, and direct advertising of pharmaceutical products to consumers. The use of some new drugs reduces

hospital costs, but not enough to offset the increase in expenditures for drugs (17)." In the year 2000, 86% of health care plans will have an annual limit on brand and generic drugs, and there will be increased use of copayments for prescription drugs (25).

The budget cuts imposed by Congress in 1997 to help balance the budget have restricted the fees that caregivers receive for the elderly and disabled. When federal health programs cut funding significantly, as occurred in the Balanced Budget Act of 1997, the resulting cutbacks at the institutional and health-system level trickled down to providers' abilities to provide an acceptable level of service designed to protect patient safety and foster appropriate medication use. Partial restoration of the Balanced Budget Act in 1999 addressed the transition to an outpatient prospective payment system for hospitals, payments to skilled nursing facilities and home health agencies, payments for indirect medical education, and a number of rural health care provisions.

The dramatic shift of third parties (government, private health insurance) toward paying for a greater and greater proportion of personal health care services has led to a paradigm shift in attitudes and actions toward health care financing and cost control. Several approaches have been adopted in the governmental sector to slow the increases in costs and expenditures. The most dramatic of these has been the introduction in 1983 of the prospective payment system (PPS) to curb the growth in hospital costs and expenditures. By imposing prospective limits on Medicare payments to hospitals through a system of reimbursing average costs of specific diagnoses, hospital utilization has decreased dramatically. The average length of stay and admission rates in community hospitals of elderly patients (those covered by Medicare) dropped sharply after the introduction of PPS (13).

Because of cost-containment strategies of both the private and governmental sectors, hospital utilization has declined. This has resulted in a decline in the number of patient beds, the average length of stay, and patient bed census (7). The present predominant view is that hospitalization of any patient, regardless of revenue source, is to be avoided wherever possible. Only those patients for whom hospitalization can be fully justified are admitted.

As much as the financing of America's health care system is a major issue on the policy agenda of the nation, so too is the continuous question about the relationship between the costs and the outcomes of care. As costs increase, the numbers of policy analysts, organizations, and governmental agencies calling for a better definition of the cost-outcome relationship has sharply risen.

Cost-effectiveness and cost-benefit analyses are frequently mentioned in academic and policy-analysis circles. These notions center on careful examination of the costs and their corresponding outputs. Eisenberg (26) defines cost-effectiveness analysis as the measure of the net cost of providing service (expenditures minus savings) as well as the results obtained (e.g., clinical results measured singly or a series of results measured on some scale). Cost–benefit analysis determines whether the cost is worth the benefits by measuring both in the same units (26). Such analyses will be critical, as future policy decisions are made with regard to the collection, allocation, and utilization of finite resources in the health care system for the enhancement of health status of the American people.

Private-sector strategies and governmental plans to curb health care costs have not escaped criticism. Ginsberg, for example, argues that the notion of "for profit" hospital chains has severe limitations with respect to garnering large proportions of market share and, consequently, greater profits (27). He bases this view on the limited amount of private funding available for hospital care. On the other hand, he sees this sector as being able to grow in the area of nursing homes and other businesses related to the care of the elderly.

ACCESS TO HEALTH CARE SERVICES IN THE UNITED STATES

There are three classes of individuals who have open access to and can derive some form of services from America's health care system:

- Those who receive support from governmental sources because of specific entitlements (indigents, elderly, and veterans)
- Those who are provided with basic health insurance coverage from their employers
- Those who choose to cover their expenses from out-of-pocket payments

There are, however, those who have no specific financial support or capacity to pay for health care services and who are not eligible for any type of entitlements. These individuals must rely on some form of charity care or services. In addition, there are those who, for reasons of geographic remoteness or total inability to gain access, have no access to health care services. This group represents a complex, resource-based demand model, which also has an equally complex pattern of health care system and services-utilization requirements.

With increasing health care costs and consequent increases in insurance premium costs, gaining access to health care services without incurring personal costs has become more difficult. Not all services are covered for individuals in the federal Medicare and Medicaid programs. Moreover, there are strict limitations on the extent of services offered in these programs. A similar set of restrictions may be found in private-sector health care coverage strategies. Because few insurance programs and none of the federal programs provide coverage for unlimited long-term care, all but the very rich are at risk of financial ruin.

The health care lexicon includes two new terms to reflect these problems: underinsured and uninsured. The underinsured may include the "working poor," those individuals who have jobs and may be covered by a very limited, if any, health insurance program by their employers. They are likely low wage earners and those receiving incomes at, or slightly above, the poverty level. Typically, they do not qualify for Medicaid entitlements, do not have employer-paid health insurance benefits, and cannot afford (or choose not to purchase) third-party coverage for payment of health care services.

There are no specific policy plans available to finance uninsured and underinsured care. Whether planned as charity care or unplanned as financial loss, the "price tag" for uncompensated care in the United States was $18.5 billion in 1997, which is 6% of the total of hospital expenses (28). This percentage has remained constant since 1984, when the percentage of total expenses for uncompensated care was also 6% (8).

Reduced payments and high levels of uncompensated care have led to the closing of hospital facilities in both urban and rural blighted areas, making access to care even more difficult for some. Whiteis and Salmon (29) refer to this phenomenon as "disinvestment in the public goods." Because privately owned and not-for-profit hospitals and private clinics, pharmacies, and physician's offices must rely on their own financial soundness, any threat to that foundation may lead to closure.

The amount of uncompensated care is magnified in areas where serious social problems exist because health status is directly related to social status. Health status should be examined in broad terms by reviewing morbidity and mortality data available for the whole population. The life expectancy of people who live in the United States has grown by almost 10 years, from 68.5 years in 1936 to 76.1 years in 1996. Women were expected to live to 79.1 years in 1996, whereas the average for men was 73.1 years (11). The leading causes of death in 1996 among people living in the United States were (11):

1. Heart disease (733,361 deaths)
2. Cancer (539,533 deaths)
3. Stroke (169,942 deaths)
4. Pulmonary diseases (108,027 deaths)
5. Accidents (94,948 deaths)
6. Pneumonia and flu (63,727 deaths)
7. Diabetes (61,787 deaths)
8. AIDS (31,130 deaths)
9. Suicide (30,903 deaths)
10. Liver disease (25,047 deaths)

Infant mortality, another measure of the health status of a nation, stated as the number of deaths per live births, was 7.2 per 1000 live births in 1997 compared to 9.9 per 1000 live births for 1988. Overall, these figures are comparable to those of the major, industrialized nations of the world.

Major morbidity in the United States is currently centered on diseases of life style. These morbidities contrast sharply with disease patterns prevalent during the early part of the 20th century. Outside of AIDS and other sexually transmitted diseases, infectious diseases represent a small proportion of prevalent morbidity. Rather, life-style diseases, associated with smoking, poor nutrition, a sedentary life style, alcohol and other chemical consumption, homicides, suicides, and accidents, represent the majority of morbidity in the United States. Significant preventive strategies can markedly reduce the incidence, prevalence, and mortality associated with these health care problems.

Not surprising, in areas with high concentrations of indigent people, there are similarly high concentrations of uninsured individuals requiring intense health care services. These areas exist in both rural and urban settings. Emergency rooms have become a major resource for primary health care services in areas where physician office services or other service providers (clinics) are not available because of location, cost, or quality. Emergency rooms have also become providers of high-intensity care for victims of gun shot wounds, drug overdoses, communicable diseases, and other trauma associated with poor social conditions. Much of the care in emergency rooms is uncompensated because the quality and amount exceed the allowable reimbursement. Some trauma centers in economically blighted areas have been closed (30).

Hospitals in inner cities and blighted rural areas also care for a higher proportion of "at-risk" patients than hospitals in the for-profit sector generally located in more affluent areas (29). In fact, affluent hospitals sometimes "dump" their uncovered patients on charity care and other public hospitals in order to reduce their financial risks. This, however, increases the financial risks of public or charity hospitals. Again, the reimbursement levels under present schemes for large numbers of "at-risk" patients simply do not cover costs; thus, the United States has

witnessed hospital closings, particularly in those areas where such loss is most noticeable (30).

American health policy continues to grapple with these issues related to the underinsured and the uninsured (31). A multiple-tiered health care system based on social class and ability to pay is unacceptable in a nation that boasts incomparable riches and political agendas of democracy and rights. Ginsberg (27) notes:

> Despite all our efforts of recent years, then, health care costs continue to increase.... There is undoubtedly waste in the health care system, but no solid proposals have been advanced to recapture the $100 billion, plus or minus, that some believe can be saved. I believe that we will not reshape our national health policy agenda unless and until we achieve a broad consensus on the key issues. Do the American people, for example, desire to ensure access to health care for the entire population? In that case they must agree to pick up a sizable additional tab, which they have thus far avoided.

The issue of quality health care has become an increasing issue of concern in the face of cost constraints and limited access to health care. The *President's Advisory Commission on Consumer Protection and Quality in the Health Care Industry* (32) states that "the purpose of the health care system must be to continuously reduce the impact and burden of illness, injury and disability and to improve the health and functioning of the people of the U.S." According to the Commission, there are basic characteristics of health care that, as a nation, we should strive to achieve. The Commission has created "Guiding Principles for the Consumer Bill of Rights and Responsibilities" for the health care of people in the United States. These include the following:

- All consumers are created equal.
- Quality comes first.
- Preserve what works.
- Costs matter.

THE FUTURE OF HEALTH CARE

Suggestions for broad reform, which address the financial, access, and quality of care issues for America's health care system, have emerged during the past decade. Iglehart emphasizes the irony of the American health care system. He writes (17):

> By many technical standards, U.S. medical care is the best in the world, but leaders in the field declared recently at a national round table that there is an "urgent need to improve health care quality." The stringency of managed care and a low inflation rate have slowed the growth of medical spending appreciably, but a new government study projects that health care expenditures will soon begin escalating again and will double over the next decade. In short, the American system is a work in progress, driven by a disparate array of interests with two goals that are often in conflict: providing health care to the sick, and generating income for the persons and organizations that assume the financial risk.

The President's Commission (32) outlines areas in which the American health care system could be improved in light of the reality that many individuals receive substandard care and 44.3 million individuals are without health insurance coverage. This commission outlines several types of quality problems including avoidable errors, underutilization of services, overuse of services, and variation in services. Based on the reality of these quality problems, the Commission recommended that the initial set of national aims should include (32):

- Reducing the underlying causes of illness, injury and disability
- Expanding research on new treatments and evidence on effectiveness
- Ensuring the appropriate use of health care services
- Reducing health care errors
- Addressing oversupply and undersupply of health care resources
- Increasing a patient's participation in his or her care

The President's Commission engages a broad consumer advocacy movement in public and private sectors calling for a major reform of the U.S. health care system to improve access to care for more individuals living in America. Consistent with previous patterns, however, these calls have only led to incremental adjustments in policy and slight quality changes in direction. The major problems, for the most part, remain unaffected. Although broad based health care reform efforts have been unsuccessful, market forces and more targeted legislation and regulatory efforts have changed the face of health in the 1990s.

The 1993–94 Clinton health care reform plan, in its ideology, provided an ambitious plan to eliminate the enormous problem of lack of access to health care. It proposed to guarantee comprehensive health benefits for all American citizens and legal residents, regardless of health or employment status. The proposal was unsuccessful due to a number of factors, including its vast scope, the complicated nature of the plan, and an underestimation of the politics involved with radically reforming health care. The failure of the Clinton administration health care reform agenda and the subsequent events to revise the American health care system are important lessons of health-care-system related politics.

Unfortunately, since the failure of the Clinton Administration plan in 1994, the number of uninsured individuals in America has grown. According to the Census Bureau, 44.3 million people are uninsured, comprising about 16.3% of the population. Of those uninsured, 15.4% are under 18 years of age, and the largest percentage is among individuals between 18 and 24 years of age. People of Hispanic origin make up 35.3% of those uninsured and 43% of the total uninsured population are not citizens of the United States (6).

The number of uninsured persons is expected to continue to grow. Proposals for health care reform to combat this problem include President Clinton's proposal for Medicare buy-in proposals for "middle aged" adults and House Majority Leader Dick Armey's (R-TX) proposal for a refundable tax credit to pay for insurance for the uninsured. The 2000 presidential campaign opened the debate for legislation that will improve health care coverage for the uninsured. This public debate on how to enhance access to care will stimulate creative ways to improve the U.S. health care system. However, rhetoric is not enough; it needs to be translated into programs that attack the problem.

The essence of the health care financing dilemma is related to how much a nation wishes to spend, on whom these funds are to be expended, and by what methods a relationship among cost, quality, and outcomes might be determined. In a time when advancing science and technology is flourishing in the health care field, "high tech" medicine will continue to evolve with an ever-increasing price tag. Furthermore, the costs of unanticipated and complex disease problems (e.g., HIV/AIDS) add to the unpredictability of health care system costs. This is all to say that most policy makers understand what needs to be done. They are in a quandary, however, in finding the appropriate and acceptable solution. Hence, it is likely that costs and expenditures will continue to rise (and, thereby, increase the percentage of GNP that will be spent for health care) and that solutions may become even more elusive.

Although some might argue that the available resources for expenditures on health care are ultimately limited, few are able to say exactly where that limit is or should be. In the United States, there has been an expansion of technologies and procedures based on scientific advancements without a concomitant development of a moral and ethical policy for determining who might be best served by such advancements. Rationing of health care services or otherwise limiting access to high cost services, for example, has resulted from political policy rather than from deliberated public policy and rational decision making. This is most notably evidenced in the Medicaid component of the U.S. health care system.

As cost pressures continue to mount, there will likely be a return to having patients pay more of the health care expenditure dollar from their own resources. This will take the form of higher deductibles and co-insurance payments. Perhaps returning the burden of health care financing to the individual will raise the collective consciousness of American society that "there is no such thing as a free lunch" insofar as using and paying for health care services is concerned. Certainly, this phenomenon has occurred in social welfare "reform" in which the programs that have had mixed success have been restructured to "roll" participants off of welfare to work.

On the other hand, there are perhaps no solutions forthcoming on some of the problems represented in the arena of health care financing. As Hardin suggests, there is indeed a class of human problems that have no technical solution (33). In using Hardin's analogies, Hiatt (34) suggests that "nobody would quarrel with the proposition that there is a limit to the resources any society can devote to medical care, and few would question the suggestion that we are approaching such a limit. The dilemma confronting us is how we can place additional stress on the medical commons without bringing ourselves closer to ruin."

CONCLUSION

These are the principal contemporary features of the U.S. health care system. A massive societal structure is at once saviour, behemoth, juggernaut, and question mark. It certainly will be in a constant state of flux and gradual change. It therefore bears constant vigilance and careful guidance by those who derive their livelihoods from it and those who are the beneficiaries of its caring. Most importantly, it will require significant pressure from those who are disenfranchised from it.

REFERENCES

1. Donabedian, A. *Aspects of Medical Care Administration: Specifying Requirements for Health Care*; Harvard University Press: Cambridge, 1973.
2. Dougherty, C.J. *American Health Care: Realities, Rights, and Reforms*; Oxford University Press: New York, 1998.

3. Bella, R.N.; Madsen, R.; Sullivan, W.M.; Swidler, A.; Tipton, S.M. *Habits of the Heart*; Harper & Row: New York, 1985.
4. Relman, A. Reforming the Health Care System. New Engl. J. Med. **1990**, *323* (14), 991–992.
5. www.census.gov (accessed Oct 1999).
6. www.doc.gov (accessed Oct 1999).
7. *American Hospital Association, Hospital Statistics*, Health Forum: Chicago, IL, 2000.
8. *Modern Healthcare*, By the Numbers 1999 Edition (Supplement to Modern Healthcare), July 19, 1999.
9. www.who.gov (accessed Nov 1999).
10. Fuchs, V.R. Health Care for the Elderly: How Much? Who Will Pay for it? Health Aff. **1999**, *18* (1), 11–21.
11. www.cdc.gov/nchs (accessed Oct 1999).
12. www.nahc.org (accessed Dec 1999).
13. www.hcfa.gov (accessed Dec 1999).
14. Quality First: Better Health Care for all Americans. *The President's Advisory Commission on Consumer Protection and Quality in the Health Care Industry*, 45.
15. www.NCQA.org (accessed Oct 1999).
16. Landis, N. Am. J. Health Syst. Pharm. **1999**, *56*, 1392.
17. Iglehart, J. The American Health Care System: Expenditures. New Engl. J. Med. **1999**, *340* (1).
18. New Health Care Model Balances Cost, Quality, Shifts Power to Patients. PR news wire. Nov. 1999.
19. www.hcfa.gov (accessed Oct 1999).
20. www.bea.doc.gov (accessed Oct 1999).
21. Schieber, G.J. Health Expenditures in Major Industrialized Countries, 1960-87. Health Center Financing Review **1990**, *11* (4), 159–167.
22. www.phrma.org (accessed Oct. 1999).
23. Smith, S.; Freeland, M.; Heffler, S.; McKusick, D. Health Tracking: Trends. Health Aff. **1998**, *17* (5), 128–140.
24. Klarman, H.E. *The Economics of Health*; Columbia University Press: New York, 1965.
25. www.hcfa.org (accessed Oct. 1999). Medicare + Choice: Changes for the Year 2000 Executive Summary.
26. Eisenberg, J.M. Clinical Economics: A Guide to the Economic Analysis of Clinical Practice. JAMA **1989**, *262*, 2879–2886.
27. Ginsberg, E.N. For Profit Medicine: A Reassesment. New Engl. J. Medicine **1988**, *319* (12), 757–761.
28. www.aha.org (accessed Nov 1999).
29. Whiteis, D.G.; Salmon, J.W. *The Corporate Transfer of Health Care*; Baywood Publishing Company: Amityville NY, 1990.
30. Christianson, J.B. Institutional Alternatives to the Rural Hospital. Health Care Financing Review **1990**, *11* (3), 87–97.
31. McLennan, K. Care and Cost. *Current Issues in Health Policy*; Westview Press: Boulder CO, 1989.
32. Quality First: Better Health Care for all Americans, The President's Advisory Commission on Consumer Protection and Quality in the Health Care Industry.
33. Hardin, G. The Tragedy of the Commons. Science **1968**, *162*, 1243–1248.
34. Hiatt, H.H. Protecting the Medical Commons: Who is Responsible. NEJM **1975**, *293*, 235–241.
35. Quality First: Better Health Care for All Americans, Final Report to the President of the United States. The President's Advisory Commission on Consumer Protection and Quality in the Health Care Industry.
36. To Err is Human: Building a Safer Health System, Institute of Medicine. National Academy of Sciences, 1999.
37. Kronick, R.; Todd, G. Explaining the Decline in Health Insurance Coverage, 1979–1995. Health Aff. **Mar/April, 1999**, 30–47.
38. Merrill, R.A. Modernizing the FDA. *An Incremental Revolution. Health Affairs*, Mar/April, 1999; 96–111.
39. Robinson, J.C. *The Future of Managed Care Organization. Health Affairs*, Mar/April, 1999; 7–24.
40. Grunback, K. Primary Care in the United States—The Best of Times, The Worst of Times. New Engl. J. Med. **1999**, *341* (26).
41. Smith, B.M. Trends in Health Care Coverage and Financing and their Implications for Policy. New Engl. J. Med. **1997**, *337* (14).
42. Vladeck, B.C. *The Political Economy of Medicare. Health Affairs*, Jan/Feb, 1999; 22–36.

Health Services Research

Teresa J. Hudson
Marisue Cody
Veterans Affairs Medical Center,
North Little Rock, Arkansas, U.S.A.

INTRODUCTION

Health services research (HSR) is a relatively new and evolving field. As the organization and financing of healthcare has changed, the need for information about the type and level of care and the effectiveness and quality of care provided in the healthcare system has increased. This chapter provides a definition of HSR, a historical perspective of the development of the field and the relationship between HSR and public health policy, and a discussion of the profession of pharmacy and its relationship to HSR. The chapter concludes by highlighting some of the institutions that commonly fund HSR and journals that publish manuscripts on HSR topics.

DEFINITION OF HEALTH SERVICES RESEARCH

Most fields of research can be identified by the academic degree of the investigators in that area of research. However, health services researchers are identified more by the work than by the particular degree of the investigator. This diversity of degrees reflects the many disciplines that work in the field. As seen in Fig. 1, many disciplines may be involved in a given HSR project, making it difficult to succinctly define the field of HSR. In 1995, the Institute of Medicine (IOM) developed a comprehensive definition that characterized HSR as a

> multidisciplinary field of inquiry, both basic and applied, that examines the use, costs, quality, accessibility, organization, delivery, financing, and outcomes of health care services to increase knowledge and understanding of the structure, processes, and effects of health services for individuals and populations.[1]

One way to understand HSR is to examine the differences between HSR and clinical research. Although the two areas are certainly related as described here, there are differences that distinguish the two. For the purposes of this discussion, comparisons are made using three categories: 1) study setting, 2) subject selection and sampling, and 3) data sources and measures.

Study Setting

In general, clinical trials study the efficacy of a medication or other treatment under defined conditions. HSR studies evaluate the effectiveness of care under usual conditions. Clinical trials are generally conducted in some type of clinical laboratory. That is, the intervention takes place in a controlled clinical setting where the process of care is dictated by the protocol (i.e., how often patient is followed, how often and which tests are performed at each visit). Any additional healthcare is provided external to the study setting. In contrast, data in a HSR study are gathered from the setting where routine clinical care is provided. Subject sampling and data collection follow a strict protocol, but the process of care continues according to the usual clinical practice in that setting.

Subject Selection and Sampling

Subject selection for clinical trials generally consists of a convenience sample of a specified number of subjects that exhibit the particular syndrome or disease of interest. There are generally very strict inclusion and exclusion criteria that determine eligibility for the study. However, HSR uses population-based sampling such that all the subjects who meet a set of criteria are identified and then a sampling plan is developed to enroll a study sample that is representative of the population of interest. The population to be studied may be defined by a specific geographic location (e.g., people living in the Mississippi delta), by a specific disease (e.g., veterans with schizophrenia), or by the health care payer (e.g., Medicare recipients). It is imperative that sampling occurs in a way that allows for comorbid diseases, differences in demographic variables, and other natural variations among

Encyclopedia of Clinical Pharmacy
DOI: 10.1081/E-ECP 120006347

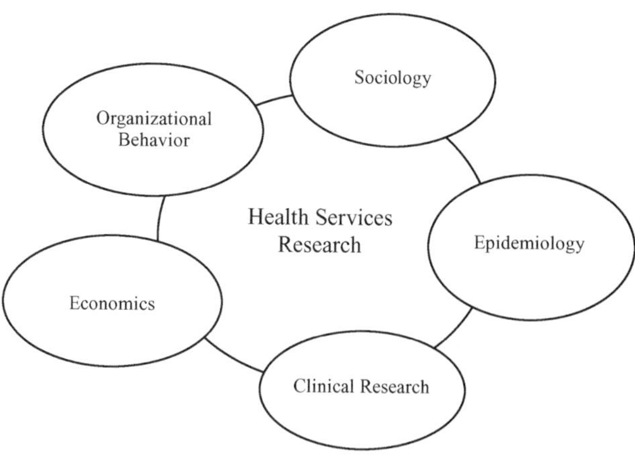

Fig. 1

subjects. Due to the extensive sampling strategies required by HSR, these projects generally require much larger sample sizes than those used in clinical research.

Data Sources and Measures

Data for clinical trials are generally collected directly from the patient and include detailed clinical information collected by trained clinicians. These data are often lengthy and disease specific, and are designed to detect small variations among subjects. In HSR projects, lay interviewers may collect data or they may be gathered by self-report. Secondary data sources such as pharmacy refill records, paid claims data, or secondary analyses of data from national surveys are used to address HSR questions. Given the large samples in HSR data collection, instruments for primary data collection tend to be shorter and do not detect small clinical differences. Instead, they offer a broad assessment of the patients health, level of function, and well-being. Some instruments are rather ''generic'' and are designed for use in any population (SF-36). Others may be designed for a specific disease or population (e.g., asthma quality-of-life questionnaire, toddler quality-of-life survey).

RELATIONSHIP BETWEEN HSR AND HEALTH POLICY

Ginzberg provides an excellent review of the history of HSR and health policy.[2] The IOM 1995 report, ''Health Services Research: Work Force and Educational Issues,'' also provides an excellent summary of HSR and health policy.[1] Because the funding and conduct of HSR has been driven by changes in health policy, a brief overview is provided.

Using a very broad definition, HSR activities can be found as far back as the late 1800s, consisting primarily of descriptive surveys of the prevalence of disease and the number and type of health care personnel and services. Federal and state government funded most projects, although professional organizations such as the American Medical Society also paid for and conducted some studies. One of the earliest true HSR projects began in the 1920s. Between 1928 and 1932, this landmark study, known as the Committee on the Cost of Medical Care (CCMC), produced 27 field studies and final reports that provided recommendations on many aspects of health care, including hospital planning, enhancing public health, and improving professional medical education. One report suggested the use of group medical practice in association with hospitals as a means to provide comprehensive health care. Some CCMC members strongly supported a system using a federally funded health care program, although most favored a system of voluntary health insurance. A significant minority, mostly physicians, firmly opposed any insurance initiative.

Despite the recommendations of the CCMC, the federal government did not intervene in the health care system until the end of World War II (WWII).[2] During WWII, most of the United States' resources were diverted to the war effort. By the time the war was over, it was clear that U.S. hospitals had suffered due to a lack of resources. Not only were the hospitals lacking modern amenities, but also as the nation moved into suburban areas, the number and location of hospital beds were inadequate. In the late 1940s, the federal government began to enact subsidies that encouraged the expansion of the stock of technology and biomedical knowledge, hospitals, and health care personnel. This was accomplished through funding for the National Institutes of Health (NIH) and programs to increase education of nurses, ancillary personnel, and physicians. In 1946, the Hill-Burton Act provided funding for construction of many new hospitals and renovation of old ones. This was the first federally mandated health planning initiative and one of the first efforts to reduce or eliminate shortages of health care facilities in rural and relatively poor regions of the United States.

From the 1940s to the 1960s, federal healthcare initiatives focused primarily on supply issues with limited efforts to improve funding of healthcare. The 1960s marked some of the most significant changes in organization and financing of health care and, therefore, in the development and funding of HSR. In 1965, the legislation that funded Medicare and Medicaid was passed, and in 1966, the Office of Economic Opportunity (OEO) was funded. The OEO opened several large community health centers throughout the United States. The federal governments new responsibility for funding and providing healthcare focused attention on the need to evaluate strengths, weaknesses, and consequences of these programs. In 1967, Congress enacted a bill that made it possible for the Secretary of the Department of Health Education and Welfare (DHEW) to establish the National Center for Health Services Research and Development (NCHSR). This new center consolidated a variety of research activities in the DHEW, and established other centers for health services research through contractual arrangements with academic institutions and other organizations. At about this same time, other federal organizations began funding HSR projects. These included the Health Care Financing Administration (HCFA) and the Department of Veterans Affairs. Ultimately, NCHSR was absorbed into the Agency for Health Care Policy and Research (AHCPR), which has been renamed the Agency for Healthcare Research and Quality (AHRQ).

By the 1970s, it was clear that costs for both Medicaid and Medicare were increasing more rapidly that anyone expected. Of particular concern was the scope of coverage provided by the Medicaid program and the cost-based reimbursement policies for Medicare. Despite only limited evidence that health maintenance organizations (HMOs) decreased costs of health care, support was increasing for use of HMOs in both the public and private sectors. During the mid-1970s, Congress passed a variety of legislation that made changes to both Medicaid and Medicare, but was unable to enact any proposal for a national healthcare program. The concern over costs for these national programs and the increasing healthcare costs in the private sector were the basis of several HSR projects to examine many aspects of health insurance and the costs incurred in these plans. However, there were problems in most of the studies, so no definitive answer was available. Because of the need for better information in this area, the OEO agreed to sponsor an extensive experiment in health insurance.

This controlled trial in healthcare financing, known as the Health Insurance Experiment (HIE), was one of the largest and longest running HSR projects ever conducted. Enrollment of a pilot sample began in 1973, and the last families completed participation in the project in 1982. The HIE randomly assigned families to four health insurance plans that varied the amount of copayment incurred by the family from 0–95% or to a staff-model HMO. One of the most notable findings of the HIE was that free care did not decrease total health care costs, as some proposed. Rather, plans in which patients received essentially free care resulted in higher total costs and higher costs for hospitalization. The HIE also suggested that cost-sharing and enrollment in HMOs was most likely to have a deleterious effect on the health of the poorest and sickest groups of patients who were enrolled. As the results of the HIE were published, many private medical insurance plans were changed to increase the amount of out-of-pocket expenses incurred by patients.

Despite the changes in the design of healthcare plans in the 1980s, the costs of medicaid and medicare and the costs to employers for private health insurance plans increased steadily. By the 1990s, there was growing interest in healthcare reform. Despite considerable effort by the Clinton Administration to propose these types of reforms, opposition in Congress prevented anything except further tweaking of federally funded plans. In both the public and private sectors, payers began to emphasize the value obtained for the dollars spent for healthcare. This emphasis on cost versus value of services is the focus of many current HSR efforts.

ROLE OF THE PHARMACY PROFESSION IN HEALTH SERVICES RESEARCH

The pharmacy profession can offer considerable expertise to the field of HSR. Furthermore, recent changes in the healthcare system mandate that, as a profession, pharmacy must broaden its focus from individual clinical interventions to include population and system-level interventions and evaluations. The techniques used in HSR provide vital tools for pharmacy to influence health policy and, ultimately, delivery of care. Furthermore, because of their unique skills and perspectives, pharmacists can offer a distinctive knowledge base that can inform HSR.

When thinking about pharmacy's involvement in HSR, it is helpful to use seven areas of HSR outlined in the IOM's 1995 report.[1] These are 1) organization and financing of health services; 2) access to health care; 3) quality of care; 4) clinical evaluation and outcomes research; 5) informatics and clinical decision making; 6) practitioner, patient, and consumer behavior; and 7) health professions work force. By thinking about the type of research questions considered in each of these categories, it is possible to better understand both the contribution pharmacists can make to HSR and what HSR has to offer the profession of pharmacy.

Organization and Financing of Healthcare

HSR has contributed to many proposals for healthcare reform since the 1960s. These include managed care,

consumer choice, and outcomes and performance monitoring. The field has devised tools and techniques that have facilitated the development of alternative methods of paying for health services, such as resource-based relative value scales for physician services. The recent emphasis on the provision of clinical pharmacy services has spurred an increase in studies evaluating the effect of these services on the costs of care.[3–6] McCombs and colleagues conducted an extensive study of the impact of pharmacists services on costs of care.[7] This study compared the effects of three models of pharmacy consultation services on hospital admissions, total healthcare costs, and medication costs. When compared with usual care, the consultations were associated with a lower likelihood of hospital admission and with lower total healthcare costs for high-risk patients. The consultations that focused on high-risk patients were associated with lower costs for office visits but with higher costs for medication.[7] This study is an excellent example of using HSR tools to provide the kind of evidence necessary to improve both the organization and financing of pharmaceutical care.

Access to Healthcare

Projects that study access to healthcare evaluate factors that influence the timely receipt of appropriate care. In the pharmacy profession, access to care implies access to both dispensing and clinical pharmacy services. Examining access can highlight the effects of changes in payments for prescription medications on access to dispensing services. For example, in 1999, Straub and Straub[8] studied access to retail pharmacies in rural Illinois. This survey showed that, although current access to pharmacies was good, changes in reimbursement from third-party payers, demands of managed care, and expanded competition provided threats to access in rural pharmacies. This type of information is vital to develop health policy that will maintain access to retail pharmacy services throughout the United States.

Access to clinical pharmacy services is a somewhat more difficult issue because barriers to these services are both geographic and financial. Because pharmacists cannot obtain a provider number to directly bill for clinical pharmacy services, it is difficult for them to receive the financial incentives necessary to make this service widely available. Furthermore, because financial incentives exist only for dispensing services, patients may have access to clinical pharmacists only in special environments such as inpatient hospitalization or in systems such as the Veterans Healthcare Administration (VHA) or HMOs.

Quality of Care

Donabedian describes quality of care in terms of the structure, process, and outcomes of the healthcare system.[9] Structure refers to the availability, organization, and financing of health care programs and the characteristics of the targeted populations. Process encompasses the transactions between patients and providers during actual care delivery. Equity, efficiency, and effectiveness serve as intermediate outcomes of the medical care delivery process, which has the ultimate goal of enhancing the populations' health and well-being. Major quality initiatives have adopted this approach. These include the National Committee on Quality Assurances, Health Plan Employer Data and Information Set, the Foundation for Accountability, the Medical Outcomes Trust, the Health Outcomes Institute, and the Joint Commission for Accreditation of Healthcare Organizations ORYX system. Relatively little work has explicitly addressed the interactive effects of organizational factors on care delivery and client outcomes. Most HSR has focused on broad structural organizational variables without studying the mechanisms that may account for differences in outcomes.

Measurement of quality is an area in which pharmacists could play a key role. Pharmacists in hospitals and other institutions are often charged with the task of evaluating quality of medication use in the form of drug use evaluation. Furthermore, many efforts at assessing quality of care rely on use of computerized prescription records.[10,11] Because of pharmacists' knowledge of clinical aspects of care along with the details of dispensing of medications and, therefore, the development of prescription records, pharmacists are poised to offer insight and leadership in quality assessment efforts in a variety of settings, particularly as it pertains to use of medication.

Clinical Evaluation and Outcomes Research

Ellwood described outcomes management as a way to help patients, payers, and providers make rational medical choices based on better insight into the effect of these choices on a patient's life.[12] Clinical evaluation and outcomes research studies include evaluation of the impact of severity of illness on clinical and economic outcomes, the effect of patient participation in care, the role of patient preferences in medication adherence, and the relationship between quality of life and satisfaction. Many studies address the impact of pharmacist activities on economic outcomes.[4,13] Some studies also evaluate pharmacists impact on clinical outcomes.[14–16] However, in a literature review of studies that examined the impact of pharmacists in ambulatory care or community practice, Tully and Seston found few studies that clearly demonstrated improvement of economic outcomes.[17] Only a minority of the literature reviewed included assessment of quality of life, patient satisfaction, or functional outcomes and, when assessed, these parameters were rarely significantly different after the pharmacists intervention. The authors believed these results were due, in part, to small sample sizes and problems with research design. It is imperative that studies evaluating the impact of pharmacists activity pay particular attention to study design, sample size, and outcomes measurement.

Informatics and Clinical Decision Making

Studies of informatics and clinical decision making concentrate on the benefits of using computerized decision support systems in clinical practice and in research to measure outcomes, efficiency, and effectiveness of care. Decision analysis in clinical research employs probability analysis to express uncertainty and utility theory to express patient preferences for health outcomes.

Computers have been used as a routine part of pharmacy practice for many years. Pharmacists use this technology in many ways. For example, computers can be used to interact with physician colleagues, track patient behaviors, or as tools to evaluate cost and effectiveness of medication regimens.[18,19] Pharmacists have also investigated concordance between traditional and computerized patient records, and are now incorporating computers into patient assessment and educational activities.[20,21] This familiarity with the technology positions pharmacists to provide leadership in using informatics to examine and improve health services.

Practitioner, Patient, and Consumer Behavior

Many HSR studies focus on the behaviors of practitioners, patients, and consumers, and the relationship between behaviors and the outcomes of care. The study of practitioner behavior includes identifying and understanding the impact of factors that influence the way providers make decisions. These factors can include training, experience, confidence level, peer pressure, patient preferences, financial incentives, and organization constraints. Strategies for changing practitioner behavior are also examined in this type of study. Investigation of patient behaviors includes treatment and medication adherence, preferences for types and delivery of services, and perceived barriers to receipt of health care services. The goal is to provide information to many types of consumers (e.g., patients, policy makers, payers, etc.) with scientifically based, understandable information that will guide decisions about many aspects of healthcare.

Clinical pharmacy has a long history of influencing provider behavior through medication formularies, clinical recommendations, drug utilization review, and provision of drug information. Pharmacists influence consumer behavior through patient education and clinical management strategies designed to optimize clinical outcomes. This experience provides a knowledge base that can be used to enhance HSR studies that seek to understand and influence patient, provider, and consumer behaviors. Pharmacists can also use the information gained by researchers in this area to improve the effectiveness of pharmacy practice.

Health Professions Work Force

HSR also asks questions about the education and supply of health care workers. For example, a project evaluating care for rural patients with cancer identified both a shortage of pharmacists to deliver pharmaceutical care and raised questions about how well pharmacy curricula prepare pharmacists to meet the needs of rural patients.[22] Other studies evaluate the number of professionals needed to optimize costs of care or evaluate the supply and demand of health care workers.[23–25] Unfortunately, these efforts have not proven particularly successful.[1,26–28] Given the increasing concern over shortages of pharmacists, it is important for the profession to work with health service researchers to develop models to better address this issue and to study the consequences of personnel shortages.[29–32]

FUNDING

Funding for HSR projects can be divided into four broad categories: 1) self-funding, 2) consultation, 3) contracts, and 4) grants.[33] Self-funding is fairly self-explanatory in that the project is conducted by using resources already available to the researcher. Data processing and library services are common examples of this type of resource. Some institutions may also allow junior researchers to use datasets already in existence at that institution. An important issue in self-funded projects is that the researchers time is already paid for, usually by an academic institution. Consultation provides funds that pay for the professional expertise of the researcher. The researcher combines this expertise with research activities to address the client's specific question or problem. Rand Corporation in Los Angeles, California, is one of the most prominent examples of this type of activity. Grants and

contracts provide funds to conduct a specific research proposal. Both may be funded by public or private institutions. The major difference between contracts and grants has to do with control of the project and the project funds. In a grant, the principal investigator generally controls all aspects of the project including study design, project implementation, disbursement of funds, and modifications to the original protocol. In a contract, the funding entity has much more input into all aspects of the project.

Among federal programs, the Agency for Healthcare Research and Quality (AHRQ), formerly the Agency for Healthcare Policy and Research (AHCPR), is the leader in research to improve quality of care. However, others such as the Government Accounting Office, Health Resources Service Administration, HCFA, Office of the Assistant Secretary for Planning and Evaluation, and Centers for Disease Control and Prevention also perform HSR. Individual institutes within the NIH also support services research. For example, there is a Services Research and Clinical Epidemiology Branch within the National Institute of Mental Health (NIMH) that funds services research.

The VHA has had a department of Health Services Research and Development (HSR&D) since 1976.[34] Since the mid-1980s, this department has emphasized use of health services research as a tool to improve the health of veterans. Funding from HSR&D is primarily for investigators in the Veterans Administration system, and consists of both investigator initiated research proposals and solicited proposals for specific research.

A number of large centers in the private sector are now offering grant funding for HSR. These institutions may also conduct HSR. These include the Center for Studying Health Systems Change, funded by the Robert Wood Johnson Foundation, and affiliated with Mathematica Policy Research Inc., RAND Health, the Health and Policy Center of the Urban Institute, and Emergency Care Research Institute. Key HSR academic centers include the Cecil G. Sheps Center For Health Services Research at the University of North Carolina at Chapel Hill, the Center for Health Services Research and Policy in the George Washington University Medical Center School of Public Health and Health Services, and Michigan Health Services Research at the University of Michigan.

There are also resources for identifying funding sources. The Federal Information Exchange provides an electronic service that uses e-mail to send grant information to researchers. The Directory of Research Grants and The Foundation Directory are also excellent sources for information regarding federal, state, and private funding institutions. A more complete listing is available in *Health Services Research Methods* written by Leiyu Shi.[33]

HSR PUBLICATIONS

The specific educational disciplines of health service researchers are reflected in the wide variety of journals that now publish health research manuscripts. Journals that are devoted entirely to HSR include *Medical Care*, published by the American Public Health Association; *Health Services Research*, the official journal of the Academy for Health Services Research and Health Policy (formerly known as the Association for Health Services Research); and the *Journal of Health Services Research and Policy*, published by the Royal Society of Medicine, Ltd. Many specialized journals are beginning to publish work based on health services research. For example, *Journal of Health Economics*, *American Journal of Public Health*, *American Journal of Medical Quality*, and *American Journal of Epidemiology* will accept manuscripts based on HSR projects. Prominent clinical journals such as the *New England Journal of Medicine* and *Journal of the American Medical Association* are also increasing the number of publications of HSR projects. Pharmacy journals also contain some HSR articles, including the *American Journal of Health-System Pharmacy*, *Annals of Pharmacotherapy*, and *Pharmacotherapy*.

REFERENCES

1. Committee on Health Services Research Training and Work Force Issues. *Health Services Research: Work Force and Educational Issues*; Field, M.J., Tranquada, R.E., Feasley, J.C., Eds.; National Academy Press: Washington, D.C., 1995.
2. Ginzberg, E. *Health Services Research: Key to Health Policy*; Harvard University Press: Cambridge, Massachusetts, 1991.
3. Mamdani, M.M.; Racine, E.; McCreadie, S.; Zimmerman, C.; O'Sullivan, T.L.; Jensen, G.; Ragatzki, P.; Stevenson, J.G. Clinical and economic effectiveness of an inpatient anticoagulation service. Pharmacotherapy **1999**, *19* (9), 1064–1074.
4. McMullin, S.T.; Hennenfent, J.A.; Ritchie, D.J.; Huey, W.Y.; Lonergan, T.P.; Schaiff, R.A.; Tonn, M.E.; Bailey, T.C. A prospective, randomized trial to access the cost impact of pharmacist-initiated interventions. Arch. Intern. Med. **1999**, *159* (19), 2306–2309.
5. Galt, K.A. Cost avoidance, acceptance, and outcomes associated with a pharmacotherapy consult clinic in a Vet-

erans Affairs Medical Center. Pharmacotherapy **1998**, *18* (5), 1103–1111.
6. Harrison, D.L.; Bootman, J.L.; Cox, E.R. Cost-effectiveness of consultant pharmacists in managing drug-related morbidity and mortality at nursing facilities. Am. J. Health-Syst. Pharm. **1998**, *55* (15), 1588–1594.
7. McCombs, J.S.; Liu, G.; Feng, W.; Cody, M.; Parker, J.P.; Nichol, M.B.; Hay, J.W.; Johnson, K.A.; Groshen, S.L.; Nye, M.T. The Kaiser Permanente/USC patient consultation study: Change in use and cost of health care services. Am. J. Health-Syst. Pharm. **1998**, *55*, 2485–2499.
8. Straub, L.A.; Straub, S.A. Consumer and provider evaluation of rural pharmacy services. J. Rural Health **1999**, *15* (4), 403–412.
9. Donabedian, A. *Explorations in Quality Assessment and Monitoring*; Health Administration Press: Ann Arbor, 1980.
10. Every, N.R.; Fihn, S.D.; Sales, A.E.; Keane, A.; Ritchie, J.R. Quality enhancement research initiative in ischemic heart disease: A quality initiative from the Department of Veterans Affairs QUERI IHD Executive Committee. Med. Care **2000**, *38* (6 Suppl. 1), I49–I59.
11. Rust, G.S.; Murray, V.; Octaviani, H.; Shcmidt, E.D.; Howard, J.P.; Anderson-Grant, V.; Willard-Jelks, K. Asthma care in community health centers: a study by the southeast regional clinicians' network. J. Natl. Med. Assoc. **1999**, *91* (7), 398–403.
12. Ellwood, P.M. Shattuck lecture—outcomes management: A technology of patient experience. N. Engl. J. Med. **1988**, *318* (23), 1549–1556.
13. Blumenschein, K.; Johannesson, M. Use of contingent valuation to place a monetary value on pharmacy services: An overview and review of the literature. Clin. Ther. **1999**, *21* (8), 1402–1417.
14. Dager, W.E.; Branch, J.M.; King, J.H.; White, R.H.; Quan, R.S.; Musallam, N.A.; Albertson, T.E. Optimization of inpatient warfarin therapy: Impact of daily consultation by a pharmacist-managed anticoagulation service. Ann. Pharmacother. **2000**, *34* (5), 567–572.
15. Self, T.H.; Brooks, J.B.; Lieberman, P.; Ryan, M.R. The value of demonstration and role of the pharmacist in teaching the correct use of pressurized bronchodilators. Can. Med. Assoc. J. **1983**, *128*, 129–131 (Jan. 15).
16. Hanlon, J.T.; Weinberger, M.; Samsa, G.P.; Schmader, K.E.; Uttech, K.M.; Lewis, I.K.; Cowper, P.A.; Landsman, P.B.; Cohen, H.J.; Feussner, J.R. A randomized, controlled trial of a clinical pharmacist intervention to improve inappropriate prescribing in elderly outpatients with polypharmacy. Am. J. Med. **1996**, *100*, 428–437 (April).
17. Tully, M.P.; Seston, E.M. Impact of pharmacist providing a prescription review and monitoring service in ambulatory care or community practice. Ann. Pharmacother. **2000**, *34*, 1320–1331.
18. Demakis, J.G.; Beauchamp, C.; Cull, W.L.; Denwood, R.; Eisen, S.A.; Lofgren, R.; Nichol, K.; Wooliscroft, J.; Henderson, W.G. Improving residents' compliance with standards of ambulatory care: Results from the VA Cooperative Study on computerized reminders. J. Am. Med. Assoc. **2000**, *284* (11), 1411–1416.
19. Monane, M.; Matthias, D.M.; Nagle, B.A.; Kelly, M.A. Improving prescribing patterns for the elderly through an online drug utilization review intervention: A system linking the physician, pharmacist, and computer. J. Am. Med. Assoc. **1998**, *280* (14), 1249–1252.
20. Hudson, T.J.; Owen, R.R.; Lancaster, A.E.; Mason, L. The feasibility of using automated data to assess guideline-concordant care for schizophrenia. J. Med. Syst. **1999**, *23* (4), 299–307.
21. Miller, L.G. Use of patient education and monitoring software in community pharmacies. J. Am. Pharm. Assoc. **1997**, *NS37* (5), 517–521.
22. Gangeness, D.E. Pharmaceutical care for rural patients: Ominous trends. J. Am. Pharm. Assoc. **1997**, *NS37* (1), 62–65, 84.
23. Bond, C.A.; Raehl, C.L.; Franke, T. Clinical pharmacy services, pharmacy staffing, and the total cost of care in United States hospitals. Pharmacotherapy **2000**, *20* (6), 609–621.
24. Institute of Medicine. *Summary. The Nation's Physician Workforce*; Lohr, K.N., Vanselow, N.A., Detmer, D.E., Eds; National Academy Press: 1996; 1–14.
25. Knapp, K.K.; Paavola, F.G.; Maine, L.L.; Sorofman, B.; Politzer, R.M. Availability of primary care providers and pharmacists in the United States. J. Am. Pharm. Assoc. **1999**, *39* (2), 127–135.
26. Feil, E.C.; Welch, H.G.; Fisher, E.S. Why estimates of physician supply and requirements disagree. J. Am. Med. Assoc. **1993**, *269* (20), 2659–2663.
27. Kindig, D.A. Counting generalist physicians. J. Am. Med. Assoc. **1994**, *271* (9), 1505–1507.
28. Capilouto, E.; Capilouto, M.L.; Ohsfeldt, R. A review of methods used to project future supply of dental personnel and the future demand and need for dental services. J. Dent. Educ. **1995**, *59* (1), 237–257.
29. Mott, D.A.; Kreling, D.H. An internal rate of return approach to investigate pharmacist supply in the United States. Health Econ. **1994**, *3* (6), 373–384.
30. Knapp, K.K. Building a pharmacy work force mosaic: New studies help to fill in the gaps. J. Am. Pharm. Assoc. **2000**, *40* (1), 13–14.
31. Magill-Lewis, J. Reverse trend. Drug Topics **2000**, *144*, 44 (Feb. 21).
32. Beavers, N. Feeling the weight. Drug Topics **2000**, *144* (Jan. 3), 38–40, 42, 47–48.
33. Shi, L. *Health Services Research Methods*; Delmar Publishers: 1997.
34. www.va.gov/resdev.

PHARMACY PRACTICE ISSUES

Health Status Assessment

Kathleen M. Bungay
The Health Institute, Boston, Massachusetts, U.S.A.

INTRODUCTION

For many pharmacists, their first encounter with the terminology ''quality of life'' was in the 1986 New England Journal of Medicine article by Croog et al.[1] entitled ''The effects of hypertensive therapy on the quality of life.'' The authors found that antihypertensive agents had different effects on the quality of life and that these differences can be meaningfully assessed with available psychosocial measures. Currently, the clinical community is more aware of patient-based measures and the potential uses of health status assessments. Curriculum of many schools of pharmacy now includes some information on outcomes of patient care beyond just the traditional biological measures.

This article discusses selected milestones in the evolution of health status assessments, the health status/quality of life conceptual framework(s), an introduction to the scientific basis and evaluation of patient health status self-assessment questions, and potential future research and application of health status measures to patient care, with special emphasis on its role in clinical pharmacy practice.

DISCUSSION

> The act of measurement is an essential component of scientific research, whether in the natural, social, or health sciences.[2]

Although measurement has always played an essential role in health sciences, measurement in laboratory disciplines rarely presented difficulty. As with other natural sciences, measurement was a fundamental part of the discipline and was approached through the development of appropriate instrumentation. Subjective judgment played a minor role in the measurement process; any issue of reproducibility or validity was therefore amenable to a technological solution.

Since the 1990s, the situation in clinical research has become more complex. The effects of new drugs or surgical procedures on quantity of life are likely to be marginal.[2] Conversely, there is an increased awareness of the impact of health care on the quality of human life. Therapeutic efforts in many disciplines of medicine, especially those increasing numbers who care for patients with chronic, long-term disease states, are directed equally if not primarily to improvement of the quality of life,[3] not the quantity of life.

With therapeutic efforts focusing more on improving patient function and well-being, the need increases to understand the relationships between traditional clinical and health-related quality of life (HRQOL[a]), especially because it is increasingly used as an outcome in clinical trials, effectiveness research, and research on the quality of care. Factors that have facilitated this increased usage include the accumulating evidence that measures of health status are valid and reliable. In an effort to promote a better understanding of linking clinical variables to HRQOL, Drs. Wilson and Cleary[4] published a valuable distinction between basic clinical medicine and social science approaches to patients' health. They also propose a model to link both, which is discussed later in this article.

> In the clinical paradigm, the ''biomedical'' model, the focus is on etiologic agents, pathological processes, and biological, physiological, and clinical outcomes. The principal goal is to understand causation to guide diagnosis and treatment. Controlled experiments are its principal methodology, and current biomedical research is directed at fundamental molecular, genetic, and cellular mechanisms of disease. Its intellectual roots are in biology, biochemistry, and physiology.
>
> In contrast, the social science paradigm, or quality of life model, focuses on dimensions of functioning and overall well-being, and current research examines ways to accu-

[a]Because quality of life represents the broadest range of human experiences, use of this general term in the health field has led to considerable confusion, particularly because of the overlap with the more specific concept, health status. To make the meaning more specific and retain the important aspects of life quality, the term ''health-related quality of life'' is both useful and important.

Encyclopedia of Clinical Pharmacy
DOI: 10.1081/E-ECP 120006248

rately measure complex behaviors and feelings. Experimental research designs are rarely possible[5] because the focus of social science is on the way that numerous social structures and institutions influence individuals. These models have their foundation in sociology, psychology, and economics, and use concepts and methods often foreign to clinicians and clinical researchers.[4]

EVOLUTION OF HEALTH STATUS OUTCOME MEASURES

During the 1940s, physicians first began to measure patient functioning; the Karnofsky Functional Status for Patients with Cancer[6] and the New York Heart Association Classification[7] were among the instruments developed during that period. The first health status measures distinguished among functional states and included symptoms, anatomic findings, occupational status, and daily living activities. Studies began in the 1950s when clinicians examined the functional status of patients with severe disabilities. When social science methods and clinical expertise came together in the 1970s, the first modern health status questionnaires emerged. Typical measures of this period include the Quality of Well Being Scale,[8] the Sickness Impact Profile,[9] the Health Perceptions Questionnaire,[10] and the OARS[11] for use in health services and clinical research as outcome measures. The next generation of measures developed in the 1980s and 1990s were the Health Insurance Experiment (HIE) health surveys,[12] the Duke–UNC Health Profiles,[13] the Nottingham Health Profile,[14] and the Medical Outcomes Study health surveys,[15] including the SF-36 Health Survey.[16]

For a more detailed discussion of the history and development of health status assessment, see Refs. [17–19]. Also, for a more exhaustive list of questionnaires, readers are directed to Spilker.[17,18,20–22]

Variations in Medical Care in Small Areas

The impetus for research on rationality of processes in health care delivery, an issue that the field of outcomes research and guidelines development are meant to address, is typically traced to the work of John Wennberg,[23] who uncovered a phenomenon known as small area variation. In brief, Wennberg and colleagues noticed large disparities in the rates of various medical procedures in different geographic areas. The differences could not be attributed to differences in the populations, but instead appeared to indicate differences in physician cultures of different regions, where certain treatment strategies became the norm. For example, a 10-fold difference in rates of tonsillectomy was observed just within the six New England states.

The Rand HIE

In 1990, when it became apparent in the United States that health expenditures accounted for 12.4% of the gross national product, whereas that proportion was 4% in 1980 and that the rate of growth of health care expenditures was exceeding the rate of inflation as well as growth in our economy,[25] questions surfaced. Does spending more buy better health? In individual cases, the answer may be an obvious yes or no, but in the population as a whole as of 1983, the point of diminishing (or absent) returns was difficult to identify.[12] This quandary prompted the federal government to support a large-scale controlled trial, now known as the Rand HIE.[24]

One purpose of the HIE was to learn whether the direct cost of medical care, when borne by consumers, affects their health. First, the researchers found that the more people had to pay for medical care, the less of it they used. Free care had no effect on major health habits associated with cardiovascular disease and some types of cancer. Second, the study detected no effects of free care for the average enrollee on any of the five general self-assessed health measures.

In addition to these remarkable findings, the HIE presented one of the first major challenges for measuring health status. A consequence of this challenge resulted in one of the most extensive applications of psychometric theory and methods (long used in educational testing) to the development and refinement of health status surveys. Researchers developed or adapted measures to evaluate the effect of cost sharing on health status. At that time, the comprehensive set included four distinct categories—general health, health habits, physiological health, and the risk of dying from any cause related to risk factors. General health was operationally defined as physical functioning, role functioning, mental health, social contacts, and health perceptions.[24]

The measurement goal in the HIE was to construct the best possible scales for measuring a broad array of functioning and well-being concepts; it demonstrated the potential of scales, constructed from self-administered surveys, as reliable and valid tools for assessing changes in health status. It, however, left two questions unanswered: Can methods of data collection and scale construction work in sick and elderly populations? In addition, could scales that are more efficient be constructed? The answer to these questions was the challenge

accepted by the Medical Outcomes Study (MOS) investigators.[26]

MOS

The MOS was a 2-year observational study designed to help understand how specific components of the health care system affected the outcomes of care. One of the two original purposes of the MOS was to develop more practical tools for monitoring patient outcomes, and their determinants, in routine practice using state-of-the-art psychometric techniques. The study and its many implications and conclusions are discussed in detail elsewhere[15] and mentioned here for completeness.

Agency for Health Care Policy and Research (AHCPR)/Agency for Healthcare Research and Quality (AHRQ)

To enhance the quality, appropriateness, and effectiveness of health care services, and access to these services the federal government in the Omnibus Budget Reconciliation Act of 1989 (Public Law 101-239)[27] established the AHCPR. The act, sometimes referred to as the Patient Outcome Research Act,[28] called for the establishment of a broad-based, patient-centered outcomes research program. In addition to the traditional measures of survival, clinical endpoints and disease- and treatment-specific symptoms and problems, the law mandated measures of "functional status and well-being and patient satisfaction." In 1999, then President Clinton signed the Healthcare Research and Quality Act, reauthorizing AHCPR as the AHRQ until the end of fiscal year 2005. Presently, its mission is to improve the outcomes and quality of health care, reduce its costs, address patient safety and medical errors, broaden access to effective services, and improve the quality of health care services.

Summary

Now that we briefly reviewed the history of some of the origins of health status assessment research and a few of the important stimuli, we proceed with a brief discussion of some of the underlying theories and assumptions.

The design of health surveys, consisting of scales measuring attributes of a person or a population's health, are supported by underlying theory known as psychometric theory.[29] Health status scales development can also be viewed as a unique application of the design and theory that support the creation of educational measurements (e.g., Standardized Achievement Tests). A person who studies these theories and conducts research or measurement of such attributes as intelligence, pain, mental well-being, or functioning is usually a doctorate-level research psychologist and can be known as a psychometrician.

PSYCHOMETRIC THEORY

Any type of measurement can be boiled down to two fairly simple concepts: "measurement" consists of rules for assigning numbers to objects so as to 1) represent quantities of attributes numerically (scaling) or 2) define whether objects fall into categories with respect to a given attribute (classification). The objects from a psychological perspective are usually people. The rules indicate that the assignment of numbers must be explicitly stated. The term "attribute" in the definition indicates that measurement always concerns some particular feature of the objects.

Scales can be created based on a number of different theories or models. Three commonly referenced scales are the Guttman scale, Thurstone scale, and Likert-type scale. Developing questions and scales using any of these theories requires some assumptions be made. To reduce error, one measures the extent that the assumptions[30] are met. For example, with the Likert-type scales, one needs to test that summated rating assumptions are met, or that the scale achieves maximum reliability and validity with a minimum number of questions.[31] Other examples of assumptions are that each item can discriminate itself from a different concept (measured by a different scale) and that its properties converge with other like scale items with its own concept. One might also address the reliability of the scale scores and the features of the scale distributions. For a much more extensive discussion, see Nunnaly (1994)[29] and for examples see papers written by Bayliss et al.,[30] Mc Horney et al.,[32] and Wagner.[33]

Basic Concepts of Measurement

One readily apparent feature of health sciences literature devoted to measuring health status is the daunting array of already available scales. Paradoxically, if you proceed a little further to find an instrument for your intended purpose, you may conclude that none of the existing scales is quite right. Many researchers tend to magnify the deficiencies of existing measures and underestimate the effort required to develop an adequate new measure. Perhaps the most common error committed by clinical researchers is to dismiss the existing scales too lightly, and embark on the development of a new instrument with an unjustifiably optimistic and naïve expectation that they can do better. The development of scales requires considerable investment of both mental and fiscal resources. A comprehensive set of standards, widely used in the assessment of psychology and education, is the manual called Standards for Educational and Psychological Tests, published by the American Psychological Association (1974).[34] In addition to these standards, there are a number of compendia of measuring scales.[2]

Face and Content Validity

These terms are technical descriptions of the judgment that a scale looks reasonable. Face validity simply indicates whether, on the face of it, the instrument appears to be assessing the desired qualities. Content validity is a closely related concept, consisting of a judgment as to whether the instrument samples all the relevant or important content of domains. Nevertheless, a researcher should be cautions not to dismiss existing measures based on a judgment of face validity—for example, if they did not like some of the questions or the scale was too long. This judgment of face and content validity comprises only one of several used to decide on the usefulness, and will need to be balanced with other evaluations of the measure.

Reliability[b]

On the surface, the concept of reliability is deceptively simple. Before one can obtain evidence that an instrument is measuring what it is intended (validity); it is first necessary to gather evidence that the scale is measuring something in a reproducible fashion. The basic idea behind the concept is an index of the extent to which measurements of individuals obtained under different circumstances yield similar results.

Validity

Reliability assesses that a test is measuring something reproducibly; it says nothing about what is being measured; it is necessary, but it is not sufficient, to establish the usefulness of measures. To determine that the test is measuring what was intended requires some evidence of validity. Many variables in health sciences are physical quantities, such as height, serum cholesterol level, or potassium. As such, they are readily observable, either directly or with the correct instruments. The situation is different when it changes to range of motion or responsibility of a physician.

Presently, validity is represented as a process whereby we determine the degree of confidence we can place on inferences we make about people based on their scores from that scale. Some different types of validity, a discussion of which is beyond the scope of this chapter, are called content validity, criterion validity, and construct validity.

DEFINITIONS

Health

Defining health is vital to developing a strategy for measuring it. Concepts of health[35] can lack clarity yet commonly hold their dimensionality as a fundamental feature. Terms used to define health include positive states—wellness and normal—and negative states—disability and illness.[35] Clues to what dimensions comprise health are found in the definition of health offered by the World Health Organization (WHO). The WHO defines health as a ''state of complete physical, mental and social well-being and not merely the absence of disease or infirmity.''[36] Dictionaries also identify both physical and mental dimensions of health. Two features of these definitions are crucial; namely, the many dimensions of health and the range of health states from disease to well-being.

Quality of Life

Quality of life is a global concept with many meanings. It is generally advisable to understand the domains included when the term is used. Quality of life, it has been suggested, involves highly subjective value judgments and is equated with ''profound satisfactions from the activities of daily life.''[37] Research and measurement of quality of life have encompassed both objective and subjective indicators involving a wide array of experiences, states, and perceptions. Cultural, psychological, interpersonal, spiritual, financial, political, temporal, and philosophical dimensions may be incorporated into various definitions.[35] In 1981, Campbell[38] defined 12 dimensions or domains of quality of life: community, education, family life, friendships, health, housing, marriage, nation, neighborhood, self, standard of living, and work. Health is but one domain or one aspect of life or the quality of one's life.

Health-Related Quality of Life

Because quality of life represents the broadest range of human experiences, use of this general term in the health

[b]For more in-depth information on this topic, the reader is directed to Refs. [2] and [29].

field has led to considerable confusion, particularly because of the overlap with the more specific concept, health status. To make the meaning more specific and to retain the important aspects of life quality, HRQOL is both a useful and important term.

MEASURING HEALTH

Conceptual Framework/Models

Researchers have proposed a number of conceptual models of the relationships among the components of HRQOL.[15,16,39–44] Wilson and Cleary, who proposed a model linking clinical variables with HRQOL, argued that "the ultimate promise of the ability to measure HRQOL will not be fulfilled until it has clear applications to clinical care."[4] Their pursuit of this goal sets their model apart from previously published models. Their model includes five levels or subdivisions: biological and physiological variables, symptom status, functional status, general health perceptions, and overall quality of life (Fig. 1).

A comparison of different conceptual models is beyond the scope of this chapter. Because the conceptual model informs the measurement, each may be slightly different although some commonly agreed upon and frequently measured general health concepts can be identified and discussed. These concepts are: 1) physical functioning, 2) mental functioning, 3) social and role functioning, and 4) general health perceptions. By denoting a measure as a general health status measure, it is understood that the questions are not disease or disorder specific, and that they cover a range of health states from life-threatening conditions to an overall sense of well-being. General measures evaluate aspects of health relevant to all ages, races, genders, and socioeconomic backgrounds. The measures also permit the examination of the benefits of treatments in comparable units.

Using general health measures, Stewart et al.[45] compared the functional status and well-being of patients with chronic conditions. They reported the usefulness of generic (non-disease-specific) health measures for monitoring progress and for use as outcomes in studies of patients with chronic conditions. The authors maintained that there are several advantages of general measures of functional status and well-being over disease-specific measures. Among these, they noted, first, they are useful for monitoring patients with more than one condition, and second, for comparing patients with different conditions by providing a common yardstick. Last, the same measures can be appropriately applied to both general (well) and patient (sick) populations, with the advantage of comparing patient groups (sicker) with the healthy standard of a general population (Fig. 2).

Commonly Measured Domains of Health: General Health Status Assessment

Physical function

Physical health is commonly measured in terms of limitations in the performance of or ability to perform self-care activities (e.g., eating, bathing, dressing), mobility, moderate or more strenuous physical activities, and bodily pain. Responses to questionnaire items in this category

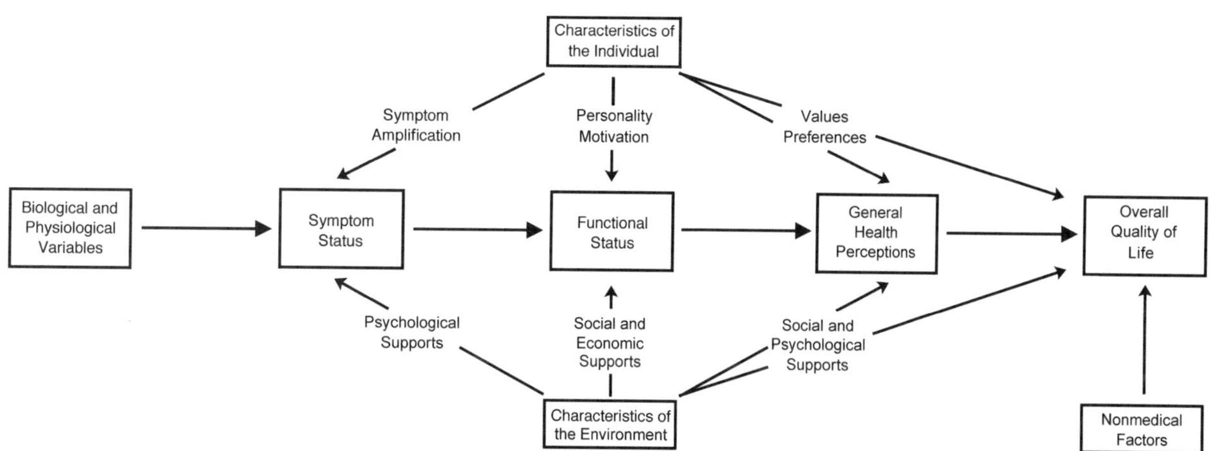

Fig. 1 Relationships among measures of patient outcome in a HRQOL conceptual model. (From Ref. [4].)

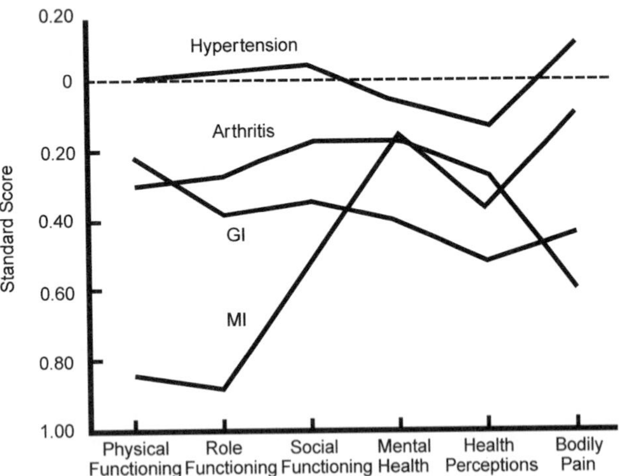

Fig. 2 Health profiles for patients with four conditions. Dotted line indicates patients with no chronic conditions. GI, gastrointestinal disorder; MI, myocardial infarction. (From Ref. [45].)

generally focus on limitations due to physical health as opposed to some other cause.

Assessments of physical health often vary in the range of functioning. Questionnaires that assess self-care often concentrate on the negative end of the continuum of physical functioning; some determine different physical health states, and others determine individual differences in the level of effort, pain, difficulty, or need for assistance in performing physical activities.[46,47] Thus, individual patient responses to question may be identical or may different substantially from one test to another. To be comprehensive as measures of physical health status for a population, measures for general use are recommended[46] to elicit information concerning activities of daily living, energy level, satisfaction with physical shape or condition, and ability to perform vigorous activities. Questions in those areas are sometimes phrased positively to extend the measurement scale into the positive range (how we feel as opposed to the limitations we experience). This allows measurement only of differences in physical well-being among those free of limitations.

Mental health

These measures often focus on the frequency and intensity of psychological distress (e.g., anxiety or depression), and include the individual's perception of psychological well-being and life satisfaction and an assessment of cognitive functioning.[48] Measures of this domain also cover the broad range of differences possible in the mental health continuum. Health disturbances commonly manifest themselves on behavioral and physical levels; however, mental health can change long before observable changes in behavior. Furthermore, clinical and social changes in mental health do not always manifest as distress or cognitive dysfunction. Disease or illness may cause a loss of zest for life or the feeling that life is less enjoyable. Capturing such a change requires the presence of questions that assess psychological well-being; therefore, general measures should encompass the full range of states in the continuum.[49]

Social and role functioning

Functioning in interpersonal relationships and in role and other daily activities are commonly lumped together as measures of social functioning. However, for evaluating HRQOL it is better to consider them separately.

Social functioning is defined as the ability to develop, maintain, and nurture mature social relationships. Our group has concluded that measures of social functioning 1) reflect physical and mental health status, 2) serve to indicate the need for health care, and 3) reflect appropriate outcomes of health care.

Usually, social well-being is separated into two areas: 1) whether and with what frequency social contacts are occurring and 2) the nature of the person's social network or community. The frequency and number of contacts, as well as personal satisfaction with those contacts, vary a great deal amount individuals. Depending on the person, quantitative data may offer no insight or may offer the wrong insight. A person's evaluation of the adequacy of the social network to which he or she belongs may be more valuable. People who consider themselves part of a community, family, or neighborhood often have a strong sense of being wanted, loved, or valued.[50] Measures of personal resources often overlap the mental dimension of health. The feeling that one is loved or cared for may assess mental health even more than it measures social well-being. Some research reviewed by Wortman[51] suggested that social circumstances are linked to both physical and mental health outcomes.[52]

Role functioning describes whether a person can meet the demands of their normal role in life (e.g., formal employment, schoolwork, housework). Persons not working in outcomes research often use the terms role function and social function interchangeably; however, in terms of measuring HRQOL, they are distinct. A role function measure seeks to identify situations in which an individual's health problem directly interferes with the performance of their everyday role, in contrast to participation in the social interaction and network to which they belong. For example, arthritis strikes a professional organist who has a loving and supportive family and a

network of friends. He may not feel isolated or unloved, and his relationships with his wife and children continue to be positive. However, to the man for whom music is fuel for mental, financial, and spiritual well-being, the loss of his professional role may be devastating. That devastation is a role function loss. If he then lost the friendships he developed and maintained through his music, then that would be loss of social function.

For many people, physical health problems limit role performance. Occasionally, mental disruption can impinge on role functioning, but measures of this health dimension seldom detect mental or emotional problems because patients seldom consider role limitations unless they are asked about them explicitly. Some questionnaires ask specifically about limitations of role function due specifically to personal or emotional problems, in addition to those due to physical health problems.

Measurement of the impact of health on role activities has grown, owing in part to legislation and information provided by the passage of the Americans with Disabilities Act.[53] Approximately 55 million working-age individuals (18–65 years of age) have chronic illnesses and/or impairments. Disabilities are a potential consequence of health problems and signify a partial or total inability to perform social roles in a manner consistent with norms or expectations.[53] National survey data suggest that 32% of employed adults have ongoing health problems that interfere with their ability to perform their job demands.[54] Historically, although limited, studies have used outcomes of "work loss" or amount of time missed from work due to illness or treatment.

Role functioning scales usually measure global, role-level disability indicators to capture disability in paid work and/or other activities. However, for some applications, these may be relatively course, distinguishing a limited range of disability levels. Recently, researchers introduced a more detailed measure of work productivity assessing on-the-job impact of chronic conditions and treatment.[54]

General health perceptions

The beliefs and evaluations of a person's over health in general, rather than a particular mental or physical aspect, constitute their general health perceptions. Questions in this area reflect each person's own health preferences, values, needs, and attitudes, and thereby discriminate between individuals whose objective levels of physical and mental health, as defined by other measures, appear identical. Such self-perceptions are important for two reasons: 1) reports of behavioral performance do not capture important subject manifestations of differences in health such as pain, difficulty, level of effort required, or worry and concern about health; and 2) questions in this domain inquire about positive feelings or a positive frame of reference, for example, a favorable health outlook in contrast to questions from other domains that focus on measures of limitation, pain, or dysfunction, and are usually stated in a negative way. Responses in the general health perception domain are subjective and evaluative. They are typically ratings rather than reports, for example, a rating of health from "excellent" to "poor."

Disease-Specific Health Status Instruments

Often, it is necessary to focus on the particular impact that a certain disease has on patients. In such cases, general health status tools are inadequate for providing the information needed. To overcome this limitation, condition- or disease-specific measures are often used instead of or along with a general health status instrument. The more narrowly focused disease-specific measure requests detailed information on the patient's perspective on the impact of a disease and its treatment. In addition, using disease-specific measures allows inclusion of domains of specific interest for the disease under study and the patients it affects. Among the specific areas previously investigated with disease-specific questionnaires are sexual and emotional functioning, nausea and vomiting, chronic pain, anxiety and depression, chronic obstructive pulmonary disease, cardiovascular disorders, hypertension, epilepsy, benign prostatic hypertrophy, end-stage renal disease, diabetes, cancers, AIDS, HIV infection, and migraine.[20,55–57] The reader is directed to Refs. [20] and [22] for a more comprehensive listing of disease-specific questionnaires.

The argument in favor of disease-specific questionnaires is twofold. The first consideration is that, if an instrument has to cover a wide range of disorders, many of the questions may be inappropriate or irrelevant for any one specific problem. The second reason is to keep the length of the questionnaire manageable. Thus, there will be fewer, relevant questions to detect real changes within patients or to detect differences among them.

On the opposite side of the argument, the cost of a greater degree of specificity is a reduction in generalizability.[58,59] That is, generic scales allow comparisons across groups of patients with different disorders, severities of disease, interventions, and perhaps even demographic and cultural groups,[60] as well as being able to measure the burden of illness of populations suffering from chronic medical and psychiatric conditions,[32] as compared with a healthy population. This is much harder to do when each study uses a different scale.

Furthermore, because any one generic scale tends to be used more frequently than a given disease-specific instrument, there are usually more data available regarding its reliability and validity.

THE INTERNATIONAL QUALITY OF LIFE ASSESSMENT (IQOLA) PROJECT

As the integration of patient-based outcome measures into all sectors of health care expands, the need arises for instruments capable of capturing data across cultures. In recent years, a rapid increase in the number of available translations of both generic and condition-specific instruments has occurred throughout the world.[61] The rise in demand for translated instruments is partially driven from the need to aggregate data from two or more cultures in clinical trials.

The IQOLA Project is translating, validating, and preparing norms for the SF-36 Health Survey for use in multinational clinical trials and for other international studies.[62–66] Based at the Health Assessment Laboratory at New England Medical Center, the project began in 1991, with sponsored investigators from 14 countries.[c] In addition, researchers from more than 30 other countries are translating and validating the SF-36 using IQOLA Project methods.[d]

The general process of translating an instrument is very complex, and is oversimplified here to give the reader a brief introduction only. The instrument is translated from the source (original) language to the target language (forward translation). Several forward translations are conducted by translators residing in the target country who are familiar with the tenets of the field of health outcomes. Then a consensus meeting with experts is convened to evaluate the efforts. A quality control step exists to ensure that the target version is equivalent to the source version, both conceptually and linguistically. This usually includes a backward translation from the target language to the source (original) language. The instrument is then pretested, which marks the final stages of the translation process.

[c]Australia, Belgium, Canada, Denmark, France, Germany, Italy, Japan, The Netherlands, Norway, Spain, Sweden, United Kingdom (English version), and United States (English and Spanish versions).

[d]Argentina, Bangladesh, Brazil, Bulgaria, Cambodia, China, Croatia, Czech Republic, Estonia, Finland, Greece, Hong Kong, Hungary, Iceland, Indonesia, Israel, Korea, Mexico, New Zealand, Poland, Portugal, Romania, Russia, Singapore, Slovak Republic, South Africa, Taiwan, Tanzania, Turkey, United Kingdom (Welsh), United States (Chinese, Japanese, Vietnamese), and Yugoslavia.

USING HRQOL ASSESSMENTS IN CLINICAL PRACTICE

When you ask a patient or any person, ''how are you?'', what type of information do you expect in response? Do you direct the patient in what units to answer? For example, how are you? . . . I am fine. If prompted, the respondent could produce a rating of how they think they are doing . . . ''On a scale of 1 to 10, I am a 2.'' Likewise, the answer could include a reference to the time span on which they are reflecting when answering your question. Such as, ''At this moment, I am just fine, in general my life is a bit unsettled.''

This example underscores and oversimplifies developers' thought processes when developing items to measure the domains of health. A measurement strategy can be defined to obtain as little or as much information. Ultimately, the amount of detail in the answers depends on what one plans to do with the information. The first, ''I am fine,'' is a global assessment. The second ''I am a 2,'' is an example of a rating scale. The third gives you and example of what one might call a recall period, or more practically, what time frame do you want the information from, yesterday, in the past 4 weeks, or in the last year. To obtain breadth in the answer to your question, you need to identify the potential extent of the answer, such as ''including both the physical and mental dimensions of health.'' Last, how much depth do you want in the reply? Parameters such as rating, breadth, depth, quantity, and frequency are the details with which measurement experts' struggle when developing patient self-administered health status questions.

Fig. 3 is an example of a mental health status subscale from the SF-36 Health Survey,[4,16,67] a popular general health status measure or instrument. The subscale is commonly referred to as the Mental Health Inventory, or the MHI-5. The five questions are each called an item stem. The balance of the item consists of the response choices; which are designed to be the same for many different items, thus making it less burdensome to the respondent.

The respondent is instructed to answer each question ''about the past 4 weeks'' (the recall period) and indicate ''how much of the time'' (quantity in amount of time). The response choices are: 1) all of the time, 2) most of the time, 3) a good bit of the time, 4) some of the time, 5) a little bit of the time, and 6) none of the time. Each of the patients' answers to the five questions are assigned a number between 1 and 6, summed and averaged, and then converted to a score between 0 and 100, with 100 being best health and 0 being worst.

Some of the questions require an endorsement such as how often have you been a happy person. In that case, the

These questions are about how you feel and how things have been with you during the past 4 weeks. For each question, please give the one answer that comes closest to the way you have been feeling. How much of the time during the past 4 weeks?

	[1] All of the time	[2] Most of the time	[3] A good bit of the time	[4] Some of the time	[5] A little of the time	[6] None of the time
Have you been a very nervous person?	☐	☐	☐	☐	☐	☐
Have you felt so down in the dumps that nothing could cheer you up?	☐	☐	☐	☐	☐	☐
Have you felt calm and peaceful?	☐	☐	☐	☐	☐	☐
Have you felt downhearted and blue?	☐	☐	☐	☐	☐	☐
Have you been a happy person?	☐	☐	☐	☐	☐	☐

Fig. 3 Example of a mental health scale from the SF-36. (From Ref. [5].)

response "all of the time" indicates better health. However, other items are stated so that "none of the time" indicates a better health. Such items are reversed before scoring. An example is "how often have you felt so down in the dumps that nothing could cheer you." Persons answering "none of the time" need to be assigned a scoring indicating better health, or the opposite direction of the endorsed items.

Advantage of Health Status Assessment Information to a Health Care Professional

Self-administered surveys allow the patient to have a voice in their care. It permits the patient to communicate to the health care professionals who are caring for them about what matters most. This may be information that you need to know but do not have time to elicit. Analogously to providing a common language for patients and health professionals, the general HRQOL information can also provide a standard or common language for different disciplines of health care professionals. For example, a nephrologist and a psychiatrist can use a common metric to discuss a dialysis patient's emotional health. A standardized method of asking patients about their functioning and well-being can be efficiently used in treatment decisions and as a monitoring parameter for efficacy and toxicity. The information may also be a tool or indicator for compliance assessments.

In addition, HRQOL can be used to add important information to the evaluation of the effectiveness of an intervention, in terms that matter most to the patient. For example, does the 34-year-old otherwise healthy woman diagnosed with depression who just started an antidepressant feel better or worse? One could just simply ask her that question when you see her 4 weeks after the start of her therapy. As pharmacists, we commonly use the question, "Are you having any side effects?" If the patient tells you she has diarrhea, you may form an impression of that diarrhea—seems like a mild side effect. However, having her answer survey questions about her functioning can reveal how trivial or nontrivial the impact of her diarrhea is to her everyday activities. What would happen if her diarrhea limits her ability to function as the checkout person in the grocery store? She cannot leave her post frequently to go to the bathroom and, if she does, she could be fired and not be able to provide for her two young children that she is raising alone. The patient sees the limitation imposed by diarrhea as considerable, and knowing more about her functioning conveys a different message to us than just knowing she is having diarrhea. A discussion employing information from a patient self-administered health status survey could also lead to the patient revealing that she has decided to stop taking her medication. She did not think it was working and the diarrhea was not worth the hassle.

Pharmacists Use of Health Status Assessment Information

As pharmacists, we can use evidence from patient self-administered health status surveys in caring for patients.[68] A common model used in teaching students to monitor therapy is to first create a problem list and, for every problem on the list, develop an assessment and plan. The diagram in Fig. 4 breaks down the assessment process. It requires one to write a potential inventory of all monitoring parameters. It reminds and guides us to monitor both the efficacy and the toxicity using subjective and objective parameters appropriate for the disease and the treatment.

We can easily incorporate the information from health status surveys in any of these boxes. Examples are bolded in Fig. 4. Now, instead of just monitoring clinical parameters of efficacy and toxicity, we can extend our

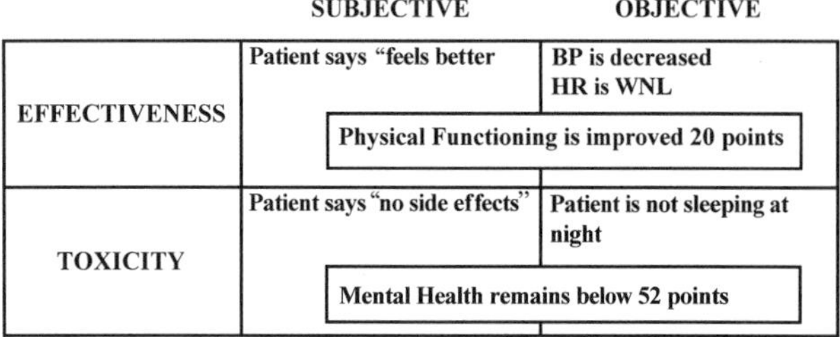

Fig. 4 Pharmacist assessment and monitoring therapy with health status assessment information included into typical paradigm.

monitoring parameters to include how a patient is functioning and feeling. For instance, is the patient able to go to the store and buy food? Do the patient's medications allow them or prevent them from socializing? Is their role in life supported by the therapy, or is it harder to do what they normally consider their job, whether home with kids or in the office.

Imagine that you get a number back just as you would a laboratory test from a scored survey. Where would you place it in the existing paradigm? What will do with it? If the patient's physical functioning after cardiac surgery and a new medication regimen is up 20 points, that could be a quantifiable part of your subjective information, or it could be considered in the objective category of effectiveness. However, if that same person still had a mental health score below 52, thus indicating a high probability of depression, then that could be reported as toxicity. Although it could be a consequence of the seriousness of his treatment (45-year-old man who just suffered a myocardial infarction), it could also be related to his medications. This is just a start; there is much more to be done to develop the use of these measures in the care of individual patients. However, it seems that the information on patients' functioning and well-being at a minimum can help pharmacists to better assess compliance and reasons for noncompliance. It also presents an opportunity to be better informed about the patient and tailor education strategies to fit the individual.

Controversies in Using Health Status Assessments for Individual Patient Care Decisions

Standardized measures capturing patient perspectives on their physical functioning, social and role functioning, mental health, and general health perceptions are likely to become more acceptable as an additional piece of evidence on which providers and their patients can make decisions about treatment and the treatment's efficacy. Mature theoretical models,[4,69] sophisticated measurement techniques,[70,71] and enhanced technology for use in measurement make the routine use of individual patient results in their own care more promising than ever before.

Two practical concerns of the critics of use of HRQOL assessments in individual patient care are: 1) respondent burden and 2) reliability of scores obtained from shorter questionnaires. Current researchers struggle with the competing demands invoked by everyday use requiring shorter forms and the reliability of a result obtained from fewer questions. Specifically, concerns are raised about the reliability of the result and the interpretation. With popular outcomes measures, the standard error around a single person estimate is large and not satisfying enough to ensure stable conclusions.

Modern test theory offers the potential for individualized, comparable assessments for the careful examination and application of different health status measures.[69] One such theory is item response theory (IRT). Researchers report that IRT has a number of potential advantages over the currently employed classical test theory in assessing self-reported health outcomes. Applications of the IRT models are ideally suited for implementing computer adaptive testing.[55] IRT methods are also reported to be helpful in developing better health outcome measures and in assessing change over time.[70]

Patients increasingly have more access to computer technology. It is becoming more practical to employ assessments using a computer. Patients answering questions about a health status concept using dynamic assessment technology are requested only to complete the number of questions needed (minimizes response burden) to establish a reliable estimate. The resulting scores for an individual are estimated to meet the clinical measures of precision.

CONCLUSION

The study of HRQOL requires a multidimensional approach. Assessments must include components that evaluate, at a minimum, the health concepts of physical functioning, social and role functioning, mental health, and perception of general health. In addition, the full continuum of these concepts must be included, from the most limited to the healthiest. Approaches to capture HRQOL data include the self-administered questionnaire, personal interview, telephone interview, observation, and postal survey. The assessment instruments must possess acceptable reliability, validity, and sensitivity, and the investigators and the participants must accept them. Psychometrics

is an essential part of HRQOL research, especially in today's research environment that requires shorter, more focused measures.

Existing health outcomes measures drawn from classic test theory and emerging approaches based on item response theory offer exciting opportunities for appreciably expanded applications in biomedical and health services research, clinical practice and decision making, and policy development. The research agenda of measurement scientists includes challenges to: 1) refine and expand measurement techniques that rely on IRT; 2) improve measurement tools to make them more culturally appropriate for diverse populations, and more conceptually and psychometrically equivalent across such groups; 3) address long-standing issues in preference- and utility-based approaches, particularly in the elicitation of preference responses and scoring instruments; and 4) enhance the ways in which data from outcomes measurement tools are calibrated against commonly understood clinical and lay metrics, are interpreted, and are made useable for different decision makers.[19]

With the advances in measurement that promise to continue, knowledgeable clinicians will become the transportation for these measures to inclusion in patient care. Interpretation, it is suggested, is partially an issue of familiarity and repeated applications of the measures would lead to a better understanding. Ideally, a better understanding of what a patient tells their provider about their health status can be used for decision making that requires the patient to more actively and routinely participate in their own care.

REFERENCES

1. Croog, S.H.; Levine, S.; Testa, M.; Brown, B.; Bulpitt, C.J.; Jenkins, C.D.; Klerman, G.L.; Williams, G.H. The effects of antihypertensive therapy on the quality of life. N. Engl. J. Med. **1986**, *314*, 1657–1664.
2. *Health Measurement Scales: A Practical Guide to Their Development and Use*; Streiner, D.L., Norman, G.R., Eds.; Oxford Medical Publications, 1999.
3. Ware, J.E. *Measuring Functioning, Well-Being, and Other Generic Health Concepts*; Osoba, D., Ed.; CRC Press: Boca Raton, Florida, 1991; 7–23.
4. Wilson, I.B.; Cleary, P.D. Linking clinical variables with health-related quality of life. A conceptual model of patient outcomes. JAMA, J. Am. Med. Assoc. **1995**, *273*, 59–65.
5. Cook, C. *Quasi-Experiments: Nonequivalent Control Group Designs*; Fankhauser, G., Ed.; Rand McNally: Chicago, 1979; 137–146.
6. Karnofsky, D.; Burchenal, J. The Clinical Evaluation of Chemotherapeutic Agents in Cancer. In *Evaluation of Chemotherapeutic Agents*; Ferrer, M.I., Macleod, C., Eds.; Columbia Press: New York, 1949.
7. Criteria Committee, N.Y.H.A. *Nomenclature and Criteria for Diagnosis of Diseases of the Heart and Great Vessels*; Little Brown and Company: Boston, 1979.
8. Fanshel, S.; Bush, J. A health-status index and its application to health-services. Operations Res. **1970**, *18*, 1021–1066.
9. Bergner, M.; Bobbitt, R.; Kressel, S.; Pollard, W.; Gilson, B.; Morris, J. The sickness impact profile: Conceptual formulation and methodology for the development of a health status measure. Int. J. Health Serv. **1976**, *6*, 393–415.
10. Ware, J. Scales for measuring general health perceptions. Health Serv. Res. **1976**, *11*, 369–415.
11. *Multidimensional Functional Assessment: The OARS Methodology*; Pfeiffer, E., Ed.; Duke University Press, Center for the Study of Aging and Human Development: Durham, North Carolina, 1975.
12. Brook, R.H.; Ware, J.E., Jr.; Rogers, W.H.; Keeler, E.B.; Davies, A.R.; Donald, C.A.; Goldberg, G.A.; Lohr, K.N.; Masthay, P.C.; Newhouse, J.P. Does free care improve adults' health? Results from a randomized controlled trial. N. Engl. J. Med. **1983**, *309*, 1426–1434.
13. Parkerson, G.; Gehlback, S.; Wagner, E.; James, S.; Clapp, N.; Muhlbaier, L. The Duke–UNC health profile: An adult health status instrument for primary care. Med. Care **1981**, *19*, 806–828.
14. Hunt, S.M.; McKenna, S.P.; McEwen, J.; Williams, J.; Papp, E. The Nottingham health profile: Subjective health status and medical consultations. Soc. Sci. Med. **1981**, *15A*, 221–229.
15. *Measuring Functioning and Well-Being: The Medical Outcomes Study Approach*; Stewart, A.L., Ware, J.E., Eds.; Duke University Press: Durham, 1992; ix-447.
16. Ware, J.E., Jr.; Sherbourne, C.D. The MOS 36-item short-form health survey (SF-36). I. Conceptual framework and item selection. Med. Care **1992**, *30*, 473–483.
17. *Proceedings of the Advances in Health Assessment Conference, Palm Springs, California, February 19–21, 1986*; J. Chronic. Dis., 1987; Vol. 40; 1S–191S.
18. Lohr, K.N. Applications of health status assessment measures in clinical practice. Overview of the Third Conference on Advances in Health Status Assessment. Med. Care **1992**, *30*, MS1–MS14.
19. Lohr, K. Health outcomes methodology symposium: Summary and recommendations. Med. Care **2000**, *38*, II-194–II-208.
20. *Quality of Life and Pharmacoeconomics in Clinical Trials*; Spilker, B., Ed.; Lippincott-Raven Publishers: Philadelphia, Pennsylvania, 1996.
21. Patrick, D.L.; Erickson, P. *Types of Health Related Quality of Life Assessments*; Oxford University Press, Inc.: New York, 1993; 113–142.
22. McDowell, I.; Newell, C. *Measuring Health: A Guide to Rating Scales and Questionnaires*; Oxford University Press: New York, New York, 1996.

23. Wennberg, J.; Gittelsohn, A. Variations in medical care among small areas. Sci. Am. **1982**, *246*, 120–134.
24. Newhouse, J.P.; Manning, W.G.; Morris, C.N.; Orr, L.L.; Duan, N.; Keeler, E.B.; Leibowitz, A.; Marquis, K.H.; Marquis, M.S.; Phelps, C.E.; Brook, R.H. Some interim results from a controlled trial of cost sharing in health insurance. N. Engl. J. Med. **1981**, *305*, 1501–1507.
25. Thorpe, K.E. *Health Care Cost Containment: Results and Lessons From the Past 20 Years*; Shortell, S.M., Reinhardt, U.E., Eds.; 1992; 227–274.
26. Tarlov, A.R.; Ware, J.E., Jr.; Greenfield, S.; Nelson, E.C.; Perrin, E.; Zubkoff, M. The medical outcomes study. An application of methods for monitoring the results of medical care. JAMA, J. Am. Med. Assoc. **1989**, *262*, 925–930.
27. *Omnibus Reconciliation Budget Act 1989 Public Law 101-239*; Government Printing Office: Washington, DC, 1989.
28. Ware, J.E., Jr. The status of health assessment 1994. Annu. Rev. Public Health **1995**, *16*, 327–354.
29. Nunnally, J.C.; Bernstein, I.H. *Psychometric Theory,* 3rd Ed.; McGraw-Hill Inc.: New York, 1994.
30. Bayliss, M.S.; Gandek, B.; Bungay, K.M.; Sugano, D.; Hsu, M.A.; Ware, J.E., Jr. A questionnaire to assess the generic and disease-specific health outcomes of patients with chronic hepatitis C. Qual. Life Res. **1998**, *7*, 39–55.
31. Thurstone, L.; Chave, E. *The Measurement of Attitude*; University of Chicago Press: Chicago, 1929.
32. McHorney, C.A.; Ware, J.E., Jr.; Lu, J.F.; Sherbourne, C.D. The MOS 36-item short-form health survey (SF-36): III. Tests of data quality, scaling assumptions, and reliability across diverse patient groups. Med. Care **1994**, *32*, 40–66.
33. Wagner, A.K.; Bungay, K.M.; Kosinski, M.; Bromfield, E.B.; Ehrenberg, B.L. The health status of adults with epilepsy compared with that of people without chronic conditions. Pharmacotherapy **1996**, *16*, 1–9.
34. American Psychological Association standards for educational and psychological testing. APA **1985**.
35. Patrick, D.L.; Erickson, P. *Health Status and Health Policy: Allocating Resources to Health Care*; Oxford University Press, Inc.: New York, New York, 1993.
36. Constitution of the World Health Organization Basic documents: World Health Organization, Geneva, Switzerland. WHO **1948**.
37. Dubos, R. The state of health and the quality of life. West. J. Med. **1976**, *125*, 8–9.
38. Campbell, A. *The Sense of Well-Being in America: Recent Patterns and Trends*; McGraw-Hill: New York, 1981.
39. Bergner, M. Measurement of health status. Med. Care **1985**, *23*, 696–704.
40. Nagi, S. Some conceptual issues in disability and rehabilitation. Soc. Rehabil. **1965**.
41. Read, J.; Quinn, R.; Hoefer, M. Measuring overall health: An evaluation of three important approaches. J. Chronic. Dis. **1987**, *40*, 7S–21S.
42. Patrick, D.L.; Bergner, M. Measurement of health status in the 1990s. Annu. Rev. Public Health **1990**, *11*, 165–183.
43. Verbrugge, L. Physical and social disability in adults. Primary Care Res.: Theory and Methods **1991**, 31–53.
44. Johnson, R.J.; Wolinsky, F.D. The structure of health status among older adults: Disease, disability, functional limitation, and perceived health. J. Health Soc. Behav. **1993**, *34*, 105–121.
45. Stewart, A.L.; Greenfield, S.; Hays, R.D.; Wells, K.; Rogers, W.H.; Berry, S.D.; McGlynn, E.A.; Ware, J.E. Functional status and well-being of patients with chronic conditions: Results from the medical outcomes study. JAMA, J. Am. Med. Assoc. **1989**, *262*, 907–913.
46. Stewart, A.L.; Ware, J.E., Jr.; Brook, R.H. Advances in the measurement of functional status: Construction of aggregate indexes. Med. Care **1981**, *19*, 473–488.
47. Jette, A.M. Functional capacity evaluation: An empirical approach. Arch. Phys. Med. Rehabil. **1980**, *61*, 85–89.
48. Bungay, K.M.; Ware, J.E., Jr. Measuring and monitoring health-related quality of life. Upjohn **1998**, 1–39.
49. Veit, C.T.; Ware, J.E., Jr. The structure of psychological distress and well-being in general populations. J. Consul. Clin. Psychol. **1983**, *51*, 730–742.
50. Donald, C.A.; Ware, J.E., Jr. The measurement of social support. Res. Community Ment. Health **1984**, *4*, 325–370.
51. Wortman, C. Social support and the cancer patient: Conceptual and methodologic issues. Cancer **1984**, *53*, 2339–2362.
52. Sherbourne, C.; Stewart, A.; Wells, K. *Role Functioning Measures*; Stewart, A.L., Ware, J.E., Eds.; Duke University Press: Durham, North Carolina, 1992; 205–219.
53. Pope, A.M.; Tarlov, A.R. *Disability in America: Toward a National Agenda for Prevention*; National Academy Press: Washington, DC, 1991.
54. Lerner, D.; Amick, B.C.; Rogers, W.H.; Malspeis, S.; Bungay, K.; Cynn, D. The work limitations questionnaire. Med. Care **2001**, *39*, 72–85.
55. Ware, J.E., Jr.; Bjorner, J.B.; Kosinski, M. Practical implications of item response theory and computerized adaptive testing: A brief summary of ongoing studies of widely used headache impact scales. Med. Care **2000**, *38*, II73–II82.
56. Patrick, D.; Deyo, R. Generic and disease-specific measures in assessing health status and quality of life. Med. Care **1989**, *27* (Suppl.), S217–S232.
57. Wu, A.W.; Rubin, H.R.; Mathews, W.C.; Ware, J.E., Jr.; Brysk, L.T.; Hardy, W.D.; Bozzette, S.A.; Spector, S.A.; Richman, D.D. A health status questionnaire using 30 items from the medical outcomes study. Preliminary validation in persons with early HIV infection. Med. Care **1991**, *29*, 786–798.
58. Aaronson, N. K. Quantitative issues in health-related quality of life assessment. Health Policy **1988**, *10*, 217–230.
59. Aaronson, N.K. Quality of life assessment in clinical trials: Methodologic issues. Control. Clin. Trials **1989**, *10*, 195S–208S.

60. Deyo, R.A.; Patrick, D.L. Barriers to the use of health status measures in clinical investigation, patient care, and policy research. Med. Care **1989**, *27*, 254–268.
61. Anonymous Approaches to instrument translation: Issues to consider. Med. Outcomes Trust Bull. **1997**, *5*, 2.
62. Aaronson, N.K.; Acquadro, C.; Alonso, J.; Apolone, G.; Bucquet, D.; Bullinger, M.; Bungay, K.; Fukuhara, S.; Gandek, B.; Keller, S. International quality of life assessment (IQOLA) project. Qual. Life Res. **1992**, *1*, 349–351.
63. Garratt, A.M.; Ruta, D.A.; Abdalla, M.I.; Russell, I.T. SF 36 health survey quetionnaire: II. Responsiveness to changes in health status in four common clinical conditions. Qual. Health Care **1994**, *3*, 186–192.
64. Ware, J.E., Jr.; Gandek, B. The IQOLA project group. The SF-36 health survey: Development and use in mental health research and the IQOLA project. Int. J. Ment. Health **1994**, *23*, 49–73.
65. Ware, J.E.; Gandek, B.; Keller, S. *Evaluating Instruments Used Cross-Nationally: Methods From the IQOLA Project*; 1996; 681–692.
66. Ware, J.; Keller, S.; Gandek, B. Evaluating translations of health status questionnaires: Methods from the IQOLA project. Int. J. Technol. Assess. Health Care **1995**, *11*, 525–551.
67. Ware, J.; Snow, K.; Kosinski, M.; Gandek, B. *SF-36 Health Survey: Manual and Interpretation Guide*; The Health Institute, New England Medical Center: Boston, Massachusetts, 1993.
68. Tietze, K. *Taking Medication Histories*; Mosby-Year Book, Inc.: St. Louis, Missouri, 1997; 39–55.
69. Patrick, D.; Chiang, Y. Measurement of health outcomes in treatment effectiveness evaluations: Conceptual and methodological challenges. Med. Care **2000**, *38*, II-14–II-25.
70. Hays, R.; Morales, L.; Reise, S. Item response theory and health outcomes measurements in the 21st Century. Med. Care **2000**, *38*, II-28–II-42.
71. McHorney, C.; Cohen, A. Equating health status measures with item response theory: Illustrations with functional status items. Med. Care **2000**, *38*, II-43–II-59.

PROFESSIONAL DEVELOPMENT

Health-Systems, Clinical Pharmacy Careers in

William E. Smith
Virginia Commonwealth University, Richmond, Virginia, U.S.A.

INTRODUCTION

Health systems evolved from a hospital into multiple facilities and levels of care during the 1980s and 1990s in the United States. A health system can include more than one hospital, ambulatory care clinics, physician office buildings, long-term care facilities, and home care services. The economic forces, both internal and external to the hospital, led to the development of health systems. From a pharmacy perspective, the scope (range) of pharmacy services expanded from acute care to ambulatory care, to home care, to long-term care, and to other diversified pharmacy services. Consequently, positions for clinical pharmacists expanded from acute care to the other care settings within the health system.

RANGE OF CAREER ACTIVITIES WITHIN THIS FIELD

A simple description of the range of career activities within health system pharmacy would be acute care, ambulatory care, and home care to administrative pharmacist positions. A more detailed description of pharmacist positions within the health system is as follows.

Acute Care

Acute care relates to hospitalized patients. Patient care areas within a hospital include internal medicine, general surgery, pediatrics, obstetrics and gynecology, critical care, cardiac care, pulmonary critical care, psychiatry, oncology, and geriatrics.

Ambulatory Care

A health system can own and operate its own ambulatory clinics and physician offices. The physician component can be either by staff physicians employed by the health system or by contract for physician services.

Community Pharmacy

A health system can own and operate its own licensed community pharmacies. Prescription services and clinical services can be provided.

Geriatrics and/or Long-Term Care

A health system can own and operate its own licensed long-term care facilities or license beds within the hospital for long-term care.

Home Care Services

A health system can own and operate its own home care services for nursing care, prescription drug products, and pharmacist services.

Drug Information Service

A health system can own and operate its own drug information service (DIS) to serve the drug information needs of pharmacists, physicians, nurses, and other professional staffs within the system. In addition to drug information, the DIS focuses on drug formulary and pharmacoeconomic issues of drug products and drug use within the system.

Therapeutic Drug Monitoring Service

A health system can own and operate a centralized therapeutic drug monitoring service (TDMS) to focus on the application of clinical pharmacokinetics to the care of patients within the system.

Management of Pharmacy Services

Depending on the size and complexity of the health system, pharmacy management positions will range from the

Encyclopedia of Clinical Pharmacy
DOI: 10.1081/E-ECP 120006175

H

director of pharmacy services to supervisor of a segment of the pharmacy services within the health system. Some examples of pharmacy manager positions include supervisor of the drug information service, supervisor of the therapeutic drug monitoring service, supervisor of ambulatory care services, supervisor of community pharmacies, supervisor of clinical services, and assistant director of pharmacy services.

WORK SETTINGS AND JOB ACTIVITIES

The typical work settings for clinical pharmacists in a health system include acute care hospital, ambulatory clinic, outpatient pharmacy, home care pharmacy, and community pharmacy.

Clinical practice in the hospital could be in the central hospital pharmacy, a satellite pharmacy, a pharmacist's office, or a patient care area. The hospital pharmacy is usually located on a lower floor of the facility, which places the pharmacist physically remote from the patient, physician, nurse, and other personnel. Communications are often by telephone, fax, or information technology rather than in person. A satellite pharmacy is a pharmacy area located in the patient care area where drug distribution and clinical services are provided. A satellite pharmacy places the pharmacist in the patient care area where drug distribution and clinical services are provided. A satellite pharmacy facilitates the placement of pharmacists in close proximity to the patients, physicians, and nurses. A pharmacist's office space is often provided as a location for the pharmacist to provide clinical services that is in close proximity to patients, physicians, and nurses. Clinical services can be provided in a drug information center, often located in the hospital pharmacy, but it may be located in the medical library. Therapeutic drug-monitoring services may be provided from a pharmacist's office location.

Clinical practice in an ambulatory clinic may be provided from an office area within the clinic. The patient, patient medical record, physician, nurse, and other practitioners are in close proximity to the pharmacist's office area. Examples of clinics in which pharmacists have provided clinical services include family practice, OB-GYN, anticoagulation, prescription refill, pain therapy, nutrition, and internal medicine.

Outpatient Pharmacy

The health system may own one or more community pharmacies. Clinical services can be provided relating to patient drug therapy counseling for prescription and nonprescription medications, management of drug therapy via physician-approved guidelines, monitoring of drug therapy, and screening tests for hypertension, diabetes, and hypercholesterolemia.

General Clinical Practice Model

The following list of pharmacist practice activities describes a general clinical practice model:

- Clarify prescription orders.
- Question inappropriate prescription orders.
- Answer drug information requests from patients.
- Answer drug information requests from physicians, nurses, and other health professionals.
- Monitor patient drug therapy for safety and efficacy using a comprehensive patient medication record:

    — Drug–drug interactions.
    — Concomitant drug therapies.
    — Appropriate drug, dose, and dosage form.
    — Patient allergies.
    — Drug–laboratory test interactions.
    — Drug–food interactions.
    — Abnormal laboratory tests that are drug induced.
    — Clinical pharmacokinetics.

- Provide patient medication counseling.
- Provide screening tests.
- Participate in collaborative practice agreements for managing drug therapy.
- Participate in clinical research.

EDUCATION, TRAINING, AND EXPERIENCE

The preferred education for a health system pharmacist is the doctor of pharmacy degree. A general practice residency is also preferred. Some clinical pharmacist practices prefer pharmacists with a specialty residency. The American Society of Health System Pharmacists for the past 25 years has adopted policies and provided programs to support these preferred education and training programs. When the criteria can be met for board certification, many health systems support clinical pharmacists in becoming board certified.

Pharmacist clinical expertise requires practice, practice, and more practice. Years, usually three to five, are often acceptable to health systems in lieu of some residency training. The challenge is to get appropriate clinical practice experience without a residency.

For supervisory positions, three to five years of practice experience is often required. During the practice experience, the pharmacist should demonstrate the ability to achieve results, complete objectives on a timely basis, possess good communication skills, and demonstrate good working relationships with coworkers, physicians, and nurses.

For director of pharmacy services, five to seven years of experience are often required in a similar health system. Additional education and training, such as an advanced residency in pharmacy management or a masters degree in business administration, are often preferred or required. Ability to manage resources, personnel, planning, financial, and interprofessional relationships with good communication skills are often required.

CAREER LADDERS AND GROWTH

Career ladders and growth within a health system can be viewed as longitudinal and/or lateral. Longitudinal would be from staff pharmacist to director of pharmacy services. Lateral would be clinical pharmacist from acute to home care.

The usual longitudinal path is staff pharmacist to clinical pharmacist, to supervisor of clinical services, to assistant or director of pharmacy services. Each practice along this path requires demonstrating knowledge, skill, and the ability to learn more; assuming new responsibilities; and successfully performing the duties and responsibilities of each position. Additional education and training often will speed the time line for the longitudinal career path.

The lateral career path relates to clinical practice at different patient care levels or settings. Acute care to ambulatory and/or home care was often required in the 1990s as health systems expanded ambulatory and home care services and reduced acute care services.

Directors of pharmacy may be asked to assume the management of other departments and programs within the health system. The pharmacy director may continue as director or may give up the management responsibility for pharmacy services.

SITES OF PHARMACIST CLINICAL PRACTICE

Pharmacist clinical services can be provided at any site or location of patient care. These services are provided directly to patients or indirectly to patients through the nurse and/or physician.

The following sites are examples within health systems where pharmacist clinical services are provided:

- Acute care hospital in the patient care area(s).
- Critical care unit.
- Pediatrics hospital.
- Neonatal intensive care unit.
- Long-term care facility.
- Family practice physician office.
- Ambulatory care clinic.
- Home care services pharmacy.
- Community pharmacy.
- Outpatient pharmacy.
- Drug information services.
- Therapeutic drug monitoring service.
- Oncology.
- Hospice.

ADVANTAGES OF WORKING IN THE HEALTH-SYSTEM ENVIRONMENT

Several of the obvious advantages for working as a pharmacist in a health system include:

- Direct access to patients and patient information.
- Availability of physicians and nurses.
- Patient care environment.
- Levels of care—primary, secondary, tertiary.
- Resources to support pharmacist clinical services.
- Patient care quality assurance activities for pharmacist participation.
- Hospital and medical staff committees for pharmacist participation.
- Opportunities for clinical research.
- Opportunities for participation in education programs for physicians, nurses, and patients.
- Provision of drug information on a daily basis.
- Participation in therapeutic drug-monitoring services.
- Collaboration with pharmacist colleagues in clinical practice.
- Participation in teaching programs for pharmacy students and residents.
- Demand to know acute care pharmacotherapy.

These examples translate into a demand for the pharmacist to know pharmacotherapy and a requirement to update clinical therapeutics knowledge and expertise; to collaborate and work effectively and efficiently with physicians, nurses, and pharmacist colleagues in providing services and care to patients; and to participate in the

many varied activities to provide quality care to patients at different levels of care.

DIFFERENT TYPES OF HOSPITALS AND PHARMACY SERVICES

A broad categorization of hospitals is government and nongovernment. Government hospitals are federal, state, and local. Nongovernment hospitals can be categorized into nonprofit and proprietary (for-profit). A teaching hospital is one that provides a postgraduate education program for physicians. All hospitals exist to provide services and care to the patients being served. Some of the key differences between hospitals include the management decision-making process, type of medical staff, scope of patient services to be provided, size, and financial objectives and strength of each hospital. There is not an existing method to determine which types of hospitals provide more pharmacy and pharmacist clinical services as there are too many variables that determine the existing scope of pharmacist services. In general, every hospital needs more clinical services from the pharmacy department and staff than currently exist.

Some questions to consider when looking at a health system for possible employment include the following:

- What is the existing scope of pharmacist clinical services? What types of services? How long have they been provided?
- Is the hospital a teaching hospital?
- Does the pharmacy have an affiliation with a school of pharmacy?
- Does the health system provide pharmacy residencies?
- What is the pharmacy director's philosophy regarding pharmacist clinical services?
- How does the pharmacy facilities look regarding to size, organization, cleanliness, automation, and drug information resources?
- Does the medical staff and health-system administration support pharmacy services and pharmacist clinical practice activities?
- What is the job satisfaction and morale of the pharmacist staff?
- How are pharmacy technicians used in the pharmacy operations?
- What is the compensation and benefit package?
- What is the strategic plan for the health system and for the pharmacy services?

The answers to these and similar questions should convey whether the health system being considered will provide an environment for clinical practice, job satisfaction, and opportunities for growth and career advancement.

WHY SEEK EMPLOYMENT IN A HEALTH SYSTEM?

Some key factors for seeking pharmacist employment in a health system relate to the opportunities to provide clinical services directly to patients, to collaborate with physicians and nurses, to cope with the personal challenge to maintain and expand clinical pharmacotherapy knowledge and expertise, to change practice settings for the different levels of care and job satisfaction, and to be viewed as an essential health care practitioner by the institution, physician, and nurse colleagues. The personal satisfaction from providing clinical services that benefit patients is the best reward for working in a health-system environment.

BIBLIOGRAPHY

ACCP Guideline, Practice guidelines for pharmacotherapy specialists. Pharmacotherapy **2000**, *20*, 487–490.

ACCP Position Statement. Position paper on critical care pharmacy services. Pharmacotherapy **2000**, *20*, 1400–1406.

ACCP White Paper. Clinical pharmacy practice in the noninstitutional setting. Pharmacotherapy **1992**, *12*, 358–364.

ACCP White Paper. Establishing and evaluating clinical pharmacy services in primary care. Pharmacotherapy **1994**, *14*, 743–758.

Healthy People 2010: Objectives for Improving Health

Carl J. Tullio
Pfizer, Inc., Yorktown, Virginia, U.S.A.

INTRODUCTION

Healthy People 2010 is a national health promotion and disease prevention program aimed at improving the health of all Americans. Progress in reaching these goals will be measured using 467 objectives organized under 28 focus areas. Healthy People 2010 is important to pharmacists because many of the objectives involve the use, or the need for proper use, of medications. This article provides a short history of this program. Using diabetes as an example, it explains the content of the focus areas, then reviews the goals and how progress toward them is assessed. Finally, the implications for pharmacists are presented. (All of the information in this article is from the Healthy People 2010 Web site, http://www.health.gov.healthypeople/.)

HISTORY OF HEALTHY PEOPLE 2010

Healthy People 2010: Objectives for Improving Health, the third decade-long national initiative, builds on the achievements of the past two decades. In 1979, the first report, *Healthy People: The Surgeon General's Report on Health Promotion and Disease Prevention*, put forth national goals for preserving independence for the elderly and reducing premature deaths. A second report, in 1980, *Promoting Health/Preventing Disease: Objectives for the Nation*, provided more than 200 health objectives for the United States to achieve over the next 10 years. *Healthy People 2000: National Health Promotion and Disease Prevention Objectives*, released in 1990, continued this program and identified health improvement goals and objectives to be attained by the year 2000. The Healthy People 2010 initiative continues in this tradition as a tool to improve our nation's health into the first decade of the 21st century.

One of the most encouraging lessons learned from the Healthy People 2000 program was that we, as a nation, can make dramatic progress in improving the nation's health in a relatively short period of time. For example, during the last decade, significant reductions were achieved in infant mortality. ''Childhood vaccinations are at the highest levels ever recorded in the United States. Fewer teenagers are becoming parents. Overall, alcohol, tobacco, and illicit drug use is leveling off. Death rates for coronary heart disease and stroke have declined.'' Significant advances have been made in the diagnosis and treatment of cancer and in reducing unintentional injuries.

But there is still much progress to be made. ''Diabetes and other chronic conditions continue to present a serious obstacle to public health. Violence and abusive behavior continue to ravage homes and communities across the country. Mental disorders continue to go undiagnosed and untreated. Obesity in adults has increased 50% over the past two decades. Nearly 40% of adults engage in no leisure time physical activity. Smoking among adolescents has increased in the past decade. And HIV/AIDS remains a serious health problem, now disproportionately affecting women and communities of color.'' The development and implementation of Healthy People 2010 will be the guiding instrument for addressing these health issues, reversing unfavorable trends, and expanding on past achievements.

DEVELOPING OBJECTIVES

Suggestions for Healthy People 2010 objectives were gathered from a variety of diverse organizations and people using a series of national and regional meetings. On two different occasions in the late 1990s, the American public was given the opportunity to express its views and opinions. More than 11,000 comments were received from every state in the Union, plus the District of Columbia and Puerto Rico. Using this input, the final Healthy People 2010 objectives were developed by teams of experts from various federal agencies under the direction of Health and Human Services Secretary Donna Shalala, Assistant Secretary for Health and Surgeon General David Satcher, and former Assistant Secretaries for Health. The Office of Disease Prevention and Health

Encyclopedia of Clinical Pharmacy
DOI: 10.1081/E-ECP 120006190

Promotion, U.S. Department of Health and Human Services, coordinated and oversaw the entire process.

GOALS AND ASSESSMENT

The two overarching goals of Healthy People 2010 are the elimination of disparities in health status among racial and ethnic groups and the improvement in the years and the quality of life for people of all ages. Progress in attaining these goals will be measured using the 467 objectives in the 28 Focus Areas (Table 1). Each focus area contains its own overarching goal. For example, the goal of the diabetes section states, "Through prevention programs, reduce the disease and economic burden of diabetes, and improve the quality of life for all persons who have or are at risk for diabetes." After listing the goal, an overview of the issues, trends, disparities, and opportunities for action is presented. If the topic was included in the previous program, Healthy People 2000, interim progress toward the objectives is detailed. Using the diabetes example, there are five objectives in the previous initiative. As Healthy People 2000 draws to a close, one objective is trending toward the goal, while the other four are trending away from the goal.

Next, the focus area objectives for 2010 are presented. Each focus area contains varying numbers of objectives. Many of the objectives are aimed at "interventions designed to reduce or eliminate illness, disability, and premature death among individuals and communities. Others focus on broader issues, such as improving access to quality health care, strengthening public health services, and improving the availability and dissemination of health-related information." In the diabetes example, the number of objectives was increased from 5 in the 2000 program to 17 for the current program. Each objective (e.g., increase the proportion of persons with diabetes who receive formal diabetes education.) lists a target (e.g., 60%) for the year 2010, the rationale behind its focus, and the national data tables from which the measurements will be extracted. Each focus area ends with a listing of related objectives from other focus areas, an explanation of the terminology used, and the references employed.

Table 1 Focus areas for Healthy People 2010

Access to quality health services
Arthritis, osteoporosis, and chronic back conditions
Cancer
Chronic kidney disease
Diabetes
Disability and secondary conditions
Educational and community-based programs
Environmental health
Family planning
Food safety
Health communication
Heart disease and stroke
HIV
Immunization and infectious disease
Injury and violence prevention
Maternal, infant, and child health
Medical product safety
Mental health and mental disorders
Nutrition and overweight
Occupational safety and health
Oral health
Physical activity and fitness
Public health infrastructure
Respiratory diseases
Sexually transmitted diseases
Substance abuse
Tobacco use
Vision and hearing

(Ref. http://www.health.gov.healthypeople/.)

LEADING HEALTH INDICATORS

In order to periodically assess the health of the nation, a set of leading health indicators was developed for the first time (Table 2). These indicators, which address major public health concerns, "were chosen based on their ability to motivate action, the availability of data to measure progress, and their relevance as broad public health issues." For each of the leading health indicators, specific objectives from Healthy People 2010 were selected and will be used to track progress. This small subset of measures will provide a snapshot of the health of the nation. Even though the leading health indicator may have the same name as a focus area, the indicator may

Table 2 Leading health indicators for Healthy People 2010

Access to health care
Environmental quality
Immunization
Injury and violence
Mental health
Overweight and obesity
Physical activity
Responsible sexual behavior
Substance abuse
Tobacco use

(Ref. http://www.health.gov.healthypeople/.)

contain only a few of the focus area's objectives and may even contain objectives from a related focus area. For example, the tobacco use focus area has 21 objectives, while the tobacco use leading indicator follows only 2 of the objectives. The indicators will highlight achievements and challenges throughout the next decade while serving as a link to the 467 objectives of the Healthy People 2010 program. The leading health indicators are intended to help the populace more easily understand the importance of health promotion and disease prevention. They are also aimed at encouraging wide participation in improving health in the next decade.

IMPLICATIONS FOR PHARMACISTS

Healthy People 2010 is important to pharmacists in all areas of practice. The focus areas and their objectives address not only clinical issues but also social issues. As pharmacists push forward into true pharmaceutical care, the entire patient must be considered, not just the medical management. Healthy People 2010 provides the information needed to help pharmacists develop services that are aligned with national goals.

For detailed information, the full text of *Healthy People 2010 Conference Edition* (Volumes 1 and 2) is available online at http://www.health.gov.healthypeople/. A CD-ROM version (B0071) can be purchased from ODPHP Communication Support Center, P.O. Box 37366, Washington, D.C. 20013–7366, (301) 468–5960. Limited numbers of the print version (B0074) are also available from the ODPHP Communication Support Center.

BIBLIOGRAPHY

Healthy People 2010: Objectives for Improving Health. http://www.health.gov.healthypeople/ (accessed October 10, 2000).

PROFESSIONAL DEVELOPMENT

Home Care, Clinical Pharmacy Careers in

H

Donald J. Filibeck
Mt. Carmel Home Infusion, Columbus, Ohio, U.S.A.

INTRODUCTION

The practice of clinical pharmacy in the home care/home infusion setting is a challenging, but rewarding practice site. The pharmacist is a vital member of the home care team, which includes the patient and/or their caregiver, the physician, the home care nurse, and various other support personnel (e.g., pharmacy technicians, customer service personnel, billing personnel). The practice sites vary greatly, and many clinical, operational, and marketing opportunities exist.

BACKGROUND INFORMATION

Home infusion therapy involves the administration of medications using the intravenous, subcutaneous, or epidural routes. Therapies administered at home include antiinfectives, chemotherapy, pain management, parenteral or enteral nutrition, and immunologic or biological agents. Many different diagnoses are treated at home, including many infectious diseases (bacterial, fungal, or viral), gastrointestinal diseases, immunologic disorders, and cardiac diseases (e.g., congestive heart failure).[1–7]

Home infusion therapy has proven to be a safe and effective alternative to patients receiving care in hospital settings. For most patients, receiving treatment in the home (or in an outpatient clinical setting) is preferable to being kept in a hospital.

Whenever a patient starts on home infusion therapy, a prescription from a qualified physician responsible for the care of the patient is needed. Home nursing services are also generally provided to ensure that the proper patient education and training occurs, and to provide ongoing clinical monitoring of the patient in the home, along with the pharmacist's clinical interventions.

From a business perspective, the home infusion market is projected to have annual revenues approaching $4.5 billion (year ending 2000). The market continues to experience cost-containment pressures (as does the entire healthcare market); however, the future for providing care at home looks good, as it is approximately one-third as costly as providing care in the hospital.

As an alternative, some infusion pharmacies also provide infusion therapies in an ambulatory cliniclike setting. This arrangement has the advantage of providing services to patients in a supervised setting. It allows several patients to receive their infusions concurrently, therefore making more efficient use of the organization's staff, particularly nursing.

THE PHARMACIST IN HOME CARE

The staff pharmacist may or may not have an advanced degree (i.e., Doctor of Pharmacy degree). Although a PharmD degree is not required, it does ensure that the pharmacist has a good, sound clinical education. More important is the person's ability to think quickly when asked difficult questions or when in difficult situations; to interact professionally with a wide range of individuals (both clinical and nonclinical); and to be able to work with little supervision in an often unstructured environment.

As a manager, when hiring, the person's previous work history should be evaluated for these abilities. However, experience working in the home care environment is not an absolute requirement. There are pros and cons to hiring someone with experience. The person must be licensed in the state in which they are practicing and must meet all continuing education requirements.

WORK ENVIRONMENTS

Typical work environments are office-type settings where the pharmacist is working alongside many different individuals. The sites may be free standing (located in light industrial or suburban office parks) or located on a health system campus. Many health systems provide home care/home infusion services as part of the for-profit arm of the system. In those cases, the home infusion provider pro-

Encyclopedia of Clinical Pharmacy
DOI: 10.1081/E-ECP 120006263

vides service for only those patients being discharged from the hospital.

For-profit home infusion providers range from single-site, private companies to multiple-site, million-dollar companies. All home care providers are licensed by the state and can chose to become accredited by several accrediting bodies (e.g., Joint Commission on Accreditation of Healthcare Organizations (JCAHO), Accreditation Commission for Health Care, Inc. (ACHC), Community Healthcare Accreditation Program (CHAP)). Accreditation is a requirement for many insurance companies to serve as a provider for their members.

An advantage of working in home care is the flexibility in hours and activities that the home care environment offers. Positions are available that range from PRN or, to as needed, to part and full time. As needed positions are often used to help cover vacations and scheduled time off or on-call activities. Part-time positions can range from 1 or 2 days per week to 4 or 5. Average hours per day are 8 or 10 depending on the home care company.

Because home care personnel must be available 24 hours per day, on-call related activity may be required. Depending on the organization and workload activities, afternoon, evening, or weekend shifts may be used.

ACTIVITIES OF THE HOME CARE PHARMACIST

Activities vary greatly, depending on the services provided and the size of the operation. In small offices, the pharmacist may wear many different hats. In large offices, the pharmacist may do only one task on a given day.

Table 1 Home care preadmission criteria

- Patient/caregiver agrees to receive services in the home.
- Patient/caregiver are willing to learn the necessary steps to administer their drug(s) in the home.
- The home environment is acceptable (clean, access to telephone and running water).
- The patient is readily accessible to the home care provider.
- The patient has adequate family support, both physically and psychologically.
- A physician is readily available in the event of an emergency, ongoing clinical updates, and/or order changes.
- The medication ordered is appropriate to be given in the home environment.
- The indication, dosage, and route of administration of the medication(s) ordered is appropriate.
- Labs etc., are ordered to access the effectiveness of the therapy ordered.

Table 2 Home care patient database

- Patient's name, address, phone number, date of birth.
- Alternate contact information in the event of an emergency.
- Information on the status of any advance directive.
- Height, weight, gender.
- Diagnoses.
- Location and type of intravenous access and date of placement.
- Pertinent laboratory test results.
- Pertinent medical history and physical findings.
- Accurate history of allergies.
- A detailed medication profile, including all prescription and nonprescription medications, home remedies, and investigational and nontraditional therapies.
- Other agencies involved in patient care.
- Prescriber's name, address, phone number, etc.
- A plan of care.
- Patient education activities.
- Any functional limitations.
- Any pertinent social history.

Tasks include dispensing-related functions; technician oversight; obtaining orders from physicians and then assessing the orders for appropriateness; assessing the patient and caregiver for the appropriateness of providing care in the home; patient and/or caregiver education; providing education for nursing agencies, discharge planners, etc; answering drug information questions; sales support; and so on. No one day is ever the same. The following explains these activities in greater detail.

One of the primary roles of the pharmacist is the preadmission assessment. This role ensures that each patient is assessed for appropriateness using predetermined admission criteria. Common criteria are outlined in Table 1.

In conjunction with other members of the home care team and with the patient's physician, a decision is made to either accept the patient for home care services or refer them back to the hospital discharge planner or referral source. Once accepted, an assessment is completed and an initial patient database established. Table 2 lists some of the items that are part of this database. Again, the pharmacist is an integral part of this process. Much of this information is obtained via the telephone in conversations with the physician, hospital personnel, or patient. Information may also be received via fax or from the home care agency nurse. Pharmacists working in a hospital-based home care pharmacy may be able to go up to the floor and obtain this information directly from the medical record, floor nurse, and/or patient.

One of the documents that is part of this patient database is the care plan or plan or care. The plan of care should indicate the treatment goals and indicators of de-

sired outcomes, any interventions that need to be done, and the frequency of those interventions. Any drug-related problems that occur or have the potential to occur should be addressed by the pharmacist, along with other members of the patient care team. When multiple providers are involved with the patient, the pharmacist is in an ideal position to coordinate the information flow and care of the patient.

The plan of care should be developed initially and updated as needed. Based on the drug(s) used and the potential for side effects and adverse drug reactions, the pharmacist should determine what type of monitoring is needed (e.g., labs, physical findings) and the frequency at which it is to occur. The pharmacist must communicate this plan to others involved and provide updates as needed.

Another role of the pharmacist is the selection of products, infusion devices (i.e., pump), and ancillary supplies. Many factors need to be considered when choosing the administration method, infusion device to use, and what ancillary supplies are needed.

The stability, compatibility of the drug, and volume of the drug are important considerations when determining what method (IVPB, IV push, continuous infusion) or infusion device (elastomeric, electronic infusion device, etc.) will be used. Nursing agency knowledge and ability of the patient to learn the methodology are all important considerations. Patient convenience, prescriber preference, and cost must also be considered. Again, the pharmacist is able to weigh the pros and cons of any method and help the patient care team make appropriate decisions.

The ongoing clinical monitoring is the hallmark of the pharmacist's involvement. By having regular, ongoing conversations with the patient/caregiver, physician, and home care nurse, the pharmacist is able to make an objective evaluation of the therapy(ies), make appropriate recommendations for changes, and effectively communicate those changes to the patient/caregiver and all involved health care providers.

On an initial and ongoing basis, the pharmacist should be providing education to the patient and/or caregiver. Some of this information may be provided verbally, although most is provided in writing. Table 3 lists some of these education-related issues. The pharmacist should be involved in the development of all educational material.

Table 3 Education-related issues

- Medication related, including dose, route of administration, dosage interval, duration, side effects, adverse reactions (and their management).
- Proper aseptic technique.
- Precautions and directions for administering the medication.
- Equipment use, maintenance, and troubleshooting techniques.
- Proper care of the vascular access device and site (if applicable).
- Home inventory management, how to contact help, emergency issues (what to do if something goes wrong).
- Special precautions and directions for the preparation, storage, handling, and disposal of drugs, supplies, and biomedical waste.

CAREER OPTIONS

The range of careers is very diverse. Pharmacists may choose to remain clinically focused, providing hands-on care to the patient. Opportunities exist to do research on the delivery and use of drugs in the home environment. Extended stability studies are one area where the pharmacist can become involved. If the pharmacist gets involved in clinical research, they should ensure that all appropriate policies and procedures are followed, that the patient and health care providers have appropriate information concerning the drug(s), and that all required record-keeping requirements are met.

Many sites offer clinical clerkships for undergraduate pharmacy students and several post-PharmD residencies in home care exist.

From an operational perspective, pharmacists who have a business background can progress from a staff-level position to branch, regional, or corporate management positions. It is not unusual for a mid- to high-level manager to have started out as a staff pharmacist.

The pharmacist should be actively involved in the organization's performance improvement activities.[8] Aspects of care that can be monitored include, but are not limited to, patient satisfaction, unscheduled admissions, medication errors, adverse drug reactions, infection control-related issues (e.g., line infections), unscheduled deliveries, and so on.

The pharmacist must also take an active role in the development, implementation, and review of an organization's policies, procedures, and protocols. The pharmacist should ensure that all aspects of care are addressed, including patient care, drug preparation and dispensing, quality control, infection control, and equipment maintenance. Involvement in such activities can have far-reaching effects on efficiency and financial outcomes.

As a manager, the pharmacist's responsibilities include: 1) setting the goals (both short- and long-term) of the pharmacy, based on the needs of the patients and mission/goals of the organization; 2) developing plans to achieve those goals; 3) implementing those plans; 4) assessing whether the goals are being met; and 5) instituting corrective actions when necessary.

The pharmacy manager will have multiple areas of responsibility, such as managing the pharmacy (including compliance with laws, regulations, and accreditation standards), financial resources (drugs, budgets, reimbursement), and pharmaceutical care and human resources (scheduling, hiring, education and training, staffing needs).

Pharmacists that have a sales/marketing nature can pursue this career tract if so desired. As mentioned previously, positions range from branch sales/marketing to corporatewide strategic sales management.

CONCLUSION

The practice of pharmacy in the home care environment presents many opportunities for professional and personal growth. The practice continues to evolve and will continue to offer pharmacists multiple opportunities (both clinically and management related), as well as continuing to provide sound pharmaceutical care to the patients receiving home care services.

REFERENCES

1. American Society of Health-System Pharmacists. ASHP guidelines on minimum standards for home care pharmacies. Am. J. Health-Syst. Pharm. **1999**, *56*, 629–638.
2. American Society of Health-System Pharmacists. ASHP guidelines on the pharmacist's role in home care. Am. J. Health-Syst. Pharm. **2000**, *57*, 1252–1257.
3. National Home Infusion Association. *White Paper: Home Infusion Services, Payment Modes and Operational Costs*; www.nhianet.org.
4. National Home Infusion Association. *Resources for Payers, Physicians & Providers: Overview of Home Infusion Therapy*; www.nhianet.org.
5. National Home Infusion Association. *Resources for Payers, Physicians & Providers: Patient Care Process in Home Infusion Therapy*; www.nhianet.org.
6. www.nhianet.org.
7. www.ashp.org.
8. Winiarski, D. Performance improvement in action: Reducing unscheduled deliveries of infusion supplies. Infusion **2000**, *6*, 19–23.

Home Care Pharmacy Practice (Spain)

Ana Clopes
Hospital de la Sta. Creu i Sant Pau, Barcelona, Spain

INTRODUCTION

The introduction of home care is unavoidably bound to the changes that have been taking place in most health systems over the last 30 years. The financial pressure to reduce the hospital length of stay has a direct relationship on the acceptance of home care, and on the growth of other activities such as nursing homes and outpatient clinics.

Home care or *hospital at home* is defined as a service that provides active treatment by healthcare professionals in the patient's home of a condition that otherwise would require acute hospital in-patient care, always for a limited time period.[1]

This definition is the same for all models of health systems but the application and focus of care differ. However, in most systems home care implies the application of high technology in the patient's home for a limited period, rather than care for chronic patients. For this reason, in most systems, the referral centers are the hospitals.

The concept of home care originated in the university hospitals in the forties. In 1947 the Montefiori Hospital in New York planned to extend the hospital to the patient's home. But home care was in fact first applied in the sixties with "Hospitalisation a Domicile" in France in 1961.[2] It has been implemented in a number of other countries, including the United States,[3] Canada, and the Netherlands.[4] Home care coverage within the Medicare program in the United States was implemented in 1966.

The acceptance of home care has been faster in North America than in European countries where there is no direct cost to the patient or an insurer when a patient is admitted to a hospital.[5]

ADVANTAGES OF HOME CARE

The advantages of home care are:

- *Reduction in hospital length of stay.*[1] This is reflected in the decrease in costs [from 30 to 85% according to different studies][1,6–8] without loss of effectiveness of treatment. A meta-analysis carried out by Hughes et al.[9] studied the impact of home care hospital days (22 studies) and demonstrated a significant reduction in hospitalization days across studies due to home care, with a cumulative effect size of -0.38 (CI, -0.42 to -0.34, $p=0.001$).
- *The patient's maintenance in his/her family environment.* This implies an improvement in the quality of life[10] and patient satisfaction.[11]
- *The patient's involvement in his/her own care.* This is not typical in conventional health care and should be considered to improve the effectiveness of treatment. At the same time it breaks the bonds of nonpositive dependence that sometimes exist between the patient and the hospital.
- *Avoidance of the risk of nosocomial infections.* Patient care in a nonhospital environment avoids contact with hospital organisms, which are usually more resistant to antibiotic treatment.
- *Development of health models which integrate the different areas (basically hospital and community care).* The separation between the different areas of patient care is artificial, while integration implies a higher quality and more individualized care.

ORGANIZATION

The way home care is organized depends more on the type of care within each country than on the kind of care provided. Home care can be classified according to the type of reference center or according to the type of structure.

Classification According to the Type of Reference Center

Hospital-based home care services

The hospital is responsible for the patients and is the decision-making center. A hospital team organizes,

Encyclopedia of Clinical Pharmacy
DOI: 10.1081/E-ECP 120006376

stimulates, and assumes the leadership of the inclusion of patients in home care. High technology such as long-term ventilation,[12] intravenous antibiotics administration,[13] chemotherapy administration,[14] or parenteral nutrition[15] is included. The length of hospital stay is reduced by early discharge of patients following elective surgery with a home-based rehabilitation program.[16–19]

Community-based home care services

These usually include patients with chronic diseases requiring low technology. The community center medical team visits the patient at home. Examples of programs applying such schemes are home care programs for diabetes, hypertension, terminally ill patients, physiotherapy at home, and care of elderly.

Programs may also consist of mixed care with collaboration between the different areas of the health system.[20]

Classification According to the Type of Structure

- *External provider.* The health care team (physicians, nurses, and pharmacists) and the drugs and ancillary supplies proceed from a commercial provider who has a contract with hospital or the reference center.
- *A mixed structure of external provider and the reference center.* The hospital may provide the medical team and pharmacy services, for example, and the external provider supplies the nurses and drugs.
- *Reference center structure.* The physicians, nurses and the pharmacy services depend on the reference center, hospital or community centers.

PATIENT SELECTION

Selection Criteria

Selection criteria for patients who are candidates for home care are adapted to each environment, geographical area, and type of patient. These criteria can be divided into medical condition and psychosocial and family support. They will be described in each protocol of patients' inclusion defined for each diagnosis. But some general environments should be evaluated in all cases: home and family environment.

Home environment

A series of home requirements must be met and in all cases assessment of the following is needed:

- Geographic access to the reference center that each home care team will define according to the characteristics of the area.
- A telephone is imperative for continued contact between the patient and the home care team.
- The home should be clean and have electricity and running water. Based on this information, the pharmacist, in conjunction with the other team members, will assess the patient's appropriateness. Other requirements such as a refrigerator will also be necessary in some cases if the patient requires medication that has to be stored at low temperatures.

Family environment

The presence of a caregiver is mandatory in most of the home care protocols, although this will depend on the therapy administered and also on the medical situation. The home care team should assess the patient's or caregiver's capacity to be involved in the care.

Patient's Origin

Patients evaluated for inclusion in a home care program may proceed to a hospital, emergency room, or community care center.

Procedure for the Patient's Admission

The whole home care team is involved in patient inclusion and care planning although each member will play a specific role in the activities.

The steps in the admission procedure are:

1. The physician in charge of the patient considers whether he/she will be a candidate for home care according to the clinical assessment described in the previously defined protocol. In the detection of patient candidates, the pharmacist and the nurse who are working in the home care team can also participate.
2. Family support and home environment are evaluated by the social worker or by the nurse together with the pharmacist, also according to the previously defined protocols.
3. The entire home care team plans the care.

Table 1 Infections most frequently included in home care programs

Skin and soft-tissue infections
Cellulitis
Abscess
Postoperative wound infection
Posttrauma wound infection
Diabetic foot
Decubitus ulcer
Bone and joint infections
Acute and chronic osteomyelitis
Septic arthritis/bursitis
Prosthetic joint infections
IV line infection
Infective endocarditis
Ear and sinus infections (sinusitis/otitis/mastoiditis)
Acute exacerbation of pulmonary symptoms in cystic fibrosis
Lung infection (hospital- or community-acquired pneumonia)
Gastrointestinal infections (abscess/peritonitis)
Kidney, bladder, and prostate infections (pyelonephritis/ perinephric abscess)
Systemic febrile syndromes
Cytomegalovirus infection
Febrile neutropenia
Brain abscess

4. The patient or the caregiver is trained in and informed about the therapy to be carried out in the home. The information has to be oral and in writing and the pharmacist and the nurse can provide it.
5. The patient goes home and therapy begins.

One option to facilitate the coordination among the different steps is periodic meetings to discuss the cases with the participation of all the members of the home care team.

TYPE OF INTERVENTIONS

Home Parenteral Antibiotics

In general, all types of infection and all organisms are susceptible to home IV antibiotic therapy. The treatment of patients with bone and joint infections has proven highly effective and is now well accepted.[21] Other bacterial infections that have been studied extensively are skin and soft tissue infections and lung infections. The reason is that these infections fulfill two important criteria: patients are clinically stable and require prolonged IV antibiotic therapy (>7 days).[22] But home care can be extended to great number of infections: bacterial, viral, and fungal (Table 1). The patient's admission to home care should be considered from the beginning of the infection or should be wait until the patient is clinically stable, depending on the infection.

A large number of cost-effectiveness studies have been carried out (Table 2), all with positive results.

AIDS

The maintenance therapies of opportunistic disease in acquired immune deficiency syndrome (AIDS) are an-

Table 2 Studies of cost savings from home IV antibiotic therapy

Study	*n*	Infection	Average savings/day/ patient ($)	Average savings/day/ patient (Euros)
Antoniskis A 1978[38]	20	NA	165	196
Stiver 1978[39]	23	NA	97	115
Kind 1979[40]	15	NA	95	113
Swenson 1981[41]	8	Osteomyelitis, pyelonephritis and others	148	176
Poretz 1982[42]	150	Osteomyelitis	142	169
Stiver 1982[43]	95	NA	135	160
Rehm 1983[44]	48	Bone and joint infections	305	362
Kind 1985[45]	315	NA	350	416
Corby 1986[46]	36	NA	345	410
Chamberlain 1988[47]	6	Osteomyelitis	265	316
Kane 1988[48]	27	Cystic fibrosis	618	735
Tice 1991[49]	290	Osteomyelitis	303	360
Williams 1993[50]	56	Cellulitis, osteomyelitis and others	262	312
Williams 1994[51]	58	Pneumonia	252	300
Clopes 1998[29]	13	Several	152	180

Table 3 Opportunistic disease in AIDS candidates for home care

Infection	Antimicrobial therapy
Cytomegalovirus infection	Maintenance and induction therapies: Ganciclovir IV, Foscarnet, Cidofovir
Acyclovir-resistant *Herpes simplex*	Foscarnet
Acyclovir-resistant *Herpes zoster*	Foscarnet
Pneumocystis carinii pneumonia	Pentamidine IV Pentamidine aerosol
Cryptococcosis	Amphotericin B
Histoplasmosis	Amphotericin B
Coccidiomycosis	Amphotericin B
Drug-resistant mycobacterium	Amikacin
Pneumonia	3rd Generation Cephalosporins Aminoglycosides

tibiotic therapy candidates for home care. The reasons are that the patient is clinically stable and requires long-term therapy. In some cases the induction can also be considered to be treated at home. These infections and treatments are described in Table 3.

Other support therapies for AIDS patients that can be given at home are nutrition support, parenteral and enteral, IV immunoglobulins, chemotherapy in lymphoma or Kaposi's sarcoma, and care of terminally ill patients.

Cystic Fibrosis

The majority of antibiotics needed for the treatment of infectious complications of cystic fibrosis have to be administered intravenously for several weeks; until recently these treatments were given on an in-patient basis. As the lung disease progresses, patients may require more frequent hospitalizations. This greatly increases health care costs and adversely affects the patient's quality of life.[23]

Home intravenous therapy in cystic fibrosis may also cut costs by avoiding hospital admissions and may improve family life and psychological well-being.

Palliative Care

Some trials have evaluated the effectiveness of hospital at home for terminally ill patients.[20,24] Patients and caregivers receiving hospital-at-home care reported greater satisfaction than those in the hospital group.

One of the fundamental pharmacological treatments in this group of patients is the opioid continuous infusion with devices adapted to outpatient treatments as patient-controlled analgesia pump.

Oncology Patients

The administration of chemotherapy at home has demonstrated that it is feasible and that it produces a decrease of adverse effects and an improvement of the quality of life and a monetary savings.[25]

However, home care can also give support to the patient with cancer in other areas: parenteral antibiotics in febrile neutropenia, nutrition and fluid support, or pain support.

Hematology Patients

In the support of hematology patients, the therapy candidates for home care may be chemotherapy, IV antibiotics in febrile neutropenia, blood products, IV immunoglobulins, fluid/electrolyte replacement, central line maintenance, and specific treatments such as deferoxamine administration.

In the support of hematopoietic stem cell transplantation there are programs developed to permit treatment with chemotherapy at home and treatment of complications.[26]

Cardiology Patients

In cardiology patients some home-based interventions have been published on the treatment of heart failure patients and heart transplant patients. In our center we also have experience with patients with pulmonary hypertension.

Increased survival of cardiovascular disease has resulted in a significant rise in the number of patients with chronic, refractory heart failure requiring intensive medical management and follow-up. In a controlled study,[27] among a cohort of high-risk patients with congestive heart failure, beneficial effects of a postdischarge home-based intervention were sustained for at least 18 months, with a significant reduction in unplanned readmissions, total hospital stay, hospital-based costs, and mortality.

Cardiac patients receiving inotropic therapy can be successfully treated in the home using specific admission criteria and monitoring guidelines,[28] and home dobutamine infusions can improve functional status and quality of life of patients with severe heart failure.

Home care can also give support to heart transplant patients. In our center we have the experience of a program for organ rejection therapy, antilymphocyte immunoglobulin, and high dosage of methylprednisolone at home. After the experience, we can say that it is feasible to carry out this treatment at home and the satisfaction of the patients is high.[29]

There are experiences of ambulatory treatment of patients with pulmonary hypertension, with inhaled nitric oxide and with prostaglandins, in both cases using an ambulatory delivery system. In our center we have an outpatient treatment program of pulmonary hypertension with inhaled iloprost leading in some patients to significant improvement in pulmonary hypertension and in the quality of life with no adverse effects.

Nutrition Support

The candidates for home nutrition support should be clinically stable patients that require enteral or parenteral nutrition for a long term. Before initiation of home nutrition support, a nutrition assessment and a care plan should be performed and after initiation nutrition status should be monitored on a regular basis.[30]

The indications included in a study of incidence of home nutrition support made by the American Society for Parenteral and Enteral Nutrition were:

Pareteral nutrition:

- Short bowel disease.
- Crohn's disease and ulcerative colitis.
- Gastric or duodenal fistula.
- Radiotherapy damage.
- Congenital disorders.
- Disorder of the GI motility.

Enteral nutrition:

- Neuromuscular diseases:
    - Amyotrophic lateral sclerosis
    - Myasthenia gravis
    - Parkinson's disease
    - Alzheimer's disease
    - Cerebral palsy
    - Cerebral vascular accident
    - Brain tumors
- Oral and GI diseases:
    - Secondary to surgical procedure:
        - Head and neck
        - Esophagus or stomach
    - Malabsorption
    - Disorder of GI motility
    - Crohn's disease and ulcerative colitis

Elderly Patients

Home care in elderly patients can help with the geriatric assessment of disability and functional status and the prevention of complications related or not related to drugs. Stuck et al. conducted a three-year, randomized, controlled trial of the effect of annual in-home comprehensive geriatric assessment and follow-up for people who were 75 years of age or older.[31] The results showed that this intervention can delay the development of disability and can reduce permanent nursing home stays among elderly people living at home.

Pediatric Patients

Some programs of home care have been applied to pediatric patients. Home care for cystic fibrosis and oncology patients is previously described. Other examples of home care programs in children are patients with asthma that require high technology at home,[32] children with newly diagnosed diabetes,[33] and infants who require neonatal special care and a family support program.[34]

Surgery and Obstetric Patients

Because of developments in surgery, early discharge after surgery is becoming popular. These programs sometimes need support at home, for example, with rehabilitation.[35,36] In obstetric patients there have been experiences of home care giving support to the woman before or after childbirth.[37]

Others

Other home care programs with smaller pharmacist implications are long-term mechanical ventilation and renal dialysis.

PHARMACIST'S ROLE

The pharmacist's role in home care should include the following functions:

1. Development of protocols or practice guidelines for each diagnosis candidate for inclusion in the home care program in collaboration with other home care team members.
2. Preadmission assessment. The pharmacist, together with the other team members, should assess

the patient's suitability for home care on the basis of criteria described in the protocol (home environment, psychosocial factors, and clinical condition).

3. Coordination of preparation and delivery of drugs to the patient or caregiver. Together with the medication, the pharmacist should provide the ancillary supplies and drug delivery systems. The pharmacist should also ensure appropriate disposal of cytotoxic products.
4. Planning home treatment and care, also with an interdisciplinary approach, involving the patient and in collaboration with other team members. This home treatment and care plan should be reviewed and updated periodically and outcome should be assessed.
5. Therapy monitoring using parameters previously defined in the protocol. The pharmacist should carry out this monitoring from the medical records and verbal exchange with the patient and/or the caregiver, nurse, physician, and other family members.
6. Patient and caregiver education about the treatment. The information should be both oral and written and include:

   - Description of the therapy (drug, dose, route of administration).
   - Goal of the therapy.
   - Administration technique.
   - Special precautions regarding storage, handling, and disposal of drugs.
   - Emergency procedure.

7. Information for home care team members regarding:

   - Drug stability and compatibility.
   - Adverse effects.
   - Administration technique.

8. Early detection and reporting of adverse drug effects.
9. Monitoring pharmacokinetic laboratory data for evaluation of efficacy and prevention of adverse effects of the specific drugs (vancomicyn, aminoglycosides, cyclosporin, etc.).
10. Selection of drug delivery systems for parenteral and inhaled drugs. This selection should be carried out in cooperation with the physician and nurse taking into account safety features, ease in handling, and cost. It should be individualized according to the patient's characteristics.
11. Participation in performance improvement activities. Patient satisfaction and outcome should be monitored to detect and resolve problems. Quality of life should also be considered.

WEB SITES OF INTEREST

- Section of Home Care Practitioners of American Society of Health-System Pharmacists (ASHP)(USA): www.ashp.org/homecare.
- Home Care Highlights of American Society of Health-System Pharmacists (USA): www.ashp.org/public/news/newsletters/homecare/index.html.
- ASHP Guidelines (USA): www.ashp.org/bestpractices.
- American Society of Parenteral and Enteral Nutrition (with the Standards of Practice for Home Nutrition Support)(USA): www.clinnutr.org.
- Joint Comission on Accreditation of Healthcare Organizations (USA): www.jcaho.org.
- American Academy of Hospice and Palliative Medicina (USA): www.aahpm.org.
- Edmonton Palliative Care Program (Canada): www.palliative.org.
- National Council for Hospice and Specialist Palliative Care Services (United Kingdom): www.hospice-spc-council.org.uk.
- Agence Nationale d'Accreditation d'Evaluation en Sante, France (with the recommendations for the medical records of the home care patients for ambulatory nursing professionals) (France): www.anaes.fr.
- American College of Chest Physicians (ACCP) (Patient Education Guides: Mechanical Ventilation at Home) (USA): www.chestnet.org/health.science.policy.

REFERENCES

1. Sheppard, S.; Iliffe, S. Hospital-At-Home Versus In-Patient Hospital Care (Cochrane Review). In *The Cochrane Library, Issue 3*; Update Software: Oxford, 2000.
2. Morris, D.E. Sante Service Bayonne: A French approach to home care. Age Ageing **1983**, *12*, 323–328.
3. Frasca, C.; Christy, M.W. Assuring continuity of care through a hospital based home heath agency. Qual. Rev. Bull. **1986**, *12*, 167–171.
4. Bosna, E. KITTZ: Innovation in Home Care. In *Capital Conference*; King's Fund Centre: London, 1993.
5. Loader, J.; Sewell, G.; Gamme, S. Survey of home infusion care in England. Am. J. Health-Syst. Pharm. **2000**, *57*, 763–766.

6. Coast, J.; Richards, S.H.; Peters, T.J.; Gunnell, D.J.; Darlow, M.A.; Pounsford, J. Hospital at home or acute hospital care? A cost minimisation analysis. BMJ **1998**, *316*, 1802–1806.
7. Chamberlain, T.; Lehman, M.; Groh, M.; Munroe, W.; Reinders, T. Cost analysis of a home intravenous antibiotic program. Am. J. Hosp. Pharm. **1988**, *45*, 2341–2345.
8. Sarabia, J. Análisis económico de la hospitalización a domicilio y su comparación con los costes hospitalarios. Rev. Seg. Soc. **1983**, *17*, 131–140.
9. Hughes, S.L.; Ulasevich, A.; Weaveer, F.M.; Henderson, W.; Manheim, L.; Kubal, J.D.; Bonarigo, F. Impact of home care on hospital days: A meta analysis. Health Serv. Res. **1997**, *32*, 415–432.
10. Shepperd, S.; Harwood, D.; Jenkinson, C.; Gray, A.; Vessey, M.; Morgan, P. Randomised controlled trial comparing hospital at home care with inpatient hospital care. I: Three month follow up of health outcomes. BMJ **1998**, *316*, 1786–1791.
11. Richards, S.H.; Coast, J.; Gunnell, D.J.; Peters, T.J.; Pounsford, J.; Darlow, M.A. Randomised controlled trial effectiveness and acceptability of an early discharge, hospital at home scheme with acute hospital care. BMJ **1998**, *316*, 1796–1801.
12. Goldberg, A.I.; Faure, E. Home care for life supported persons in England: The responaut preogram. Chest **1984**, *86*, 910–914.
13. Jennings, P. *An Evaluation of a Hospital at Home Service for Children*; NHS Trust and Department of Health: South Buckinghamshire, 1993.
14. Holdsworth, M.T.; Raisch, D.W.; Chavez, C.M.; Duncan, M.H.; Parasuraman, T.V.; Cox, F.M. Economic impact with home delivery of chemotherapy to pediatric oncology patients. Ann. Pharmacother. **1997**, *31*, 140–148.
15. Mughal, M.; Irving, M. Home parenteral nutrition in the United Kingdom and Ireland. Lancet **1995**, *16*, 383–387.
16. Goldthorpe, P.; Hodgson, E.; Evans, E.; Bradley, J.G. How we set up an audit of hospital at home in orthopaedic surgery. Med. Audits News **1994**, *4*, 40–43.
17. O'Caithain, A. Evaluation of a hospital at home scheme for the early discharge of patients with fractured neck of femur. J. Public Health Med. **1995**, *16*, 205–210.
18. Pryor, G.A.; Williams, D.R. Rehabilitation after hip fractures. Home and hospital management compared. J. Bone Jt. Surg., Br. Vol. **1989**, *71*, 471–474.
19. Hollingsworth, W.; Todd, C.; Parkr, M.; Roberts, J.A.; Williams, R. Cost analysis of early discharge after hip fracture. BMJ **1993**, *307*, 903–906.
20. Hughes, S.L.; Cummings, J.; Weaver, F.; Manheim, L.; Braun, B.; Conrad, K. A randomized trial of the cost effectiveness of a VA hospital-based home care for terminally ill. Health Serv. Res. **1992**, *26*, 801–817.
21. Poretz, D.M. Introduction. Outpatient use of intravenous antibiotics. Am. J. Med. **1994**, *97*, 1–2.
22. Hazas, J.; Sampedro, I.; Fernández-Miera, M.F.; García de la Paz, A.M.; Sanroma, P. Un programa de antibioterapia intravenosa domiciliaria. Enferm. Infecc. Microbiol. Clin. **1999**, *17*, 463–469.
23. Marco, T.; Asensio, O.; Bosque, M.; de Gracia, J.; Serra, C. Home intravenous antibiotics for cystic fibrosis (Protocol for a Cochrane Review). In *The Cochrane Library, Issue 3*; Update Software: Oxford, 2000.
24. Grande, G.E.; Todd, C.J.; Barclay, S.I.; Farquhar, M.C. Does hospital at home for palliative care facilitate death at home? Randomised controlled trial. BMJ **1999**, *319*, 1472–1475.
25. Holdsworth, M.T.; Raisch, D.W.; Chavez, C.M.; Duncan, M.H.; Parasuraman, T.V.; Cox, F.M. Economic impact with home delivery of chemotherapy to pediatric oncology patients. Ann. Pharmacother. **1997**, *31*, 140–148.
26. Herrmann, R.P.; Leather, M.; Leather, H.L.; Leen, K. Clinical care for patients receiving autologous hematopoietic stem cell transplantation in the home setting. Oncol. Nurs. Forum **1998**, *25*, 1427–1432.
27. Stewart, S.; Vandenbroek, A.J.; Pearson, S.; Horowitz, J.D. Prolonged beneficila effects of a home-based intervention on unplanned readmissions and mortality among patients with congestive heart failure. Arch. Intern. Med. **1999**, *159*, 257–261.
28. Mayes, J.; Carter, C.; Adams, J.E. Inotropic therapy in the home care setting: Criteria, management and implications. J. Intraven. Nurs. **1995**, *18*, 301–306.
29. Clopes, A. *Hospitalización Domiciliaria: Participación del Servicio de Farmacia*; Reunión Internacional Farmacia Clínica y Atención Farmacéutica: Barcelona, Spain, June 4–6, 1998.
30. American Society for Parenteral and Enteral Nutrition (ASPEN). Standars of practice for home nutrition support. NCP **1999**, *14*, 151–162.
31. Stuck, A.E.; Aronow, H.U.; Steiner, A.; Alessi, C.A.; Bula, C.J.; Gold, M.N.; Yuhas, K.E.; Nisenbaum, R.; Rubenstein, L.Z.; Beck, J.C. A trial of annual in-home comprehensive geriatric assessments for elderly people living in the community. N. Engl. J. Med. **1995**, *333*, 1184–1189.
32. Ryan, C.A.; Willan, A.R.; Wherrett, B.A. Home nebulizers in childhood asthma. Clin. Pediatr. **1988**, *27*, 420–424.
33. Dougherty, G.E.; Soderstrom, L.; Schiffrin, A. An economic evaluation of home care for children with newly diagnosed diabetes: Results from a randomized controlled trial. Med. Care **1998**, *36*, 586–598.
34. Rieger, I.D.; Henderson-Smart, D.J. A neonatal early discharge and home support programme: Shifting care into the community. J. Pediatr. Child Health **1995**, *31*, 33–37.
35. Worland, R.L.; Arredondo, J.; Angles, F.; Lopez-Jimenez, F.; Jessup, D.E. Home continous passive motion machine versus professional physical therapy following total knee replacement. J. Arthroplast. **1998**, *13*, 784–787.
36. Donald, I.P.; Baldwin, R.N.; Bannerjee, M. Gloucester hospital-at-home: A randomized controlled trial. Age Ageing **1995**, *24*, 434–439.

37. Brooten, D.; Knapp, H.; Borucki, L.; Jacobsen, B.; Finkler, S.; Arnold, L.; Mennuti, M. Early discharge and home care after unplanned cesarean birth: Nursing care time. J. Obstet. Gynaecol. Neonat. Nurs. **1996**, *25*, 595–600.
38. Antoniskis, A.; Anderson, B.C.; Van Volkinburg, E.J.; Jackson, J.M.; Gilbert, D.N. Feasibility of outpatient self-administration of parenteral antibiotics. West J. Med. **1978**, *128*, 203–206.
39. Stiver, H.G.; Telford, G.O.; Mossey, J.M. Intravenous antibiotic therapy at home. Ann. Intern. Med. **1978**, *89*, 690–693.
40. Kind, A.C.; Williams, D.N.; Persons, G.; Gibson, J.A. Intravenous antibiotic therapy at home. Arch. Intern. Med. **1979**, *139*, 413–415.
41. Swenson, J.P. Training patients to administer intravenous antibiotics at home. Am. J. Hosp. Pharm. **1981**, *38*, 1480–1483.
42. Poretz, D.M.; Eron, L.J.; Goldenberg, R.I. Intravenous antibiotic therapy in an outpatient setting. JAMA, J. Am. Med. Assoc. **1982**, *248*, 336–339.
43. Stiver, H.G.; Trosky, S.K.; Cote, D.D.; Oruck, J.L. Self-administration of intravenous antibiotics: An efficient, cost-effective home care program. J. Can. Med. Assoc. **1982**, *127*, 207–211.
44. Rehm, S.J.; Weinstein, A.J. Home intravenous antibiotic therapy: A team approach. Ann. Intern. Med. **1983**, *99*, 388–392.
45. Kind, A.C.; Williams, D.N.; Gibson, J. Outpatient intravenous antibiotic therapy: Ten year's experience. Postgrad. Med. **1985**, *77*, 105–111.
46. Corby, D.; Schad, R.F.; Fudge, J.P. Intravenous antibiotic therapy: Hospital to home. Nurs. Manage. **1986**, *17*, 52–61.
47. Chamberlain, T.M.; Lehman, M.E.; Groh, M.J.; Munroe, W.P.; Reinders, T.P. Cost analysis of a home intravenous antibiotic program. Am. J. Hosp. Pharm. **1988**, *45*, 2341–2345.
48. Kane, R.E.; Jennison, K.; Wood, C.; Black, P.G.; Herbst, J.J. Cost savings and economic considerations using home intravenous antibiotic therapy for cystic fibrosis patients. Pediatr. Pulmonol. **1988**, *4*, 84–89.
49. Tice, A.D. An office model of outpatient parenteral antibiotic therapy. Rev. Infect. Dis. **1991**, *13* (S2), 184–188.
50. Williams, D.N.; Bosch, D.; Boots, J.; Schneider, J. Safety, efficacy, and cost savings in an outpatient intravenous antibiotic program. Clin. Ther. **1993**, *15*, 169–179.
51. Williams, D.N. Reducing costs and hospital stay for pneumonia with home computerized cefotaxime treatment results with a computerized ambulatory drug delivery system. Am. J. Med. **1994**, *97* (2A), 50–55.

Hospice and Palliative Care

Arthur G. Lipman
University of Utah, Salt Lake City, Utah, U.S.A.

INTRODUCTION

Hospice programs have existed in the United States for more than 25 years, providing symptom control-based palliative care for patients with advanced, life-limiting disease. In fact, more than 3000 hospice programs are now operating or are in formation in the United States. The World Health Organization (WHO) defines and comments on palliative care as follows: ''Palliative care is active total care of patients whose disease is not responsive to curative treatment. Control of pain, of other symptoms, and of psychological, social and spiritual problems is paramount. The goal of palliative care is achievement of the best possible quality of life for patients and their families.''[1] Pharmacists often serve as key members of hospice interdisciplinary teams, and many opportunities for pharmacists to provide valuable clinical services exist in hospice programs. Most hospice care is provided in the patients' homes. Palliative care units are increasingly being integrated into hospitals and long-term care facilities.

OVERVIEW

The explosive growth of interdisciplinary hospice and palliative care programs for patients with terminal illnesses has created excellent opportunities for pharmaceutical care. In the United States, the number of hospice programs has grown from 1 less than 30 years ago to approximately 3000 today. The importance of pharmacists providing care to terminally ill patients appeared in the American pharmaceutical literature over 25 years ago.[2] In addition, rapidly expanding opportunities for pharmacists in hospice care were defined in the 1990s.[3] All pharmacists should know about the availability and quality of hospice care in their communities, and should be able to refer patients to programs appropriate for their needs.

In the 1970s and 1980s, hospices provided care primarily for patients with advanced cancer. Today, hospice care is common for patients with cancer; acquired immunodeficiency syndrome (AIDS); degenerative neurological diseases, such as multiple sclerosis and amyotrophic lateral sclerosis; end-stage organ system failure, including congestive heart failure, hepatic disease, pulmonary disease, and renal disease; and patients with dementia and other progressive, irreversible disorders.

The word ''hospice'' is derived from a medieval French term for resting places established for Crusaders on their journeys to the Holy Land. It was revived in the last century by a Catholic order that provided resting places for terminally ill patients in Ireland and England. By the mid-1900s, several such hospice programs existed in the United Kingdom. However, the modern hospice movement based on comprehensive symptom control only began in 1967, with the opening of St. Christopher's Hospice in London. The first American hospice—originally called simply Hospice, Inc., now The Connecticut Hospice—was started in the early 1970s in New Haven, CT. That program became the National Cancer Institute Demonstration Project of Hospice Care from 1974 to 1977. More than 1000 American pharmacists are now estimated to provide hospice pharmaceutical care as integral parts of their practices. Many more are needed.

A hospice is a program of care, not necessarily a facility, per se. In the United States, most hospice care is provided in patients' homes. Some dedicated inpatient hospice facilities exist, as do hospice wings of long-term care facilities and hospice beds in hospitals. These inpatient hospices commonly provide support for the home care programs, respite care (admission of patients to allow their families to rest so that they can resume home care), admissions for difficult symptom control problems, and admissions for care in the last hours or days, when necessary.

The term ''palliative care'' was used initially to define the provision of symptom relief for patients who were no longer considered to be candidates for cure or remission. Today, the need for palliative care throughout the course of life-threatening disease, including patients for whom cure will be achieved, is becoming more widely accepted. Palliative medicine is a recognized medical specialty in the United Kingdom and several other countries. In 1997,

Encyclopedia of Clinical Pharmacy
DOI: 10.1081/E-ECP 120006340

the report of the Committee on Care at the End of Life of the Institute of Medicine of the National Academy of Sciences concluded that ''palliative care should become, if not a medical specialty, at least a defined area of expertise, education and research.''[4] Hospice and palliative care are often used interchangeably.

Hospices provide care for patients with advanced, irreversible disease and a life expectancy measurable in weeks to months as opposed to years. The defined unit of care is the patient and their family. The focus of care spans physical, psychological, social, and spiritual domains. This requires an interdisciplinary team. Nurses usually coordinate home visits and serve as team leaders. The patient's primary care physician normally continues to provide care, often in consultation with the hospice medical director. Other key members of the team are social workers, nursing assistants/home health aides, chaplains, volunteers, and pharmacists. Persons from several other disciplines support the hospice team (Fig. 1). As shown in Fig. 1, the central focus of care is the patient, family, and primary care person(s), who is usually a family member. They continue to work with their primary physician. The interdisciplinary hospice team listed in the next concentric circle from the center provides direct support. This team may include both health care professionals and other persons who are equipped to deal with issues that are complicating the lives of the patients/families (e.g., financial counselors). The third concentric circle from the center includes persons who support the team. Pharmacists have both direct patient care and supportive roles in hospice teams as described in the following paragraphs.

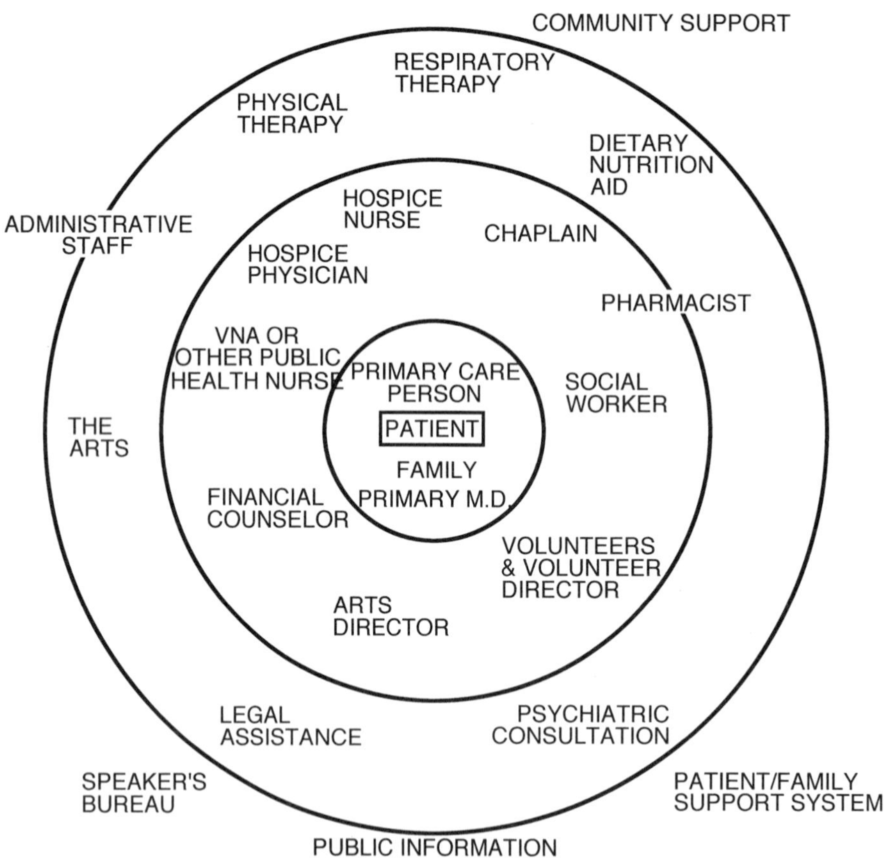

Fig. 1 The hospice interdisciplinary team. The patient, primary caregiver, and family are the focus of the hospice team's efforts in collaboration with the patient's primary physician. The core team is represented by the next circle away from the center. The support team is indicated by the outer circle. Community resources that support hospice care are listed outside that circle. Pharmacists serve on both the core team (second circle from the center) by providing direct pharmaceutical care to patients and families, and on the support level (next circle out from the center) by providing professional and public education about drug therapy in the care of terminally ill patients. (From Lipman AG, Berry JI. Pharmaceutical care of terminally ill patients. *Journal of Pharmaceutical Care Pain and Symptom Control*, 1996; *3*(2):31–56.)

There is a need for elimination of artificial barriers between the time when a cure is sought and the inevitability of death is accepted. This barrier exists, at least in part, due to the requirement for documenting life expectancy by the Medicare Hospice Benefit. The U.S. Congress defined this benefit in the 1980s through which Medicare beneficiaries can assign their medicare part B benefits to any Medicare-certified hospice program. That program then receives a daily fee from Medicare in return for assuming responsibility for the patient's total care, including drugs and pharmaceutical care. To be eligible for this benefit, the patient's physician must certify a probable life expectancy of 6 months or less. This arbitrary time limit has created psychological barriers for physicians, patients, and their families, resulting in many patients not being referred, seeking, or receiving the hospice care to which they are entitled. Because pharmacists commonly have long-standing relationships with families they serve and enjoy their patients' trust, pharmacists are often in the best position to advise patients about the importance of developing relationships with a hospice program as soon as possible after determination that the disease has a probability of being life ending.

Hospice and palliative care are becoming much more widely recognized by healthcare providers. The American Academy of Hospice and Palliative Medicine (founded in 1988 as the Academy of Hospice Physicians) and the Association of Hospice Nurses are respected national organizations of health care professionals who provide palliative care. The National Hospice and Palliative Care Organization (NHPCO, formerly known as the National Hospice Organization [NHO]) includes the National Council of Hospice Professionals (NCHP). The 15 membership sections of the NCHP include an active pharmacist section.

PHARMACEUTICAL CARE OPPORTUNITIES

It is unfortunate that many physicians and families remain unaware of the benefits that modern hospice care provides. As a result, referrals to hospice programs often do not occur, or occur when the patient has only days to live. Hospice care is most efficacious and cost effective when referrals are made early, while the patient still has months to live and is reasonably active. Relationships between the hospice team and the patient/family that are established before crises occur are most effective. Such relatively early relationships permit the hospice team to provide more effective and efficient care when it is actively needed. As the most accessible and trusted healthcare professionals (Gallup surveys), pharmacists are often in an excellent position to recommend hospice care and to refer families to appropriate programs.

Services provided by pharmacists in American hospices have only been qualitatively and quantitatively documented twice, in 1979[5] and 1991.[6] Many more pharmacists provide these services today than when the latter survey was completed, but the observed types and mix of services do not appear to have changed much since the 1990s.

Although many pharmacists serve as hospice volunteers, about three-fourths are paid for their services. The majority of pharmacists who provide services to hospice programs are not employed directly by the hospices, but by a provider of pharmaceutical services such as a home health pharmacy or hospital. Many are employees of pharmacies that have contracts with hospices to provide drugs and services. In recent years, specialized hospice pharmacy service providers have been developed in several parts of the United States.

The 1991 survey[6] reported that dispensing fees accounted for about one-half of the reimbursement received by pharmacists. In the past few years, payment for cognitive services has become more common. Some pharmacists provide only consulting or dispensing services, but many provide drug products, home health supplies and equipment, and pharmaceutical care. Pharmaceutical services other than prescription dispensing services are not usually required by licensing or certifying agencies. Payment for cognitive services is at the discretion of the hospice administration. The experience of many hospices has been that integration of pharmacists directly into planning and provision of patient care both improves the quality of symptom control and lowers costs. In many hospices, pharmacists are now active participants in weekly or biweekly interdisciplinary team (IDT) meetings at which patients' progress is discussed and care plans are refined.

HOW TO GET STARTED

Most pharmacists possess many of the skills needed to provide pharmaceutical care to terminally ill patients. In the last few years, pharmacy curricula have placed increased emphasis on pain management and symptom control.[7]

Many pharmacists increase their knowledge of drugs and dosing regimens for symptom control in seriously ill patients through consultation and visits with experienced hospice pharmacists. Pharmacists can gain a valuable perspective on hospice care by taking hospice volunteer training. Continuing pharmaceutical education directly

Table 1 Selected palliative care resources

Journals

Journal of Pain and Palliative Care Pharmacotherapy (incorporating the former *Journal of Pharmaceutical Care in Pain and Symptom Control* and *The Hospice Journal*)
Pharmaceutical Products Press, an imprint of The Haworth Press, 10 Alice Street, Binghamton, New York; (800) HAWORTH; e-mail: getinfo@haworth.com

Journal of Pain and Symptom Management
Elsevier Science, Inc.; (888) 437-4636

Pain
Journal of the International Association for the Study of Pain (IASP); (206) 547-6409

The Journal of Pain
Official Journal of the American Pain Society; (800) 654-2452; e-mail: info@ampainsoc.org

Newsletters

IASP (International Association for the Study of Pain) Newsletter
(206) 547-6409

American Pain Society Bulletin
American Pain Society; (847) 375-4715; e-mail: info@ampainsoc.org

Texts

Berger AM, Portenoy RK, Weissman DE. *Principles and Practice of Supportive Oncology*. Philadelphia, Lippincott-Raven, 1998. Doyle D, Hanks GWC, MacDonald N, editors. *Oxford Textbook of Palliative Medicine*, 2nd edition. New York and Oxford, Oxford University Press, 1997. Berger AM, Portenoy RK, Weissman DE. *Principles and Practice of Supportive Oncology*, 2nd Ed.; Philadelphia, Lippincott-Raven, in press 2002.

Web sites

National Hospice and Palliative Care Organization
www.nho.org

PDQ (Physician Data Query)
www.cancernet@icii.nci.nih.gov/

Talarian Map Cancer Pain
www.stat.washington.edu/TALARIA/TALARIA.html

Open Society Institute: Project Death in America
www.cyberspy.com/~webster/death.html

The Palliative Medicine Program
www.mcw.edu/pallmed

Hospice Foundation of America
www.hospicefoundation.org

Information about hopsice with links
www.hopsiceweb.com

Hospice Hands web site
http://hospice-cares.com

Purdue Pharma Pain and Palliative Care Information
http://www.partnersagainstpain.com

Additional web references can be found in Ref. [9].

relevant to hospice care and symptom control is often provided at meetings of the American Society of Consultant Pharmacists, American Society of Health-System (previously Hospital) Pharmacists, American Pharmaceutical Association, the National Hospice and Palliative Care Organization, and at some state and local professional associations. Several journals, newsletters, and web sites focus on pain and symptom control. Examples are listed in Table 1. Because hospice care is interdisciplinary by definition, most programs are open to suggestions of additional ways in which any discipline can contribute to the program's overall objectives.

Hospice programs are always recruiting and training new volunteers. Therefore, most readily welcome calls from persons in the community interested in learning more about the program or becoming involved in patient care. Any pharmacist can simply call a local hospice and make an appointment to meet with the staff to discuss unmet pharmaceutical care needs. These include a range of activities, including administrative responsibilities, provision of medications, and outcome-oriented pharmaceutical care.

Common administrative functions include the following:

- Managing program or facility (if applicable)
- Serving on the hospice board or professional advisory committee
- Negotiating contracts with provider pharmacies
- Reviewing and ensuring compliance with state and federal laws and regulations that relate to the provision of hospice pharmaceutical care and services
- Developing drug-related policies and procedures
- Participating in continuous quality improvement and quality assurance activities, including drug-use evaluations and cost-avoidance and cost-effectiveness studies
- Procuring medications for indigent patients through pharmaceutical industry patient assistance programs
- Managing the hospice formulary

Common clinical functions include the following:

- Developing pharmaceutical care plans, including assessment and monitoring for therapeutic and toxic outcomes
- Participating in hospice interdisciplinary team meetings (chart rounds)
- Performing drug regimen reviews
- Providing pain and symptom management consultations to team members and to patients' primary physicians
- Preparing routine admission orders
- Developing drug-use protocols
- Making home visits as needed to assess medication needs and use, and to educate patients/families about medication use

Common educational functions include:

- Providing staff education in drug therapy for symptom control and other indications
- Providing education to patients and their families on medication use
- Providing physician education to hospice patient primary physicians
- Providing public education on drug use in terminal care
- Educating hospice volunteers about desired and achievable outcomes from medication use
- Providing clerkships for pharmacy students

Common dispensing functions include the following:

- Dispensing prescription and over-the-counter medications, including therapeutic interchange
- Providing for delivery of medications to patients' homes
- Extemporaneous compounding of dosage forms that are not commercially available
- Providing home infusion service
- Maintaining patient medication profiles

MARKETING PHARMACEUTICAL CARE SKILLS

The broad range of relevant pharmaceutical services needed by a progressive hospice program nearly always requires more than one provider. Simple, informal needs assessments of programs with which pharmacists want to affiliate is an effective way to market their services. Hospice Medicare payments and most other insurance reimbursement is capitated (i.e., a flat daily fee is paid to the program for all aspects of care). Therefore, the full range of care must be provided within a defined cost structure. Efficient formulary management, including generic and therapeutic interchange and elimination of unneeded drug therapy, can improve both patient care and the fiscal health of the program. Patient and family satisfaction are also important considerations for every hospice program. Medication-related education provided by a pharmacist can markedly increase satisfaction. Hospice nurses often work relatively independently from their patients' physicians. Therefore, by providing nursing education and consultation about patient assessment for responses to therapy and about drug use, pharmacists can increase their perceived need on the hospice team.

DOCUMENTATION FROM CARE PROVIDERS

Most hospice referrals come to the programs from patients' primary physicians. Some come directly from families who have heard about hospice from other families that used the service or from presentations made in the community. Most hospice programs send a nurse to the patient's home (hospital or nursing home) to assess the patient and to perform an intake evaluation. This evaluation requires a detailed history, including a medication history.

Pharmacists need to know patients' prescription and nonprescription medication intake; use of nutritional supplements that may be pharmacologically active, physical, and psychiatric diagnoses; and relevant laboratory test data when they are available. Usually, that information is available from the primary physician's referral and the documentation from the intake interview. Frequently, laboratory test data are not available because of the hospice philosophy of only doing tests that will directly affect patient outcomes. Renal function often can be estimated from the quantity and quality of the patient's urinary output balanced against intake. Careful dose titration is often needed in the absence of laboratory test data as patients' metabolic and elimination capabilities decline. Sometimes, pharmacists make home visits to get more complete medication histories, and to ascertain the family's understanding of medications and ability to administer them correctly.

CONCLUSION

Most pharmacists will interact with terminally ill patients or their family members at some time. Many pharmacists will provide services to dying patients and hospice programs. An increasing number of pharmacists will work with hospice programs as a substantial part of their practices.

Effective management of pain and other symptoms associated with life-threatening disease is usually attainable with the proper combination of pharmacological and nonpharmacological interventions.[8] Pharmacists can, and should, play an important role in ensuring that their patients receive this care when it is needed.

REFERENCES

1. WHO Expert Committee. *Cancer Pain and Palliative Care*; Technical Report Series, World Health Organization: Geneva, 1990; Vol. 804.
2. Lipman, A.G. Drug therapy for terminally ill patients. Am. J. Hosp. Pharm. **1975**, *32*, 270–276.
3. Arter, S.G.; Lipman, A.G. Hospice care; a new opportunity for pharmacists. J. Pharm. Pract. **1990**, *3*, 28–33.
4. *Approaching Death: Improving Care at the End of Life*; Field, M.J., Casell, C.K., Eds.; National Academy Press: Washington, 1997.
5. Berry, J.I.; Pulliam, C.C.; Caiola, S.M.; Eckel, F.M. Pharmaceutical services in hospices. Am. J. Hosp. Pharm. **1981**, *38*, 1010–1014.
6. Arter, S.G.; Berry, J.I. The provision of pharmaceutical care to hospice patient: Results of the national hospice pharmacist survey. J. Pharm. Care Pain Symptom Control **1993**, *1* (1), 25–42.
7. Lipman, A.G. Curriculum on pain for pharmacy students. IASP Newsl. **1992 May/June**, 2–4.
8. Jacox, A.; Carr, D.B.; Payne, R., et al. *Management of Cancer Pain*, Clinical Practice Guideline. AHCPR Publication Number 94-0592, Rockville, MD. Agency for Health Care Policy and Research; U.S. Department of Health and Human Services, Public Health Service, 1994.
9. Gavrin, J.R. An annotated guide to pain and palliative care on the World Wide Web. J. Pain Palliat. Care Pharmacotherap. **2002**, *16* (2), 37–48.

Hospital Pharmacy Practice in Spain

Joaquin Giraldez
Ana Ortega
Antonio Idoate
Azucena Aldaz
Carlos Lacasa
Clinica Universitaria de Navarra, Pamplona, Spain

INTRODUCTION

''Hospital pharmacy service'' refers to the pharmacy that is inside a hospital to serve inpatients and outpatients who receive care in the hospital or require drugs that are only delivered in hospitals. ''Hospital pharmacy practice'' makes reference to all activities carried out by hospital pharmacy service personnel to serve those patients.

In Spain, by law, there must be a hospital pharmacy service in every hospital with 100 beds or more.[1] This service must be under the supervision of a hospital pharmacist. The total number of pharmacists depends on different factors such as number of beds, services provided to patients, and type of hospital. All hospital pharmacists working in the service must be hospital pharmacy specialists.

Activities common to all hospital pharmacy services in Spain are pharmacy management, dispensing of drugs, drug information, and drug manufacture. Many other activities are also conducted in many hospital pharmacies such as centralized parenteral admixture preparation, design and preparation of parenteral and enteral nutrition as well as follow-up of patients under this kind of nutrition, therapeutic drug monitoring, pharmacoeconomics, drug surveillance, research, activities related to medical devices, radiopharmaceutical activities, clinical pharmacy activities, pharmaceutical care, participation in committees, and so on.

In what follows, hospital pharmacy practice in Spain will be described. As an introduction, a brief history and description of the evolution of this discipline and the Spanish hospital pharmacists training program will be presented. Then, activities currently conducted by hospital pharmacy service personnel will be described and clinical pharmacy opportunities will be indicated. And finally, future trends will be outlined. Useful references will be given throughout the report.

BRIEF HISTORY OF HOSPITAL PHARMACY IN SPAIN

Pharmacists have always worked with doctors and nurses, in and outside hospitals, but it was not until 1955 that a National Association of Pharmacists from Civil Hospitals was created in Spain. In 1967, the Spanish Public Health Service created its own hospital pharmacy services in state hospitals. In 1977, hospital pharmacy services were regulated as was the training of hospital pharmacists. In 1988 the name of the association changed to the Spanish Society of Hospital Pharmacists, as it is known today.[31] In 1990, the Spanish Parliament approved the ''Medicine Law'',[1] which consolidated hospital pharmacy services as the basic structure for the rational use of drugs and specified the residency program as the training needed to work in those services.

EVOLUTION/TRANSITION

Hospital pharmacy is a discipline in permanent transition. In Fig. 1, activities conducted by hospital pharmacists as well as the number of hospital pharmacists in Spain from 1955 are presented. Originally, hospital pharmacists were responsible for management and delivery of stocks of drugs to the wards. Since then the role of the pharmacist has evolved to include a more rational dispensing system (unit-dose delivery) and clinical activities and pharmaceutical care.

Hospital pharmacists in Spain, as in other countries, are increasing their direct communication with patients, nurses, and doctors and at the same time are transferring some activities to others, such as drug manufacture to the pharmaceutical industry. The therapeutic role of drugs is increasing; furthermore, the responsibility of pharmacists

Encyclopedia of Clinical Pharmacy
DOI: 10.1081/E-ECP 120006377

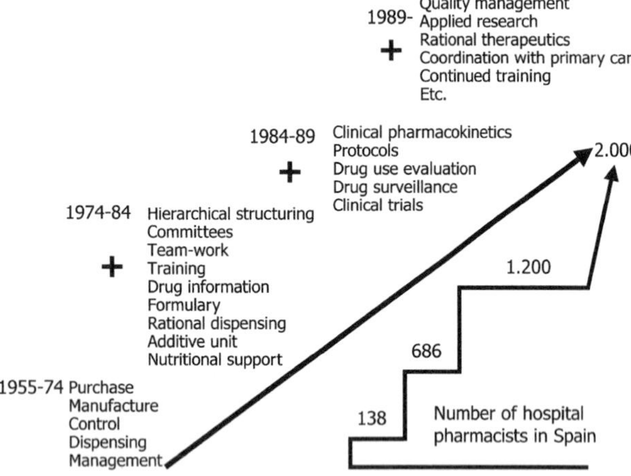

Fig. 1 Activities conducted by hospital pharmacists and the number of hospital pharmacists in Spain since 1955.

goes beyond simple delivery of prescriptions. Clinical pharmacy is appearing as a new culture for professional practice. Clinical pharmacy can be defined as a compound of beliefs, rules, and values that constitute the foundation of the pharmacy practice, cooperation in the health care team, and direct pharmacist interventions.[2] The objective is better patient care. Spanish health care organizations are incorporating this new role of the pharmacist in different ways. An example is the addition of activities involving direct contact with patients (clinical pharmacy and pharmaceutical care) to the hospital pharmacist training program.

Clinical pharmacy is founded on three basic activities: drug selection, drug information, and rational distribution (unit-dose). If these activities are not present, other clinical activities cannot be developed. Drug information, the unit-dose delivery system, parenteral and enteral nutrition programs, therapeutic drug monitoring, participation in clinical trials, and other activities developed by Spanish hospital pharmacists are modest examples of what is known as clinical pharmacy.[2] Clinical pharmacy is slowly changing society's idea of hospital pharmacy in Spain.

THE HOSPITAL PHARMACIST TRAINING PROGRAM IN SPAIN (RESIDENCY)

In Spain, a hospital pharmacy training (residency) is mandatory in order to work as a hospital pharmacist. This specialization has been regulated by law since 1982.[3,29] Until 1999 the residency program lasted for three years; it is now four years. The reason for this extension is that activities conducted by hospital pharmacists have increased considerably and training had to adapt to these changes. Activities outside the pharmacy service and in proximity to the patient and health care team are necessary. In the fourth year, residents are supposed to take their knowledge to the bedside and be with the patient and health care team. They have to take responsibility for the pharmacotherapy given to each patient, work as part of a team, and develop a critical ability to solving all pharmacotherapeutic problems.[4] The residency program is regulated, practice-based, and can be done only in certain accredited hospitals, and students have to first pass a national exam.[3,5,6] Currently, approximately 100 pharmacists per year can be admitted to the residency program, which includes all activities conducted in hospital pharmacy services.

ACTIVITIES CONDUCTED IN HOSPITAL PHARMACY SERVICES: CURRENT SITUATION

Activities developed in a hospital pharmacy service can be conducted either from inside the service or outside of it. In the latter case, activities are obviously connected to centralized activities.

In order to completely understand the situation in Spain, it is important to know first how the Spanish health system works. In Spain, there are public and private hospitals (around 795 hospitals; see Fig. 2).[30] Every Spanish person has the right to free public health care; however, if patients prefer, they can go to a private hospital and pay for the health care that they receive. In addition, some

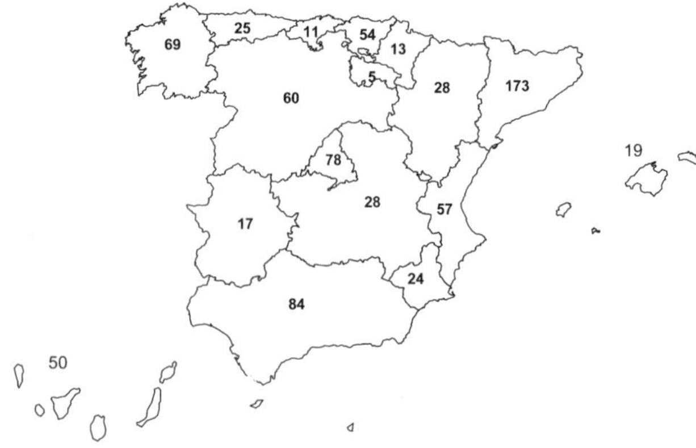

Fig. 2 Number of hospitals per region in Spain (From Ref. [30]).

private hospitals have agreements with the public sector or with insurance companies.

In what follows, some activities conducted by Spanish hospital pharmacists will be briefly described. Three activities are considered the foundation of hospital pharmacy in Spain: adequate drug selection, drug information, and drug delivery. Some Spanish references include most of the activities developed at Spanish hospitals[7,8] as well as statistics on hospital activity.[9] Recommendations of the Spanish Society of Hospital Pharmacists (SEFH) for some of the activities can be found at www.sefh.es/normas/normasy.htm. In addition, there are now many possibilities for networking. The SEFH facilitates interhospital communication and interest groups have been created.[31] Some other international organizations provide the same opportunities for their specific topics (e.g., www.senpe.com/Gtrabaj/textos2.htm for parenteral and enteral nutrition). Statistical data will not be presented here but can be obtained from a survey conducted by the SEFH in 1995;[10] more up-to-date figures will become available from the year 2000 survey.

Management

There are two important areas in management, clinical and purchasing management. In every hospital pharmacy service it is necessary to establish basic procedures for drug selection, acquisition, reception, storage, and distribution with the least cost and risk for patients.

Clinical management refers to an efficient and safe use of drugs according to pharmaceutical criteria. To achieve this goal there are many possible courses of action; however, the most basic one, which is conducted in all Spanish hospital pharmacy services, is the definition of a hospital-specific drug formulary that lists all the drugs approved by the hospital's Pharmacy and Therapeutics (P&T) committee. In Spain a hospital pharmacist is one of the members of P&T committee, frequently the president or the secretary. The P&T committee has the following tasks: to select drugs; to recommend a drug use policy; to educate about correct drug use; to set drug use protocols and establish the means of ensuring compliance; to introduce a program for the detection, follow-up, and evaluation of adverse drug reactions; and to cooperate in a quality control program. Criteria used by the P&T committee for drug selection are, in order of importance: efficacy, safety, cost-effectiveness, therapeutic contribution, and incidence.

Regarding purchasing management, the main responsibility of the purchasing unit of a hospital pharmacy service is to have available the necessary drugs to treat hospital patients. Almost all purchasing units in Spanish hospital pharmacies are computerized and all have the following tasks: to define requested drugs, to establish purchasing procedures according to Spanish law, to place orders, to inform hospital directors of acquisitions, and to develop a quality control program. Spanish references to management techniques are given in the bibliography.[7,8,11]

Drug Dispensing/Distribution

Drug dispensing/distribution is one of the main clinical activities of Spanish hospital pharmacists. Many studies have shown that the unit-dose distribution system has reduced drug errors, and it is one of the main contributions of the hospital pharmacy[12] to patient care. Pharmacist participation in medical rounds and presence at the time of prescription can result in even better patient care and a prompter detection of treatment failures.[13,14] Such "clinical pharmacy" activity is being conducted with some groups of patients in some Spanish hospitals[15] and is becoming more frequent.

Most Spanish hospitals have a unit-dose drug distribution system (Fig. 3). The main objectives of such a system are the following: knowledge of patient pharmacotherapeutic profile, which encourages pharmacist intervention before drug dispensation and administration; decrease of drug errors, interactions, and adverse reactions; reduction in treatment costs; decrease of drug manipulation by nurses on the wards; and billing or economic assignment according to each patient's real expenses.

In a unit-dose system, the pharmacy service delivers drugs to be directly administered to the patient without need of further intervention by others. In hospitals, the distribution of some drugs (e.g., narcotics, compassionate-use drugs, research drugs, and drugs for emergencies) requires a special control and distribution procedure. Normally, these drugs are not sent with the rest of the medication; and the procedure for these drugs will be presented later on. The following is a description of the unit-dose system as it is applied in most Spanish hospital pharmacy services.

Medical orders are handwritten by doctors and a copy is sent to the hospital pharmacy service, where it is recorded in the computer system. However, in some hospitals, doctors enter the medical order directly into the computer; few proceed in this way at the moment but the number is increasing. In a few hospitals, with some types of patients, pharmacists are present at the time of prescription. Prescriptions may specify generic or brand names depending on the hospital's policy, and pharmacists can choose bioequivalent drugs depending on what is available.

Fig. 3 Unit-dose area in a Spanish hospital pharmacy service.

Medical orders are checked by pharmacists, and doctors or nurses are consulted if necessary. At this point, pharmacists have a good opportunity for intervention. To prove the appropriateness of the prescription for a specific patient, patient data must be checked. The unit-dose system is computerized in all hospital pharmacies. Computer programs may be in-house or standard. Some information can be checked on the computer; in some cases programs even make suggestions.[16] Subsequently, lists are created for auxiliary personnel to prepare the delivery trolleys to take the medications to the wards. In a few hospitals, for some specific units, automated delivery (e.g., Pyxis®, Suremed®, Omnicell®) is used. In this case, pharmacists, or someone under their supervision, have to check the accuracy of the delivery content. Quality and security in delivering medication must be fully guaranteed. These systems require a medical order, and information regarding patient name, doctor, and quantity of drug dispensed must be recorded.

In most Spanish hospitals, there is just one delivery a day, in the afternoon, because in many hospitals doctors see patients between 8 A.M. and 3 P.M. However, the number of visiting hours is increasing and pharmacy working procedures may have to adapt to the new situation. Parenteral admixtures and nutritional preparations, if chemically stable, are generally prepared for each patient in a centralized unit (described later), labeled, and then delivered with the rest of the medication. Cytotoxic drugs require special control and handling and are not normally sent with the rest of the medication. Sometimes, in intensive care units and other acute care settings, drug delivery is not based on a unit-dose system but on stocks kept on the wards.

Some outpatient services are provided by the inpatient pharmacy, but discharged patients in Spain cannot receive drugs from the inpatient pharmacy. At discharge, patients may receive drug information and a copy of their medication administration record for reference. Computer software (InfoWin®) has been developed by a Spanish group (with a Spanish drug database) for this purpose.

Drugs that require a special delivery procedure are:

1. *Drugs for compassionate use.* Hospital pharmacists have to control the ordering, dispensing, and use of compassionate-use drugs. These are drugs for nonauthorized indications and/or research drugs not included in a clinical trial. In Spain, activities in relation to these drugs are regulated.[1,17] In order to use a drug for compassionate care, the pharmacy service of the hospital applies to the Dirección General de Farmacia y Productos Sanitarios with the following documents: a clinical report in which the doctor justifies the application for the drug, a consent form signed by the patient,

and a form signed by the hospital medical director who is responsible for drug use. It is common practice for the pharmacy service to prepare a technical report with relevant references to support the application and to inform hospital directors of the process.

2. *Research drugs.* Regarding drugs for clinical trials conducted at the hospital, the pharmacy service is responsible for their reception, storage, dispensing, distribution, and return of unused drugs. Spanish requirements are that clinical trials be regulated.[17] A copy of the clinical trials committee approval must be kept at the pharmacy service, and dispensing is done only after a written and signed prescription is received.

3. *Foreign drugs.* Drugs marketed in a foreign country but not available in Spain may, according to Spanish law, be obtained but only for the specific indications for which the drug is approved in that foreign country.[1] The hospital pharmacy service applies to the Dirección General de Farmacia y Productos Sanitarios with the necessary documentation for use with an individual patient or according to a protocol.

4. *Stocks in wards.* There are some drugs (e.g., urgent medications, PRN, drugs dispensed as needed) and medical devices that have to be in stock on the ward. These are normally sent to the floor on a regular basis, according to a fixed schedule. These stocks are periodically checked by pharmacists (with regard to composition, expiry date, correct identification), and the results of the control are filed. The nurse supervisor of each ward is responsible for the safekeeping of the stock; the pharmacist is responsible for control and supervision.

Manufacture

Manufacture implies the manipulation of active substances and drugs in order to make them suitable for direct administration to patients. Separate areas are needed for the manufacture of intravenous admixtures and parenteral nutrition, cytotoxic drugs, and sterile preparations. No separate areas or biological security are needed for other, nonsterile preparations or drug repackaging. Following Spanish regulation,[18] written protocols and procedures for manufacturing processes must exist in every pharmacy service. A sterile area is normally achieved with a vertical or horizontal airflow hood.

Central Intravenous Additive Service

Centralized units of intravenous therapy (or CIVAS, for "central intravenous additive service") were created in Spanish hospital pharmacy services as both a consequence of the growing importance in the hospital of intravenous drugs, and parenteral nutrition and fluids and as a consequence of the clinical and technical progress in this area of the pharmacy. In recent years, Spanish hospital pharmacy services are almost obligated to have a CIVAS,[19] and it is now considered, along with the unit-dose distribution system, one of the main units in the pharmacy service.[19] The main objectives of the CIVAS are preparation of products therapeutically and pharmaceutically appropriate for the patient (right dose, administration route, chemically compatible, stable); preparation of admixtures free of particles, microorganisms, or toxins; preparation of admixtures with the correct drug in the exact amount; labeling, identification, storage, and distribution of admixtures according to good drug control principles; cost control of intravenous fluids; monitoring and clinical follow-up of patients; drug use evaluation studies; and participation in the

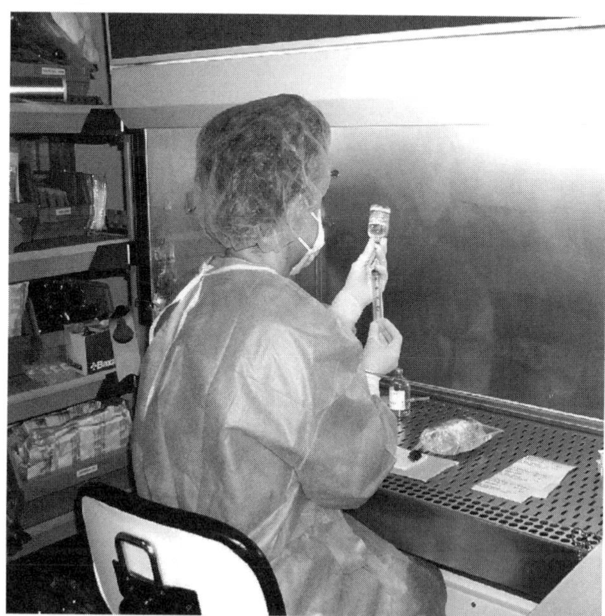

Fig. 4 Nurse preparing an anticancer drug in a central unit in a Spanish hospital pharmacy service.

intravenous therapy policy of the hospital (indications, selection, preparation, administration, etc.).

Most hospitals have a computerized CIVAS that is integrated into the unit-dose distribution system. Preparations handled in these units include cytotoxic drugs, antibiotics, parenteral nutrition, other drugs, and therapy with fluids (Fig. 4). References to Spanish articles dealing with recommendations for managing these units can be found in the bibliography.[7,8,19,20] Protocols must include every procedure carried out in the unit, from preparation to identification, hazard handling, waste treatment, and so on. In Spain, admixtures and nutritional preparations are normally prepared by pharmacy nurses supervised by pharmacists.

Centralized units have some advantages, such as less investment in equipment, better use of multidose vials, recycling of unused preparations, better working conditions, a good opportunity for clinical intervention by pharmacists, and improvement in the quality of patient care when the CIVAS is well coordinated with the unit-dose system.

Enteral and Parenteral Nutrition

In Spain, preparation of enteral and parenteral nutrition is carried out in the centralized units of the pharmacy services. Hardly any hospitals obtain their parenteral nutrition preparations from an external company. In most hospitals there are some standard nutritional preparations as well as others designed for specific patients. Nutrition design and patient follow-up is done by hospital pharmacists or by a team of various professionals (doctors, dieticians, nutritionists, pharmacists), depending on the hospital. Normally, laboratory data, clinical results, and patient progression are observed by pharmacists and nutrition support is changed accordingly, which gives pharmacists another opportunity for clinical intervention. References on how to manage such a service are given in the bibliography.[7,8,19] Computer software is used to make this task easier, permitting data entry (general patient data, lab results, prognosis, nutritional status, diet) and preparation of working sheets, reports, and labels for nutrition identification. Complications or incidents can also be registered; some software programs include Spanish products for nutrition support (e.g., NutriData®, Nutri2000®).

Drug Information

Drug information is one of the main responsibilities of pharmacists in hospitals and one of their most important contributions to a rational use of drugs and better patient care. In 1973, the first drug information center was established in Spain, and today activities related to drug information are part of every hospital pharmacy service. Drug information is another area where clinical intervention by pharmacists could be increased.

Information provided by pharmacists can be classified as passive or active. The former includes answering questions and preparing the requests/controls for foreign, compassionate-use, and research drugs. Active information includes providing support to the P&T committee (drug formulary preparation, diffusion of main decisions), establishment of protocols, writing of the drug information bulletin, sessions, adverse drug reactions programs, advising in- and outpatients, health education activities, information management, and so on. Some hospitals make their drug formulary and other information available on the Internet (e.g., www.hsanmillan.es/farma/index.htm) and some participate in the dissemination of drug information, in the Spanish language, to patients.[32] Additional sources of information and recommendations for the management of drug information centers that have been proposed by the SEFH and others are given in the bibliography.[7,8,11,21,22]

Clinical Pharmacokinetics and Therapeutic Drug Monitoring

Clinical pharmacokinetics is a multidisciplinary field that has been growing in importance over the last 20 years. Its main objective is therapy optimization by achieving drug concentrations in the therapeutic range and thereby obtaining maximum efficacy with minimum adverse effect. The concentration–effect relationship of many drugs is better than the dose–effect relationship. This is due to high interindividual variability. In these drugs, therapeutic drug monitoring is justified.

To assure the best efficacy, the pharmacist designs a pharmacotherapy that is specific to each individual patient. This is achieved by obtaining blood samples, gathering patient data (clinical situation, laboratory results, physiopathology, progression, therapy), applying pharmacokinetic principles, and applying knowledge of drug behavior in the population in which the patient is included. Even though drug concentration is an important piece of information, it is not enough on its own and patient follow-up is required. Times of sample collections must be carefully established in order to obtain maximum information from the minimum number of samples.

The usefulness of therapeutic drug monitoring has been demonstrated for some drugs (e.g., some antibiotics, cardiovascular agents and antiepileptics, theophylline, inmunosupressants, litium, methotrexate),[8,23] and these are the drugs that are included in clinical pharmacokinetic programs in Spanish hospital pharmacy units. The be-

nefits of therapeutic drug monitoring of other drugs, such as some anticancer drugs, are now being studied in some centers.[24]

Sample analysis requires specific techniques, such as fluorescence polarization immunoassay (FPIA) and high-performance liquid chromatography (HPLC). These techniques are not always available in the pharmacy and so sample analysis is not always done in Spanish hospital pharmacy services but in laboratories. However, it is a pharmacist who interprets results, makes recommendations, and follows up on patients, all as part of clinical activities to pursue better patient care. In all hospitals with such a pharmaceutical service, doctors and other members of the health team welcome the contribution of pharmacists, with their pharmacokinetic knowledge, to the rational use of drugs.

Drug Surveillance

Drug surveillance includes drug follow-up with the purpose of observing, evaluating, and communicating any adverse reactions that a drug can produce when used in clinical practice. A drug surveillance program must be established in every hospital in order to detect these reactions, and the drug information center must support this activity technically. Observed events are communicated to the regional center for drug surveillance, either directly or through the SEFH. The Spanish Drug Agency[33] facilitates drug surveillance activities and the diffusion of information among professionals. Spain has an organized drug surveillance system—a national committee reporting to the Ministry of Health was constituted for this purpose in 1987. Spontaneous communication of adverse drug reactions is voluntary in Spain and is conducted through an official form known as the ''yellow card.''[8]

Radiopharmacy

In Spain, pharmacy practice is also applied to the study, manufacture, control, and distribution of radiopharmaceuticals. Radiopharmaceuticals must be isolated from other drugs and personnel, and devices must follow Spanish regulations.[25] Radiopharmacy is part of the hospital pharmacy service; however, it is recommended that the unit be located close to the nuclear medicine department and supervised by a pharmacist specialist in radiopharmacy.[7]

Pharmacoeconomics

Pharmacoeconomic evaluations consist of comparing different alternatives in terms of costs and benefits. In Spain, pharmacoeoconomics is becoming more important due to increased pressure to make the best use of limited resources. Furthermore, advances in the methodology[26] have increased the scientific rigor of pharmacoeoconomics. Pharmacoeconomics is used by Spanish hospital pharmacists as a tool for decision making regarding drugs, medical devices, or related activities. Studies are conducted and pharmacists adapt published studies to each unique hospital setting.

Activities Related to Medical Devices

Many Spanish hospital pharmacies participate in the selection, ordering, storage, distribution, and provision of information relating to medical devices. Such hospital pharmacies are also involved in rational use programs. A guide to medical devices used in Spanish hospitals has been published, which gives a classification to each device.[27]

FUTURE TRENDS

The number of activities conducted by hospital pharmacy services is continually increasing as the needs of doctors, personnel, and patients evolve. This gives the pharmacist the opportunity to develop a range of activities (clinical roles, management, administrative duties) that are of interest to and positive for the hospital. Pharmacists must continue to focus on the impact that technological and professional changes may exert on the efficacy and safety of medications as well as on patient care.

The role to be played by hospital pharmacists should be determined by all health care professionals, not just by pharmacists themselves. The 1999 meeting of the Spanish Society of Hospital Pharmacists took this into account and a roundtable was held incorporating representatives of all health team members as well as a representative of patient opinion based on a survey of patients. Better information for patients, more integration of pharmacists in the health team, and more direct contact with patients seem to be the activities to be developed in the future.[28]

Any activity that contributes to patient care must be nurtured, no matter who suggests it. Pharmacists as well as other professionals know that teamwork is the key to improving results for the patient.

REFERENCES

1. Ley 25/1990, de 20 de Diciembre, Boletín Oficial del Estado (B.O.E.) del 21, del Medicamento. (Also at www.msc.es/farmacia/legislacion/home.htm).

2. Bonal, J. Management en Farmacia Hospitalaria. In *Farmacia Hospitalaria,* 2nd Ed.; Editorial Médica Internacional: Madrid, Spain, 1992; 30–55.
3. Suñé, J.M.; Bel, E. *Compilaciónde Legislación en Farmacia Hospitalaria,* 2nd Ed.; Sociedad Española de Farmacia Hospitalaria, Ed.; International Marketing and Communications: Madrid, Spain, 1994; 1290.
4. *Guía de Formación de Especialistas. Farmacia Hospitalaria*; Ministerio de Sanidad y Consumo, Ministerio de Educación, Cultura, Consejo Nacional de Especializaciones Farmacéuticas: Madrid, Spain, 1999; 31.
5. Suñé, J.M.; Bel, E. Legislación. In *Farmacia Hospitalaria,* 2nd Ed.; Editorial Médica Internacional: Madrid, Spain, 1992; 172–268.
6. Simó, R.M. Docencia. In *Farmacia Hospitalaria,* 2nd Ed.; Editorial Médica Internacional: Madrid, Spain, 1992; 136–157.
7. *Guía de Gestión de los Servicios de Farmacia Hospitalaria*; Insituto Nacional de la Salud: Madrid, Spain, 1997, (Also at www.sef.es/guiafarmacia/home.htm).
8. *Farmacia Hospitalaria,* 2nd Ed.; J. Bonal, A. Dominguez-Gil, Dir.; Editorial Médica Internacional: Madrid, Spain, 1992, 1717 pp.
9. *Guía para la Evaluación y Mejora de los Servicios de Farmacia Hospitalaria*; Insituto Nacional de la Salud: Madrid, Spain, 1998, (Also at www.sef.es/guia/index.htm).
10. Situación de la Farmacia Hospitalaria. Encuesta-1995. SEFH Bol. Inf. **1996**, *XX* (76), 100.
11. *Curso de Administración de Servicios Hospitalarios de la Clínica Universitaria. El Servicio de Farmacia*; Universidad de Navarra: Pamplona, Spain, 1992; 509.
12. Ferrandiz, J.R. Distribución unidosis de Medicamentos en Hospitales. In *XIX Asamblea Nacional de Farmacéuticos de Hospitales*; Gráficas Orión: Madrid, Spain, 1975; 71–88.
13. Leape, L.L.; Cullen, D.J.; Clapp, M.D.; Burdick, E.; Demonaco, H.J.; Erickson, J.I.; Bates, D.W. Pharmacist participation on physician rounds and adverse drug events in the intensive care unit. JAMA, J. Am. Med. Assoc. **1999**, *282* (3), 267–270.
14. Scroccaro, G.; Alós Almiñana, M.; Floor-Schreudering, A.; Hekster, Y.A.; Huon, Y. The need for clinical pharmacy. Pharm. World Sci. **2000**, *22* (1), 27–29.
15. Atención Farmacéutica. In *XLIV Congreso de la SEFH, Pamplona, Sept. 1999*; Farm. Hosp., 1999; Vol. 23, 3–6, (Num. Especial).
16. Codina Jane, C. In *Sistemas Expertos y Aplicaciones Informáticas*, Ponencias XLIV Congreso Nacional de Farmacia Hospitalaria, Pamplona, Spain, Sept.1999; Sociedad Española de Farmacia Hospitalaria: Madrid, Spain, 2000.
17. Real Decreto 561/1993, del 16 de abril, Boletín Oficial del Estado de 13 de Mayo, por el que se establecen los requisitos para la realización de ensayos clínicos con medicamentos.
18. Normas de correcta fabricación de fórmulas magistrales y preparados oficinales. SEFH Bol. Inf. **1994**, *XVIII* (69), 15–28.
19. Jiménez Torres, N.V. *Mezclas, Intravenosas y Nutrición Artificial,* 4th Ed.; C.E.E. Convaser: Godella, Valencia, Spain, 1999; 722.
20. *Manejo de Medicamentos Citostáticos,* 2nd Ed.; Asociación Española de Farmacéuticos de Hospitales: Madrid, Spain, 1987; 56.
21. Proyecto de recomendaciones de la SEFH: Información de medicamentos. SEFH Bol. Inf. **1996**, *XX* (77), 15–20.
22. Tordera, M.; Magraner, J.; Fernández, M.J. Informaciónde medicamentos e internet. Estrategias de búsqueda farmacoterapéutica en la World Wide Web. Farm. Hosp. **1999**, *23* (1), 1–13.
23. *Applied Pharmacokinetics. Principles of Therapeutic Drug Monitoring,* 3rd Ed.; Evans, W.E., Schentag, J.J., Jusko, W.J., Eds.; Applied Therapeutics, Inc.: Vancouver, Washington, 1992.
24. Aldaz, A. Farmacocinética de Citostáticos. Conceptos Generales. In *El Paciente Oncohematológico y su Tratamiento. Módulos de Actualización Multidisciplinar*; SEFH, Editores Médicos, S.A.: Madrid, Spain, 1997; 32–37.
25. Real Decreto 479/1993, de 2 de Abril, por el que se regulan los medicamentos Radiofármacos de uso humano (1). Boletín Oficial del Estado num 109 de 7 de Mayo de 1993.
26. Drummond, M.F.; O'Brien, B.; Stoddart, G.L.; Torrance, G.W. *Methods for Economic Evaluation of Health Care Programmes,* 2nd Ed.; Oxford University Press, Inc.: New York, 1997; 305.
27. Giráldez, J.; Idoate, A.; Romero, B.; Ursúa, C.; Errea, M.T.; Lacasa, C.; Aldaz, A. *Guía de Productos Sanitarios,* 2nd Ed.; EUNSA: Barañain, Navarra, Spain, 1998; 508.
28. In *El Medicamento en el Proceso de Atención al Paciente. Análisis Multidisciplinar*, Ponencias XLIV Congreso Nacional de Farmacia Hospitalaria, Pamplona, Spain, Sept. 1999; Sociedad Española de Farmacia Hospitalaria: Madrid, Spain, 2000; 9–64.
29. Text available at www.msc.es/farmacia/legislacion/home.htm.
30. www.msc.es/centros/catalogo/home.htm.
31. See www.sefh.es.
32. See www.viatusalud.com.
33. See www.msc.es/agemed/.

PROFESSIONAL DEVELOPMENT

Hyperlipidemia Pharmacy Practice

Theresa M. Bianco
Oregon Health Sciences University, Portland, Oregon, U.S.A.

H

INTRODUCTION

Hyperlipidemia is a disorder that is widely prevalent in the U.S. population. Elevations of total and low-density lipoprotein (LDL) cholesterol have been documented to increase the risk of coronary heart disease (CHD). The Third National Health and Nutrition Evaluation Survey (NHANES III) estimated that 52 million Americans have cholesterol elevations that require intervention, of which 12.7 million may require drug therapy.[1] A number of studies have shown a reduction in cardiovascular mortality or morbidity with lipid-lowering therapy in subjects with CHD (secondary prevention)[2–5] and in some patients without known CHD (primary prevention).[6,7] Despite this, the use of lipid-lowering agents in patients who have had a prior coronary event is disturbingly low.[8] When drug therapy is initiated, compliance may be poor and adherence to therapy may be as low as 35% in some series.[9,10] Other data indicate that even where cholesterol-lowering drugs are prescribed, many patients do not reach the goals of therapy recommended by the National Cholesterol Education Program (NCEP).[11]

Hyperlipidemia is a disease particularly suitable for pharmacist management for a number of reasons. It is a disorder that can be diagnosed and monitored primarily by laboratory testing. There are accepted guidelines for LDL goals. The drugs that are used vary in their effectiveness for altering the different lipoproteins and require someone skilled in this knowledge to select them for use. The rate of adherence to drug therapy is low, possibly in part because patients do not feel elevated cholesterol and therefore do not understand the need to take medication. These drugs are in some cases unpalatable or difficult to tolerate and require much patient education to initiate therapy and maintain compliance. Drug interactions with cholesterol-lowering agents can be clinically significant. These include inhibition of absorption of drugs such as levothyroxin or warfarin given concurrently with bile acid binding resins, or inhibition of the metabolism of statin drugs resulting in myopathy or even rhabdomyolysis.

JUSTIFICATION OF PHARMACIST INTERVENTION

Pharmacist intervention was effective in maintaining compliance and achieving LDL goals in patients treated with colestipol.[12] In a small study at a Veterans Administration (VA) Medical Center, patients received 1 hr of education and assessment by pharmacists before initiating colestipol therapy. They also were telephoned at 2-week inervals until an 8-week follow-up appointment. They were contacted by telephone again at 26 and 52 weeks. When compared at 52 weeks with patients receiving usual care, the pharmaceutical care group had greater persistence with colestipol therapy, were taking higher doses, had lower mean LDLs, and had a higher rate of reaching goal.

The effect of weekly contacts with patients initiated on combination lipid-lowering therapy of lovastatin and colestipol was investigated.[13] Patients from a university-affiliated tertiary care center were enrolled if they had undergone coronary artery bypass graft surgery or percutaneous transluminal coronary angioplasty (PTCA). In addition to instructions on appropriate drug use prior to hospital discharge, patients were telephoned at home weekly for 12 weeks at which time ''emphasis was placed on the importance of therapy in reducing the risk of cardiac events.'' Interestingly, when these patients were compared with a control group at the end of the 12 weeks compliance with therapy was high in both groups and not significantly different. However, when refill history was obtained from the patients' pharmacies at 1- and 2-year intervals, the patients in the intervention group had significantly higher rates of compliance.

Provision of patient education in combination with bimonthly cholesterol testing in a community pharmacy resulted in a significant reduction in cholesterol values over a 6-mo study.[14] Changes in patient-reported behaviors such as dietary habits and exercise were also noted. Although the lack of control group made this study less than definitive, it indicated that a combination of cholesterol monitoring and education could result in lower cholesterol concentrations. A second study demonstrated that screen-

Encyclopedia of Clinical Pharmacy
DOI: 10.1081/E-ECP 120006308

ing in combination with education and referral to a primary care physician when appropriate resulted in a significant number of patients receiving follow-up for cholesterol concentrations that were higher than the NCEP goals.[15]

DEVELOPING A HYPERLIPIDEMIA PHARMACY PRACTICE

Hyperlipidemia management can exist wherever pharmacists practice, including community pharmacies, institution-based or free-standing ambulatory clinics, or inpatient services. Despite these different settings, some universal requirements need to be addressed.

The nature of the practice may be influenced by the availability of space in which to provide patient care. For example, the lack of facilities in which to meet privately with the patient may result in a telephone-based practice. Offering lipid management in the community pharmacy may require an investment in infrastructure. Some remodeling of the pharmacy may be needed to provide an area where confidential communications can occur. A lipid analyzer, as well as a dedicated clean area, must be supplied if blood lipid monitoring is to be offered.

Staffing must be adequate. A redistribution of duties among pharmacists and technicians, possibly in addition to hiring additional pharmacists, may be necessary to allow pharmacists time to provide the service.[16]

Most pharmacists will need to justify their provision of this service, whether it be in the form of a business plan for an independent pharmacist or a proposal demonstrating benefit to an institutional employer. If the pharmacist will be relying on referrals to the service or will be collaborating with physicians to implement therapy, the pharmacist must first determine whether physicians will use the service and be accepting of input. An evaluation of a cholesterol screening program found that a significant number of physicians in the geographic area were resistant to their patients directly receiving the results of their cholesterol tests from the pharmacy. These physicians were less likely to contact patients with the results of elevated cholesterol values obtained at the screening.[17] Patients may also be surveyed as to acceptance of pharmacist management, particularly if they are going to be expected to pay part or all the costs of the service.

In all models, a scope of practice agreement or protocol is recommended, if not required. This should outline the following:

1. The hours of operation.
2. The pharmacists who are responsible for providing the service.
3. The supervising physician, if applicable. This may be especially needed if the pharmacist has prescriptive authority.
4. The population to be managed. For example, in the case of limited resources, the service may be restricted to secondary prevention patients, patients requiring combinations of drugs, or those with mixed lipid disorders versus those with only elevated LDL, or other parameters as determined by the needs of the facility.
5. The means of identification of patients. This could vary from seeing potentially low-risk patients, such as any patient followed in a general medicine clinic or referred by a primary care provider, to identifying high-risk patients, such as anyone discharged from the hospital with a diagnosis of myocardial infarction or after a revascularization procedure or with other evidence of CHD risk.
6. The goals of the clinic and methods for achieving them. Explain how patients will be evaluated and how the need and type of therapy will be determined. Describe any protocols for deciding on drug therapy or the rationale for allowing clinical decision making instead of following an algorithm. Will patients be seen once for evaluation and recommendations, as often as necessary to achieve control, or indefinitely? How frequently will they be seen? Will all contacts be by visit, or will telephone calls be routinely used?

PRACTICE MODELS

Community Pharmacy Practice

The functions of a pharmacist in lipid management in a community setting may include screening for elevated cholesterol and/or low HDL cholesterol, providing patient education and counseling to enhance adherence with drug and nondrug therapy, monitoring of lipid profiles for assessment of efficacy, and making recommendations to providers for drug therapy management.

Screening programs

The accessibility of community pharmacists to both patients and physicians makes them an ideal resource for identifying the presence of lipid abnormalities. Screening may consist of offering to measure cholesterol levels to the general population, or may involve targeted screening of patients at high-risk for CHD, also called case finding. In either case, screening should involve more than pro-

vision of a laboratory value. The total and HDL cholesterol values should be evaluated and interpreted in the light of the patient's risk factors for CHD. Education about cholesterol and cholesterol-lowering strategies should be provided, and the pharmacist should be prepared to refer the patient to their primary care provider if warranted. Failure to interpret these values may result in unnecessary concern on the part of the patient or, potentially more damaging, result in a patient not seeking care when needed.

Gardner and colleagues[18] demonstrated that a community pharmacy prescription database can be used to identify patients at risk for CHD. This is important because it targets those individuals most likely to benefit from lipid-lowering interventions. They identified four clinical indicators that were believed to be likely to identify patients at risk for CHD: prescription for sublingual nitroglycerin, prescription for beta-adrenergic blocking agents or thiazide diuretics, males with a prescription for nicotine gum or patch, or those receiving oral hypoglycemic agents or insulin therapy and who were greater than 50 yr of age. A search of the pharmacy database was performed to identify individuals prescribed at least one of these agents, and the pharmacy profiles were screened to ensure the age and sex met the criteria. These subjects, who were invited to a free cholesterol screening, were compared with an unselected population who self-referred to the screening. Twenty-one percent of those identified as high risk responded to the invitation. A significantly greater percentage of the screened patients had cholesterol values that were higher than desired. In addition, two-thirds to three-fourths of the patients with a clinical indicator had cholesterol values over 200 mg/dl, indicating that these indicators may be predictive of the need for cholesterol-lowering intervention.

Einarson et al.[19] reported the financial feasibility of a pharmacy-based cholesterol screening program. Subjects were asked how much they would be willing to pay for a cholesterol measuring service in a pharmacy. Patients who completed a pharmacy service questionnaire indicated they would be willing to pay a mean of $11.54. Patients who received the service were surveyed afterward, and indicated a willingness to pay $14.47 (1987 dollars). Of note, it does not appear that these patients received pharmacist education as part of their testing but were reacting to the value of obtaining cholesterol results at a pharmacy.

Lipid management practices

Shibley and Pugh[20] described the provision of pharmaceutical care in independent community pharmacies. Patients were recruited by the investigator and included in the study if their primary physicians agreed to allow them to do so. The physicians were recruited by letter and by meetings with the pharmacists. Pharmacists provided basic education about lipid disorders, the relationship to coronary artery disease, and diet and exercise. Lipoproteins were measured at the pharmacy using the Cholestech® analyzer. If warranted, drug therapy recommendations were provided to the physician via telephone or letter; if accepted, the patient was seen at 2 months to assess efficacy and adverse effects. All patients were also seen by a certified dietician. Significant reductions in LDL cholesterol were observed, although it is not clear how many patients reached their therapeutic goal. Given choices ranging from $15 to $55, patients indicated they would be willing to pay $23.75 ± $11.42 for each encounter with the pharmacist.

Project ImPACT: Hyperlipidemia was a multicenter community pharmacy-based demonstration project that aimed to demonstrate the benefits of a pharmacist on patient adherence and compliance with lipid-lowering therapy.[16] The pharmacists used cholesterol analyzers at their sites to enhance their interactions with patients and their physicians. Emphasis was placed on patient education and communication with the physicians to bring patients to their NCEP cholesterol goals. Of interest, 62.5% of the patients, who were predominantly primary prevention, did reach and maintain their goals by the end of the study. Persistence with therapy was excellent, with 93.6% remaining on the prescribed cholesterol-lowering agent throughout the study. Compliance with therapy, defined as fewer than five missed doses or refills obtained within 5 days of when due, was 90.1%. Physician acceptance of pharmacist interventions was high, with 76.65% of recommendations resulting in a change. These interventions involved coordination of care, adverse drug reactions, drug interactions, drug dosing, drug selection, and side effects.

The participating pharmacies were primarily independents, with some chains, clinic pharmacies, health maintenance organizations, and home health/home infusion pharmacies. All pharmacies scheduled patients for appointments with the pharmacist. Most used time before the regular pharmacy hours or on weekends, as well as during usual business hours. Seventy-two percent of sites changed the pharmacist's duties to accommodate this new role, and 59% changed technician duties. Increasing pharmacist overlap was also a commonly used strategy. Fewer than one-third added pharmacist staff to implement the program. The average amount of time spent on patient encounters was about 45 min for a new patient and 22 min for a follow-up appointment.

Clinic Models

Development of lipid management practices in the institutional or free-standing clinic settings may take many forms. The types of practice can range from provision of consultative services by pharmacists in conjunction with patients' appointments with their primary care provider, to free-standing pharmacist-managed clinics in which the pharmacist has prescriptive authority to initiate, discontinue, and change drug therapy.

Pharmacists in a consultative role improved management of lipid disorders in an ambulatory internal medicine clinic.[21] In this study, the pharmacist met with patients prior to their physician appointment. Medication histories were taken, compliance encouraged, drug costs were tracked, and the least costly recommendation made to the physician. The pharmacist reviewed laboratory data and recommendations with the physician and attached a copy of these to the front of the chart. Decisions to accept or decline the recommendations were made by the physician. The majority of recommendations were accepted. When compared with usual care where pharmacists were not involved, significantly more patients reached LDL goals.

Furmaga[22] described the structure of a pharmacist-managed lipid clinic at a VA Medical Center outpatient clinic. Initially patients were identified using the hospital computer database to identify those with a total cholesterol of greater than 260 mg/dl. These patients were invited to a general educational seminar and subsequently scheduled into the lipid clinic, if needed. As this resulted in more patients identified than could be reasonably accepted into the clinic, the system was changed so that patients were referred from outpatient clinics. Patients were scheduled for 30-min appointments. The activities of the pharmacist included patient education, identification of secondary causes of hyperlipidemia with subsequent referral to other clinics as indicated, compliance assessment, and intervention and recommendation of addition of drug therapy to diet therapy. Clinical judgment was used in lieu of a protocol for drug selection. The pharmacist did not have prescriptive authority but was responsible for monitoring of drug therapy for efficacy and adverse events, and determining when changes were needed. Activities were documented in the medical record.

Shectman and colleagues[23] demonstrated that use of physician extenders resulted in improved LDL cholesterol concentrations when compared with usual care. In this model, also at a VA hospital clinic, the pharmacist or nurse used an algorithmic stepwise approach to assist in drug selection and optimization in reaching NCEP LDL goals. More patients reached their LDL goals in the physician extender group. The total costs of the physician extender care was higher, primarily due to higher drug costs. The cost per unit of LDL lowering, however, was significantly less.

Institutional Pharmacy Model

Inpatient pharmacists can also provide care by helping to initiate lipid-lowering therapy. In addition to the data that support treatment to lower cholesterol in patients who have had a coronary event, the National Committee for Quality Assurance is instituting a new Health Plan Employer Data and Information Set (HEDIS) indicator for cholesterol management in patients who have experienced an acute cardiovascular event. This will provide a challenge to identify and treat patients with coronary artery disease. A program by which pharmacists identified patients through acute myocardial infarction/percutaneous transluminal coronary angioplasty orders has been detailed.[24] Pharmacists placed a standardized note on the outside of the patient chart that included the goals of therapy and recommended that a lipid panel be obtained. The proportion of patients receiving lipid-lowering therapy at discharge was significantly increased after initiation of the program.

USEFUL TOOLS FOR PROVISION OF SERVICES

Pharmacist Education and Training

Regardless of the practice setting, a pharmacist needs certain tools to provide lipid management services. The first tool is an in-depth understanding about the disease and antihyperlipidemic drugs. Understanding of the disease includes knowledge about lipid metabolism, the influence of lipids on atherogenesis and vascular function, the risk of dyslipidemia and CHD mortality and morbidity, and the benefits of lipid-lowering as demonstrated in clinical outcome trials. Knowledge of the drugs includes pharmacology, pharmacokinetics (especially as pertains to the potential for drug interactions), adverse effects that are most often experienced or most severe, and how to manage these effects. The influence of each drug on the various lipoproteins, the effects of dose on lipoprotein lowering, the risks and benefits of combination therapy, and the goals to be targeted should be known. This type of education can be obtained through self study or by attending certificate programs, conferences, or other training programs.

The American Pharmaceutical Association's "Pharmaceutical Care for Patients with Dyslipidemias" is a

2-day training session that includes material on evidence of cholesterol-lowering and use of antihyperlipidemic agents, training on the Cholestech® analyzer, discussions on preparing the pharmacy practice site to provide the service, marketing to patients and physicians, communication with physicians, and reimbursement and billing. The National Pharmacy Cardiovascular Council offers a comprehensive three-tiered educational program. The Lipid Managers Training Program begins with the basics of lipid disorders and progresses to on-site training in a lipid clinic.[32] Many state organizations offer certificate programs in lipid management.

The NCEP was initiated by The National Heart, Lung, and Blood Institute (NHLBI) of the National Institutes of Health (NIH) in 1985. The goal of this program is to promote cholesterol awareness in the U.S. population as a risk factor for CHD and provide guidelines for cholesterol-lowering to physicians, patients, and the community, thus reducing CHD mortality and morbidity. The program consisted of five panels that are responsible for evaluation of the evidence and establishing guidelines in their specific areas: the Expert Panel on Detection, Evaluation, and Treatment of High Blood Cholesterol in Adults (Adult Treatment Panel or ATP) develops guidelines for the detection, evaluation, and treatment of high blood cholesterol in adults; and the Expert Panel on Blood Cholesterol Levels in Children and Adolescents developed recommendations for healthy diets for children and adolescents, and for detection and treatment of high blood cholesterol in children and adolescents from high-risk families.

The guidelines for treatment recommended by the ATP are considered the standard for dietary and drug therapy in adults. The most recent guidelines were released in May 2001;[25] the panel is currently revising these and updated guidelines are anticipated after Spring 2001. The pediatric guidelines were released in 1992.[26] The American Diabetes Association clinical practice guidelines make recommendations for managing hyperlipidemia in persons with diabetes that are more aggressive than the current NCEP guidelines, as well as more specific to this population.[27]

PATIENT NAME　　ID#　　DOB　　HT:
REFERRING CLINIC/ PROVIDER　　PRIOR DIETARY CONSULT / INTERVENTION?
DX:
SMOKER?　　ETOH? (QUANTITY):
RISK FACTORS:

MALE > 45 YR	FEMALE > 55 YR	CHD	HDL > 60?
DIABETES	SMOKING	FAMILY HX	
HTN	CVD	PVD	

LDL GOAL　　TG GOAL

DATE	LIPID DRUG / DOSE	WT	TC	TG	HDL	LDL	HBA1C	LFTS	OTHER LABS	COMMENTS

Fig. 1 Sample lipid monitoring form.

Patient Education

The second set of tools involves imparting some of this information to the patient. This can be done verbally, with written educational materials, with videotapes, or a combination thereof. The level of the material should be adjusted for the educational level of the patient population. The information should include definitions of cholesterol, triglycerides, and lipoproteins; factors that increase or decrease these values; and the goals for the patient. A risk calculator that can be used to illustrate to patients how their individual factors increase or decrease their risk of a coronary event should also be included.[28] In patients without physical limitations, handouts and counseling about beginning an exercise program may be provided, although patients with known vascular disease should be referred to their primary care provider for guidance on appropriate activity. Providing a diary in which patients can document their activity, heart rate, notes on dietary changes, and weights can be helpful, especially in the initial stages of making lifestyle changes. Information sheets about the individual drugs should also be distributed. The American Heart Association (AHA) web site provides a variety of tools for the health care provider to order for a fee or to download at no charge.[29]

Pharmacists who practice lipid management should be familiar with dietary factors that influence lipids. If referrals to a dietician are allowed by law, the pharmacist should have a referral base from which to guide the patient. Handouts that describe the goals of fat content, specific foods to choose and avoid, and how to read and interpret food labels should be available for distribution. These are available from a variety of sources. Patients may be referred to the AHA web site, which offers information about recommended diets as well as recipes. Drug companies that market cholesterol-lowering medications often provide free patient information materials that may not be product specific but will contain a company logo and product brand names.

Documentation

The third set of tools involves the pharmacist's documentation of interventions and results. If lipids are to be measured and followed, the use of a monitoring flow sheet is extremely useful (Fig. 1). Flow sheets may be on paper files, created on computer spreadsheets, or use special software programs.

Initial demographic data including height should be collected. The information obtained at each visit should include weight, exercise, lipid values, drug therapy (if any), and compliance. If available, other pertinent labs such as glucose or hemoglobin A1C, liver transaminases, or measures of renal function should be noted. A comments section is useful to document items such as adverse drug effects, noncompliance, or other issues that can affect lipid control.

Communication

The fourth set of tools regards communication with physicians or other primary care providers. Interventions made by the pharmacist or recommendations to the physician may be made by telephone, letter, fax, or personal contact, depending on the practice setting. These communications are important in both obtaining and maintaining provider buy-in as well as demonstrating the active role the pharmacist is playing in the care of the patient. In addition, there is less likelihood for misunderstanding than if all information is provided by the patient.

Lipid Measurement Devices

Lipid analyzers are not necessarily a needed tool for providing lipid management services but can be very

Table 1 Cholesterol monitoring tests granted CLIA waived status

System	Manufacturer	Lipoprotein measured
Advanced Care	Johnson & Johnson	Cholesterol
Cholestech LDX	Cholestech	Total cholesterol, HDL, triglycerides, glucose
Accu-Chek InstantPlus	Boehringer Mannheim	Cholesterol
ENA.C.T Total Cholesterol Test	ActiMed Laboratories	Cholesterol
Lifestream Technologies Cholesterol Monitor	Lifestream Technologies	Cholesterol
Polymer Technology Systems (PTS) MTM Bioscanner 1000 (for OTC use)	Polymer Technology Systems, Inc.	Cholesterol, HDL

(Adapted from Ref. [31].)

helpful. They allow the pharmacist to provide information and make recommendations for dietary and drug therapy at the time of the interaction, instead of having to schedule another time or attempt to reach patients by phone. It allows reenforcement of the information provided at the last visit as the patient can see the results of the intervention, and the implications of adherence or nonadherence to therapy can be demonstrated and discussed, and strategies for improvement can be presented.

Measuring cholesterol in the practice setting requires both the equipment and the legal authority to perform testing. The 1988 Clinical Laboratory Improvement Amendments (CLIA) established quality standards for accuracy, reliability, and timeliness in all laboratory testing. Certain devices are considered to be of low complexity and are therefore regarded as CLIA waived, which means that the site where they are used must be enrolled in the CLIA program but that routine on-site visits and monitoring are not required.

The cholesterol measuring devices that are in the CLIA waived category are listed in Table 1. At this time, the only waived analyzer that measures total and HDL cholesterol and triglycerides is the Cholestech LDX®. State law will also need to be followed because some states, for example, do not permit pharmacists to act as laboratory directors or to obtain blood via finger stick. Information about obtaining CLIA certification, a list of waived devices, and contact information for state survey agencies may be found on the CLIA web site.[30]

Reimbursement Strategy

Although obtaining reimbursement is beyond the scope of this chapter, a number of studies have tried to assess what patients will pay or perceive to be the value of provision of cholesterol monitoring and lipid management. The data range from assessing the value of a cholesterol level without attendant counseling to how much patients believe insurance companies should pay for each visit to the pharmacist where education and medication review and management are provided.

Project ImPACT is one of few studies that reports actual billing and reimbursement results. Both patients and insurers were billed for services. On average, pharmacists billed $28 for counseling services and $27 for lipid profiles. Seventy-five percent of patients billed paid an average of $35 per visit, and 53% of third-party payers paid an average of $30. Reimbursement by third-party payers was more frequent, however, for lipid profiles than for counseling.

CONCLUSION

Pharmacist practices in hyperlipidemia management have been shown to be effective in improving compliance, adherence to therapy, and LDL lowering. Studies that establish cost effectiveness are limited and are needed to support efforts to expand pharmacist involvement and justify reimbursement.

REFERENCES

1. Sempos, C.T.; Cleeman, J.I.; Carroll, M.D.; Johnson, C.L.; Bachorik, P.S.; Gordon, D.J.; Burt, V.L.; Briefel, R.R.; Brown, C.D.; Lippel, K.; Rifkind, B.M. Prevalence of high blood cholesterol among U.S. adults: An update based on guidelines from the second report of the National Cholesterol Education Program adult treatment panel. JAMA, J. Am. Med. Assoc. **1993**, *269* (23), 3009–3014.
2. Sacks, F.M.; Pfeffer, M.A.; Moye, L.A., et al. The effect of pravastatin on coronary events after myocardial infarction in patients with average cholesterol levels. N. Engl. J. Med. **1996**, *335*, 1001–1009.
3. Anonymous. Prevention of cardiovascular events and death with pravastatin in patients with coronary heart disease and a broad range of initial cholesterol levels. The Long-Term Intervention with Pravastatin in Ischaemic Disease (LIPID) study group. N. Engl. J. Med. **1998**, *339*, 1349–1357.
4. Anonymous. Randomised trial of cholesterol-lowering in 4444 patients with coronary heart disease: The Scandinavian Simvastatin Survival Study (4S). Lancet **1994**, *344*, 1383–1389.
5. Rubins, H.B.; Robins, S.J.; Collins, D.; Fye, C.L.; Anderson, J.W.; Elam, M.B., et al. Gemfibrozil for the secondary prevention of coronary heart disease in men with low levels of high-density lipoprotein cholesterol. Veterans Affairs High-Density Lipoprotein Cholesterol Intervention Trial study group. N. Engl. J. Med. **1999**, *341*, 410–418.
6. Frick, M.H.; Elo, O.; Haapa, K.; Heinonen, O.P.; Heinsalmi, P.; Helo, P., et al. Helsinki heart study: Primary-prevention trial with gemfibrozil in middle-aged men with dyslipidemia. Safety of treatment, changes in risk factors, and incidence of coronary heart disease. N. Engl. J. Med. **1987**, *317*, 1237–1245.
7. Shepherd, J.; Cobbe, S.M.; Ford, I.; Isles, C.G.; Lorimer, A.R.; MacFarlane, P.W., et al. Prevention of coronary heart disease with pravastatin in men with hypercholesterolemia. West of Scotland Coronary Prevention study group. N. Engl. J. Med. **1995**, *333*, 1301–1307.
8. Grundy, S.M.; Balady, G.J.; Criqui, M.H.; Fletcher, G.; Greenland, P.; Hiratzka, L.F.; Houston-Miller, N.; Kris-Etherton, P.; Krumholz, H.M.; LaRosa, J.; Ockene, I.S.; Pearson, T.A.; Reed, J.; Smith, S.C., Jr.; Washington, R.

When to start cholesterol-lowering therapy in patients with coronary heart disease: A statement for healthcare professionals from the American Heart Association Task Force on Risk Reduction. Circulation **1997**, *95*, 1683–1685.

9. Avorn, J.; Monette, J.; Lacour, A.; Bohn, R.L.; Monane, M.; Mogun, H., et al. Persistence of use of lipid-lowering medications: A cross-national study. JAMA, J. Am. Med. Assoc. **1998**, *279*, 1458–1462.
10. Andrade, S.E.; Walker, A.M.; Gottleib, L.K.; Hollenberg, N.K.; Testa, M.A.; Saperia, G.M.; Platt, R. Discontinutation of antihyperlipidemic drugs—do rates reported in clinical trials reflect rates in primary care settings? N. Engl. J. Med. **1995**, *332*, 1125–1131.
11. Pearson, T.A.; Laurora, I.; Chu, H.; Kafonek, S. The Lipid Treatment Assessment Program (L-TAP). Arch. Intern. Med. **2000**, *160*, 459–467.
12. Konzem, S.L.; Gray, D.R.; Kashyap, M.L. Effect of pharmaceutical care on optimum colestipol treatment in elderly hypercholesterolemic veterans. Pharmacotherapy **1997**, *17*, 576–583.
13. Faulkner, M.A.; Wadibia, E.C.; Lucas, B.D.; Hilleman, D.E. Impact of pharmacy counseling on compliance and effectiveness of combination lipid-lowering therapy in patients undergoing coronary artery revascularization: A randomized, controlled trial. Pharmacotherapy **2000**, *20* (4), 410–416.
14. Ibrahim, O.M.; Catania, P.N.; Mergener, M.A.; Supernaw, R.B. Outcome of cholesterol screening in a community pharmacy. DICP, Ann. Pharmacother. **1990**, *24*, 817–821.
15. McKenney, J.M. An evaluation of cholesterol screening in community pharmacies. Am. Pharm. **1993**, *NS33*, 34–40.
16. Bluml, B.M.; McKenney, J.M.; Cziraky, M.J. Pharmaceutical care services and results in Project ImPACT: Hyperlipidemia. J. Am. Pharm. Assoc. **2000**, *40* (2), 157–165.
17. Madejski, R.M.; Madejski, T.J. Cholesterol screening in a community pharmacy. J. Am. Pharm. Assoc. (Wash) **1996**, *NS36*, 243–248.
18. Gardner, S.F.; Skelton, D.R.; Rollins, S.D.; Hastings, J.K. Community pharmacy data bases to identify patients at high risk for hypercholesterolemia. Pharmacotherapy **1995**, *15*, 292–296.
19. Einarson, T.R.; Bootman, J.L.; McGhan, W.F.; Larson, L.N.; Gardner, M.E.; Donohue, M. Establishment and evaluation of a serum cholesterol monitoring service in a community pharmacy. Drug Intell. Clin. Pharm. **1988**, *22*, 45–48.
20. Shibley, M.C.; Pugh, C.B. Implementation of pharmaceutical care services for patients with hyperlipidemias by independent community pharmacy practitioners. Ann. Pharmacother. **1997**, *31*, 713–719.
21. Bogden, P.E.; Koontz, L.M.; Williamson, P.; Abbott, R.D. The physician and pharmacist team. An effective approach to cholesterol reduction [see comments]. J. Gen. Intern. Med. **1997**, *12*, 158–164.
22. Furmaga, E.M. Pharmacist management of a hyperlipidemia clinic. Am. J. Hosp. Pharm. **1993**, *50*, 91–95.
23. Schectman, G.; Wolff, N.; Byrd, J.C.; Hiatt, J.G.; Hartz, A. Physician extenders for cost-effective management of hypercholesterolemia. J. Gen. Intern. Med. **1996**, *11*, 277–286.
24. Birtcher, K.K.; Bowden, C.; Ballantyne, C.M.; Huyen, M. Strategies for implementing lipid-lowering therapy: Pharmacy-based approach. Am. J. Cardiol. **2000**, *85* (3A), 30A–35A, Feb. 10.
25. National Cholesterol Education Program. Executive summary of the third report of the expert panel on detection, evaluation, and treatment of high blood cholesterol in adults (adult treatment panel III). JAMA **2001**, *285*, 2487–2502.
26. Expert Panel. National Cholesterol Education Program. Reprot of the expert panel on blood cholesterol levels in children and adolescents. Pediatrics **1992**, *89* (Suppl. 2), 525–584.
27. American Diabetes Association. Management of dyslipidemia in adults with diabetes. Diabetes Care **2000**, *23* (Suppl. 1).
28. Grundy, S.M.; Paternak, R.; Greenland, P.; Smith, S., Jr.; Fuster, V. Assessment of cardiovascular risk by use of multiple-risk-factor assessment equations. J. Am. Coll. Cardiol. **1999**, *34*, 1348–1359.
29. www.americanheart.org/CAP/pro/prof_tctools.html (accessed October 2000).
30. www.hcfa.gov/medicaid/clia/cliahome.htm (accessed October 2000).
31. www.hcfa.gov/medicaid/clia/waivetbl.htm (accessed November 2000).
32. www.npccnet.org/ (accessed November 2000).

PROFESSIONAL DEVELOPMENT

Infectious Diseases Specialty Pharmacy Practice

Steven C. Ebert
Meriter Hospital, Inc., Madison, Wisconsin, U.S.A

INTRODUCTION

Since the 1980s, the specialty area of infectious diseases within pharmacy practice has evolved into a distinct discipline that is directed at providing optimum antimicrobial therapy to patients. The pharmacist is uniquely qualified to apply therapeutic, pharmacokinetic, and pharmacodynamic principles to antimicrobial therapy. These skills serve to complement rather than compete with the roles of infectious diseases physicians. Infectious diseases pharmacists are employed in private and teaching hospitals, clinics, academia, and industry. Literature that document the positive impact of the infectious diseases pharmacist on patient outcomes is now being published.

EDUCATION/TRAINING

Education and Postgraduate Training

Infectious diseases pharmacists have typically been awarded either a postbaccalaureate or entry-level Doctor of Pharmacy degree. In addition, most have completed 2 to 3 years of postdoctoral training that consists of a 1-year residency in pharmacy practice, followed by either 1 year in an infectious diseases specialty residency or a 2-year infectious diseases fellowship. It should be noted, however, that a select number of motivated practitioners have not completed these postgraduate training programs, but instead have become proficient in infectious diseases through ''on-the-job training.''

Board Certification

Pharmacists who have been practicing for more than 3 years and/or have completed postgraduate training may become certified in pharmacotherapy (BCPS) through the Board of Pharmaceutical Specialties (BPS). This certification is achieved via examination. In addition, as of the year 2000, BCPS awardees could be granted Added Qualifications in Infectious Diseases Pharmacotherapy by submitting an application to BPS. The application consists of a portfolio that describes the applicant's practice in infectious diseases pharmacotherapy. The portfolio includes[1]

1. A letter from applicant requesting review of portfolio for purpose of granting Added Qualifications in Infectious Diseases Pharmacotherapy.
2. Current curriculum vitae (CV).
3. A detailed summary of each of the following elements (if not included in CV):
   a. Any special training or professional development programs in the area of infectious diseases pharmacotherapy.
   b. Work experience in the area of infectious diseases pharmacotherapy.
   c. Specific professional responsibilities for care of patients with infectious diseases in outpatient and inpatient settings.
   d. Any professional awards, honors, or special achievements relative to infectious diseases pharmacotherapy.
   e. Bibliography of applicant's relevant professional publications.
   f. List of applicant's past and present research or other scholarly activities in the area of infectious diseases pharmacotherapy.
   g. Summary of past and current educational/in-service activities for health care professionals in infectious diseases pharmacotherapy.
   h. List of memberships in professional organizations relative to infectious diseases, with specific notation of any service or leadership activities to the organization.

At this time, Board Certification in Pharmacotherapy with Added Qualifications in Infectious Diseases is a means for recognizing outstanding practitioners. It is not a means of licensure or a prerequisite for practicing in the area of infectious diseases pharmacotherapy.

Encyclopedia of Clinical Pharmacy
DOI: 10.1081/E-ECP 120006229

VOCATIONS FOR INFECTIOUS DISEASES PHARMACISTS

Hospital Practice

Pharmaceutical care of the hospitalized patient with infection is the most traditional role for infectious diseases pharmacists. Numerous opportunities for proactive interventions in antimicrobial selection, dosing, route of administration, and monitoring of patients with changing clinical status make this a popular practice setting for many individuals.

Practice solely in infectious diseases

Infectious diseases pharmacists typically practice in a hospital setting that allows them to devote all their time to managing antimicrobial therapy. All aspects of infectious diseases pharmacotherapy, including interventions on antimicrobial selection, antimicrobial dosing, and intravenous-to-oral conversion are the responsibility of the infectious diseases pharmacist. In addition, the pharmacist is usually responsible for analyzing new antimicrobials for formulary inclusion, medication use evaluations, and antimicrobial restriction or therapeutic interchange policies.

Some infectious diseases pharmacists may collaborate with infection control practitioners to reduce nosocomial infections and control antimicrobial resistance. Others may work closely with clinical microbiologists to design institution-specific susceptibility testing and reporting methods, and to generate periodic antibiotic susceptibility reports.

In hospitals with a significant pharmacy influence on antimicrobial therapy, it may be impossible for the infectious diseases pharmacist to perform all the functions described here. Instead, the pharmacist may need to delegate the responsibility for conducting standardized antimicrobial ''protocols'' (therapeutic interchange, intravenous to oral, aminoglycoside pharmacokinetics) to other pharmacists, while maintaining accountability for the quality of these programs.

Although the salary for many hospital pharmacists who practice exclusively in the area of infectious diseases comes from the hospital in which they practice, a substantial number are cofunded by hospitals and schools of pharmacy or medicine.

Combined with other responsibilities

In hospital pharmacy departments with limited resources or incomplete antimicrobial management programs, pharmacists trained in infectious diseases may practice in a clinical setting that requires not only expertise in antimicrobial therapy, but also other therapeutic areas. For example, an infectious diseases pharmacist may practice as a clinical coordinator who is charged with developing practice areas such as cardiology, nutrition support, etc., in addition to antimicrobial management programs. Other practice sites that require a working knowledge of infectious diseases include clinical pharmacokineticists, critical care pharmacists, and transplant pharmacists. These practice positions are typically funded 100% by the hospital in which they are located.

Outpatient Practice

An increasing number of infectious diseases pharmacists practice in outpatient settings. These individuals usually practice in one of two areas. One area is in outpatient clinics, where they are directly involved in patient care. This is particularly true for pharmacists who specialize in treatment of patients infected with human immunodeficiency virus (HIV) or other chronic infectious diseases (e.g., leprosy). Pharmacists take medication histories, counsel patients about their medications, assess response to antimicrobial therapy, and make adjustments in the-rapy, as necessary.

Infectious diseases pharmacists also make valuable contributions to patient care in the managed care setting. By evaluating antimicrobial prescribing patterns, creating drug treatment protocols, directing formulary decisions, and ''counterdetailing'' prescribers, infectious diseases pharmacists help to curtail inappropriate antibiotic prescribing that may lead to increased antibiotic resistance.

Pharmaceutical Industry

An increasing number of infectious diseases pharmacists have found a career in the pharmaceutical industry. Some initially take positions in pharmaceutical sales. Others may be hired as research associates, where they assist in the collection and analysis of data for clinical studies. More often, they are hired as ''medical science liaisons.'' These individuals interact with physician and pharmacist practitioners, where they provide drug information, grant support for research and educational efforts, assist in medication use evaluations, and give in-services to medical and pharmacy staff.

Promotions within industry have lead many of these pharmacists into advanced positions such as Director of Medical Affairs, Associate Director for Research, or Associate Director for Education.

Contract Research Organization

Some infectious diseases pharmacists join contract research organizations. These organizations work primarily with pharmaceutical companies to test the in vitro activity of new antimicrobials, assess their efficacy in intro and animal infection models, and conduct clinical trials. Pharmacists may be hired into positions ranging from researcher to director.

Government

Some infectious diseases pharmacists have been hired into government positions. These individuals direct government-initiated studies, care for patients in clinics, and formulate policies regarding medication use. Infectious diseases pharmacists currently hold positions in the Food and Drug Administration, National Institutes of Health, and World Health Organization.

Independent Consultant

Many infectious diseases pharmacists devote some time to work as consultants. In most cases, they serve as ad hoc consultants for pharmaceutical companies, where they assess the likely impact of a newer antimicrobial and/or providing advice on direction of future studies. They may also educate pharmaceutical sales staff or write review articles.

Other infectious diseases pharmacists work full time as consultants. Usually, they are employees of larger consulting firms that are hired by hospitals or other health care institutions to detect inefficiencies in process and to improve financial success.

MODEL CLINICAL PRACTICE SETTINGS

Hospital Setting

Rounding with an infectious diseases consult service

Most infectious diseases pharmacists who practice in a hospital setting round with an infectious diseases consult service. This service usually consists of an infectious diseases physician, an infectious diseases medical fellow, medical students, an infectious diseases pharmacist, and (possibly) pharmacy students, residents, or fellows. Patients are usually identified through infectious diseases "consults." The pharmacist usually acts to "optimize" the antimicrobial regimen by adjusting antibiotic doses and apprising the service members of any imminent drug interactions or adverse effects. The pharmacist also monitors patients followed by the service, to assess therapeutic response and/or adverse events. Finally, the pharmacist serves as a resource for drug information for service members.

The advantages of rounding with an infectious diseases consult service include a sense of "teamwork" and camaraderie; the backing of an infectious diseases physician, which means that most recommendations will be followed; direct interaction with only a limited number of (infectious diseases) physicians, which will quickly establish mutual trust and respect; and the potential for collaboration in research. Disadvantages include limited patient exposure (usually only patients involved in "consults" are followed) and, potentially, limited usefulness if the infectious diseases attending physician is knowledgeable in antimicrobial pharmacology.

Pharmacist–infectious diseases physician collaboration

Another common practice model for hospital-based pharmacists is a one-on-one collaboration between an infectious diseases pharmacist and an infectious diseases physician. Under this model, the infectious diseases physician is generally responsible for standard infectious diseases "consults." The pharmacist acts as an extension of the infectious diseases physician's clinical practice clinical practice, rather than competition or duplication. The pharmacist identifies patients in whom antimicrobial therapy is suboptimal (i.e., wrong drug, wrong dose, questionable indication, potential for IV-to-oral conversion). After conferral with the infectious diseases physician, an intervention is recommended or implemented. These interventions usually follow predefined criteria established by the Pharmacy and Therapeutics Committee.

Some advantages of this model are the establishment of a close relationship between infectious diseases physicians and pharmacists, the backing of the infectious diseases service and the Pharmacy and Therapeutics Committee on interventions, and the potential for pharmacists to bill for clinical pharmacy services through a physician provider. Potential disadvantages exist if the infectious diseases physician and pharmacist do not interact well.

Independent practice

Under a third practice model in the hospital setting, infectious diseases physicians and pharmacists conduct separate services: the physician handles infectious diseases consults, and the infectious diseases pharmacist identifies patients with inappropriate antimicrobial therapy and makes interventions. Under this system, the Pharmacy and Therapeutics Committee will ideally grant the pharmacist some authority to automatically order modifications in therapy. This model is used when infectious diseases physicians are either unwilling or unable to become involved in interventions concerning antimicrobial therapy. A potential disadvantage is the perceived ''competition'' between infectious diseases physicians and pharmacists for consults. Indeed, the Infectious Diseases Society of America (IDSA) has issued a statement condemning the independent practice of a pharmacist to advise physicians on selection of antimicrobial therapy.[2] In hospitals that have limited or no infectious diseases physician presence, this model may be the only viable option.

Outpatient Setting

As mentioned previously, some infectious diseases pharmacists have established effective clinical practices in the outpatient setting. The most common example of this is the presence of a pharmacist in an HIV clinic. The myriad of antimicrobial drug interactions and adverse effects associated with antiretroviral therapy, the need to periodically assess antiretroviral efficacy, and the considerable potential for noncompliance literally necessitate the need for a pharmacist in any established HIV clinic. Infectious diseases pharmacists work with infectious diseases and/or immunology physicians. Pharmacists conduct medication histories and answer drug information questions. In some settings, they may act under protocol to assess patient response to antiretroviral therapy based on virologic and immunologic measures, and to make appropriate modifications in therapy.

IMPACT OF INFECTIOUS DISEASES PHARMACISTS ON PATIENT CARE

The original published reports of the impact of infectious diseases pharmacists' interventions on patient outcomes were limited to therapeutic drug monitoring of aminoglycosides. Therapeutic drug monitoring of aminoglycosides by pharmacists resulted in more appropriate utilization of serum aminoglycoside concentrations, more serum concentrations within the therapeutic range, and reduced nephrotoxicity when compared with monitoring by physicians (Destache et al.).[3]

Subsequent reports of the impact of interventions by infectious diseases pharmacists have focused more on improving the antimicrobial therapy process. Specifically, reports of ''antibiotic streamlining'' (narrowing the spectrum of therapy based on culture and susceptibility reports)[4–6] and intravenous-to-oral conversion of antibiotics[7,8] have shown that interventions by pharmacists can reduce costs and lengths of stay without adversely effecting quality of patient care. However, more research and publications are necessary to fully document the beneficial impact of infectious diseases pharmacist interventions.

TOOLS/MATERIALS USED BY INFECTIOUS DISEASES PHARMACISTS

Journals

A number of published journals specifically directed toward infectious diseases and antimicrobial therapy are available as resources for infectious diseases pharmacists:

Clinical Infectious Diseases—This journal, formerly named *Reviews of Infectious Diseases*, is an official publication of the IDSA. Articles are primarily directed at the diagnosis and treatment of infectious diseases, including clinical trials. Frequently, ''State of the Art'' articles are published that summarize current therapy of a particular infection. In addition, IDSA guidelines for the treatment of infectious diseases are published in this journal.

The Journal of Infectious Diseases—This journal is also published by IDSA. The contents of this journal are generally directed at the cellular mechanisms of pathogenesis and immunity of infection. From a pharmacist practitioner standpoint, it is of less usefulness than *Clinical Infectious Diseases*.

Antimicrobial Agents and Chemotherapy—One of the official journals of the American Society for Microbiology, this journal focuses on characterizing and quantifying the activity of antimicrobial agents against various pathogens. Many papers are directed at mechanisms for antimicrobial resistance and activity of newer antimicrobials in vitro. Studies of the efficacy of antimicrobials, as measured via in vitro pharmacokinetic and animal infection models, are published frequently. Studies of drug treatment in humans are also published, but less frequently.

Journal of Antimicrobial Chemotherapy—This British publication addresses all aspects of infectious diseases pharmacotherapy and therapeutics. Both American and European authors contribute to this journal. A review article at the beginning of each issue addresses a pertinent clinical issue. Supplements are published regularly that focus on new antimicrobial agents. More so than other journals, this journal regularly addresses pharmacokinetic/pharmacodynamic issues. Although it is very popular, it occasionally suffers from lack of relevance, in that position papers are usually European rather than American organizations.

Infectious Diseases Clinics of North America—This quarterly, hardbound journal focuses on a single infectious disease topic in each issue. Experts in the field of infectious diseases author state-of-the-art articles that are useful for review or for teaching purposes. Although the articles usually do not present breaking information, they are useful in defining current practice in infectious diseases.

Infectious Diseases in Clinical Practice—This is a very pragmatic journal with topics that clearly state that this journal is authored ''by practitioners, for practitioners.'' Although it is not currently referenced in MEDLINE, this journal offers special insights into practice-related issues that are authored by eminent infectious diseases practitioners.

Journal of Infectious Diseases Pharmacotherapy—This journal, still in its infancy, is the first attempt by clinical pharmacy practitioners to author a journal devoted entirely to infectious diseases pharmacotherapy. It is also not referenced in MEDLINE. Although it has suffered from ''identity crisis,'' the articles are excellent, well referenced, and pertinent to current practice. Hopefully, this journal will continue to grow in stature over the next few years.

In addition to those described here, a number of journals devoted to internal medicine and/or pharmacology topics will from time to time publish articles concerning infectious diseases. They are not discussed further in this article.

Books/Texts

Although they are republished less frequently than journals and therefore may contain dated material, some books and texts have stood the test of time and remain valuable resources:

Mandell, Douglas, and Bennett, eds., *Principles and Practice of Infectious Diseases* (Churchill Livingstone, Philadelphia, 2000)—This ''bible'' of infectious diseases is a must for every infectious diseases practitioner's bookshelf. This book addresses virtually all infectious diseases topics from both a disease- and a pathogen-related perspective. Although the antimicrobial pharmacology component represents only a small portion of the text, it is well written and is a useful resource for those looking for information on antimicrobial pharmacokinetics, interactions, and adverse effects.

Kucers, Crowe, Grayson, and Hoy, eds., *The Use of Antibiotics* (Butterworth–Heinemann, Oxford, 1997)—''Kucers'' remains the standard as a reference text for antimicrobial pharmacology. Arranged by drug classes, this text describes the clinical pharmacology of all known antimicrobials, and has the advantage of its long publication history to include ''classic'' references in antimicrobial pharmacokinetics, adverse effects, and interactions. Because of the evolving nature of antimicrobial resistance, the sections on in vitro activity are often dated and of limited usefulness when comparing antibiotics.

Yu, Merigan, and Barriere, eds., *Antimicrobial Therapy and Vaccines* (Williams and Wilkins, Baltimore, 1998)—This text, first published in 1998, is a laudable attempt to create a comprehensive reference of pathogenic organisms and antimicrobials. Unlike *Principles and Practice of Infectious Diseases*, the text does not specifically address infectious diseases syndromes. Approximately half of the chapter authors are infectious diseases pharmacists. Chapters are addressed by pathogen and by corresponding therapeutic agent(s). They are relatively short but extremely well referenced and clinically relevant. Each chapter devoted to a pathogen is divided into sections on general description, microbiology, susceptibility, and treatment of infections caused by that pathogen. Chapters on antimicrobial agents are divided into sections on chemistry, antimicrobial activity, pharmacodynamics, pharmacokinetics, and adverse effects. Overall, this is a very useful text. Hopefully, it will continue to a second (and subsequent) edition.

Guidelines

Guidelines or consensus statements are important for infectious diseases pharmacists because they identify the ''state of the art'' on paper, which creates a template by which they may conduct their practice. For the most part, pharmacists are relatively content to follow and adhere to clinical guidelines, as long as they are logical and well written. Unfortunately, many guidelines written by physicians address diagnosis and patient assessment much more than the specifics of drug therapy, an area that is of outmost importance to infectious diseases pharmacists. Nevertheless, these guidelines are invaluable for pharmacists who seek ''reinforcement'' when developing treatment protocols for their own institutions.

Infectious Diseases Society of America (IDSA)—The IDSA is the single most common source of guidelines for treatment of infectious diseases in the United States. These guidelines are created largely on an ad hoc basis and are published in *Clinical Infectious Diseases*. Examples of such guidelines include treatment of community-acquired pneumonia, urinary tract infections, and febrile neutropenia. Unfortunately, no systematic process for generating and routinely updating these guidelines appears to be in place.

Centers for Disease Control and Prevention (CDC)—The CDC has begun to exert itself in the area of treatment guidelines for infectious diseases.[9] Well known for their reports and recommendations that are published in *Morbidity and Mortality Weekly Report*, the CDC has now also issued guidelines for treatment of community-acquired pneumonia. Although these guidelines are generally more conservative than those issued by other groups, they do carry the weight of the CDC and the U.S. government. Hopefully, the issuance of guidelines by the CDC will continue.

American Thoracic Society (ATS)—The ATS has issued guidelines for the treatment of community- and hospital-acquired pneumonia. These guidelines have also proven valuable as a reference point for clinicians want to establish treatment guidelines in their own institution. Although ATS guidelines carry considerable political influence, they are typically more consensus driven than evidence based in nature.

Other organizations—Other organizations may occasionally publish guidelines for the use of antimicrobials in selected clinical settings. For example, the American Society of Health-System Pharmacists published guidelines for surgical and nonsurgical prophylaxis in 1999. When such guidelines are published, most infectious diseases pharmacists will obtain them and use them as a resource. However, the sporadic publication of guidelines from these sources means that practitioners are often left without specific guidance in many therapeutic areas.

PROFESSIONAL NETWORKING OPPORTUNITIES

Professional Societies/Meetings

A number of professional societies exist that are either entirely devoted to infectious diseases or support subgroups directed toward infectious diseases. Infectious diseases pharmacists are active members of all the organizations.

Society of Infectious Diseases Pharmacists (SIDP)—SIDP, formed in 1990, is the only organization entirely devoted to practice and research by infectious diseases pharmacists. Currently, more than 400 infectious diseases pharmacists are members of SIDP. An elected Board of Directors and active standing committees conduct the majority of SIDP's business. SIDP provides grants to members to conduct research, and also awards funds to support three infectious diseases residencies annually. SIDP also cosponsors two to three educational symposia with other societies each year.

A 1-day annual meeting is held in conjunction with ICAAC (see below). In addition, members receive a quarterly newsletter and may visit SIDP's web site (www.sidp.org).

Interscience Conference for Antimicrobial Agents and Chemotherapy (ICAAC)—The annual meeting, sponsored by the American Society for Microbiology (ASM), is the largest meeting devoted to infectious diseases in the world. Infectious diseases physicians and pharmacists, as well as microbiologists and infection control practitioners, comprise the majority of ICAAC attendees. More than 15,000 people gather at ICAAC to review the most recent research on antimicrobials and attend state-of-the-art symposia. The sheer volume of information presented at this meeting makes time management a priority. Numerous infectious diseases pharmacists attend this meeting every year and are able to network over the 4-day meeting.

Infectious Diseases Society of America (IDSA)—IDSA is an organization primarily composed of infectious diseases physicians. However, more than 50 infectious diseases pharmacists are members of IDSA. Topics at IDSA's annual meeting parallel the content of their two journals, *Journal of Infectious Diseases* and *Clinical Infectious Diseases*, and include cellular and biochemical mechanisms of infectious diseases, and the epidemiology, diagnosis, and management of infectious diseases. The limited presence of infectious diseases pharmacists at IDSA makes networking more difficult.

International Society of Antiinfective Pharmacology (ISAP)—ISAP is a small but influential organization of infectious diseases physicians and pharmacologists whose focus is infectious diseases pharmacokinetics/pharmacodynamics. The society is truly international in scope, with members from the United States, Canada, and Europe. ISAP's 1-day annual meeting is held immediately after ICAAC and consists of state-of-the-art lectures on current concepts in antimicrobial pharmacodynamics. The timing of the ISAP meeting and its relatively ''low profile'' limit the number of infectious diseases pharmacists that attend this meeting, but those who do attend remain avid supporters of ISAP.

American College of Clinical Pharmacy (ACCP)—The ACCP is an organization devoted to the promotion of clinical pharmacy practice and research. ACCP holds two meetings annually. The content of material presented at these meetings spans the scope of clinical pharmacy practice. However, ACCP has created specialized practice and research networks (PRNs), one of which focuses on infectious diseases. For an additional $10, an ACCP member can join a PRN. PRN members hold business and scientific sessions at ACCP meetings, which allows for networking among members. ACCP also supports PRN listservs on its web site. Finally, ACCP supports a personnel placement service at its fall meeting, where members can recruit residents and fellows.

International Conference of Chemotherapy (ICC)—The ICC is a biannual conference that is similar in size and scope to ICAAC, but is usually held outside the United States. Although the content is excellent, the travel, registration, and housing expenses make this meeting cost prohibitive for most American infectious diseases pharmacists. Those who do attend enjoy the opportunity to interact with practitioners and researchers from around the globe.

International Conference on Macrolides, Azalides, Streptogramins, and Ketolides (ICMASKO)—ICMASKO is a biannual conference that is attended primarily by infectious diseases researchers who present their research on macrolides, azalides, streptogramins, and ketolides. This is a relatively small, intimate meeting that allows for networking for those in attendance.

American Society of Health-System Pharmacists (ASHP) Midyear Clinical Meeting—ASHP's midyear clinical meeting is one of the largest annual meetings of pharmacists in the world. The scope of topics presented at the meeting is very diverse, ranging from clinical topics to reimbursement issues. No specific subgroup of pharmacists devoted to infectious diseases exists within the ASHP. However, numerous infectious diseases satellite symposia and research papers are presented during the meeting. Many infectious diseases pharmacists use this meeting to recruit residents and fellows.

International Conference on Retroviruses—The International Conference on Retroviruses focuses specifically on the treatment of patients with HIV infection. This conference attracts pharmacists who care for these patients. A variety of papers dealing with new antiretroviral agents and combinations are presented, in addition to presentations outlining the best means for caring for these patients. Unfortunately, infectious diseases pharmacists currently are not fully ''recognized'' at this meeting, and problems with registration have occurred. Many other pharmacist societies are lobbying to correct this problem.

Industry Consultantships

From time to time, infectious diseases pharmacists are invited to serve as consultants at small meetings held by pharmaceutical manufacturers. Typically, 6 to 12 consultants are invited to give their opinions about the likelihood of success of a new antimicrobial, or to suggest new marketing or research strategies. These meetings serve an additional purpose in that they allow an additional opportunity for interaction and networking between the pharmacists in attendance.

REFERENCES

1. Board of Pharmaceutical Specialties. *Statement #9903, Added Qualifications for Infectious Diseases*; March 6, 1999.
2. Infectious Diseases Society of America Hospital pharmacists and infectious diseases specialists. Clin. Infect. Dis. **1997**, *25*, 802.
3. Destache, C.J.; Meyer, S.K.; Bittner, M.J.; Hermann, K.G. Impact of a clinical pharmacokinetic service on patients treated with aminoglycosides: A cost-benefit analysis. Ther. Drug Monit. **1990**, *12*, 419–426.
4. Gentry, C.A.; Greenfield, R.A.; Slater, L.N.; Wack, W.; Huycke, M.M. Outcomes of an antimicrobial teaching program in a teaching hospital. Am. J. Health-Syst. Pharm. **2000**, *57*, 268–274.
5. Berman, J.R.; Zaran, F.K.; Rybak, M.J. Pharmacy-based antimicrobial monitoring service. Am. J. Hosp. Pharm. **1992**, *49*, 1701–1706.
6. Briceland, L.L.; Lesar, T.S.; Lomaestro, B.M.; Lombardi, T.P.; Gailey, R.A.; Kowalski, S.F. Streamlining antimicrobial therapy through pharmacists' review of order sheets. Am. J. Hosp. Pharm. **1989**, *46*, 1376–1380.
7. Paladino, J.A.; Sperry, H.E.; Backes, J.M.; Gelber, J.A.; Serrianne, D.J.; Cumbo, T.J.; Schentag, J.J. Clinical and economic evaluation of oral ciprofloxacin after an abbreviared course of intravenous antibiotics. Am. J. Med. **1991**, *91*, 460–472.
8. Ramirez, J.A.; Vargas, S.; Ritter, G.; Brier, M.E.; Wright, A.; Smith, S.; Newman, D.; Burke, J.; Mushtaq, M.; Huang, A. Early switch from intravenous to oral antibiotics and early hospital discharge: A prospective observational study of 200 consecutive patients with comminuty-acquired pneumonia. Arch. Intern. Med. **1999**, *159*, 2449–2454.
9. www.cdc.gov.

Institute for Safe Medication Practices

Michael R. Cohen
Institute for Safe Medication Practices, Huntington Valley, Pennsylvania, U.S.A.

INTRODUCTION

The Institute for Safe Medication Practices (ISMP) is a nonprofit organization devoted entirely to safe medication use and to the prevention of medication errors. A broad-based and interdisciplinary organization, ISMP aims to provide impartial, timely, and accurate medication safety information at all times. As an independent watchdog organization, ISMP receives no advertising revenues and depends entirely on volunteer efforts, educational project, grants, and donations to pursue its work. ISMP's work has been acknowledged and honored by a number of organizations worldwide. For example, in 2000 ISMP received the prestigious Award of Honor from the American Hospital Association (AHA). In addition, in 1998 the Institute received the highly regarded Pinnacle Award from the American Pharmaceutical Association (APhA) and the Healthcare Quality Alliance (HQA).

MISSION

ISMP's mission is:

- To continually expand knowledge about medication errors and prevention methods through system analysis in an interdisciplinary and cooperative manner.
- To collaboratively develop and implement effective error-prevention strategies to reduce the risk of medication errors.
- To educate healthcare policymakers about legislative and regulatory steps that can help prevent medication errors.
- To communicate broadly and to educate healthcare professionals and the public about the nature of medication errors, how to prevent them, and how to manage errors that do occur.
- To provide professional support for healthcare practitioners in preventing and handling medication errors.

OBJECTIVES

ISMP's work, which focuses primarily on improving the safety of medication distribution and use, naming, packaging, and labeling, falls into five key areas. These areas are knowledge, analysis, education, cooperation, and communication. All efforts are built on a non-punitive approach and systems-based solutions.

Knowledge-Based ISMP Initiatives

- Independent review of all errors reported to the USP–ISMP Medication Errors Reporting Program (MERP) and acting partner in the FDA's MedWatch Program.
- Comprehensive collection/analysis of error information through the organization's global information-sharing network.
- Original and impartial research and practitioner surveys on medication errors and prevention.

Analysis-Based ISMP Initiatives

- Comprehensive use of failure mode and effects analysis (FMEA) to learn where or when errors are most likely to occur and to help prevent them.
- A thorough review process, using an innovative computer software program, to study and prevent product name- and packaging-related errors.
- Site visits and confidential consultations in various healthcare delivery settings and throughout the healthcare industry.
- A wholly owned subsidiary called Medical Error Recognition and Revision Strategies (Med-E.R.R.S.™), which works confidentially with pharmaceutical companies to predict error potential and thereby avoid problems that might stem from proposed drug names, labels, and packaging.

Encyclopedia of Clinical Pharmacy
DOI: 10.1081/E-ECP 120006207

Education-Based ISMP Initiatives

ISMP provides the following products and services that are targeted to healthcare professionals, pharmaceutical industry, insurance industry, and regulatory agencies:

- Slide programs, CD-ROMs, posters, books, and videotapes on medication safety including such topics as Failure Mode and Effects Analysis (FMEA); timely and accurate educational sessions and conferences, slide programs, and presentations.
- Knowledgeable and articulate speakers including nurses, pharmacists, and physicians.
- Free, Web-based access to archived issues of the biweekly *ISMP Medication Safety Alert!* newsletter.
- Safe Medication Management Fellowship that provides postgraduate training to healthcare practitioners in safe medication management.

Cooperation-Based ISMP Initiatives

- Official partnership in the U.S. Food and Drug Administration's (FDA) MedWatch program.
- Collaborative work toward error prevention with the American Hospital Association (AHA), the Joint Commission on Accreditation of Healthcare Organizations (JCAHO), the National Coordinating council on Medication Error Reporting and Prevention (NCCMERP), the National Patient Safety Foundation (NPSF), the United States Pharmacopeia (USP), and dozens of other consumer and professional organizations.
- Highly effective educational efforts with legislative and regulatory bodies to improve the safety of medication use.

Communication-Based ISMP Initiatives

- *ISMP Medication Safety Alert!*—the nation's only biweekly publication that reaches nearly every U.S. hospital and tens of thousands of healthcare professional with warnings about medication errors and practical prevention strategies.
- Special electronic and direct mail hazard warnings targeted to healthcare professionals.
- Scholarly and practical articles and continuing education columns that reach practitioners in virtually every healthcare field.
- Web site with timely and accurate information on errors and prevention recommendations for healthcare professionals and consumers.
- Media relations campaigns that reach millions of healthcare professionals and the public every month by placing error-prevention information in healthcare publications and with the nation's most prestigious news organizations.

Contact Information: Martin S. Goldstein, Director of Program Development, Institute for Safe Medication Practices, 1800 Byberry Road, Huntingdon Valley, PA 19006, 215.947.7797, mgoldstein@ismp.org.

FURTHER READING

Institute for Safe Medication Practices; www.Ismp.org.

ISMP, Medication Safety Alert! Huntington Valley, Pennsylvania.

Institute for Safe Medication Practices—Spain

María-José Otero
Alfonso Domínguez-Gil
Hospital Universitario de Salamanca, Salamanca, Spain

INTRODUCTION

The Institute for Safe Medication Practices—Spain (ISMP—Spain) is an independent, nonprofit Spanish organization committed to the prevention of medication-system errors and adverse drug events.

Called *Instituto para el Uso Seguro de los Medicamentos* in Spanish, the ISMP—Spain was established in Salamanca in October 1999, at the Clinic Hospital of the University of Salamanca. It is an interdisciplinary organization, currently under the direction of Alfonso Domínguez-Gil and María-José Otero.

As an independent organization, ISMP—Spain depends upon volunteer efforts, grants, and donations. It works in cooperation with healthcare practitioners, professional organizations, the pharmaceutical industry, and the government in an effort to enhance patient safety. The ISMP—Spain is also a partner of the ISMP—U.S.A., which has agreed to provide a full range of logistic and scientific support for ISMP—Spain, including consultative support and access to its resources and published recommendations for use in error prevention.

MISSION

It is the mission of ISMP—Spain to enhance the safety of the medication-use system and to improve the quality of patient healthcare. The most important goal is to reduce the risk of medication errors and preventable adverse drug events.

OBJECTIVES

Specific objectives of ISMP—Spain include:

- To maintain a national medication error reporting program to collect observations and experiences of healthcare practitioners and to analyze this information with a systems approach in order to draw valid conclusions. This program shares data with the ISMP—U.S.A. and is important for the exchange of information and the coordination of efforts in the prevention of medication-system errors worldwide.
- To increase awareness of the importance of medication errors among healthcare professionals, institutions, organizations, the pharmaceutical industry, the patients themselves, and throughout the entire healthcare system, and to build a safety culture of commitment to medication error reporting and prevention.
- To develop recommendations and effective strategies for preventing medication errors and reducing adverse drug events and to educate healthcare professionals and patients to ensure that these recommendations and practices are implemented appropriately.

MEDICATION ERRORS REPORTING PROGRAM

The key to reducing medication errors lies in learning effectively from failures. Since its founding in October 1999, ISMP—Spain has maintained a national notification error reporting program. The principal objective of this program is to obtain information on medication errors and their causes in order to establish and transmit practical recommendations to prevent the recurrence of the errors.

This program has three main characteristics: it is voluntary, confidential, and independent. It collects observations and experiences concerning those potential or actual medication errors that healthcare professionals voluntarily report. The information is independently analyzed, with no conflicts of interests nor administrative pressure, and all information is treated confidentially.

Healthcare professionals can either complete a report form or contact the ISMP—Spain directly by e-mail, fax, or telephone to report medication errors with complete confidentiality. The types of medication errors submitted include confusion over look-alike or sound-alike drug names, ambiguity or similarity in packaging or labeling,

Encyclopedia of Clinical Pharmacy
DOI: 10.1081/E-ECP 120006373

misinterpretation of handwritten orders, errors in prescribing and monitoring, or errors in drug administration.

ISMP—Spain carefully reviews and analyzes all reported errors, and depending on the characteristics, sends a copy of the report to the Spanish Medicines Agency (AEM) and to the pharmaceutical companies whose products are mentioned in the reports. This information is also shared with the ISMP—U.S.A.

ISMP—Spain publishes information about submitted reports on their web site www.usal.es/ismp and includes safety recommendations designed to help reduce the probability of such errors recurring. The goal is to make this information readily available to healthcare professionals.

OTHER ACTIVITIES AND PROJECTS

It is the belief of ISMP—Spain that the first step in the long road to reducing adverse drug events is getting everyone to recognize this problem and to become committed to combating it. To achieve this goal it will be necessary to effectively quantify the extent and cost of this problem on a national basis. The next step would then be to identify solutions appropriate for the Spanish healthcare system, solutions that will lead to the reduction of medication errors. For this reason, a key initiative of ISMP, Spain is to carry out and to promote research on the incidence, nature, and cause of adverse drug events in different settings, as well as their impact on patients and healthcare costs. This effort will compile important reference material to help improve safety in Spanish healthcare.

The education and dissemination of information is another primary objective of ISMP—Spain: If everyone understands the nature and causes of medication errors, there is a much greater possibility of improving patient safety. In this sense, ISMP—Spain makes educational presentations and holds conferences at healthcare professional meetings to provide information about adverse drug events. ISMP—Spain also publishes opinion articles and practical articles in Spanish healthcare journals in an effort to broadly disseminate a culture of safety and error prevention.

Another important goal of ISMP—Spain is to encourage the development of local medication error reporting and analysis systems in individual healthcare organizations. ISMP—Spain has coordinated a project with the Spanish Society of Hospital Pharmacy to create a standard terminology and classification system of medication errors so that healthcare organizations can design databases and analyze medication error reports using the same system.

FUTURE GOALS

ISMP—Spain believes in the importance of coordinating efforts to enhance patient safety in countries all over the world. It is open to the creation of a work platform in Europe and also to cooperating with Spanish-speaking countries with any initiatives they may undertake to improve their medication systems.

Institute of Medicine

Kenneth I. Shine
Institute of Medicine, Washington, District of Columbia, U.S.A.

INTRODUCTION

The Institute of Medicine (IOM) was chartered by the National Academy of Sciences in 1970 "to improve the health of the American people and peoples of the world" by advancing the health sciences and by providing analysis of important issues in health and health policy for government, the professions, and the public and private sectors. The Institute is an independent, nongovernmental organization. It carries out its work largely through committees of pro bono experts who employ an evidence-based deliberative process to produce scientifically valid nonpartisan reports. Studies originate in several ways: by Congress mandating that an Executive Branch agency contract with the IOM; by direct request of Executive Branch agencies, foundations, or other private organizations; or as self-initiated projects when the Institute determines that an important or highly sensitive issue might not be the subject of a request from an outside organization. In addition to committee studies, IOM plays a unique convening role by sponsoring workshops, roundtables, symposiums, forums, and other activities that enable parties on all sides of an issue to come together and discuss problems and solutions in a neutral, unbiased setting.

The Institute also has an honorific function. Each year it elects 60 regular members, five senior members, and five foreign associates. Elected members include distinguished individuals whose expertise and leadership cover the broad range of biomedical sciences, public health, nutrition, environmental sciences, and social and community medicine, as well as pharmacy, the development of new drugs and biologics, and vaccine and drug safety. The Institute's charter stipulates that at least one-quarter of IOM members be from professions other than those primarily concerned with medicine and health. Thus, the membership includes leading ethicists, economists, and social and behavioral scientists, among others.

ACTIVITIES

The Institute's portfolio of activities is extensive, ranging from issues of scientific integrity to the future of specific areas of health sciences research. The Institute has undertaken studies on the processes of innovation, including new drug, vaccine, and biologics development. Its Roundtable for the Development of Drugs and Vaccines Against AIDS was an important example of IOM's ability to convene disparate opinions around important issues. In this case, the roundtable included representatives of the National Institutes of Health, the Food and Drug Administration, the pharmaceutical industry, AIDS activists, and other interested parties. Other roundtables have considered such issues as the development of drugs for new and emerging infections, antibiotic resistance, and the development of biologics and devices.

The Institute's studies have had a profound effect on public health. The range of topics that have been the subject of IOM studies is wide and varied. Some examples follow:

- In addition to recommending priorities for new vaccine development, IOM committees have also provided analysis and advice regarding complications associated with vaccines, barriers to immunization, immunization finance, and appropriate uses of vaccines.
- An IOM committee examined the role of women in clinical trials—a complicated issue involving scientists, industry, and ethicists who evaluated the participation of women, particularly women of childbearing age, in drug trials.
- The 1999 report entitled *To Err is Human: Building a Safer Health Care System* emphasized that medication errors were an important contributor to morbidity and mortality. Yet this committee also noted that the elimination of such errors will require a comprehen-

Encyclopedia of Clinical Pharmacy
DOI: 10.1081/E-ECP 120006290

sive systematic approach involving physicians, nurses, hospitals, pharmacists, patients, and others in health care working together. Better information technology, computerized data entry, and nonpunitive reporting of near misses were a few of the elements recommended as an approach to accomplish this goal.

- Substance abuse has been the subject of several IOM studies, which have covered a whole host of issues related to the topic, including federal regulation of methadone treatment, the development of medications for the treatment of opiate and cocaine additions, and community-based research to find better ways to treat people who abuse drugs.
- Fluid replacement has been examined both in relation to conditions such as heat stress and when used to resuscitate and treat combat casualties and civilian injuries.
- The FDA Advisory Committee process and other FDA roles and functions have been the subject of IOM studies that have led to significant changes in the agency's function. In one important study, *Halcion: An Independent Assessment of Safety and Efficacy Data*, an IOM committee addressed the difficult problem of the criteria and procedures for withdrawing a previously approved drug from the market.
- In 1997, the report *Pharmacokinetics and Drug Interactions in the Elderly and Special Issues in Elderly African–American Populations* considered the special challenges confronted by proper use of agents in these groups.
- A landmark 1988 report, *The Future of Public Health* (soon to be updated), identified many of the critical challenges to education, practice, and applications of public health. In 1998, the Institute helped develop the prototype leading indicators for ''Healthy People 2010,'' the nation's blueprint for prevention. The importance of community organization and partnerships in furthering the public health has been underscored by a number of other Institute reports.
- A series of studies on health and behavior and the role of the social and behavioral sciences in health have had important implications for public health, as well as for other aspects of medicine. The 1992 report *Emerging Infections: Microbial Threats to Health in the United States* was among the earliest warnings with regard to this issue and also focused on the development of antibiotic-resistant organisms. It has been followed by a number of efforts in both public and private sectors to respond to these threats.
- The Institute is also responsible for establishing ''dietary reference intakes''—quantitative estimates of nutrient intakes to be used for planning and assessing diets for healthy people—which update and replace the recommended dietary allowances.
- The 1986 report *Confronting AIDS: Direction for Public Health, Health Care, and Research* was an IOM-initiated project that addressed seriously what had been to that time a largely ignored epidemic. Subsequent reports have addressed needle exchange, the behavioral and mental health aspects of HIV infection and AIDS, and the prevention of perinatal transmission of HIV. The most recent IOM report on the subject, *No Time to Lose: Getting More from HIV Prevention*, provides a comprehensive review of current HIV-prevention efforts in the United States, as well as a framework for future activities.
- Several studies have focused on environmental issues, including the concept of environmental justice, environmental and occupational education and training in medicine and nursing, and the role of environmental factors in illness (e.g., asthma).
- Among the reports issued by the Institute on tobacco and tobacco control, the 1994 report, *Growing Up Tobacco Free: Preventing Nicotine Addiction in Children and Youths*, was particularly influential in establishing national policy.
- The Institute also conducts a significant program in international health, including efforts to control hepatitis and diarrheal diseases in the Middle East, that are being conducted through collaborations involving American, Israeli, Egyptian, and Palestinian scientists. More recently, Jordan has joined the academies from these countries in addressing problems of water conservation and micronutrients in the region.

The full text of Institute of Medicine publications is available on-line at the National Academy Press' web site, www.nap.edu. Additional information about the Institute and its activities, as well as a list of all publications, can be found at www.iom.edu.

PHARMACY PRACTICE ISSUES

Integrative Medicine

Kathryn L. Grant
University of Arizona, Tucson, Arizona, U.S.A.

William Benda
Big Sur, California, U.S.A.

INTRODUCTION

Although the terminology ''integrative medicine'' has been used synonymously with such terms as complementary, alternative, and unconventional medicine, it in fact represents the emergence of a totally new model of healthcare delivery.[1–3] In the broadest sense, integrative medicine seeks to combine the best therapies of Western (conventional) medicine with the best alternative modalities to provide each individual patient with an optimal treatment plan for their specific situation.[4] In addition, this synergistic approach has as its foundation an assumption of the innate healing power of each human organism and a belief in the centrality of the healing relationship between doctor and patient.[1]

BACKGROUND

Practitioners of conventional medicine are justifiably proud of the achievements of their profession, showcased by the pharmacological, radiological, and surgical advances of the 20th century. Fundamental reliance on such technologies, however, has led to the dismissal of numerous therapies developed outside the conventional medical model.[5] Exclusionary attitudes escalated when studies revealed that an estimated one-third of the U.S. population was using complementary and alternative modalities.[6–9] The result was increasing divisiveness between proponents and opponents of unconventional therapies, even though practitioners on both sides had received identical training during their medical education.[10,11]

Perhaps the greatest controversy of interest to the pharmacist has involved the increasing use of botanicals and nutritional supplements. Despite objections from conventional physicians and pharmacists, the U.S. Congress responded to pressure from the public and nutraceutical industry by passing the Dietary Supplement Health and Education Act of 1994 to ensure broad access to these products.[12] As further evidence of governmental intervention, the National Institutes of Health (NIH) created the National Center for Complementary and Alternative Medicine (NCCAM) and provided a current budget of $68.3 million.[13] More recently, the Office of Dietary Supplements (ODS) within the NIH has established four centers for phytomedicinal research, ensuring that selected botanicals will be evaluated by the same or similar processes expected for prescription medications.[14]

Why is the public turning toward complementary and alternative medicine (CAM)? Surveys showing that CAM use tends to be higher in patients with diseases often inadequately treated by conventional therapies (e.g., arthritis, cancer).[15,16] This may seem to suggest an inherent dissatisfaction with conventional medicine. However, studies specifically addressing this issue have demonstrated that patients are actually turning toward CAM in keeping with their individual philosophical values and belief systems.[17,18] Reasons for patient use not withstanding, it would be beneficial for conventional practitioners to investigate the attraction of CAM, and how it can be used to strengthen and enhance their individual practices. This perspective appears even more appealing in light of the fact that public interest and usage of CAM continues to increase.[7]

Although the breadth of CAM therapies range from traditional systems developed in other cultures (e.g., traditional Chinese medicine [TCM]) to recently developed practices loosely based on science (e.g., functional medicine), most tend to share certain underlying perspectives.[19,20] Such philosophies include belief in the interconnectedness of mind and body, preference for innate rather than artificial (e.g., pharmaceutical) sources of healing, and recognition of an ultimate meaning underlying each individual's illness.[20] Consequently, many

Encyclopedia of Clinical Pharmacy
DOI: 10.1081/E-ECP 120006402

CAM practitioners spend far more time with their patients than conventional practitioners, listening attentively and attempting to truly understand what the patient wants and who they are as an individual.[21] A patient can thus leave a session with a feeling of empowerment and a belief that they can be well again.

DEFINITION OF INTEGRATIVE MEDICINE

The basic tenet of integrative medicine is that it neither rejects conventional medicine nor uncritically accepts CAM.[1] Just as various alternative practices are as yet unproven and some do carry significant risk, practitioners of integrative medicine recognize that it is also important to be just as analytical of conventional medicine. For example, the fact that adverse reactions to prescription medications represent between the fourth to sixth leading cause of inpatient deaths came as a shock to many in the healthcare field.[22] An Institute of Medicine survey found iatrogenic illnesses to be the eighth leading cause of death, exceeding the deaths attributable to motor vehicle accidents, breast cancer, and acquired immunodeficiency syndrome.[23,24] Consequently, in weighing the risks inherent to any therapy, conventional or CAM, the integrative practitioner seeks the least invasive, least toxic, and least costly interventions whenever possible.

Another cornerstone of the integrative model is the assertion that healing optimally occurs when all factors that influence health, as well as illness, are addressed. Beyond the patient's physical condition lie the mental, emotional, and spiritual influences on quality and duration of life.[2,20] Sir William Osler is quoted as saying, "It is more important to know what sort of patient has a disease than what sort of disease a patient has."[2] Whether searching for relief from illness or for promotion of health, patients choosing the integrative model are offered more tools than just drugs or surgery.[1] The most common recommendations made are modification in diet and increase in level of physical activity. Stress reduction techniques (e.g., biofeedback, yoga, tai chi) are encouraged to be used in place of negative coping activities (e.g., tobacco, alcohol, recreational drugs) during difficult times. Positive coping skills can also be based in spiritual practices such as prayer, church attendance, meditation, or even such common activities as nature walks. Community involvement including volunteer work can also help to increase an individual's positive outlook. These cornerstones of a healthy life (good nutrition, physical activity, and stress reduction) are considered the primary means by which illness can be both prevented and treated. Health, therefore, becomes more than just the absence of disease. Other tools that are available to the integrative medicine practitioner include hypnosis, guided imagery, TCM, Ayurvedic medicine, homeopathy, osteopathy, chiropractic treatments, and a host of others. Discussion of such complex systems is beyond the scope of this article.[25]

The actual foundation upon which the integrative model rests is the creation of a deeply respectful and understanding relationship between patient and practitioner.[21,26] The practitioner recognizes that they are merely the means by which patients can discover, or actually rediscover, their innate capacity to restore health. This process begins with asking open-ended questions and listening patiently to the answers with a nonjudgmental attitude.[21] Patients are allowed to make choices about therapies or lifestyle changes that are consistent with their values and philosophical beliefs, which reconfirms one of the primary reasons they are using CAM.[17,26] Medical decision making is shared rather than dictated by informing the patient of all possible alternatives, whether conventional or CAM. This requires that the practitioner becomes knowledgeable about a wide array of potential interventions that may be useful to their patient, as well as which interventions should be avoided.[26]

Essential to a partnership in medical decision making is the necessity that each patient realigns his or her own approach toward health and the healthcare system. Currently, our Western culture advertises and promotes poor choices in lifestyle that ultimately result in burgeoning healthcare expenditures. Fundamental to this new paradigm of healthcare is the requirement that individuals take responsibility for their own wellness, making healthy choices concerning nutrition, physical activity, and response to stressful situations.

Finally, it is recognized that to ensure patient trust and compliance, health professionals must develop and model a healthy lifestyle for themselves. To this end, integrative practitioners are encouraged to examine and remedy the indoctrination of the current system that rewards the overworked, driven, and harried healthcare professional.

As part of the analysis and critique of both CAM and conventional modalities, integrative medicine practitioners search for accurate information in texts, primary literature, and the Internet to counsel their patients on the relative efficacy, safety, and appropriate use of these therapies. To ensure validity of research findings, it is becoming increasingly evident that other methodologies, in addition to the randomized controlled trial, need to be conceived and completed to fully evaluate these unconventional therapies as well as the integrative medical model itself.

AN INTEGRATIVE CLINICAL MODEL

The past 10 years have witnessed the emergence of a large number and wide variety of integrative clinical models. Configurations have ranged from providing mostly conventional therapies with a smattering of complementary modalities to groups of alternative practitioners simply sharing both office space and patient populations. A feasible model that most accurately reflects the definition of integrative medicine incorporates practitioners who have been trained in both conventional and alternative therapeutic modalities. In support of this approach, a patient survey by the University of Arizona's Program in Integrative Medicine reports that their primary desire was to be treated by a physician who was knowledgeable in both conventional medicine and CAM.[27]

In the ideal integrative clinic, the initial visit entails an in-depth interview lasting from 60 to 90 minutes. During this visit, the practitioner concentrates on understanding the patient as an individual, as well as delineating the medical history and determining desired outcomes. After the initial interview and review of prior records, a comprehensive treatment plan is formulated and presented to the patient, not as a directive but as a series of options to be chosen under the guidance of the experienced practitioner. This approach provides the greatest potential of adherence and success due to the patient's sense of participation and empowerment. Often the treatment plan results in patient referral to other specialized practitioners. For example, a dietician might be suggested for nutritional counseling; a pharmacist for medication/dietary supplement counseling; a psychologist for hypnosis or guided imagery; or an osteopathic, homeopathic, or TCM practitioner to address specific issues. The patient is then followed either by return visits or by telephone to determine either success or need for treatment plan modification.

CHALLENGES TO INTEGRATIVE MEDICINE

Because the integrative model incorporates practices that are often deemed ''quackery'' and unsafe by many conventional practitioners, there has been significant vocal and written opposition to the growth of CAM, in general, and integrative medicine, in particular.[11] One such avenue of opposition has been to block or strongly discourage incorporation of CAM teachings into medical school curricula.[10] Despite this resistance, 60% of medical schools and 72% of pharmacy schools have incorporated CAM into their curriculum.[28,29] One significant endeavor to close the gap between consumer demand and professional response has been the formation of the Consortium of Academic Health Centers for Integrative Medicine.[30] Their mission is to become a significant voice addressing the importance of reevaluating and restructuring medical education, and to have integrative medicine programs in one-fifth of the United States' medical schools within the next few years. Table 1 lists the current members of the consortium.

Table 1 Current members of the Consortium of Academic Health Centers for Integrative Medicine

Albert Einstein/Yeshiva University College of Medicine, New York
Duke University School of Medicine, North Carolina
Georgetown University School of Medicine, Virginia
Harvard Medical School, Massachusetts
Jefferson Medical College, Pennsylvania
Stanford School of Medicine, California
University of Arizona College of Medicine, Arizona
University of California–San Francisco School of Medicine, California
University of Maryland School of Medicine, Maryland
University of Massachusetts School of Medicine, Massachusetts
University of Minnesota Medical School, Minnesota

Another challenge to the growth of integrative medicine lies in the realm of reimbursement policies. Although third-party payers are increasing coverage of some CAM modalities, this tends to be a slow and erratic process.[31] Because CAM therapies tend to be less costly than conventional interventions, reimbursement should increase as research validates efficacy and cost effectiveness. If hypnosis decreases pain and eliminates the need for lifelong use of addicting medications or if techniques such as Feldenkrais prevent or postpone orthopedic surgery, the cost savings could be great for a healthcare system currently in economic crisis.

Difficulties have also arisen in relation to funding of CAM research. This has been partially addressed by the establishment within the NIH of the NCCAM and the ODS. However, NCCAM and ODS grants tend to be awarded to those following the conventional reductionistic research model when investigating a CAM modality (e.g., studying a single acupuncture point rather than acupuncture or TCM as a whole). Although this approach may add another tool to the conventional medicine arsenal, it does not address the fundamental questions of whether a different medical system (e.g., TCM) or a new medical model (integrative medicine) is appropriate and effective.

Research methodologies must be developed that allow for studying systems as a whole (e.g., using systems theory or multidimensional outcome measures) to determine whether component parts are truly synergistic.[32] A small step in the right direction has been the creation of four botanical centers by ODS.[14] These centers are charged with studying the effects of marketed dietary supplements currently purchased and used by the public, rather than the reductionistic search for a solitary active principle.

Another roadblock on the path to credibility is an apparent bias on the part of mainstream medical journals when reviewing research or position papers on CAM or integrative medicine. Journals may appear to be more willing to publish negative results for CAM modalities than those that indicate a positive outcome.[33] Although editors insist that the review process is objective, it may be that the bar is set higher for CAM, and such bias has been recognized by such authorities as the editors of the *New England Journal of Medicine*.[34] As research methodology in the CAM field improves and a biased editorial policy is corrected, the system of peer review may be more fairly applied to all scholarly submissions.

INTEGRATIVE MEDICINE AND THE FUTURE OF HEALTHCARE

Given the current state of healthcare in the United States, most efforts by the political, academic, and corporate entities have been in the direction of repair of the system rather than creation of a new model. The results of these labors have been equivocal at best. A growing number of people planning the future of healthcare now envision the need for a totally new system, one that retains the beneficial technologies of conventional medicine while returning the focus of the profession toward health rather than illness.[35,36] The result would be a system of healthcare in which the public once again trusts and respects the healthcare practitioner as an ally, rather than as part of what often appears to be a mechanistic, profit-driven industry. Integrative medicine as a new medical model may fill the need for a totally new system, such that in the future the term ''integrative'' will no longer be needed, and we will all simply practice good medicine.[30]

REFERENCES

1. Gaudet, T.W. Integrative medicine: The evolution of a new approach to medicine and to medical education. Integrative Med. **1998**, *1* (2), 67–73.
2. Maizes, V.; Caspi, O. The principles and challenges of integrative medicine. West J. Med. **1999**, *171* (3), 148–149.
3. Guerrera, M.P.; Ryan, A.H. Integrative medicine: An old concept with new meaning. Conn. Med. **1999**, *63* (1), 50–51.
4. Weil, A. The significance of integrative medicine for the future of medical education. Am. J. Med. **2000**, *108* (5), 441–443.
5. Anonymous. Quackery persists. JAMA, J. Am. Med. Assoc. **1972**, *221* (8), 914.
6. Eisenberg, D.M.; Kessler, R.C.; Foster, C.; Norlock, F.E.; Calkins, D.R.; Delbanco, T.L. Unconventional medicine in the United States: Prevalence, costs, and patterns of use. N. Engl. J. Med. **1993**, *328* (4), 246–252.
7. Eisenberg, D.M.; Davis, R.B.; Ettner, S.L.; Appel, S.; Wilkey, S.; Van Rompay, M.; Kessler, R.C. Trends in alternative medicine use in the United States, 1990–1997: Results of a follow-up national survey. JAMA, J. Am. Med. Assoc. **1998**, *280* (18), 1569–1575.
8. Skolnick, A.A. FDA petitioned to ''stop the homeopathy scam.'' JAMA, J. Am. Med. Assoc. **1994**, *272* (5), 1154–1155.
9. Barrett, S.; Tyler, V.E. Why pharmacists should not sell homeopathic remedies. Am. J. Health-Syst. Pharm. **1995**, *52* (9), 1004–1006.
10. Dalen, J.E. Is Integrative medicine the medicine of the future? A debate between Arnold S. Relman, MD, and Andrew Weil, MD. Arch. Intern. Med. **1999**, *159* (18), 2122–2126.
11. Relman, A.S. A trip to stonesville: Andrew Weil, the boom in alternative medicine and the retreat from science. New Repub. **1998**, http://www.tnr.com/archive/1298/121498/relman121498.html.
12. Dietary Supplement Health and Education Act of 1994 (DSHEA). In *The Complete German Commission E Monographs: Therapeutic Guide to Herbal Medicine,* 1st Ed.; Blumenthal, M., Ed.; Integrative Medicine Communications: Boston, 1998; 12.
13. Nahin, R.L.; Straus, S.E. Research into complementary and alternative medicine: Problems and potential. BMJ **2001**, *322* (7279), 161–164.
14. Stern, M. *NIH Announces Two Additional Centers for Dietary Supplement Research. Office of Dietary Supplements*; National Institutes of Health, 2000, Sep. 20, http://www.nih.gov/news/pr/sep2000/ods-20.htm (accessed January 2001).
15. Boisset, M.; Fitzcharles, M.A. Alternative medicine yse by rheumatology patients in a universal healthcare setting. J. Rheumatol. **1994**, *21* (1), 148–152.
16. Nam, R.K.; Fleshner, N.; Rakovitch, E.; Klotz, L.; Trachtenberg, J.; Choo, R.; Morton, G.; Danjoux, C. Prevalence and patterns of the use of complementary therapies among prostate cancer patients: An epidemiological analysis. J. Urol. **1999**, *161* (5), 1521–1524.
17. Astin, J.A. Why patients use alternative medicine: Results

of a national study. JAMA, J. Am. Med. Assoc. **1998**, *279* (19), 1548–1553.

18. Donnelly, W.J.; Spykerboer, J.E.; Thong, Y.H. Are patients who use alternative medicine dissatisfied with orthodox medicine? Med. J. Aust. **1985**, *142* (10), 539–541.
19. Anonymous. *What are the Major Types of Complementary and Alternative Medicine?*; http://nccam.nih.gov/health/whatiscam/index.htm#4 (accessed July 2002).
20. Kaptchuk, T.J.; Eisenberg, D.M. The persuasive appeal of alternative medicine. Ann. Intern. Med. **1998**, *129* (12), 1061–1065.
21. Gianakos, D. Alternative healer. Ann. Intern. Med. **2000**, *133* (7), 559.
22. Lazarou, J.; Pomeranz, B.H.; Corey, P.N. Incidence of adverse drug reactions in hospitalized patients: A meta-analysis of prospective studies. JAMA, J. Am. Med. Assoc. **1998**, *279* (15), 1200–1205.
23. *To Err is Human: Building a Safer Health System*; Kohn, L.T., Corrigan, J.M., Donaldson, M.S., Eds.; National Academy Press: Washington, D.C., 2000. http://www.nap.edu/openbook/0309068371/html/index.html (accessed January 2001).
24. Leape, L.L. Institute of Medicine medical error figures are not exaggerated. JAMA, J. Am. Med. Assoc. **2000**, *284* (1), 95–97.
25. Benda, W.; Grant, K.L. Integrative medicine and the search for health. Support Line **2000**, *22* (5), 23–28.
26. Sugarman, J.; Burk, L. Physician's ethical obligations regarding alternative medicine. JAMA, J. Am. Med. Assoc. **1998**, *280* (18), 1623–1625.
27. Gaudet, T.W., Unpublished data, The University of Arizona Program in Integrative Medicine.
28. Berman, B.M. Complementary medicine and medical education. BMJ **2001**, *322* (7279), 121–122.
29. Rowell, D.M.; Kroll, D.J. Complementary and alternative medicine education in United States pharmacy schools. Am. J. Pharm. Educ. **1998**, *62* (4), 412–419.
30. Rees, L.; Weil, A. Integrated medicine. BMJ **2001**, *322* (7279), 119–120.
31. Tracey, E. *Delivery of Complementary Medicine Needs Improvement*; Reuters Health Newsservice: Washington DC, 2000, December 11.
32. Bell, I. **2000**, Personal Communication, October 18.
33. Ezzo, J.; Berman, B.M.; Vickers, A.J.; Linde, K. Complementary medicine and the Cochrane collaboration. JAMA, J. Am. Med. Assoc. **1998**, *280* (18), 1628–1630.
34. Tye, L. Journals breaks with past: New editor cleans slate at medical publication. Boston Globe Online. Archives **2000**, September 12.
35. Carlson, R.J.; Ellwood, P.M.; Etzioni, A.; Goldbeck, W.B.; Gradison, B.; Johnson, K.E.; Kitzhaber, J.; Koop, C.E.; Lee, D.R.; Lewin, L.S.; McNerney, W.J; Ray, R.D.; Riley, T.; Schmoke, K.L.; Thier, S.O.; Tilson, H.H.; Tuckson, R.V.; Warden, G.L. *The Belmont Vision for Healthcare in American*; Institute for Alternative Futures: Alexandria, Virginia, 1992.
36. Anonymous. *The Future of Complementary and Alternative Approaches (CAAs) in US Healthcare: Executive Summary*; Institute for Alternative Futures: Alexandria, Virginia, 1998.

International Pharmaceutical Abstracts (ASHP)

Carol Wolfe
American Society of Health-System Pharmacists, Bethesda, Maryland, U.S.A.

INTRODUCTION

International Pharmaceutical Abstracts (*IPA*) is the official abstracting and indexing service of the American Society of Health-System Pharmacists. *IPA* is unique in its international coverage of pharmacy and related health journals.

AIMS AND SCOPE

An expert panel selects, abstracts, and indexes the most pertinent articles from over 600 journals published throughout the world. Translators with pharmacy expertise review the major pharmacy journals published in languages other than English and prepare abstracts in English. In addition, *IPA* publishes abstracts of meetings conducted by the American Society of Health-System Pharmacists and the American Association of Colleges of Pharmacy. Over 15,000 abstracts per year are added to the *IPA* database covering the clinical and scientific literature related to pharmacy, as well as legal aspects, professional practice issues, and trends in education and research. *IPA*'s unique structure supports searching by drug class, disease state, MeSH heading, generic or trade drug name, and chemical registry number.

Examples of topic areas targeted for coverage are drug evaluations, pharmacology, investigational drugs, toxicity, therapeutic advances, new technology, herbals, and adverse drug reactions. Although a printed version of *IPA* is available from ASHP, use is now primarily electronic. Online versions are available by subscription through several database vendors, including SilverPlatter, Ovid, Dialog, Cambridge Scientific Abstracts, STN International, DataStar, DIMDI, and EBSCO Publishing.

Individuals may also economically purchase online search blocks by selecting "IPA PharmSearch" at www.ashp.org.

HISTORY

The inaugural issue of *IPA* was published in January 1964 under the leadership of *IPA*'s founding editor, Donald E. Francke. The seminal concept for *IPA* was created in 1957 when Dr. Francke established a section called ''Selected Pharmaceutical Abstracts'' in the *Bulletin*, the forerunner of the *American Journal of Health-System Pharmacy*. Dr. Francke's early objective for *IPA* was to ''serve as an alerting service to keep the busy pharmacy practitioner, professor, researcher, and student keenly aware of a wide variety of information to permit him to do a better professional job.''

In late 1966, Dwight R. Tousignaut succeeded Donald Francke as editor of *IPA*. *IPA*'s production process was computerized in 1970, and *IPA*'s first electronic licensing arrangement was formalized in 1971 in an agreement to supply magnetic tape to the National Library of Medicine for use in its ToxLine database.

The International Pharmaceutical Abstracts (Coden: IPMAAH; ISSN: 0020-8264) is published by the American Society of Health-System Pharmacists twice each month, on the 1st and 15th. Its circulation is primarily electronic. The print and electronic subscription rates are available upon request. Internet access to *IPA PharmSearch* is available at $90 per 100 searches. The editorial and subscription offices are located at 7272 Wisconsin Avenue, Bethesda, Maryland 20814, U.S.A. (Telephone: 301-657-3000, extension 1241; Fax: 301-664-8857; E-mail: ipa@ashp. org; Web site: http://www. ashp.org; Managing Editor: Victoria Ferretti-Aceto).

The first edition of the *IPA Thesaurus and Frequency List* was published in 1981. Now in its eighth edition, the *Thesaurus* provides a controlled vocabulary designed to support robust searching of pharmaceutical literature.

Encyclopedia of Clinical Pharmacy
DOI: 10.1081/E-ECP 120006388

International Society for Pharmacoeconomics and Outcomes Research

Elizabeth Tracie Long
ISPOR—International Society for Pharmacoeconomics and Outcomes Research, Lawrenceville, New Jersey, U.S.A.

INTRODUCTION

The International Society for Pharmacoeconomics and Outcomes Research (ISPOR) is a nonprofit, 501(c)(3) member organization governed by an annually elected board of directors, with administrative headquarters located in the heart of central New Jersey's pharmaceutical corridor. (3100 Princeton Pike, Building 3, Suite D, Lawrenceville, New Jersey 08648; Tel: 609-219-0773; Fax: 609-219-0774. Its mission is to translate research and outcomes/results in pharmacoeconomics into practice to ensure fair and efficient distribution of healthcare resources.

HISTORY

The International Society for Pharmacoeconomics and Outcomes Research (ISPOR), formerly the Association for Pharmacoeconomics and Outcomes Research (APOR), was founded in 1995 by a group of 32 healthcare professionals and researchers with the goal of developing research practice standards for assessing the value of healthcare therapies to consumers, healthcare systems, and societies. In 1997, APOR merged with the International Society for Economic Evaluation of Medicines to form ISPOR, the first and largest international pharmacoeconomics society. In 1998, ISPOR launched *Value in Health*, the definitive, peer-reviewed journal for this scientific discipline. In that same year, two regional ISPOR chapters were established in Russia and Poland, and collegiate chapters were organized across the United States and Canada for students in the field. The student chapters include the University of Texas at Austin, University of Arizona, University of Toronto, University of Louisiana at Monroe, University of Michigan, University of Washington, University of Southern California, and the University of Maryland.

With a steadily growing membership of 2000 representing 29 countries, ISPOR remains steadfast in its mission to translate pharmacoeconomics and outcomes research into practice to ensure that society allocates scarce healthcare resources wisely, fairly, and efficiently.

The Society serves the public interest by:

- Providing a forum that fosters the interchange of scientific knowledge in pharmacoeconomics and patient health outcomes.
- Facilitating and encouraging communications among the research community, healthcare professionals, governmental, educational groups, the media, and the general public.
- Educating public and private agencies on the usefulness of research in pharmacoeconomics and patient outcomes assessment.
- Acting as a resource in the formation of public policy relevant to pharmacoeconomics, healthcare outcomes assessment, and related issues of public concern.
- Promoting this area of scientific research by providing services and educational activities that advance it.
- Representing the discipline before public and governmental bodies.

To implement these objectives, ISPOR has created several steering committees and task forces to lead the initiatives, formulate strategies, and promote good research practices. They are:

- Health Science Steering Committee.
- Prospective Studies.
- Modeling Studies.
- Retrospective Database Studies.
- Outcomes Assessment Using Quality of Life Indicators.
- Use of Pharmacoeconomic/Health Economic Information in Health Care Decision making.

Encyclopedia of Clinical Pharmacy
DOI: 10.1081/E-ECP 120006292

- Medical Information Access.
- Code of Ethics.
- Education Steering Committee.
- Fellowship.
- Short Course Development and Quality Assurance.
- Pharmacoeconomics and Outcomes Research Curriculum Development, North America.
- Pharmacoeconomics and Outcomes Research Curriculum Development, Europe.

Currently, ISPOR initiatives include developing standards of research practices to guide the activities of those conducting pharmacoeconomics and outcomes research, and developing educational programs to communicate those research results to healthcare decision makers who could greatly benefit from it.

ISPOR members are scientists, economists, and healthcare practitioners from 29 different countries and four different work environments: academia, the pharmaceutical and biotechnology industry, research and consulting organizations, and healthcare practice environments (hospitals, clinics, private practice, managed care, pharmacy benefit management, clinicians, and government). Their educational backgrounds reflect degrees in statistics, nursing, accounting, economics, business administration, public health, and other health sciences. A number possess doctoral degrees in medicine, philosophy, pharmacy, public health, and jurisprudence.

Through the ISPOR administrative staff, members are provided with a variety of services to support their research and professional growth, including:

- www.ispor.org—the Society's web site which provides key research, educational, and public information on the discipline, the organization, and its members.
- *LEXICON*—a unique dictionary of pharmacoeconomics and outcomes research terminology.
- *Value in Health*—the official peer-reviewed research journal of the Society published bimonthly.
- *ISPOR NEWS*—a bimonthly newsletter featuring the latest research activities and employment opportunities.
- PRAP (Professional Recruitment Assistance Program)—provides a specialized job placement service to members.
- The Annual International Meeting (generally held in May each year) and Annual European Conference (generally held in November or December each year)—promote the discipline of pharmacoeconomics and outcomes research through networking and education.
- Student chapters, special interest groups, steering committees, and task forces allow members to target and affect issues of particular interest to them and to the discipline.

The current ISPOR 2000–2001 Board of Directors is as follows: President: Jon C. Clouse, M.Ph., M.S., Ingenix Pharmaceutical Services; President-Elect: Eva Lydick Ph.D., SmithKline Beecham; Past President: Bryan R. Luce Ph.D., M.B.A, MEDTAP International; Directors: A. Mark Fendrick M.D., University of Michigan Medical Center; Karen Rascati R.Ph., Ph.D., University of Texas; Joan Rovira Ph.D., University of Barcelona and SOIKOS; Kent H. Summers Ph.D., Eli Lilly & Company; Adrian Towse M.S., Mphil, Office of Health Economics (United Kingdom); and Executive Director: Marilyn Dix Smith R.Ph., Ph.D.

Janus Commission (AACP)

Richard P. Penna
American Association of Colleges of Pharmacy,
Alexandria, Virginia, U.S.A.

INTRODUCTION

The Janus Commission was established by, 1995–96, AACP President Mary-Anne Koda-Kimble to scan the healthcare environment and to identify, analyze, and predict those changes within the environment likely to profoundly influence pharmacy practice, pharmaceutical education, and research, and to alert the academy to both threats and opportunities that such environmental changes present. The Commission took its name from the Roman god Janus, which had one head with two faces capable of looking in opposite directions at the same time.

HISTORY

The Commission chose to define "the environment" as encompassing the healthcare delivery system, health professions education, the academic health center and research enterprise, and the related social, economic, and political forces which impact these three areas. It has did so by recognizing that this did not constitute the entire environment which could be considered in such a discussion.

The Commission convened via face-to-face meetings and conference calls throughout 1996 and 1997 and extensively discussed several issues. Throughout its deliberations, the Commission was drawn repeatedly to the pervasive influence of a changing, more intensively integrated and managed healthcare system on both pharmaceutical education and pharmacy practice. This fundamental change in healthcare was the primary influence in the Commission's thinking and recommendations.

MISSION

The Commission believes that a revised model for pharmaceutical education is needed to meet the challenges presented by the changing healthcare system. In particular, schools and colleges of pharmacy must become true "activists" in healthcare policy, services delivery, and research in order to effectively achieve their missions in professional education. Employing the analogy of the pharmaceutical industry, the Commission suggests the following fundamental areas of emphasis for schools and colleges:

- The need for enhanced research and development activities related to the provision of, compensation for, and outcomes of pharmaceutical care.
- Sustained curricular reform efforts that ensure successful "manufacturing" of competent and caring pharmaceutical care providers.
- Aggressive "marketing" programs, working in collaboration with the profession of pharmacy, that promote the delivery of pharmaceutical care and foster enhanced practice/education partnerships.
- Enhanced interaction with the "product" of professional education programs—the pharmacist—to ensure that graduates can and will continue to provide effective clinical, humanistic, and economic outcomes in the course of their professional careers.

Finally drawing upon the previous works of the Pew Health Professions Commission and the AACP Commission to Implement Change in Pharmaceutical Education, the Janus Commission believes that fundamental actions must be taken immediately within the academy to ensure the future success of our professional programs and graduates within the evolving healthcare system. Among these are:

- Completion of the evolution of the "values system" of pharmaceutical education toward producing graduates who are patient-centered providers of pharmaceutical care.
- Active involvement by administrators and faculty of colleges and schools of pharmacy in healthcare sys-

Encyclopedia of Clinical Pharmacy
DOI: 10.1081/E-ECP 120006192

tem decision making, policy determinations, and research activities.
- Creation by colleges and schools of action-oriented business plans that result in effective partnerships with practice organizations and healthcare delivery systems.

The Commission encourages a thorough and critical reading of its report and looks forward to a very healthy, lively, and productive dialogue within the Academy.

APPENDIX

Members of the Janus Commission

J. Lyle Bootman (Chair)
University of Arizona, Tucson, Arizona

Robert H. Hunter
Merck & Company, West Point, Pennsylvania

Robert A. Kerr
University of Maryland, Baltimore, Maryland

Helene L. Lipton
University of California, San Francisco, California

John W. Mauger
University of Utah, Salt Lake City, Utah

Victoria F. Roche
Creighton University, Omaha, Nebraska

Joint Commission for the Accreditation of Health-Care Organizations

Kathryn T. Andrusko-Furphy
Atascadero, California, U.S.A.

INTRODUCTION

The Joint Commission on Accreditation of Healthcare Organizations (JCAHO) is a private, voluntary, not-for-profit organization that serves as a gatekeeper for health care quality and safety. Currently, more than 19,000 health care organizations in the United States and many other countries are JCAHO accredited.[1] Since its earliest inceptions almost 50 years ago, JCAHO has striven to establish health care standards and performance measures. The mission of JCAHO is to continuously improve the safety and quality of care provided to the public through the provision of health care accreditation and related services that support performance improvement in health care organizations.

BACKGROUND

The organization takes its origins from a peer-to-peer-based practice starting first in the early 1900s with the American College of Surgeons, which established minimum hospital standards. The majority of early hospitals participating could not pass even these minimum standards. Ernest Codman, MD, is credited as the father of JCAHO. He proposed the end-result system of hospital standardization. Under this system, a hospital would track every patient long enough to determine whether the treatment was effective. If not effective, a determination would be made as to why not.

In 1951, the American College of Surgeons were joined by the American College of Physicians, the American Hospital Association, the American Medical Association, and the Canadian Medical Association to create JCAHO as it is known today. The Canadian Medical Association has since withdrawn from JCAHO;[1] however, they continue a process similiar to the current JCAHO. The governing body is made up of 28 members with a diverse background in health care, business, and public policy. Members include nurses, physicians, consumers, medical directors, administrators, providers, employers, labor representation, health plan leaders, quality experts, ethicists, health insurance administrators, and educators. Their role is to provide policy and leadership oversight.

Currently, JCAHO accredits the following types of organizations:

- General, psychiatric, children's, and rehabilitation hospitals.
- Health care networks, including health plans, integrated delivery networks, and preferred provider organizations.
- Home care, including infusion, home health, hospice, personal care and support, and home medical equipment.
- Nursing homes and long-term care, including long-term care pharmacies.
- Assisted living.
- Behavioral health, including mental health, chemical dependence, mental retardation, and foster care.
- Ambulatory care, including outpatient surgery centers, rehabilitation centers, infusion centers, group practices, and specialty centers, such as birthing centers, endoscopic centers, and pain centers.
- Clinical laboratories.

ACCREDITATION STATUS

Even though the accreditation process is voluntary, much emphasis is placed by health care organizations to successfully achieve accreditation status. Hospitals who receive federal funding from the Centers for Medicare and Medicaid Services, in particular, must adhere to the rigorous preparation and survey process. Because 80% of the approximate 5000 hospitals or health care systems receive federal funding, the JCAHO accreditation process, although voluntary, takes on a lot more meaning if the organization wants to continue to receive federal funds. The Social Security Amendment passed in 1965 granted

Encyclopedia of Clinical Pharmacy
DOI: 10.1081/E-ECP 120006293

hospitals deemed status in the 1960s if they were JCAHO accredited. Organizations who were JCAHO accredited were deemed compliant with the Medicare Conditions of Participation for hospitals. Consequently, today the majority of hospital systems must participate in the JCAHO accreditation process and adhere to its standards.

The long-term care (LTC) accreditation process is actually the second oldest JCAHO program and has been accrediting LTCs since 1966.[1] More than 2800 organizations participate in this program. However, unlike hospitals, it has not received deemed status to date. Consequently, many skilled nursing home facilities continue to not opt for JCAHO accreditation to the same level as hospitals. Instead, they adhere to state licensing requirements alone. In the late 1990s, a resurged interest in the LTC accreditation process occurred precipitated by managed care's interest in JCAHO accreditation. Many managed care organizations were requiring JCAHO accreditation to service their patient population and LTC beds. However, the advent of the Prospective Payment System (PPS) in 1999, fueled by the balanced Budget Act of 1996, has more recently left a mixed signal as to the value and cost of being JCAHO accredited.

The home care accreditation program[2] is one of the more diverse processes. The reasons for the 6400 organizations to participate in accreditation are actually four separate and distinct submarkets: home infusion pharmaceuticals; home medical equipment; home health, personal care, and support; and hospice. Collectively, these entities are often referred to as alternate site. That is, alternate to hospitalization. First, the home infusion or home intravenous (IV) segment. As an organized form of health care delivery, home infusion has only been in use since the early 1980s. Early patients requiring home infusion services were generally those with chronic disease in which lifetime hospitalization was not an option (e.g., short gut syndrome requiring parenteral or enteral nutrition, hemophiliacs requiring factor VIII). The changes in the Medicare reimbursement structure or diagnosis-related groups (DRGs) in the mid-1980s had only an indirect impact on the growth in patients serviced in the home care setting because Medicare does not pay for prescription medications, including most IV medications in the outpatient or ambulatory setting. Mainly, the growth in home infusion occurred due to the financial pressures of managed care influences to release patients more quickly from the hospital when therapies were considered safe, yet required prolonged use of IV medications. For example, the use of IV antibiotics for the treatment of such conditions as osteomyelitis and other soft-tissue infections, such as cellulitis or status post-surgical infections. The human immunodeficiency syndrome epidemic of the late 1980s and early 1990s also fueled the growth in the home infusion industry. However, this industry's interest in the home care accreditation program that started in 1998 stemmed more from competitive marketing reasons and financial reasons more from the pressures of managed care and other third-party payers instead of federal funding. About this time, JCAHO changed its name from the Joint Commission on Hospital Accreditation to the Joint Commission on Healthcare Organizations, thus acknowledging the changing shape of health care delivery.

The home health agency or skilled visiting nurse component of home care also has diverse reasons to seek JCAHO accreditation. JCAHO received deemed status in 1993 for its Medicare-certified organizations. However, non-Medicare agencies also exist across the United States. In fact, not all states require agency licensing. However, the movement in the last decade has been for states to require agencies to be licensed. More often, similar marketing and financial motives to attract third-party payers have also fueled the motivation for non-Medicare home health agencies to seek JCAHO accreditation. However, the Balanced Budget Act of 1996 has had profound effects on home health agencies and forced the closure and consolidation of agencies as PPS took effect on October 1, 2000.

The hospice segment of home care usually has a federal funding component as well. JCAHO just received deemed status in 1999.[1] Originally, JCAHO had a separate accreditation program for hospice; however, the hospice standards have been folded into the home care standards manual.

Of this diverse segment of health care, home medical equipment (HME) is probably the most diverse group within this segment. Participants range from HME dealers with little to no clinical expertise providing ambulatory aids such as wheelchairs, walkers, and intense clinical services and equipment, such as clinical respiratory therapists providing oxygen, ventilators, and apnea monitors. Unlike hospitals and home health care that receive federal funds through Medicare part A benefits, HME receives federal payment through Medicare part B benefits. As of yet, deemed status or the requirement to be JCAHO accredited has not been a motivating factor for seeking accreditation in this market. However, there are some reports that it may be required in the future. Marketing and financial reasons from third-party payers and managed care predominantly have been the motivating force to voluntarily seek JCAHO accreditation. In addition, other healthcare organizations, such as hospitals and LTC facilities that are JCAHO accredited, require their vendors to be accredited. This exempts these HME organizations

from being included in the hospital, home health, or LTC organization's accreditation process.

With the changes in health care delivery, ambulatory care is also a growing segment of accreditation. Currently, more than 1000 centers are accredited. These include ambulatory surgery centers, community health centers, group medical practices, Native American health clinics, and other specialty centers. In addition, they received deemed status in 1996.[3]

ACCREDITATION PROCESS STANDARDS

When not motivated to participate for involuntary reasons such as the withholding of federal funding, most health care organizations find the JCAHO accreditation process to be an investment into risk management. Organizations agree to be measured against national standards set by health care professionals. No longer is the accreditation process a set of minimum standards. Once the standards emphasis were structural in nature. Structural standards assumed that care should be good because the opportunity existed for such, such as the right physical plant or the right number of staff. In 1987, the Agenda for Change was launched. Its emphasis is based on actual organizational performance and processes performed; that is, is the care provided actually safe and quality oriented? Organizations during the last 10 years have had to show that the care provided is in fact efficacious, efficient, cost effective, and safe. JCAHO has almost gone full circle with Dr. Codman's origin idea, except the current JCAHO process continues to be standards based, not just outcomes based. Organizations must show that access to care is timely, staff is oriented and competent, and sentinel events and complaints are incorporated into the organization's assessment processes.

However, JCAHO is also striving for a more continuous and data-driven process. The precursor to their ORYX Initiative or outcomes performance measurement was the IMS system first launched in 1988.[4] In 1998, hospitals and LTC facilities began to participate in the collection of outcomes data via select clinical measures and began to submit that data to JCAHO for analysis. Home care has followed suit in the year 2000. However, as of this writing, JCAHO is looking to modify its home care requirements for collecting outcomes data for ORYX.

The goal of the JCAHO accreditation process is that it will continue to be standards driven with an increased emphasis on continuous data. In the last few years, the JCAHO accreditation process—in particular, for hospitals—has been challenged. Payers continue to challenge the value of the accreditation process cost versus quality. As health care organizations continue to evolve, change, and consolidate, so must JCAHO evolve, change, and consolidate.

In 1999, the Board of Commissions established an Oversight Task Force for the Accreditation Process Improvement (API) Initiative. The purpose is to oversee the continuous improvement in the accreditation process.[1] The resulting changes are intended to enhance the evaluation of critical patient safety and patient care functions and to achieve an accreditation process that remains consultative and focused on performance improvement. A white paper was published outlining a future operational model that will continue to build and expand on technology, performance data, presurvey self-assessments, a fully automated interface with JCAHO, increased surveyor development, and a more continuous accreditation process. Instead of a once every 3 years site visit, two 18-month site visits would occur that evaluate select standards. In addition, since health care entities are so diverse, there is a desire to create a model that is more data driven, less predictable, and more customized to an individual organization.

REFERENCES

1. www.jcaho.org.
2. 1999–2000 Comprehensive Accreditation Manual for Home Care. Joint Commission on Accreditation of Healthcare Organizations. One Renaissance Boulevard Oakbrook Terrace, Illinois 60181.
3. 2000–2001 Comprehensive Accreditation Manual for Ambulatory Care. Joint Commission on Accreditation of Healthcare Organizations. One Renaissance Boulevard Oakbrook Terrace, Illinois 60181.
4. ORYX: The Next Evolution in Accreditation. Joint Commission on Accreditation of Healthcare Organizations. One Renaissance Boulevard Oakbrook Terrace, Illinois 60181.

Joint Commission of Pharmacy Practitioners (JCPP)

Robert M. Elenbaas
American College of Clinical Pharmacy, Kansas City, Missouri, U.S.A.

INTRODUCTION

Because of the diversity of practice locales and specialties found within pharmacy, there is no one professional association of which all pharmacists are members. The existence of multiple professional associations, each with its own unique focus, is an effective way to meet many of the professional needs of a highly differentiated practitioner population. However, it is important that these organizations have a means to collaborate effectively on major professional, regulatory, and legislative issues that confront pharmacy. The Joint Commission of Pharmacy Practitioners (JCPP) fulfills this need.

HISTORY

The JCPP was established in 1977 and serves as a discussion forum on matters of common interest and concern to the profession. As a discussion forum, JCPP facilitates effective representation of pharmacists on professional, educational, legislative, and regulatory issues. Although JCPP has no intrinsic authority to speak on behalf of its member organizations, discussions within the JCPP forum have identified many areas of agreement and have allowed participating organizations to adopt identical positions on a variety of professional issues. The significance of this is found in the fact that, collectively, the member organizations of JCPP represent virtually all pharmacy practitioners in all practice environments.

MEMBER ORGANIZATIONS

The membership of JCPP is not composed of individual pharmacists. Instead, JCPP's membership is composed of national pharmacy organizations who themselves represent either individual pharmacists or entities critical in some way to the practice of pharmacy (e.g., schools of pharmacy, state boards of pharmacy). The member organizations of JCPP are as follows:

Academy of Managed Care Pharmacy (AMCP)
100 N. Pitt Street, Suite 400, Alexandria, Virginia 22314; Phone: (703) 683-8416; Fax: (703) 683-8417; Judith A. Cahill, Executive Director.

American Association of Colleges of Pharmacy (AACP)
1426 Prince Street, Alexandria, Virginia 22314; Phone: (703) 739-2330; Fax: (703) 836-8982; Lucinda L. Maine, Executive Vice-President.

American College of Apothecaries (ACA)
2830 Summer Oaks Drive, Bartlett, Tennessee 38134; Phone: (901) 383-8119; Fax: (901) 383-8882; D.C. Huffman, Jr., Executive Vice-President.

American College of Clinical Pharmacy (ACCP)
3101 Broadway, Suite 650, Kansas City, Missouri 64111; Phone: (816) 531-2177; Fax: (816) 531-4990; Robert M. Elenbaas, Executive Director.

American Council on Pharmaceutical Education (ACPE)
20 N. Clark St., Suite 2500, Chicago, Illinois 60602; Phone: (312) 664-3575; Fax: (312) 664-4652; Peter H. Vlasses, Executive Director.

American Pharmaceutical Association (APhA)
2215 Constitution Avenue NW, Washington, DC 20037; Phone: (202) 628-4410; Fax: (202) 429-6300; John A. Gans, Executive Vice-President.

American Society of Consultant Pharmacists (ASCP)
1321 Duke Street, Alexandria, Virginia 22314; Phone: (703) 739-1300; Fax: (703) 739-1321; R. Timothy Webster, Executive Director.

Encyclopedia of Clinical Pharmacy
DOI: 10.1081/E-ECP 120006294

American Society of Health-System Pharmacists (ASHP)
7272 Wisconsin Avenue, Bethesda, Maryland 20814; Phone: (301) 657-3000; Fax: (301) 652-8278; Henri R. Manasse, Executive Vice-President.

National Association of Boards of Pharmacy (NABP)
700 Busse Highway, Park Ridge, Illinois 60068; Phone: (847) 698-6227; Fax: (847) 698-0124; Carmen A. Catizone, Executive Director.

National Community Pharmacists Association (NCPA)
205 Daingerfield Road, Alexandria, Virginia 22314; Phone: (703) 683-8200; Fax: (703) 683-3619; Bruce Roberts, Executive Vice-President.

National Council of State Pharmacy Association Executives (NCSPAE)
4041 Devlin Court, Tallahassee, Florida 32308; Phone: (850) 906-0779; Fax: (850) 893-1845; Steven E. Glass, Administrative Manager.

MAJOR PROGRAMS

JCPP meets four times each year, and those organizations whose membership is composed of individual pharmacy practitioners chair and host the meeting on a rotational basis (i.e., AMCP, ACA, ACCP, APhA, ASCP, ASHP, and NCPA). Participants in the quarterly meetings include the chief elected officers and staff of the member organizations.

One or more major topics for in-depth discussion are identified for each meeting. These topics represent issues confronting the profession at that point in time where there is value in exchanging information or seeking collaboration among the member organizations. These discussions may result in a consensus position among all eleven organizations, or allow those organizations who share a common point of view to identify each other and work collaboratively. Over the years, JCPP has facilitated collaboration among its member organizations on issues dealing with pharmacy education, practitioner licensure, professional and technical manpower, federal legislation or regulations, and programs developed by pharmaceutical companies that directly impact on pharmacy practice.

One initiative that is now organized and guided by JCPP is a series of strategic planning conferences for the profession—originally titled ''Pharmacy in the twenty-first Century'' (P-21). Although the first P-21 conference in 1984 was organized through other means,[1] JCPP was responsible for the P-21 conferences held in 1989 that focused on ''Opportunities and Responsibilities in Pharmaceutical Care,'' and in 1994 that dealt with ''The Changing Healthcare System: Implications for Pharmaceutical Care.''[2,3] In 1999, with the imminent dawn of the twenty-first century, the P-21 moniker was dropped, and the conference was titled ''Re-engineering the Medication Use System.''[4]

REFERENCES

1. *Pharmacy in the 21st Century. Planning for an Uncertain Future*; Bezold, C., Halperin, J.A., Binkley, H.L., Ashbaugh, R.A., Eds.; American Association of Colleges of Pharmacy: Alexandria, Virginia, 1985.
2. Proceedings. Conference on pharmacy in the 21st century. Am. J. Pharm. Educ. **1989**, *53*, 1S–78S.
3. Proceedings. Pharmacy in the 21st century. The third strategic planning conference for pharmacy practice. Am. Pharm. **1995**, 1–51, Suppl.
4. Proceedings. Re-engineering the medication-use system: An interdisciplinary conference. Am. J. Hosp. Pharm. **2000**, *57*, 537–601.

PROFESSIONAL DEVELOPMENT

Long-Term Care, Clinical Pharmacy Careers in

Diane B. Crutchfield
Pharmacy Consulting Care, Knoxville, Tennessee, U.S.A.

INTRODUCTION

The role of the clinical pharmacist in long-term care, also called a consultant pharmacist, continues to expand with the recognition of drug-related problems that affect the quality of life of seniors. In addition to the potential adverse outcomes associated with medication-related problems, the cost of these problems is staggering. In fact, the total annual direct medical cost of medication-related problems in nursing homes exceeds $7 billion.[1]

ROLE OF CONSULTANT PHARMACISTS

Consultant pharmacists closely monitor each nursing home resident's medication regimen, monitoring for potential drug or disease interactions, dosing irregularities or side effects, and for medication use that may be considered inappropriate for the geriatric resident. This review is multifaceted and takes into account the information provided in the written medical record, as well as input from the staff, resident, and family. Various services are provided to the facility (Table 1). Inservice education programs are provided to the nursing staff and to family or resident meetings, if requested. Consultant pharmacists frequently participate on interdisciplinary care plan teams that focus on individual resident needs, such as fall risk assessment, weight loss evaluations, or pain management programs. Through ongoing monitoring of quality assurance programs, the consultant assists the facility in providing quality improvement for better resident care and in meeting state and federal regulations.

The success of consultant pharmacists' drug regimen reviews was reported in the Fleetwood study, which demonstrated that the reviews improved therapeutic outcomes by 43% and save as much as $3.6 billion annually in costs associated with medication-related problems.[2]

HISTORY

The concept of consultant pharmacy was first born in the early 1970s with the formation of the American Society of Consultant Pharmacists (ASCP), which has since also become known as America's Senior Care Pharmacists. About the same time, the centers for Medicare and Medicaid Services (formerly the Health Care Financing Administration) recognized the importance of medication management for the frail elderly in nursing homes, and in 1974, required that all medications for Medicare residents of nursing homes be reviewed on a monthly basis. In 1987, the requirement was amended so that all Medicare and Medicaid residents are required to have a monthly medication review by a pharmacist. In addition to the need for therapeutic drug monitoring, pharmacists recognized the need for unit dose packaging and specialized delivery services for nursing home residents in an effort to reduce medication errors and increase the efficiency of the systems.[3]

A pharmacy career in long-term care is typically associated with the provision of consultant pharmacy services to nursing home facilities. More recently, however, the opportunities for pharmacists in this practice setting have expanded far beyond the role of providing medications and medication reviews to only nursing homes.

Table 1 Types of consultant pharmacist services

- Drug regimen review
- Drug information
- Quality assurance programs
- Inservice education programs
- Therapeutic drug monitoring
- Patient counseling
- Pain management
- Interdisciplinary care planning
- Pharmaceutical care planning
- Pharmacokinetic dosing services
- Drug research programs

Encyclopedia of Clinical Pharmacy
DOI: 10.1081/E-ECP 120006262

Table 2 Job settings for consultant pharmacists

- Nursing facilities (skilled/intermediate care)
- Assisted living facilities/board and care homes
- Correctional institutions
- Hospitals (subacute/transitional care)
- Home health agencies
- Industry
- Mental institutions
- Alcohol or drug rehabilitation centers
- Mental retardation facilities
- Community health centers
- Retail pharmacies
- Health maintenance organizations
- Hospice
- Retirement communities
- Adult day care
- Physician offices
- Ambulatory care centers

(From Ref. [1].)

Terms such as consultant pharmacist and senior care pharmacist are used interchangeably to describe the type of pharmacist, rather than focusing on the physical setting. In addition to providing services for 1.8 million residents in more than 16,000 nursing homes in the United States, consultant pharmacists also provide medications and medication reviews for jails and prisons, home health agencies, and assisted living facilities, among others (Table 2).

TRAINING AND CERTIFICATION REQUIREMENTS

The usual training for a pharmacist in long-term care, or senior care, is a background or strong interest in providing care for the elderly. Beyond the pharmacy degree, there are now more than 800 certified geriatric pharmacists (CGPs) practicing in all areas of the world. The certification for recognition as a geriatric pharmacist first became available in 1997, and is administered by the Commission for Certification in Geriatric Pharmacy (CCGP). The CGP is not required for a career in long-term care pharmacy but is an asset for someone seeking a position in the field related to geriatric pharmacy. The CGP has demonstrated knowledge in the specific clinical areas that are required for the provision of consultant services to the elderly. In addition to educational training, the pharmacist must possess excellent verbal and written communication skills specific to the needs and considerations of the elderly. It is imperative to have the ability to work effectively with physicians, nurses, and other healthcare providers because the interdisciplinary approach is key to providing positive patient outcomes by improving the quality of life for residents in the variety of available long-term care settings. Postgraduate training programs, or short-term clinical rotations, are available in a variety of geriatric-related topics, including dementia, Alzheimer's disease, geropsychiatry, and Parkinson's disease, through the ASCP Research and Education Foundation.

PROFESSIONAL OUTLOOK

As mentioned previously, careers are not limited to those serving the elderly who reside in a nursing home. Some states require consultant pharmacist services in assisted living facilities. As the senior population continues to grow, so does the need for pharmacists trained specifically in the area of geriatrics.

Within the job settings, consultant pharmacists may be self-employed or work for a provider pharmacy. Provider services typically include the operation and management of the medication distribution system and consultant services typically refer to the clinical services. Some consultants provide only the consulting services and others provide consulting services, in addition to dispensing the medications.

Innovation has been the key word in the evolution of the practice of consultant practice in long-term care settings. Thus, it will continue to be a key to the future success of pharmacists willing to step out of the traditional pharmacy practice settings and provide much needed services to the growing senior population.

REFERENCES

1. www.ascp.com. Senior Care Pharmacy Facts. accessed February 24, 2002.
2. Bootman, J.L.; Harrison, D.L.; Cox, E. The healthcare cost of drug-related morbidity and mortality in nursing facilities. Arch. Intern. Med. **1997**, *157*, 2089–2096.
3. Webster, R.T. A perspective on consultant pharmacy's future: Changing information into dollars. Consult Pharm. **1989**, *4*, 8–12.

BIBLIOGRAPHY

http://www.ascpfoundation.org/.
www.aacp.org/students.
http://www.ccgp.org/.

Managed Care, Clinical Pharmacy Careers in

Barbara Zarowitz
Henry Ford Health System, Bingham Farms, Michigan, U.S.A.

INTRODUCTION

When care is managed, health systems are accountable for providing high-quality clinical care, while balancing the economic factors, to deliver cost-efficient health outcomes.[1] The application of business practices to health care requires accountability for financial and clinical outcomes. Financial accountability is ensured by applying fixed reimbursement rates to the delivery of care and services. Reimbursement rates are set by the government [Centers for Medicare and Medicaid Services (CMS)], insurance companies, or health maintenance organizations (HMOs). Fixed reimbursement rates require that the delivery system or physician group provide care for their assigned patients at a set rate or premium. If the cost of managing assigned patients exceeds the reimbursement rate, the delivery system loses money. The system or group of physicians is at risk for, or capitated for, the management of their assigned patients.

Accountability for clinical outcomes is negotiated by managed care organizations with independent practice associations and preferred provider organizations (PPOs) for a discounted fee for service. Physician providers accept a discounted fee for services rendered and agree to abide by certain quality and utilization requirements in return for access to a defined membership group. Quality standards for medical care have been established by the National Committee for Quality Assurance (NCQA), whose mission it is to improve health care in the United States by providing information about the quality of care and service delivered by managed health care systems.[2] The Health Plan Employer Data and Information Set (HEDIS) measures focus on specific quality issues defined by NCQA and provides employers with standardized performance measures to compare managed care organizations. Representative NCQA/HEDIS measures include the use of appropriate medications in asthma, chlamydia screening in women, controlling high blood pressure, beta-blocker treatment post myocardial infarction, and antidepressant medication management. Employers evaluate plans based on these medical quality indicators, as well as effectiveness of care, use of services, cost of care, informed health care choices, and health plan stability.

Managed care applies to a wide range of settings including HMOs, PPOs, integrated delivery systems (IDSs), physician hospital organizations (PHOs), and point-of-service plans.

CAREER CHOICES AND SETTINGS

Managed care has penetrated most geographic areas within the United States and essentially all health care settings. As a result, career opportunities for clinical pharmacists are vast and diverse. While previously pharmacists interested in managed care positions were limited to settings such as pharmacy benefit management (PBM) companies or insurance companies, career opportunities in managed care have expanded and are available in hospitals, clinics, and private practices in both urban and rural areas.

Traditional managed care pharmacy roles were those that placed the pharmacist in the PBM or insurance company setting to provide administrative, formulary management, prescription claims management, and contracting services. With increasing managed care penetration, IDSs, physician practices, PPOs, PHOs, and HMOs are hiring pharmacists to provide clinical services involving both quality of care and utilization management functions. Representative pharmacy opportunities in managed care are outlined in Table 1.

PREPARING FOR A CLINICAL PHARMACY CAREER IN MANAGED CARE

Managed care needs qualified clinical and financial pharmacy managers, grounded in solid business principles and clinical pharmacy skills. The doctor of pharmacy degree is strongly desired in pharmacists who graduated prior to 1995, and mandatory for recent graduates. Depending on the position focus, advanced training in pharmacoeconomics or managed care as either a 1-year residency or 2-year fellowship is recommended. Doctor

Encyclopedia of Clinical Pharmacy
DOI: 10.1081/E-ECP 120006259

Table 1 Clinical pharmacy careers in managed care

Role	Typical settings[a]	Typical functions	Typical prerequisites[a]
Retail network manager	HMO PBM	Contracting with retail pharmacies Audit and compliance functions Pharmacy reimbursement	Pharmacy degree Retail experience Business background
Claims manager	HMO PBM	Processing prior authorizations Developing criteria for new drugs	Pharmacy degree Managed care experience
Account manager	HMO PBM Drug company	Managing each business ventures Tracking financial performance Marketing new programs	MBA Pharmacy degree
Quality initiatives manager	HMO PBM IDS	Compliance programs for NCQA/HEDIS Coordinating data management	PharmD
Clinical specialist	HMO IDS PPO PHO PBM Drug company	Drug expert functions for providers, members, and formulary functions	PharmD Res
Data/population manager	HMO PBM Employer	Management and analysis of data for quality and cost-reduction initiatives	PharmD Res or Fel Pharmacoeconomics
Group practice pharmacist	PPO PHO IDS HMO	Managing the group's patient population and pharmacy costs	PharmD
Formulary manager	HMO IDS PBM	Identifying new strategies to manage drug entries, formulary reviews, and process	PharmD
Manager of clinical initiatives	HMO PBM IDS Employer	Design and implement cost-reduction and quality initiatives	PharmD
Utilization/case manager	IDS HMO	Manage high-risk populations as part of a team	PharmD
New business developer	HMO PBM	Identifies new client opportunities and develops business plans	MBA Pharmacy degree
Contracting	PBM HMO IDS	Contracts with drug companies, pharmacy providers, and/or physician's providers for services and drugs	Pharmacy degree MBA
Call center pharmacist	PBM IDS HMO.com	Provide telepharmacy services to patients and providers Compliance programs	Pharmacy degree
Home care pharmacist	IDS HMO	Prepare, deliver, administer, and monitor intravenous and enteral nutrition or drug products to patients in their homes	Pharmacy degree
Outcomes researcher	IDS HMO PBM	Measure the value of therapeutic interventions on the quality of life	Pharmacy degree Fel PhD or PharmD
Drug information specialist	IDS PBM	Research and develop answers to questions related to drugs and drug policy	Pharmacy degree Res
Information technology/ database manager	IDS HMO PBM	Integrate and manage database functions, informatics, and technology	Pharmacy degree

(Continued)

Table 1 Clinical pharmacy careers in managed care (*Continued*)

Role	Typical settings[a]	Typical functions	Typical prerequisites[a]
Outpatient pharmacist	IDS HMO	Counseling and dispensing medications Reinforce compliance and disease management programs	Pharmacy degree Res
Direct patient care	IDS PHO PPO	Condition/disease-focused clinics	PharmD Res Board certification

[a]Employer group, Employer; HMO, health maintenance organization; PPO, physician provider organization; PHO, physician hospital organization; IPA, independent practice association; IDS, integrated delivery system; PBM, pharmacy benefit manager organization; Res, residency training; Fel, fellowship training;.com, Internet commerce sites for medications.

of pharmacy students interested in a managed care career should pursue external rotations in progressive managed care settings. Experience with claims and benefit management, formulary management, disease management, quality certification, and large relational databases is very important. Pharmacists interested in account management may benefit from formal training in business administration and finance. Table 1 summarizes career opportunities that may require advanced training as a prerequisite.

Desirable skills and knowledge are summarized in Table 2. In addition to academic training, managed care pharmacists are called to respond to a rapidly changing, often unstable financial and clinical environment. Individual behavioral characteristics that are helpful in these circumstances may include flexibility and ease of response to change, open-mindedness, ''pioneerism,'' and respect for the sense of urgency surrounding health care business evolution. Many managed care organizations are willing to hire pharmacists with desirable behavioral and clinical characteristics, realizing that knowledge related specifically to managed care practice can be acquired. Large organizations, whether managed care organizations, HMOs, IDSs, or PBM firms, are often willing to provide on-the-job training to advance pharmacists' skills in managed care. However, prospective employers may preferentially select candidates with previous managed care experience, thus underscoring the importance of selecting elective rotations in managed care settings.

Table 2 Desirable skills, knowledge, and behaviors of clinical pharmacists in managed care

Problem-solving skills
Exemplary oral and written communication skills
Knowledge of population management and pharmacoeconomic principles
Facility with large relational databases and information systems
Business and financial expertise
Comfort with the measurement science
Solid clinical skills, knowledge, and behaviors
A sense of urgency
Comfort in rapidly changing, unstable environments
Motivated by challenge
Teamwork
Interpersonal skills

PROFESSIONAL GROWTH IN MANAGED CARE PHARMACY

Many opportunities exist for career growth in managed care practice settings, just as they do in other practice environments. New graduates can usually qualify for staff pharmacist careers in benefit design, clinical, data analyst, or case management positions. Advancement can follow experience at entry-level positions, with or without advanced skills enhancement through external degrees or training programs. Much of the management of pharmacy benefits, irrespective of the site of practice, is science and can be learned didactically. However, a significant component of success in a managed care pharmacy career, like other pharmacy career paths, depends on facility with the art of the application of managed care skills, knowledge, and strategies—the so-called ''craft'' of the managed care pharmacist. The development of ''craft'' skills requires experience through application, which can only be learned during practice. Entry-level pharmacists may move on to clinical manager or team leader positions, or transition into higher business or administrative positions.

A parallel track is available within PBM companies for pharmacists with retail pharmacy experience. Pharmacy network management and contracting have become important mechanisms for managing costs and enhancing revenue to health plans and PBMs. Retail pharmacy experience and business skills are desirable characteristics for pharmacists desiring advancement in the more traditional roles of pharmacists within managed care organizations.

Board certification and other postgraduate credentials may be useful to individuals, depending on their practice setting. However, in general, these credentials have not attained widespread acceptance as prerequisites for advancement within managed care organizations. The paucity of qualified individuals to deliver exemplary pharmacotherapy services, while balancing business and financial prerequisites has led to a limited candidate pool for managed care pharmacy positions. Over time, as a greater number of pharmacists enter and advance in managed care clinical pharmacy careers, board certification, certificate programs, and association affiliations or recognition may become more important delineators of high-quality pharmacists. In the interim, it is more important that pharmacists gain experience in the field and determine their compatibility with a career in managed care.

Managed care pharmacists are moving up the ranks within their organizations. In a survey conducted among 200 managed care pharmacists in 1997, the five most frequent job titles were director of pharmacy, regional director of pharmacy, pharmacy manager, vice president of pharmacy, and director of pharmaceutical division.[3] Many clinical pharmacists report directly to vice presidents or chief executive officers within their organizations.

CASE STUDIES OF CLINICAL PHARMACY PRACTICE IN MANAGED CARE

Hospice

Hospice services are often offered as part of an IDS or managed care organization. Typically, when patients and their families determine that end-of-life measures are indicated, they search for options to make the patient maximally comfortable, with appropriate care, and at an affordable cost. Hospice care can be offered as a purchased service or a covered benefit. In most cases, hospice care is capitated and must operate within a budget or lose money.[4] Pharmacists practicing in hospice settings may be called on to optimize rational pharmacotherapy and help to discontinue medications deemed no longer necessary for patient comfort. In this regard, the pharmacist and the rest of the patient's care team are managing care within a capitated limit.

All-Inclusive Care for the Elderly (PACE) Programs

There are several PACE programs across the United States that serve as day care and/or home care opportunities for elderly patients, referred to as participants.[5] PACE programs offer an alternative to nursing home placement, enabling frail elderly people to remain independent in their homes within the community. When elders enroll in a PACE program, their care becomes the responsibility of the program. PACE programs are funded through medicare and medicaid. The PACE program is responsible for provision of both drug therapy and medical care. Pharmacists can serve an important role in optimizing and simplifying drug therapy, at lowest cost. Given that many of the participants are seniors with multiple concurrent medications, careful individualization of dose is essential to minimize adverse events.

Utilization Management

Clinical pharmacists have been redeployed in ambulatory managed care settings to work with primary care providers to enhance both the quality and cost effectiveness of care delivered.[3,6] They are highly integrated within the system of care delivery. Population management strategies are used to identify the high-cost or low-quality providers with the greatest need for pharmacy care management. Provider profiles are used to continuously provide feedback to physicians, identifying cost-reduction and quality improvement opportunities. Clinical pharmacists can work with individual physicians or groups, such as a PPO or PHO, to ensure that the highest quality of care is provided within the capitation limits of the plan or group. Improvements in quality are accomplished through the pharmacists'' role in the development of clinical guidelines and pathways, while helping physicians understand the patients'' pharmacy benefits. Pharmacists may also work with individual high-risk patients to streamline drug therapy, decrease cost, and improve patient medication safety.

Specialty Clinics

In managed care, it is often a small percentage of patients (approximately 20%) who are responsible for the majority of the cost (80%)—often referred to as the 80/20 rule. It has been shown to be cost effective to manage these high-risk patients in specialty clinics or programs. Disease management programs offer many patients with specific conditions, enhanced management strategies to improve

the likelihood that they will achieve desired therapeutic outcomes as outpatients. Managed care organizations typically offer disease management programs in diabetes, asthma, depression, and other conditions in which the quality of follow-up and intervention has been shown to vastly improve patient outcomes, such as anticoagulation. In addition, specialty clinics are an approach to improving the management of patients with specific disorders, such as asthma, hypertension, and dyslipidemias.[7] Pharmacists may provide individualized care or participate as team members for case management or disease management of high-risk patients.

CONCLUSION

There are numerous opportunities for clinical pharmacists to contribute to high-quality patient care, business results, and drug benefit administration in a wide variety of managed care settings. Pharmacists are no longer limited to a narrow range of managed care opportunities within PBMs. Managed care settings offer intensity and challenge for clinical pharmacists with opportunity for upward mobility and career growth.

REFERENCES

1. Blissenbach, H.; Penna, P. Pharmaceutical Services in Managed Care. In *The Managed Care Handbook*; Kongstvedt, P., Ed.; Aspen: Gaithersburg, 1996; 367–387.
2. National Committee of Quality Assurance, www.ncqa.org (accessed Aug. 2000).
3. Sardinha, C. Managed care pharmacists: Leading the way for a new millennium. J. Man Care Pharm. **1997**, *3* (4), 383–388.
4. Virnig, B.A.; Persily, N.A.; Morgan, R.O.; DeVito, C.A. Do medicare HMOs and medicare FFS differ in their use of the medicare hospice benefit? Hospice J. **1999**, *14* (1), 1–12.
5. Eng, C.; Pedulla, J.; Eleazer, G.P.; McCann, R.; Fox, N. Program of all-inclusive care for the elderly (PACE): An innovative model of integrated geriatric care and financing. J. Am. Geriatr. Soc. **1997**, *45* (2), 223–232.
6. Barreuther, A. Academic detailing to influence prescribing. J. Man Care Pharm. **1997**, *3* (6), 631–638.
7. Alsuwaidan, S.; Malone, D.C.; Billups, S.J.; Carter, B.L. Characteristics of ambulatory care clinics and pharmacists in Veterans Affairs Medical Centers. IMPROVE investigators. Impact of managed pharmaceutical care on resource utilization and outcomes in Veterans Affairs Medical Centers. Am. J. Health-Syst. Pharm. **1998**, *55* (1), 68–72.

Managed Care Pharmacy Practice

Beverly L. Black
American Society of Health-System Pharmacists, Bethesda, Maryland, U.S.A.

INTRODUCTION

Health system integration has been a dominant phenomenon since the early 1980s, and the trend is likely to continue. Understanding health system integration and its impact on pharmacy practice is challenging for many reasons, one of which is the continued rapid pace of change.[1]

Responding to the challenges of system integration is also complicated by the various types of organizations that are characterized as integrated health systems. In broad terms, an integrated health system is one that brings together hospital, medical, pharmaceutical, and other health services into a single organization and thereby offers the prospect of lower cost, an improved and more consistent quality of care, and greater marketing power. Managed care organizations (MCOs), such as health maintenance organizations (HMOs), preferred provider organizations (PPOs), and independent practice associations (IPAs), are considered integrated health systems because they manage healthcare services for a defined population, even though they may not be fully vertically integrated.[2]

Regardless of their size, structure, or service mix, integrated health systems have the same goal: to deliver high-quality care to a defined population at a competitive cost. If desired patient outcomes and standards of fiscal performance are to be achieved, pharmaceutical care must be delivered efficiently and consistently, and drug products must be used appropriately.[3] This is where the clinical pharmacist plays a crucial role.

The growth of managed care in the United States has fostered new and innovative roles for pharmacists in this area, such as consulting, disease management, wellness program development, technology assessment, and outcomes research. The skills necessary for these roles include communication and mediation, assertiveness, assimilation of drug information, evaluation of clinical and economic data, and the ability to work in teams. Demonstrating and documenting the value of these roles will be essential to the future of the profession of pharmacy.[4] As in many other areas of pharmacy, there is currently a tremendous demand for well-qualified clinical pharmacists in managed care settings.

OPPORTUNITIES AVAILABLE IN MANAGED CARE FOR CLINICAL PHARMACISTS

Core Strengths of Pharmacy

Managed care pharmacists have identified the following core strengths that pharmacists should have to enable them to thrive in a managed care setting. Pharmacists should:

- Have an established reputation and competency as drug therapy experts.
- Have established skills in automation and computer technology.
- Have a demonstrated understanding of, and appreciation for, quantitative and qualitative information.
- Be committed to helping patients.
- Be accessible to patients.
- Be focused on drugs and knowledge related to medications.
- Be experienced in adapting skills for surviving in a changing healthcare environment.
- Possess a wealth of online information and databases.
- Possess a strong clinical knowledge base and training that can support case management.
- Be experienced in applying data to business applications.
- Have access to automated dispensing technologies, which have increased pharmacy's capacity to take on additional responsibilities.
- Have good organizational skills.
- Possess leadership skills and be passionate about furthering the success of the profession.

Essential Skills

The changing marketplace and different patient needs are creating innovative roles for pharmacists in integrated

Encyclopedia of Clinical Pharmacy
DOI: 10.1081/E-ECP 120006395

health systems. However, unique skills are necessary for effectiveness in managed care; some of these skills will require additional training and education. Also, a different attitude is essential; for example, pharmacists must be willing to take risks and accept new responsibilities—ones that may be unlike those of a traditional position.[5]

There are certain skills that clinical and dispensing positions have in the managed care setting. These include drug information knowledge, communication and mediation skills, assertiveness, and the ability to work in teams.

Drug information knowledge

The trends that are occurring in managed care are those of disease state management, outcomes, wellness program emphasis, technology, and pharmacoeconomics. The important role of the pharmacist with regard to drug information is to assimilate and combine information with other components for use in the decision-making processes. These include not only patient-specific decisions, but also formulary decisions, for the entire managed population.

Pharmacists have the specific background to make the appropriate recommendations for product selection; however, is no longer the product's pharmacologic advantage over other medications the only factor in selection. Pharmacoeconomics and population characteristics, as well as many other factors, must be considered.[6]

Other essential skills for pharmacists in managed care include the following.

Communication and mediation skills

Understanding what other healthcare team members are looking for and communicating effectively is vital. As pharmacists increase their involvement in direct patient care, interacting with patients will require the ability to present information in terms that the patient will understand. With respect to being a mediator, the goals of cost containment must be clearly understood and accepted by all parties in order for success to be achieved; this responsibility often falls to the pharmacist.

Assertiveness

Healthcare and pharmacy are not exempt from market pressures, and pharmacists must accept the fact that, unless they assert their role and importance in the services that they offer, they may not be needed.

Ability to work in teams

The pharmacist must understand role definitions, develop empathy for other team members, and acquire the skills necessary to surpass expectations.[7] Team relationships are essentially balancing acts that can be upset easily if someone does not understand the practice philosophy or the role definitions within the team.

Innovative positions

Managed care systems offer innovative pharmacy practice positions in such areas as pharmacoeconomics, disease state management, outcomes research, wellness program management, and technology assessment. For individuals willing to expand their pharmacy practice and develop new skills, these positions can offer unique opportunities and growth.

Care coordinators and clinic coordinators

Some organizations have created interdisciplinary rounding teams for inpatient care. The responsibility for coordinating the transition from hospital to ambulatory care is delegated to these teams. For example, pharmacists from the HMO's home IV service attend rounds within the contract hospitals, focusing on monitoring drug therapy with respect to quality and cost effectiveness and facilitating a smooth transition from hospital to home care.

Many MCOs are using pharmacists to influence physician prescribing and to conduct clinics promoting appropriate use of pharmaceuticals. In addition, these pharmacists participate in direct patient care activities that relate to drug therapy, such as ordering and monitoring laboratory tests and providing effective patient follow-up to ensure patient satisfaction and high-quality care. Many of the pharmacist-coordinated clinics providing this patient care have been shown to improve patient outcomes and satisfaction and to support the efficient provision of services from other members of the healthcare team.[8]

Disease management and the patient

Disease management has been defined as quantifying, tracking, and controlling the unique set of cost drivers in any disease and the interactions of these drivers over the course of the disease across all elements of the healthcare system.[9] Studying the distribution and effects of medications in populations (pharmacoepidemiology), as well as the costs and comparisons of alternative courses of therapy (pharmacoeconomics), helps determine where disease management can change practice the most.[10] These conditions are generally the most common, expensive, and treatable (e.g., asthma, diabetes, peptic ulcer disease, depression, cardiovascular disease) and provide the greatest opportunity for economic savings when disease management programs focus on them.[11]

Disease management requires that pharmacists add a number of skills to their professional role. These skills include managing information systems to collect and organize data, total quality management to improve processes, teamwork with patients and other professionals, budgeting to obtain compensation, teaching clinical care and appropriate drug use, and the ability to help move pharmacy from compartmentalized care to integrated care.[12] Most models of disease management use the pharmacist in a clinical role that goes beyond dispensing. Pharmacists are ideally qualified to participate in disease management because of their knowledge of drug therapy, communication, computer applications, and marketing, as well as sales, finance, and basic wellness issues. Pharmacists have a great opportunity in these situations, given their ability to communicate complex concepts in easily understood terms to help employee groups take advantage of disease management options.[13]

In the future, as healthcare systems make the transition from stage 1 (primarily fee for service) to stage 4 (primarily capitation), opportunities for disease management will increase.[14] For many MCOs, shifting from formulary management to disease management will be a means of improving the overall quality of care, reducing costs, and expanding patient involvement in care.[15] Decisions will not be based on individual components of care, such as the cost of drugs or other treatments; rather, a disease will be viewed in its entirety. As customized cost-management techniques are applied to each disease and patient, the pharmacist will assume a greater role in, and responsibility for, patient outcomes.[16]

Managing health

Wellness programs have been described as the community outreach component of disease management. These programs can link with disease management programs to provide follow-up support.

Pharmacists are ideally positioned to play a major role in wellness programs. Combining their outstanding clinical skills acquired during formal education, internships, externships, and work experiences with significant transferable skills, especially communication and interpersonal skills, makes pharmacists ideal candidates for counseling patients and producing the outcome of better overall healthcare at lower costs. Application of principles of disease prevention and management in individual or group counseling sessions can help patients learn how to care for themselves properly, thus reducing healthcare costs and improving their quality of life.[17]

In the future, wellness programs will most likely be a joint effort among payers, employers, patients, and providers. In these programs, pharmacists could serve as coordinators to ensure appropriate utilization of resources, effective communication, and optimal patient outcomes.[18]

Evaluating clinical and economic data

The focus on disease state management and outcomes will require that data on a patient's drug therapy be accessible to both hospital and ambulatory care providers. The pharmacist can play an integral role in the coordination and dissemination of these data. For example, institutional pharmacists will need to establish communication links with pharmacy providers in other settings to ensure optimal and seamless care.[19]

Pharmacists can provide a valuable service in evaluating the data resulting from treatments and helping patients make the most of their prescribed therapies.[20] A pharmacist could help select the appropriate formulary drug within a drug class and then teach the patient how to take the medication correctly.

Pharmacists also have a role as pharmacoeconomic analysts in comparing medications. Sometimes the investment in a more expensive medication reduces other costs to the HMO, but the data must be collected and tracked for these assumptions to be proved. To perform these analyses successfully, the pharmacist must be able to evaluate and interpret statistics and research articles. A sound understanding of pharmacoeconomic principles is also essential.

As more information about a given treatment is received, the pharmacist must always be prepared to perform the analysis again and incorporate the new information. Depending on new research or changes in the HMO-covered population and its needs, the pharmacoeconomic evaluation may lead to different results at a later date.

Because of the financial pressures to reduce healthcare costs without compromising quality, the pharmacist must have the ability to apply basic quantitative skills to evaluate options and then to blend those results with qualitative information to make decisions and recommendations.[21] Pharmacists will have to work directly with prescribers, helping them interpret and use phar-

macy claims data to achieve the best patient outcomes. Having a basic understanding of business and knowing when to seek additional information are increasingly necessary to operate successfully within the managed care environment.

Measuring outcomes

Measuring outcomes shifts the emphasis from products to patient results, which could eliminate the need for formularies.[22] Simply put, outcomes measurements evaluate systems and decide what works and what does not.

Healthcare providers are increasingly relying on pharmacists to perform outcomes research and quality-of-life studies. Pharmacists can apply basic quantitative skills in evaluating options and combine the results with qualitative information to make decisions and recommendations. For disease state management programs, measuring outcomes can be the key to success.[23]

As health delivery systems move toward total managed care, the need for outcomes studies will increase.[24] Issues to be addressed by outcomes research will include clinical efficacy of interventions, health-related quality of life, patient satisfaction, employee productivity, and resource utilization. Pharmacists can either participate in or direct outcomes research. Outcomes studies may also be used to support a pharmacy position for interventions.

Outcomes research must always be patient focused and useful for improving patient care. Many pharmacists already have the data to do their own outcomes research. Questions about appropriate prescribing and compliance can be answered by using pharmacy claims data. That information could be a tool in providing positive feedback to the patient, as well as to the prescriber.

MODEL CLINICAL PRACTICES IN MANAGED CARE

The American Society of Health-System Pharmacists (ASHP) and the Academy of Managed Care Pharmacists developed the *Accreditation Standard and Learning Objectives for Residency Training in Managed Care Pharmacy Practice* in 1997. This standard outlines specific requirements and principles that managed care pharmacies should have in place for training residents. The relevant practice areas include direct patient care, drug information, population-based pharmaceutical care, business administration and management activities, and practice management. Regular accreditation surveys and visits ensure that each site maintains the practice standards. The *Managed Care Pharmacy Practice Standard* can be found on the ASHP web site.[25]

In addition to the *Accreditation Standard and Learning Objectives for Residency Training in Managed Care Pharmacy Practice*, ASHP has published the following standards and guidelines, which provide guidance on model clinical practices to managed care pharmacies. (Note: The citations provided are from the *American Journal of Health-System Pharmacy*, but each of these standards or guidelines can also be located on ASHP's web site.[36])

- *ASHP Guidelines on Pharmaceutical Services for Ambulatory Patients.*[26]
- *ASHP Guidelines: Minimum Standard for Pharmacies in Institutions.*[27]
- *ASHP Guidelines on a Standardized Method for Pharmaceutical Care.*[28]
- *ASHP Guidelines for Obtaining Authorization for Documenting Pharmaceutical Care in Patient Medical Records.*[29]
- *ASHP Statement on the Pharmacist's Responsibility for Distribution and Control of Drug Products.*[30]
- *ASHP Statement on the Pharmacist's Role with Respect to Drug Delivery Systems and Administration Devices.*[31]
- *ASHP Technical Assistance Bulletin on Drug Formularies.*[32]
- *ASHP Technical Assistance Bulletin on the Evaluation of Drugs for Formularies.*[33]
- *ASHP Guidelines on Medication-Use Evaluation.*[34]
- *ASHP Guidelines on Adverse Drug Reaction Monitoring and Reporting.*[35]

Other organizations that provide guidance on model clinical practices are those that measure pharmacy quality and utilization in the managed care area. The primary organizations are as follows:

The Joint Commission on Accreditation of Healthcare Organizations (JCAHO)—JCAHO began providing accreditation services by focusing on hospitals; more recently, its services have been expanded to provide accreditation services for a wide continuum of healthcare providers. In fact, JCAHO shifted its accreditation focus toward a continuous quality improvement (CQI) process and incorporated outcomes measures into the standards. Information on JCAHO and its standards can be found at www.jcaho.org.

The National Committee for Quality Assurance (NCQA)—NCQA was founded in 1979, in an effort to establish a comprehensive quality measurement process for MCOs. The focus of the NCQA accreditation process is the effective implementation of a CQI process into the medical services provided by the managed care organization. Information on NCQA and its standards can be found at www.ncqa.org.

The Health Plan Employer Data and Information Set (HEDIS)—The HEDIS reporting system is a series of specific performance measures designed to provide healthcare consumers with the information they need to reliably compare MCOs. HEDIS measures are self-reported by participating MCOs on a quarterly basis. Information on HEDIS, which is a joint project with NCQA, can be found at www.ncqa.org/pages/programs/hedis.

References to Published Material

References to published materials documenting the benefit of pharmacists in managed care settings include the following:

- Mistry SK. Helping primary care providers with appropriate, cost-effective prescribing. *Am J Health-Syst Pharm.* 2000; 57:1575.
- Knapp KK, Blalock SJ, O'Malley CH. ASHP survey of ambulatory responsibilities of pharmacists in managed care and integrated health systems—1999. *Am J Health-Syst Pharm.* 1999; 56:2431–43.
- Carroll NV. Formularies and therapeutic interchange: healthcare setting makes a difference. *Am J Health-Syst Pharm.* 1999; 56:467–72.
- Knowlton CH. Pharmaceutical care in 2000: engaging in a moral covenant in turbulent times. *Am J Health-Syst Pharm.* 1998; 55:1477–82.
- Kay B, Crowling GH, Kershaw VI et al. Perspectives on pharmacy's role in managed care. *Am J Health-Syst Pharm.* 1998; 55:1482–8.
- Reeder CE, Kozma CM, O'Malley CH. ASHP survey of ambulatory care responsibilities of pharmacists in integrated healthcare systems—1997. *Am J Health-Syst Pharm.* 1998; 55:35–43.
- Hawkins PR. Pharmacist as health education coordinator. *Am J Health-Syst Pharm.* 1997; 54:1497–9.
- Hepler CD. Where is the evidence for formulary effectiveness? [Letter] *Am J Health-Syst Pharm.* 1997; 54:95.
- Additional citations validating the benefits of clinical pharmacists' participation in managed care are listed in the Bibliography.

Professional Networking Opportunities

There are many professional networking opportunities for clinical pharmacists who practice in the managed care setting. Some of the more prominent organizations include the following:

- The American Society of Health-System Pharmacists Center on Managed Care Pharmacy (www.ashp.org).
- The American Association of Health Plans (www.aahp.org).
- The American Medical Informatics Association (www.amia.org).
- The American Telemedicine Association (www.atmeda.org).
- The Academy of Managed Care Pharmacy (www.amcp.org).
- The National Business Coalition on Health (www.nbch.org).
- SCRIPT (http://scriptproject.org).

REFERENCES

1. Knapp, K.K.; Blalock, S.J.; O'Malley, C.H. *Survey of Managed Care and Ambulatory Care Pharmacy Practice in Integrated Health Systems, 1999*; ASHP: Bethesda, MD, 1999; 5.
2. Knapp, K.K.; Blalock, S.J.; O'Malley, C.H. *Survey of Managed Care and Ambulatory Care Pharmacy Practice in Integrated Health Systems, 1999*; ASHP: Bethesda, MD, 1999; 5.
3. *Survey of Managed Care and Ambulatory Care Pharmacy Practice in Integrated Health Systems, 1999*; ASHP: Bethesda, MD, 1999; 5.
4. ASHP Center for Managed Care. *Innovative Roles for Managed Care Pharmacists*; ASHP: Bethesda, MD, 1998; 13.
5. Knapp, K.K.; Blalock, S.J.; O'Malley, C.H. *Survey of Managed Care and Ambulatory Care Pharmacy Practice in Integrated Health Systems,* 1999; ASHP: Bethesda, MD, 1999; 3.
6. Knapp, K.K.; Blalock, S.J.; O'Malley, C.H. *Survey of Managed Care and Ambulatory Care Pharmacy Practice in Integrated Health Systems,* 1999; ASHP: Bethesda, MD, 1999; 4.
7. Knapp, K.K.; Blalock, S.J.; O'Malley, C.H. *Survey of Managed Care and Ambulatory Care Pharmacy Practice in Integrated Health Systems,* 1999; ASHP: Bethesda, MD, 1999; 4.
8. Knapp, K.K.; Blalock, S.J.; O'Malley, C.H. *Survey of Managed Care and Ambulatory Care Pharmacy Practice in Integrated Health Systems,* 1999; ASHP: Bethesda, MD, 1999; 6.
9. Toscani, M.R. Introduction to Disease Management. In *Paper Presented at ASHP Midyear Clinical Meeting. New Orleans, LA, 1996, Dec. 10.*
10. Larson, L.N.; Bjornson, D.C. Interface between pharmacoepidemiology and pharmacoeconomics in managed care pharmacy. J. Manage. Care Pharm. **1996**, *2* (3), 282–289, May/Jun.

11. Marcille, J. Is there a place for pharmacists in the new world of disease management? J. Manage. Care Pharm. **1996**, *3* (1), 16–21, Jan./Feb.
12. Campbell, W.H.; Newsome, L.A.; Ito, S.M. The Evolution of Managed Care and Practice Settings. In *A Pharmacist's Guide to Principles and Practices of Managed Care Pharmacy*; Ito, S.M., Blackburn, S. Eds.; Foundation for Managed Care Pharmacy: Alexandria, Virginia, 1995; Vol. 5, 13.
13. *Innovative Roles for Managed Care Pharmacists; 7–8.*
14. Toscani; 7.
15. Rosenburg, G.B. Opportunities for alliances between industry and pharmacy. Am. J. Hosp. Pharm. **1994**, *51*, 3061–3065.
16. Marcille, J. Is there a place for pharmacists in the new world of disease management? J. Manage. Care Pharm. **1996**, *3* (1), 20, Jan./Feb.
17. Boysen, H. **1996**, Personal communication, Oct.
18. ASHP Center for Managed Care. *Innovative Roles for Managed Care Pharmacists*; ASHP: Bethesda, MD, 1998; 8.
19. Bond, W.E. Direct contracting: The next purchaser strategy. J. Manage. Care Pharm. **1996**, *2* (1), 11–16, Jan./Feb.
20. Kostoff, L. **1996**, Personal communication, Nov.
21. Boysen, L. **1996**, Personal communication, Oct.
22. Marcille, J. Is there a place for pharmacists in the new world of disease management? J. Manage. Care Pharm. **1996**, *3* (1), 18, Jan./Feb.
23. Buchner, D. Role and Importance of Outcomes Measurement in Maximizing the Success of Disease Management Initiatives. In *Abstract Presented at ASHP Midyear Clinical Meeting, New Orleans, LA, 1996; Dec. 11.*
24. Gross, T. Considerations in Developing and Selecting Disease Management and Outcomes Programs in Managed Care. In *Paper Presented at ASHP Midyear Clinical Meeting, New Orleans, LA, 1996; Dec. 10.*
25. www.ashp.org-public-rtp-MCPHSTND.html.
26. American Society of Hospital Pharmacists. ASHP guidelines on pharmaceutical services for ambulatory patients. Am. J. Hosp. Pharm. **1991**, *48*, 311–315.
27. American Society of Hospital Pharmacists. ASHP guidelines: Minimum standard for pharmacies in institutions. Am. J. Hosp. Pharm. **1985**, *42*, 372–375.
28. American Society of Health-System Pharmacists. ASHP guidelines on a standardized method for pharmaceutical care. Am. J. Heath-Syst. Pharm. **1996**, *53*, 1713–1716.
29. American Society of Hospital Pharmacists. ASHP guidelines for obtaining authorization for documenting pharmaceutical care in patient medical records. Am. J. Hosp. Pharm. **1989**, *46*, 338–339.
30. American Society of Hospital Pharmacists. ASHP Statement on the Pharmacist's Responsibility for Distribution and Control of Drug Products. In *Practice Standards of ASHP 1996–97*; Deffenbaugh, J.H., Ed.; American Society of Health System Pharmacists: Bethesda, Maryland, 1996.
31. American Society of Hospital Pharmacists. ASHP statement on the pharmacist's role with respect to drug delivery systems and administration devices. Am. J. Hosp. Pharm. **1989**, *46*, 2342–2343.
32. American Society of Hospital Pharmacists. ASHP technical assistance bulletin on drug formularies. Am. J. Hosp. Pharm. **1991**, *48*, 791–793.
33. American Society of Hospital Pharmacists. ASHP technical assistance bulletin on evaluation of drugs for formularies. Am. J. Hosp. Pharm. **1988**, *45*, 386–387.
34. American Society of Health-System Pharmacists. ASHP guidelines on medication-use evaluation. Am. J. Heath-Syst. Pharm. **1996**, *53*, 1953–1955.
35. American Society of Hospital Pharmacists. ASHP guidelines on adverse drug reaction monitoring and reporting. Am. J. Hosp. Pharm. **1995**, *52*, 417–419.
36. www.ashp.org.

BIBLIOGRAPHY

Following are some selected references to published materials validating the benefits of pharmacists' participation in managed care:

Butler, C.D. Practical approaches to meaningful outcomes analysis. ASHP Home Care Meet. **1995**, *2*, HC-24, Aug.

Goss, T.F. Overview of outcomes studies conducted in managed care organizations. ASHP Midyear Clin. Meet. **1996**, *31*, PI-58, Dec.

Grasela, T.H. Pharmacoepidemiology: Scientific basis for outcomes research. Ann. Pharmacother. **1996**, *30*, 188–190, Feb.

Hedblom, E. Process of identifying medication-related outcomes. ASHP Annu. Meet. **1996**, *53*, PI-85, Jun.

Horn, S.D. Clinical practice improvement model and how it is used to examine the availability of pharmaceuticals and the utilization of ambulatory healthcare services in HMOs. PharmacoEconomics **1996**, *10* (Suppl. 2), 50–55.

Horn, S.D.; Sharkey, P.D.; Phillips, H.C. Formulary limitations and the elderly: Results from the managed care outcomes project. Am. J. Managed Care **1998**, *4*, 1105–1113, Aug.

Horn, S.D.; Sharkey, P.D.; Levy, R. Managed care pharmacoeconomic research model based on the managed care outcomes project. J. Pharm. Pract. **1995**, *8*, 172–177, Aug.

Hughes, T. Research in managed care: Outcomes research in disease prevention and management. J. Managed Care Pharm. **1996**, *2*, 212; 214; 216; 221–222, May–June.

Leon, J.; Neumann, P.J. Cost of Alzheimer's disease in managed care: A cross-sectional study. Am. J. Managed Care **1999**, *5*, 867–877, Jul.

Lindgren, D.L. Issues for managed care pharmacy. Am. Pharm. **1995**, *NS35*, 51–58, Aug.

Parker, D.A. Advancing outcomes research in managed care pharmacy. J. Managed Care Pharm. **1998**, *4*, 257–258, 261–271; 266–267, May–June.

Robison, R.N. Prioritizing managed care organization needs for outcomes studies and identifying disease Management programs and outcomes studies worth implementing: Managed care medical director's perspective. ASHP Midyear Clin. Meet. **1996**, *31*, PI-59, Dec.

Wernsing, D. Health outcomes and managed care. Drug Benefit Trends **1996**, *8* (Suppl. C), 29–35, Mar.

Medicaid and Medicare Pharmaceutical Programs

Albert I. Wertheimer
Stephen H. Paul
Temple University, Philadelphia, Pennsylvania, U.S.A.

INTRODUCTION

There are two major government supported healthcare programs in the United States today. They are completely different in structure, purpose, and financing. One is Medicaid, which is operated by the state governments with financial support from the federal government. This support varies but is generally in the range of 53–80% of total expenditures, depending upon the state's per capita income. Medicaid is Title 19 of the Social Security laws and was enacted in 1965. It is intended for medically indigent persons. Such low-income persons must pass a means test of income and wealth maximum criteria. In 1998, about 40.6 million persons received benefits at an expense of $142 billion. Medicaid can be seen as a welfare program to replace the very different programs operated by states and counties before 1965, which had different eligibility criteria, benefit structures, and waiting/residency requirements impacting persons who moved residences. In essence, it standardized welfare programs.[1]

Medicare, on the other hand, was established also in 1965 to provide assistance for medical expenses for the aged and disabled. It is not a welfare program, but rather an insurance program, as beneficiaries have their premiums deducted from their monthly social security checks. All persons 65 years of age and above are eligible for hospital insurance (called Part A). Part B is a supplemental health insurance plan covering physician's and surgeon's fees, laboratory work, and other outpatient services. In 2000, 39.33 million persons were enrolled in Part A and 37.4 million of these were enrolled in Part B. This is an insurance program funded by enrollee payments of about $50.00 per month. Part A is funded by a tax on incomes of all persons of 7.65%.[2] Medicare is operated directly by the federal government through the Social Security Administration.

The role of drugs and pharmaceutical services could not be any different. In Medicaid, every state program has an outpatient drug benefit that covers virtually all prescription and OTC drugs available in the United States. There are some differences among the states, which will be discussed later. However, there has never been an outpatient drug benefit in Medicare. It has been discussed, and legislative initiatives have been planned, but it has not been brought to fruition in the 35-year history of the program. Medicare covers drugs used while a patient is hospitalized. Estimates of the likely cost of such an outpatient program are massive, and there is controversy as to the need.

MEDICARE: DESCRIPTION AND CONTACT POINTS

Medicare *is the health insurance program for people 65 years of age or older, certain younger individuals with disabilities, and citizens with End-Stage Renal Disease (permanent kidney failure with dialysis or a transplant, sometimes called ESRD).* This program is administered by the Department of Health and Human Services through the Health Care Financing Administration (HCFA) of the United States federal government.

- Contact with HCFA can be accomplished over the Internet at: http://www.hcfa.gov or http://www.medicare.gov.
- Via the telephone by calling: 1-800-Medicare (1-800-633-4227).
- Or by mailing the Medicare agency at: U.S. Department of Health and Human Services; Health Care Financing Administration; 7500 Security Boulevard; Baltimore, Maryland 21244-1850.

History

The quest for universal healthcare in the United States started during the depths of the great depression in the 1930s. President Franklin D. Roosevelt wanted it, however, American medicine vehemently opposed it. From that point, the protagonists and antagonists have been fighting the battle. The goal appeared to be in reach numerous times during the last 70 years; however, an inci-

Encyclopedia of Clinical Pharmacy
DOI: 10.1081/E-ECP 120006184

dent always came along to derail comprehensive reform. In 1965, the ''Great Society'' under President Lyndon B. Johnson moved forward with legislation providing a level of care for the elderly (Medicare). Medicaid was also enacted at the same time to cover low-income aged, blind, and disabled individuals, and parents and their dependent children on welfare. Piecemeal health coverage for select populations was continuing, and comprehensive reform was thwarted. These two federally initiated health proposals did not include mandatory ambulatory drug coverage. In 1988, the Medicare Catastrophic Act was signed into law by President Ronald W. Reagan and promptly vetoed by President George H. W. Bush before it could be implemented. This legislation would have provided a prescription drug benefit and a cap on patient liability. The law was rescinded, because high-income elderly felt they would pay more in premiums than they would receive in benefits.[3]

The passage of the Omnibus Budget Reconciliation Act (OBRA) of 1990 provided poor seniors who qualified for Medicare and Medicaid a level of mandatory patient care provided by pharmacists in addition to Medicaid prescription coverage.

The 1990s witnessed the creation of optional non-federal prescription benefits for Medicare beneficiaries enrolled in Health Maintenance Organizations (HMOs), and in 1997, Medicare + Choice was established by President William J. Clinton which created numerous choices for recipients. The start of this millennium has witnessed the commitment of President George W. Bush to create a prescription benefit program for the elderly. This program could take several different paths. There could be comprehensive Medicare reform with prescription coverage and pharmaceutical care becoming a mandatory or optional benefit. There could also be just a prescription benefit passed for categorically defined individuals. Another option could be a block grant program enabling states to receive federal funds. States could have the flexibility in choosing to establish a drug benefit or enhance pharmaceutical coverage with existing voluntary plans. Failure of a political consensus among the President, Congress, the pharmaceutical industry, pharmacies, and patients could result in no additional prescription and pharmaceutical care coverage for consumers, at least until the next presidential election in 2004.

The Medicare Program

The basic Medicare insurance program, which provides benefits to consumers who are 65 or older, consists of two components.

The hospital portion is ''Part A.'' It receives its name from the fact that it is ''Part A'' of Title XVIII of the Social Security Act.[4] This section of the legislation covers inpatient hospitalization, critical access hospitals, skilled nursing facilities, hospice care, and limited home healthcare.[5] Critical access hospitals are small facilities that provide limited outpatient care and inpatient services to individuals in rural areas.

Most people pay for ''Part A'' during their working years. They therefore receive this insurance benefit automatically when the appropriate time comes. Employees and employers each pay 1.45% tax on all wages and salaries.

The medical portion is ''Part B,'' and it is included in the legislation as ''Part B'' of Title XVIII. It covers medical services including physician care, outpatient hospitalization services, and selected medical activities not covered in ''Part A,'' such as occupational and physical therapists. This section will pay for diabetic supplies when medically necessary.

The cost to the Medicare patient changes each year. During the year 2001, it was $50.00 per month. This charge is deducted from the recipients' monthly Social Security check before it is received. The premiums paid by participants represent 25% of the program's cost. The remaining 75% is paid directly by a federal budget appropriation.

Healthcare Coverage Options

Eligibility for Medicare enables individuals to select one of a myriad of choices for receiving care.

The Original Medicare Plan

The first Medicare health insurance is now called the ''Original Medicare Plan.''[6] This is a fee-for-service plan. The provider charges a fee to the patient and or the government each time the service is provided. This arrangement is offered nationwide.

The federal government has authorized supplemental insurance polices to aid in paying for services not covered in the Original Medicare Plan. Up to ten supplemental plans can be marketed by private insurance carriers. These plans must be labeled ''Medicare Supplement Insurance.'' There are high-cost and lower-cost policies. They differ in the scope of coverage, deductibles, and copayments. Several of these optional programs have limited outpatient prescription coverage. In addition to these supplements that are also called Medigap policies, the standardized benefits may also be sold as ''Medicare Select'' policies. The Medicare Select benefits should be less expensive, because the freedom-of-choice to choose providers is limited to selected physicians and hospitals. Insurance carriers may not add or subtract benefits to the ''Medicare

Supplement Insurance'' policies. The only variable allowed is the premium charged.

Medicare + Choice

In 1997, legislation was passed to aid in diminishing the growth of Medicare expenditures and provide more options to participants in the healthcare plans. This legislative modification to a section of the Social Security Act is popularly known as Medicare+C or Medicare+Choice. These arrangements include the offering of managed care plans (health maintenance, provider sponsored, preferred provider, and point of service option), private fee-for-service plans, and medical savings accounts.

THE PRESENT STATUS OF DRUG COVERAGE

Pharmaceuticals are covered for patients who are admitted to hospitals for acute or chronic care. The law does not provide for ambulatory drug coverage. Managed care plans and some supplemental policies may offer a prescription benefit to its senior citizens. Medicare does not directly pay for this coverage. Some Internet pharmacies may attempt to lure unsuspecting patients to their sites with special Medicare prescription plans. It is imperative to recognize that the potential for scams on the elderly exists because of their tendency to be trusting.[7]

The issue of pharmacy benefits becomes more complicated and challenging to implement as each year passes. Medicare will provide coverage for diabetic supplies of glucose monitors, test strips, and lancets. Diabetic drugs are not covered. The present official status is even difficult for providing a simple answer to drug coverage.

> Generally, Original Medicare does not cover prescription drugs. However, Medicare does cover some drugs in certain cases such as immunosuppressive drugs (for transplant patients) and oral anti-cancer drugs.[8]

> There are some Medicare Health Plans that cover prescription drugs. You can also check into getting a Medigap or supplemental insurance policy for prescription drug coverage. Medicaid may also help pay for prescription drugs for people who are eligible.[9]

Many former employers of retired workers provide a level of voluntary prescription benefits in addition to other retiree benefits.

Presently, pharmacists cannot directly bill for providing pharmaceutical services to patients.

Extending Drug Coverage to All Medicare Beneficiaries

Extending the Medicare benefit package to include a pharmaceutical care benefit could close a significant gap in program coverage. The issues revolve around drug costs and services. The various protagonists and antagonists all recognize the cost will be substantial to the taxpayers and participants.[10]

Recently, prescription drug spending has far surpassed growth in spending for other types of healthcare. The rising expenditures have had a significant impact on Medicare beneficiaries, employers who offer retiree health coverage, and on state governments providing drug coverage to the elderly.[11]

The issues involved in providing drug and pharmaceutical care are numerous. They can be examined from the financial and patient care perspectives, and by using management tools.

Medicare participation

Most beneficiaries use drugs. The distribution of the use of drugs is slanted toward patients with chronic conditions of diabetes, hypertension, and cardiovascular diseases. Adverse patient selection to participate is an important issue, because there is an uneven distribution of drug use, and patients must utilize medications over a long period of time.

The financial conundrum. There are a myriad of choices for funding the prescription benefit. Costs could be financed as they are right now with ''Part B'' participation. Beneficiaries and the government contribute to the benefit. Workers and retirees enrolled in employer-sponsored heath plans will have to be taken into account. Beneficiaries who are on Medicaid will present issues that the state and federal governments will have to resolve. Recent drug benefit proposals for the low-income participants have indicated a strong preference for full or partial subsidies for the premium payments and cost sharing for the prescriptions and pharmaceutical services utilized.

Drug benefit coverage

Legislation involving medical and hospital care has always achieved a commitment to comprehensive care. This obligation has not been honored when it comes to providing drugs to patients. Employers and legislators have fashioned creative reasons to provide less than comprehensive care. Since the past is considered prologue, drugs will have to be considered from the vantage point of false economy.

There are various ways to contain the drug benefit. The drug benefit could be restricted by placing a cap on the value of the benefits provided for a benefit period. Another restraint would be to provide drug coverage after a drug benefit deductible was exceeded or only provide the benefit for catastrophic situations. This type of action would save drug money; however, the overall costs for healthcare would probably increase.

The pharmacy profession has recognized that the benefits to society greatly exceed the cost of the drugs. The benefits of the drugs and pharmacy care center around an improved quality of life, a decrease in overall healthcare expenditures, and an increased life span.[12]

Administration of the drug benefit. Proposals being discussed utilize the private-sector model of the Pharmacy Benefit Manager (PBM). Some proposals utilize multiple PBMs for a geographic area, while others rely on a single PBM to manage the program. The issue of whether they will be merely a claims processor or have additional responsibilities is varied. The remaining proposals require government management of the drug benefit. The managerial control would occur at the federal or state level.

Additional tools to manage the pharmacy benefit. Coordination of the Medicare benefit by implementing a multidisciplinary approach to patient care should produce a quality outcome for the beneficiary. From the pharmacy perspective, numerous issues should be addressed.[13] Some of the concerns include:

- *Formulary coverage.* A drug formulary is an instrument that contains safe, effective, and affordable medications designed to improve or maintain patient health. The breadth and depth of coverage and the ability to add drugs are imperative to maintaining a workable drug formulary.
- *Pharmaceutical care and disease management.* The collaborative efforts of the interdisciplinary team of pharmacists and other health professionals help to ensure patients are utilizing medications appropriately.
- *Critical (clinical) pathways and drug step therapy.* These concepts can be effectively used to create a synergistic impact on improving the health of the elderly. Critical pathways are designed to provide continuity of care and decrease the fragmentation of services. Their use helps guide the patient and family through the expected treatments and progress. It also increases the satisfaction of patients, families, health professionals, and the various payers for healthcare services.[14]

Step therapy for the drug treatment of patients creates a ''road map'' to be used with various medications in order to control the disease or medical condition. The initial step is usually the most common one used in this situation. More complex steps are not attempted to correct the patient's situation until the earlier steps have failed.

Drug evaluation. Drug evaluation is an ongoing, systematic process designed to maintain the appropriate and effective use of medications. It involves the review of the physicians' prescribing relationships, review of the pharmacists' dispensing patterns, and patients' use of medications. This evaluation goes by several names in different healthcare settings.[15] The names include DUR (drug utilization review), DUE (drug use review), and MUE (medical use evaluation).

Resource-Based Relative Value System

Reimbursement for health services is extremely complicated. Providers want higher reimbursement, and payers desire to reduce, maintain, or limit increases paid each year. HCFA developed a methodology to deal with physician reimbursement and allow for an increase in payments for physician services. It is a system based on approximately 7500 relative value service codes. These codes are more complex than the approximate 450 Diagnosis Related Groups (DRG) used by hospitals in their Medicare reimbursement. In addition to the service codes, the formula has a relative value unit (RVU) for practice expenses and a separate one for malpractice insurance. Added to these components is a geographic practice cost index (GPCI) for each defined work service area. The GPCI is designed to take into account high- as well as low-cost practice expenses and physician services as compared to the national average for each constituent of the model. A conversion factor (CF) is also part of the model. This variable is designed to maintain fiscal budget neutrality in the event total payments exceed a certain monetary sum, determined by Congress, each year. The complexities involved can be understood more completely by checking out the designated HCFA web site.[16]

The model used to compute physician payment can be expressed as:

$$\begin{aligned}\text{Physician Payment} &= [((\text{RVU service activity} \times \text{GPCI service activity}) \\ &\quad + (\text{RVU practice expense} \\ &\quad \times \text{GPCI practice expense}) \\ &\quad + (\text{RVU malpractice expense} \\ &\quad \times \text{GPCI malpractice expense})) \times \text{CF}]\end{aligned}$$

The physician payment model is not perfect; however, it represents a quantum leap forward compared to how pharmacy practitioners are reimbursed. The reimbursement in many Medicare+Choice plans, in most state Medicaid programs, and private PBM (third party) arrangements fails to recognize that the costs for providing the pharmacy benefit changes at the practitioner level. PBMs only recognize that the pharmaceutical cost changes.

Conclusion

Medicare represents a federal entitlement program for healthcare services that will continue to expand in coverage of beneficiaries and benefits provided. Providers will be continually challenged to maintain and increase productivity, while payers explore innovative ways to pay for services. A modification of the resource-based relative value scale could be utilized in this endeavor. The drug benefit is exceeding Congressional Budget Office estimates each year. Drug spending for beneficiaries is projected to be $1,500,000,000,000 over the time period 2002–2011. The House of Representatives Budget Resolution allocated $153,000,000,000 for the drug benefit for the same period.[17] The monetary difference is approximately seven times greater than the major political organizations have proposed. The gap between the estimate and the budget amount will require one of several responses. Creative spin masters will have to explain the continued lack of drug coverage, or there will be a final recognition that drugs and pharmaceutical care represent a cost-effective mechanism to aid in controlling healthcare expenditures. "The key to an effective pharmaceutical benefit is management," (and) "a well-trained, well-equipped pharmacist is critically important to the smooth operation of a drug benefit."[18]

MEDICAID

The 50 states, District of Columbia, Puerto Rico, Guam, and the Virgin Islands and other territories all have medical assistance (Medicaid) programs that vary somewhat but are within federal guidelines. States qualify for federal reimbursement by agreeing to provide benefits to certain categories of needy persons who meet the requirements of the block grant for (TANF), temporary assistance to needy families, and the subsequent aid to families with dependent children (AFDC) programs, and for blind and disabled persons receiving social security income.

The program covers children under six years whose family income is no more than 133% of the federal poverty level definition, pregnant women up to 133% of the poverty level, some Medicare beneficiaries, and recipients of adoption assistance and foster care programs. Medicaid covers virtually all outpatient and inpatient health and rehabilitative services, home health, long-term care, dental, prosthetic, pharmacy, and optical services and goods. Pharmacy benefit rules state:

> Prescribed drugs are simple or compound substances or mixtures of substances prescribed for the cure, mitigation or prevention of disease, or for health maintenance, which are prescribed by a physician or other licensed practitioner of the healing arts within the scope of their professional practice, as defined and limited by Federal and State law (42 CFR 440.120). The drugs must be dispensed by licensed authorized practitioners on a written prescription that is recorded and maintained in the pharmacists or practitioners records.[19]

While there are over 30 million beneficiaries enrolled in Medicaid, not all of these persons were recipients of services during any given year. Table 1 presents the utilization of various Medicaid services in fiscal year 1998.

Reimbursement

The Health Care Financing Administration (HCFA) establishes the policies that individual State Medicaid programs must adhere to. In the realm of prescription drug reimbursement, rules were established in 1987 for multisource drugs. Upper payment limits based upon estimated

Table 1 Utilizing number of individuals of Medicaid services in FY 1998

Pharmaceuticals	19,337,543
Physicians	18,554,746
Hospital outpatient	12,157,729
Lab/X-ray	9,380,689
EPSDT	6,174,628
Clinic	5,285,415
Dental	4,965,202
Hospital inpatient	4,408,162
Other practitioners	4,341,915
Personal support services	3,108,432
Family planning	2,011,124
Nursing facility	1,645,728
Home healthcare	1,224,714
ICF-mentally retarded	126,490

(From Ref. [20].)

acquisition costs (EAC) were established. For drugs certified by the FDA as being interchangeable, if the prescriber writes on the face of the prescription: ''brand necessary'' or ''medically necessary,'' the patient can receive the branded product instead of the generic equivalent product. For 1998, HCFA spent $13.52 trillion for 19.3 million recipients which is about $700 per recipient that year.

The top ten states in prescription expenditures for 1998 (in descending order) were California, New York, Florida, Texas, Ohio, Illinois, Pennsylvania, Massachusetts, North Carolina, and New Jersey. The ten lowest expense states were Oklahoma, Arizona. Tennessee, Wyoming, North Dakota, South Dakota, Alaska, Nevada, Hawaii, and District of Columbia. For all states, drugs and related services consumed 9.5% of the total Medicaid budget. Because of this more than $13 trillion expenditure, in 1990, Congress considered alternative means to reduce this expense. The result was a compromise where in exchange for Medicaid formularies to be open to all drugs, manufacturers agreed to agree to a rebate program with HCFA in the OBRA 1990 legislation. Rebates were to be a minimum of 10% of that state's purchases from a company. OBRA was amended in 1992, and today, manufacturers pay 15.1% of the average manufacturer's price back to the state for innovator (single-source) products, and 11% is returned for generic, multi-source products.[21]

To give one a feeling for the quantities involved, the total rebate for 1998 was $2.5 billion. While all drugs should be available, state Medicaid agencies may restrict availability of certain drugs of limited value, regarding safety, effectiveness, or clinical outcome if the drug may be obtained through the prior approval procedure. Other drugs may be excluded completely if they are:

- For anorexia, weight gain, fertility, hair growth, cosmetic effect, smoking cessation, or symptomatic relief of cough or cold.
- Vitamins or minerals or OTC drugs (fluorides and prenatal vitamins excluded).
- Drugs requiring monitoring to be obtained from the manufacturer.
- Barbiturates or benzodiazepines.

A significant component of the Medicaid drug program is a Drug Utilization Review (DUR) activity, which is defined as a structured and continuing program that reviews, analyzes, and interprets patterns of drug usage in a given healthcare environment against predetermined standards. This is conducted for two purposes: to improve the quality of care and to assist in containing costs. OBRA requires that states provide prospective DUR and retrospective DUR programs. The prospective DUR activity is performed at the time of dispensing.[22]

As is the case of an HMO patient presenting a card at the pharmacy, the Medicaid patient does the same thing. Each state decides whether it will have a patient copayment, and if so, its amount. About 15 States have no copayment requirement, and the others charge between 50 cents and $3.00 per prescription.

The pharmacy is paid a dispensing fee that ranges between $3.00 and $5.50 per prescription, depending upon the state. The pharmacy is reimbursed the wholesale price of the drugs minus a discount, which is a percentage reduction from the ''sticker'' price (called AWP or average wholesale price) which is higher than the actual price paid by most pharmacies due to quantity discounts, direct purchases from manufacturers, and the taking advantage of ''deals.'' The discount averages about 11 or 12% of the average wholesale price. This brings the ingredient reimbursement more in line with the actual price paid by the pharmacy.

For an example, let us consider a drug where the AWP is $60.00. The patient paid $3.00, and the pharmacy will be reimbursed $60.00 less 12%, which equals $52.80 less the $3.00 patient copayment or $49.80 by that state Medicaid agency. In addition, the pharmacy will receive $4.00 as a dispensing fee.

Because of budget problems in some states from time-to-time, limitations have been implemented on occasion. Some states have limited the number of prescriptions per month for limited periods or established caps on the value of the benefits. Usually, these have been lifted when the budget situation improved, especially because there is no evidence that such restrictions are cost-effective overall, and, in fact, there is considerable suspicion that patients might not get needed drugs, resulting in potentially massive hospital or other costs.

With such a huge price tag, HCFA administrators and legislators are always searching for means to reduce costs. Some relief has come from the prior authorization program as well as from a mandatory generic dispensing policy in many states, but additional savings are still desired. There has been discussion about placing recipients in managed care plans that are capitated and having the practitioners control utilization with actual incentives. Some states have asked for supplemental rebates that provide a discount well beyond the OBRA 1990 dictated amount.[23]

Health economists continue to advocate greater emphasis on prevention, screening, patient education, wellness education with emphasis on nutrition, smoking, and alcohol use reduction, avoidance of substance abuse, and prenatal care as examples of strategies to reduce illness trauma and cost throughout life. Perhaps enrollment in capitated managed care organizations can provide the environment and suitable incentives for greater cost savings in the future.

REFERENCES

1. Wright, J.W. *2001 New York Times Almanac*; Penquin: New York, 2000; Vol. 153.
2. *Pharmacy and the US Health Care System,* 2nd Ed.; Fincham, J., Wertheimer, A., Eds.; Pharm. Products Press: Binghamton, New York, 1998; Vol. 34.
3. http://www.MEDICARE.GOV/35/milestones.asp (accessed April 2001).
4. Social Security Act. http://www.ssa.gov/OP_Home/ssact/comp-toc.htm, accessed April 2001.
5. *Medicare and You 2001, HCFA-10050*; Health Care Financing Administration: Maryland, 2000; Vol. 5; 1–73.
6. *Medicare and You 2001, HCFA-10050*; Health Care Financing Administration: Maryland, 2000; Vol. 14; 1–73.
7. http://www.MEDICARE.GOV/publications/pubs/pdf/2049fina.pdf File: /publications/pubs/pdf/2049fina.pdf.
8. Medicare 2000: 35 Years of Improving Americas' Health and Security, pp. 2, July 2000, Medicare fact10.pdf http://www.medicare.gov/FAQs/Top20.asp, April 15, 2001.
9. Medicare 2000: 35 Years of Improving Americas Health and Security, pp. 2, July 2000, Medicare fact10.pdf http://www.medicare.gov/FAQs/Top20.asp, April 15, 2001.
10. Laying the Groundwork for a Medicare Prescription Drug Benefit, Statement of Dan L. Crippen, Committee on Ways and Means, Subcommittee on Health, U.S. House of Representatives, March 27, 2001.
11. Laying the Groundwork for a Medicare Prescription Drug Benefit, Statement of Dan L. Crippen, Committee on Ways and Means, Subcommittee on Health, U.S. House of Representatives, March 27, 2001.
12. Rx price controls likely unless manufacturers help curb costs, Vt. Gov says; McCaughan, M., Ed.; The Pink Sheet **2001**, *63* (14), 6.
13. *Where We Stand: Medicare Prescription Drug Coverage*; Stables, C., Ed.; Academy of Managed Care Pharmacy: Virginia, 1999; 1–12.
14. Sapienza, A.; Broeseker, A. Health Care Professionals and Interprofessional Care. In *Introduction to Health Care Delivery A Primer for Pharmacists*; McCarthy, R., Ed.; Aspen Publishers, Inc.: Maryland, 1998; 48–59.
15. *Drug Use Evaluation*; The Academy of Managed Care Pharmacy. http://amcp.org/public/pubs/concepts/druguse.html (accessed March 2001).
16. http://www.hcfa.gov/stats/pufiles.htm#rvu; http://www.MEDICARE.GOV/publications/pubs/pdf/2049fina.pdf File: /publications/pubs/pdf/2049fina.pdf.
17. Medicare PBMs could offer ''loose'' and ''tight'' Rx options—Rep. Johnson; McCaughan, M., Ed.; The Pink Sheet **2001**, *63* (14), 9–10.
18. http://www.healthnewsdaily.com/rs/channels/fdc/HND/Current + Articles/hnd + pink/stories/0425p2.asp.
19. *Pharmaceutical Benefits Under State Medical Assistance Programs, 1999*; National Pharm. Council: Reston, Virginia, 2000; Vol. 4.8.
20. *Pharmaceutical Benefits Under State Medical Assistance Programs, 1999*; National Pharm. Council: Reston, Virginia, 2000; Vol. 4.12.
21. *Pharmaceutical Benefits Under State Medical Assistance Programs, 1999*; National Pharm. Council: Reston, Virginia, 2000; 4.32.
22. Wertheimer, A.; Navarro, R. *Managed Care Pharmacy: Principles and Practice*; Pharm Products Press: Binghamton, New York, 1999; Vol. 232.
23. Sultz, H.; Young, K. *Healthcare USA*; Aspen: Gaithersburg, Maryland, 2001; 272.

Medical Communications, Clinical Pharmacy Careers in

Lara E. Storms
Virginia Commonwealth University School of Pharmacy, Richmond, Virginia, U.S.A.

Cindy W. Hamilton
Hamilton House, Virginia Beach, Virginia, U.S.A.

INTRODUCTION

Pharmacists rely on medical communications to acquire the information needed to practice pharmacy, but they may not be familiar with the scope of medical communications, much less the potential for an attractive career opportunity in this field. In fact, a popular guide to careers for pharmacists has only one paragraph on the subject.[1] Medical communications refers to the written, audiovisual, or oral dissemination of medical information with a goal of informing audiences about health issues.[2,3] These materials are prepared by medical communicators who are educated in the humanities or sciences, including pharmacy. This monograph describes job activities, work settings, and training and skills for medical communicators. Finally, this monograph explains how pharmacists can become medical communicators.

METHODS

To identify information on careers for pharmacists in medical communications, a database at Virginia Commonwealth University, which includes MEDLINE and 212 other databases, was searched. Because this search did not reveal many useful citations, the director of a biomedical writing program at a university that has an established school of pharmacy (Table 1) was interviewed, as were eight pharmacists who had at least 10 years of experience in medical communications and who represented different work settings; the mean duration of experience was 15 years (range: 10–21 years). Approximately 30 questions were developed, consent was obtained to record and quote the interviewees, telephone interviews were conducted in June 2000, and interviewees were invited to review the first draft. The information in this chapter is based on these interviews, unless otherwise stated.

Two additional interviews were conducted to obtain missing data. To supplement salary information, we contacted Beth Rhoads of Pharmaceutical Search Professionals, Inc. (North Wales, PA). To determine where pharmacists find employment as medical communicators, membership data were collected from the American Medical Writers Association (AMWA).

WHAT DO MEDICAL COMMUNICATORS DO?

Medical communicators prepare many types of projects for health care professionals, consumers, regulatory agencies, and other audiences. To convey information to different target audiences, medical communicators use traditional media, such as books, journals, slides, and other newer media, such as compact discs and the Internet (Table 2). Most of these projects begin as written manuscripts, but medical communicators perform other types of activities, such as editing, interviewing, and moderating focus and other discussion groups.

Writing or editing a manuscript for publication in medical journals is a popular type of project. A medical communicator can collaborate with clinical investigators to transform a clinical trial report into an original research article. Hamilton said, ''Medical writing is perceived as a difficult task, so investigators are happy to delegate the work.'' The medical communicator can identify references to be cited in the introduction and discussion, develop an outline, write the first draft, incorporate author's changes, and style the manuscript for submission to the target journal.

Additional examples of popular projects are preparing monographs and regulatory documents. A monograph describes a topic, such as a drug or disease, for the purpose of

Encyclopedia of Clinical Pharmacy
DOI: 10.1081/E-ECP 120006381

Table 1 Medical communicators interviewed in June 2000[a]

Interviewee	Title and affiliation
Freelance setting	
Donna K. Curtis, PharmD	Freelance writer, Curtis Consulting & Communications
Cindy W. Hamilton, PharmD	Freelance writer, Hamilton House® Medical and Scientific Communications
John Russo, Jr., PharmD	Freelance writer, Medical Communications Resources
Medical communications or pharmaceutical company	
Karen M. Overstreet, EdD, MS, RPh	President, Meniscus Educational Institute, a division of Meniscus Limited
Gary E. Pakes, PharmD	Senior Medical Publications Scientist II, Glaxo Wellcome, Inc.
Nancy C. Phillips, BS, RPh	President, The Phillips Group
Publisher, professional association, or both	
Eugene M. Sorkin, PharmD	Associate Editor, *The Annals of Pharmacotherpy*, Harvey Whitney Books
Cheryl A. Thompson, MA, RPh	Editor, American Society of Health-System Pharmacists
University	
Lili F. Vélez, PhD	Director, Biomedical Writing Program, University of the Sciences in Philadelphia
Other	
Beth Rhoads, BS	Recruiter, Pharmaceutical Search Professionals, Inc.

[a]Some titles and affiliations have changed since the interviews were conducted.

providing continuing education or promoting a new drug. Regulatory documents are required for drug development; examples include clinical investigator brochures, clinical trial reports, new drug applications, and package inserts. With these projects, the medical communicator can research the literature, collaborate with other team members to analyze and interpret data, write an outline and then the first draft, edit it, and coordinate revisions. In the case of a monograph, the medical communicator can work with a graphic artist to design and produce a printed product.

Table 2 Scope of medical communications

Types of projects		
Abstracts	Grant applications	Policy writing for professional practice
Bibliographies	Journal articles	Practice guidelines
Book reviews	Marketing and promotional material	Product monographs
Case histories and reports	Medical letters	Proposals for new business
Clinical investigator brochures	New drug applications	Sales training
Clinical trial reports	Package inserts	Study protocols
Continuing education (CE) programs	Patient education material	
Activities		
Critically evaluating the literature	Moderating focus groups	Writing
Editing	Researching	
Interviewing	Reporting	
Media		
Audio	Journals	Radio
Books and book chapters	Monographs	Slides
Compact discs	Newsletters	Television
Internet	Newspaper articles	Video

(From interviews and Ref. [2].)

M

Table 3 Estimated annual income for pharmacists employed as medical communicators in different settings in June 2000

Employment setting	Annual income ($)	
	Starting	Maximum
Contract research organization	45,000–65,000	85,000–90,000
Freelance setting	50,000–80,000[a]	90,000–200,000+
Medical communications company	55,000–70,000	100,000–130,000
Pharmaceutical company	55,000–70,000	100,000+
Publishing company	45,000–50,000	100,000
Professional association	Not available	Not available
University	Not available	Not available

[a]Based on hourly rate × 1000 billable hours/year.
(From interviews and Ref. [3].)

WHERE DO MEDICAL COMMUNICATORS WORK?

Medical communicators work in many settings, such as pharmaceutical companies, medical communication companies, freelance situations, professional associations, publishing companies, and universities. Pharmacists can also work in these settings (Table 3), but the salaries are not always competitive with those in traditional pharmacy practice sites. One way to estimate where pharmacists work as medical communicators is to compare the primary sections chosen by pharmacists who join AMWA with those chosen by the entire membership. The primary sections are pharmaceutical; freelance; editing/writing; public relations, advertising, and marketing (PRAM); and education. AMWA membership data confirm that pharmacists ($n = 137$) choose the same primary sections as the entire membership ($n = 4637$), but there are some differences in distribution (Fig. 1). For example, pharmacists were most likely to choose the pharmaceutical section, whereas the entire membership was most likely to choose the editing/writing section. Unfortunately, our survey was not designed to determine the reasons for these differences, and only limited salary information was available for each setting. Additional limitations of our survey were small sample size and potential for overlap between sections. Finally, our survey did not capture membership data for pharmacists who do not have the PharmD degree because this was the only relevant searchable term.

Interviewees described advantages and disadvantages associated with their positions, some of which recurred in different work settings. For example, Pakes enjoys his position in international medical publications at a

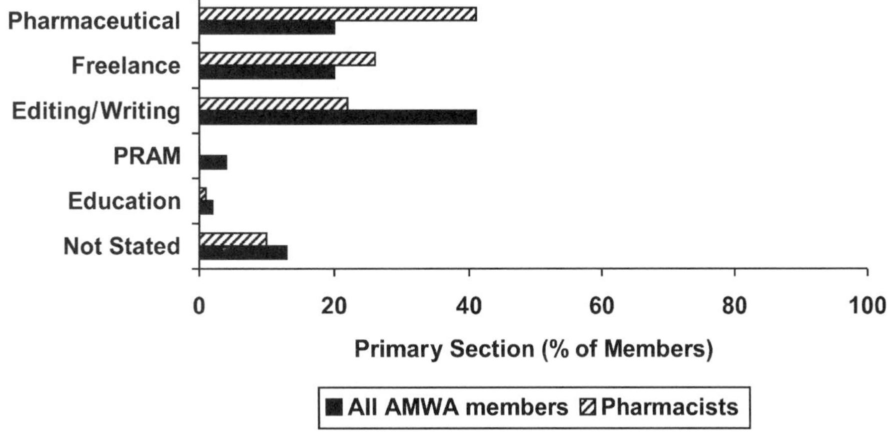

Fig. 1 Primary section in American Medical Writers Association (AMWA) chosen by all members ($n = 4637$) and pharmacists ($n = 137$) in July 2000. PRAM = public relations, advertising, and marketing.

pharmaceutical company because it is intellectually challenging and, with the finished product, ''there is something tangible.'' Pharmaceutical companies offer opportunities for advancement and comprehensive benefits such as a retirement package; paid vacation, holidays, and sick leave; and medical and dental insurance. Entry-level salaries can be higher than in some settings because pharmaceutical companies typically require some experience.

Medical communication companies prepare educational, promotional, and regulatory materials, usually for pharmaceutical companies. Contract research organizations, which oversee clinical trials for pharmaceutical companies, may also prepare regulatory materials. Phillips said that her job at a medical communications company ''allows me to work with some of the leading physicians and pharmacists in the world.'' Overstreet added that her position allows her ''the flexibility to do a variety of different things every day, and the opportunity to learn about new techniques and new therapies to supplement my pharmacy background.'' Benefits vary depending on the size of the organization. There are opportunities for advancement to positions such as editorial director or even president, but Phillips lamented that she has less time to write now that she is president of her own company.

Freelance was the second most popular AMWA section chosen by pharmacists. Advantages include working in a home office and being self-employed, independent, and flexible. Russo explained, ''As a freelancer, I'm able to choose the projects that I'm interested in doing.'' Curtis, Hamilton, and Russo agreed that freelance writing has some disadvantages, especially regulating the workflow. Curtis explained, ''Often you are juggling projects with very tight deadlines for multiple clients, and every client is important, so things can get really stressful.'' The earnings potential for freelancers appears to be one of the highest in medical communications, but this can be misleading because self-employed people must pay for their own equipment, benefits, and the part of social security that is otherwise paid by employers. In addition, freelancing can be very solitary because of minimal interaction with professional colleagues and patients.

Pharmacists who work at professional associations and publishing companies solicit manuscripts for journals, organize the peer review process, and edit manuscripts. Pharmacists also write news, editorials, and monographs for their journals, newsletters, web sites, and other publications. Thompson likes being an editor at a professional organization because her job encourages creativity, but she admitted that ''being an editor is not a 40-hour a week job and sometimes I feel overwhelmed.'' Salary information is not available for pharmacists who work in these settings, but Thompson said, ''Medical communicators should expect lower wages from non-profit organizations like ASHP.'' The benefits, however, are comparable to those offered by pharmaceutical companies, and there are opportunities for advancement. Sorkin said, ''There is a trade-off and people just have to determine what works best for them.''

WHAT TRAINING AND SKILLS ARE NEEDED?

The minimum training and skill requirements for a medical communicator are knowledge of medicine and ability to communicate. Hamilton explained, ''By their training, pharmacists know and understand medicine and the drug development process.'' Curtis added, ''Having the clinical knowledge about drugs and therapeutic areas really helps to bring some expertise to the writing.'' Another relevant aspect of pharmacy training is drug literature evaluation, because medical communicators

Table 4 Resources for pharmacists interested in learning about career options in medical communications

Type of resource	Example[a] (Internet address)	Comments on example
Professional organization	American Medical Writers Association (www.amwa.org)	Annual conference, regional meetings, workshops, networking
Colleges and universities	University of the Sciences in Philadelphia (www.usip.edu/graduate/biomedical_writing.htm)	Degree in biomedical writing, article on career opportunities (www.home.earthlink.net/~rhetrx/bmw/)
Practical experience	Hamilton House (www.hamiltonhouseva.com)	Clerkship for pharmacy students, list of recommended books

[a]The examples were selected because they provide information available on the Internet, including links to other useful web sites.

are expected to research new therapeutic categories. It is also helpful to have good computer skills, including familiarity with word processing and presentation software.

All but two of the interviewees agreed that good writers are trained, not born. Vélez said, ''A writer is simply someone who is willing to keep at it.'' Pakes disagreed and said, ''I don't think that you can take just any old pharmacist and turn them into a writer. A person is either a writer or they're not. What you do is you hone your skills and become a better writer.''

Regardless of whether good writers are trained or born, good medical communicators must be organized, accurate, flexible, curious, self-motivated, disciplined, and able to meet deadlines. Good pharmacists also have many of these qualities. Hamilton explained, ''Pharmacists are typically very well organized because we are trained to put the right medication into the right bottle and administer it at the right time. These skills are transferable to medical writing.''

Another skill that is transferable from some pharmacy settings to medical communications is the ability to manage a business. This is especially important for the individual freelance writer who must balance the time required to manage the business with the time required to write and generate income. Russo said, ''You need to have an appreciation and an understanding of marketing'' to thrive in the freelance business.

HOW CAN PHARMACISTS BECOME MEDICAL COMMUNICATORS?

There are many ways for pharmacists to obtain training and improve their communication skills (Table 4). Professional organizations offer workshops on writing and other relevant topics. For example, AMWA, an educational organization that promotes advances in biomedical writing, offers a certificate for completing their core and advanced curricula.[4] Workshops are available at the popular annual meeting in late autumn, at regional meetings throughout the year, and through an onsite option. AMWA membership also provides credibility and a venue for networking, which can lead to writing assignments or even full-time employment.

Colleges and universities offer courses and degrees in technical writing, journalism, and communications. The University of the Sciences in Philadelphia offers a master's degree in biomedical writing, the only degree program of its kind in the United States. According to Vélez, about one-fourth of the students in the master's program are pharmacists, and many other pharmacists take classes as nonmatriculating students or via the Internet.

A degree in biomedical writing is not likely to become a prerequisite for a career in medical communications, according to the interviewees. Vélez explained, ''There is a lot of lost creativity and innovation when you force people to have a particular credential before they can participate in a certain field.'' However, Vélez continued,

> Traditional pharmacy education does not stress written communication abilities. This means pharmacists interested in biomedical writing could probably benefit from additional coursework. Knowing *why* a text needs to be written in a particular way gives you more confidence and flexibility than only knowing *what* needs to be written.

Practical experience is also available. Pharmaceutical companies, professional associations, and publishers offer residencies and fellowships in medical communications or related areas, such as drug information. Schools of pharmacy offer clerkships for students under the direction of medical communicators. Finally, there is on-the-job training. Overstreet said, ''PharmDs are a hot commodity right now, even without experience. Some employers are willing to accept a new PharmD graduate and train them themselves.''

Interviewees offered advice for making the transition from traditional pharmacy practice to medical communications. To learn how to differentiate between good and bad writing, pharmacists should read and critically evaluate the published literature. To gain insight about career opportunities, pharmacists can interview a medical communicator[5] or search the Internet; some web sites[3,6] offer links to other useful sites. To develop a writing portfolio, pharmacists should write for publication; an example of a popular target for new writers is a newsletter published by an institution or local professional association. In fact, pharmacists can try medical writing on a part-time basis while maintaining a traditional pharmacy career.[2] Sorkin concluded, ''If you're interested in medical writing, you just have to get into it and do it. You won't know how good you are and if you like it until you do it.''

CONCLUSIONS

Medical communications is an attractive career option for pharmacists because jobs are available and pharmacists are well suited for the work. Employment is attainable in many settings, some of which offer competitive salaries. Sorkin attributed the abundance of career opportunities to the growing demand for medical information because medical communicators are the conduits for that information. For example, the Internet is creating a surge in the demand for direct-to-consumer access to complex medical issues. Thompson predicted, ''There is going to be a greater need for people who can explain medical information to consumers.''

Finally, pharmacists who work as medical communicators generally like their jobs often because the work is intellectually challenging, and medical communications is a natural extension and application of the education and skills characteristic of pharmacists.[2] They have a great advantage compared with many other professionals who want to redirect their careers. As Vélez explained, ''Pharmacists don't have to walk away from their years of specialized education. Pharmacists can bring that expertise right into their new lives as writers, and employers will love them for it!''

REFERENCES

1. *The Pfizer Guide: Pharmacy Career Opportunities*; Merritt Communications, Inc.: Old Saybrook, Connecticut, 1994.
2. Price, K.O. Medical communications. Pharm. Stud. **1992,** Dec. 13–15.
3. Vélez, L.F.; Bicknese, J.; Connelly, P.; Lantz, K.; MacKay, P.; Snyder, A. Biomedical writing. http://www.home.earthlink.net/~rhetrx/bmw/ (accessed June 22, 2000).
4. *American Medical Writers Association Membership Directory*; American Medical Writers Association: Rockville, Maryland, 2000.
5. Barry, D.M. Six months out: The ongoing conversion of a bench scientist turned science writer. AMWA J. **2000**, *15* (2), 22–25.
6. Hamilton, C.W. Resources for medical writers. http://www. hamiltonhouseva.com (accessed September 28, 2000).

Medical Information, Industry-Based

Deena Bernholtz-Goldman
Fujisawa Healthcare, Inc., Deerfield, Illinois, U.S.A.

INTRODUCTION

Medical information departments in the pharmaceutical industry vary greatly from one company to another in size, scope, level of responsibility, and even placement in the company. Typically, these departments can be located under the umbrella of research and development, sales and marketing, or medical affairs. In a survey conducted by Ernst and Young,[1] 75% of the industry' s medical affairs and drug information departments report to clinical (or medical) groups versus sales and marketing. The group may also have several different names such as medical information, professional services, scientific information, or medical communications. For the purposes of this discussion, the designation medical information department is used. It is important to note that the group exists to support the company in its efforts to provide quality pharmaceutical products in respective therapeutic areas.

DISSEMINATION OF DRUG PRODUCT INFORMATION

Guidelines and Limitations

The primary responsibility of any medical information department is the provision of product-specific information to healthcare professionals and/or consumers in response to specific inquiries. If a response to a question involves information outside the Food and Drug Administration (FDA)-approved package insert/labeling, the information is considered to be ''off label,'' and several factors limit the way in which information is communicated to the customer.

The FDA guidance that addresses the dissemination, by pharmaceutical manufacturers, of information on drugs and devices exists in response to the issue of off-label information. This type of information can be provided to a caller from the medical information department; however, it does not serve as a recommendation from the company. The FDA views the support of a medical information department as the exchange of scientific information among healthcare professionals and allows that information to be exchanged only in response to a spontaneous request from a caller. In other words, by providing information that the caller has not requested, promotion is assumed. The FDA Modernization Act of 1997 states that a manufacturer can disseminate information in response to an unsolicited request from a healthcare practitioner.[2] A disclaimer that establishes that the data provided are to be used by the caller only for their scientific information is always stated either verbally or in writing, depending on which is applicable to the given situation. In contrast, sales representatives are not at liberty to discuss off-label topics.

The information disseminated must be scientifically balanced; however, it must also be ''fair balanced.'' This refers to the inclusion of negative as well as positive information. If the indications of a drug are being described, so too must the contraindications be described; if the efficacy is described, so too must the safety of the product.

In general, a medical information staff member can answer an off-label question if the request is of a spontaneous and unsolicited nature, if the response contains both scientific and fair balance, and if an appropriate disclaimer is included. Each company must interpret the FDA guidances and develop policies with respect to the details of how information is disseminated to healthcare professionals. The FDA does not provide any guidelines that delineate how to address consumer inquiries. Each company must again develop policies regarding the dissemination of information to consumers. The majority of companies in the United States limit discussions with consumers, if any occur, to data contained in the product's package insert (PI) or the patient package insert, if available.

Drug Information Resources

Package labeling

In the realm of the pharmaceutical industry, the FDA-approved package insert/labeling is the document used

Encyclopedia of Clinical Pharmacy
DOI: 10.1081/E-ECP 120006344

to answer questions with regards to company recommendations. The PI, as it is known in the United States, is a summary of the new drug application (NDA) provided to the FDA from the manufacturer for the approval of a new product. This summary reflects in-depth discussions between the FDA and the manufacturer, and generally represents certain compromises on the part of both parties. The information within the PI becomes the foundation for approved promotional materials and is used by healthcare professionals as general guidelines for the use of the product, although the entire wealth of information known about any given product is not included in the PI. The PI should be the first resource used to answer an inquiry. Even if additional off-label information on a specific topic is provided, the PI data/recommendations should be described first.

Primary, secondary, and tertiary references

As in university or academic-based drug information centers, published literature used to develop an answer to a drug information inquiry can be divided into primary, secondary, and tertiary sources. Primary sources are the most desirable type of data to use and can be subdivided, depending on the type of data collected. Information gleaned from randomized, prospective, double-blind trials result in data more useful than just one case report on a given topic. Ideally, the most rigorous data available are used; however, case reports and abstracts may be the only available literature published. Peer-reviewed, well-respected journals should be used.

Secondary resources such as pharmacopeias and reference texts can be used for very basic information, but the health professional customer usually has these types of resources available. Data described in a review article should be traced to their original reference if said data are to be used to answer a query.

Tertiary references, such as textbooks, may be used for background information necessary to research a given topic such as a rare disease state.

Data on file

Manufacturers retain considerable data on file that are not found in the public literature. These data can include such information as extended stability studies, compatibility studies, drug safety reports, and comprehensive information in the NDA. The guidelines regarding dissemination of this data are subject to company policies and are typically used when no published literature is available. If the data are used and referenced in any communications with healthcare professionals, the data must be easily retrieved, if required.

Summary

In summary, the development of a well-balanced reply to an inquiry involves the use of published materials in the same manner as that conducted in an academic or clinical setting. In an industry setting, PI information and unpublished data on file with the company are used. In addition, the information communicated must be scientifically accurate, balanced, supported by appropriate literature, and should not include the author's editorial perspectives.

MANAGING THE DAILY WORK FLOW

Customer Base

Medical information staff receive questions from a varied source of customers, including healthcare professionals practicing in clinical/academic settings employed by managed care organizations, pharmacy buying groups, and health maintenance organizations; and consumers. These callers are considered external customers. The mix of these calls will depend on the type of products manufactured by the company. A company whose products are all prescription intravenous drugs will have more calls from health professionals, whereas a company where the majority of products are oral, over-the-counter (OTC) medications will receive more calls from the consumer market. In addition, the type of health professional calling depends on the types of products produced.

The services of the medical information group are also used by ''internal customers,'' that is, other departments within the pharmaceutical company. The group supports internal customers in the same fashion as external ones. The communication with the field sales force is extremely important. It is imperative to impress upon sales representatives the limits of their ability to answer questions. Questions that include information outside the package insert should be referred to the medical information group.

Volume and Type of Requests

The numbers of questions received varies greatly based on several factors. Although companies with more products usually receive more calls, they may also have more mature drugs in their catalog. Drugs that are further along

in the life cycle may generate less interest in the medical community and therefore fewer calls. Along similar lines, the type of inquiry depends on the type of product. Intravenous products used in an acute setting may generate questions of a clinical nature that require more in-depth research than that of a consumer call regarding an OTC product and its proper storage.

Triage Procedures

The efficiency with which requests are handled reflects on the level of service that a medical information department provides. The organization of the medical information group depends on the size of the company and the department, as well as the number of products the department supports. The phone triage may begin with an administrative representative who can refer the call to the appropriate healthcare professional. In larger groups, there may be a two-tiered system whereby a trained nonhealth care professional service representative may answer standard questions with scripted answers. If a question cannot be appropriately answered with the information available to them, the call is referred to the next tier of health care professionals, typically pharmacists, nurses, or physicians. Calls may also be received via the Internet through a company's web site, by fax, or by mail. Procedures for the efficient routing of these inquiries must also be in place.

Third-party companies who specialize in the provision of medical information may also be used. This outsourcing of responsibilities may be done in anticipation of a large increase in call volume (e.g., the launch of a new product) or may be used by companies who choose not to maintain a call center as part of the organization. In both cases, there are usually medical information staff employed by the pharmaceutical companies who develop and maintain a library of standard information (or scripts) used by the first tier of company representatives or by third-party organizations to answer queries. After-hours coverage may be provided through a pager or an answering service.

Documentation

Regardless of the type of product, every call must be documented. Documentation serves several purposes, including to store data for report generation; to create a database that can be repeatedly used to answer questions; and to meet legal needs in case of litigation. Most industry-based medical information departments maintain a library of ''standard answers'' that consist of replies to frequently asked questions. Commercially available electronic databases are available specifically for this setting. These types of databases enable the medical information representative to document all calls; store, generate, and maintain standard replies; and generate reports as needed. Systems of this nature can be customized to a degree for the particular needs of the company. Alternatively, similar systems can be developed by a company's own information systems department.

INTERNAL SUPPORT FUNCTIONS

Services provided by any given medical information group vary greatly. In general, however, the group is a service provider for the company at large and those services can be employed by all departments including, but not limited to, the sales and marketing, regulatory affairs, and clinical research departments.

Sales

The medical information group can support the sales group in many ways. Invariably, the sales organization is the most frequent user of the medical information department, with respect to internal customers. The sales organization relies on the medical information department to answer or clarify any questions they have regarding the products they sell and related issues. They also refer questions from their customers to the medical information department who in turn communicates with the customers. Additional support may include product training and educational newsletters to equip the sales force with new information that has been published regarding the products they sell or competitors' products. More often than not, this information is off label and serves to educate the sales force and prevent them from hearing news (positive or negative) about their products from their customers. Although they cannot discuss this information with the customer, they will still be aware that the medical information group has developed a reply to queries regarding new published data.

Marketing

The marketing department works not only on the ''dollars and cents'' of selling products, but also on how to best position products in the marketplace. In the development of advertising campaigns and key messages, the medical information department can be very helpful in ensuring the accuracy of the data and the fair balance with which the data are used.

Regulatory Affairs

Safety reporting

All pharmaceutical manufactures have a group devoted to safety issues and the reporting of adverse events to the FDA according to specific regulations. However, the report of an adverse event (AE) may initiate within the medical information group. A caller may contact the company to inquire about the incidence of a specific AE or to ascertain how to treat a specific event. The caller may have no intention of reporting an AE; however, it may become clear from the conversation that an event has actually occurred. If four specific pieces of information exist—an identifiable patient, identifiable reporter, an identifiable drug, and an identifiable event—the company is obligated to report the AE. Because many of the regulations are time sensitive, medical information groups must work hand in hand with safety groups to assure that AE reporting is done in a timely fashion. Some pharmaceutical companies combine the medical information and safety departments, typically in the setting of a small company.

Quality assurance

As part of the regulatory affairs department, the quality assurance groups often call upon the medical information department to assist in ensuring that data used in any formal documentation that they review (from NDA submissions to changes in the package insert to new advertising campaigns) are accurate and appropriately used.

Clinical Research

Clinical research teams work on investigational trials from the time a product reaches phase II clinical development through FDA approval. Medical information staff use their expertise to assist in protocol development, medical writing, and researching drug or disease state questions that may arise in the course of these trials.

COMMITTEE INVOLVEMENT

As in any organization, pharmaceutical manufacturers develop committees to make decisions, initiate new programs, and maintain specific procedures. Many of these committees include representation from many different groups in the organization, each contributing their expertise in a given area. The following is meant to exemplify several types of committees that may have medical information representation, but is not meant to be all inclusive.

Labeling and Promotional Review Committee

All the information included in a product's package labeling is under careful scrutiny by the FDA. The development of the insert is usually done with the input of clinical research, sales and marketing, medical information, regulatory affairs, legal, pharmaceutical sciences, and other appropriate groups, depending on the nature of the product. All these departments are represented on this type of committee. The drafts of the initial PI are reviewed, edited, and agreed upon by the members of the Labeling and Promotional Review Committee prior to submission to the FDA. Even after final development and FDA approval of the initial PI, any change in the labeling (from storage changes to new indications) must be reviewed by the Labeling and Promotional Review Committee. All advertisements and promotional materials must be reviewed by this committee. A representative from the medical information group well versed in the literature for a given product serves as a valuable contributor to this committee, lending an objective view of the information presented.

Publications Committee

When research projects sponsored by pharmaceutical companies are complete, investigators often write reports of the results and submit them for presentation and/or publication. The writing of these reports is often completed with the assistance of professionals from the medical information department and other involved departments in conjunction with the investigators. In addition, most companies have a publications committee to map out where and when reports should be submitted.

Global Medical Information

The pharmaceutical industry has become a global business. Companies have headquarters in a specific part of the globe, with subsidiaries in many other locations. It becomes important to communicate with medical information colleagues in those subsidiaries to ensure that consistent messages are being communicated globally regarding the same products. Information must always be

tailored for the specific location because government-approved indications and labeling for the same products may vary by country.

EDUCATION

Because a medical information group is considered to be a support service for the company and its customers, a logical extension is to provide education for those customers.

Continuing Education (CE)

Decreasing budgets and staffing have made it increasingly difficult for health professionals to regularly attend large off-site meetings. Therefore, the provision of high-quality CE at local presentations or as home study programs is a well-received service that a medical information department can initiate and develop. Guidelines set by the FDA and the individual accreditation organizations for each type of professional CE vary greatly. Accreditation must be received by each institution separately and care must be taken to abide by all respective guidelines. In general, educational pieces are given as live presentations or self-study programs authored by outside experts that should not be influenced by the company.

Employee Education

As described previously, the medical information department can provide learning tools to enable the sales organization to remain current on their product line. These communications can be made available to all employees to educate them on the products they support.

Academic Rotation Site

As opportunities for pharmacists in nontraditional roles continue to grow, the medical information department of a pharmaceutical company provides a unique student clerkship site. Pharmacy students experience a different medical information practice while learning the organizational makeup of a corporation versus an academic, hospital, or retail setting. In addition, students can be exposed to many opportunities afforded to pharmacists in departments outside the medical information departments.

CONCLUSION

The primary objective of an industry-based medical information department is to provide accurate, well-balanced, and timely responses to inquiries regarding a company's products. However, the department can be involved in every aspect of the life cycle of a pharmaceutical product. The level of that involvement depends on the size and organization of the company, as well as the initiative and creativity of the medical information staff.

REFERENCES

1. Hopkins, F.; Galligher, C.; Levine, A. Pharmaceutical medical affairs and drug information practices. Drug Inf. J. **1999**, *33*, 69–85.
2. Food and Drug Administration Modernization Act of 1997 sec 401 [FFDCA sec 557; 21 USC 360aaa-6].

Medicare Benefits and Improvement Act of 2000 for Outpatient Immunosuppressive Agents

Marie A. Chisholm
University of Georgia College of Pharmacy, Athens, Georgia, U.S.A.

INTRODUCTION

Immunosuppressive therapy is life long, or is needed for as long as the individual has a functional transplanted organ, and the cost of therapy often exceeds $10,000 annually. Since the mid-1970s, the question of how to help patients pay for immunosuppressive therapy has received congressional attention. The Medicare Benefits Improvement and Protection Act of 2000 (BIPA 2000) addresses Medicare time limitation for outpatient immunosuppressive coverage.

COVERAGE UNDER MEDICARE

Outpatient immunosuppressive medications—most commonly, azathioprine, cyclosporine, mycophenolate, prednisone, sirolimus, and tacrolimus—are covered by Medicare Part B. Currently, medicare pays for 80% of immunosuppressive drug therapy. The other 20% is the patient's responsibility. To qualify for Medicare immunosuppressive coverage, the patient must have the following:

1. Medicare benefits due to *age, disability, or end-stage renal disease (ESRD).*
2. Medicare part B.
3. Received a solid organ transplant that was medicare "covered" at the time of transplant.
4. Surgery performed at a medicare-approved transplant facility.
5. A viable transplanted organ.

Prior to January 2000, medicare provided outpatient immunosuppressive therapy for 36 months posttransplant to patients who were Medicare eligible due to age or disability and met criteria in items 2–5 listed previously. As of January 2000, eligible beneficiaries whose Medicare coverage for immunosuppressants expired during the year 2000 received an additional 8 months of Medicare coverage beyond the 36-month period. In December 2000, President Bill Clinton signed the Medicare BIPA 2000, which eliminated the time limitation for Medicare benefits for immunosuppressive drugs for patients who receive Medicare benefits due to age or disability. A program memorandum from the Health Care Financing Administration stated:

> Effective with immunosuppressive drugs furnished on or after December 21, 2000, there is no longer any time limit for Medicare benefits. This PM supersedes PM AB-99-98, which describes the method for determining the former time limit for this benefit that applied to drugs furnished prior to December 21, 2000. This policy applies to all Medicare entitled beneficiaries who meet all of the other program requirements for coverage under this benefit. Therefore, for example, currently entitled beneficiaries who had been receiving benefits for immunosuppressive drugs in the past, but who immunosuppressive drug benefit was terminated solely because of the time limit described in PM AB-99-98, would now resume receiving that benefit for immunosuppressive drugs furnished on or after December 21, 2000.[1]

Therefore, BIPA 2000 expanded Medicare benefits to lifetime coverage, or for as long as the patient needs immunosuppressive therapy, for patients who have Medicare due to age or disability. The lifetime coverage benefits of BIPA 2000 do not include patients who have Medicare based solely on ESRD. Currently, those patients who have Medicare due to ESRD only and meet criteria in items 2–4 will receive immunosuppressive coverage for 36 months posttransplant.

REFERENCE

1. *Department of Health and Human Services (DHHS) Health Care Financing Administration. January 24, 2001*, HCFA-Pub. 60AB.

Encyclopedia of Clinical Pharmacy
DOI: 10.1081/E-ECP 120006415

Medication Assistance Programs, Pharmaceutical Company-Sponsored

Marie A. Chisholm
University of Georgia College of Pharmacy, Athens, Georgia, U.S.A.

INTRODUCTION

Pharmaceutical company-sponsored medication assistance programs are a philanthropic effort promoted by the pharmaceutical companies to provide medications free or at reduced prices to eligible patients who cannot afford prescription medications or have no other means of obtaining them.[1–3] Although more than 2.7 million prescriptions costing over $500 million were filled through medication assistance programs in 1999, every prescription medication cannot be obtained through medication assistance programs. The existence of an assistance program is determined by the manufacturer of the medication.[3]

MEDICATION ASSISTANCE PROGRAMS

Pharmaceutical companies offer many different types of medication assistance programs to help qualifying patients obtain prescription medications. One type of assistance program includes providing free or reduced prices for medications to patients who financially qualify. For eligible patients, the medication may be received by mail, from a local pharmacy, or from the prescribing physician. Another type of assistance program involves helping to determine the patient's healthcare coverage, identifying billing claims for medications, and attempting to resolve denied medication claims. Some manufacturers offer payment limitation programs, where the companies set an annual drug cost limit, and when costs exceed the limit, they are absorbed by the manufacturer. For all of these programs, eligibility criteria are established by each company's assistance program and may require that the patient have no prescription coverage or limited prescription coverage, that the patient has reached a certain medication cost, and/or has limited income and assets to afford the medication.

The references cited in this text provide a listing of some medications that can be obtained through medication assistance programs. For the most current information concerning pharmaceutical company-sponsored medication assistance programs, contact the manufacturer of the medication. Additional sources of information include Web sites (Table 1), publications including the *Reimbursement Assistance Programs Sponsored by the Pharmaceutical Industry* and the *Directory of Prescription Drug Patient Assistance Programs* published by the Pharmaceutical Research and Manufacturers of America, and healthcare professionals such as social workers and pharmacists.[3]

Enrollment

Patient enrollment in these programs is usually initiated by the patient's physician, although enrollment may occur by the patient, a patient advocate, or other healthcare professionals. Information frequently needed to process the application is listed in Table 2. Regardless of who initiates enrollment, the prescribing physician has to participate in the process.

Companies have their own processes for patient enrollment that typically require the completion of an application and a reporting of the patient's health insurance coverage, assets, income, and liabilities. A prescription for the medication is usually required. Many programs allow enrollment over the phone, while others require written correspondence. In order to determine eligibility, many companies require verification of a patient's financial status by documents such as income tax returns, W-2 forms, social security and other benefit

Table 1 Internet sites providing information concerning medication assistance programs

www.themedicineprogram.com
www.needymeds.com
www.phrma.org/patients
www.familyvillage.wisc.edu

Encyclopedia of Clinical Pharmacy
DOI: 10.1081/E-ECP 120006326

Table 2 Information frequently requested by medication assistance programs

- Patient and physician's names and contact information
- Physician's state license number and drug enforcement number
- Name, strength, and amount of medication requested (prescription)
- Patient's income and other financial disclosures
- Patient's insurance coverage
- Patient and physician signatures

reports, and asset statements. Once the information is received by the company, it is then evaluated and the patient's acceptability is determined.

Medication

Once enrolled in the program, the medication may be provided free, or patients may be required to pay a co-payment or shipment charge. Medications are supplied by direct delivery to the patient or physician, or the patient may be issued a benefit card or voucher to be presented at a pharmacy. The amount of medication supplied and the enrollment period varies according to the program.

Pharmacists' Role

Pharmacists can play an important role in this process. Pharmacists can inform patients and other healthcare professionals about the availability of pharmaceutical company medication assistance programs, initiate the enrollment process, and develop strategies that use medication assistance programs to reduce institutional financial losses.[4,5]

ETHICAL ISSUES

Two ethical issues concerning pharmaceutical company-sponsored medication assistance programs include drug diversion and the reporting of false information to obtain medications. Drug diversion and the reporting of false information may lead pharmaceutical manufacturers to consider ending the assistance program or creating more stringent eligibility criteria. Therefore, it is important to ensure that patients report accurate information during the enrollment process and that the intended party receives the medication. By adequately screening and monitoring information, healthcare professionals will be able to ensure that patients in need receive the intended benefits of these programs.

CONCLUSION

Many patients cannot afford to purchase needed prescription medication. Approximately 16% of the U.S. population does not have any health insurance, and a greater percentage has health insurance without prescription medication benefits.[6] Pharmaceutical company-sponsored medication assistance programs provide valuable options to both patients and healthcare providers by providing drugs to patients who have limited means of purchasing their medications.

REFERENCES

1. Chisholm, M.A.; Reinhardt, B.O.; Vollenweider, L.J.; Kendrick, B.D.; DiPiro, J.T. Medication assistance programs for uninsured and indigent patients. Am. J. Health-Syst. Pharm. **2000**, *57*, 1131–1136.
2. Needymeds. www.needymeds.com (accessed 2000 Dec. 29).
3. Pharmaceutical Research and Manufacturers of America. *Facts and Figures and Publications*; www.phrma.org/facts and www.phrma.org/patients (accessed 1999 Oct. 1).
4. Decane, B.E.; Chapman, J. Program for procurement of drugs for indigent patients. Am. J. Hosp.-Syst. Pharm. **1994**, *51*, 669–671.
5. Hotchkiss, B.D.; Pearson, C.; Listano, R. Pharmacy coordination of an indigent care program in a psychiatric facility. Am. J. Health-Syst. Pharm. **1998**, *55*, 1293–1296.
6. Bureau of the Census. Current Population Reports. In *Health Insurance Coverage, 1998*; Government Printing Office: Washington, DC, 1999. Available at http://www.census.gov/prod/99pubs/p60-208.pdf.

Medication Errors and Adverse Drug Events Prevention

Vicki S. Crane
Parkland Health and Hospital System, Dallas, Texas, U.S.A.

INTRODUCTION

Around 1995, the U.S. public began to become much more aware of the risks associated with medications as there was a series of highly publicized medication errors and adverse drug events (ADEs), such as the fatal Betsey Lehman chemotherapy overdose.[1] In November 1999, the Institute of Medicine (IOM) Report concluded that medication errors and ADEs were in the top 10 causes of American deaths. The report detailed many strategies to reduce medication errors and ADEs, and called upon healthcare organizations and professionals to give a higher priority to safe medication systems.[2] In interviews with hospital patients, a survey found the most common fear of patients was being give two or more medications that interacted to produce a negative outcome.[3]

Health care providers and organizations have accelerated the search for ways to make the medication-use process safer for patients. Numerous prevention strategies have been developed.[2,4–6] Although well intended, some of the strategies have introduced different complexities and workload into a medication-use process that had all ready demonstrated its fragility. More effective methods for decreasing medication errors and associated ADEs have to be designed, implemented, and evaluated post-implementation. This article describes how changes in the medication-use cycle can significantly reduce medication errors and preventable ADEs. It is not intended to replace comprehensive references already available,[2,4–6] but to introduce key ideas into the body of knowledge about medication error and ADE prevention.

THE MEDICATION-USE CYCLE

A medication error happens when there is general agreement that a person should have done other than what they did and so the person is labeled as having committed a medication error.[7] An ADE occurs when patient harm occurs as a result of a medication error (e.g., wrong medication given) or a patient idiosyncrasy (e.g., a previously unidentified allergy to penicillin). Preventable ADEs usually result from medication errors, but not all medication errors produce ADEs.

Fig. 1 lists the typical steps and personnel involved in the medication-use cycle in the acute care setting. In the ambulatory setting, the patient or family instead of the nurse would be receiving, administering, and initially monitoring the effects of the medication. The typical acute or ambulatory care medication cycle contains 10–15 independent steps in the process. If the personnel and systems operating within the cycle perform at 99% accuracy, this means that in a 10 to 15-step process, there is a 10–14% error rate. Based on the literature, the Advisory Board Company projects that medication errors and ADEs extract a terrible toll. In an average 400-bed hospital with 15,500 admissions per year and a 6.5% ADE rate, there will be 575 patients per year with significant ADES, 302 patients with serious ADEs, 120 patients with life-threatening ADEs, and 10 patients with fatal ADEs as well as additional costs of $2,318,400 and additional lengths of stay of 2218 days per year.[8]

Any successful reform strategy to decrease medication errors and ADEs must begin with understanding the sources of these medication errors and ADEs within the medication-use cycle as presented in Fig. 1. There are four discrete activities in the medication-use cycle that need reform—physician prescribing, order processing, drug delivery, and reporting and event capture. Forty nine percent of medication errors occur at the ordering stage. Twenty five percent of medication errors occur at the medication processing stage. Twenty six percent of medication errors occur at the medication administration. The largest percentage of preventable ADEs occurs at the physician prescribing stage. Fortunately, the largest percentage of intercepted ADEs also occur at the physician prescribing stage because the physician has the pharmacist, nurse, and patient who can potentially intercept their errors. The lowest percentage of intercepted ADEs occurs at the administration stage of the process because

DOI: 10.1081/E-ECP 120006327

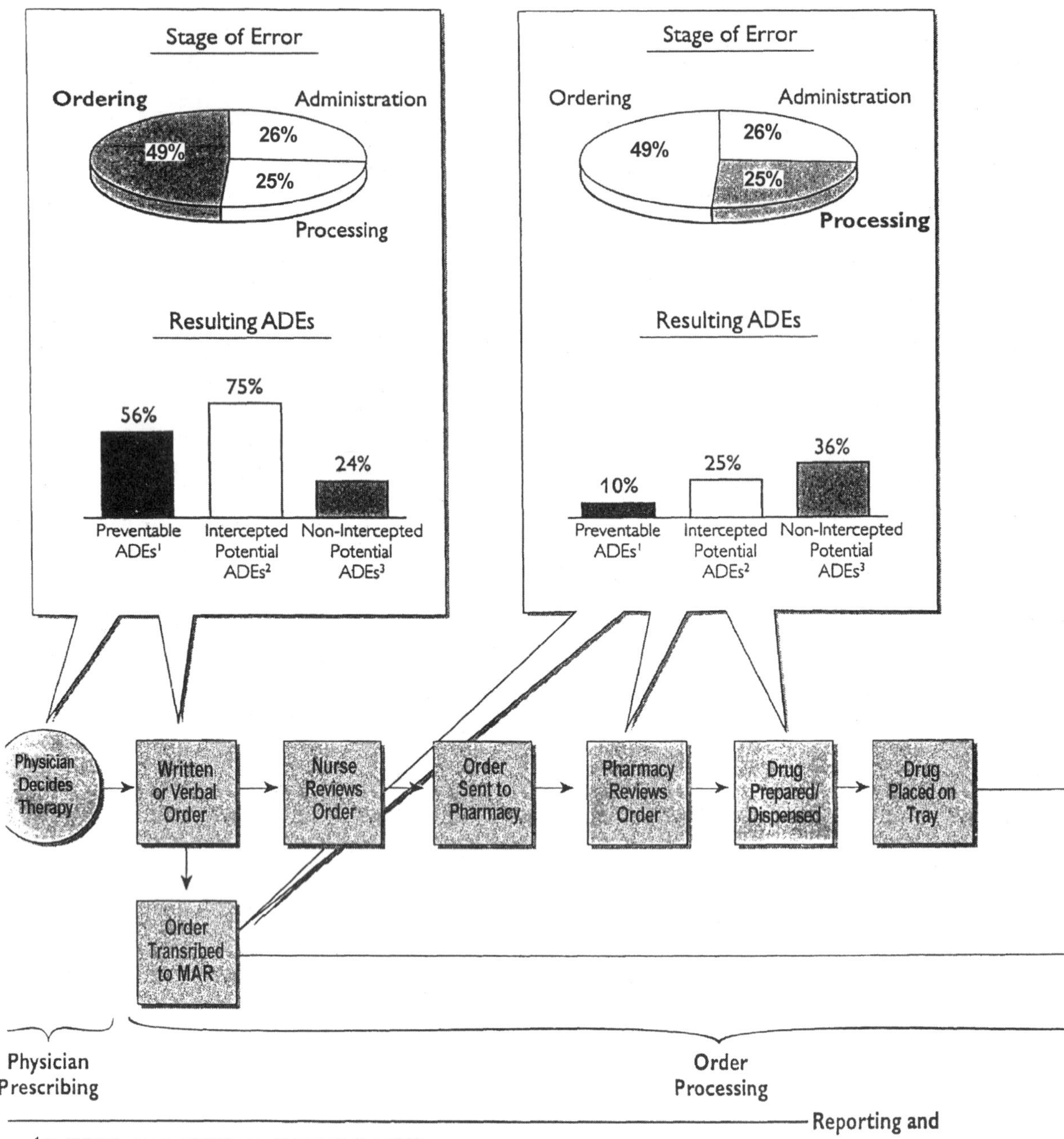

[1] An ADE that was preventable by any means currently available.

[2] A medication-related incident with the potential for injury that was intercepted before it reached the patient.

[3] A medication-related incident with the potential for injury that reached the patient and did not result in an ADE.

Fig. 1 Incidence of ADEs lends insight to reform strategy. (© 1999 The Advisory Board Company.)

the nurse (or the ambulatory patient by analogy) has no other health care professional to check the final step and possibly catch errors.

Within the medication cycle, there are two basic sources of medication errors and ADEs: systems and humans. It is essential to understand how systems and humans interact to create or prevent medication errors and ADEs and to identify and intervene at potential fail points. Historically, health care organizations have relied heavily on the individual accountability approach

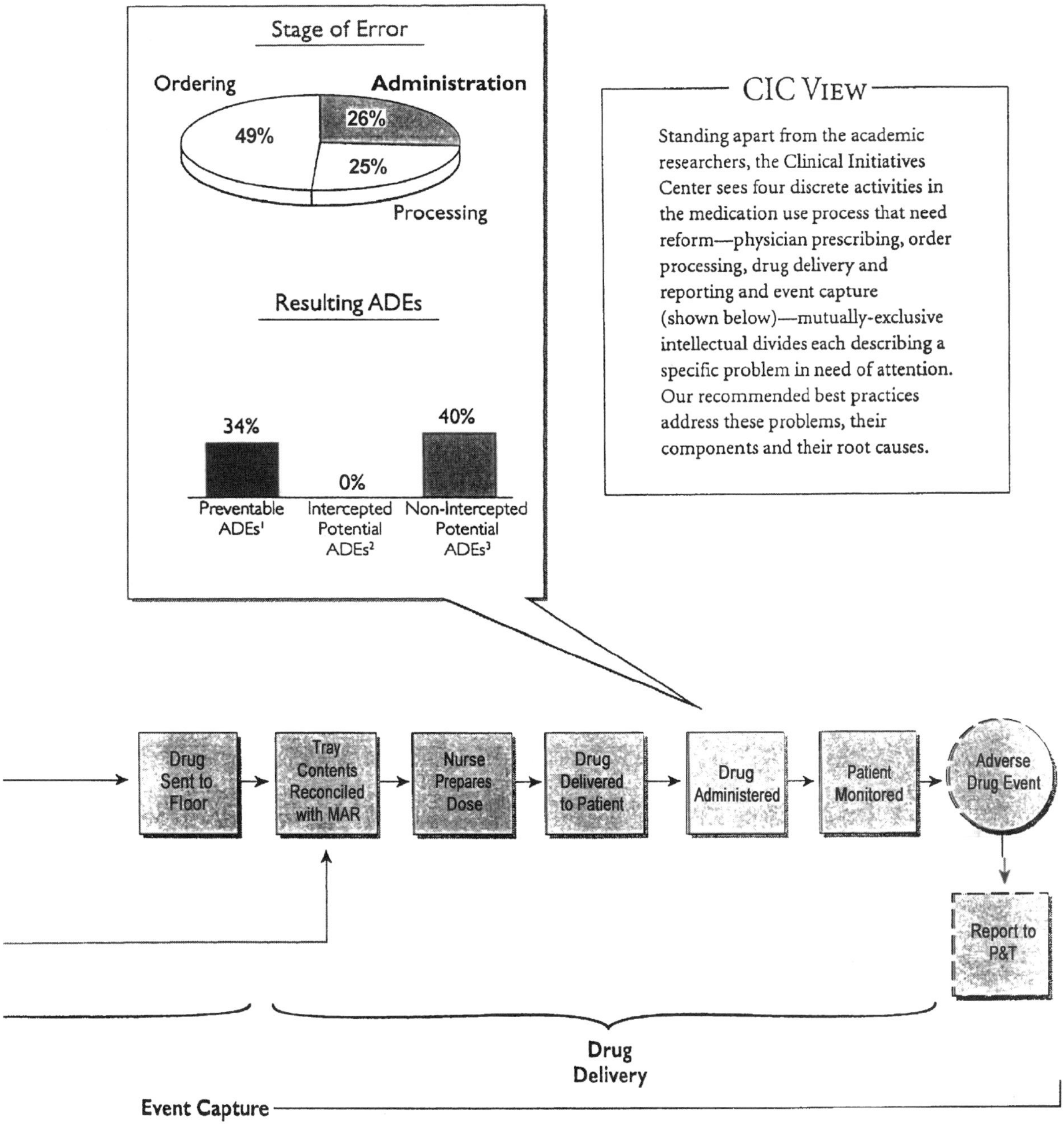

Stage of Error

Ordering 49%

Administration 26%

Processing 25%

Resulting ADEs

Preventable ADEs[1]	Intercepted Potential ADEs[2]	Non-Intercepted Potential ADEs[3]
34%	0%	40%

CIC VIEW

Standing apart from the academic researchers, the Clinical Initiatives Center sees four discrete activities in the medication use process that need reform—physician prescribing, order processing, drug delivery and reporting and event capture (shown below)—mutually-exclusive intellectual divides each describing a specific problem in need of attention. Our recommended best practices address these problems, their components and their root causes.

Drug Sent to Floor → Tray Contents Reconciled with MAR → Nurse Prepares Dose → Drug Delivered to Patient → Drug Administered → Patient Monitored → Adverse Drug Event → Report to P&T

Drug Delivery

Event Capture

Fig. 1 Incidence of ADEs lends insight to reform strategy. (© 1999 The Advisory Board Company.) (*Continued*).

with elaborate systems for scoring the severity of medication errors and threats of severe punishment. The individual accountability approach focuses on the individual who generated the error or ADE as being fully responsible. The systems' approach focuses on the system failures that set up individuals to generate errors. What is needed is an approach that appropriately integrates individual accountability with systems' enhancements on a continuum that provides reliable, safe care to each patient. The recommended best practices within this article address these problems, their components, and their root causes, as well as point out strategies for identifying and preventing medication errors and ADEs.

SCIENTIFIC INVESTIGATION OF PREVENTABLE PATIENT HARM

Data about preventable medication errors from organizational experience and the literature should be used to improve medication-use systems to enhance patient safety.[9] Each health care organization has a different culture, resources, and priorities, and is at a different evolutionary stage in reaching ''best practices.'' Rather than proposing a rigid, stepwise process for investigating preventable patient harm, readers are encouraged to investigate patient harm in new ways appropriate to their own practice environments. There are no easy solutions, and there is no consensus about how best to prevent medication errors and ADEs. Two inductive approaches to prevention that offer a broader scientific, investigation of patient safety are the process-improvement approach and the outcomes-measurement approach. Other similar techniques, such as root-cause analysis and failure-mode and effects analysis have been discussed elsewhere.[2,4]

The Process-Improvement Approach

The process-improvement approach to system safety evolved from an expert workshop and continuing dialogue on patient safety sponsored by the National Patient Safety Foundation (NPSF).[10] This line of thinking says that study of the ultimate (rather than just the proximate) source of errors and ADEs is needed. Reason[11] first developed the idea that harmful patient outcomes are the result of latent, small degradations within a system com-

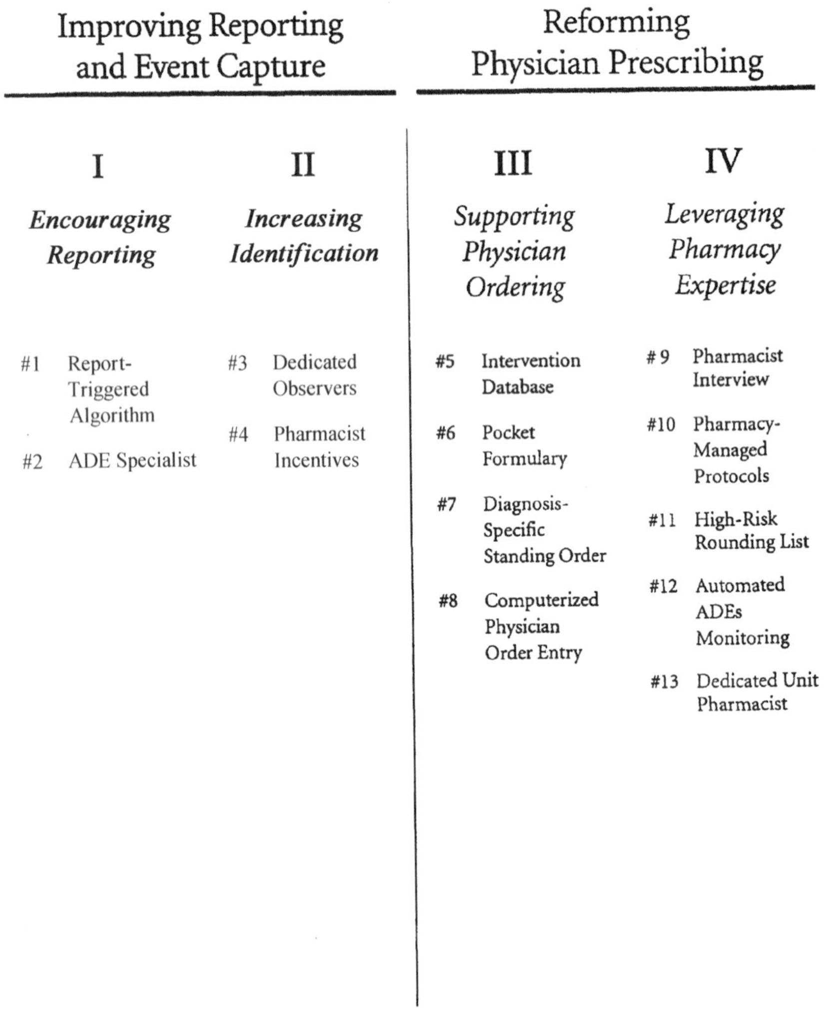

Improving Reporting and Event Capture		Reforming Physician Prescribing	
I	II	III	IV
Encouraging Reporting	*Increasing Identification*	*Supporting Physician Ordering*	*Leveraging Pharmacy Expertise*
#1 Report-Triggered Algorithm	#3 Dedicated Observers	#5 Intervention Database	#9 Pharmacist Interview
#2 ADE Specialist	#4 Pharmacist Incentives	#6 Pocket Formulary	#10 Pharmacy-Managed Protocols
		#7 Diagnosis-Specific Standing Order	#11 High-Risk Rounding List
		#8 Computerized Physician Order Entry	#12 Automated ADEs Monitoring
			#13 Dedicated Unit Pharmacist

Fig. 2 Identifying and preventing ADEs. (© 1999 The Advisory Board Company.)

bining at some point to cause a harmful patient outcome. The harmful outcome was long in coming and was a natural result of the way the system was designed and operated. According to the process-improvement approach, medication-use is a complex system with many potential points of failure. Each latent point of failure within the system is in itself insufficient to cause an accident (e.g., the sterile-products pharmacist catches an error made by a technician in preparing an intravenous admixture). The pattern of these potential failures is dynamic and depends on the individual elements of the medication-use process.

There is a "sharp end" and a "blunt end" to the medication-use process; both may harbor latent points of failure. The sharp end of the system is where the health care providers are. They interact directly with the medication-use process and are directly exposed to its hazards (e.g., heavy workload, drug supply shortages, paperwork associated with new Medicare compliance regulations). The blunt end of the system is where regulations, new drug technologies, and reimbursement policies are structured, and where many of the competing demands within the system are created. For example, there is a constant competing demand not only to produce care very quickly but also to have it be error free. Despite this competing demand, safe operations are the rule and harmful outcomes are infrequent.

When a visible failure occurs, people focus on evaluating the performance of health care providers. Because the negative outcome is known, this results in a bias toward hindsight in which the events leading to the negative outcome are seen as obvious, even though they may not

Reforming Order Processing		Reforming Drug Delivery	
V	VI	VII	VIII
Writing Orders Accurately	*Dispensing Drugs Precisely*	*Reinforcing Nursing Staff*	*Supplementing Nursing Staff*
#14 Zero-Tolerance Ordering Standards	#17 "HAZMAT" Dispensing Protocols	#19 Drug Administration Tests	#22 Dedicated Medication Personnel
#15 Preprinted Order Forms	#18 Automated Dispensing Systems	#20 Dosing Crib Sheets	
#16 Pharmacist Order Entry		#21 Bar Code Reconciliation	

Fig. 2 Identifying and preventing ADEs. (© 1999 The Advisory Board Company.) (*Continued*).

have seemed obvious to the provider at the time of the adverse drug event. Hindsight bias is the greatest obstacle to making personnel performance more robust and resilient in providing patient safety. Retelling of these ''celebrated accident'' or ''first stories'' reinforces the view that the provider acted alone in creating the error.

The immediate reaction to celebrated accidents in many health care organizations is to blame the provider and institute new training, rules, technology, and disciplinary actions so that this type of error will never reoccur. Investigation of the harmful outcome then stops. Many of these new changes are not tested thoroughly, add complexity to the already fragile system, and introduce new types of failures; thus, the cycle of harmful outcomes merely repeats itself at different points in the medication-use process. In the case of medication use, the classic celebrated accident occurs when a potentially lethal drug is misused (e.g., a patient receive a dose 10 times too high) and the patient dies as a result. What typically happens next is that restrictions are placed on the use of the offending drug and on the provider's related functions. Unfortunately, these restrictions can keep patients who may benefit from the drug from receiving it in a timely and accurate manner, and can prompt the provider to focus on getting through cumbersome mechanics rather than paying attention to the drug's safe use. Restrictions should be thought of in the context of the entire medication-use process, not just one drug. The thought process should be expanded to include safe handling of other potentially lethal drugs, not just the one that harmed a patient.

Because each step in the medication-use cycle can be associated with preventable medication errors and ADEs, systems must be assessed and enhanced, in their entire scope and each component. Fig. 2 lists 22 examples of strategies for identifying and preventing medication errors and adverse drug events by reforming various components of the medication-use cycle. In addition, in their standards for medication use, the Joint Commission of Accreditation of Healthcare Organizations (JCAHO) recognizes that the multidisciplinary teams must be responsible for all components of the medication-use cycle. Keys to improving reporting and event capture usually result when reporting is nonpunitive or easy to do, or when dedicated resources such as an ADE specialist or observers are in place or incentives are used.[12] Although reporting and event capture can be improved, they will never be perfect. A related and important strategy is to proactively learn from the experience of other organizations. Comparisons can be made with other organizations through such reporting mechanisms as the Institute for Safe Medication Practices (ISMP) Newsletter, FDA Alerts, USP's MedMARx national medications errors database, JCAHO's Sentinel Event Alerts (high-alert medications), and the media. Regardless of whether there is medication error or ADE specific to a health care practice environment or not, if the medication error or ADE is happening elsewhere, it could likely happen at the individual practice site. Examples of preemptive strikes to protect patients from harm include instituting protocols for use, restrictions on the duration or route of therapy, goal setting, education, and feedback procedures, and sometimes even turning down requests for a drug to be added to the formulary.

Because prescribing errors have been identified as a leading cause of preventable medication errors and ADEs, reforming physician prescribing should also be given a high priority. Although computerized physician order entry with clinical decision support would appear ideal, many organizations may not have the information systems infrastructure or funds to accomplish such a goal in the near future. It is generally recognized that leveraging pharmacy expertise is in many cases the most cost-effective way to support physicians in reforming prescribing through a variety of initiatives. Examples of leveraging pharmacy expertise include dedicated unit pharmacists, pharmacy admissions interviews, pharmacy-managed protocols, and pharmacist-led ambulatory care clinics.[2,12,13]

Thus, the process-improvement approach to the safety of the medication-use cycle goes beyond the celebrated cases and first stories to scientifically investigate the system as a whole. Data on near-misses and uncelebrated errors should be analyzed to find hidden flaws and strengths, and to better understand the dynamics of our medication-use system. Scientific investigation of the whole cycle—peeling away the layers of the onion—will reveal latent points of failure and facilitate a redesign that substantially reduces the occurrence of harmful outcomes.[11]

The Outcomes-Measurement Approach

The second scientific approach to prevention focuses on negative patient outcomes (ADEs) rather than on process (medication errors). The outcomes-measurement approach pioneered by Classen[14] views the process-improvement angle as flawed because it does not distinguish between medication errors and ADEs. Although dissecting medication errors may improve patient outcomes, only 1% of all medication errors result in ADEs and 50% of ADEs can be prevented.[14] The extensive studies of the drug-use process and of error prevention have often been done at the expense of demonstrating ways of identifying and preventing ADEs that produce actual patient harm. Improving the system (preventing medication errors) does not necessarily improve patient outcomes or reduce

costs. The outcomes-measurement approach advocates the rigorous study of the epidemiology of actual ADEs, combined with continuous quality improvement techniques and nonpunitive communication to achieve a true impact on patient safety through improved outcomes.

Specifically, the outcomes-measurement approach asserts that rigorous tracking of actual ADEs, coupled with an epidemiologic investigation of their sources, is the foundation for improving patient outcomes through prevention, research, and education. Computerized surveillance systems have already shown great potential to prevent ADEs through such mechanisms as alerts about drug allergies, drug interactions, and inappropriate dosages. Many ADEs result from a lack of integrated information at the point of clinical decision making rather than from carelessness or lack of knowledge.

Measurement of the outcomes of ADEs cannot wait until the health care organization has an electronic medication record or a comprehensive, computerized disease-management surveillance system. An ADE outcomes-prevention project should be selected and implemented. Good tools certainly make measurement faster and easier, but in their absence there are still many outcomes-measurement strategies that can be successfully used to prevent ADEs. Various researchers have studied the causes of ADEs by tracking them to their root causes, even though this was done through an entirely manual system of chart reviews and drug usage evaluations.[13] Common findings of major causes of preventable ADEs were related to dosage errors (e.g., with renally excreted drugs), drug allergies, and initial use of new drugs. Pharmacists can use many drugs with potentially severe adverse effects safely with appropriate monitoring, intervention, and education. Typical examples from the literature or an organization's own ADE-tracking or medical records system might include analyzing the use of antimicrobials in association with ADEs, admissions due to ADEs or visits to ambulatory clinics due to ADEs.[15–19]

Systematic and timely feedback are keys to generating cooperation so that ADEs in progress do not become more severe and can be prevented in the future. Feedback should be communicated in a nonpunitive way to encourage cooperation with and enthusiasm for a program of research, education, and prevention.[20]

Summary

In the process-improvement or outcomes-measurement approaches, medication-use system redesign should 1) identify and assess lynchpin safety components of the medication-use cycle as well as improving event capture and reporting, 2) develop a medication system safety plan through a multidisciplinary team approach with the support from executive management, 3) set priorities and timelines for corrective actions in accordance with the organization's strategic planning and budgeting processes taking into account the cost effectiveness and feasibility of the strategy for the organization, and 4) work through local constraints or risks to provide a safer medication-use system for patients. The process-improvement approach may be more appropriate in a health care practice environment where systems are clearly not providing the safety net they should be. In other organizations, a focus on preventing negative outcomes may provide the best investment in patient safety. Both strategies have validity because processes and outcomes overlap.

ENHANCING HUMAN RELIABILITY

The objective of this section is to introduce readers to the concepts of human reliability in the complex medication-use cycle. It is not intended to replace comprehensive treatments of the discipline of human factors research already available but rather to provide a starting point for readers to view human reliability and specifically human errors in the medication-use cycle from a different and more dynamic perspective.[6,7,11] The study of human reliability gives insights into how health care providers operate under competing demands and time limits in caring for patients. There are many factors affecting human reliability. Some factors move human reliability toward 100%, while others drive it away[7] (Fig. 3). Human reliability never reaches 100% all the time but knowing what influences human behavior to become more reliable can be used to develop added strategies to

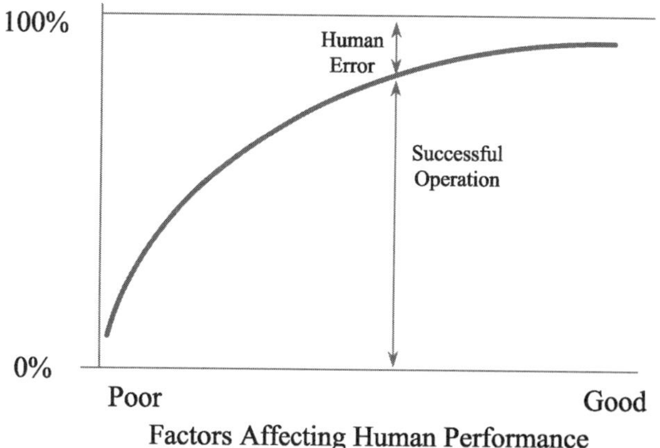

Fig. 3 Human reliability curve. (From Ref. [7].)

Imperfect Behavior	At-Risk Behavior	High-Culpability Behavior
Product of Current System Design	*Unintentional Risk Taking*	*Intentional Risk Taking*
Manage through nonpunitive changes in • Processes • Procedures • Training • Design • Environment	Manage through non punitive changes in •Understanding our at-risk behaviors •Removing incentives for at-risk behaviors •Creating incentives for healthy behavior •Increasing situational awareness	Manage through • Disciplinary (punitive) action

Fig. 4 Three manageable behaviors. (From Ref. [7].)

reduce medication errors and ADEs. Three types of human behavior influence human reliability and thus can be used to expand the approaches to reducing medication errors and associated ADEs[7] (Fig. 4).

Imperfect Behavior

The first aspect of human reliability focuses on situations where humans can not produce 100% reliability all the time. Although we cannot change our nature as humans, we can change the medication-use system in which we work. Imperfect behavior is the product of systems design. The classic example is physician handwritten medication orders that are often illegible or incomplete. This is a system designed to produce errors. It is natural or normal for errors to occur with this ''as-designed'' system. For the average physician, pharmacist, nurse, or patient, quality and safety fail when the medication-use system fails. The approach to decreasing imperfect behavior and increasing human reliability is to identify and change those system factors that negatively affect human reliability. These system factors fall into four categories: people, environment, actions, resources. Human system analysis includes such factors as knowledge, skills, physical capability, adaptability, and fallibility. Environmental system analysis includes such factors as temperature, noise, light, peer communication, facility layout and location, and management/labor relations. Action systems analysis include factors such as task design, technology and information management, work measurement and methods improvement, and policies and procedures. Resource system analysis includes such factors as tools, equipment, support from others, and time. In exhibiting imperfect behavior, health care providers are trying to do their best, are not engaging in risk-taking behavior, and are in a situation where their human reliability or performance is shaped by the system in which they operate. Thus, solutions to imperfect behavior decreasing human reliability are not disciplinary or punitive but are systems analysis and redesign oriented.

At-Risk Behavior

The second aspect of human reliability focuses on situations where humans unintentionally engage in risk taking. At-risk behavior means that intentional conduct unintentionally results in creating a significant and unjustifiable risk that may cause an undesirable outcome. The classic example of this is the need for the health care provider to provide patient care simultaneously, both quickly and accurately under conditions of competing demands and time limits. There are three at-risk cognitive factors affecting human reliability (both expertise and error) within the medication-use cycle: knowledge factors, attentional dynamics, and strategic factors.[21]

Knowledge factors are the types of knowledge that are available to solve problems as they arise in the medi-

cation-use process. It is normal for health care providers to adopt a limited number of strategies to deal with processes or routine variations within the medication-use process. However, when an unusual or surprising situation presents itself, the relevant knowledge may be inadequate, imprecise, or too simplified. For example, risks of new drug therapy are not always perfectly predictable, because drugs are tested in a small subset of the general population. Known adverse effects do not occur in all patients given the new drug. In many cases, it is easier to see the benefits the patient could receive from the drug than to determine the risks associated with the drug.

Attentional dynamics are how humans control and manage the mental workload as circumstances develop and change over time. The mind can pay attention to only so many situations at a time, so attention shifts as events occur. If a situation comes to light that does not agree with what is expected, attention usually shifts to investigate. An ICU patient often receives numerous intravenous medications. If the wrong one is infused (e.g., a vasopressor), the patient would receive an incorrect drug and suffer the consequences. The responding health care providers might not immediately discern that the wrong drug was infusing. The providers would probably focus their attention on whatever seemed necessary to address the patient's adverse reaction (e.g., lowering the blood pressure) even though the source of the event was not known. Humans are great at recovery after an event, even when their lack of attention might have been the source of the problem.

Strategic factors are how humans deal with situations when conflicting goals are present and the best course of action is not readily apparent (because, for example, of limited resources and the need to act immediately). The health care provider must decide whether to act immediately with inadequate information, to wait for more information to be become available, or to consider other alternatives. This may result in the misapplication of usually good guidelines. In hindsight (after all the strategic factors are known), the decision made as the situation evolves often appear to be reckless. If a pharmacist receives five critical orders simultaneously, then he/she must decide which one to fill first or to call for help. The pharmacist must decide whether to dispense the drugs immediately or first check a suspiciously high dose in relationship to other variables when the values are not immediately available. If the pharmacist dispenses the suspiciously high doses and the patient's outcome is positive, the dispensing will be reviewed as uneventful even though the safety margin was reduced. However, if the outcome is negative, the pharmacist may be seen as reckless and having made an error.

Thus, attributes of at-risk behavior are judged against a specific outcome, judged risky by an external observer, involve a choice between different behaviors, are not understood to be of significant or unjustifiable risk by the health care provider taking the risk, and are generally considered more blameworthy than system-induced or imperfect behavior errors.

The approach to decreasing at-risk behavior and inreasing human reliability is to change incentives so those safe behaviors are chosen over at-risk behaviors. In health care systems where mediation-use systems already have appropriate checks and balances, changing at-risk behaviors represents the largest opportunity for safety and quality gains since medication errors and ADEs are often side effects of weak cognitive functions. The analysis and corrective actions should include 1) creating a tally sheet of incentives for comparison of safe behaviors with at-risk behaviors for a specific function or area in the medication-use cycle so that safe behavior can be compared with at-risk behavior, 2) rate each incentive as strong or weak, and 3) modify incentives to change behavior. It is essential to recognize that the strength of safe or at-risk behaviors is influenced by the following factors: 1) certain consequences are stronger than uncertain consequences, 2) consequences are stronger than policies and procedures or other environmental factors, and 3) immediate consequences are stronger than delayed consequences. In exhibiting at-risk behavior, the health care provider's conduct is intentional but they do not understand that the conduct itself significantly increases the risk of a negative consequence, and is a situation where their human reliability or performance is shaped by expediency and at-risk incentives rather than safe practices and potential consequences. The health care provider may be negligent in that they should have been aware of a substantial and unjustifiable risk and under the law may be required to provide compensation. However, for the health care organization, the solutions to at-risk behavior decreasing human reliability are not disciplinary or punitive but are incentive oriented. Punitive sanctions are an obstacle to other initiatives such as event reporting.[7] Correct solutions to at-risk behaviors include such things as enhanced management–front line worker communication and direction, increased health care team training versus just technical or professional training, and consistent errors management awareness.

High-Culpability Behavior

The third aspect of human reliability focuses on situations where humans intentionally engage in risk taking. High-culpability behavior means that the person consciously disregarded the fact that their conduct would significantly and unjustifiably increase the risk that negative consequences would occur. The risk must be of such a nature and degree that, considering the circumstances known to the healthcare provider, its disregard involves a gross deviation from the standard of care that a reasonable employee would observe in his situation. The approach to decreasing high-culpability behavior is disciplinary or punitive because high-culpability behavior must be changed or the person cannot continue to function as a healthcare provider. High-culpability behavior is the single behavior where punitive sanctions should be considered as a solution.[7] Punitive strategies may include such things as progressive disciplinary actions, time off, punitive training or duties, and even termination.

In summary, human nature can not be changed, but the changing the circumstances under which it operates in the medication-use process can enhance human reliability. Traditional responses to a medication error and resultant patient harm include discipline, new rules, retraining, technology, or environmental changes. These responses have a place but they address symptoms rather than trying to understand human reliability. Unfortunately, most of the time traditional responses have assumed that human recklessness or high-culpability behavior was the root cause of the error when it was not. Traditional approaches have sought to regiment, rather than enhance human reliability. Preventable medication errors and ADEs happen because of reliance on weak human cognition and reliability. Strategies that seek to minimize human error (e.g., adding additional procedures and steps) can often make the system more prone to medication errors and ADEs at other points, even though the intention was exactly the opposite. Strategies designed to enhance human reliability have a higher probability of making the medication-use cycle safer for patients. Much effort has been put into improving processes and outcomes in the medication-use system, but human reliability has been largely ignored. Successful preventive strategies must take into account human abilities and limitations, and will use these strategies to correct for the limitations, and to enhance the abilities.

THE ROLE OF THE PHARMACIST

The IOM Report has given health care a clear mandate to make the medication-use cycle safer and has given pharmacists a clear opportunity to shine. Pharmacists must have an unwavering commitment to the priority of preventing medication errors and ADEs. Now is a significant time to be in the pharmacy profession and pharmacists should act the part. The following practice management guidelines are presented as a practical approach by which pharmacists can shape their own practice environments to decrease medication errors and ADEs and increase patient safety. Medication quality only begins when patient safety begins. Therefore, medication safety strategies should be an essential part of evidence-based medicine practices.

Practice management guideline #1: Make medication safety improvement a pharmacy leadership priority. Every recent treatise on error management recognizes leveraging pharmacy expertise as an essential key in preventing medication errors and ADEs. A multidisciplinary medication safety team and a 5-year medication safety plan should be part of every pharmacy department or health care team's strategic plan. As the drug therapy experts, pharmacists should be in leadership positions to create medication cultures of safety.

Practice management guideline #2: Stop using the medication error reporting system primarily for punishment and shift the emphasis to using reporting for improvement. Instead, help to create an environment in which fear may be alleviated in large part by creating an open and safe environment for detecting, reporting, and investigating how errors and ADEs occur so they can be reduced in the future. Extend the collegial peer review process of morbidity and mortality conferences to the context of improving medication safety. There has been a great deal of erosion of trust, and it is exceedingly difficult to have a high-performance organization without having a high-trust organization.

Practice management guideline #3: Initiate and maintain ongoing self-assessments of the safety of individual clinical practices as well as with the health care team, the pharmacy department, and the organization. Use the learning experiences of others and self to proactively implement known "best practices" locally to prevent medication errors and to develop your own safety audit tools. Be dissatisfied with the status quo and seek to develop an error management vision and then take the practical first steps to implement it.

Practice management guideline #4: Be accountable for medication safety by designing, implementing, and maintaining safe practices in the medication-use cycle. This might include such things as educating patients and their families regarding appropriate use of their medications, using protocols for the use of potentially lethal or

narrow therapeutic index drugs, insisting that pharmacy (not nursing) prepare all intravenous admixtures, and implementing technology, automation, and information support systems that enhance the frontline pharmacist's ability to care for patients.

Practice management guideline #5: Integrate medication safety understanding into all pharmacist, technician, resident, student training continuing education, and evaluations. Ensure that everyone understands that medication safety is everyone's business and everyone has a role in keeping patients safe. If the messenger aide does not understand the importance of delivering the needed intravenous admixture to the right patient care area on time, everything that the clinical specialist did in customizing the drug therapy and the staff pharmacist in reviewing the medication profile and preparing the order can become academic.

Practice management guideline #6: Realize that competing priorities and severe financial constraints are the environment in which health care organizations operate today. If the organization needs to improve revenues or margins by a certain percentage, can the pharmacist quantify how medication error and ADE prevention contribute to the financial goal? Further, can the pharmacist demonstrate why the health care organization should shift limited resources to medication error management over some other worthy and cost-effective goal? Demonstration of a positive impact on patient outcomes and health care costs is a prerequisite to getting needed resources for medication safety, but it is not a guarantee. Financial constraints and competing priorities are perhaps the most daunting challenge, but they are also simultaneously the greatest opportunity.

Practice management guideline #7: Make medication safety research a priority. How can technology and information management be deployed to make the medication-use cycle safer and position pharmacists to better care for patients? How will gene therapy affect the demand for traditional drug therapies, and will pharmacists be in the forefront of safely managing gene therapy? Intuitively, medication error and ADE prevention should result in better financial, clinical, and humanistic outcomes, but added meaningful outcomes measurement and research are needed to validate that business costs rise significantly with poor quality and safety. If pharmacists do not make medication safety research a priority, it is likely that this role will be filled by other health care providers, regulatory agencies, or even sophisticated information systems.

The urgent need to enhance the safety of drug therapy shows that the medication-use cycle must change. Pharmacists as the drug therapy experts are positioned to shape their practice environments to focus attention on a critical set of priorities to enhance the medication-use cycle to better serve patient welfare and safety.

CONCLUSION

Much of the knowledge developed about medication errors and ADEs has depended on the ability of individual pharmacists to detect problems and take an active part in resolving them. The increasing use of complex medication regimens has drawn attention to the number of iatrogenic medication errors and ADEs, as well as their associated costs. Pharmacists must work to reduce predisposing factors so that safety can be enhanced and costs reduced. A new practice model as an adjunct to evidence-based medicine practices must be created to prevent medication errors and ADEs, and to let others outside the pharmacy know that we are ready to lend our expertise and energy to this critical endeavor.

REFERENCES

1. Voelker, R. 'Treat systems, not errors,' experts say. JAMA, J. Am. Med. Assoc. **1996**, *276*, 1537–1538.
2. *To Err is Human: Building a Safer Health System*; Kohn, L.T., Corrigan, J.M., Donaldson, M.S., Eds.; National Academy Press: Washington, DC, 1999.
3. Medication errors rank as a top patient worry in hospitals, health systems. ASHP Press Release. September 7, 1999.
4. *Medication Errors*; Cohen, M.R., Ed.; American Pharmaceutical Association: Washington, DC, 1999.
5. *Error Reduction in Health Care: A Systems Approach to Improving Patient Safety*; Spath, P.L., Ed.; Jossey-Bass Publishers: San Francisco, 2000.
6. *Human Error in Medicine*; Bogner, M.S., Ed.; Lawrence Erlbaum: Hillsdale, New Jersey, 1994.
7. Marx, D. *Managing Human Error in High-Risk Industries Seminar. Houston, Texas*; December 1999.
8. The Advisory Board Company. *The Year 2000 Engagements*; The Advisory Board Company: Washington, DC, 2000.
9. Runkle, D.C. The Scientific Investigation of Avoidable Patient Injury: Two Approaches. In *Proceedings of Enhancing Patient Safety and Reducing Errors in Health Care*; Schettler, A.L., Zipperer, L.A., Eds.; National Patient Safety Foundation: Chicago, 1998; 51–52.
10. Cook, R.I.; Woods, D.D.; Miller, C. *A Tale of Two Stories: Contrasting Views of Patient Safety*; National Patient Safety Foundation: Chicago, 1998.

11. Reason, J. *Human Error*; Cambridge Univ. Press: New York, 1990.
12. The Advisory Board Company. *Prescription for Change Seminar*; The Advisory Board Company: Washington, DC, September 30–October 1, 1999; Dallas, Texas.
13. Crane, V.S. New perspectives on preventing medication errors and adverse drug events. Am. J. Health-Syst. Pharm. **2000**, *57*, 690–697.
14. Classen, D.C. Adverse Drug Events and Medication Errors: The Scientific Perspective. In *Proceedings of Enhancing Patient Safety and Reducing Errors in Health Care*; National Patient Safety Foundation: Chicago, 1998; 56–60.
15. Classen, D.C.; Evans, R.S.; Pestonik, S.L.; Horn, S.D.; Menlove, R.L.; Burke, J.P. The timing of prophylactic administration of antibiotics and the risk of surgical wound infection. N. Engl. J. Med. **1992**, *326*, 281–286.
16. Dennehy, C.E.; Kishi, D.T.; Louis, C. Drug-related illness in emergency department patients. Am. J. Health-Syst. Pharm. **1996**, *53*, 1422–1426.
17. Prince, B.S.; Goetz, C.M.; Rhin, T.L.; Olsky, M. Drug-related emergency department visits and hospital admissions. Am. J. Hosp. Pharm. **1992**, *49*, 1696–1700.
18. Aparasu, R.R.; Helgeland, D.L. Visits to hospital outpatient departments in the United States due to adverse effects of medications. Hosp. Pharm. **2000**, *35*, 825–831.
19. Leape, L.L.; Cullen, D.V.; Clapp, M.D.; Burdick, E.; Demonaco, H.J.; Erickson, J.I.; Bates, D.W. Pharmacist participation on physician rounds and adverse drug events in the intensive care unit. JAMA, J. Am. Med. Assoc. **1999**, *282*, 267–270.
20. Clemmer, T.P.; Spuhler, V.J.; Berwick, D.M.; Nolan, T.W. Cooperation: The foundation of improvement. Ann. Intern. Med. **1998**, *128* (12, part 1), 1004–1009.
21. Cook, R.I.; Woods, D.D. Operating at the sharp end: The complexity of human error. In *Human Error in Medicine*; Bogner, M.S., Ed.; Erlbaum: Hillsdale, New Jersey, 1994; 258–287

Medication Use Evaluation

Marjorie A Shaw Phillips
Medical College of Georgia Hospitals and Clinics, Augusta, and
University of Georgia College of Pharmacy, Athens, Georgia, U.S.A.

INTRODUCTION

A Medication Use Evaluation (MUE), or Drug Use Evaluation (DUE) program is a planned, criteria-based systematic process for monitoring, evaluating, and continually improving medication use, with the ultimate aim of improving medication-related outcomes for a group of patients or consumers. The improvement process that is MUE has application in all settings where pharmaceutical care is provided. Its focus is on assessing and improving one or more of the steps involved in medication use (patient assessment, prescribing, preparation and dispensing, administration, patient monitoring for medication safety and efficacy, and patient education). Practitioners in different settings may use a variety of terms to describe interventions that fall under the lexicon of MUE, including drug utilization review, drug usage review, drug use review, drug use evaluation, and drug utilization evaluation. MUE overlaps with the provision of pharmaceutical care when a pharmacist uses criteria, therapeutic guidelines or clinical pathways to identify, resolve, and prevent medication-related problems and guide the care for an individual patient.

DEFINITIONS

An audit (or study) is a single evaluation (data collection and assessment) focusing on a particular medication, drug class, disease state, or process. Typically, information is collected, organized, analyzed, and reported to measure current practice. Changes (''corrective actions'') are then made, as needed, to address identified concerns and improve care. Individual audits are sometimes erroneously referred to as ''an MUE'' or ''a DUE.'' Because a one-time project will have minimal impact, audits are only one small component of a comprehensive MUE program.[1]

Criteria are used to measure the quality of medication use. Criteria have been defined as ''predetermined elements against which aspects of the quality of a medical service may be compared.''[2] Criteria reflect the standard of practice and, therefore, should be current, literature-based, and objective (explicit) rather than subjective (implicit).

Drug regimen review (DRR) is the concurrent review of an individual patient's medication regimen at regular intervals, designed to identify and resolve potential medication-related problems. This pharmaceutical care process is typically guided by the pharmacist's personal judgement, rather than explicit criteria. Monthly drug regimen review by a pharmacist is required for patients in U.S. nursing homes. Reviewing the appropriateness of a new prescription at the time of dispensing (in the context of the patient's overall medication profile) is a DRR activity. This type of drug regimen review is referred to in the Federal Omnibus Budget Reconciliation Act of 1990 (OBRA '90) regulations[3] and certain state pharmacy practice acts as prospective drug use review (DUR). When the pharmacist uses organizationally sanctioned criteria to guide individual patient evaluations and collects data on the results, this care becomes part of the MUE process.

Drug utilization review or drug-use review (DUR) has been defined as ''a formal program for assessing data on drug use against explicit, prospective standards, and, as necessary, introducing remedial strategies to achieve some desired end.''[4] Traditionally, many DUR audits have been retrospective in approach, such as computerized evaluations of large databases maintained by insurers or pharmacy benefit organizations. While most experts agree there is little difference among the processes when implemented fully, a number of professional groups now prefer to use the term ''Medication Use Evaluation''[5] or ''Drug Use Evaluation,''[6] rather than DUR.

Indicators provide a ''quantitative measure of an aspect of patient care that can be used in monitoring, evaluating, and improving the quality and appropriateness of healthcare delivery.''[7] An indicator may serve as a screen or ''red flag'' to identify a potential problem (postoperative infection rate, number of serious medication errors) or measure progress toward an established goal (percent of patients with atrial fibrillation who are anticoagulated).

DOI: 10.1081/E-ECP 120006323

GOALS OF MUE

Within the overall goal of improving medication-related outcomes, an MUE program can have a number of more specific aims. These include improving the efficiency and effectiveness of medication use, promoting medication safety, increasing patient satisfaction, supporting a formulary system, and controlling treatment costs while optimizing care.

ROLE OF PHARMACIST AND OTHERS

Because the medication-use process involves complex steps and interactions among a number of healthcare professionals and caregivers, an MUE program must be the responsibility of the entire organization and an interdisciplinary approach must be used, involving physicians and other prescribers, nurses, pharmacists, administrators, and (ideally) patients or consumers. Other healthcare personnel should contribute their unique perspectives and skills when the evaluation and improvement process addresses their area of responsibility. Pharmacists, by virtue of their expertise and their mission of ensuring optimal patient outcomes,[8] should exert leadership and work collaboratively with other members of the health care team in the ongoing process of medication use improvement through MUE. Having physician champions for the improvement process can make or break the success of an MUE program.[9,10]

SCOPE

An MUE program must be comprehensive, with the goal of impacting all the patients served by an organization or institution. The target population, therefore, could be all the patients at a single ambulatory pharmacy or hospital, within a healthcare system, the ''covered lives'' managed by a health maintenance organization or insurer, or enrollees in a government program such as Medicaid. To benefit the overall population, one initial step should be to define the types of patients (ages, most prevalent disease states, medication utilization, and unmet needs) and settings (acute care, ambulatory, long-term care, home care, etc.) under the purview of an MUE program.

PRIORITY SETTING

Once the scope of an MUE program has been determined, priorities should be set for evaluation and improvement activities. This simply involves identifying where the greatest positive impact can be achieved by allocating resources, either to correct identified problems or to identify and improve less-than-optimal practices. To make the best use of existing resources, an MUE program may be limited to three to five major initiatives at one time or could encompass a large number of concurrent initiatives. Selected topics could vary widely based on the characteristics of the organization and patients involved.

Some common ways of setting priorities are to focus on medications, drug classes, disease states, or processes that are:

1. *High risk*: Medications, procedures, processes, and disease states for which there is a greater chance for adverse outcomes (e.g., chemotherapy administration, intravenous potassium use, intrathecal injections, anticoagulation, medications with a narrow therapeutic index, and care of immunocompromised patients).
2. *High use/high volume*: Those medications used most frequently, as well as the most common procedures and disease states. Incremental improvements made in these areas will impact the greatest number of patients (e.g., antibiotic use, pain management, acetaminophen, community-acquired pneumonia, otitis media, and patient counseling).
3. *High cost*: High-cost medications and procedures are often targeted for intervention, because wasted resources divert funds from more beneficial patient care and have a negative financial impact on the organization.
4. *Associated with high risk patient groups*: These include infants and children, the elderly, and patients with organ dysfunction or cancer. By targeting populations that are more prone to adverse effects or treatment problems, a number of important issues or medications can be addressed at the same time.
5. *''Problem-prone''*: These are areas identified either by local surveillance or based on medical literature findings. An organization may label an issue ''problem-prone'' if outcomes are worse than national benchmarks or standards (e.g., a postoperative infection rate of 5%) or prior efforts have not reached their goals. Thought leaders and national publications frequently highlight issues that are expected to be problematic in most organizations (such as antimicrobial resistance, overuse of vancomycin, or inadequate pain management).

It is helpful to get organization-wide consensus on the priorities for medication use improvement. Issues that fall into more than one category, for example, frequent and high risk or problem-prone, should be addressed or evaluated first.

STEPS IN THE MUE PROCESS

In its simplest form, the MUE process is an ongoing improvement cycle with no set beginning or end. For each identified high-priority issue, there are four major action steps:

1. *Identify or determine the standard of care or optimal use*: Therapeutic guidelines and consensus documents and standards established by national professional organizations or governmental groups often define ideal processes and outcome goals. Healthcare organizations may adopt national guidelines, or adapt them for local use. An evidence-based medicine approach can also be used to develop guidelines or clinical pathways to meet local needs. It is important to gain consensus from practitioners who will utilize guidelines as part of the implementation process. Ideally, guidelines and protocols should be shared across an organization to encourage implementation before current practices are evaluated.
2. *Data collection (compare actual to optimal)*: This step is the ''evaluation,'' where performance is measured against objective criteria for the processes or outcomes of care. Criteria may relate to indications (for selection of a particular medication or procedure), processes (drug dosing and administration, patient assessment and monitoring) or outcome measures (cure, relief of symptoms, treatment failure, adverse drug event, patient satisfaction).
3. *Intervene*: If indicated, corrective actions should be taken to improve processes, practices, or medication use. Possible actions include the development of consensus statements or guidelines when they do not already exist, providing feedback on performance, education of health professionals (seminars, newsletters, feedback letters, one-on-one communication), creation of tools to support use of guidelines and treatment protocols (such as standard order sets, computerized pathways, pocket cards), redesigned processes, or patient education.
4. *Evaluate the impact of the intervention*: Additional data collection or ongoing monitoring to ensure corrective actions had the desired impact (document improvement) and to guide subsequent incremental steps if further improvement is needed.

To be most effective, a majority of MUE program resources need to be devoted to education and process improvement, rather than measurement and simple data collection. An additional important component is regular (at least annual) evaluation of the overall MUE program to assess its value and improve the MUE process.

APPROACHES TO EVALUATION

Assessments of medication use can be performed through a concurrent, retrospective, or prospective study design (or a combination of these approaches). Each type has advantages and disadvantages.

Retrospective review is conducted after the patient has received a medication, and the treatment is complete. Medical charts or computerized records are screened to identify patterns and trends. This may be the simplest type to perform, because large amounts of data can be reviewed quickly, and the outcomes are known. Information gained from large retrospective reviews is helpful for problem identification and initiating actions that will benefit future patients. However, inaccurate or incomplete documentation could limit the usefulness of information collected.

Concurrent review is performed during the course of treatment and often involves ongoing monitoring of drug therapy. Concurrent evaluation gives pharmacists the opportunity to intervene if potential problems are detected, and thus, current patients can benefit. This type of review also allows additional information to be collected if documentation is incomplete (such as querying prescribers, observing behavior, or interviewing patients). The concurrent approach can be integrated with case management or ongoing pharmacist monitoring activities. It may be more time-consuming if data is collected on individual patients over time. Furthermore, if outcome data is required, it may be months or years before outcomes are known and data collection can be completed.

Prospective review takes place before medication use, and thus, its usual focus is on patient assessment and prescribing. This approach to evaluation can detect and resolve potential problems in individual patients before they occur. Pharmacists routinely perform prospective reviews when assessing new medication orders for possible drug interactions, accurate dosing, or duplicate therapy. Data collection occurs at only one point in time and can easily be integrated with daily pharmacist care activities. However, while data collection for a single patient may not be very time-consuming, it can take a long time for data to be collected on enough patients to draw conclusions and guide improvement efforts.

TOOLS AND RESOURCES

The professional literature, national treatment guidelines, and locally developed protocols provide key resources for the development of MUE criteria. Various professional groups as well as private businesses also produce medication-specific, disease-specific, or process-oriented criteria that can be adapted for local use.[11–13]

Performance measurement systems, such as the HEDIS® (Health Plan Employer Data and Information Set) managed by the National Committee for Quality Assurance (NCQA)[14] and the ORYX Initiative of the Joint Commission on Accreditation of Healthcare Organizations (JCAHO)[15] provide indicators for identifying concerns and monitoring the impact of improvement initiatives. In Australia, the government is developing a searchable database of major quality use of medications (QUM) initiatives as a means of sharing resources and ideas.[16]

Computerized databases (such as insurance billing files or electronic medical records) greatly improve the ability to gather and evaluate information compared to manual data collection (e.g., chart reviews). Computer software programs, including proprietary software designed specifically for MUE functions, may be helpful in managing data and reporting.

VALUE OF MUE

The key to MUE is not the ''drug use review'' or ''evaluation'' but the improvement process that is guided by the information learned about medication use. The MUE program can be an important tool for gaining professional consensus and focusing organizational energy on activities that will improve medication-related outcomes. An MUE program can be used to integrate a number of important medication improvement initiatives, such as the formulary system, adverse drug reaction reporting, medication error prevention, pharmacist interventions, and other clinical pharmacy services.

LIMITATIONS AND PITFALLS OF MUE

According to critics, the way DUE was practiced in many institutions might better be described as ''Don't Use Expensively'' or ''Don't Use Ever,''[17] because the focus became punitive actions against individual practitioners with a focus on cost control rather than quality improvement. To have legitimacy within the organization, the emphasis must be on actions that truly improve outcomes for patients. The MUE program must have both the administrative authority to promote positive change and support (''buy-in'') from practitioners at every level. One important consideration is to avoid using words that imply control or a loss of professional autonomy, because these can be threatening to practicing physicians. Rather than labeling an action ''appropriate'' (with the connotation of condescending approval or disapproval), state that it is ''consistent with the published literature'' or ''meets Medical Staff-approved criteria''; use ''evaluation'' rather than ''enforcement'' and ''consult'' or ''inform'' instead of planning an ''intervention.''[18]

Because evidence-based medicine is an evolving discipline, and the state of knowledge is constantly changing, guidelines must be frequently updated. Indicators, criteria, and even process improvements need to be reassessed to ensure they remain valid.

Often, an MUE program commits to too many initiatives, resulting in insufficient time and energy to complete any of the projects. Follow-through is essential when corrective actions are implemented-the impact must be assessed and adjustments made if needed. If an MUE program becomes authoritarian, bureaucratic, emphasizes analysis of data rather than action, or loses sight of its mission to improve care, it will be ineffective. MUE must be seen as a tool or means to foster improved medication use processes and safer, more effective medication use, rather than an end in itself. Achieving and maintaining improvement requires not only streamlining systems and processes, but also changing the behavior of healthcare practitioners and patients—a difficult proposition.

REFERENCES

1. Stolar, M.H. Drug-use review: Operational definitions. Am. J. Hosp. Pharm. **1978**, *35* (1), 76–78.
2. Todd, M.W. Drug Use Evaluation. In *Handbook of Institutional Pharmacy Practice,* 3rd Ed.; Brown, T.R., Ed.; American Society of Hospital Pharmacists: Bethesda, Maryland, 1992; 261–271.
3. Medicaid program: Drug use review program and electronic claims management system for outpatient drug claims. Fed. Regist. **1992**, *57*, 49397–49412 (Nov. 2).
4. Principles of Drug Utilization Review (DUR) adopted by American Medical Association, American Pharmaceutical Association, and Pharmaceutical Manufacturers Asso-

ciation, 1991. (based on a quotation from Rucker TD. Drug-utilization review: Moving toward an effective and safe model. In *Society and Medication: Conflicting Signals for Prescribers and Patients*; Morgan, J.P., Kagan, D.C., Eds.; Lexington Books: Lexington, MA, 1983; 25–51).

5. American Society of Health-System Pharmacists. ASHP guidelines on medication-use evaluation. Am. J. Health-Syst. Pharm. **1996**, *53* (8), 1953–1955, May also be accessed at http://www.ashp.org/bestpractices/formulary/guide/medication.pdf (accessed October 2000).
6. The Academy of Managed Care Pharmacy. Drug Use Evaluation. In *Concepts in Managed Care Pharmacy Approved by the AMCP Board of Directors*; June 12, 1999 (accessed at http://www.amcp.org/professional_res/concepts/index_concepts.asp, July, 2002).
7. Nadzam, D.M. Development of medication-use indicators by the Joint Commission on Accreditation of Healthcare Organizations. Am. J. Hosp. Pharm. **1991**, *48* (9), 1925–1930.
8. Angaran, D.M. Quality assurance to quality improvement: Measuring and monitoring pharmaceutical care. Am. J. Hosp. Pharm. **1991**, *48* (9), 1901–1907.
9. Dzierba, S. Limited resource approach to drug-use review. Top Hosp. Pharm. Manage. **1991**, *11* (1), 87–91.
10. Phillips, M.S. Collaborating with physicians in the drug-usage evaluation process. Top Hosp. Pharm. Manage. **1991**, *11* (1), 38–45.
11. *Criteria for Drug Use Evaluation*; American Society of Hospital Pharmacists: Bethesda, Maryland, 1993; Vol. 4.
12. *APhA Guide to Drug Treatment Protocols: A Resource for Creating and Using Disease-Specific Pathways*; Manolakis, P.G., Ed.; American Pharmaceutical Association: Washington, DC, 1996.
13. *Drug Regimen Review: A Process Guide for Pharmacists,* 3rd Ed.; American Society of Consultant Pharmacists: Alexandria, Virginia, 1998.
14. http://www.ncqa.org and http://www.ncqa.org/Pages/Communications/News/somcrel00.htm (assessed Oct 2000).
15. http://www.jcaho.org/oryx_frm.html (assessed Oct. 2000).
16. http://www.qummap.health.gov.au (assessed Oct. 2000).
17. Enright, S.M.; Flagstad, M.S. Quality and outcome: Pharmacy's professional imperative. Am. J. Hosp. Pharm. **1991**, *48* (9), 1908–1911.
18. Teagarden, J.R. A-E-I and you. Three words you can't use in drug usage evaluation programs. Hosp. Pharm. **1991**, *26* (7), 590, 615.

Medication Use for Unapproved Indications

Kimberly J. La Pointe
Children's Hospital of The King's Daughters, Norfolk, Virginia, U.S.A.

INTRODUCTION

Prescribing drugs for "unapproved" uses is common. The American Medical Association (AMA) estimates that 40%–60% of all prescriptions in the United States are written for drugs being used for something other than their approved purpose.[1] The term *unapproved uses* actually has no legal meaning. An unapproved use merely indicates a lack of Food and Drug Administration (FDA) approval and does not imply an improper or illegal use. Unlabeled use is a more appropriate term and is defined as the use of a drug product in doses, patient populations, routes of administration, or for indications that are not included in FDA-approved product labeling.[2]

With each new prescription or physician order, the pharmacist must decide if the chosen therapy is appropriate for the patient and if there is evidence to support the use of the prescribed drug for the specific indication. How does the pharmacist make this decision when the drug has been prescribed for an unlabeled indication? What is the pharmacist's liability? Will this affect reimbursement?

To address these questions, a Medline search and literature review was performed using the terms unapproved, nonapproved, not approved, unlabeled, off-label, unindicated, and not indicated. A large number of articles, many of which centered around changes in legislation, were found written by or for physicians. Only a handful of articles specifically addressed unlabeled medication use from a pharmacist perspective. This article begins with a brief review of contributing factors to unlabeled medication use and then highlights its frequent occurrence. The remaining sections focus on patient safety and product efficacy, pharmacist liability, and reimbursement issues for unlabeled medication uses. For each section, information found in the literature review is summarized, followed by implications for clinical pharmacy practice.

BACKGROUND AND FREQUENCY

The FDA approves drugs based on large, randomized, well-designed studies that show a product is safe and effective for a specific indication. Once a product is approved and marketed, many other uses can emerge from postmarketing experience. The manufacturer may apply for FDA approval of these additional uses if the patient population is large enough to support the time and expense of more clinical trials. Regardless of labeling, however, the Food, Drug, and Cosmetic Act does not limit the manner in which a physician may use an approved drug. Therefore, once a drug is approved for any indication, a physician may prescribe it for uses or in treatment regimens or patient populations that are not included in approved labeling.[3]

Surveys indicate that 88%–100% of physicians prescribe drugs for unlabeled indications with 25% prescribing for an unlabeled indication daily.[4,5] Published reports, although few in number, reflect a similar occurrence of unlabeled medication use in pharmacy practice. Based on the published studies, chart reviews, and surveys, drug use for unlabeled indications is especially common in the areas of obstetrics, oncology, pediatrics, psychiatry, and infectious diseases (Table 1). Depending on the practice area, patient population, and drug being evaluated, 9%–100% of patients received a medication for an unlabeled use.

PATIENT SAFETY AND PRODUCT EFFICACY

There is a danger that many unlabeled uses are derived from inadequately controlled or uncontrolled studies and anecdotal reports. Two articles reviewing the unlabeled uses of intravenous immunoglobulin (IVIG) evaluated the level of evidence supporting these uses. In the first study,[6] only one-third of the unlabeled uses of IVIG was supported by randomized controlled trials. The other two-thirds were derived from case reports and uncontrolled trials. The second study showed that a substantial proportion of the unlabeled use of IVIG was not even supported by evidence-based guidelines.[7] The authors concluded that it would be prudent to discourage indiscriminate unlabeled uses of IVIG because of the product's unproven clinical effectiveness in unlabeled situations. As Dr. Arnold Relman, editor emeritus of the

Encyclopedia of Clinical Pharmacy
DOI: 10.1081/E-ECP 120006328

Table 1 Frequency of drug use for unlabeled indications

Practice area	Drug(s) evaluated	Unlabeled Use (%)	
		Patients	Prescriptions
Family Practice [18]	Various	9	NR
Hospital [19]	Metoclopramide	100	100
Infectious Disease [20]	HIV*	81	40
Infectious Disease [7]	IVIG	52	29–100
Obstetrics [21]	Various	23	NR
Oncology [16]	Antineoplastics	56	35
Pediatrics [22]	Ondansetron	NR	12–73
Pediatrics [23]	Various	92	66
Pediatrics [24]	Various	36	18
Pediatrics [25]	Various	NR	7
Psychiatry [26]	Antipsychotics	67	NR

*Drugs used to treat human immunodeficiency virus and its clinical manifestations.
NR = not reported, IVIG = intravenous immunoglobulin.

New England Journal of Medicine has pointed out, ''publication of an article on the unapproved use of a drug in a peer-reviewed journal is no guarantee of safety or efficacy.''[8] However, in some clinical situations, such as pediatrics, unlabeled uses may be rational and represent the most appropriate treatment for optimal patient care.

When the unlabeled use of an approved drug endangers the public health, the FDA is obligated to investigate thoroughly and to take whatever action is warranted to protect the public. The FDA can require a change in the labeling to warn against the unlabeled use, seek evidence to substantiate the use, restrict the channel of distribution, and even withdraw approval of the drug and remove it from the market. In such instances, the official labeling may include a prominently displayed ''box warning'' to alert practitioners that certain uses are hazardous and, in effect ''disapproved.''[9]

Implications for Pharmacists

The American Society of Health-System Pharmacists states that pharmacist should serve as patient advocates and drug information specialists concerning unlabeled uses of medications.[10] Pharmacists must evaluate the available information to assess product efficacy and patient safety. Therefore, pharmacists must have access to accurate and unbiased information about unlabeled uses of prescription drugs. The following textbooks are recognized as authoritative sources on unlabeled uses by the FDA, Blue Cross/Blue Shield, and the Health Insurance Association of America: 1) *The American Hospital Formulary Service Drug Information* (AHFS DI); 2) *The American Medical Association Drug Evaluations* (AMA DE); and 3) *The United States Pharmacopeial Drug Information* (USP DI). The information in AHFS DI is derived from published medical literature, as well as scientific meetings, educational programs, professional interactions, and comments provided by editorial reviewers. Not all unlabeled uses included in AHFS DI are well established. Some unlabeled uses may be acceptable but are considered investigational because of limited experience. Unlabeled uses listed in the USP DI are selected by USP Advisory Panels based on current literature, current prescribing, and utilization practices. The AMA DE is no longer published, but its contents have been incorporated into USP DI. None of these authoritative textbooks include references.

Additional resources that pharmacists can use for information about unlabeled indications include medical and pharmacy textbooks, handbooks, journals, electronic databases, and the Internet. Although the Internet can be a useful resource for finding information, many sites offer unreferenced information, often with unsubstantiated claims. Drug information services staffed by pharmacists trained in the retrieval, interpretation, and dissemination of medical literature are excellent resources for determining if an unlabeled use of a drug is appropriate.

Another concern for pharmacists is manufacturers' distribution of information on unlabeled drug uses. The FDA Modernization Act passed by Congress in 1997 allows pharmaceutical companies to distribute reprints of peer-reviewed journal articles or reference textbooks that include information about unlabeled uses of their products to health care practitioners.[11,12] The FDA does not include pharmacists as ''health care practitioners'' thus limiting access to information about unlabeled indications disseminated by the pharmaceutical companies to physicians. Pharmacists can and should gain access to this and other information using the drug information resources discussed previously. It is important to perform a comprehensive review of all information available on the unlabeled uses of a medication and to not determine product efficacy or patient safety on the basis of a single source.

LIABILITY

Dispensing a medication for an unlabeled use, dose, or dosage form does not violate federal law. In fact, to date, no pharmacist is known to have been prosecuted for dispensing a product for an unlabeled use. However, a deviation from the safe and effective guidelines provided in the product labeling does expose the pharmacist to some liability. A pharmacist may be subject to malpractice action if patient injury results from the administration of a medication for an unlabeled use.[13,14]

Lack of criminal prosecution is not a predictor of future actions or exposure to civil litigation. Therefore, it is important for a pharmacist to be familiar with labeled indications and unlabeled uses of a drug and to be prepared to defend the appropriateness of a drug's use in the event of litigation.[3] Liability will be minimized if the pharmacist, in the exercise of sound professional judgment, concludes that the use is rational, safe, and reasonable.[13]

In addition, a patient is entitled to know when a drug is used for an unlabeled use. It is reasonable for the physician or pharmacist to inform the patient that the drug's use is not FDA approved, it is safe and effective for other uses, and some data exist to support its unlabeled use. Such discussions should be documented in the patient's medical record.[12] A consent form signed by the patient is highly recommended but is not currently necessary if the drug is used in the normal practice of medicine.[9,12]

REIMBURSEMENT

Reimbursement for unlabeled uses of drugs has become an important medical and financial issue. Decisions on coverage for drug therapy are complicated because it is often difficult to differentiate between an accepted standard of practice, an evolving standard of practice, and investigational therapies. Consequently, many insurance carriers and managed care providers have elected to cover only those indications included in FDA-approved product labeling. Coverage has frequently been denied for unlabeled or non–FDA-approved indications, citing these as ''investigational'' uses.[15]

The Health Insurance Association of America has created guidelines to assist insurers in assessing coverage for unlabeled drug uses. A growing number of insurance carriers are following these guidelines, which encourage the use of references such as the three authoritative drug sources, peer reviewed literature, and consultation with experts in research and clinical practice to make specific drug coverage decisions.[10]

Implications for Pharmacists

At this time, there is minimal financial impact on pharmacists when insurance coverage is denied for an unlabeled use of a medication. However, lack of FDA approval for drug reimbursement can affect patient safety. Results of a 1991 General Accounting Office survey of medical oncologists showed denials for unlabeled drug use by insurance companies were having adverse effects on the treatment decisions of one of every eight cancer patients.[16] When an oncologist learns that a service, procedure, or drug is not covered by insurance, they often choose another less effective therapy that will not jeopardize the patient's financial situation.[17] Consequently, pharmacists must consider if the chosen therapy is financially appropriate for the patient. If the patient can not afford the unlabeled medication, the pharmacist should assist in selecting the best alternative therapy.

CONCLUSION

FDA-approved product labeling is not intended to set the standard of medical practice, but to ensure consumer safety. However, product labeling often reflects the manufacturer's regulatory or financial concerns rather than current literature-supported information. Pharmacists must possess good drug information skills, and have access to accurate and unbiased information to assess therapeutic appropriateness when dispensing medications for unlabeled indications.

Pharmacists are faced with product efficacy, patient safety, liability, and reimbursement issues when medications are used for unlabeled indications. Lack of FDA approval for a specific use should not prevent a pharmacist from dispensing an available drug. Rather, drugs should be used in a manner consistent with good medical practice and in the patient's best interest. It is good pro-

fessional practice for a pharmacist to be familiar with a drug's unlabeled uses, to evaluate the literature supporting its use, and to keep the patient informed.

REFERENCES

1. Gorman, C. Double-duty drugs. Time **1995**, *146* (12), 96–97.
2. Cranston, J.W.; Williams, M.A.; Nielsen, N.H.; Bezman, R.J. For the council on scientific affairs. Unlabeled indications of Food and Drug Administration-approved drugs. Drug Inf. J. **1998**, *32*, 1049–1061.
3. Department of Health and Human Services. Use of approved drugs for unlabeled indications. FDA Drug Bull. **1982**, *12* (1), 4–5.
4. Serradell, J.; Galle, B. Prescribing for unlabeled indications. HMO Pract. **1993**, *7* (1), 44–47.
5. Li, V.W.; Jaffe, M.P.; Li, W.W.; Haynes, H.A. Off-label dermatologic therapies: Usage, risks, and mechanisms. Arch. Dermatol. **1998**, *134* (11), 1449–1454.
6. Ratko, T.A.; Burnett, D.A.; Foulke, G.E.; Matuszewski, K.A.; Sacher, R.A.; University Hospital Consortium Expert Panel for Off-Label Use of Polyvalent Intravenously Administered Immunoglobulin Preparations. Recommendations for off-label use of intravenously administered immunoglobulin preparations. JAMA **1995**, *273* (23), 1865–1870.
7. Chen, C.; Danekas, L.H.; Ratko, T.A.; Vlasses, P.H.; Matuszewski, K.A. A multicenter drug use surveillance of intravenous immunoglobulin utilization in US academic health centers. Ann. Pharmacother. **2000**, *34* (3), 295–299.
8. Marwick, C. Implementing the FDA Modernization Act. JAMA **1998**, *279* (11), 815–816.
9. Nightingale, S.L. Use of drugs for unlabeled indications. Am. Fam. Physician **1986**, *34* (3), 269.
10. American Society of Health-System Pharmacists. Issue: Dissemination of information on unlabeled uses of medications, Available at: http://www.ashp.org/public/proad/Dissemination.html (accessed January 23, 2001).
11. Landow, L. Off-label use of approved drugs. Chest **1999**, *116* (3), 589–591.
12. Henry, V. Off-label prescribing. Legal implications. J. Leg. Med. **1999**, *20* (3), 365–383.
13. Podell, L.B. Legal implications of preparing and dispensing approved drugs for unlabeled indications. Am. J. Hosp. Pharm. **1983**, *40* (1), 792–794.
14. McKenzie, C.A.; Barnes, C.L.; Boyd, J.A.; Poinsett-Holmes, K. Unlabeled uses for approved drugs. U.S. Pharm. **1993**, *18* (2), 63–76.
15. Laetz, T.; Silberman, G. Reimbursement policies constrain the practice of oncology. JAMA **1991**, *266* (21), 2996–2999.
16. *Off-Label Drugs: Reimbursement Policies Constrain Physicians in Their Choice of Cancer Therapies*, GAO/PEMD-91-14; U.S. General Accounting Office: Washington, DC 1991; 1–59.
17. Mortenson, L.E. Health care policies affecting the treatment of patients with cancer and cancer research. Cancer **1994**, *74* (S7), 2204–2207.
18. Erickson, S.H.; Bergman, J.J.; Schneeweiss, R.; Cherkin, D.C. The use of drugs for unlabeled indications. JAMA **1980**, *243* (15), 1543–1546.
19. Schwinghammer, T.L. Prescribing of oral metoclopramide for nonapproved indications. Drug Intell. Clin. Pharm. **1984**, *18* (4), 317–319.
20. Brosgart, C.L.; Mitchell, T.; Charlebois, E.; Coleman, R.; Mehalko, S.; Young, J.; Abrams, D.D.I. Off-label drug use in human immunodeficiency virus disease. J. Acquired Immune Defic. Syndr. Hum. Retrovirol. **1996**, *12* (1), 56–62.
21. Rayburn, W.F.; Farmer, K.C. Off-label prescribing during pregnancy. Obstet. Gynecol. Clin. North Am. **1997**, *24* (3), 471–478.
22. McQueen, K.D.; Milton, J.D. Multicenter postmarketing surveillance of ondansetron therapy in pediatric patients. Ann. Pharmacother. **1994**, *28* (1), 85–92.
23. 't Jong, G.W.; Vulto, A.G.; de Hoog, M.; Schimmel, K.J.M.; Tibboel, D.; Van den Anker, J.N. Unapproved and off-label use of drugs in a children's hospital. N. Engl. J. Med. **2000**, *343* (15), 1125.
24. Turner, S.; Longworth, A.; Nunn, A.J.; Choonara, I. Unlicensed and off label drug use in paediatric wards: Prospective study. BMJ **1998**, *316* (7128), 343–345.
25. Thompson, D.F.; Heflin, R.N. Frequency and appropriateness of drug prescribing for unlabeled uses in pediatric patients. Am. J. Hosp. Pharm. **1987**, *44* (4), 792–794.
26. Weiss, E.; Hummer, M.; Koller, D.; Ulmer, H.; Fleischhacker, W.W. Off-label use of antipsychotic drugs. J. Clin. Psychopharmacol. **2000**, *20* (6), 695–698.

Millis Commission—Study Commission on Pharmacy

William A. Miller
University of Iowa, Iowa City, Iowa, U.S.A.

INTRODUCTION

Appointment of the Study Commission on Pharmacy culminated after a variety of conferences and studies in the 1960s and early 1970s to examine the changing role of pharmacy practice. Clinical pharmacy was emerging during the 1960s; during this period, academic pharmacy was also debating how pharmaceutical education should be changed to prepare pharmacists for the future. In 1972, American Association of Colleges of Pharmacy President Arthur E. Schwarting called for the Association to conduct an evaluation of pharmacy service within the overall structure of healthcare and an evaluation of recent and current changes in pharmacy education. His call led to the formation of the Study Commission on Pharmacy in 1992.

ORGANIZATION

To assure objectivity, Schwarting required that the Commission include other health professionals and healthcare planners in addition to pharmacists. Dr. John Millis, President of the National Fund for Medical Education, was selected to Chair the Commission. Eleven distinguished individuals from pharmacy, medicine, nursing, and the pharmaceutical industry were selected to serve on the Commission. The Commission spent two years studying the practice of pharmacy and the process of pharmacy education before publishing its report in 1975, and utilized 80 consultants to help formulate opinions, observations, and recommendations. A record of the events that led to the formation of the Commission is detailed in a historical record of the American Association of Colleges of Pharmacy.[1]

Change is most often evolutionary and not revolutionary. Reading the report of the Study Commission on Pharmacy[2] today reinforces how slow change occurs when it involves changing the purpose of a profession or the basis of an educational program to produce a professional. Reading key historical documents such as the Commission report also demonstrates that today's practice and education trends forecast the future. Many of the recommendations and opinions offered in the report of the Commission have been reiterated in subsequent reports and addresses by leaders within and outside the profession. Nevertheless, over time, many of the recommendations made by the Commission have been implemented, while other recommendations have only been partially achieved or have been discarded and not achieved at all.

The following includes a review of each of the major concepts, findings and recommendations of the Study Commission on Pharmacy; observations on the changes that have occurred since the report was published; and commentary on influence of the report on clinical pharmacy practice.

Pharmacists as Providers of Drug Information

The study commission indicated that one of the major deficiencies of the healthcare system was the lack of adequate information for those who consume, prescribe, dispense, and administer drugs. The Commission saw pharmacists as health professionals who could make an important contribution to the healthcare system of the future by providing information about drugs to consumers and health professionals.[3] The Commission judged pharmacy as a knowledge system as only being partially successful in delivering its full potential as a healthcare service. The report stimulated further emphasis on the role of pharmacists as providers of drug information. Since the report was published, tremendous progress has been made to advance the role of pharmacists as providers of drug information to patients and other health professionals. The clinical pharmacy education and practice movement emphasized the role of pharmacists as providers of drug information to patients and other health professionals. Today, organized healthcare settings clearly value and utilize pharmacists as providers of objective drug information for patients, interdisciplinary patient care, and drug use policy making. Today, a greater number of physicians in solo or group practice rely upon pharmacists for objective drug information, rather than relying solely on some texts such as the *Physician's Drug Reference*[4] or

Encyclopedia of Clinical Pharmacy
DOI: 10.1081/E-ECP 120006193

pharmaceutical representatives. Today's Doctor of Pharmacy curriculum prepares pharmacy graduates to retrieve, analyze, and provide drug information to other health professionals, patients, and committees concerned with health policy. In the last decade, there has been an explosion of information available to providers and patients using information technology and the Internet. It is clear that the provision of drug information, as suggested by the Commission, remains central to the future of pharmacy practice. It is also clear that our educational programs must prepare pharmacists to have more advanced drug information skills to enable them to continue to be a primary source of drug information.

Defining Pharmacy and Pharmacists

The study commission indicated that a pharmacist must be defined as an individual who is engaged in one of the steps of a system called pharmacy. The Commission asserted that they could not define a pharmacist simply as one who practices pharmacy. Rather, a pharmacist must be defined as a one who practices a part of pharmacy, which is determined by the activities carried on in the subsystems of pharmacy. Further, the Commission stated that a pharmacist is characterized by the common denominator of drug knowledge and the differentiated additional knowledge and skill required by his particular role. The Commission predicted that the future role of pharmacists would emphasize dispensing of drug information and not drugs.[5] The Commission noted that although the concept of clinical pharmacy was still evolving, it should be regarded as a powerful force internal to pharmacy that would produce change in the system of pharmacy and the practice of pharmacists.[6] The Commission opined that differentiation did not equate to specialization. The Commission defined differentiation based upon differences in place of practice, function, compensation arrangements, and patients served. The Commission stated that specialization should be based upon uniqueness of knowledge and not place or function of patient service. Further, the Commission offered cautionary comments about potential negatives in specialization and stated that specialization could only be justified by improved quality and effectiveness of patient care.[7] Although the Commission report influenced the thinking of clinical pharmacy leaders about differentiation and specialization, clinical pharmacy leaders rejected the notion that all pharmacists have the same knowledge and ability to improve the quality and effectiveness of patient care, and thus, pursued recognition of clinical pharmacy as a specialty rather than as a differentiated area of pharmacy practice as defined by the Commission. After many years, the American College of Clinical Pharmacy (ACCP) was successful in getting the Board of Pharmaceutical Specialties to recognize pharmacotherapy as a specialty practice. The terms pharmacotherapy and pharmacotherapist were used in the final specialty petition, because these terms more clearly connoted a different type of pharmacist than the terms clinical pharmacy and clinical pharmacist.[8,9] ACCP organizational leaders and individuals involved with the petition to recognize clinical pharmacy and eventually pharmacotherapy did so with a belief that patient care quality would be enhanced by board certification of pharmacists and thus meet the criterion set by the Commission. Since the report of the Commission, other knowledge-based specialties have also been approved as a part of the clinical pharmacy movement. Today, the profession continues to debate the concept of differentiation and specialization as evidenced by the current debate concerning the credentialing of pharmacists to provide disease management services to patients and the need to recognize advanced practice pharmacists in community settings who have greater knowledge and clinical skills than many community pharmacists. Thus, the observations of the Commission on differentiation and specialization live on.

Design of Pharmacy Curricula

The Commission opined that curricula of colleges of pharmacy should be based upon the competencies desired for their graduates rather than upon the basis of knowledge available in the several relevant sciences. The Commission suggested that one of the first steps to review a curriculum for a college of pharmacy would be to weigh the relative emphasis given to the physical and biological sciences against the behavioral and social sciences. The Commission said we should start with the outcomes of the curriculum and work backwards to determine the coursework needed to achieve the desired outcomes.[10] Major changes in curricula for producing pharmacists have occurred since the report of the Commission. The work of the American Association of Colleges of Pharmacy (AACP) Commission to Implement Change in Pharmaceutical Education[11,12] and the revision of standards by the American Council on Pharmaceutical Education (ACPE) led to adoption of the Doctor of Pharmacy degree as the sole entry-level degree for the profession. The clinical component of most pharmacy curricula is now the single most significant component of the curriculum. The Commission report was criticized for being too general and not specific about changes that should be made to improve the curricula of colleges of pharmacy.[2] The report of the Commission to Implement Change in Pharmaceutical Education and revised ACPE standards provided the specific detail needed to guide colleges to revise their curricula. Regretfully, some college curricular reviews have ignored the recommendation of the Commission by approaching curriculum revision by starting with the existing curriculum and working forward to improve the outcomes of the curriculum. This approach as noted by the Commission leads to less innovative and satisfactory approaches to achieving desired curricular outcomes.

The Concept of Clinical Scientists

The Commission opined that the greatest weakness of colleges of pharmacy is a lack of an adequate number of clinical scientists who can relate their specialized scientific knowledge to the development of the practice skills required to provide effective, efficient, and needed patient services.[13] The concept of a ''clinical scientist'' was discussed and debated in many forums after the Commission offered this opinion. Since the report, colleges of pharmacy with a research mission have hired faculty into tenured track faculty positions that require significant clinical research for promotion in addition to teaching and service through clinical practice. These faculty are now considered clinical scientists who contribute to the advancement of scientific knowledge to enhance patient care. As tenure track clinical pharmacy faculty developed as scientists, the amount of time they spend in practice has declined due to the time demands required to be a successful researcher and teacher. In addition to tenured clinical faculty, full-time clinical track faculty positions have been developed at most colleges to teach therapeutics and other clinical courses and to serve as practitioner role models for students. Clinical pharmacy practitioners in various settings also frequently hold adjunct faculty appointments at colleges of pharmacy and serve as role models and teachers for clinical clerkships. Clinical faculty now significantly contribute to the scholarship and research conducted by colleges of pharmacy.

The Role of Graduate and Advanced Professional Educational Programs

The Commission stated that colleges of pharmacy with adequate resources should offer programs at the graduate and advanced professional level for more differentiated roles of pharmacy practice. The Commission recognized the need for advanced professional education and stated that colleges without adequate resources and outside the environment of a health science center should not consider offering advanced professional degree programs beyond the baccalaureate degree.[14] The Commission did not define the terms graduate and advanced professional level. However, the report stimulated discussion of the role of Master of Science, Ph.D., and fellowship programs as methods to develop clinical scientists, further development of Doctor of Pharmacy degree programs as advanced professional degree programs, and further development of postgraduate residency programs to produce differentiated pharmacists for advanced professional roles. The Commission considered the Doctor of Pharmacy as an advanced professional degree and warned against colleges offering this degree without adequate resources.[14] The debate on the preparation of clinical scientists or tenured track faculty continues today. A major issue likely to be confronted by academic pharmacy and the practice community is the future role of residency programs in the preparation of pharmacy graduates for general practice.

Environment for the Preparation of Pharmacists

The Commission opined that the optimal environment for pharmacy education is the university health science center but felt that adequate arrangements of colleges not located in health science centers could be made to provide an acceptable environment for the education of students at the baccalaureate level.[14] This recommendation was debated within academic pharmacy. It caused colleges of pharmacy to examine whether they had adequate resources and the environment to offer a postbaccalaureate Doctor of Pharmacy degree program. It helped spur colleges to develop relationships with hospitals and other healthcare organizations to provide clinical education and to add sufficient clinical faculty to deliver a quality postbaccalaureate Doctor of Pharmacy degree program. Over time, colleges with sufficient clinical faculty offered postbaccalaureate Doctor of Pharmacy degree programs. Soon, all colleges will offer the Doctor of Pharmacy degree as their sole professional degree program. Clinical pharmacy practice has evolved beyond the walls of the university hospital and other tertiary care hospitals to community hospital, community pharmacy, ambulatory care settings, and other innovative practice sites. Today's curricula are designed to prepare pharmacists to provide clinical pharmacy services to patients in all practice settings. All colleges now seek and have arrangements with multiple organizations to provide an acceptable environment for the preparation of pharmacists to provide clinical services.

Credentialing of Pharmacists and Pharmacy Education

The Commission opined that all aspects of credentialing of pharmacists and pharmacy education should be brought together under an umbrella organization such as a National Board of Pharmacy Examiners.[15] Over the years, this recommendation has been debated through many forums. Today, many leaders in pharmacy believe the creation of a new organization to provide oversight for all credentialing and accreditation programs in pharmacy would benefit the profession and society. A new credentialing and accreditation organization could better coordinate all programs and minimize interorganizational competition to allow a more unified vision and direction. Perhaps with time, the right political and organizational circumstances will occur to bring this recommendation of the Commission to fruition.

SUMMARY

The work of the Study Commission on Pharmacy has had an important impact on the development of clinical pharmacy and pharmaceutical education. It, coupled with other significant reports and individual and organizational leadership, have advanced the profession. As a result, pharmacists are now better educated and trained and play a more significant role in concert with other health providers to help ensure drugs are used safely and effectively in patient care.

REFERENCES

1. Buerki, R.A. In search of excellence: The first century of the American Association of Colleges of Pharmacy. Am. J. Pharm. Educ. **1999**, *63* (Suppl.), 167–172.
2. *Pharmacists for the Future: The Report of the Study Commission on Pharmacy*; Health Administration Press: Ann Arbor, Michigan, 1975.
3. *Pharmacists for the Future: The Report of the Study Commission on Pharmacy*; Health Administration Press: Ann Arbor, Michigan, 1975; 139.
4. *Physicians' Desk Reference*; Medical Economics Company: Montvale, New Jersey, 2000.
5. *Pharmacists for the Future: The Report of the Study Commission on Pharmacy*; Health Administration Press: Ann Arbor, Michigan, 1975; 11–14, 139–140.
6. *Pharmacists for the Future: The Report of the Study Commission on Pharmacy*; Health Administration Press: Ann Arbor, Michigan, 1975; 92–94.
7. *Pharmacists for the Future: The Report of the Study Commission on Pharmacy*; Health Administration Press: Ann Arbor, Michigan, 1975; 104–106.
8. Definition of clinical pharmacy as a specialty in clinical practice. Committee on Clinical Pharmacy as a Specialty. American Pharmacy Association. Drug Intel. Clin. Pharm. **1985**, *19* (2), 149–150.
9. A petition to the Board of Pharmaceutical Specialties requesting recognition of pharmacotherapy as a specialty. The American College of Clinical Pharmacy. Am. Pharm. **1988**, *NS28* (5), 44–50.
10. *Pharmacists for the Future: The Report of the Study Commission on Pharmacy*; Health Administration Press: Ann Arbor, Michigan, 1975; 109, 121–128.
11. Miller, W.A. Planning for pharmacy education in the 21st century: AACP's leadership role. Am. J. Pharm. Educ. **1989**, *53*, 336–339.
12. Papers from the Commission to implement change in pharmaceutical education. Am. J. Pharm. Educ. **1993**, *57*, 366–399.
13. *Pharmacists for the Future: The Report of the Study Commission on Pharmacy*; Health Administration Press: Ann Arbor, Michigan, 1975; 123–126, 142.
14. *Pharmacists for the Future: The Report of the Study Commission on Pharmacy*; Health Administration Press: Ann Arbor, Michigan, 1975; 129–132, 143.
15. *Pharmacists for the Future: The Report of the Study Commission on Pharmacy*; Health Administration Press: Ann Arbor, Michigan, 1975; 135–138, 143.

National Childhood Vaccine Injury Act of 1986

Rachel Crafts
Idaho Drug Information Service, Pocatello, Idaho, U.S.A.

INTRODUCTION

Vaccination against infectious illnesses provides unseen protection against contagious diseases—afflictions causing permanent disability or even death. Vaccines have been responsible for dramatic decreases in morbidity and mortality secondary to infectious disease, and in the case of smallpox, has globally eradicated a once life-threatening illness.[1–7] However, while true adverse consequences of vaccination have never exceeded the level of adverse consequences of infection in the absence of vaccination, the public perception of harm secondary to vaccine administration has threatened to overshadow the victory of disease prevention.[1–8] With the inception and continued evolution of immunization, the number of individuals protected against diseases has steadily increased. Unfortunately, the number of vaccine-related adverse events has also increased proportionally to vaccine use.[2]

Historical Advancements

Immunization against infectious disease has saved more lives than any other health intervention in the history of modern medicine.[2,8] The success of continued disease prevention, however, directly correlates to continued immunobiological research, vaccine availability, and those clinicians who administer vaccinations.

Vaccination Against Childhood Diseases: The Controversy

The National Childhood Immunization Program has proven to be one of the most successful health promotion practices in medicine.[8,9] The production of disease-specific vaccines has been lauded as one of the second millennium's greatest accomplishments.[10] In spite of this, however, more than 20% of U.S. two-year-olds remain susceptible to preventable illness by not receiving primary immunizations on schedule.[11,12] This is due in part to socioeconomic and health care accessibility barriers, but also to a loss of parental confidence in immunization programs. Religious or philosophical exemptions account for roughly 0.58% of U.S. children who do not receive immunizations.[13] The risk-to-benefit issue has been argued since the introduction of smallpox vaccine and remains a subject of contention for some. In 1904, in the case *Jacobson v. Massachusetts*, the U.S. Supreme Court ruled in favor of routine smallpox innoculation despite ''quite often'' serious, and sometimes, fatal side effects.[1] The basis for this ruling was predicated upon the idea that overall societal good outweighs individual rights when prevention of contagious disease is at stake.

Vaccine Adverse Events

Untoward vaccine effects are variable, manifesting from injection site irritation, fever, and irritability to encephalopathy, paralysis, and even death.[14–23] Reporting of certain vaccine adverse events to the Vaccine Adverse Event Reporting System (VAERS) is mandatory. VAERS was established by the Centers for Disease Control and Prevention (CDC) and the Food and Drug Administration (FDA) to facilitate the process of gathering postmarketing surveillance data on vaccine-related adverse events.[2] VAERS reports record adverse events temporally but not necessarily causally associated with vaccination.[24]

Manufacturer Liability

Adverse events secondary to vaccine products have inspired lawsuits against vaccine manufacturers.[1] Two theories of tort (duties created by law regarding professional conduct) are reviewed when litigation against a vaccine manufacturer ensues. The first regards safe vaccine development, with a vaccine manufacturer strictly liable for defective vaccine products. Liability cases involving defective vaccines rarely occur, and manufac-

Encyclopedia of Clinical Pharmacy
DOI: 10.1081/E-ECP 120006197

turers are not held liable for inherent or unavoidable adverse events under the theory of strict liability. Second, a manufacturer is responsible for providing adequate warning of possible side effects. Prior to 1986, a manufacturer had to prove that despite adequate warning, the plaintiff would still have received the necessary innoculation.[1]

THE VACCINE INJURY ACT OF 1986

Escalating numbers of lawsuits, primarily aimed at DTP vaccine manufacturers during the 1970s and 1980s, forced many manufacturers from the market. The remaining manufacturers increased DTP vaccine prices, making the limited products available very expensive.[2,5] Widespread vaccine shortages combined with exorbitant increases in vaccine costs placed mandatory vaccination protocols for school-aged children in serious threat of discontinuation. A battle fought largely by the American Academy of Pediatrics called for government reform to ensure the future of the U.S. vaccine supply and continuation of mass immunization programs.[8] In 1986, Congress passed the National Childhood Vaccine Injury Act (NCVIA), administered by the Department of Health and Human Services. This legislation was aimed at the "prevention of human infectious disease through immunization and to achieve optimal protection against adverse reactions to vaccines."[25] The NCVIA program responsibilities include continuation of vaccine research and development, licensing, distribution, and monitoring.[26]

The NCVIA is a no-fault, nontort compensation program for persons injured by certain covered vaccines. The purpose of a no-fault compensation program is to resolve the controversy surrounding government-mandated administration of vaccines for school-aged-children.[27] The NCVIA applies to any vaccine containing diphtheria, tetanus, pertussis, measles, mumps, rubella, poliovirus, hepatitis B, *H. influenzae* type b, and varicella or conjugate pneumococcal antigens.[28] With the probability that rare instances of injury will ensue during mass innoculation practices, the act provides insurance coverage against adverse vaccine events.[4]

The National Vaccine Injury Compensation Program

To effectively address the vaccine litigation crisis, the Vaccine Injury Compensation Program (VICP) was established by the NCVIA and became effective on October 1, 1988.[29] The directive of the VICP is to bring together information about vaccinations (VAERS was officially adopted in 1990), collaborate with the CDC for developing vaccine administration information forms, promote the development of safer vaccines, and establish a federal compensation program for those injured by covered vaccines. The compensation program is funded via an excise tax on vaccines compensable under the

Table 1 Vaccine Injury Table (effective date: December 18, 1999)

Vaccine	Adverse event and time period for manifestation
Tetanus toxoid	Anaphylaxis within 4 hours; brachial neuritis within 2–28 days
Whole-cell pertussis bacteria, partial cell pertussis bacteria, specific pertussis antigens	Anaphylaxis or anaphylactic shock within 4 hours; encephalopathy or encephalitis within 72 hours
Measles, mumps, or rubella virus	Anaphylaxis or anaphylactic shock within 4 hours; encephalopathy or encephalitis within 5–15 days (not less than 5 days, not more than 15 days)
Rubella virus	Chronic arthritis within 7–42 days
Measles virus	Thrombocytic purpura within 7–30 days; vaccine-strain measles viral infection in an immunodeficient recipient within 6 months
Live poliovirus (OPV)	Paralytic polio or vaccine strain polio viral infection within 30 days in an immunocompetent recipient or within 60 days in an immunodeficient recipient.
Inactivated poliovirus (IPV)	Anaphylaxis or anaphylactic shock within 4 hours
Unconjugated Hib vaccine	Early onset Hib disease within 7 days
Conjugated Hib vaccine	No condition specified
Varicella	No condition specified
Pneumococcal conjugate vaccines	No condition specified

(Adapted from Ref. [31,32].)

VICP.[4,30] The VICP largely replaces traditional tort law for deciding vaccine-related injuries and was designed to stabilize the vaccine market by decreasing liability costs of manufacturers and health care providers and to ease reward recovery by eligible claimants.[4,29]

The Vaccine Injury Table

The VICP established the Vaccine Injury Table (VIT) (Table 1) to facilitate the process of justifying compensation to injured vaccinees. The VIT lists specific vaccines with corresponding injuries to be compensated if injury manifests within the predetermined time frame. Therefore, the table eliminates the need for determining a causal relationship between vaccine and injury.[4,5,28,31]

PHARMACY-BASED IMMUNIZATION PRACTICE

Since 1971, the National Center for Health Services Research has recognized the clinical role of pharmacy to include immunization.[28] Pharmacists administered more than 100,000 immunizations in 1997 alone, and as of July 1999, 30 U.S. states officially recognized pharmacists as having the authority to vaccinate.[28,33] As defined by the Model Pharmacy Practice Act of the National Association of Boards of Pharmacy, the practice of pharmacy includes services not limited to compounding, dispensing, labeling, interpreting, and evaluating prescriptions, but also the *administration* and distribution of drugs and devices.[33]

Liability

The liability risk surrounding vaccine providers is relatively low.[33] However, pharmacists administering vaccines can be held liable for negligent administration of vaccines and should take precautions to prevent adverse events. Several important duties surround the administration of vaccines. Health care professionals wishing to immunize should have adequate training in vaccine protocols (including proper management of rare anaphylactic reactions) and administration techniques and maintain current CPR certification.[33] Pharmacists requiring up-to-date immunization information should be familiar with the Web site www.immunofacts.com for vaccine-related information and current standards of practice. Pharmacists must also thoroughly review indications and contraindications to vaccine use and provide adequate consent information defining possible untoward events. Failure to warn represents the bulk of litigation suits. Thus, pharmacists should be aware of vaccine contraindications and keep abreast of information regarding possible adverse outcomes.[1]

Protection Under the NCVIA

Clinicians administering vaccines covered by the NCVIA are protected against litigation surrounding possible vaccine-related adverse events.[33] Pharmacists are commonly involved with the administration of pneumococcal 23-valent and influenza virus vaccines—vaccines not currently covered by the NCVIA.[28] Despite this, pharmacists should be familiar with this far-reaching legislation and understand both the ramifications and limitations on vaccine coverage.

Required Documentation

Health care practitioners administering vaccines covered by the NCVIA are required to record pertinent information in either a permanent patient record or a vaccine log. The patient's name, date of administration, vaccine name, manufacturer, lot number, provider name, address, and title are all required.[33–35] Administration of a NCVIA-covered vaccine also requires practitioners to give vaccine information statements (provided by state health departments and the CDC at www.cdc.gov/nip/publications/VIS/default.htm) to patients or guardians outlining product use, benefits, risks, and warnings.[33,34,36,37] Pharmacists involved with the administration of pneumococcal, influenza, and other vaccines should also adopt this practice.

Mandatory Reporting

In the event of an adverse vaccine event, it is the practitioner's responsibility to contact the VICP for information on how to file a claim, eligibility, and required documentation. Also, the NCVIA mandates reporting to the VAERS for any adverse event covered by the VICP, some of which are specifically listed in the vaccine injury table.[2,34,38]

SUMMARY

Society has a vested interest in routine immunization practices.[2,39] Although many vaccine-preventable diseases are controlled and even eliminated in some regions of the world, complete eradication (with the exception of polio) is unlikely in the near future.[2] Although the conrred benefit received from mass immunization is among the greatest of public health achievements, the fact remains that vaccine products are not completely safe. Under the NCVIA, research and development of new vaccine products will also foster the production of safer and more efficacious vaccines. Further attributes of the NCVIA include the VICP along with VAERS and their capability for accumulating data regarding the safety of vaccines.

REFERENCES

1. Prins-Stairs, J.C. The National Childhood Vaccine Injury Act of 1986. Can congressional intent survive judicial sympathy for the injured? J. Leg. Med. **1989, Dec.**, *10* (4), 703–737.
2. P. Chen, R.T.; Rastogi, S.C.; Mullen, J.R.; Hayes, S.W.; Cochi, S.L.; Donlon, J.A.; Wassilak, S.G. The Vaccine Adverse Event Reporting System (VAERS). Vaccine **1994, May**, *12* (6), 542–550.
3. Bedford, H.; Elliman, D. Concerns about immunisation. BMJ **2000, Jan. 22**, *320* (7229), 240–243.
4. Ridgway, D. No-fault vaccine insurance: Lessons from the National Vaccine Injury Compensation Program. J. Health Polit. Policy Law **1999, Feb.**, *24* (1), 59–90.
5. Mariner, W.K. The National Vaccine Injury Compensation program. Health Aff. (Millwood) **1992, Spring**, *11* (1), 255–265.
6. Evans, G. National Childhood Vaccine Injury Act: Revision of the vaccine injury table. Pediatrics **1996, Dec.**, *98* (6 Pt. 1), 1179–1181.
7. General recommendations on immunization. Guidelines from the immunization practices advisory committee. Centers for Disease Control. Ann. Intern. Med. **1989, Jul. 15**, *111* (2), 133–142.
8. Smith, M.H. National Childhood Vaccine Injury Compensation Act. Pediatrics **1988, Aug.**, *82* (2), 264–269.
9. Evans, G. Vaccine liability and safety revisited. Arch. Pediatr. Adolesc. Med. **1998, Jan.**, *152* (1), 7–10.
10. Ten great public health achievements—United States, 1900–1999. MMWR, Morb. Mortal. Wkly. Rep. **1999 Apr. 2**, *48* (12), 241–243.
11. National, state, and urban area vaccination coverage levels among children aged 19–35 months—United States, 1998. Morb. Mortal. Wkly. Rep. CDC Surveill. Summ. **2000 Sep. 22**, *49* (9), 1–26.
12. Mayer, M.L.; Clark, S.J.; Konrad, T.R.; Freeman, V.A.; Slifkin, R.T. The role of state policies and programs in buffering the effects of poverty on children's immunization receipt. Am. J. Public Health **1999, Feb.**, *89* (2), 164–170.
13. Rota, J.S.; Salmon, D.A.; Rodewald, L.E.; Chen, R.T.; Hibbs, B.F.; Gangarosa, E.J. Processes for obtaining nonmedical exemptions to state immunization laws. Am. J. Public Health **2001, Apr.**, *91* (4), 645–648.
14. Meszaros, J.R.; Asch, D.A.; Baron, J.; Hershey, J.C.; Kunreuther, H.; Schwartz-Buzaglo, J. Cognitive processes and the decisions of some parents to forego pertussis vaccination for their children. J. Clin. Epidemiol. **1996, Jun.**, *49* (6), 697–703.
15. Weibel, R.E.; Benor, D.E. Reporting vaccine-associated paralytic poliomyelitis: Concordance between the CDC and the National Vaccine Injury Compensation Program. Am. J. Public Health **1996, May**, *86* (5), 734–737.
16. Weibel, R.E.; Caserta, V.; Benor, D.E.; Evans, G. Acute encephalopathy followed by permanent brain injury or death associated with further attenuated measles vaccines: A review of claims submitted to the National Vaccine Injury Compensation Program. Pediatrics **1998, Mar.**, *101* (3 Pt. 1), 383–387.
17. Griffin, M.R.; Ray, W.A.; Mortimer, E.A.; Fenichel, G.M.; Schaffner, W. Risk of seizures after measles-mumps-rubella immunization. Pediatrics **1991, Nov.**, *88* (5), 881–885.
18. Cody, C.L.; Baraff, L.J.; Cherry, J.D.; Marcy, S.M.; Manclark, C.R. Nature and rates of adverse reactions associated with DTP and DT immunizations in infants and children. Pediatrics **1981, Nov.**, *68* (5), 650–660.
19. Howson, C.P.; Fineberg, H.V. The ricochet of magic bullets: Summary of the Institute of Medicine Report, Adverse Effects of Pertussis and Rubella Vaccines. Pediatrics **1992, Feb.**, *89* (2), 318–324.
20. Duclos, P.; Ward, B.J. Measles vaccines: A review of adverse events. Drug Saf. **1998, Dec.**, *19* (6), 435–454.
21. Gale, J.L.; Thapa, P.B.; Wassilak, S.G.; Bobo, J.K.; Mendelman, P.M.; Foy, H.M. Risk of serious acute neurological illness after immunization with diphtheria-tetanus-pertussis vaccine. A population-based case-control study. JAMA, J. Am. Med. Assoc. **1994, Jan. 5**, *271* (1), 37–41.
22. Miller, D.; Madge, N.; Diamond, J.; Wadsworth, J.; Ross, E. Pertussis immunisation and serious acute neurological illnesses in children. BMJ **1993, Nov. 6**, *307* (6913), 1171–1176.
23. Hoffman, H.J.; Hunter, J.C.; Damus, K.; Pakter, J.; Peterson, D.R.; van Belle, G.; Hasselmeyer, E.G. Diphtheria-tetanus-pertussis immunization and sudden infant death: Results of the National Institute of Child Health and Human Development Cooperative Epidemiological Study of Sudden Infant Death Syndrome risk factors. Pediatrics **1987, Apr.**, *79* (4), 598–611.
24. Braun, M.M.; Ellenberg, S.S. Descriptive epidemiology of adverse events after immunization: Reports to the Vaccine Adverse Event Reporting System (VAERS), 1991–1994. J. Pediatr. **1997, Oct.**, *131* (4), 529–535.
25. 42 U.S.C.A. §300, aa-1.

26. 42 U.S.C.A. §300, aa-2.
27. Stratton, W.T. Physicians' obligations under the National Childhood Vaccine Injury Act. Kans. Med. **1989, Feb.**, *90* (2), 32–33.
28. Grabenstein, J.D.; Bonasso, J. Health-system pharmacists' role in immunizing adults against pneumococcal disease and influenza. Am. J. Health-Syst. Pharm. **1999, Sep. 1**, *56* (17 Suppl. 2), S3–S22.
29. Clayton, E.W.; Hickson, G.B. Compensation under the National Childhood Vaccine Injury Act. J. Pediatr. **1990, Apr.**, *116* (4), 153–508.
30. http://frwebgate3.access.gpo.gov (accessed January 2001).
31. 42 U.S.C.A. §300, aa-14.
32. http://bhpr.hrsa.gov/vicp/table.htm (accessed January 2001).
33. Grabenstein, J.D.; Baker, K.R. Legal and liability issues related to immunizations. Hosp. Pharm. **1998, Jan.**, *33* (1), 97–115.
34. 42 U.S.C.A. §300, aa-25.
35. Current trends national childhood vaccine injury act: Requirements for permanent vaccination records and for reporting of selected events after vaccination. MMWR, Morb. Mortal. Wkly. Rep. **1988, April 08**, *37* (13), 197–200.
36. Publication of vaccine information pamphlets. MMWR, Morb. Mortal. Wkly. Rep. **1991, Oct. 25**, *40* (42), 881–882.
37. 42 U.S.C.A. §300, aa-26.
38. Current trends vaccine adverse event reporting system—United States. MMWR, Morb. Mortal. Wkly. Rep. **1990, Oct. 19**, *39* (41), 730–733.
39. Evans, G. Vaccine injury compensation programs worldwide. Vaccine **1999, Oct. 29**, *17* (Suppl. 3), S25–S35.

PROFESSIONAL ORGANIZATIONS

National Committee for Quality Assurance

Kevin M. Jarvis
Biocentric, Inc., Berkley, Michigan, U.S.A.

INTRODUCTION

Since the early 1980s, there has been a more than fivefold increase in the number of Americans enrolled in HMOs, which has had a dramatic impact on the delivery of healthcare in the United States. As a result of this trend, HMOs realized the need to demonstrate the quality of care they provided to their members compared with fee-for-service health plans; however, at the time the industry lacked established standards. Subsequently, various independent review processes evolved, which often did not have consistent assumptions about the parameters that defined quality. It was not until the late 1980s that a group of HMO industry leaders recognized that an organization called the National Committee for Quality Assurance (NCQA) had the potential to act as an independent authority on quality control in managed care.[1]

The NCQA is a private, not-for-profit organization that was originally founded in 1979 by two managed care trade associations; however, for 8 years, the organization remained relatively inactive.[1–3] In 1988, a series of meetings were held with the benefits managers of Fortune 500 companies that had experience in external quality assessment, and the evolution of the NCQA was mandated. With a grant from the Robert Wood Johnson Foundation and financial support from the managed care industry, funding was provided for the development of the NCQA, and in the spring of 1990, the NCQA was officially launched.[1,4]

The primary activities of the NCQA are assessing and reporting on the quality of the managed health care plans; the mission is to provide information that enables purchasers and consumers of managed health care to distinguish among plans based on quality, thereby allowing them to make more informed healthcare purchasing decisions.[5] The primary method by which the NCQA evaluates managed care health plans (i.e., managed care organizations, managed behavioral healthcare organizations, preferred provider organizations) involves accreditation and reporting on Health Plan Employer Data and Information Set (HEDIS) measures.[2] The NCQA also has developed or participated in other benchmark programs, including the Consumer Assessment of Health Plans Study (CAHPS, a comprehensive survey of health plan member satisfaction developed by the Agency for Healthcare Quality and Research) and accreditation/certification programs.[2]

NCQA ACCREDITATION SURVEY

The accreditation process for managed care organizations is strictly voluntary, and before accreditation can be pursued, several criteria must be met. The care provided must include adult and pediatric medical/surgical services, obstetrics, mental health care, and preventive care. Furthermore, the plan must provide ambulatory and inpatient services, have been in operation for 2 or more years, have a review process in place to continually improve care, and have access to patients' clinical records.[2] The accreditation survey consists of on- and off-site components, which are conducted by a team that typically consists of physicians and experts in managed care. More than 60 different standards are used to evaluate the clinical and administrative systems of the health plan during the accreditation survey (Table 1).[2] In particular, the survey is used to evaluate the efforts of health plans to continuously improve the quality of care provided and the service it delivers. The findings of the survey teams are reported to a national oversight committee of physicians that assigns one of five possible accreditation levels (excellent, commendable, accredited, provisional, denied). Other accreditation categories include appealed by plan, in process, revoked, scheduled, suspended, and under review by NCQA. The NCQA views accreditation as an ongoing learning process; therefore, it does not assign pass or fail grades. Instead, the accreditation process offers health plans an opportunity to identify how they can improve their services. Initially, only the NCQA and the health plan had access to the accreditation results; now the accreditation status of most health plans can be accessed via the Internet.[2] Historically, HMOs are the principal groups that have pursued accreditation; however, more recently, the NCQA launched a preferred provider organization accreditation program.[2]

Encyclopedia of Clinical Pharmacy
DOI: 10.1081/E-ECP 120006213

Table 1 Classification of accreditation survey standards

Category	Example
Access and service	Do health plan members have access to the care and service they need? • Are providers in the health plan free to discuss all available treatment options? • Do patients report problems getting the necessary care? • How well does the health plan follow up on grievances?
Qualified providers	Does the health plan assess each provider's qualifications and what health plan members say about their providers? • Does the health plan regularly check the licenses and training of providers? • How do health plan members rate their personal doctor or nurse?
Staying healthy	Does the health plan help people maintain good health and avoid illness? • Does it give its doctors guidelines about how to provide appropriate preventive health services? • Are members receiving tests and screenings as appropriate?
Getting better	How well does the health plan care for people when they become sick? • How does the health plan evaluate new medical procedures, drugs, and devices to ensure that patients have access to safe and effective health care?
Living with illness	How well does the health plan care for people with chronic conditions? • Does the plan have programs in place to assist patients in managing chronic conditions such as asthma? • Do diabetics, who are at risk for blindness, receive eye exams as needed?

(From Ref. [2].)

HEDIS

Another means by which the NCQA evaluates health plans is via HEDIS reporting, which differs from the accreditation process in that HEDIS measures the performance and not the quality of systems or processes in the health plan. The primary objective of HEDIS reporting is to provide purchasers of health plan services (i.e., employers) with objective information about the relative value and accountability of health plans by focusing on basic areas of performance (Table 2).[6] Essential services evaluated with HEDIS include preventive medicine (e.g., immunization and screening programs), prenatal care, acute and chronic disease management, and mental health programs. Other areas of performance that are included in HEDIS measures include member access and satisfaction, health plan utilization, and financial stability of the health plan.[6] The Committee on Performance Measurement identifies areas of priority for which measures should be developed.[2] The responsibility of developing the actual measure is left up to the Measurement Advisory Panels, which identify the best measures and methodologies for assessing healthcare performance. Via the Measurement Advisory Panels, measures are updated on a regular basis to reflect advancements in performance measurement and information systems technology, as well as changes in the managed care industry.

ACCREDITATION AND CERTIFICATION PROGRAMS

Beginning in 2002, the NCQA will implement a flexible health plan evaluation program that applies to various

Table 2 Examples of effectiveness of care measures in HEDIS 2000

Childhood/adolescent immunization status	Breast cancer screening
Chlamydia screening in women	Cervical cancer screening
Prenatal care in first trimester/check-ups after delivery	Controlling high blood pressure
Cholesterol management after heart attack	Comprehensive diabetes care
Use of appropriate medications for people with asthma	Beta blocker treatment after a heart attack
Follow-up after hospitalization for mental illness	Antidepressant medication management
Advising smokers to quit	Flu shots for older adults
Medicare health outcomes survey	

(From Ref. [6].)

services provided by healthcare organizations.[7] Examples include the following:

- Healthcare organizations that offer disease-management services to patients or providers can apply for disease-management accreditation (health plans that provide comprehensive services) or certification (organizations that provide clinical systems to support disease management).
- Physician groups or organizations can apply for NCQA certification, which is similar to the managed care accreditation process and allows the physician groups to demonstrate the quality of their healthcare services.
- Healthcare organizations or physician groups/organizations that are not eligible for managed care accreditation programs can apply for certification in utilization management and credentialing, which allows them to demonstrate the quality of their healthcare services in a manner consistent with managed care organizations.
- Organizations or departments that verify the credentials of healthcare providers can apply for NCQA certification.

NCQA QUALITY COMPASS AND THE STATE OF MANAGED CARE QUALITY REPORT

The NCQA Quality Compass is a compilation of accreditation and HEDIS information from hundreds of health plans. Contained in this document are national and regional averages and benchmarks, which can be used to identify areas for improvement in a particular health plan. Based on the information compiled in the Quality Compass, the NCQA produces an annual State of Managed Care Quality Report, which is a broad assessment of the overall performance of managed care throughout the United States. Similar to the Quality Compass, the Report includes national and regional averages for key performance measures and identifies the levels of performance that are attainable. The Report also uses the information to extrapolate the benefits on the nation's health if all health plans performed at the benchmark level.[2]

Table 3 2000 NCQA Board of Directors

Barbara A. Brickmeier (IBM Global Benefits)
Mark Chassin, MD (Mt. Sinai Medical Center)
James Cubbin (General Motors Corporation)
Robert S. Galvin, MD (General Electric Company)
Charles M. Gayney (UAW Social Security Department)
Alice G. Gosfield, Esq. (Alice G. Gosfield & Associates, PC)
Alan Hoops (PacifiCare Health Systems)
David Kessler, MD, JD (Yale School of Medicine)
Trudy Lieberman (Consumer Reports)
John M. Ludden, MD (Harvard Medical School)
Dorothy H. Mann, PhD, MPH (Consumer Representative)
Robert J. Margolis, MD (HealthCare Partners Medical Group)
Ira Millstein, JD (Weil, Gotshal & Manges)
Margaret E. O'Kane (NCQA)
Thomas R. Reardon, MD (American Medical Association)
Neil Schlackman, MD (Aetna, US Healthcare)
I. Steven Udvarhelyi, MD (Independence Blue Cross)
Andrew M. Wiesenthal, MD (Permanente Foundation)

Table 4 NCQA contact information

National Committee for Quality Assurance
2000 L Street, NW, Suite 500, Washington, DC 20036
Telephone: 202/955-3500
Fax: 202/955-3599
E-mail: customersupport@ncqa.org; TIL@ncqa.org (for technical questions regarding accreditation); HEDIS@ncqa.org (for technical questions regarding HEDIS auditing)
Web site: http://www.ncqa.org

NCQA/HEDIS AND PHARMACY PRACTICE

Pharmacy performance measures have been included in the managed care health plan accreditation process as well as in HEDIS. Therefore, pharmacists can play an important role in the process by understanding and utilizing the NCQA/HEDIS indicators and performance measures as a guide to improving patient care.[8] Pharmacists also play an important role in the accreditation/certification process of other healthcare organizations by providing input and participating in disease-management services as well as utilization management.

NCQA INFORMATION

Members of the 2000 NCQA Board of Directors are listed in Table 3. NCQA contact information is shown in Table 4.

REFERENCES

1. Venable, R.S.; Pagano, R. The long and winding road to NCQA accreditation. Drug Benefit Trends **1996**, *8*, 32–34, 36.

2. National Committee for Quality Assurance (NCQA) Web site. Available at http://www.ncqa.org. Accessed July 12, 2000; September 10, 2000.
3. Bell, D.; Brandt, E.N., Jr. Accreditation by the National Committee for Quality Assurance (NCQA): A description. J. Okla. State Med. Assoc. **1999**, *92* (5), 234–237.
4. O'Kane, M.E. A new approach to accreditation of HMOs. J. Health Care Benefits **1991**, 48–53, Sept./Oct.
5. Sennett, C. An introduction to the National Committee for Quality Assurance. Pediatr. Ann. **1998**, *27* (4), 210–214.
6. Health Plan Employer Data and Information Set (HEDIS) Web site. Available at: http://www.hedis.org. Accessed September 10, 2000.
7. National Committee for Quality Assurance (NCQA). Disease Management Fact Sheet. Available at: http://www.ncqa.org/Programs/Accreditation/Certification/DM%20Fact%20Sheet.pdf. Accessed March 30, 2002.
8. American Society of Health-Systems Pharmacists. Action guide for pharmacists in the managed care world. Practice-level strategies. Available at: http://www.ashp.org/managedcare/ashpactionguide.pdf. Accessed March 30, 2002.

National Community Pharmacists Association

Calvin J. Anthony
Robert A. Malone
National Community Pharmacy Association, Arlington, Virginia, U.S.A.

INTRODUCTION

The National Community Pharmacists Association (NCPA) was established on October 20, 1898 in St. Louis, Missouri, as the National Association of Retail Druggists (NARD), a name it held until 1996. The organization was conceived by a small group of independent community drugstore owners from major metropolitan areas in the East and Midwest, who chose to band together for the express purpose of protecting their professional practice and economic livelihood from a stamp tax on proprietary medicines to fund the Spanish-American war. From this modest organizational effort (predictably over a legislative matter) to coalesce a national group representing both the professional and proprietary interests of independent retail pharmacy grew one of the most powerful national associations—pharmacy or otherwise—ever to grace the halls of Congress. In 1937, in the wake of two extraordinary congressional victories that ensured the enactment of the landmark Robinson–Patman Anti-Discrimination Act and the Tydings–Miller Fair Trade Law—both of which remain foundations of protections for small business— *Business Week* Magazine labeled NARD "the most powerful trade association today."

HISTORICAL OVERVIEW

Due to its tenacity, singularity of purpose (promoting the cause of the independent community pharmacist), and the grassroots strength of its active and vocal membership, NARD/NCPA earned the respect of its adversaries as well as of its allies, including several U.S. presidents. The issues confronting independent pharmacists at the turn of the 20th century have a familiar ring to today's practitioners—price cutting, price protection, price discrimination, fair trade, mail order, physician dispensing, and encroachments by the federal government on the practice of the pharmacy profession. Legislative accomplishments of NARD/NCPA during the latter half of the 20th century sought to address these issues, particularly as they relate to discriminatory pharmaceutical pricing and other practices (such as the Prescription Drug Marketing Act of 1988, designed to stop illegal prescription drug diversion, and "best price" legislation in 1990 requiring manufacturers to offer their best prices in the Medicaid drug program); retaining consumers' "freedom of choice" to patronize the pharmacy of their choice through enactment of state laws inspired by NARD/NCPA model legislation; preserving practices traditionally reserved for the pharmacy profession (such as compounding); and allowing pharmacists to collectively negotiate third-party contracts (in the 106th Congress as the Quality Health Care Coalition Act).

Concurrent with these legislative concerns, NARD/NCPA also provided leadership in advancing professional issues affecting its membership and the public it serves. Examples of this leadership included NARD/NCPA-supported "Ask Your Pharmacist" labeling on over-the-counter medicines, a 20-year campaign that culminated in 1999 with a Food and Drug Administration (FDA) directive permitting such labeling; the creation of the National Institute for Pharmacist Care Outcomes (NIPCO), an organization devoted to advancing pharmacist care services through a disease management certification program for pharmacists; and coalition-building to create the National Institute for Standards in Pharmacist Credentialing, a national model and process for credentialing pharmacists in disease state management.

NCPA is entering its 103rd year as the sole organization representing the interests of independent community pharmacists. New challenges for the organization and its members lie ahead in the new millennium—surviving in a managed care environment, fine-tuning

Encyclopedia of Clinical Pharmacy
DOI: 10.1081/E-ECP 120006212

pharmaceutical delivery services to ensure fair compensation, enhancing professional knowledge through credentialing to deliver the highest levels of personalized pharmacist care, and continuing to refine the mission of the profession.

ORGANIZATIONAL STRUCTURE AND GOVERNANCE

The current organizational and governance structure of NCPA is essentially unchanged from that created by the organization's founding fathers in 1898, and has served as its enduring strength in the intervening 102 years. The structure consists of a full-time salaried Executive Vice-President, who serves as the Chief Executive Officer of the Association, as well as eight officers and six members of the Executive Committee, elected to staggered terms of office once a year. The officers and members of the Executive Committee (OEC), which also includes the Executive Vice-President as an ex-officio member, are responsible for the management of association priorities and activities.

The OEC are named at the annual meeting of the association's House of Delegates, which meets once a year at NCPA's Annual Convention. Officers and Executive Committee members typically serve a period of 12 to 14 years in association leadership as each moves up the ladder of responsibility (for example, from fifth Vice-President to fourth Vice-President, service on the Executive Committee, President-Elect to President).

The House of Delegates is comprised of representatives, who are appointed by their respective state or local pharmacy associations, to serve in this policy-making body. The House of Delegates provides policy guidance to the association's OEC through deliberations and passage of resolutions and policy positions, which are generated by eight standing committees that meet each year to formulate policy directives.

The eight committees are consumer affairs and public relations, home healthcare pharmacy services, long-term care pharmacy services, independent pharmacy chains, management, national legislation and government affairs, third-party payment programs, and professional relations. One NCPA member from each state is nominated by his or her state or local pharmacy association and appointed by the NCPA president to serve on a committee. The committees review and evaluate current NCPA policies and activities, develop recommendations for existing and new projects, and offer suggestions on how best to respond to the emerging needs and challenges of the pharmacy marketplace.

The House of Delegates and standing committees structure ensures that the NCPA membership at large has the opportunity to provide maximum input into association policies and programs.

The leadership of the NCPA can be found at ncpanet.org.

Executive Vice President:
Calvin J. Anthony, P.D.

MISSION

The mission of the association is quoted as follows:

- We are dedicated to the continuing growth and prosperity of independent community pharmacy in the United States.
- We are the national pharmacy association representing the professional and proprietary interests of independent community pharmacists and will vigorously promote and defend those interests.
- We are committed to high-quality pharmacist care and to restoring, maintaining, and promoting the health and well-being of the public we serve.
- We believe in the inherent virtues of the American free enterprise system and will do all we can to ensure the ability of independent community pharmacists to compete in a free and fair marketplace.
- We value the right to petition the appropriate legislative and regulatory bodies to serve the needs of those we represent.
- We will utilize our resources to achieve these ends in an ethical and socially responsible manner.

CURRENT MAJOR INITIATIVES OR DIRECTIONS

The year 1999–2000 ushered in the second century of independent pharmacy service to the nation's communities and NCPA's existence as the only national association representing independent pharmacy. The new millennium finds the association's membership robust and thriving, thus stemming a decade-long decline in independent store count brought about by a difficult managed care environment and aggressive chain drugstore and mail order incursions in the marketplace.

There are nearly 25,000 single-store independent pharmacies, independent chains, independent franchises, and independent pharmacist-owned supermarket pharmacies in the United States or nearly half of the 52,600 total stores in the pharmacy sector. Independent pharmacy today represents a $49 billion marketplace, where independents' prescription sales are $41 billion or nearly one-half of the retail prescription market. Independent pharmacies dispense 1.1 billion prescriptions annually.

The average independent's pharmacy sales are $1.97 million, up 37% in the last 3 years; average prescription sales are $1.64 million, up 43% in the last 3 years.

Independents are rated the best pharmacies in America by consumers because of their personal attention, the speed and efficiency of dispensing, and the medication information provided. According to the *1999 Ortho Biotech Retail Pharmacy Digest*, consumers using independent pharmacies reported the highest level of satisfaction, and customers of independent pharmacies were the most likely to recommend their pharmacy to a friend or relative.

Independents are leaders in providing pharmacist care services: Nearly 42% of independents are certified to provide disease management services, and more than 12,000 community pharmacists have completed disease management programs offered by NCPA's National Institute for Pharmacist Care Outcomes (NIPCO).

Highlights of NCPA's 1999–2000 year were forged on both the legislative and technology fronts as the association grappled with the age-old challenges of third-party issues and saw many of its initiatives in the great healthcare reform debate of the early 1990s become standard fare for public embrace. At the same time, NCPA defined a brand new frontier for independents in the realm of the Internet and the World Wide Web.

It was no small achievement that NCPA marshaled widespread support in the House of Representatives for the Campbell/Conyers Bill (the Quality Health Care Coalition Act), calling for collective negotiations by pharmacists and other healthcare providers, free of the shadow of antitrust. By the time of a key vote on the measure, more than one-half of the House of Representatives (220 members), showing broad bipartisan support, signed on as cosponsors of the bill—a clear signal that this was an issue whose time had come. The bill passed the House in late June 2000 by a resounding vote of 276 to 136—a 2 to 1 margin.

Another bill that garnered widespread bipartisan support was the patient bill of rights, which riveted national attention on remedies proposed to improve the quality of health care and eliminate managed care abuses. Chief among the bill's provisions is the restoration of patient rights to have better access to the provider of their choice and prescriptions selected by their physicians.

Another legislative victory for independents was resounding support in both the House and Senate for landmark legislation that enables pharmacists, wholesalers, pharmacy buying groups, independent chains, and other licensed distributors to purchase Food and Drug Administration (FDA)-approved medicines from other countries, including Canada, the United Kingdom, and the European Union. Current law restricts imports to drug manufacturers. The Senate's 74 to 21 vote in July 2000 placed the import legislation before a House–Senate conference committee, where a vote was imminent in late-September.

On the regulatory front, a 20-year campaign waged by NCPA for federal recognition of community pharmacists as vital sources of medication information for consumers earned Food and Drug Administration (FDA) sanction in spring 1999. Pharmacists will now be included on the labeling for over-the-counter (OTC) medications—as in "Ask your doctor or pharmacist." NCPA was prominently featured at the White House ceremony announcing the FDA OTC label recognition.

On the professional development front, the NCPA-created NIPCO continued to lead the way in developing programs that help establish pharmacist care services in community pharmacies. To date, more than 12,000 community pharmacists have completed NIPCO-accredited disease management programs in pharmacist care skills, cardiovascular care, respiratory care, diabetes care, immunization skills, osteoporosis, and mental health.

On the business development front, NCPA and the Chain Drug Marketing Association (CDMA) announced an alliance, which bore fruit as a combined Expo 2000 midyear meeting sponsorship and the promise of other joint marketing programs. CDMA represents more than 100 small and regional drugstore chains.

NCPA was instrumental in the formation and funding of two major foundations that will benefit community pharmacy in 1999. The first is the Institute for the Advancement of Community Pharmacy, which received a $27.5 million grant from Knoll Pharmaceutical Company to benefit community pharmacy. The board of directors is composed of the CEOs of NCPA and the National Association of Chain Drug Stores (NACDS). The second is the result of the distribution of funds from a $723 million

settlement approved by a U.S. district court judge of a price-fixing case involving 20 major brand name pharmaceutical manufacturers. Approximately $18.5 million of settlement proceeds has been set aside for a new foundation, "the Foundation for the Advancement of Retail Pharmacy."

The Internet and World Wide Web became a reality for many of NCPA's members. In August 1999, CornerDrugstore.com was launched as the first national Internet solution for independents. Community pharmacists signed up in droves to become part of the largest virtual network of independents anywhere in the nation. Independents could build their own web sites by using CornerDrugstore.com templates, or join their web site in the national network hosted by CornerDrugstore.com. To date, more than 4000 independent pharmacies have enrolled in the program, which has earned the endorsement of more than 50 pharmacy organizations.

Lastly, NCPA and TheraCom, Inc., a leading provider of specialty drugs, formed the Specialty Drugs Network in early 2000. The Specialty Drugs Network represents the largest and most comprehensive network in the United States for the distribution, administration, and care of patients in need of specialty drugs. The network will enable independent pharmacists to dispense a growing array of specialty pharmaceuticals for those drugs they choose to dispense and with "just-in-time" inventory control that offers cost-effective management of these medications. The network also offers professional education to build clinical competency through a specialty pharmaceuticals certificate training program.

In all, 1999–2000 was a good start to the second century—the new millennium—for the independent. As the figures indicate, independent community pharmacy has become a formidable market force, one to be reckoned with in the years ahead.

National Institute for Standards in Pharmacist Credentialing

Walter J. Morrison
Provider Payment Solution, Inc., Little Rock, Arkansas, U.S.A.

INTRODUCTION

The National Institute for Standards in Pharmacist Credentialing (NISPC) was formed in 1998 by the American Pharmaceutical Association, the National Association of Boards of Pharmacy (NABP), the National Association of Chain Drug Stores, and the National Community Pharmacists Association. These organizations developed NISPC to satisfy the need for a nationally recognized credentialing process that establishes appropriate national standards of care for, and facilitates recognition of the value of, disease state management (DSM) services provided by pharmacists.

As the scope of pharmacy practice continues to expand to meet the needs of a more health-conscious, knowledgeable consumer and the cost-containment concerns of managed care organizations, pharmacists are working with patients, physicians, and other providers to provide disease-specific primary care services. NISPC-credentialed pharmacists have demonstrated via a national examination process that they possess the competencies required to provide disease-specific services that merit inclusion in public and private health plans. Consequently, the NISPC credential facilitates prudent managed care plan design and payment decisions. Because of the focus on cognitive services for which payment can be expected, the NISPC credential also can be expected to facilitate the transition from practices which are primarily distributive to those that have a better balance between distributive and cognitive services. The NISPC credentialing process provides the most viable approach to achieve the acceptance of pharmaceutical care in the national marketplace.

Pharmacists are providing DSM services to help patients manage such chronic diseases as asthma, diabetes, and dyslipidemia, and drug therapies such as anticoagulation. Hundreds of pharmacists, including those serving on college faculties, have made the decision to seek NISPC credentialing. They agree that the NISPC credential provides the national recognition and assurance that patients and payers will require of pharmacists who choose to provide disease-specific patient management services.

ORGANIZATIONAL STRUCTURE/MEETINGS

NISPC is an autonomous and independent agency with a four-member Board of Directors, one from each of the founding organizations, and an Executive Director. In April 2000, Dr. Walter J. Morrison was selected to serve as the first Executive Director by the NISPC Board of Directors. The NISPC board includes Calvin J. Anthony, Executive Vice President, National Community Pharmacists Association; Carmen A. Catizone, Executive Director/Secretary, National Association of Boards of Pharmacy; John A. Gans, Executive Vice President/CEO, American Pharmaceutical Association; and Kurt A. Proctor, Senior Vice President, National Association of Chain Drug Stores. Mr. Anthony has served as president since incorporation. The NISPC Board of Directors has scheduled quarterly meetings, usually on the day immediately preceding the quarterly meetings of the Joint Commission of Pharmacy Practitioners (JCPP).

To provide assurance that the competencies, standards, and objectives on which the respective examinations are based remain current, NISPC has established a Standards Review Board. A Payers Advisory Panel has been established to provide recommendations related to the ambulatory care credentialing needs of firms paying for those primary health care services provided by pharmacists. NISPC is advised by a Standards Review Board and Payers Advisory Panel, which provide a broad coalition of

Encyclopedia of Clinical Pharmacy
DOI: 10.1081/E-ECP 120006295

professional and industry representation, including boards of pharmacy, colleges of pharmacy, interdisciplinary health professionals, national health care associations, payers, pharmaceutical manufacturers, practitioners, public/consumer advocates, quality assessment organizations, and state pharmacy associations. These advisory groups are scheduled to meet annually.

NISPC DSM EXAMINATIONS

Under an agreement with NISPC, the NABP has developed examinations that are psychometrically valid and legally defensible. NISPC currently offers four disease-specific examinations that assess a pharmacist's competency in the management of diabetes, dyslipidemia, asthma, and anticoagulation therapy. Pharmacists can pursue credentialing in one, two, three, or all four areas of practice. NISPC DSM examinations are administered daily in a computer-based format through LaserGrade, Inc.'s, nationwide system of testing centers. Each exam consists of multiple-choice questions and takes approximately 3 hours to complete. Results are made available in 7 to 10 business days.

Pharmacists who earn a passing score on the NISPC DSM examinations are credentialed and issued a certificate by NISPC. The certificate documents to employers, peers, and the public that they possess the knowledge and skills to provide disease-specific services to patients. NISPC-credentialed pharmacists are recognized on the NISPC and NABP web sites.

Information about the NISPC DSM examinations is available in the *Disease State Management (DSM) Examination Registration Bulletin*. In addition to a registration form and information about the testing procedures, the *Bulletin* contains the competency statements, or blueprint, upon which the DSM examinations are developed. Also included in the *Bulletin* are the standards and objectives for each DSM examination. Cross-referenced to the competency statements, the standards and objectives delineate the knowledge base expected of pharmacists providing disease-specific primary care services.

Examination Eligibility Requirements

Pharmacists seeking the NISPC credential must be licensed in good standing with their state board of pharmacy to sit for the examinations. The NISPC DSM exams are designed to assess the competencies of practicing pharmacists with 2 or more years of experience; however, there is no traditional practice experience requirement. Demonstrated competence, rather than years of experience, is the factor being evaluated. Also, unless the exams are being used as personal assessment tools, continuing professional education programs designed to prepare pharmacists for NISPC examinations are strongly recommended. During the first 2 years, the exams were offered, 1649 pharmacists took 1973 examinations with a success rate of 56%.

Exam Registration Process

The cost of a single NISPC DSM examination is $135. If a person registers for two or more exams on the same registration form, there is a charge of $135 for the first and $110 for each additional exam. To register for an examination, contact the NISPC Testing Center to request a *DSM Registration Bulletin*. Also available is the electronic testing *DSM Examinations Candidate's Review Guide* (CRG), which contains information about the examinations and sample questions. The CRG may be purchased for $10 through the NISPC Testing Center or downloaded for free from the NISPC web site (www.nispcnet.org). A CRG order form is also included in the *DSM Registration Bulletin*. The NISPC Testing Center address is 700 Busse Highway, Park Ridge, IL 60068. The Center phone number is (847) 698-6227, the fax number is (847) 698-0124, and the e-mail address is dsm@nabp.net.

ACTIVITIES AND ISSUES

NISPC will continue to support efforts to revise state and federal laws and regulations that limit a pharmacist's ability to provide DSM services. State boards of pharmacy will be encouraged to adopt the NISPC competencies, standards, and objectives, and to provide for inspectors that are credentialed and competent to evaluate the quality of care that is being provided. The preceding will position boards to address any complaints brought to their attention regarding the quality of care provided by the pharmacists they have licensed. Colleges of pharmacy will be encouraged to make available the knowledge and skills that will adequately prepare their graduates to pass the NISPC exams immediately after licensure. Colleges and pharmacy associations will be encouraged to provide continuing professional education with the depth and breadth of content required for success on the NISPC exams. They will also be encouraged to develop implementation manuals and ''safety nets'' to facilitate decisions regarding the provision of DSM services. NISPC will be encouraging those conducting studies to adopt standardized data collection procedures to address the need for significant studies with documented results.

In addition to the annual critique of the respective competency statements, standards, and objectives, the following outstanding issues will require consideration by the NISPC advisory groups and disposition by the Board of Directors:

- The need for practical and written exams to meet credentialing requirements.
- Recertification requirements.
- Establishment of an appropriate approach for pharmacists to indicate that they have satisfied NISPC credentialing requirements.
- Expansion of the current dyslipidemia exam to cover all coronary vascular-related diseases, except coronary heart failure.
- Consideration of other pharmacist/pharmacy ambulatory care credentialing needs, if recommended by the Payers Advisory Panel.

Walter J. Morrison, the former Executive Director, National Institute Standards in Pharmacist Credentialing prepared this document.

National Institutes of Health

Karim Anton Calis
Charles E. Daniels
National Institutes of Health, Bethesda, Maryland, U.S.A.

INTRODUCTION

Founded in 1887, the National Institutes of Health (NIH) is one of the world's foremost biomedical research centers, and the federal focal point for medical research in the United States. The NIH, comprised of 27 separate institutes and centers, is one of eight health agencies of the Public Health Service, which, in turn, is part of the U.S. Department of Health and Human Services[1] (For information on the history of NIH, visit http://www.nih.gov/od/museum/.)

The goal of NIH research is to acquire new knowledge to help prevent, detect, diagnose, and treat disease and disability, from the rarest genetic disorder to the common cold. The NIH mission is to uncover new knowledge that will lead to better health for everyone and works toward that mission by conducting research in its own laboratories; supporting the research of nonfederal scientists in universities, medical schools, hospitals, and research institutions throughout the country and abroad; helping in the training of research investigators; and fostering communication of medical and health sciences information.

Scientific progress depends mainly on the work of dedicated scientists. About 50,000 principal investigators—working in every state and in several foreign countries, from every specialty in medicine, every medical discipline, and at every major university and medical school—receive NIH extramural funding to explore unknown areas of medical science. Supporting and conducting both NIH's extramural and intramural programs are about 15,600 employees, more than 4000 of whom hold professional or research doctorate degrees. The NIH staff includes intramural scientists, physicians, dentists, veterinarians, nurses, pharmacists, and laboratory, administrative, and support personnel, plus an ever-changing array of research scientists in training.

The rosters of those who have conducted research, or who have received NIH support over the years include the world's most illustrious scientists and physicians. Among them are 97 scientists who have won Nobel Prizes for achievements as diverse as deciphering the genetic code and learning what causes hepatitis. Five Nobelists made their prize-winning discoveries in NIH laboratories: Drs. Christian B. Anfinsen, Julius Axelrod, D. Carleton Gajdusek, Marshall W. Nirenberg, and Martin Rodbell.

CONTRIBUTIONS TO PUBLIC HEALTH

NIH is a key element in a partnership that has thrived for decades and includes universities and academic health centers, independent research institutions, and private industry, where research programs and product development activities help make federally funded research findings more widely available. The partnership, which has produced many of the medical advances that benefit Americans today, also includes voluntary and professional health organizations, and the Congress, which consistently has supported this vast enterprise.

NIH research has played a major role in making possible the following achievements of the last few decades: Mortality from heart disease, the number one killer in the United States, dropped by 36% between 1977 and 1999. Death rates from stroke decreased by 50% during the same period. Improved treatments and detection methods increased the relative five-year survival rate for people with cancer to 60%. Paralysis from spinal cord injury is significantly reduced by rapid treatment with high doses of a corticosteroid. Treatment given within the first 8 h after injury increases the likelihood of recovery in severely injured patients who have lost sensation or mobility below the point of injury. Long-term treatment with anticlotting medicines cuts stroke risk by 80% from atrial fibrillation. In schizophrenia, where patients suffer frightening delusions and hallucinations, new medications can reduce or eliminate these symptoms in 80% of patients. Chances for survival increased for infants with respiratory distress syndrome, an immaturity of the lungs, due to development of a substance to prevent the lungs from collapsing. In general, life expectancy for a baby born today is almost three decades longer than one born at the be-

Encyclopedia of Clinical Pharmacy
DOI: 10.1081/E-ECP 120006186

ginning of the century. With effective medications and psychotherapy, the 19 million Americans who suffer from depression can now look forward to a better, more productive future. Vaccines protect against infectious diseases that once killed and disabled millions of children and adults. Dental sealants have proved 100% effective in protecting the chewing surfaces of children's molars and premolars, where most cavities occur. In 1990, NIH researchers performed the first trial of gene therapy in humans. Scientists are increasingly able to locate, identify, and describe the functions of many of the genes in the human genome with the ultimate goal of developing screening tools and gene therapies for cancer and many other diseases.

INSTITUTES, CENTERS, AND OFFICES

The NIH is an amalgam of many diverse research and support organizations (Fig. 1). A list of NIH's 27 Institutes and Offices (along with links to their web sites) is available at http://www.nih.gov/icd/. The missions and activities of the Institutes and Centers are described at http://www.nih.gov/icd/programs.htm. The Institutes include the National Cancer Institute, the National Heart, Lung, and Blood Institute, the National Institute of Mental Health, the National Institute of Allergy and Infectious Diseases, and the Human Genome Research Institute. Centers include the Vaccine Research Center, the Fogarty International Center, the Center for Scientific Review, and the National Center for Complementary and Alternative Medicine. Other notable components of the NIH are the Warren Grant Magnuson Clinical Center, Office of Social and Behavioral Research, and the National Library of Medicine.

Campus and Facilities

NIH intramural scientists conduct their research in laboratories located on a 300-acre campus in Bethesda, and in several field units across the country and abroad. Maps of the campus and of the local area are located at http://www.nih.gov/about/maps.html. The Intramural Research Programs, although representing only a small part of the total NIH budget, are central to the NIH scientific effort. First-rate scientists are key to NIH intramural research. They collaborate with one another regardless of institute affiliation or scientific discipline, and have the intellectual freedom to pursue their research leads in NIH's own laboratories. These explorations range from basic biology, to behavioral research, to studies on the diagnosis and treatment of major diseases. Through clinical research, promising discoveries in the laboratories are translated into new therapies and treatments for patients. The NIH also has facilities in the Rockville, Maryland, area and the NCI Frederick Cancer Research and Development Center (FCRDC) at Fort Detrick in Frederick, Maryland. The National Institute of Environmental Health Sciences, which studies the adverse effects of environmental factors on human health, operates from its main facility in Research Triangle Park (RTP), North Carolina. Other laboratory facilities include the NIH Animal Center in Poolesville, Maryland; the National Institute on Aging's Gerontology Research Center in Baltimore, Maryland; the Division of Intramural Research of the National Institute on Drug Abuse, also in Baltimore; the National Institute of Allergy and Infectious Diseases' Rocky Mountain Laboratories in Hamilton, Montana, and several smaller field stations.

The 14-story Warren Grant Magnuson Clinical Center (http://www.cc.nih.gov) is the NIH's hospital and center for clinical research. Patients come from all over the world to participate in clinical studies here. Each year, the Clinical Center admits about 7000 inpatients. The ambulatory clinical research unit accounts for nearly 68,000 outpatient visits per year. Patients at the NIH participate in clinical trials that span a wide range of diseases and conditions. The Clinical Center also houses the Visitor Information Center, which is NIH's information liaison and host to thousands of visitors each year. Construction began in 1997 for the new Mark O. Hatfield Clinical Research Center. The Hatfield Center will house the research hospital's 250 beds for inpatient and outpatient care, outpatient facilities, and research laboratories. It will connect to the current building, which opened its doors in 1953. The hallmark of the original building is the proximity of the patient care to scientific labs. The new facility will amplify the bench-to-bedside tradition, providing a crucial link from the rapidly moving biomedical findings of the laboratory into the mainstream of medical practices. The Clinical Center Pharmacy Department (http://www.cc.nih.gov/phar) provides extensive clinical and research services and conducts several highly regarded training programs. Pharmacists also contribute directly to the NIH intramural and extramural research programs.

Funding

From a total of about $300 in 1887, the NIH budget has grown to more than $20.3 billion in 2001. The NIH is funded through direct appropriations from the U.S. Congress. Approximately 82% of the investment is made through grants and contracts supporting research and

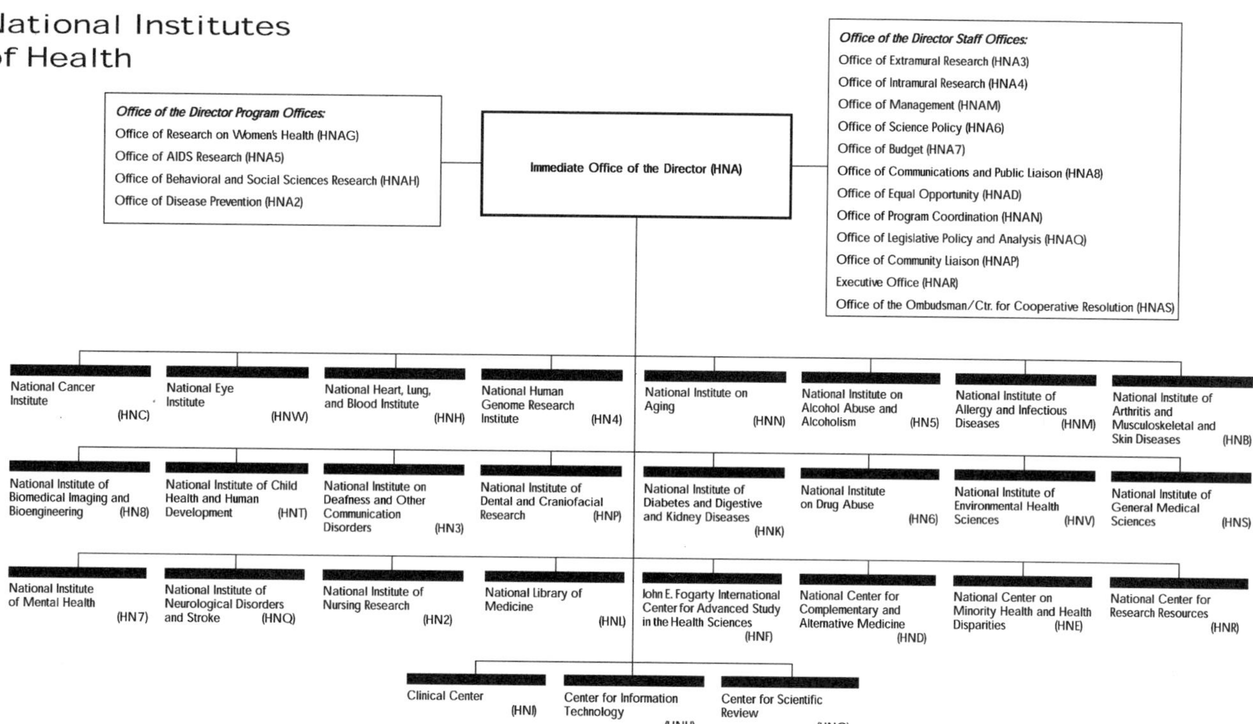

Fig. 1 The Mission of the National Institutes of Health is science in pursuit of knowledge to improve human health. This means pursuing science to expand fundamental knowledge about the nature and behavior of living systems; to apply that knowledge to extend the health of human lives; and to reduce the burdens resulting from disease and disability. The National Institutes of Health seeks to accomplish its mission by: ① Fostering fundamental discoveries, innovative research, and their applications in order to advance the Nation's capacity to protect and improve health; ② Developing, maintaining, and renewing the human and physical resources that are vital to ensure the Nation's capability to prevent disease, improve health, and enhance quality of life; ③ Expanding the knowledge base in biomedical, behavioral, and associated sciences order to enhance America's economic well-being and ensure a continued high return on the public investment in research; and ④ Exemplifying and promoting the highest level of scientific integrity, public accountability, and social responsibility in the conduct of science.

training in more than 2000 research institutions throughout the United States and abroad. In fact, NIH grantees are located in every state in the country. These grants and contracts comprise the NIH Extramural Research Program. Approximately 10% of the budget goes to NIH's Intramural Research Programs, the more than 2000 projects conducted mainly in its own laboratories. About 8% of the budget is for both intramural and extramural research support costs.

Clinical Studies

Information regarding clinical trials conducted at the NIH Clinical Center in Bethesda, Maryland, can be found at http://clinicalstudies.info.nih.gov/. ClinicalTrials.gov (http://clinicaltrials.gov) provides information on federal and private medical studies at thousands of locations nationwide. Information regarding the types and conduct of clinical trials is also provided.

Training

Clinical research training is an integral part of the NIH mission. Training opportunities are available for students, physicians, other health care professionals, and postdoctoral researchers. These training programs vary in scope and can range from several weeks to several years in duration. Information regarding NIH research training opportunities is available at http://www.training.nih.gov. Several remote-access resources have been developed for training researchers and clinicians. For example, the course on Clinical Research Training (http://www.cc.nih.gov/ccc/cr/index.html) was developed to assist those with an interest in conducting clinical research. Other training programs are described at http://www.cc.nih.gov/profscientists.cgi.

GRANTS AND CONTRACTS

Information regarding NIH grants and funding opportunities may be found at http://www.grants.nih.gov/grants/index.cfm. The grants page provides information about NIH grant and fellowship programs, policy changes, administrative responsibilities of awardees, the Computer Retrieval of Information on Scientific Projects (CRISP) database, and the numbers and characteristics of awards made by the NIH. CRISP is a searchable database of federally funded biomedical research projects conducted at universities, hospitals, and other research institutions. The database, maintained by the Office of Extramural Research at the National Institutes of Health, includes projects funded by the National Institutes of Health (NIH), Substance Abuse and Mental Health Services (SAMHSA), Health Resources and Services Administration (HRSA), Food and Drug Administration (FDA), Centers for Disease Control and Prevention (CDCP), Agency for Health Care Policy Research (AHCPR), and Office of Assistant Secretary of Health (OASH). Users, including the public, can access the CRISP interface to search for scientific concepts and emerging trends and techniques or to identify specific projects and/or investigators. This grants page also contains information regarding research contracts and research training opportunities and provides access to the NIH Guide for Grants and Contracts. The Guide is the official document for announcing availability of NIH funds for biomedical and behavioral research and research training.

Extramural Research

Final decisions about funding extramural research are made by senior scientific staff at the NIH headquarters. But long before this happens, the process begins with an idea that an individual scientist describes in a written application for a research grant. The project might be small, or it might involve millions of dollars. The project might become useful immediately as a diagnostic test or new treatment, or it might involve studies of basic biological processes with practical value that may not be apparent for many years. Each research grant application undergoes a peer review process. A panel of scientific experts, primarily from outside the government, who are active and productive researchers in the biomedical sciences, first evaluates the scientific merit of the application. Then, a national advisory council or board, comprised of eminent scientists as well as public members who are interested in health issues or the biomedical sciences, determines the project's overall merit and priority in advancing the research agenda of the particular NIH funding institute. Altogether, about 38,500 research and training applications are reviewed annually through the NIH peer review system. At any given time, the NIH supports 35,000 grants in universities, medical schools, and other research training institutions both nationally and internationally.

SCIENTIFIC RESOURCES

Scientific resources available from the NIH are described at http://www.nih.gov/science/. These include research sourcebooks, documents relating to molecular biology and molecular modeling, research labs on the Web, library and literature resources, model organisms for biomedical research, reagent programs, data banks, animal care and use guidelines, NIH-supported human specimen resources, bioethics resources, and lab safety and waste management resources, to name a few. Protomechanics, a guide for intramural scientists on preparing and conducting a research study, is available at http://www.cc.nih.gov/ccc/protomechanics/.

The National Library of Medicine is another principal unit of the NIH. The 10-story Lister Hill Center houses the Lister Hill National Center for Biomedical Communications and the National Center for Biotechnology Information. Both are components of the National Library of Medicine. The Library produces and publishes Index Medicus, a comprehensive monthly listing of articles appearing in the world's leading medical journals. The Library also operates a computerized Index Medicus, known as MEDLINE (PubMed), and has pioneered the introduction of large medical bibliographic databases.

NIH Image is a public domain image processing and analysis program for the Macintosh. It was developed at the Research Services Branch (RSB) of the National Institute of Mental Health (NIMH), part of the National Institutes of Health (NIH). The software is located at http://rsb.info.nih.gov/nih-image/Default.html.

HEALTH INFORMATION RESOURCES

A wide array of health information is provided by the NIH to consumers and healthcare professionals. The NIH Health Information Index, available at http://www.nih.gov/health/InformationIndex/HealthIndex/Pubincov.htm, provides information on diseases currently under investigation by NIH or NIH-supported scientists and major NIH research areas with links to the appropriate institute(s), center(s), or other component(s) to call for information, along with the appropriate phone numbers. The NIH web site does not offer personalized medical advice to individuals about their condition or treatment. The resources on this site should not be used as a substitute for professional medical care, and patients are urged to work with their medical care providers for answers to personal health questions. For general medical questions, patients are encouraged to visit the Health Information (http://www.nih.gov/health/) section of the NIH web site. MEDLINEplus at http://www.nlm.nih.gov/medlineplus/ is another useful source for obtaining information on various conditions, diseases, and wellness programs. The site also has a medical encyclopedia, a medication guide for consumers (*USP DI Advice for the Patient*), a glossary, medical dictionaries, health-related directories, special NIH programs, and other resources. For questions relating to specific foods, or prescription, or over-the-counter drugs, the Food and Drug Administration's (FDA) web site http://www.fda.gov/ is recommended. Healthfinder at http://www.healthfinder.gov/ contains valuable information on choosing quality medical care. Also, toll-free medical Information Hotlines may be found at http://www.nih.gov/health/infoline.htm.

Extensive information resources are available directly from the Institutes and Centers that comprise the NIH. For example, the National Cancer Institute provides extensive cancer-related information through its CancerNet program (http://cancernet.nci.nih.gov/) and the Cancer Information Service (1-800-4-CANCER). This includes cancer statistics, cancer information for patients, and access to a bibliographic cancer database (Cancerlit) and the cancer information database PDQ® (Physician Data Query). Health information and clinical guidelines are also available from other Institutes and Centers. The National Heart, Lung, and Blood Institute, for example, provides information for patients, healthcare professionals, and the general public about various vascular, heart, lung, blood, and sleep disorders (http://www.nhlbi.nih.gov/health/ index.htm). Authoritative national clinical guidelines pertaining to asthma, cholesterol, chronic obstructive pulmonary disease, hypertension, obesity, and others are also available (http://www.nhlbi.nih.gov/guidelines/ index.htm). Most of these NIH guidelines and those from other reputable organizations may also be accessible from the National Guideline Clearinghouse (http://www.guideline.gov/). Information regarding dietary supplements and herbal products is available from the National Center for Alternative and Complementary Medicine (http://nccam.nih.gov/). These include fact sheets, consensus reports, and the Complementary and Alternative Medicine Databases (CAM).

NEWS AND EVENTS

The News and Events page (http://www.nih.gov/news/) is the source for NIH news, calendars of events, press releases, special reports, and information about NIH-sponsored events—including events of interest to the media and those available through NIH VideoCasting. Information regarding NIH Consensus Development Conferences and NIH State-of-the-Science Conferences can be obtained via http://odp.od.nih.gov/consensus/default.html. NIH clinical alerts and advisories (http://www.nlm.nih.gov/databases/alerts/clinical_alerts.html) are provided to expedite the dissemination of findings from NIH-funded clinical trials where such release could significantly affect morbidity and mortality. The NIH Calendar of Events, or "Yellow Sheet," is located at http://www.nih.gov/news/calendar/nihcalendar.htm. It lists NIH-sponsored meetings and other meetings of interest to scientists, healthcare professionals, and the general public. It is updated daily, and the listed meetings are free and open to the public.

GENERAL INFORMATION

The e-mail and phone directory for NIH employees may be found at http://directory.nih.gov/.

Information about employment opportunities and summer internships at NIH may be found at http://www.nih.gov/about/index.html#employ.

The emergence of several abuses of the research process has generated some confusion about which office handles what type of abuse within the Department of Health and Human Services (HHS). The attached document (http://ori.dhhs.gov/) was prepared to help clarify the matter.

Information regarding Freedom of Information Act (FOIA) requests may be found at http://www.nih.gov/icd/od/foia/index.htm.

REFERENCE

1. National Institutes of Health Home Page (http://www.nih.gov).

PROFESSIONAL RESOURCES

National Library of Medicine

Joan K. Korth-Bradley
JB Ashtin Group Inc., Canton, Michigan, U.S.A.

INTRODUCTION

The National Library of Medicine (NLM) is the world's largest biomedical library containing nearly 6 million books, reports, journals, photographs, manuscripts, and computer images.[1,2] Located on the campus of the National Institutes of Health in Bethesda, Maryland, this national resource houses material on the topics of medicine, healthcare, and biomedical technology. Information on physical, life, and social sciences is also a part of the NLM. In addition to storing this immense collection of medical literature, the NLM operates more than 40 online databases,[3] the most popular of which is MEDLINE, the world's largest up-to-date online collection of biomedical citations.[1,4]

HISTORY

The NLM dates back to 1836 as a collection of medical texts in the office of the United States Army Surgeon General.[1] Publishing of the *Index Medicus* (a printed, monthly subject/author guide to medical journals) commenced in 1879 under the direction of John Shaw Billings. By 1910, this once-modest library of the Surgeon General's office had grown to be the largest medical library in the world. The National Library of Medicine became the official name of this collection in 1956 when the United States Army transferred the oversight of the library to the U.S. Public Health Service.

The goal of the NLM has remained constant—to increase the availability of medical literature worldwide. In the past 50 years the NLM, working to achieve this goal, has contributed tremendously to the biomedical field (Table 1). In 1971, the NLM launched MEDLINE, an index to biomedical articles, to support the information needs of healthcare professionals.[4,5] Originally used only by trained librarians, MEDLINE became directly accessible to healthcare professionals in the 1980s.[5] During the past two decades, the NLM expanded the topics of its databases and now includes a variety of databases on AIDS, cancer, research, clinical trials, chemical information, toxicology, and ethics (Table 2).

With the arrival of the World Wide Web and its ability to facilitate global communication came the opportunity to expand the services of the NLM. In 1997, the cumbersome methods of searching the MEDLINE database were replaced with user-friendlier search systems, namely Internet Grateful Med and PubMed. Moreover, searching via these resources was now free. These search systems fueled a 10-fold increase in the use of MEDLINE to 75 million searches annually.[6] Today, more than 250 million MEDLINE searches are performed each year.[1] In addition to MEDLINE, users can access almost any of the NLM databases from the NLM health information page, http://www.nlm.nih.gov/hinfo.html.

MEDLINE

MEDLINE is a bibliographic database that contains more than 11 million references to journal articles from 4300 national and international biomedical journals, covering the period from 1966 to the present.[1] Types of journals indexed in MEDLINE include pharmacy, allied health, nursing, medicine, veterinary medicine, dentistry, preclinical science, and life science journals. Each week, nearly 8000 reference citations, created by the NLM and its partners, are added to the MEDLINE database.

As Yahoo! and Excite are to the World Wide Web, PubMed and Internet Grateful Med are to MEDLINE (see Tables 2 and 3). Although both of these latter search services access MEDLINE, the two differ in the way searches are conducted.[5] (Search strategies and instructions for PubMed and Internet Grateful Med will not be included within this article because this information can be found in detail on their respective Web sites). Each service offers access to different databases. Accessing each database is simple and there is a direct link on the PubMed home page to Internet Grateful Med and vice versa. Results from searches in both services offers direct links to full-text articles from more than 800 participating

Encyclopedia of Clinical Pharmacy
DOI: 10.1081/E-ECP 120006185

Table 1 Highlights of the National Library of Medicine since 1956

Computerization of the *Index Medicus* and the development of MEDLINE
Provision of free access to the MEDLINE database
Development of
The NLM Web site (www.nlm.nih.gov), which is a portal to information on its healthcare databases, news, research programs, and library programs
PubMed, a free NLM database search service
MEDLINE*plus*, a Web site for consumer health information
the National Network of Libraries of Medicine
Development and implementation of an integrated library catalog (LOCATOR*plus*) that can be searched through the Internet
Creation of
A grant program, to support research in medical informatics and in biotechnology and health information
The Toxicology and Environmental Health Information Program, to facilitate governmental and nongovernmental literature (this program operates TOXNET, http://toxnet.nlm.nih.gov/)
The Lister Hill National Center for Biomedical Communications (http://lhncbc.nlm.nih.gov), which oversees NLM research and development
The National Center for Biotechnology Information (NCBI, www.ncbi.nlm.nih.gov), which is responsible for the databases that contain DNA and protein sequences, genome mapping data, and 3-D protein structures.
The National Information Center on Health Services Research and Health Care Technology
PubMed Central

(Adapted from Ref. [1].)

journals. The services offered by PubMed and Internet Grateful Med are updated continually, and the reader is encouraged to visit their Web sites often to check out new features.

The NLM developed PubMed as a user-friendly search tool to explore MEDLINE. For each citation listed in a search result, the user has the option to click on ''Related Articles,'' which takes the user to a listing of additional citations that are related to the search topic. Additionally, PubMed offers access to publisher Web sites for full-text articles, a citation matcher (service that assists a user in finding a complete citation), and access to the molecular biology databases of the NCBI. One of the newest features of PubMed is called ''Cubby,'' a page on the site that allows the user to store a search and update it in the future (users need to register for this free service). AIDSLINE soon will be added to PubMed.[7] A user can search MEDLINE and 14 other NLM databases (e.g., AIDSLINE, AIDSDRUGS, TOXLINE) via Internet Grateful Med. Since 1998, Internet Grateful Med searches MEDLINE using the PubMed retrieval engine.[3]

Once a MEDLINE search has been done via either PubMed or Internet Grateful Med, the user has the option of ordering listed articles from Loansome Doc®, an international, online, article-ordering service from the National Network of Libraries of Medicine.[1] Before users can place an order, however, they must identify an ''Ordering Library,'' then register using the Loansome Doc registration page. Although there is no fee for using Loansome Doc, there is a charge for copies of articles.

IMPLICATIONS FOR PHARMACY PRACTICE AND RESEARCH

The NLM has brought about important changes to pharmacy practice and research. By searching MEDLINE and its other databases, pharmacists, pharmacy researchers, and pharmacy students can find answers to clinical questions, seek help in decision making, obtain information to support research, expand their knowledge on a certain drug topic, and explore and obtain available data to prepare educational tools for colleagues, students, patients, and consumers.

When a pharmacist or pharmacy student has a clinical question, a good place to begin seeking an answer is the medical literature. MEDLINE is used often to solve clinical questions;[10,11] rare clinical conditions have been diagnosed from the results of a MEDLINE search.[4] Some hospitals or medical libraries offer search systems that will search medical literature from MEDLINE and other sources, such as *Best Evidence* and the Cochrane Library.[10] Other sources for clinical decision making by pharmacists include clinical practice guidelines. The NLM offers the Health Services/Technology Assessment Text (HSTAT),[12] a free resource that allows users to access full-text evidence reports, clinical practice guidelines and consensus statements (such as those from the NIH and the AHCPR), technology assessments, and other documents useful in healthcare decision making. One other interesting resource made available online by the NLM is the Online Mendelian Inheritance in Man™ Web site.[13] This site, primarily for healthcare professionals, is a searchable database of human genes and genetic disorders.

Aiming to bring the benefits of advanced technology to the healthcare profession, other NLM programs, like the Visible Human Project, offer new opportunities to study human anatomy and options to further medical research. The Visible Human Project[14,15] is essentially a database of electronic transverse CT, MRI, and cross section images from an entire female and male human cadaver.

Table 2 Primary databases of the National Library of Medicine

NLM database	URL	Description
MEDLINE	Search via PubMed or via Internet Grateful Med (below)	Can be searched using NLM's controlled vocabulary, MeSH, or by author name, title word, text word, journal name, phrase, alone or in combination. The result of a search is a list of citations (including authors, title, source, and often an abstract) to journal articles.
PubMed	www.ncbi.nlm.nih.gov/PubMed	Launched in 1997 to provide free access via the World Wide Web to MEDLINE. Also links to molecular biology databases.[a]
Internet Grateful Med	www.igm.nlm.nih.gov	Launched in 1997 to provide free access via the World Wide Web to MEDLINE. Also provides access to 14 other NLM databases.[a]
PubMed Central	www.pubmedcentral.nih.gov	System launched in 2000 to make available life science research results online. Provides free full-text articles from its own journals; also provides links to free full-text articles from peer-reviewed journals (*nih.gov*).
MEDLINE*plus*	http://medlineplus.gov	Site aimed at answering health-related questions of the consumer. This site contains information on hundreds of diseases, conditions, and wellness issues and includes directories of physicians, hospitals, and medical libraries.
ClinicalTrials.gov	http://clinicaltrials.gov	Site targeted toward consumers and healthcare professionals that provides easy access to information on nearly 5000 clinical trials for a wide range of diseases and conditions.
TOXNET	http://toxnet.nlm.nih.gov	User-friendly Web site that allows searching of a cluster of bibliographic and factual databases with information on toxicology and environmental health. Includes the Hazardous Substances Databank (HSDB), with detailed information about 4500 chemicals, and TOXLINE, containing >3 million citations to journal articles and technical reports.

[a]See Table 3.
(Adapted from Refs. [1, 3, 7–9].)

Table 3 Choosing PubMed versus Internet Grateful Med

Both search services provide functions for helping create and refine searches, including access to hundreds of thousands of biomedical terms from the NLM Unified Medical Language System.
In general, the choice of which Web site to use is personal.
Use PubMed when you need:
  An exhaustive search (i.e., 1996 to present)
  Genetics research information
  Information from a textbook[a]
Use Internet Grateful Med when you need:
  Access to any of the NLM AIDS databases or databases other than MEDLINE
  Articles from 1963 to 1965 (OLDMEDLINE)

[a]This is a "future feature." At the time of this printing, only one book is available on the PubMed Web site. PubMed is working with book publishers to be able to offer a larger selection of textbooks for the user to search online.
(Adapted from Refs. [5] and [7].)

Several applications are offered on the Web site to allow the user to view the images from the project. Although these data sets are free, the user must have a license from the NLM to view and use images from the site.

Pharmacists, with their knowledge of drugs and diseases, may have a tremendous impact on the healthcare team if they possess the necessary skills required to search NLM databases and interpret search results. Even with the amazing technology of the Internet and with the enormity of accessible medical databases, research shows that physicians still do not make use of the information retrieval systems like MEDLINE.[16] With the arrival of high-speed Internet connections, Loansome Doc, and the increased number of journal publishers offering full-text articles online, pharmacists quickly can obtain articles from the results of their searches. This new technology saves mounds of time spent in the library searching for articles—which potentially translates into more time to assist other members of the healthcare team and for patient care.

TELEMEDICINE AND ELECTRONIC PUBLISHING: CURRENT FOCI OF THE NLM

Telemedicine and electronic publishing are top priorities for the NLM. Providing timely, quality healthcare to residents in remote and rural locations via telemedicine is a technological advance that we would not have without the Internet or without funding by organizations like the NLM. Telemedicine is the ''transfer of electronic medical data (i.e., high-resolution images, sounds, live video, and patient records) from one location to another.''[17] Currently, the NLM is one of six strategic partners that provide support to the Telemedicine Information Exchange,[17] an organization that keeps track of all that is happening worldwide in the area of telemedicine, including operating a database of nearly 10,000 citations regarding this topic. The NLM is funding at least 19 projects under the National Telemedicine Initiative to determine the feasibility, costs, and efficacy of telemedicine compared with conventional care.[18]

In terms of electronic publishing, the NLM launched PubMed Central this year in an effort to reduce the time from article submission to publication and to increase the dissemination of new information in the life sciences.[19,20] PubMed Central eventually will contain two types of data depositories: The first will be peer-reviewed content from journals that will partner with NLM to make their content available free to the public; and the second depository will contain articles submitted directly by authors. Ideally, PubMed Central would increase the options for researchers and academicians to publish their research. This is a very new concept in medical publishing and it will be exciting to watch how it takes shape in terms of the quality of research articles submitted and the way in which the healthcare community embraces this new source of information.

SUMMARY

The NLM is an organization devoted primarily to improving the access by health professionals, researchers, students, and consumers to the worldwide biomedical literature. Although MEDLINE is the most recognized and most frequently used[4] service of the NLM by the pharmacy community, the NLM serves also to create and expand databases and databanks to explore new communication technologies in order to improve the organization and use of biomedical information, to support a national network of local and regional medical libraries and to educate users about available sources of information so that they may conduct their own research concerning medical topics.[1]

CONTACTING THE NLM

The NLM is located at 8600 Rockville Pike, Bethesda, Maryland, U.S.A., 20894. To contact the NLM, send an e-mail to custserv@nlm.nih.gov; call (888) FIND-NLM, (888) 346-3656, (301) 594-5983; or view its Web site at http://www.nlm.nih.gov. For information about the holdings of the NLM, visit LOCATOR*plus* at www.nlm.nih.gov/locatorplus/.

REFERENCES

1. www.nlm.nih.gov (accessed Oct. 2000).
2. Ball, M.J. International efforts in informatics. Creating a global village for healthcare. MD Comput. **2000**, 50–53, May/June.
3. www.igm.nlm.nih.gov (accessed Aug. 2000).
4. Lindberg, D.A.B. The NLM and Grateful Med: Promise, public health, and policy. Pub. Health Rep. **1996**, *111*, 552–555.
5. DeGeorges, K.M. Help! I need a MEDLINE search! AWHONN Lifelines **1998**, *2* (5), 38–40.
6. Lindberg, D.A.B.; Humphreys, B.L. Medicine and health on the Internet: The good, the bad, and the ugly. JAMA, J. Am. Med. Assoc. **1998**, *280* (15), 1303–1304.
7. www.ncbi.nlm.nih.gov/PubMed (accessed Oct. 2000).
8. Stephenson, J. National Library of Medicine to help consumers use online health data. JAMA, J. Am. Med. Assoc. **2000**, *283* (13), 1675–1676.
9. www.clinicaltrials.gov (accessed Oct. 2000).
10. Hunt, D.L.; Jaeschke, R.; McKibbon, K.A. For the evidence-based medicine working group. Users' guides to the medical literature XXI. Using electronic health information resources in evidence-based practice. JAMA, J. Am. Med. Assoc. **2000**, *283* (14), 1875–1879.
11. Westberg, E.E.; Miller, R.A. The basis for using the Internet to support the information needs of primary care. JAMA, J. Am. Med. Assoc. **1999**, *6*, 6–25.
12. http://text.nlm.nih.gov/ (accessed Oct. 2000).
13. www.ncbi.nlm.nih.gov/entrez/query.fcgi?db=OMIM (accessed Oct. 2000).
14. McCray, A.T. Improving access to healthcare information: The Lister Hill National Center for Biomedical Communications. MD Comput. **2000**, 29–34, March/April.
15. http://www.nlm.nih.gov/research/visible/visible_human.html (accessed Oct. 2000).
16. Hersh, W.R.; Hickham, D.H. How well do physicians use electronic information retrieval systems? A framework for investigation and systematic review. JAMA, J. Am. Med. Assoc. **1998**, *280* (15), 1347–1352.
17. http://tie.telemed.org (accessed Oct. 2000).
18. Larkin, M. Telemedicine gets a chance to prove itself. Ann. Intern. Med. **1997**, *127* (12), 1149–1150.
19. Caelleigh, A.S. PubMed Central and the new publishing landscape: Shifts and tradeoffs. Acad. Med. **2000**, *75* (1), 4–10.
20. www.pubmedcentral.nih.gov (accessed Oct. 2000).

PROFESSIONAL DEVELOPMENT

Neurology Specialty Pharmacy Practice

Susan C. Fagan
University of Georgia College of Pharmacy, Athens, Georgia, U.S.A.

Melody Ryan
University of Kentucky, Lexington, Kentucky, U.S.A.

INTRODUCTION

Neurology specialty pharmacy practice involves the provision of pharmaceutical care to patients with neurological diagnoses in collaboration with neurologists and other members of the neurology team. Although it has many similarities with other clinical pharmacy specialties, there are unique aspects and strategies employed that warrant discussion. The neurological deficits themselves make it sometimes more challenging to obtain information from the patient, and a relative paucity of well-studied treatments for many neurological illnesses poses significant challenges. Consensus statements, whether national or regional, are important sources of treatment information in this specialty.

Historically, neurology pharmacy specialists have concentrated on the most medication-intense subspecialty within neurology, which is epilepsy. However, the 1990s saw an explosion in the number of new therapies for neurological illness, both for previously untreatable conditions (e.g., amyotrophic lateral sclerosis, acute stroke) and as additions to the therapeutic armamentarium in already treatable conditions (e.g., epilepsy, Parkinson's disease). The twenty-first century ushers in a continued intense interest in the development of novel therapies for nervous system diseases and steadily broadening opportunities for the pharmacy specialist in neurology.

APPROACH TO THE NEUROLOGIC PATIENT

Pharmacotherapy History

The importance of an accurate pharmacotherapy history to the neurology pharmacy specialist cannot be overemphasized. When integrated with information from the community pharmacy, the outpatient clinic, and the laboratory, a careful assessment of the patient's recollection, together with that of family members, can often expedite implementation of effective therapy and prevent adverse drug effects. The pharmacotherapy history includes assessments of the following:

1. Current problems (History of present illness? Relationship to drug therapy?).
2. Past medical history (Validate chart, probe in more depth regarding relevant history.)
3. Current prescribed medication and indication and duration for each, including physician office samples.
4. Current nonprescription medication and nutritional/herbal products with indications and durations.
5. Assessment of current therapy efficacy (Frequency and severity of headaches? Number of seizures? Improvement of tremor?).
6. Assessment of adverse effects (Ataxia? Drooling? Dyskinesias?).
7. Past medications, efficacy assessment, and reasons for discontinuation, with dates.
8. Known adverse reactions to medication, including allergies and sensitivities, with descriptions of each.
9. Adherence assessment.
10. Insurance assessment (Does patient pay for medication?).
11. Targeted questions depending on suspected diagnosis (e.g., aspirin use in stroke patients; grapefruit juice in elderly patient with hypotension and nifedipine, antipsychotic or antiemetic medication in Parkinson's disease).
12. Name of community pharmacy where patient obtains medications.

REVIEW OF SYSTEMS

A complete discussion of the review of systems is included elsewhere.[1] In reviewing the systems of a

Encyclopedia of Clinical Pharmacy
DOI: 10.1081/E-ECP 120006230

neurological patient, depending on the patient's diagnosis, some systems should receive special attention. Some examples of potential neurology-associated problems that can be detected by a review of systems are as follows:

1. *Vital signs*: Are there any abnormalities that could be drug related? For example, sympathomimetics or stimulants may cause hypertension that could increase stroke risk. Temperature, heart rate, and blood pressure are important therapeutic monitoring parameters for medications such as antihypertensives, corticosteroids, and stimulants. Weight gains or losses can be attributed to some medications, such as valproic acid and tricyclic antidepressants for weight gain or topiramate for weight loss.
2. *Cardiovascular*: Does the patient have atrial fibrillation, a coronary artery disease history, or hyperlipidemia increasing stroke risk? Could orthostatic hypotension be due to medications for Parkinson's disease treatment? Is a medication such as amantadine causing peripheral edema?
3. *Pulmonary*: Does the patient require medications for pulmonary disease that may affect neurological functioning such as beta agonist-induced tremor? Could a medication such as beta-blockers prescribed for essential tremor cause deterioration of pulmonary function? Could an acute pulmonary illness such as pneumonia be contributing to delirium? Does the patient have a history of pulmonary embolism that may indicate a predisposition to thromboembolic disorders? Is the patient a current or past smoker, contributing to stroke risk or migraines?
4. *Fluid/electrolyte/nutritonal status*: Does the patient require nutritional supplementation as is the case for folic acid during pregnancy to prevent neural tube defects, a vitamin B_{12} deficiency-associated peripheral neuropathy, or a disease-associated inability to ingest adequate nutrition (amyotrophic lateral sclerosis or debilitating stroke)? Does the patient have adequate hydration (dehydration affects cerebral blood flow)? Are feeding tubes present, requiring special dosing form considerations? Is the patient hypoalbuminemic, which may impact dosing adjustments for highly protein-bound medications? Is the patient on the ketogenic diet for seizure control that requires careful examination of all carbohydrate sources, including medications?
5. *Renal*: Is it necessary to adjust doses for poor renal function?
6. *Hepatic*: Is there evidence of hepatic dysfunction that may require dosing adjustment of antiepileptic agents or statins?
7. *Endocrine*: Diabetes increases risk of stroke, may worsen stroke outcomes, and causes peripheral neuropathy. Thyroid disorders may contribute to mental status changes. Is the patient postmenopausal, increasing stroke risk?
8. *Hematology*: Does the patient have a blood dyscrasia attributable to antiepileptic drugs or other medications? Is the patient hypercoagulable, causing an increased risk of stroke? Does the patient have any particular bleeding risk that limits antiplatelet drug use?
9. *Gastrointestinal*: Does the patient have nausea, vomiting, diarrhea, or constipation caused by a medication or neurological disease state?
10. *GU/reproductive*: Because of neurological disease, some patients must use a urinary catheter, predisposing them to urinary tract infections and requiring frequent antibiotic use. What form of contraceptive does the patient use, if any? Oral contraceptives may be less effective with antiepileptic drugs or may contribute to headaches or stroke risk in a smoker. Is the patient pregnant or breastfeeding, thus prompting close examination of medication use? Stroke risk is also increased in the postpartum period.
11. *Musculoskeletal*: Does the patient have muscular pain that could be caused by statins? Does the patient have muscular spasms or central spasticity requiring an antispasmodic? Does the patient complain of arthritic joint pain? Are tension headaches a problem? Does the patient have a tremor that could be attributed to medication use such as lithium, valproic acid, or beta agonists? Does the patient have weakness that could be due to medication use, underuse, or overuse (e.g., pyridostigmine underuse, antispasmodic overuse)?
12. *Neurological*: Discussed later.
13. *Psychological*: Does the patient have undertreated depression, schizophrenia, or bipolar disorder that may complicate neurological disease treatments? Could psychotic symptoms be caused by neurological illnesses such as diffuse Lewy body disease or Huntington's chorea? Could hallucinations be caused by medication use such as dopaminergic agents? Does the patient require medication for attention-deficit disorder?

14. *Skin*: Check for drug related skin abnormalities (rashes). Know the patient's baseline skin conditions before a new medication is started.
15. *EENT*: Does the patient have gingival hyperplasia caused by phenytoin? Does the patient have thrush or stomatitis that could be caused by a medication? Does the patient have dry eyes caused by an anticholinergic medication effect? Has the patient noted hair loss that could be due to valproic acid? Is the patient able to swallow solid dosage forms. Does the patient complain of vertigo that could be medication induced?

Pharmacist's Targeted Neurological Exam

The pharmacist need not rely on the physician's record of the neurological examination to monitor the safety and effectiveness of the prescribed therapy. A brief (5-minute), targeted neurological exam can be performed on each patient and is described in Table 1. This examination is derived from the complete neurological examination, performed by a physician in the diagnosis of neurological disease, and is designed to provide the pharmacist with data necessary for the design, implementation, and monitoring of pharmaceutical care plans.

INPATIENT NEUROLOGY SPECIALTY PRACTICE

The clinical pharmacist specialist performs many functions for the hospitalized neurological patient. In addition to performing the pharmacotherapy history, review of systems, and neurological exam on new admissions to formulate a care plan, the pharmacist monitors the patient's progress and documents the therapeutic outcomes. Table 2 lists the common inpatient neurological diagnoses with the medications and monitoring parameters used by clinical pharmacists.

One of the most important contributions the pharmacist can make to the care of the patient is the detailed instructions and medication education provided prior to discharge to the patient, family, or caregiver. Identifying barriers to optimal use of outpatient medications and creating a viable circumvention plan (e.g., indigent programs from pharmaceutical companies, prescription of formulary items only, streamlining regimen) begins at admission and ends with the final discharge interview.[2] Follow-up phone calls at 48 hours to answer questions are helpful to the patient/caregiver who is often overwhelmed during hospitalization and thinks of questions when faced with the challenge of correctly following the prescribed medication after discharge.

On a more global scale, the neurology pharmacy specialist working in an inpatient setting is involved in

Table 1 Pharmacist's targeted neuro examination

Domain	Assessment	Diseases/therapies monitored
Mental status	Greeting the patient daily: Does the patient remember you? Does the patient know where he/she is? Assess alertness, speech, memory, and ability to communicate and compare with baseline. Mini Mental Status Examination may be required for insurance approval of some therapies for Alzheimer's disease.	Dementias, delerium, stroke, encephalopathies, agents that cause sedation and/or confusion
Cranial nerves	Note facial tone and symmetry, gaze preference, and ptosis; check for nystagmus by having patient follow object across visual field; engage in conversation to assess for slurred speech.	Myasthenia gravis (ptosis), Parkinson's disease (masked facies), stroke (palsy), antiepileptic drug therapy (nystagmus, slurred speech)
Motor function	Assess handshake for ability to lift arm against gravity and motor control; move toward target (your hand); assess tremor (resting, action, and postural) of hand; assess grip strength. Ask patient to lift leg off bed. Observe for dyskinesias.	Stroke (arm and leg weakness), Parkinson's disease (tremor, bradykinesia, rigidity, dyskinesias), essential tremor
Sensory function	Touch patient on both arms/legs and assess symmetry; inquire about paresthesias.	Stroke, neuropathy (diabetes, HIV, drug-induced)
Gait	Do not test in hospitalized patient (ask the patient if he/she has been walking); observe patient walking normally, heel-to-toe, and turning.	Stroke, vertigo, Parkinson's disease, antiepileptic drug toxicity

Table 2 Common inpatient neurology diagnoses

Diagnosis	Interventions	Monitoring
Ischemic stroke	TPA, heparin, ASA, clopidogrel, ticlopidine, dipyridamole and aspirin, warfarin, surgery	Neurological worsening, bleeding, PT/APTT, INR, CBC
Back pain	Analgesics, surgery, antibiotics if infectious	Pain, sedation, cultures/sensitivites
Bell's palsy	Steroids, acyclovir, artificial tears and lubricants	Weakness, paresthesias
Guillain barre	Plasmapheresis, IVIG	Hemodynamics, respiratory status, renal function
Headache	Surgery if SAH, IV DHE, analgesics, triptans	Pain, sedation, neurological status
Multiple sclerosis	Methylprednisolone, beta interferons, glatiramer acetate, antispasmodics, anticholinergics	Weakness, paresthesias, vision, injection technique
Seizures	Antiepileptic agents	Seizure activity, CNS side effects, respiration, EEG
Subarachnoid hemorrhage	Surgery, BP control, nimodipine, volume expansion, sedation, seizure prophylaxis, stool softeners, ventricular drainage	Neuro status, BP, hydration status, EKG, ICP
TIA/vertebrobasilar insufficiency	Surgery, antiplatelets, heparin/warfarin	Neuro status, bleeding, platelets, BP/hydration, aPTT, INR
Weakness, acute generalized	Depends on final diagnosis	Drug induced? Infectious?

Key: TPA, tissue plasminogen activator; ASA, aspirin; PT, prothrombin time; APTT, activated partial thromboplastin time; INR, international normalized ratio; CBC, complete blood count; IVIG, intravenous immune globulin; SAH, subarachnoid hemorrhage; IV DHE, intravenous dihydroergotamine; CNS, central nervous system; EEG, electroencephalogram; BP, blood pressure; EKG, electrocardiogram; ICP, intracranial pressure; TIA, transient ischemic attack.

guideline development (e.g., treatment of hypertension in stroke patients), critical pathway/protocol development,[3] adverse drug reaction reporting, and Pharmacy and Therapeutics Committee new drug evaluations.

OUTPATIENT NEUROLOGY SPECIALTY PRACTICE

The clinical pharmacist positioned in the outpatient setting can also significantly affect patient care. In contrast to the inpatient neurological specialist, the outpatient specialist usually cares for more chronic neurological problems and may develop a long-standing relationship with their patients. The elements of the pharmacotherapy history, review of systems, and neurological exam are the same as previously described for the inpatient setting. Table 3 lists common outpatient neurological diagnoses with the medications used and monitoring parameters used by clinical pharmacists.

Formulating the pharmaceutical care plan is a very important part of the outpatient neurological pharmacists' duties. Special care must be taken to remove barriers to the patient obtaining the required medications. This may involve facilitating refill requests, obtaining insurance or health maintenance organization approvals for nonformulary medications, or requesting patient assistance from pharmaceutical companies for medically indigent patients. Patients determined to be nonadherent to their medication regimen should also be investigated for the cause. This etiology may be as diverse as lack of understanding of the disease state to difficulty tolerating the medication to inability to afford the medication. In most cases, the pharmacist can assist the patient with these barriers to adherence.

In the neurology population, patient medication education is very important as regimens can be very complex and the patient may have communication or comprehension difficulties. The pharmacist may develop educational materials and clear, concise directions for medication adjustments to help patients. Specialized calendars or dosing charts can provide a visual reminder to patients in addition to verbal counseling.

From these basic clinical pharmacy functions, the ambulatory neurological pharmacist may choose to subspecialize in any of the common disease states seen in neurology clinics. In disease states where medication plays an important role in patient care and close monitoring is necessary, the pharmacist can be an essential part of patient management. As previously mentioned, the majority of clinical pharmacist involvement in neurology has been in epilepsy clinics. There are several good re-

Table 3 Common outpatient neurology diagnoses

Diagnosis	Interventions	Monitoring
Epilepsy	Antiepileptic drugs, surgery	Medication serum concentrations, sedation, cognitive abilities, liver function tests, blood dyscrasias, bleeding abnormalities, CNS toxicities, rashes, seizure counts, other drug-specific adverse effects
Headaches (abortive treatment)	Triptans, ergotamines, NSAIDS, opiates, oxygen, Midrin®	Sedation, chest pain, tingling/numbness, pain relief
Headaches (prophylactic treatment)	Beta blockers, calcium channel blockers, tricyclic antidepressants, divalproex sodium, gabapentin	Disease-specific cautions (i.e., beta-blockers and diabetes), blood pressure, heart rate, tremor, medication-specific adverse effects, headache count
Parkinson's disease	Levodopa/carbidopa, dopamine agonists, amantadine, selegiline, COMT inhibitors, anticholinergics, surgery	Dyskinesias, orthostatic hypotension, hallucinations, sedation, anticholineric effects, relief of symptoms
Multiple sclerosis	Beta interferons, glatiramer acetate, glucocorticoids, antispasmodics, amantadine, stimulants, immunosuppressants	Number of exacerbations, self-injection technique, injection site reactions, amount of spasticity, amount of fatigue, CBC and infections (immunosuppressants), medication-specific adverse effects
Alzheimer's disease	Cholinesterase inhibitors, antipsychotics	Mini Mental Status Examination, gastrointestinal complaints, number of hallucinations, liver function tests (tacrine)
Myasthenia gravis	Cholinesterase inhibitors, immunosuppressants	Weakness, fatiguability, diarrhea, drooling, CBC and infections (immunosuppressants)
Ischemic stroke prevention	Antiplatelets, anticoagulation, lipid-lowering agents, antihypertensives, smoking cessation	Signs/symptoms of stroke/TIA or bleeding, INR, serum lipid concentrations, blood pressure, smoking status
Amyotrophic lateral sclerosis	Riluzole, antisecretory medication, supportive care	Weakness, drooling
Peripheral neuropathy	Tricyclic antidepressants, antiepileptic drugs, mexilitine, capsacian	Sedation, anticholinergic effects, blood dyscrasias, arrhythmias, medication-specific adverse effects, pain relief
Chronic pain	Surgery, NSAIDs, opiates, tricyclic antidepressants, antiepileptic drugs	Sedation, pain relief

Key: CNS, central nervous system; NSAIDs, nonsteroidal anti-inflammatory drugs; COMT, catechol-O-methyltransferase; CBC, complete blood count; TIA, transient ischemic attack; INR, international normalized ratio.

ports of these clinics in the literature.[4–8] The reported responsibilities of the clinical pharmacists included taking medication histories, performing directed neurological examinations, ordering laboratory evaluations for monitoring medications, assessing adverse drug reactions, providing medication counseling, and providing pharmacokinetic consultations.

The role of the clinical pharmacist in a specialty headache clinic has been described by Adelman and Von Seggem.[9] This individual obtains medication histories and extensively counsels patients on their medication regimens. In addition, the clinical pharmacist determines changes to the patient's therapeutic regimen to enhance effectiveness and minimize toxicity.

Pharmacists may be involved in a multidisciplinary stroke program with an outpatient component. Here the pharmacist may monitor and adjust anticoagulation treatment and provide safety monitoring for ticlopidine through collaborative care agreements with the prescriber and the patient.[10] Vigilant monitoring of disease states that may contribute to stroke risk such as diabetes, hyperlipidemia, or hypertension with appropriate medication adjustments can also be the realm of the pharmacist. Extensive patient education materials may be necessary, and the clinical pharmacist may want to participate in community education efforts such as stroke screenings.[11]

Other medication-intensive neurological disorders include Parkinson's disease and multiple sclerosis. Al-

though specific pharmacist interventions have not been reported in the literature for these disease states, some opportunities for involvement can be envisioned. For Parkinson's disease, the pharmacist may develop a collaborative relationship with the neurologist and make necessary adjustments to medication regimens in response to patient needs. In this case, the pharmacist would likely develop a specialized set of physical assessment skills, including detailed motor examination. Parkinson's disease-specific diaries and medication regimen handouts could be provided to this population and be used for dosage adjustments. The multiple sclerosis population can also benefit from pharmacist involvement. When disease-modifying therapies such as beta interferons, glatiramer acetate, or immunosuppressants are prescribed, a great deal of patient education must be performed. Patients must understand that these medications are not curative, but slow progression and/or decrease multiple sclerosis exacerbations. In addition, the patient must be instructed in self-injection technique for the parenteral products and monitoring regimens must be established for the immunosuppressants to ensure patient safety. Other issues that the pharmacist may address include therapy for muscle and bladder spasticity, fatigue, sexual dysfunction, pain, and urinary tract infections.

OUTCOMES

Very few examples of outcomes research in neurological pharmacy exist. However, one historical control study determined that the implementation of a pharmacokinetics consultation service in an epilepsy clinic decreased seizure frequency and number of adverse effects compared with the baseline frequency in the 4 months prior to offering the service.[7] A second study conducted in the pediatric epilepsy population described the effect of establishment of a specialty pediatric epilepsy clinic with clinical pharmacy services. Compared with patients seen before the beginning of the clinic, patients seen in the specialty clinic had decreased numbers of antiepileptic drugs and decreased doses of these medications. Frequency of seizures was not examined in this report.[8]

OTHER ACTIVITIES

Most specialists are involved in both clinical and didactic teaching of pharmacy students, residents, and fellows as well as trainees in other programs (resident physicians, nurses, etc.). Many hold faculty appointments at Colleges of Pharmacy and/or Medicine. In addition, pharmacy specialists participate in and conduct clinical and/or basic research in the neurosciences.

GUIDELINES USED IN NEUROLOGY PHARMACY PRACTICE

The following guidelines may be helpful in providing care to neurology patients:[12]

1. Practice advisory: Thrombolytic therapy for acute ischemic stroke. American Academy of Neurology, 1996 (5 pages).
2. Intravenous immunoglobulin preparations. University Health System Consortium, 1999 Mar (216 pages).
3. Second report of the Expert Panel on the Detection, Evaluation, and Treatment of High Blood Cholesterol in Adults (Adult Treatment Panel II). National Heart, Lung, and Blood Institute (U.S.), 1993 Sept (Reviewed 1998) (169 pages).
4. Sixth ACCP consensus conference on antithrombotic therapy. American College of Chest Physicians, 2001 (20 pages).
5. Practice advisory on selection of patients with multiple sclerosis for treatment with Betaseron. American Academy of Neurology, 1994 (reviewed 1998) (4 pages).
6. Sixth report of the Joint National Committee on the Prevention, Detection, Evaluation and Treatment of High Blood Pressure. National Heart, Lung, and Blood Institute (U.S.), 1997 Nov (33 pages).
7. Practice guideline for the treatment of patients with Alzheimer's disease and other dementias of late life. American Psychiatric Association, 1996 Dec (93 pages).
8. An Algorithm (Decision Tree) for the Management of Parkinson's Disease: Treatment Guidelines. Neurology 1998; 50 (suppl 3) S1–S57.
9. PRN Opinion Paper: The Ketogenic Diet. *Pharmacotherapy* 1999; 19:782–786.

NETWORKING AND EDUCATIONAL OPPORTUNITIES

Neurology pharmacists have formal networking and educational opportunities at professional pharmacy organizations (American Society of Health-System Pharmacists, American College of Clinical Pharmacy) and within the confines of neurology subspecialty groups (American Epilepsy Society). In addition, stand-alone groups of clinical specialists, usually in collaboration with psychiatry specialists (College of Psychiatric and Neurologic Pharmacists), develop and offer high-quality, intensive continuing education several times per year in the United States. The introduction of electronic communication has facilitated the exchange of clinical experiences and information via list servs and e-mail directories available on the Internet.

REFERENCES

1. Cippole, R.J.; Strand, L.M.; Morley, P.C. The Patient Care Process. In *Pharmaceutical Care Practice*; Cippole, R.J., Strand, L.M., Morley, P.C., Eds.; McGraw-Hill: New York, New York, 1998; 121–176.
2. Nelson, W.J.; Edwards, S.A.; Roberts, A.W.; Keller, R.J. Comprehensive self-medication program for epileptic patients. Am. J. Hosp. Pharm. **1978**, *35*, 798–801.
3. Gonzaga-Camfield, R. Developing an emergency department team for treatment of stroke with recombinant tissue plasminogen activator. Crit. Care Nurs. Clin. North Am. **1999**, *11*, 261–268.
4. Kootsikas, M.E.; Hayes, G.; Thompson, J.F.; Perlman, S.; Brinkman, J.H. Role of a pharmacist in a seizure clinic. Am. J. Hosp. **1990**, *47*, 2478–2482.
5. Hixon-Wallace, J.A.; Barham, B.; Miyahara, R.K.; Epstein, C.M. Pharmacist involvement in a seizure clinic. J. Pharm. Pract. **1993**, *6* (6), 278–282.
6. Allen, J.P.; Ludden, T.M.; Walton, C.A. The role of clinical pharmacists in a epilepsy clinic. Drug Intel. Clin. Pharm. **1978**, *12*, 242–244.
7. Ioannides-Demos, L.L.; Horne, M.K.; Tong, N.; Wodak, J.; Harrison, P.M.; McNeil, J.J.; Gilligan, B.S.; McLean, A.J. Impact of a pharmacokinetics consultation service on clinical outcomes in an ambulatory-care epilepsy clinic. Am. J. Hosp. Pharm. **1988**, *45*, 1549–1551.
8. Summers, B.; Summers, R.S.; Rom, S. The effect of a specialist clinic with pharmacist involvement on the management of epilepsy in paediatric patients. J. Clin. Hosp. Pharm. **1986**, *11*, 207–214.
9. Adelman, J.U.; Von Seggem, R.L. Office Rx: A renewed professional relationship. N. Carolina Med. J. **1995**, *56* (6), 262–264.
10. Ryan, M. In *Collaborative Care Agreements to Facilitate Pharmaceutical Care. International Pharmaceutical Abstracts*, 32nd American Society of Health-System Pharmacists Midyear Clinical Meeting, Atlanta, Georgia, Dec. 7–11, 1997; American Society of Health-System Pharmacists: Washington, DC, 1997, 3412273, 2164.
11. Ryan, M.; Kammer, R. In *Report on a Multidisciplinary Stroke Screening. International Pharmaceutical Abstracts*, 33rd American Society of Health-System Pharmacists Midyear Clinical Meeting, Las Vegas, Nevada, Dec. 6–10, 1998; American Society of Health-System Pharmacists: Washington, DC, 1998, 3513328, 2359.
12. see www.guidelines.gov.

BIBLIOGRAPHY

Graves, N.M. The role of the pharmacist in investigational AED trials. Epilepsy Res., Suppl. **1993**, *10*, 201–209.

McCauley, J.W.; Mott, D.A.; Schommer, J.C.; Moore, J.L.; Reeves, A.L. Assessing the needs of pharmacists and physicians in caring for patients with epilepsy. J. Am. Pharm. Assoc. **1999**, *39*, 499–504.

Mort, J.T.; Tasler, M.K. Managing dementia-related behavior in the community. J. Am. Pharm. Assoc. **1996**, *NS36* (4), 249–256.

Pitterle, M.E.; Sorkness, C.A.; Wiederholt, J.B. Use of serum drug concentrations in outpatient clinics. Am. J. Hosp. Pharm. **1985**, *42*, 1547–1552.

Turner, C.J.; Pryse-Phillips, W.Can. J. Clin. Pharmacol. **1999**, *6* (2), 113–117.

Nontraditional Pharm.D. Programs

Vaughn L. Culbertson
Catherine A. Heyneman
Idaho State University, Pocatello, Idaho, U.S.A.

INTRODUCTION

In 1989, the American Council on Pharmaceutical Education (ACPE) announced its intent to accredit only Doctor of Pharmacy (Pharm.D.) degree programs beginning in the year 2000. Not surprisingly, the creation of a new entry-level degree generated significant controversy. Initial concern focused on the potential impact that two levels of pharmacy practitioner (i.e., baccalaureate vs. Pharm.D.) would have on the profession. Understandably, many baccalaureate pharmacists were particularly concerned about their ability to successfully compete against future graduates in the professional marketplace. Early attempts to resolve the issue included discussions of potential degree transfer mechanisms for baccalaureate pharmacists.[1] For example, the National Community Pharmacists Association (NCPA), formerly known as the National Association of Retail Druggists, responded by issuing a nonacademic Pharmacy Doctor (PD) degree to any association member for a nominal fee. However, as practitioners began to understand the paradigm change espoused by pharmaceutical care and to accept their new roles as vital health care providers, the need for programs that could significantly enhance knowledge and skills gradually overrode the impulse to grandfather degrees.

BACKGROUND INFORMATION

The nation's three largest pharmacy practitioner organizations again addressed the issue in 1991.[2] In a consensus statement released by the American Pharmaceutical Association, the American Society of Health-Systems Pharmacy, and the National Community Pharmacists Association, an entry-level Pharm.D. degree that would prepare pharmacists as generalist practitioners rather than specialist practitioners was strongly endorsed. In addition, the joint statement endorsed the concept of a degree transfer process for current baccalaureate degree practitioners to avoid even the appearance of two distinct practitioner levels. Although never accomplished, the organizations further proposed developing an institute for granting ''Pharm.D. equivalency'' certificates for baccalaureate practitioners whose colleges of pharmacy do not develop a degree transfer process.

Finally, in 1996, a National Association of Boards of Pharmacy (NABP) Task Force released their report entitled, *Development of an Equitable Degree Upgrade Mechanism.* The task force defined general criteria for pharmacy schools intending to develop a uniform approach to the delivery of nontraditional Pharm.D. (NTPD) programs. These consisted of four essential characteristics for NTPD programs:

- Assessibility—a competency-based process (anchored to NAPLEX endpoint competencies) to be conducted by a committee of faculty and practitioners.
- Accessibility—defined as a practical, nondisruptive program that will not require the applicant to relocate or significantly interfere with his or her practice.
- Academic soundness—defined as a documented evaluation that does not disrupt or compromise accreditation standards.
- Affordability—a program that can be offered to applicants at a reasonable cost and may be completed in a timely manner (e.g., 36 semester hours).

It is within this general context that schools and colleges of pharmacy have attempted to balance the feasibility issues as voiced by the profession with the academic standards and endpoint competencies expected of doctoral-level practitioners. Although NTPD programs existed prior to the landmark ACPE decision to mandate entry-level Pharm.D. practice requirements, this decision and its resulting controversy have directly influenced the rapid growth and innovative approaches to NTPD curriculum delivery. Experimentation in curricular delivery methods has emerged and continued efforts are essential.[3–8] This report will attempt to efficiently summarize the emerging educational approaches and provide an overview of the NTPD programs currently in operation.

Encyclopedia of Clinical Pharmacy
DOI: 10.1081/E-ECP 120006358

Table 1 Admission requirements for nontraditional Pharm.D. programs

School/college	Enrollment	Maximum time to complete[b] (yr)	Web site URL	Cost[c]
Samford (Alabama)	Discontinued in 2001[a]	3	http://www.Samford.Edu/schools/pharmacy/ntpd/ntpd.htm	$$$$
Auburn (Alabama)	Open	6	http://pharmacy.auburn.edu/nontrad/	$$
Arkansas	Open	5	www.uams.edu/ntpdp	(R) $ (NR) $$$
Arizona	Open[a]	4	No web site	$$$$
Colorado	Open	5–6	http://www.uchsc.edu/sp/sp/programs/prostudies.htm	(R) $$$ (NR) $$$$
Nova Southeastern (Florida)	Open	5	http://pharmacy.nova.edu/	$$$
Florida	Open	7	http://www.cop.ufl.edu/wppd/www.intelicus.com	(R) $$$ (NR) $$$
Georgia (Mercer and University of Georgia)	Restricted	5	http://www.rx.uga.edu/main/home/ntpharmd/index.htm	$$$
Idaho State	Open	6	http://rx.isu.edu/	(R) $ (NR) $$$$
Midwestern (Illinois and Arizona)	Open	4	http://www.midwestern.edu/CCP/nontrad.html	$$$
Illinois–Chicago	Open	4.5	http://www.uic.edu/pharmacy/offices/cco/cco.html	(R) $ (NR) $$$
Purdue	Restricted	7	www.pharmacy.purdue.edu/~ntdpp/	(R) $$ (NR) $$$$$
Iowa	Restricted[a]	5	No web site	$$
Kansas	Open	5	http://www.pharm.ukans.edu/nontrad/ntpd.htm	(R) $$
Kentucky	Restricted	6	www.uky.edu/pharmacy/npo/	(R) $$ (NR) $$$$$
Louisiana at Monroe	In development		http://rxweb.ulm.edu/pharmacy/	
Xavier (Louisiana)	Not admitting	5	http://www.xula.edu/PharmPostBac.html	$$
Maryland	Open	3.5	No web site found	
Massachusetts COP	Discontinued			
Minnesota	Restricted[a]	3–5	http://www.pharmacy.umn.edu/dp4/	$$
Mississippi	Open	5	http://www.olemiss.edu/depts/pharm_school/NTPharmD.pdf	(R) $ (NR) $$$
Missouri–Kansas City	Restricted	5	http://www.umkc.edu/pharmacy/CE.HTML	$$
Montana	Restricted	6	http://www.umt.edu/ccesp/distance/pharmd/	$$
Creighton (Nebraska)	Open	8	http://pharmacy.creighton.edu/spahp/non_traditional/curriculum/pharmacy/curr_async-pprof.asp	$$$
New Mexico	Restricted	6	http://hsc.unm.edu/pharmacy/Nontraditional.htm	$
Albany (New York)	Open	5.5	http://www.acp.edu/Academics/catalog/programs.htm#non-trad	(R) $$$
Arnold and Marie Schwartz (New York)			No web site found	
Buffalo (New York)	Open	5	http://pharmacy.buffalo.edu/programs/ntpharmd/ntprogram.html	(R) $$$$ (NR) $$$$$

(Continued)

Table 1 Admission requirements for nontraditional Pharm.D. programs (*Continued*)

School/college	Enrollment	Maximum time to complete[b] (yr)	Web site URL	Cost[c]
North Carolina	Open	3	http://www.pharmacy.unc.edu/xpharmd/index.html	(R) $$
North Dakota State	Restricted	6	http://www.ndsu.nodak.edu/instruct/hanel/pharmacy/ntp_let.htm	$$$
Ohio Northern	Open	5	http://www.onu.edu/pharmacy/ntpd/default.asp	$$$$$
Ohio State	Open	6	http://www.pharmacy.ohio-state.edu/ntpd/	$$$
Oklahoma (Southwestern Oklahoma State University and Oklahoma University)	Not admitting[a]		No web site found	
Duquesne (Pennsylvania)	Open	5	http://www.duq.edu/pharmacy/Programs/non-traditional.html	$$$
Philadelphia (Pennsylvania)	Open	5	http://www.usip.edu/graduate/flex	$$$$$
Wilkes (Pennsylvania)	Start Fall 2001		No web site found	
Temple (Pennsylvania)	Discontinued			
Rhode Island	Discontinued			
South Carolina	Not admitting[a]		No web site found	
South Dakota	see Univ. Minnesota			
Tennessee	see Univ. Kentucky			
Texas Schools	Restricted	5	http://www.txpharm.org/	$$$
Utah	see Idaho State			
Virginia Commonwealth	Open	2	http://www.pharmacy.vcu.edu/nontrad/index.html	(R) $$$
Shenandoah (Virginia)	Open	4	http://pharmacy.su.edu/Ntdp/index.html	$$$$
Washington (University of Washington and Washington State University)	In-state preferred	5	http://depts.washington.edu/expharmd	(R) $
West Virginia	Restricted	6	http://www.hsc.wvu.edu/sop/academic/	$$$$
Wisconsin	Open	6	www.pharmacy.wisc.edu/ntpd	(R) $$$

(R) = Resident; (NR) = Nonresident.
Relative cost scale: $ = $6–10,000; $$ = $11–15,000; $$$ = $16–20,000; and $$$$ = $21–25,000; $$$$$ > $26,000.
[a]Indicates programs that did not continue after 2000–2001.
[b]Usual time frame for completion—where applicable, maximum time requirement (max.) is presented.
[c]Costs are subject to change and may vary depending on credit load taken or length of time to complete degree requirements.

Only those schools providing nontraditional approaches (i.e., alternative curricular pathway to an on-campus program) are included here, although it is worth noting that several schools still offer traditional on-campus postbaccalaureate Pharm.D. programs.

A general comparison of the admission requirements, didactic coursework, and clerkship curricula is provided for current NTPD programs in Tables 1–3. This information was collected from a review of each program's web site and/or written information. To ensure current and accurate information, verification was attempted using the AACP's Pharmacy School Admission Requirements for 2001–2002[9] and a telephone call to the program, if needed. Nonetheless, many programs are still under development and others are continuing to experiment with new techniques and methodologies to enhance nontraditional learning. Therefore, program requirements and curricula may change quickly and frequently. A web site address (URL) is provided where available, and readers are encouraged to contact the schools for more specific information.

HISTORY AND GOALS

Although the number of programs has grown dramatically in recent years, nontraditional delivery of Pharm.D. curricula began in the early 1980s, long before the issue of entry-level degree requirements was debated. Purdue University began offering off-campus coursework in 1981. Shortly thereafter, the University of Kentucky (1986), the University of Illinois–Chicago (1986), and Idaho State University (1989) implemented nontraditional programs. These early pioneers represent a small group of highly experienced and well-established programs with successful graduation records.

Nontraditional pharmacy education originated because of the recognized need to enhance the knowledge and clinical abilities of practicing pharmacists while allowing them to continue full-time employment. As is discussed in more detail later in this article, the methods used are diverse and have engendered innovative and at times somewhat controversial educational approaches. However, at this point, it is worth emphasizing that doctoral education generally implies and probably demands some level of self-sacrifice. NTPD programs can only hope to minimize this disruption.

The guiding principle in the development of NTPD programs has been the insistence of ACPE that the educational goals and endpoint competencies are identical to those established for traditional on-campus programs. In other words, the use of the term ''nontraditional Pharm.D.'' does not imply a different degree because no true distinction is recognized by ACPE. Only the delivery of that curriculum differs, and institutions are encouraged to formulate different pathways by which accomplishment of curricular outcomes fulfills requirements for the Doctor of Pharmacy degree. Nevertheless, perceptions of a ''correspondence course'' or ''mail order degree'' may exist within some sectors of the profession as evidenced by recent commentaries in the pharmacy literature.[10,11] Hopefully, this overview of current NTPD programs will help to dispel any misconceptions or perceived differences between traditional and nontraditional pharmacy education.

ADMISSION REQUIREMENTS

Although admission requirements vary among NTPD programs, the major differentiating factor is open vs. closed enrollment. As indicated in Table 1, open enrollment refers to programs that will accept applications from practicing pharmacists holding licensure in the United States or its territories. In 1990, one-half of the existing nontraditional NTPD programs restricted admission to licensed practitioners within their state or alumni of the school. Currently, only 14 programs have similar admission restrictions.

Programs offering open enrollment often require foreign applicants to provide proof of English language proficiency either through examination (e.g., TOEFL) or on-campus interview. Canadian pharmacists present a unique situation. Only two Canadian colleges of pharmacy offer a postbaccalaureate Pharm.D. option: the University of British Columbia and the University of Toronto. Because of the high demand, many NTPD programs have opened admission to Canadian pharmacists.[12]

As part of the admission process, NTPD programs generally require applicants to provide the following information: proof of licensure, a completed application form, application fee, official college transcripts, and several letters of recommendation. Other admission criteria vary substantially between programs but may include an on-campus interview, satisfactory achievement on minimum competency assessments, satisfactory completion of prerequisite coursework, and/or minimum GPA criteria.

Recently, much attention has been devoted to prior learning assessment (PLA) as a method to award academic credit or advanced level placement for competencies acquired through previous work experience. As is discussed in more detail later in this article, PLA is a useful methodology for decreasing coursework and time requirements. However, it is not a simple process and requires considerable effort by both students and faculty.

DIDACTIC INSTRUCTIONAL METHODS AND COURSE DELIVERY

Virtually all NTPD programs are structured to allow practicing pharmacists to continue working while taking didactic courses on a part-time basis. This is accomplished using a variety of instructional methods, and a description of the relative merits of each is summarized here. These terms are also used in Table 2 to identify the methods used by each NTPD program.

Videotaped Lectures

In addition to facilitating course delivery in the student's own home, the videotape delivery of didactic courses provides several advantages that nontraditional student-practitioners often find desirable. Studying can be tailored to individual work schedules (i.e., rotating shifts, weekends, or on-call schedules), thereby allowing much greater flexibility and a self-directed study program for

those individuals with irregular schedules or other conflicts. In addition, it allows students to "fast forward" quickly through information that may have been acquired previously while allowing "replay" to review new information when needed.

The major disadvantage of videotaped coursework is the inability to interact directly with the instructor or other classmates. However, most programs currently employ the use of supplemental or web-based interactions with instructors or online small group discussions to overcome, or at least minimize, these deficiencies. Occasionally, poor technical quality may substantially impact course delivery and practitioners considering these programs may want to review sample videotapes or discuss this issue with current students.

Distance Learning

Using a variety of available technologies, transmission of live didactic lectures to remote classrooms remains a common vehicle for course delivery. For many practitioners, this format is probably easiest to assimilate because it closely resembles more traditional teaching methods and affords at least some direct student and instructor interaction. A regular class schedule and close geographic proximity to one of the regional transmission sites are the major limitations to distance learning programs.

Web-Based Coursework

Dramatic improvements in computer technology have fostered the development of unique web-based approaches to curriculum delivery. Several programs offer virtually all didactic coursework via the Internet, and nearly all NTPD programs are gravitating toward this new educational tool. Some programs have found it necessary to contract with private corporations to assist with the rather daunting task of redesigning their entire curricula to an online, interactive format.

Web-based didactic coursework generally allows students greater flexibility in determining their own study schedule and may allow easier incorporation of supplemental educational materials (e.g., Internet links to additional information and/or demonstrations). The positive quality of web-based flexibility must be tempered by recognition of technological limitations, however. Many students find themselves feeling isolated and have difficulty coordinating group projects over the Internet.

Even with recent technological improvements, not all web-based coursework measures up to advertised benefits, and significant problems may occur. Slow transmission rates, server malfunctions, and poor utilization of computer capabilities can result in student dissatisfaction and frustration. As with other educational mediums, students should carefully investigate the quality of web-based materials and attempt to validate program claims via inquiries to active students whenever possible. In a speech, Dr. Victoria Roche cautioned against becoming enamored with technology for its own sake.[13] Studies comparing cybercourses and traditional didactic formats are essential to apply the most effective learning strategies in the future.

NTPD program didactic requirements are remarkably similar; the marked differences reported in Table 2 can be accounted for primarily by differences in prerequisite coursework or delivery methods used. Few programs provide a self-paced format and, despite the widespread potential for self-paced curricula, most NTPD programs follow a traditional semester sequence. Thus, flexibility may be an important distinguishing characteristic depending on the self-directed nature, motivation, or family and work demands of individual students. Conversely, self-paced programs tend to be less interactive because students are generally at different points in the curriculum at any given time. They are less conducive to small-group learning activities (e.g., case discussions and interactive projects) and are best suited for practitioner-students that are highly self-directed.

ASSESSMENT OF PRIOR LEARNING

One of the more difficult issues facing nontraditional programs has been the development of methodologies for assessing learning acquired through previous experience. Obviously, daily practice affords a rich environment for continuing education, yet the extent to which individual practitioners use this varies considerably. To address this issue and provide a mechanism for assessing prior learning, many schools and colleges have developed innovative evaluation processes. Portfolio review has become an increasingly popular mechanism in this regard. Under this process, student-practitioners are given academic credit if they can document their ability to perform specific learning objectives. Frequently, this involves describing in significant detail work situations or patient cases that demonstrate the student-practitioner's ability to understand and perform the objective appropriately. Common examples include the following types of information and documentation:

- A statement of career goals.
- A detailed description of current practice activities, including services provided and patient population

Table 2 Nontraditional Pharm.D. didactic coursework

School/college	Didactic coursework (semester credits)[a]	Challenge exams[b]	Portfolio review[b]	Other[b,d]	Self-paced	Delivery methods[c]
Samford (Alabama)	56 (includes clerkships)	No	No			Distance learning by videoconferencing
Auburn (Alabama)	44	✓(20)	✓(20)			Internet, CDs, traditional lectures
Arkansas	21				✓	Internet, videotapes
Arizona	24				✓	AACP Consortium, videotapes
Colorado	35	✓(3)	✓(4)			All Internet-based course delivery
Nova Southeastern (Florida)	31		✓(15)			Distance learning to regional sites, Internet
Florida	27					Videotapes, Internet, 3 days on-campus/semester
Georgia Schools (Mercer and University of Georgia)	29	✓(21)				Distance learning, videotapes, teleconferencing, Internet
Idaho State	28	✓(7)		BCPS (10)	✓	Videotapes, Internet
Midwestern (Illinois and Arizona)	35 quarter hours	✓	✓(30)			Evening/weekend classes
Illinois–Chicago	24		✓(3)	BCPS, DI, and Statistics		Internet, web conferencing
Purdue	28			Yes	✓	Videotapes, Internet, regional workshop sites
Iowa	29			Recent grads only		Videotapes, Internet, weekend
Kansas	24	✓	✓			All Internet-based course delivery
Kentucky	26	✓(26)	✓(16)	Yes (4)		Videotaped case studies, Internet, interactive patient activities
Louisiana at Monroe		✓	✓			Distance learning, videotapes, Internet
Xavier (Louisiana)	24	✓			✓	Videotapes, Internet
Maryland	24	✓(6)	✓(10)			On-campus and distance learning, Internet
Massachusetts	Discontinued					
Minnesota	26		✓(15)	(15) total curriculum		Internet with 1 weekend/year on campus
Mississippi	33	✓(32)	✓(2)	BCPS (32)	✓	Problem-based learning format via chat room groups
Missouri–Kansas City	22				✓	Videotapes
Montana	25	✓		BCPS (14)		Internet, one course videotaped
Creighton (Nebraska)	34	✓(9)	✓(9)	BCPS	✓	Converting to Internet-based courses
New Mexico	Two courses (see delivery methods)			BCPS (all didactics)	✓	ASHP Clinical Skills Program and the ACCP Pharmacotherapy Self-Assessment Program (PSAP)

(Continued)

Table 2 Nontraditional Pharm.D. didactic coursework (*Continued*)

School/college	Didactic coursework (semester credits)[a]	Challenge exams[b]	Portfolio review[b]	Other[b,d]	Self-paced	Delivery methods[c]
Albany (New York)	26	✓	✓(8)	Occasionally	✓	Audiotapes
A and R Schwartz (New York)		✓(9)				Weekend/evening classes
Buffalo (New York)	30		✓(12)		✓	Videotapes, Internet, CD and DVD teleconferences
North Carolina	25	✓	✓			Videotapes, web bulletin board
North Dakota State	37		✓			Uses Colorado didactic program
Ohio Northern	39 quarter hours	✓(16)	✓(16)	16 quarter hours total	✓	Videotapes, computer-based programs
Ohio State	40 quarter hours					Internet-based course delivery
Oklahoma Schools (Southwestern Oklahoma State University and Oklahoma University)	16	✓(4)				Videotapes, weekend/evening classes
Duquesne (Pennsylvania)	28		✓	(4) previous graduate work	✓	Self-directed study, Internet, and 1–3 day workshops
Philadelphia (Pennsylvania)	24	✓(9)				On campus
Wilkes (Pennsylvania)	27					Weekend/evening classes
Temple (Pennsylvania)	Discontinued					
Rhode Island	Discontinued					
South Carolina	Not admitting					
South Dakota	26					Uses Minnesota didactic program
Tennessee	26					Uses Kentucky didactic program
Texas Schools	19–22	✓(7)	✓(7)			Statewide distance learning and Internet-based technologies
Utah	28					Uses Idaho State didactic program
Virginia Commonwealth	18	✓(9)				Internet
Shenandoah (Virginia)	33		✓(8)			Internet
Washington (University of Washington and Washington State University)	Individualized (Max. 66 quarter hours)		✓(8)		✓	Self-study and workshops
West Virginia	30 (26 from Kentucky)					Uses part of Kentucky didactic program—videotapes, CDs, Internet
Wisconsin	24	✓		(4)	✓	Distance learning, Internet workshops, audio-conferencing

[a]Indicates semester credit hours, except where noted.
[b]Maximum credit hours allowable—not necessarily additive for determining total program waivers.
[c]See text for description of delivery methods.
[d]BCPS, Board of Pharmaceutical Specialties Certification (can be used to waive requirements); ASHP—American Society of Health System Pharmacy; ACCP, American College of Clinical Pharmacy.

Table 3 Clerkship requirements for nontraditional Pharm.D. programs

School/college	Required clerkship (week)	Part-time allowed	Credit for prior experience (max. weeks waived)	Completed at student's worksite (max. weeks)
Samford (Alabama)	None, but patient monitoring required for disease modules	Yes	None	Yes, but not all
Auburn (Alabama)	8 (typically)	Yes	Clerkship requirements are individualized	All
Arkansas	16 (4 weeks on-campus)	Yes	None	12 Weeks self-directed study
Arizona	26	Yes	None	Under certain circumstances
Colorado	30	Yes	Portfolio waiver (12)	(15)
Nova Southeastern (Florida)	16	Yes		
Florida	No, patient monitoring required for disease modules	Yes		Yes
Georgia Schools (Mercer and University of Georgia)	18	Yes	Precepted clerkships (6)	(14)
Idaho State	28	Yes	Substitution only	(16)
Midwestern (Illinois)	20 (32 quarter hours)	Yes	Yes	Under certain circumstances
Illinois–Chicago	30 (20 semester credits)	No	ASHP residency (12)	Clerkships must be completed in Chicago
Purdue	28	Yes	(8)	Under certain circumstances
Iowa	35 (28 semester credits)	Yes	(10)	Under certain circumstances
Kansas	20	Yes	(12)	Yes
Kentucky	32	Yes	(4–16)	Yes, but not all
Louisiana at Monroe	TBA			
Xavier (Louisiana)	15 (9 semester credits)	No	(10)	Yes
Maryland	4.5 (total approx. 180 hours)	Yes	Yes	All
Massachusetts	Discontinued			
Minnesota	25	Yes	Yes, maximum 15 credit hr total for didactic and clerkships	One practice enhancement clerkship
Mississippi	24	Yes	30% of clerkship requirement	Optional case management
Missouri–Kansas City	28	Yes	(8)	(4)
Montana	28	Yes	No	All
Creighton (NE)	24	Yes	(8)	(12)
New Mexico	36	Yes	(16)	No
Albany (New York)	25	Yes	(5)	One self-directed rotation
A & R Schwartz (New York)	16			
Buffalo (New York)	36	Yes	(12)	Yes
North Carolina	28	Yes	(12)	Under certain circumstances
North Dakota State	36			
Ohio Northern	24 (30 quarter hours)	Yes	(8) 10 quarter hours	(8) 10 quarter hours
Ohio State	TBA			

(Continued)

Table 3 Clerkship requirements for nontraditional Pharm.D. programs (*Continued*)

School/college	Required clerkship (week)	Part-time allowed	Credit for prior experience (max. weeks waived)	Completed at student's worksite (max. weeks)
Oklahoma Schools	16	Yes		(8–12)
Duquesne (Pennsylvania)	25 (10 credit hours)	Yes	(10) 4 credit hours	Yes
Philadelphia (Pennsylvania)	24	Yes	(8)	
Temple (Pennsylvania)	Discontinued			
Wilkes (Pennsylvania)	16			
Rhode Island	Discontinued			
South Carolina	Not admitting			
South Dakota State	32	Yes		(1)
Tennessee	32	Yes	(16)	
Texas Schools	19–21	Yes	(3)	(9)
Utah	28			
Virginia Commonwealth	30	Yes	(20)	Yes
Shenandoah (Virginia)	24	No	16	Yes
Washington Schools (University of Washington and Washington State University)	Portfolio review determines clerkship requirements	Yes	Yes	Yes, strongly encouraged
West Virginia	28	Yes	3 Clerkships	2 Rotations
Wisconsin	24	Yes	Substitution only	(4)

served; generally, this includes examples of written patient work-ups or care plans that relate to specific educational objectives.

- Documentation (i.e., dates, location, and copies of handouts, etc.) of presentations at pharmacy or other professional meetings.
- Previous educational activities including disease state management programs, certifications, etc.
- Other community or preventative health care activities.

CLERKSHIP REQUIREMENTS

The experiential component presents special challenges for both NTPD programs and students, particularly in rural areas.[14] Although credit for work-related experience and some flexibility in designing the clerkship schedule are offered, the majority of programs require a significant clerkship component, typically 24 to 28 weeks. Because many programs require full-time clerkship training, a temporary leave of absence, use of vacation time or other work arrangements are necessary. Although this frequently imposes financial and personal difficulties, it is perhaps the most important aspect of Pharm.D. education and should be investigated thoroughly to ensure that career goals and objectives are satisfied.

The types of clerkship experiences and time commitment vary considerably among NTPD programs (Table 3). Internal medicine and ambulatory care constitute the core clerkship requirements of many programs, with electives in mental health, pediatrics, geriatrics, and drug information commonly offered. Some programs, such as Idaho State University's NTPD program, offer a capstone rotation during which the student implements and documents the resultant benefits of some aspect of pharmaceutical care at their original place of employment. This serves not only to further the goals of the profession, but also rewards employers for their flexibility and patience by upgrading the level of care at the employment site.

Most programs allow for part-time completion of clerkships. This can be arranged on a half-time basis (e.g., 4 hours per day) or, more frequently, students may complete one 4- or 6-week rotation, followed by returning to full-time work for a period of time. This allows students significant financial flexibility.

Restrictions vary with respect to fulfillment of clerkship requirements at the student's place of employment. Some programs require students to complete as many clerkships as possible outside their normal workplace, reasoning that exposure to different practice environments and styles will ultimately benefit the student-practitioner. Other programs have taken the opposite approach, and encourage students to complete all experiential rotations in their normal workplace under the construct that this provides an opportunity to enhance the practice setting by practicing pharmaceutical care skills on their own patient population.

Many NTPD programs have simply funneled their nontraditional students through available traditional clerkship sites, whereas other programs have combined traditional sites with the identification and approval of new clerkship venues. The older NTPD programs have the advantage of expanding clerkship opportunities by using nontraditional graduates as preceptors.

Evaluation of clerkship outcomes and competencies are, by nature, much more subjective than their didactic counterparts. This element, combined with the fact that clerkship requirements of traditional, entry-level Pharm.D. candidates often take priority over nontraditional clerkship needs,[15,16] results in the potential for inclusion of clerkship sites with less than minimal qualifications. Some programs have attempted to sidestep the costly and time-consuming process of evaluating and approving off-site clerkships by creating regional sites (satellites) where students congregate at predetermined intervals (usually monthly or at certain times during the semester) for clinical practice assessments via paper case presentations and group interaction. Students are then expected to accumulate direct patient contact experience independently by following patients identified from their workplace. Students from hospital pharmacy environments may encounter no difficulty finding interesting cases to follow; those with pharmaceutical industry or community pharmacy backgrounds will have more difficulty, and run the risk of sacrificing their capstone clerkship experience for the sake of convenience.

UNDERLYING MOTIVATIONAL FACTORS

Pursuit of a doctorate through the nontraditional pathway requires a significant commitment in terms of time, money, and determination on the part of the candidate.[17–19] The motivations behind obtaining a postbaccalaureate Pharm.D. are multifactorial, but can be separated into two broad categories—external or internal.[20,21] External motivations for application to a nontraditional Pharm.D. program include earning a promotion, employer pressure, and concerns about future job security as a result of the shift from BS to Pharm.D. as the sole entry degree. Internal motivation, however, derives from a desire to enhance personal job satisfaction. Examples of internal motivations include a desire to improve clinical skills to optimize patient care, as well as the desire for greater responsibility in making decisions regarding drug therapy. It has been our experience that students who are internally motivated tend to progress faster through the nontraditional Pharm.D. curriculum and have a lower attrition rate compared with those students who are externally motivated.

The motivations of pharmacists interested in pursuing a nontraditional Pharm.D. degree have been surveyed.[20–22] The most important reasons given for seeking a postbaccalaureate degree were improvement of clinical skills, enhanced work quality, and increased personal satisfaction. Kelly et al. surveyed Pennsylvania pharmacists and discovered that the degree of satisfaction with current employment was inversely correlated with likelihood of enrolling in an NTPD program; more than 73% of pharmacists who found their jobs to be ''very unfulfilling'' stated they would probably apply.[23]

Two surveys of nontraditional Pharm.D. program graduates have assessed the effect of the degree on practice patterns.[6,24] Respondents from both surveys reported spending a significantly greater percentage of their time performing pharmaceutical care (e.g., making therapeutic recommendations, developing treatment plans) and less time dispensing or processing prescriptions compared with their last position as a BS-trained pharmacist. The respondents also reported significantly greater levels of job satisfaction after graduation from their respective NTPD programs.

Dunn et al. surveyed nontraditional Pharm.D. students currently enrolled at the University of Arkansas.[25] The majority of respondents from this survey reported enhancements in self-esteem, pride in their profession, and an elevation in the overall level of pharmaceutical care at their practice site. More than 85% of the Arkansas nontraditional Pharm.D. students stated that they had gained confidence in their ability to provide pharmaceutical care as a result of the program.

DISCUSSION

The information presented in Tables 1–3 is evidence of the commitment, dedication, and concerns of academic pharmacy regarding the facilitation and enhancement of clinical skills and abilities of active practitioners. This is impressive, particularly in light of the limited lifespan anticipated for NTPD programs. As colleges of pharmacy begin to confer the Doctor of Pharmacy degree exclusively, the demand for NTPD programs will inevitably

wane. Given the high costs of starting a NTPD program (both in terms of faculty time and distance learning infrastructure)[4] and the limited time frame to recoup any investment, some universities have developed unique partnerships. For example, the University of Florida has partnered with a private corporation to facilitate delivery of course materials. However, Idaho State University (ISU) has taken a different approach. ISU has entered into a collaborative clinical training agreement with the University of Utah. As a result, ISU benefits from this agreement by bringing additional out-of-state students into its established NTPD program, whereas the University of Utah benefits by avoiding the high startup and maintenance costs of repackaging their curriculum into a distance learning format. Utah pharmacists also benefit from the agreement by having access to ISU's NTPD program at a reduced tuition rate. As the pool of prospective NTPD program applicants dries up, many more universities may consider such collaborations to minimize costs, while still providing this essential educational service to practitioners.

Considerable resources have been allocated toward the development and implementation of NTPD programs, generally without any corresponding increase in appropriated budgets or overall institutional support. Hence, programs have been largely developed and are solely dependent on practitioner-derived tuition fees to maintain and support their existence. Total tuition costs for in-state residents of NTPD programs range from approximately $5000 to $15,000, excluding the costs of textbooks, computer use, etc. Nonresident tuition rates are usually 2–3 times higher.

Financial cost, unfortunately, is a major impediment to participation. Because these programs are designed for full-time working practitioners, demonstration of financial need is usually difficult and most practitioner-students will not qualify for federal financial aid programs. Thus, for those individuals who do not work for institutions or corporations that offer employee educational benefits, financing their return to school will require careful consideration. Even with financial assistance, a substantial burden of the cost will accrue to the individual. Although a few programs may offer scholarship assistance specifically for nontraditional students, these are extremely rare.

Clearly established goals, strong self-discipline, and an extremely supportive family and employer are essential attributes of successful nontraditional students. Although schools/colleges have expended significant efforts to design and implement user-friendly programs to allow full-time practitioners to return to school, prospective candidates must approach this process with the same intensity and rigor expected of other doctoral-level programs. Thus, the goal of nontraditional education is not to make it easy to obtain the degree, but rather to make it easier to return to school. Substantial time, effort, and self-sacrifice can be expected of nearly all programs, and these generally will directly correspond to enhancement of clinical skills and abilities. At this juncture, the evolution of NTPD programs has virtually assured that practitioners will be required to *earn their doctorate* and that grandfathering of the Pharm.D. degree will not occur.

Considerable effort has gone into designing NTPD programs that achieve the same educational endpoints as their on-campus entry-level counterparts. The use of standardized patients as an assessment method has been proposed as a mechanism to document parity between nontraditional and traditional tracks.[26] Evidence supports the equivalency of nontraditional education;[24] yet, the issue is still somewhat controversial.[11] Hopefully, this will quickly disappear since NTPD programs are now required to demonstrate equivalent educational outcomes as part of their ACPE accreditation process.[27]

Perhaps more controversial is the relative quality and equivalency between various NTPD programs, especially with respect to major differences in clerkship requirements for some programs. The ability to take time off from work to complete full-time or even half-time clerkship experiences is certainly a major impediment to pursuing a NTPD, and flexibility in designing the clerkship program is imperative. This has resulted in development of truly unique experiential components in an attempt to overcome this critical element. The major issue is whether NTPD programs requiring primarily self-directed clerkships (i.e., no or very limited onsite preceptor supervision) are able to provide the breadth and depth of clinical educational experiences compared to with more traditional clerkship programs (i.e., student-practitioner progress directly monitored by onsite preceptor). Given the significant differences, debate will likely continue until more extensive experience accumulates or an objective comparison of NTPD programs and/or graduates can be made.

CONCLUSION

Final resolution and adoption of the Pharm.D. as the sole entry degree have provided the impetus for development of innovative nontraditional programs. Practitioners are now afforded a variety of equivalency-based academic degree programs from which to select. Consideration of career goals must be carefully weighed against the potential financial and personal costs.

ACKNOWLEDGMENT

The authors acknowledge Ms. Kathleen Moe for her assistance in collecting the individual NTPD program information.

REFERENCES

1. Task Force on the Development of an Equitable Degree Upgrade Mechanism. In *Guidelines for a Uniform Method for a Baccalaureate Degreed Pharmacist to Earn a Doctor of Pharmacy Degree*, NABP 91st Annual Meeting Proceedings, Dec. 10, 1994.
2. American Pharmaceutical Association; American Society of Hospital Pharmacists; NARD. *Joint Statement on the Entry-Level Doctor of Pharmacy Degree*; July 7, 1992.
3. Piascik, M.M.; Lubawy, W.C. Demographic comparisons of on-campus and off-campus post-BS Pharm.D. students. Am. J. Pharm. Educ. **1992**, *56*, 1–7.
4. Carter, R.A. Rationale, strategies and cost analysis of a nontraditional Pharm.D. program. Am. J. Pharm. Educ. **1993**, *57*, 211–215.
5. Vanderbush, R.E.; Goad, D.L.; Udeaja, J.O.; Rollins, S.; Morrison, W.J. Nontraditional clerkships at the University of Arkansas. Am. J. Pharm. Educ. **1994**, *58*, 294–299.
6. Fjortoft, N.F.; Engle, J.P. Effect of the nontraditional Pharm.D. on individual practice patterns. Am. J. Pharm. Educ. **1995**, *59*, 223–227.
7. Kaplan, I.P.; Patton, L.R.; Hamilton, R.A. Adaptation of different computerized methods of distance learning to an external Pharm.D. degree program. Am. J. Pharm. Educ. **1996**, *60*, 422–425.
8. Stratton, T.P.; Bartels, C.L.; Miller, S.J.; Carter, J.T. Distance learning via Lotus Notes Learning Space in a nontraditional Pharm.D. program: A preliminary report. Am. J. Pharm. Educ. **1999**, *63*, 328–333.
9. The American Association of Colleges of Pharmacy Admissions Report. In *Pharmacy School Admission Requirements 2001–2002*; AACP: Alexandria, Virginia, 2000; 85–95.
10. Ukens, C. It's not over yet: The great Pharm.D.-B.S. race still going strong. Drug Top. **1995**, *139* (Apr. 24), 48–49, 51, 53, 55–56.
11. Gebhart, F. Is there a bias against nontraditional Pharm.D.s? Drug Top. **1999**, *143*, 40 (Nov. 15).
12. Pugsley, J.A.; Einarson, T.R.; Spino, M. Demand for a Canadian Doctor of Pharmacy program. Am. J. Pharm. Educ. **1991**, *55*, 331–334.
13. Roche, V.F. In *Challenges and opportunities for schools and colleges of pharmacy in the next decade*, Speech Presented at the NAPB/AACP Combined District 7 and 8 Meeting in Tucson, Arizona, September 22, 2000; http://www.aacp.org/About_AACP/speeches.html (accessed 10/29/2000).
14. Scott, D.M.; Miller, L.G.; Letcher, L.A. Assessment of desirable pharmaceutical care practice skills by urban and rural Nebraska pharmacists. Am. J. Pharm. Educ. **1998**, *62*, 243–252.
15. Carter, R.A. In nontraditional education—assuring quality is job one. Am. J. Pharm. Educ. **1994**, *58*, 411–413.
16. Vanderbush, R.E.; Dunn, E.B.; Morrison, W.J. Nontraditional Pharm.D. program at the University of Arkansas for Medical Sciences. Am. J. Health-Syst. Pharm. **1995**, *52*, 620–623.
17. Halley, H.J. Nontraditional doctor of pharmacy program: One graduate's experience. Am. J. Hosp. Pharm. **1994**, *51*, 2717–2720.
18. Beavers, N. Taking a nontraditional approach. Am. Drug. **1996**, *213*, 20, (Aug.).
19. Milito, D.A. Taking a nontraditional path to the Doctor of Pharmacy degree. Am. J. Health-Syst. Pharm. **1995**, *52*, 1168, 1171.
20. Garst, W.C.; Ried, L.D. Motivational orientations: Evaluation of the education participation scale in a nontraditional doctor of pharmacy program. Am. J. Pharm. Educ. **1999**, *63*, 300–304.
21. Zgarrick, D.P.; MacKinnon, G.E., III. Motivations and practice-area preferences of pharmacists interested in pursuing a Pharm.D. degree through a nontraditional program. Am. J. Health-Syst. Pharm. **1998**, *55*, 1281–1287.
22. Joyner, P.U.; Pittman, W.; Campbell, W.H.; Dennis, B.H. A needs assessment for an external Doctor of Pharmacy degree program. Am. J. Health-Syst. Pharm. **1997**, *61*, 292–296.
23. Kelly, W.N.; Chrymko, M.M.; Bender, F.H. Interest and resources for nontraditional Pharm.D. programs in Pennsylvania. Am. J. Pharm. Educ. **1994**, *58*, 171–176.
24. Fjortoft, N.F.; Weigand, L.; Lee, M. Effect of the nontraditional Pharm.D. on practice patterns based on a survey of graduates from six programs. Am. J. Pharm. Educ. **1999**, *63*, 305–309.
25. Dunn, E.B.; Vanderbush, R.E.; McCormack, J.; Morrison, W.J. Survey results of nontraditional Pharm.D. students at the University of Arkansas for Medical Sciences. Hosp. Pharm. (Saskatoon, Sask.) **1996**, *31*, 955–958.
26. Monaghan, M.S.; Vanderbush, R.E.; Gardner, S.F.; Schneider, E.F.; Grady, A.R.; McKay, A.B. Standardized patients: An ability-based outcomes assessment for the evaluation of clinical skills in traditional and nontraditional education. Am. J. Pharm. Educ. **1997**, *61*, 337–344.
27. American Council on Pharmaceutical Education. *Standards 2000 Self-Study: A Self-Study Guide for Accreditation Standards and Guidelines for the Professional Program in Pharmacy Leading to the Doctor of Pharmacy Degree,* Adopted June 14, 1996 and effective July 1, 2000. http://www.acpe-accredit.org/frameset_Pubs.htm (accessed 11/2/00).

Nutraceuticals and Functional Foods

John Russo, Jr.
The Medical Communications Resource, Mahwah, New Jersey, U.S.A.

INTRODUCTION

Functional foods and nutraceuticals are assuming a middle ground between food and drugs due to a growing body of evidence that supports their role in maintaining health and contributing to the treatment of diseases. Since Hippocrates advised ''Let food be thy medicine and medicine be thy food,'' we have defined medicines and foods based on what is known about each substance in terms of efficacy, safety, and the significance of its perceived contribution to health. Over time, we tend to redefine these substances as our experience and expectations change. The ancient Greeks, for example, looked upon garlic as a performance-enhancing drug and officially sanctioned it for this use during the first Olympic games. During the age of sailing ships, lemons were dispensed to sailors to prevent and treat scurvy. John Woodall, the father of naval hygiene and a ''Master in Chirurgerie,'' published ''The Surgeon's Mate'' in 1636, in which he wrote, ''The juyce of lemmons is a precious medicine .... It is to be taken each morning two or three teaspoonfuls, and fast after it two hours.''

Modern functional foods became available in the 1920s, when iodine was added to salt to prevent goiter. This was followed by vitamin D milk. Today, many Americans start their day with calcium-fortified orange juice (to strengthen their bones). Then, they spread a margarine that lowers cholesterol on folate-enriched toast (to protect their hearts and prevent birth defects).

DISTINGUISHING BETWEEN FUNCTIONAL FOODS AND NUTRACEUTICALS

It is not surprising that the terminology to define and promote these products is escalating (Table 1). For practical purposes however, two terms (*functional food* and *nutraceutical*) can be used to distinguish among nutritional products that make health claims beyond basic nutrition.

Functional foods, as defined by the American Dietetic Association, are products whose nutritional value is enhanced by the addition of natural ingredients. Functional foods may provide specific health benefits beyond basic nutrition when consumed as part of a varied diet.

Nutraceuticals, according to the American Nutraceutical Association, are functional foods with potentially disease-preventing and health-promoting properties. They also include naturally occurring dietary substances in pharmaceutical dosage forms. Thus, they include ''dietary supplements'' as defined by the Dietary Supplement Health and Education Act of 1994 (DSHEA).

From the FDA regulatory perspective, all substances that influence our health can be divided into two groups: food and drugs, with food further divided into conventional food and dietary supplements (Fig. 1). Despite the fact that many terms are used to describe nutritional substances that influence health, each of these terms fits into one of these categories. When viewed in this way, healthcare professionals will be less susceptible to the vague claims that tend to characterize this growing field.[1]

EXAMPLES OF FUNCTIONAL FOODS

All foods are functional; however, the term *functional* refers to an additional physiological benefit beyond meeting basic nutritional needs (Table 2). For example, epidemiological data tend to show that cancer risk in people consuming diets high in fruits and vegetables is about one-half that seen among people consuming few of these foods.[2] Certain components from animal sources make similar contributions to health, including omega-3 fatty acids found in fish, calcium in dairy products, and the anticarcinogenic fatty acid known as conjugated linoleic acid in beef.[3]

Once the benefits of a key component in food are documented, the challenge is to increase its concentration, and presumably its benefits, while maintaining safety. For example, isoflavones in soy are phytoestrogens with a chemical structure similar to estrogen. Isoflavones may reduce cholesterol, but what is the risk of increasing the intake of a compound that may modulate estrogens? Knowledge of the toxicity of functional food components is crucial to improve their benefit–risk ratio. The efforts

DOI: 10.1081/E-ECP 120006312

Table 1 Categories of healthful foods and dietary supplements

Category/example	Description
Dietary supplements Products containing one of more of the following: • Vitamin, mineral, herb or other botanical • An amino acid or metabolite • An extract • Any combination of the above	• Products (other than tobacco) intended to supplement the diet. • May be marketed in a food form if not "represented" as a conventional food and labeled as a dietary supplement. • Specific health or structure/function claims[a] can be made if the FDA deems adequate scientific substantiation exists.
Fortified foods • Breakfast cereal • Vitamin B added to baked goods	• Foods enriched with vitamins and minerals, usually up to 100% of the DRI. • Often mandated by law to replace nutrients lost during processing.
Functional foods • Soy • Salad dressing with omega-3 polyunsaturated fatty acids • Carrots with 170% of daily requirement of vitamin A	• A food or ingredient that may provide a health benefit beyond the "traditional nutrients" it contains.[b] • Specific health or structure/function claims can be made if the FDA deems adequate scientific substantiation exists. • Super fortified foods have more than 100% of the DRI and/or foods with added botanicals or other supplements.
Medicinal foods • Food to treat diabetes, obesity, or heart disease, sold through physicians, not by conventional retailers	• Food formulated to be consumed or administered internally while under the supervision of a physician. • Intended for specific dietary management of a disease or condition for which distinctive nutritional requirements are established.
Nutraceuticals • Orange juice with calcium • Dietary/herbal substances in pharmaceutical dosage forms	• Dietary supplements and fortified foods enriched with nutrients not natural to the food.

DRI = dietary reference intake.
[a]Structure/function claims state that a product may affect the structure or function of the body (e.g., calcium builds strong bones, antioxidants maintain cell integrity, fiber maintains bowel regularity), but may not claim that a therapy can prevent or cure a disease (e.g., alleviates constipation).
[b]"Traditional nutrients" refers to vitamins and minerals considered essential to the diet and/or to correct a classical nutritional deficiency disease. For example, foods containing vitamin C to correct scurvy or vitamin D to alleviate rickets are not functional foods. However, soy, which contains soy protein and is associated with a reduced cardiovascular risk, is a functional food.
(From Ref. [8].)

that go in to making these determinations are costly. As a result, companies that are successful can be expected to market products that will be branded and extensively promoted (Table 3).

CHALLENGES FACING NUTRACEUTICALS

Some of the most popular nutraceutical products marketed today are botanicals such as St. John's wort, echinacea, ginkgo biloba, saw palmetto, and ginseng. Unfortunately, manufacturers are not required to prove their safety or efficacy before marketing them. Dosages are not standardized. The quality of the raw source and the plant parts used are not regulated. And, unlike prescription drugs or over-the-counter medicines, there is no federal quality control standard to ensure that the label reflects what is in the bottle.

The problem is illustrated by reports that labeled concentrations of active ingredients often significantly overestimate the content in the dosage form. In one report, nearly one-third of the brands tested did not contain what their manufacturers claimed.[4] It is a serous problem that undermines the nutraceutical market, encourages skeptics who criticize the value and role of nutraceuticals, and (most importantly) is potentially harmful to the public.

SIGNIFICANCE FOR PHARMACISTS

The issues surrounding nutraceuticals and functional foods are important to pharmacists for two reasons. It is

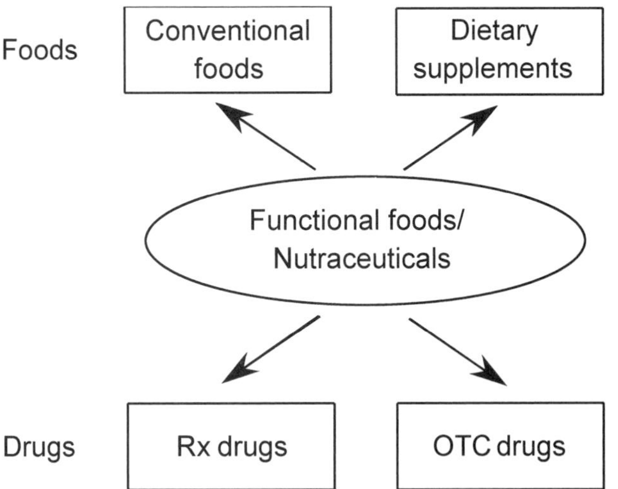

Fig. 1 Options for categorizing functional foods and nutraceuticals.

important to their patients and is certain to be the source of questions. Also, the growing field of nutraceuticals and functional foods is likely to change the practice of pharmacy over the next decade.

More than 100 million Americans regularly use dietary supplements, and in the year 2000, consumers spent about $16 billion on them.[5] Furthermore, 39% of Americans say they have made dietary changes to reduce their risk of getting cancer, 43% claim to take a daily multivitamin for cancer protection, while 21% take another form of nutritional or dietary supplement (i.e., concentrated doses of single vitamins, minerals, or herbal substances) to lower their cancer risk.[6]

Americans are concerned about their health and they view their pharmacist as an important resource for health matters. Almost two-thirds of respondents in one sample stated that they regularly talk with their pharmacist when choosing an OTC product, and 58% have come to think of their local pharmacist in the same terms as their family doctor.[6]

In the future, according to Dr. Benadette Marriott, who is vice president of programs and communications at the Burroughs Wellcome Fund, the need to guide consumers through the maze of functional foods and nutraceuticals may lead to significant changes in pharmacy practice. Food stores will continue to evolve into ''one-stop wellness centers,'' where consumers go for basic wellness screening activities, nutritional counseling, and medication advice. At the center will be the pharmacy, where pharmacists will work with nutritionists and others to help consumers recognize their options and select among several sources of an ingredient in order to safely treat or lower their risk for disease.[1]

CASE HISTORY: YEAR 2010

Mr. D is a 50-year-old man with a family history of prostate cancer and a personal history that is remarkable for a diet high in saturated fats. He comes to the pharmacy knowing that his family and dietary history place him at risk. He has no symptoms or laboratory values consistent with prostate cancer but is committed to making changes in his life that will lower his risk. Part of his strategy includes eating more tomatoes to increase his intake of the antioxidant lycopene. However, he is confused because of a news story in which researchers reported that lycopene administered to mice as a supplement acted as a pro-oxidant and encouraged tumor growth.[7]

In addition to wanting to take an effective source of lycopene, Mr. D wants to know which is safest and most

Table 2 Examples of functional foods, their key components, and potential health benefits

Functional food	Key component	Potential health benefits
Black and green tea	Catechins	Reduce risk for cancer
Broccoli and other cruciferous vegetables	Sulforaphane	Reduce risk for cancer
Citrus fruits	Limonoids	Reduce risk for cancer
Fish	Omega-3 fatty acids	Reduce risk for heart disease
Fruits and vegetables	Many different phytochemicals	Reduce risk for cancer and heart disease
Garlic	Sulfur compounds	Reduce risk for cancer and heart disease
Oats and oat-containing foods	Soluble fiber beta glucan	Reduce cholesterol
Purple grape juice and red wine	Polyphenolic compounds	Support normal, healthy cardiovascular function
Soy foods	Soy protein	Reduce cholesterol
Tomatoes and tomato products	Lycopene	Reduce risk for cancer
Yogurt and fermented dairy products	Probiotics	Improve gastrointestinal health

Table 3 Examples of products marketed or planned to be marketed as functional foods by pharmaceutical companies

Novartis: Aviva product line	• Breakfast bars, cereals, and beverages. • Claims: Benefits the heart, bones, and digestion. • Marketed in United Kingdom and Switzerland. • U.S. launch planned.
McNeil Consumer Health, Division of Johnson & Johnson	• Benecol brand margarine, salad dressing, and health bars in U.S. • Claim: Reduce LDL cholesterol up to 14% within two weeks of product use. • In first few months, Benecol margarine captured ~2% of U.S. margarine sales.
Mead Johnson, Division of Bristol-Myers Squibb	• EnfaGrow nutrient-enriched oatmeal and snack line for toddlers, marketed primarily to physicians. • Allergy alert: EnfaGrow Nutritional Oatmeal for Toddlers in maple brown sugar and cinnamon and strawberry flavors may contain trace amounts of milk protein, not listed in the ingredients.[a] • Viactiv Soft Calcium Chews sold in U.S. grocery and drug stores. • Claim: To meet women's special nutritional needs.

[a]From Ref. [9].

economical. In the year 2010, there are no fewer than five sources of lycopene, listed here in order of increasing cost:

- Nutritional lycopene: vine-ripened tomatoes (and other foods).
- Organically grown sources of lycopene: organically grown tomatoes.
- Lycopene-enhanced functional foods: genetically enhanced tomatoes with guaranteed high levels of lycopene.[a]
- Nutraceuticals: dietary supplements that isolate and contain high levels of lycopene.
- Prescription drugs: pharmaceutical-grade lycopene containing the highest concentrations of lycopene and clinically tested to meet FDA standards for safety and efficacy.*

GUIDELINES

The challenge facing the pharmacist is to guide this patient through the maze of study results and conflicting claims. There are few absolute answers, but there are guidelines that can help consumers make intelligent decisions.[7] And it is likely that the guidelines for today will be valid a decade from now.

* Currently, not marketed.

Regard advertising and articles about supplements with caution:

- Be cautious of any product that claims to ''boost'' the immune system or ''rejuvenate'' health. Advise consumers to look for specific findings, not vague claims.
- Nutraceuticals can have side effects under certain circumstances and should be thoroughly tested before being used by the public.
- Remember, if it doesn't have side effects, it probably hasn't been thoroughly tested.

Natural first, supplement second:

- When there is a choice between a vegetable and a pill, recommend to eat the vegetable.
- Nutraceuticals are products that isolate recognized active ingredients. It is often not clear whether the active ingredient is as effective when taken in the absence of other nutrients found in food.
- When taking supplements, consumers should take them with food to aid absorption and minimize the risk of GI upset.

Read the label; then, get a second opinion:

- Before buying anything, read the label.
- Confirm the value of the supplement with a health care advisor.

Moderation is best:

- More is not always better; vitamin C, selenium, and vitamin D are examples where too little has no effect, but too much has adverse effects.

- Start with low doses and work up. A commitment to taking nutraceuticals is a long-term strategy. Starting doses that are too high increase the risk of side effects, which will dampen the consumer's resolve to continue treatment.

Keep everything in perspective:

- Supplements are only part of the picture. They are not substitutes for a healthy diet, stress reduction, exercise, weight control, and (when needed) prescription drug therapy.
- Nutraceuticals do not offset the negative effects of smoking.

Finally, remind consumers not to become obsessed by one disease. For example, men concerned about prostate cancer and women concerned about breast cancer should not ignore their risk factors for heart disease.

REFERENCES

1. Lewis, **2000**. http://www.nytimes.com/library/national/science/health/062000hth-lab-tests.html.
2. Block, G.; Patterson, B.; Subar, A. Fruit, vegetables, and cancer prevention: A review of the epidemiological evidence. Nutr. Cancer **1992**, *18*, 1–29.
3. Hasler, C.M. Functional foods: Their role in disease prevention and health promotion. Food Technol. **1998**, *52* (2), 57–62.
4. Hemphill, C. *Putting Dietary Supplements to the Test*; The New York Times, June 20, 2000 [Accessed 20 June 2000.] Available at URL: http://www.nytimes.com/library/national/science/health/062000hth-lab-tests.html.
5. Augsburger, L.L. In *Introduction and Welcoming Remarks*, AAPS Dietary Supplements Forum: Exploring the Science of Nutraceuticals, Washington, DC, June 28–30, 2000.
6. AICR. *Seniors Prefer Taking Supplements to Changing Diet*; American Institute for Cancer Research (AICR), 31 August 2000 [Accessed 12 November 2000.] Available at http://www.newswise.com/articles/2000/8/SUPSURV.CR1.html.
7. Moyad, M.A. *The ABC's of Nutrition and Supplements for Prostate Cancer*; Sleeping Bear Press: Chelsea, Michigan, 2000.
8. Mulry, M. Functional foods and nutraceuticals. Nutr. Sci. News **July 2000**, *5* (7), 292–296.
9. SafetyAlert.com. *Mead Johnson Issues Allergy Alert for Two Flavors of EnfaGrow Nutritional Oatmeal for Toddlers*; 10 January 2000 [Accessed 13 November 2000]. Available at URL: http://www.safetyalerts.com/recall/f/00/enfgrw.htm.
10. Marriott, B.M. In *Dietary Supplements: Clinical Studies and Trials to Evaluate the Efficacy and Safety of Nutraceuticals and Dietary Supplements: Approaches, Stumbling Blocks and the Consumer Quandary*, AAPS Dietary Supplements Forum: Exploring the Science of Nutraceuticals, Washington, DC, June 28–30, 2000.

Oddis, Joseph A.

C. Richard Talley
American Society of Health-System Pharmacists, Bethesda, Maryland, U.S.A.

INTRODUCTION

Joseph A. Oddis served as the Executive Vice-President of the American Society of Health-System Pharmacists (ASHP, formerly the American Society of Hospital Pharmacists) for 37 years (1960–1997). During that period ASHP emerged as pharmacy's strongest membership association, with 31,000 members, a staff of 180, an annual budget of $30 million, and net assets of $30 million. Oddis's leadership was a major force behind establishing pharmacy as a clinical profession, establishing ASHP as the accrediting body for pharmacy residency and technician-training programs, organizing the world's largest, ongoing meeting of pharmacists (the ASHP Midyear Clinical Meeting), and creating the ASHP Research and Education Foundation. Oddis was also seminal to the publication of a periodical devoted to clinical practice (*Clinical Pharmacy*), as well as books considered essential to pharmacy practice (e.g., *The Handbook on Injectable Drugs*), and patient-oriented sources of drug information (e.g., *Consumer Drug Digest*).

EDUCATION AND TRAINING

Oddis graduated from Duquesne University in 1950 with a degree in pharmacy. He served as a staff pharmacist at Pittsburgh's Mercy Hospital before serving in the U.S. Army. He returned to Mercy Hospital as assistant director of pharmacy and later served as chief pharmacist at Western Pennsylvania Hospital. In recognition of his leadership, Oddis received five honorary doctor of science degrees: from the Massachusetts College of Pharmacy and the Philadelphia College of Pharmacy and Science in 1975, the Albany College of Pharmacy in 1976, Duquesne University in 1989, and Mercer University in 1995. He received an honorary doctor of humane letters degree from Long Island University in 1991.

MAJOR POSITIONS AND PROFESSIONAL INVOLVEMENT

In 1956 Oddis joined the American Hospital Association (AHA) in Chicago as the staff representative for hospital pharmacy. In 1960 he was recruited by ASHP and the American Pharmaceutical Association (APhA) as secretary of ASHP and director of the Division of Hospital Pharmacy at APhA. When the APhA position was dissolved a few years later, Oddis became full-time chief executive officer at ASHP, a position he held until his retirement in 1997.

For the International Pharmaceutical Federation (FIP), Oddis served as President, Section on Hospital Pharmacy (1977–1982); FIP Vice-President (1984–1986); President (1986–1990), and immediate Past President (1990–1994).

For the American Association for the Advancement of Science, Oddis was a Fellow; Secretary, Pharmaceutical Scientists Section (1968–1970); and Vice President and Chairman, Section on Pharmacy (1961).

For the American Society of Association Executives Foundation, Oddis was a member of the Board of Directors (1977–1978); Treasurer (1982–1983); Chairman (1984–1985), and immediate Past Chairman (1985–1986.)

ACCOMPLISHMENTS

Oddis's leadership at ASHP yielded a long list of achievements. The *American Journal of Health-System Pharmacy* (*AJHP*) grew from modest beginnings to the profession's most highly regarded peer-reviewed scientific journal. The *American Hospital Formulary Service* (*AFHS*) became a world-reknowned drug information system. Educational meetings grew in number, size, and quality; ASHP's Midyear Clinical Meeting became the largest meeting of pharmacists in the world. ASHP

Encyclopedia of Clinical Pharmacy
DOI: 10.1081/E-ECP 120006256

created and published more than 100 practice standards, as well as numerous guidelines, technical assistance bulletins, and position papers, many of which dealt with clinical pharmacy issues. In 1962, ASHP produced the first Statement on Accreditation of Hospital Pharmacy Internship Training Programs; in the Oddis era more than 10,000 residents graduated from 375 programs accredited by ASHP. In 1968, ASHP created the ASHP Executive Residency Program, which produced 26 graduates under Oddis's stewardship. In 1969, the ASHP Research and Education Foundation was created. In addition to the numerous honorary doctor of science degrees bestowed on him, Oddis received the Harvey A.K. Whitney Lecture Award, pharmacy's highest award, presented by ASHP in 1970; the Hugo H. Schaefer Award, presented by APhA in 1983; the Donald E. Francke Medal, presented by ASHP in 1986; the Remington Honor Medal, presented by APhA in 1990; and the Andre Bedat Award, presented by FIP in 1994.

INFLUENCE ON CLINICAL PHARMACY PRACTICE

Oddis, more than any other pharmacy association executive, advanced the role of pharmacists in healthcare. He was a key facilitator and motivator in the 1960 creation of the ASHP/AHA joint statement of the legal basis of the hospital formulary system, widely regarded as a crucial catalyst for the development of clinical pharmacy services in hospitals. The quality and quantity of publications created by ASHP during the Oddis years stimulated pharmacy's development as a clinical profession. As the sole accrediting body for residency and technician-training programs, ASHP during the Oddis administration contributed substantially to the skill levels of clinical pharmacy practitioners. Oddis was one of the profession's strongest advocates for making the doctor of pharmacy degree the entry-level standard. As much as anything else, it was Oddis's persistence—behind the scenes, in a gentlemanly and nonconfrontational manner—that crystallized these achievements for pharmacy.

BIBLIOGRAPHY

McConnell, W.E.; Harris, R.R. The American Society of Hospital Pharmacists: A history. Am. J. Hosp. Pharm. **1993**, *50* (suppl. 2), S1–S45.

Talley, C.R.; Reilly, M.J. Joseph A. Oddis: Influences and achievements. Am. J. Health-Syst. Pharm. **1997**, *54*, 1815–1825.

PROFESSIONAL DEVELOPMENT

Oncology Specialty Pharmacy Practice

Lisa E. Davis
Philadelphia College of Pharmacy, Philadelphia, Pennsylvania, U.S.A.

INTRODUCTION

Oncology pharmacy specialists have been recognized for their expertise and fundamental role in cancer care programs for more than 30 years. Oncology specialty pharmacy practice requires a diverse group of knowledge and skills and the ability to apply these principles to highly patient-specific clinical situations. Skilled practitioners must possess knowledge of malignant processes and how they are treated optimally with antineoplastic agents and ancillary medications. In addition, a solid foundation of general pharmacotherapeutics is essential, since oncology patients may require integrated care plans to address concomitant nonmalignant diseases.

In the clinical setting, the oncology specialty pharmacist is responsible for reviewing medication orders to ensure completeness and accuracy, and to minimize drug-related adverse effects; he/she also serves as a resource for drug and patient education information. Clinical practitioners actively participate in prospective therapeutic decision-making processes for treatment of individual patients with malignancy and serve as integral members of team efforts to evaluate and identify specific agents, clinical indications, and treatment regimens for their use within a healthcare system. Although the types of settings in which oncology pharmacy specialists practice may be quite different, their clinical practice roles share a common set of knowledge, skills, and responsibilities.

CLINICAL PHARMACY OPPORTUNITIES

Clinical pharmacy opportunities for oncology pharmacy specialty practitioners are available in every type of pharmacy practice setting including patient care, research, the pharmaceutical industry, medical communications, and governmental agencies (Table 1). Direct patient care settings include adult, pediatric, and specialty oncology practices; palliative care and hospicare; and home care patient care services. Since the majority of oncology patient care is conducted in outpatient clinics, tremendous practice opportunities exist in these typically busy care settings. In addition, the growth of stem cell transfusion protocols has led to wider use of these often-complex and toxic treatment programs, leading to increased need for specialty practitioners to work with oncology transplant services.

Several factors have driven the demand for oncology pharmacist involvement in patient care, including disease-management diagnosis-related group (DRG) reimbursement policies, high-cost oncology drugs, increasing complex technology, high risk for medication errors, the need for outcomes-based research, and public and health professional concerns regarding patient safety.[1] The type and extent of pharmacists' responsibilities are influenced by institutional needs and the individual pharmacist's professional strengths and training. The focus of one's professional responsibilities may also change over time. It is therefore important to recognize that, for most practitioners, no single clinical practice model represents the entire scope of their professional activities during their career.

Pharmacists who practice in oncology patient care settings are responsible for reviewing medication orders, monitoring patients' therapies, and providing drug information to patients and healthcare providers. In an early survey of pharmacists who were members of the American Society of Health-System Pharmacists (ASHP) Oncology Special Interest Group (SIG), the bulk of practice activities described include patient-oriented services (performing medication histories and medication discharge counseling), drug distributive activities (determining appropriate volumes and diluents for admixture solutions and dispensing medications), participating in drug studies as a clinical investigator, providing in-service education programs, assuming management responsibilities (determining pharmacy budget appropriations, performing inventory maintenance and management, and assuming managerial responsibilities for pharmacy personnel), and providing therapeutics consultations.[2] Regardless of whether the pharmacist is directly responsible for dispensing or overseeing the release of medications for patients, medication order review is a fundamental responsibility shared by all clinical oncology

Encyclopedia of Clinical Pharmacy
DOI: 10.1081/E-ECP 120006234

Table 1 Oncology specialty practice roles

- Oversee cancer drug preparation and dispensing in patient care settings
- Direct patient care
- Investigational drug services
- Research
- Drug information
- Educational instruction
- Outcomes assessment
- Pharmaceutical industry
- Medical communications

pharmacists. Preventing medication errors is a vital goal of this process.

Medication Order Review

Oncology pharmacists are responsible for ensuring that medication orders for antineoplastic agents that are released for patient care administration are complete, accurate, and appropriate for each patient. The increased use of cytotoxic agents for patients with nonmalignant diseases, such as rheumatologic and dermatologic disorders, has led to more complex therapies in patient care settings where healthcare professionals are less familiar with these agents and their potential toxicities. While nonspecialists assume this role in many settings, institutions that provide care to large numbers of patients with cancer most often involve pharmacists with specialty training in oncology to minimize medication errors and optimize pharmacotherapy.

Strategies for preventing medication errors in oncology include using of standardized chemotherapy order forms, conducting computer-assisted chemotherapy order screening, allowing only pharmacists to prepare cytotoxic agents, and double-checking drug orders.[1] Several organizations have developed processes for uniform medication order reviews; in addition, institution-specific processes may be developed through the pharmacy and therapeutics and other interdisciplinary committees. Although a number of computerized medication order-screening software programs are increasingly utilized, the flagging of excessive drug doses is not the only basis for order review. Also, if the patient has received prior courses of chemotherapy, a comparison between the new and past orders should be made; significant discrepancies are addressed with the prescriber.

Key recommendations for preventing chemotherapy medication errors are well recognized and include: education of healthcare providers, an established dose-verification process, established dose limits, use of standardized prescribing vocabulary, patient education, interdisciplinary team review of medication errors and communication issues, and medication error reporting with cooperation from drug manufacturers.[3] The use of standardized chemotherapy order forms improves completeness of physician ordering, prevents potential medication errors, and reduces the time spent by pharmacists clarifying orders.[4] The oncology pharmacist is instrumental in the development of such forms. Chemotherapy checklists can also be used in the drug order verification process. Some institutions include these on the product patient label, as well.

In an effort to identify opportunities to reduce chemotherapy medication errors, a multidisciplinary committee reviewed practices of more than 100 hospitals' processes to prevent chemotherapy errors.[5] A summary of the authors' recommendations is presented in Table 2. The nurse and pharmacist should both be independently responsible for checking each item.

Drug Handling, Preparation, and Dispensing

Pharmacists assume overall responsibility for the acquisition, preparation, handling, and distribution of antineoplastic agents within their healthcare system. Pharmacists are necessary to ensure optimal safety and minimal expo-

Table 2 Recommendations for chemotherapy order verification

- Check entire set of chemotherapy orders against an acceptable reference.
- Verify current body surface area, height, and weight.
- Verify the final dose of each drug.
- Check the rate of administration, amount, and type of admixture solution.
- Check the antiemetic regimen and prehydration and posthydration orders for omissions and additions of ancillary medications.
- Compare current chemotherapy orders with previous orders and discuss significant discrepancies with prescriber.
- Check that an x-ray has been performed and read to confirm central venous access for newly placed lines for vesicant chemotherapy infusions.
- Confirm parameters with standard references or the research protocol to determine that any dose modifications that have been made are appropriate; discuss significant discrepancies with prescriber.
- Determine that pertinent (hematologic, organ-specific tests based on the chemotherapeutic agents used) laboratory values are within normal parameters; discuss significant abnormalities with prescriber.

(Adapted from Ref. [5].)

sure risks for themselves, other pharmacy staff, individuals who administer cytotoxic drugs, and patients who receive those agents. Although clinically focused practice positions typically do not require pharmacists themselves to prepare antineoplastic agents, in many settings, those activities often comprise part of their responsibilities. Regardless of whether those pharmacists actually prepare the agents themselves, the overall responsibility for drug handling, preparation, and appropriate dispensing falls to the pharmacist. He or she is typically involved in developing training, recertification, and employee-monitoring policies. Determining technician competencies for drug preparation and handling and documenting technician training are also important components of this process.[6] Anyone who is involved with any aspect of this process needs to understand the risks and protective measures that are available to them. The Occupational Safety and Health Administration (OSHA) Technical Manual on Hazardous Drug Exposure[41] and the ASHP Technical Assistance Bulletin on Handling Cytotoxic and Hazardous Drugs[42] are easily accessible resources. The ASHP Technical Assistance Bulletin,[7] last reviewed in 1996, is currently under review and updated guidelines will likely be released in the near future. Criteria for facilities and personnel for administration of parenteral antineoplastic agents have been developed by the American Society of Clinical Oncology.[8]

Drug Administration Policies and Guidelines

Oncology pharmacy specialists provide guidance and assistance to address drug administration issues and provide input toward the development of guidelines for administration of certain medications within their healthcare system. The rationale for such guidelines may involve the potential to reduce medication errors, infusion-related reactions, or treatment-related toxicities or to optimize drug delivery of time-consuming treatment regimens in busy outpatient treatment settings. Oncology pharmacists, through their efforts in reviewing primary literature and participating as investigators in clinical studies, facilitated the recognition that certain agents that were initially approved by the FDA for administration over 24 hours (i.e., pamidronate, paclitaxel) could be safely administered over shorter time periods.

Patient Education

Patient education is standard practice and crucial for patient well-being and compliance. As such, patients are typically required to provide written informed consent prior to receiving antineoplastic agents and other anticancer therapies. Pharmacists frequently collaborate with other healthcare practitioners to develop patient education materials to address general complications of their cancer treatment as well as drug-specific information sheets. Cancer.gov, a National Cancer Institute Web site for current and accurate cancer information, makes some of these materials accessible online.[43] Chemotherapy and You: A Guide to Self-Help During Treatment, a patient-focused booklet that addresses common treatment-related side effects and coping strategies, is also available online.[44] Many commercially available print and online drug information resources provide patient cancer medication education sheets; however, practitioners often develop their own patient education information materials for use in their specific patient care settings.

Patient Monitoring

Patient monitoring functions comprise a key component of clinical pharmacists' activities in oncology practice. Examples of common patient care problems addressed by clinical oncology pharmacists are listed in Table 3. Based on their understanding of the nature and time course of specific drug-related toxicities and an individual patient's characteristics, pharmacists may anticipate certain treatment toxicities and provide input in the design of the regimen to minimize adverse outcomes. Drug dose reductions and use of hematopoietic growth factors, chemoprotective agents, certain antiemetic agents, or other ancillary medications may be recommended. For patients who develop significant toxic effects, pharmacists assess their current and previous drug therapy history to assist in managing these problems.

Table 3 Common patient care problems in oncology pharmacy practice

- Medication education and compliance
- Dose verification
- Adverse drug effects
- Nausea
- Mucositis
- Cytopenias
- Infection
- Fatigue
- Pain
- Nutrition
- Therapeutic drug monitoring
- Practice guideline development and adherence
- Medication error prevention and tracking
- Quality improvement
- Pharmacoeconomic assessment

Therapeutic drug monitoring is another important aspect of pharmacists' patient-monitoring responsibilities. Target blood concentrations of cyclosporin have been established for minimizing graft versus host disease and renal dysfunction whereas optimal busulfan blood concentrations are associated with minimizing graft versus host disease and venous occlusive disease. Serum methotrexate level determinations are important for establishing appropriate leucovorin rescue treatment regimens. The pharmacist ensures that the levels are obtained at the appropriate time intervals and are processed in a timely manner. When patients are discharged prior to having completed methotrexate serum level monitoring, pharmacists may continue to estimate methotrexate levels to determine an appropriate dosing schedule and duration of leucovorin therapy. Research investigations of therapeutic drug monitoring offer further opportunities to improve patient care. At St. Jude Children's Research Hospital, investigations by oncology pharmacy specialists demonstrate that, by individualizing doses of certain chemotherapeutic agents based on drug elimination, survival rates can be increased for children with acute lymphocytic leukemia (ALL).[9]

Therapeutic drug monitoring and pharmacokinetic dosing adjustments are also important for nononcologic agents as well. Aminoglycosides, vancomycin, and anticonvulsant agents are frequently administered to oncology patients; the patients may be followed by an institutional pharmacokinetics service or by the oncology pharmacy specialist. In selected situations, requests for drug levels may be initiated for agents that are not routinely monitored in general clinical practice.

Drug Information

Clinical pharmacists in all practice settings routinely provide drug information to other healthcare providers; drug queries are particularly common in oncology practice due to the specialized nature of these agents and complex treatment regimens. The use of oncologic agents for non-FDA-approved indications is not uncommon and many patients receive drugs for an off-label indication. Pharmacists are responsible for ensuring that the literature supports such uses and that therapy is delivered safely in conjunction with other agents that the patient may be receiving.

When new agents become available for use within a healthcare system, the oncology pharmacy specialist is usually the individual responsible for providing drug information to the general pharmacy staff, nursing staff, house staff, and prescribers. The provision of this information can be disseminated through memorandums, newsletters, e-mail notices, chemotherapy administration guideline updates, and/or interactive in-services. Pharmacists who spend a significant portion of their day in patient care areas may find that their presence stimulates even more requests for drug information from other healthcare providers.

At the National Institutes of Health Clinical Center, oncology clinical pharmacists participate as members on the NCI Physician's Data Query online cancer information service supportive care and drug information panels and serve on the Board of Pharmaceutical Specialties Council on Oncology Pharmacy Practice.[10] These pharmacists, in conjunction with the NIH Division of Safety, developed the first recommendations for safe handling and disposal of antineoplastic drug products and waste.

Education

Providing education to healthcare providers and trainees, in addition to patients, is an important responsibility of clinical practitioners. The interdisciplinary nature of oncology patient care practice settings makes them desirable experiential training rotations for medical, nursing, pharmacy, physician's assistant, social work, physical therapy, occupational therapy, and psychology students. Oncology clinical pharmacists likely interact with most of these disciplines in their practice setting and serve as primary preceptors for pharmacy students, residents, and/or fellows during their training program. Salary support for many practitioners is, in part, based on their responsibilities for pharmacy student education.

Cancer centers typically require nursing staff who administer cytotoxic agents to participate in chemotherapy certification programs. The oncology pharmacy specialist, in conjunction with oncology nurse specialists, generally participates in the administration of these programs. Pharmacists also participate in ongoing education programs for house staff and medical hematology-oncology fellows.

MODEL CLINICAL PRACTICES

It is not unusual that an individual patient, during the course of treatment for his or her malignancy, will receive care in both an outpatient and inpatient environment. Although most chemotherapy is administered to patients in ambulatory treatment clinics, some patients will be hospitalized to receive certain complex treatment regimens or for complications due to the treatments or underlying disease process. Those patients who are undergoing induction treatment regimens for an acute leukemia, re-

ceiving high-dose chemotherapy in preparation for stem cell transplantation, or receiving salvage treatment regimens for refractory tumors are most likely to be hospitalized sometime during the course of their disease. Due to the nature of patient problems and heavy practice workloads, busy cancer treatment programs require multiple clinical practitioners to cover ambulatory and inpatient care settings. Regardless of whether the patient care setting is inpatient or outpatient, the basic functions of the medication review and patient monitoring processes, as well as provision of drug information and patient education by oncology pharmacy specialists, are similar.

Inpatient Care Practice Roles

Typically, pharmacists attend and participate in patient care rounds and discussions with the medical staff. Even a small portion of time devoted to this activity can yield tremendous information that can provide the pharmacist with an opportunity to discuss intended therapeutic interventions or suggest modification or discontinuation of current therapies. This can improve the efficiency of the therapeutic decision-making process through facilitating immediate interventions and decrease unnecessary drug wastage by previewing medication orders before they are delivered to and prepared by the pharmacy. Examples may include initiating appropriate patient-specific antibiotic dosage regimens, ensuring appropriate first-line antiemetic prophylaxis, or recommending changes in analgesic therapy. In addition, the presence of the pharmacist in this setting generally stimulates questions by the medical staff, thereby providing opportunity to improve drug knowledge. Moreover, pharmacists may be alerted to certain patient-specific issues that may not be obvious through a review only of hospital orders or the patient's medical chart.

At the University of California, San Francisco, approximately seven clinical pharmacists provide oncology services to adult inpatients.[1] These practitioners perform admission medical histories, attend rounds with medical house staff, attend bone marrow transplant rounds, make therapeutic interventions, provide drug information to staff and patients, perform medication order reviews, perform ongoing drug utilization reviews and evaluation, follow practitioner adherence to institutional drug therapy guidelines, and coordinate medication discharge planning and counseling.

At Cedars-Sinai Medical Center, oncology pharmacists provide services to hospitalized patients on a specialized inpatient unit and in an ambulatory infusion center.[11] In these settings, targeted areas to improve patient outcomes, prevent adverse drug reactions, and reduce costs include the chemotherapy process (ordering, dispensing, and administration), treating patients with febrile neutropenia, and managing peripheral blood stem cell patients. A clinical pathway for treating febrile neutropenia, which includes the use of hematopoietic growth factors, was implemented and reevaluated. In patients undergoing peripheral blood stem cell transplantation, pharmacists screen medication orders and participate in patient care rounds to ensure that inappropriate medications are avoided, antiemetic therapy is safe and cost-effective, and electrolyte supplementation therapy is appropriate and well-tolerated.[11] Professional activities that are common in the inpatient treatment setting include pharmacotherapeutics monitoring, pharmacokinetic dosing, managing clinical protocols and critical pathways, maintaining patient profiles, conducting patient monitoring, and providing adverse drug reaction (ADR) reporting and documentation.

Depending on the institution, the nurses also attend patient care rounds with the medical staff or discuss patients' status with other healthcare professionals. Through these discussions, additional patient care issues are frequently identified. Discharge planning rounds typically involve social workers, patient representatives, and nurse case managers. The pharmacists may assist in this process by helping to initiate therapy that is appropriate for the discharge setting and that will be covered adequately financially. The knowledge that imminent discharge is pending may help the pharmacist to suggest converting parenteral drug therapy (e.g., antibiotics, analgesics, antiemetics) sooner during the hospitalization to get the patient stabilized on the new therapy prior to discharge. Pharmacists also assist in arranging for qualifying patients to enroll in indigent drug assistance programs. The Pharmaceutical Research and Manufacturers of America provides a directory of patient medication assistance programs.[45]

Inpatient cancer patient care settings differ among institutions. In some settings, patients are managed directly by internal medicine house staff; the oncologist may be an attending physician for that service or a hematology-oncology specialty service may interact with the medical house staff in a consultative arrangement. Patients may be admitted for direct care by a hematology-oncology specialty service or an oncology nurse practitioner service. Although patients' needs are the same, regardless of their primary care team, the pharmacists' interactions may vary, depending on those individuals. Furthermore, the pharmacists' position may be entirely based within the pharmacy department or supported by the hematology-oncology department. If the institution has a pharmacokinetics service, pharmacokinetics issues

Table 4 Specialty oncology patient care services

- Stem cell transplantation
- Pediatric oncology
- Gynecologic oncology
- Solid tumor services
- Surgical oncology
- Radiation oncology
- Palliative care
- Hematology
- Hospice
- Home infusion
- Outpatient clinics

may be addressed by that service or by the oncology pharmacist, or may be shared. The oncology pharmacist also interacts with other clinical pharmacists that are involved in a patient's care (e.g., infectious diseases, nutritional support, pain, etc.).

Institutions that care for large numbers of cancer patients also have subspecialty oncology services to which specific pharmacists may be assigned. Examples of these services are listed in Table 4. At M.D. Anderson Cancer Center, for example, clinical pharmacists provide direct patient care in concert with the leukemia, lymphoma, bone marrow transplantation, medical breast cancer, gastrointestinal, critical care, and pediatrics services. The clinical pharmacists participate in drug regimen design and monitoring; provide pharmacokinetic dosing recommendations, drug information, and comprehensive patient education; and precept experiential training rotations for residency candidates and pharmacy students.

Transitional Patient Care Practice Roles

In the transitional patient care setting, patients may be receiving follow-up from pharmacists for chemotherapy that has extended from the inpatient to the outpatient environment. Pharmacists may also expedite chemotherapy orders for patients who are being admitted.[1] An example of such a treatment setting where continuous pharmaceutical care is delivered is the autologous bone marrow transplant (ABMT) outpatient program at Duke University Medical Center.[12] This program required months of planning with input from nurses, physicians, and pharmacists to deliver pharmaceutical services to meet individual patients' needs. Patients undergo 5 days of hospitalization to receive high-dose chemotherapy, continuous hydration, and antiemetic therapy after which time they are discharged with hematopoietic growth factors, prophylactic oral antibiotics, antiemetics, and electrolyte supplements. Approximately half of patients require readmission to the inpatient unit during this period of care. Antibiotic use guidelines for prevention and management of infection were developed and a clinical pharmacist is responsible for following pharmacokinetic parameters, toxicities, and microbiologic results for all patients. A system was developed in which drugs could be supplied to patients in the clinics and drug usage could be tracked for patient profile records and reimbursement purposes. A bone marrow transplant clinical pharmacist is available on-call to answer questions and to meet with patients at the clinic site if necessary. Delivery of pharmacy services through the entire treatment program is achieved through coordination efforts between the bone marrow transplant clinical pharmacists, the outpatient pharmacy, and the inpatient pharmacy satellite and facilitates a cost-effective and safe approach to administering high-dose chemotherapy to patients with solid tumors.

In an evaluation of system distress in this treatment setting, the most frequent symptoms involved loss of appetite, fatigue, insomnia, and intermittent nausea.[13] Medication compliance was determined to be greater than 90% and is believed attributable in part to patient and caregiver education by pharmacy and nursing personnel.

Outpatient Care Practice Roles

Clinical practice opportunities for oncology pharmacy specialists in outpatient treatment settings are significant as patients continue to receive increasingly complex anticancer therapies outside of the inpatient unit. Pharmacists conduct medication and patient profile reviews, perform medication histories, provide patient counseling and drug information, and participate in therapeutic decision-making. Within Veterans Affairs Medical Centers and some other practice settings, pharmacists see patients as primary care providers and initiate prescriptions for analgesics, growth factors, antiemetics, and other ancillary medications by protocol. The scope of clinical pharmacy practitioners' activities in oncology clinics is not well documented in the literature, however.

Raehl et al. reviewed results of questionnaires that were sent to directors of pharmacy in one-half of acute-care general medical-surgical hospitals in the United States that had at least 50 licensed beds to determine the extent that pharmacists provide ambulatory clinical pharmacy services.[14] Overall, pharmacists performed non-dispensing activities in ambulatory clinics in 19% of responding hospitals. Pharmacist involvement in oncology clinics occurred in 9% of institutions, second only to diabetes clinics (10%), and was more common than in cardiology (6%) or geriatrics, infectious disease, or pain clinics (4%).

In a prospective study to evaluate the impact of a clinical pharmacist in outpatient hematology-oncology clinics, patient charts and profiles were reviewed and patient interviews were conducted to obtain medication histories.[15] Within a 36-day time period, 211 interventions were documented; the majority of these interventions were not related to chemotherapy. The most frequent activities consisted of patient counseling and therapeutic recommendations. Of the problems identified, most were of high and moderate significance and pharmacy interventions yielded positive clinical outcomes. The physician acceptance rate to pharmacists' interventions was 94.5%.

In a unique practice role at the University of California, San Francisco (UCSF), a clinical pharmacist has been responsible for setting up chemotherapy and heparin administration via external or implanted infusion pumps.[1] The pharmacist also monitors the patient for treatment toxicities and efficacy, and provides pump maintenance. Thus, this practice setting enables the pharmacist to manage patients' outpatient chemotherapy first-hand, identify potential treatment-related problems, and initiate any necessary changes to optimize the treatment regimen. In addition, UCSF has recently implemented a distribution and clinical pharmacy service 40 hours a week using 5 pharmacists (4 with specialized training in oncology). These pharmacists work out of the Cancer Infusion Center and service 60 to 70 patients weekly.

Oncology pharmacists who care for patients in outpatient practice settings typically review patients' medication and medical profiles in advance of seeing the patients. This enables them to identify those individuals who are most likely to have medication-related issues or poor symptom management, and who will benefit from the pharmacist's interventions.

Investigational Drug Use and Management

Investigational drug use is an important aspect of pharmacotherapy for patients with malignancies. Oncology pharmacy specialists have an important role throughout the investigational drug use process; their responsibilities may involve directly maintaining drug-dispensing records and investigational protocols, participating in protocol development, having membership on institutional review committees, developing and disseminating information about the protocols and drugs to the pharmacy and nursing staff, and coordinating protocol management between study investigators, institutional staff, and study sponsors. Protocols for the use of investigational agents are developed and made readily available to all involved personnel. Included in these protocols are drug-specific characteristics and information that would not otherwise be readily available from typical drug information resources. At many institutions, these activities comprise the sole responsibilities of one or more pharmacists dedicated to research protocols and investigational drugs. Cancer centers also often have a clinical trials review committee that is separate from the Human Subjects Review Board. Pharmacist participation in any initial trial review process is crucial to ensure that the project is feasible, coordination and dispensing practices are worked out, and any clinical trial involving chemotherapeutic agents does not have a significant negative financial impact on the institution.

Situations may arise when a patient is in need of a drug that is not available commercially or in an ongoing clinical trial at the institution. The oncology pharmacy specialist may arrange to obtain the drug which may only be available from the pharmaceutical company or the National Cancer Institute for nonresearch (compassionate) use.[16] The NCI Cancer Therapy Evaluation Program (CTEP) Treatment Referral Center provides guidelines and contact information for this process.[46]

Drug Information

While provision of drug information comprises a portion of an oncology pharmacy specialist's daily activities, some practitioners devote their entire career to the practice setting in drug information. Large comprehensive cancer centers such as Memorial Sloan-Kettering and M.D. Anderson are examples of institutions where care is devoted to patients with cancer and workload justifies having an oncology clinical pharmacy specialist in drug information. Pharmacists resolve general oncology drug-related queries and provide key information to solve a specific patient problem. However, their responsibilities extend beyond these activities.

Pharmacists serve as members of the healthcare system's Pharmacy and Therapeutics Committee which directs the review process for considering medication additions to and deletions from the drug formulary. The committee may review and manage data from the ADR reporting program and from medication usage evaluations and reviews, and may monitor policies for medication use guidelines.

Other institutional programs that typically involve pharmacists include pharmacoeconomics, outcomes research and management, quality improvement, and reimbursement assistance programs.

Clinical Research

Patient participation in clinical research trials is a significant component of cancer care therapy, particularly at comprehensive cancer centers and their affiliated medical practices and institutions. Whereas all oncology pharmacy specialists facilitate these activities, some practitioners focus their entire clinical practice in clinical research. Examples of their responsibilities include developing and reviewing research and funding proposals, screening and recruiting patient participants, overseeing medication disbursement, evaluating patient response and treatment outcomes, managing adverse effects, providing patient education, overseeing study records and report forms, reviewing and interpreting study data, and summarizing and reporting study results.

Clinical pharmacists participate as principal and coinvestigators on studies that are purely clinical in nature, basic and translational research, pharmacoeconomic studies, and outcomes-based research. Oncology pharmacy specialists have had a significant impact on oncology therapeutics through their participation in trials in drug development (pharmacokinetics, pharmacodynamics, and pharmacogenetics), supportive care, pharmacoeconomics, and outcomes research. Several oncology pharmacy specialists receive NIH support for their research activities.

DOCUMENTED BENEFITS OF ONCOLOGY SPECIALTY PHARMACY PRACTICE

The benefits of oncology specialty pharmacist practitioners are evidenced by the high demand for skilled clinical practitioners in oncology patient care settings. Furthermore, the Board of Pharmaceutical Specialties (BPS) recognized oncology as a pharmaceutical specialty in 1996. Information regarding the specialty certification process can be accessed at the BPS Web site.[47] Pharmacists are examined on their ability to collect and interpret clinical data to recommend, design, implement, monitor, and modify pharmacotherapeutics plans and optimize drug use for patients with cancer; basic knowledge of malignant disease processes and their management; ability to provide education and medication-related counseling, and ability to address public health issues related to oncology pharmacy practice; and knowledge of the drug development process. The petition requesting specialty recognition of oncology pharmacy practice to the BPS in 1992 delineated the societal need and demand for highly skilled oncology pharmacy practitioners.[17]

Ignoffo and King have identified potentially important patient outcomes related to anticancer therapy[18] (Table 5). By documenting the incidence of specific outcomes associated with patient care services, pharmacists can establish a threshold at which corrective measures can be instituted. Quality-of-life issues related to cancer patient care are a particularly active area of investigation. Both disease and symptom-specific assessment tools are available and can be used to document quality-of-life outcomes for individual patients and within a treatment population.

Table 5 Outcomes related to anticancer therapy

Desirable outcomes	Undesirable outcomes
Cure	Mortality
Improved survival	Progression of disease
Decreased mortality rate	Worsening of symptoms
Slowing of disease progression	Side effects
Decreased symptoms	Severe organ toxicity
Prevention of disease symptoms	Increased cost of ancillary therapies
Decreased cost of treatment	
Low incidence of adverse effects	Drug resistance

(From Ref. [18].)

Although the impact of this has not been formally ascertained, a method for designing and implementing toxicity outcome indicators for specific chemotherapy protocols has been described.[19] For each clinical protocol, clinical monitoring criteria were established; pharmacists are able to use predetermined agent-specific monitoring and toxicity ratings to document and improve pharmaceutical care services for oncology patients.

Practice Guideline Development and Implementation

Oncology pharmacists develop drug and use practice guidelines; the impact of these guidelines has been reported extensively at professional meetings and to some degree in the professional literature. Hirsh et al. report that pharmacist involvement in the development and implementation of guidelines for managing extravasations from antineoplastic agents resulted in prompt and uniform management of these drug effects.[20] However, the magnitude of this impact was not quantified.

Pharmacists who have been responsible for the design and implementation of antiemetic dosing guidelines have resulted in significant cost savings at many institutions.[21,22] In addition, patients who are treated according to established practice guidelines have satisfactory or improved antiemetic control.[22,23] Moreover, institutional drug use guidelines serve as educational tools and help to improve the drug-ordering and -administration

process. Significant cost savings and improved health outcomes have resulted from practice guidelines for hematopoietic growth factors and intravenous-oral drug interchange programs.

Publications by NIH Clinical Center clinical oncology pharmacists have received national attention and improve pharmacy practice processes; examples include ways to standardize expression and nomenclature of cancer treatment regimens[24] and pharmacy procedures for dealing with gene therapy products.[25]

Through a pharmacist's development of a clinical pathway for a methotrexate infusion protocol in pediatric patients, prechemotherapy hydration parameters were reached sooner and length of stay decreased from an average of 3.27 days to 2.4 days.[26] This was associated with a savings of approximately $600 per patient. At the same institution, a change in empiric antibiotic therapy in penicillin-allergic patients and other changes resulted in a decrease in length of stay by approximately one day (average 5.48 days versus 4.66 days) that was associated with a savings of approximately $1500 in direct costs per patient. Criteria for developing, implementing, and evaluating critical pathways in oncology practice have been published.[27–29]

Oncology Pharmacy Specialist Interventions

McClennan et al. conducted a prospective 2-month intervention study to determine clinical outcomes associated with pharmacist interventions at a tertiary referral cancer institute in Australia.[30] Clinical pharmacists documented clinical interventions and relevant patient information which were assessed by an independent pharmacist. Members of the healthcare team reviewed and discussed medical progress notes to determine outcomes. Of 674 total interventions that were documented during the study, outcomes were assessed for 10% of the interventions reported; 90% of the interventions led to documented clinical benefit. Pharmacist interventions most frequently consisted of initiating changes in drug therapy associated with antiemetic, antimicrobial, and analgesic agents.

Pfeiffer documented the impact of an oncology pharmacist's review of chemotherapy administrations.[31] The pharmacist reviewed chemotherapy admissions on a monthly basis for 20 months. Pertinent patient demographic data, chemotherapy administrations information, and admissions stay data were collected and were compared to an ideal that was predetermined based on the institution's admissions standards and chemotherapy care maps. The pharmacist identified opportunities for education for cases where length of stay exceeded the institution's standards or where problems were identified following a discussion by the Oncology Care Committee. The pharmacist provided education in the form of in-services or written material to the appropriate staff, resulting in a reduction in the average length of stay for chemotherapy administration, the number of variances in timing of chemotherapy, and the number of problem cases.

An analysis of self-reported interventions by hematology-oncology pharmacists and staff was also performed to document pharmaceutical care interventions over a period of approximately 8 months at the Walter Reed Army Medical Center.[32] The interventions were analyzed to determine the types of interventions that are most frequently performed, prescribing errors encountered, medication cost avoidance that resulted from the interventions and types of interventions that are associated with medication cost interventions, and intervention acceptance rate by physicians. Interventions were entered into a personal computer and analyzed using CliniTrend Web Support System software (ASHP). Medication cost avoidance was determined if less medication was used, an equally effective but less costly medication was used, or a medication that could not be reused was not prepared.

A total of 503 interventions were reported; clinical consultations, correction of prescribing errors, and patient treatment procedures accounted for approximately two-thirds of the interventions (Table 6). In this study, all clinical interventions were related to chemotherapy-related toxicities or drug-dosing and -administration issues. The 167 documented clinical consultations were related to nausea or vomiting (29.3%), chemotherapy dosing or

Table 6 Summary of interventions reported by hematology-oncology pharmacists at Walter Reed Army Medical Center (October 1, 1995–May 31, 1996)

Intervention	Number	%
Clinical consultation	167	33.2
Correction of prescribing error	85	16.9
Patient treatment procedure	65	12.9
Clarifying/verifying prescription with physician	38	7.5
Other	33	6.6
Enrolling/monitoring patient in clinical trial	28	5.6
Formulary or therapeutic interchange	26	5.1
Compatibility consultation	23	4.6
Verifying patient's chemotherapy schedule	20	4.0
Drug information counseling for patient	10	2.0
Alerting physician of need for a test dose	5	1.0
Patient counseling (other than drug information)	3	0.6

(From Ref. [32].)

administration (22.7%), general clinical consultations (18.0%), hematologic toxicity (9.0%), pain control (7.8%), other toxicities (7.2%), and mucositis (3.0%), or were allergy-related (3.0%). The impact of these consultations on health outcomes was not determined but likely contributed to improve patient care. Physicians accepted 488 (97%) of the pharmacy-initiated interventions. The total medication cost avoidance was $23,091; $19,696 (85.3%) was associated with correction of prescribing errors.

At the Baltimore Cancer Research Center, an NCI cancer treatment unit located in a U.S. Public Service hospital, pharmacists demonstrated that their counseling interventions improved patient adherence to medications. After pharmacist counseling, the patient compliance rate to oral antibiotic medication regimens for gut sterilization increased to 90% compared to 65% before counseling. A subsequent discontinuation and reinstitution of pharmacist counseling resulted in changes in compliance rates back to 65% and 90% again, respectively.[10]

Reimbursement Assistance and Investigational Drug Services

Through pharmacist involvement with other healthcare providers, identifying opportunities for medication reimbursement from drug manufacturers with indigent drug access programs can result in significant cost savings. In a program for procuring medications at no cost for indigent patients, a full-time pharmacist was responsible for identifying qualifying patients, assisting in the application process, and coordinating receipt and distribution of the medications.[33] In 1992, during the first year of the program, 200 patients were served and the potential cost avoidance resulting from obtaining these free medications was estimated at $150,000. The medications obtained included immune globulin, antiemetics, growth factors, cancer chemotherapy, antimicrobials, and interferons.

Pharmacy-managed investigational drug services represent an additional opportunity to reduce costs, particularly in oncology patient care settings where there is high-volume use of investigational drugs. In a study of cost avoidance associated with investigational drug services at two institutions in Washington State, drug costs associated with acquired immunodeficiency syndrome and oncology accounted for the largest figures.[34]

Patient Care Services

Oncology pharmacist participation in home delivery of chemotherapy for selected patients with good home care support can result in cost savings with antiemetic control.[35] Pharmacist involvement with infusion pump use and patient education services for drug reconstitution and self-injection have resulted in reduced medical costs that were recognized more than 20 years ago.[36,37] At many centers where patients are treated using ambulatory drug infusions, pharmacists decide how the drug will be prepared for delivery based on patient factors, drug stability data, and what type of pump is available. The pharmacist may also determine what type of pump will be used, how often the patient will return to the infusion center for follow-up during the treatment, and whether the patient will change the medication bags at home. During the treatment period, the patient may return to the infusion center to see the pharmacist to have the pump refilled or checked, receive additional patient education, or undergo monitoring for drug toxicities; follow-up with the oncologist may not occur until the infusion has been completed.

RESOURCES FOR ONCOLOGY PHARMACY PRACTICTIONERS

Oncology Practice Information and Guidelines

Oncology pharmacy specialists frequently participate in developing and monitoring adherence to health system guidelines for patient care. Several professional organizations have developed guidelines which serve as valuable resources for practitioners (Table 7). Information and treatment guidelines for cancer pain management have been developed by the American Pain Society and the Agency for Healthcare Policy and Research (AHCPR). Furthermore, the Joint Commission on Accreditation of Healthcare Organizations (JCAHO)[48] has developed standards for assessing and managing pain in accredited healthcare organizations. Standards are available for ambulatory care, behavioral healthcare, home care, healthcare network, hospital, long-term care, and long-term care pharmacy practice settings. Internet Web sites for pharmaceutical manufacturers of analgesics such as Roxane Laboratories[49] and Purdue Pharma L.P.[50] are also sources for professional and patient-oriented information and resources. Talaria is a cancer pain management resource for healthcare professionals that provides multimedia instructive and interactive tools regarding pain management, in addition to general technical information.[51] The National Comprehensive Center Network (NCCN), American Society of Clinical Oncology (ASCO), and the American Society of Health-System Pharmacists (ASHP) are among several groups that have produced practice guidelines for cancer or supportive care treatments.

Table 7 Resources for oncology practice guidelines

Organization	Web site URL	Applicable contents
Agency for Healthcare Research and Quality (AHRQ)	http://www.ahcpr.gov/clinic/cpgonline.htm	Online clinical practice guideline for management of cancer pain. References include a clinical guide, quick-reference guide, and consumer version (English and Spanish language available). Note: This was published by the AHCPR in 1994 (Publication No. 94-0592: March 1994) and is recognized as being in need of revision.
American Cancer Society	http://www.cancer.org	Guidelines for breast and prostate cancer treatment (developed in conjunction with the National Comprehensive Cancer Network (NCCN); recommendations for the early detection of breast, prostate, colon and rectum, and uterine cancers.
American Pain Society	http://www.ampainsoc.org	Consensus and quality improvement guidelines for treatment of acute and chronic pain.
American Society of Clinical Oncology (ASCO)	http://www.asco.org	Clinical practice guidelines for chemotherapy and radiotherapy protectants, hematopoietic colony-stimulating factors, antiemetics, bisphosphonates in breast cancer, and treatment of selected neoplastic diseases.
American Society of Health-System Pharmacists	http://www.ashp.org	Guidelines for managing postoperative and chemotherapy- and radiation-therapy-induced nausea and vomiting.
BC Cancer Agency	http://www.bccancer.bc.ca/default.htm	Online Cancer Management Manual with disease management guidelines.
Cancer Care Ontario Program in Evidence-Based Care and the Cancer Care Ontario Practice Guidelines Initiative	http://hiru.mcmaster.ca/ccopgi/index.html	Clinical practice treatment guidelines for numerous malignancies and use guidelines for supportive care agents.
National Comprehensive Cancer Network (NCCN)	http://www.nccn.org	Approximately 17 cancer centers across the United States are members of this organization. The NCCN develops cancer screening, prevention, detection, treatment, and surveillance guidelines for numerous malignancies and supportive care therapies.
National Cancer Institute	http://ctep.cancer.gov/index.html	Gateway to National Cancer Institute information and resources for: cancer information, clinical trials, statistics, research programs, and research funding.
PDQ®–NCI's Comprehensive Cancer Database	http://www.cancer.gov/cancer_information/pdq/	Cancer information summaries, including screening and detection, prevention, treatment, genetics, and supportive care information.

Table 8 Resources for drug handling, drug development, and investigational drug management

American Society of Health-System Pharmacists	http://www.ashp.org/	ASHP technical assistance bulletin on handling cytotoxic and hazardous drugs.
Cancer Therapy Evaluation Program (CTEP)	http://www.cancer.gov/clinical_trials/	Information for investigators, healthcare professionals, and patients related to cancer drug development and clinical trials; guideline and policy information for clinical trials and investigation drug use and accountability.
Cancer Trials Support Unit (CTSU) of the National Cancer Institute	http://cancertrials.nci.nih.gov	Investigator's Handbook for participants in clinical studies of investigational agents; directory of research tools and services for cancer researchers; clinical trials information; links to cooperative study group Web sites.

Table 9 Selected oncology-specific reference texts

- *Cancer Principles and Practice of Oncology*, 6th Ed.; DeVita, V.T.; Hellman, S.; Rosenberg, S.A., Eds.; Lippincott Williams and Wilkins: Philadelphia, 2000.
- *Cancer Chemotherapy and Biotherapy: Principles and Practice*, 3rd Ed.; Chabner, B.A.; Longo, D.L., Eds.; Lippincott Williams and Wilkins: Philadelphia, 2001.
- *Principles and Practice of Supportive Oncology*; Berger, A.M.; Portenoy, R.K.; Weissman, D.E., Eds.; Lippincott-Raven: Philadelphia, 1998.
- *The Chemotherapy Sourcebook*, 3rd Ed.; Perry, M.C., Ed.; Lippincott Williams and Wilkins: Philadelphia, 2001.
- *Cancer Chemotherapy Handbook*, 2nd Ed.; Dorr, R.T.; Von Hoff, D.D., Eds.; Appleton and Lange: East Norwalk, 1994.
- *A Clinician's Guide to Chemotherapy Pharmacokinetics and Pharmacodynamics*; Grochow, L.B.; Ames, M.M., Eds.; Williams and Wilkins: Baltimore, 1998.
- *Cancer Chemotherapy Pocket Guide*; Ignoffo, R.J; Viele, C.S.; Damon, L.E.; Venook, A., Eds.; Lippincott Williams and Wilkins: Philadelphia, 1997.
- *William's Hematology*, 6th Ed.; Beutler, E.; Lichtman, M.A.; Coller, B.S.; Kipps, T.J.; Seligsohn, U., Eds.; McGraw-Hill: New York, 2000.
- *Hematology: Basic Principles and Practice*, 3rd Ed.; Hoffman, R.; Benz, E.J.; Shattil, S.J.; Furie, B.; Cohen, H.J.; Silberstein, L.E.; McGlave, P., Eds.; Churchill Livingstone: New York, 1999.
- *Concepts in Oncology Therapeutics*, 2nd Ed.; Finley, R.S.; Balmer, C., Eds.; American Society of Health-System Pharmacists: Bethesda, 1998.
- *Clinical Oncology*. Abeloff, M.D.; Armitage, J.O.; Lichter, A.S., Eds.; Churchill Livingstone: New York, 2000.
- *Principles and Practice of the Biologic Therapy of Cancer*, 3rd Ed.; Rosenberg, S.A., Ed.; Lippincott Williams and Wilkins: Philadelphia, 2000.
- *Principles and Practice of Gynecologic Oncology*, 3rd Ed.; Hoskins, W.J., Ed.; Lippincott Williams and Wilkins: Philadelphia, 2000.

Oncology Practice Tools

The FDA Oncology Tools Web site[52] is a general access point for oncology treatment and drug information for healthcare professionals. The site contains links to information regarding approved oncology drugs with approved use indications; oncology drugs advisory committee meeting calendar; meeting transcripts; disease summaries; FDA policies and publications regarding cancer and clinical research; oncology reference tools (toxicity grading criteria, performance status scales, disease staging information, dose calculator, human fluid and calorie calculator); the cancer liaison program; cancer information for patients; and contact information for FDA review divisions that are responsible for oncology products.

Investigational Drug Development and Drug Use Procedures

The home page for the Cancer Therapy Evaluation Program (CTEP) division of the National Cancer Institute is a key reference for information on investigational drug use and development. (Table 8). CTEP provides information on policies and guidelines for investigation agents, and includes information, forms, and handbooks for policies regarding investigational drug trials of NCI-sponsored and cooperative group programs.[53]

Publications

Any practitioner who is involved with anticancer drug therapy should have access to standard reference texts as a source of detailed oncology drug information. Examples of standard anticancer drug pharmacology and hematology-oncology disease management textbooks are provided in Table 9. In addition, the oncology pharmacist should periodically scan or review several important medical

O

Table 10 Selected medical journals of interest to oncology pharmacists

Blood
Cancer Practice (Chemotherapy update column, edited by R. Ignoffo and R. King)
Current Opinion in Oncology
European Journal of Cancer
Gynecologic Oncology
Hospital Pharmacy (Cancer chemotherapy update, by Solimando and Waddell)
Journal of Clinical Oncology
Journal of Oncology Pharmacy Practice
New England Journal of Medicine
Oncology
Seminars in Oncology
The Lancet

journals. A library that contains access to these journals is recommended (Table 10). Since many publishers provide e-mail notification of the current table of contents for selected journals at no cost, a subscription to these services can help the practitioner keep up with new developments.

Online publications

Highlights in Oncology Practice, CA-A Cancer Journal for Clinicians, and Cancer Control Journal are available online free of charge. These high-quality publications provide timely reviews of oncology-related topics that many oncology and general practice practitioners find useful (Table 11). Practitioners will also find that online publications from several academic and government centers can be valuable resources for timely and accurate information regarding alternative and complementary medicines (Table 11).

Treatment updates

The American Society of Hematology (ASH),[54] an organization of clinicians and scientists who are involved in care and research related to hematologic disorders, provides online access to meeting abstracts and its meeting educational booklet, among other educational materials. Practitioners will find that the focus of these materials is primarily in hematologic malignancies and stem cell transplantation, immunology, and hemostasis. The American Society of Clinical Oncology (ASCO),[55] an organization of oncology professionals, also posts searchable meeting abstracts. The ASCO Online Journal Club provides online reviews of key oncology-related articles in the medical literature and can be accessed at the organization Web site. The American Association of Cancer Research[56] allows searchable access to proceedings abstracts and other association-related symposia. Medscape Hematology-Oncology[57] is an outstanding online resource for hematology-oncology practice guidelines, news, conference summaries, and treatment updates. Many full-text reviews and clinical updates are available as well. Registration is required but access to all materials is free. These online sites are a source of new research and drug research findings in hematology-oncology.

PROFESSIONAL NETWORKING OPPORTUNITIES

Education and Training Programs

Specialized oncology pharmacy practice residencies are minimum 12-month postgraduate programs that are designed to develop competencies for practitioners to par-

Table 11 Online cancer-related publications

Publication title	Web site URL	Source
CA-A Cancer Journal for Clinicians	http://www.ca-journal.org/	Peer-reviewed journal from the American Cancer Society
Cancer Control Journal	http://www.moffitt.usf.edu/pubs/ccj/	Published by the H. Lee Moffitt Cancer Center and Research Institute
Highlights in Oncology Practice	http://www.meniscus.com/web/publications/hl/index.htm	Serial publication for healthcare professionals from Meniscus Limited
Other Online Resources	http://nccam.nih.gov/	National Center for Complementary and Alternative Medicine, National Institutes of Health
	http://www.cancer.gov/occam/index.html	Office of cancer complementary and alternative medicine, National Cancer Institute
	http://dietary-supplements.info.nih.gov/	Office of Dietary Supplements, National Institutes of Health

ticipate in care of oncology patients and oncology drug distribution within oncology healthcare systems. The American Society of Health-System Pharmacists has developed accreditation standards for oncology specialty residency programs;[58] individuals who are considering such programs, accredited or nonASHP accredited, should refer to these educational objectives. Specialized pharmacy residency training in oncology will provide the practitioner with the skills and knowledge to work effectively as a multidisciplinary team member to improve patient care.[38] Residency programs provide excellent opportunities for the resident to develop professional relationships with other oncology practitioners.

Oncology specialty review courses are offered by the American Society of Health-System Pharmacists[59] and the American College of Clinical Pharmacy.[60] Some organizations offer mini-traineeships (ASHP, ACCP) to enable practitioners to develop clinical expertise in specialized practice settings; Roxane Laboratories has sponsored a Scholar Program for healthcare practitioners in pain and palliative care for several years. Participation in these review courses and training programs allows practitioners to network with other similarly interested individuals outside their practice setting or institution.

Professional Organizations

The Section of Clinical Specialists of the American Society of Health-System Pharmacists[61] provides networking opportunities for ASHP members who practice or conduct clinical research in oncology pharmacy practice. Members also have input for specialty practice educational programming and receive a networking directory to assist in identifying other practitioners with similar practice and/or research interests.

The Hematology/Oncology Practice and Research Network (PRN) of the American College of Clinical Pharmacy[62] is organized to facilitate dissemination of clinical practice, research, and teaching-related information in hematology and oncology and is also a resource for hematology and oncology expertise among its membership. Members direct and participate in educational programming and network to establish collaborative relationships.

The International Society of Oncology Pharmacy Practitioners[63] was formed at the IV International Symposium on Oncology Pharmacy Practice held in 1995. The mission of this organization is to promote and enhance oncology pharmacy practice worldwide for the purpose of improving cancer patient care. Annual meetings of ISOPP provide national and international networking opportunities for oncology pharmacists.

Several professional oncology-focused organizations are available for pharmacists to join. In particular, oncology pharmacist membership and attendance at annual meetings of the American Society of Clinical Oncology (ASCO) and the American Society of Hematology (ASH)[54] are common. There are numerous other organizations related to pain and palliative care, as well as specialty oncology areas. Pharmacists are active members in their institutions' cooperative oncology groups and participate in and attend those meetings as well.

INTERNATIONAL PERSPECTIVES

Trends and issues regarding the safety and efficiency of delivery of patient care services in oncology are similar worldwide, although the structures of these services vary among different healthcare systems and in their maturity. The need for standards for pharmacist training and knowledge in oncology in these treatment settings is widely recognized. Since European hospitals typically have fewer pharmacists than do American hospitals, the provision of comprehensive clinical services in some countries has yet to be accomplished.[39] Some chemotherapy use guidelines have been developed by oncology pharmacy groups. Guidelines for pharmaceutical care of cancer patients, developed by the London oncology pharmacy group, specify that chemotherapy given on nononcology wards should be dispensed by pharmacists who possess some basic clinical training and pharmacists with overall oncology responsibility should oversee such therapy in London.[40]

Policy guidelines for pharmacy services have also been established and adopted by the Society of Hospital Pharmacists in Australia.[41] These guidelines suggest specific areas for which pharmacists should place special emphasis: potential chemotherapy drug–drug interactions, emetic control, analgesia, and gastrointestinal preparations.

SUMMARY AND RECOMMENDATIONS

Oncology clinical pharmacy specialists are well-recognized for their expertise and valued contributions toward improving the efficiency, safety, and outcomes of care provided to patients with malignancies. Formal programs such as oncology specialty residencies offer exceptional opportunities for practitioners to gain the specialized knowledge and skills needed to meet challenging practice

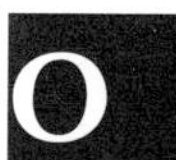

responsibilities in oncology pharmacy practice. Numerous available practice resources such as practice guidelines and reference and online publications facilitate practitioners' use of new knowledge to improve patient care. Despite the wide recognition and appreciation of oncology pharmacy specialists' services, there remains a tremendous need for documentation of the benefits of their services in the medical literature. There is also a need for the public to learn more about the value of oncology pharmacy services. This may be enhanced by having pharmacy organizations develop better public relations and promote the importance of oncology pharmacist certification as a requirement for all institutions that treat patients with cytotoxic chemotherapy.

REFERENCES

1. Wong, W.M.; Ignoffo, R.J. If there are expert systems and dose checks, why do we still need the clinical pharmacist? Pharm. Pract. Manage. Q. **1996**, *16* (1), 50–58.
2. Whitmore, C.K.; Hasdall, R.S. Professionalism and activity profiles in oncology pharmacy practice. J. Soc. Admin. Pharm. **1986**, *3* (3), 102–111.
3. Cohen, M.R.; Anderson, R.W.; Attilio, R.M.; Green, L.; Miller, R.J.; Pruemer, J.M. Preventing medication errors in cancer chemotherapy. Am. J. Health-Syst. Pharm. **1996**, *53* (7), 737–746.
4. Thorn, D.B.; Sexton, M.G.; Lemay, A.P.; Sarigianis, J.S.; Melita, D.D.; Gustafson, N.J. Effect of a cancer chemotherapy prescription form on prescription completeness. Am. J. Hosp. Pharm. **1989**, *46*, 1802–1806.
5. Fischer, D.S.; Alfano, S.; Knobf, M.T.; Donovan, C.; Beaulieu, N. Improving the cancer chemotherapy use process. J. Clin. Oncol. **1996**, *14* (12), 3148–3155.
6. Waddell, J.A.; Hannan, L.C.; Stephens, M.R. Pharmacy technician competencies for practice in an oncology pharmacy. J. Pharm. Technol. **1998**, *14*, 191–201.
7. American Society of Hospital Pharmacists. ASHP technical assistance bulletin on handling cytotoxic and hazardous drugs. Am. J. Hosp. Pharm. **1990**, *47*, 1033–1049.
8. American Society of Clinical Oncology. Criteria for facilities and personnel for the administration of parenteral systemic antineoplastic therapy. J. Clin. Oncol. **1997**, *15* (11), 3416–3417.
9. Evans, W.E.; Relling, M.V.; Rodman, J.R.; Crom, W.R.; Boyett, J.M.; Pui, C.H. Conventional compared with individualized chemotherapy for childhood acute lymphoblastic leukemia. N. Engl. J. Med. **1998**, *338* (8), 499–505.
10. Wright, L.J. United States Public Health Service bicentennial 1798-1998: A focus on oncology pharmacy practice at the National Institutes of Health. J. Oncol. Pharm. Pract. **1998**, *4* (3), 139–142.
11. Pon, D. Service plans and clinical interventions targeted by the oncology pharmacist. Pharm. Pract. Manage. Q. **1996**, *16* (1), 18–30.
12. Gilbert, C.J.; Morgan, R.T.; May, D.M. Inpatient high-dose chemotherapy with outpatient autologous stem cell support: A model of continuous pharmaceutical care. J. Pharm. Pract. **1995**, *8* (6), 290–296.
13. Lawrence, C.C.; Gilbert, C.J.; Peters, W.P. Evaluation of symptom distress in a bone marrow transplant outpatient environment. Ann. Pharmacother. **1996**, *30*, 941–945.
14. Raehl, C.L.; Bond, C.A.; Pitterle, M.E. Ambulatory pharmacy services affiliated with acute care hospitals. Pharmacotherapy **1993**, *13* (6), 618–625.
15. Wong, S.W.; Gray, E.S. Clinical pharmacy services in oncology clinics. J. Oncol. Pharm. Pract. **1999**, *5* (1), 49–54.
16. Montello, M.J.; Greenblatt, J.J.; Fallavollita, A.; Shoemaker, D. Accessing investigational anticancer agents outside of clinical trials. Am. J. Health-Syst. Pharm. **1998**, *55* (Apr. 1), 651–652, 660.
17. Task Force on Specialty Recognition of Oncology Pharmacy Practice. Executive summary of petition requesting specialty recognition of oncology pharmacy practice. Am. J. Hosp. Pharm. **1994**, *51*, 219–224.
18. Ignoffo, R.J.; King, R. Pharmaceutical care and the cancer patient. J. Oncol. Pharm. Pract. **1995**, *1* (4), 7–19.
19. Broadfield, L. Pharmaceutical care in oncology pharmacy practice: A method for using outcome indicators. J. Oncol. Pharm. Pract. **1995**, *1* (1), 9–14.
20. Hirsh, J.D.; Conlon, P.F. Implementing guidelines for managing extravasation of antineoplastics. Am. J. Hosp. Pharm. **1983**, *40*, 1516–1519.
21. Nolte, J.; Lucarelli, C.; Baltzer, L.; Berkery, R.; Grossano, D.; Yoo, S.; Tyler, E.; Pizzo, B.; Kris, M. In *Development, Implementation, and Assessment of Antiemetic Guidelines*, Proceedings of the American Society of Clinical Oncology, Los Angeles, CA, May 20–23, 1995; Perry, M.C., Ed.; American Society of Clinical Oncology: Alexandria, Virginia, 1995; 1735, 525.
22. Valley, A.W.; Holle, L.M.; Weiss, G.R. In *One-year Outcomes from an Institutional Antiemetic Guideline*, Proceedings of the American Society of Clinical Oncology, New Orleans, LA, May 20–23, 2000; Perry, M.C., Ed.; American Society of Clinical Oncology: Alexandria, Virginia, 2000; 2383, 605a.
23. Wood, A.M.; Pisters, K.; Chi, M.; King, K.M.; Arbuckle, R.; Rubenstein, E.B.; Oholendt, M.; Rodriguez, M.A.; Hagemeiser, F.; Cabanillas, F. In *Evaluation of Antiemetic Guidelines and Outcomes in Patients Receiving Highly Emetogenic Chemotherapy Utilizing the Palm Pilot*, Proceedings of the American Society of Clinical Oncology, New Orleans, LA, May 20–23, 2000; Perry, M.C., Ed.; American Society of Clinical Oncology: Alexandria, Virginia, 2000; 2512, 636a.
24. Kohler, D.R.; Montello, M.J.; Green, L.; Huntley, C.; High, J.L.; Fallavolita, A.; Goldspiel, B.R. Standardizing the ex-

pression and nomenclature of cancer treatment regimens. Am. J. Health-Syst. Pharm. **1998**, *55* (2), 137–144.
25. DeCederfelt, H.J.; Grimes, G.J.; Green, L.; DeCederfelt, R.O.; Daniels, C.E. Handling of gene-transfer products by the National Institutes of Health Clinical Center pharmacy department. Am. J. Health-Syst. Pharm. **1998**, *54* (14), 1604–1610.
26. Acosta, M. Improving disease management in pediatric oncology. Hosp. Pharm. Rep. **1996**, *10* (5), 59.
27. Goodin, S.; Torma, J.A. Development of critical pathways for chemotherapy agents. Hosp. Pharm. Rep. **1997**, *11* (3), 52–61.
28. Finley, R.S. The influence of clinical practice guidelines on professional practice and decision-making. Highlights Oncol. Pract. **1999**, *17* (2), 28–37.
29. Wood, A.M. Oncology clinical practice guidelines. Highlights Oncol. Pract. **1999**, *17* (2), 38–42.
30. McLennan, D.N.; Dooley, M.J.; Brien, J.E. Beneficial clinical outcomes resulting from pharmacist interventions. J. Oncol. Pharm. Pract. **1999**, *5* (4), 184–189.
31. Pfeiffer, C.M. In *Impact of a Clinical Pharmacist's Review of the Chemotherapy Administration Process*, ASHP Midyear Clinical Meeting, Las Vegas, NV, Dec. 3–7, 2000; American Society of Hospital Pharmacists: Bethesda, Maryland, 2000; 94E.
32. Waddell, J.A.; Solimando, D.A.; Strickland, W.R.; Smith, B.D.; Wray, M.K. Pharmacy staff interventions in a medical center hematology-oncology service. J. Am. Pharm. Assoc. **1998**, *38* (4), 451–456.
33. Decane, B.E.; Chapman, J. Program for procurement of drugs for indigent patients. Am. J. Hosp. Pharm. **1994**, *51* (5), 669–671.
34. McDonagh, M.S.; Miller, S.A.; Naden, E. Costs and savings of investigational drug services. Am. J. Health-Syst. Pharm. **2000**, *57* (1), 40–43.
35. Holdsworth, M.T.; Raisch, D.W.; Chavez, C.M.; Duncan, M.H.; Parasuraman, T.V.; Cox, F.M. Economic impact with home delivery of chemotherapy to pediatric oncology patients. Ann. Pharmacother. **1997**, *31*, 140–148.
36. Schneider, P.J. How pharmacists can reduce hospital costs. Pharm. Times **1980**, *46* (1), 70–76.
37. Kubica, A.J.; Coarse, J.F. Reducing medical costs through clinical pharmacy intervention and applied medical technology. Top. Hosp. Pharm. Manage. **1982**, *2* (1), 22–31.
38. LeRoy, M.L.; Pruemer, J. Impact of an oncology residency training program on quality improvement initiatives in an oncology center. Pharm. Pract. Manage. Q. **1996**, *16* (1), 59–65.
39. Bingham, J.M. Issues in pharmaceutical care and quality assurance. J. Oncol. Pharm. Pract. **1995**, *1* (Suppl. 1), 13–17.
40. London Oncology Pharmacy Group. Guidelines for the pharmaceutical care of cancer patients. Pharm. J. **1995**, *255*, 841–842.
41. The Society of Hospital Pharmacists of Australia Specialty Practice Committee for Oncology. SHPA policy guidelines for pharmacy services to oncology units. Aust. J. Hosp. Pharm. **1985**, *15* (1), 29–30 (http://www.osha-slc.gov/dts/osta/otm/otm_vi/otm_vi_2.html).
42. (http://www.ashp.org/bestpractices/TABs.html).
43. (http://www.cancer.gov/cancer_information/).
44. (http://www.cancer.gov/cancer_information/coping/).
45. (http://www.phrma.org/patients/).
46. (http://ctep.cancer.gov/requisition/index.html/).
47. (http://www.bpsweb.org/).
48. (http://www.jcaho.org/).
49. (http://www.roxane.com).
50. (http://www.partnersagainstpain.com/).
51. (http://www.talaria.org/).
52. (http://www.fda.gov/cder/cancer/index.htm).
53. (http://ctep.cancer.gov/index.html).
54. (http://www.hematology.org/).
55. (http://www.asco.org).
56. (http://www.aacr.org/).
57. (http://www.medscape.com/hematology-oncologyhome).
58. (http://www.ashp.org/public/rtp/ONCSUPPS.html).
59. (http://www.ashp.org/).
60. (http://www.accp.com/).
61. (http://www.ashp.org/).
62. (http://www.accp.com/).
63. (http://www.isopp.org/).

Orphan Drugs

Carolyn H. Asbury
University of Pennsylvania, Philadelphia, Pennsylvania, U.S.A.

INTRODUCTION

As drug development costs began to rise in the late 1960s and 1970s with new Food and Drug Administration (FDA) requirements for demonstrating relative safety and efficacy, manufacturers of drugs, biologicals, and diagnostic agents faced a dilemma. How could they consider developing drugs that were important medically but not likely to be profitable? Drugs for diseases and conditions that were rare in the United States were one of several—and the most visible—of such drug types. Rarity meant few patients would be available to participate in clinical trials of safety and efficacy under the sponsor's Investigational Exemption of a New Drug (IND) application. Slow patient recruitment into trials increased development time, with the 17-year patent-protection clock ticking. And once the pharmaceutical sponsor received approval for the New Drug Application (NDA) required for interstate marketing, there would be few customers. This generally translated into a low return on investment (ROI). Moreover, the time devoted to testing and gaining approval for a drug for a rare disease was an opportunity cost. Manufacturers would otherwise have devoted the time and resources to products with more favorable market returns.

Six other drug categories were of limited commercial interest. They included products for (1):

1. Chronic diseases, requiring extended testing phases to assess long-term effects, using up critical patent-protection time;
2. Single administration, such as vaccines, requiring separate production facilities, used by many but on a one-time basis, and carrying high liability risks; and diagnostic agents that had relatively low sales volume;
3. Women of childbearing age, presenting unparalleled liability risks;
4. Children and the elderly, both of whom were excluded from clinical testing at the time. Children presented high liability risks and could be difficult to enroll in clinical trials. Elderly people, who often had multiple conditions, were excluded from trials by the FDA because the multiple conditions and drugs used to treat them would confound trial results. (Once a drug was marketed, physicians could prescribe it for these two populations. However, optimal dosing had not been determined, and likely drug interactions in elderly patients had not been identified);
5. Diseases, rare or common, for which the intended drugs were not patentable including shelf chemicals, drugs known to exist, natural chemicals, and drugs for which the patent had expired;
6. Substance abuse or relapse prevention, intended for a population considered to be a high liability risk, uncooperative with treatment trials, and requiring extensive records once marketed; and
7. Developing countries, with the related problems of distributing and paying for products.

Observing that certain drugs could be in common use but were not potentially profitable enough to invite commercial introduction, Provost referred to them in the *American Journal of Hospital Pharmacy* in 1968 as homeless or ''orphan'' drugs (2). This chapter focuses on orphan drugs for rare diseases. It provides background on the issues and on attempts to deal with them—a description of the Orphan Drug Law, enacted in 1983, to address these issues—and an accounting of progress and problems to date emanating from the law.

BACKGROUND

The 1938 Food, Drug and Cosmetics Act and the 1962 Kefauver-Harris Amendments to it substantially altered drug development in this country. They strengthened the safety and efficacy determinations of drugs, but at cost, one of which was industry disinterest in drugs for rare diseases. The 1938 law required proof of safety after diethylene glycol, used as a sulfanilamide vehicle was found to form lethal quantities of oxalic acid in the body. This resulted in 100 fatalities, mostly children (3). The law required manufacturers to provide evidence that the drugs were relatively safe. For the FDA to keep a drug off the market, however, the onus was on the *agency* to prove that the drug was not safe. The thalidomide disaster changed that situation. Tragic birth defects were reported

Encyclopedia of Clinical Pharmacy
DOI: 10.1081/E-ECP 120006169

in infants born to women in Europe and Canada who had taken the drug while pregnant. At the time the news broke, Congress was working on amendments to the 1938 law that required sponsors to provide proof of drug efficacy before receiving market approval. The thalidomide catastrophe propelled Congress to require sponsors also to show proof of safety. To implement these two requirements, the FDA created the IND process requiring sponsors to test for and establish relative safety before embarking on clinical trials to determine drug efficacy. The FDA set up the NDA approval process for ruling on safety and efficacy before the drug could be marketed interstate (4).

An essential interplay developed between patent protection and regulatory requirements. Patent protection became essential for emerging products developed by research-intensive, vertically integrated firms that looked to a few major market winners to survive in this era of increased development costs (5). Patent time used in the IND and NDA premarket stages reduced the time remaining for manufacturers to protect their products, once marketed, against less costly generic drug competition. Drug development also became more expensive. Development costs in the late 1960s and early 1970s, before the FDA amendments became operationally implemented, were estimated to range from a low of $2.7 million to a high of $16.9 million per new chemical entity (NCE) (6, 7). By the late 1970s, the estimate had risen to $54 million (8). Since then, estimates have continued to leap upward: $124 million in the late 1980s, $231 million in 1991, and $500 million today (9, 10).

By the late 1960s and early 1970s, market winners generated most of a company's profits and also helped encourage brand loyalties by prescribing physicians. Other drugs needed to at least break even. Although not blockbusters, these drugs generally were for large markets. Drugs for rare diseases usually did not break even and became therapeutic orphans. They became wards of government and university-sponsored development efforts.

Cancer treatment drugs were among the first wards of government. Even before the 1962 amendments to the Food, Drug and Cosmetics Act, the federal government had developed and maintained a role in stimulating the development of cancer drug treatment that was not being addressed by industry. The National Cancer Institute (NCI) created the Cancer Chemotherapy Program in 1955 with a $5 million Congressional authorization. Congress decided that the NCI should take on the challenge. Impressed with industry's spectacular antibiotics development, Congress recognized that a low ROI was precluding industry's interest in exploiting the early successes in antitumor drugs (antifol aminopterin for acute childhood leukemia and methotrexate for uterine choriocarcinoma). Beginning as a small grant-oriented program to develop antileukemia agents, the program was based on the use of transplantable tumors in syngeneic rodents as a system for testing new drugs. After the NCI drafted an agreement allowing for the trade secret status of data, industry submitted compounds for screening of bioactivity. Within three years, the Cancer Chemotherapy Program had grown into a $35 million industrial contract effort. After a few missteps, when the FDA would not accept the cancer-funded research results as provided, the FDA and NCI developed a clear understanding of how data needed to be provided. From that point until at least the early 1980s, the NCI was involved in the development and/or clinical testing of every antineoplastic drug available in the United States (11–13).

In 1966, the National Institute of Neurological and Communicative Disorders and Stroke, as it was then named, established the Antiepileptic Drug Development Program to develop clinical trial methodologies and then to conduct clinical trials of antiepileptic agents already available in other countries. The program later developed a screening program similar to that of the NCI. The Institute filed INDs for drugs that entered clinical testing, and by 1981, four drugs had commercial sponsorship by the NDA stage (14, 15).

Although most of the federal funding for drug programs by the National Institutes of Health (NIH) was for the development of drugs for rare diseases—cancers and epilepsy—the NIH also devoted funding to development of other categories of orphan drugs. These included drugs to prevent and treat substance addiction and relapse, contraceptives for women of childbearing age, and some vaccines being developed by NIH and by the Centers for Disease Control (now the Centers for Disease Control and Prevention). Before 1983, when the Orphan Drug Act was signed into law, NIH drug development programs and grant-supported research had resulted in 13 drugs for the treatment of rare diseases being approved and on the market. In the 17 years before the Orphan Drug Law, the pharmaceutical industry had developed and marketed 34 drugs or biologicals for use in rare diseases or conditions (16). Ten of these marketed orphan drugs and biologicals had been developed solely by industry without government or university support (17).

The Gathering Storm

Nonetheless, several articles published in medical journals by academic researchers chronicled their plight in formulating new dosages for available drugs or

encapsulating chemical ingredients for clinical tests in patients with rare diseases, because no pharmaceutical sponsor had come to their rescue (18–20). One of these researchers, Cambridge University's John Walshe, proclaimed that this "do-it-yourself" problem had gone far enough and should be placed on a sound commercial basis (21). The FDA published a list of potentially promising therapeutic agents that were under development primarily by academic or government-based scientists who had failed to find commercial sponsors to undertake the costly phase III clinical trials for efficacy, file the NDA, and market the product (22). In 1979, the FDA convened a Task Force on Drugs of Little Commercial Value to determine how to find commercial sponsors. That same year, Louis Lasagna, who was then at the University of Rochester, invoked Provost's notion in an article in the journal *Regulation* by asking who will adopt the orphan drugs (23)?

NIH-funded academic researchers were reporting failed attempts to interest industry in taking on and securing NDA approval for drugs it had not developed. Industry had cited three major problems. First, NIH-funded scientists may not have gathered and analyzed data according to FDA requirements. Second, the trade secret status of data could not be preserved if the researcher had previously published information in the scientific literature. And third, many of the NIH-funded studies were of drugs that were not (or no longer) patentable. The FDA task force recommended that incentives be provided to industry to develop drugs of limited commercial value, but that any profits made from those incentives be returned, in whole or in part, to the government. (24). This arbitration-type approach was based on the assumption that industry would be willing to make trade-offs with the FDA on behalf of these drugs. But this assumption was made in the absence of any indication from industry that this was the case (25).

Although the task force was the first official policy response to the orphan drug situation (apart from the NIH-sponsored research responses), there were several private-sector efforts underway to promote the development and availability of orphan products. The Pharmaceutical Manufacturers Association (PMA, now called the Pharmaceutical Research and Manufacturers of America, or PhRMA) had been seeking sponsors among its member companies for promising orphan therapeutics. And a consumer group, the National Organization for Rare Disorders (NORD), was growing in number (now representing more than 20 million patients and their families) and undertaking efforts to raise public awareness and to help link patients to researchers conducting clinical studies of their diseases or conditions.

Lightning Strikes

A researcher at the Mount Sinai School of Medicine New York, who was seeking a pharmaceutical company sponsor for the drug L-5HTP for myoclonus, appealed to his Congressperson, Elizabeth Holtzman (D-New York) for a legislative remedy to the plight of orphan drug research. She introduced a bill in 1980 to establish the Office of Drugs of Limited Commercial Value to assist the NIH in developing drugs for the treatment of rare diseases (26). Although no action was taken on the bill, the committee heard testimony from a young California man suffering from Tourette's syndrome, a genetically determined neurological condition causing its victims to twitch, tic, and have uncontrollable verbal (and often abusive) outbursts. When haloperidol proved unsuccessful in controlling his symptoms, the patient had tried desperately to obtain and try pimozide, a drug available for the condition in Canada and Europe, but not in the United States.

A Los Angeles newspaper carried an account of the testimony, which caught the eye of the producer of the television series *Quincy*. Before long, a *Quincy* episode was devoted to dramatizing the conundrum of patients trying to cope with Tourette's syndrome and of industry trying to contend with the commercial disincentives to develop drugs to treat these patients. The episode demonstrated that there were no villains and no remedies. In a compelling scene before a Congressional committee, *Quincy* delivered an impassioned appeal to Congress to find a remedy. Shortly thereafter, *Quincy* star Jack Klugman was asked by Congressman Henry Waxman (D-California) to appear at a hearing on an orphan drug bill that was similar to Holtzman's, introduced by her colleague Representative Ted Weiss (D-New York). Klugman testified at the hearing, using the identical appeal he had delivered on the show (27). A *Wall Street Journal* editorial, entitled "Leave of Reality," likened Klugman's appearance before the Waxman subcommittee as an orphan drug expert to having Leonard Nimoy (Mr. Spock on *Star Trek*) testify as an expert on the nation's space program (28). By the end of 1981, Congressman Waxman had introduced H.R. 5238, the Orphan Drug Act.

The orphan drug survey

To find out the current status of development of drugs for the treatment of rare diseases, the subcommittee surveyed the industry, the NIH, and the FDA on three groups of drugs:

1. Products listed by PMA (now PhRMA) member companies as drugs for the treatment of rare diseases

that industry sponsors had marketed or had made available on a compassionate basis to specialists;
2. Drugs and biologicals listed by the FDA as under development, but needing a commercial sponsor for FDA approval and marketing; and
3. Drugs under development by NIH scientists or grantees.

From this survey, the Subcommittee learned that industry had marketed 34 drugs for the treatment of rare diseases extending back to 1965 and had made an additional 24 drugs (under development) available to physicians on a compassionate basis for treating patients with rare diseases. Most (82%) of the marketed drugs were for conditions affecting fewer than 100,000 people in the United States; 10% were for 100,000 to 500,000 people, and the remaining 8% were for up to 1 million people. Industry respondents indicated that substantial federal funding had been provided for research and/or development (R&D) for all but 10 of these marketed drugs.

The picture of disincentives confirmed claims by industry spokespeople sponsors indicated that the ROI for 83% of the drugs was lower than the sponsors' average return for marketed drugs, whereas development costs were higher than average for 12% of these drugs. Other issues cited by industry were the lack of clarity of FDA clinical testing guidelines when small numbers of patients were available and involved. This further eroded the sponsors' ability to estimate the length of clinical testing, and therefore the length of remaining patent protection time, once the drug was approved. Survey data indicated that industry-sponsored marketed orphan drugs took an average of 5.75 years for clinical testing (from filing the IND to filing the NDA). Unpatented orphan products were increasingly unlikely to be submitted for NDA approval, suggesting the importance of having at least some period of market protection. Whereas nearly two-fifths (39%) of industry-sponsored drugs for the treatment of rare diseases had been marketed in the 1960s, even though they were not protected by patent, this fell to 29% by the 1970s. Liability claims had been filed against the manufacturers of one-fifth of the marketed orphan drugs. The promising finding was that one in four industry-sponsored marketed orphan drugs also had a common indication. This suggested that orphan drugs were a relatively good market gamble (29).

Orphan Drug Act's Passage Provides Market Incentives and Regulatory Assistance

Based primarily on Congressional testimony marshaled by NORD revealing that millions of people and their families were profoundly affected by rare diseases and conditions, and on survey data revealing market and regulatory disincentives to therapeutic progress, Congress passed the Orphan Drug Act during a lame duck session in December 1992. Called "the golden egg of the lame duck Congress" by the head of the Generic Pharmaceutical Industry Association (GPIA), the bill was signed into law the first week of January 1983 (30). The act (Public Law 97–414) was supported initially by the GPIA, which had begun seeking sponsors for developed orphan products, and eventually by the PMA after certain provisions were deleted or modified. Provisions in the law, and in subsequent amendments in 1984, 1985, and 1988, addressed market and regulatory issues and provided incentives for pharmaceutical industry development of drugs for the treatment of rare diseases.

The law, as amended, defines a rare disease or condition as one that affects fewer than 200,000 persons in the United States. Alternatively, the disease or condition can affect more than 200,000 persons in the United States, if there is no reasonable expectation that the cost of developing and making a drug available for such disease or condition in the United States will be recovered from U.S. sales of the drug. Therefore, a drug can be designated by the FDA as an orphan by demonstrating applicability of the law's financial criteria, regardless of the total number of people affected in the United States. Major provisions are the availability of two market incentives and the reduction of regulatory barriers. The FDA Office of Orphan Products Development administers nearly all the law's provisions.

One incentive is seven years of market exclusivity granted by the FDA for a specific indication of a product. Exclusivity begins on the date the FDA approves the marketing application for the designated orphan drug and applies only to the indication for which the drug has been designated and approved. An application for designation as an orphan drug for a specific indication must be made before submission of the NDA (or biologic product license application, PLA) for market approval (31). Other sponsors can receive approval for a different drug to treat the same rare disease or can receive approval to market an identical drug for some other orphan or common indication. Market exclusivity, therefore, only precludes a second sponsor from obtaining approval to provide an identical drug for the identical orphan indication for which the first sponsor received exclusive market approval. Initially, market exclusivity pertained only to unpatented products. A 1985 amendment allowed exclusive approval for all orphan drugs, whether patented or not. This change was designed to provide incentives for sponsors of products for the treatment of rare diseases whose patents would expire before or soon after approval, or in cases in which prior publication (usually by an academic or government scientist) had precluded issuance of a patent.

A second incentive provides tax credits equal to 50% of the costs of human clinical testing undertaken in any given year by a sponsor to generate data required for obtaining FDA market approval through successful completion of the NDA process. The Internal Revenue Service administers the tax credit provisions.

The act provides for the FDA to award grants to support clinical studies of designated orphan products under-development. FDA grant funding as of March 2000 from the FDA totaled $126.3 million. From initial funding in 1983 of $500,000 in grants, the grant program peaked in 1994 at $12.3 million and has declined slightly but steadily thereafter, totaling $11.1 million in 1999 (32). Applications are reviewed by outside experts and are funded according to a priority score. The FDA Office of Orphan Product Development provides information at its website (www.fda.gov/orphan/GRANTS/patients) on investigators seeking research subjects. Listed by the name of the disease or condition, information includes a description of the study, criteria for inclusion in clinical trials (such as age, stage of disease, etc.), and contact information on the clinical investigator seeking participants for clinical trials. Patients, their families, or their physicians are able to follow up with the clinical investigator.

To address regulatory barriers the FDA also provides formal protocol assistance when requested by the sponsor of an orphan drug. Although formal review of a request for protocol assistance is the direct responsibility of the FDA Centers for Evaluation and Research (one for drugs, the other for biologicals), the Office of Orphan Products Development is responsible for ensuring that the request qualifies for consideration. A sponsor need not have obtained orphan drug designation to receive protocol assistance.

Table 1 Designated and approved orphan drug products

	Pre-1983	1989	1991	3/2000
Cumulative total approved orphan products	34[a]	36	54	235
Total sponsors of approved orphan drugs[b]	17			110

[a]At the time, these were called drugs for rare diseases, not orphan drugs.
[b]Sponsors identified only for the two endpoints.

Finally, the FDA is required to encourage sponsors to design open protocols for drug availability to patients not included in clinical trials.

The Current Status of Orphan Drugs

The FDA approved 201 orphan products between 1983 and March 2000 (33). In the 17 years since passage of the act, therefore, the number of approved orphan drugs has increased sixfold from the number approved in the 17 years before the act (Table 1). The drugs are classified within 16 therapeutic categories, primarily for the treatment of cancer, infectious disease, AIDS and related conditions, and central nervous system conditions (Table 2). Included in the 201 approved drugs are 24 (12%) that received FDA grant support for clinical trials (34). A list of these grant-supported products is available at the FDA Office of Orphan Products Development website (www.fda.gov/orphan/GRANTS).

The number of sponsors of approved orphan drugs (nearly all of them produced at pharmaceutical or biotechnology companies) has increased from 17 (for the 34 drugs marketed before the act) to 110 as of March 2000. The number of approved drugs per sponsor ranges from one to eight, with a preponderance of one-drug sponsors (Table 3). A total of 813 products have received orphan designations as of March 2000. Sponsors of approximately 25% of these designated products have filed INDs for the products, indicating that they are under active development (35).

Many of the designated and approved orphan products are developed at biotechnology companies. Beginning in the 1970s, molecular biology had begun to spur the creation of biotechnology research companies. These companies reportedly recognized early on that market exclusivity, and lack of competition to develop products for the treatment of rare diseases, provided protection essential to raising venture capital. As a result, orphan drugs are now among biotechnology's most prevalent, and, according to some, most lucrative products. Between 1988 and 1992, biotechnology product designations increased by 31%, from 8 to 39% of total orphan designations (36). In addition to industry, however, sponsors have included a university, an individual researcher, and a state public health unit, which in aggregate sponsored six orphan drugs at the time of approval. Lists of approved and designated orphan drugs can be obtained from the FDA Office of Orphan Products Development. These data suggest that orphan drug development has consistently risen over the years since the law was enacted. Aggregate sales of

Table 2 Approved orphan drugs: Number per therapeutic category[a]

Category	Number of approved orphan products $N = 201$
Cancer	49
Infectious disease	23
Central nervous system	22
Hematopoetic	21
AIDS, AIDS-related	21
Endocrine	18
Inborn errors of metabolism	12
Renal	8
Cardiovascular disease	7
Respiratory	5
Gastrointestinal	5
Bone	3
Immunological	2
Dermatological	2
Antidote	2
Urinary tract	1

[a]Does not include drugs for rare diseases marketed before the 1983 Orphan Drug Law.

orphan drugs have been reported to be more than $1 billion a year (37).

A Resulting Issue: High Pricing for Some Products

By the early 1990s, U.S. sales data collected for 41 of the approved orphan drugs that had been on the market for a year or more indicated that 75% of the products had generated earnings of less than $10 million per drug. Three products had earnings between $10 and $25 million, six drugs between $26 and $100 million, and two products more than $100 million. Biotechnology firms produced four of the 11 drugs with relatively high sales.

Table 3 Approved orphan drugs per sponsor, March 2000[a]

Drugs per sponsor	Sponsors
1	69
2	19
3	6
4	2
5	5
6	2
7	1
8	2

[a]Does not include sponsors of drugs for rare diseases approved before the Orphan Drug Act of 1983.

Those orphan drugs that command high prices have generated intense controversy over whether the market exclusivity provision is creating an unnecessary monopoly, keeping prices artificially high. One example is recombinant human erythropoietin (r-EPO), intended for patients with chronic renal failure-related anemia. EPO eliminates the need for frequent blood transfusions by patients with end-stage renal disease who are undergoing kidney dialysis. These patients are covered under the federally financed Medicare program. Both Amgen Inc. and Genetics Institute applied to the FDA for market exclusivity for their r-EPO products. Amgen was the first to receive FDA approval and market exclusivity. Genetics Institute was the first to receive a patent. On appeal, the court ruled that Amgen had exclusive marketing rights. Sales of r-EPO exceeded $100 million in the first six months of marketing, paid for by Medicare. By 1991, sales totaled $400 million (38).

Human growth hormone (r-hGH), another example, generated sales of $150 million in 1991. Intended to treat approximately 12,000 children in the United States with retarded growth caused by a lack of endogenous pituitary hormone, two companies provide r-hGH. Genentech received FDA market exclusivity in 1985. Eli Lilly received market approval two years later, based on the determination that the two products differed by one amino acid (39,40). But the shared market did not lead to price competition: each company was earning approximately $20,000 per child annually, depending on the dosage needed.

A third example, aerosol pentamidine, generated sales of $130 million in 1991. Helping to prevent *Pneumocystis carinii* pneumonia associated with the human immunodeficiency virus (HIV), the increasing number of users resulting from a rapidly escalating HIV prevalence rate was expected to soon exceed the 200,000 population figure specified in the law. This example, along with the other two, prompted efforts to seek a legislative remedy. A series of amendments were introduced and passed by Congress in 1990 that sought to eliminate orphan status for products intended for use in epidemics and to allow shared market access for identical products developed simultaneously for the same indication, in the hope that this would lead to price competition. The amendments were vetoed (41). As Arno, Bonuck, and Davis present in *Milbank Quarterly*, AIDS treatment drugs exemplify the policy dilemma of how to use the Act to meet the legislative intent of stimulating development of drugs for small patient populations without resulting in prices that make such drugs inaccessible (42).

Another, more recent, example is enzyme therapy for Gaucher's disease, an inborn error of metabolism, treated with Ceredase. The therapy, which requires more than a ton of placenta annually to extract and make the drug, can cost as much as $500,000 per year per person, depending on the dosage needed. A 1996 National Institutes of Health technology assessment panel addressed issues in diagnosis and treatment of the disease and concluded that despite the success of enzyme therapy, treatment is limited by the cost. The panel reported that it was imperative to define the appropriate clinical indications for treatment and to determine the lowest effective initial and maintenance doses (43).

Although the survey undertaken prior to the Act found that retail costs of drugs for the treatment of rare diseases were reportedly higher than the manufacturer's average for 60% of the products marketed before the act, the pricing for some orphan drugs after the act, as exemplified by these examples, is a critically important consequence of the law. This consequence merits continued public scrutiny. It is a sign of progress that for some orphan drugs, accessibility rather than availability is now the challenge requiring creative solutions.

REFERENCES

1. Asbury, C.H. Medical Drugs of Limited Commercial Interest: Profit Alone is a Bitter Pill. Intern. J. Health Serv. **1981**, *11* (3), 451–462.
2. Provost, G.P. Homeless or Orphan Drugs. Am. J. Hosp. Pharm. **1968**, *25*, 609.
3. Silverman, M.; Lee, P.R. *Pills, Profits and Politics*; University of California Press: Berkeley, 1974; 2.
4. Harris, R. *The Real Voice*; Macmillan: New York, 1964; 21–25.
5. Temin, P. Technology, Regulation and Market Structure in the Modern Pharmaceutical Industry. Bell J. Econ. **1979**, *10* (2), 427–446.
6. Report of the Panel on Chemicals and Health of the President's Science Advisory Committee, 73–500, NSF, U.S. Government Printing Office: Washington, DC, 1973.
7. Schwartzman, D. *Innovation in the Pharmaceutical Industry*; Johns Hopkins University Press: Baltimore, 1976; 106–107.
8. Hansen, R.W. The Pharmaceutical Development Process: Estimates of Development Costs and Times and Effects of Proposed Regulatory Changes. *Issues in Pharmaceutical Economics*; Chien, R., Ed.; Lexington Books: Lexington, MA, 1980; 151–181.
9. DiMasi, J.; Hanson, R.; Grabowski, H.; Lasagna, L. The Cost of Innovation in the Pharmaceutical Industry and Health Economics. J. Health Econ. **1991**, *10*, 107–142.
10. Rosenbaum, D.E. The Gathering Storm Over Prescription Drugs. The New York Times **Nov. 14, 1999**, *1*, Section 4 (Week in Review).
11. Devita, V.; Oliverio, V.T.; Muggia, F.M.; Wiernick, P.W.; Ziegler, J.; Goldin, A.; Rubin, D.; Henney, J.; Schepartz, S. The Drug Development and Clinical Trials Programs of the Division of Cancer Treatment, National Cancer Institute. Cancer Clin. Trials **1979**, *2*, 195–216.
12. Zubrod, C.G.; Schepartz, S.; Leiten, J.; Endicott, K.M.; Carrese, L.M.; Baker, C.E. *Cancer Chemotherapy Reports October 1996, 50 (7)*; DHEW: Washington, DC, 1968.
13. Rate of Development of Anticancer Drugs by the National Cancer Institute and the U.S. Pharmaceutical Industry and the Impact of Regulation. *Final Report for the National Cancer Institute*; University of Rochester Medical Center: New York, 1981; 40.
14. Krall, R.A. Anti-Epileptic Drug Development. I. History and a Program for Progress. Epilepsia **1978**, *19* (4), 398–408.
15. Program Performance Summary. *National Institute of Neurological and Communicative Disorders and Stroke Antiepileptic Drug Development Program*; NINCDS: Washington, DC, 1982.
16. Orphan Drugs in Development, 1992 Annual Survey. Pharmaceutical Manufacturers Association: Washington, DC, 1992.
17. Orphan Drug Act Report, 97th Congress, 2nd Session. Washington, DC, Sept 17; 1982.
18. Van Woert, M. Profitable and Non-Profitable Drugs. N. Engl. J. Med. **1978**, *298* (16), 903–906.
19. Rawlins, M. No Utopia Yet. Br. Med. J. **1977**, *2*, 1076.
20. Rawlins, M. Editorial. Lancet **1976**, *2*(7970), 835–836.
21. Walshe, J.M. Treatment of Wilson's Disease with Trientine (Triethylene Tetramine) Dihhydrochloride. Lancet **1982**, *1*, 643–647.
22. Asbury, C.H. *Medical Drugs of Limited Commercial Interest: The Development of Federal Policy*; Johns Hopkins School of Hygiene and Public Health Thesis, Baltimore, 1981; 157–175.
23. Lasagna, L. Who Will Adopt the Orphan Drugs? Regulation **1979**, *3* (6), 27–32.
24. U.S. Food and Drug Administration Report. *Interagency Task Force Report to the Secretary of DHEW*; FDA: Washington, DC, 1979;, 1–82.
25. Asbury, C.H.; Stolley, P Orphan Drugs: Creating Federal Policy. Ann. Intern. Med. **1981**, *95* (2), 221–224.
26. Holtzman, E. Rep. H.R. 7089, 96th Congress, 2nd Session. Washington, DC, April 17, 1980.
27. Klugman, J.; Seligman, A. Testimony, Serial No. 97–17, U.S. Government Printing Office: Washington, DC, 1981; 10–16.
28. Leave of Reality (Editorial). The Wall Street Journal **March 12, 1981**.
29. Asbury, C.H. *Orphan Drugs: Medical vs. Market Value*; D.C. Heath: Lexington, MA, 1985; 136–155.
30. Waxman, H.R. Rep. H.R. 5328, 97th Congress, 1st Session. Washington, DC, 1981.
31. FDA Office of Orphan Products Development, www.fed.gov/orphan/about/progovw.htm (accessed Feb 2000).
32. Personal Communication, *FDA OOPD*, March 2000.
33. *FDA List of Approved Orphan Products through 02/24/2000*, FDA: Washington, DC, 2000; Provided on Request from the FDA OOPD.

34. Grant-Supported Products with Marketing Approval, *FDA OOPD*, Washington, DC, www.fda.gov/orphan/grants/magrants (accessed Feb 2000).
35. Personal Communication, *FDA OOPD*, March 2000.
36. Schuman, S.R. Implementation of the Orphan Drug Act. Food Drug Law J. **1992**, *47* (4), 363–403.
37. Grady, D. In Quest to Cure Rare Diseases, Some Get Left Out. *The New York Times*; Nov. 16, 1999.
38. Coster, J.M. Recombinant Erythropoietin: Orphan Product with a Silver Spoon. Intern. J. Technol. Assess. Health Care **1992**, *8* (4), 635–646.
39. Hilts, P.J. Seeking Limits to a Drug Monopoly. *The New York Times*; May 14, 1992, D1; 7.
40. Asbury, C.H. Evolution and Current Status of the Orphan Drug Act. Intern. J. Technol. Assess. Health Care **1992**, *8* (4), 573–582.
41. Asbury, C.H. The Orphan Drug Act: The first 7 years. J. Am. Med. Assoc. **1991**, *265*, 893–897.
42. Arno, P.S.; Bonuck, K.; Davis, M. Rare Disease Drug Development and AIDS: The Impact of the Orphan Drug Act. Milbank Q. **1995**, *73* (2), 231–252.
43. NIH Technology Panel on Gaucher Disease. Gaucher Disease. Current Issues in Diagnosis and Treatment. J. Am. Med. Assoc. **1996**, *275* (7), 548–553.
44. Asbury, C. *Orphan Drugs: Medical vs Market Value*; D.C Heath and Company: Lexington, MA, 1985.
45. Scheinberg, I.H., Walshe, J.M. Eds. *Orphan Diseases and Orphan Drugs;* Manchester University Press: Manchester, UK, 1986.
46. Wagner, J. Orphan Technologies. Int. J. Technol. Assessment **1992**, *8* (4), www.fda.gov/orphan.
47. Arno, P.S.; Bonuck, K.; Davis, M. Rare Diseases, Drug Development, and AIDS: The Impact of the Orphan Drug Act. Milbank Q. **1995**, *73* (2), 241–252.
48. Henkel, J. Orphan Drug Law Matures into Medical Mainstay. FDA Consumer **1999**, *33* (3), 29–32.

PROFESSIONAL DEVELOPMENT

Pain Management, Pharmacy Practice in

P

Theresa A. Mays
Cancer Therapy and Research Center, San Antonio, Texas, U.S.A.

INTRODUCTION

All people will experience pain at some point in their lives. On average, most Americans experience one form of pain an estimated three to four times per year. As our population continues to age in the United States, pain management will become an even larger issue. The annual cost of pain to our society was estimated at $79 billion in 1990 with more than 50 million Americans being either partially or totally disabled because of pain.[1]

The lack of appropriate pain management strategies has been significantly documented since the early 1980s. In 1979, the International Association for the Study of Pain (IASP) defined pain as "an unpleasant sensory and emotional experience arising from actual or potential tissue damage or described in terms of such damage."[2] In 1982, the World Heath Organization (WHO) brought together a panel of cancer pain experts to discuss this worldwide crisis. This group reached the consensus that relief from cancer pain is an appropriate goal for most patients. The WHO placed a high priority on the relief of cancer pain, and another meeting was held in 1984 that resulted in the first publication of *Cancer Pain Relief* in 1986.[3]

This publication was groundbreaking, not only in its discussion of the lack of appropriate cancer pain management worldwide, but in its overall recommendations for managing cancer pain. The concept of the three-step analgesic ladder is introduced, which divides pain ratings into three categories mild, moderate, or severe, with medication selection based on a step-up approach. When this step approach to pain management is employed, up to 90% of all cancer pain can be relieved.

In 1990, the WHO published a report from their expert committee on cancer pain relief and active supportive care.[4] The committee's purpose was to evaluate the current state of cancer care and pain relief, and to provide recommendations and an action plan to improve the overall care of these patients. This publication not only addresses appropriate pain management, but also discusses symptom management for other problems experienced by cancer patients. Since the early 1990s, significant strides have been made in pain management; yet, unfortunately, the WHO goal that every cancer patient has the right to be free from pain has not yet been met.

Ten years after the original publication, an updated version of *Cancer Pain Relief* was published.[5] The update included information from the 1989 publication, as well as any new advances in the management of cancer pain. This publication also included sections on the availability of opioids worldwide and the regulation of health care workers. Even though these publications are specific to cancer pain management, the basic principles can be applied to all chronic pain conditions.

Recently, the Joint Commission on Accreditation of Healthcare Organizations (JCAHO) announced the development and approval of new standards for pain management. These standards affect the management of both acute and chronic pain in all health care settings. These standards are included in the 2000–2001 accreditation manuals and are highlighted in this article because they offer numerous opportunities for pharmacists to participate in the pain management process.

Numerous opportunities exist for pharmacists to become involved in pain management. The management of pain can be divided into two major categories: acute and chronic. Examples of acute pain management include the surgical setting or acute injury. Examples of chronic pain management include malignant (cancer) and nonmalignant pain [human immunodeficiency virus (HIV)/acquired immunodeficiency syndrome (AIDS) or rheumatoid arthritis]. This article discusses the opportunities existing for pharmacists in both acute and chronic pain management settings. The goal of this article is not to review the appropriate management of pain, but to provide the reader with an overview of clinical pharmacy involvement in the area of pain management. Therefore, different types of pain management practices are presented and discussed. Important clinical guidelines in these areas are briefly presented. Finally, a comprehensive reference list, of both print media and

Encyclopedia of Clinical Pharmacy
DOI: 10.1081/E-ECP 120006309

Internet resources, is also included to facilitate your understanding of clinical pharmacy in pain management.

BARRIERS TO APPROPRIATE PAIN CONTROL

Numerous barriers exist that lead to suboptimal management of pain, and these barriers affect patients, health care professionals, and the health care system. The Agency for Healthcare Policy and Research (AHCPR) published a guideline for the management of cancer pain in March 1994.[6] This publication discusses the major barriers that currently exist to appropriate pain managment. Patients can be reluctant to report their pain, and they may be concerned that pain means worsening disease or that it may distract their physician from treating the underlying cause. Patients may not take their pain medication due to concerns of addiction, worries of opioid side effects, or concerns about becoming resistant to pain medications. Patients and their families can be misinformed regarding pain management. We live in a "just say no" society where opioids are confused with illicit drug use. Patients and their families need to understand the difference between using a medication for a medical reason and using one to maintain a high. A useful analogy is comparing insulin use in diabetics with the use of opioid medications for a painful condition. Both are recognized disease states requiring appropriate management of their disease process.

The health care system has given low priority to pain management due to inadequate reimbursement for provision of services, the legal regulation of controlled substances, and problems of treatment availability or lack of those trained to provide these services. The most significant barrier to pain management is lack of knowledge of the basic principles for appropriate pain management. Several surveys have been conducted in health care professionals, including physicians, nurses, and pharmacists. These studies have revealed a lack of knowledge and education in pain management leading to fear in both prescribing, dispensing, and administering of opioid medications.[7–15]

A survey of 59 questions sent to physicians practicing in the state of Texas revealed many areas of concern when prescribing opioids.[14] Ten percent of those surveyed would withhold opioid medications until a patient in severe pain had either a prognosis of less than 1 year or was terminal. Forty percent were extremely concerned regarding potential addiction. More than one-half falsely believed that opioid addiction is a common occurrence with legitimate prescribing. The physicians surveyed also fear regulatory scrutiny when opioid medications are prescribed for pain control.

Several studies have been published regarding pharmacists' attitudes and knowledge about opioid medication.[8,11,12] These studies have revealed pharmacists as barriers due to insufficient inventories of opioid analgesics, and inadequate counseling of both patients and their families about the importance of using these medications. In the study by Bressler and colleagues, 79% of pharmacists surveyed falsely believed that a patient using sustained release morphine 150 mg every 12 hours would become "addicts" if this dosing level was continued.[11]

Numerous barriers still exist to pain management. These barriers may come from the patient themselves, their physician, or other health care professionals, including pharmacists. One of the most important roles a pharmacist working in pain management can fill is the provision of education, not only to the patient and their family, but also to pharmacists, physicians, nurses, and other health care professionals.

TYPES OF PAIN AND IMPLICATIONS FOR CARE

The exact mechanisms of pain transmission are unknown. Pain is a complex interaction involving receptor stimulation, pain transmission, and response. Pain is generally divided into two categories: acute and chronic. These categories have different characteristics, which affect the approach to treatment. Patients may suffer from both acute and chronic pain at the same time.

Acute pain may manifest itself as a warning when we do something harmful to ourselves. Postoperative pain is also classified as acute pain. Acute pain is considered to be meaningful, linear, and reversible. The therapeutic goal is pain relief, and sedation is often a desirable effect. Medications chosen for treatment of acute pain should have a rapid onset of action and a short duration of action. These opioids are administered on a prn, or as needed basis, usually through an IV with the doses being standardized.[16]

Chronic pain is usually defined as pain that persists for months to years. This may be due to a specific disease state, such as arthritis, diabetes, AIDS, or cancer. Chronic pain is considered to be meaningless, cyclical, and irreversible. The therapeutic goal is pain prevention and sedation is an undesirable adverse effect. Two forms of medication are needed, a long-acting formulation to prevent the pain and a rapid-acting formulation to cover episodes that "breakthrough" the long-acting

opioid. Therefore, the patient takes an opioid on a scheduled basis around the clock and uses supplemental quick-acting opioids to manage additional pain. Dosages are titrated for the individual patient and can vary significantly.[16]

In general, chronic pain is more difficult to manage than acute pain due to the psychological problems caused by unrelenting pain. Chronic pain is usually divided into two categories: malignant and nonmalignant pain. Malignant pain is associated with a diagnosis of cancer and, more recently, HIV/AIDS. The nonmalignant category includes back pain, neuropathic pain, migraines, and many other diagnoses of chronic pain.

Pain Assessment

The first step in managing a patient's pain, whether it is acute or chronic in nature, is assessment. There are numerous methods available to assess pain, including verbal, visual analog, categorical, and faces scales. These methods are discussed in detail in several publications.[6,17–21] The choice of pain assessment scale will depend on the environment in which pain will be assessed, the patient's ability to comprehend the scale, and how detailed the assessment needs to be. Most pharmacists will find a simple verbal scale of 0 to 10, with 0 = no pain and 10 = worst pain imaginable useful.

A component of the new JCAHO standard is documentation of pain assessment and intensity in all patients. In addition, the interventions that are performed based on the pain assessment, such as administration of a pain medication, need to be documented along with the reassessment of the effect of these interventions. Pharmacists can play a key role in developing standardized assessment forms that include recommendations for pain medications based on the results.

Nonpharmacologic Measures for Pain Management

Patients suffering from either acute or chronic pain problems may benefit from nonpharmacologic measures. These measures may also be combined with pain medication. The AHCPR guidelines for the management of cancer pain contains a chapter outlining these options.[6] Nonpharmacologic interventions are divided into behavioral (psychosocial) interventions and mechanical (physical) interventions. Examples of behavioral interventions include biofeedback, self-hypnosis, and relaxation or imagery training. Examples of mechanical interventions include exercise, cutaneous stimulation, and acupuncture.

Behavioral interventions help patients gain a sense of control over their pain management and should be introduced early in the course of their illness. When deciding on the appropriate behavioral intervention, consideration must be given to the intensity of the pain, the expected duration of pain, and the patient's mental clarity, past experience with technique, physical ability, and desire to employ active or passive techniques. Relaxation techniques are easily taught to patients and include focused breathing, progressive muscle relaxation, music-assisted relaxation, and meditation. These techniques are more useful when combined with pleasant images.

The use of physical modalities to manage pain may lead to a decreased requirement for pain-relieving drugs and, just as with behavioral interventions, should begin early in the disease process. Cutaneous stimulation techniques include the application of heat or cold, massage, pressure, and vibration. Exercise techniques should be aimed at preventing immobilization and may include range of motion and stretching. Counterstimulation techniques include the use of transcutaneous electrical nerve stimulation (TENS) devices or acupuncture. These last two techniques have not been formally studied in cancer-related pain.

Use of Nonopioids in Pain Management

Nonopioid agents available for pain management include acetaminophen, nonsteroidal antiinflammatory agents, muscle relaxants, antidepressants, anticonvulsants, corticosteroids, bisphosphonates, and topical preparations, such as lidocaine patches. Depending on a patient's presenting complaints of pain, one of these agents may be an acceptable medication to start. All these medications can also be used in combination with opioid agents. Nonsteroidal agents are useful for pain due to muscle injury or bone metastases. Corticosteroids may be useful in patients suffering from pain due to inflammation or edema, such as spinal cord compression. Bisphosphonates are useful in the management of pain due to bone metastases. The other agents are discussed more in-depth under the section entitled, "Neuropathic Pain Management."

ACUTE PAIN SERVICES

The American Pain Society (APS) published quality assurance standards for relief of acute pain and cancer pain in 1991.[22] These standards were updated and republished in 1995.[17] In February 1992, the AHCPR published a guideline on *Acute Pain Management: Operative or Medical Procedures and Trauma*.[23] In 1995, the American Society of Anesthesiologists published guidelines for acute pain management in the perioperative setting.[24]

The AHCPR guideline has information about management strategies for various surgeries, including dental, radical head and neck, neurosurgery, chest and chest wall, abdominal, perianal, and musculoskeletal. Other sections of the guideline address pediatric and geriatric acute pain management. A discussion of handling patients with potential addiction problems and those with concomitant disease states is also included. Another important section is recommendations for handling patients with shock, trauma, or burns.

The AHCPR guideline is well referenced and an excellent beginning point for developing acute pain services. The guideline ends with several useful appendices, including a summary of the existing scientific evidence for pain interventions, potential nonpharmacologic interventions, relaxation techniques, and adult and pediatric dosing for nonsteroidal antiinflammatory agents and opioids. Several tools are also included, such as an initial pain assessment form and a monitoring form.

The APS, the American Society of Anesthesiologists, and the AHCPR guidelines outline the basic patient right of being pain free. Improving the quality of pain management includes several key areas. These areas are 1) defining levels of pain that trigger a review of the patient's care plan, 2) providing reference information on the basic analgesic principals that is readily available and near the area where orders are written, 3) educating patients to report pain and provide assurance that patients will receive attentive analgesic care, 4) developing and implementing policies for the use of analgesic technologies, 5) coordinating and assessing these measures, 6) charting pain assessments with a display of the assessment used and relief provided by the intervention, 7) surveying patient satisfaction with pain management, 8) providing specialized analgesic technologies and nonpharmacologic interventions, and 9) periodic monitoring of the efficacy of pain treatments. Unfortunately, even with the publication of these guidelines, data still show that postoperative patients continue to suffer from significant pain.[25]

In 1994, Smythe and colleagues[26] published a study evaluating pharmacy and nursing time requirements, quality of pain control, and cost of patient-controlled analgesia (PCA) versus intramuscular (IM) analgesic therapy. This study showed the PCA therapy provided better pain control than IM therapy after gynecological surgery at an increased cost, but with minimal increases in pharmacy and nursing time.

An article report by Triller and colleagues[27] details their development of a clinical pharmacy program at an existing home health care agency. The pharmacy resident undertook a pain management initiative in this patient population. Patients who had reported pain were identified and, at admission, 53% reported pain interfering with activities of daily living. Twenty-nine percent of these patients were still experiencing pain at discharge from the home health care service. Charts of the next 20 patients admitted to the service who had complaints of pain were reviewed. This review revealed that pain assessments were routinely occurring; however, documentation of interventions was lacking. A committee was formed to develop documentation methods and staff education programs. This article identifies several critical areas for pharmacy involvement in the home care setting, including the management of pain.

A publication by Blau and colleagues focuses on the organization of a multidisciplinary hospital-based acute pain management program.[28] This article stresses that appropriate management of postoperative pain may improve the outcome of surgery, decrease time in the surgical intensive care unit, and therefore allow for earlier discharge of the patient. The first step to initiate a pain management program is to compile an interdisciplinary team. The critical members of this team include physicians, nurses, pharmacists, social workers, and psychologists. The involved physicians in acute pain management are usually anesthesiologists. The role of the pharmacist is as an educational resource, consultant, and technical support. This article does a superb job of describing the basic steps to developing an acute pain management service, and the authors include a copy of the preprinted physician's order form they developed.

Another article describing the implementation of a pain management service at a U.S. Army medical center was published.[29] Even though this project was developed and implemented by clinical nurses, it provides useful information for pharmacists for developing recommendations for an epidural and PCA service. The author also includes copies of their institution's pain assessment flow sheet and pain assessment form. An article describing the application of the American Pain Society's guideline in the management of pain in surgical, oncology, and hospice inpatients in Taiwan was also published.[30] This article contains useful information on performing patient satisfaction surveys on patients undergoing pain management in an inpatient setting, which is another requirement of the new JCAHO standards.

The JCAHO has identified pain management as an import focus for accreditation visits starting in 2000 and 2001. Standards have been developed for ambulatory care, behavioral health care, home care, health care networks, hospitals, long-term care, and long-term care pharmacies.[31] These new standards should help improve acute pain management in these settings.

There are numerous additional helpful references for acute pain management. These include the 1999 *AHSP Therapeutic Position Statement on the Safe Use of Oral Nonprescription Analgesics*,[32] and a review of cyclooxygenase-2 enzyme inhibitors and their role in pain management.[33] For patients with renal or gastrointestinal problems, a helpful discussion of the nonopioid considerations can be found in *Pain*.[34] A useful reference to support the development of acute pain management services is the 1994 article by Lewis and colleagues.[35] This article discusses the physiological consequences of acute pain and the importance of appropriate analgesic therapy.

There are many roles for pharmacists in the management of acute pain. The changes in the JCAHO standards provide an excellent opportunity for pharmacists to become active in the pain management movement. Policies, procedures, and institution-specific guidelines need to be developed and implemented. Pharmacists have essential knowledge of opioid medications, their half-lives, potency, and management of adverse effects. Unfortunately, there have not been many publications by pharmacists discussing their pain management practices. Therefore, pharmacists who successfully implement acute pain services should make every effort to publish their efforts and results.

Pharmacists should be involved in all steps of pain management. This includes the initial development of pain management policies and procedures; development of standardized pain assessment and documentation forms; development of standardized treatment orders, including recommendations to manage adverse effects; education of staff on the implementation of the new pain management standards; and quality assurance efforts to evaluate the policies and procedures.

Some institutions use pharmacists to round on all patients receiving PCA therapy. The patient's total usage over 24 hours is reviewed and changes are recommended when necessary. Other institutions have pharmacists on their pain management teams. These pharmacists see a mixture of both acute and chronic pain patients. Their roles are varied, but include recommendations of adjuvant medications to facilitate in pain relief, patient counseling, health care professionals, and patient education. Patient's who are being discharged should have pain management counseling and a plan developed for home management when necessary.

CHRONIC PAIN SERVICES

Malignant Pain

In March 1994, the AHCPR released a clinical practice guideline for the management of cancer pain.[6] This guideline builds on the previous WHO report published in 1990.[4] This guideline is an excellent reference containing complete information on patient work-up and assessment. Numerous tools for assessment are provided, including the Brief Pain Inventory (Short Form), visual analog scales, and the faces of pain scale. Discussions of nonopioid and opioid doses are included, in addition to conversion tables for switching a patient from one opioid agent to another. Other chapters focus on pediatric and geriatric therapy, and the use of these principles in HIV/AIDS patients.

The AHCPR guidelines recommend using the WHO three-step method for managing cancer pain. Patients should be started on a nonopioid pain medication, such as acetaminophen or a nonsteroidal antiinflammatory agent. If this does not relieve their pain, proceed to the next level of a weak opioid agent, such as codeine or hydrocodone, with or without a nonopioid. If this does not relieve their pain, proceed to the final level of a strong opioid, such as morphine, with or without a nonopioid.

The WHO published an update to their 1990 report on cancer pain relief in 1996.[5] This update presents the advances in pain management since the mid-1980s. This includes new drug entities and dosing recommendations. In addition, it includes a guide to opioid availability from around the world. The American Society of Anesthesiologists published their guidelines for cancer pain management in 1996.[36]

Numerous other publications are available in the area of cancer pain management.[6,37–43] The AHCPR guidelines are an excellent starting place for pharmacists interested in chronic pain management, regardless of it being due to malignant or nonmalignant conditions. These guidelines built on the original recommendations by WHO published in 1990.

Recently, the National Comprehensive Cancer Network (NCCN) published their practice guidelines for cancer pain.[18] This publication has numerous useful figures and tables outlining the process for cancer pain management and serves as an update to the 1994 AHCPR guidelines. The NCCN guidelines provide the best illustration of today's management of cancer pain.

The biggest impact of these guidelines is changing the basic approach to pain management. Instead of starting all patients out on the first level of pain management (nonopioid), the NCCN guidelines recommend matching the pain medication to the patient's pain level. The patient should undergo a comprehensive pain assessment, including rating their pain using the verbal 0 to 10 scale discussed previously. If their rating is between 1 and 3, the NCCN recommends starting them on a nonopioid; for a rating between a 4 and 6, then begin a short-acting opioid and titrate as needed; and for a rating between 7 and 10, rapidly titrate opioid. At any point, an adjunct analgesic may be added to the regimen.[18]

In addition, the NCCN guideline stresses the importance of beginning a bowel regimen in all patients started on opioids, the use of antinausea medications as required, and psychosocial support and educational activities. In addition, the effectiveness of a pain regimen can be reassessed at 24 to 72 hours, depending on the severity of their initial complaint. This guideline also provides a discussion of surgical or interventional pain management strategies.

A 1999 article discussing a pharmacist-based analgesic dosing service for cancer pain management provides useful information for pharmacists interested in starting an inpatient pain management service in oncology.[44] Northwestern Memorial Hospital has a 33-bed dedicated oncology unit that admits 15 to 20 patients per week from both attending and private physicians. On average, 44% of admitted patients have pain as a continuing problem. A bedside analgesic dosing service (ADS) was developed to educate rotating interns, residents, and fellows on appropriate pain management techniques. This service uses 40% (16 hours) of a full-time clinical pharmacist's hours with rounds being performed each weekday afternoon. The pharmacist is available throughout the day if a patient's pain requires immediate intervention. After-hours and weekend coverage is provided by the pharmacist staffing the oncology satellite pharmacy.

Upon admission, the cancer patient is asked to complete a survey to assess their current pain situation. A copy of the form is provided in the article. Every 8 hours, a nurse records three pain-intensity values on the bedside flow sheet. After a patient is seen and evaluated, recommendations are conveyed orally to the prescribing physician or documented in the progress notes. Verbal orders are obtained when rapid changes are needed to analgesic orders. An education series is provided every month by the ADS attending physician or pharmacist to cover basic approaches to management of cancer-related pain, equianalgesic dosage conversions, use of adjunct medication, and barriers to pain management.

During the first 3 years that this service was in place, 1029 patients with pain were seen by the ADS and 941 (91%) had some kind of interaction with the service's pharmacist, with an average of 3.5 recommendations per patient. These recommendations ranged from correcting medical omissions or incorrect dosing, to changing the route of administration or dosing schedule, to providing agents to manage constipation.

Bonomi and colleagues[45] published an excellent pharmacist guide to quality-of-life assessment in acute, chronic, and cancer pain. The various quality-of-life assessment tools are presented and discussed to help guide the pharmacists in documenting appropriate pain management of these patients. This information can also be used to document the importance of pharmacy involvement in pain management services. Additional helpful tools include two recently published scales, one of which assess symptoms related to malignancy[46] and the scale rating the amount of symptom distress[47] in cancer patients.

Several reviews have also been published regarding the management of pain caused by bone metastases. The most recent advances include the utilization of the bisphosphonates and new radioactive compounds.[48] The American Society of Clinical Oncology (ASCO) published clinical guidelines for the use of bisphosphonates in breast cancer in 2000.[49]

Nonmalignant Pain

The American Society of Anesthesiologists published their guidelines for chronic pain management in 1997.[50] The purpose of these guidelines is to optimize pain control, while minimizing adverse effects and costs, and enhancing functional abilities, physical and psychological well-being, and quality of life. The guidelines also review adjuvant analgesics that are available for pain management, in addition to opioid therapy.

A key component for management of chronic pain, whether malignant or nonmalignant, is the development of a multimodality and integrated treatment approach involving medical, physical, and behavioral therapies. This approach should include physicians (anesthesiologists, neurologists, rehabilitation), pharmacists, social workers, phycologists, and physical therapists.[51] When conventional therapies fail to provide adequate pain control, the patient may require interventional pain management.[52]

More recently, there has been a treatment shift in the management of chronic nonmalignant pain. Increasingly, physicians are including opioids in their management

plans for patients.[53–58] Studies have shown that nociceptive pain responds favorably to opioids, neuropathic pain responds reasonably well, and idiopathic pain syndromes do not seem to respond.[58] This change has led to increased concerns regarding opioid addiction. However, once publication showed that increased medical use of opioid analgesics to manage pain does not appear to contribute to increases in the health consequences of opioid analgesic abuse.[59]

Patients who have a past addiction history need to be closely followed when opioid medications are prescribed. In addition, the health care team may decide to employ an opioid contract in these patients. Whenever possible, these patients should be managed by a pain management clinic and only one physician should be allowed to prescribe opioids.[60]

Model guidelines for the use of controlled substances for the treatment of pain are available.[61] These guidelines recommend the following steps when evaluating the use of controlled substances for pain control: 1) evaluate the patient, 2) develop a treatment plan, 3) sign an informed consent/agreement for treatment, 4) review patient periodically, 5) seek consultation when appropriate, 6) maintain medical records, and 7) comply with controlled substance laws and regulations.

There are numerous publications of recommendations to treat various chronic pain problems. The purpose of this chapter is not to review the various treatments in pain management, but to provide an overview and references. Therefore, included in the following paragraphs is a brief listing of chronic pain conditions and treatments. These articles provide basic reference information for pharmacists interested in pain management.

Treatment recommendations for post herpetic neuralgia were published in 1996 by Kost and colleagues.[62] Several reviews on the treatment of back pain have been recently published.[56,63,64] Treatment guidelines for migraine were published in 1999.[65] In 1998, a review of the treatment of burn pain was published by Ulmer and colleagues.[66]

The APS published a comprehensive new guideline entitled *Guideline for the Management of Acute and Chronic Pain in Sickle Cell Disease* in August 1999. For more information about the guideline, write to the APS at 4700 W. Lake Avenue, Glenview, IL 60025-1485, or call 847-375-4715. The cost is $15 for each guideline with a discount on 10 or more copies.

The management of chronic, nonmalignant pain is another area ready for increased participation and education of pharmacists. This involvement should also include community pharmacists who see these patients on a monthly basis. Understanding the appropriate conditions for prescribing opioid analgesics is essential in preventing discrimination against this patient population. In addition, counseling patients on appropriate regimens to prevent opioid-induced constipation should occur in all patients receiving a opioid prescription.

ISSUES IN THE MANAGEMENT OF PAIN IN THE ELDERLY

There are special considerations in the management of elderly patients with malignant and nonmalignant pain. A major concern is the undertreatment of elderly patients in pain, based on a fear of increased toxicity. In 1998, Bernabei and colleagues[67] reported a retrospective study evaluating the adequacy of pain management in the elderly and minority cancer patients being admitted to nursing homes. Thirty eight percent of patients reported daily pain; however, only 26% of these patients were receiving morphine. Patients older than the age of 85 were even less likely to receive weak or strong opioids. In addition, 26% of the patients experiencing daily pain were receiving no medication for this complaint.

This is another area primed for active pharmacy involvement as many of these patients are in nursing homes. These patients are usually undergoing monthly medication reviews by pharmacists who can suggest appropriate dosages and combinations for the management of pain. An article by Shimp[68] estimated that 45–80% of nursing home residents and 20–50% of elderly in the community suffer from pain. This article discusses appropriate selection of therapy and the risks for adverse effects.

The elderly patient usually has multiple medical problems, in addition to pain. A helpful article was published in 1998 by Ruoff for practicing physicians.[69] This article discusses issue with the use of nonsteroidal antiinflammatory and opioid medications in the elderly and patients with comorbid conditions.

NEUROPATHIC PAIN MANAGEMENT

One of the most complex pain states to manage is neuropathic pain, whether it is due to cancer or other chronic illnesses including diabetes, multiple sclerosis, or prior nerve damage. Neuropathic pain may be constant or intermittent and is usually described as burning, stabbing, or lancinating. Patients may exhibit hyperalgesia (increased pain response to noxious stimuli) and allodynia (pain elicited by a non noxious stimuli;[43]). An example of hyperalgesia is an exaggerated pain response to a poke by a wooden cotton swab. An example of allodynia is an exaggerated pain response elicited by a gentle stroking of the skin or by clothing touching the area.

Treatment usually involves a combination of opioid and nonopioid agents, including antidepressants and anticonvulsants. Several excellent reviews have been published on this topic, and the reader is encouraged to read them.[39,43,69–81] These classes of agents are briefly discussed below.

Chronic pain patients tend to have concurrent depression; however, the antidepressants chosen may not have any pain-relieving properties. Antidepressants that affect one neurotransmitter in the brain, such as selective serotonin reuptake inhibitors have not appeared to be effective in the management of pain in clinical trials. Antidepressants that affect multiple neurotransmitters—namely, serotonin and norepinephrine—have been shown to be effective pain relievers.[82] Two published meta-analyses have shown that tricyclic antidepressants amitriptyline, desipramine, imipramine, and nortriptyline are the most effective treatment for the management of neuropathic pain.[70,78] These publications review the published clinical trial data for all agents available for the management of neuropathic pain.

For patients who are unable to tolerate a tricyclic antidepressant, the choice of agent should consider the patients underlying mental state. If the patient has concomitant depression, another antidepressant such as venlafaxine, nefazodone, or mirtazapine may be the appropriate choice. If the patient does not have concomitant depression, an anticonvulsant agent such as gabapentin may be an acceptable alternative.[73,74,76]

Pharmacists are in possession of unique knowledge regarding these adjunct medications and their mechanisms of action. They can serve as excellent references and educators for health care professionals. Pharmacists who are active in the management of pain can ensure the proper choice of adjunct medications.

CLINICAL PHARMACY PRACTICES IN PAIN MANAGEMENT

Practices in pain management vary greatly among pharmacists. The clinical practice depends greatly on the practice setting. Pharmacists working in acute care or hospital settings may be involved in developing guidelines and standard orders for the appropriate use of PCA pumps. Additional responsibilities may include daily monitoring of all patients on PCA pumps and making recommendations for treatment adjustments. Pharmacists may also round with an inpatient pain management team that provides consultations for pain patients.

Pharmacists in ambulatory settings may work in chronic pain clinics as part of the treatment team. These clinicians perform extensive medication reviews and provide alternatives when treatments fail. Pharmacists who work with diabetic patients may be responsible for managing neuropathic pain.

In the oncology setting, pain management practices vary greatly. The pharmacist may assist in the management of an acute pain crisis with a PCA pump. After an adequate analgesic level is reached, the patient will need to be converted to oral therapy. In other patients, initial regimens can be developed as outpatients. The patient is then followed closely and monitored with medication adjustments when necessary.

In home health care or hospice settings, the pharmacist may be involved in compounding special admixtures for patients. The use of subcutaneously continuous infusions of opioids is also increasing. Pharmacists who review patients' medical administration records in nursing homes can recommend appropriate therapies for pain management.

Regardless of the practice setting, pharmacy opportunities exist in the area of pain management. In most of these settings, patients can be followed and managed over the telephone. Other important issues for pharmacists include the treatment of adverse effects of pain medications, especially constipation.

Patient Assistance Programs

Another excellent area for pharmacists to become involved in pain management is in assisting those patients who are unable to afford their medications. All the pharmaceutical companies have patient assistance programs. These programs require a close relationship with a prescribing physician because the medications are usually shipped directly to the physician's office. There are several web sites devoted to patient assistance programs that are excellent places to start:

- http://www.needymeds.com/MainPage.html (accessed October 2001).
- http://www.cancersupportivecare.com/drug_assistance.html (accessed October 2001).

ORGANIZATIONS FOR PAIN MANAGEMENT

There are numerous organizations devoted to pain management. The decision on which one to join will be affected by the area of pain management in which you are involved. The American College of Clinical Pharmacy (ACCP) and the American Society of Health-System

Pharmacists (ASHP) both have pharmacy-specific groups devoted to pain management. Descriptions of these groups and enrollment procedures can be found at the organization web sites:

- ACCP: http://www.accp.com/index.html (accessed October 2001).
- ASHP: http://www.ashp.org/ (accessed October 2001).

In addition to pharmacy organizations for pain management, there are several national societies devoted to this cause. These organizations are multidisciplinary and provide an excellent resource for pharmacists who want to specialize in pain management. The area of pain management in which you want to specialize will once again affect the organization you choose to join:

- American Academy of Pain Management: http://www.aapainmanage.org/ (accessed October 2001).
- American Academy of Pain Medicine: http://www.painmed.org (accessed October 2001).
- American Chronic Pain Association: http://www.theacpa.org/ (accessed October 2001).
- American Pain Foundation: http://www.painfoundation. org/ (accessed October 2001).
- American Pain Society: http://www.ampainsoc.org/ (accessed October 2001).
- American Society of Addiction Medicine: http://www.asam.org/ (accessed October 2001).
- American Society of Anesthesiologists: http://www.asahq.org/homepageie.html (accessed October 2001).
- International Association for the Study of Pain: http://www.halcyon.com/iasp/ (accessed October 2001).
- National Chronic Pain Outreach Association: http://neurosurgery.mgh.harvard.edu/NCPAINOA.HTM (accessed October 2001).
- National Foundation for the Treatment of Pain: http://www.paincare.org/ (accessed October 2001).

WEB SITES FOR ADDITIONAL INFORMATION AND TOOLS FOR PAIN MANAGEMENT

Purdue Fredrick: http://www.partnersagainstpain.com/ (accessed October 2001).

Abbott Total Quality Pain Management: http://www.abbotthosp.com/prod/pain/pain.htm (accessed October 2001).

City of Hope Palliative Care Resource Center: http://www.cityofhope.org/mayday/ (accessed October 2001).

Mayday Upper Peninsula Project: http://www.painandhealth.org (accessed October 2001).

Medical College of Wisconsin Palliative Care Medicine Program: http://www.mcw.edu/pallmed/ (accessed October 2001).

OncoLink: http://cancer.med.upenn.edu/ (accessed October 2001).

Pain Net: http://www.painnet.com/ (accessed October 2001).

Project on Death in America: http://www.soros.org/death/index.htm (accessed October 2001).

Robert Wood Johnson Foundation: http://www.rwjf.org/main.html (accessed October 2001).

University of Texas Caner Pain Page: http://palliative.mdanderson.org/ (accessed October 2001).

Wisconsin Cancer Pain Initiative: http://www.wisc.edu/wcpi/ (accessed October 2001).

World Health Organization's Cancer Pain Release: http://www.medsch.wisc.edu/WHOcancerpain/ (accessed October 2001).

The web sites of pharmaceutical companies who manufacture medications to relieve pain should be searched regularly for continuing education programs for pharmacists and physicians in both acute and chronic pain management.

CONCLUSION

Even in the twenty-first century the adequate treatment of pain is lacking. There are several reasons for this, including lack of education of health care professionals on the basic principles and knowledge of pain management. Pharmacists can play a vital role in overcoming the undertreatment of pain. Roles include education of health care professionals, patients, and their families; development of clinical treatment pathways for acute and chronic pain; and monitoring and documenting outcomes of pain management, and providing guidance for improvement.

Pain management is a significant expectation of the JCAHO in their year 2000 surveys. Expectations not only include established treatment pathways, but also documentation of outcomes and patient satisfaction. Pharmacists, whether in acute or home health care, community, or ambulatory settings should strive to become leaders in the area of pain management.

REFERENCES

1. Baumann, T.J. Pain Management. In *Pharmacotherapy*, 4th Ed.; Dipiro, J.T., Talbert, R.L., Yee, G.C., Matzke, G.R., Wells, B.G., Posey, L.M., Eds.; Appleton and Lange: Stamford, Connecticut, 1999; 1014–1026.
2. Mersky, H. Pain terms: A list with definition and notes

on usage. IASP Subcommittee on Taxonomy. Pain **1979**, *6*, 249–252.

3. World Health Organization (WHO). *Cancer Pain Relief*, 1st Ed. World Health Organization: Geneva, 1986; 1–80.
4. World Health Organization (WHO). *WHO Expert Committee on Cancer Pain Relief and Active Supportive Care*, 1st Ed., 1990; 7–73.
5. World Health Organization (WHO). *Cancer Pain Relief*, 2nd Ed.; World Health Organization: Geneva, 1996; 1–63.
6. AHCPR. Management of cancer pain: Adults. Am. J. Health Syst.-Pharm. **1994**, *51*, 1643–1653, (July 1).
7. Weissman, D.E.; Joranson, D.E.; Hopwood, M.B. Wisconsin physicians' knowledge and attitudes about opioid analgesic regulations. Wis. Med. J. **1991**, *90* (12), 671–675.
8. Holdsworth, M.T.; Ralsch, D.W. Availability of narcotics and pharmacists' attitudes toward narcotic prescriptions for cancer patients. Ann. Pharmacother. **1992**, *26*, 321–326, (March).
9. Von Roenn, J.H.; Cleeland, C.S.; Gonin, R.; Hatfield, A.K.; Pandya, K.J. Physician attitudes and practice in cancer pain management. Ann. Intern. Med. **1993**, *119* (2), 121–126.
10. Elliott, T.E.; Murrary, D.M.; Elliott, B.M.; Braun, B.; Oken, M.M.; Johnson, K.M.; Post-White, J.; Lichtblau, L. Physician knowledge and attitudes about cancer pain management: A survey from the Minnesota Cancer Pain Project. J. Pain Symptom Manage. **1995**, *10* (7), 494–504.
11. Bressler, L.R.; Geraci, M.C.; Feinberg, W.J. Pharmacists' attitudes and dispensing patterns for opioids in cancer pain management. J. Pain Symptom Manage. **1995**, *3* (2), 5–20.
12. Doucette, W.R.; Mays-Holland, T.; Memmott, H.; Lipman, A.G. Cancer pain management: Pharmacist knowledge and practices. J. Pain Symptom Manage. **1997**, *5* (3), 17–31.
13. Furstenberg, C.T.; Ahles, T.A.; Whedon, M.B.; Pierce, K.L.; Dolan, M.; Roberts, L.; Silberfarb, P.M. Knowledge and attitudes of health-care providers toward cancer pain management: A comparison of physicians, nurses, and pharmacists in the state of New Hampshire. J. Pain Symptom Manage. **1998**, *15* (6), 335–349.
14. Weinstein, S.M.; Laux, L.F.; Thornby, J.I.; Lorimor, R.J.; Hill, C.S., Jr.; Thorpe, D.M.; Merrill, J.M. Physicians' attitudes toward pain and the use of opioid analgesics: Results of a survey from the Texas Cancer Pain Initiative. South. Med. J. **2000**, *93* (5), 479–487.
15. Weinstein, S.M.; Laux, L.F.; Thornby, J.I.; Lorimor, R.J.; Hill, C.S., Jr.; Thorpe, D.M.; Merrill, J.M. Medical student's attitudes toward pain and the use of opioid analgesics: Implications for changing medical school curriculum. South. Med. J. **2000**, *93* (5), 472–478.
16. Levy, M.H. Pain management in advanced cancer. Semin. Oncol. **1985**, *12* (4), 394–410.
17. American Pain Society Quality of Care Committee. Quality improvement guidelines for the treatment of acute pain and cancer pain. JAMA, J. Am. Med. Assoc. **1995**, *274* (23), 1874–1880.
18. Grossman, S.A.; Benedetti, C.; Brock, C.; Cleeland, C.S.; Coyle, N.; Dubé, J.E.; Ferrell, B.; Hassenbusch, S., III; Janjan, N.A.; Lema, M.J.; Levey, M.H.; Loscalzo, M.J.; Lynch, M.; Muir, C.; Oakes, L.; O'Neill, A.; Payne, R.; Syrjala, K.L.; Urba, S.; Weinstein, S. NCCN practice guidelines for cancer pain. Oncology (Huntingt) **2000**, *14* (11A), 135–150.
19. American Society of Anesthesiologists Task Force on Pain Management. Practice guidelines for acute pain management in the perioperative setting. Anesthesiology **1995**, *82* (4), 1071–1081.
20. American Society of Anesthesiologists Task Force on Pain Management. Practice guidelines for cancer pain. Anesthesiology **1996**, *84* (5), 1243–1257.
21. American Society of Anesthesiologists Task Force on Pain Management. Practice guidelines for chronic pain management. Anesthesiology **1997**, *86* (4), 995–1004.
22. American Pain Society. American Pain Society Quality Assurance Standards for Relief of Acute Pain and Cancer Pain. In *VIth World Congress on Pain*; Bond, M.R., Charleton, J.E., Woolf, C.J., Eds.; Elsevier Science Publishers: 1991; 185–189.
23. Acute Pain Management Guideline Panel. *Acute Pain Management: Operative or Medical Procedures and Trauma. Clinical Practice Guideline, Ed.*; Agency for Health Care Policy and Research Public Health Service, U.S. Department of Health and Human Services: Rockville, Maryland, 1992.
24. American Society of Anesthesiologists Task Force on Pain Management. Practice guidelines for acute pain management in the perioperative setting. Anesthesiology **1995**, *82* (4), 1071–1081.
25. Warfield, C.A.; Kahn, C.H. Acute pain management: Programs in US. hospitals and experiences and attitudes among U.S. adults. Anesthesiology **1995**, *83*, 1090–1094.
26. Smythe, M.; Loughlin, K.; Schad, R.F.; Lucarroti, R.L. Patient-controlled analgesia versus intramuscular analgesic therapy. Am. J. Health-Syst. Pharm. **1994**, *51*, 1433–1440, (June 1).
27. Triller, D.; Hamilton, R.; Briceland, L.; Waite, N.; Audette, C.; Furman, C. Home care pharmacy: Extending clinical pharmacy services beyond infusion therapy. [In Process Citation]. Am. J. Health Syst.-Pharm. **2000**, *57* (14), 1326–1331.
28. Blau, W.; Dalton, J.; Lindley, C. Organization of hospital-based acute pain management programs. South. Med. J. **1999**, *92* (5), 465–471.
29. Ruzicka, D.L.; Daniels, D. Implementing a pain management service at an Army medical center. Mil. Med. **2001**, *166* (2), 146–151.
30. Lin, C.-C. Applying the American Pain Society's QA standards to evaluate the quality of pain management among surgical, oncology, and hospice inpatients in Taiwan. Pain **2000**, *87*, 43–49.

31. http://www.jcaho.org/standards/stds.2001_mpfm.html (accessed October 2001).
32. American Society of Health-System Pharmacists. ASHP therapeutic position statement on the safe use of oral nonprescription analgesic. AJHP **1999**, *56*, 1126–1131 (June 1).
33. Noble, S.; King, D.; Olutade, J. Cyclooxygenase-2 enzyme inhibitors: place in therapy. Am. Fam. Physician **2000**, *61* (12), 3669–3676.
34. Cyer, B.; Dalgrn, P.H.; Ruoff, G.E.; Whelton, A. Non-narcotic analgesics: Renal and GI considerations. Pain **1998** (September). Available at www.pain.com.
35. Lewis, K.S.; Whipple, J.K.; Michael, K.A.; Quebbeman, E.J. Effect of analgesic treatment on the physiological consequences of acute pain. Am. J. Health-Syst. Pharm. **1994**, *51*, 1539–1554 (June 1).
36. American Society of Anesthesiologists Task Force on Pain Management. Practice guidelines for cancer pain. Anesthesiology **1996**, *84* (5), 1243–1257.
37. Levy, M.H. Pharmacologic treatment of cancer pain. N. Engl. J. Med. **1996**, *335* (15), 1124–1132.
38. Breitbart, W. Psychotropic adjuvant analgesics for pain in cancer and AIDS. Psychooncology **1998**, *7*, 33–45.
39. Grond, S.; Radbruch, L.; Meuser, T.; Sabatowski, R.; Loik, G.; Lehmann, K. Assessment and treatment of neuropathic cancer pain following WHO guidelines. Pain **1999**, *79* (1), 15–20.
40. Portenoy, R.; Lesage, P. Management of cancer pain. Lancet **1999**, *353* (9165), 1695–1700.
41. Chang, H.M. Cancer pain management. Med. Clin. North Am. **1999**, *83* (3), 711–736.
42. Cleary, J.F. Cancer pain management. Cancer Control **2000**, *7* (2), 120–131.
43. Regan, J.M.; Peng, P. Neurophysiology of cancer pain. Cancer Control **2000**, *7* (2), 111–119.
44. Lothian, S.T.; Fotis, M.A.; von Gunten, C.F.; Lyons, J.; von Roenn, J.H.; Weitzman, S.A. Cancer pain management through a pharmacist-based analgesic dosing service. Am. J. Health-Syst. Pharm. **1999**, *56*, 1119–1125, (June 1).
45. Bonomi, A.E.; Shikiar, R.; Legro, M.W. Quality-of-life assessment in acute, chronic, and cancer pain: A pharmacist's guide. J. Am. Pharm. Assoc. **2000**, *40* (3), 402–416.
46. Chang, V.T.; Hwang, S.S.; Feuerman, M.; Kasimis, B.S.; Thaler, H.T. The memorial symptom assessment scale short form (MSAS-SF) validity and reliability. Cancer **2000**, *89* (5), 1162–1171.
47. Cleeland, C.S.; Mendoza, T.R.; Wang, X.S.; Chou, C.; Harle, M.T.; Morrissey, M.; Engstrom, M.C. Assessing symptom distress in cancer patients. Cancer **2000**, *89* (7), 1634–1646.
48. Diener, K.M. Bisphosphonates for controlling pain from metastatic bone disease. Am. J. Health-Syst. Pharm. **1996**, *53*, 1917–1927, (Aug. 15).
49. Hilner, B.E.; Ingle, J.N.; Berenson, J.R.; Janjan, N.A.; Albain, K.S.; Lipton, A.; Yee, G.C.; Biermann, J.S.; Chlebowski, R.T.; Pfister, D.G. American Society of Clinical Oncology Guidelines on the role of bisphosphonates in breast cancer. J. Clin. Oncol. **2000**, *18* (6), 1378–1391.
50. American Society of Anesthesiologists Task Force on Pain Management. Practice guidelines for chronic pain management. Anesthesiology **1997**, *86* (4), 995–1004.
51. Gallagher, R.M. Treatment planning in pain medicine: Integrating medical, has, and behavioral therapies. Med. Clin. North Am. **1999**, *83* (3), 823–849.
52. Krames, E.S. Interventional pain management: Appropriate when less invasive therapies fail to provide adequate analgesia. Med. Clin. North Am. **1999**, *83* (3), 787–808.
53. Belgrade, M.J. Opioids for chronic nonmalignant pain. Postgrad. Med. **1999**, *106* (6), 115–124.
54. Savage, S.R. Opioid use in the management of chronic pain. Med. Clin. North Am. **1999**, *83* (3), 761–786.
55. McQuay, H. Opioids in pain management. Lancet **1999**, *353* (i9171), 2229–2234.
56. Marcus, D.A. Treatment of nonmalignant chronic pain. Am. Fam. Physician **2000**, *61*, 1331-1335, 1345-1336.
57. Portenoy, R. Current pharmacotherapy of chronic pain. J. Pain Symptom Manage. **2000**, *19* (Suppl. 1), S16–S20.
58. Collett, M.-J. Chronic opioid therapy for non-cancer pain. Br. J. Anaesth. **2001**, *87* (1), 133–143.
59. Joranson, D.E.; Ryan, K.M.; Gilson, A.M.; Dahl, J.L. Trends in medical use and abuse of opioid analgesics. JAMA, J. Am. Med. Assoc. **2000**, *283* (13), 1710–1714.
60. Fishman, S.M.; Bandman, T.B.; Edwards, A.; Borsook, D. The opioid contract in the management of chronic pain. J. Pain Symptom Manage. **1999**, *18* (1), 27–37.
61. Gilson, A.M.; Joranso, D.E. Controlled substances and pain management: Changes in knowledge and attitudes of state regulators. J. Pain and Symptom Manage. **2001**, *21* (3), 227–237.
62. Kost, R.; Straus, S.E. Postherpetic neuralgia-pathogenesis, treatment, and prevention. N. Engl. J. Med. **1996**, *335* (1), 32–42.
63. Biewen, P.C. A structured approach to low back pain. Postgrad. Med. **2000**, *106* (6), 102–114.
64. Rosomoff, H.L.; Rosomoff, R.S. Low back pain. Med. Clin. North Am. **1999**, *83* (3), 643–662.
65. Weitzel, K.W.; Thomas, M.L.; Small, R.E.; Goode, J.-V.R. Migraine: A comprehensive review of new treatment options. Pharmacotherapy **1999**, *19* (8), 957–973.
66. Ulmer, J.F. Burn pain management: A guideline-based approach. J. Burn Care Rehabil. **1998**, *19* (2), 151–159.
67. Bernabei, R.; Gambassi, G.; Lapane, K.; Landi, F.; Gastsonis, C.; Dunlop, R.; Lipsitz, L.; Steel, K.; Mor, V. Management of pain in elderly patients with cancer. JAMA, J. Am. Med. Assoc. **1998**, *279* (23), 1877–1882.
68. Shimp, L.A. Safety issues in the pharmacologic management of chronic pain in the elderly. Pharmacotherapy **1998**, *18* (6), 1313–1322.
69. Ruoff, G. Management of pain in patients with multiple health problems: A guide for the practicing physician. Am. J. Med. **1998**, *105* (1B), 53S–60S.

70. McQuay, H.J.; Tramèr, M.; Nye, B.A.; Carroll, D.; Wiffen, P.J.; Moore, R.A. A systematic review of antidepressants in neuropathic pain. Pain **1996**, *68*, 217–227.
71. Portenoy, R.K. Adjuvant analgesic agents. Hematol. Oncol. Clin. North Am. **1996**, *10* (1), 103–119.
72. MacFarlane, B.V.; Wright, A.; O'Callaghan, J.; Benson, A.E. Chronic neuropathic pain and its control by drugs. Pharmacol. Ther. **1997**, *75* (1), 1–19.
73. Wetzel, C.H.; Connelly, J.F. Use of gabapentin in pain management. Ann. Pharmacother. **1997**, *31*, 1082–1083, (Sept.).
74. Backonja, M.; Beydoun, A.; Edwards, K.R.; Schwartz, S.L.; Vivian, F.; Hes, M.K.; LaMoreaux, L.; Garofalo, E. Gabapentin for the symptomatic treatment of painful neuropathy in patients with diabetes mellitus: A randomized controlled trial. JAMA, J. Am. Med. Assoc. **1998**, *280* (21), 1831–1836.
75. Weinstein, S. Phantom limb pain and related disorders. Neurol. Clinic **1998**, *16* (4), 919–935.
76. Morello, C.M.; Leckband, S.G.; Stoner, C.P.; Moohouse, D.F.; Sahagian, G.A. Randomized double blind study comparing the efficacy of gabapentin with amitriptyline on diabetic peripheral neuropathy pain. Arch. Intern. Med. **1999**, *159* (16), 1931–1937.
77. Schwartzman, R.J.; Maleki, H. Post injury neuropathic pain syndromes. Med. Clin. North Am. **1999**, *83* (3), 597–626.
78. Sindrup, S.; Jensen, T.S. Efficacy of pharmacological treatments of neuropathic pain: An update and effect related to mechanism of drug action. Pain **1999**, *83*, 389–400.
79. Vaillancourt, P.D.; Langevin, L.H. Painful peripheral neuropathies. Med. Clin. North Am. **1999**, *83* (3), 527–542.
80. Woolf, C.J.; Mannion, J. Neuropathic pain: Aetiology, symptoms, mechanisms, and management. Lancet **1999**, *353* (9168), 1959–1964.
81. McCleane, G. The pharmacological management of neuropathic pain: A review. Anaesthesia On-Line **2000**.
82. Mays, T.A. Antidepressants in the management of cancer pain. Curr. Pain Headache Rep. **2001**, *5* (3), 227–236.

PHARMACY PRACTICE ISSUES

Patient Education/Counseling

P

Matthew Perri III
University of Georgia College of Pharmacy, Athens, Georgia, U.S.A.

Jamie Cristy
Solvay Pharmaceuticals, Atlanta, Georgia, U.S.A.

INTRODUCTION

Patient education and medication counseling are an integral part of the healthcare process. The two-way flow of information in these processes is important to improve the quality of care and patient outcomes, and to build patient–practitioner relationships. This article discusses how, through the processes of patient education and medication counseling, pharmacists can ensure that patients take their medications correctly and achieve the most beneficial outcomes of medication use. A variety of methods of communication and counseling skills are presented.

PATIENT EDUCATION

Patient education is a broad term that describes the process through which healthcare professionals attempt to increase patient knowledge of healthcare issues. Patient education can occur in a variety of environments from hospitals and long-term care institutions to physicians' offices, community pharmacies, and other ambulatory care facilities. Patient education may be verbal or written, performed on an individual basis or in groups, and provided directly to the patient or caregiver. Although there are many different types of patient education, the process uses basic communication and educational techniques to achieve its goals of better health and improved health outcomes.

Patient education should be initiated with the first patient contact and is developed with each patient interaction. The process provides for the exchange of information between the patient and the health practitioner. The information gathered is needed to assess the patient's medical condition to further design, select, implement, evaluate, and modify health interventions. The information provided to patients can assist them in making better health decisions. With this two-way flow of information, the process of patient education can assist health practitioners in improving patient outcomes as well as helping to establish or build patient–practitioner relationships.

MEDICATION COUNSELING

Medication counseling is a type of patient education strategy. It is designed to increase patient awareness and comprehension of information related to medication and its use. In general, good patient medication counseling should help the patient to understand why a medication is important in treating their condition, how to use the medication appropriately, and what to expect from the treatment. It is a process that should be highly individualized to meet the patient's specific informational and healthcare needs. Patient medication counseling can be accomplished in a variety of ways, including oral instructions, written information, and/or other forms of audiovisual teaching aids. Counseling may occur between the practitioner and the patient directly, or between the practitioner and the patient's caregiver.

Similar to patient education, medication counseling should be viewed as a process that can improve patient and practitioner knowledge of essential information used to improve health outcomes and quality of life for the patient. The kinds of information conveyed can range from providing very basic levels of information about medicines to very detailed patient-specific information designed to assist a patient with managing a specific medical condition.

Current medical practice relies more heavily on medications than in the past. For these medications to work to their full potential, patients need to take them correctly. Pharmacists are the final health professional contact for most patients receiving prescription medications. Because of this, the community pharmacist is charged with the duty to ensure that patients take their

Encyclopedia of Clinical Pharmacy
DOI: 10.1081/E-ECP 120006329

medications correctly. Pharmacists and other health practitioners can use patient medication counseling strategies to ensure that patients: 1) take the medication correctly; 2) deal effectively with problems or adverse effects that might arise from taking the prescription; 3) remain compliant with recommended therapies; and 4) know how to assess if the medication is working.

Patient Counseling Methods and Skills

The value and importance of patient medication counseling has been recognized in the literature[1] and by federal legislation (OBRA 90),[2] which has mandated the provision of certain information to patients. When this medication information is combined with the specific needs and perspectives of the patient (e.g., the patient's health beliefs; education level; emotional, physical, psychological, social, and cultural needs), effective medication counseling should result. The responsibility of the health practitioner (pharmacist) is to ensure that the information is appropriate to the patient and is effective in helping the patient to modify or maintain appropriate medication-taking behaviors. *Communication Skills in Pharmacy Practice* provides a complete overview of counseling skills, issues, and methods.[6]

In 1994, the United States Pharmacopeia (USP) established an Ad Hoc Panel on Medication Counseling Behavior Guidelines. This panel was a subgroup of the USP Consumer Interest/Health Education Advisory Panel. The work of the panel resulted in the development of patient medication counseling inventory delineating 35 behaviors that could be part of a patient counseling session.[3] The behaviors are divided into four groups that structure the counseling session into: 1) the introduction of the session; 2) the content of the session; 3) the process followed; and 4) the conclusion of the session. The behaviors (detailed descriptions are available at the USP web site[3]) include the following:

- Conducts appropriate counseling by identifying self and the patient or the patient's agent
- Explains the purpose of the counseling session
- Reviews patient records prior to counseling
- Obtains pertinent initial drug-related information (e.g., allergies, other medication, age)
- Warns patient about taking other medication, including over-the-corner drugs, herbals/botanicals, and alcohol, which could inhibit or interact with the prescribed medication
- Determines whether the patient has any other medical conditions that could influence the effects of this drug or enhance the likelihood of an adverse reaction
- Assesses the patient's understanding of the reason(s) for therapy
- Assesses any actual and/or potential concerns of problems of importance to the patient
- Discusses the name and indication of the medication
- Explains the dosage regimen, including scheduling and duration of therapy when appropriate
- Assists the patient in developing a plan to incorporate the medication regimen into their daily routine
- Explains how long it will take for the drug to show and effect
- Discusses storage recommendations and ancillary instructions (e.g., shake well, refrigerate)
- Tells patient when they are due back for a refill
- Emphasizes the benefits of completing the medication as prescribed
- Discusses potential (significant) side effects
- Discusses how to prevent or manage the side effects of the drug if they occur
- Discusses precautions (activities to avoid, etc.)
- Discusses significant drug–drug, drug–food, and drug–disease interactions
- Explains in precise terms what to do if the patient misses a dose
- Explores with the patient potential problems in taking the medication as prescribed (e.g., cost, access)
- Helps patient to generate solutions to potential problems
- Provides accurate information
- Uses language that the patient is likely to understand
- Uses appropriate counseling aids to support counseling
- Responds with understanding/empathic responses
- Presents facts and concepts in a logical order
- Maintains control and direction of the counseling session
- Probes for additional information
- Uses open-ended questions
- Displays effective nonverbal behaviors
- Verifies patient understanding via feedback
- Summarizes by acknowledging and/or emphasizing key points of information
- Provides an opportunity for final concerns and questions
- Helps patient to plan follow-up and next steps

This inventory of behaviors was designed to be relevant to a wide variety of counseling situations and can provide a framework for pharmacists and other health professionals in providing patient medication education and counseling.

It should be noted that, for any given patient counseling encounter, any combination of these items

may be included to focus attention on the specific needs of the patient and may also include verbal instruction, written information, or demonstration. The practitioner should keep in mind that the patient's ability to use written information effectively may be limited, depending on the nature of the written information. The USP Drug Information Advice for the Patient[7] provides an excellent resource for patient-oriented material, including pictograms for use with hearing- or language-impaired patients. Also, demonstration, such as with the use of inhalation aerosols or insulin injections, may be used to either educate or verify how a patient is using one of these products.

In addition to pharmacist–patient education, there are a variety of other resources available for use in patient teaching. These include company-prepared materials such as videotapes and audiotapes for patients, booklets, leaflets, and even periodic newsletters. For some disease states, pharmaceutical companies have formed virtual support groups where questions can be discussed online with various health professionals. In addition, there are texts prepared to assist the pharmacist in determining the most appropriate information for a patient encounter such as the *Patient Counseling Handbook*.[8]

Communication Skills and Counseling Methods

There are many different techniques that pharmacists and other health professionals may use to provide patient medication counseling. In addition to the information requirements of patient medication counseling, the communication skills employed by pharmacists are important to good patient education. In particular, skills that build rapport (proper introductions, learning and using patient names, talking in lay terms, etc.), good listening skills (summarizing, paraphrasing, and the use of empathic responses), appropriate body language (tone of voice, eye contact, posture, etc.), and interactive techniques (use of open-ended questions) are key to good pharmacist–patient communication. Structured counseling strategies that have been developed for pharmacists to enable them to provide better care are presented in the following paragraphs.

Interactive patient counseling

Interactive patient counseling methods[4,5] use open-ended questions to determine what the patient knows about the medical condition and its treatment. For example, by asking ''What did your doctor tell you this medication was for?'', the pharmacist can discern if the patient knows what condition is being treated. Other questions can be employed to assess the patient's knowledge of the medication directions, ''How did your doctor tell you to use this medication?'' and expected outcomes, ''What did your doctor tell you to expect.'' Once the existing knowledge of the patient is determined, the pharmacist can fill in any information gaps that are identified. Verifying the patient's understanding of the information communicated will then complete the counseling. Counseling in this manner can improve the efficiency of medication counseling because only information that the patient does not already know is provided. It also ensures and reinforces patient knowledge through the verification process.

The PAR technique

In medication counseling there are often times when patients can present with challenging or difficult to handle situations. These situations can arise from behavior on the part of the patient or the pharmacist. Some patient examples might include when a patient is in a hurry to get back to work, has a sick child, has a hearing impairment, or does not speak English well. On the pharmacists side of the process, challenging situations could arise as a result of a busy pharmacy, interruptions, or lack of confidence about the ability to counsel. These challenging situations may be the result of either functional (e.g., disability such as loss of hearing, a patient who is a wheelchair user) or emotional (e.g., a patient who is angry, in a hurry, or who is confused by their problem) barriers. Any barriers to communication that are present must be resolved before effective counseling can occur.

When confronted with a potential or apparent communication barrier, the goal of the pharmacist should be to react and respond to the patient in ways that defuse tension to effectively remove barriers. The PAR (Prepare, Assess, and Respond) technique,[4] originally developed in the field of psychology, involves preparing for each patient encounter, assessing situations before and after they arise, and responding to patients in a manner that removes any barrier to communication.

To effectively prepare for a patient encounter, a pharmacist examines the environment in the pharmacy setting, the prescription, the patient profile, as well as any personal knowledge of the patient to identify potential barriers to communication before they arise. Once a problem has emerged or been identified during a counseling session, the pharmacist must assess it accurately. To assess, pharmacists must identify possible emotional or functional barriers. This may require careful observation and attention to the patient to determine the exact nature of the problem.

Once the pharmacist has identified and assessed a problem that has been identified, it is necessary to respond to the patient in an appropriate manner. The PAR technique suggests that the use of empathic responses, reflecting patient feelings, and the use of probing questions to encourage patients to talk will help in resolving communication barriers. Pharmacists should keep in mind that the ultimate goal is to ensure that the patient takes the medication correctly. This can be accomplished through medication counseling only after any communication barriers have been removed. Acting in this manner, rather than reacting defensively to the patient, can open the lines of communication and facilitate patient education.

Counseling for compliance

A process referred to as the RIM (Recognize, Identify, and Manage) technique[4] structures a process for pharmacists to use to help patients solve compliance problems through patient education. To maximize patient compliance with recommended therapies, pharmacists can learn to recognize objective and subjective forms of evidence to determine if the patient has a potential problem with medication compliance. Some examples of objective evidence might include reviewing the patient profile to identify problems with the refill record, the presence of complex regimens, or patients with high-risk disease states. Subjective evidence, for example, might involve comments from patients expressing uncertainty about the effectiveness of their therapy, questions about the dose prescribed, or hesitation when asked how the medication may be working.

Using various forms of empathic or reflecting responses, universal statements, and probing questions, pharmacists can learn to identify the specific causes of noncompliance. For example, questions such as ''It sounds like you are unsure about how to take this medication'' (reflecting response), ''Tell me more about how this medication makes you feel dizzy'' (probing question), and ''Many patients will feel sleepy after taking this medication'' (universal statement) can be useful in determining the nature of the patient's problem.

Once a compliance problem has been identified, pharmacists can use patient education to help patients to manage their medication regimen better. For example, providing information to alleviate patient fears about side effects, affirming the need for a medication, assurance about benefits, or even helping the patient to plan a typical day to include taking their medication on schedule.

CONCLUSION

Patient education and medication counseling are an integral part of the healthcare process. The two-way flow of information in these processes is important to improve the quality of care and patient outcomes, and to build patient–practitioner relationships. Through medication counseling, pharmacists can ensure that patients take their medications correctly. Knowledge of a wide variety of communication and counseling skills is important to successful patient education. Specific methods are available to assist pharmacists in providing effective counseling and patient education.

REFERENCES

1. Hatoum, H.T.; Akhras, K. 1993 bibliography: A 32 year literature review on the value and acceptance of ambulatory care provided by pharmacists. Ann. Pharm. **1993**, *27*, 1106–1118.
2. Public Law 101-5008, S4401, 1927(g) November 5, 1990 and Fed. Regist., November 2, 1992; 57FR(212): 49397–49401.
3. www.usp.org.
4. Pfizer Corporation, New York, New York, USA and the US Indian Health Service.
5. Ekedahl, A. Open-ended questions and 'show and tell' a way to improve pharmacist counseling and patient's understanding of their medications. J. Clin. Pharm. Ther. **1996**, *21*, 95–99.
6. Tindal, W.N.; Beardsley, R.S.; Kimberlin, C.L. *Communication Skills in Pharmacy Practice,* 2nd Ed.; Lea and Febiger: USA, 1994.
7. *USP DI Advice for the Patient,* 16th Ed.; United States Pharmacopeial Convention: Rand McNally, Rockville, Maryland, 1996.
8. *Patient Counseling Handbook,* 3rd Ed.; American Pharmaceutical Association: Washington, DC, 1998.

Patient Satisfaction

Jon C. Schommer
University of Minnesota, Minneapolis, Minnesota, U.S.A.

Suzan E. Kucukarslan
Henry Ford Hospital, Detroit, Michigan, U.S.A.

INTRODUCTION

Patient satisfaction is a commonly used measure to help assess how well healthcare providers are meeting the needs of their patients. Donebedian states that satisfying patients is one important indicator of quality care, because it demonstrates the ability of the provider to meet expectations and values of the patient.[1] Industry reports on managed care organizations include patient satisfaction as one measure of service quality.[2–4] However, the science of measuring patient satisfaction is yet in its infancy.

WHAT IS SATISFACTION?

Oliver defines satisfaction as ''an individual's *judgment* about the extent to which a product or service provides a pleasurable level of consumption-related fulfillment.''[5] Stated simply, satisfaction results from an *evaluation* of a product or service that nets some emotional reaction. A judgment is made by an individual as to how well the service was provided, and this judgment results in pleasure if satisfaction occurs or displeasure if dissatisfaction occurs.

Satisfaction is not to be confused with dimensions of quality such as 1) quality of conformance or 2) quality of design. *Quality of conformance* is the consistency of how the product is manufactured or the service is delivered. For example, pharmaceutical tablets are manufactured to fit within FDA standards for dissolution. Each batch is tested to determine if tablets within that batch fit standards. *Quality of design* is how well the product or service meets the needs of consumers or how well the product or service compares with those of competitors in the eyes of the consumer. In the pharmaceutical tablet example, the manufacturer would evaluate the quality of design by comparing the clinical outcomes provided by the product to the desired outcomes of the patients. Such factors as returning to ''normal'' health or avoiding side effects would be used in the evaluation of quality of design. Finally, a pharmaceutical company must compare its product with those of competitors from the viewpoint of the consumer. The question to ask would be, ''Is my product perceived by consumers as safer and more effective than my competitor's product?''

In contrast to how quality is measured and evaluated, satisfaction is specific to the individual and is based on the perceptions and judgments of individuals. It follows, then, that *patient satisfaction* is an individual's judgment about the extent to which a healthcare product or service provides a pleasurable level of consumption-related fulfillment. In the recent literature,[6] satisfaction has been conceptualized in four ways: 1) performance evaluation; 2) disconfirmation of expectations; 3) affect-based assessment; and 4) equity-based assessment (Table 1).

SATISFACTION AS PERFORMANCE EVALUATION

Patient satisfaction can be viewed as a ''personal evaluation of healthcare services and providers.''[7] Ware and colleagues developed the Patient Satisfaction Questionnaire (PSQ) using a performance evaluation approach. In such an approach, service characteristics are divided into categories (e.g., interpersonal characteristics, accessibility). Then, individuals are asked to respond to each statement about the service on a Likert-type scale usually ranging from ''strongly agree'' to ''strongly disagree.''

Ware and colleagues viewed patient satisfaction as a multidimensional construct in which distinct features of care are assumed to influence individuals' attitudes toward providers and services, with each of these features possibly having a different effect on satisfaction. Much of the literature on patient satisfaction has focused on identifying salient characteristics of healthcare services for evaluation by individuals.[8]

Viewing patient satisfaction as performance evaluation is most useful for services that have characteristics in-

Encyclopedia of Clinical Pharmacy
DOI: 10.1081/E-ECP 120006349

Table 1 Conceptualizations of satisfaction

Conceptualization	Focus	Strengths	Weaknesses
Performance evaluation	Salient characteristics of a service.	Can evaluate specific characteristics of a service.	Characteristics are selected by the researcher and might not be reflective of individuals' views.
Disconfirmation of expectations	Cognitive appraisal of a service experience.	Provides an understanding of the psychological process of service evaluation.	Results are sensitive to the type and level of expectations used for the study.
Affect-based assessment	Emotional response to a service and resultant actions by the individual.	Allows the investigation of the emotional responses to services, especially in the absence of prior expectations.	Might be applicable to short-term evaluations but not to long-term evaluations.
Equity-based assessment	Fairness in what is gained compared with what it cost the individual.	Allows the investigation of the relationship between inputs and outputs of the individual and the service provider.	Assumes that fairness is the key determinant of satisfaction.

(From Ref. [6].)

dividuals can identify and understand. However, individuals might not have the expertise to assess unfamiliar or ambiguous services. In these instances, individuals might base their evaluations on their expectations of whether or not such a service should be offered to them or on how the service experience makes them feel.

SATISFACTION AS DISCONFIRMATION OF EXPECTATIONS

In his conceptual article, Oliver described satisfaction as a result of one comparing an expectation to the actual service experience. This gap between the expectation and service experience then can impact how one feels about the service experience or the individual's satisfaction with the service.[9] The disconfirmation of expectations model that Oliver describes in his paper has been tested and validated by various researchers.[10–14] This model states that individuals compare their experiences with the service to expectations. These expectations serve as a reference point.[9] Although many researchers have accepted this view of satisfaction, they have held different opinions about which expectations, or comparison standards, are most pertinent and about which interrelationships among variables are most important in the satisfaction process.[9,15]

SATISFACTION AS AFFECT-BASED ASSESSMENT

Satisfaction also has been viewed as a pleasurable response to a service encounter. This type of conceptualization takes into consideration that individuals with moderate expectations of a healthcare service will likely be indifferent if those expectations are met. However, individuals with high expectations of a healthcare service might be delighted if a healthcare provider met their high expectations. Thus, the focus in this conceptual framework is the degree of affective response and not only how well expectations were or were not met. Also, Kucukarslan, Pathak, and Segal showed that this conceptual view of satisfaction can be very useful in cases when individuals have no prior expectations of a healthcare service.[16] In the healthcare domain, individuals often do not know what to expect when experiencing a healthcare service. In these cases, how a person feels would be an appropriate way to gauge patient satisfaction because knowledge about the service and prior expectations about the service are absent.

SATISFACTION AS EQUITY-BASED ASSESSMENT

A fourth conceptualization of patient satisfaction is an equity-based assessment. This is a comparison of one's outcomes versus inputs with respect to someone else's outcomes versus inputs.[5,17] The resulting perception of fairness has a direct relationship to satisfaction. Individuals who perceive that a service provider has gained more than they have are likely to be less satisfied. Studies show that perceptions of fairness increased as the differences between inputs and outputs increased in favor of the individual.[18,19]

Such an approach to understanding patient satisfaction assumes that fairness is the key determinant of patient satisfaction. An understanding of the key inputs and outputs of a healthcare service are required as well. There are only a few examples of this approach in the healthcare literature,[16,19] but it could be quite useful in situations where equity is a primary consideration (e.g., satisfaction with insurance coverage for healthcare services).

TOOLS TO MEASURE PATIENT SATISFACTION

No single, standard conceptualization of patient satisfaction is applicable to all situations. When selecting a conceptualization of satisfaction, the investigator must identify the research question. For example, if the primary research objective is to compare service providers in terms of how well individuals rate specific aspects of service performance, an experience-based evaluation of performance would be appropriate. If the research is focused instead on the relationship between individuals' evaluation of a service and their future commitment to the service provider, a satisfaction measure that reflects evaluative and affective reactions by the individual would be appropriate.

Also, one should remember that patient satisfaction might not be the most relevant outcome for study. Performance, quality, loyalty, complaining behavior, trust, or some other outcome might be more useful for decision making, planning, implementation, and evaluation related to healthcare products and services. It is important to link the construct being measured to the problem or decision that the study results will help address.

Two popular patient satisfaction measures for which psychometric properties have been investigated and reported include 1) Ware et al.'s Patient Satisfaction Questionnaire (PSQ)[7,20,21] and 2) MacKeigan and Larson's Patient Satisfaction with Pharmacy Services Questionnaire (PSPSQ).[22,23]

Ware et al.'s Patient Satisfaction Questionnaire (PSQ),[7,20,21] and related versions, has become one of the most widely used measures of satisfaction with medical care.[24] The most recent version, called the Short-Form Patient Satisfaction Questionnaire (PSQ-18), can be obtained from RAND Corporation, Santa Monica, California.[24] This version contains 18 items that focus on seven distinct areas: 1) general satisfaction; 2) technical quality; 3) interpersonal manner; 4) communication; 5) financial aspects; 6) time spent with doctor; and 7) accessibility/convenience. Acceptable reliability and validity have been reported for this measure.[24] It constitutes a widely used and accepted measure of patient satisfaction with medical care.

A widely accepted example of patient satisfaction measurement that is specific to pharmacy services is MacKeigan and Larson's Patient Satisfaction with Pharmacy Services Questionnaire (PSPSQ).[22,23] Introduced in 1989,[22] MacKeigan and Larson patterned their measure after Ware et al.'s Patient Satisfaction Questionnaire (PSQ).[7] Eight dimensions of pharmacy services were used for the questionnaire: 1) explanation; 2) consideration; 3) technical competence; 4) financial aspects; 5) accessibility; 6) drug efficacy; 7) nonprescription products; and 8) quality of the drug product dispensed. Over time, this measure has been tested and revised. In 1994, Larson and MacKeigan[23] further refined their questionnaire into a 33-item measure reflective of seven dimensions: 1) explanation; 2) consideration; 3) technical competence; 4) finance; 5) accessibility; 6) product availability; and 7) general. Reliability and validity of this measure are supported,[22,23] and it has been used in the pharmacy services domain.[25,26]

CONSIDERATIONS FOR MEASURING PATIENT SATISFACTION

Because a number of conceptualizations of patient satisfaction can be used, the measurement of patient satisfaction must fit the context of the overall research process. A research process proposed by Churchill[27] involves six stages: 1) formulate the problem; 2) determine the research design; 3) design data collection method and forms; 4) select a sample and collect the data; 5) analyze and interpret the data; and 6) prepare the research report. Each stage is linked, and decisions made at one stage will affect decisions made at other stages.

When investigating patient satisfaction, one must also consider the requirements and limitations imposed by the research process itself. For example, the purpose of the research, potential users of the data, time and money available for the study, the population of interest, and the data collection methods will all influence the type of satisfaction measure determined to be the most appropriate. Limitations in these factors must be recognized, whether a new measure of patient satisfaction is being developed or an existing measure is being used.

Because the meaning attributed to a satisfaction measure by the investigator may not be the same as the meaning imputed to it by the respondents, a systematic process for developing measures should be followed that includes: 1) specifying the domain of the construct; 2) generating a pool of items and determining the format of the measure; 3) having the initial pool of items reviewed by experts; 4) considering inclusion of validation items;

5) administering items to a development sample; 6) purifying the measure; and 7) optimizing the practicality of the measure (see Ref. [28] for a useful summary of this process).

When using existing measures of satisfaction or when developing new measures for a specific research purpose, acquiescent responding can be a problem. Acquiescent responding is the tendency to agree with statements of an attitudinal nature regardless of the statement's content.[24] To overcome this problem, the use of a balanced multiple-item measure, asking several positively worded questions and several negatively worded questions, is crucial to the measurement of patient satisfaction. This method tends to cancel out any systematic tendency to simply agree with items contained in the questionnaire.[24]

Finally, when comparing satisfaction among groups of individuals who belong to different health plans or access different care providers within a health plan, it is important to control for case-mix. *Case-mix* refers to the different patient attributes that might be apparent in different settings.[24] To the extent that patients self-select or are assigned to different comparison groups in non-random ways, comparisons of satisfaction ratings may be biased by factors other than the provision of the health-care services in the compared groups. For example, relationships between health status and patient satisfaction have been reported.[24] Thus, it is often prudent to control for the effects of health status when comparing patient satisfaction among or between groups. Other factors might account for differences in satisfaction scores depending on the question being addressed in the study. Case-mix control variables such as age, gender, socio-economic status, disease severity, physical functioning, service expertise, or service provider familiarity also should be considered for inclusion in the study. Methods used to control or adjust for case-mix variables can then be used (e.g., restriction, stratification, matching, statistical adjustment). Johnson[24] has written a useful summary of these methods.

CONCLUSION

Patient satisfaction is an individual's judgment about the extent to which a healthcare product or service provides a pleasurable level of consumption-related fulfillment. In the recent literature, patient satisfaction has been conceptualized in four ways: 1) performance evaluation; 2) disconfirmation of expectations; 3) affect-based assessment; and 4) equity-based assessment.[6] Commonly used patient satisfaction measures include Ware et al.'s Patient Satisfaction Questionnaire (PSQ)[7,20,21] and MacKeigan and Larson's Patient Satisfaction with Pharmacy Services Questionnaire (PSPSQ).[22,23] However, no single standard measure of patient satisfaction is applicable to all situations. We suggest that an existing measure of patient satisfaction, with demonstrated reliability and validity, should be used if it fits the construct domain of the study. Whenever a new satisfaction measure is needed, its conceptualization should be defined carefully and a systematic process for developing a measure should be followed.

REFERENCES

1. Donebedian, A. The Definition of Quality and Approaches to its Assessment, Vol. 1. In *Explorations in Quality Assessment and Monitoring*; Health Administration Press: Ann Arbor, 1980.
2. *1997 Novartis Report on Member Satisfaction within Managed Care*; Novartis Pharmaceuticals Corporation: 59 Route 10, East Hanover, NJ 07936-1080, 1997.
3. Carey, R.G.; Seibert, J.H. A patient survey system to measure quality improvement: Questionnaire reliability and validity. Med. Care **1993**, *31* (9), 834–845.
4. Dearmin, J.; Brenner, J.; Migliori, R. Reporting on QI efforts for internal and external customers. J. Qual. Improv. **1995**, *21* (6), 277–288.
5. Oliver, R.L. *Satisfaction: A Behavioral Perspective on the Consumer*; McGraw-Hill: New York, 1997.
6. Schommer, J.C.; Kucukarslan, S.N. Measuring patient satisfaction with pharmaceutical services. Am. J. Health-Syst. Pharm. **1997**, *54*, 2721–2732.
7. Ware, J.E.; Snyder, M.K.; Wright, W.R.; Davies, A.R. Defining and measuring patient satisfaction with medical care. Eval. Program Plan. **1983**, *6*, 291–297.
8. Hall, J.A.; Dornan, M.C. What patients like about their medical care and how often they are asked: A meta-analysis of the satisfaction literature. Soc. Sci. Med. **1988**, *27*, 935–939.
9. Oliver, R.L. A cognitive model of the antecedents and consequences of satisfaction decisions. J. Mark. Res. **1980**, *17*, 460–469.
10. Woodruff, R.B.; Cadotte, E.R.; Jenkins, R.L. Modeling consumer satisfaction processes using experienced-based norms. J. Mark. Res. **1983**, *20*, 296–304.
11. Cadotte, E.R.; Woodruff, R.B.; Jenkins, R.L. Expectations and norms in models of consumer satisfaction. J. Mark. Res. **1987**, *24*, 305–314.
12. Tse, D.K.; Wilton, P.C. Models of consumer satisfaction formation: An extension. J. Mark. Res. **1988**, *25*, 204–212.
13. Yi, Y. The determinants of customer satisfaction: The moderating of role ambiguity. Adv. Cons. Res. **1993**, *20*, 502–506.
14. Schommer, J.C. Roles of normative and predictive expectations in evaluation of pharmacist consultation

services. J. Consum. Satisfac., Dissatisfac., Complain. Behav. **1996**, *9*, 86–94.

15. Spreng, R.A.; MacKenzie, S.B.; Olshavsky, R.W. A reexamination of the determinants of consumer satisfaction. J. Mark. **1996**, *60*, 15–32.
16. Kucukarslan, S.N.; Pathak, D.S.; Segal, R. The Relationship of Expectations and Perceived Equity with Consumers' Evaluation of Pharmaceutical Information Services. In *Presented at American Marketing Association 1996 Winter Educators' Conference. Hilton Head, South Carolina*; 1996; Feb. 5.
17. Oliver, R.L.; Swan, J.E. Equity and disconfirmation perceptions as influences on merchant and product satisfaction. J. Consum. Res. **1989**, *16*, 372–383.
18. Oliver, R.L.; Swan, J.R. Consumer perceptions of interpersonal equity and satisfaction in transactions: A field survey approach. J. Mark. **1989**, *53*, 21–35.
19. Pathak, D.S.; Kucukarslan, S.N.; Segal, R. Explaining patient satisfaction/dissatisfaction in the high blood pressure prescription drug market: An application of equity theory and disconfirmation paradigm. J. Consum. Satisfac., Dissatisfac., Complain. Behav. **1994**, *7*, 53–73.
20. Tarlov, A.R.; Ware, J.E., Jr.; Greenfield, S.; Nelson, E.C.; Perrin, E.; Zubkoff, M. The medical outcomes study: An application of methods for monitoring the results of medical care. JAMA, J. Am. Med. Assoc. **1989**, *262* (7), 925–930.
21. Marshall, G.N.; Hays, R.D.; Sherbourne, C.D.; Wells, K.B. The structure of patient satisfaction with outpatient medical care. Psychol. Assess. **1993**, *4*, 477–483.
22. MacKeigan, L.D.; Larson, L.N. Development and validation of an instrument to measure patient satisfaction with pharmacy services. Med. Care **1989**, *27*, 522–536.
23. Larson, L.N.; MacKeigan, L.D. Further validation of an instrument to measure patient satisfaction with pharmacy services. J. Pharm. Mark. Manage. **1994**, *8*, 125–139.
24. Johnson, J.A. Patient Satisfaction. In *Pharmacoeconomics and Outcomes: Applications for Patient Care*; American College of Clinical Pharmacy, 1997; 111–154.
25. Malone, D.C.; Rascati, K.L.; Gagnon, J.P. Consumers' evaluation of value-added pharmacy services. Am. Pharm. **1993**, *NS33* (3), 48–56.
26. Johnson, J.A. A Comparison of Satisfaction with Pharmacy Services between Mail and Traditional Pharmacy Patrons and an Evaluation of the Relationship of Health Status. PhD Dissertation; University of Arizona, 1996.
27. Churchill, G.A. *Marketing Research: Methodological Foundations*; Dryden Press: Orlando, Florida, 1995.
28. DeVellis, R.F. *Scale Development Theory and Applications*; Sage: Newbury Park, California, 1991.

Pediatric Dosing and Dosage Forms

Rosalie Sagraves
University of Illinois, Chicago, Illinois, U.S.A.

INTRODUCTION

The administration of medications to pediatric patients is in many ways difficult because health care providers and parents are faced with many challenges not experienced, or experienced to a lesser degree, than when medications are prescribed for and taken by adults. First, less information is available about the use of most medications for pediatric patients. In fact only about 20% of drugs marketed in the United States have labeling for pediatric use.[1] Milap Nahata, in a 1999 article on pediatric drug formulations, stated that ''only five of the 80 drugs most commonly used in newborns, and infants are approved for pediatric use''.[1] Second, many drugs that are used for some pediatric patients are not in appropriate dosage forms for use by children. This includes even some medications approved for use in pediatric patients. These issues have resulted in many questions that need to be answered about drug administration to pediatric patients. For example, is the drug approved for use in pediatric patients and in what age groups? If not approved, is there scientific information that enables us to determine whether the drug is safe and effective for pediatric patients of various ages? If the drug is available commercially for pediatric use, what dose should be administered and how frequently? What route should be used for administration, and what dosage form selected? If the drug is not available in an appropriate dosage form for childhood use, can it be prepared extemporaneously? Are there stability studies, palatability tests, clinical data in children, etc. that pertain to the extemporaneous formulation? How should the drug be monitored for effectiveness as well as for adverse effects? Information determined in adult medication studies may not be applicable to pediatric patients because of pharmacokinetic and pharmacodynamic differences as well as differences in disease states for which a particular drug might be used. Many questions about the use of particular drugs in various age groups of pediatric patients can only be answered through well-designed, randomized controlled studies in pediatric patients who need certain medications for particular health problems.

HISTORICAL BACKGROUND

In 1997, the Food and Drug Administration (FDA) proposed new regulations for how pharmaceutical manufacturers would access safety and efficacy of certain new drugs that could have pediatric indications.[1,2] Thereafter, the FDA and the American Association of Pharmaceutical Scientists (AAPS) held a conference with academicians, pharmaceutical industry representatives, and U.S. Pharmacopeia (USP) representatives to discuss these proposed FDA regulations.

The FDA Modernization Act (FDAMA) of 1997 contains within it financial incentives for the development and marketing of drugs that could be used for pediatric patients.[3] Some of these incentives include an extension of 6 months on market exclusivity and waiving fees for supplemental applications needed for receiving the approval of drugs for pediatric use that are already approved for adult use. In addition, the FDA published a list of drugs approved in adults for which additional pediatric data may produce health benefits for pediatric patients.[4] For drugs on this list, FDA may ask a pharmaceutical manufacturer why it has not sought approval of a particular drug for pediatric use.

Various medical and pharmacy organizations have worked hard throughout the years in their efforts to better educate children, parents, educators, and health care providers about the medications and their appropriate use. Indeed, individuals who help care for children may not be adequately trained to educate children about medications that they need to use. In June 2000, the USP developed three target initiatives that related to pediatric medication use: principles for educating children about their medications, guidelines for developing and evaluating information for children, and developing specific curricular information in a modular format.[5] The USP position

Encyclopedia of Clinical Pharmacy
DOI: 10.1081/E-ECP 120006418

about educating children about medications may be found on their website (www.usp.org). The following information pieces have been developed by the USP:[5]

- *Guide to Developing and Evaluating Medicine Education Programs and Materials for Children and Adolescents* (joint publication of the American Health Association and USP).
- *A Kid's Guide to Asking Questions about Medicines.*
- *Teaching Kinds about Medicines.*
- *Talking to Children about Their Medicines* (pamphlet developed jointly by Pfizer and USP to be disseminated to pediatricians and children's families).
- An Annotated Bibliography of Research and Programs Relating to Children and Medications.

The USP worked with the National Center for Health Education in New York to develop educational materials that can help school systems nationwide to know more about medications that students may need to take. The USP also adopted the following resolution to address the work that needs to be done in the area of health education:[5]

> Facilitate and contribute to the development of a rational school medicines policy, including guidelines for student, faculty, and staff medicine education, for acquisition, transport, storage, administration, use, and disposal of medicines; for protection of privacy; and for record-keeping in primary and secondary schools. Initiative should be undertaken in collaboration with appropriate partners.

The USP worked with the National Institute of Child Health and Human Development (NICHD) to develop a list of drugs for which more pediatric information is needed to insure proper use in children.

This overview of pediatric dosing and dosage forms covers issues that peditric health care providers face daily, such as age-related drug pharmacokinetic and pharmacodynamic changes that occur secondarily to physiologic changes in maturing neonates, infants, children, and adolescents that can affect drug absorption from various routes of administration as well as drug distribution, metabolism, and elimination. To be more knowledgeable about pharmacokinetic changes, therapeutic drug monitoring (TDM) must be undertaken for drugs with narrow therapeutic indexes and for those for which pharmacodynamic data (i.e., pharmacologic response that correlates to the drug concentration at the receptor site) correlates with pharmacokinetic information. Also addressed will be drug administration by various routes including intravenous (i.v.), oral (p.o.), intramuscular (i.m.), subcutaneous (s.c.), percutaneous, rectal, otic, nasal, ophthalmic, and inhalation. Another issue discussed is product selection for pediatric patients.

To better understand changes in drug disposition, the pediatric population needs to be categorized into various groups (Table 1) because children vary markedly in their absorption, distribution, metabolism, and elimination of medications. This occurs because neonates, infants, children, adolescents, and adults have different body compositions (i.e., as to their percentages of body water and fat) and have their body organs in different stages of development.

PEDIATRIC PHARMACOKINETICS AND PHARMACODYNAMICS

Effect of Developmental Physiologic Changes on Pharmacokinetics and Pharmacodynamics of Drugs

Rational pediatric pharmacotherapy is primarily based on the knowledge about a particular drug, including its pharmacokinetics and pharmacodynamics, that may be

Table 1 Pediatric age groups terminology

Terms	Definition
Gestational age	Time from the mother's last menstrual period to the time the baby is born; at birth, a Dubowitz score in weeks gestational age is assigned, based on the physical examination of the newborn
Postnatal age	Age since birth
Postconceptional age	Age since conception, i.e., gestational plus postnatal age
Neonate	First 4 weeks or first month of life
Premature neonates	Born at less than 37-weeks gestation
Fullterm neonates	Born between 37- and 42-weeks gestation
Postterm neonates	Born after 42-weeks gestation
Infant	1 month to 1 year of age
Child	1–12 years of age
Adolescent	12–18 years of age

modified by physiologic maturation of the child from birth through adolescence. Physiologic changes that occur can affect drug absorption, distribution, metabolism, and elimination. The most dramatic changes occur during the neonatal period.

Oral Absorption

Drug absorption from the gastrointestinal (GI) tract is dependent on patient factors, physicochemical properties of the orally administered drug, and the drug formulation. Patient factors that affect GI absorption include absorptive surface area, maturation of the mucosal membrane, gastric and duodenal pH, gastric emptying time, GI motility, enzyme activity, bacterial colonization of the GI tract, and dietary intake, including the specific gastric content status at the time when a medication is ingested.[6–8] Patient factors are influenced by rapid maturational changes that occur throughout early childhood, but which occur primarily during the first few months of life.

Most drugs are absorbed across the GI tract by passive diffusion, but a variety of drug physicochemical factors influence the extent of absorption. These factors include molecular weight, lipid solubility, ionization as well as disintegration and dissolution rates.[7] In addition, drug absorption may be dependent on the dosage form selected (e.g., a liquid, a tablet that may need to be crushed, or a sustained-release product), and the particular brand selected. For time-release preparations, the release characteristics must also be taken into consideration.

Gastric pH

When examining patient-specific factors such as gastric pH, which affect oral absorption, it should be noted that infants born vaginally who are at least 32-weeks gestation, usually have gastric pHs between 6 and 8 at birth.[7,8] Gastric pH then falls rapidly within a few hours after delivery to a pH of less than 3.[7,8] The initial gastric pH is alkaline compared to that of adults and results from the presence of amniotic fluid in the infant's stomach.[9,10] Thereafter, gastric pH remains acidic until approximately day 10, then a nadir in acid production occurs between days 10 and 30 of life. Then gastric acid production begins to increase, but gastric pH and maximal gastric output may not mirror that of adults on a per kilogram basis until after the neonatal period.[7]

Gastric emptying and gastrointestinal motility

Gastric emptying time in neonates, especially those less than 24 h of age, may be variable.[7] It may not reach adult levels until 6–8 months of age and may be associated with diet.[11,12] Gastrointestinal transit time may be prolonged and peristaltic activity unpredictable in young infants;[8,13] both appear affected by the feeding.[13] Lebenthal and colleagues noted that breast-fed infants, older than 45 days of age, had gastric transit times longer than 10 h while formula-fed infants had transit times less than 10 h.[14] It should also be noted that young infants have a propensity to reflux their gastric contents because of GI immaturity. All these factors affect the extent to which a drug may be absorbed.

Enzyme activity and microflora in the gastrointestinal tract

Pancreatic enzyme activity may be low at birth, but enzymes such as amylase, lipase, and trypsin develop to adult levels within the first year of life.[15] Premature infants appear to have lower amylase levels than do full-term infants. Low concentrations of pancreatic enzymes may be the reason why newborns have a decreased ability to cleave prodrug esters such as chloramphenicol palmitate.[7] Lipid-soluble drugs may not be well absorbed by neonates because of low lipase concentrations and bile acid pool.[8]

More information is needed about the microflora of the GI tract and its effect on drug absorption. In addition, the effects of various diets and antibiotic use can alter the microflora of the GI tract.[7]

Absorptive surface area

The surface area of the small intestine in young infants is proportionately greater than in adults. This physiologic difference may allow for increased drug absorption from the GI tract.

Intramuscular Absorption

When a child is unable to take a medication orally or the drug is unavailable for oral use, there may be a need to administer a drug parenterally by either the i.v. or i.m. route. Of these, the latter may be less desirable because of pain, irritation, and decreased drug delivery as compared to i.v. administration. Drug absorption after i.m. administration depends on various physicochemical and patient factors. Physicochemical factors to be considered include lipid or water solubility, drug concentration, and surface area. When addressing drug solubility, it should be noted that lipophilic drugs readily diffuse through the capillary walls of endothelial cells

whereas water-soluble drugs diffuse at fairly rapid rates from interstitial fluid to plasma via pores in capillary membranes.[16] A lipid-soluble drug may be more rapidly absorbed i.m., but a water-soluble drug may be more desirable because the drug must be stable in an aqueous solution until administered. After administration, the drug must then be water soluble at physiologic pH until absorption occurs.[16]

Drug absorption may be dependent on concentration, but available data do not allow us to determine whether an increased or decreased drug concentration results in better absorption. An increase in the osmolality of a pharmaceutical preparation secondary to the addition of another substance such as an excipient may decrease or slow down i.m. adsorption.[16] Absorption occurs more rapidly when diffusion involves a large area of muscle or the drug spreads over a large muscle mass. The massaging of an injection site after i.m. administration increases the rate of absorption.[16]

A physiologic determination of i.m. drug absorption is dependent on the adequacy of blood flow to muscle groups used for drug administration. Absorption rates differ at injection sites because blood flow varies among different muscle groups. For example, the absorption of a drug-administered i.m. in the deltoid muscle is faster than from the vastus lateralis that, in turn, is more rapid than from the gluteus.[16,17] This occurs because blood flow to the deltoid muscle is 7% higher than to vastus lateralis and 17% higher than to gluteal muscle groups.[18] Physiologic conditions that reduce blood flow to a muscle group may adversely alter the rate and/or extent of a drug-administered i.m. Decreased perfusion or hemostatic decompensation, frequently observed in ill neonates and young infants, may reduce i.m. drug absorption. Drug absorption may be adversely affected in neonates who receive a skeletal muscle-paralyzing agent such as pancuronium[16] because of decreased muscle contraction. A small muscle mass in neonates and young infants may also reduce the ability of a drug to be adequately absorbed.

The injection technique used may alter i.m. absorption. This was noted when needles of different lengths were used. The use of a longer needle (38 vs. 31 mm, 1 1/2 vs. 1 1/4 in.) for i.m. administration in adult patients resulted in higher diazepam serum concentrations.[18] This probably occurred because the drug administered with the shorter needle was actually administered s.c. rather than i.m.

Some drugs are absorbed more slowly after i.m. than oral administration; examples include diazepam, digoxin, and phenytoin. This probably occurs because these drugs require a mixture of alcohol, propylene glycol, and water for solubility, and they are insoluble in the muscle after i.m. administration.[18]

Complications associated with i.m. administration include nerve injury, muscle contracture, and abscess formation.[19] Less common problems include intramuscular hemorrhage, cellulitis, skin pigmentation, tissue necrosis, muscle atrophy, gangrene, and cyst or scar formation. In addition, injury may occur from broken needles and inadvertent injection into a joint or vein.[19]

Subcutaneous Absorption

The s.c. route is used for the administration of drugs such as insulin that require slow absorption. Injection technique and patient factors, such as fluid status and physical build, are important.[18] Exercise, elevation or warming of the injection site, or inadvertently administrating a drug i.m. rather than s.c. can increase absorption and be dangerous in some situations, such as hypoglycemia occurring in a diabetic patient from excessive insulin absorption.[18] Adverse effects that can occur secondarily to s.c. administration include tissue ischemia, sterile and nonsterile abscesses, lipodystrophy, cysts, and granulomatous formation.

Intraosseous Drug Absorption

If an i.v. line cannot be placed, the intraosseous drug administration route can be used for pediatric patients during, for example, cardiopulmonary resuscitation (CPR) because drug delivery by this route is similar to that for i.v. administration.[20] If drug or fluid delivery by this route is sluggish, a saline flush can be used to clear the needle. Intraosseous administration is used to deliver medications such as epinephrine, atropine, sodium bicarbonate, dopamine, diazepam, isoproterenol, phenytoin, phenobarbital, dexamethasone, and various antibiotics.[20]

Percutaneous or Transdermal Absorption

The percutaneous (transdermal or topical) route for systemic drug delivery is used infrequently for pediatric patients. Medications are typically applied to the skin for their local effect. In the future, this route may be used more frequently for systemic effects as more transdermal systems are developed for drug delivery.

The percutaneous absorption or the transdermal delivery of a drug occurs in the following manner. Initially a topically applied drug is absorbed into the stratum corneum and diffuses through that layer of skin into the epidermis and then into the dermis where drug molecules reach capillaries and enter the circulatory system. Diffusion through the stratum corneum is the rate-determining step unless skin perfusion is decreased. If the latter case, diffusion is controlled by the transfer of drug molecules into capillaries rather than by the diffusion process previously explained. Percutaneous or transdermal absorption[21,22] is affected by

- Patient age.
- Application site.
- State of hydration of the stratum corneum.
- Thickness and intactness of the stratum corneum.
- Physical characteristics of the solute, and
- Physical characteristics of the vehicle or solvent.

Drug diffusion may be explained by Eq. 1:

$$J = \frac{K_m \times D_m \times C_s}{\ell} \tag{1}$$

where J is flux, K_m is the partition coefficient, D_m is the diffusion constant under specific conditions such as temperature and hydration, C_s is the concentration gradient, and l is the length or thickness of stratum corneum.[21]

Lipid-soluble drugs are better absorbed into the stratum corneum than are water-soluble drugs, but the latter do not easily traverse the stratum corneum. Thus, lipid-soluble drugs are more likely to be stored in the stratum corneum, whereas water-soluble drugs are more likely to diffuse across the stratum corneum to the epidermis and dermis.[21]

Patient age

Drug absorption transdermally is not appreciably different in various age groups of patients except for neonates less than 32-week gestation at birth.[23] Drug absorption is increased in premature neonates, because the stratum corneum is not completely formed at birth. An example of increased drug absorption occurred in two premature neonates who were repeatedly washed with 3% hexachlorophene and developed encephalopathy secondary to drug absorption.[21] The absorption of the corticosteroid betamethasone valerate after topical application in children resulted in hypothalamus-pituitary-adrenal axis suppression. Children may have increased drug absorption from the percutaneous application of drugs not because of higher absorption rate but because of a greater topical application or a larger dose per kilogram. Examples of deaths in children from percutaneous drug absorption include those caused by salicylic acid and phenol absorption.[21] Toxicity has also been noted with the topical application of iodine and alcohol-containing products.[23]

Application site

The ability of a drug to be absorbed transdermally depends on the thickness of the stratum corneum. For example, absorption occurs more readily through abdominal skin than through skin on the plantar surface of the foot. Topical absorption may be enhanced from a particular site by the application of an occlusive dressing.

Status of the stratum corneum

Percutaneous absorption of a drug is enhanced by the hydration of the stratum corneum. Such hydration affects the absorption of hydrophilic drugs more than lipophilic drugs. Drugs will penetrate damaged skin more than intact skin. Skin damaged because of dryness will allow for increased drug penetration through areas where the skin is cracked or broken.

Solute

The penetration of the solute (or drug) depends on its polarity and on the polarity of the delivery vehicle.

Vehicle or solvent

Drug-delivery vehicles typically used for topical application include lotions, ointments, creams, emulsions, and gels. Substances such as emulsifiers may be added to the drug and vehicle to improve the texture of an emulsion, a stabilizer to preserve drug stability, the vehicle, or both, a thickening agent to increase viscosity, or a humectant to draw moisture into the skin.[21] It is particularly important to consider the vehicle and other additives when selecting a topical drug preparation for a neonate, especially when premature, because of the greater possibility of absorption of not only the drug but also other product ingredients. Toxic reactions have occurred in neonates from ingredients considered ''inactive.''

Transdermal Drug-Delivery Systems

Drugs chosen for delivery via a transdermal drug-delivery system must adequately penetrate the skin in such a way that the system determines the delivery rate that should be fairly constant.[21] In addition, the drug must not irritate or

sensitize the skin. It is hoped that in the future more drugs will be developed for transdermal delivery. This could become an alternative route for drug delivery to children who have difficulty with oral administration.

Endotracheal Absorption

The endotracheal (ET) route has been used to administer medications during CPR when other routes, such as the i.v. route, are unavailable. It provides rapid access as well as rapid drug absorption and distribution.[24] Some studies have shown that the time to reach peak absorption is similar to that for the i.v. route, but serum concentrations achieved were 10–33% of that achieved with i.v. administration, resulting in a weaker response. A depot effect has also been demonstrated for drugs such as epinephrine. Work needs to be done to determine the optimal dose by this route, drug-delivery vehicle, and the most effective delivery technique.

Rectal Absorption

The rectal route is used for local and systemic therapy for the following reasons:[25]

- Nausea or vomiting.
- Rejection of oral administration because of its taste, texture, etc.
- Upper GI disease that might affect absorption.
- Medication absorption affected by food or gastric emptying.
- Medication is readily decomposed in gastric fluid but may be stable in rectal fluid, and
- First-pass effect of high-clearance drug may be partially avoided.

Absorption from the rectum depends on various physiological factors such as surface area, blood supply, pH, fluid volume, and possible metabolism by microorganisms in the rectum. The rectum is perfused by the inferior and middle rectal arteries, whereas the superior, the middle, and the inferior rectal veins drain the rectum.[25] The latter two are directly connected to the systemic circulation; the superior rectal vein drains into the portal system. Drugs absorbed from the lower rectum are carried directly into the systemic circulation, whereas drugs absorbed from the upper rectum are subjected to hepatic first-pass effect.[25] Therefore, a high-clearance drug should be more bioavailable after rectal than oral administration. The volume of fluid in the rectum, the pH of that fluid, and the presence of stool in the rectal vault may affect drug absorption. Because the fluid volume is usually low compared to that in other areas of the GI tract, a drug may not be completely soluble. In addition, a variety of organisms colonize the rectum, and it is debated whether these organisms are involved in drug metabolism.[25] Absorption is also influenced by the dosage form used. For example, drugs are rapidly absorbed rectally from aqueous or alcoholic solutions, whereas absorption from a suppository depends on its base, the presence of a surfactant, particle size of the active ingredient(s), and drug concentration.[25] The following problems may be associated with the rectal route for drug administration:[25]

- Decreased absorption secondary to defecation of the rectally administered pharmaceutical product.
- Less adsorption rectally than orally because the absorbing surface area of the rectum is smaller.
- Dissolution problems for rectally administered medications because of lower fluid volume in the rectum than in the stomach, duodenum, etc.
- Microorganisms in rectum may cause degradation of some medications.
- Patient or parent acceptance.

Distribution

A drug is distributed by moving from a patient's systemic circulation to various compartments, tissues, and cells. Distribution depends on patient factors, drug physiochemical properties, and the route of drug administration. Patient factors that influence drug distribution or the volume of drug distribution (V_d) include body composition, perfusion, protein- and tissue-binding characteristics, and permeability.[7,8] Many of these characteristics are age dependent. Drug physiochemical properties that may influence distribution include molecular weight, pK_a, and partition coefficient.

Differences in body composition

Age-related changes in body composition can alter the V_d of a drug. At birth, 85% of the weight of a premature infant may be water, compared to approximately 75% as total body water (TBW) in a full-term infant.[13] Neonates have the highest percentage of extracellular water (65% of TBW in premature infants as compared to 35–44% in full-term neonates and 20% in adults). The intracellular water (ICW) is more stable throughout life (i.e., 25% in premature neonates, 33% in full-term neonates, and 40% in adults).[7] An infant's percentage of TBW approaches that of an adult male by 1 year of age (60% TBW); it reaches the same about the time of puberty or 12 years of age.[8] Women have a lower percentage of TBW (50%)

than men do because they have a higher concentration of body fat. Thus, neonates, because of their high TBW, have a higher V_d for water-soluble drugs such as aminoglycosides than older children or adults. For example, the V_d for an aminoglycoside such as gentamicin approximates that of extracellular cellular fluid volume, 0.5–1.2 L/kg for a neonate, but only 0.2–0.3 L/kg for an older child or an adult.[8]

Adipose tissue increases from as little as 0.5% in a premature infant to approximately 16% of body weight for a full-term infant.[26,27] Boys experience a spurt in body fat between the ages of 5 and 10 years, and then a gradual decrease in fat content until about 17 years of age; girls usually have a rapid increase in adipose tissue at puberty.[9] Thus, one would expect neonates and young infants to have a decreased V_d for lipid-soluble drugs. This has been noted for diazepam in neonates who have exhibited an apparent V_d of 1.4–1.8 L/kg compared to 2.2–2.6 L/kg in adults.[28]

Protein binding

Neonates have lower concentrations of various plasma proteins (e.g., albumin concentrations about 80% of those in adults) for drug binding, but the albumin present may also have a lower affinity for binding drugs than noted for adults who are receiving the same medications. This lower affinity for binding drugs may result in a competition for various albumin-binding sites with substances such as bilirubin. Plasma protein binding noted in adults is usually achieved in children by the age of 1 year.[13]

In neonates drugs such as various penicillins, phenobarbital, phenytoin, and theophylline have lower protein-binding affinity than in adults. This may increase the concentration of free or pharmacological active drug in neonates, and may also change the apparent volume of distribution. Thus, neonates may require different doses on a mg/kg basis compared to that for adults for these drugs to achieve appropriate therapeutic serum concentrations.

In addition to binding to plasma proteins in the neonate, some drugs such as sulfonamides may displace plasma bilirubin from binding sites. This may increase an infant's risk for developing kernicterus. The significance of drugs displacing bilirubin is controversial because bilirubin may have a greater affinity for albumin than drugs have.[29]

Tissue binding

The binding of drugs to various body tissues appears to vary with age; for example, digoxin binding to erythrocytes is higher in neonates than in adults. This may be due to the increased number of binding sties on neonatal erythrocytes.[30]

Drug penetration into the central nervous system

A drug is more likely to cross into the central nervous system (CNS) of a neonate rather than an older child or an adult. This most likely occurs because its CNS is less mature and the blood-brain barrier is less formed. This is an important consideration when antimicrobial therapy is needed for the treatment of bacterial meningitis or anticonvulsant for seizures.

Metabolism

Although drug metabolism can occur in various body organs including the lungs, GI tract, liver, and kidneys as well as in the blood, the liver is the primary organ for metabolism. Most drugs are metabolized from lipid-soluble parent compounds to more polar, less lipophilic metabolites that are more readily eliminated renally.[9]

Hepatic metabolism

Most drug metabolism occurs in the liver by phase I or phase II metabolic processes. Phase I reactions primarily biotransform an active drug to a more water-soluble compound that typically is inactive or has less activity than the parent compound. Oxidation, reduction, hydralysis, and hydroxylation are examples of phase I reactions.[6,8,16] Oxidation is primarily catalyzed by the cytochrome (CYP) P450 system that has a multitude of isozymes (at least 13 primary enzymes) with a multitude of isozymes of specific gene families.[6] It appears that isozymes CYP450 1A2, 2D6, 2C19, and 3A3/4 are involved in drug metabolism in humans.[6] Oxidizing enzyme systems appear to mature after birth so that by the age of 6 months, activity is similar to or even exceeds adult levels. More information about drugs affected by phase I reactions may be found in Ref. [6].

Phase II reactions (glucuronidation, sulfation, acetylation, and glutathione conjugation) usually involve the conjugation of active drugs with endogenous molecules to form metabolites that are more water soluble;[16] glucuronidation is the most thoroughly studied reaction. It is postulated that maternal glucocorticoids inhibit the development of glucuronyltransferase, the enzyme involved in glucuronidation in utero. After birth this metabolic system matures rapidly and reaches adult levels by the age of 2 years.[29]

Sulfate conjugation appears to be fully developed immediately prior to or at the time of birth. Infants and

young children readily sulfate acetaminophen; in adults the major metabolic route is glucuronidation.[31] Little is known about acetylation in neonates or infants. It is believed that neonates have an extremely low capacity for acetylation at birth, but this pathway matures at approximately 20 days of age.[29]

Theophylline is an example of a drug that is readily metabolized in neonates by N-methylation to caffeine (process not relevant clinically in older infants, children, and adults). It is also a compound that has pharmacologic activity versus apnea (like theophylline), but which may have toxicity when it is not readily metabolized by the liver, and its elimination is slowed by immature kidneys.[6]

Neonates require close monitoring if their mothers received enzyme inducers such as phenytoin, phenobarbital, carbamazepine, or rifampin during pregnancy or if they need one of these drugs themselves.[16] Examples of drugs that inhibit the metabolism of other medications include cimetidine, erythromycin, and ketoconazole.[16]

Renal Elimination

The kidneys are the major route for drug elimination, especially for water-soluble compounds or the metabolites of lipid-soluble drugs. Renal drug elimination is dependent on renal blood flow, glomerular filtration, and tubular secretion and reabsorption. These functions appear to mature at different rates in the neonate and infant. Full-term infants achieve renal blood flow similar to that of adults by the age of 5–12 months; glomerular filtration approaches adult values by the age of 3–5 months.[6] Premature neonates exhibit lower rates for glomerular filtration at birth than do full-term neonates, and more time is required of them postnatally to develop filtration ability.[32] This is probably due to their lack of as many functional nephrons at birth. Tubular function is less mature in the neonate at birth than is glomerular filtration, and it matures at a slower rate. Tubular function begins to approach adult values by 7 months of age. Renal function is equal to that of adults by 1 year of age.

Aminoglycosides (e.g., gentamicin, tobramycin, amikacin) and digoxin are drugs whose eliminations are affected by renal maturation. The renal elimination of aminoglycosides in neonates and young infants parallels the maturation of glomerular function and correlates with creatinine clearance.[33] The renal elimination of digoxin parallels kidney maturation. Dosage adjustment for this drug is necessary as renal function matures in neonates and young infants. In addition, older infants and children require higher mg/kg doses of digoxin than do adults to achieve the same serum concentrations. This may be due to decreased digoxin absorption or increased renal elimination.[8]

THERAPEUTIC DRUG MONITORING

Therapeutic drug monitoring should encompass the entire drug-use process including drug selection, product selection, administration route, patient age, appropriate dosing on a mg/kg or mg/m^2 basis, and monitoring serum concentrations when appropriate and observing the patient for optimal drug effect(s) and possible adverse drug events.

Important Differences in Pediatric Serum Drug Concentrations

For many drugs, especially for those with narrow therapeutic indexes, serum concentration ranges have been determined that correlate to minimum and maximum therapeutic effects as well as to the development of toxicity. Therapeutic serum concentration ranges for various drugs have been developed for adults, and these data have been applied to pediatric patients including neonates. Such data may be appropriate to monitor drug therapy in children, but possibly not in children of all ages or possibly not in children at all. For example, Painter et al.[34] noted that neonates need higher serum phenobarbital concentrations than do older children and adults to terminate seizures. Gilman et al.[35] observed that higher phenobarbital loading doses were needed to achieve serum concentrations in neonates that would reduce the occurrence of seizures. Thus, there may be a need for different serum concentration ranges for various drugs needed by different age groups of patients for a similar pharmacodynamic or therapeutic outcome.

Free serum concentrations, rather than total concentrations, of some drugs such as phenytoin may need to be monitored in some patients, including neonates, who have low serum albumin. Gilman has advocated the possibility of using individualized dosing and serum concentration range for pediatric patients because children, especially neonates, have rapidly maturing functions of various organs and changes in albumin for drug binding.[36]

Serum concentration monitoring of various drugs administered to pediatric patients may appropriately give information about the drug but not its metabolites. This may be a problem when children metabolize specific drugs differently than adults with resulting differences in metabolite concentrations or the presence of different metabolites. This has been noted when premature infants have been administered theophylline for central apnea. A major metabolite of theophylline in neonates is caffeine, although only small concentrations of this metabolite are noted in older children and adults.[37,38] Caffeine is effective in treating apnea, and thus may add to the effectiveness of theophylline. This may help explain why lower theophylline serum concentrations may be needed for apnea rather than asthma. In addition, the presence of the 4-en metabolite of valproic acid noted in the serum of infants and young children, but not adults, receiving this medication for seizures may be responsible for the hepatoxicity of this drug in young pediatric patients.[36,39]

Serum Drug Concentrations

Because of the cost associated with therapeutic drug monitoring, serum drug concentrations must be drawn appropriately to provide useful information. Drugs typically followed pharmacokinetically are those with narrow therapeutic indexes for which there is an association between pharmacokinetic and pharmacodynamic data or toxicity. For many drugs, especially for those administered orally, the determination of trough concentrations (serum concentrations obtained prior to administration) may be most appropriate. This eliminates differences in absorption rates that could influence peak concentrations (e.g., orally administered phenobarbital, phenytoin, carbamazepine, or valproic acid). Trough concentrations may be important for drugs such as digoxin that take time to distribute to tissue receptors in such a way that serum concentrations reflect pharmacodynamic effects. Peak concentrations are best used for determining toxicity and therapeutic effects of drugs with short half-lives.

Table 2 gives therapeutic serum concentrations and pharmacokinetic information for some drugs administered to pediatric patients.

Technical Factors

Sample size and timing of blood drawing for serum concentration determination

Because of the small blood volume and the small size of veins, it is technically difficult to draw blood from neonates, infants, and young children for therapeutic drug monitoring, and it is therefore important to determine the best drawing schedule. For example, when are peak and trough data needed compared to trough data only? For anticonvulsants administered orally or i.v., trough concentrations are needed, whereas for aminoglycosides it may be important to obtain both peaks and troughs.

DOSING REGIMENS

Drugs for pediatric patients should be dosed on a mg/kg or a mg/m^2 basis using information available for the patient's age group. In addition, the patient's renal and hepatic functions must be considered. The route for administration must be determined based on the severity of the illness, the availability of the medication for a particular route of administration, and whether the patient is able to take a medication orally.

The Bibliography succeeding the References at the end of this chapter contains a list of handbooks and other references that are useful sources of dosing information for neonatal and/or pediatric pediatrics. In addition, drug information centers in pediatric hospitals or university settings are another excellent resource for pediatric drug information.

EXCIPIENTS OR ADDITIVES IN MEDICATIONS

Pharmaceutical products may contain, in addition to the active or therapeutic agent(s), a variety of other ingredients that are termed inactive or inert that are categorized as excipients or additives (flavorings, sweeteners, preservatives, stabilizers, diluents, lubricants, etc.). The words inert or inactive are misnomers for some excipients because some have been shown to cause adverse effects. Neonates and young children are at risk for such adverse effects, because they may not be able to metabolize or eliminate an ingredient in a pharmaceutical product in the same manner as an adult. In addition, patients of various ages have experienced allergic reactions to excipients such as tartrazine dyes.

Benzyl alcohol is a preservative that may be present in multidose vials of bacteriostatic sodium chloride and bacteriostatic water for injection and pharmaceuticals available in multidose vials for parenteral use. An association between the presence of benzyl alcohol in solutions used for flushing intravascular catheters and to reconstitute medications and a gasping syndrome and deaths in neonates was first reported in the early 1980s.[40,41] The neonates also displayed clinical findings such as an elevated anion gap, metabolic acidosis, CNS depression, seizures, respiratory failure, renal and hepatic failures, cardiovascular collapse, and death. Those at highest risk were premature infants who weighted less

Table 2 Pediatric pharmacokinetic data of some medications[a]

Drug	Therapeutic serum concentration (μg/ml)	Bioavailability (for oral drugs) (%)	Plasma protein binding (%)	V_d (L/kg)	$t_{1/2}$ (h)
Carbamazepine	4–12	>70	40–90	1.5 (neonate) 0.8–1.9 (child)	8–25 (child) $t_{1/2}$ varies with multiple dosing
Clonazepam	20–80 ng/mL	>85	47–80	3.2 (child)	20–40 (child)
Ethosuximide	40–100	~100	0	0.6–0.7 (child)	24–36 (child)
Gentamicin	trough ≤2 peak 4–10	Not available	<30	0.4–0.6 (neonate) 0.3–0.35 (child)	3–11.5 (<1 wk) 3–6 (1 wk-6 mo) 1.2 (child)
Phenobarbital	15–40	80–100	40–60	0.6–1.2 (neonate) 0.7–1 (child)	45–173 (neonate) 37–72 (child)
Phenytoin	10–20	85–95	>90	1–1.2 (premature neonate) 0.8–0.9 (full-term neonate) 0.7–0.8 (child)	6–140 (<8 days)[b] 5–80 (9–21 days)[b] 2–20 (21–36 days)[b] 5–18 (child)[b]
Theophylline	5–15	Up to 100%, depending on the formulation	32–40 (neonate) 55–60 (child)	0.4–1 (premature neonate) 0.3–0.7 (child)	19.9–35 (neonate) 3.4 ± 1.1 (1–4 yrs)
Valproic acid	40–100 (150)[c]	100	>90[d]	0.2 (child)	23–35 (neonate) 4–14 (child)

[a]Age or stage of life in parentheses.
[b]Michaelis–Menton pharmacokinetics; $T_{1/2}$ varies with serum concentration.
[c]Upper end of the serum concentration range is not definitely established.
[d]May vary with serum concentration.
(Adapted from Sagraves, R. Epilepsy and other Convulsive Disorders. *In Pediatric Pharmacotherapy*, 2nd Ed.: Kuhn, R.J. (Ed.) University of Kentucky: Lexington, 1993; Taketomo, C.K.; Hodding, J.H.; Kraus, D.M. *Pediatric Dosage Handbook*; LEXI-COMP, Inc.: Hudson, OH, 2000; Kauffman, R.E. Drug Therapeutics in the Infant and Child. In *Pediatric Pharmacology: Therapeutic Principles in Practice*. Yaffe, S.J.; Aranda, J.V., (Eds.); W.B. Saunders Co.: Philadelphia, 1992, 212–219; A. Rane, Drug Disposition and Action in Infants and Children. In *Pediatric Pharmacology: Therapeutic Principles in Practice*. Yaffe, S.J.; Aranda, J.V.; (Eds.); W.B. Saunders Co.: Philadelphia, 1992, 10–19.)

than 1250 g at birth.[40–42] In a study by Benda et al., premature neonates who survived benzyl alcohol administration were compared to neonates born after the use of benzyl alcohol-containing flush solutions was discontinued.[43] They noted that survivors had a higher incidence of cerebral palsy (50%) compared to infants who did not receive benzyl alcohol flushes (2.4%) ($P < 0.001$). In addition, the incidence of cerebral palsy and developmental delay was 53.9% versus 11.9% in the two populations ($P < 0.001$). The cause is probably associated with benzyl alcohol use and the inability of neonates, especially those who are premature, to adequately metabolize benzyl alcohol.[44] The American Academy of Pediatrics,[45] the Centers for Disease Control,[46] and the FDA[47] recommend that the administration of products containing benzyl alcohol be avoided in infants. Preservative-free i.v. flush solutions are recommended.[45–47]

Initially, it was believed that benzyl alcohol was only toxic in neonates who received doses greater than 99 mg/kg,[42] but it has been suggested that lower doses may be toxic, resulting in kernicterus and intraventricular hemorrhages.[48,49] Therefore, pharmaceutical preparations and fluids containing benzyl alcohol should be avoided in premature neonates.

Benzoic acid and sodium benzoate are added in low concentrations to various pharmaceutical preparations as bacteriostatic and fungistatic agents. Hypersensitivity reactions to benzoates have occurred when administered to allergic patients, such as those with asthma, those who do not tolerate aspirin, and those with a history of urticaria.[44] Hyperbilirubinemia and systemic effects attributed to benzyl alcohol may occur in premature neonates because benzyl alcohol is metabolized to benzoic acid.[44]

Propylene glycol is found as a solvent in some i.v. multiple vitamin preparations and a variety of pharmaceutical preparations for parenteral administration including phenytoin, digoxin, and diazepam. MacDonald et al.[50] noted that neonates who received MVI-12 (propylene glycol dose of approximately 3 g/day) versus those who received MVI concentrate (propylene glycol dose of approximately 300 mg/day) had a significant increase in seizures. In addition, infants in the first group suffered from hyperbilirubinemia and renal failure. (Although MVI concentrate is no longer on the market in the United States, it was used in the early 1980s.)

Serum hyperosmolality has been reported in infants who received vitamin preparations,[51] and in burn patients due to the topical absorption of propylene glycol-containing products.[52,53] In addition, burn patients have experienced metabolic acidosis with a high anion gap, decreased ionized calcium concentrations, acute renal failure, and death from topical propylene glycol absorption.[44,54] Problems associated with the oral ingestion of propylene glycol-containing products by children include CNS depression, seizures, and cardiac dysrhythmias.[55] Hypotension, cardiac dysrythmias, respiratory depression, and seizures have occurred after the rapid administration of phenytoin that may be associated with the propylene glycol.[56]

The American Academy of Pediatrics Committee on Drugs recommends that medications intended for pediatric use be ethanol free.[57] If, because of stability or solubility problems with the active ingredients(s), liquid medications need ethanol as an ingredient, but they should not contain more than 5% v/v ethanol.[57] The Academy also recommends that the ingestion of a single dose of an ethanol-containing product by a pediatric patient should not result in blood ethanol concentrations greater than 25 mg/100 mL, the volume of a packaged liquid medication should be of a minimal amount so that its entire ingestion would not result in a lethal dose and safety closures should be on all medicinals containing greater than 5% v/v ethanol. In addition, the Academy suggests that children under 6 years of age who need an ethanol-containing OTC preparation be under medical supervision and that doses of any ethanol-containing product be spaced at intervals to avoid ethanol accumulation.[57]

The Academy of Pediatrics made their recommendations concerning ethanol exposure from medications based on potential acute and chronic ethanol-related problems. Acutely, the coadministration of ethanol may alter drug adsorption or metabolism, and may result in drug interactions (e.g., increased sedation when taken with sedatives). Disulfiram-like reactions have occurred after the ingestion of an alcohol-containing medication or when an ethanol-containing product is used in conjunction with medications such as metronidazole, sulfonamides, or chloramphenicol.[57] The CNS effects (muscle incoordination, a longer reaction time, behavioral changes) are the most commonly reported acute adverse reactions associated with ethanol ingestion. Such reactions have occurred with blood ethanol concentrations in the range of 1–100 mg/100 ml.[57] Lethal ethanol doses in children occur at approximately 3 gm/kg although deaths due to ethanol-induced hypoglycemia have occurred at lower doses or because of interactions with other medications.[57,58] Chronic ethanol exposure may induce hepatic enzymes, and may thus alter the clearance of drugs such as phenytoin, phenobarbital, and warfarin.[59] Examples of other additives that have been problematic in pediatric patients include lactose,[55] tartrazine dyes[44,55] and sulfites.[55]

It is therefore important for health care professionals, and especially those who are responsible for selecting and administering medications to premature neonates, to examine pharmaceutical preparations for the presence of inactive ingredients as well as for the active drug. The provision of medications should be based on choosing the safest preparations possible. Various brands of medications should be compared to ensure that products without hazardous excipients. In the hospital setting, pharmacy and therapeutics committees and the pharmacy department play important roles in this process because they compare pharmaceutical preparations for formulary selection. In the outpatient setting, physicians and pharmacists must responsibly select the most appropriate brand of a particular medication. Kumar et al.[60] recommend that labeling for pharmaceutical products should include the names and the amounts of excipients as well as active ingredients to help health care professionals select appropriate drug products for neonates.

INTRAVENOUS ADMINISTRATION

Without being properly instructed about methods used for administering i.v. medications to pediatric patients, health care personnel may give a medication incorrectly, resulting in an inappropriate or unexpected therapeutic response. Therefore, it is important that health care personnel (nurses, physicians, pharmacists) understand how medications are administered by this route.

The i.v. route is most frequently chosen for medication delivery when a patient's clinical condition requires that a medication be administered by the most expeditious and complete method possible. In addition, some drugs are only available for i.v. administration. Although this route is the most reliable for drug delivery to the systemic circulation, problems can occur that reduce and/or delay medication delivery because of the product selected, dosage volume needed, or frequency of administration, but problems can also be associated with the i.v. delivery system used. The latter occur most frequently when a small medication volume is administered at a slow rate as is often needed for a neonate or young infant. A brief discussion of problems associated with i.v. drug delivery to pediatric patients is given here. A thorough overview of i.v. drug administration to pediatric patients is provided in Refs. [61] and [62].

Disposable IV Equipment, Effects on Drug Delivery

Infusion rates and location of injection sites

The first article to explore problems that can occur with i.v. drug delivery to pediatric patients was published in 1979 by Gould and Roberts.[63] They demonstrated in their study using an in vitro system for drug administration (Fig. 1) that infusion rates as well as the location of the injection sites in the i.v. infusion system influence the infusion profile of i.v. administered medications. Fig. 1 shows the effects of different i.v. fluid rates on the length of time to infuse 95% of a gentamicin dose administered at various sites in the infusion system.[63] It was reported that at a slow infusion rate of 3 ml/h, a drug takes longer to be infused and that the time for infusion time depends on the site of administration (i.e., the further the drug injection from a patient, the longer to administer 95% of the medication, see Fig. 2). Thus, it took approximately 400 min to infuse gentamicin at an infusion rate of 3 ml/h via a Y-site in the administration system. However, the same drug administered at a butterfly injection site at the same fluid flow rate reduced the length of time to administer 95% of the drug to less than 20 min. Gould and Roberts also stated that the time needed for drug administration in their i.v. system was longer than what had been expected.[63]

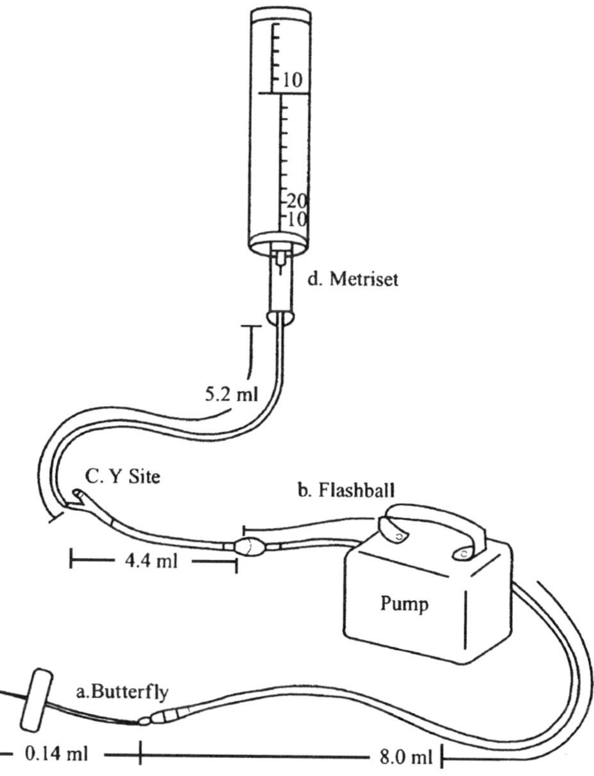

Fig. 1 Intravenous administration system used by Gould and Roberts. (From Ref. [63].)

Type of injection site

Leff and Roberts[61] demonstrated that the amount of drug received by a pediatric patient and the drug-delivery rate are influenced by the type of injection site (Y-site, T-type, T-connector, stopcock, etc.) and the volume (dead space) contained in the particular site. For the delivery of small dosage volumes (less than 1 ml) i.v. tubing should have microinjection sites that prevent a drug from being sequestered in the injection site. In addition, the amount of i.v. fluid needed to adequately flush microinjection sites to clear the medication would be less than needed to flush injection sites found on tubing used to administer drugs to adults.

Fluid flow dynamics

Another characteristic of i.v. tubing that affects drug delivery is fluid flow dynamics. It appears that flow in i.v. tubing is best characterized by laminar flow, and the radius of the tubing. Poiseuille's law describes flow in i.v. tubing as

$$q_v = \frac{Pr^4}{4nL} \tag{2}$$

where q_v is the volumetric flow rate, P is the pressure change in the i.v. tubing, r is the radius of the i.v. tubing, n is the viscosity of the fluid, and L is the length of the i.v. tubing.

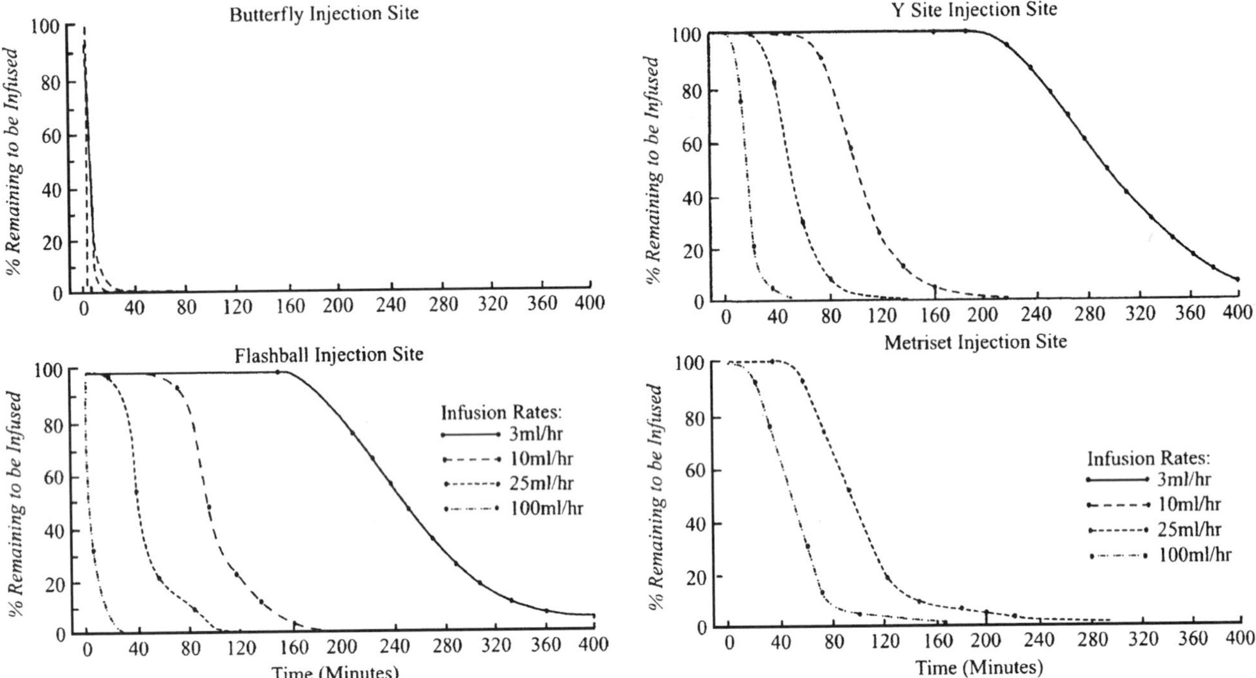

Fig. 2 Influence of i.v. flow rate on the infusion profile of gentamicin. (From Ref. [61].)

Thus, for i.v. delivery to pediatric patients, microbore tubing with an intraluminal diameter of <0.06 in. should be used rather than macrobore tubing. The use of microbore tubing allows the use of longer tubing lengths without significantly increasing delivery time.

Filters

A filter, especially one with a large reservoir volume, may prolong and/or reduce drug delivery. This occurs if the drug and its diluent are of different densities and there is a layering out of the drug in the filter.[64] Therefore, a filter with a smaller reservoir volume should be selected.

Drug and Fluid Considerations for Intravenous Drug Administration

Characteristics of the drug and the fluid such as drug volume, osmolality, pH, and density may affect i.v. drug delivery. The frequency and duration of drug administration is also important as is the need for the infusion system to handle multiple drugs. This may lead to drug incompatibilities and problems in medication scheduling.

Osmolality and pH

Osmolality and pH must be considered when preparing a drug solution for i.v. administration to pediatric patients. Problems such as tissue irritation, pain on injection, phlebitis, electrolyte shifts, and even intraventricular hemorrhages in neonates have been associated with the administration of drug solutions with high osmolalities.[65] Drug solutions should have osmolalities similar to serum osmolality, if possible. To control the osmolality, a drug can be diluted with a vehicle selected for i.v. infusion via a syringe infusion system[65] or the i.v. flow rate can be adjusted to achieve a particular drug-vehicle osmolality.[61]

Density

If the density of a drug is significantly different from that of the diluent, the drug may layer out on the filter or in the i.v. tubing. The latter occurs more frequently if macrobore tubing, a low flow rate, the i.v. system, or if the tubing is in a particular position. A density problem can be avoided by using microbore tubing which promotes mixing; this is especially important when i.v. flow rates are low, as are needed for neonates or young infants.

Frequency and Duration of Drug Administration; Multiple Drugs

To ensure that frequent doses are administered at appropriate intervals or that multiple drugs are administered to avoid drug incompatibilities, a syringe infusion pump can be used to administer drug volumes over a specific length of time. This helps avoid a situation where part of a drug dose is left in the tubing when the i.v. set is changed, as has been reported for manual administration techniques. More than one syringe pump can be used to simultaneously administer compatible drugs in a parallel system into a micro-Y-site or stopcock.

Types of Intravenous Administration

Drugs may require i.v. administration as continuous infusions or at intervals (q4h, q6h, q12h, etc.). Manual methods require the administration of the drug into the i.v. system at an injection site (Y-site, T-connector, stopcock, etc.), added to the i.v. solution in a mixing chamber, or added to an i.v. bag to be administered via gravity. A syringe pump or another mechanical device may be used for drug administration.

Manual administration

Manual administration is not as accurate as using a syringe pump for drug administration. It has been used for small volumes of medication (<3 ml), a low flow rate (<20 ml/h), or if the antegrade (forward toward the patient) injection of a drug bolus is safe.[61] If a medication is to be administered antegrade, it should be administered slowly into a microinjection site toward the patient; microbore tubing should be placed between the injection site and the patient to reduce the time to get the drug to the patient. Leff and Roberts[61] recommended that the volume of the drug to be injected by the antegrade technique should be a smaller volume than the tubing fluid volume between the injection site and the patient. If the medication volume is too large to be safely given by antegrade administration, but the i.v. fluid flow rate is low (<20 ml/h), the drug may be administered by a retrograde technique (Fig. 3).

Mechanical system for drug administration

If a mechanical system is chosen for drug administration, the appropriate infusion device must be selected based on its operating mechanism, flow accuracy, flow continuity, and ability to detect occlusions. Other important factors include an alarm system, ease of operation, ability to be cleaned easily, and safety from children inadvertently trying to change pump settings. A syringe pump is best for delivering small dosage volumes and when intermittent intervals are needed for medications. It is the mechanical device most often selected for medication administration because it can be used for intermittent administration of small and large doses, or for the continuous infusion of medications at low rates. A drug can be administered separately from the primary i.v. fluid flow rate, with the drug and the fluid mixing for a short distance therefore in microbore tubing (see Fig. 4) before reaching the patient. In addition to being able to more accurately deliver medications than by manual methods, syringe pump systems have the advantage of being able to separate the administration of incompatible drugs, reduce difficulties associated with the administration of

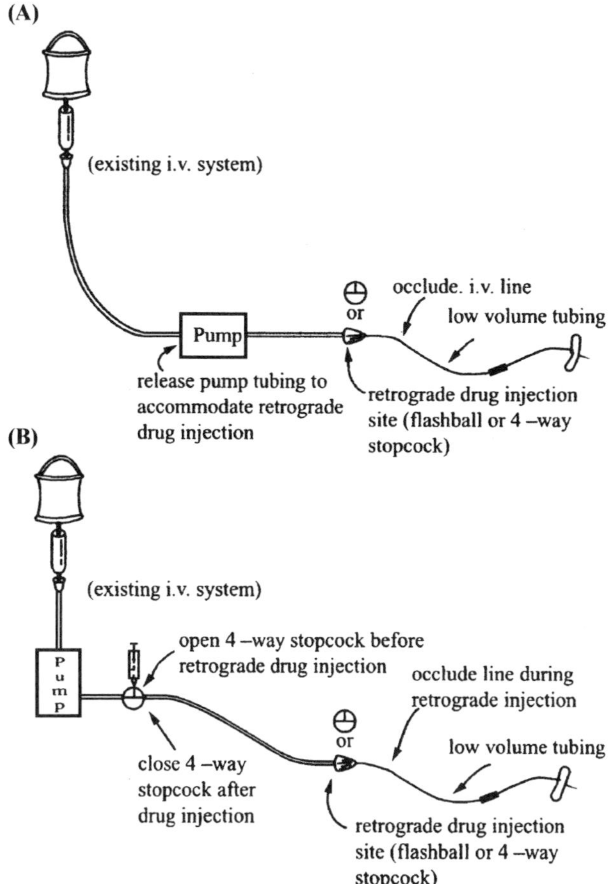

Fig. 3 Examples of retrograde system setups. (From Leff, R.D.; Roberts, R.J. Methods for intravenous drug administration in the pediatric patient. J. Pediatr. 1981, *98*, 631–635.)

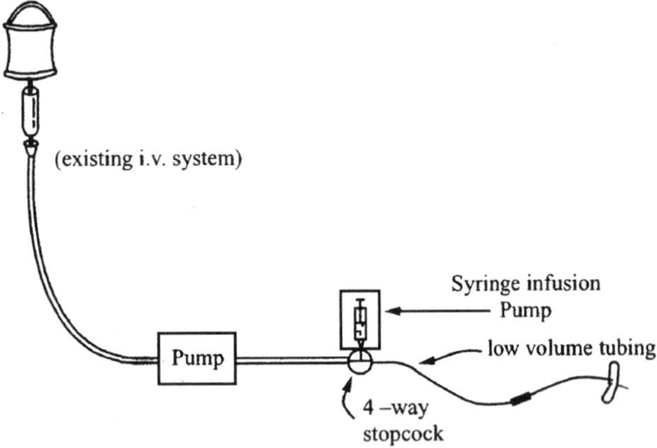

Fig. 4 Syringe pump setup with drug administered separately of the primary i.v. fluid flow rate. Mixing of the drug and i.v. fluid occurs at a stopcock and for a short distance in microbore tubing. (From Leff, R.D.; Roberts, R.J. Methods for intravenous drug administration in the pediatric patient. J. Pediatr. 1981, *98*, 631–635.)

multiple doses, and shorten the time required to administer medications.

Additional Comments About IV Drug Administration

Reed and Gal[6] recommended the following steps to decrease problems associated with i.v. drug administration to pediatric patients:

1. Standardize and document total time for drug administration.
2. Document the volume of any solution used to flush an i.v. dose.
3. Standardize infusion techniques for drug administration, especially for those with a narrow therapeutic index.
4. Use the largest gauge cannula that can be used.
5. Standardize dilution and infusion volumes for drugs given by intermittent i.v. injection, and avoid attaching lines for drug infusion to a central hub with solutions infused at widely disparate rates, and
6. Use low-volume i.v. tubing and use the most distal sites for drug administration.

In addition, one must remember that for infants the amount of fluid required for drug administration may take away from the amount of fluid available for nutrition. Thus, with medication administration, the fluid volume must be as restrictive as possible so that the bulk of the daily fluid intake can be saved for nutrition. Health care providers must closely monitor daily fluid intake from all sources to prevent fluid overload and must also watch the osmolality of medications with diluents.

ADMINISTRATION OF ORAL MEDICATIONS

The oral route is typically the preferred route for medication administration to pediatric patients. Other routes may be used, if the patient cannot take a medication orally because of vomiting, being unable to swallow, or the medication is unavailable for oral use. In addition, for specific problems it may be better to deliver the medication directly to the area being treated, for example, inhalation, ophthalmic administration, or otic administration.

Dosage Forms

Oral liquids

Liquid medications are the most commonly administered oral medications to pediatric patients because of the ease of swallowing by infants and young children who cannot swallow solid dosage forms. However, availability of some medications as liquid formulations may be limited. If not available in liquid form, a solid dosage form may need to be modified by the pharmacist, other health care provider, or by the parent. If a solid dosage form is modified, for example a suspension is prepared, will the drug be stable and for how long, and will it be absorbed differently than the original dosage form? These are just a few questions that must be answered about the extemporaneous preparation of a drug product for a pediatric patient.

Alcohol-free products should be selected for pediatric patients whenever possible. Furthermore, the inactive ingredients or excipients contained in an oral preparation should be identified. This is especially important if the patient is known to have had an adverse reaction to a particular excipient or there is another reason to avoid a particular additive in a medication. The Committee on Drugs of the American Academy of Pediatrics recommended that pharmaceutical products contain a qualitative listing of inactive ingredients in order that products containing these substance could be avoided in patients who had problems with specific adjuvants.[55] Kumar et al.[60] contains lists of inactive ingredients (sweeteners, flavorings, dyes, and preservatives) found in many liquid medications such as analgesics, antipyretics, antihistamine decongestants, cough and cold remedies, antidiarrheal agents, and theophylline preparations. The authors of the

Table 3 Osmolalities of liquid pharmaceuticals for oral administration

Liquid preparation	Osmolality (mean ± sEM)
Propylene glycol	8326 ± 1467
Saccharin-containing drugs	
Albuterol	65
Haloperidol	47
Suspensions	2500 ± 246
Sugar-containing[a] drugs	5574 ± 594

[a]Sucrose, mannitol, glucose, and others.
(From Ref. [66].)

previous article and Golightly et al. have reviewed adverse effects associated with many inactive ingredients.[44,60]

Liquid medications, taken orally, can cause diarrhea and other GI symptoms, or they may aggravate GI distress that a patient is already experiencing. These GI effects can be associated with the high osmolality of some oral liquids. Osmolalities have been determined for various oral liquids.[66] For preparations containing propylene glycol, or various sugars (e.g., sucrose, mannitol, glucose), osmolalities were noted to be high (see Table 3). It is important to compare various brands of liquid medications because they may contain different excipients and may have different osmolalities.

Sustained-Release Preparations

Most medications have shorter half-lives in children than in adults, and therefore children may need sustained-release products to maintain serum concentrations in the therapeutic range. For example, a sustained-release theophylline product may be needed for a child with asthma. It may need to be administered every 8 h to the child as compared to every 12 h for a healthy, nonsmoking adult to maintain therapeutic serum concentrations. When choosing a sustained-release theophylline preparation for a child, it must be remembered that because of differences in release properties, theophylline sustained-release products are not interchangeable. A product selected for the pediatric asthma patient should be reliably absorbed with a minimal serum concentration variation and not a preparation that has exhibited a difference in bioavailability when administered with or without food.[67–69]

Extemporaneous liquid preparations

Because many medications are not available as liquid preparations, there are times when powder papers or suspensions must be prepared. An excellent information source about the preparation of liquid dosage forms for pediatric patients has been published by Nahata and Hipple (Nahata, M.C., Hipple, T.F. Pediatric Drug Formulations, 4th Ed. Harvey Whitney Books: Cincinnati, 2000).

Product selection

Products for oral administration should be in a dosage form most readily taken by the child. If the child is old enough to participate in the decision-making process, he or she may state a preference for a liquid, chewable tablet, tablet, or capsule, if the needed drug is available in a variety of dosage forms and appropriate dosage. If a liquid medication is needed, a product should be chosen based on texture, taste, and ease of administration. Other factors that must be considered are the absence of alcohol and dyes, and an osmolality that is close to physiologic (280–290 mOsm/kg). Are there excipients or adjuvants in the product, and if so, what are they and what is their concentration? Is there bioavailability information or pharmacokinetic information for the oral medication in pediatric patients, and if so, in what age groups? Is there information about the extemporaneous product that is to be prepared?

Rebecca Chater, a North Carolina pharmacist, recommends that pediatric patients be involved in medication counseling in order to improve their understanding of why a medication is needed. In the counseling process, the word medication should be used and not drug because of the connotation associated with the latter in today's society.[70,71] Chater recommends that, when possible, a product be selected that requires the fewest number of doses administered per day, for example, every 12 h dosing rather than every 8 h, so the medication does not need to be taken to school or day care for administration. If a medication must be given outside of the home, she recommends that two small labeled bottles be dispensed or one large bottle with a small empty bottle labeled to be used for medication administration at day care or school.

Health care providers including nurses, pharmacists, and physicians should demonstrate to parents and older children how medications should be administered and offer appropriate dosing devices (oral syringe, dropper, cylindrical medication spoon, or a small-volume doser with attachable nipple) to enable parents to accurately measure liquid products. A household teaspoon or tablespoon should not be used for medication administration because they are inaccurate. Kraus and Stohlmeyer[72] explain the use of a new oral liquid medication delivery system that can be used for infants and young children who still use a bottle for feeding.

Administration Techniques

The following information is presented to help health care providers counsel parents and older children on how medication should be administered by various routes.

Oral liquids

An oral liquid medication needed for an infant or young child should be shaken well, if required, and then administered accurately using an appropriate device. If a dropper or an oral syringe is used, the liquid should be administered toward the inner cheek. Administration in the front of the mouth may allow the child to spit out the medication, whereas administration toward the back of the mouth may result in gagging or choking. The oral syringe should be of an appropriate size to allow for administration into the inner cheek.

Oral solid dosage forms

A medication available only as a solid dosage form, may be prepared as an extemporaneous liquid (e.g., suspension) or it may be modified for oral use, for example, by crushing. As mentioned previously, a sustained-release product should not be crushed or chewed. For a solid, nonsustained-release medication, the product can be crushed and mixed with a small amount of food just prior to administration. Examples of foods that may be used for mixing include applesauce, yogurt, or instant pudding, but the medication should not be added to an entire dish of food or to infant formula, because the infant

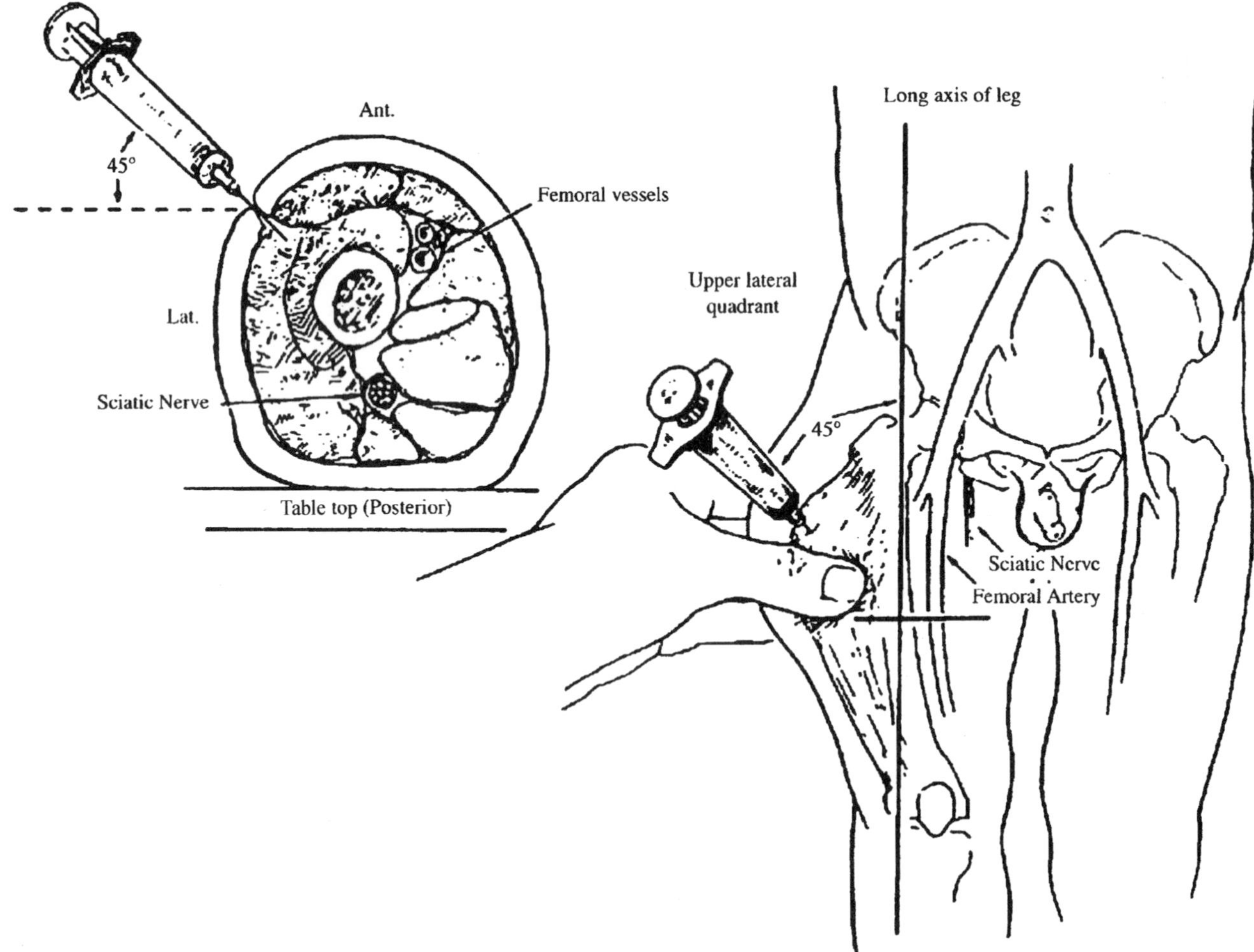

Fig. 5 A technique for anterior lateral-thigh intramuscular injection. (From Ref. [19].)

or child may not eat/drink the entire portion and thus not receive the total amount of medication.

OTHER ROUTES

Intramuscular Administration

Absorption of i.m. administered medications depends on the injection site because perfusion of individual muscle groups differs. For example, drug absorption from the deltoid muscle is faster than that from the vastus lateralis that is more rapid than from the gluteus.[16,17] In addition, lower perfusion or hemostatic decompensation, frequently observed in ill neonates and young infants, may reduce i.m. absorption. It may also be decreased in neonates who receive a skeletal muscle-paralyzing agent such as pancuronium because of decreased muscle contraction. In addition, the smaller muscle mass of neonates and young infants provide a small absorptive area.

The injection technique and the length of the needle used may affect drug absorption, and thus serum concentrations. For example, using a longer needle (1 1/2 vs. 1 1/4 in. or 3.8 vs. 3.1 cm.) for i.m. administration resulted in higher diazepam serum concentrations in adults.[18] Therefore, it is important to select the appropriate site for drug administration as well as the appropriate length and needle bore. Sites that can be used for i.m. administration include anterior thigh and vastus lateralis, gluteal area and deltoid.

The midanterior thigh (rectus femoris) and the middle third of the vastus lateralis are used for i.m. administration to young infants as well as to older children.[19] These sites are better developed and larger than other muscle groups that are used for drug administration to older children or adults. The technique is shown in Fig. 5.[19] With the patient lying supine, the "needle should be inserted in the upper lateral quadrant of the thigh, directed inferiorly at an angle of 45° with the long axis of the leg and posteriorly at a 45° angle"[19] to the surface on which the patient is lying. The person administering the injection should compress the tissues of the injection site to help stabilize the extremity. A 1-in. (2.5-cm) needle has been recommended for pediatric patients by Bergeson et al.[19] while Newton et al.[18] recommend a 23–26 gauge 11/2-in. (3.8-cm) needle. The volume of drug that can be administered in this manner is 0.1–1 ml in infants and 0.1–5 ml in older children and adults.[18]

The gluteal musculature develops as the infant or child increases his or her mobility; it becomes a more suitable injection site in children who are walking.[18,19] Damage to the sciatic nerve is the major problem associated with this injection site, and it occurs more commonly in infants because of their lack of gluteal muscle mass.[73] Injury to the gluteal nerve, resulting in muscle atrophy, has occurred even when the injection technique was appropriately performed.[19] Other nerves including the pudendal, posterior femoral cutaneous, and the inferior cluneal nerves have been damaged because of poor injection technique.[19] Additional adverse effects associated with this drug administration route are discussed by Bergeson et al.[19]

A technique for gluteal administration is shown in Fig. 6 although other techniques are also used.[19] All techniques involve the determination of the upper outer quadrant (see Fig. 6 for anatomical landmarks). After the location of the upper outer quadrant is determined, the needle should be inserted at a 90° angle to the surface on which the patient is lying.[19] This site can be used for older children. A 1-in. (2.5-cm) needle has been recommended.[19] The volume of drug that can be administered in this manner is 0.1–5 ml for older children and adults.[18]

The ventrogluteal (gluteus medius and minimus) site may be less hazardous for i.m. administration than the dorsogluteal (gluteus maximus) site.[19] The technique is shown in Fig. 7.[19] The person administering a drug i.m. ventrogluteally should first note the anatomical landmarks (anterior superior iliac spine, tubercle of the iliac crest, and upper border of the greater trochanter). The needle is

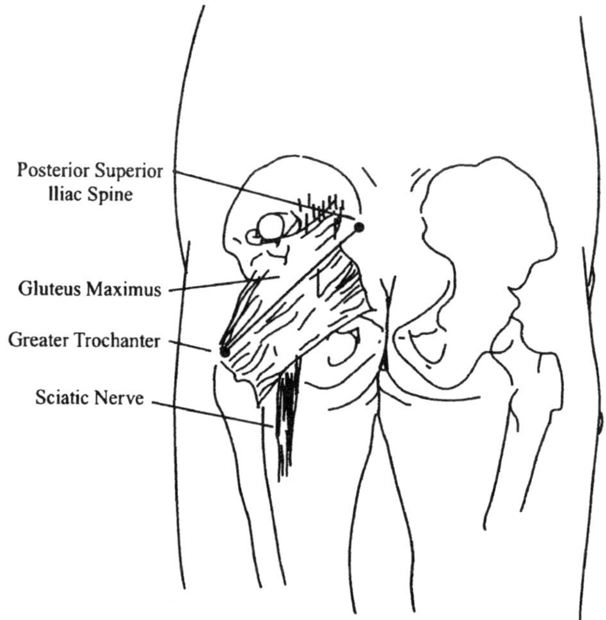

Fig. 6 A technique for gluteal-area intramuscular injection. (From Ref. [19].)

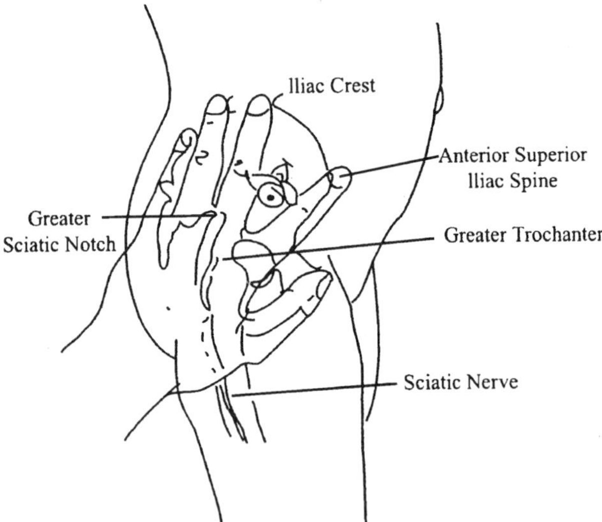

Fig. 7 von Hochstetter technique for ventrogluteal intramuscular injection. (From Ref. [19].)

inserted into a triangular area bounded by these landmarks while the patient is in the supine position. The location for this injection can be determined "by placing the palm over the greater trochanter, the index finger over the anterior superior iliac spine, and spreading the index and middle fingers as far as possible".[19]

The deltoid muscle can be used for i.m. injections in older children, but it is not an option for young infants and children because of their limited muscle mass. Although there are few complications associated with this administration route, nerve injury can occur.[21] The technique is shown in Fig. 8.[19] The area for deltoid administration should be fully visible so that the anatomical landmarks can be visualized. Then the needle for deltoid injection should enter the muscle halfway between the acromium process and the deltoid tuberosity to avoid hitting the underlying nerves.[19] The drug volume that can be administered by this route to older children and adults is 0.1–2 ml.[18] The recommended needle length for older children is 1 in. (2.5 cm).

Subcutaneous Administration

The s.c. route is used for drug administration, such as insulin, that requires slow absorption. It is not commonly employed for medication administration for pediatric patients but is used for specific drugs. Typically a 1/2- or 1-in. (1.25- or 2.5-cm) needle is used with the volume of drug that can be administered by this route ranging from 0.1 to 1 ml (drug volume administered depends on patient size).

Percutaneous Administration

The skin should be thoroughly cleaned prior to applying a topical ointment, cream, etc. A thin layer of ointment or cream should be applied to the prescribed area to reduce the possibility of a toxic reaction. The area of the skin where the medication is applied should not be covered or occluded unless instructed to do so by the physician because this procedure may increase drug absorption. Specific information should be given on how to cover the area.

Rectal Administration

Before the administration of a rectal suppository, the child's rectal area should be thoroughly cleaned. The infant or child should be placed on his or her side or stomach. The wrapper should be removed from the suppository and its pointed end should be inserted into the rectum above the anal sphincter. (If only half a suppository is prescribed, the suppository should be cut lengthwise before administration.) A finger cot or finger wrapped in plastic can be used for administering the suppository.[74] Because an infant or small child cannot adequately retain a suppository in the rectum, the buttocks can be held together firmly for a few minutes after rectal administration to hold the suppository in place.[74]

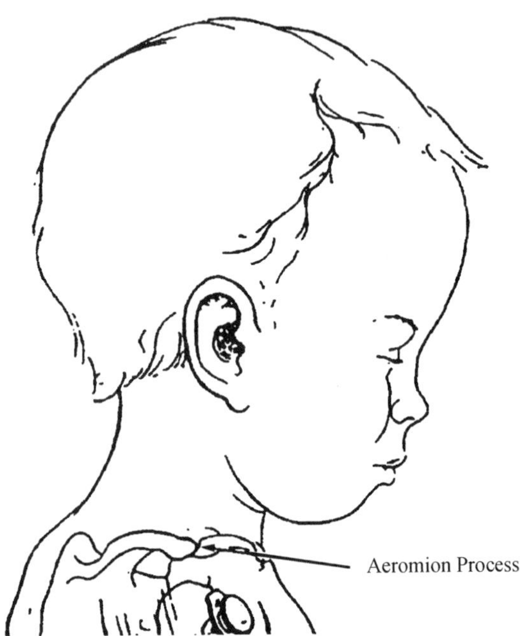

Fig. 8 A technique for deltoid intramuscular injection. (From Ref. [19].)

Otic Administration

Otic preparations should be at room temperature prior to administration. If the otic product is a suspension, it should be gently shaken for approximately 10 sec. before administration.[74] Prior to otic administration, the individual administering the medication should wash their hands thoroughly. The child should be lying on his or her side, and the earlobe should be gently pulled down and back to straighten the outer ear canal (for adults the earlobe is pulled up and back). Then the prescribed number of drops should be instilled into the ear without placing the dropper in the ear canal. The patient should be kept in a position with the ear tilted for approximately 2 min to help keep the drops in the ear.[75] This procedure may be repeated for the treatment of the other ear, if needed. The tip of the dropper should wiped with a clean tissue after use.

Nasal Administration

For adults, the first step in administering nose drops or a nasal spray is blowing the nose to clear the nasal passages of mucus and other secretions, but infants and young children are unable to do this. Therefore, the nasal passages may need be cleared with a bulb syringe prior to medication administration. A child should lie down on his or her back, or a young infant or child should be placed in a lying position, and the head should be tilted slightly backwards. An appropriate amount of medication should then be placed in each nostril. Thereafter, the infant or child should remain quiet for a few minutes to allow the medication to be absorbed. The dropper should be rinsed with hot water before it is returned to the medication container.

Ophthalmic Administration

An ophthalmic medication should be at room temperature prior to administration. If the eye drops are in a suspension, the container should be gently shaken before administration. A child old enough to follow directions should tilt his or her head slightly backward and to the side so that the eye drops will not drain into the tear ducts near the nose. The eyelids should be separated and the patient should be asked to look up. The appropriate amount of medication is instilled into the lower eyelid, using the medication dropper, which should be accomplished without touching the eyelids. The patient should look downward for a few seconds after drug administration. The eye(s) should then be closed for several minutes in order to spread the medication across the eyeball and be absorbed if the effect is to be systemic.[76] In addition, it has been recommended to gently put pressure on the inside corner of the eye for 3–5 minutes to retard drainage of the medication.[74] Do not use the eyedrops if they have changed color or if there is any particulate matter in the ophthalmic solution. The dropper should not be rinsed after use because this could lead to contamination of the dropper and the medication. The package insert should be reviewed for specific information.[76]

Another method for administering eyedrops to children recommends that the drops be applied to the inner canthus of the eye while the patient keeps his/her eyes closed until told to open them after medication administered.[77] Approximately 66% of the medication administered in this fashion was absorbed. In addition, this method may increase compliance and make children more cooperative.

For the administration of an ophthalmic ointment to a child who can cooperate, the child should tilt his or her head backwards and look up. After the hands have been washed, the person to administer the medication should squeeze out and discard the first 0.25 inches of ointment and discard it. Thereafter, the eyelid(s) should be gently pulled down for drug administration. A thin layer of ointment (0.25–0.5 inches) should then be placed in the lower eyelid(s). Afterwards, the eyelid(s) should be closed for 1 to 2 min to allow for the spreading of the medication and absorption. During this process, the tip of the ophthalmic applicator should not touch the eye. After administration is completed, the tip of the applicator tube should be cleaned with a clean tissue and be tightly capped.[76] The package insert should also be reviewed for specific information.

Inhalers

For the use of an inhaled medication (e.g., β2-agonists, corticosteroids, antivirals, cromolyn, etc.), it is crucial for the child and parents to understand the mechanism of the metered dose inhaler (MDI) or nebulizer, if used. The package insert should also be reviewed for information about the specific drug product. A decision may also need to be made as to whether a spacer may be needed for use with the medication canister.

REFERENCES

1. Nahata, M.C. Pediatric drug formulations: Challenges and potential solutions. Ann. Pharmacother. **1999**, *33*, 247–249.

2. Food and Drug Administration. *Regulations Requiring Manufacturers to Assess the Safety and Effectiveness of New Drugs and Biological Products in Pediatric Patients, 21 CFR Parts 201, 312, 314, and 601*, Docket No. 97N-0165; U.S. Department of Health and Human Services; Washington, DC, August 17, 1997.
3. FDA Modernization Act of 1997 Public Law 105–115 105th Congress, Washington, DC, 1997.
4. Food and Drug Administration. *List of Approved Drugs for Which Additional Pediatric Information may Produce Health Benefits in the Pediatric Population*, Docket No. 98N-0056; U.S. Department of Health and Human Services: Washington, DC, May 20, 1998.
5. USP Member Memorandum, U.S. Pharmacopoeia, Rockville, MD, June 22, 2000.
6. Reed, M.D.; Gal, P. Principles of Drug Therapy. In *Nelson Textbook of Pediatrics,* 16th Ed.; Behrman, R.E., Kliegman, R.M., Jenson, H.B., Eds.; W.B. Saunders Company: Philadelphia, 2000; 2229–2234.
7. Reed, M.D. The ontogeny of drug disposition: Focus on drug absorption, distribution, and excretion. Drug Inf. J. **1996**, *30*, 1129–1134.
8. Milsap, R.L.; Hill, M.R.; Szefler, S.J. Special Pharmacokinetic Considerations in Children. In *Applied Pharmacokinetics: Principles of Therapeutic Drug Monitoring,* 3rd Ed.; Evans, W.E., Schentag, J.J., Jusko, W.J., Eds.; Applied Therapeutics: Vancouver, Washington, 1992; 10.1–10.32.
9. Stewart, C.F.; Hampton, E.M. Effect of maturation on drug disposition in pediatric patients. Clin. Pharm. **1987**, *6*, 548–564.
10. Christie, D.L. Development of Gastric Function During the First Month of Life. In *Infancy*; Lebenthal, E., Ed.; Raven Press: New York, 1981; 109–120.
11. Cavell, B. Gastric emptying in preterm infants. Acta Paediatr. Scand. **1979**, *68*, 725–730.
12. Cavell, B. Gastric emptying in infants fed human milk or infant formula. Acta Pediatr. Scand. **1981**, *70*, 639–641.
13. Blumer, J.L.; Reed, M.D. Principles of Neonatal Pharmacology. In *Pediatric Pharmacology: Therapeutic Principles in Practice*; Yaffe, S.J., Aranda, J.V., Eds.; W.B. Saunders Co.: Philadelphia, 1992; 164–177.
14. Lebenthal, E.; Lee, P.C.; Heitlinger, L.A. Impact of the development of the GI tract on infant feeding. J. Pediatr. **1983**, *101*, 1–9.
15. Kearns, G.L.; Reed, M.D. Clinical pharmacokinetics in infants and children: A reappraisal. Clin. Pharmacokinet. **1989**, *17* (Suppl. 1), 29–67.
16. Greenblatt, D.J.; Koch-Weser, J. Intramuscular injection of drugs. N. Engl. J. Med. **1976**, *295*, 542–546.
17. Evans, E.F.; Proctor, J.D.; Frantkin, M.J.; Velandia, J.; Wasserman, A.J. Blood low in muscle groups and drug absorption. Clin. Pharmacol. Ther. **1975**, *17*, 44–47.
18. Newton, M.; Newton, D.; Fudin, J. Reviewing the big three injection routes. Nursing **Feb. 1992**, *92*, 34–42.
19. Bergeson, P.S.; Singer, S.A.; Kaplan, A.M. Intramuscular injections in children. Pediatrics **1982**, *70*, 944–948.
20. Sagraves, R.; Kamper, C. Controversies in cardiopulmonary resuscitation: Pediatric considerations. DICP, Ann. Pharmacother. **1991**, *25*, 760–772.
21. Ghadially, R.; Shear, N.H. Topical Therapy and Percutaneous Absorption. In *Pediatric Pharmacology: Therapeutic Principles in Practice*; Yaffe, S.J., Aranda, J.V., Eds.; W.B. Saunders Co.: Philadelphia, 1992; 72–77.
22. American Academy of Pediatrics Committee on Drugs. Alternative routes of drug administration—advantages and disadvantages (Subject review). Pediatrics **1998**, *100*, 143–153.
23. McRorie, T. Quality drug therapy in children: Formulations and delivery. Drug Inf. J. **1996**, *30*, 1173–1177.
24. Ward, J.T., Jr. Endotracheal drug therapy. Am. J. Emerg. Med. **1983**, *1*, 71–82.
25. de Boer, A.G.; Moolenaar, F.; de Leede, L.G.J.; Breimer, D.D. Rectal drug administration: Clinical pharmacokinetic considerations. Clin. Pharmacokinet. **1982**, *7*, 285–311.
26. Iob, V.; Swanson, W.W. Mineral growth of the human fetus. Am. J. Dis. Child. **1934**, *47*, 302–306.
27. Widdowson, E.; Spray, C.M. Chemical development in utero. Arch. Dis. Child. **1951**, *26*, 205–214.
28. Morselli, P.L. Clinical pharmacokinetics in neonates. Clin. Pharmacokinet. **1976**, *1*, 81–98.
29. Reed, M.D.; Besunder, J.B. Developmental pharmacology: Ontogenic basis of drug disposition. Pediatr. Clin. North Am. **1989**, *36*, 1053–1074.
30. Kearin, M.; Kelly, J.G.; O'Malley, K. Digoxin "receptors" in neonates: An explanation of less sensitivity to digoxin in adults. Clin. Pharmacol. Ther. **1980**, *28*, 346–349.
31. Levy, G.; Khanna, N.N.; Soda, D.M.; Tsuzuki, O.; Stern, L. Pharmacokinetics of acetaminophen in the human neonate: Formation of acetaminophen glucuronide and sulfate in relation to plasma bilirubin concentration and d-glucaric acid excretion. Pediatrics **1975**, *55*, 818–825.
32. Siegel, S.R.; Oh, W. Renal function as a marker of human fetal maturation. Acta Paediatr. Scand. **1976**, *65*, 481–485.
33. Siber, G.R.; Smith, A.L.; Levin, M.J. Predictability of peak serum gentamicin concentration with dosage based on body surface area. J. Pediatr. **1978**, *94*, 135–138.
34. Painter, M.J.; Pippenger, C.E.; Wasterlein, C.; Barmada, M.; Pitlick, W.; Carter, G.; Aberin, S. Phenobarbital and phenytoin in neonatal seizures: Metabolism and tissue distribution. Neurology **1981**, *31*, 1107–1112.
35. Gilman, J.T.; Gal, P.; Duchowny, M.S.; Weaver, R.L.; Ransom, J.L. Rapid sequential phenobarbital treatment of neonatal seizures. Pediatrics **1989**, *83*, 674–678.
36. Gilman, J.T. Therapeutic drug monitoring in neonate and

paediatric age group. Problems and clinical pharmacokinetic implications. Clin. Pharmacokinet. **1990**, *19*, 1–10.

37. Tserng, K.Y.; Takieddine, F.N.; King, K.C. Developmental aspects of theophylline metabolism in premature infants. Clin. Pharmacol. Ther. **1983**, *33*, 522–528.
38. Lonnerholm, G.; Lindstrom, B.; Paalzow, L.; Sedin, G. Plasma theophylline and caffeine and plasma clearance of theophylline during theophylline treatment in the first year of life. Eur. J. Clin. Pharmacol. **1983**, *24*, 371–374.
39. Dreifuss, F.E.; Santilli, N.; Langer, D.H.; Sweeney, K.P.; Moline, K.A.; Menander, K.B. Valproic acid hepatic fatalities: A retrospective review. Neurology **1987**, *37*, 379–385.
40. Gershanik, J.J.; Boecler, B.; George, W.; Sola, A.; Leitner, M.; Kapadia, C. The gasping syndrome: Benzyl alcohol (BA) poisoning. Clin. Res. **1981**, *29*, 895A, Abstract.
41. Brown, W.J.; Buist, N.R.M.; Gipson, H.T.C.; Huston, R.K.; Kennaway, N.G. Fatal benzyl alcohol poisoning in a neonatal intensive care unit. Lancet **1982**, *1*, 1250, Letter.
42. Gershanik, J.J.; Boecler, B.; Ensley, H.; McCloskey, S.; George, W. The gasping syndrome and benzyl alcohol poisoning. N. Engl. J. Med. **1982**, *307*, 1384–1388.
43. Benda, G.I.; Hiller, J.L.; Reynolds, J.W. Benzyl alcohol toxicity: Impact on neurologic handicaps among surviving very low birth weight infants. Pediatrics **1986**, *77*, 507–512.
44. Golightly, L.K.; Smolinske, S.S.; Bennett, M.L.; Sutherland, E.W., III; Rumack, B.H. Pharmaceutical excipients. Adverse effects associated with inactive ingredients in drugs products (Part I). Med. Toxicol. **1988**, *3*, 128–165.
45. American Academy of Pediatrics Committee on Fetus and Newborn and Committee on Drugs. Benzyl alcohol: Toxic agent in neonatal units. Pediatrics **1983**, *72*, 356–358.
46. Centers for Disease Control. Neonatal death associated with use of benzyl alcohol—United States. MMWR **1982**, *31*, 290–291.
47. Benzyl alcohol may be toxic to newborns. FDA Drug Bull. **1982**, *12*, 10–11.
48. Hiller, J.L.; Benda, G.I.; Rahatzad, M.; Alen, J.R.; Culver, D.H.; Carlson, C.V.; Reynolds, J.W. Benzyl alcohol toxicity: Impact on mortality and intraventricular hemorrhage among very low birth weight infants. Pediatrics **1986**, *77*, 500–506.
49. Jadine, D.S.; Rogers, K. Relationship of benzyl alcohol to kernicterus, intraventricular hemorrhage, and mortality in preterm infants. Pediatrics **1989**, *83*, 153–160.
50. MacDonald, M.G.; Getson, P.R.; Glasgow, A.M.; Miller, M.K.; Boeckx, R.L.; Johnson, E.L. Propylene glycol: Increased incidence of seizures in low birth weight infants. Pediatrics **1987**, *79*, 622–625.
51. Glascow, A.M.; Boeckx, R.L.; Miller, M.K.; MacDonald, M.G.; August, G.P. Hyperosmolality in small infants due to propylene glycol. Pediatrics **1983**, *72*, 353–355.
52. Berkeris, L.; Baker, C.; Fenton, J. Propylene glycol as a cause of an elevated serum osmolality. Am. J. Clin. Pathol. **1979**, *72*, 633–636.
53. Fligner, C.L.; Jack, R.; Twiggs, G.A.; Raisys, V.A. Hyperosmolality induced by propylene glycol. JAMA, J. Am. Med. Assoc. **1985**, *253*, 1606–1609.
54. Kulick, M.I.; Lewis, N.S.; Bansal, V.; Warpeha, R. Hyperosmolality in the burn patient: Analysis of an osmolal discrepancy. J. Trauma **1980**, *20*, 223–228.
55. American Academy of Pediatrics Committee on Drugs. ''Inactive'' ingredients in pharmaceutical products. Pediatrics **1985**, *76*, 635–643.
56. Louis, S.; Jutt, H.; McDowell, F. The cardiocirculatory changes caused by intravenous dilantin and its solvent. Am. Heart J. **1967**, *74*, 523–529.
57. American Academy of Pediatrics Committee on Drugs. Ethanol in liquid preparations intended for children. Pediatrics **1984**, *73*, 405–407.
58. Rumack, B.H.; Spoerke, D.G. *Poisindex Information System*; Micromedex, Inc.: Denver, 2000.
59. Hoyumpa, A.M.; Schenker, S. Major drug interactions. Effect of liver disease, alcohol, and malnutrition. Annu. Rev. Med. **1982**, *33*, 113–149.
60. Kumar, A.; Rawlings, R.D.; Beaman, D.C. The mystery ingredients: Sweeteners, flavorings, dyes, and preservative in analgesic/antipyretic, antihistamine/decongestant, cough and cold, antidiarrheal, and liquid theophylline preparations. Pediatrics **1993**, *91*, 927–933.
61. Leff, R.D.; Roberts, R.J. *Practical Aspects of Intravenous Drug Administration*; American Society of Hospital Pharmacists: Bethesda, MD, 1992.
62. Nahata, M.C. Intravenous infusion conditions: Implications for pharmacokinetic monitoring. Clin. Pharmacokinet. **1993**, *24*, 221–229.
63. Gould, T.; Roberts, R.J. Therapeutic problems arising from the use of the intravenous route for drug administration. J. Pediatr. **1979**, *95*, 465–471.
64. Rajchgot, P.; Radde, I.C.; MacLeod, S.M. Influence of specific gravity on intravenous drug delivery. J. Pediatr. **1981**, *99*, 658–661.
65. Santeiro, M.L.; Sagraves, R.; Allen, L.V. Osmolality of small—Volume I. V. Admixtures for pediatric patients. Am. J. Hosp. Pharm. **1990**, *47*, 1359–1364.
66. Bloss, C.S.; Sybert, K. Osmolality of commercially available oral liquid drug preparations. ASHP Annu. Meet. **June 1991**, *48*, p-35, Abstract.
67. Maish, W.; Sagraves, R. Childhood asthma. U.S. Pharmacist **1993,** Jan., *36–58*, 105.
68. Hendeles, L.; Weinberger, M.; Szefler, S. Safety and efficacy of theophylline in children with asthma. J. Pediatr. **1992**, *120*, 177–183.
69. Hendeles, L.; Weinberger, M. Selection of a slow-release

theophylline product. J. Allergy Clin. Immunol. **1986**, *78*, 743–751.

70. Martin, S. Catering to pediatric patients. Am. Pharm. **1992**, *32*, 47–50.
71. Chater, R.W. Pediatric dosing: Tips for tots. Am. Pharm. **1993**, *33*, 55–56.
72. Kraus, D.M.; Stohlmeyer, L.A.; Hannon, P.R. Infant acceptance and effectiveness of a new oral liquid medication delivery system. Am. J. Health-Syst. Pharm. **1999**, *56*, 1094–1101.
73. Gilles, F.H.; Matson, D.D. Sciatic nerve injury following misplaced gluteal injection. J. Pediatr. **1970**, *76*, 247–254.
74. Administration guides. Drug Store News **1993,** Dec., *13*, 7–14.
75. Proper Use of Ear Drops. In *Facts and Comparisons*; Burnham, T.H., Ed.; Walters Kluwer Co.: St. Louis, January 2000; 1802.
76. Recommended Procedures for Administration of Solutions or Suspensions (Ophthalmics). In *Facts and Comparisons*; Burnham, T.H., Ed.; Walters Kluwer Co.: St. Louis, January 2000; 1725.
77. Smith, S.E. Eyedrop instillation for reluctant children. Br. J. Ophthalmol. **1991**, *75*, 480–481.

BIBLIOGRAPHY

Harriett Lane Handbook, 15th Ed.; Siberry, G.K., Iannone, Eds.; Mosby: St. Louis, 2000.

Leff, R.D.; Roberts, R.J. *Practical Aspects of Intravenous Drug Administration*; American Society of Health-System Pharmacists: Bethesda, 1992.

Nahata, M.C.; Hipple, T.F. *Pediatric Drug Formulations,* 4th Ed.; Harvey Whitney Books: Cincinnati, 2000.

Nelson, J.; Bradley, J. *Nelson's 2000–2001 Pocket Book of Pediatric Antimicrobial Therapy,* 14th Ed.; Lippincott Williams & Wilkins: Philadelphia, 2000.

Phelps, S.; Hak, E. *Guidelines for Administration of Intravenous Medications to Pediatric Patients,* 6th Ed.; American Society of Health-System Pharmacists: Bethesda, 1999.

Taketomo, C.K.; Hodding, J.H.; Kraus, D.M. *Pediatric Dosage Handbook*, 8th Ed.; Lexi-Comp., Inc.: Hudson, OH, 2002.

The Pediatric Drug Handbook, 3rd Ed.; Benitz, W.E., Tatro, D.S., Davis, S., Eds.; Harcourt Health Sciences Group: St. Louis, 1995.

Young, T.; Mangum, B. *Neofax 2000,* 13th Ed.; Blackwell Science: Oxford, 2000.

Pediatric Pharmacy Specialty Practice

Marcia L. Buck
University of Virginia Medical Center, Charlottesville, Virginia, U.S.A.

INTRODUCTION

Providing pharmacy services to children can be challenging but also rewarding. The changes in body size and function that occur with normal growth and development result in a dynamic transition in drug disposition and therapeutic response throughout childhood and adolescence. In the past, there was little attention paid to these physiologic differences. Little information was available about the use of drugs in children; as a result, dosing and monitoring information were often extrapolated from studies conducted in adults. In fact, although nearly all drugs on the market in the United States were being used in pediatric patients, less than one-half carried a pediatric indication from the Food and Drug Administration (FDA).[1,2]

Since the 1980s, the lack of attention to pediatric-specific medication information has slowly changed as more data on dosing, pharmacokinetics, monitoring, and adverse effects in children have begun to appear in the medical literature. However, much research remains to be done. Many of the reports currently being published involve only small numbers of patients and often lack controls. Health care providers, including pharmacists, must be adept at drawing appropriate therapeutic information from these limited resources. In addition to the relative lack of medication information, pharmacists are frequently also faced with a lack of an appropriate dosage formulation. Most drug preparations on the market in the United States are designed for adults. These dosage forms must be tailored for use in children, often by compounding tablets or capsules into oral solutions, syrups, or suspensions.[3] Pediatric practice has evolved as a specialty within pharmacy to meet unique needs such as these in infants and children.

PRACTICE MODELS

There are many different types of pediatric clinical pharmacy positions. Although most pediatric clinical pharmacists work in a traditional hospital setting, many are branching out into ambulatory pediatric practice sites.[4,5] Clinic and retail pharmacy-based pediatric practitioners are finding a variety of ways to provide patient care. For example, many pharmacies now offer age-appropriate written and verbal patient/parent counseling and suggestions for improving medication compliance.[6] In either the acute or ambulatory setting, pediatric clinical pharmacists may be generalists, seeing children of many different ages with a variety of disease states, or specialize in a particular age or condition, such as neonates or children with pulmonary illnesses. Pediatric clinical pharmacists may also move into administrative and managerial roles. Other opportunities exist in the pharmaceutical industry, where pediatric clinical pharmacists may assist in the development or marketing of drugs for children.

The role of the pediatric pharmacist in organized health care has become well established. As early as 1971, reports of pediatric specialization in pharmacy were available.[7] Since that time, many more papers have been published describing pediatric practice models in both the hospital and retail settings.[4,6,8–10] In 1994, the American Society of Hospital (now Health-System) Pharmacists published guidelines for providing pediatric pharmaceutical services. Among the activities that the society recommended were the preparation of individualized doses, through the use of unit dosing and intravenous admixture systems, the provision of drug information and family education, and the maintenance of therapeutic drug monitoring and pharmacokinetic services. In addition, the group suggested that programs to provide ongoing evaluation of medication use and to orient and train pharmacy staff be established at each institution. The guidelines also call for the participation of pharmacists in clinical research involving children, both as investigators and as members of institutional review boards to ensure patient safety in drug trials.[5]

Whether hospital-, clinic-, retail-, or industry-based, it is becoming increasingly likely that the pediatric pharmacist will be involved in drug research. The need for pediatric-specific drug information and pediatric dosage

Encyclopedia of Clinical Pharmacy
DOI: 10.1081/E-ECP 120006235

formulations has been targeted by the FDA.[11] In October 1992, the FDA took the first steps toward improving the amount of pediatric information available on drug labeling as part of their ruling entitled, "Specific Requirements on Content and Format of Labeling for Human Prescription Drugs: Revision of 'Pediatric use' Subsection in the Labeling." These new regulations were designed to promote the inclusion of information gained from new clinical trials as well as information gained from previously published studies and case reports in children. In an effort to support this research, a network of pediatric pharmacology units (PPRUs) were created at medical centers throughout the United States. Many of the PPRUs include pharmacists as investigators. Further progress was made in December 1994, when the FDA announced plans to mandate labeling information on pediatric use for all pertinent new drug applications. Manufacturers also were required to examine their existing products to determine if there was adequate cause to modify their pediatric sections to include more specific pediatric dosing and safety information.

As anticipated, this ruling generated considerable concern on the part of manufacturers who saw this as a potential roadblock to drug approval for adults patients and an additional expense. The FDA took a stronger stance on November 21, 1997, with the enactment of the Modernization Act. This ruling was the first major amendment of the Food, Drug, and Cosmetic Act and was designed to incorporate changes for the twenty-first century, including advances in technology and trade practices, as well as changing public health concerns. The Modernization Act covers many different aspects of the drug approval process, such as fast track policies, industry guidance, and postmarketing studies, but one of the most significant changes has been the tightening and enforcement of regulations relating to pediatrics. The complete Modernization Act became effective on April 1, 1999, and an increase in industry-sponsored research has already been observed. It is anticipated that the implementation of the Modernization Act will usher in a new era in pediatric drug research and provide opportunities for many pediatric pharmacists to contribute to the expansion of our knowledge of pediatric therapeutics.

DOCUMENTING BENEFIT

The benefits of having a clinical pharmacist within a pediatric practice setting have been documented in several different ways. Reducing medication errors and patient costs have been the primary methods by which new or additional pediatric pharmacy positions have been justified.[12,13] In 1987, Folli and colleagues published their study of the impact of pharmacy intervention on medication errors in two children's hospitals.[13] The pharmacists in this study detected, and prevented, mistakes at a rate of 5 errors per 1000 medication orders. The majority of the errors (64%) occurred in children younger than 2 years of age. This study received considerable attention in both the medical and lay press. It continues to serve as a landmark citation for the support of pharmacist participation in pediatric acute care. In addition to error prevention and cost savings, pediatric pharmacists have also documented their value in their contributions to our knowledge of medication use in children. It is now common for studies of new drugs in children to have at least one pharmacist author.[14–17] Pharmacists have also added to our knowledge of pediatric medicine by publishing case reports describing drug interactions and toxicities.[18] In addition, pharmacists have conducted studies of extemporaneous oral liquid formulations and intravenous drug compatibility, which have alleviated many of the problems with pediatric drug administration.[19,20]

PEDIATRIC PHARMACY RESOURCES

Pediatrics, unlike most other areas of pharmacy practice, encompasses a wide variety of disease states. Clinical pharmacists must not only have an understanding of the effects of growth and development on drug disposition, but also a general appreciation for therapies related to disease in any organ system. Despite the complexity of this patient age group, relatively little emphasis is given to pediatrics in most pharmacy school curricula. In a survey of pharmacy schools, an average of 16.7 hours was devoted to pediatric didactic content.[21] Most pediatric pharmacists develop their knowledge base in the clinical setting, beginning with experiential training during postgraduate programs and continuing throughout their careers. There are a number of resources available to provide information on pediatric medications for those beginning their study of pediatric pharmacy. Table 1 provides a list of topics that may serve as a guide for developing a basic knowledge of pediatric therapeutics.

Reference Texts

Fortunately for students and new practitioners, the pharmacotherapy texts most frequently required for didactic therapeutics courses contain introductory chapters on pediatrics.[22–24] These general chapters can be very helpful as brief overviews to pediatric specialty practice. General pediatric texts, such as the *Nelson Textbook of Pediatrics*, are also useful to provide a basic understand-

Table 1 Pediatric topics for clinical pharmacists

Growth and development
Drug dosage calculations and dosage formulations
Pediatric pharmacokinetics
Fluid and electrolyte management
Nutrition
Immunizations
Asthma
Cystic fibrosis
Infectious diseases
Management of fever
Otitis media
Sepsis and meningitis
Pneumonia
Diarrhea/dehydration
Urinary tract infections
Pediatric seizure disorders
Pediatric cardiology
Pediatric oncology
Pediatric endocrinology
Neonatology
Behavioral disorders
Pain management in infants and children
Cardiopulmonary resuscitation in infants and children

ing of pathophysiology in infants and children.[25] Although it has not been updated recently, the 1992 text *Pediatric Pharmacology: Therapeutic Principles in Practice* remains one of the best general references for pediatric drug information.[26] Another valuable reference is the Pediatric Module contained in the third edition of the *Pharmacotherapy Self-Assessment Program* (PSAP).[27] The PSAP program is designed to provide the reader with information on current and innovative clinical pharmacy practices. In addition to a basic chapter on pediatric pharmacotherapy, the module contains chapters on neonatology, critical care issues, respiratory infections, gastroenterology, and epilepsy. Additional pediatric topics are included in the first two editions. Each PSAP chapter also contains an annotated bibliography and self-test questions. More information on the PSAP program can be obtained through its publisher, the American College of Clinical Pharmacy, on their web site at www.accp.com.

Dosing Handbooks

All pediatric clinical pharmacists need access to current dosing handbooks. There are many dosing references available, but the most useful for pharmacists are those that also include information on administration, pharmacokinetics, and adverse effects, in addition to dosage and dosing frequency. Two examples of these references are the *Pediatric Dosage Handbook* and the *Handbook of Pediatric Drug Therapy*.[28,29] Both of these references contain thorough drug monographs as well as appendices with additional information on drugs for resuscitation, immunization schedules, and unit conversions. With the rapidly expanding information available on pediatric medication use, it is important to select dosing references that are frequently updated. A good rule of thumb is to choose those pediatric dosing handbooks that are revised at least every 2 to 3 years to incorporate new drugs and dosing information.

In addition to general dosing handbooks, the reference shelf for a pediatric clinical pharmacist may include several other more specific texts. Many clinicians use the report of the American Academy of Pediatrics Committee on Infectious Diseases as their standard reference for antibiotic and immunoglobulin therapy, as well as immunizations.[30] This report, often referred to as the ''Red Book,'' is published approximately every 3 to 4 years. The *Pediatric Pain Management and Sedation Handbook* is also useful for pediatric pharmacists. This reference is filled with tables, nomograms, and algorithms to help clinicians to determine the appropriate drug, dose, and monitoring techniques for sedation and analgesia in a variety of settings.[31] For the preparation of intravenous medications, *Teddy Bear Book. Pediatric Injectable Drugs* is considered a standard reference.[32] This book contains single-page monographs with information on dosing, medication preparation, and infusion rates. Like the ''Red Book,'' it is extensively referenced with citations from the original medical literature.

Pharmacists involved in patient/family teaching may also want to have written medication instructions available. The *Pediatric Medication Education Text* contains single-page instructions in English and in Spanish that can be photocopied for families of young patients.[33] These instructions meet the current FDA recommendations for written information and are designed to serve as a supplement to verbal medication counseling. The information in this book is targeted at patients younger than 12 years of age. For example, instead of containing warnings against driving for those drugs causing dizziness or sedation, these sheets instruct parents to watch children closely when they are going up or down stairs. Another frequently used reference for family counseling is *Drugs in Pregnancy and Lactation*.[34] This text, also available on CD-ROM, contains brief summaries of the literature regarding the potential for teratogenicity during pregnancy and toxicity during breastfeeding. In addition to presenting the results of case reports and small-scale studies, this reference also includes the outcomes of surveillance studies conducted in the United States and foreign countries, such as Canada's MotherRisk program. Like many of the handbooks and references described earlier, this text lists drugs by generic name in alphabetical order, making it easy to use for quick consultations. A separate quarterly update subscription is also available from the publisher.

Journals

There are many pediatric medical, nursing, and pharmacy journals that include articles on pediatric drug therapies (Table 2). *Pediatrics*, the journal of the American Academy of Pediatrics (AAP), and the *Journal of Pediatrics* are considered by most pediatric practitioners to be the top in the field. *Pediatrics* is of particular use to clinicians because it includes the policy statements developed by the AAP. These statements are considered to represent standards of practice by pediatricians. Many of these statements are also of interest to clinical pharmacists, such as the yearly schedule for routine childhood immunizations.[35] Other AAP policy statements of note include recommendations on the administration of medications during breastfeeding,[36] the ethical treatment of children enrolled in clinical research trials,[37] and methods to reduce medication errors in the pediatric inpatient setting.[38] The *Journal of Pediatrics* has also published useful practice recommendations, such as the guidelines for antithrombotic therapy.[39] The pediatric journal for the American Medical Association, *Archives of Pediatrics and Adolescent Medicine*, often contains large-scale surveillance studies such as a retrospective review of methylphenidate toxicity.[40] These policy statements, guidelines, and reviews will make useful additions to the files of most pediatric clinical pharmacists.

Table 2 Examples of peer-reviewed journals containing articles on pediatric therapeutics

American Journal of Health-System Pharmacy
Annals of Pharmacotherapy
Archives of Diseases in Childhood
Archives of Pediatrics and Adolescent Medicine
Clinical Pediatrics
Clinical Pharmacokinetics
Current Pediatric Therapy
Drug Safety
Drugs
Journal of Pediatrics
Journal of Pediatric Pharmacology and Therapeutics (formerly *The Journal of Pediatric Pharmacy Practice*)
Paediatric Drugs
Pediatric Clinics of North America
Pediatric Emergency Care
Pediatric Infectious Disease Journal
Pediatric Nursing
Pediatrics
Pharmacotherapy

Web Sites

A growing number of web sites have been developed for pediatric medication information. Most university teaching hospitals and children's medical centers have web sites that include educational programs. For example, the University of Virginia Children's Medical Center web site offers tutorials for health care providers on attention deficit/hyperactivity disorder and cerebral palsy, as well as the institution's *Pediatric Pharmacotherapy* newsletter.[41] In addition, many professional organizations for pediatric health care practitioners have web sites, as well as mailing lists, chat rooms, or electronic bulletin boards. Two other useful resources on the Internet are PEDINFO and Harriet Lane Links, the latter sponsored by The Johns Hopkins University. These two noncommercial web sites provide a large number of links to pediatrics-related sites, as well as some feedback on the quality of information contained on many of the linked sites. Table 3 lists these web sites and several others that may be of interest to pediatric clinical pharmacists.

PROFESSIONAL ORGANIZATIONS FOR PEDIATRIC PHARMACY PRACTITIONERS

There are several organizations that provide educational programs and networking opportunities for pediatric clinical pharmacists. Two of the most active groups are the Pediatrics Practice and Research Network (PRN) within the American College of Clinical Pharmacy (ACCP) and the Pediatric Pharmacy Advocacy Group (PPAG). The ACCP Pediatrics PRN currently has approximately 200 members and meets twice yearly at the ACCP Spring Forum and annual meetings. The focus of the Pediatrics PRN is to support the practice, teaching, and research of its members by fostering collaboration and sharing information among colleagues. For more information about this group, visit the ACCP web site at www.accp.com.

PPAG was also founded by pharmacists practicing in pediatrics. Although this group has traditionally been focused on issues related to hospital practice, with an emphasis on pharmacy administration, the past several years have seen a renewed interest in expanding the organization and incorporating clinicians from a variety of practice settings. PPAG holds an annual meeting and also sponsors a meeting in conjunction with the American Society of

Table 3 Websites of interest to pediatric clinical pharmacists[a]

Source	URL
Agency for Healthcare Research and Quality	www.ahrq.gov
American Academy of Pediatrics	www.aap.org
American College of Clinical Pharmacy	www.accp.com
Centers for Disease Control and Prevention	www.cdc.gov
Food and Drug Administration (FDA)	www.fda.gov
FDA Pediatric Medicine Page	www.fda.gov/cder/pediatric/index.htm
Harriet Lane Links	www.med.jhu.edu/peds/hll
Institute for Safe Medication Practices	www.ismp.org
Kids Meds	www.kidsmeds.org
MedWatch (Adverse Event Reporting)	www.fda.gov/medwatch
Neonatal and Paediatric Pharmacists Group	www.nppg.demon.co.uk
Pediatric Critical Care Medicine	www.pedsccm.org
Pediatric Pharmacy Advocacy Group	www.ppag.org
PEDINFO	www.pedinfo.org
U.S. Pharmacopeia: Children and Medicine	www.usp.org/information/programs/children
Vaccine Adverse Event Reporting System	www.vaers.org

[a]Accessed June 2002.

Health-System Pharmacists' Clinical Midyear Meeting. PPAG also organizes the Pediatric Adverse Drug Reaction Program (PADR), a central database of adverse effects in children. For more information about this group, visit their web site at www.ppag.org. Both the ACCP Pediatrics PRN and PPAG offer a wide array of member services, including mailing lists for communication among their members. An international group, the Neonatal and Paediatric Pharmacists Group (NPPG), may also be of interest. This organization was formed in 1994 to advance the quality of pharmacy services for children in the United Kingdom. Their web site (www.nppg.demon.co.uk), offers links to a variety of resources and bulletin boards for subgroups in oncology, as well as pediatric and neonatal intensive care.

CONCLUSION

Pediatric pharmacy practice has come a long way since Shirkey first coined the term ''therapeutic orphans'' to describe the lack of information and research support for children requiring medications.[42] As we begin the twenty-first century, there are a growing number of pediatrics drug resources available to the health care provider. We have seen the benefits of increased information in the medical literature, the publication of several pediatric-specific dosage handbooks, and the advent of the Internet as a means of sharing information with colleagues. These changes, along with a renewed interest from the FDA in pediatric drug research, are coming together to allow pediatric practitioners to expand our knowledge of drug disposition in children and improve the quality of their health care.

REFERENCES

1. Buck, M.L. Pediatric Pharmacotherapy. In *Pharmacotherapy Self-Assessment Program,* 3rd Ed.; Carter, B.L., Ed.; American College of Clinical Pharmacy: Kansas City, Missouri, 2000; 179–202, Module 9.
2. Zenk, K.E. Challenges in providing pharmaceutical care to pediatric patients. Am. J. Hosp. Pharm. **1994**, *51*, 688–694.
3. Nahata, M.C. Lack of pediatric drug formulations. Pediatrics **1999**, *104* (3 Supplement Part 2), 607–609.
4. Buck, M.L.; Connor, J.J.; Snipes, C.J.; Hopper, J.E.H. Comprehensive pharmaceutical services for pediatric patients. Am. J. Hosp. Pharm. **1993**, *50*, 78–84.
5. American Society of Hospital Pharmacists ASHP guidelines for providing pediatric pharmaceutical services in organized health care systems. Am. J. Hosp. Pharm. **1994**, *51*, 1690–1692.
6. Martin, S. Catering to pediatric patients. Am. Pharm. **1992**, *1* (NS32), 47–50.
7. Klotz, R. Is the pediatric pharmacist needed? Hosp. Pharm. **1971**, *6* (7), 18–20.
8. Cupit, G.C. An approach to pediatric pharmacy practice. Hosp. Pharm. **1974**, *9* (10), 399–400, 402.
9. Oakley, R.S. Pharmacy practice in a small pediatric hospital. Am. J. Hosp. Pharm. **1984**, *41*, 694–697.
10. Scott, S.A.; Smith, R.S.; Mason, H.L. Pharmaceutical

services in hospitals treating pediatric patients. Am. J. Hosp. Pharm. **1985**, *42*, 2190–2196.
11. Buck, M.L. Impact of new regulations for pediatric labeling by the Food and Drug Administration. Pediatr. Nurs. **2000**, *26* (1), 95–96.
12. Tisdale, J.E. Justifying a pediatric critical-care satellite pharmacy by medication-error reporting. Am. J. Hosp. Pharm. **1986**, *43*, 368–371.
13. Folli, H.L.; Poole, R.L.; Benitz, W.E.; Russo, J.C. Medication error prevention by clinical pharmacists in two children's hospitals. Pediatrics **1987**, *79* (5), 718–722.
14. Findling, R.L.; Preskorn, S.H.; Marcus, R.N.; Magnus, R.D.; D'Amico, F.; Marathe, P.; Reed, M.D. Nefazodone pharmacokinetics in depressed children and adolescents. J. Am. Acad. Child. Adolesc. Psych. **2000**, *39* (8), 1008–1016.
15. Erenberg, A.; Leff, R.D.; Haack, D.G.; Mosdell, K.W.; Hicks, G.M.; Wynne, B.A. Caffeine citrate study group. Caffeine citrate for the treatment of apnea of prematurity: A double-blind, placebo-controlled study. Pharmacotherapy **2000**, *20* (6), 644–652.
16. Avent, M.L.; Gal, P.; Ransom, J.L.; Brown, Y.L.; Hansen, C.J. Comparing the delivery of albuterol metered-dose inhaler via an adapter and spacer device in an in vitro infant ventilator lung model. Ann. Pharmacother. **1999**, *33* (2), 141–143.
17. Hovinga, C.A.; Chicella, M.F.; Rose, D.F.; Eades, S.K.; Dalton, J.T.; Phelps, S.J. Use of intravenous valproate in three pediatric patients with nonconvulsive or convulsive status epilepticus. Ann. Pharmacother. **1999**, *33* (5), 579–584.
18. Zell-Kanter, M.; Toerne, T.S.; Spiegel, K.; Negrusz, A. Doxepin toxicity in a child following topical administration. Ann. Pharmacother. **2000**, *34* (3), 328–329.
19. Nahata, M.C.; Morosco, R.S.; Hipple, T.F. Stability of mexiletine in two extemporaneous liquid formulations stored under refrigeration and at room temperature. J. Am. Pharm. Assoc. **2000**, *40* (2), 257–259.
20. Nahata, M.C.; Morosco, R.S.; Hipple, T.F. Stability of lamotrigine in two extemporaneously prepared oral suspensions at 4 and 25°C. Am. J. Health-Syst. Pharm. **1999**, *56*, 240–242.
21. Low, J.K.; Baldwin, J.N.; Greiner, G.E. Pediatric pharmacy education for U.S. entry-level doctor of pharmacy programs. Am. J. Pharm. Educ. **1999**, *63*, 323–327.
22. Nahata, M.C.; Taketomo, C. Pediatrics. In *Pharmacotherapy: A Pathophysiologic Approach,* 5th Ed.; DiPiro, J.T., Talbert, R.L., Yee, G.C., Matzke, G.R., Wells, B.G., Posey, L.M., Eds.; McGraw-Hill: New York, 2002; 69–77.
23. Bolinger, A.M.; Chan, C.Y.J.; Zanwill, A.B. Pediatrics. In *Applied Therapeutics: The Clinical Use of Drugs,* 7th Ed.; Koda-Kimble, M.A., Young, L.Y., Eds.; Lippincott Williams & Wilkins: Baltimore, 2000; 91-1–91-27.
24. Levin, R.H. Pediatric and Neonatal Therapy. In *Textbook of Therapeutics: Drug and Disease Management,* 6th Ed.; Herfindal, E.T., Gourley, D.R., Eds.; Williams & Wilkins: Baltimore, 1996; 1687–1696.
25. Behrman, R.E.; Kliegman, R.M.; Jenson, H.B. *Nelson Textbook of Pediatrics,* 16th Ed.; W.B. Saunders: Philadelphia, 1999; 2414 pp.
26. *Pediatric Pharmacology: Therapeutic Principles in Practice,* 2nd Ed; Yaffe, S.J., Aranda, J.V., Eds.; W.B. Saunders: Philadelphia, 1992; 663 pp.
27. Module 9, Pediatrics. In *Pharmacotherapy Self-Assessment Program,* 3rd Ed; Carter, B.L., Ed.; American College of Clinical Pharmacy: Kansas City, Missouri, 2000; 277 pp.
28. Taketomo, C.K.; Hodding, J.H.; Kraus, D.M. *Pediatric Dosage Handbook,* 8th Ed.; Lexi-Comp, Inc.: Hudson, Ohio, 2001; 1302 pp.
29. *Handbook of Pediatric Drug Therapy,* 2nd Ed.; Springhouse Corporation: Springhouse, Pennsylvania, 2000; 807 pp.
30. *2000 Red Book: Report of the Committee on Infectious Diseases,* 25th Ed.; Pickering, L.K., Ed.; American Academy of Pediatrics: Elk Grove Village, Illinois, 2000; 855 pp.
31. Yaster, M.; Krane, E.J.; Kaplan, R.F.; Cote, C.J.; Lappe, D.G. *Pediatric Pain Management and Sedation Handbook*; Mosby: St. Louis, 1997; 674 pp.
32. Phelps, S.J. *Teddy Bear Book. Pediatric Injectable Drugs.* 6th Ed.; American Society of Health-System Pharmacists, Inc.: Bethesda, Maryland, 2002; 423 pp.
33. Buck, M.L.; Hendrick, A.E. *Pediatric Medication Education Text,* 3rd Ed.; American College of Clinical Pharmacy: Kansas City, Missouri, 2001; 276 pp.
34. Briggs, G.G.; Freeman, R.K.; Yaffe, S.J. *Drugs in Pregnancy and Lactation,* 6th Ed.; Williams & Wilkins: Baltimore, 2002; 1595 pp.
35. Committee on Infectious Diseases, American Academy of Pediatrics Recommended childhood immunization schedule—United States, January–December 2001. Pediatrics **2001**, *109*, 162.
36. Committee on Drugs, American Academy of Pediatrics The transfer of drugs and other chemicals into human milk. Pediatrics **2001**, *108*, 776–789.
37. Committee on Drugs, American Academy of Pediatrics Guidelines for the ethical conduct of studies to evaluate drugs in pediatric populations. Pediatrics **1995**, *95*, 286–294.
38. Committee on Drugs and Committee on Hospital Care, American Academy of Pediatrics Prevention of medication errors in the pediatric inpatient setting. Pediatrics **1998**, *102*, 428–430.
39. Andrew, M.; Michelson, A.D.; Bovill, E.; Leaker, M.; Massicotte, M.P. Guidelines for antithrombotic therapy in pediatric patients. J. Pediatr. **1998**, *132*, 575–588.
40. White, S.R.; Yadao, C.M. Characterization of methylphenidate exposures reported to a regional poison control center. Arch. Pediatr. Adolesc. Med. **2000**, *154*, 1199–1203.
41. University of Virginia Children's Medical Center webpage at www.med.virginia.edu/medicine/clinical/pediatrics/CMC/cmchome.html (accessed June 2002).
42. Shirkey, H. Editorial comment: Therapeutic orphans. J. Pediatr. **1968**, *72*, 119–120.

Pew Health Professions Commission Reports

Carl E. Trinca
Edward H. O'Neil
Western University, Pomona, California, U.S.A.

INTRODUCTION

The Pew Commission Reports are a series of forward-looking documents that critically examine the education and training of health professionals in the context of the most dynamic decade ever faced by the nation's healthcare providers. Anchored by four "Commission" reports, the series is complemented by public policy papers, task force recommendations, and research reports on such topics as accreditation of educational programs,[1] regulation of the healthcare workforce,[2] and federal graduate medical education.[3]

ABOUT THE COMMISSION

Established in 1989 by the Pew Charitable Trusts[4] and administered initially at Duke University and later at the University of California–San Francisco's Center for the Health Professions, the Commission was "inspired by the belief that the education and training of health professionals is out of step with the evolving health needs of the American people." Since its inception, five values have guided the Commission's work:

- Universal access to basic healthcare services coordinated by the states, but implemented and supported by combined federal, state, and private resources.
- More efficient use of resources to better manage capacity and the outcomes of care while using best practices and reducing duplication.
- Socially desired outcomes based on empirical evidence.
- Creativity and innovation through experimentation with new providers, technologies, administrative structures, and ways to involve the patient within a modern regulatory framework.
- Informed choice and greater public responsibility through better use of information, decision making, and incentives for behavioral change.

The Commission Reports

The first Commission Report, *Healthy America: Practitioners for 2005*,[5] set the stage for the Commission's ten-year agenda by establishing the premise that reform of the system that prepares health professionals is central to crafting good healthcare policy for the nation. It described a framework for change within its view of evolving healthcare trends, laid out specific challenges to health profession educators and administrators, and proposed an agenda for action. The Commission made it quite clear that the success of its work was dependent on broad-based input and participation from all stakeholders, and it established advisory panels and health and public policy task forces, and launched research activities supporting its work.

The following three reports built on this initial framework. Each provided an updated snapshot of the evolving healthcare system in the United States including sections on access, quality, financing, and satisfaction. These sections were generally followed by observations and general recommendations to the healthcare and health profession education systems, followed by specific recommendations to individual professions.

Perhaps the most thoughtful and valuable contribution of the Commission was its 1993 presentation *Health Professions Education for the Future: Schools in Service to the Nation*[6] and subsequent revisions of competencies (Table 1) for successful practice in the 21st century. This effort has been used uniformly across the health professions by schools, professional associations, accrediting agencies, and regulatory bodies.

Critical Challenges: Revitalizing the Health Professions for the Twenty-first Century,[7] the Commission's third report published in 1995, noted that while the federal government was unable or unwilling to reform the healthcare system, a number of market-driven forces were at work directly challenging the high costs of the system. The most controversial recommendations in this Report centered on the belief that an oversupply of health professionals was imminent: The numbers of health pro-

Encyclopedia of Clinical Pharmacy
DOI: 10.1081/E-ECP 120006194

Table 1 Twenty-one competencies for the twenty-first century

Embrace a personal ethic of social responsibility and service.
Exhibit ethical behavior in all professional activities.
Provide evidence-based, clinically competent care.
Incorporate the multiple determinants of health in clinical care.
Apply knowledge of the new sciences.
Demonstrate critical thinking, reflection, and problem-solving skills.
Understand the role of primary care.
Rigorously practice preventive healthcare.
Integrate population-based care and services into practice.
Improve access to healthcare for those with unmet health needs.
Practice relationship-centered care with individuals and families.
Provide culturally sensitive care to a diverse society.
Partner with communities in healthcare decisions.
Use communication and information technology effectively and appropriately.
Work in interdisciplinary teams.
Ensure care that balances individual, professional, system, and societal needs.
Practice leadership.
Take responsibility for quality of care and health outcomes at all levels.
Contribute to continuous improvement of the healthcare system.
Advocate for public policy that promotes and protects the health of the public.
Continue to learn and to help others to learn.

fessionals entering practice should be reduced by closing schools in medicine, nursing, and pharmacy.

The Commission's final recommendations were presented in its fourth report. *Recreating Health Professional Practice for a New Century*[8] built on previous work and, as with earlier reports, suggested a strategic guide for changing schools, regulations, and professional practice.

THE COMMISSION AND PHARMACY

From the outset, pharmacy practice and pharmaceutical education figured in the Commission's agenda. In 1990, preceding its first report, *Perspectives on the Health Professions*[9] was commissioned and the chapter on pharmacy authored by Jack R. Cole and Jere E. Goyan. It described pharmacy's evolution as a profession, and presented the contemporary challenges facing pharmacy education and practice. Like other chapters, it served to identify common themes and initiate a dialog between the Trusts, professional leadership, and Commission staff.

Dr. Goyan also participated in the deliberations resulting in the first and second Commission Reports. In the first Report, little about pharmacy was specifically published except for the results of survey research quantifying pharmacists' perceptions of their education. Rather, a continuing case was being built for the inclusion of medication use within the context of responsible healthcare. Clearly the trends affecting healthcare in general (i.e., coordinated care, diversity and aging, expansion of science and technology, consumer empowerment) would have a profound impact on pharmacy. The second Report dedicated a chapter to pharmacy and introduced the term "pharmaceutical care" as a "responsible and participatory" process to a broader audience outside of the profession. It was no easy task to suggest that pharmacists "contribute to the drug therapy decision-making process, including the determination of dose, schedule, form, and delivery method; provide the medication; and monitor patient compliance, progress, and outcomes."

Specific strategies recommended in the second Report included curricular reform, documentation of pharmacist–patient–physician interactions, medication-use information systems, clinical training in non-institutional sites, and faculty development. These strategies provided guidance and supported the work of the American Association of Colleges of Pharmacy's Commission to Implement Change in Pharmaceutical Education[10] and its Center for Advancement of Pharmaceutical Education.

As previously described, the third Report offered a radical recommendation for pharmacy: "Reduce the number of pharmacy schools by 20–25% by the year 2005." The rationale for this recommendation, supported with preliminary data, was several-fold:

- Managed care would require a knowledge base and skill set different from contemporary pharmacy practice, more aligned with clinical pharmacy and pharmaceutical care.
- Technology, robotics, and technicians could do the manipulation-intensive work of pharmacists more efficaciously.
- Schools were perceived to be moving too slowly to get in step with the evolving healthcare needs of the American people. A less controversial recommendation suggested that pharmaceutical education focus even more on issues of clinical pharmacy, system management, and working with other healthcare providers.

As one of its five case studies, the third Report highlighted pharmaceutical care in education and practice. The Commission specifically chose to offer pharmaceutical education's experience in the decade:

> The process of educational evaluation and reform that has accompanied this professional evolution can serve as an

example as other health professions re-examine their missions in light of ongoing changes in the health care delivery system.

The fourth Report set forth three recommendations for pharmacy, each supporting the views of previous Commission's work:

- Continue to orient pharmaceutical education to reflect pharmacists' changing practice roles and settings under managed care and in clinical drug therapy. Referenced within this recommendation are various elements of the 21 competencies and more emphasis on residency training.
- Embrace an interdisciplinary approach to healthcare. Again the competencies are referenced as well as collaboration with pharmacy technicians and other allied health workers.
- Provide opportunities for re-training and continuing education for practitioners to develop skill sets for expanded clinical roles beyond dispensing pharmaceuticals. Included in this recommendation is greater emphasis on distance education and collaboration with managed care and chain pharmacy settings in developing and delivering re-training programs.

The Pew Charitable Trusts also supported other pharmacy-based initiatives during the 1990s, which were stimulated by the work of the Commission. The Pharmacy Manpower Project[11] worked with state boards of pharmacy to establish a method for monitoring supply and determining base-line numbers of and demographics for practicing pharmacists. The Scope of Pharmacy Practice Project[12] provided a contemporary update to an earlier study that described what pharmacists do in various practice settings. The results of these efforts, and their respective databases, are available to researchers for continuing analysis.

REFERENCES

1. Taskforce on Accreditation of Health Professions Education. *Strategies for Change and Improvement*; The Pew Health Professions Commission: San Francisco, California, June 1999.
2. Finocchio, L.J.; Dower, C.M.; Blick, N.T.; Gragnola, C.M.; Taskforce on Health Care Workforce Regulation. *Strengthening Consumer Protection: Priorities for Health Care Workforce Regulation*; The Pew Health Professions Commission: San Francisco, California, October 1998.
3. Pew Health Professions Commission Federal Policy Taskforce. *Beyond the Balanced Budget Act of 1997: Strengthening Federal GME Policy*; The Pew Health Professions Commission: San Francisco, California, October 1998.
4. The Pew Charitable Trusts support nonprofit activities in the areas of culture, education, the environment, health and human services, public policy and religion. Based in Philadelphia, The Trusts make strategic investments that encourage and support citizen participation in addressing critical issues effecting social change. In 1999, with more than $4.9 billion in assets, The Trusts awarded $250 million to 206 nonprofit organizations.
5. *Healthy America: Practitioners for 2005, an Agenda for Action for U.S. Health Professional Schools*; Shugars, D.A., O'Neil, E.H., Bader, J.D., Eds.; The Pew Health Professions Commission: San Francisco, California, 1991.
6. O'Neil, E.H. *Health Professions Education for the Future: Schools in Service to the Nation*; The Pew Health Professions Commission: San Francisco, California, 1993.
7. Pew Health Professions Commission. *Critical Challenges: Revitalizing the Health Professions for the Twenty-First Century*; UCSF Center for the Health Professions: San Francisco, California, 1995.
8. O'Neil, E.H.; Pew Health Professions Commission. *Recreating Health Professional Practice for a New Century*; The Pew Health Professions Commission: San Francisco, California, December 1998.
9. *Perspectives on the Health Professions*; O'Neil, E.H., Hare, D.M., Eds.; The Pew Health Professions Commission: Duke University, Durham, North Carolina, 1990.
10. *Papers from the Commission to Implement Change in Pharmaceutical Education*; The American Association of Colleges of Pharmacy: Alexandria, Virginia, 1992.
11. *The Scope of Pharmacy Practice Project: Final Report*; Greenberg, S., Smith, I.A., Eds.; The American Association of Colleges of Pharmacy: Alexandria, Virginia, 1994.
12. National Census of Pharmacists. The Pharmacy Manpower Project, Inc.: Alexandria, Virginia, 1992.

Pharmaceutical Benefits Scheme (Australia)

John Jackson
Integrated Pharmacy Services, Victoria, Australia

INTRODUCTION

The Pharmaceutical Benefits Scheme (PBS) is the Australian Federal Government program that provides public subsidy of a range of medications determined to meet safety, quality, effectiveness, and cost-effectiveness criteria. The PBS is available to every Australian resident and visitors from countries that have reciprocal healthcare agreements with Australia. PBS medicines can be prescribed for patients in ambulatory and aged care settings and in private hospitals. An alternative funding program is utilized in government funded hospitals. The Repatriation Pharmaceutical Benefits Scheme (RPBS) is an enhanced version of the PBS available to eligible war veterans and war widows.[1]

Under the program the government: 1) negotiates a price for a medicine with the supplier of the product; 2) controls the mark-ups applied at all stages of the supply chain; and 3) subsidizes the cost of the medicine to patients where the price exceeds certain limits. The subsidies are linked to the welfare needs of the patient and a safety net provides protection for high users of medication.

A number of programs have been introduced to the PBS that are aimed at improving the prescribing and utilization of medication. An example relevant to pharmacy is the Medication Management Program under which accredited pharmacists are remunerated for conducting medication reviews in both aged care and domiciliary settings. Under the Pharmacy Development Program, funding grants are provided to pharmacies achieving quality accreditation, for research into professional pharmacy services and, in the near future, for expanding the provision of medication information to consumers.

APPROVAL OF MEDICINES

Two separate branches of the Department of Health and Family Services determine which medicines are approved for use in Australia and which will be subsidized via the PBS. The Therapeutic Goods Administration is responsible for assessing medicines for quality, safety, and efficacy in order to approve them for marketing. Once approved for marketing, the sponsor (manufacturer) of the medication may apply to the Pharmaceutical Benefits Advisory Committee (PBAC) of the Pharmaceutical Benefits Branch to list the medicine on the Schedule of the PBS. The PBAC is an independent statutory body that applies cost-effectiveness criteria in determining if a medicine will be included on the PBS and any restrictions that are to apply to its use.[2]

On its inception in 1948 the PBS funded 139 ''life-saving and disease preventing drugs.'' By May 2000 the Schedule included 581 generic drugs available as 1,417 dosage forms and strengths marketed as 2,102 branded products.[3] It is estimated that 75% of prescriptions in the ambulatory care setting are for medicines listed on the PBS. The PBS has wide spread public support and bipartisan political support.

PBS PRESCRIBING AND DISPENSING

PBS listed medicines are eligible for subsidy if they are prescribed by a registered medical practitioner or, for certain medicines, a registered dental practitioner, in accordance with the regulations governing the scheme. The prescription must be dispensed by an approved pharmacist or, in limited cases, by an approved medical practitioner. The vast majority of medical practitioners, dentists, and pharmacists involved with the scheme are private practitioners.[4]

The PBS includes a set of general prescribing and dispensing regulations aimed at standardizing practices, simplifying processes, and minimizing waste and abuse. For example, a prescriber can not write more than one prescription for the same medicine for the same patient on the same day. Similarly, there are specified periods that must elapse before a pharmacist can dispense a second prescription of the same medicine for the same patient.[4]

In addition to the general regulations, there are specific regulations that apply to particular medicines. The maximum quantity of a medicine and maximum number of times a medicine can be dispensed per prescription are

Encyclopedia of Clinical Pharmacy
DOI: 10.1081/E-ECP 120006365

specified depending on the type of medicine. In general terms, medicines for chronic diseases can be prescribed in quantities adequate for one month of treatment at standard doses with five repeat supplies providing a total of six months of therapy per prescription. The maximum PBS quantities for medicines for acute conditions reflect standard treatment periods. Manufacturers supply most medicines in packs corresponding to the PBS maximum quantities.[4]

Medicines listed in the PBS Schedule fall into three categories of restriction: 1) Unrestricted medicines can be prescribed at the discretion of the medical practitioner; 2) restricted medicines should only be prescribed for specific therapeutic uses; and 3) authority medicines are also restricted to specific therapeutic uses but can only be prescribed following approval from a government agency.[4]

PRICES AND SUBSIDIES

The federal government has used its position of dominant funder in the prescription market to negotiate prices for medication that historically have been 30–40% lower than in the United States and Western Europe. Wholesale distributors and pharmacists are allowed to add specified mark-ups, and pharmacists also receive a professional dispensing fee resulting in the ''PBS dispensed price.'' In 1999, the average PBS dispensed price was $26.35, which equated to 3.65% of average weekly earnings.[5,6]

Most patients pay a co-payment toward the PBS dispensed price. Neither the price nor the therapeutic use of the medicine is a factor in determining the level of the co-payment. There is one level of co-payment for general patients and a lesser one for patients with concessional status. General patients pay the cost of the dispensed medicine up to a maximum of $21.00. All people of retirement age, on pension, receiving unemployment or welfare benefits, and all Veterans pay the concessional co-payment of $3.50 per medication (year 2001 rates). After dispensing a prescription, the pharmacist submits it to the Federal Government's Health Insurance Commission, which pays to the pharmacist the PBS dispensed price less the applicable patient co-payment. If a family unit requires chronic medication and reach expenditure thresholds on PBS prescriptions approximating either 30 or 50 prescriptions per calendar year, a stepped safety net scheme provides a higher level of subsidy. At maximum safety net benefit, the patient co-payment is eliminated.[7]

In 1999, the concessional patient contribution (then $3.20) represented 14% of the average PBS dispensed price of concessional prescriptions, and the general patient contribution ($20.30) represented 40% of the average PBS dispensed price of general prescriptions. Each Australian has on average six prescriptions for PBS medicines each year; however concessional patients have an average of 20 prescriptions per year, resulting in 79% of government expenditure on the PBS being directed to concessional patients.[6]

In the mid-1990s, expenditure on pharmaceuticals—largely expenditure on PBS listed medicines—accounted for 15% of total healthcare expenditure in Australia, which was 8.5% of gross domestic product. Of the total outlays on the PBS, manufacturers receive 67%, pharmacists 26%, and wholesalers 7%.[7]

Although the government's dominant position has enabled it to maintain relatively low costs for individual medicines, the PBS is an uncapped government program in which the level of overall government subsidy is largely driven by doctors' prescribing habits. With the introduction of new drugs that have benefits over older, cheaper medicines and with a population that is aging, the government outlay has increased rapidly. In the seven years from 1992 to 1998, the cost to the government of PBS prescriptions increased by 107%. An increase in prescription volumes of 6.5% in calendar 1999 over the previous year resulted in a further increase in government expenditure of 14.5%.

Brand Price Premiums and Brand Substitution

Different brands of the same drug may be listed in the PBS Schedule at different prices and the federal government subsidizes up to the price of the lowest priced brand. As a result of this policy, patients are required to pay a surcharge on their patient co-payment if they choose the higher priced brand of a drug when a lower priced generic exists.

This policy promotes generic substitution and pharmacists are allowed to substitute a bioequivalent medicine without reference to the prescriber if the patient agrees and the prescriber has not vetoed such substitution.[4]

Therapeutic Group Premiums

This policy extends the principle of brand price premiums to four therapeutic groups of medicines that are clinically similar. The groups are angiotensin converting enzyme inhibitors, calcium channel blockers, HMG CoA reductase inhibitors, and H2 receptor antagonists. There is at least one drug in each group available at the basic patient co-payment without supplementary charge.[4]

Both the Brand Price Premium and Therapeutic Group Premium policies are aimed at establishing benchmark prices within equivalent groups of medicines and promoting use of the cheapest of the options available.

REVIEW OF PBS LISTING OF MEDICINES

The PBS gives priority to listing of medicines for the treatment of conditions not amenable to self diagnosis and treatment. Consequently if a particular medicine or a group of medicines no longer requires prescriptions, it may be delisted from the PBS schedule. Similarly if medicines are deemed to be no longer an acceptable or cost-effective therapeutic choice, they may be delisted. Examples of delisting include all antihistamines, all preparations containing dextropropoxyphene, and all nasal sprays.

MEDICATION MANAGEMENT PROGRAM

The Medication Management Program (MMP) provides federal government funding for accredited pharmacists to undertake medication reviews. The aims of the reviews are to:

- Contribute to optimizing the therapeutic effectiveness and manageability of the medication regimen.
- Facilitate a cooperative working relationship between pharmacists and other members of the healthcare team in order to benefit the health and well being of the patient.
- Provide an information resource for colleagues in relation to patients' medications and the medication review process.[8]

A medication review involves the systematic evaluation of the patient's complete medication regimen including clarification of the indications for use and administration details of both prescription and nonprescription medicines plus the outcome of therapy. The process results in the generation of a report by the pharmacist documenting interventions and recommendations to other healthcare providers to optimize therapeutic outcomes.

The MMP replaced the Medication Review Program. Under the initial program pharmacists were funded to undertake systematic medication reviews for residents of aged care facilities (nursing homes and supported care hostels). The MMP increases the level of funding per review episode and introduces funding for medication reviews for domiciliary patients. While reviews in the aged care facility setting are at the initiation of pharmacists, reviews in the domiciliary setting are on referral from a medical practitioner.

Under this program pharmacists are being remunerated for their clinical expertise rather than for the supply of medicines. The federal government has determined that the cost of this program will largely be met out of savings in expenditure on PBS medicines as a result of rationalization of prescribing following the medication reviews.[9]

Other clinically focused services that are to be funded via the MMP include case discussions between pharmacists and medical practitioners and support for pharmacists to work within divisions of general practice as part of an integrated, geographically based primary healthcare service.

PHARMACY DEVELOPMENT PROGRAM

The Pharmacy Development Program (PDP) provides federal government support for programs of quality accreditation of pharmacies, for the provision of medicine information for consumers, and for research programs to identify cost effective professional pharmacy initiatives that can demonstrate net benefits for the Australian health system.

With this federal government support, professional pharmacy bodies have established practice standards, measurable performance indicators, and pharmacy accreditation programs. The standards address health promotion, drug dispensing, dose administration aids, patient counseling, compounding, medication reviews, comprehensive pharmaceutical care, drug information services, liaison pharmacy, and pharmacy services in residential care facilities.[10]

Funding arrangements for the accreditation and medicine information services are currently under negotiation but will take the form of block grants to accredited pharmacies.

The MMP and PDP are funded partly from the general revenue of the federal government and partly by the transfer of funds from projected growth in remuneration to pharmacists for dispensing of PBS prescriptions. As such, pharmacists are being remunerated for services aimed at improving the use of medication at the expense of growth in their remuneration from dispensing medication.

The total of the funds available for pharmacists' clinical practice via the MMP and PDP is small in comparison with the total PBS expenditure; however, both

government and professional organizations support continued expansion in these areas.[11]

REFERENCES

1. *Pharmaceutical Benefits Scheme (2002)*; Health Access and Financing, Australian Department of Health and Ageing: Canberra, Australia. www.health.gov.au/pbs/aboutus.htm (accessed July 2002).
2. Hall, R. *Medication in Australia—An Overview of the Key Decision Makers*; MediScene Newsletter, Commonwealth Department of Health and Aged Care: Canberra, Australia, 2000; 16, Issue 14.
3. Hall, R. *Pharmaceutical Benefits Scheme*; MediScene Newsletter, Commonwealth Department of Health and Aged Care: Canberra, Australia, 2000; 15, Issue 14.
4. *Schedule of Pharmaceutical Benefits for Approved Pharmacists and Medical Practitioner*; Commonwealth Department of Health and Aged Care: Canberra, Australia, August 2000; 25–48.
5. Davey, P.; Lees, M.; Aristides, M.; Ziss, S. *Report on the Australian System of Pharmaceutical Financing and Delivery*; Medical Technology Assessment Group: Chatswood, Australia, 1999; Vol. 1; 12–14.
6. *PBS Statistics (2002)*; Health Access and Financing, Australian Department of Health and Ageing: Canberra, Australia. www.health.gov.au/pbs/stats.htm (accessed July 2002).
7. *Who Pays? Where does the money go? (2002)*; Health Access and Financing, Australian Department of Health and Ageing: Canberra, Australia. www.health.gov.au/pbs/pbs/whopays.htm (accessed July 2002).
8. National Policy Committee Comprehensive Medication Reviews in Residential Aged Care Facilities. In *Pharmacy Practice Handbook 2000*; Pharmaceutical Society of Australia: Canberra, Australia, 2000; 80–82.
9. Elliot, R.; Thomson, W. Assessment of a nursing home medication review service provided by hospital based clinical pharmacists. Aust. J. Hosp. Pharm. **1999**, *29* (5), 255–260.
10. Cameron, N. Pharmacy standards ensure professional consistency. Pharm. Rev. **1999**, *23* (2), 58–59.
11. Wooldridge, M. Federal health minister: Pharmacy's key role in nation's health care structure. Pharm. Rev. **2000**, *24* (1), 8–9.
12. Contact for the Pharmaceutical Benefits Branch: telephone +61 2 62 89 70 99; URL www.health.gov.au/haf/branch.

PROFESSIONAL DEVELOPMENT

Pharmaceutical Care

Terry L. Schwinghammer
Grace D. Lamsam
University of Pittsburgh School of Pharmacy, Pittsburgh, Pennsylvania, U.S.A.

INTRODUCTION

As defined by Hepler and Strand in 1990, ''pharmaceutical care is the responsible provision of drug therapy for the purpose of achieving definite outcomes that improve a patient's quality of life.''

A revised definition provided in 1998 states that ''pharmaceutical care is a practice in which the practitioner takes responsibility for a patient's drug-related needs and is held accountable for this commitment. In the course of this practice, responsible drug therapy is provided for the purpose of achieving positive patient outcomes.''

The desired outcomes of pharmaceutical care are as follows: 1) cure of a patient's disease; 2) reduction or elimination of disease symptoms; 3) arresting or slowing progression of a disease; and 4) preventing a disease or symptoms. These outcomes are achieved through the identification and resolution of drug therapy problems.

HISTORY OF LANDMARK DEVELOPMENTS

In their seminal paper on pharmaceutical care, Hepler and Strand described the evolution of the pharmacy profession from traditional apothecary functions in the early twentieth century to dispenser of medications in the 1950s, and through the clinical pharmacy movement beginning in the mid-1960s. The concept of pharmaceutical care represents the culmination of this professional maturation from a primary focus on the dispensing and administration of drug products to the acceptance of a social responsibility for the health care needs of individual patients.

This new practice is required to meet an unfilled need in the health care system that has developed in part because of rapid increases in the number and complexity of available drug products. These therapeutic advances can improve health and may prolong life, but their use is also associated with a commensurate increase in the risk of preventable drug-related morbidity and mortality. In a 1995 study of the U.S. health care system, the estimated yearly costs associated with managing drug-related morbidity and mortality exceeded $76 billion in outpatient settings. The same study estimated that approximately 28% of hospital admissions resulted from drug-related problems. In 1999, the National Institute of Medicine estimated that 7000 patients die each year from medication errors that occur both within and outside hospitals.

To reduce medical problems associated with drug use, a rational decision-making process for drug therapy must be established, and a health care practitioner is needed to systematically apply this approach. The pharmacy profession recognizes the need to improve drug therapy outcomes, and since 1990 it has gradually moved toward adopting pharmaceutical care as the mission of pharmacy practice. The American Pharmaceutical Association (APhA), American Society of Health-System Pharmacists (ASHP), American Association of Colleges of Pharmacy (AACP), and Federation Internationale Pharmaceutique (FIP) have embraced the principles of pharmaceutical care. In 1998, the Alliance for Pharmaceutical Care was founded and now consists of 10 pharmacy organizations. Its purpose is to demonstrate to state legislators that patient care services provided by pharmacists can improve medication use and reduce health care costs associated with drug-related problems.

Pharmaceutical care is very different from other cognitive patient care services such as clinical pharmacy and disease state management programs. It is not a separate service that is superimposed on an existing practice but a comprehensive professional practice much like medicine and nursing. Consequently, it has a clearly defined philosophy of practice, patient care process, and practice management system. The remainder of this article describes the essential principles and processes involved in the practice of pharmaceutical care.

Encyclopedia of Clinical Pharmacy
DOI: 10.1081/E-ECP 120006330

PRINCIPAL CONCEPTS

Pharmaceutical Care as a Generalist Practice

Pharmaceutical care is a *generalist* practice in which the practitioner takes responsibility for providing consistent, ongoing, comprehensive care to a population regardless of age, gender, medical problem, drug class, or organ system disease. Care is not provided to some patients but not others because of time constraints, convenience, or simply personal preference. The practitioner is responsible for ensuring the appropriateness of *all* aspects of a patient's drug therapy. Moreover, each patient receives care that meets the same standard of quality. The generalist practitioner seeks the expertise of a specialist only when complex drug therapy problems exist that exceed the generalist's knowledge, skills, or experience.

Application to All Practice Settings

The principles and processes of pharmaceutical care practice are needed and applicable to patients in all pharmacy settings, including (but not limited to) inpatient, outpatient, community pharmacy, and academic sites. Practice is not restricted to select pharmacists based on years of experience, degree, specialty practice certificate, board certification, residency experience, or academic appointment. It is not a function of professional credentials or place of work but rather the desire and competence to take responsibility for the outcomes of each patient's drug therapy.

Creation of Collaborative Partnerships

Pharmaceutical care is a necessary component of the existng health care system and must be integrated with other aspects of health care. The pharmacist cooperates with the patient and other health care professionals but does not have sole responsibility for the medication use process. Other health professionals (e.g., physicians, nurses) also have clearly defined roles related to proper medication use. The concept of pharmaceutical care does not imply a reduction or elimination of those roles. In fact, qualified health care providers other than pharmacists can provide pharmaceutical care. However, pharmacists possess knowledge and skills that make them uniquely suited to assume the responsibility for reducing drug-related morbidity and mortality.

Consistency with Other Health Care Practices

Pharmaceutical care is a professional *practice* that addresses a unique set of health care needs not currently addressed by other practitioners. A practice is the application of a specialized set of knowledge to resolve specific patient problems in accordance with standards accepted by both the profession and society. These standards require that the patient care process be completed in its entirety for each patient. Pharmaceutical care practice shares three attributes common to all health care practices: 1) a single philosophy of practice; 2) one patient care process; and 3) a practice management system.

Philosophy of Pharmaceutical Care Practice

The practice of pharmaceutical care has a clearly articulated philosophy that defines values and explains *what* all practitioners must do. According to this philosophy, the practitioner performs the following: 1) takes responsibility for meeting society's need to reduce drug-related morbidity and mortality; 2) employs a patient-centered approach that addresses *all* the patient's drug-related needs; 3) establishes a caring therapeutic relationship with individual patients; and 4) assumes a clearly defined set of responsibilities that directs patient care activities. These responsibilities are to ensure that patients receive the most appropriate, effective, safe, convenient, and economical therapy; to identify, resolve, and prevent drug therapy problems; and to ensure that optimal patient outcomes are achieved.

Patient Care Process

The patient care process (Fig. 1) describes *how* an individual practitioner fulfills the responsibilities delineated in the philosophy of practice. This process includes three distinct elements that must be completed for each patient: 1) patient assessment; 2) creation of a pharmaceutical care plan; and 3) follow-up evaluation. The success of the process depends on the quality of the therapeutic relationship established with the patient.

The *Pharmacist's Workup of Drug Therapy* (PWDT) is a tool available to practitioners that serves as a guide through the steps of the patient care process. It offers a standardized format for efficient documentation of patient-specific information needed for a financially viable pharmaceutical care practice.

Patient assessment

The practitioner begins by assessing the patient's understanding, expectations, concerns, and behaviors relative to drug therapy to determine if *drug-related needs* exist. Other subjective and objective information is also gathered from review of patient records and other sources. The pharmacist then interprets that information to ensure all current therapy is appropriate and to iden-

Fig. 1 The patient care process within a pharmaceutical care practice. Adapted from Cipolle, R.J.; Strand, L.M.; Morley, P.C. *Pharmaceutical Care Practice*; McGraw-Hill: New York, 1998, p. 129.

tify *drug therapy problems* that may be interfering with the goals of therapy or placing the patient at future risk.

A *drug therapy problem* is an undesirable event involving drug therapy that actually or potentially interferes with a desired patient outcome. Seven major types of drug-related problems have been identified that relate to the appropriateness, effectiveness, safety, and convenience of drug therapy (Table 1). There are multiple possible causes of each type of problem.

Drug therapy problems drive subsequent steps in the patient care process; therefore, they are stated clearly and with a high level of specificity. Moreover, they are prioritized according to their degree of urgency as determined by the severity of potential harm to the patient.

Pharmaceutical care plan

The purposes of a care plan are as follows: 1) to solve the existing drug therapy problems that were identified during the assessment; 2) to achieve the desired outcomes for each medical condition; and 3) to prevent the development of future drug therapy problems.

In creating a care plan, the practitioner:

- Specifies the desired therapeutic goal for each problem identified.
- Considers all feasible actions or choices that may be made to achieve the stated goals.
- Outlines interventions designed to resolve drug therapy problems.
- Designs a monitoring plan to assess effectiveness and minimize undesirable adverse effects of drug therapy.

These functions require ongoing collaboration with the patient and other clinicians to develop and implement a plan that is agreed upon and understood by everyone involved. The individual providing pharmaceutical care is then responsible for ensuring that optimal therapeutic outcomes are achieved.

Follow-up evaluation

The purposes of follow-up evaluations are as follows: 1) to determine the actual outcomes that were achieved by the plan; 2) to assess the extent to which the plan has achieved the desired outcomes; 3) to determine if there are new or changing drug therapy problems that must be addressed; and 4) to determine if anything has occurred that increases the patient's risk for developing new drug therapy problems. The patient's progress is documented accurately in the pharmacy record and communicated effectively to the patient and other health care providers.

The status of each medical condition can be described according the following categories:

- *Resolved*—the goals have been achieved and therapy is completed.
- *Stable*—the goals have been achieved, but continue the same therapy.
- *Improved*—progress is being made toward achievement of the goals, so continue the same therapy.
- *Partial improvement*—progress is being made, but minor adjustments in the therapy are required.
- *Unimproved*—there is no measurable progress yet, but continue the same therapy.
- *Worsened*—there is a decline in health, so revise the therapy accordingly.

Table 1 Relationship of drug-related needs to drug therapy problems

Drug-related needs	Drug therapy problems
Appropriate indication for drug therapy	1. *Need for additional drug therapy*: Patient has a medical condition that requires initiation of new or additional drug therapy. 2. *Unnecessary drug therapy*: Patient is taking drug therapy that is not needed given their present medical condition.
Effective drug therapy	3. *Wrong drug*: Patient has medical problem treated with therapy that is less effective, more costly, or more hazardous than alternative therapies. 4. *Dose too low*: Patient is taking correct drug for medical condition, but too little of drug is being taken.
Safe drug therapy	5. *Adverse drug reaction*: Patient has medical problem caused by an adverse drug effect, which may include a side effect as well as an allergic reaction; idiosyncratic reaction; and a drug–drug, drug–food, or drug–laboratory test interaction.[a] 6. *Dose too high*: Patient is taking correct drug for medical condition, but too much of drug is being taken.
Convenient drug therapy	7. *Nonadherence*: Patient has medical problem resulting from not taking or receiving drug prescribed.

[a]In early descriptions of drug therapy problems, drug–drug, drug–food, and drug–laboratory test interactions were included as a separate eighth type of problem.

- *Failure*—the goals are not achievable with the present therapy, so initiate new therapy.
- *Expired*—the patient died while receiving drug therapy.

If changes in the plan are required to maintain or improve its efficacy, safety, or economy, the clinician coordinates these changes and communicates them to the patient and other health care providers.

Practice Management System

A practice management system provides the support necessary for effective and efficient practice. The system must allow for the addition of new patients and ensure the long-term financial viability of the practice. The system generally includes the following elements: 1) a statement of the mission of the practice; 2) the physical, financial, and human resources needed to support the practice; 3) a documentation system to evaluate the practice; and 4) reimbursement mechanisms to compensate the clinician and support the continuation of the practice.

MEASURING OUTCOMES OF PHARMACEUTICAL CARE PRACTICE

Cipolle and colleagues reported the outcomes achieved by 50 pharmacists who had completed the pharmaceutical care training program of the Peters Institute of Pharmaceutical Care at the University of Minnesota. Between 1993 and 1999, 14,357 patients received pharmaceutical care in 30 different ambulatory practice settings. The pharmacists assessed 97,511 drug therapy regimens and identified, resolved, and prevented 19,140 drug therapy problems. In a clinical and financial analysis of 1500 patients conducted between January 1998 and August 1999, pharmacist interventions avoided 193 unnecessary clinic visits, 72 multiple office visits, 36 emergency department visits, 28 urgent care visits, and 14 hospital admissions. Drug costs were reduced on 177 occasions. An estimated cost savings of $144,626 was realized, which reflects a savings of approximately $60 per patient. This represents a savings-to-cost ratio of 2:1 for the pharmaceutical care provided. When clinical outcomes of patients were evaluated, problems were resolved in 10%, stable in 38%, improved in 17%, and partially improved in 17%. Fourteen percent of patients were unimproved, 3% were worse, and 1% failed therapy. The authors concluded that the quality of care provided had a substantial positive impact on the clinical well-being of patients as well as on economic outcomes.

OBTAINING COMPENSATION FOR SERVICES PROVIDED

For a patient care practice to be recognized by the federal government and other third-party payers, four elements must be present: 1) a defined patient care practice;

2) a documentation system; 3) a formal evaluation system; and 4) reimbursement systems based on the level of patient needs.

Pharmacists participating in collaborative patient care environments (including disease state management and clinical pharmacy service programs) have experienced varying degrees of success in obtaining payment for the care provided. Although direct payment from patients is infrequently sought, a mail survey of 2500 American adults indicated that 56% of respondents were willing to pay for comprehensive pharmaceutical care services after they were provided with a description of what such care entailed.

Several different types of billing mechanisms have been used to gain compensation from third parties for services that are not tied directly to dispensing a drug product. Examples include fee-for-service, capitation payment, and the Health Care Financing Administration (HCFA) 1500 claim form. Each method has inherent advantages and disadvantages and may not be a suitable method for compensation for a comprehensive pharmaceutical care practice.

Payment for the majority of the patient care services provided by physician and nonphysician practitioners is based on the *resource-based relative value scale (RBRVS)*. The RBRVS ranks services according to the relative costs of the resources needed to provide them. The resulting relative value scale is then multiplied by a dollar figure to convert the service into a payment schedule. This payment model was successfully applied to pharmaceutical care practice in the Minnesota Pharmaceutical Care Project. Five levels of patient need were created based on the following: 1) the number of the patient's medical conditions; 2) the number of medications the patient is taking; and 3) the number of drug therapy problems identified. At 10 different community pharmacy practices in 1994, the average payment for a patient encounter was $12.14.

CHALLENGES TO IMPLEMENTATION OF PHARMACEUTICAL CARE

To date, actual and perceived barriers have impeded widespread acceptance and implementation of pharmaceutical care practice. Some factors affecting individual pharmacists include inadequate education and training in the required skills, absence of suitable role models and mentors, lack of time due to dispensing pressures, unwillingness to change to an entirely new philosophy of practice, fears of legal liability, fear of failure, and the absence of a tangible reward system. Pharmacy systems may pose impediments related to insufficient support personnel, lack of automated dispensing systems, limited physical facilities or poor organization of the available space, lack of adequate pharmacist compensation for cognitive functions, limited communication due to physical isolation from the patient and other health professionals, and inability to access necessary patient medical information. Other real or perceived barriers may include the resistance of physicians, administrators, and others; outdated and overly restrictive state board of pharmacy regulations; and lack of a market-driven demand for pharmaceutical care by the public.

CONCLUSION

Pharmaceutical care is a practice in which the clinician, working in concert with the patient and other health care providers, takes shared responsibility for the outcomes of all aspects of a patient's drug therapy. The goal of the practice is to ensure that every patient seen by the practitioner receives the most appropriate, effective, safe, economical, and convenient drug therapy possible. This is accomplished through the identification and resolution of actual and potential drug therapy problems. The essential components of a pharmaceutical care practice include a philosophy of practice, a patient care process, a documentation and evaluation system, and a method for obtaining financial compensation. All these elements must be in place for the practitioner to become recognized and compensated as a health care provider within the U.S. health care system. The future widespread adoption of pharmaceutical care practice by the pharmacy profession has great potential to reduce morbidity and mortality due to preventable drug therapy problems.

ACKNOWLEDGMENTS

The authors gratefully acknowledge Linda M. Strand, PharmD, PhD, and Robert J. Cipolle, PharmD, for their thoughtful review of this article.

BIBLIOGRAPHY

American College of Clinical Pharmacy. A vision of pharmacy's future roles, responsibilities, and manpower needs in the United States. Pharmacotherapy **2000**, *20*, 991–2000.

American Society of Hospital Pharmacists. ASHP statement on pharmaceutical care. Am. J. Hosp. Pharm. **1993**, *50*, 1720–1723, http://www.ashp.org (accessed Oct. 2000).

Cipolle, R.J.; Strand, L.M.; Morley, P.C. A New Professional Practice. *Pharmaceutical Care Practice*; McGraw-Hill: New York, 1998; 1–35.

Cipolle, R.J.; Strand, L.M.; Morley, P.C.; Frakes, M. *The Outcomes of Pharmaceutical Care Practice*, Presented in part at the 1st National Congress of Pharmaceutical Care, Donostia-San Sebastian, Spain, Oct. 28–30, 1999.

Commission to Implement Change in Pharmaceutical Education. *Background Paper I. What is the Mission of Pharmaceutical Education?* The American Association of Colleges of Pharmacy: Alexandria, Virginia, 1994; 3–9. http://www.aacp.org (accessed Oct. 2000).

FIP Statement of Professional Standards: Pharmaceutical Care (Adopted September 4, 1998); http://www.fip.nl (accessed Nov. 2000).

Hepler, C.D.; Strand, L.M. Opportunities and responsibilities in pharmaceutical care. Am. J. Hosp. Pharm. **1990**, *47*, 533–543.

Institute of Medicine Division of Health Care Services Committee on Quality of Health Care in America. *To Err is Human: Building a Safer Health System*; National Academy Press: Washington, DC, 1999. http://www.books.nap.edu (accessed Oct. 2000).

Johnson, J.A.; Bootman, L.J. Drug-related morbidity and mortality. Arch. Intern. Med. **1995**, *155*, 1949–1956.

Larson, R.A. Patients' willingness to pay for pharmaceutical care. J. Am. Pharm. Assoc. **2000**, *40*, 618–624.

Manasse, H.R., Jr. Medication use in an imperfect world: Drug misadventuring as an issue of public policy, part 1. Am. J. Hosp. Pharm. **1989**, *46*, 929–944.

Manasse, H.R., Jr. Medication use in an imperfect world: Drug misadventuring as an issue of public policy, part 2. Am. J. Hosp. Pharm. **1989**, *46*, 1141–1152.

McCallian, D.J.; Carlstedt, B.D.; Rupp, M.T. Elements of a pharmaceutical care plan. J. Am. Pharm. Assoc. **1999**, *39*, 82–83.

Strand, L.M.; Cipolle, R.J.; Morley, P.C. Documenting clinical pharmacy services: Back to basics. Drug Intell. Clin. Pharm. **1988**, *22*, 63–67.

Strand, L.M.; Cipolle, R.; Morley, P.C.; Ramsay, R.; Lamsam, G.D. Drug-related problems: Their structure and function. DICP **1990**, *24* (11), 1093–1097.

The American Pharmaceutical Association. *APhA Mission, Vision, Values, and Strategic Goals*; http://www.aphanet.org (accessed Oct. 2000).

The American Society of Hospital Pharmacists. Implementing pharmaceutical care. Proceedings of an invitational conference conducted by the American Society of Hospital Pharmacists and the ASHP Research and Education Foundation. Am. J. Hosp. Pharm. **1993**, *50*, 1585–1656.

Vaneslow, N.A.; Donaldson, M.S.; Yordy, K.D. From the Institute of Medicine: A new definition of primary care. JAMA **1995**, *273*, 192.

Pharmaceutical Care Spain Foundation

Joaquin Bonal de Falgas
Pharmaceutical Care Spain Foundation, Barcelona, Spain

INTRODUCTION

In 1998, a group of Spanish community pharmacists created an organization to promote educational, scientific, and professional activities in pharmaceutical care. After several contacts and meetings, a nonprofit foundation was constituted in 1998 named the Pharmaceutical Care Spain Foundation and was established in Barcelona (Muntaner Street 560 pal 1ª, 08022 Barcelona, Spain; phone and fax: 0034-93-2113108; e-mail: vade1@nacom.es).

The aims of this Foundation are:

- To implement, promote, and develop pharmaceutical care in Spain and Latin American countries, according to the concept formulated by Hepler and Strand in 1990.[1]
- To support health institutions and researchers interested in pharmaceutical care, to achieve a high quality of the research projects.
- To promote research projects on pharmaceutical care.
- To offer educational programs and counseling on subjects related to pharmaceutical care.
- To spread the results of the research studies and progress on pharmaceutical care through the organization of congresses, courses, meetings, workshops, and publications.

The Foundation is constituted by patrons that can be individuals or institutions. Most individuals are pharmacists, but there are also some physicians. Among the institutions that are patrons of the Foundation are medical organizations such as the Academy of Medical Sciences of Catalonia and Balearic Islands (ACMCB); Spanish primary healthcare net (REAP); Group CESCA, a scientific organization of primary care; and the Gerontology International Foundation (FGI). The support of these medical societies is essential for the credibility and progress of pharmaceutical care practice.

The Foundation also has a growing number of "collaborators"—individual pharmacists who are interested in the objectives of the Foundation. The number of such collaborators is constantly growing—at the end of May 2001, it was 400.

WHY THE FOUNDATION WAS CREATED

In Spain, during the 1970s, some hospital pharmacists initiated an important change in hospital pharmacy practice. Pharmacists began to leave their pharmacies, to offer their knowledge to physicians and nurses in hospitals, to integrate themselves into the healthcare team, and to reorient their professional practice by having the patient as the center of their activity. This process became the introduction of clinical pharmacy practice in hospital pharmacies that later demonstrated its effectiveness by improving the quality of care as well as its economic efficiency.[2]

However, the concept and philosophy of clinical pharmacy is not yet implemented in all hospitals and is found even less among community pharmacists. Most community pharmacists saw clinical pharmacy linked to hospital practice and hardly adapted to the needs of the community practice setting. That is probably the reason why the pharmaceutical care concept was formulated by Hepler and Strand. But with the concept of pharmaceutical care, we carry the same risk as with clinical pharmacy. Many people are talking about the concept, but few are putting in the effort and means required to put it into practice.

Pharmaceutical Care Spain Foundation was created with the desire to influence the society in general—starting with the community of pharmacists—in order to disseminate pharmaceutical care practice to improve healthcare quality and to provide better service to the patients and to the healthcare system. Consequently, the Foundation has to extend pharmaceutical care practice among practitioners. The Foundation works with officially established pharmaceutical organizations in order to clarify the direction in which pharmacy practice should be oriented and to determine which services are offered to the citizens. These services have to go further than simply drug dispensing. The Foundation must also convince healthcare authorities that pharmaceutical care is adding a significant

Encyclopedia of Clinical Pharmacy
DOI: 10.1081/E-ECP 120006384

value to the drug itself, making the prescription and use of drugs more rational and consequently more cost effective. Also, in relation to Spanish universities, the Foundation has to influence the reorientation of its educational programs and teaching methods in order to create the culture, attitudes, knowledge, and skills required among pharmacy students to provide pharmaceutical care to society. Pharmaceutical care should be the main educational objective for all schools of pharmacy. Finally, another reason for the Foundation is the need to inform and to show the benefits that people can obtain from the pharmaceutical care practice in terms of ensuring the safety and effectiveness of their pharmacotherapy. No service will be demanded by society if people ignore its existence and the benefits that can be obtained from it.

ACTIONS OF PHARMACEUTICAL CARE SPAIN FOUNDATION

In two years of existence, the Foundation has developed some activities in order to achieve the above-mentioned objectives:

- *Journal Pharmaceutical Care Spain*: This publication was initiated in January 1999,[3] with six issues per year. The contents of the journal are original papers, reviews, international news, drug information, etc. Also, a continuing education program is offered through the journal to all subscribers.
- Vade mecum of the official technical information about drugs commercialized in Spain: this is offered via the Internet to all physicians and pharmacists of the country through the Web at www.saludaliafarma.com or www.saludaliamedica.com.
- Procedures manual on pharmaceutical care: Because many colleagues plan to initiate provision of pharmaceutical care but do not know how to start, the Foundation published this manual to guide them.[4]
- First National Congress of Pharmaceutical Care: Held in San Sebastián with more than 1000 participants, this congress included several distinguished speakers, including Charles Hepler, Linda Strand, J.W. Foppe van Mil, and Robert Cipolle. The Ministry of Health also attended the meeting and 77 communications on Pharmaceutical Care experiences were presented.
- Consensus on pharmaceutical care: Due to the differing conceptions that Spanish professional groups have on pharmaceutical care, and the difficulties in translating English terminology into Spanish, a working party with all the interested groups was established in order to achieve a consensus document on the subject. This working group had been coordinated by the General Director of Pharmacy of the Ministry of health.[5]
- Competencies document: A competencies document of the pharmacy profession is planned with strong participation by the Pharmaceutical Care Spain Foundation. The document is intended to define the areas in which pharmacists are competent within the healthcare practice.[6]

RECENT AND FUTURE DEVELOPMENTS

During 2001, the Foundation offered different courses through the Internet. Topics include:

- Methodology to implement pharmaceutical care into practice.
- Concepts on pharmacoepidemiology.
- Pharmacoeconomics.
- Evaluation of the scientific biomedical literature.

Other courses will be more oriented to pharmacotherapy:

- Treatment of blood coagulation disorders.
- Arterial hypertension.
- Peripheral vascular pathologies.
- Congestive heart failure.

The Foundation held the Second National Congress on Pharmaceutical Care November 2001 in Barcelona. The main subject of this congress was ''communication with patients and with the other healthcare practitioners.'' The Foundation believes that communication skills are the main deficiencies of community pharmacists, and these are essential for pharmaceutical care implementation and practice.[7]

Some research projects promoted or initiated by the Foundation are currently ongoing. A European research consortium, with the participation of different countries, is proposed to study the influence of pharmaceutical care in the control and treatment of minor diseases from community pharmacists, and it has submitted to the European Community in Brussels asking for economic support. Such research will be coordinated by the School of Pharmacy of the University of Manchester, United Kingdom. A project promoted by the European Society of Clinical Pharmacy is currently studying the number of drug related problems (DRP) identified by community pharmacists on patients discharged from the hospital. A research study on the number of DRP identified by pharmacists in patients admitted to emergency services of hospitals was initiated to find out how many patients are admitted in emergency rooms because they are suffering some DRP, and how many of these DRPs could be prevented through pharmaceutical care practice.

The Pharmaceutical Care Spain Foundation is also cooperating in an organization called Pharmaceutical Care Network Europe (PCNE) that is promoting pharmaceutical care practice and research. One of the difficulties we have in Europe is the diversity of pharmacy organization system and practice, as well as languages and healthcare structure. Such differences, between European countries, are well discussed by Foppe van Mil.[8]

REFERENCES

1. Hepler, C.D.; Strand, L.M. Opportunities and responsibilities in the Pharmaceutical Care. Am. J. Hosp. Pharm. **1990**, *47*, 533–543.
2. Bonal, J. Porque se ha creado una Fundación de Pharmaceutical Care? Editorial Pharm. Care Esp. **1999**, *1*, 1–2.
3. Fernandez-Llimós, F.; Herrera, J. Por qué creamos una revista? Editorial Pharm. Care Esp. **1999**, *1*, 3.
4. Alvarez de Toledo, F.; Barbero, J.A.; Bonal, J.; Dago, A., et al. *Manual de Procedimientos en Atención Farmacéutica,* 1st Ed.; Fundación Pharmaceutical Care España: Barcelona, Septiembre 1999.
5. Consenso sobre Atención Farmacéutica. Grupo de Consenso de AF (2000). Dirección General de Farmacia y Productos Sanitarios. Ministerio de Sanidad y Consumo. Madrid.
6. Competencias de la profesión de Farmacia. CCECS borrador 2000. Barcelona.
7. Bonal, J. El segundo Congreso Nacional de Atención Farmacéutica. Editorial Pharm. Care Esp. **2000**, *2*, 387–389.
8. Van Mil, J.W.F. Pharmaceutical care in community pharmacy in Europe, challenges and barriers. Pharm. Care Esp. **2000**, *2*, 42–56.

Pharmaceutical Outcomes

Neil J. MacKinnon
Ingrid S. Sketris
Dalhousie University, Halifax, Nova Scotia, Canada

INTRODUCTION

Pharmaceuticals are among the most widely used interventions in health care today. They are used in both the prevention and treatment of disease and amelioration of symptoms to improve patient outcomes. The relative importance of pharmaceuticals in health care is growing in most industrialized nations, as the percentage of health care expenditures for pharmaceuticals continues to increase. Spending on prescription drugs in the United States is increasing at a rate of 12% annually and will continue to increase approximately 10% per year over the next 8 years.[1]

Not all outcomes from the use of pharmaceuticals are positive. Adverse drug reactions are the fourth to sixth leading cause of death,[2] and in the United States, medication errors are the largest category of the 44,000–98,000 deaths each year due to medical errors.[3] The total annual cost of drug-related morbidity and mortality in the ambulatory setting in the United States has more than doubled in 5 years, increasing from $76.7 billion in 1995 to $177.4 billion in 2000.[4]

Many drug-related adverse outcomes are preventable. For example, a significant percentage of hospital admissions are due to drug-related morbidity,[5] and approximately one-half of these admissions are preventable.[6,7] Preventable adverse outcomes related to drug use—often referred to as preventable drug-related morbidities (PDRMs)—are typically not due to the inherent pharmacological properties of medicines, but rather to problems such as a lack of systematic approach to monitoring of patients receiving drug therapy by health professionals. A Canadian study estimated that the annual cost of PDRM in older adults in that country is $10.9 billion (Canadian) CND.[8]

One framework for looking at pharmaceutical outcomes—whether these outcomes are positive or negative—is the ECHO (Economic, Clinical, and Humanistic Outcomes) model, proposed by Kozma, Reeder and Schulz.[9] According to this model, three main types of health-related outcomes should be considered: economic (drug-related and non-drug-related), clinical (clinical events, physiologic and metabolic measures), and humanistic outcomes [such as patient satisfaction and health-related quality of life (HRQoL)].[9] Economic and clinical (and fairly often humanistic) outcomes are incorporated into pharmacoeconomic evaluations.

Patient satisfaction is growing in importance and has become a key part of most health care report cards. Perceived satisfaction with medical treatment is key in determining the overall well-being and health of patients, and is recognized as being one of the main components of quality. HRQoL relates to those aspects of a patient's life dominated by personal health. Although differing opinions exist, there is some consensus that there are four main dimensions (or domains) constituting "health": a physical dimension, a functional dimension, a psychological dimension, and a social dimension.[10] Increasingly, HRQoL is being used in clinical drug trials and in pharmacy program evaluations, and to monitor patients.[10]

Although this article focuses on pharmaceutical outcomes, it is worth noting the relationship of outcomes to processes and structures. Donabedian has argued that understanding health care structures, process, and outcomes is critical.[11] Structure and process performance measures are commonly used by health care organizations. However, they are of limited value because few structures and processes have been proven to be directly associated with patient outcomes.[12] Still, an understanding of processes, and even structures, is important to improve the efficiency of health care services.[13] Lipowski recommends that outcomes be evaluated first and then processes, but cautions that determining a definite causal relationship between the two is often difficult.[14] Monitoring process and outcomes are interrelated activities.[15–18] Improving outcomes is the goal; studying processes can point the way to achieving that goal, and outcomes must be linked to processes to improve the quality of health care.[19] Global outcomes such as

Encyclopedia of Clinical Pharmacy
DOI: 10.1081/E-ECP 120006332

morbidity and readmission are the results of multiple factors, and it is difficult to attribute those outcomes to a specific encounter or element of care without additional information about key factors.[16,20–22] Although outcome studies are increasingly being undertaken, there is much to be learned about the relationships between processes and outcomes.

EVALUATING PHARMACEUTICAL OUTCOMES

Misuse, underuse, and overuse of medical care and pharmaceuticals can lead to suboptimal patient outcomes. These are severe problems that require urgent attention.[19] Indeed, measurement of quality of medical care is a daunting but essential task.[23] Most organizations such as pharmacy regulatory agencies and hospitals, which evaluate the medication use process, focus on structural elements (such as checking whether narcotics are safe and secure, and whether pharmacists' licenses are displayed) that have questionable relationships to patient outcomes. However, the evaluation of pharmaceutical-related patient outcomes is becoming more important as employers and health care payers are demanding more accountability and transparency with health care spending.

Performance Measures

Performance measures, or indicators, are frequently used to measure and compare pharmaceutical outcomes. Performance measures are used to monitor and evaluate important governance, management, clinical, and support functions that affect patient outcomes.[24] A performance measure can be used in one of three ways: to warn of potential quality problems, to measure the result of process improvement, and to monitor continuing performance.[25] They are typically expressed as ratios.

These measures can define practice norms when absolute standards or the prevalence of conditions and practices are unknown. Findings can be compared over time and space, across sites of care and health care organizations. Patterns, trends, and gross variation from practice norms are discernible.

The two major types of performance measures are: 1) rate-based performance measures, which measure an event for which a certain proportion of the events are expected to occur even with quality care (such as mistimed prescription refills); and 2) sentinel-event performance measures, which measure a serious event that requires an indepth review for each occurrence of the event (such as a medication error).[24] To provide meaningful information, performance measures must be reliable, valid, and feasible. To ensure this, they must be carefully selected, thoroughly tested, evaluated, and used within the context of a comprehensive performance measurement system.[26] In isolation, no single measure serves as a direct measure of quality but as an indicator of the need to direct attention to a potential problem.

Performance Measurement Systems

Some of the more common performance measurement systems used to assess pharmaceutical outcomes include report cards, balanced scorecards, clinical value compasses, profiling, performance-based evaluation systems, and others. The goals of pharmaceutical performance measurement systems are to: 1) compare treatment modalities fairly; 2) recognize and promote good care; 3) identify and eliminate substandard care; and 4) improve the level of care overall.[27] Because performance measures can include data over the course of treatment, the outcomes of alternative therapies and practices may be detected. The end goal of any performance measurement system should not be cost containment only; improving patient outcomes must be a primary concern, keeping in mind the cost effectiveness of the therapy and sustainability of the system.

All performance measurement systems must consider the level of outcomes data collection and aggregation. Assessment can occur at multiple levels: by individual practitioner, institution, geographic region or state/province; by symptoms, diseases, or organ system; by encounter, episode of care, or continuous monitoring; and by the setting or settings in which care was delivered. The size of the population will affect the level of aggregation that is reasonable. Large samples are required to detect differences in processes and outcomes of care. Moreover, the time span covered by the data in conjunction with the size of the population will determine the level of aggregation possible.

Uses of Performance Measurement Systems

A survey of stakeholders of provider profiling of pharmaceutical use in the United States revealed that the top five uses for these systems are (in descending order of use): 1) educate and give information to the physician; 2) change physician behavior and influence prescribing patterns; 3) monitor and improve the quality of care; 4) identify potential problems; and 5) improve patient outcomes.[28] A Canadian survey had similar results.[29]

Performance measures can reveal the impact of provider practice patterns on medical care, resource use, and cost of care. Often, this information is provided back

to the physician or other health care practitioner so the individual can review their own data. Indeed, when interventions have shown positive changes in medical practice, one consistent finding is that providers were given information on how their practices and outcomes compared with others in the community.[30,31] Feedback is successful when used alone or in combination with education, rewards, or the support of opinion leaders.

Pharmaceutical outcome data are also used to improve the quality of care, identify potential problems, and improve patient outcomes. These data are often used within a continuous quality improvement (CQI) cycle, where rate-based performance measures are tracked over time and used in conjunction with control charts to show changes in quality and assess the impact of programs or changes in process. Information can be fed back to front line health care practitioners, areas for possible improvement identified, appropriate changes made, and reassessments initiated.

Examples of Specific Performance Measurement Systems

The United States has several major performance measurement systems in place. Two of the most important systems are those directed by the National Committee for Quality Assurance (NCQA) for managed care organization (MCO) accreditation and The Joint Commission on Accreditation in Healthcare Organizations (JCAHO) for health care organization accreditation. Both NCQA's performance measurement system, HEDIS, and JCAHO's IMSystem include several pharmaceutical outcomes indicators.

Several other projects in the United States are developing further pharmaceutical outcome measures. SCRIPT (Study of Clinically Relevant Indicators for Pharmacologic Therapy), initiated by the Health Care Financing Administration (HCFA), is a multidisciplinary coalition representing 52 public and private sector organizations that is currently beta-testing several pharmaceutical performance measures.[32] The Performance Indicators for Continuous Quality Improvement in Pharmacy (PICQIP) study, a partnership between the University of Florida and the American Pharmaceutical Association Foundation Quality Center, is an example of new pharmacy indicators being used in a CQI cycle.[33,34]

The Canadian Institute for Health Information (CIHI) is actively working on developing drug performance indicators for uptake and application by the Canadian Council on Health Services Accreditation (CCHSA). In Great Britain, the National Health Service (NHS) has established a National Performance Framework that has developed primarily process indicators to assess the impact of health services. The National Prescribing Service (NPS) was launched by the Australian Government in 1998 and is undertaking work in Quality Use of Medicines (QUM). It has also done work in the area of measurement of pharmaceutical outcomes.

Limitations of Evaluating Pharmaceutical Outcomes

David Eddy at Duke University has written extensively on the problems and potential solutions related to pharmaceutical performance measurement systems.[22] According to a U.S. survey, the most commonly perceived problems with pharmaceutical performance measurement systems are limitations with billing and administrative databases, lack of time to review summary data by physicians, and incomplete data.[28] Other limitations include risk adjustment (what if my practice has ''sicker'' patients), overreliance on administrative (claims) data rather than clinical data (therefore lacking key patient outcomes), patient individuality and variation in medical practice, and lack of capacity for taking into account a discipline-specific rather than a whole programs-oriented CQI approach. There has also been some debate on the reliability of performance measurement systems to assess the true impact of physician care on the quality of health care.[35,36]

DATABASES FOR PHARMACEUTICAL OUTCOMES

Difficulty in obtaining the data necessary for measurement hinders pharmaceutical outcomes assessment. Both clinical and administrative data sources have limitations. Medical chart review is time consuming and expensive, and not all findings and services rendered by providers are documented faithfully.[37] Administrative records, often in the form of paid claims for reimbursement, are more readily retrievable, but are collected for administrative purposes and need validation to be used for assessing pharmaceutical outcomes.

Available sources from the public system include census data, health expenditure data, physician clinical record billing data, prescription drug claims data, population health surveys, hospital and nursing home medical records, and disease- or organ-specific registries. Private sector data include data from the pharmaceutical industry, private insurance industry, drug benefit management companies, and individual private firms.[38] Developments in health care information technology (IT) should facilitate the collection of pharmaceutical outcomes data in the future.

ETHICS, PRIVACY, CONFIDENTIALITY, AND DATA SECURITY

It is generally accepted that persons have a right to control their health information so they are not harmed. Many jurisdictions are developing guidelines or legislation related to privacy, confidentiality, and security of personal health information. For example, in the United States, the Health Insurance Portability and Accountability Act (HIPAA) of 1996 and the associated final Privacy Rule of December 28, 2000 (which took effect on April 14, 2001) gave patients more control over how their personal health information is used, and addressed the duties of health care plans and providers to protect health information.[39] Research ethics committees need to determine when patients need to give explicit consent for the use of their personal health information for new research questions when it has been collected for other research questions or administrative purposes. The soundness of scientific methodology needs to be assured. The potential public health benefit of the research needs to be weighed against any risk to the individual. Guidelines for data confidentiality and anonymity have been developed in many jurisdictions and methods for masking or delinking personal identifiers or reporting only aggregate data have been developed and implemented. Conflict of interest and financial relationships in the research need to be disclosed. There should also be vigilance on the technical side of ensuring security of data including controlled access to and storage, maintenance and transmission of, personal health information.[40–46]

ROLE OF VARIOUS STAKEHOLDERS IN PHARMACEUTICAL OUTCOMES MEASUREMENT

Improving drug use outcomes requires the participation and cooperation of everyone involved in the case process. Physicians and other prescribers should specify the therapeutic objective in writing for each prescription they write, and this should be communicated to the patient and other members of the health care team. Pharmacists, nurses, and other health care professionals should monitor patients, documenting their interactions with patients, and communicate this information back to the health care team. ASHP's Statement on the Pharmacist's Role in Primary Care advocates the measurement of pharmaceutical care outcomes: ''High-quality, coordinated, and continuous medication management for patients should be measurable as a result of the provision of pharmaceutical care within a primary care delivery model,'' and ''Pharmacists should evaluate all components of the medication-use process to optimize the potential for positive patient outcomes.''[47] Health administrators need to implement systematic ways of measuring the processes, outcomes, and costs of medication use. Health policy makers should take note of policy synthesis documents, evidence of the effectiveness of drug policies implemented in other jurisdictions and, when possible, evaluate policies and programs to determine their effect on patient outcomes, health care resource utilization and costs. Pharmaceutical companies need to help health practitioners proactively identify patients at risk for PDRM from their products and help to implement systems to improve drug use. They need to continue to participate in methodologically sound pharmaceutical outcomes research. Finally, each time they receive a prescription, patients should question what is the intended outcome of this drug and who will be working with me to ensure that I achieve this outcome?

USEFUL RESOURCES

Textbooks

International Society for Pharmacoeconomics and Outcomes Research, ISPOR Lexicon Pashos, Klein, Wanke, L.A., Eds.; ISPOR: New Jersey, 1998.

Spilker, B. In Quality of Life and Pharmacoeconomics in Clinical Trials; 2nd Ed.; Spilker, B., Eds.; Lippincott-Raven: Philadelphia, PA, 1996.

Organizations/Web Sites

Academy of Managed Care Pharmacy (www.amcp.org).

American College of Clinical Pharmacy (Outcomes and Economics PRN) (www.accp.com).

Academy for Health Services Research and Health Policy (www.academyhealth.org).

Canadian Coordinating Office for Health Technology Assessment (www.ccohta.ca).

International Health Economics Association (www.healtheconomics.org).

International Society for Pharmacoeconomics and Outcomes Research (www.ispor.org).

International Society for Pharmacoepidemiology (www.pharmacoepi.org).

International Society for Quality of Life Research (www.isoqol.org).
National Committee for Quality Assurance (www.ncqa.org).
Society for Medical Decision Making (www. smdm.org).

ACKNOWLEDGMENTS

The authors acknowledge the contributions of Sharon Boriel, Susan Bowles, Murray Brown, Chole Campbell, and Rick Manuel.

REFERENCES

1. Mehl, B.; Santell, J.P. Projecting future drug expenditures—2001. Am. J. Health-Syst. Pharm. **2001**, *58*, 125–133.
2. Lazarou, J.; Pomeranz, B.H.; Corey, P.N. Incidence of adverse drug reactions in hospitalized patients. A meta-analysis of prospective studies. JAMA, J. Am. Med. Assoc. **1998**, *279* (15), 1200–1205.
3. *To Err is Human. Building a Safer Health System. Institute of Medicine*; Kohn, L.T., Corrigan, J.M., Donaldson, M.S., Eds.; National Academy Press: Washington, 1999.
4. Ernst, F.R.; Grizzle, A.J. Drug-related morbidity and mortality: Updating the cost-of-illness model. J. Am. Pharm. Assoc. **2001**, *41* (2), 192–199.
5. Grymonpre, R.E.; Mitenko, P.A.; Sitar, D.S.; Aoki, F.Y.; Montgomery, P.R. Drug-associated hospital admissions in older medical patients. J. Am. Geriatr. Soc. **1988**, *36* (12), 1092–1098.
6. Nelson, K.M.; Talbert, R.L. Drug-related hospital admissions. Pharmacotherapy **1996**, *16* (4), 701–707.
7. Hallas, J.; Worm, J.; Beck-Nielsen, J.; Gram, L.F.; Grodum, E.; Damsbo, N.; Brøsen, K. Drug related events and drug utilization in patients admitted to a geriatric hospital department. Dan. Med. Bull. **1991**, *38*, 417–420.
8. Kidney, T.; MacKinnon, N.J. Preventable drug-related morbidity and mortality in older adults: A Canadian cost-of-illness model. Geriatr. Today **2001**, *4* (3), 120 (presentation abstract).
9. Kozma, C.M.; Reeder, C.E.; Schulz, R.M. Economic, clinical, and humanistic outcomes: A planning model for pharmacoeconomic research. Clin. Ther. **1993**, *15*, 1121–1132.
10. MacKinnon, N.J.; Chaudhary, D.A.N. Pharmacy practice research: An introduction to generic health-related quality of life instruments. Can. Pharm. J. **2000**, *133* (4), 40–42.
11. Donabedian, A. The quality of care: How can it be assessed? JAMA, J. Am. Med. Assoc. **1988**, *260* (12), 1743–1748.
12. Holdford, D.A.; Smith, S. Improving the quality of outcomes research involving pharmaceutical services. Am. J. Health-Syst. Pharm. **1997**, *54*, 1434–1442.
13. Campbell, A.B. Benchmarking: A performance intervention tool. Jt. Comm. J. Qual. Improv. **1994**, *20* (5), 225–228.
14. Lipowski, E.E. Evaluating the outcomes of pharmaceutical care. J. Am. Pharm. Assoc. **1996**, *NS36* (12), 726–739.
15. Jencks, S.F.; Wilensky, G.R. The health care quality improvement initiative. A new approach to quality assurance in medicare. JAMA, J. Am. Med. Assoc. **1992**, *268* (7), 900–903.
16. McNeil, B.J.; Pedersen, S.H.; Gatsonis, C. Current issues in profiling quality of care. Inquiry **1992**, *29*, 298–307.
17. Wray, N.P.; Ashton, C.M.; Kuykendall, D.H.; Petersen, N.J.; Souchek, J.; Hollingsworth, J.C. Selecting disease-outcome pairs for monitoring the quality of hospital care. Med. Care **1995**, *33* (1), 75–89.
18. Siu, A.L.; McGlynn, E.A.; Morgenstern, H.; Beers, M.H.; Carlisle, D.M.; Keeler, E.B.; Beloff, J.; Curtin, K.; Leaning, J.; Perry, B.C.; Selker, H.P.; Weiswasser, W.; Wiesenthal, A.; Brook, R.H. Choosing quality of care measures based on the expected impact of improved care on health. Health Serv. Res. **1992**, *27* (5), 619–650.
19. Hannan, E.L. The continuing quest for measuring and improving access to necessary care. JAMA, J. Am. Med. Assoc. **2000**, *284* (18), 2374–2376.
20. Tompkin, C.P.; Bhalotra, S.; Garnick, D.W.; Chilingerian, J.A. Physician profiling in group practices. J. Ambulatory Care Manage. **1996**, *9* (4), 28–39.
21. Heineken, P.A.; Charles, G.; Stimson, D.H.; Wenell, C.; Stimson, R.H. The use of negative indexes of health to evaluate quality of care in a primary-care group practice. Med. Care **1985**, *23* (3), 198–208.
22. Eddy, D.M. Performance measurement: Problems and solutions. Health Aff. (Millwood) **1998**, *17* (4), 7–25.
23. Fihn, S.D. The quest to quantify quality. JAMA, J. Am. Med. Assoc. **2000**, *283* (13), 1740–1742.
24. Joint Commission on Accreditation of Healthcare Organizations. *Primer on Indicator Development and Application: Measuring Quality in Health Care*; JCAHO: Chicago, 1992.
25. Angaran, D.M. The performance-based evaluation system. Pharmaguide Hosp. Med. **1993**, *6* (3), 4–7.
26. MacKinnon, N.J. Performance measures in ambulatory pharmacy. Pharm. Pract. Manage. Quart. **1997**, *17* (1), 52–62.
27. Lee, C.; Rogal, D. *Risk Adjustment: A Key to Changing Incentives in the Health Insurance Market*; Alpha Center: Washington, DC, 1997; 1–26.
28. MacKinnon, N.J.; Lipowski, E.E. Opinions on provider profiling: Telephone survey of stakeholders. Am. J. Health-Syst. Pharm. **2000**, *57*, 1585–1591.
29. MacKinnon, N.J.; MacDonald, N.L. The State of Performance Measurement for Pharmacy in Canada. In *A Report for the Pharmacy Endowment Fund*; Dalhousie University College of Pharmacy, January 2000.

30. Braham, R.L.; Ruchlin, H.S. Physician practice profiles: A case study of the use of audit and feedback in an ambulatory care group practice. Health Care Manage. Rev. **1987**, *12* (5), 11–16.
31. Palmer, R.H.; Louis, T.A.; Peterson, H.F.; Rothrock, J.K.; Strain, R.; Wright, E.A. What makes quality assurance effective? Results from a randomized, controlled trial in 16 primary care group practices. Med. Care **1996**, *34*, SS29–SS39.
32. Landis, N.T. Coalition continues work on medication-use performance measures. Am. J. Health-Syst. Pharm. **1999**, *56* (20), 2022, 2024.
33. MacKinnon, N.J.; Hepler, C.D.; Bennett, L. Performance Indicators for Continuous Quality Improvement in Pharmacy (PICQIP): A Study Update. In *American Pharmaceutical Association Annual Meeting, San Antonio, TX*; Mar. 1999; Posey, L.M., Ed.; (Presentation abstract).
34. Hepler, C.D. Observations from the conference: A pharmacist's perspective. Am. J. Health-Syst. Pharm. **2000**, *57* (6), 590–594.
35. Hofer, T.P.; Hayward, R.A.; Greenfield, S.; Wagner, E.H.; Kaplan, S.H.; Manning, W.G. The unreliability of individual physician ''report cards'' for assessing the costs and quality of care of a chronic disease. JAMA, J. Am. Med. Assoc. **1999**, *281* (22), 2098–2105.
36. Bindman, A.B. Can physician profiles be trusted? [editorial]. JAMA, J. Am. Med. Assoc. **1999**, *281* (22), 2142–2143.
37. Brand, D.A.; Quam, L.; Leatherman, S. Medical practice profiling: Concepts and caveats. Med. Care Res. Rev. **1995**, *52* (2), 223–251.
38. Sanchez, L.A.; Lee, J.T. Applied pharmacoeconomics: Modeling data from internal and external sources. Am. J. Health-Syst. Pharm. **2000**, *57* (2), 146–155.
39. Health Insurance Portability and Accountability Act (HIPAA) http://www.hcfa.gov/medicaid/hipaa/adminsim/privacy.htm, accessed June 24, 2002.
40. Cahill, N.E. Caveats in interpreting and applying pharmacoeconomic data. Am. J. Health-Syst. Pharm. **1995**, *52* (4 Suppl.), S24–S26.
41. Lincoln, T.L. Privacy: A real world problem with fuzzy boundaries. Methods Inf. Med. **1993**, *32*, 104–107.
42. Meux, E. Encrypting personal identifiers. Health Serv. Res. **1994**, *29* (2), 247–256.
43. Westrin, C.G.; Nilstun, T. The ethics of data utilization: A comparison between epidemiology and journalism. BMJ [Br. Med. J.] **1994**, *308*, 522–523.
44. Prince, L.H.; Carroll-Barefield, A. Management implications of the health insurance portability and accountability act. Health Care Manager **2000**, *19* (1), 44–49.
45. Casarett, D.; Karlawish, J.; Andrews, E.; Caplan, A. Bioethical Issues in Pharmacoepidemiology Research. In *Pharmacoepidemiology,* 3rd Ed.; Strom, B.L., Ed.; John Wiley and Sons Ltd., 2000; 417–431.
46. Else, B.A.; Armstrong, E.P.; Cox, E.R. Data sources for pharmacoeconomic and health services research. Am. J. Health-Syst. Pharm. **1997**, *54* (22), 2601–2608.
47. American Society of Health-System Pharmacists. ASHP statement on the pharmacist's role in primary care. Am. J. Health-Syst. Pharm. **1999**, *56* (16), 1665–1667.

Pharmaciens Sans Frontières (Pharmacists without Borders)

Myrella T. Roy
The Ottawa Hospital, Ottawa, Ontario, Canada

INTRODUCTION

Pharmaciens Sans Frontières is a humanitarian nongovernmental organization conducting emergency and development missions in medically underprivileged countries. Its contributions range from the provision of essential drugs and medicosurgical supplies to the renovation of healthcare facilities and the education of resident healthcare professionals.

GOAL

Pharmaciens Sans Frontières (PSF) strives to promote and establish economical and geographical access to quality healthcare for all, through adapted pharmaceutical means.

OBJECTIVES

Pharmaciens Sans Frontières is pursuing the following objectives:

- To analyze and itemize the hygiene and public health (sanitary) needs of populations.
- To provide selected drugs and medicosurgical supplies to recipient countries.
- To ascertain the best supplying strategies for a prompt, careful, and secure delivery of goods.
- To support the establishment of community clinics, hospitals, and public pharmacies in collaboration with local authorities, international institutions, and other nongovernmental organizations.
- To give technical assistance and train resident staff working in pharmaceutical facilities.
- To enhance drug knowledge of resident healthcare professionals by providing information material, organizing educational programs, and encouraging professional associations and committees.
- To promote rational drug use through training programs and public education campaigns.
- To collaborate with health care professionals from other humanitarian agencies.
- To apply pharmaceutical expertise to the safe management of healthcare waste.

HISTORY

Pharmaciens Sans Frontières was founded in 1985 by five French pharmacists: Philippe Bon, Michel Camus, Jean-Louis Machuron, Bernard Toureet, and Daniel Tsantucci. The original ambition was to collect unused drugs, then to sort and distribute those still valid to the most disadvantaged nations of the world. Through this first initiative, the pharmaceutical profession secured its legitimacy within the field of humanitarian aid. From the outset, PSF's action was aimed at improving the sanitary situation of developing countries. In 1987, the first interventions commenced in Mali and in Mauritania.

These early efforts highlighted the unsuitability of unused drugs for the needs of medically underprivileged countries. Since then, the World Health Organization has issued guidelines for drug donations (www.drugdonations.org) and a list of essential drugs (www.who.int/medicines/edl.html) endorsed by PSF. Intended to meet the needs and resources of third world and developing countries, the list of essential drugs is used to guide purchasing decisions and procurement schemes both during acute emergencies and development aid. PSF then created a wholesaler for the provision of essential drugs, medicosurgical supplies and biological substances.

In late 1989, PSF had acquired the necessary credibility with French and European authorities to establish an emergency mission in Romania. Soon after, programs of a similar magnitude were launched in Burkina Faso and in Bulgaria.

Encyclopedia of Clinical Pharmacy
DOI: 10.1081/E-ECP 120006302

The number of missions has continued to grow since 1992. The association conducts programs in all facets of the pharmaceutical profession, namely goods distribution, management, training, technical assistance, and facility renovation. Consequently, PSF's organizational structure has expanded markedly. An International Committee has been formed that currently includes Belgium, Canada, France, Italy, Luxembourg, the Netherlands, Poland, and Switzerland. Through its sustained efforts, PSF has now gained worldwide recognition.

MISSIONS

Pharmaciens Sans Frontières administers three different types of international missions: emergency, development, and technical assistance.

Emergency Missions

During these missions, PSF supplies drugs and medicosurgical materials to countries afflicted by natural, economic, or human catastrophes. These interventions are rapidly launched to respond to a critical situation. Soon after, a postemergency program should be planned in order to extend the aid beyond the initial punctual contribution.

Development Missions

These missions foster the health care autonomy of populations through more long-term aid. They result in establishing village pharmacies and in renovating hospital facilities and medical laboratories.

Technical Assistance Missions

When the situation is favorable, missions can be accomplished that involve expertise and training and affect the whole pharmaceutical field. Examples of such missions include restoration of industrial drug production capabilities and reinstatement of a national laboratory for drug quality control with subsequent training of technicians.

Current Missions

Pharmaciens Sans Frontières is currently operating an emergency mission in Tadjikistan and development or technical assistance missions in the following countries: Albania, Bosnia, Cambodia, Equator, Honduras, Kenya, Kosovo, Mali, Mauritania, Moldovia, Montenegro, Niger, Serbia, and Soudan.

FUNDING

Like all nongovernmental organizations, PSF relies on two sources of funding: institutional sponsors and private contributions. These funds ensure the self-reliance of the association.

Institutional Sponsors

Whether from the European Union, the United Nations High Commissioner for Refugees, the French Ministry of Foreign Affairs, or the United States Agency for International Development, these funds are intended exclusively for the establishment or the continuation of missions in the field. Granting of these funds entails a detailed application, following the assessment of the pharmaceutical needs in a given region or country, and a thorough account at the completion of the program. Most funds make provision for administrative charges, which partially cover the operating cost of these programs.

Private Contributions

Donations, subscriptions, and industry partnerships mainly serve to fund development missions, transitional programs between emergency and development missions, logistic material not otherwise covered by the institutional sponsors, and the operations of the association.

Financial Transparency

Close to 90% of the funds are earmarked for the carrying out and direct management of missions in the field. In order to assure transparency, PSF's finances are controlled by independent agents and are stratified by mission.

HUMAN RESOURCES

Board of Directors

The Board is formed of approximately 20 volunteer directors, elected by the members of the association. Claudi M. Cuchillo currently serves as President.

Staff

Some 30 employees work at the international headquarters in Clermont-Ferrand and at the marketing office in Paris, France. The key managers are as follows:

- Executive Director: Gérard Plantin
- Director of Finance: Raphaël Chaize
- Director of Human Resources: Gérard Pourreaux
- Chief of Marketing: Isabelle Leroi
- Chief of Accounting: Françoise Chargueraud
- Communication Officer: Samuel Falgoux

National Associations

- PSF Belgium: Mrs. Solange Gossieaux, Forêt d'Houthulst 27, 1000 Brussels, Belgium.
- PSF Canada: Mr. Hubert Brault, a/s A.Q.P.P, 4378 Pierre de Coubertin, Montreal, Qc, Canada H1V 1A6.
- PSF Netherlands: Mr. Tammes, Apothekers Zonder Grenzen, PO Box 10493, 7301 GL Apeldoorn, Netherlands.
- PSF Italia: Mrs. Elena Gandini, Farm- Piazza Vittorio Emanuele III no. 4, Monzambano, Italia.
- PSF Luxembourg: Mr. Camille Groos, 13 G. Diderich, 1420 Luxembourg.
- PSF Poland: Mr. Zygmund Ryznerski, Farmaceuci Bez Granic, 2 PL W. Slkorslkiego, 31115 Krakow, Poland.
- PSF Switzerland: Mr. Pierre Alain Jotterand, Route de Crochy 2, Case Postale 229, 1024 Ecublens, Switzerland.

Missionaries

Over 1000 volunteers are presently members of PSF around the world. At any point in time, between 90 to 100 expatriated of diverse nationalities and trades (such as pharmacists, technicians, nurses, logisticians, and administrators) and some 400 local staff combine forces to improve health care in the field.

MEETING

The annual general meeting of the members is held in May.

PROFESSIONAL DEVELOPMENT

Pharmacist Managed Vaccination Programs

David S. Ziska
Joseph K. Bonnarens
Applied Health Outcomes, Tampa, Florida, U.S.A.

INTRODUCTION

In the United States, the concept of pharmacists as vaccine providers is new; however, a small number of pharmacists have been administering vaccines for decades. Pharmaceutical care initiatives, spearheaded by state and national pharmacy organizations, have rapidly expanded the number of pharmacists who are qualified to administer vaccines. Based on the increased number of certificate programs offered each year, pharmacists are expanding their role in the health of American citizens by administering vaccines. The role of vaccine prescriber and provider is also new to school of pharmacy curricula and continues the expansion of community and institutional practitioners providing pharmacy-based clinical services. If embraced, pharmacist-managed vaccination programs will enhance other clinical services (e.g., diabetes pharmaceutical care) by providing one-stop shopping for basic disease state prevention and management services.

CERTIFICATE PROGRAMS

Pharmacists must be trained to provide and administer immunizations to their patients. The principle method of training has been through the development of immunization certificate programs. The first formal certificate program for pharmacists as providers of immunization services occurred in the state of Washington in 1994.[1] The American Pharmaceutical Association (APhA) organized a national Pharmacy-Based Immunization Delivery program that was first offered to Mississippi pharmacists in 1996. State pharmacy associations, many in conjunction with APhA and other national pharmacy associations, have sponsored live educational programs to help encourage and train interested pharmacists.[2] Since the mid-1990s, thousands of pharmacists have received specialized knowledge and training to appropriately select and administer vaccines through the completion of various certificate programs in eight states: Alabama, Arkansas, Iowa, Michigan, Mississippi, Texas, Tennessee, and Washington.[3] Once successfully completed, this credential indicates that a pharmacist has achieved a high level of competence in the field of immunizations. Twenty-four other states allow pharmacists to administer drugs; thus, the potential exists for pharmacists in these states with their current pharmacy practice acts, or slightly amended ones, to establish pharmacy-based immunization delivery services. If you are not sure about your state pharmacy practice act, call your board of pharmacy and ask if your state's practice act includes wording that allows for pharmacists to administer medications.

BENEFITS OF PHARMACISTS ADMINISTERING VACCINES

With each passing year, pharmacists are becoming more involved in the development of immunization programs and are accepting additional responsibilities by becoming active immunizers. The presence of pharmacists in most communities highlights the immense potential for the development of immunization programs. Although these opportunities are exciting for pharmacists, there are larger issues that are positively affected by these opportunities. In conjunction with physicians, nurses, public health clinics, and others, pharmacists fill a unique role for all those attempting to increase immunization rates. Compared with all other health care professionals, pharmacists are consistently more accessible in most geographic areas and have been regarded by the public as one of the most respected professions for more than 20 years. Most physicians and other health professionals in their local area see the emergence of this service as a positive development. Although there will always be some tension regarding such turf issues, the overall benefit of increasing the immunization rates nationally and decreasing morbidity and mortality far outweigh them.

Encyclopedia of Clinical Pharmacy
DOI: 10.1081/E-ECP 120006240

Health Economic Benefits

Vaccine administration is economically beneficial. Several studies have demonstrated the cost effectiveness of various vaccines. One study suggested that advising 100,000 at-risk patients to accept influenza vaccine would increase the number of vaccinations and could avert 139 hospitalizations and 63 deaths.[4] A study in the Journal of the American Medical Association analyzed pneumococcal vaccination of people 65 years of age and older, and when their findings were extrapolated to 23 million senior citizens, immunizing this population would save $194 million and gain 78,000 years of healthy life.[5]

With the addition of pharmacists to the role of immunizer, the immunization rate continues to increase. The public health benefits, as well as the professional opportunities for practitioners, increase. As patients become accustomed to receiving immunizations from their local pharmacist, the opportunities to develop other patient care services, such as disease management programs, increases significantly because patients become more comfortable with pharmacists in a clinical role.

Pharmacists have provided a needed boost in advocating the importance of vaccinations. Beginning in the early 1980s, pharmacists began to get more involved in vaccine advocacy by hosting influenza immunization teams at their pharmacies, evolving to present-day efforts that entail pharmacists in various practice settings—independent, chain, institutional, and consulting—actually administering immunizations.[6] Over this period of time, the addition of pharmacists to the campaign increased immunizations in all settings, including physicians' offices and pharmacies. With the development of pharmacist-managed vaccination programs, pharmacies have also experienced an increased patient traffic. Overall, the development of pharmacist-managed vaccination programs has demonstrated that participants—patients, other health care professionals, pharmacists, and society—benefit greatly from these programs.

Profit

Pharmacist-managed vaccination programs should be viewed as revenue generators, not necessarily as a profit center. According to the *2001 NCPA-Pharmacia Digest*, responding independent pharmacists that offer immunizations in their practices reported that, on average, 79% charged a separate fee for the service.[7] It is unlikely that any pharmacist will be able to make a living exclusively administering vaccines, but this service would be a good way to begin or expand clinical services. Even the most prolific pharmacist-immunizers have only been reimbursed $5000 to $10,000 per year. Immunizing is a way to get closer to clients as well as increase the status, in their eyes, of a pharmacist as a caregiver. In furthering one's status as a caregiver, one would expect an increase in spin-off business. For example, the client comes in for an influenza vaccine and while there the client takes the opportunity to ask the pharmacist's recommendation on purchasing an over-the-counter product.

IMMUNIZATION OPPORTUNITIES FOR PHARMACISTS

Pediatrics

Infant and childhood immunization opportunities abound, but many practitioners shy away from this practice because of liability concerns and because of being inexperienced at injecting infants and small children. If one can overcome these challenges, there are federally and state-funded programs designed to encourage adherence to complex pediatric vaccination regimens and to reimburse qualified clinicians providing these services.[8] The most widely recognized programs include the Vaccines For Children (VFC) program, and state-operated low-cost or free health insurance for children (e.g., Children's Health Insurance Program [CHIP], KidCare programs). One of the many services covered by most state children's health insurance programs includes immunizations, whereas the VFC was designed exclusively to, "provide FREE vaccine to children between the ages of birth and 18 years"[9] (Table 1).

Qualifications for the VFC program are as stated here. Each state's children's health insurance qualifications are unique but can be found at their web site.[10] Children through 18 years of age qualify to receive federally purchased vaccines under the VFC program if they are medicaid-enrolled, uninsured, or Native Americans/Native Alaskans. Children who have insurance that does not cover immunizations also qualify to receive VFC vaccine at federally qualified health centers and rural health clinics. In addition, states may use their funds to purchase vaccines for additional groups of children. Contact the immunization program in your state for more information.

Adult

Pharmacists certified to administer vaccines typically begin their clinical practices by immunizing adults against influenza and pneumococcal pneumonia. This

Table 1 Vaccines covered by Vaccines for Children program

Vaccine name(s): Common (brand)	Disease(s) prevented	Required for K–12 in most states
DTaP (Tripedia®, etc.)	Diphtheria, tetanus, whooping cough	Yes
Hib (PedvaxHIB™, etc.)	Haemophilus influenzae type b	No, DC
Hepatitis A (Havrix®, VAQTA®)	Hepatitis A	No
Hepatitis B (Recombivax HB®, Engerix-B®)	Hepatitis B	Yes
Flu (Fluogen®, etc.)	Influenza	No
MMR (M-M-R® II)	Measles, mumps, and rubella	Yes
Pneumonia (Pneumovax®, Pnu-Imune® 23)	Pneumococcal diseases	No
Pneumococcal conjugate vaccine (Prevnar®)	Infant pneumococcal	No
Poliovirus (IPOL™)	Paralytic poliomyelitis	Yes
Varicella (Varivax®)	Chicken pox	No

Key: DC, day care; DTaP, diphtheria, tetanus, and acellular pertussis; Hib, haemophilus influenza type b; MMR, measles, mumps and rubella.

broad population and these vaccines target the most undervaccinated population and can affect the two most vaccine preventable diseases in the United States. Combined, these two diseases kill between 50,000 and 80,000 people per year. Other vaccine-preventable diseases that are common in adult populations include hepatitis B, hepatitis A, and varicella.[11] Unlike the vaccines for influenza and pneumococcal pneumonia, these other diseases require multiple doses and the vaccines cost $25 to $60 or more per dose.[12] Yet another vaccine that many are not given because providers simply forget to offer it to their clients is the tetanus/diphtheria vaccine or Td. This vaccine requires a booster every 10 years after the age of 11 or 12.[13]

Elderly/Long-Term Care

The elderly suffer most in terms of increased morbidity and mortality when recommended vaccines are not administered. Pneumococcal diseases are the vaccine preventable diseases that cause, by far, the most mortality in the elderly.[14] Unfortunately, the vaccine for pneumococcal-induced pneumonia has one of the lowest administration rates. Other vaccines that most elderly and long-term care residents should receive include the influenza and tetanus/diphtheria vaccines. Consultant pharmacists or pharmacists providing services to long-term care facilities are in excellent positions to identify those who need vaccines.[15]

MODEL PHARMACIST-MANAGED PRACTICES

There are a number of opportunities for pharmacists to develop an immunization service in their current practice. However, the degree of development for a vaccination program depends on numerous factors, including, but not limited to, location of the pharmacy (both the state in which the pharmacy exists, as well as the physical location in the community), staff size, physical layout of the pharmacy, and other community health professionals currently involved in vaccinations (i.e., physicians, nurse practitioners, nurses, state health departments). Based on guidelines approved by the APhA Board of Trustees in 1997, there are three levels of involvement for pharmacists in vaccine advocacy:

1. Pharmacist as educator (motivating people to be immunized).
2. Pharmacist as facilitator (hosting others who immunize).
3. Pharmacist as immunizer (providing and administering immunizations).[16]

A logical role for all pharmacists is in the area of educating the public about the benefits and importance of immunizations. Pharmacists regularly speak at local civic, service, and school organization meetings to discuss health-related issues, and these serve as ideal settings to speak about the importance of immunizations. By keeping individual patient profiles, pharmacists also have a unique opportunity to identify and directly inform patients with special needs, such as those with a high risk of infection and the elderly, about the importance of receiving their pneumococcal and annual influenza immunizations. Vast amounts of educational material are available from local, state, and national organizations to aid in educating children and adults (see Table 2 for specific contact information). All pharmacists should meet this level of advocacy.

In daily dealings with physicians, nurses, insurance companies, and patients, the role of facilitator is com-

Table 2 Internet-based vaccine and immunization resources

Resource	Web address
AAP	http://www.aap.org/family/parents/immunize.htm
ACIP publications	http://www.cdc.gov/nip/publications/acip-list.htm
Adobe Acrobat Reader	http://www.adobe.com/products/acrobat/readstep2.html
CDC	www.cdc.gov
FDA web site for biological products	http://www.fda.gov/cber/efoi/approve.htm
Immunization Gateway: Your Vaccine Fact-Finder	http://www.immunofacts.com
Medicare Part B	http://www.medicare.gov/Basics/PartAandB.asp
Minnesota Dept. of Health phone numbers for vaccine companies	http://www.health.state.mn.us/divs/dpc/adps/manual/pgphonef/pgvacmfg.htm
MMWR	http://www.cdc.gov/mmwr
State health insurance programs for children	http://www.insurekidsnow.gov/childhealth/states/states.asp
Vaccine catch-up schedule	http://www.cdc.gov/nip/publications/pink/AccSch00.pdf
Vaccine information statements	http://www.cdc.gov/nip/publications/VIS
The Vaccine Page	http://www.vaccines.com
Vaccine price list	http://www.cdc.gov/nip/vfc/cdc_vaccine_price_list.htm
VFC	http://www.cdc.gov/nip/vfc/

Key: AAP, American Academy of Pediatrics; ACIP, Advisory Committee on Immunization Practices; CDC, Centers for Disease Control and Prevention; FDA, Food and Drug Administration; MMWR, Morbidity and Mortality Weekly Report; VFC, Vaccines For Children.

monplace for a pharmacist. As a facilitator, pharmacists continue to educate and motivate the public about immunizations; however, they assume a more active role by ensuring that individuals actually receive their immunizations. This has been accomplished by the development of immunization programs that are offered in community pharmacies. As early as 1982, ambulatory care pharmacists in Georgia instituted a program to identify and immunize patients at risk for pneumonia and influenza.[17] In 1985, community pharmacies in Denver, Colorado, began to invite nurses to administer vaccines to the public. The early 1990s witnessed the development of similar programs in numerous chain pharmacies.[18] As facilitators, pharmacists have also played an active role in referring individuals in need of immunization to the county health clinic, a local physician, or nurse practitioner. Even though all pharmacies do not offer immunizations, the referral of patients continues to be an active area of immunization advocacy.

In 1994, the University of Washington School of Pharmacy and the Washington State Pharmacists Association developed the first coordinated program to train pharmacists to administer immunizations.[19] With this development, the need to host others, (e.g., nurses) in the pharmacy to provide immunizations began to decrease in states that allowed pharmacists to administer drugs. As of June 2001, 32 states explicitly authorize pharmacists to vaccinate.[20] Although the role of facilitator is important in connecting those in need of immunizations to those who can provide the vaccine, there are opportunities for pharmacists to expand their level of advocacy by becoming an immunizer.

For pharmacists to achieve the highest level of immunization advocacy, a number of responsibilities and decisions must be finalized. These responsibilities have been broken into the following categories: legal/regulatory, training, practice requirements, compensation, and documentation.

Legal/Regulatory

If a pharmacist is interested in developing an immunization program, the principle legal issue that must be addressed is how that state views the administration of drugs by a pharmacist. As mentioned previously, more than 30 states currently authorize pharmacists to vaccinate. If you are not sure, check with your state board of pharmacy to determine your options. If your state currently does not allow pharmacists to administer drugs, this would be an opportune time to get involved with your state pharmacy association and state board to attempt to change your state pharmacy regulations. If your state allows pharmacists to administer medications, additional legal and regulatory issues should be addressed. Depending on a state's practice regulations, initiating a patient care service, such as an immunization program, requires a pharmacist to establish a collaborative practice agreement or protocol with a prescriber. As an immunizer, a pharmacist is at risk of needle sticks and exposure to blood-borne pathogens. Employers and employees must become familiar with the regulations of the Occupational Safety and Health Administration to minimize potential harm to all employees.[21]

Training

As mentioned previously, certificate programs represent the primary method used to educate and train interested pharmacists in how to develop a pharmacy-based immunization program. The training programs available provide individuals with the knowledge, tools, and encouragement to achieve the highest level of advocacy. A list of the organizations offering such programs is provided in Table 3. The training programs are offered at various times of the year and in various locations. Contact the listed organizations or your state pharmacy association to identify a program near you.

Although many of the training programs focus on practicing pharmacists, pharmacy students have also become active in advocating immunizations. Students in all 81 APhA Academy of Students of Pharmacy chapters and 40 Student National Pharmaceutical Association chapters have been involved with *Operation Immunization: The Nation's Pharmacy Students and Practitioners Protecting the Public Health.* Since its inception in 1997, *Operation Immunization* has helped provide more than 125,000 vaccinations.[22]

Practice Requirements

Although there are levels of advocacy, there are also levels of program development. The range of possible levels could start in an apothecary-style pharmacy with a single pharmacist providing immunizations to individuals in between filling all the prescriptions to a large pharmacy practice (independent or chain store) where the immunizations are provided to patients in a separate, private room with pharmacists and support personnel dedicated entirely to this service.

Table 3 Organizations providing standardized training in pharmacy-based immunizations

Organization	Internet address
American Pharmaceutical Association	www.aphanet.org
Centers for Disease Control and Prevention	www.cdc.gov
National Community Pharmacists Association	www.ncpanet.org
Washington State Pharmacists Association	www.PharmCare.org

In the *2000 NCPA-Pharmacia Digest*, responding independent pharmacists reported that on average, 21% of all respondents offered immunizations. Based on the pharmacy's sales volume, the range of offerings was 9% for $500,000 to $750,000 per year in sales to 42% in pharmacies with more than $4,000,000 per year in sales.[23] Generally, vaccine program offerings increased as sales volume increased. Similarly, the percentage of pharmacies offering an immunization service tended to increase as square footage increased. Based on the pharmacy's total square footage, the range of offerings was 11% for pharmacies with less than 1000 square feet to 32% in pharmacies larger than 4000 square feet.[24]

There are numerous examples of immunization programs being offered in independent pharmacies, as well as in chain pharmacies. Although there are a number of differences among these practice sites—including, but not limited to, staff size, physical space, and facility location—in all cases, the development of an immunization program is possible. Staff size is a critical issue for a pharmacy developing an immunization service. Some questions that should be asked are "Who will be the immunizers?", "How will this service effect work flow?", and "Will staff be dedicated to the immunization program, or will immunizations be administered by staff in between regular scheduled duties?" In regards to physical space needs, the minimum requirements for providing immunizations should include space with privacy for administering injections (with room to sit and possibly to lay down in case of an emergency), a dose-preparation area, and a place for patients to wait after the immunization.[25,26] The physical space used for immunization programs has varied from a chair at the end of the counter separated from the waiting area by a partition or curtain, to stock rooms converted into administration areas, to using the owner's/manager's office on scheduled days, to having separate counseling facilities already available. When asked, most immunizers shared that the service was easier to set up than expected and most facilities host such a service. Also, future immunizers will obtain a large amount of information regarding staff and physical needs through the respective training programs offered. Another requirement for this service is a refrigerator with freezer for storage of the vaccine supplies. Vaccine products have specific storage requirements, with standard temperature ranges, so the presence of a quality refrigerator/freezer is critical to the service. A back-up

plan, such as a cooler or alternative power source for the refrigerator, is also important in case of power interruptions of greater than 24 hours.[27]

Compensation

The critical aspect of any service is the covering of costs incurred to provide the service and profit to sustain not only the service, but also the business as a whole. For immunization programs, as with any patient care service developed in a pharmacy, there will be direct and indirect costs associated with the service. Direct costs are those that would not be incurred by the business if the service were not performed, and indirect costs are those incurred even if the service in question was not developed.[28] Although each pharmacy is unique, with its own needs and requirements, the following examples are areas to consider in determining potential costs of such a program. As for direct costs, areas such as service personnel costs, specialty training, additional liability protection, supplies and equipment, service workspace, service-specific advertising and promotion, and additional information resources must be analyzed to help determine the overall costs of this service. To calculate overall costs, indirect costs must also be added to the formula, such as personnel costs of nonservice personnel, rent, utilities, taxes, insurance, general advertising and promotion, and general repair and maintenance. Pricing of each immunization will depend on a number of variables. By having the expected costs of the service, coupled with the cost of the product and the other factors such as competition, demand, standard rates, and image, a price for each immunization may be determined.

For an immunization service, there are a number of methods for compensation. The first method is patients paying cash for the immunization. Patients are familiar with this payment method based on the model of county/state health clinics that have long charged a fee for immunization administration. The second method relates to billing a patient's third-party payer. In some cases, vaccines are paid for as a prescription would be, sent electronically, and the pharmacy is reimbursed for the vaccine as well as an administration fee. The inclusion of an administration fee varies with groups, so an administration fee may also be collected from the patient.[29] In other cases, patients pay cash for the immunization and receive a detailed receipt, listing the type of vaccine, appropriate billing code, and amount paid. The patients are then responsible for submitting the claim to their own insurance company for reimbursement. The third method is billing medicare. To claim reimbursement under medicare, the pharmacist, pharmacy, or both must apply for a medicare provider identification number as an immunizer from the local medicare part B carrier.

Both pharmacists and pharmacies use Health Care Finance Administration (HCFA) Form 855, Provider/Supplier Enrollment Application, to request a provider number. This form may be obtained from medicare part B carriers or provided by the previously mentioned training programs. This is a cursory review of medicare policies. Training programs will provide additional details, or contact your local medicare part B carrier, HCFA regional office, or other pharmacy reimbursement publications.[30] A fourth method of compensation is direct contracts with local area employers. By contracting with an area employer, pharmacists have administered influenza and pneumococcal vaccines to all employees. This type of on-site program can be offered to various businesses, such as plants, office-based facilities, and health care-related facilities, such as assisted living facilities and adult family homes.[31] One popular place for a traveling immunization program has been schools. This type of service primarily focuses on the adults (teachers, staff members); however, some pharmacies have also established child immunization programs. The final method involves special programs that provide funding to specific populations. One such program is VFC. Begun in 1994, children eligible for medicaid, uninsured, Native American, or Alaska Natives may receive vaccinations. Vaccine providers are supplied the vaccines for free. Although some immunization programs charge a small administration fee, all eligible children will receive an immunization, regardless of whether a fee can be paid. Additional details for the VFC program are available by contacting your state immunization coordinator or state department of health.

Documentation

Documentation of vaccination is critical to an immunization service. In many cases, reimbursement is not obtained without proper documentation. However, with a large portion of immunizations paid for in cash and the need to know which patient received which vaccination in case of a recall, there are some documentation issues that must be addressed. The use of software, either specifically designed for vaccines or general patient management, is available from several sources. Free software is available from the Center for Disease Control and Prevention's (CDC's) National Immunization Program (www.cdc.gov).[32] If patient profile software is used to track immunizations, it is critical for an easily accessible, permanent record to be established for immunizations because prescription records can be purged and even deleted over time, based on state board regulations. Depending on the requirements of individual states, immunizers must determine if there is a requirement for the use and filing of consent forms, protocol agreements, and other related documentation. The various training programs provide necessary information regarding documentation requirements, or your respective state board of pharmacy can be contacted directly with specific questions.

RESOURCES FOR VACCINE INFORMATION

A variety of resources are available for pharmacists interested in beginning a pharmacy-based immunization practice. Some of the best resource people include state health officers; county health clinic physicians; state department of health vaccine coordinators, especially those that administer the CHIP or KidCare-like immunization programs; and local, board-certified primary care and infectious disease physicians. There are also a number of excellent web sites (see Table 2). The primary and most reliable source for vaccine and immunization information is the CDC web site. Most reliable infor mation about vaccines and immunization on other web pages is based on information first published in the public domain by the CDC in their journal, *Morbidity and Mortality Weekly Report* (MMWR). For those interested in keeping tabs on up-to-the-minute vaccine and immunization changes and recommendations, one should access the CDC listserv. This service forwards all issues of MMWR to your e-mail in portable document format (pdf) for viewing on freeware—Adobe Acrobat Reader.

Table 4 Immunization networks, organizations and listservs

Resource	Web address
Allied Vaccine Group	http://vaccine.org
APhA	http://www.aphanet.org (Pharmaceutical Care)
APhA-Pharmacist Immunizer Listserv	apha-immpharm-subscribe@egroups.com
Immunization action coalition	http://www.immunize.org
MMWR subscription	listserv@listserv.cdc.gov
National immunization program	http://www.cdc.gov/nip
NNII	http://www.immunizationinfo.org
VISI	http://www.cdc.gov/nip/visi

APhA, American Pharmaceutical Association; MMWR, Morbidity and Mortality Weekly Report; VISI, The Vaccine Identification Standards Initiative; NNII, National Network for Immunization Information.

NETWORKS AND ORGANIZATIONS

There are a number of immunization and vaccine networks designed to help inform professionals and the public about immunization efforts regionally and nationally. Listservs can also be helpful to those having specific questions in need of expert commentary. One should be cautious, as there is as much helpful information about immunizations and vaccines as there is unsubstantiated and erroneous information. Look carefully at who sponsors and writes the information that you are reading. If the information seems inflammatory or extremist, it is probably not scientifically valid or presents a skewed interpretation of the facts. Also, whenever reading tertiary literature, pull the references and interpret the information yourself if you are in doubt of its validity. This is the best method of ensuring that you are making decisions based on first-hand information. Of course, the referenced material must come from a refereed, reputable source or else it too is suspect. Table 4 lists some reliable networks, organizations, and listservs that pharmacists can access to augment their knowledge base.

REFERENCES

1. Grabenstein, J.D. *Pharmacy-based Immunization Delivery: Self-Study Learning Program,* 5th Ed.; American Pharmaceutical Association: Washington, DC, 2000; 5.
2. Grabenstein, J.D.; Bonasso, J. Health-system pharmacist's role in immunizing adults against pneumococcal disease and influenza. Am. J. Health Syst.-Pharm. **1999**, *56* (17 Suppl. 2), S1.
3. Grabenstein, J.D. *Pharmacy-based Immunization Delivery: Self-Study Learning Program,* 5th Ed.; American Pharmaceutical Association: Washington, DC, 2000; 6–7.
4. Grabenstein, J.D.; Hartzema, A.G.; Guess, H.A.; Johnston, W.P.; Rittenhouse, B.E. Community pharmacists as immunization advocates: cost-effectiveness of a cue to influenza vaccination. Med. Care **1992**, *30* (6), 503.
5. Sisk, J.E.; Moskowitz, A.J.; Whang, W.; Lin, J.D.; Fedson, D.S.; McBean, A.M.; Plouffe, J.F.; Cetron, M.S.; Butler, J.C. Cost effectiveness of vaccination against pneumococcal bacteremia among elderly people. JAMA, J. Am. Med. Assoc. **1997**, *278*, 1333.
6. Grabenstein, J.D. *Pharmacy-based Immunization Delivery: Self-Study Learning Program,* 5th Ed.; American Pharmaceutical Association: Washington, DC, 2000; 6–7.
7. NCPA. *2001 NCPA-Pharmacia Digest*; National Community Pharmacists Association: Alexandria, Virginia, *in press.*

8. Hoeben, B.J.; Dennis, M.S.; Bachman, R.L.; Bhargava, M.; Pickard, M.E.; Sokol, K.M.; Vu, L.; Rovers, J.P. Role of the pharmacist in childhood immunizations. J. Am. Pharm. Assoc. **1997**, *NS37*, 557–562.
9. www.cdc.gov/nip/vfc/ (accessed September 2000).
10. http://www.insurekidsnow.gov/childhealth/states/states.asp (accessed September 2000).
11. Grabenstein, J.D. *Pharmacy-Based Immunization Delivery: Self-Study Learning Program*, 5th Ed.; American Pharmaceutical Association: Washington, DC, 2000; 3.
12. http://www.cdc.gov/nip/vfc/cdc_vaccine_price_list.htm (accessed September 2000).
13. Grabenstein, J.D. *Pharmacy-Based Immunization Delivery: Self-Study Learning Program,* 5th Ed.; American Pharmaceutical Association: Washington, DC, 2000; 3–4.
14. Spruill, W.J.; Cooper, J.W.; Taylor, W.J.R. Pharmacist-coordinated pneumonia and influenza vaccination program. Am. J. Hosp. Pharm. **1982**, *39*, 1904–1906.
15. Weitzel, K.W.; Goode, J.R. Implementation of a pharmacy-based immunization program in a supermarket chain. J. Am. Pharm. Assoc. **2000**, *40*, 252–262.
16. http://www.aphanet.org/pharmcare/immguide.html (accessed Sept 2000).
17. Spruill, W.J.; Cooper, J.W.; Taylor, W.J.R. Pharmacist-coordinated pneumonia and influenza vaccination program. Am. J. Hosp. Pharm. **1982**, *39*, 1904–1906.
18. Grabenstein, J.D. *Pharmacy-Based Immunization Delivery: Self-Study Learning Program,* 5th Ed.; American Pharmaceutical Association: Washington, DC, 2000; 5–6.
19. Grabenstein, J.D.; Bonasso, J. Health-system pharmacist's role in immunizing adults against pneumococcal disease and influenza. Am. J. Health Syst.-Pharm. **1999**, *56* (17 Suppl. 2), S1.
20. http://www.immunize.org/laws/pharm.htm (accessed June 2001).
21. Gardner, J.S. Establishing a vaccine administration program in the community pharmacy. J. Am. Pharm. Assoc. **2000**, *40* (1), 111.
22. English, T. Immunization project lauded at APhA2000. Pharm. Today **2000**, 25, (May).
23. Huffman, D.C., Jr. *2000 NCPA-Pharmacia Digest*; National Community Pharmacists Association: Alexandria, Virginia, 2000.
24. Huffman, D.C., Jr. *2000 NCPA-Pharmacia Digest*; National Community Pharmacists Association: Alexandria, Virginia, 2000.
25. Grabenstein, J.D. *Pharmacy-Based Immunization Delivery: Self-Study Learning Program,* 5th Ed.; American Pharmaceutical Association: Washington, DC, 2000; 61–74.
26. Gardner, J.S. A practical guide to establishing vaccine administration services in community pharmacies. J. Am. Pharm. Assoc. **1997**, *NS37* (6), 683.
27. Grabenstein, J.D. *Pharmacy-Based Immunization Delivery: Self-study Learning Program,* 5th Ed.; American Pharmaceutical Association: Washington, DC, 2000; 89–104.
28. Rupp, M.T. *Pricing Pharmacist Care Services*; National Community Pharmacists Association: Alexandria, Virginia, 1999.
29. Grabenstein, J.D. *Pharmacy-Based Immunization Delivery: Self-Study Learning Program,* 5th Ed.; American Pharmaceutical Association: Washington, DC, 2000; 61–74.
30. Grabenstein, J.D. *Pharmacy-Based Immunization Delivery: Self-Study Learning Program,* 5th Ed.; American Pharmaceutical Association: Washington, DC, 2000; 61–74.
31. Beavers, N. Vaccinations, pharmacy style. Drug Top. **2000**, *144*, 56 (Feb 21).
32. Grabenstein, J.D. *Pharmacy-Based Immunization Delivery: Self-Study Learning Program,* 5th Ed.; American Pharmaceutical Association: Washington, DC, 2000; 61–74.

Pharmacist Prescribing

Deborah E. Boatwright
Veterans Affairs Medical Center, San Francisco, California, U.S.A.

INTRODUCTION

The modern practice of pharmacy is in a state of evolution. One important element of that evolutionary state is the inclusion within the scope of practice of some pharmacists in some jurisdictions of the right to prescribe medications to a patient. A pharmacist who undertakes the new skill of prescribing must acquire the appropriate skills consistent with the sophistication of the task or risk liability on various levels. This article will summarize critical legal and ethical issues associated with the evolving function of the pharmacist prescriber.

THE EVOLVING ROLE OF THE PHARMACIST AND EXPANSION OF SCOPE OF PRACTICE

Scope of practice is defined as those duties and limitation of duties placed on pharmacists by applicable laws. Traditionally, the scope of practice of a pharmacist was limited to accurately filling and dispensing valid prescriptions as directed by the treating physician.[1,2] This was a largely nonjudgmental role. However, in the past 20 years, the role of pharmacists has evolved from technical to more judgmental, broadening their scope of practice.[3] Generic substitution[4] and therapeutic substitution, usually through a formulary system,[5] represented initial expansive tasks undertaken by pharmacists. The Omnibus Budget Reconciliation Act of 1990 (P.L. 101-508) (OBRA 90)[6] further expanded the responsibility of the pharmacist to include counseling and drug utilization review.

Clinical pharmacy[7] evolved into pharmaceutical care, which included therapeutic outcome monitoring and disease-state management to improve quality of life and minimize long-term health costs.[8] Pharmacists are now receiving compensation from third-party payers for such cognitive services and have expanded their clinical skills.[9–11] There is a move by more and more states to seek prescriptive authority for pharmacists, enabling seamless and effective monitoring and treatment of patients, especially those with chronic diseases on maintenance therapy.[12]

PRESCRIPTIVE AUTHORITY

Dependent Versus Independent Prescribing Authority

Physicians exercise plenary independent prescribing authority. Plenary authority refers to the ability to prescribe all drugs, treatments, and devices, including controlled substances, without supervision, control, or oversight by another profession.[13] Independent prescribing authority means that the prescriber has the sole authority to make treatment decisions and is wholly responsible for the resultant outcomes.[14]

Limited independent prescribing authority permits independent prescribing within the bounds of the prescriber's scope of practice or within the restrictions of a certain formulary of drugs. Doctors of dental surgery, doctors of veterinary medicine, and doctors of podiatric medicine have limited independent prescribing authority in every state; this authority is limited to their scope or course of practice.[15]

Prescribers with dependent prescribing authority are dependent on a prescriber with independent authority for their own authority, and the discretion involved in this authority may be limited by written guidelines, a protocol, or supervisory approval. Pharmacists with dependent prescribing authority have prescribing authority delegated by a physician on the basis of that physician's belief that the pharmacist has the professional judgment and skills necessary to perform the delegated duties. The pharmacist shares with the collaborating physician the risks and responsibility for the patient's overall outcome.[16]

In Collaborative Drug Therapy Management (CDTM), the pharmacist and physician work in collaboration to manage a patient's drug therapy. Through the use of written guidelines or protocols, the pharmacist may be delegated authority to collect and review patient drug histories, initiate and/or modify drug therapy, order and evaluate laboratory tests relating to drug therapy, and order and/or perform drug-related patient assessments. The written protocol is usually specific to a certain pharmacist and a certain physician and may also define the

Encyclopedia of Clinical Pharmacy
DOI: 10.1081/E-ECP 120006331

individual and shared responsibilities of the physician and pharmacist. CDTM protocols authorizing pharmacists to initiate or modify drug therapy in effect give the pharmacist dependent prescribing authority.[17]

State Perspective

States have the ''police power'' to take regulatory actions to protect the public health, welfare, and safety of their citizens.[18] Thus, states have the authority to license health care professionals and to dictate which health care professionals can prescribe medications. State statutes called Pharmacy Practice Acts dictate what duties pharmacists may perform.

Many states have enacted legislation that expands the practice of pharmacy to include CDTM. As of June 2000, there were at least 27 states granting the statutory authority for the practice of CDTM.[19] The nature of CDTM prescriptive authority for pharmacists varies from jurisdiction to jurisdiction and its definition is in flux. In some jurisdictions, the pharmacist will be required to acquire advanced training and experience in order to participate in such a role. In other jurisdictions, delegated prescriptive authority may be restricted to certain settings, such as health care institutions or nursing homes.

At least one author has defined pharmacist prescriptive authority as ''the practice of pharmacy whereby a pharmacist has jointly agreed, on a voluntary basis, to work in conjunction with one or more practitioners under protocol whereby the pharmacist may perform certain functions authorized by the practitioner or practitioners under certain specified conditions or limitations.''[20] This approach does not contemplate prescriptive authority as currently practiced by physicians. Another article in this volume will expand on CDTM.

Regarding independent prescriptive authority, no pharmacists in any state have plenary independent prescriptive authority. Florida is the only state where pharmacists enjoy true independent prescriptive authority; however, this authority extends only to a limited formulary of drugs, listed in 21 categories. There are other restrictions as well. For example, a pharmacist cannot prescribe any injectable drugs nor any oral drugs for a pregnant woman or nursing mother. The pharmacist cannot exceed the manufacturer's recommended dosage or in any case give more than a 34-day supply.[21]

Federal Perspective

The sale and distribution of drugs in the United States is primarily controlled by two legislative acts. They are the Federal Food, Drug and Cosmetic Act (FDCA)[22] and the Federal Controlled Substances Act (CSA).[23] The FDCA involves the establishment of rules and regulations by which drugs are imported, manufactured, distributed, and sold in the United States, and the CSA involves the prevention and control of the abuse of controlled substances. Federal law does not dictate ''who'' may prescribe and does not state provisions for licensure.[24,25] The CSA also fails to establish specific criteria regarding who may prescribe, merely stating that the prescriber must be licensed to prescribe under state law and be registered with the Drug Enforcement Administration (DEA) to prescribe controlled substances.[26] The supremacy clause (U.S. Constitution, Article VI, Paragraph 2) establishes supremacy of federal law over state law; federal agencies enjoy freedom from state or local regulation where Congress has not affirmatively made federal agencies subject to state law.

Within federal services and agencies, such as the Armed Services, Indian Health Service (IHS), and the Veterans Administration (VA), there may exist directives or regulations that confer the authority to prescribe on certain classes of practitioners. Federal facilities require that their health care professionals be licensed by at least one state, but the scope of practice of these professionals may be broadened or limited by written policies.[27]

The federal government has experimented with various models of pharmacist prescribing.[28–32] The VA and the IHS appear to have the most liberal policies toward pharmacist prescribing.[33,34] In the VA, clinical pharmacy specialists are required to have an advanced degree or have completed an accredited residency (e.g., an American Society of Health Systems Pharmacy, ASHP, accredited residency) or specialty board certification (e.g., a Board of Pharmaceutical Specialties, BPS, certification) before they may prescribe medications within their scope of practice. The scope of practice is established within the local VA facility. Once this criterion is satisfied, the clinical pharmacy specialist may function as an independent health care provider. In the IHS, pharmacists can be credentialed to provide primary care and use their prescriptive authority to evaluate and manage the care of certain patients.[35]

LIABILITY AND ETHICAL CONSIDERATIONS

Traditionally, the learned intermediary doctrine, shielded drug manufacturers and pharmacists from liability by imposing on the physician the duty to explain and to warn the patient about the effects of specific medications. Courts have remained reluctant to impose a ''duty to warn'' on pharmacists dispensing prescriptions, unless pharmacists voluntarily undertake such a duty, e.g., by advertising that they will undertake this duty.[36,37] However, mandated requirements for patient counseling will likely soon impact court decisions. As the scope of practice expands and judgmental functions increase, so does the potential for liability.

As pharmacists gain prescriptive authority, as well as responsibility for primary management of certain medical conditions, they must acquire new skills and a new awareness of the associated legal and ethical pitfalls. Pharmacists will be held accountable for negligence in performing new tasks or for practicing outside the scope of their practice. Pharmacists must also be aware of potential conflicts of interest, particularly in the managed care setting. Areas that have historically been problems for physicians may also become problem areas for pharmacists.[38–40]

Civil Actions

A patient alleging harm by a prescribing pharmacist can allege that the pharmacist, acting within the scope of practice of the profession of pharmacy, has performed in a fashion that is substandard and that, as a direct result of that substandard performance, the patient has suffered harm. This patient would allege that the pharmacist was negligent. A plaintiff can establish negligence by establishing that a duty to conform to a certain standard of conduct existed, that the duty was breached, that damages occurred, and that there was a reasonably close causal connection between the conduct and the resulting injury or damage.[41]

Traditionally, malpractice actions against pharmacists have stemmed from dispensing errors.[42] Dispensing errors can be distinguished from prescribing errors. Under a dispensing error standard, if a pharmacists fills a prescription in the manner in which it was ordered by the prescriber and there are no other obvious contraindications to the prescription, the pharmacist is not liable for any harm that comes to the patient. In short, courts have viewed dispensing as a nonjudgmental technical task that should be undertaken with great care and accuracy. Even alleged errors associated with faulty therapeutic interchanges (e.g., incorrect dosage adjustments made when one drug is dispensed in place of another) are dispensing errors, because pharmacists follow a protocol, decided on institutionally, when substituting one drug for another. Hence, the liability applies to the institution, which exercised the judgment, rather than to the pharmacist, who executed the decision. There is substantial case law on systems errors, based on the theory of respondeat superior, a doctrine in which the employer accepts responsibility for the tort liability of the employee, and corporate or institutional negligence, where protocols and procedures were inadequate to prevent errors.[43]

Pharmacists assuming additional clinical duties are at increased risk for malpractice lawsuits because patient assessment, treatment, and prescribing are more complex functions than making sure a prescription is filled correctly or counseling patients about adverse reactions. There is potential for negligence in any aspect of patient care, including taking a history, performing a physical examination, interpreting a laboratory test, or prescribing or renewing a drug.

Administrative Actions

The voluntary assumption of different or more sophisticated clinical duties by pharmacists can involve administrative action.[44,45] Administrative actions are brought by the state board and involve the statutorily authorized revocation or suspension of a license upon administrative finding of unprofessional conduct. Examples of unprofessional conduct for physicians are prescribing a drug without medical justification, prescribing a drug in excessive amounts, and unlawfully dispensing a controlled drug to a known addict. Unprofessional conduct might also include ethical violations like supplying a patient with drugs in return for sex and performing inappropriate physical examinations for sexual purposes.

State boards of pharmacy will have to develop guidelines and procedures to address the expanded clinical functions of pharmacists so that adequate oversight can be maintained. While tightly drawn CDTM agreements and protocols can eliminate some problems, they cannot eliminate all the problems.

Criminal Actions

Criminal sanctions may be imposed on health care professionals in certain situations.[46,47] These situations include improper prescribing of controlled substances, Medicaid and Medicare fraud, sexual abuse of patients, and even negligent care of patients.

Prescribers must be careful in prescribing controlled substances; it has been held that a licensed physician who prescribes controlled substances outside the bounds of professional medical practice is no different from a drug "pusher" subject to prosecution under the Controlled Substances Act. (21 U.S.C.S.§841[a][1]). Criminal liability may extend to the prescribing of controlled substances.[48]

The position of the Drug Enforcement Administration (DEA) is that if a state recognizes the authority of a pharmacist to prescribe controlled substances, then the DEA will register pharmacists as midlevel practitioners. Spe-

cifically, in 21 CFR §1304.02(f), the DEA establishes a new category of registration, midlevel practitioners, under which advanced-practice nurses, physician assistants, and others (e.g., pharmacists) will receive individual DEA registration granting controlled substance privileges consistent with authority granted them under state law. Other DEA regulations, 21 CFR §1301.24(b) and 1301.24(c), exempt agents and employees of a registered individual practitioner, hospital, or institution from the requirement of individual registration when they administer, dispense, or prescribe controlled substances in the course of their official duties or business. Thus, individual registration may not be necessary in order for pharmacists to prescribe controlled substances in an organizational setting.

It has long been clear that pharmacists would be liable for exchanging controlled drugs for sexual favors from either patients or nonpatients. A pharmacist may now also be liable for writing a prescription in exchange for sexual favors. Furthermore, pharmacists doing physical examinations must avoid behaviors that could be construed as inappropriate touching or sexual assault, both of which could result in criminal or administrative sanctions.

Smith[49] stated that there is currently an ''enthusiasm for professional accountability, especially when human life is lost.'' Criminal prosecution of negligence involving deliberate disregard of patient safety effectively serves the dual purpose of deterrence and punishment.[50] Like physicians, pharmacists will have to be sensitive to this added potential liability as they undertake the clinical care of patients, including prescribing medications.

Standard of Care

In analyzing whether a duty has been breached in a medical malpractice civil action, an administrative action, or a criminal case, the court considers whether the health care professional met the standard of care. If a lawsuit were brought because of damages caused by an incorrectly filled prescription, the court would look at the care that a reasonably prudent pharmacist in the community would have provided in the same or similar circumstances. Because prescribing drugs has traditionally been a physician function, a pharmacist undertaking this function may be held to the standard of care of a physician rather than the traditional standard of care of a pharmacist. This may be a more stringent standard because most physicians have more clinical training and experience than most pharmacists. If pharmacists hold themselves out as competent to physically assess patients and prescribe medications, citing additional degrees and board certification, the court may decide to hold them to the higher standard.

Managing Risk: Boundaries, Conflicts of Interest, Attitudes, and Skills

To minimize the risk of liability, pharmacists must know their scope of practice and not exceed it. Phillips[51] stated that an adequate legal definition of the scope of practice of a health care professional, as well as realistic standards for that profession, is necessary in order for the courts to correctly apportion liability for wrongful conduct and correctly apply malpractice theory.

CDTM or prescribing protocols must be written carefully, and they should be reexamined and updated regularly to ensure that they are not based on a standard too high to be reasonably met at all times and in all settings. In the pharmacist's judgment, a patient's condition or treatment requirements may seem only slightly inconsistent with the written protocol; however, pharmacists must be aware that protocols may leave little room for discretion. Pharmacists must be careful not to overstep the boundaries of their scope of practice, for this could result in a charge of practicing medicine without a license.

Another issue to be considered is the inadvertent misrepresentation of one's profession to patients. If a pharmacist performs a medical procedure or writes a prescription for a patient while the patient is mistaken about the professional status of the pharmacist, the pharmacist could potentially be charged with breach of the duty to obtain informed consent.[52]

The pharmacist who obtains patient histories, conducts physical examinations, orders laboratory tests, and prescribes medications must adopt new attitudes and develop new judgment skills. Communication is paramount, both with patients and with other health care professionals, particularly physicians. It has been said that poor communication and emotional issues are at the heart of nearly all medical malpractice actions.[53] Interdisciplinary communication is vital to ease the passage of pharmacists into expanded prescribing roles; good communication will enhance collaboration in patient care and diminish perceived threats to physicians' autonomy and authority. Obviously, expanded clinical roles for pharmacists will require increased patient contact and strong communication skills. If the patient and the pharmacist fail to understand each other, the pharmacist's risk of liability is increased.

POLICY ISSUES

The expansion of pharmacist prescribing may be affected by marketplace factors such as the abundant availability of other nonphysician prescribers,[54] the current focus of the health policy arena on medication errors,[55] and the nationwide shortage of pharmacists.[56] CDTM is a politically acceptable stepping stone to independent prescriptive authority and is supported by many leading national pharmacy organizations.[57] On the other hand, the American Medical Association has taken a strong stand against nonphysician prescribing, and it will take time for pharmacists to establish themselves as prescribers.[58] Traditionally, pharmacists have been gatekeepers and patient advocates. However, if suddenly the pharmacist becomes a physician surrogate, the pharmacist's role will be seen differently by the patient. The pharmacist will be acting in a more cognitive discretionary manner and asking for more compensation; from the patient's perspective, the pharmacist will also become less available or approachable. Given the complexity of medications and the volume of medication errors, there will be an interest by the public in retaining pharmacists in their traditional roles as gatekeepers unless these matters can be addressed effectively.

CONCLUSION

This article has summarized critical legal and ethical issues associated with the evolving function of the pharmacist prescriber. The purpose of this chapter is not to discourage pharmacists interested in prescribing, but to review critical issues associated with a new role. With proper training, communication with the patient and the patient's physician, adherence to the appropriate standard of care, avoidance of ethical and criminal breaches, the modern pharmacist will be able to contribute even more to the health care team.

REFERENCES

1. Fleischer, L. From pill counting to patient care: Pharmacists' standard of care in negligence law. Fordham Law Rev. **1999**, *LXVIII* (1), 165–187.
2. Green, T.C. Licking, sticking, counting, and pouring-is that all pharmacists do? McKee v. American Home Products Corp. Creighton Law Rev. **1991**, *24*, 1449–1477.
3. Gianutsos, G. *Pharmacy Law: Prescriptive Authority*; Power-Pak C.E. Program No. 424-000-98-008-H03 at http://www.powerpak.com/CE/PharmLaw (accessed April 22, 2001).
4. Fink, J.L., III; Vivian, J.C.; Reid, K.K. Civil Liability. In *Pharmacy Law Digest*, 35th Ed.; Facts and Comparisons, Wolters Kluwer, Co.: St. Louis, Missouri, 2000; 235–305.
5. Gianutsos, G. *Pharmacy Law: Prescriptive Authority*; Power-Pak C.E. Program No. 424-000-98-008-H03 at http://www.powerpak.com/CE/PharmLaw (accessed April 22, 2001), at II.
6. Omnibus Budget Reconciliation Act of 1990, 42 U.S.C. §1396 et seq.
7. Penna, R.P. Pharmacy: A profession in transition or a transitory profession? Am. J. Hosp. Pharm. **1987**, *44* (9), 2053–2059.

8a. Hepler, C.D. Pharmaceutical care. Pharm. World Sci. **1996**, *18* (6), 233–235.

8b. Hepler, C.D.; Strand, L.M. Opportunities and responsibilities in pharmaceutical care. Am. J. Hosp. Pharm. **1990**, *47* (3), 533–543.

9. Rosenthal, W.M. Establishing a pharmacy based laboratory service. J. Am. Pharm. Assoc. **2000**, *40* (2), 146–148, 150–152, 154–156.
10. National Association of Boards of Pharmacy. *1999–2000 National Association of Boards of Pharmacy Survey of Pharmacy Law*; National Association of Boards of Pharmacy: Park Ridge, Illinois, 1999; 94.
11. Ukens, C. Iowa pharmacists win Medicaid payments. Drug Topics **1999**, *143* (22), 37.
12. Magill-Lewis, J. More states seek prescriptive authority for pharmacists. Drug Topics **2000**, *144* (6), 57.
13. Conway, A.; Beister, D. Independent plenary prescriptive authority. J. Pediatr. Nurs. **1995**, *10* (3), 192–193.
14. Galt, K.A. The key to pharmacist prescribing: Collaboration. Am. J. Health-Syst. Pharm. **1995**, *52*, 1696–1698.
15. National Association of Boards of Pharmacy. *1999–2000 National Association of Boards of Pharmacy Survey of Pharmacy Law*; National Association of Boards of Pharmacy: Park Ridge, Illinois, 1999; 67.
16. Boatwright, D.E. Legal aspects of expanded prescribing authority for pharmacists. Am. J. Health-Syst. Pharm. **1998**, *55*, 585–594.
17. Fink, J.L., III; Vivian, J.C.; Reid, K.K. Regulation of Pharmaceuticals. In *Pharmacy Law Digest*, 35th Ed.; Facts and Comparisons, Wolters Kluwer, Co.: St. Louis, Missouri, 2000; 27–122.

18a. *Dent v. West Virginia* 129 U.S. 114 (1889).

18b. Gianutsos, G. *Pharmacy Law: Prescriptive Authority*; Power-Pak C.E. Program No. 424-000-98-008-H03 at http://www.powerpak.com/CE/PharmLaw (accessed April 22, 2001).

19. States authorize collaborative practice. ASHP News Views **2000**, *33* (6), 3.
20. Riley, K. *Collaborative Prescribing Authority for Pharmacists Gains Momentum*; At www.ascp.com/public/pubs/tcp/1996/sep/collab.html (accessed April 27, 2001).
21. Specific Authority 465.186(2) FS, Regulation 64B16-27.210, 64B16-27.220, 64B16-27.230, 64B16-27.300.
22. 21 U.S.C. § 301, et seq.
23. 21 U.S.C. § 801, et seq.
24. Gianutsos, G. *Pharmacy Law: Prescriptive Authority*; Power-Pak C.E. Program No. 424-000-98-008-H03 at http://www.powerpak.com/CE/PharmLaw (accessed April 22, 2001), at 1A.

25. Nielsen, J.R. *Handbook of Federal Drug Law*, 2nd Ed.; Mundorff, G., Ed.; Lea and Febiger: Philadelphia, 1992; 206.
26. Nielsen, J.R. Regulations of Pharmaceuticals. In *Pharmacy Law Digest*; Fink, III, J.L., Vivian, J.C., Reid, K.K., Eds.; Lea and Febiger: Philadelphia, 1992; 54.
27. Nielsen, J.R. *Handbook of Federal Drug Law*, 2nd Ed.; Mundorff, G., Ed.; Lea and Febiger: Philadelphia, 1992; 206, Boatwright at 587.
28. Ogden, J.E.; Muniz, A.; Patterson, A.A. Pharmaceutical services in the Department of Veterans Affairs. Am. J. Health-Syst. Pharm. **1997**, *54*, 761–765.
29. Paavola, F.G.; Dermanoski, K.R.; Pittman, R.E. Pharmaceutical services in the United States Public Health Service. Am. J. Health-Syst. Pharm. **1997**, *54*, 766–772.
30. Williams, R.F.; Moran, E.L.; Bottaro, S.D., II. Pharmaceutical services in the United States Army. Am. J. Health-Syst. Pharm. **1997**, *54*, 773–778.
31. Bayles, B.D.; Hall, G.E.; Hostettler, C. Pharmaceutical services in the United States Navy. Am. J. Health-Syst. Pharm. **1997**, *54*, 778–782.
32. Young, J.H. Pharmaceutical services in the United States Air Force. Am. J. Health-Syst. Pharm. **1997**, *54*, 783–786.
33. VHA Directive 10-95-019. In *General Guidelines for Establishing Medication Prescribing Authority for Clinical Nurse Specialists, Nurse Practitioners, Clinical Pharmacy Specialists and Physician Assistants*; 1995 (Mar. 3).
34. VHA Directive 96-034. In *Scope of Practice for Clinical Pharmacy Specialists*; 1996 (May 7).
35. www.ihs.gov (accessed April 30, 2001), specifically at www2.ihs.gov/Pharmacy/index.asp.
36. See *Baker v. Arbor Drugs, Inc.* 544 N.W. 2d 727 (Mich. Ct. App. 1996).
37. Termini, R.B. The pharmacist duty to warn revisited: The changing role of pharmacy in health care and the resultant impact on the obligation of a pharmacist to warn. Ohio N.U. Law Rev. **1998**, *4*, 551–566.
38. Krohn, L. Cause of Action Against Physician for Negligence in Prescribing Drugs or Medicine. In *9 Causes of Action 1.*
39. Federal Criminal Liability of Licensed Physician for Unlawfully Prescribing or Dispensing ''controlled substance'' or drug in violation of the Controlled Substances Act (21 U.S.C.S.§§801 et seq). 33ALR Fed 220.
40. Gulbus, V.M. Wrongful or excessive prescription of drugs as ground for revocation or suspension of physician's or dentist's license to practice, Lawyers Cooperative Publishing Co., Rochester, New York, 22ALR4th Supplement, 1983 and 1995; 669.
41. Keeton, W.P. *Prosser and Keeton on The Law of Torts*, 5th Ed.; West Publishing: St. Paul, 1984; §30.
42. Abood, R.R.; Brushwood, D.B. *Pharmacy Practice and the Law*; Aspen: Gaithersburg, Maryland, 1994; 355.
43. *See e.g. Bookman v. Ciolino,* 684 So.2d 982 (La.App. 5th Cir. 1994), *Harco Drugs v. Holloway,* 669 So.2d 878 (1995), *Dunn v. Dept. of Veterans Affairs,* 98 F.3d 1308 (1996).
44. Altman, L.K. Overdoses of cancer drug revealed by patient's death. N.Y. Times **1995**, A9, Mar. 24.
45. Cancer center begins disciplinary process in overdose cases. N.Y. Times **1995**, A7, Apr. 3.
46. Barrett, D.C. Homicide predicated on improper treatment of disease or injury, 45 ALR3d 114.
47. Annas, G.J. Medicine, death and the criminal law. N. Engl. J. Med. **1995**, *333*, 527–530.
48. California Health and Safety Code §11153(a), 21 U.S.C. §841.21, 21 C.F.R. §1306.04.
49. Smith, A.M. Criminal or merely human? The prosecution of negligent doctors. J. Contemp. Health Law Policy **1995**, *12* (1), 131–146.
50. See e.g. *People v. Einaugler,* 618 NYS2d 414 (App. Div.2d Dept 1994).
51. Phillips, R.S. Nurse practitioners: Their scope of practice and theories of liability. J. Leg. Med. **1985**, *6*, 391–414.
52. Cook, J.H. The legal status of physician extenders in Iowa. Review, speculations and recommendations. Iowa Law Rev. **1986**, *72*, 215–239.
53. Fitzgerald, M.A.; Jones, P.E. The midlevel provider: Colleague or competitor? Patient Care **1995**, *29* (1), 20 (13).
54. NABP Appendix on Prescribing Authority. In *Pharmacy Law Digest*, 35th Ed.; Facts and Comparisons, Fink, III, J.L., Vivian, J.C., Reid, K.K., Eds.; Wolters Kluwer, Co.: St. Louis, Missouri, 2000; 794–796.
55. Continuing efforts to prevent medication errors. Ipsa Loquitur **2000**, *27* (9), 1.
56. The Pharmacist Shortage. In *California Medicine*; December/January 1999-2000 at www.healthcarebusiness.com/archive...rniamedicine/current/pulsepharmacist.html (accessed April 27, 2001).
57. *See e.g.* Organization and Delivery of Services: Collaborative Drug Therapy Management (9812) and Collaborative Drug Therapy Management Activities (9801) at www.ashp.org/bestpractices/MedTherapy/org/position/position.pdf (accessed April 30, 2001) or Pharmacists Finding Solutions Through Collaboration at www.accp.com/postition/paper10.pdf (accessed April 30, 2001).
58. Breu, J. M.Ds still give thumbs down to nonphysician prescribing practice. Drug Topics **2000**, *144* (1), 21.

Pharmacotherapy, The Journal of Human Pharmacology and Drug Therapy

Richard T. Scheife
American College of Clinical Pharmacy, Boston, Massachusetts, U.S.A.

INTRODUCTION

Pharmacotherapy, The Journal of Human Pharmacology and Drug Therapy, is the official journal of the American College of Clinical Pharmacy (ACCP) and is published monthly. *Pharmacotherapy* is peer reviewed and is indexed in *Index Medicus, Current Contents, Exerpta Medica, Biological Abstracts, Chemical Abstracts, Current Awareness in Biological Sciences*, and *Referativnyi Zhurnal*.

Pharmacotherapy seeks out and publishes original, clinically relevant research, evidence-based reviews about drugs and therapeutics, and pharmacoeconomic and population-based therapeutic outcomes studies. Instructive case reports are also published. Editorial departments within the journal include Original Research Articles, Brief Reports, Reviews of Therapeutics, Mini-Reviews, Practice Insights, and Case Reports. The American College of Clinical Pharmacy publishes all of its official papers (e.g., official position statements, white papers) as well as its meeting abstracts in *Pharmacotherapy*. With some regularity, *Pharmacotherapy* publishes peer-reviewed supplements.

AVAILABILITY OF *PHARMACOTHERAPY* ARTICLES, ABSTRACTS, AND INSTRUCTIONS FOR AUTHORS ONLINE

Pharmacotherapy (1999 and later issues) is available online in a searchable, full-text, full-graphics format, (downloadable HTML files) free of charge through Medscape.com. You must sign up once for Medscape's free service. Thereafter, you need only to use a password to gain access to this valuable online library and information source. To navigate directly to the ''*Pharmacotherapy* reading room'' on Medscape, the direct URL is http://www.medscape.com/viewpublication/132_index. Alternatively, you may go to medscape.com and utilize the many resources available on this site while navigating to the ''*Pharmacotherapy* reading room.''

Abstracts of all *Pharmacotherapy* articles (from 1997 onward) as well as a downloadable/printable version of *Pharmacotherapy's* ''Instructions for Authors'' are posted in the *Pharmacotherapy* section of the ACCP Web site (www.accp.com).

Pharmacotherapy's Staff

Richard T. Scheife, Pharm.D., FCCP
  Editor-in-Chief
Wendy R. Cramer, B.Sc., FASCP
  Associate Editor
Sarah Jefferies
  Associate Editor
Nancy Gerber, M.Sc.
  Assistant Editor
Ann Chella-Nigl
  Business Manager
Minhoai Lam, B.Sc.
  Production Manager and Information Technology
Steve Cavanaugh, B.A.
  Production Assistant and Information Technology

Contact Information

Pharmacotherapy is located at NEMCH/806, 750 Washington Street, Boston, Massachusetts 02111; Phone: (617) 636-5390; Fax: (617) 636-5318; e-mail: rscheife@accp.com.

EDITORIAL BOARD AND SCIENTIFIC EDITORS

The Editor-in-Chief enjoys the counsel and support of a multidisciplinary Editorial Board of over 40 respected clinicians, researchers, and other scientists. In addition, a small, elite core of Scientific Editors, composed of nationally and internationally recognized scientists and thought leaders, help set editorial policy, help determine the journal's direction, and actively participate as Field

Encyclopedia of Clinical Pharmacy
DOI: 10.1081/E-ECP 120006276

Editors, inviting important papers and directing the pivotal elements of their editorial life.

A BRIEF HISTORY OF *PHARMACOTHERAPY*

Pharmacotherapy, The Journal of Human Pharmacology and Drug Therapy, was founded in 1981 by Russell R. Miller, Pharm.D., Ph.D., a young, fiery visionary who conceptualized a journal whose embrace would include all scientifically anchored aspects of clinical pharmacy and clinical pharmacology. For several years previously, Dr. Miller had been a Column Editor for the *American Journal of Hospital Pharmacy*, captaining a column prophetically named by him, ''Pharmacotherapy.'' Armed with this experience, his vision, and his famous energy and laser focus, Dr. Miller developed and brought *Pharmacotherapy* to market nearly single-handedly. Early on, the value of the journal was recognized by William A. Gouveia, Director of Pharmacy at Tufts-New England Medical Center in Boston, who instrumentally and graciously supported the journal and has provided hospital space for the journal to nurture and grow.

The journal grew steadily under Dr. Miller's steady hand and in 1985, he appointed Richard T. Scheife, Pharm.D., FCCP, as Deputy Editor of *Pharmacotherapy*. Since 1986, following the untimely death of Russell Miller, Richard Scheife has served as Editor-in-Chief.

Because of their shared commitment to improving the science and practice of pharmacotherapy, and because of their common long-term goals regarding the professional and scientific direction of the journal, *Pharmacotherapy* and ACCP initiated an affiliation agreement in 1988 whereby *Pharmacotherapy* became the official journal of the American College of Clinical Pharmacy. This relationship continued to strengthen and, in 1994, *Pharmacotherapy* became part of the ACCP family. Today, *Pharmacotherapy* continues to flourish and has become one of the preeminent scientific journals in the area of pharmacy and clinical pharmacology.

FUTURE DIRECTIONS

The future directions for *Pharmacotherapy* can be summed up in the preamble of our vision statement: ''*Pharmacotherapy* will become the preeminent journal in the broad field of pharmacotherapy and clinical pharmacology by serving as a knowledge source and publication venue for all health disciplines committed to optimizing drug therapy outcomes.'' Further, *Pharmacotherapy* will be the source of first choice by faculty when teaching health professionals about drugs and therapeutics. *Pharmacotherapy* will fully utilize all relevant forms of media to allow the broadest accessibility to *Pharmacotherapy's* important and high-impact knowledge.

PROFESSIONAL RESOURCES

Pharmacotherapy Self-Assessment Program (ACCP)

John D. Benz
American College of Clinical Pharmacy, Kansas City, Missouri, U.S.A.

INTRODUCTION

The *Pharmacotherapy Self-Assessment Program* (PSAP), published by the American College of Clinical Pharmacy, is a series of books that provides information on the therapeutic use of drugs. A broad range of therapeutic topics is covered. The topics included in PSAP are presented in a modular, curricular format (Table 1). Books, each containing two to three modules, are distributed every three months throughout the life of each edition.

DESCRIPTION

The focus of PSAP is on gathering data to make informed decisions and to support complex therapeutic decision making and problem solving. Information is presented and summarized through the use of tables, diagrams, figures, and algorithms to assist in decision making. Each chapter includes an update of pharmacotherapy with a focus on the provision of pharmaceutical care to improve patient outcomes, annotated references that direct health care professionals to additional resources for further study, and multiple-choice self-assessment questions in a case study format. A detailed rationale for the answers is provided, with references, to enhance the learning experience. The detailed rationale explains the approach that the author would take in evaluating and treating the patient described in the case. The program is designed to provide an update of the drug therapy of diseases with information from the three years previous to publication. PSAP emphasizes an extremely short time frame between when the text is written until it is printed and distributed. This resource provides a professional development tool that permits pharmacists to work at an individual pace.

Table 1 Topics by edition

First Edition	Second Edition	Third Edition	Fourth Edition
Cardiovascular	Cardiovascular	Cardiovascular	Cardiovascular
Infectious diseases	Critical care	Critical care	Systems of care
Neurology	Infectious diseases	Infectious diseases	Sites of care
Psychiatry	Biostatistics	Neurology	Respiratory
Respiratory	Research design and literature evaluation	Psychiatry	Endocrinology
Gastroenterology	Drug information	Research, biostatistics, and drug information applications	Rheumatology
Nephrology	Technologies and strategies	Consensus-driven pharmacotherapy	Infectious diseases
Endocrinology	Drug regulatory process	Respiratory	The science and practice of pharmacotherapy
Immunology	Respiratory	Endocrinology	Critical care
Oncology	Nephrology	Nephrology	Urgent care
Nutrition	Gastroenterology	Immunology	Neurology
Fluid and electrolytes	Endocrinology	Gastroenterology	Psychiatry
	Immunology	Nutrition	Gastroenterology
	Nutrition	Pediatrics	Pediatrics
	Oncology	Oncology	Nephrology
		Women's health	Hematology
			Oncology
			Women's health
			Men's health

Encyclopedia of Clinical Pharmacy
DOI: 10.1081/E-ECP 120006390

HISTORY OF PSAP

PSAP was initiated in 1991 with the goal of developing and distributing a high-quality, intensive, structured program for advanced-level clinical practitioners. Other objectives of the program included: 1) assisting other structured programs (e.g., staff development or residencies) in upgrading the clinical competence of clinical pharmacy generalists; 2) assisting advanced-level clinical practitioners in maintaining clinical competence; 3) emphasizing decision-making skills for complex clinical problems.

The goals and objectives of this program have been carried out consistently through the dedication and direction of the editorial boards (Table 2), the quality writing of leaders in the field of pharmacotherapeutics, and the input of skilled reviewers. Since the second edition of PSAP, the Board of Pharmaceutical Specialties acknowledges its quality by recognizing participation in PSAP as a method of obtaining recertification as a Board Certified Pharmacotherapy Specialist.

CONTINUING FOCUS OF PSAP

PSAP is continuing to focus on innovations in the drug treatment of diseases. Additional emphasis is being placed on the integration of alternative medicine into the treatment plan and the role of genomics in the prevention and treatment of diseases. Information is being provided on new ways that patients may choose to access the health care system, including home health care and telemedicine. In addition, new sources of drug information are discussed and described.

Consistent with the progressive nature of the PSAP series, innovative formats for delivery of PSAP information are being evaluated. These will include a version available on the Internet, starting with the fourth edition. The online version of PSAP provides the same information as the print version but with additional functionality, including the ability to take the exams online. The online version also permits access by pharmacists in remote areas of the world.

Table 2 Editorial boards by edition

First Edition	Second Edition	Third Edition	Fourth Edition
Barry L. Carter, Pharm.D., FCCP (Chair) David M. Angaran, MS, FCCP Thomas Sisca, Pharm.D., FCCP	Barry L. Carter, Pharm.D., FCCP (Chair) David M. Angaran, MS, FCCP Kathleen D. Lake, Pharm.D., BCPS Marsha A. Raebel, Pharm.D., FCCP, BCPS	Barry L. Carter, Pharm.D., FCCP, BCPS (Chair) Kathleen D. Lake, Pharm.D., FCCP, BCPS Marsha A. Raebel, Pharm.D., FCCP, BCPS Karen E. Bertch, Pharm.D. Marc K. Israel, Pharm.D., BCPS, CDE Donna M. Jermain, Pharm.D., BCPP H. William Kelly, Pharm.D., FCCP, BCPS Nancy P. Lam, Pharm.D. BCPS Wendy L. St. Peter, Pharm.D., FCCP, BCPS Bruce A. Mueller, Pharm.D., FCCP, BCPS Dennis F. Thompson, Pharm.D., FCCP, FASHP	Bruce A. Mueller, Pharm.D., FCCP, BCPS (Chair) Karen E. Bertch, Pharm.D. Teresa S. Dunsworth, Pharm.D., BCPS Susan C. Fagan, Pharm.D., FCCP, BCPS Mary S. Hayney, Pharm.D., BCPS Mary Beth O'Connell, Pharm.D., FCCP, BCPS Glen T. Schumock, Pharm.D., MBA, BCPS Dennis F. Thompson, Pharm.D., FCCP James E. Tisdale, Pharm.D., FCCP, BCPS Daniel M. Witt, Pharm.D., BCPS Barbara J. Zarowitz, Pharm.D., FCCP, BCPS

PHARMACY PRACTICE ISSUES

Pharmacotherapy Specialists, Practice Guidelines for (ACCP)

American College of Clinical Pharmacy
Kansas City, Missouri, U.S.A.

INTRODUCTION

The purpose of these practice guidelines is to describe the level of clinical pharmacy practice, knowledge, specialized skills, and unique functions that characterize the pharmacotherapy specialist.

Pharmacotherapy is that area of pharmacy practice that ensures the safe, appropriate, and economical use of medications. To function in this capacity requires specialized education and/or structured training in the clinical sciences. The pharmacotherapy specialist possesses unique professional knowledge and skills gained through advanced training in the biomedical, pharmaceutical, and clinical sciences, as well as through practice experiences. The pharmacotherapy specialist is skilled in the use of sound judgment in the collection, interpretation, and application of patient-specific data that are used to assist with the design, implementation, and monitoring of therapeutic regimens. Knowledge and skills of the pharmacotherapy specialist may be acquired through primary academic curricula, postgraduate residency or fellowship training, or innovative and extensive pharmacy practice experiences. The skills of the pharmacotherapy specialist, when applied across the entire continuum of the health care system, benefit individual patients, health care organizations, and society in general. The pharmacotherapy specialist is a licensed pharmacist and graduate of an accredited college or school of pharmacy.

The pharmacotherapy specialist is recognized by the pharmacy, medical, and allied health professions and by society in general as an expert in the area of applied pharmacotherapeutics and as an integral member of the health care team. Regardless of whether the specialist practices in acute or ambulatory care, the pharmacotherapy specialist meets the guidelines outlined in this document by the routine performance of the activities described in the assessment factors as part of daily pharmacy practice responsibilities.

GUIDELINE I

The pharmacotherapy specialist designs, implements, monitors, evaluates, and modifies patient pharmacotherapy to ensure effective, safe, and economical patient care.

Rationale

Using specialized knowledge of pharmacology, pharmacokinetics, pathophysiology, pharmacoeconomics, and therapeutics, the pharmacotherapist takes responsibility for patient outcomes. The pharmacotherapy specialist manages pharmacotherapy by evaluating therapeutic agents in the context of patients and populations and working collaboratively to assure their effective, safe, and economical use. The pharmacotherapy specialist makes clinical observations and incorporates them with information gained from other health care providers to optimize therapeutic decisions. The pharmacotherapy specialist participates in the evaluation of drug efficacy; identifies, reports, prevents, and participates in the management of adverse reactions; and initiates appropriate changes in the pharmacotherapeutic management of patients. This may include the discontinuation of drug treatment regimens, the prevention of unnecessary or potentially harmful treatment regimens, individualized adjustment of therapy in patients having unstable disease entities, management and moni-

This document was approved by the ACCP Board of Regents on January 17, 2000. It replaces "Practice Guidelines for Pharmacotherapy Specialists" (*Pharmacotherapy* 1990, 10:308–311).
From *Pharmacotherapy* 2000, 20(4):487–490, with permission of the American College of Clinical Pharmacy.

Encyclopedia of Clinical Pharmacy
DOI: 10.1081/E-ECP 120006408

toring of chronic drug therapy, or management of the drug formulary.

Assessment factors

1. Collaborates with other health professionals to make therapeutic decisions such as drug and drug product selection, therapeutic drug monitoring, and drug dosing.
2. Participates in the planning and development of patient treatment.
3. Investigates therapeutic alternatives and recommends or initiates the management of patient-related problems based on interpretation of relevant literature and clinical experience. Communicates the results of these investigations to health care practitioners in a manner appropriate to the training, skill, and need of that health professional.
4. Assists in the management, monitoring, and modification of drug therapy in patients with chronic disease.
5. Reviews patient records and orders regarding drug therapy and recommends and initiates changes as appropriate.
6. Evaluates patients by means of interview and, when appropriate, physical assessment to determine past medical history, previous medication use, present medical history, present medication use, present medical condition, and response to therapy. The pharmacotherapy specialist performs accurate and reproducible physical examination in accordance with their formal training and experience.
7. Interprets laboratory and other patient-specific data to aid in determining treatment plans and monitoring response to therapy.
8. Solves therapeutic queries posed by physicians and other health care providers.
9. Identifies complications resulting from drug therapy and recommends or initiates the necessary treatment alternatives to minimize or negate them.
10. Utilizes available state-of-the-art knowledge and technology to assess, improve, and monitor drug therapy regimens.
11. Establishes procedures for detecting significant drug-drug, drug-laboratory, drug-food, and drug-herbal interactions, and develops the necessary means to minimize adverse patient consequences that might result from such interactions.
12. Effectively communicates oral and/or written therapeutic recommendations or other aspects of drug therapy to health professionals, peers, patients, the public, and health care managers.
13. Assesses and participates in the management of patients with drug overdose and patients exposed to poisons.
14. Performs basic cardiac life support, and assesses and participates in drug therapy management during medical emergencies.
15. In conjunction with licensed medical practitioners, develops, manages, and assists in the implementation of pharmacotherapeutic protocols.
16. Works with other health care providers and relevant committees to develop programs for improving drug use and quality of patient care.
17. Documents the economic impact of clinical pharmacy activities for use by organized health care managers, practitioners, institutions, and providers.
18. Identifies therapeutic categories or individual therapeutic agents warranting drug utilization evaluation. Develops and conducts drug utilization evaluation in these targeted areas.
19. Coordinates the timely, accurate delivery of medications to patients in conjunction with other pharmacy practitioners.
20. Utilizes pharmacokinetic principles in the formulation of therapeutic drug regimens.
21. Coordinates the timing and collection of drug concentration samples in biologic fluids, interprets drug concentration results, makes recommendations to physicians regarding dosage adjustments, and monitors response to recommended dosage regimens.
22. Interprets patient-specific data, physical findings, medical history, and other pertinent information to aid in designing treatment plans, and monitors the patient's response to the recommended dosage regimen.
23. Evaluates the biomedical literature to determine optimal therapeutic drug-monitoring strategies and population pharmacokinetic parameters.
24. Interprets and applies population pharmacokinetic data to the design of patient-specific drug dosage regimens.
25. Determines patient-specific pharmacokinetic parameters on the basis of measured drug concentrations and prospectively applies these data to dosage regimen design.
26. Educates health professionals, students, patients, and the public regarding the utility of clinical pharmacokinetics.

GUIDELINE II

The pharmacotherapy specialist retrieves, analyzes, evaluates, and interprets the scientific literature as a means of providing patient- and population-specific drug information to health professionals and patients.

Rationale

Having expertise in literature evaluation and specialized training in therapeutics enables the pharmacotherapy specialist to retrieve and apply drug treatment information. The pharmacotherapy specialist will interpret the primary literature, evaluate its applicability to given patient care situations, and apply it in the synthesis of a solution to patient-specific drug therapy problems.

Assessment factors

1. Identifies and retrieves the best available information about pharmacotherapy by searching appropriate tertiary, secondary, and primary sources. The pharmacotherapist is adept at using computer technology to collect information.
2. Evaluates biomedical and pharmacoeconomic literature to determine criteria for optimal use and monitoring of therapeutic agents.
3. Routinely reviews biomedical and pharmacoeconomic literature relevant to the pharmacotherapeutic management of patient populations.
4. Evaluates the literature with regard to study design and methodology, statistical analysis, and significance of reported data so that appropriate assessments and conclusions may be applied to the solution of drug therapy problems.

GUIDELINE III

The pharmacotherapy specialist participates in the generation of new knowledge relevant to the practice of pharmacotherapy, clinical pharmacy, and medicine.

Rationale

The pharmacotherapy specialist has responsibility for the continuing development and refinement of knowledge regarding the appropriate use of medications. This knowledge may be generated by conducting research and clinical experimentation of therapeutic agents, by devising new and innovative approaches to pharmacotherapy practice, or by evaluating outcomes of new and innovative roles of pharmacotherapy specialists. The pharmacotherapy specialist has the responsibility to effectively convey the results of pharmacotherapeutic research to health professionals, patients, the public, and health care managers.

Assessment factors

1. Identifies pharmacotherapeutic questions to be studied or problems to be solved within the realm of pharmacotherapy practice.
2. Develops, implements, evaluates, and participates in scientifically valid and ethically designed studies.
3. Supports internal and external mechanisms for review of research protocols with regard to study design and protection of human subjects.
4. Collects data regarding the outcomes of patients managed by the pharmacotherapist.
5. Presents research results at scientific meetings and publishes results in the scientific literature.

GUIDELINE IV

The pharmacotherapy specialist educates health care professionals and students, patients, and the public regarding rational drug therapy.

Rationale

Experience and training in education combined with specialized knowledge and skills in pharmacotherapeutics render the pharmacotherapy specialist uniquely qualified to educate health care providers and consumers regarding effective, safe, and economical use of drugs. The specialist may teach in either the classroom or the clinical environment (e.g., instruction during hospital rounds, patient conferences, primary care settings, ambulatory clinics, emergency departments, etc.). The pharmacotherapy specialist is experienced in behavior modification and communication with patients, peers, and students. When applied appropriately, these skills may result, for example, in improved patient compliance with drugs and altered physician prescribing habits.

Assessment factors

1. Assumes responsibility for the education of all members of the health care team involved in patient pharmacotherapy.

2. Participates in continuing education programs concerning pharmacotherapeutics.
3. Develops patient education materials and participates in patient instruction programs to facilitate appropriate medication therapy and compliance.

GUIDELINE V

The pharmacotherapy specialist continually develops his/her knowledge and skills in applicable practice areas and demonstrates a commitment to continued professional growth by engaging in a lifelong process.

Rationale

The frequent introduction of new pharmacotherapeutic agents into practice, the increasing complexity and technicality of new drugs and biologic products, and the evolution of pharmacotherapy practice necessitate that the pharmacotherapy specialist continually refine, improve, and expand the unique, advanced skills which he/she possesses.

Assessment factors

1. Participates in professional organizations related to areas of expertise, thereby nurturing and enhancing personal knowledge and leadership skills.
2. Increases personal level of knowledge and skills by reading professional journals, and attending or participating in professional seminars, professional symposia, and national and international conferences.
3. Obtains board certification.

ACKNOWLEDGMENTS

Prepared by the 1998–1999 ACCP Clinical Practice Affairs Committee, Subcommittee B: Terry L. Seaton, Pharm.D., BCPS, St. Louis College of Pharmacy, St. Louis, MO; Aileen Bown-Luzier, Pharm.D., Department of Pharmacy Practice, SUNY at Buffalo, Buffalo, NY; Catherine E. Cooke, Pharm.D., BCPS, School of Pharmacy, University of Maryland, Baltimore, MD; John F. DeConinck, B.S., Owen/Cardinal Healthcare Corp., Desert Springs Hospital Pharmacy, Las Vegas, NV; Robert E. Dupuis, Pharm.D., BCPS, School of Pharmacy, University of North Carolina, Chapel Hill, NC; Rex W. Force, Pharm.D., BCPS, Department of Family Medicine, Idaho State University, Pocatello, ID; Peter Gal, Pharm.D., BCPS, FCCP (Chair), Greensboro Area Health Center, Greensboro, NC; Darrell Hulisz, Pharm.D., BCPS, Department of Family Medicine, University Hospitals of Cleveland, Cleveland, OH; Nathan L. Kanous, II, BCPS, School of Pharmacy, University of Wisconsin-Madison, Madison, WI; Dannielle C. O'Donnell, Pharm.D., BCPS, College of Pharmacy, The University of Texas at Austin, Austin, TX; Eric Racine, Pharm.D., Pharmacy Department, Harper Hospital, DMC, Detroit, MI; Theresa Salazar, Pharm.D., School of Pharmacy and Pharmacal Sciences, Purdue University, West Lafayette, IN; Allen F. Shaughnessy, Pharm.D., FCCP, BCPS, Harrisburg Hospital Family Practice; Harrisburg, PA; Ronald Taniguchi, Pharm.D., Pharmacy Services, Kaiser Permanente, Hawaii Region, Honolulu, HI; and Kay M. Uttech, Pharm.D., BCPS, College of Pharmacy, University of Illinois at Chicago, Chicago, IL.

Significant contributions to the development of this article were made by Karen S. Oles, Pharm.D., BCPS, Wake Forest University School of Medicine, Winston-Salem, NC.

PROFESSIONAL DEVELOPMENT

Pharmacotherapy Specialty Practice

Jack J. Chen
Western University of Health Sciences, Pomona, California, U.S.A.

INTRODUCTION

Pharmacotherapy is that area of pharmacy practice that ensures the safe, appropriate, and economical use of medications.[1] The pharmacist who practices as a pharmacotherapy specialist is considered to be an expert in the area of applied pharmacotherapeutics and is viewed as an essential member of the health care team. Similar to the practice of medicine, pharmacotherapy specialists may practice as generalists or may align themselves by specialties (e.g., geriatric medicine, infectious disease medicine, pediatric medicine, oncology medicine, nutrition support, psychiatric medicine). In addition, just as physician hospitalists—specialists in inpatient medicine—are responsible for managing the care of hospitalized patients and primary care physicians are responsible for managing the care of outpatients, a similar analogy may be applied when comparing the pharmacotherapy specialist with the ambulatory care pharmacist. This article focuses on the pharmacotherapy specialist who possesses the set of knowledge, skills, and attitudes specific to the care of hospitalized adult patients and compatible with the diversity of diseases and drugs encountered in the practice of internal medicine. Information on pharmacy practice in outpatient settings or in other types of specialties (e.g., infectious disease, oncology, pediatrics, psychiatry) are not discussed in this article.

FUNCTIONS AND ACTIVITIES

The function and activities of a pharmacotherapy specialist are well described by guidelines published by the American College of Clinical Pharmacy.[1]

Guideline One: *The pharmacotherapy specialist designs, implements, monitors, evaluates, and modifies patient pharmacotherapy to ensure effective, safe, and economical patient care.* The pharmacotherapy specialist takes responsibility for patient outcomes. The pharmacotherapy specialist will use his or her specialized knowledge, clinical observations, analytical skills, and communication skills to work collaboratively with other health care professionals to make decisions that facilitate optimal pharmacotherapy-related outcomes.

Guideline Two: *The pharmacotherapy specialist retrieves, analyzes, evaluates, and interprets the scientific literature as a means of providing patient- and population-specific drug information to health professionals and patients.* The pharmacotherapy specialist should be capable of using their literature evaluation skills to complement their knowledge base and problem-solving skills. The pharmacotherapy specialist optimizes patient outcomes by correctly interpreting the biomedical literature, evaluating its applicability to given patient care situations, and applying when solving drug therapy problems for a given patient.

Guideline Three: *The pharmacotherapy specialist participates in the generation of new knowledge relevant to the practice of pharmacotherapy, clinical pharmacy, and medicine.* The pharmacotherapy specialist has a responsibility to contribute to the generation of new biomedical knowledge and information through research and to share the research results through publication or presentation at scientific meetings.

Guideline Four: *The pharmacotherapy specialist educates health care professionals and students, patients, and the public regarding rational drug therapy.* The pharmacotherapy specialist is uniquely qualified to educate others on the optimal use of drugs. For example, the specialist may provide teaching to health care professionals and students in a classroom setting (e.g., pharmacy curriculum, scientific conference, educational symposium) or a clinical setting (e.g., hospital rounds, hospital clerkships). Education to hospitalized patients usually takes place prior to hospital discharge and may

Encyclopedia of Clinical Pharmacy
DOI: 10.1081/E-ECP 120006236

also focus on nonpharmacologic (e.g., dietary or lifestyle changes), in addition to pharmacologic (e.g., medication education), issues. The pharmacotherapy specialist may also be invited to speak at community health symposiums to educate the public on a variety of pharmacotherapy-related issues (e.g., preventative health medications used to treat various conditions, medication safety).

Guideline Five: *The pharmacotherapy specialist continually develops his or her knowledge and skills in applicable practice areas, and demonstrates a commitment to continued professional growth by engaging in a lifelong process.* As new pharmacotherapeutic agents are introduced into the marketplace and as the practice of pharmacotherapy evolves, the specialist must be capable of continually refining, improving, and expanding their advanced knowledge and skills.

As conceptualized by Holland and Nimmo, pharmacy practice can be categorized into five interrelated practice models (see Appendix).[2] The five interrelated practice models can be simplistically delineated as clinical pharmacy, distributive pharmacy, drug information, pharmaceutical care, and self-care. The value-added functions of a pharmacotherapy specialist often depend on the successful utilization of knowledge and skills derived from three of the five basic practice models—clinical pharmacy, drug information, and pharmaceutical care. Therefore, the pharmacotherapy specialist often has to acquire additional training and education, beyond the level of general licensed pharmacists, that will allow them to provide specialized, value-added functions.

QUALIFICATIONS

The pharmacotherapy specialist is a licensed pharmacist who has received specialized education and/or advanced training in the biomedical, pharmaceutical, and clinical sciences. The pharmacotherapy specialist must also be capable of using sound judgment when practicing pharmacotherapy. The appropriate knowledge, skills, and attitudes of a pharmacotherapy specialist are acquired through academic curricula (e.g., Doctor of Pharmacy program) at an accredited college or school of pharmacy, postgraduate training (e.g., pharmacy residency or fellowship), and/or extensive pharmacy practice experiences.

The pharmacotherapy specialist is generally intelligent and capable of resolving problems that may not have a clear right or wrong answer. They possess a strong sense of responsibility, are confident and conscientious, exercise practicality and logic, exhibit emotional stability and persistence, are able to convey expertise, are socially adept, and possess mature interpersonal skills that are required to successfully participate in collaborative decision making.

Certification

According to the Board of Pharmaceutical Specialties,

> *the primary purpose of specialization in any health care profession is to improve the quality of care individual patients receive, to increase the chances of positive treatment outcomes, and ultimately, to improve the patient's quality of life. Specialties evolve in response to the development of new knowledge or technology that can affect patient care, and the resulting changes in patient-care needs. The rapid, dramatic advancement in drug therapy in recent decades has created a clear need for pharmacy practitioners who specialize in specific kinds of treatment and aspects of care. Specialty certification is a responsible, progressive initiative from the profession to ensure the best possible patient care.*[3]

Although not required, a pharmacotherapy specialist may become a Board Certified Pharmacotherapy Specialist (BCPS) through a process established by the Board of Pharmaceutical Specialities (BPS). In addition to pharmacotherapy, the BPS certifies pharmacists in several other specialties (e.g., nuclear pharmacy, nutrition support pharmacy, psychiatric pharmacy, oncology pharmacy).

Recall that licensure to practice as a pharmacist is attained through a government agency after an individual demonstrates a *minimum degree of competency*. It does not indicate that the pharmacist has attained any additional skills or knowledge that is required to practice as a pharmacotherapy specialist. However, successful completion of BPS certification in pharmacotherapy indicates that the pharmacist possesses a level of education, experience, knowledge, and skills that surpass those required to obtain pharmacy licensure.

Additional information on BPS certification can be obtained from the Board of Pharmaceutical Specialties, 2215 Constitution Avenue, NW, Washington, DC, 20037-2985 or online at www.bpsweb.org.

VALUE

A growing body of literature has emerged that supports the value of pharmacist-related activities in improving and promoting the safe, effective, and economical use of drugs.

Researchers from the Harvard School of Public Health, Massachusetts General Hospital, and the Brigham and

Women's Hospital conducted a study designed to measure the effect of a single pharmacist's participation on medical rounds on the rate of preventable adverse drug events in an intensive care unit.[4] The participation of an experienced BCPS was associated with a 66% reduction in preventable adverse drug events and approximately $270,000 in annual cost savings to the hospital. The BCPS was accepted as a member of the multidisciplinary team, and the researchers noted that mutual cooperation and positive interpersonal relationships between the pharmacy and medical staff were vital criteria for successful pharmacist interventions.

In another study, services provided by a pharmacotherapy specialist (e.g., participation on rounds and reviewing of patients' medications) in internal medicine and intensive care units was associated with significant cost savings without negatively affecting the quality of care provided to patients. The drug cost savings alone was estimated at approximately $400,000 per year.[5]

On a daily basis, pharmacotherapy specialists are often involved with many types of clinical pharmacy services and functions associated with added value to patients and the health care system. For example, involvement in cardiopulmonary resuscitation teams, clinical research, drug information services, and medication admission histories is associated with a significant number of lives saved in hospitals within the United States (see Table 1). Six types of clinical pharmacy services—adverse drug reaction reporting, drug information services, drug protocol management, drug use evaluation, medication admission histories, and participation on medical rounds—are associated with an estimated health care cost savings of billions of dollars annually (see Table 2).[6,7]

SUPPLEMENTAL INFORMATION

For additional reading, the American College of Clinical Pharmacy has published a position paper on pharmacotherapy practice.[8] This document may also be viewed online at http//:www.accp.com.

Table 1 Clinical pharmacy services associated with reduced mortality rates in United States hospitals

Clinical pharmacy service	Estimated reduction in deaths per year*
Clinical research	201,274
Medication admission histories	131,815
Drug information services	45,428
Cardiopulmonary resuscitation participation team	18,416

*Based on 1029 hospitals that offer these clinical services. (From Ref. [6].)

Table 2 Hospital-based clinical pharmacy services associated with reductions in health care costs

Clinical pharmacy service	Estimated annual cost savings*
Adverse drug reaction monitoring	$1,636,614,476
Drug information services	$5,309,746,272
Drug protocol management	$1,757,282,145
Drug use evaluation	$1,137,727,143
Medication admission histories	$7,075,571,493
Participation on medical rounds	$8,107,395,977

*Based on 1016 hospitals that offer these clinical pharmacy services. (From Ref. [7].)

The American Society of Health-System Pharmacists (ASHP) supplemental standard and learning objectives for residency training in internal medicine and pharmacotherapy are published annually by the Accreditation Services Division of the ASHP in the Residency Directory, Volume Two. These documents may also be viewed online at:

http://www.ashp.org/public/rtp/IMSUPPST.html for internal medicine and at
http://www.ashp.org/public/rtp/PHTHSUPP.html for pharmacotherapy practice

APPENDIX

Clinical Pharmacy Practice

The clinical pharmacy specialist requires strong skills in designing, recommending, monitoring, and evaluating drug therapy. The ability to convey expertise and persuade prescribers is important. Knowledge of prescription medications and of chronic and acute disease states are important. Their primary role is to help prescribers make appropriate drug therapy decisions. A successful clinical pharmacy specialist must enjoy complex technical problem solving or solving ill-defined problems. In the acute care setting, patient contact may be minimal. However, a high degree of interaction with other members of the health care team (e.g., nurses, dieticians, social worker) is required. In the ambulatory setting, a high degree of patient contact is required.

Distributive Pharmacy Practice

The distributive pharmacy specialist requires strong technical skills in the preparation and dispensing of pres-

criptions, and communication skills to counsel patients on the use and administration of medications. They enjoy technical problem solving and routine activities. Their primary role is to ensure that the right medication gets to the right patient at the right time. Common practice settings include the community pharmacy and the hospital pharmacy.

Drug Information Pharmacy Practice

The drug information specialist requires strong literature evaluation skills, analytical skills, group communication skills, and the ability to persuade prescribers. Their primary role is to provide analytical and educational support to groups of health care providers or the public as a whole. A successful drug information specialist must enjoy researching the scientific literature. This practice does not necessarily involve a high degree of social interaction. Common practice settings include hospitals, managed care groups, and pharmaceutical companies.

Pharmaceutical Care Practice

The ability to provide pharmaceutical care is the profession's ultimate goal. In this practice, the pharmacist shares responsibility and accountability for the patient and drug therapy outcomes. The principal goal of pharmaceutical care is to achieve definite outcomes from medication use that improve the patient's quality of life. These outcomes include cure of a disease, elimination or reduction of disease symptoms, slowing a disease process, prevention of disease, and/or diagnosis of disease. Strong communication skills are required to establish a collaborative relationship with the patient, prescriber, and/or other members of the health care team.

Self-Care Pharmacy Practice

The self-care specialist requires strong skills in health screening, design and recommendation of drug therapy, use of judgment for referrals, and the ability to establish trust with the patient, to convey expertise, and to communicate clearly. Knowledge of nonpres-cription medications and therapy of minor diseases is important. The primary role of the self-care specialist is to help patients make informed decisions about their own care. A successful self-care specialist must enjoy technical problem solving and a high degree of social interaction. A common practice site is the community pharmacy.

REFERENCES

1. ACCP Clinical Practice Affairs Committee, Subcommittee B. Practice guidelines for pharmacotherapy specialists. A position statement of the American College of Clinical Pharmacy. Pharmacotherapy **2000**, *20*, 487–490.
2. Holland, R.W.; Nimmo, C.M. Transitions, part I: Beyond pharmaceutical care. Am. J. Health-Syst. Pharm. **1999**, *56*, 1758–1764.
3. Anonymous. Board of Pharmaceutical Specialties: Over 20 years of service to the public and the pharmacy profession. Career Manager **1997,** 1–7.
4. Leape, L.L.; Cullen, D.J.; Clapp, M.D.; Burdick, E.; Demonaco, H.J.; Erickson, J.I.; Bates, D.W. Pharmacist participation on physician round and adverse drug events in the intensive care unit. JAMA **1999**, *282*, 267–270.
5. McMullin, S.T.; Hennenfent, J.A.; Ritchie, D.J.; Huey, W.Y.; Longergan, T.P.; Schaiff, R.A.; Tonn, M.E.; Bailey, T.C. A prospective, randomized trial to assess the cost impact of pharmacist-initiated interventions. Arch. Intern. Med. **1999**, *159*, 2306–2309.
6. Bond, C.A.; Raehl, C.L.; Franke, T. Clinical pharmacy services and hospital mortality rates. Pharmacotherapy **1999**, *19*, 556–564.
7. Bond, C.A.; Raehl, C.L.; Franke, T. Clinical pharmacy services, pharmacy staffing, and the total cost of care in United States hospitals. Pharmacotherapy **2000**, *20*, 609–621.
8. ACCP Clinical Practice Affairs Committee, Subcommittee B, 1998–2000. Practice guidelines for pharmacotherapy specialists. Pharmacotherapy **2000**, *20*, 487–490.

PHARMACY PRACTICE ISSUES

Pharmacovigilence

James Buchanan
Tularik, Inc., South San Francisco, California, U.S.A.

INTRODUCTION

The primary goal of pharmacovigilence is to ensure that medications are used to maximal benefit while minimizing the risks of treatment. While clinicians directly involved in patient care seek to provide drug therapy with these same goals, healthcare professionals involved in the field of pharmacovigilence actively monitor the safety profile of medications so as to provide their colleagues directly involved in patient care with the information necessary to use medications in the safest possible manner. The scope of activities that comprise pharmacovigilence is broad, extending from preclinical safety pharmacology work through clinical studies to the ''postmarketing'' arena. The richest source of safety experience arises once a drug is marketed, yet it remains the most challenging from an analysis perspective. Due to the breadth of sources of safety information, the pharmacovigilence professional draws on a variety of disciplines, including pharmacology, pharmacokinetics, statistics, and pharmacoepidemiology.

REGULATORY ENVIRONMENT

The regulatory requirement that drugs be shown to be safe prior to marketing first became codified in the United States with the Federal Food, Drug and Cosmetic Act of 1938.[1] Prior to that time there was no mandate for an evaluation of drug safety. The Pure Food and Drug Act of 1906 prohibited interstate commerce of adulterated and mislabeled drugs. The Sherley Amendment four years later made it illegal to make therapeutic claims that are false and fraudulent. However, there was no legal basis for establishing that drugs were safe to use. The Food and Drug Administration in 1933 advocated for new legislation that would improve drug safety by illustrating the many shortcomings of the 1906 law. The need for revised legislation became acutely apparent in 1937 after 107 people died as a result of a sulfanilamide product that contained diethylene glycol. The Food, Drug and Cosmetic Act the next year gave the FDA new regulatory authority to require that drugs be demonstrated to be safe prior to commercialization. Since then, the scope of regulations and guidelines applicable to safety surveillance in the Unite States has become extensive.[2,3]

Other countries faced with similar circumstances also enacted similar legislation. Regulations requiring registration of drug products in Japan were enacted in the 1950s. The thalidomide tragedy in 1962 prompted regulatory action in Europe.[4] Across Europe, North America, and Japan this resulted in a panoply of regulations that, while based on the same fundamental goals of safety and efficacy, diverged in their technical requirements. Pharmaceutical companies engaged in the global marketplace found themselves facing duplicative and expensive work in order to meet the various regulations. Regulators in different countries also found it a challenge to exchange safety information when the methods of information collection were not consistent.

The European Community initiated efforts at regulatory harmonization in the 1980s. However, it was during the World Health Organization (WHO) Conference on Drug Regulatory Authorities in 1989 when international harmonization efforts were conceived.[5] The International Conference on Harmonization (ICH) was founded the following year, with industry and regulatory representatives from the United States, Europe, and Japan. The members were given the task to make more efficient the process for developing and registering new medicinal products in the three member areas.

The scope of the resulting ICH guidelines includes safety, efficacy, quality, and multidisciplinary issues.[5] Within the sections on safety, guidelines have been prepared for testing carcinogenicity, genotoxicity, and safety pharmacology, for example. Additional safety issues were addressed in other sections dealing with the expedited and periodic reporting of adverse events, a common nomenclature for describing adverse events, and the extent of population exposure to adequately assess safety. An indication of the success of this organization has been the extent that its recommendations have been adopted by

Encyclopedia of Clinical Pharmacy
DOI: 10.1081/E-ECP 120006345

various countries worldwide, further improving the steps toward global harmonization.

A number of the ICH guidelines were based on proposals offered by the Council of International Organizations for the Medical Sciences (CIOMS).[6,7] Originally founded by the WHO and UNICEF to produce scientific conferences, CIOMS has since matured into a ''think tank'' for addressing safety issues faced by both industry and regulators. Like ICH, the CIOMS Working Groups are comprised of members from the pharmaceutical industry and national regulators. Examples of the work of the CIOMS group have been development of a common form for reporting individual cases and periodic summaries of adverse events, elements for electronic data transmission of adverse events, a framework for determining attributability of an adverse event to a drug, and a process for weighing the risks and benefits of a drug product.

INTERNATIONAL ORGANIZATIONS

Additional organizations have developed to foster drug safety. The WHO Department of Essential Drugs and Medicines (EDM), created in response to the thalidomide disaster, established the Drug Safety Programme to communicate information among member states on drug safety and efficacy, including the prompt transmission of serious adverse event experiences.[8] The Uppsula Monitoring Centre, through an agreement with WHO, oversees a collaborative effort among 64 countries, as of January 2000, to monitor drug safety worldwide.[9,10] The Centre's goals are to identify safety risks as soon as possible and communicate them to healthcare professionals and the public in order to facilitate risk–benefit assessments. The WHO program also supports the European pharmacovigilence Research Group (EPRG). The EPRG, established in 1993, is comprised of members from academia and regulatory agencies from 10 European Union member states.[11] The group conducts research and investigations into specific drug safety issues.

CLINICAL DEVELOPMENT

The maturing of the field of pharmacovigilence has led industry and regulators to understand the evolutionary nature of safety surveillance. Whereas much of the impetus for promulgating regulations to ensure public safety was based on problems with commercialized products, a greater appreciation has developed that understanding the safety profile of a medication begins before it reaches the clinic. Safety surveillance begins with preclinical toxicology studies, evolves through clinical testing, and matures with postmarketing experience. Pharmacovigilance groups within pharmaceutical companies take this ''long view'' of drug safety and put into place systems to address the monitoring and evaluation needs at each phase.

The potential for a new drug or biologic to elicit untoward effects is first examined in preclinical studies.[12] Single-dose and repeated-dose toxicity studies examine the pharmacodynamic effects in at least two mammalian species. Genotoxicity studies examine the propensity of the drug to cause mutations and chromosomal damage. Carcinogenicity studies assess the ability of the compound to produce cancerous changes in cells. The toxic effects on reproductive organs and fertility is examined in reproductive toxicity studies. Untoward effects may first be identified during these initial pharmacodynamic studies. Safety pharmacology studies then seek to further identify adverse effects across a series of organ systems and to define the dose–response curve for these adverse effects. This information forms the initial basis for the scope of adverse effects that might be anticipated in humans.

The first clinical trials of a new compound generally take place in healthy volunteers. These studies seek to establish the initial safety profile in humans and obtain pharmacokinetic information. Up until this point, the extrapolation of the adverse events identified from preclinical studies to humans remains theoretical. Even adverse effects associated with other members of the drug class remain to be demonstrated for the product under study. Emphasis is placed on the clinical investigator to closely monitor the subjects for the development of any adverse effects and to attempt to determine whether a causal relationship exists between the event and the study drug.

The essential elements of pharmacovigilence during clinical trials of an investigational compound span from adverse event identification through characterization and analysis.[13,14] Identification is focused on ''treatment-emergent'' adverse events. These events can then be characterized by incidence, prevalence, severity, seriousness, and relationship to the study drug (causality). The data can then be examined to identify potential risk factors and markers with which to anticipate the occurrence of an adverse event.

Determination of a causal relationship between the use of a drug and the development of an adverse event can be difficult. A number of algorithms have been developed in an attempt to standardize this process.[15–22] Although each employs elements that are generally accepted, none has been widely adopted in its entirety due to limitations in reproducibility and predictive value. Over time, emphasis has shifted from attempting to assign causality at the individual case level and instead applying determinations of attributability to aggregate data.

The CIOMS Working Group published a guideline that included considerations of how to approach this assessment. The guideline describes how to create a source of the minimum drug safety information that should be communicated by manufacturers to clinicians, that is, information most needed to help clinicians balance a product's risks against its benefits and thus make good therapeutic decisions. The guideline identifies and ranks 39 factors that can be used to determine which adverse events to include in this ''core'' safety information document, events that are most likely to be causally due to the use of the drug. Although initially developed from a postmarketing perspective, the CIOMS V and revised CIOMS III report extended the concept of core safety information to the clinical development period prior to marketing.

The accumulating information in the core safety information document serves a number of participants in clinical trials. The investigators brochure, a document detailing the properties of the investigational drug, is augmented by the core document to provide investigators with a sense of the important safety information that can guide their management of clinical trial subjects. The adverse event data in the core document also serves as the basis of the risk section in the informed consent document, the essential adverse events that a subject would need to know about to make an informed decision regarding participation in the clinical trial. This information is also of relevance to ethics review committees (institutional review boards) when assessing the benefits and risks to participants in clinical trials.

Phase III studies are the pivotal trials by which efficacy is established for gaining marketing approval. The larger patient numbers, and often longer duration of drug treatment, in such trials allow for a more extensive evaluation of known adverse events and an opportunity to detect new events present at a lower frequency. The presence of a control group gives pharmacovigilence personnel the opportunity to estimate the baseline frequency rates for adverse events that previously may have been attributed to the drug. Certain adverse events may be secondary to the disease state under study or due to concomitant medication. From the background rate and the observed frequency of the adverse event in the treated group, an estimate can be made of the attributable rate, i.e., the extent to which the drug is causing the adverse event above the intrinsic rate. This information is useful not only in causality assessment but also in determining the magnitude of the effect produced by the study drug and, hence, in making a benefit–risk assessment.

POSTMARKETING ENVIRONMENT

The adverse event experience accumulated during the conduct of clinical trials, in the form of the core safety information document, serves as the basis for product labeling (prescribing information) upon marketing licensure. The safety profile of the drug at this point is characterized by an experiential database numbering only in the hundreds of patients, with limited long-term exposure data. Invariably, adverse events that occur at a low frequency or upon extended use will only emerge when larger numbers of patients are treated. Idiosyncratic drug reactions may also require greater numbers of exposed patients to be detectable.[23] One example of the failure of clinical trials to detect an important adverse event was the cough produced by angiotensin-converting-enzyme (ACE) inhibitors.

The sources of safety information once a drug is marketed can be many. Healthcare professionals using the medication in clinical practice become the frontline for detecting adverse reactions. Spontaneous reports of adverse events to the pharmaceutical manufacturer and/or regulatory authorities is an extremely important source of data for detecting emerging safety signals. Recognizing the importance of this experience, in the United States the Joint Commissions for the Accreditation of Health Organizations (JCAHO) mandates that healthcare organizations maintain a process for the identification, reporting, analysis, and presentation of sentinel events (any unexpected occurrences involving death or serious physical or psychological injury, or the risk thereof).[24] In France and Denmark, reporting adverse reactions that occur in clinical practice is mandated by law. In most other countries, reporting of postmarketing adverse events is voluntary.

The scientific literature is an important source of adverse event information, particularly in the form of case reports. Pharmacovigilence departments routinely scan the published literature for references to their products in order to identify adverse experiences. Case reports and case series can give clues to emerging safety issues. Because of the importance of these experiences, clinicians should publish case reports as soon as possible rather than waiting to assemble a case series.

Pharmaceutical manufacturers may be required to conduct postmarketing safety studies as a condition of drug approval. These may be warranted when licensure is based on limited data submitted for expedited approval or when the safety profile of the drug necessitates additional research. Such investigations can take the form of formal studies or registries, programs where limited but targeted information is collected on a large group of patients.

A significant limitation when analyzing spontaneous reports of adverse events is the inability to calculate frequency rates. In a controlled clinical trial, the number of events (numerator) and the number of trial subjects (denominator) are well known. A frequency rate for the adverse event can be calculated. However, in the postmarketing setting it must be assumed that all adverse events are underreported; hence the numerator value is poorly known and the total number of patients exposed to the drug cannot be definitively established, rendering the denominator value unclear. At best, postmarketing data can be described as reporting rates that cannot be directly compared to the frequency rates arrived at in the clinical trials.

A significant advance in the analysis of safety information for marketed drugs is the availability of large-scale, record-linked databases. These are databases of patient demographics, prescription history, laboratory data, and clinical data that are linked. Health maintenance organizations and academic programs have been instrumental in the construction of these databases, e.g., Group Health of Puget Sound, Kaiser Permanente, and General Practice Research Database. Epidemiologists utilize these systems for case-control studies that can provide estimates of relative and absolute risk. Based on a signal derived from spontaneous reports, for example, pharmacoepidemiologists can use these record-linked databases to explore the validity of this signal in a large patient experience.

The application of epidemiological principles has been a valuable development in the practice of pharmacovigilence.[25] The detection of new adverse reactions is made difficult by their unanticipated nature, delayed presentation, infrequent occurrence, presence of multiple concomitant drug treatments and/or other confounding risk factors. The science of epidemiology brings to pharmacovigilence the concept of risk assessment, tools for making comparisons, and a vocabulary for describing risk. Relative risk provides a measure of the size of an association between drug exposure and an adverse event, which is helpful for establishing associations. The absolute or excess risk is an indicator of the extent to which an adverse event can be attributed to drug use. The epidemiologic principles of relative and absolute risk provide a basis on which these risks can be balanced against anticipated benefits.

RISK–BENEFIT ASSESSMENT

Once a safety signal is identified and analyzed, pharmacovigilence personnel must determine whether it signifies a shift in the relationship between risks and benefits. Weighing the risks and benefits of drug treatment is a fundamental process in clinical medicine. Unfortunately, what appears to be a straightforward process has no agreed-on procedures or quantifiable statistics. Approaches to risk–benefit assessment during development have been published,[26–29] but little work has been directed toward analysis of marketed drugs.

The CIOMS Working Group addressed the task of developing a common approach to risk–benefit analysis in its fourth guidance—CIOMS IV.[7] In addition to providing a standardized methodology for risk assessment, the guidance describes concepts and procedures for determining the magnitude of a safety problem and deciding on appropriate actions. Although not a definitive set of algorithms, the guidance offers pharmacovigilence personnel not only a common schema for approaching risk–benefit assessment but also a recommendation that such common approaches be utilized by industry and regulators alike.

FUTURE DIRECTIONS

The momentum of international harmonization efforts will continue to improve the ability with which adverse events are classified, analyzed and communicated. Paper-based system will be replaced by electronic transmission of adverse event information. The WHO monitoring program will extend the Bayesian artificial neural networks for analysis of the large amounts of adverse event data at its disposal.[30] In the United States, efforts are underway at the FDA to improve the content and format of product prescribing information.[31]

Pharmacogenomics will likely play an increasingly important role in pharmacovigilence as genetic polymorphisms become better understood. Genetic variation can play a significant role in the safety profile of a drug. The science of pharmacogenomics—the study of genetic factors in drug response—has only recently become a viable tool with the advent of sufficiently sophisticated analysis techniques.[32,33] Knowledge of genetic variations in known polymorphic enzymes can help identify genotypes at particular risk. An appreciation that terfenadine metabolism was dependent on a cytochrome P450 isoenzyme (CYP2D6) that was effectively absent in 6–10% of the Caucasian population may have altered the development decisions for the compound or the target population.[32] Such examples of genetic polymorphisms of genes associated with drug metabolism are among the best studied to date. Genetic variants can give rise to unexpected drug effects, e.g., hemolysis in glucose-6-phosphate dehydrogenase deficiency, and variants in the drug target can alter the response and frequency of adverse effects, e.g., variants in the beta-adrenergic receptor can alter the response to beta-agonists in asthmatics.[34] Mutations in membrane transporters involved in drug absorption can lead to substantial changes in bioavailability. Transporter pharmacogenomics is a rapidly developing field in its own right. As the field of pharmacogenomics matures, it promises to afford a mechanism to improve the safety profile of a large number of drugs.

The process of pharmacovigilence requires the contribution of many healthcare professionals in industry, government, and clinical practice. With improved harmonization, more sophisticated analysis tools, and clearer communication, more patients will be able to realize the benefits of drug therapy while minimizing the attendant risks.

REFERENCES

1. http://www.fda.gov/cder/about/history/time1.htm (accessed Dec. 2000).
2. Curran, C.F.; Engle, C. An index of United States Federal Regulations and Guidelines which cover safety surveillance of drugs. Drug Inf. J. **1997**, *31*, 833–841.
3. Curran, C.F.; Sills, J.M. An index of United States Federal Regulations and Guidelines which cover clinical safety surveillance of biological products. Drug Inf. J. **1997**, *31*, 843–847.
4. http://www.eudra.org/humandocs/humans/PhVWP.htm (accessed Dec. 2000).
5. http://www.ifpma.org/ich1.html (accessed Dec. 2000).
6. http://www.who.int/ina-ngo/ngo/ngo011.htm (accessed Dec. 2000).
7. http://www.who.int/dsa/cat98/zcioms8.htm (accessed Dec. 2000).
8. http://www.who.int/medicines/ (accessed Dec. 2000).
9. http://www.who.pharmasoft.se/umc.html (accessed Jan. 2000).
10. Olsson, S. The role of the WHO programme on international drug monitoring in coordinating worldwide drug safety efforts. Drug Saf. **1998**, *19* (1), 1–10.
11. http://www.vsms.nottingham.ac.uk/vsms/talks/pharmacov/eprg.html (accessed Dec. 2000).
12. *ICH M3(M)—Non-Clinical Safety Studies for the Conduct of Human Clinical Trials for Pharmaceuticals.* http://www.ich.org/ich5s.html#multi.
13. Gait, J.E.; Smith, S.; Brown, S.L. Evaluation of safety data from controlled clinical trials: The clinical principles explained. Drug Inf. J. **2000**, *34*, 273–287.
14. Nowak, H. A systematic approach for handling adverse events. Drug Inf. J. **1993**, *27*, 1001–1007.
15. Kramer, M.S.; Leventhal, J.M.; Hutchinson, T.A.; Feinstein, A.R. An algorithm for the operational assessment of adverse drug reactions. I. Background, description and instructions for use. JAMA, J. Am. Soc. Assoc. **1979**, *242* (7), 623–632.
16. Hutchinson, T.A.; Leventhal, J.M.; Kramer, M.S.; Karch, F.E.; Lipman, A.G.; Feinstein, A.R. An algorithm for the operational assessment of adverse drug reactions. II. Demonstration of reproducibility and validity. JAMA, J. Am. Soc. Assoc. **1979**, *242* (7), 633–638.
17. Leventhal, J.M.; Hutchinson, T.A.; Kramer, M.S.; Feinstein, A.R. An algorithm for the operational assessment of adverse drug reactions. III. Results of tests among clinicians. JAMA, J. Am. Soc. Assoc. **1979**, *242* (18), 1991–1994.
18. Louik, C.; Lacouture, P.G.; Mitchell, A.A.; Kauffman, R.; Lovejoy, F.H.; Yaffe, S.J.; Shapiro, S. A study of adverse reactions algorithms in a drug surveillance program. Clin. Pharmacol. Ther. **1985**, *38* (2), 183–187.
19. Pere, J.C.; Begaud, B.; Haramburu, F.; Albin, H. Computerized comparison of six adverse drug reaction assessment procedures. Clin. Pharmacol. Ther. **1986**, *40* (4), 451–461.
20. Naranjo, C.A.; Shear, N.H.; Lanctot, K.L. Advances in the diagnosis of adverse drug reactions. J. Clin. Pharmacol. **1992**, *32*, 897–904.
21. Lanctot, K.L.; Naranjo, C.A. Computer-assisted evaluation of adverse events using a Bayesian Approach. J. Clin. Pharmacol. **1994**, *34*, 142–147.
22. Lanctot, K.L.; Naranjo, C.A. Comparison of the Bayesian Approach and a simple algorithm for assessment of adverse drug events. Clin. Pharmacol. Ther. **1995**, *58*, 692–698.
23. Knowles, S.R.; Uetrecht, J.; Shear, N.H. Idiosyncratic drug reactions: The reactive metabolite syndromes. Lancet **2000**, *356*, 1587–1591.
24. http://www.jcaho.org/ptsafety_frm.html (accessed Jan. 2000).
25. Kaufman, D.W.; Shapiro, S. Epidemiological assessment of drug-induced disease. Lancet **2000**, *356*, 1339–1343.
26. Spilker, B. Incorporating benefit-to-risk determinations in medicine development. Drug News & Perspect. **1994**, *7* (1), 53–59.
27. Chuang-Stein, C.A. New proposal for benefit-less-risk analysis in clinical trials. Controlled Clin. Trials **1994**, *15*, 30–43.
28. Methods and Examples for Assessing Benefit/Risk and Safety for New Drug Applications. In *Proceedings of a DIA Workshop*; Drug Inf. J.; 1993; Vol. 27, 1011–1049.
29. Cocchetto, D.; Nardi, R.V. Benefit/risk assessment of investigational drugs: Current methodology. Limitations and alternative approaches. Pharmacotherapy **1986**, *6*, 286–303.
30. Edwards, I.R.; Aronson, J.K. Adverse drug reactions: Definitions. Diagnosis, and management. Lancet **2000**, *356*, 1255–1259.
31. http://www.fda.gov/OHRMS/DOCKETS/98fr/122200a.htm (accessed Jan. 2000).
32. Kurth, J.H. Pharmacogenomics: Future promise of a tool for identifying patients at risk. Drug Inf. J. **2000**, *34*, 223–227.
33. Meyer, U.A. Pharmacogenetics and adverse drug reactions. Lancet **2000**, *356*, 1667–1671.
34. Ligget, S.B. The pharmacogenetics of beta2-adrenergic receptors: Relevance to asthma. J. Allergy Clin. Immunol. **2000**, *105*, 487–492.

PHARMACY PRACTICE ISSUES

Pharmacy Benefit Management

Chip Robison
BeneScript Services, Inc., Norcross, Georgia, U.S.A.

Kimberly Vernachio
Canton, Georgia, U.S.A.

INTRODUCTION

Pharmacy benefit managers (PBMs) are business entities that provide processes and services related to the acquisition and utilization of drugs by customers. They are the technical and clinical conduits for benefit payers (e.g., employers, employer groups, government) to develop, administer, and maintain a pharmacy benefit on behalf of the patients receiving the benefits.

A prescription benefit program is often regarded by consumers as a privilege that accompanies employment. Presentation of a prescription drug card to a pharmacy allows members of a PBM to have prescriptions filled at substantial discount. When filling these prescriptions, most pharmacies interface electronically with the database of the PBM. They then charge the member a copay based on the specific benefit design of their ''company-sponsored'' prescription program.

Employers share the prescription cost by charging the member a copay. The amount of the copay is typically based on whether the drug is a generic or brand name drug. The copay may also be based on whether the drug is found on a preferred list of medications known as a formulary. Formulary drugs typically are priced more favorably, due to rebate agreements made with the pharmaceutical company supplier. This is reflected in a lower copay when a PBM member's physician writes a prescription for a formulary drug.

PBMs in effect bring about competitive drug pricing. By their management of formularies, PBMs cause competition within the pharmaceutical industry marketplace. They leverage a large client base to negotiate price discounts from drug manufacturers in return for formulary inclusion of the manufacturer's product(s).

Managed care organizations that use PBMs include health maintenance organizations (HMOs), preferred provider organizations (PPOs), government agencies (primarily medicare, medicare plus choice, and state medicaid), and third-party administrators (TPAs). Third-party administrators are organizations outside the insuring organization that handle claims processing, administrative duties, and sometimes utilization review. They are used by organizations that fund the prescription (or total health) benefits, but do not find it cost effective to administer the health plan benefits themselves.

In addition to organized healthcare, private self-insured corporations may also contract directly with a PBM for services. Depending on the size of the PBM and its willingness to work with small groups, contracts may be written for prescription benefit management for small companies. However, many small companies enlist the services of a TPA for representation when it comes to purchasing the services of a PBM.

ORIGIN OF PBMs

Some of the first PBMs originated from managed care, as in the case of Express Scripts.[1] Prescription Card Service (PCS) evolved from an innovative network of community chain and independent pharmacies. The services of PCS were originally the processing of prescription claims for employer groups seeking discounts, and is responsible in large part for the birth of the National Council for Prescription Drug Programs (NCPDP).[2] The NCPDP is an industry group responsible for setting electronic adjudication and prescription submission standards nationwide. In 2000, PCS merged with Advance Paradigm to become Advance PCS the largest PBM at this writing.

The NCPDP helps PBMs standardize online pharmacy claim adjudication. Online adjudication of claims is a key revenue generator for many PBMs. When prescriptions are filled at the pharmacy level, the claim for payment is transmitted electronically to the PBM. At the PBM, computers interface with eligibility data, plan design, and a set of rules that process the claim and send notification back to the pharmacy that the claim is processed. In the case of a rejection, the reason for rejection in the form of a rejection code is sent to the pharmacy. The NCPDP helped the PBMs standardize the data files used to adju-

DOI: 10.1081/E-ECP 120006334

dicate claims and the codes used to submit utilization data for formulary drug rebates.

Diversified Pharmaceutical Services, Inc., with roots going back to the early 1970s, is credited for establishing maximum allowable cost (MAC) programs and mandatory generic drug programs.[3] It has been associated with SmithKline Beecham Healthcare Services and was once a subsidiary of the United HealthCare Corporation, but is now part of the Express Scripts network.

Actual ownership of the PBMs varies. Some companies are privately held corporations (e.g., National Prescription Administrators). Some are publicly traded (e.g., Express Scripts). Some are now or have been owned by managed care organizations (e.g., PacifiCare's ownership of Prescription Solutions or Prudential's ownership of Integrated Pharmacy Solutions). They have also been owned by retail pharmacy chains, as in the case of Rite Aid's ownership of PCS prior to its evolution into Advance PCS. Pharmaceutical manufacturers have purchased PBMs (e.g., Merck owns Medco and Lilly at one time owned PCS).

INITIAL MANAGED CARE CHALLENGE: COST

In the 1990s, both utilization and overall cost of pharmaceuticals increased drastically. This allowed cost containment programs and the PBM business to flourish. The annual percent change in retail prescription prices rose significantly higher than the consumer price index. Marketplace entry of new, higher-priced medications and direct to consumer marketing of medicines contributed significantly to increased patient utilization. Analysis of the recent explosion of drug prices is beyond the scope of this chapter; however, a sobering fact to note is that national health care expenditures for prescription drugs totaled $91 billion in 1998. This is expected to reach about $243 billion in 2008.[4]

The annual percentage increases in prescription expenditures have surpassed all other components of personal health care expenditures in the past decade. Yet, growth in expenditures and prescription drugs still comprise a relatively small proportion of total health care costs.

ONGOING MANAGED CARE CHALLENGE: BUILD OR BUY YOUR PBM

Typically, PBMs charge a managed care organization by the number of prescription claims administered. Plans with low to moderate levels of membership can realize substantial savings by outsourcing the services of a PBM. For example, a plan with 100,000 members who use their benefit averaging 0.7 prescription claims per member per month at 60 cents per claim would be billed $42,000 per month. Compare this with the cost of building a plan with an in-house pharmacy program staff, which could run into the hundreds of thousands of dollars.

The question that must ultimately be answered before contracting with a PBM is whether doing so will yield drug cost savings. Managed care organizations considering contracting for PBM services must compare the cost of the potential PBM contact with the cost of building an internal pharmacy program. For many organizations, purchasing the services of an established PBM (with its attached clinical expertise) makes better business sense than building one internally.

In addition, PBMs can assist with many of the challenges facing managed care organizations in the administration of a prescription drug program. These challenges may include

1. *Administering multiple plan designs and copayments*. Managed care organizations will inherently differ in the way their plans are designed. For example, a step therapy program for the treatment of hypertension may be part of one organization's design, but not anothers. An assertive PBM can set both prescribing protocols for the physicians as well as produce policies for prescription processing with regard to copay levels.
2. *Performing audits of contracted pharmacy providers (e.g., retail pharmacies contracted to provide pharmacy services)*. This is done to ensure that both the PBM and the managed care organization (payer) are being charged properly by the dispensing pharmacy for what is actually dispensed to the patients.
3. *Adjudicating claims electronically*. This refers to the actual transmission of a prescription claim via telephone lines from the pharmacy to the PBM database. Various benefit design rules are applied to the electronic claim as it is processed. This processing or adjudication step is an expensive item and can be administered on a cost per claim basis. Depending on the number of covered patient lives within a prescription plan, economies of scale may prevail, allowing for a low cost per claim adjudication fee to be offered to large plans. As the reader may imagine, cost per claim is a competitive area among marketers of PBM services.

4. *Providing Drug Utilization Review (DUR) edits.* These are electronic messages transmitted back to the dispensing pharmacist when filling a prescription. These edits provide messages concerning many aspects of drug therapy, and can flag nonformulary drugs when attempts are made to fill them. The DUR message from most PBMs alert the pharmacist to preferred drugs in lieu of the nonformulary drug being considered for dispensing. The National Drug Code (NDC) number flagging or blocking at the point of dispensing is a common function of DUR edits. Other parameters may be programmed into DUR edits by a PBM. For example, age limit parameters may be used for certain NDC numbers. Cosmetic (e.g., retinoic acid) products or attention-deficit disorder drugs may be excluded from coverage in patients over a certain age. Entire therapeutic category exclusions can be managed by DUR edits as well. For example, the oral contraceptive or fertility drug class prescriptions can be programmed for DUR flags. However, DUR edits can be patient friendly. Drugs that require a prior authorization can be logged into a patient's drug coverage profile. Once authorization is complete for the initial filling, subsequent refills can be approved electronically by means of DUR edits.
5. *Introduce initiatives on behalf of drug manufacturers to support the goals of a managed care organization.* This may take the form of unrestricted grants, promotion of internal health fairs, and wellness programs. For example, the manufacturer of cholesterol medicine may support health initiatives of a managed care organization aimed at reducing heart disease. A PBM can often be a valuable conduit for this type of support.
6. *Communicating prescription benefits, formulary drug lists, utilization management, and disease management initiatives to physicians and to patients.*
7. *Negotiating the contracted drug reimbursement rates for network and mail service pharmacies.* Some PBMs use a MAC pricing schedule for generic drugs, limiting the pharmacist's reimbursement rates, and thereby containing cost for the health plan payor.
8. *Maximizing cost savings by sharing savings from pharmaceutical contracts with the managed care organization.*
9. *Ensuring competitive administrative fees associated with prescription claims and management reports.* Competition in the marketplace puts pressure on the PBM to offer a managed care organization a fair and equitable rate on the prescription claims processed by the PBM. As a service not unlike others in health care, fees charged by PBMs are negotiable and should provide value to the managed care organization.

Once hired by a managed care organization, the PBM must also report utilization trends back to the organization. These reports must be meaningful in a way that will provide the managed care organization with data from which to draw informed conclusions as to how it is spending its prescription drug dollar.[5]

PBM ACTIVITIES

A number of products and services beyond the collection and processing of pharmacy prescriptions are offered to customers by PBMs. The following paragraphs, although not complete, provide a listing of some of the many services offered by PBMs.

Claims Adjudication/Processing

A claim is a bill, generally submitted electronically in a standardized NCPDP format (NCPDP), for products and services rendered by retail pharmacies. Once received, the claim is audited via a series of programs that make up the adjudication process. The information is scrutinized for appropriate content, correctness, eligibility, pharmaceutical, and benefit appropriateness. Real-time electronic point-of-sale claims are adjudicated according to the plan design of the patient's benefit. The technology used allows for the data to be merged with eligibility and pharmacy data pertaining to each member and plan. The adjudication process also provides electronic messaging for ease of prescription filling by the pharmacist. (The message received is guided by the results of the various adjudication programs.) The goal of this messaging is to foster plan/formulary compliance and ultimately bring about cost efficiency for payors.

There are 12 basic audit steps in the adjudication process (Fig. 1). They are as follows:

1. *Submit claim*—The pharmacy sends the claim electronically to the PBM responsible for processing, as indicated by the patient's benefit card.

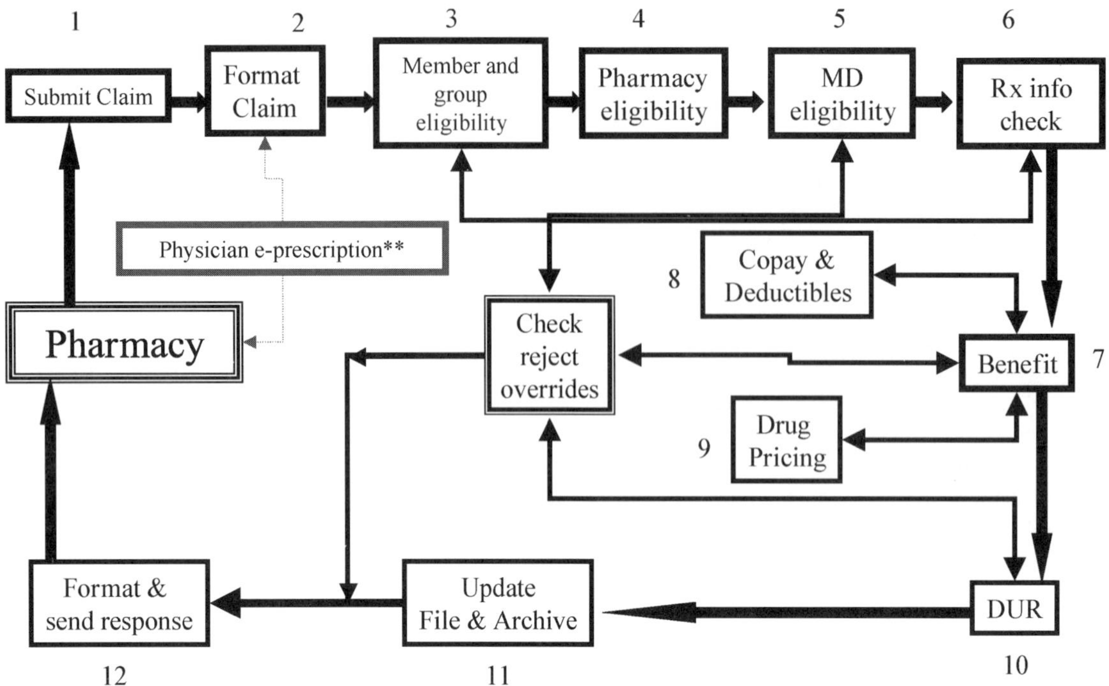

Fig. 1 PBM claim adjudication process.

2. *Format claim*—Claim information is inserted into appropriate fields of the NCPDP standard format.
3. *Member and group eligibility*—The patient/member information is checked against a current list of patients eligible to receive a medical benefit. The type and extent of the benefit received is indicated by the ''group'' to which the individual belongs (e.g., all employees of a single company usually make up a ''group'').
4. *Pharmacy eligibility*—Not all medical benefits include coverage of pharmaceuticals. The member/patient information must be checked against a list of individuals eligible for a pharmacy benefit (e.g., currently medicare beneficiaries do not receive a pharmacy benefit; thus, no assistance is offered in the coverage of pharmaceutical costs).
5. *MD eligibility*—The prescribing physician is checked against a list of approved health plan providers.
6. *Prescription information check*—Once the prescription information is placed in the proper format, each piece of information is checked by a program for agreement with the other components of the prescription information (e.g., drug selection agrees with NDC number, drug agrees with the dose, dose agrees with the dosing schedule, quantity agrees with day's supply).
7. *Benefit*[6]—This indicates what medications are covered and under what circumstances. The benefit is defined by the degree to which the patient will contribute to products and services via a copay. The entity that provides the benefit quantitates and determines the benefit limits (e.g., if the benefit is provided to the patient by the employer, then the employer dictates the conditions of the benefit—for example, ''lifestyle drugs'' that offer cosmetic improvement may not be covered in favor of covering critical medications that treat disease).
8. *Copays and deductibles*[6]—These are the prescribed dollar contributions that the patient must pay to receive a particular pharmacy product or service. The prevailing logic for copays is that increasing the patient's share of the cost will encourage the patient to influence prescribing. If the patient must pay a higher copay, then the patient may influence the physician to seek

lower copay (less costly to the health plan) drugs. Thus, copays are used indirectly to steer a physician's prescribing habits. Copays are also used to encourage patients to request the generic version of medications when available (e.g., a patient benefit may state a low copay of $5 for a generic medication, $10 for a brand name medication, and/or $25 for a nonformulary medication).

9. *Drug pricing*—The prescription information is matched with a listing of per unit price, as dictated by PBM purchase contracts with pharmaceutical companies.

10. *DUR*—This process can take many forms depending on the goal to be accomplished. The drug can be scrutinized against the existing patient profile for drug interactions, accuracy of dosing, and/or prescribing instructions or duplicity. The purpose is to ensure appropriate prescribing for the most common or majority use. It is not intended to check appropriateness for clinical exceptions to ''the rule.'' Actions to be taken, if any, are relayed back to the pharmacy by an edit message. The edit message may indicate that the prescription will not be covered until discrepancies are corrected or further information is given. The DUR message is one of the vital communications between the PBM and the dispensing pharmacy. The actual DUR message can suggest a preferred formulary drug for a particular nonformulary drug submitted. The DUR message may be in the context of an interaction warning or overdosage warning supplied by the logic of the PBM's internal software. Discrepancies between quantity and day's supply violations may be communicated back to the dispensing pharmacist as well. For example, a pharmacy may enter correct information for a birth control prescription. If the patient file indicates the patient is female and that birth control is part of the allowable benefit, then the prescription will be covered with a copay. If the patient's file indicates the patient is male, a rejection code will be generated with an appropriate message sent back to the pharmacy.

11. *Update file and archive*—The patient's drug profile and the PBM's data archive are updated to reflect the latest information. The information necessary to complete the transaction is sent electronically back to the pharmacy. Updating a patient file ensures a continuous, accurate prescription record. This record can then be checked for compliance, drug interactions, compared with medical claims to check for appropriateness of treatment, etc., all of which offer opportunity for the pharmacist, physician, or PBM to avoid an adverse event or improve a clinical outcome.

12. *Format and send response*—The adjudicated claim is returned to the dispensing pharmacy. The industry standard for claims communication is National Council for Prescription Drug Program (NCPDP) formatting. With this format, the various field containing characters and text are standardized and easily recognizable to pharmacies. This final link will clarify such parameters as:

    - patient identification
    - ingredient cost(s)
    - days supply of medication
    - copay due from patient
    - dispensing fee
    - final amount reimbursed to the dispensing pharmacy

Retail Network Access

PBMs have contractual relationships with retail pharmacies and share a common electronic network. Through this common network, prescription claims can be sent to a common point of collection (the PBM). Large PBM–retail networks offer customers easy access to prescriptions throughout the United States. Network relationships are less costly to PBM customers due to volume discount and greater billing/payment efficiencies (e.g., all pharmacies that contract with the PBM are part of that PBM's pharmacy network).

Mail Order Services

Providing volume discounts and less administrative cost to the payer, by virtue of making payments to only one business, makes mail order a cost-efficient alternative. Patients are often incentivized to use mail order services with offerings of home delivery, discontinued multiple-month supplies of medication, and discounted copayments.[6,7]

Account Management

PBM representatives act as liaisons with the purchaser of pharmacy benefits for the purposes of exchanging information on benefit utilization, collaboration on health education or interventional programs, maintaining and updating the benefit as needed, and addressing patient needs and satisfaction issues. PBM representatives and purchasers also collaborate to ensure that quality pharmaceuticals and services are delivered at minimal cost. Good disease management translates into greater productivity and less absenteeism for the employer. Thus employers have a vested interest in improving the health of their employees. An employer with a young covered population—young men, women, and children—would have concerns that medications for diseases predominant in the young, like asthma, would be adequately covered by the benefit and used appropriately by patients. Account managers work with employers to create ways to meet these health goals.

DUR/DUE (Drug Utilization Evaluation)

PBMs offer analysis of drug utilization patterns at both individual and group levels. From these analyses, they may recommend an intervention, education, or course of action to ensure appropriate utilization and outcome. DUR/DUE edits also occur online as the prescription is processed. If the prescription information does not match program criteria for appropriateness, an edit message is generated to the filling pharmacy.

Formulary Management

PBMs undergo a continual process of evaluating individual drugs, drug classes, and their application in disease. The process includes the use of internal and external medical and pharmaceutical expertise to evaluate the efficacy, safety, and cost of pharmaceuticals in the context of their use in a variety of disease states. This process provides for continual improvement of the pharmacy benefit as new products come to market. In addition to an ongoing evaluation of the value and application of formulary drugs, PBMs negotiate contracts for these drugs. Contracts are negotiated in consideration for a PBM placing a partciular particular manufacturer's medication on a preferred drug formulary. Contracting with the pharmaceutical industry allows revenue to flow into the PBM and ultimately back to the PBM's clients. This in effect brings about discounting of pharmaceuticals.

Disease Management

PBMs evaluate prescription data to identify drug/disease patterns of use, and develop programs targeted to ensure correct dosing, compliance, and maintenance of treatment standards are maintained. Disease management programs may originate from within the PBM or may be outsourced to a specialty company with this expertise. Generally, pharmaceutical manufacturers assist PBMs with their disease management initiatives.

Utilization Reporting

PBMs offer aggregate reports from their prescription data archives. These data are stripped of all patient-identifying information and are disseminated to physicians or employers for the purpose of enhancing patient outcomes. PBMs offer the expertise of clinical pharmacists to assist in the interpretation of aggregate reports. Clinical pharmacists help identify trends and goals to set for maximization of patient outcomes.

Patient Education

PBMs offer educational services to patients in many forms. These may include newsletters, specialty programs, advice lines, consultation services, and internet-based information services, and so on.

Customer Support

PBMs offer services to both customers and benefit purchasers to ensure that questions are answered and problems are resolved in a timely manner.

Prior Authorization

The purpose of prior authorization is to safeguard the appropriate use of specialized or niche drugs, drugs with potentially serious toxicities or narrow therapeutic index, and very costly medications. This step requires that additional information be given before the drug in question will be paid for by insurance. Information requirements necessary to receive authorization for coverage of a particular drug may include previously failed drug therapies, requirement of certain diagnostic tests, and complications of disease.

Benefit Development

PBMs offer expertise in designing the most appropriate and cost-efficient benefit based on a number of demo-

graphic parameters to include prevalence of disease, age of the insured population, and cost limitations.

TRENDS

The business of medicine and of managed care is ever evolving. The PBM industry is continually challenged with issues surrounding the development of new treatments and technologies. Listed here are a few of the PBM trends of the future.

Four-Tier Formulary Design

A tiered benefit is a type of pharmacy benefit that has grown in popularity in recent years. Rather than follow the ''traditional'' formulary process of selecting ''best in class'' medications for formulary inclusion, a tiered benefit simply stratifies medications according to some basic parameters, usually efficacy, safety, and cost. This design of benefit usually does not preclude the coverage of any given medication (as with the traditional closed formulary design) but rather allows for a greater cost sharing by the patient. Medications in lower tiers are usually available as generic, possess a clinical advantage of some type, or are more cost-efficient. Medications placed in higher tiers are usually those that offer no clinical advantage, are more expensive, or are ''lifestyle enhancing'' in nature (e.g., medications for sexual performance, cosmetic drugs, fertility agents, hair restoration drugs, appetite suppressants, oral contraceptives).

Traditionally, pharmacy benefits have been of three-tier design; each tier associated with an ever-increasing copay (e.g., 1st tier, $5; 2nd tier, $10; 3rd tier, $25) for common medications. Under this design a few select medications, such as ''lifestyle-enhancing'' drugs, were either not included in the benefit or were severely restricted.

Consumer demands for greater access to medications has prompted the development of a fourth tier. This differs in that it introduces the concept of coinsurance, or percentage copay to consumers. Medications in this tier would be covered as a percentage of total medication cost as opposed to a set copay amount. For example, fertility medications might be covered at 50% of their total cost. The patient would pay 50% and the pharmacy benefit would pay 50%. Thus providing the consumer and the physician with an incentive to discuss considerations of cost.

Electronic Prescribing

Electronic prescribing has been a much sought after goal by the health care industry. Electronic prescribing would close many of the loopholes through which medication errors often pass. By eliminating the handwritten word, e-prescribing would significantly reduce errors in prescription interpretation; can allow for real-time drug screening, formulary, and DUR checks; and streamline the communication process between the HMO, pharmacist, and physician.

Barriers to electronic prescribing generally circulate around issues of connectivity and ease-of-use issues with physicians, and the cost of implementing this technology in both physician practice and pharmacies. For a seamless system to work, the prescribing physician must have

1. *Ready access to benefit information.* What is the formulary, and what if any conditions must be met for coverage?
2. *The Benefit provider information.* Who is the HMO/PBM benefit provider?
3. *A Common electronic interface.* Will a single program interface with all benefit providers? Will it interface with all retail pharmacy providers? How will the prescription information get to the correct provider? What devices will physicians use to interface with the system? Computers, handheld devices?

Although the benefits to the patient and the health care marketplace would be tremendous, there are many obstacles to overcome. One market development holds great promise for bringing the health care system one step closer to making e-prescribing a reality: On February 22, 2001, three of the largest PBMs, Advance PCS, Express Scripts, and Merck–Medco announced an agreement to establish a common utility for the collection of prescription information. The company, known as RxHub LLC, will act as the central ''conduit of information between all parties'' and ''will provide a single standardized channel of communication to link physicians through electronic prescribing software on their handheld computer or practice management system to pharmacies, PBMs and health plans.''[8] The success of e-prescribing and of ventures like RxHub holds great promise for increasing efficiency.

Potential Regulation of PBMs

Due to their inherent activities, PBMs are alleged to be involved in the practice of pharmacy by some lobbying groups. As a result, the National Association of Boards of Pharmacy has advised each state board of pharmacy to evaluate and license PBMs in their area. The Academy of

Managed Care Pharmacy disagrees with PBM licensure, claiming that a very small percentage of PBM activity could be defined as practicing pharmacy. Legislation has been proposed by various state legislatures to license PBMs as pharmacies. Licensing legislation most probably will take the form of a separate class of pharmacy licensure. Separate licensing would in turn lead to potential state regulation of PBMs. Georgia, for instance, has signed into law a bill that mandates the licensure of pharmacy benefit managers doing business within the state. This licensure requirement allows for inspection of pharmacy benefit managers by state inspectors.

Another regulatory issue is therapeutic interchange of medications. PBM clinical departments often set up internal protocols for alternate product substitution within a therapeutic class. The National Association of Boards of Pharmacy contends that therapeutic substitution is practicing pharmacy. In one of its position papers, the Academy of Managed Care Pharmacy supports the use of therapeutic interchange programs. The Academy of Managed Care Pharmacy contends that this decision is ultimately that of the physician.[9]

Another form of regulation now facing PBMs is in the form of the Health Information Portability and Accountability Act. This enormous set of regulations, which is expected to have full legal effect in the near future, would potentially affect not only the processing of confidential patient information, but also the administration of other clinical programs (e.g., disease management, formulary management, drug compliance programs, patient and clinician education programs).

ACKNOWLEDGMENTS

Paul Hueseman, PharmD, Medical Information Scientist, National Accounts, AstraZeneca Pharmaceuticals, and Ben Sloan, National Account Director, Salix Pharmaceuticals.

REFERENCES

1. Navarro, R.P. *Managed Care Pharmacy Practice*; Aspen Publishers: 1999; 223–231.
2. The National Council for Prescription Drug Programs website: www.ncpdp.org.
3. *Diversified Pharmaceutical Service, Internal Promotional Document*; 1998.
4. Smith, S., et al. The next decade of health spending: A new outlook. Health Aff. **1999**, 94, July/August.
5. Sawyers, T. Test prospective PBM before signing contract. Managed Care **2000**, March.
6. Stern, C. The history, philosophy and principles of pharmacy benefits. J. Managed. Care Pharm. **1999**, *5* (6), 525–531.
7. Copland. Employee Benefit Research Institute, 1999, April, no. 208 (www.ebri.org).
8. http://www.express-scripts.com/other/rxhub/rxhub.html, Press Release: February 22, 2001; New Venture to Jumpstart Electronic Link with Physicians and Pharmacies.
9. *Where We Stand, Therapeutic Interchange, Position Paper. Revised by the AMCP Board of Directors*; Academy of Managed Care Pharmacy: 100 North Pitt Street, Suite 400, Alexandria, Virginia 22314, 1999, November (www.amcp.org).

BIBLIOGRAPHY

Web Sites

Academy of Managed Care Pharmacy web site: www.amcp.org

Employee Benefit Research Institute web site: www.ebri.org

Medscape web site: www.medscape.com

National Council for Prescription Drug Programs web site: www.ncpdp.org

Pharmaceutical Care Management Association web site: www.pcmanet.org

Pharmacy Benefit Management Institute web site: www.pbmi.com

Pharmscope news, articles, and editorials: www.pharmscope.com

RxHub LLC web site: www.rxhub.net

Journals/Text

Drug Benefit Trends is published monthly by SCP/Cliggott Communications, Inc., 330 Boston Post Road, Darien, CT 06820-4027.

Healthplan is published bimonthly by the American Association of Health Plans (AAHP). 1129 20th Street, NW Suite 600, Washington, DC 20036-3421.

Journal of Managedcare Pharmacy is peer-reviewed and published bimonthly by the Academy of Managed Care Pharmacy, 100 North Pitt Street, Suite 400 Alexandria, VA 22314.

Managed Care is published monthly by Managed Care, Medimedia USA, 275 Phillips Boulevard, Trenton, NJ 08618.

Navarro, Robert P., *Managed Care Pharmacy Practice*; Aspen Publishers: 1999.

Pharmacy in the 21st Century

Mae Kwong
American Society of Health-System Pharmacists, Bethesda, Maryland, U.S.A.

INTRODUCTION

Since 1984, four Pharmacy in the 21st Century conferences have occurred. Each conference sought to provide a current and future assessment of the pharmacy profession and healthcare.

1984: THE FIRST CONFERENCE

The first Pharmacy in the 21st Century Conference[1] was held March 25–28, 1984, in Millwood, Virginia. The conference was sponsored by several pharmacy and trade associations, pharmaceutical companies, the Institute for Alternative Futures, and the Institute for Health Policy of Project HOPE. Over 50 representatives from the sponsoring associations, the pharmaceutical industry, and other healthcare professions participated.

The conference focused on the future of pharmacy and its role in healthcare over the next 25 years. The mission of the conference was to consider such issues as the nature and quantity of healthcare and pharmaceutical services, total drug volume, classes of drugs, pharmacy functions, practice locations, and pharmacy personnel needs. Fourteen leaders in healthcare gave presentations on different issues affecting pharmacy's future. The participants discussed the changes in pharmacy practice and imagined what pharmacy would be like in 2010.

Work groups identified several trends for the future. Major factors expected to affect pharmacy included a growing elderly population, more health promotion and wellness activities, increased efforts to prove the cost effectiveness of drugs, a decline in healthcare expenditures as a proportion of the Gross National Product, and an increase in nontraditional therapies. There would be more sophisticated drug delivery systems, and more prescription drugs would become available on a nonprescription basis. Finally, it was believed that pharmacists would play a greater role in monitoring medications and that most pharmacists would practice within large healthcare systems or chain pharmacies.

1989: THE SECOND CONFERENCE

The second Pharmacy in the 21st Century Conference[2] was held October 11–14, 1989, in Williamsburg, Virginia. This invitational conference was sponsored by seven pharmacy associations and nine other groups. The conference had 108 participants, including pharmacy leaders, pharmacy practitioners, and representatives from consumer groups, government, and healthcare administration.

The objective of the conference was to identify the critical issues facing the pharmacy profession in the next 15 to 20 years. Keynote speakers described trends relating to public policy, technology, healthcare economics, and healthcare delivery. C. Douglas Hepler, Ph.D. presented an address coauthored by Linda M. Strand, Pharm.D., Ph.D. on pharmacy's opportunities and responsibilities in healthcare. This presentation was a milestone for pharmacy and defined "pharmaceutical care" as "the responsible provision of drug therapy for the purpose of achieving definite outcomes that improve a patient's quality of life."

The consensus statements that were drafted identified several needs in pharmacy: Create a mission statement, develop standards for pharmacists to apply in managing pharmaceutical care, demonstrate and communicate the value of pharmaceutical care to healthcare, measure the quality and outcomes of services, strengthen political action, and convince the public and payers of the benefits of pharmaceutical care. The obstacles facing practice were also noted. The overarching conclusion was that pharmacy had to actively demonstrate and communicate its value to healthcare.

Encyclopedia of Clinical Pharmacy
DOI: 10.1081/E-ECP 120006195

1994: THE THIRD CONFERENCE

The third Pharmacy in the 21st Century Conference[3] was held October 7–10, 1994, in Lansdowne, Virginia. This invitational conference was organized and sponsored by the Joint Commission of Pharmacy Practitioners, a consortium of national pharmacy associations and liaison organizations. Members of JCPP included the Academy of Managed Care Pharmacy, the American Association of Colleges of Pharmacy, the American College of Apothecaries, the American College of Clinical Pharmacy, the American Pharmaceutical Association, the American Society of Consultant Pharmacists, the American Society of Health-System Pharmacists, the National Community Pharmacists Association, the National Association of Boards of Pharmacy, and the National Council of State Pharmacy Association Executives. The conference had 131 participants, including pharmacists and representatives from the sponsoring organizations, medicine, nursing, and public and private agencies.

The objective of the conference was to understand and overcome the barriers to delivering pharmaceutical care in all settings. The two keynote speakers were David A. Zilz, who presented "The Changing Health System: Implications for Pharmaceutical Care," and Carl E. Trinca, who presented "The Pharmacist's Progress Toward Implementing Pharmaceutical Care." A panel of five pharmacists representing hospital pharmacy, long-term and specialty care, managed care, and independent and chain practice described the services they provided and how they overcame obstacles to providing pharmaceutical care. These presentations set the stage for work groups to identify barriers to implementing pharmaceutical care, the reasons for those barriers, and strategies for overcoming the barriers and implementing pharmaceutical care in community and ambulatory care, hospital and institutional care, long-term care, home healthcare, and managed care.

Work groups classified the barriers as economic, educational, political and legal, technological, and social, interprofessional, and intraprofessional. The work groups also proposed strategies for addressing these problems in each practice setting. It was concluded that pharmacists should direct their attention to patients, acquire the skills needed to provide pharmaceutical care, collect and share patient information, document the services and care provided, perform and participate in outcomes research on pharmacists' interventions to demonstrate the clinical and economic value of pharmaceutical care, and recognize that new credentials will be developed for pharmacists providing high-quality patient care.

1999: THE FOURTH CONFERENCE

The fourth Pharmacy in the 21st Century Conference,[4] "Reengineering the Medication-Use System," was held October 1–3, 1999, in Baltimore, Maryland. This invitational, interdisciplinary conference was organized and sponsored by the Joint Commission of Pharmacy Practitioners and had 108 participants from pharmacy, medicine, and nursing. Also included were benefit managers and representatives from federal regulatory agencies, consumer advocacy groups, the pharmaceutical industry, the information systems industry, and the health insurance industry.

Instead of focusing strictly on the pharmacy profession, the fourth conference addressed the quality of the entire medication-use system. The objectives were to describe—and publicize—the extent of preventable morbidity, mortality, and excess costs resulting from suboptimal medication use; to stimulate discussion of the social and economic impact of dysfunction in the medication-use system; to develop a reengineered system for medication use that would optimize clinical, economic, and humanistic outcomes; to identify strategies that could be used to evaluate and implement the models developed; and to foster greater interprofessional collaboration and a shared commitment to optimal medication use.

Several keynote presentations set the stage for workshops focusing on opportunities for improving the medication-use system; recommendations for applying technology and preventing human error; recommendations for medicine, nursing, and pharmacy; and strategies for improving the medication-use system. The final exercise was the development of recommended strategies for reengineering the medication-use system. These included improving the clinical decision-making process related to medication use; improving communication among health professionals and between health professionals and patients; ensuring safety in drug distribution, dispensing, and administration; and educating individual patients and the public about responsible medication use.

DISCUSSION

Each Pharmacy in the 21st Century Conference had a different focus yet each had a positive impact on the profession and on healthcare. For instance, the concept of pharmaceutical care—a major milestone for pharmacy—was widely introduced as a result of the second conference. Since that time, pharmacists have been increasingly recognized for what they do: helping people make the best use of medications.

The series of conferences helped the profession to evaluate pharmacy practice and move forward. In addition, they provided a platform for pharmacy to work with other healthcare professions, healthcare administrators, consumer groups, and public and private agencies.

REFERENCES

1. Bezold, C.; Halperin, J.A.; Binkley, H.L., et al. *Pharmacy in the 21st Century—Planning for an Uncertain Future*; Institute for Alternative Futures and Project HOPE: Virginia, 1985.
2. *Conference on Pharmacy in the 21st Century, October 11–14, 1989, Williamsburg, Virginia*; Am. J. Pharm. Educ., 1989; 53 (winter supplement).
3. The third strategic planning conference for pharmacy practice—a conference 1994 dedicated to understanding and overcoming the obstacles to delivering pharmaceutical care (undated supplement to Am. Pharm.).
4. *Re-engineering the Medication-Use System—Proceedings of a National Interdisciplinary Conference Conducted by the Joint Commission of Pharmacy Practitioners, October 1–3, 1999, Baltimore, Maryland*; Am. J. Health-Syst. Pharm., 2000; 57, 537–601.

PHARMACY PRACTICE ISSUES

Placebo Effects

Renee F. Robinson
Milap C. Nahata
The Ohio State University, Columbus, Ohio, U.S.A.

INTRODUCTION

The definition of placebo and its role in medicine have changed dramatically over the past 50 years. Prior to 1945, *placebo* was defined as an inactive substance given to appease the patient rather than for medical purposes.[1] Lack of a definitive diagnosis or treatment often resulted in the use of an inert preparation (e.g., sugar pills, colored water) to treat the patient. These agents were given for peace of mind without medical consequence and their efficacy separated imaginary symptoms from those secondary to disease. In the 1950s, widespread acceptance of clinical trials to determine safety and efficacy of drug therapy led to the need to assess the therapeutic effectiveness of the *placebo*. The double-blind, randomized, placebo-controlled, clinical trial design was developed to standardize medical research, decrease physician and patient bias, and account for changes secondary to the natural history of disease. Clinical outcome was considered significant only if response in the experimental group was superior to that in the placebo group. It was confirmed that a certain percentage of patients improved with placebo therapy.

In 1955, Beecher attempted to quantify the placebo response in 15 randomly selected clinical trials encompassing a wide variety of ailments (e.g., pain, cough, behavior changes).[1] Of the 1082 patients receiving placebo, 35.2% experienced therapeutic benefits.[1] Response was noted regardless of the intelligence of the person. Further, patient beliefs, habits, and educational background appeared to be consistent within those who responded to placebo.[1]

The ethical use of placebos as the gold standard to establish efficacy and safety of new agents in clinical trials has recently come into question. In order to establish efficacy of antiretroviral therapy in decreasing placental transmission of HIV, the HIV-infected pregnant females in Thailand and Uganda received placebo therapy, despite the known efficacy of zidovudine in decreasing placental transfer. Concern regarding the use of placebos when effective antiretroviral therapies are available led to the recent revision of The Declaration of Helsinki. The Declaration of Helsinki, produced in 1964 by the World Medical Association, has been the primary document describing ethical practice in medical research. The original doctrine was developed to set international standards for human experimentation after World War II. The Declaration of Helsinki has no legal authority; therefore, it is the role of the investigational review board (IRB) to set and enforce the standards for the ethical participation of humans in medical research at each institution.

PLACEBO EFFECT/RESPONSE

Placebo therapy encompasses any therapeutic intervention or treatment for which no scientific theory including mechanism of action, the effect of administration, and/or a medical act has a specific effect on the condition for which it is prescribed.[1–3] It is distinct from the natural history of disease. When used appropriately, it can idetify the variation due to nonspecific effects such as physical setting, informed consent process, cultural background and equivalence of participants, differences in healing rates, patient or provider bias about treatment, characteristics and communication skills, and perception of the practitioner by the patient.[4,5] Placebo use is further divided based on the agent or therapy administered into two groups: pure and impure placebos.[2] Pure placebos do not possess pharmacologic activity (i.e., contain lactose and normal saline), while impure placebos possess pharmacologic activity irrelevant to the disease under study.[2]

Placebo effect or placebo response has been referred to as the lie that heals.[2] It is the change in patient condition due to a symbolic intervention and increased awareness (e.g., consultation, procedure) rather than pharmacologic or physiologic intervention. It is used to eliminate observational bias in many clinical trials.[2] Elements of placebo response exist in almost every patient encounter. Variables that have been shown to affect patient outcomes

Encyclopedia of Clinical Pharmacy
DOI: 10.1081/E-ECP 120006393

include method of recruiting, the informed consent and consenting process, procedures for blinding, vehicle delivery (brand of medication, capsule color, size, preparation, consistency of the product, and equivalence to the medication being compared), and patient-specific factors (preexisting conceptions and beliefs).[6–8]

The method of patient recruiting, consenting, and blinding has been shown to influence patient beliefs and subsequent reporting of efficacy and safety of the medication.[4] Means and type of vehicle delivery appear to affect patient perception of intervention strength, and adverse events.[8] One study found that increased confidence in the manufacturer appeared to supplement relief obtained in both placebo and active medication groups.[6]

Patient beliefs and attitudes have the most influence on outcomes, and reporting of efficacy and adverse events. When patients are provided with information, emotional support, and an intervention consistent with preexisting conceptions, a sense of control over their illness and positive response to therapy were achieved.[2] The expectation from the drug had an influence on the magnitude of the response.[9]

PLACEBO RESPONSE IN VARIOUS DISEASES

Over the past 50 years, placebo-controlled trials have been used by practitioners and researchers to determine the efficacy and safety for a variety of therapeutic interventions. In all clinical trials regardless of disease state (e.g., peptic ulcer disease, ulcerative colitis, dental pain management, hypertension, hypercholesterolemia, migraine) or therapy received (e.g., antacids, clofibrate) patients exhibited placebo response (Table 1). The extent of placebo response (subjective and objective measures) varies between trials and disease states. Evaluation of the literature demonstrates the extent and variety of factors implicated in response. Faizallah et al. studied the effects of high- and low-dose placebo versus high- and low-dose antacids in the treatment of peptic ulcer disease.[10] Patients with more significant disease received high-dose placebo versus high-dose antacid therapy. Those with less severe disease were given low-dose placebo versus low-dose antacid therapy. Response appeared to be dependent on the agent received and extent of the underlying disease. Patients in the low-dose placebo group reported a 35% response while those receiving high-dose placebo reported a 13% response to therapy.[10] This suggested that patients with more severe disease were less likely to demonstrate a placebo response. However, poor compliance (2/3 of the desired used volume) among the patients receiving high-dose placebo may account for the decreased response.

Placebo therapy in the treatment of ulcerative colitis has consistently shown a measurable benefit. Ilnyckyj et al. reviewed the literature to determine the extent of placebo response in the treatment of acute ulcerative colitis.[11] Analysis of all double-blind, placebo-controlled trials of patients with active ulcerative colitis by multiple reviewers ensured appropriate data extraction and a series of analysis were conducted on the various clinical, endoscopic, and histological endpoints. Remission (10%) occurred less frequently than response (30%) with placebo therapy. Number of patient visits appeared to be related to placebo response (clinical, endoscopic, and histologic) but not remission. Therefore, the authors concluded that conditioned response with active intervention may have a significant effect in placebo response.

Understanding endogenous mechanisms may account in part for the placebo response in patients with pain. The release of endorphins (endogenous opiate-like substances) is believed to mediate the response to pain.[12] Naloxone, a pure opiate antagonist, should block the effect of endorphin analgesia that is often quantified as placebo response.[12] Patients who received naloxone post dental extraction experienced more pain compared to placebo responders. The placebo response was attributed to endorphin release.[12]

The mechanism of placebo response in blood pressure is not well understood. In a study of 1292 stage I and stage II hypertensive females, 30% responded to placebo therapy.[13] Response varied with age and ethnicity. The response in elderly Caucasian females was greater than in other age and race subgroups, 38% versus 23–27%, respectively.[13] Factors such as natural history of the disease and regression to the mean have been suggested as the basis for this placebo response.

Placebos have been useful in determining the efficacy of various therapies in decreasing mortality and morbidity. The Coronary Drug Project Research Group evaluated the efficacy and safety of several lipid-lowering drugs such as clofibrate on the long-term morbidity and mortality of patients with coronary artery disease.[14] Subgroups defined by patient response were determined after randomization. Patient adherence of 80% or greater appeared to substantially lower five-year mortality in both the clofibrate and placebo groups.[14] Factors such as patient adherence, social and behavioral characteristics, and/or biochemical response appeared to influence patient outcomes.

Route of administration may have a significant effect on patient response. A meta-analysis of 22 acute migraine treatment trials showed greater effect associated with a

Table 1 Placebo response rates from various disease states

Ref.	Medication/ disease state	Therapy received by patient	No. of patients	Responders
[22]	Caffeine	SBP in caffeine	20	13 (65%)
		SBP in placebo	20	7 (35%)
[13]	Hypertension	One of 6 active drugs	1105	643 (58%)
		Placebo	187	58 (30%)
[12]	Analgesia	1st dose naloxone, 2nd placebo	11	2 (18%)
		1st dose placebo, 2nd placebo	14	5 (36%)
[23]	Parkinson's disease	Placebo	198	41 (20.7%) 67 (33.8%) mixed response
[10]	Peptic Ulcer disease	High-dose antacids	19	6 (35%)
		Placebo for high dose	15	2 (13%)
		Low-dose antacid	18	12 (67%)
		Placebo for low dose	17	6 (35%)
[24]	Neuropathic pain	Fentanyl	26	(50%)
		0.9% sodium chloride injection	24	(8.3%)
[25]	Congestive heart failure	Amrinone	47	27 (57.4%)
		Placebo	52	41 (78%)
[26]	Schizophrenia	Fluphenaine deconate	27	2 (7%)
		Placebo	27	8 (30%)

parenteral than an oral route.[15] Of those receiving oral placebo therapy, 25.7% experienced relief compared to 32.4% response in the subcutaneous placebo group.[15] Future trials are necessary to determine the effect of other routes of administration on patient response.

FACTORS AFFECTING PLACEBO RESPONSE

Numerous factors may explain the reason for placebo response and identify patients most likely to respond to placebo. These include preconceived patient views of disease severity, available treatment options, desire of the patient to please the provider, biochemical release, and stress.[16]

Preconceived patient views of illness, presence of an emotional support network, and practitioner response and intervention influence the patient attitude about the disease and its treatment.[16] Patients with acute severe symptoms are more likely to have improvement in symptoms, due to natural or spontaneous recovery, which may be attributed to the placebo. Placebo response in the treatment of acute ulcerative colitis was related to the number of study visits. Greater than three patient visits demonstrated greater placebo response.[11] Response to placebo in individuals suffering from panic disorder was directly related to the severity of the underlying disease. Those with extensive pathologic conditions responded only to active therapy.[17]

Placebo response may be related to the Hawthorne effect.[18] As the practitioner provides more attention to the patient's health, there is often an increased desire of the patient to please the caregiver and to express appreciation for an effective intervention.

Many hypotheses have been proposed about the underlying cause of placebo response such as the release of endogenous endorphins in response to painful stimuli. Patients experiencing postoperative pain after molar extraction responded to placebo due to the release of endorphins.[12]

The percentage of patients responding to placebo increased when stress was associated with pain.[1] Patient anxiety appears to be a factor in determining response to placebo.[18]

On reanalysis of 14 of the 15 clinical trials, the factors accounting for the improvements in the placebo group

included spontaneous improvement, fluctuation of disease symptoms, effects of previous and/or concurrent treatment and medication, conditional switching of placebo treatment, irrelevant response variables, answers given to be polite but not truthful, experimental subordination, conditioned answers, and false assumption of toxic placebo effect.[19] Practitioner and patient expectations directly influence the results and may distort the data.[20] Recall bias, parental assessment, and context influence expectations may affect reporting of subjective outcomes and adverse events.[20] It is difficult to determine the effects related to the treatment versus those related to the supportive, confirming environment.

Placebos have been found to have effects similar to the active medication being compared.[21] Many investigators have attempted to quantify the placebo response. Scales such as the placebo effect scale (PES) and the side effect scale (SES) have been used to identify patients likely to exhibit placebo effects. These scales determine possible predictors of placebo response such as demographic variables, illness characteristics, and diagnostic and symptom variables to identify patients at increased benefit or risk of demonstrating a placebo response.[18] The need for routine use of such scales prior to inclusion in clinical trials and implications of identifying those most susceptible to placebo response require further study.

MISCONCEPTIONS

There are numerous misconceptions or generalizations about placebo response. Common misconceptions include but are not limited to the following: One-third of all patients respond to placebo therapy; response is brief; certain demographic or personality factors predict response; those who respond to placebo therapy do not have medical problems; and giving placebo is equivalent to no intervention.[21] In addition, placebo response in one clinical trial or disease state does not ensure equivalent placebo response in other disease states.[10,12,13,22–26]

USE OF PLACEBOS IN CLINICAL PRACTICE

Placebos employed in the course of general medical practice may be given to patients believed to be experiencing exaggerated symptoms or to those who were unresponsive to current medical therapies. Such patients are likely to receive placebo therapy when practitioners are unable to determine the cause of disease or if inadequate treatment is available.[27,28]

Is it ethical for physicians to prescribe placebos? This question has been debated in the medical literature since the 19th century. Arguments for placebo use have focused on the positive effects noted in clinical trials.[2,10,12,13,22–26]

The primary arguments against placebo use without consent focus on the deception and manipulation involved, and violation of the a priori moral rule that a person deserves respect and should be treated as expected with an active drug.[2] It eliminates the patient's right to make an informed decision about his/her care, diminishes the patient's dignity, and results in decreased trust in the physician/patient relationship. It also advocates the use of medication for every ailment including those without medical justification or for which no treatment is available. Placebo administration may further increase requests for treatment of symptoms, delay appropriate diagnosis and treatment, and result in the withholding of established therapies.

CURRENT DECLARATION OF HELSINKI

The current Declaration of Helsinki was revised in October 2000 to provide guidance to physicians and other medical researchers on the issue of placebo controls when effective agents are available. The changes reflect the ethical concerns raised by the use of placebo controls in pregnant HIV-infected females in Uganda and Thailand. Available agents such as zidovudine have been shown to significantly decrease the incidence of transmission of HIV to the fetus when taken throughout the pregnancy. The revisions stress that the well-being of the human subject takes precedence over the interests of medical progress. Medical research is justified only if the benefits to the patient outweigh the risks and the informed patient consents to participate. The efficacy and safety of the investigational product should be tested against the best current treatment available. If no treatment exists, placebo therapy or no intervention is acceptable. It is important to realize that the Declaration of Helsinki does not have legal power to enforce these changes. These are guidelines for ethical conduct of medical research. Institution-specific IRBs are responsible for enforcing the rules and guidelines for human subject safety and research conduct.

SUMMARY

Placebos are used in clinical practice and research. Placebo response has been demonstrated in a variety of disease states. Despite the demonstrated efficacy of placebo therapy, the use of placebos in clinical practice may be considered unethical. Attempts by researchers to determine the patient-specific characteristics responsible for placebo response have been unsuccessful. Patient, practitioner, and environmental factors make it difficult to quantify placebo response. The Declaration of Helsinki has been revised to address the issue of placebo use in clinical trials. If an effective agent is available, the new agent should be tested for efficacy against the available effective therapy, not the placebo. Further studies are needed to clearly understand the reasons for placebo response in patients of various demographic characteristics.

REFERENCES

1. Beecher, H.K. The powerful placebo. JAMA, J. Am. Med. Assoc. **1955**, *159*, 1602–1606.
2. Brody, H. The lie that heals: The ethics of giving placebos. Ann. Intern. Med. **1982**, *97*, 112–118.
3. Gotzsche, P.C. Is there logic in the placebo? [see comments]. Lancet **1994**, *344*, 925–926.
4. Kleijnen, J.; de Craen, A.J.; van Everdingen, J.; Krol, L. Placebo effect in double-blind clinical trials: A review of interactions with medications. Lancet **1994**, *344*, 1347–1349.
5. Moerman, D.E. Cultural variations in the placebo effect: Ulcers, anxiety, and blood pressure. Med. Anthropol. Q **2000**, *14*, 51–72.
6. Branthwaite, A.; Cooper, P. Analgesic effects of branding in treatment of headaches. Br. Med. J. (Clin. Res. Ed.) **1981**, *282*, 1576–1578.
7. Kaptchuk, T.J. Powerful placebo: The dark side of the randomised controlled trial. Lancet **1998**, *351*, 1722–1725.
8. Buckalew, L.W.; Coffield, K.E. An investigation of drug expectancy as a function of capsule color and size and preparation form. J. Clin. Psychopharmacol. **1982**, *2*, 245–248.
9. Gracely, R.H.; Dubner, R.; Deeter, W.R.; Wolskee, P.J. Clinicians' expectations influence placebo analgesia [letter]. Lancet **1985**, *1*, 43.
10. Faizallah, R.; De Haan, H.A.; Krasner, N., et al. Is there a place in the United Kingdom for intensive antacid treatment for chronic peptic ulceration? Br. Med. J. (Clin. Res. Ed.) **1984**, *289*, 869–871.
11. Ilnyckyj, A.; Shanahan, F.; Anton, P.A.; Cheang, M.; Bernstein, C.N. Quantification of the placebo response in ulcerative colitis [see comments]. Gastroenterology **1997**, *112*, 1854–1858.
12. Levine, J.D.; Gordon, N.C.; Fields, H.L. The mechanism of placebo analgesia. Lancet **1978**, *2*, 654–657.
13. Preston, R.A.; Materson, B.J.; Reda, D.J.; Williams, D.W. Placebo-associated blood pressure response and adverse effects in the treatment of hypertension: Observations from a Department of Veterans Affairs Cooperative Study. Arch. Intern. Med. **2000**, *160*, 1449–1454.
14. Influence of adherence to treatment and response of cholesterol on mortality in the coronary drug project. N. Engl. J. Med. **1980**, *303*, 1038–1041.
15. de Craen, A.J.; Tijssen, J.G.; de Gans, J.; Kleijnen, J. Placebo effect in the acute treatment of migraine: Subcutaneous placebos are better than oral placebos. J. Neurol. **2000**, *247*, 183–188.
16. Bernstein, C.N. Placebos in medicine. Semin. Gastrointest. Dis. **1999**, *10*, 3–7.
17. Uhlenhuth, E.H.; Matuzas, W.; Warner, T.D.; Thompson, P.M. Growing placebo response rate: The problem in recent therapeutic trials? Psychopharmacol. Bull. **1997**, *33*, 31–39.
18. Shapiro, A.K.; Struening, E.L.; Barten, H.; Shapiro, E. Correlates of placebo reaction in an outpatient population. Psychol. Med. **1975**, *5*, 389–396.
19. Kienle, G.S.; Kiene, H. The powerful placebo effect: Fact or fiction? J. Clin. Epidemiol. **1997**, *50*, 1311–1318.
20. Swartzman, L.C.; Burkell, J. Expectations and the placebo effect in clinical drug trials: Why we should not turn a blind eye to unblinding, and other cautionary notes. Clin. Pharmacol. Ther. **1998**, *64*, 1–7.
21. Turner, J.A.; Deyo, R.A.; Loeser, J.D.; Von Korff, M.; Fordyce, W.E. The importance of placebo effects in pain treatment and research [see comments]. JAMA, J. Am. Med. Assoc. **1994**, *271*, 1609–1614.
22. Flaten, M.A.; Blumenthal, T.D. Caffeine-associated stimuli elicit conditioned responses: An experimental model of the placebo effect. Psychopharmacology (Berlin) **1999**, *145*, 105–112.
23. Shetty, N.; Friedman, J.H.; Kieburtz, K.; Marshall, F.J.; Oakes, D. The placebo response in Parkinson's disease. Parkinson study group. Clin. Neuropharmacol. **1999**, *22*, 207–212.
24. Dellemijn, P.L.; Vanneste, J.A. Randomised double-blind active-placebo-controlled crossover trial of intravenous fentanyl in neuropathic pain. Lancet **1997**, *349*, 753–758.
25. Massie, B.; Bourassa, M.; DiBianco, R., et al. Long-term oral administration of amrinone for congestive heart failure: Lack of efficacy in a multicenter controlled trial. Circulation **1985**, *71*, 963–971.
26. Jolley, A.G.; Hirsch, S.R.; McRink, A.; Manchanda, R. Trial of brief intermittent neuroleptic prophylaxis for selected schizophrenic outpatients: Clinical outcome at one year. BMJ **1989**, *298*, 985–990.
27. Goodwin, J.S.; Goodwin, J.M.; Vogel, A.V. Knowledge and use of placebos by house officers and nurses. Ann. Intern. Med. **1979**, *91*, 106–110.
28. Berger, J.T. Placebo medication use in patient care: A survey of medical interns. West. J. Med. **1999**, *170*, 93–96.

Poison Information Pharmacy Practice

Edward P. Krenzelok
Children's Hospital of Pittsburgh, Pittsburgh, Pennsylvania, U.S.A.

INTRODUCTION

Poison information centers, sometimes referred to as poison control centers, provide an important emergency hot-line service to the lay public, medical professionals, and individuals from every aspect of society. The primary mission of poison information centers is to provide emergent information about the identification and recognition of actual and potential poisons, the diagnosis of poisoning, and the treatment of poisoning to improve patient care. Most poison information centers provide service 24 hours/day, every day of the year. Pharmacists provide an important function in poison centers by serving as the individuals who respond to poison hot-line emergency calls and by providing overall administrative and clinical management of the poison information center.

EVOLUTION OF POISON INFORMATION CENTERS

In 1950 the American Academy of Pediatrics appointed a committee to explore the problem of accidental injury.[1] The committee disseminated a national survey and the results revealed that a significant percentage of the reported injuries were the consequence of unintentional poisonings. In response to those findings, the Illinois Chapter of the American Academy of Pediatrics, in cooperation with the Illinois Department of Health and seven hospitals, pooled their resources leading to the establishment of the first poison information center (PIC) in Chicago in 1953. The roots of pharmacy in this evolutionary process are quite evident—the director of the first poison center was a pharmacist, Louis Gdalman.[2] The textbooks, scientific literature, and electronic databases of today could only have been dreams of Gdalman and his colleagues as they labored intensively to obtain and compile product ingredient information from manufacturers and to develop basic protocols for the treatment of poisonings.

The first PIC was so successful that other centers evolved and it became apparent quite rapidly that they were replicating the labor-intensive efforts of each other as they compiled product information. Furthermore, they had no way to share the information with each other. In response, the surgeon general instructed the U.S. Public Health Service to develop the National Clearinghouse for Poison Control Centers.[3] The Clearinghouse was charged with tabulating the collective experience of PICs and with providing information on the diagnosis and treatment of poisonings. The first poison information was provided to PICs on cards. Pharmacists served as the primary staff of the National Clearinghouse.

As PICs proliferated the American Association of Poison Control Centers (AAPCC) was founded in 1958.[3] The AAPCC provided an interdisciplinary venue for poison information center staff—that included nurses, pharmacists, physicians, and veterinarians—to gather at an annual scientific meeting and for the eventual development of poison prevention materials and criteria for poison center standardization and regionalization. The American Academy of Clinical Toxicology (AACT) was established in 1968 as the interest in clinical toxicology grew beyond those working in the traditional PIC. Subsequently, as the PIC movement spread, a variety of related international organizations evolved [e.g., Canadian Association of Poison Control Centres (CAPCC), European Association of Poisons Centres and Clinical Toxicologists (EAPCCT), and World Federation of Poison Control Centres (no longer in existence)] that were multidisciplinary in nature.

From the humble beginnings of a single poison center in 1953, the number of centers exploded to 661 by 1978, including over 100 in the state of Illinois alone![1] There were no federal guidelines that regulated the number of or quality of PICs. Some centers operated during daylight hours and not on weekends, others were simply a telephone in an emergency department or pharmacy (answered by the unlucky individual standing closest to the telephone!), and some actually provided 24-hour/day coverage. In about 1980 the AAPCC developed voluntary guidelines to standardize minimal care by certifying qualified centers as Regional Poison Information Centers.

Encyclopedia of Clinical Pharmacy
DOI: 10.1081/E-ECP 120006237

However, the large number of centers flourished until healthcare reform (and associated cost containment) occurred in the late 1980s and 1990s, resulting in significant attrition among poison centers. Today there are only 69 PICs, 54 of which are AAPCC Certified Regional Poison Information Centers.[14]

THE CONTEMPORARY PIC AS A PRACTICE SITE FOR CLINICAL PHARMACISTS

The PIC provides the best career opportunity for the pharmacist who is pursuing an interest in the discipline of clinical toxicology. While there are opportunities to practice clinical toxicology in acute-care settings such the emergency department or critical care unit, the poison center provides a variety of experiences from frontline work as a specialist in poison information to that of a poison center director. To best describe the opportunities for pharmacists, it is important to characterize U.S. poison centers.

Poison Center Certification

There are two types of PICs in the United States—AAPCC Certified Regional Poison Information Centers (RPICs) and nonregional centers. All centers that chose not to participate in the voluntary certification process or fail to become certified are defined as nonregional centers. (See Appendix 1 for the complete description of the AAPCC certification criteria.) In summary, RPICs do the following:

- Provide service 24 hours/day, every day of the year.
- Cover a defined region with a minimum population of 1 million.
- Utilize properly trained nurses or pharmacists to provide telephone consultation.
- Feature administrative/clinical direction by a pharmacist or physician that is board certified in clinical toxicology.
- Feature medical direction by a physician.
- Maintain toxicosurveillance by participating in the AAPCC Toxic Exposure Surveillance System.
- Provide professional and lay education.

Specialists in Poison Information

The typical RPIC responds to approximately 60,000 exposure and information inquiries annually[4] (Fig. 1). Consultations about pharmaceutical agents comprise 43.1% of PIC calls, cleaning substances 9.9%, personal care products 9.3%, and plants 5.2%.[5] Therefore, the clinical and basic science education of the pharmacist in therapeutics and pathophysiology, pharmacology, pharmacognosy, and chemistry makes them well-suited to serve as a specialist in poison information.

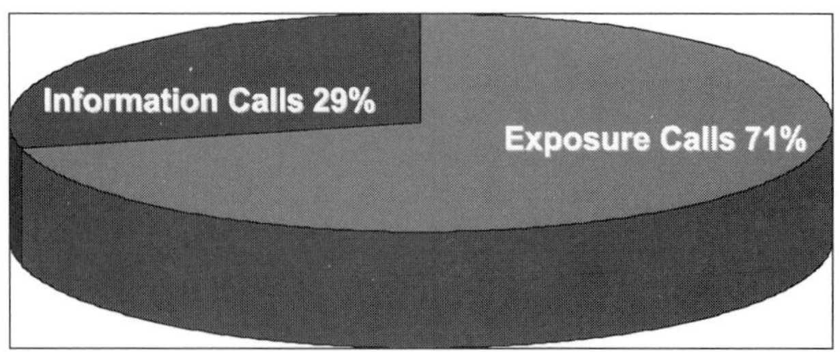

Fig. 1 RPIC call types.

The specialist in poison information (SPI) is the foundation of a PIC. The SPIs respond to incoming poison information calls and document each inquiry on a medical record. Nearly all of the RPICs utilize one of the electronic patient management/medical documentation software programs (e.g., Dotlab, Toxicall). Medical documentation customarily follows the S.O.A.P. approach to medical documentation:

- Subjective complaint.
- Objective findings.
- Assessment.
- Plan.

A variety of sophisticated online databases, such as Poisindex, as well as a multitude of toxicology references and papers from the medical literature, are utilized to provide a toxicology profile of the substance in question and to help the SPI formulate an opinion about the potential morbidity and mortality of an exposure. (Poisindex identifies ingredients for hundreds of thousands of commercial, pharmaceutical, and biological substances. Each substance is linked to one or more management documents providing information on clinical effects, range of toxicity, and treatment protocols for exposures involving the substances.)

Among RPICs there is a pool of 531 SPIs: Pharmacists account for 35% of the SPIs, nurses 58%, and others 7% (Fig. 2).[5] Within nonregional centers there are an additional 58 full-time equivalent pharmacist staff positions. Therefore, it is apparent that there is a career opportunity for pharmacists within this discipline, albeit a somewhat limited one. Nurses dominate RPIC staffing. During the formative years of poison center evolution,

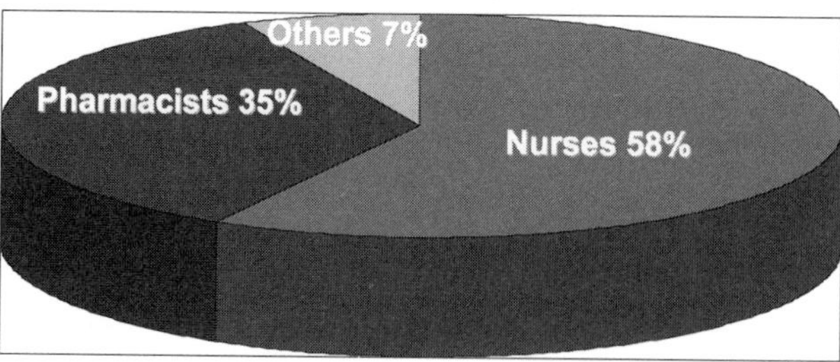

Fig. 2 RPIC staffing.

nurses were utilized more extensively due to their clinical training. However, the clinical training in the pharmacy curriculum and the extensive pharmacology, therapeutics, and chemistry background make the pharmacist well-prepared for a career as a specialist in poison information. Despite these attributes, it is unlikely that pharmacists will erode the current staffing patterns where nurses dominate. The decision to hire nurses instead of pharmacists is based largely on financial motivation, since nursing salaries are generally lower than those of pharmacists. The average RPIC has 7.7–17.0 Certified SPIs on the staff.[4] On an annual basis, the average SPI in an RPIC responds to 3,650 human poisoning exposures or approximately 14 human exposures/8 hour shift.[4] It is important to realize that PICs are open 24 hours/day and that call volume drops dramatically after 11:00 p.m. and does not increase appreciably until approximately 9:00 a.m.[5] (Fig. 3). SPIs working during the peak utilization hours of operation will actually manage 30–45 exposures per 8-hour shift.

Specialists in poison information have the opportunity to become Certified Specialists in Poison Information (CSPI) by qualifying for the AAPCC CSPI Certification Examination (Appendix 2). RPICs must have at least one CSPI on staff at all times. Most RPICs mandate that their SPIs become certified within one to two years of beginning employment at the center. Successful matriculation of the examination is often associated with pay grade enhancement. A CSPI is required to recertify only once every seven years.

Poison Center Director

Each RPIC must have a director who is responsible for the daily administrative and clinical operation of the poison center. Additionally, the RPIC must have a medical director who is responsible for the medical oversight and direction of the center. The majority of RPIC directors are pharmacists (generally Pharm.D.s) and in a limited number of centers a physician fulfills the role of both director and medical director. Pharmacists who aspire to becoming a director of a poison center should consider serving a postgraduate fellowship in an RPIC. Since there are 69 poison centers in the United States, the director position provides a limited but viable and rewarding career opportunity for a pharmacist with postgraduate training in clinical toxicology and poison center management. The AAPCC mandates that nonphysician directors must be eligible for or certified by the American Board of Applied Toxicology (ABAT). The ABAT examination is a rigorous two-day clinical toxicology examination that is administered on an annual basis just prior to the North American Congress of Clinical Toxicology. (See Appendix 3 for more information on the ABAT.)

The Consulting Clinical Toxicologist

The traditional role of the clinical pharmacist in this discipline has been in the PIC as a CSPI or director. A number of pharmacists practice clinical toxicology on a daily basis as part of their clinical pharmacy responsibilities. Clinical pharmacists working in critical care environments such as the emergency department or the critical care unit must have a substantial appreciation of clinical toxicology since unintentional and intentional overdoses are common reasons for hospitalization. There is also a substantial opportunity for the clinical toxicologist to become engaged in consultation with industry through the analysis and interpretation of toxicity issues. The legal arena also provides an opportunity for the clinical toxicologist to serve as an expert witness in the resolution of litigation.

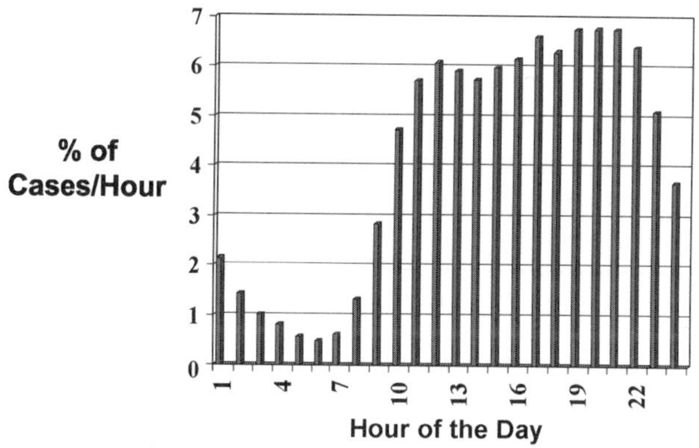

Fig. 3 Poison center call volume by hour.

P

POISON CENTER AND STAFF COST JUSTIFICATION

There has been significant attrition among poison centers over the last two decades. This is due to the fact that poison centers are generally financial liabilities to their sponsoring institutions and do not generate sufficient revenue to offset operation expenses. This has resulted in the closure of nearly 600 poison centers since 1978. Mistakenly, poison centers have been perceived by the public, medical establishment, and government to be part of the healthcare infrastructure. Each of these entities has had the erroneous perception that one of the others was responsible for funding. This misperception has left many poison centers on the brink of financial disaster. However, poison centers must shoulder some of the blame for this precarious position by failing to diversify poison center revenue sources. The successful poison center should have support from local, regional, and/or state government; from the hospitals that they serve; and from business and industry (serving as a company after-hours medical/poison information service).

Since 1998 there has been intense activity at the federal level to legislate stabilization money from the government and to ensure the future viability of PICs. The Poison Center Enhancement and Awareness Act of 1999 (Appendix 4) will provide funding to: 1) stabilize poison center services, 2) improve access to poison center services by providing a single toll-free number to be shared by all poison centers, 3) provide national poison prevention education, and 4) improve the treatment of poisoning victims. This legislation will fulfill the goals of a recent report by the Department of Health and Human Services, which stated that every United States resident should have access to a certified regional poison center. However, the $20 milion appropriation (reduced from the original $25 million in the Senate and House bills) will not be adequate to totally fund a poison center operation and there will be a continued need to aggressively seek additional funding to support PICs. The clinical pharmacist who chooses a career in a PIC, especially at the director level, must be prepared to identify a variety of financial resources to ensure that the mission of the poison center to improve patient care is not compromised. The influx of federal support will not eliminate the financial scrutiny of poison centers by sponsoring institutions and government. It will be necessary to continuously justify the human and cost-effective benefits of poison centers.[6] The best strategy is to utilize validated facts and data to help strengthen the argument in favor of investing in a PIC.

Financial Benefits

- For each dollar spent on poison center activities, $6.50 is saved.[7]
- When a PIC is utilized by healthcare professionals, the cost per successful outcome is approximately 50% less than that achieved without using a PIC.[8]
- When PICs have closed, admissions secondary to poisoning increased by 16% and outpatient claims increased by 35%.[1]
- Each dollar spent on poison prevention activities saves $3.00.[9]

General Benefits[10]

- Improved healthcare of the poisoned patient.
- Reduction of unintentional poisoning in the home and the workplace.
- Reduction in unnecessary hospital visits and inappropriate use of medical facilities.
- Reduced morbidity by advising callers against the use of antiquated first aid methods and ineffective/dangerous home remedies resources.
- Decreased burden on emergency medical services and emergency transportation.
- Use of toxicosurveillance for the early detection of biochemical terrorism events and hazardous materials leaks/spills.
- Early identification of problems associated with medications and product packaging defects.
- Enhanced patient care through education of the lay public and medical professionals.

INFORMATION RESOURCES

What product does the clinical pharmacist working in a PIC provide? Pharmacist CSPIs provide contemporary information on product toxicity and patient management. Pharmacist clinical toxicologists consult with other healthcare professionals regarding patient diagnosis and management. The poison center product is an intangible entity referred to as *knowledge*. Knowledge is the sum of poison center staff experience and information.

Experience + Information = Knowledge

Toxico-informatics, or converting information to knowledge, is the most formidable challenge that clinical pharmacists in poison centers face in the near future due to the exponential growth of information and technology.[11] For example, available information doubled

between 1750 and 1900—150 years; 1900–1950 saw information double again in only 50 years. From 1960–1992, information doubled at least once every five years. By 2020, it is estimated that information will double every 73 days. Clinical pharmacists working in poison centers must proactively prevent the information explosion from incapacitating centers and must find ways to filter the information to make it usable. Half of knowledge is knowing where to find data and information and then converting them to knowledge once you find them. Twenty-five years ago, pharmacists working in a PIC had a limited number of textbooks, a few hundred papers from the medical literature, and a few pieces of Poisindex microfiche as their information resources. Since that time hundreds of textbooks have been written to address basic clinical toxicology and very specific topics such as botanical toxins. (See Appendix 5 for a list of relevant textbooks.) The advent of the electronic database era has made an incredible number of resources available on general toxicology, hazardous materials, natural toxins, and biochemical terrorism. (See appendix 5 for specific URLs.)

CLINICAL TOXICOLOGY ORGANIZATIONS

Clinical toxicology and poison information are well defined but very small disciplines when compared to mainstream health specialties. Therefore, collaboration among those who practice the specialty is critical.

American Association of Poison Control Centers

For individuals who practice in a PIC it is advisable to become a member of the American Association of Poison Control Centers (AAPCC). The AAPCC is a nationwide organization of poison centers and interested individuals. The main objectives of the AAPCC are: 1) to provide a forum for poison centers and interested individuals to promote the reduction of morbidity and mortality from poisonings through public and professional education and scientific research and 2) to set voluntary standards for poison center operations. The major activities of the AAPCC are:

- Certification of regional poison centers and poison center personnel.
- Interaction with private and government agencies whose activities influence poisoning and poison centers.
- Development of public and professional education programs and materials.
- Collection and analysis of national poisoning data.

For more information or to join the AAPCC contact: AAPCC, 3201 New Mexico Avenue, NW, Suite 310, Washington, DC 20016; Tel.: 202-362-7217, Fax: 202-362-8377, E-mail: aapcc@poison.org, Web site: www.aapcc.org.

American Academy of Clinical Toxicology

Pharmacists who practice in a PIC or in private practice would benefit from membership in the American Academy of Clinical Toxicology (AACT). The AACT was established in 1968 as a not-for-profit multidisciplinary organization uniting scientists and clinicians in the advancement of research, education, prevention, and treatment of diseases caused by chemicals, drugs, and toxins. The founders of the AACT established the Academy to:

- Promote the study of health effects of poisons on humans and animals.
- Unite into one group scientists and clinicians whose research, clinical, and academic experience focus on clinical toxicology.
- Foster a better understanding of the principles and practice of clinical toxicology.
- Encourage development of new therapies and treatment in clinical toxicology.
- Facilitate information exchange among individual members and organizations interested in clinical toxicology.
- Define the position of clinical toxicologists on toxicology-related issues.

Today, the AACT is the largest international clinical toxicology society whose membership is comprised of clinical and research toxicologists, physicians, veterinarians, nurses, pharmacists, analytical chemists, industrial hygienists, poison information center specialists, and allied professionals. The AACT is affiliated with many professional organizations and organizes the North American Congress of Clinical Toxicology and is a cosponsor of the EAPCCT International Congress, which brings clinical toxicologists together annually for scientific meetings. An additional benefit of AACT membership includes the *Journal of Toxicology-Clinical Toxicology* (the official journal of the Academy and the European Association of Poisons Centres and Clinical Toxicologists). The journal is a leading clinical toxicology journal and has published the AACT/EAPCCT cutting-edge position statements on gastrointestinal decontamination.[12,13] The Academy is also the host organization of the American Board of Applied Toxicology, which certifies nonphysicians as clinical toxicologists—a mandatory certification for PIC directors. The AACT supports career development of members involved in toxicology research through financial awards and sponsorship of fellowships and scholarships. Additionally, the Academy sponsors an electronic bulletin board to foster collaborative clinical toxicology research among its members.

For more information or to join the AACT contact: AACT, 777 East Park Drive, PO Box 8820, Harrisburg, PA 17105-8820; Tel.: 717-558-7750, ext. 1467 Fax: 717-558-7845, E-mail: aact@pamedsoc.org, Web site: www.clintox.org.

European Association of Poisons Centres and Clinical Toxicologists

The European Association of Poisons Centres and Clinical Toxicologists (EAPCCT) was founded in 1964 by a group of physicians and scientists with the specific goal of advancing knowledge and understanding of the diagnosis and treatment of all forms of poisoning. The Association's objectives are to:

- Foster a better understanding of the principles and practice of clinical toxicology in order to prevent poisoning and to promote better care for the poisoned patient, particularly through poison information centers and poison treatment centers.
- Unite into one group individuals whose professional activities are concerned with clinical toxicology whether in a poison center, university, or hospital or in government or industry.
- Encourage research into all aspects of poisoning.
- Facilitate the collection, exchange, and dissemination of relevant information among individual members, poison centers, and organizations interested in clinical toxicology.
- Promote training in, and set standards for the practice of, clinical toxicology and to encourage high professional standards in poison centers and in the management of poisoned patients generally.
- Collaborate with international and integrational organizations including the WHO and European Union.
- Establish and maintain effective collaboration with governments, governmental organizations, professional bodies, and other groups or individuals concerned with clinical toxicology.

For more information or to join the EAPCCT contact: EAPCCT General Secretary, National Poison Information Service, Guy's & St. Thomas Hospital Trust, Medical Toxicology Unit, Avonley Road, London SE14 5ER, U.K.; Tel.: 44-207-771-5310, Fax: 44-207-771-5309, E-mail: alex.campbell@gstt.sthames.nhs.uk, Web site: www. eapcct.org.

Active participation in these societies and their annual meetings provides an excellent venue to learn about contemporary issues in clinical toxicology and poison information practice. All of the societies are interdisciplinary and cater to the needs of people with an interest in clinical toxicology irrespective of the individual's health science training. Membership in one or all of the societies provides the best opportunity to network with clinical toxicologists.

THE NATION'S CERTIFIED REGIONAL POISON INFORMATION CENTERS

There are 54 RPICs serving the majority of the United States. Clinical pharmacists who are looking for a career opportunity in poison information practice should contact those centers for additional information. (A current list of all PICs in the United States may be found at www. aapcc.org.)

APPENDIX 1: AAPCC REGIONAL POISON INFORMATION CENTER CERTIFICATION CRITERIA

Introduction

The purpose of this document is to establish criteria by which poison centers and poison center systems can be recognized as possessing the qualifications needed to adequately serve their designated population.

Definitions

A *poison center* is an organization that provides the following services to a region which it has been designated to serve: 1) poison information, telephone management advice, and consultation about toxic exposures; 2) hazard surveillance to achieve hazard elimination; and 3) professional and public education in poison prevention, diagnosis, and treatment.

A *poison center system* consists of two or more poison centers functionally and electronically linked to provide poison center services.

A *certified poison center system* consists of a poison center system which either: 1) collectively meets the same criteria as a single certified poison center, or 2) consists of a number of centers serving a given region, all of which are certified individually. There cannot be any state-designated noncertified centers in the region of a certified system unless they are components of the system.

I. Determination of Region

A. Geographical characteristics

A certified poison center or system may serve a single state, a multistate area, or only a portion of a state. The region should be determined by state authorities in conjunction with local health agencies and healthcare providers. In instances where multiple states are involved, designation from each state is necessary. Documentation of state designations of poison centers and systems must be in writing and must clearly delineate the region to be served, the services to be provided, and the exclusivity of the designation. In instances where a state declines in writing to designate any poison center or system, designation by other political or health jurisdictions (e.g., county, health district) may be an acceptable alternative. In instances where more than one center or system is designated to serve the same area, evidence of cooperative arrangements must be provided.

B. Population base

The certified poison center or poison center system must provide evidence that it adequately serves its entire region. It is unlikely that a single poison center could adequately serve more than 10 million people.

C. Penetrance

The *penetrance* of a poison center or system in a region is defined as the number of human poison exposure cases handled per 1000 population per year. Penetrance is assumed to be most affected by the public's awareness of the appropriate use of the poison center. A certified poison center or system must demonstrate a minimum average penetrance of 7.0 throughout its service area. Poison centers and poison center systems should strive to achieve a penetrance of 12 to 15 throughout the region served, by increasing or maintaining awareness of poison center services.

II. Regional Poison Information Service

A. The certified poison center or system shall provide information 24 hours/day, 365 days/year to both health professionals and the public

This criterion will be considered to be met if the certified poison center has at least one Specialist in Poison Information in each center at all times, sufficient additional staff to promptly handle each center's incoming calls, and the availability of the medical director or qualified designee, on-call by telephone, at all times.

Only if Part of a System. Certified poison centers may divert calls to another certified poison center within the same state, within a contiguous region, or to the closest certified poison center, if: 1) unequivocal continuity of clinical care is achieved through functional access to all open patient records, and 2) the receiving certified poison center staff are fully trained and informed about all healthcare, EMS, and lab facility capabilities and regional toxicology variations. (The criteria relating to diversion of calls, functional access to open patient records, and knowledge of facilities and regional toxicology variations do not apply when assisting another poison center in a disaster situation.)

B. The certified poison center or system shall be readily accessible by telephone from all areas within the region

1. The certified poison center or system must maintain a direct incoming telephone system that is extensively publicized throughout the region to both health professionals and the public.
2. The certified poison center or system must maintain a telecommunications system adequate to assure ready access and must provide data verifying ready access.
3. In the absence of a toll-free system, the certified poison center or system must demonstrate that the lack of a toll-free service is not an impediment to public use of the center.
4. A certified poison center or system may not impose a direct fee to individual members of the lay public (either by direct billing or pay-for-call services) for poison exposure emergency calls received from the public within its region.
5. The certified poison center or system must be able to respond to inquiries in languages other than English as appropriate to the region using language translation services, interpreters, and/or bilingual staff.

6. Access for hearing-impaired individuals must be provided.
7. A plan to provide poison center services in response to natural and technological disasters must be in place.

C. The certified poison center or system shall maintain comprehensive poison information resources (at each site)

This criterion will be considered to be met if each center maintains:

1. One or more comprehensive product information resources, immediately available to the Specialist in Poison Information at all times.
2. Current comprehensive references covering both general and specific aspects of acute and chronic poisoning management immediately available to the Specialist in Poison Information at all times. There must be access to the most current primary information resources and ready availability of a major medical library or comparable online resources.
3. Evidence of the competency of all specialists and information providers in using texts, information resources, and primary literature.

D. The certified poison center or system shall maintain written operational guidelines which provide a consistent approach to evaluation and management of toxic exposures

This criterion will be considered to be met if the certified poison center or system provides written operational guidelines which include but are not limited to the follow-up of all potentially toxic exposures and appropriate criteria for patient disposition. These guidelines must be available in the center at all times and must be approved in writing by the medical director of the program. In addition, these guidelines must have evidence of periodic review, and the center must provide evidence of action taken to remedy problems with guideline content or guideline adherence through quality assurance programs and staff education.

E. Staffing requirements and qualifications for the certified poison center or system

1. Toxicological Supervision. Certified poison centers and each center within a certified poison center system must provide full-time toxicological supervision. This must include at least one full-time equivalent of on-site toxicological supervision and appropriate backup. These components must meet the specific criteria listed below. Each site of a certified poison center system must meet the requirements for medical and managing direction.

A) Medical direction and medical backup.

1) Medical direction may be provided by a single medical director or by more than one individual. If more than one individual provides medical direction, one individual must be designated as medical director and that person is responsible for approving other individuals involved and for coordinating their activities.
2) The medical director and all other individuals designated as providers of medical direction must be board-certified in medical toxicology or board-prepared in medical toxicology as determined by a letter from the board indicating that the candidate will be allowed to sit for the next examination. Such board-prepared physicians must successfully complete the examination within two consecutive administrations of the exam. Board certification through either the American Board of Medical Toxicology (pre-1994) or through the American Board of Medical Specialties subspecialty exam in medical toxicology (after 1994) is acceptable.
   Physicians without board certification in medical toxicology will be considered qualified as medical directors for the purpose of determining compliance with the current criteria if: 1) the physician served as medical director of a poison center certified by AAPCC as of September 14, 1998; and 2) the physician met the immediately previous AAPCC criteria for medical directors on September 14, 1998.
3) The medical director and all other individuals designated as providers of medical direction must have medical staff appointments at an inpatient treatment facility, must be involved in the management of poisoned patients, and must regularly consult with Specialists in Poison Information about the management of poisoned patients.
4) The individual or individuals providing medical direction must individually or collectively devote at least 20 hours per week of professional activity time to toxicology. An additional 10 hours per week of medical direction time must be provided for each 25,000 human poison exposure cases per year received by the certified poison center, above

Table 1 Time requirements for medical directions

No. of human poison exposures	100% PC hours	Total tox hours
25,000	10	20
50,000	20	30
75,000	30	40
100,000	40	50
125,000	50	60

the initial 25,000 human poison exposure cases. Time applied to this total should conform to the following standards:

a. Up to 10 hours per week of the total time applied to medical direction may consist of toxicology activities not directly related to certified poison center operation, such as clinical, academic, teaching, and research activities. No more than 10 percent of clinical time in emergency department, clinic, ward, or intensive care unit service will apply to this total, unless specific documentation is provided to verify that the additional time was directly related to toxicology.

b. The remainder of the total time applied to medical direction activities must consist of poison center operational activities during the time that is 100 percent dedicated to on-site medical direction at each certified poison center or site of a poison center system.

The time requirements for medical direction are summarized in Table 1.

5) Medical backup must be available, in a timely manner, at all times. If not provided by the medical director, medical backup may be provided by those providing medical direction or other individuals designated by the medical director. All medical backup must be provided by board-certified or board-prepared medical toxicologists. Other individuals identified and qualified by the medical director (e.g., fellows, managing director) may serve as immediate certified poison center backup if timely secondary backup is provided at all times by a board-certified or board-prepared medical toxicologist. Direct clinical effort as backup can be applied to item 4.A. above.

B) Managing direction.

1) The managing director provides direct toxicological supervision of poison center staff, strategic planning, and oversight of administrative functions of programs, e.g., staff training, quality assurance, budgeting.

2) Managing direction may be provided by a single managing director or may be provided by more than one individual, each with the qualifications identified below. If more than one individual is involved in providing managing direction, one individual must be designated as managing director (or comparable title), and that person is responsible for coordinating managing direction activities.

3) The managing director must be a nurse with a baccalaureate degree, associate degree, or three-year diploma; a pharmacist; or a physician or may hold a degree in a life science discipline if a diplomate of the American Board of Applied Toxicology. If the managing director is also the medical director, this person must have a full-time commitment to the poison center.

4) The managing director with toxicological supervision responsibilities must be board-certified or board-prepared, as evidenced by a letter from the appropriate board indicating that the candidate will be allowed to sit for the next examination. For physicians this board can be the ABMT (pre-1994) or the ABMS subspecialty examination in medical toxicology (post-1994). For all others, the board must be the American Board of Applied Toxicology. Candidates for the board examination must successfully complete the examination within two consecutive administrations of the examination. Individuals without board certification in applied toxicology will be considered qualified as managing directors for the purpose of determining compliance with the current criteria if: 1) the individual served as managing director of a poison center certified by AAPCC as of September 14, 1998; and 2) the individual met the immediately previous AAPCC criteria for managing directors on September 14, 1998.

2. Specialists in Poison Information. A Specialist in Poison Information must be on duty in the certified poison center, or at each functioning site of a poison center system, at all times.

A) Specialists in Poison Information must be: 1) a nurse with a baccalaureate degree, associate degree, or three-year diploma; a pharmacist; or a physician; 2) currently certified by AAPCC as a Specialist in Poison Information; 3) a diplomate of the American Board of Applied Toxicology; or

4) a board-certified medical toxicologist. Specialists in Poison Information must be qualified to understand and interpret standard poison information resources and to transmit that information in a logical, concise, and understandable way to both health professionals and the public.

B) All Specialists in Poison Information must complete a training program approved by the medical director and, unless a diplomate of the American Board of Applied Toxicology or a board-certified medical toxicologist, must be certified by AAPCC as a Specialist in Poison Information within two examination administrations of his or her initial eligibility for certification. If a Specialist in Poison Information fails to pass a certification exam within two exam administrations of his or her initial eligibility for certification, he or she may work only as a Poison Information Provider under direction as described in Section II.E.3. If an individual fails a recertification examination or does not take a recertification examination, that person reverts to the position of Specialist in Poison Information.

C) Specialists in Poison Information not currently certified by AAPCC as Specialists in Poison Information must spend an annual average of no fewer than 16 hours per week in poison center–related activities, including providing telephone consultation, teaching, public education, or in poison center operations. Specialists in Poison Information currently certified by AAPCC as Specialists in Poison Information must spend an annual average of no fewer than 8 hours per week in poison center–related activities, including providing telephone consultation, teaching, public education, or in poison center operations. Individuals who do not meet this criterion may work as Poison Information Providers with direction as described in II.E.3.

D) All Specialists in Poison Information, whether full-time or part-time, must be 100% dedicated to poison center activities during periods when they are assigned to the center. Poison center calls must be their first priority. In cases where a poison center assumes other roles, the center must demonstrate policies and safeguards that assure that poison center calls are given priority and that these other activities pose no conflict with poison exposure cases and cause no reduction of service quality or quantity within the certified poison center's region.

E) At the time of initial application for poison center certification and thereafter, at least 50% of Specialist in Poison Information full-time equivalent positions (FTEs) must be filled by Certified Specialists in Poison Information. For certified poison center systems, at least 50% of Specialists in Poison Information FTEs at each site must be Certified Specialists in Poison Information.

F) To maintain experience and expertise, on average each certified poison center must handle at least 2,000 human poison exposures per SPI/PIP full-time equivalent.

3. Other Poison Information Providers. Other poison information providers must be qualified to understand and interpret standard poison information resources and to transmit that information understandably to both health professionals and the public. This requirement will be considered to be met if the person has an appropriate health-oriented background and has specific training and/or experience in poison information sciences. While providers may be part-time staff or have a part-time commitment to the poison center, 100% of their time should be dedicated to poison center activities while assigned to the center. At all times, poison information providers must be under the on-site direction of a Certified Specialist in Poison Information, a qualified managing director, or the medical director; these individuals may provide direction for no more than two poison information providers at one time.

4. Certified Poison Center Specialty Consultants. Certified poison center specialty consultants should be qualified by training or experience to provide sophisticated toxicology or patient care information in their area(s) of expertise. These consultants should be available on-call and provide consultation on-call on an as-needed basis. The list of consultants should reflect the type of poisonings encountered in the region.

5. Administrative Staff. Certified poison center administrative personnel should be qualified by training and/or experience to supervise finances, operations, personnel, data analysis, and other administrative functions of the certified poison center.

6. Education Staff.

A) Professional education. Professional education personnel should be qualified by training or experience to provide quality professional education lectures or materials to health professionals. This role will be supervised by the medical director.

B) Public education. Public education personnel should have proven skills in communication and program plan-

ning, implementation, and evaluation, and/or an appropriate educational background with which to provide public-oriented presentations about poison center awareness and the value of the poison center, poison prevention, and first aid for poisoning. This role will be supervised by the medical and/or managing director.

F. The certified poison center or system shall have an ongoing quality improvement program

1. A certified poison center or system shall implement quality assurance activities which incorporate specific monitoring parameters and staff education programs.
2. A certified poison center or system shall demonstrate that patient outcomes are monitored regarding high-risk, high-volume, or problem-prone cases. The corrective actions taken to improve patient care shall be documented. In addition, the certified poison center should demonstrate monitoring of customer satisfaction and assessment of staff competency.

III. Regional Treatment Capabilities

The certified poison center or system shall identify the treatment capabilities of the treatment facilities of the region

At a minimum, the certified poison center or system should: 1) identify emergency and critical care treatment capabilities within the region for adults and children; 2) have a working relationship with all poison treatment facilities in its region; 3) understand the analytical toxicology facilities in the region and how to interface with them; 4) understand how the region's prehospital transportation system is structured and how to interface with it; and 5) know where critical antidotes are available within the region and how they can be transferred between facilities when necessary.

IV. Data Collection System

A. The certified poison center or system shall keep records of all cases handled by the center in a form that is acceptable as a medical record. This criterion will be considered to be met if the Center completes a record that contains data elements and sufficient narrative to allow for peer review and medical and/or legal audit, and such records are retrievable.

B. The certified poison center or system must submit all its human exposure data (except as noted in IV.B.1.) to AAPCC's Toxic Exposure Surveillance System meeting specified submission deadlines and quality requirements and including all required data elements.
   1. The submission of human exposure data derived from industry contracts is encouraged but not required for certification.
   2. Certified poison centers that withhold industry-derived human exposure data must annually submit the number of industry-derived human exposures that were withheld.

C. The certified poison center or system shall tabulate its experience for regional program evaluation and hazard surveillance on at least an annual basis. This criterion will be considered to be met if the center completes an annual report summarizing its own experience.

D. The certified poison center shall monitor the emergence of poisoning hazards and take specific actions to eliminate poisoning hazards.

V. Professional and Public Education Programs

A. The certified poison center or system shall provide information on the management of poisoning to the health professionals throughout the region who care for poisoned patients. This criterion will be considered to be met if the certified poison center offers ongoing information about poison center access and services and updates on new and important advances in poisoning management to the health professionals throughout the region.

B. The certified poison center or system shall provide a variety of public education activities targeting identified at-risk populations. The programs shall address poisoning dangers, poison prevention strategies, first aid for poisoning, and when and how to access poison center services. These programs must be implemented throughout the certified poison center's region.

VI. Association Membership

The applicant poison center and each site in a poison center system must be an institutional member in good standing of the American Association of Poison Control Centers. (Approved April 1988 by the AAPCC Board of Directors. Amended October 1991; September 1992; January 1996; September 14, 1998.)

APPENDIX 2: CERTIFICATION EXAMINATION FOR SPECIALISTS IN POISON INFORMATION

The AAPCC Certification Examination for Specialists in Poison Information is intended for poison center staff whose primary job is handling calls involving human toxic exposures. The exam is administered annually at 40–50 test sites throughout the United States. and Canada. The exam takes place each year in the afternoon of the *second Wednesday in May*. Successful completion of this examination will lead to the designation of Certified Specialist in Poison Information. To maintain certification, the examination must be passed every *seven years*. Applicants must meet the following criteria to sit for the examination:

- Applicants must be members of the AAPCC. This criterion will be considered to be met by either individual membership or by employment at a poison center which is an institutional member.
- Candidates for certification must be registered nurses, pharmacists, physicians, or previously certified specialists in poison information. Physicians are eligible only if they are primarily employed as specialists in poison information, meet the criteria of one year experience, and have handled at least 2,000 human poison exposure telephone consultations. It is intended that physicians take other qualifying exams such as that given by the American Board of Medical Toxicology instead. Center directors, medical directors, assistant directors, consultants, and toxicology fellows in training may *not* apply for certification.
- The candidate must be currently employed as a specialist in poison information. The specialist in poison information must have at least one year (2000 hours) of experience providing telephone poison center consultations and have handled an accumulated 2000 cases. The recommendation of the director of the poison center is required to be eligible to sit for the exam.
- Part-time information specialists must be able to demonstrate at least 2000 hours' experience providing poison center consultations. (For example, a staff member working half-time must work for two years to meet the equivalent part-time experience clause.) Staff of poison centers with personnel performing other duties besides poison control (nursing, pharmacy staffing) may be eligible for the Certification Examination if they can demonstrate unequivocally that they have handled at least 2000 human poison exposure calls, and spent at least 2000 hours providing poison center consultations.

Appropriate documentation will require either:

- Copies of annual call volume statistics for the center listed by human exposures, inquiries, and staff member receiving call, or
- Documentation of percent effort dedicated to poison center operation, primary department, number of hours worked in that department each week, duration of employment (in weeks), and annual call volume for the center. This documentation must be verified and signed by the poison center director. Any figures which clearly demonstrate 2000 hours of experience handling telephone poisoning consultations and sufficient time expended (based on the center's call volume) to have handled at least 2000 calls will be accepted.

Application forms are mailed to poison center directors in February. Application forms must be received by AAPCC by the stated deadline. Applications will be accepted only if:

- All items are completed.
- An application fee is enclosed.
- A photocopy of the applicant's nursing, pharmacy, or medical license; diploma or certificate; or CSPI certificate is also enclosed.
- Applications are received by the announced deadline.

Applicants should not apply if they do not meet these criteria. Application fees for unqualified applicants will not be refunded. Partial refunds will be made to qualified applicants who are unable to sit for the examination if the AAPCC office is notified no later than 2 weeks prior to the examination date.

Currently certified specialists in poison information are not required to be recertified until seven years from the time of certification. If a certified specialist seeks to be recertified before the seven years have elapsed, he/she must pass the examination in order to maintain certification.

APPENDIX 3: THE AMERICAN BOARD OF APPLIED TOXICOLOGY (ABAT)

The American Academy of Clinical Toxicology established the American Board of Applied Toxicology (ABAT) to provide special recognition to practitioners of clinical toxicology who demonstrate exceptional knowledge, experience, and competence. ABAT exists as a self-governing entity whose membership consists solely of all current diplomates of the ABAT. ABAT is governed by a board of directors elected by ABAT membership from among its diplomates.

All ABAT functions are undertaken in accordance with ABAT bylaws and the bylaws of AACT; these bylaws and all regulations and procedures are promulgated by ABAT. ABAT is a not-for-profit organization. No member of ABAT receives or derives any profit from the operation of ABAT and no part of the net income of ABAT benefits any individual member. ABAT is not involved in authorizing or designating any political lobby action. The principal office of ABAT is usually established in the city of the residence of the president.

Purpose of ABAT

ABAT was created as a not-for-profit organization for the unique purpose of fostering the development of clinical toxicology among the nonphysician, nonveterinarian members of the American Academy of Clinical Toxicology by:

- Advancing the science, study, and practice of clinical toxicology.
- Improving the quality of clinical toxicology consultation available to the public.
- Establishing and maintaining standards of excellence for nonphysician practitioners by developing and administering examinations, as well as other criteria, for the certification and recertification of these practitioners in clinical toxicology.
- Granting certificates and other forms of recognition to professionals who demonstrate exceptional ability in clinical toxicology.
- Maintaining a registry of ABAT diplomates and, upon request, furnishing lists of ABAT diplomates to the public, governmental agencies, healthcare professionals, and educational institutions.

ABAT Certification Examination Credentialling Information

Applicants must meet the following initial criteria to become a candidate for examination:

1. Applicants must be a graduate of a college or university with an earned doctoral degree in a biomedical discipline. Applicants without doctoral degrees must possess a baccalaureate degree in a health science discipline, such as pharmacy or nursing, followed by a minimum of five years of full-time professional experience in applied clinical toxicology. Scholastic course work is not considered to be professional experience.

   Because the American College of Medical Toxicology and the American Board of Veterinary Toxicology are responsible for certification in their respective areas, applicants holding the Doctor of Medicine, Doctor of Osteopathy, or Doctor of Veterinary Medicine degree are *not eligible* to sit for the ABAT examination.

2. Applicants must complete at least 12 months of postdoctoral training (i.e., residency or fellowship) in clinical toxicology or a closely related field. Applicants without postdoctoral training must have a minimum of at least three years of professional experience related to applied clinical toxicology after completion of their doctoral degree. To be prepared for the examination, candidates should have considerable clinical experience and an understanding of the clinical and environmental factors associated with various types of toxicological problems. Examples of activities related to the practice of applied clinical toxicology include consulting with medical personnel on patient care issues; having administrative responsibility for a poison control center with consultative responsibilities; rendering opinions on product toxicity; teaching clinical toxicology to students, practitioners, or colleagues; collaborating with medical toxicologists; and conducting research in applied clinical toxicology.
3. Applicants must demonstrate experience in all the areas of clinical, research, and teaching activities and leadership. An abundance of experience in one area will not substitute for lack of experience in another.
4. Applicants holding a degree in a healthcare profession in which licensing is required must be in good standing with the appropriate jurisdictional board and must be eligible for, or possess, a valid, unrestricted license to practice. A copy of the license must accompany the application.
5. Applicants must be members in good standing of the American Academy of Clinical Toxicology at the time of their application.

Credentialling Committee Review

Following receipt of the completed application and application fee, the candidate's submission is reviewed by the Credentialling Committee. The committee uses a standardized credential review document among the application reviewers. A formal letter from the president of ABAT will inform the candidate of the committee's decision, and if required, will list areas of improvement the committee felt would allow the candidate to successfully pass credential review on a subsequent submission. Once credentialled, an applicant must take the exam within two examination cycles.

Application Fees

A combined application and testing fee of $400, made payable to the American Board of Applied Toxicology, must accompany the application. If credentialling is denied for any reason, the $300 portion for the examination will be refunded.

APPENDIX 4: S.632 POISON CONTROL CENTER ENHANCEMENT AND AWARENESS ACT

One Hundred Sixth Congress of the United States of America

At the second session

Begun and held at the City of Washington on Monday, the twenty-fourth day of January, two thousand
An Act
To provide assistance for poison prevention and to stabilize the funding of regional poison control centers.

Be it enacted by the Senate and House of Representatives of the United States of America in Congress assembled,

Sec. 1. Short Title

This Act may be cited as the 'Poison Control Center Enhancement and Awareness Act.'

Sec. 2. Findings

Congress makes the following findings:

(1) Each year more than 2,000,000 poisonings are reported to poison control centers throughout the United States. More than 90 percent of these poisonings happen in the home. Fifty-three percent of poisoning victims are children younger than 6 years of age.

(2) Poison control centers are a valuable national resource that provide lifesaving and cost-effective public health services. For every dollar spent on poison control centers, $7 in medical costs are saved. The average cost of a poisoning exposure call is $32, while the average cost if other parts of the medical system are involved is $932. Over the last 2 decades, the instability and lack of funding has resulted in a steady decline in the number of poison control centers in the United States. Within just the last year, 2 poison control centers have been forced to close because of funding problems. A third poison control center is scheduled to close in April 1999. Currently, there are 73 such centers.

(3) Stabilizing the funding structure and increasing accessibility to poison control centers will increase the number of United States residents who have access to a certified poison control center, and reduce the inappropriate use of emergency medical services and other more costly health-care services.

Sec. 3. Definition

In this Act, the term 'Secretary' means the Secretary of Health and Human Services.

Sec. 4. Establishment of a National Toll-Free Number

(a) IN GENERAL—The Secretary shall provide coordination and assistance to regional poison control centers for the establishment of a nationwide toll-free phone number to be used to access such centers.

(b) RULE OF CONSTRUCTION—Nothing in this section shall be construed as prohibiting the establishment or continued operation of any privately funded nationwide toll-free phone number used to provide advice and other assistance for poisonings or accidental exposures.

(c) AUTHORIZATION OF APPROPRIATIONS—There is authorized to be appropriated to carry out this section, $2,000,000 for each of the fiscal years 2000 through 2004. Funds appropriated under this subsection shall not be used to fund any toll-free phone number described in subsection (b).

Sec. 5. Establishment of Nationwide Media Campaign

(a) IN GENERAL—The Secretary shall establish a national media campaign to educate the public and

healthcare providers about poison prevention and the availability of poison control resources in local communities and to conduct advertising campaigns concerning the nationwide toll-free number established under Section 4.

(b) CONTRACT WITH ENTITY—The Secretary may carry out subsection (a) by entering into contracts with 1 or more nationally recognized media firms for the development and distribution of monthly television, radio, and newspaper public service announcements.

(c) AUTHORIZATION OF APPROPRIATIONS—There is authorized to be appropriated to carry out this section, $600,000 for each of the fiscal years 2000 through 2004.

Sec. 6. Establishment of a Grant Program

(a) REGIONAL POISON CONTROL CENTERS—The Secretary shall award grants to certified regional poison control centers for the purposes of achieving the financial stability of such centers, and for preventing and providing treatment recommendations for poisonings.

(b) OTHER IMPROVEMENTS—The Secretary shall also use amounts received under this section to—

(1) Develop standard education programs;
(2) Develop standard patient management protocols for commonly encountered toxic exposures;
(3) Improve and expand the poison control data collection systems;
(4) Improve national toxic exposure surveillance; and
(5) Expand the physician/medical toxicologist supervision of poison control centers.

(c) CERTIFICATION—Except as provided in subsection (d), the Secretary may make a grant to a center under subsection (a) only if—

(1) The center has been certified by a professional organization in the field of poison control, and the Secretary has approved the organization as having in effect standards for certification that reasonably provide for the protection of the public health with respect to poisoning; or
(2) The center has been certified by a State government, and the Secretary has approved the State government as having in effect standards for certification that reasonably provide for the protection of the public health with respect to poisoning.

(d) WAIVER OF CERTIFICATION REQUIREMENTS—

(1) IN GENERAL—The Secretary may grant a waiver of the certification requirement of subsection (c) with respect to a noncertified poison control center or a newly established center that applies for a grant under this section if such center can reasonably demonstrate that the center will obtain such a certification within a reasonable period of time as determined appropriate by the Secretary.
(2) RENEWAL—The Secretary may only renew a waiver under paragraph (1) for a period of 3 years.

(e) SUPPLEMENT NOT SUPPLANT—Amounts made available to a poison control center under this section shall be used to supplement and not supplant other Federal, State, or local funds provided for such center.

(f) MAINTENANCE OF EFFORT—A poison control center, in utilizing the proceeds of a grant under this section, shall maintain the expenditures of the center for activities of the center at a level that is not less than the level of such expenditures maintained by the center for the fiscal year preceding the fiscal year for which the grant is received.

(g) MATCHING REQUIREMENT—The Secretary may impose a matching requirement with respect to amounts provided under a grant under this section if the Secretary determines appropriate.

(h) AUTHORIZATION OF APPROPRIATIONS—There is authorized to be appropriated to carry out this section, $25,000,000 for each of the fiscal years 2000 through 2004.

APPENDIX 5: MICROMEDEX ELECTRONIC DATABASES

CHRIS: U.S. Coast Guard chemical hazard response information.

Dolphin Software: Over 160,000 material safety data sheets.

Drugdex: Extensive pharmaceutical information.

Drug-Reax: Drug interaction information on over 8000 pharmaceuticals.

Fisher Scientific/ACROS Organics MSDS: Over 18,000 material safety data sheets for pure chemicals. Available in 5 different languages.

Hazardtext: Emergency response information for hazardous material emergencies.

HDSB from NLM: Health and environmental effects of over 4500 toxic chemicals.

Identidex: Tablet and capsule identification system.

Index Nominum: International drug directory.

Infotext: Environmental health and safety information directed toward occupational hygienists.

IRIS: U.S. Environmental Protection Agency health risk assessment information for over 450 chemicals.

LOLI: Environmental, international, and state regulations on chemicals maintained by ChemAdvisor.

NAERG: 1996 North American Emergency Response Handbook.

NIOSH Pocket Guide: Occupational information including exposure limits, incompatibilities, and reactivities.

NJ Hazardous Substances Fact Sheets: Employee-oriented information on 700 hazardous substances developed by the New Jersey Health Department.

OHM/TADS: Physiochemical and toxicological information on 1400 substances. Includes oils and other environmental hazards.

Poisindex: Ingredient and clinical toxicology management information on over 1 million substances.

RegsLink: Access to federal and state chemical regulatory information.

Reprorisk: Contains 4 databases on reproductive risk information.

RTECS from NIOSH: Registry of the toxic effects of over 142,000 substances.

Safetydex: Over 8000 hospital-specific material safety data sheets.

Tomes: Medical and hazard information for chemicals.

APPENDIX 6: STANDARD TOXICOLOGY TEXTBOOKS

General Toxicology

- *Poisoning and Toxicology Compendium with Symptoms Index*, Jerrold B. Leikin, and Frank P. Paloucek; Lexi-Comp, Hudson, Ohio, 1998.
- *Clinical Management of Poisoning and Drug Overdose*, Third Edition; Lester M. Haddad, Michael W. Shannon, and James F. Winchester; WB Saunders, Philadelphia, Pennsylvania, 1998.
- *Goldfrank's Toxicologic Emergencies*, Sixth Edition; Lewis R. Goldfrank, Neal E. Flomenbaum, Neal A. Lewin, Richard S. Weisman, Mary Ann Howland, and Robert S. Hoffman; Appleton and Lange, Stamford, Connecticut, 1998.
- *Toxicology of the Eye*, Fourth Edition; W. Morton Grant, and Joel S. Schuman; Charles C Thomas, Springfield, Illinois, 1993.
- *Casarett and Doull's Toxicology*, Sixth Edition; Curtis D. Klaassen; McGraw-Hill, New York, New York, 2001.

Hazardous Materials Toxicology

Hazardous Materials Toxicology: Clinical Principles of Environmental Health, Second Edition; John B. Sullivan, Jr. and Gary R. Krieger; Williams and Wilkins, Baltimore, Maryland, 2001.

Sax's Dangerous Properties of Industrial Materials, Eighth Edition; Richard J. Lewis, Sr.; Van Nostrand Reinhold, New York, New York, 1992.

Occupational, Industrial, and Environmental Toxicology; Michael I. Greenberg, Richard J. Hamilton, and Scott D. Phillips; Mosby, St. Louis, Missouri, 1997.

Proctor and Hughes' Chemical Hazards of the Workplace, Fourth Edition; Gloria J. Hathaway, Nick H. Proctor, James P. Hughes, and Michael L. Fischman; Van Nostrand Reinhold, New York, New York, 1996.

Plants and Mushrooms

Handbook of Mushroom Poisoning: Diagnosis and Treatment; David G. Spoerke and Barry H. Rumack; CRC Press, Boca Raton, Florida, 1994.

AMA Handbook of Poisonous and Injurious Plants; Kenneth F. Lampe and Mary Ann McCann; American Medical Association, Chicago, Illinois, 1985.

Fetal Risk and Human Lactation

Breastfeeding: A Guide for the Medical Profession, Fifth Edition; Ruth A. Lawrence and Robert M. Lawrence; Mosby, St. Louis, Missouri, 1999.

A Reference Guide to Fetal and Neonatal Risk Drugs in Pregnancy and Lactation, Fourth Edition; Georald G. Briggs, Roger K. Freeman, and Sumner J. Yaffe; Williams and Wilkins, Baltimore, Maryland, 1994.

Herpatology

Snake Venom Poisoning, Second Edition; Findlay E. Russell; Scholium International, Great Neck, New York, 1983.

Alternative Medicines

Herbal Drugs and Phytopharmaceuticals: A Handbook for Practice on a Scientific Basis; Norman Grainger Bisset; Medpharm, Stuttgart, Germany, 1989.

The Lawrence Review of Natural Products; Facts and Comparisons; Facts and Comparisons, St. Louis, Missouri, 1999.

Analytical Toxicology

Disposition of Toxic Drugs and Chemicals in Man, Fourth Edition; Randall C. Baselt and Robert H. Cravey; Chemical Toxicology Institute, Foster City, California, 1995.

Bio-chemical Terrorism

Textbook of Military Medicine Part I—Medical Aspects of Chemical and Biological Warfare; Russ Zajtchuk, Editor; Department of the Army, Bethesda, Maryland, 1997.

Chemical Warfare Agents; Timothy C. Marrs, Robert L. Maynard and Frederick R. Sidell; John Wiley and Sons, New York, New York, 2000.

Handbook of Chemical and Biological Warfare Agents; D. Hank Ellison; CRC Press, Boca Raton, Florida, 2000.

Military Chemical and Biological Agents; James A.F. Compton; Telford Press, Caldwell, New Jersey, 1987.

Chemical Warfare Agents; Satu M. Somani; Academic Press, San Diego, California, 1992.

APPENDIX 7

General Toxicology Searching

General Medical and Clinical Toxicology Guide to the Internet
http://www.swmed.edu/toxicology/toxlinks.html
This site provides a single access point to numerous PIC Web sites, journals that publish clinical toxicology papers, and other links to important clinical toxicology information.

MedLine—PubMed
http://www.ncbi.nlm.nih.gov/pubmed/
This site provides online MedLine searching capabilities.

National Academy Press
http://www.nap.edu/catalog/6035.html
Catalog site for publications from the National Academy of Sciences.

National Library of Medicine
http://www.nlm.nih.gov
Entry point for the National Library of Medicine including general information, databases, and photographic archives.

National Library of Medicine's Search Site via Internet Grateful Med
http://igm.nlm.nih.gov/
The Internet Grateful Med Web site provides a variety of database accesses to a variety of searches including MEDLINE, AIDS line, AIDS drugs, AIDS trials, Chemical Identification, NIH Clinical Alerts, and a variety of other very useful access sites.

Specialized Information Services of the National Library of Medicine
http://sis.nlm.nih.gov/ToxSearch.htm
This provides a variety of information on toxicology and environmental health with searches and links to a variety of important organizations.

The Visible Human Project
http://www.npac.syr.edu/projects/vishuman/VisibleHuman.html
The Visible Human Project is a three-dimensional representation of the male and female body. The current phase deals with transverse CT, MR, and cryosection images at 1-mm intervals.

TOXNET ToxLine
http://toxnet.nlm.nih.gov/servlets/simple-search
This site provides free access for tox line searches. These include access to HSDB (Hazardous Substances Data Bank), CCRIS (Chemical Carcinogenesis Research Information System), RTECS (Registry of Toxic Effects of Chemical Substances), GENE-TOX [Genetic Toxicology (Mutagenicity) Data], IRIS (Integrated Risk Information System), TRI (Toxicology Releases Search), Chem-info (Chemical information identification), and many others.

Toxicology Environmental Health Information Program (TEHIP)
http://sis.nlm.nih.gov/tehipl.htm
This is the toxicology section of the National Library of Medicine that has a wealth of information that is searchable in many databases.

Environmental Chemicals Data and Information Network (ECDIN)
http://ecdin.etomep.net/Ecdin/E_hinfo.html
This is a factual database created by the European Commission, Joint Research Centre at the Ispra (I) site. It contains a list of chemical information for each chemical listed. One of its sections is PHATOX (Pharmacological and Toxicological) data which includes health evaluations, toxicological data, epidemiological data, and health hazard evaluations.

Karolinska Institute—Poisoning Information
http://www.mic.ki.se/Diseases/c21.613.html
The site provides access to many links with detailed information on a variety of poison issues including: general; bites and stings; food poisoning; gas poisoning; plant poisoning; lead, iron, mercury, cadmium, nickel, and drug poisoning; and hazardous substances. There are many pictures and multiple links.

The MedNet's Extensive Toxicology Database
http://www.internets.com/mednets/stoxicology.htm
This site provides a list of links to various toxicological sites.

Reprotox—An online reproductive toxicology resource
http://www.reprotox.org
Reprotox provides current assessments on potential harmful affects of environmental exposure to chemicals and physical agents on human pregnancy, reproduction, and development. This online source requires a subscription.

Pharmacy and Pharmaceuticals

Clinical Pharmacology Drug Database
http://www.cponline.gsm.com
Search engine for commonly prescribed drugs with dosages, indications, interactions, pharmacokinetics, costs, and more.

Dietary Supplements
http://odp.od.nih.gov/ods
This Web site provides information regarding dietary supplements.

Drug Database at Pharmaceutical Information Association
http://www.pharminfo.com
Information about pharmaceutical industry drugs and research from the Internet service PharmInfoNet.

Drugs of Abuse—Street Drugs
http://www.mninter.net/~publish/
This is a comprehensive collection of drugs of abuse. Although the information is somewhat basic it is very comprehensive.

Holistic Medicine
http://www.holisticmed.com/www
This is an interesting and very useful site that provides a lot of information on holistic medicines and alternative therapies.

Hyperreal
http://www.hyperreal.org
Important site that includes drugs of abuse primarily involving the ''rave'' scene.

Martindale Health Science Guide—The Virtual Pharmacy Center
http://www-sci.lib.uci.edu/HSG/Pharmacy.html
Access to Martindales pharmacy center for drug information.

Pharmacy (Medicine, Biosciences)
http://www.pharmacy.org
This site contains many links to pharmacy-related resources, including schools of pharmacy, online journals, CME, and societies.

PharmWeb
http://www.pharmweb.net
This is a searchable site from the U.K. with information for the patient and health professional.

Physician's GenRX Drug Compendium Program
http://www1.mosby.com/Mosby/PhyGenRx
A comprehensive listing that is searchable for generic or brand names and by drug category.

RxList Internet Drug Name Category Cross Index
http://www.rxlist.com
Searchable database of 4000 prescription and OTC drug products designed for the lay public.

ScripWorld Pharmaceutical News
http://www.pjbpubs.co.l;uk/acrip/scrhome.html
Scrip is the only international, twice-weekly newsletter reporting on the pharmaceutical sector, covering prescription and OTC medicines and biotechnology news.

Natural Toxins

American Zoo and Aquarium Association
http://www.aza.org
Information and pictures of toxic animals. Order form to obtain the Antivenom Index which is $20–30 in the publications area.

Antivenom Handbook for Australia
http://www.wch.sa.gov.au/paedm/clintox/cslb_index.html
This site has the handbook for a variety of poisonous animals from Australia. See the preceding site for North America.

Foodborne Pathogenic Microorganisms and Natural Toxins Handbook
http://vm.cfsan.fda.gov/~mow/intro.html
This is a site from the U.S. FDA Center for Food Safety and Applied Nutrition. Useful information on mushroom toxidromes.

Fungi Images on the Net
http://isa.dknet.dk/~pip1833/mushimage/
This a searchable Web site that contains information and photographs of various fungi, both poisonous and nonpoisonous.

Medical Botany Library
http://www.floridaplants.com/mpois.htm
Information on poisonous plants and mushrooms from around the world.

Medicinal and Poisonous Plant Databases
http://www.inform.umd.edu/EdRes/Colleges/LFSC/life_sciences/.plant_biology/Medicinals/medicinals.htm
Information and links on poisonous and medicinal plants from the University of Maryland.

NAMA—North American Mycological Association
http://Sorex.tvi.cc.nm.us/nama/poison/poison.htm
This is a very useful site that presents various mushroom toxidromes. There is also a case registry report form that can be downloaded and reported to NAMA.

Natural Toxins FDA
http://vm.cfsan.fda.gov/~mow/toxintoc.html
Lists information on fish, shellfish, mycotoxins, and many other natural toxins from the FDA.

Poison Information Centre—Singapore's Ministry of Health
http://www.gov.sg/health/mohiss/poison/index.html
Singapore's poison information center provides great information about natural toxins in their region of the world. The site also has pictures of many creatures.

Poisonous Plant Database (Plantox)
http://vm.cfsan.fda.gov/~djw/readme.html
The Poisonous Plant Database is a set of working files of scientific information about the animal and human toxicology of vascular plants of the world. The initial files were created in 1994, and are updated periodically.

Poisonous Plants Web Page of Cornell University
http://www.ansci.cornell.edu/plants/plants.html
This site provides information and pictures of poisonous plants. This site also lists veterinary species of interest.

The Poisonous Plant Guide
http://www.ednet.ns.ca/educ/museum/poison/ppguide.htm
An illustrated guide to some common poisonous plants in Nova Scotia including algae, fungi, and leafy plants.

Biochemical Terrorism

Anthrax Vaccine Immunization Program
http://www.anthrax.osd.mil/
The U.S. military's Web site for information related to the anthrax vaccination program. Includes information about anthrax and its use as a biological weapon, Q and A about the vaccine, a newsletter, and a section on related links.

Cal Poly Chemical and Biological Warfare Page
http:/www.calpoly.edu/~drjones/chemwarf.html
A page created by the students in the 1996 spring semester class of Chemistry 405. Contains documents on the history of chemical and biological warfare in ancient and modern times. Also contains sections on the nerve and riot control agents. Even shows the stepwise process of synthesizing several of the nerve agents.

CDC Bioterrorism Preparedness and Response
http:/www.bt.cdc.gov/
The CDC has dedicated a specific page within their Web site on bioterrorism. The page has two important sections for poison centers. The first is the Biological Agents section with a FAQ (frequently asked questions) document, Fact Sheet, Case Definition, and CDC Prevention Guidelines on a specific biological agent. The other section is the Learning Resources section. Included in this area are CDC resources such as articles on bioterrorism from the CDC Journal of Emerging Diseases, official statements on bioterrorism, a video library, and news and events. The video library contains two sets of tapes, the six-tape set from the September 1998 Medical Response to Biological Warfare and Terrorism and one tape from the Public Health Ground Rounds—*Bioterrorism: Implications for Public Health*. Watching the tapes requires the use of the Real Player program that is downloadable from the site.

Chemical and Biological Defense Information Analysis Center (CBIAC)
http:/www.cbiac.apgea.army.mil/
The center is located at the Battelle Memorial Institute and serves as the Department of Defense's focal point for information related to chemical warfare/chemical and biological defense. As part of its services, it offers an excellent search engine for the CBIAC site as well as password-protected bibliographic database (you can apply for a password online). Articles identified by a search in the bibliographic database are available only through the Defense Technical Information Center and have to be special ordered, which means setting up an account with them. However, the results of a search of the CBIAC site itself leads to documents that have unrestricted access.

FEMA Rapid Response Information System (RRIS)
http://www.rris.fema.gov/
Although this site is named ''Rapid Response Information System'' and it has a wide range of resources, it usually takes multiple screens to get to the information. The site has several sections including a relatively complete list of equipment with descriptions and vendors; an extensive list of monographs on nuclear, chemical, and biological agents; and a reference library with links to other sites and documents. There is also a symptom-based search engine that will find any symptom or group of symptoms names in the monographs.

EmergencyNET
http://www.emergency.com/
Web page for the Emergency Response and Research Institute. The site has separate pages for different EMS topics including Infectious Diseases and Chemical/Biological Terrorist Attack. The latter page contains a lesson plan for EMTs and First Responders regarding NBC incidents.

Medical NBC Online Information Server
http://www.nbc-med.org
Developed by the U.S. Surgeon General to provide a learning and reference resource for medical NBC information. Although the site has been developed for U.S. Army medical personnel, the site makes available many NBC health-related resources to any practitioner with access to the Internet. Site includes a news section, medical references (e.g., Army medical field manuals such as FM8-9(B) and the July 1998 edition of Medical Management of Biological Casualties), video and audio clips, training and calendar sections, and a search engine. Also there are numerous links to other NBC sites and many governmental and nongovernmental agencies involved with NBC information.

Mitretek Systems—Chemical, Biological, and Nuclear Systems
http://www.mitretek.org/mission/envene/nbc.html
This site has excellent monographs on the chemical and biological warfare agents (see Background on Chemical and Background on Biological Warfare Agents). The monographs are detailed with either chemical structures of the chemicals or toxins or photographs of the biological organism. Many other excellent references and documents related to chem/bioterrorism can be found as either onsite documents or links.

Outbreak
http://www.outbreak.org/
Outbreak is an online information service that addresses emerging diseases for the health professional and the interested layperson. It attempts to provide a worldwide collaborative database to collect information about possible disease outbreaks. There are registered user and nonregistered user portions of the site. There is the usual list of biological agents found at other sites with fact sheets for each one, but in addition, there are reports from ProMED about any past outbreaks of the disease. ProMED (Program for Monitoring Emerging Diseases), a list server from the Federation of American Scientists, provides information on emerging diseases via e-mail to its subscribers. Outbreak maintains a library of information about past outbreaks reported in ProMED.

Sarin Nerve Gas
http://www.geocities.com/CapCanaveral/Lab/7050/
Brief but informative site on sarin nerve agent. Describes sarin, its history as a terrorist agent, protective equipment, and dosage effects. Nice bibliography and links to other sites with reference documents on sarin.

The Chemical Weapons Conventions Web Site—Organization for the Prohibition of Chemical Weapons
http://www.opcw.nl/
Although this site's primary focus is the Chemical Weapons Treaty, there is an excellent section, Fact-finding Files, that gives accurate and even illustrated information on chemical and biological weapons. The main source for this information is the FOA Briefing Book on Chemical Weapons.

The John Hopkins Center for Civilian Biodefense Studies
http://www.hopkins-biodefense.org/
The primary value of this site is its summary of the National Symposium on Medical and Public Health Response to Bioterrorism held on February 16–17, 1999, in Arlington, VA. Many nationally recognized clinicians, both military and civilian, spoke at this conference along with many top government officials. The site contains the audio (you will need RealPlayerG2 on your computer to hear it) and PowerPoint slides from the conference. There are very brief written summaries of the presentations with the official publication of the conference proceeding appearing in the CDC publication *Emerging Infectious Diseases* Vol. 5, No. 4, the July–August issue (see http://www.cdc.gov/ncidod/eid/).

U.S. Army Research Institute of Chemical Defense (USAMRICD)
http://chemdef.apgea.army.mil/
The Army's center for chemical defense, USAMRICD Web page has several sections of interest. The first in an extensive bibliography by year of all the book chapters and published scientific papers produced by the USAMRICD staff. The next section of importance is a downloadable version of the Institute's 1995 edition of the Medical Management of Chemical Casualties handbook. The links section may not be functioning, but could be a good gateway to other sites.

U.S. Army Medical Research Institute of Infectious Diseases (USAMRIID)
http://www.usamriid.army.mil/
This is the home page of USAMRIID and a good site for two sources of information. First under its publications section, you can find downloadable versions of the Army's reference books, FM8-9 Handbook on The Medical Aspects of NBC Defensive Operations, Medical Management of Biological Casualties (July 1998 edition), and Defense Against Toxin Weapons. The second section of importance is the Continuing Education section which contains the text and PowerPoint slides used in the Army's Medical Management of Biological Casualties course.

U.S. Army Soldier and Chemical and Biological Defense Command
http://www.sbccom.army.mil/
Home page for the U.S. Army's Soldier and Chemical and Biological Defense Command. Best resource for description of military devices and products used in the detection and defense of a WMD release. Also has a description of the Domestic Preparedness training programs currently being administered by the Dept. of Defense.

WHO Communicable Disease Surveillance and Response (CSR)
http://www.who.int/emc/index.html
The primary value of this site is its monographs on various tropical diseases that could be potentially used as biological agents of terrorism. The monographs are relatively short but complete. There are also brief reports of outreaks of infectious diseases in various parts of the world.

REFERENCES

1. Scherz, R.B.; Robertson, W.O. The history of poison control centers in the United States. Clin. Toxicol. **1978**, *12* (3), 291–296.

2. Botticelli, J.T.; Pierpaoli, P.G. Louis Gdalman, pioneer in hospital pharmacy poison information services. Am. J. Hosp. Pharm. **1992**, *49*, 1445–1450.
3. Robertson, W.O. National organizations and agencies in poison control programs: A commentary. Clin. Toxicol. **1978**, *12* (3), 297–302.
4. Youniss, J.; Litovitz, T.; Villaneuva, P. Characterization of U.S. poison centers: A 1998 survey conducted by the American Association of Poison Control Centers. Vet. Hum. Toxicol. **2000**, *42* (1), 43–53.
5. Litovitz, T.L.; Klein-Schwartz, W.; White, S.; Cobaugh, D.J.; Youniss, J.; Drab, A.; Benson, B. 1999 Annual report of the American Association of Poison Control Centers Toxic Exposures Surveillance System. Am. J. Emerg. Med. **2000**, *18* (5), 517–574.
6. Krenzelok, E.P. Do poison centers save money? What are the data? J. Toxicol., Clin. Toxicol. **2000**, *36* (6), 545–547.
7. Miller, T.R.; Lestina, D.C. Costs of poisoning in the United States and savings from poison control centers. A benefit–cost analysis. Ann. Emerg. Med. **1997**, *29*, 239–245.
8. Harrison, D.L.; Draugalis, J.R.; Slack, M.K.; Langley, P.C. Cost-effectiveness of regional poison control centers. Arch. Intern. Med. **1996**, *156*, 2601–2608.
9. Van Buren, J.; Fisher, L.L. *Monroe County Poison Prevention Demonstration Project: Final Report to U.S. Consumer Product Safety Commission*; New York State Department of Health: Albany, New York, 1990.
10. Leikin, J.B.; Krenzelok, E.P. *Poison Centers. Clinical Toxicology,* 1st Ed.; WB Saunders Company: Philadelphia, 2000; 111–114.
11. Krenzelok, E.P. Poison centers at the millennium and beyond. J. Toxicol., Clin. Toxicol. **2000**, *38* (2), 169–170.
12. Krenzelok, E.P.; Vale, J.A. Position statements: Gut decontamination. J. Toxicol., Clin. Toxicol. **1997**, *35* (7), 695–697.
13. American Academy of Clinical Toxicology, European Association of Poisons Centres and Clinical Toxicologists Position statements on gut decontamination. J. Toxicol., Clin. Toxicol. **1997**, *35* (7), 699–762.
14. Personal communication with the AAPCC, September 26, 2000.

Policy Documents and Laws That Guide Clinical Pharmacy Practice in Spain

Eduardo Echarri Arrieta
Sociedad Española de Farmacia Hospitalaria, La Coruca, Spain

Henar Martinez Sanz
Sociedad Española de Farmacia Hospitalaria, Madrid, Spain

INTRODUCTION

Clinical pharmacy practice in Spain is very closely linked to the concept of pharmaceutical care, as defined by Hepler and Strand:[1] "Pharmaceutical care is the responsible provision of drug therapy for the purpose of achieving definite outcomes which improve a patient's quality of life." So much so, in fact, that the term *pharmaceutical care* already implies the concept of clinical pharmacy. The evolution toward pharmaceutical care implies the step in which the pharmacist *selects* the therapy, taking responsibility for it, instead of only *suggesting* it to the doctor.[2,3]

Originally introduced in hospital pharmaceutical practice, and although lack of human staff has prevented its consolidation at all hospitals, clinical pharmacy has grown throughout Spain in recent years. Pharmaceutical care, introduced more recently, is the current trend in hospital pharmacy, and is being increasingly adopted in community pharmacy practice.

Indeed, clinical pharmacy has become so important to pharmacy as a whole that it has now been included as an obligatory subject in pharmaceutical teaching and courses. Under the new educational plan, the second cycle of the pharmacy *licenciatura* (roughly equivalent to an honors degree) includes a central subject entitled "pharmacology and clinical pharmacy."[4] In 1984, a new European Community Directive (EC Directive 84/432) established that pharmacy undergraduate curricula should contain a 6-month stage in a community or hospital pharmacy. In the last year, the training period for hospital pharmacists has been increased by a year to a total of 4 years, an extra year dedicated exclusively to in-house clinical pharmacy training (the pharmacist visits the patients with the physician at the ward).[5]

The functions of the clinical pharmacist and the documents governing such activities in Spain are listed below.

GENERAL LAWS AND POLICY DOCUMENTS

There are a few documents that include nearly all functions developed by clinical pharmacists in Spain (see Table 1). They are, in chronological order, as follows.

Laws

Orden por la que se regulan los Servicios Farmacéuticos de hospitales (Ministerio de Sanidad y Consumo, 1977).[6] This is the law that primarily regulated hospital pharmacy services, defining for the first time some of the clinical activities that the pharmacist must develop, such as drug information, clinical trials, or pharmacovigilance.

Ley General de Sanidad (Ministerio de Sanidad y Consumo, 1986).[7] This is a general law that regulates most subjects health-related, including all health service structures and requirements for drug commercialization, as well as some of the pharmacist's functions as a healthcare provider.

Ley del Medicamento (Ministerio de Sanidad y Consumo, 1990).[8] In force since 1990, it represents the most important law in the drugs area. It regulates most clinical activities developed by pharmacists at any health institutions.

Policy Documents

Coloquios de aproximación a la farmacia clínica (Asociación Española de Farmacéuticos de Hospitales, 1981).[9] This document collects the conclusions of eight debates about clinical pharmacy in which participated 200 hospital pharmacists, including subjects such as drug selection, drug information, and pharmacist integration in

Encyclopedia of Clinical Pharmacy
DOI: 10.1081/E-ECP 120006368

Table 1 Clinical content in general laws and policy documents that guide clinical pharmacy in Spain

Type of document	Title, publication year, reference	Drug selection	Drug information	Pharmaco-kinetics	Artificial nutrition	Clinical trials	Drug use	Pharmaco-vigilance	Other
Laws	Orden 1 febrero, 1977[6]	No	Yes	No	No	Yes	No	Yes	No
	Ley General de Sanidad, 1986[7]	No	No	No	No	Yes	No	Yes	No
	Ley del Medicamento, 1990[8]	Yes	Yes	Yes	No	Yes	Yes	Yes	No
Policy documents	Farmacia Hospitalaria, 1990, 1992[10]	Yes	Yes	Yes	Yes	Yes	Yes	Yes	Yes
	Manual del Residente, 1999[11]	Yes	Yes	Yes	Yes	Yes	Yes	Yes	Yes
	SEFH Practice Guidelines and Consensus Statements[13,15,17,19,30]	Yes	Yes	Yes	Yes	No	No	No	Yes

the healthcare team. The impact of this document reached Europe through the Eighth European Symposium on Clinical Pharmacy, held in Lyon, France, in 1979.

Farmacia hospitalaria (Sociedad Española de Farmacia Hospitalaria, 1990).[10] Edited in 1990, this book compiles all activities, clinical and nonclinical, developed by an hospital pharmacist. Due to the success of the first edition, two years later, the contents were reviewed, and a second edition was published, including more details about clinical contribution of the pharmacist on patient's pharmacotherapy.

Acreditación docente de servicios de farmacia hospitalaria (Sociedad Española de Farmacia Hospitalaria, 1991).[11] Elaborated by the National Speciality Commission, this document contains the acreditation requirements for a hospital pharmacy service to be able to train future specialists, as well as the program of hospital pharmacy specialization. In 2000, the program was revised to include one more year dedicated exclusively to clinical pharmacy training.[5]

Manual del residente de farmacia hospitalaria (Sociedad Española de Farmacia Hospitalaria, 1999).[12] Elaborated to by a consult handbook for future specialists, all teaching hospitals participated in its creation; it contains most activities developed by the hospital pharmacist, with theory and practical cases.

SEFH consensus statements, practice recommendations, and guidelines (Sociedad Española de Farmacia Hospitalaria). As any society, the Spanish Society of Hospital Pharmacy (SEFH) periodically publishes practice recommendations to facilitate its members' work and daily practice at hospital. They include some clinical areas such as pharmacokinetics, artificial nutrition, and drug information (see below).

THE CLINICAL PHARMACIST'S AREAS OF ACTION

Drug Selection

Law: *Ley General de Sanidad.* Ministerio de Sanidad y Consumo, 1986.[7]

Art. 95.5. ''All people who work at health services must collaborate in drug evaluation and control. . . .''

Law: *Ley del Medicamento.* Ministerio de Sanidad y Consumo, 1990.[8]

Art. 84.4. and 84.6. ''Public Administration will proceed for primary and specialized care structures to make a scientific drug selection and evaluation, through Drug and Therapeutics Committee . . .'' and ''. . . will promote the publication of drug formularies for healthcare professionals' use.''

Art. 91.2. ''Hospital pharmacy services will take part in hospital commissions to participate in drug selection and evaluation according to evidence, and drug use.''

Policy document: *SEFH Recommendations: Drug formulary edition.* Sociedad Española de Farmacia Hospitalaria, 1994.[13]

Through this document, the SEFH defines the concept of drug formulary and tries to give a guide in how to

elaborate it. It must contain the drugs selected by the Drugs and Therapeutics Committee with their basic information, institutional proceedings regarding drug prescription, and practical drug information such as drugs in pregnancy and lactation, dose adjustment in renal or hepatic impairment, drug interactions, and so on.

Further information provided by the SEFH about drug selection in hospitals is compiled in Chapter 2.1 of the book *Farmacia Hospitalaria*.[14]

Drug Information

Law: *Ley del Medicamento.* Ministerio de Sanidad y Consumo, 1990.[8]

In primary healthcare

Art. 87. "Functions that guarantee the rational use of drugs: drug information to health-care professionals; drug information to patients, drug monitoring and pharmacovigilance; promotion and participation in educating the general public about drugs, their rational use and the prevention of abuse. . . ."

In hospital and specialized care

Art. 91.2. "To achieve the rational use of drugs, the hospital pharmacy services shall: establish a drug information service for all hospital staff, organize educational activities dealing with issues in their field addressed to healtcare staff at the hospital and their patients. . . ."

Policy document: *Procedural regulations regarding drug information.* Sociedad Española de Farmacia Hospitalaria, 1996.[15]

This is a guide with recommendations on how best to perform the basic duties at a Drug Information Center (CIM). Individual CIMs adapt the recommendations in line with their specific features. CIM directors should be qualified pharmacists.

According to the SEFH, one of the basic functions of a CIM is to evaluate and transmit information concerning drugs. The following recommendations are considered particularly important:

- A regular information procedure should be established on new developments in pharmacotherapy for the pharmacy service staff and technical advice for the dispensation unit concerning any medication-related problems detected.
- Priority should be given to aspects such as efficiency, safety, and cost when technical reports to drug-selection committees are being prepared.
- Active information will be education- and training-oriented and will be addressed to healthcare staff via bulletins and to patients.
- Passive information in response to enquiries will be as objective, concise, useful, and clear as possible.

More complete details about a hospital pharmacist's activities concerning information on drugs and the functions of the CIM are given in Chapter 2.4 of the book *Farmacia Hospitalaria* (Hospital Pharmacy) by the SEFH.[16]

Clinical Pharmacokinetics

Law: *Ley del Medicamento* (Ministerio de Sanidad y Consumo, 1990).[8]

Art. 91.2. "To achieve rational use of drugs, hospital pharmacy services shall carry out clinical pharmacokinetic activities. . . ."

Policy document: *SEFH recommendations on inical pharmacokinetics* (Sociedad Española de Farmacia Hospitalaria, 1997).[17]

The SEFH recommends that a hospital's clinical pharmacokinetic unit should form part of the pharmacy service and be directed by a pharmacist specializing in hospital pharmacy. It defines the functions of this unit as the following:

- Drug selection to be included in the pharmacokinetic monitoring program, on the basis of narrow therapeutic margin and extensive pharmacokinetic variability.
- Selection of patients to benefit from this program.
- Selection of analytical methods on the basis of specificity and sensitivity.
- Interpretation of plasma levels depending on the characteristics of the drug, the patient's clinical data, and concomitant treatment.
- In cases where dosage needs to be modified, the pharmacist informs the doctor in charge of the new dosage, when it should be started, if any doses should be avoided, the expected plasmatic levels, and subsequent analytical controls, if necessary.

Chapter 3.2 of *Farmacia Hospitalaria* (Hospital Pharmacy) contains a more detailed description of these activities.[18]

Artificial Nutrition

Policy document: *SEFH recommendations on artificial nutrition* (Sociedad Española de Farmacia Hospitalaria, 1997).[19]

The SEFH considers artificial nutrition to be multidisciplinary in range and scope and, therefore, recommends the creation of Artificial Nutrition Commissions, in which the pharmacy service should be always and actively represented. The main task of these Commissions is to establish an artificial nutrition protocol that includes:

- Records of enteric nutrition, peripheral parenteral nutrition, and central parenteral nutrition.
- Assessment of nutritional state.
- Calculation of calorie requirements.
- Administration routes.
- Biochemical monitoring.
- Standard formulas and regulations governing individual prescriptions.

In the SEFH's view, the preparation and dispensation of artificial nutrition is not the pharmacy services' only activity, as they can also perform prescription and clinical monitoring. This already occurs at a number of Spanish hospitals: the pharmacist prescribes and performs daily checkups on patients being fed artificially.

Policy document: *Nutritional therapeutics* (Sociedad Española de Farmacia Hospitalaria, 1992).[20]

This chapter describes the activities performed by the pharmacist as a member of the nutritional support team, which involve a range of tasks from indications, assessment of nutritional state, and the preparation of artificial nutrition for adults and children to concepts on basic dietetics and mother–infant nutrition. Drug–nutrient interaction analysis is another duty.

Clinical Trials

Law: *Ley del Medicamento* (Ministerio de Sanidad y Consumo, 1990).[8]

Chapter 3 (Art. 59-69) is devoted entirely to clinical trials. Besides the duties of dispensation and control of clinical trial samples, the hospital pharmacist's clinical duties in this area arise from his participation as a member of the CEIC, Spain's Ethical Clinical Research Committees.

Art. 64.2. ''The committee shall consider the methodological, ethical and legal aspects of the proposed protocol, and the balance of risk and benefits expected from the test.''

Art. 64.3. ''The CEIC shall, at the very least, consist of an interdisciplinary team involving doctors, hospital pharmacists, clinical pharmacologists, nursing staff and several people unconnected with any of the healthcare professions, one of whom, at least, shall be a jurist.''

Law: *Real Decreto establishing the requirements for performing clinical trials with drugs* (Ministerio de Sanidad y Consumo, 1993).[21]

Heading 3 deals with CEIC. Art. 41.2 emphasizes that the hospital pharmacist should be a member of the CEIC, and Art. 42 describes the duties and functions of the CEIC: ''...to evaluate the protocol and the research team, evaluate the written information to be given to the subjects of the research, perform monitoring of trials....''

Policy document: *Clinical trials* (Sociedad Española de Farmacia Hospitalaria, 1992).[22]

Involving the pharmacist in the clinical trials may actually facilitate the work in some aspects, including:

- Patient monitoring and follow-up, in which the pharmacist collaborates with the researcher in the compilation of analytical parameters.
- Registering and channeling any adverse reactions observed and trying to establish the causal relation.
- Conveying information to the patients to help them comply properly with the protocol.

Drug Use Evaluation and Pharmacoepidemiology

Law: *Ley del Medicamento* (Ministerio de Sanidad y Consumo, 1990).[8]

Art. 91.2. ''To ensure the rational use of drugs the hospital pharmacy services... shall prepare reports on drug use... and perform any duties that lead to improved drug use and control.''

Policy document: *Pharmacoepidemiology and drug use evaluation* (Sociedad Española de Farmacia Hospitalaria, 1992).[23]

Types of drug use evaluation reports, postmarketing pharmacovigilance trials and economic appraisal in the context of pharmacoepidemiology are described.

Pharmacovigilance

Law: *Ley General de Sanidad* (Ministerio de Sanidad y Consumo, 1986).[7]

Art. 99. ''Importers, manufacturers and healthcare professionals are obliged to inform on all adverse reactions caused by drugs and medical devices, whenever there is a risk to the life or health of patients.''

Law: *Ley del Medicamento* (Ministerio de Sanidad y Consumo, 1990).[8]

Art. 57.1. insists on the obligation to declare unexpected or toxic effects of drugs.

Art. 58.3. "Doctors, veterinarians, pharmacists, nurses and other healthcare professionals shall be obliged to collaborate in the Spanish Pharmacovigilance System."

Pharmacovigilance is one of the functions of both the primary health care pharmacist (art. 87) and the hospital pharmacist (art. 91.2).

The SEFH has signed a Collaboration Agreement with the Spanish Ministry of Health and Consumer Affairs to promote and encourage the hospital pharmacist's participation in the Spanish Pharmacovigilance System, which includes, among other things, using the "yellow card" system to notify any adverse drug reactions.[24]

Law: *Circular 18/90 de la Dirección General de Farmacia y Productos Sanitarios*, 1990.[25]

Includes several guidelines on preparing pharmacovigilance trials.

Other Clinical Activities of the Hospital Pharmacist

Published by the SEFH in 1992, *Farmacia Hospitalaria* covers other specific duties of the clinical pharmacist in the health system, such as clinical toxicology[26] and clinical activities in hospital clinical units: infectious diseases, pediatrics, oncology, gynecology, surgery, geriatrics, clinical hematology, cardiovascular, intensive care units, psychiatry, emergencies, etc.[27]

Activities in Community Pharmacy

Law: *Ley de regulación de servicios de las oficinas de farmacia* (Jefatura de Estado, 1997).[28]

Art. 1: Enumerates the functions and duties of pharmacists working in pharmacy offices, which include "information and follow-up of pharmacological treatment of patients; collaboration on the control of individualized use of drugs to detect any possible adverse reactions and inform the relevant pharmacovigilance bodies."

Since the law came into effect, the community pharmacist has evolved from a mere "pharmaceutical adviser" to becoming involved in the launch of pharmaceutical care programs.

Policy document: *Granada Consensus on drug-related problems*, 1998.[29]

On the assumption that the three basic objectives of pharmaceutical care are to identify, solve, and anticipate medicinal drug-related problems, the Consensus of Granada established a classification of such problems based on the application of a systematic criterion of evaluation of the requirements of pharmacotherapy: to be indicated, effective, and safe.

OTHER DOCUMENTS

Spanish Code of Pharmaceutical Ethics (Commission on Bioethics and Pharmaceutical Ethics, Sociedad Española de Farmacia Hospitalaria, 1998).[30]

This is a consensus document promoted and prepared by a task force from the SEFH and accepted by other organizations, including the *Sociedad Española de Farmacéuticos de Atención Primaria* (Spanish Association of Primary Healthcare Pharmacists), the *Consejo General de Colegios Oficiales de Farmacéuticos* (General Council of the Official Colleges of Pharmacists), and the *Real Academia de Farmacia* (Royal Spanish Academy of Pharmacy). This code of ethics includes the basic principles and responsibilities of the pharmacist in his relations with the patient, other health care professionals, and society in general.

Web Sites on Clinical Pharmacy/Pharmaceutical Care

An increasing number of Web pages of interest to the clinical pharmacist are appearing on the Internet, including the following in Spain:

- Atención Farmacéutica (Pharmaceutical Care), Carlos III Health Institute, National Health School, http://www.isciii.es/unidad/Sgpcd/ens/atenfar/paginaprincipal. htm.
- *Club de Atención Farmacéutica* (Pharmaceutical Care Club), Pharmacy Faculty, University of Granada, http://www.ugr.es/~atencfar/welcome.htm.
- *Atención Farmacéutica* (Pharmaceutical Care), Cinfa laboratories, http://www.atencion-farmaceutica.com/.
- *Unidad de Farmacia Clínica y Farmacoterapia* (Clinical Pharmacy and Pharmacotherapy Unit), Pharmacy Faculty, University of Barcelona, http://www.ub.es/farcli/wp0.htm.
- *Fundación Pharmaceutical Care España*, Barcelona, http://www.pharmaceutical-care.org.

Journals on Pharmaceutical Care/Clinical Pharmacy

The following are a few journals pertaining to pharmaceutical care and clinical pharmacy:

- *Farmacia Hospitalaria (Farm Hosp).* Sociedad Española de Farmacia Hospitalaria. http://www.sefh.es/revistas/revistas.htm.
- *Atención Farmacéutica—European Journal of Clinical Pharmacy (Aten Farm).* http://www.farmaclin.com.

- *Pharmaceutical Care España (Pharm Care Esp).* Fundación Pharmaceutical Care España. http://www.pharmaceutical-care.es.

REFERENCES

1. Hepler, C.D.; Strand, L.M. Opportunities and responsibilities in pharmaceutical care. Am. J. Hosp. Pharm. **1990**, *47* (3), 533–543.
2. Foppe, J.W. Atención farmacéutica en farmacia comunitaria en Europa, retos y barreras. Pharm. Care Esp. **2000**, *2*, 42–56.
3. Martínez Sánchez, A.M. Farm. Clin. **1998**, *5*, 519.
4. www.uv.es/~farmacia/P_N_Farm.htm (accessed Oct 2000).
5. *Programa Oficial de Formación en la Especialización de Farmacia Hospitalaria*; Sociedad Española de Farmacia Hospitalaria. http://www.sefh.es/residentes/programa/programa.htm (accessed Mar 2001).
6. Orden de 1 de febrero de 1977 por la que se regulan los servicios farmacéuticos de hospitales. BOE no. 43, Feb 19, 1977.
7. Ley 14/1986, de 25 de abril, General de Sanidad. BOE no. 102, Apr. 29, 1986.
8. Ley 25/1990, de 20 de diciembre, del Medicamento. BOE no. 306, Dec. 22, 1990.
9. *Coloquios de Aproximación a la Farmacia Clínica*; Asociación Española de Farmacéuticos de Hospital, 1981.
10. Domínguez-Gil, A.; Bonal, J. *Farmacia Hospitalaria,* 1st Ed.; SEFH: Madrid, 1990.
11. *Acreditación Docente de Servicios de Farmacia Hospitalaria*; Sociedad Española de Farmacia Hospitalaria; SEFH: Madrid, 1991.
12. *Manual del Residente de Farmacia Hospitalaria*; Bermejo, T., Cuña, B., Napal, V., Valverde, E., Eds.; SEFH: Madrid, 1999.
13. Recomendaciones de la SEFH: Edición de formularios oguías farmacoterapéuticas. SEFH **1994**, *18* (67), 36–38. http://www.sefh.es/normas/normay5.htm (accessed Oct. 2000).
14. Bonafont, X.; Pla, R. Selección de Medicamentos. In *Farmacia Hospitalaria,* 2nd Ed.; SEFH: Madrid, 1992; 269–288.
15. Proyecto de recomendaciones de la SEFH. Información de medicamentos. SEFH **1996**, *20* (77), 15–20. http://www.sefh.es/normas/normay9.htm (accessed Oct. 2000).
16. Aizpuru, K.; Arrizabalaga, M.J.; Cao, C. Información de Medicamentos. In *Farmacia Hospitalaria,* 2nd Ed.; SEFH: Madrid, 1992; 349–369.
17. Recomendaciones de la SEFH sobre farmacocinética clínica. SEFH **1997**, *21* (80), 10–11. http://www.sefh.es/normas/normay7.htm (accessed Oct. 2000).
18. Calvo, M.V.; García, M.J.; Lanao, J.M. Farmacocinética Clínica. In *Farmacia Hospitalaria,* 2nd Ed.; SEFH: Madrid, 1992; 436–470.
19. Recomendaciones de la SEFH sobre nutrición artificial. SEFH **1997**, *21* (79), 7–8. http://www.sefh.es/normas/normay8.htm (accessed Oct. 2000).
20. Cardona, D.; Alastrué, A.; Clopés, J.; Llorens, E.; Massó, J.; Mira, A.; Schinca, N. Terapéutica Nutricional. In *Farmacia Hospitalaria,* 2nd Ed.; SEFH: Madrid, 1992; 850–923.
21. Real Decreto 561/1993, de 16 de abril, por el que se establecen los requisitos para la realización de ensayos clínicos con medicamentos. BOE no. 114, May 13, 1993.
22. Idoate, A.J.; Caínzos, M.D.; Idoipe, A. Ensayos clínicos. In *Farmacia Hospitalaria,* 2nd Ed.; SEFH: Madrid, 1992; 645–691.
23. Altimiras, J.; Segu, J.L. Farmacoepidemiología y Estudios de Utilización de Medicamentos. In *Farmacia Hospitalaria,* 2nd Ed.; SEFH: Madrid, 1992; 396–435.
24. Castro, I.; Altimiras, J.; Mas, P.; Napal, V.; Olalla, J.F.; Pardo, C.; Rodríguez, J.M. Farmacovigilancia. In *Farmacia Hospitalaria,* 2nd Ed.; SEFH: Madrid, 1992; 601–644.
25. Circular 18/90, de 21 de noviembre, de la Dirección General de Farmacia y Productos Sanitarios. Ministerio de Sanidad y Consumo, 1990.
26. García, E.; Agudo, M.A.; Valverde, E.; Villalba, D. Toxicología Clínica. In *Farmacia Hospitalariam,* 2nd Ed.; SEFH: Madrid, 1992; 471–543.
27. Farmacia Clínica. In *Farmacia Hospitalaria,* 2nd Ed.; SEFH: Madrid, 1992.
28. Ley 16/1997, de 25 de abril, de regulación de servicios de las oficinas de farmacia. BOE no. 100, Apr. 26, 1997.
29. Panel de consenso ad hoc. Consenso de Granada sobre problemas relacionados con medicamentos. Pharm. Care Esp. **1999**, *1*, 107–112.
30. Código de ética farmacéutica. SEFH **1998**, *22* (86), 15–17. http://www.sefh.es/codigoetico.htm (accessed Oct. 2000).

Post-Marketing Surveillance

Beth A. Lesher
Hamilton House, Virginia Beach, Virginia, U.S.A.

INTRODUCTION

The United States has one of the most rigorous drug approval processes in the world, but 51% of marketed drugs have serious adverse effects that are not detected during pre-marketing clinical studies.[1] Consequently, post-marketing surveillance is required to collect data about drugs or other medical products once they are available to the general population.[2] Post-marketing surveillance in the United States is the responsibility of the Food and Drug Administration (FDA). Pharmacovigilance expands on the concept of post-marketing surveillance and is the continuous process of evaluation accompanied by steps to improve safe use of drugs.[3] Both post-marketing surveillance and pharmacovigilance involve pharmaceutical companies, regulatory authorities, healthcare providers, and patients. This monograph provides an overview of the post-marketing surveillance process in the United States including a justification of need; the roles of the FDA, drug manufacturers, healthcare providers, and patients; and strengths and weaknesses of the current system.

JUSTIFICATION OF NEED

Pre-marketing clinical studies are only expected to detect adverse drug events (ADEs) with an incidence of 0.1% or greater; the ability to detect those with an incidence between 0.1% to 1% is unreliable.[4] Limitations of pre-marketing clinical studies include their short duration and use of surrogate endpoints rather than clinical outcomes. In addition, pre-marketing clinical studies are conducted in well-defined study populations, for one indication, and with limited concomitant medications. Finally, the number of patients exposed to the drug during pre-marketing clinical studies is generally between 3000 and 4000. An ADE with an incidence of 0.01%, may require a study population of 30,000 patients to have a 95% chance of detection.[5]

The Prescription Drug User Fee Act of 1992 (PDUFA) and the FDA Modernization Act of 1997 (FDAMA) have been credited with reducing clinical development and regulatory review times from 30 to 12 months.[6] Consequently, the percent of products first available in the United States has increased.[7] Less than one-third of new drugs were previously first introduced in the United States; drug events in other countries served as an ADE early warning system. Although a decrease in review time and an increase in first available products has raised safety concerns,[8,9] the frequency of labeling changes for serious adverse events has decreased from 52% in the 1980s to 30% in the 1990s.[10] Furthermore, the rate of safety-based market withdrawals of new molecular entities has not changed; ranging from 1–3.5% over the past several decades (Fig. 1).[6]

POST-MARKETING SURVEILLANCE PROCESS

At the center of the current U.S. post-marketing surveillance process is the FDA (Fig. 2); drug manufacturers, healthcare providers, and patients are also vital to the success of this process. Currently, only the roles of the FDA and drug manufacturers are defined by statute. The role of hospitals in reporting ADEs is defined by the Joint Commission on Accreditation of Healthcare Organizations[11] and of hospital pharmacists by the American Society of Health-System Pharmacists.[12] For an index of U.S. federal regulations and guidelines that covers safety surveillance of drugs, the reader is referred to an article published by Curran and Engle.[13]

Food and Drug Administration

The FDA has operated a post-marketing surveillance program since 1961, replacing a system operated by the American Medical Association. The FDA currently receives more than 250,000 ADE reports yearly.[14] Within the FDA, the Center for Drug Evaluation and Research (CDER) Office of Post-marketing Drug Risk Assessment (OPDRA) is responsible for post-marketing surveillance. The objectives of the post-marketing surveillance program are to detect ADEs not previously observed, to

Encyclopedia of Clinical Pharmacy
DOI: 10.1081/E-ECP 120006335

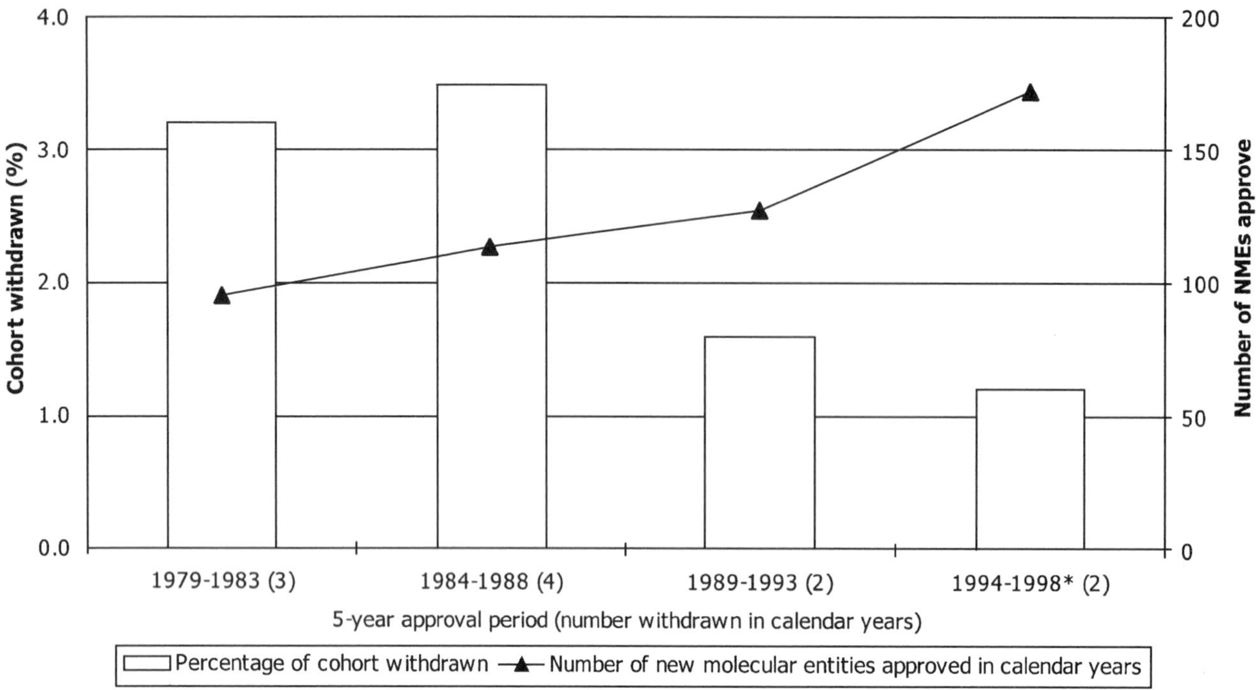

Fig. 1 New molecular entity withdrawals. NME, new molecular entity; *Prescription Drug User Free Act years. (From Ref. [6].)

improve the understanding of the potential severity of previously anticipated risks, to detect events resulting from drug interactions or drug effects in particular populations, and to assess the potential for causal relationships.[6] The FDA's primary mechanism for identifying serious ADEs is the spontaneous reporting system (SRS). Additional approaches include the use of large healthcare databases, cohort and case-control studies, and product registries.

In 1993, the FDA introduced a new streamlined SRS, MedWatch. MedWatch allows healthcare providers and patients to report ADEs to the FDA by several mechanisms, including electronic and print media (Table 1). The goals of the MedWatch program are to increase awareness of drug- and device-induced disease, clarify what should be reported to the FDA, simplify the reporting process, and provide regular feedback to the healthcare community about safety issues involving medical products.[15]

Information received through the MedWatch program is entered into the Adverse Event Reporting System (AERS) database. The AERS database is designed to permit electronic or paper submission of reports, comply with international standards and terminology, and facilitate pharmacovigilance screening. In addition to the MedWatch program, the FDA runs a program for adverse events related to veterinary products and in conjunction with the Center for Disease Control and Prevention, a program for adverse events with vaccines, the Vaccine Adverse Event Reports System (VAERS).

After determining that a problem exists through the post-marketing surveillance process, the FDA has several options. These options include medical alerts such as "Dear Health Professional" letters or safety alerts published in the FDA Medical Bulletin or on the MedWatch web site, press releases, product labeling changes, boxed warnings, and product withdrawals. The FDA may also require a drug manufacturer to conduct post-marketing safety studies. It should be noted that the FDA is only able to request that a manufacturer remove a product from the market.

Drug Manufacturers

Drug manufacturers may identify ADEs through several means, including spontaneous reports by healthcare providers or patients, clinical studies, and reviews of the medical literature. Once the company is aware of an ADE, the data are evaluated, the seriousness and ex-

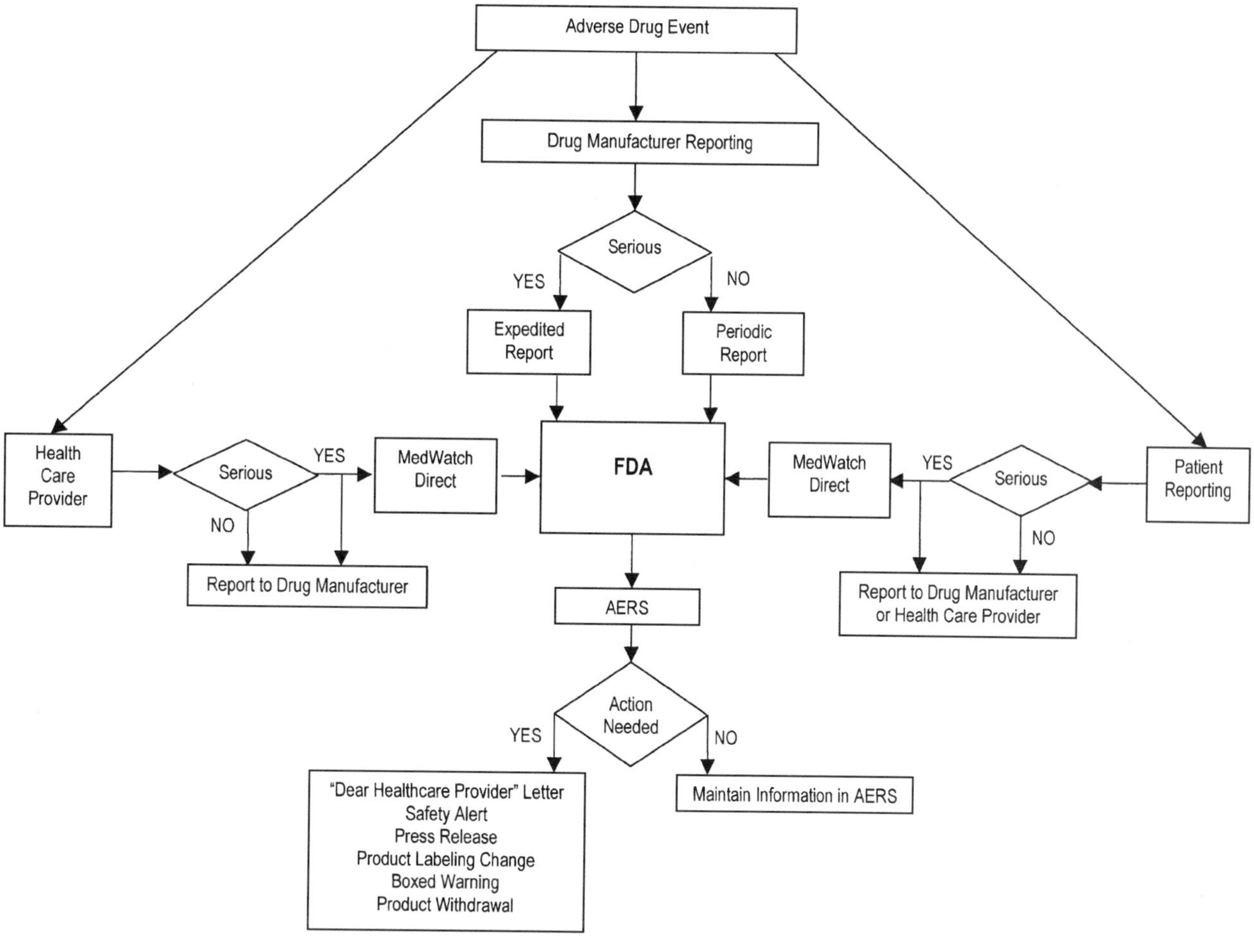

Fig. 2 Post-marketing surveillance process. An ADE is serious if the outcome is death; life-threatening; requires an intervention to prevent permanent impairment or damage; or results in hospitalization, disability, or congenital anomaly. AERS, Adverse Event Reporting System. (From Ref. [16].)

pectedness are determined, and a report is submitted to the FDA (Fig. 2). Approximately 75% of all reports received by the FDA are either expedited or periodic reports from drug manufacturers (Fig. 3). The process for notification of the FDA by a drug manufacturer is defined by the Federal Code.

The Federal Code describes two levels of ADE reporting for drug manufacturers. The first level of reporting is an expedited report for serious or unexpected events reasonably attributed to the drug. These events require submission of a 15-day alert report. The drug manufacturer is also required to submit a 15-day alert report when an increased frequency of an ADE or of drug failure is noticed. The second level of reporting is a periodic report for all ADEs attributed to the drug. These periodic reports are submitted quarterly for the first 3 years a drug is marketed and yearly thereafter.

Healthcare Providers

Voluntary reporting of ADEs by healthcare providers is invaluable to the post-marketing surveillance process (Fig. 2). These reports often provide the first signal to the FDA that a problem exists. Whereas only serious ADEs should be reported to the FDA, all ADEs can be reported to the drug manufacturer. An ADE is classified as serious if the outcome is death; life-threatening; requires an intervention to prevent permanent impairment or damage; or results in hospitalization, disability, or congenital anomaly.[16] The reporting healthcare provider needs only suspect an association between the drug and ADE; causality of the event does not need to be determined. In all cases, healthcare providers need to provide complete and accurate details of the ADE.

Table 1 Contacts for adverse event reporting

Method	Contact
MedWatch	
Direct mail	MedWatch The FDA medical products reporting program Food and Drug Administration 5600 Fishers Lane Rockville, MD 20852-9787
Facsimile	1-800-FDA-0178
Internet	https://www.accessdata.fda.gov/scripts/medwatch/
Modem	1-800-FDA-7737
Telephone	1-800-822-1088
VAERS[a]	
Internet	http://www.fda.gov/cber/vaers/vaers.htm
Telephone	1-800-822-7967
Veterinary Products	
Telephone	1-888-332-8387

[a]VAERS = Vaccine Adverse Event Reports System.

Patients

The role of the patient is also vital to the post-marketing surveillance process, particularly in identifying an ADE in an outpatient setting. Although patients may report serious ADEs directly to the FDA or any ADE to the drug manufacturer, it is suggested that they work through a healthcare provider who may be able to more accurately assess the ADE.[17]

WEAKNESSES AND STRENGTHS OF CURRENT SYSTEM

The current post-marketing surveillance system involves multiple processes and interactions. A breakdown in any one of these processes or interactions could lead to a weakness. Several weaknesses of the current system have been identified. The first is the identification of an ADE. Identification of an ADE is subjective and imprecise.[18] A study to evaluate how accurately ADEs are identified, found less than a 50% agreement between clinical pharmacologists and treating physicians.[19] A second weakness of the current system is underreporting of ADEs. The FDA estimates that only 1–10% of ADEs are reported.[20] A third weakness is the introduction of bias. ADEs reported through the SRS are not subjected to the same rigorous standards as those determined through controlled clinical studies. These reports are therefore subjected to a number of biases, including the length of time a product has been marketed, reporting environment, detailing time, and quality of data.[21] A fourth weakness is the inability to estimate population exposure to the drug. The FDA attempts to overcome this weakness by forming cooperative agreements with outside research organizations. Unfortunately, these data still need to be interpreted cautiously because drug prescribing may not equal drug usage.[22] The fifth weakness of the current system is the

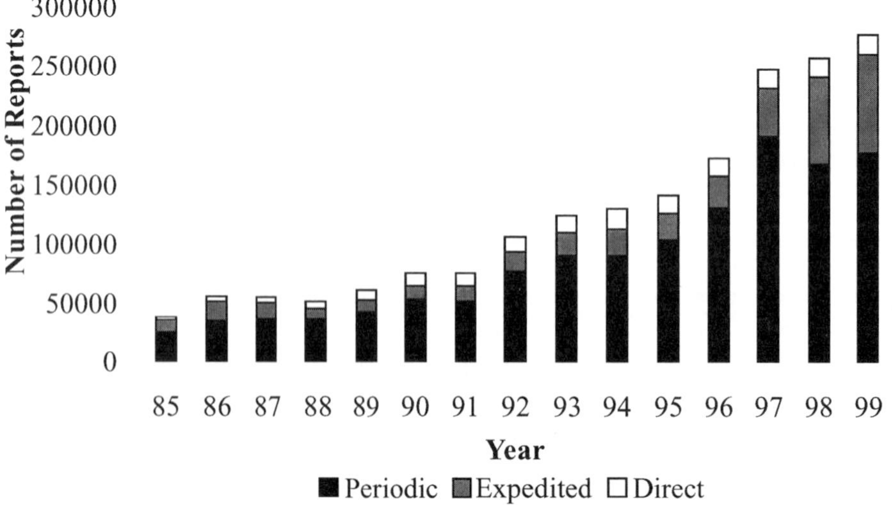

Fig. 3 Number of periodic, expedited, and direct reports received by the FDA for the years 1985–1999. (From Ref. [14].)

quality of reports submitted to the FDA. Often the poor quality of information received does not allow the FDA to adequately evaluate the report and be proactive in protecting the public.[6] The final weakness of the current system is the lack of effect of the FDA's action. A study documented that the prescribing habits of cisapride did not change after the FDA expanded the black box warning, issued a press release, and had the manufacturer distribute 800,000 ''Dear Health Professional'' letters.[23]

In spite of these limitations, the current system does have some strengths. For example, it covers large numbers of diverse patients. Reports generated through the SRS reflect everyday use of the drug. A second strength of the system is the ability to monitor ADEs in an inexpensive manner. The current system is relatively inexpensive compared with an observational event monitoring system.[24] The last strength of the current system is its ability to generate hypotheses and signals. Currently, the FDA is in the process of determining what constitutes a signal; this strength is only beginning to be fully used.

FUTURE DIRECTION OF POST-MARKETING SURVEILLANCE

The future direction of the FDA's post-marketing surveillance process is to strengthen and expand the current system. To increase the ability to identify ADEs, the FDA is piloting a ''sentinel'' system that uses a small subset of healthcare institutions and charges them with preparing frequent, detailed ADE reports. To increase the number of reported ADEs, the FDA is increasing its education programs. Improvement in physician knowledge of ADEs was reported after a 2-year program of sustained physician education. This was accompanied by a 17% increase in the number of ADE reports submitted.[25] To more rapidly investigate potentially serious ADEs, the FDA is evaluating a plan that would increase their access to usage and event data through large healthcare databases.

In addition to those items previously mentioned, the FDA is also participating in the development and use of international standards to facilitate pharmaceutical development. The goals of the International Conference on Harmonization of Technical Requirement for Registration of Pharmaceuticals for Human Use (ICH) is to facilitate the mutual acceptance of data submitted in support of drug marketing applications by the European Union, Japan, and the United States. The ICH has developed a number of guidelines and standards to harmonize post-marketing surveillance efforts on an international level.[26]

In addition to those items developed by the FDA, the creation of an independent center for assessing pharmaceutical effectiveness has been proposed.[1,27,28] This center would be an independent agency that would assist in developing answers to issues that face healthcare providers and patients after a drug has been marketed, including therapeutic differences and safety issues.

CONCLUSION

The current post-marketing surveillance system in the United States is a multifaceted, international process involving the FDA, drug manufacturers, healthcare providers, and patients. Whereas the FDA is strengthening the current system, its success will still rely heavily on the submission of spontaneous reports. Therefore, healthcare providers need to be continually reminded of the value of reporting ADEs.

REFERENCES

1. Moore, T.J.; Psaty, B.M.; Furberg, C.D. Time to act on drug safety [see comments]. JAMA, J. Am. Med. Assoc. **1998**, *279* (19), 1571–1573.
2. Goldman, S.A. *The Clinical Impact of Adverse Event Reporting [Continuing Education Article]*; October 1996, Available at: http://www.fda.gov/medwatch/articles/medcont/medcont.htm. Accessed February, 2001.
3. Waller, P.C. Pharmacovigilance: Towards the next millennium. Br. J. Clin. Pharmacol. **1998**, *45* (5), 417.
4. Bess, A.L. A successful global drug safety system: The Hoffmann–La Roche experience. Drug Inf. J. **1999**, *33*, 1109–1116.
5. Buyse, M.; Poplavsky, J.L. Clinical evaluation of virus safety and inhibitor incidence: Statistical considerations. Blood Coagulation Fibrinolysis **1995**, *6* (Suppl. 2), S86–S92.
6. Task Force on Risk Management. *Managing the Risks from Medical Product Use: Creating a Risk Management Framework*; U.S. Department of Health and Human Services Food and Drug Administration: Washington, DC, May, 1999; 35, Available at: http://www.fda.gov/oc/tfrm/riskmanagement.pdf. Accessed 2/23, 2001.
7. Kaitin, K.I.; Healy, E.M. The new drug approvals of 1996, 1997, and 1998: Drug development trends in the user fee era. Drug Inf. J. **2000**, *34*, 1–14.
8. Friedman, M.A.; Woodcock, J.; Lumpkin, M.M.; Shuren, J.E.; Hass, A.E.; Thompson, L.J. The safety of newly approved medicines: Do recent market removals mean there is a problem? [see comments]. JAMA, J. Am. Med. Assoc. **1999**, *281* (18), 1728–1734.
9. Lurie, P.; Wolfe, S.M. *FDA Medical Officers Report Lower Standard Permit Dangerous Drug Approvals*; 11/15/

00, Available at: http://www.citizen.org/hrg/publications/FDAsurvey/FDAsurvey.htm. Accessed 2/6, 2001.

10. Tufts Center for the Study of Drug Development. User Fees Credited with 51% Drop in Average Approval Times Since 1993. In *Tufts Center for the Study of Drug Development Impact Report*; June 1999, Available at: http://www.tufts.edu/med/csdd/impact12.pdf. Accessed 2/22, 2001.
11. Joint Commission on Accreditation of Healthcare Organizations. *Comprehensive Accreditation Manual for Hospitals*; Joint Commission on Accrediation of Healthcare Organizations: Oakbook Terrace, Illinois, 2001.
12. ASHP guidelines on adverse drug reaction monitoring and reporting. Am. J. Health-Syst. Pharm. **1995**, *52*, 417–419.
13. Curran, C.F.; Engle, C. An index of United States Federal Regulations and Guidelines which cover safety surveillance of drugs. Drug Inf. J. **1997**, *31*, 833–841.
14. Chen, M. *Scientific and QA/QC Safety Evaluation Initiatives in OPDRA [Slide Presentation]*; 07/03/00, Available at: http://www.fda.gov/cder/present/dia-62000/opdra/sld040.htm. Accessed 2/22, 2001.
15. US Food and Drug Administration. *MedWatch, The FDA Medical Products Reporting Program*; Available at: http://www.fda.gov/medwatch/what.htm. Accessed 2/22, 2001.
16. US Food and Drug Administration. *The FDA Desk Guide for Adverse Event and Product Problem Reporting*; Available at: http://www.fda.gov/medwatch/report/desk/tpcfinal.htm#toc. Accessed 2/22, 2001.
17. Henkel, J. *MedWatch: FDA's "Heads Up" on Medical Product Safety*; Available at: http://www.fda.gov/fdac/features/1998/698_med.html. Accessed 02/26, 2001.
18. Koch-Weser, J.; Sellers, E.M.; Zacest, R. The ambiguity of adverse drug reactions. Eur. J. Clin. Pharmacol. **1977**, *11* (2), 75–78.
19. Karch, F.E.; Smith, C.L.; Kerzner, B.; Mazzullo, J.M.; Weintraub, M.; Lasagna, L. Adverse drug reactions—a matter of opinion. Clin. Pharmacol. Ther. **1976**, *19* (5 Pt. 1), 489–492.
20. Major Management Challenges and Program Risks. In *United States General Accounting Office*; January, 2001, Available at: http://frwebgate.access.gpo.gov/cgi-bin/useftp.cgi?IPaddress=162.140.64.21&filename=d01247.pdf&directory=/diskb/wais/data/gao. Accessed 2/23, 2001.
21. Sachs, R.M.; Bortnichak, E.A. An evaluation of spontaneous adverse drug reaction monitoring systems. Am. J. Med. **1986**, *81* (5B), 49–55.
22. Serradell, J.; Bjornson, D.C.; Hartzema, A.G. Drug utilization study methodologies: National and international perspectives. Drug Intell. Clin. Pharm. **1987**, *21* (12), 994–1001.
23. Smalley, W.; Shatin, D.; Wysowski, D.K.; Gurwitz, J.; Andrade, S.E.; Goodman, M.; Chan, K.A.; Platt, R.; Schech, S.D.; Ray, W.A. Contraindicated use of cisapride: Impact of food and drug administration regulatory action. JAMA, J. Am. Med. Assoc. **2000**, *284* (23), 3036–3039.
24. Fletcher, A.P. Spontaneous adverse drug reaction reporting vs event monitoring: A comparison. J. R. Soc. Med. **1991**, *84* (6), 341–344.
25. Scott, H.D.; Thacher-Renshaw, A.; Rosenbaum, S.E.; Waters, W.J.; Green, M.; Andrews, L.G.; Faich, G.A. Physician reporting of adverse drug reactions. Results of the Rhode Island Adverse Drug Reaction Reporting Project. JAMA, J. Am. Med. Assoc. **1990**, *263* (13), 1785–1788.
26. Hamrell, M. Current regulations and practices for adverse event reporting: Implications for labeling. Drug Inf. J. **2000**, *34*, 975–980.
27. Wood, A.J.; Stein, C.M.; Woosley, R. Making medicines safer—the need for an independent drug safety board. N. Engl. J. Med. **1998**, *339* (25), 1851–1854.
28. Ray, W.A.; Griffin, M.R.; Avorn, J. Evaluating drugs after their approval for clinical use [see comments]. N. Engl. J. Med. **1993**, *329* (27), 2029–2032.

Prescriptions for Health: The Lowy Report

William McLean
Les Consultants Pharmaceutiques de l'Outaouais, Cantley, Quebec, Canada

INTRODUCTION

Prescriptions for Health, commonly referred to as the Lowy report after its chairman, was initiated by a provincial government after noting astronomical rises in its health care expenses on prescription drugs (20% annually and 500% over the past decade). The report contains 147 recommendations, many directed at government, but many also directed at the health professions, the health science centers, the pharmaceutical manufacturers, distributors, and the public. The report was produced by a committee appointed by the Ontario Ministry of Health. Its members were F. Lowy (Chair), M. Gordon, R. Moulton, R. Spunt, J. Thiessen, D. Webster, and W. Wensley. (Members Thiessen and Wensley are pharmacists.)

MAJOR CONCLUSIONS

The committee found that Ontarians receive ''excellent treatment involving prescription drugs'' and that they were safe, available, and affordable through insurance plans and government-supported programs for the aged and those receiving social welfare. However, the committee went on to note some problem areas:

- Some products paid for by the government plan were suboptimally effective or had unfavorable benefit/risk ratios.
- Physicians are not well prepared educationally for prescribing the many agents available, especially for the elderly; CE programs were found lacking.
- Pharmacists were ''not meeting their full professional potential as members of the health care team.''
- Some citizens do not have access to these drugs, because they are the ''working poor'' without insurance or have extraordinary drug costs because of catastrophic illness.

The committee found less than adequate cooperation between the federal government (responsible for approving new drugs on the market) and provincial governments (responsible for managing drug programs for seniors and those on social assistance).

Recommendations are directed at government, medicine, hospitals, nursing, and the public; the following is a synopsis of those recommendations involving pharmacy:

4.22: The continuation of drug interchangeability whereby pharmacists provide the generic equivalent (defined) of the lowest-cost product acceptable to government. This is followed by a number of other recommendations regarding drug pricing and the development of the concept of ''best available price.''

6.8: That unit-dose/IV admixture be the system of choice for institutional drug delivery in hospitals and that the Ministry aid hospitals in the costs of conversion.

7.4: That physicians and pharmacists jointly study the appropriateness of the quantity of drugs on a prescription and prepare guidelines.

7.7: That postmarketing surveillance studies be established involving large numbers of family physicians and others.

7.9: That ''Choice of Medications—1990'' be made available to all physicians and pharmacies and that it contain guidelines and objective information on drug prescribing.

7.11 and **7.12**: That the hospital formulary system be fostered and that the concept be extended to group clinics, nursing and old age homes, health maintenance organizations, and comprehensive health organizations.

7:13: That generic names be on all prescription labels.

7.15: That a pilot project be established to test the Harvard model of academic detailing.

7.18: That physicians and pharmacists establish formal mechanisms on the issues of patient counseling and communication.

8.1: That the role of pharmacy assistants be defined and ensure that they perform product-oriented tasks while

DOI: 10.1081/E-ECP 120006270

pharmacists concentrate on patient-oriented tasks such as monitoring drug therapy and providing advice to patients and other health professionals.

8.2: That information management systems be established to ensure optimal access to patient profiles and adverse drug interaction programs.

8.3 and **8.4**: That standards for original packaging dispensing be developed.

8.5: That auxiliary labels and other printed information on prescription drugs along with verbal be increased, with reinforcement by the pharmacist.

8.6: That pharmacists pay particular attention to the visually handicapped when labeling.

8.7: That pharmacists label in the language of the patient, with verbal reinforcement.

8.8: That a campaign be aimed at seniors to pick one pharmacy for all services.

8.9: That patient profiles be required in all community pharmacies.

8.10: That portable patient medication records such as the ''smart card'' be developed.

8.11: That guidelines be developed to ensure the pharmacist has access to the patient's diagnosis(es).

8.12: That a campaign be launched to make the public more aware of the services to be expected of a pharmacist, especially those related to the provision of information on proper use.

8.13: That all pharmacies have a private and comfortable consultation area.

8.14 and **8.15**: That pharmacists exert more control over the sale of nonprescription drugs.

8.16: That pilot projects be established to assess alternative reimbursement for the provision of pharmaceutical services not based on the sale of a drug.

8.17: That linkages involving the patient be developed between the medication profile in hospital and in the community.

8.18: That the clinical role of pharmacists be encouraged and expanded in monitoring and intervention as necessary.

8.19: That pilot projects of community drug therapy committees be established, modeled on institutional settings.

8.20: That drug information programs be supported and expanded.

8.22: That a second faculty of pharmacy, with a five-year curriculum, be established.

8.22: That the present faculty of pharmacy (U. Toronto) increase its curriculum to five years with more instruction on patient counseling, therapeutics, drug information, pathophysiology, and geriatrics.

8.24: That the practical training period be restructured.

8.25: That a PharmD program be established in two years.

8.26: That joint teaching occur in medicine and pharmacy on patient-oriented services, including therapeutics, monitoring techniques, and patient counseling.

8.30: That pharmacy services to long-term institutions be improved.

8.31: That pharmacy and nursing work out their roles in counseling.

8.36: That mechanisms be put in place to ensure appropriate patient education regarding treatment plans prior to discharge.

8.37: That all hospital pharmacies require patient profiles.

8.39: That mechanisms be established to compensate hospitals providing pharmacy services to outpatients.

8.41: That funding be increased for hospital pharmacy residencies.

8.42: That standards and ethically appropriate methods of cost containment be established for investigational drugs.

8.45: That community programs for patient compliance be funded.

9.3: That drug utilization review programs be integrated into medical CE.

9.4: That the provincial adverse drug reaction program be supported.

9.6: That approaches be researched to reduce use of hypnotics, sedatives, and tranquilizers.

11.1: That liquid formulations be provided to the elderly with swallowing difficulties.

11.2: That smart cards be introduced to facilitate monitoring and that pharmacy computer systems be programmed to handle problems of the elderly.

11.5: That guidelines be developed to limit hypnotics, sedatives, and tranquilizers for the shortest possible time with no repeats.

INFLUENCE

Eleven years later, many of the recommendations have been carried out, at least partially. Others have not moved significantly since the report's publication. Probably the most important aspect of the report was the laundering of the province's medication system in public, with responsibilities being levied at our government, our universities, and our professions. Yet, although the highlights of this report may have hit the lay press for a day or two, particularly the aspects of drug pricing, there is little long-term effect. The provincial pharmacy associations have taken the report seriously and have initiated many of its recommendations.

The initial concern of the government and the committee, namely, a rapidly climbing total drug bill, has only been partially addressed. Efforts have been made to control drugs such as omeprazole through the ''limited use'' requirement, which places additional (unpaid) work on pharmacists. Significant efforts have also been made in some pilot projects on antibiotic prescribing; the model involves joint educational efforts for community physicians and pharmacists and joint efforts to follow guidelines on a variety of infectious diseases. In areas where these efforts have been started, antibiotic costs have fallen by 10–20%. However, overall in the province, the drug budget continues to gallop ahead at the rate of 12% per year.

The report is as relevant today as it was a decade ago. It recognized the valuable work of clinical pharmacists in institutions and recommended that it be extended not only in institutions but into the community as well. The development of the PharmD programs at the University of Toronto (in Ontario) and the University of British Columbia (farthest western province) has provided further stimulus to this effort. Standards (hospital accreditation, hospital pharmacy, and community pharmacy) also seem to have been influenced or minimally have paralleled the recommendations of this report.

Because Ontario is Canada's most populated province, events that occur there are often considered by other provinces or in national efforts. It is therefore fair to suggest that the Lowy report probably had significant influence beyond the boundaries of Ontario.

CONCLUSION

All in all, we would do well do revisit these recommendations, particularly those dealing with seamless care and computerized records. The patient orientation urged by Lowy for pharmacists has certainly spurred the clinical movement in this part of the world.

BIBLIOGRAPHY

Prescriptions for Health/Recommandations pour la santé. Report of the Pharmaceutical Inquiry/Rapport du Comité enquête de l'Ontario sur les produits pharmaceutiques. Government of Ontario, July 1990.

PROFESSIONAL DEVELOPMENT

Preventive Medicine

Teresa Bassons Boncompte
Carmen Permanyer Munné
Pharmaceutical Association of Catalonia, Barcelona, Spain

Benet Fité Novellas
Pharmacist, Barcelona, Spain

Rafael Guayta Escolies
General Directorate of Public Health, Autonomous Govern of Catalonia, Catalonia, Spain

INTRODUCTION

Health Promotion and Disease Prevention: Definitions and Concepts

In the health sciences field, preventive medicine has traditionally been defined as the ensemble of actions and advice aimed specifically at preventing diseases[1–4] and promoting health, and the process for enabling individuals and communities to increase their control over the determining factors involved in health to improve it.[5] Besides health-restoring activities, there are also those that are fundamentally preventive such as immunization, health education for healthy people, and screenings.

As Hogarth[7] pointed out, over the last few years the meaning of the term *preventive medicine* has expanded considerably. It is being applied more and more to the health activities organized for defending and promoting the population's health at the community level. Those who accept this broad definition of preventive medicine equate it with health promotion. For them, preventive medicine would comprise all the preventive activities of the public health services that concern the individual (vaccination of a child, case finding in a healthy adult, health education through the conversation during the visit, including also certain aspects of pharmaceutical care) or collective (immunization campaigns, collective health check-ups, population screening, mass media health information and education campaigns, etc.). All these actions concern the individual and are carried out by health services, physicians, pharmacists, and qualified nurses, on the basis of knowledge furnished by medical science.

Today, what we understand as health protection is the health promotion and defense actions concerning the environment (environmental health and food hygiene) and it is carried out by public health professionals (veterinarians, pharmacists, biologists, health engineers, etc.) on the basis of scientific principles furnished not only by medicine but also by other sciences (health engineering, architecture, food hygiene and technology, etc.).[6]

In short, as Last[8] pointed out, the term *preventive medicine* in its broadest sense implies a more personal encounter (immunization, screening, health education) between the individual and the medical or pharmaceutical personnel than that entailed in health protection activities (potabilization and fluoridation of mains water, processing of milk, sewage disposal, etc.), since in the latter case the professionals have no contact at all with the user.

THE NATURAL HISTORY OF DISEASE

Any disease or morbid condition is the result of a dynamic process. The causative agents or risk factors present in the environment interact, after a variable period of incubation with the host (whose greater or lesser susceptibility to the disease is conditioned to a large extent by genetic factors) and cause the disease. Leavell and Clark differentiate three defined periods in the natural history of the disease: the prepathogenic period, the pa-

Encyclopedia of Clinical Pharmacy
DOI: 10.1081/E-ECP 120006386

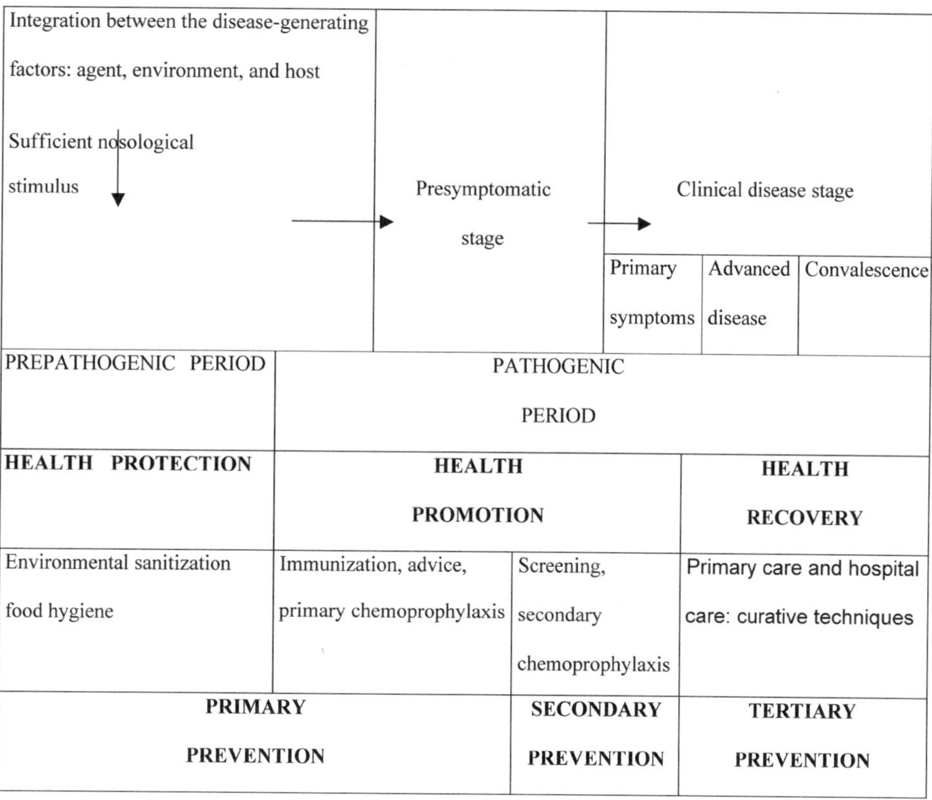

Fig. 1 Natural history of disease and levels of prevention. (From Ref. [9].)

thogenic period, and the result[9] (Fig. 1). This division is of great interest, as we see later in the article with regard to the levels of prevention.

Prepathogenic Period

The prepathogenic period is characterized by the presence of factors that favor or determine the development of the disease. These factors may be environmental (infectious, physical, chemical agents, etc.) and behavioral (overconsumption of fats or carbohydrates, sedentary lifestyle, smoking, excess consumption of alcohol, use of illegal substances, etc.), and they affect the endogenous genetic predisposition[2–4] toward developing the disease.

Some of these factors are necessary (but not sufficient) for the onset of the disease. The clearest example is that of the agents of infectious diseases (bacteria, viruses, etc.). At other times, the factors are not absolutely necessary for disease onset, as this can occur in their absence, although their presence implies an increased likelihood of the condition arising. Such is the case with the risk factors for chronic diseases (high blood pressure, obesity, smoking, hypercholesterolaemia, etc.).[2–4]

It should be noted that the existence of a close statistical relationship between a risk factor and a disease does not mean that all individuals with the risk factor will necessarily develop the disease, or that the absence of this risk factor is any guarantee that the disease will not develop.

Pathogenic Period

The pathogenic period has two stages: the presymptomatic stage and the clinical disease stage. During the presymptomatic period, there are no symptoms or clinical signs but, as a result of the causal stimulus, the anatomical and pathological changes responsible for the disease (arteriosclerosis in the coronary arteries, premalignant disorders in the tissues, etc.)[10] are already under way.

In the clinical stage, the changes in the organs and tissues are already important enough for signs and symptoms of the disease to appear in the patient.[10]

Finally, the result is the last period in the natural history of the disease and reflects the end of the process: death, disability, chronicity, or recovery.

PREVENTIVE ACTIVITIES: LEVELS OF PREVENTION

Application of the concept of ''levels of prevention'' is possible because, as mentioned earlier, all diseases in their natural history present more or less defined periods during each of which it is possible to apply some form of preventive measures.[2]

Although preventive medicine experts do not fully agree on the limits required between each of the levels that can be established, their differences are more semantic than essential.[6,7]

At present, preventive activities are classified into three levels: primary, secondary, and tertiary prevention levels,[7] as shown schematically in Fig. 1.

Primary Level of Prevention

The aim of primary prevention is to reduce the probability of disease occurring. From the epidemiological viewpoint, it aims to lower their incidence.[7] Primary prevention measures act during the prepathogenic period of the disease's natural history before the interaction of the risk factors and/or agents with the host produces the stimulus that triggers the disease.

Currently, a distinction is usually made between two types of primary prevention activities: those for health protection, which are applied to the environment, and those for health promotion and disease prevention, which are applied to people. Preventive activities, then, are really those that affect people; that is, health promotion and disease prevention activities carried out by health professionals and teams in primary health care.

Health protection is oriented toward the environment and includes activities aimed at controlling the causal factors of the disease that are present in the general environment (environmental sanitization), the work environment (safety in the workplace), or in food (food hygiene).

Health promotion and disease prevention are oriented toward people. With disease prevention, the idea is to reduce the incidence of specific diseases by means of specific actions exercised by health professionals, generally within the framework of primary care, although they can also be applied in other areas (schools, factories, etc.). With health promotion, the idea is that people themselves should adopt favorable lifestyles and give up unhealthy habits through educational intervention (health education in schools, factories, primary care centers, and pharmacies, and via the mass media) and legislative measures.

Secondary Level of Prevention

Secondary prevention takes place once the disease-triggering stimulus has occurred and acted, as the only preventive possibility is to interrupt or delay progress of the condition by detaining it with the appropriate, early treatment, with the aim of curing it or preventing it from becoming chronic and preventing the onset of sequelae and invalidity.[11–13] From the epidemiological viewpoint, secondary prevention aims to reduce the prevalence of the condition or disease. The basic assumption of secondary prevention is that early diagnosis and treatment improve disease prognosis and management.

The increase in chronic conditions in developed countries during the twentieth century has aroused a great deal of interest in the early detection of disease. In such diseases, in the majority of which primary prevention is very difficult or impossible, the strategy of the health services must be to aim for early detection to treat them as soon as possible and improve the prognosis. To solve the problem of the delay in detecting chronic diseases the application of different selection processes (screenings) has been proposed to detect them in asymptomatic persons.

Tertiary Level of Prevention

Tertiary prevention comes into play when the pathological lesions are irreversible, the disease is well established and the chronic phase has passed, and whether or not sequelae have appeared (somatic or mental functional limitation).[7] The proposed objective is to slow the course of the disease and alleviate any disability when it arises.[8]

In all this process, the pharmacist has traditionally been associated to health protection activities. More recently, however, everyone agrees on the pharmacist's role in the field of health promotion and disease prevention: the pharmacist should incorporate, in daily practice, screenings and health advice for the people attended to in the pharmacy.

ROLE OF THE COMMUNITY PHARMACIST IN HEALTH PROMOTION AND DISEASE PREVENTION

Pharmaceutical Care

Essentially, the activity of community pharmacists—that is, pharmacists who provide pharmaceutical care for the community from the pharmacy—is based on the purchase, storage, and dispensing of drugs, making up extemporaneous preparations, pharmacotherapeutic follow up of patients, and advice about drugs and health problems.[14,23]

P

Table 1 Importance of community pharmacy in health promotion and disease prevention

1. Community pharmacy is frequently the first contact with the health system: high frequency of contacts with low barriers to access (no appointments, no long waiting time, convenient opening hours, located within communities, information readily on display, etc.).
2. Community pharmacy reaches the broad population: they see a wide range of potential target groups within the patients/clients/users (people with incidental needs, people with chronic illness, healthy people, people from all social strata).
3. Community pharmacy contacts have a good potential to integrate the social environment into counseling (geographic distribution, positioning on the "high street," principal shopping areas, etc.). Contacts take place often and normally on a regular basis and include interactive, communicative, and counseling aspects.
4. There are a large number of experts, pharmacists, and other qualified staff working in community pharmacy, offering a high level of competence and knowledge of medicines.
5. Health issues are already on the agenda of the users of community pharmacies, and the pharmacists are considered experts in a variety of health issues and often actively provide a link to other health professionals, mostly to general practice doctors.
6. Community pharmacy can offer a balanced mix of contributors to curative care, prevention, rehabilitation, and health development.

(From Ref. [23].)

Over the last few years, applying the criteria of pharmaceutical care, pharmacists have extended their activity to embrace other health interventions associated with improving people's quality of life. In 1993, within the framework of the International Pharmacists Congress, FIP'93,[15] the Declaration of Tokyo was approved. In that declaration, pharmacists gave the definition of pharmaceutical care and stated that the patient is the principal beneficiary of their actions. *Pharmaceutical care* is defined as the ensemble of attitudes and actions involving medication and the role of the pharmacist as a health agent. The main implication arising from this viewpoint was the conviction that pharmaceutical care must be developed as an integral part of health services.

The objective of pharmaceutical care in the community is to participate in health promotion, disease prevention, and education. The declaration also proposes specific spheres of action: the rational use of drugs; smoking cessation; and advice on vaccination, hygiene, family planning, AIDS prevention, etc. Even now, pharmacists have still not fully developed their potential as health agents within this health system,[16] but the community pharmacy is of specific importance for health promotion (Table 1).

Specific Programs of Preventive Activities in Community Pharmacy

In practice, health promotion from the community pharmacy has been described in different studies published since 1950, although more frequently since 1980 and, specifically, in studies conducted in England and Canada.

In 1993, the Health Education Authority and the National Pharmaceutical Association published "Health Promotion and the Community Pharmacist",[17] which serves as a practical guide for all health promotion activities that can be carried out by British pharmacists from the community pharmacy.

Along the same lines, another publication to be noted is "Pharmacies and Smoking Cessation",[18] which within the WHO's European program plan of action for a tobacco-free Europe was started in 1993 in pharmacies in Denmark. The result being that, at present, 20% of Danish pharmacies routinely offer a smoking cessation service for their population.

In Catalonia (Spain), in 1997, a consensus among the Department of Health's experts in preventive medicine, nursing, and pharmaceutical scientific associations was attained. It was decided to initiate a process to implement[13] different activities in the community pharmacies in Catalonia (Table 2). These activities have been chosen according to the health plan priorities. The most important point is that these activities are carried out with collaboration with the health care teams of primary health care centers. It was first implemented as a "training trainers" procedure and now these activities are carried out in most pharmacies in Catalonia since 1997.[1,19,20]

Also in Catalonia—Barcelona to be specific—the Pla Farmacèutic d'Educació Sanitària[19] (Pharmaceutical Plan for Health Education), drawn up by the Barcelona

Table 2 Specific preventive activities in Catalonia's community pharmacies

1. Screening for high blood pressure.
2. Screening for excess weight and obesity.
3. Screening for hypercholesterolaemia.
4. Advice on drug use and self-medication.
5. Advice on active immunization.
6. Advice on preventing sexually transmitted diseases and AIDS.
7. Advice on giving up smoking.
8. Dietary advice.
9. Advice on preventing mouth and teeth diseases.
10. Advice on, and prescription of, physical exercise.
11. Advice on excess alcohol consumption.
12. Advice on screening for breast cancer and cervical cancer.

(From Ref. [1].)

Pharmaceutical Association, proved the viability of health education in both the therapeutic and health promotion spheres. Three hundred pharmacists took part in the experiment in Barcelona, the educative sessions being attended by around 10,000 people. The evaluation proved the value of exercise as it helped to significantly increase the knowledge of those attending about the matters discussed.

Also to be noted are the two health education plans drawn up in Spain by the Consejo General de Colegios Farmacéuticos (General Council of Pharmaceutical Associations), with the collaboration of the provincial associations: the Plan de Educación Nutricional (Plan for Education on Nutrition)[21] and the Plan de Educación sobre el medicamento (Plan for Education on Drugs).[22]

Requirements to Carry Out Preventive Activities in the Community Pharmacy

For the pharmacy pharmacist to be able to carry out the health promotion and disease prevention activities with due rigor and consistency, and for these activities to be well coordinated in primary health care, it is necessary to perform the following:

1. Give priority to activities that, in healthy adults, have shown to be effective in the population; that is, they have provided some health benefit for the community and constitute a framework of reference that makes it clear which preventive activities should be carried out at the pharmacy.
2. Adapt the joint development of these activities to the needs and programs established for each basic health area. The idea resulting from all this is that there has to be coordination between all the agencies and professionals involved in primary health care, and the appropriate supports and mechanisms for achieving this coordination.

In practice, to carry out the preventive activities defined, the pharmacist will have to take into account the need for the following:

1. Reinforcing knowledge, skill, and attitude through accredited, continued training.
2. Creating an atmosphere and place for these activities.
3. Having the necessary measuring equipment and resources for the activities.
4. Setting a good example, both the pharmacist and the other personnel, to reinforce the aforementioned healthy atmosphere.
5. Organizing, recording, and establishing protocol for the activities to evaluate them in accordance with the quality criteria established and divulge the results.
6. Using the communication currents established with the appropriate graphic supports for the purpose of exchanging information (Appendix 1) with the primary care team of the health areas.

Table 3 What can health promotion offer to the community pharmacy?

- Increase impact and effectivessnes of pharmacotherapy
- Increase impact and effectivessnes of counseling and health-related advice
- Increase job satisfaction by job enrichment:
  - By fostering liaison with users
  - By improving cooperation with other providers
  - By making visible ''the cognitive service'' information, advice, and counseling
- Make the contribution of community pharmacy evident
  - In the management of drug therapy
  - As an important place for health-related service
- Support the integration of community pharmacy into the primary health care team

(From Ref. [23].)

Applying health promotion principles could be a means to further develop the professional role (Table 3).

CONCLUSION

Health promotions activities in primary health care, in general practice and in community pharmacy, is not yet well documented or evaluated with sufficiently high quality standards. There seems to be a wide variance in the degree to which health-promoting services, projects, and programs are documented, evaluated, and evidence based. In particular, there seems to be need for further work concerning lifestyle counseling, health education, and interventions for health development.[23]

The need to be able to evaluate the effectiveness of the activities is obvious. Therefore, it will be essential to specify, during the process of implementing health promotion at the pharmacy, what the particular objectives are and which activities must be carried out, while ensuring that the former are attainable and the latter feasible on the basis of the resources available. It will be necessary to plan a careful implementation strategy that will ensure the participation of as many professionals as possible while also permitting evaluation of the repercussions, on the population's health, of the integration of health promotion activities into the pharmacist's practice.

P

APPENDIX 1

Servei Català de
la Salut

Consell de Col·legis
Farmacèutics de Catalunya

INTER-CONSULTATION FORM

From — pharmacy ☐ / primary care centre ☐

(full name and address)

telephone number

To — primary care centre ☐ / pharmacy ☐

(full name and address)

telephone number

Patient's data

1st surname — 2nd surname — name

Reason for consultation

Comments on the reason for consultation

Stamp and signature — Date:

Report from the department consulted (if necessary)

Stamp and signature — Date:

NCR paper

REFERENCES

1. White Paper. *Integration of Prevention in the Community Pharmacy*; Catalan Health Service and Pharmaceutical Association of Catalonia, 1997.
2. Salleras, L. Le basi scientifiche della medicina preventiva. (The scientific principles of preventive medicine). Riv. Ital. Med. Comun. **1986**, *4*, 5–11.
3. Salleras, L. La medicina preventiva en la asistencia primaria. (Preventive medicine in primary care). Rev. Sanid. Hig. Publica **1987**, *61*, 545–570.
4. Salleras, L. Salud del adulto. (The Adult's Health). In *Medicina Preventiva y Salud Pública. (Preventive Medicine and Public Health)*; Piédrola, G.G., Del Rey, C.J., Domínguez, C.M., Cortina, G.P., Gálvez, R., Eds.; Ed. Salvat: Barcelona, 1991; 1148–1162.
5. *Terminology for the European Health Policy Conference*; WHO Regional Office for Europe: Copenhagen, 1994.
6. Salleras, S.L. *Educación Sanitaria: Principios, Métodos y Aplicaciones. (Health Education: Principles, Methods and Applications)*; Díaz de Santos, Ed.; Madrid, 1985.
7. Hogarth, J. *Glossary of Health Care Terminology. Public Health in Europe*; WHO Regional Office for Europe: Copenhagen, 1978.
8. *A Dictionary of Epidemiology*; Last, J.M., Ed.; Oxford University Press: New York, 1983.
9. Leavell, H.R.; Clark, E.G. *Textbook of Preventive Medicine*; McGraw Hill: New York, 1953.
10. *Commission on Chronic Illness: Chronic Illness in the United States. Vol. I Prevention of Chronic Illness*; Harvard University Press: Cambridge, Massachusetts, 1957.
11. Fletcher, R.H.; Fletcher, S.V.; Wagner, F.H. Clinical Epidemiology. In *The Essentials*; Williams and Wilkins: Baltimore, 1982.
12. Wilson, J.M.C.; Junger, G. Principios y métodos del examen colectivo para identificar enfermedades. (Principles and Methods of Collective Screening for Identifying Diseases). In *Cuadernos de Salud Pública*; Organización Mundial de la Salud: Ginebra, 1969; Vol. 34.
13. Àlvarez, D.C.; Bolumar, F.; García, B.F. La detección precoz de enfermedades. (Early detection of diseases). Med. Clín. (Barcelona) **1989**, *93*, 221–225.
14. Decision no. 645/96/CEE of the European Parliament and Council, of 29 March 1996, for adopting a Community programme for promotion, information, education and training in health matters within the framework of action in the public health sphere (1996–2000).
15. *53è International Pharmacists' Congress, FIP'93, Tokyo.*
16. *The Role and Function of the Pharmacist in Europe*; STYX Publications: Groningen, 1989.
17. *Health Promotion and the Community Pharmacist*; Health Education Authority: London, 1994.
18. *International Guide: Pharmacies and Smoking Cessations, 1993*; WHO Regional Office for Europe: Denmark, 1993.
19. Bassons, M.T.; Gascon, M.P.; Gamundi, M.C. *Pla farmacèutic d'educació sanitària, Collegi de Farmacèutics de la Província de Barcelona, Abril 1995. (Pharmaceutical plan for health education; Pharmaceutical Association of the Province of Barcelona, April 1995).*
20. Pla de Salut de Catalunya 1996–1998. (Health Plan for Catalonia 1996–1998). In *Generalitat de Catalunya*; Departament de Sanitat i Seguretat Social: Barcelona, Abril 1997.
21. *Informe Plan de Educación Nutricional por el farmacéutico (Plenufar). (Report on the Plan for Nutritional Education Through Pharmacists)*; Consejo General de Colegios Farmacéuticos, 1993.
22. *Guía para el farmacéutico. Texto base del programa de educación sobre el medicamento. (Guide for the pharmacist. Basic text of the drug education programme)*; Consejo General de Colegios Oficiales de Farmacéuticos: Madrid, 1994.
23. Krajic, K. Health Promotion in Primary Health Care: General Practice and Community Pharmacy, A European Project. In *Opportunities, Models, Experiences*; Ludwig Boltzmann-Institute for the Sociology of Health and Medicine, April 2001.

Primary Care, Clinical Pharmacy Services in (ACCP)

American College of Clinical Pharmacy
Kansas City, Missouri, U.S.A.

INTRODUCTION

Health care reform has renewed the interest in primary care. Major problems with America's health care system include escalating health care costs, maldistribution of health care providers into urban areas, lack of health care insurance, and the excessive utilization of specialists. These issues have assured that health care reform will take place. At the time of this report, the nature and format of a reform plan has not been determined. Nevertheless, many agencies are rapidly evaluating their current health care coverage and are preparing for the inevitable reform that will take place.

There have been disproportionately high numbers of medical specialists compared with generalists since the 1960s.[1] Worldwide, there are approximately five to six generalists for every subspecialist. For various reasons, perhaps most importantly, economics, this ratio is reversed in the American health care system. For well over 20 years, governmental agencies and medical academicians have tried to increase the numbers of primary care providers without significant success. With millions of underinsured Americans, the provision of primary care has become a national priority. Clinical pharmacists must become more involved in the provision of primary patient care. Most clinical pharmacists, however, have not viewed themselves as primary care providers and, therefore, may not feel adequately prepared to become a member of an interdisciplinary primary care team.

This article is an extension of a previous American College of Clinical Pharmacy (ACCP) White Paper on Clinical Pharmacy Practice in the Noninstitutional Setting.[2] That white paper described the functions that should be expected of clinical pharmacists in ambulatory care settings. This article assists practitioners and administrators who want to establish and evaluate services in ambulatory care and primary care settings. This article presents approaches to define the scope of a pharmacist's practice and obtain clinical privileges, evaluate the process of delivering care, evaluate patient outcomes related to pharmacotherapeutic decisions, and define the legal implications of providing primary patient care.

DEFINITIONS

There is considerable confusion in pharmacy concerning current definitions of practice sites and practice philosophies. *Ambulatory care* includes all health-related services in which patients walk to seek their care.[1] These services may be provided in emergency rooms, urgent centers, private offices, primary care clinics, specialty and subspecialty clinics, and community pharmacies.

Primary care is a subset of ambulatory care with unique features and philosophies.[1] (By definition, inpatient care is never a primary level of care.) One set of definitions,[3] suggests that primary care is a form of care that includes:

1. ''First-contact'' care, serving as a point-of-entry for the patient into the health care system;
2. Continuity by virtue of caring for patients over a period of time, both in sickness and in health;
3. Comprehensive care, drawing from all the traditional major disciplines (medical specialties, nutrition, and social) for its functional content;
4. The assumption of continuing responsibility for individual patient follow-up and community health problems; and
5. Highly personalized care.

Dr. Elizabeth Short of the Veterans Administration (VA) Central Office has stated, ''Primary care is the coordinated, interdisciplinary provision of health care that consists of health promotion, disease prevention, comprehensive management of acute and chronic medical and mental health conditions, and patient education. A primary care physician coordinates access to and integration of other components of health care, such as inpatient, long-term, or subspecialty care, and psychosocial support. Un-

From *Pharmacotherapy* 1994, 14(6):743–758, with permission of the American College of Clinical Pharmacy.

DOI: 10.1081/E-ECP 120006411

der primary care, a provider or provider team is the primary source of a patient's care, and the place that a patient turns to for health care information and support.''[4]

The key feature of primary care clinicians is that they handle a wide range of medical conditions. They serve as the entry point into the health care system and decide on referral or triage to secondary or tertiary levels of care. Specialty clinics provide ambulatory care, and, in many cases, some primary care. Most specialists (e.g., cardiology, neurology, nephrology, etc.), however, are considered secondary levels of care. These secondary and tertiary levels of care should be utilized when a problem is beyond the expertise of the primary care clinician. The typical family practice physician or general internist cares for well over 90% of problems that present to them. There is a small percentage of problems that would require referral to secondary or tertiary care. Even when a patient is referred for a specific problem, the primary care clinician should maintain overall care for the patient and coordinate all other aspects of care. This continuity implies chronic care and preventive care that are more conducive to long-term assessments of patient outcomes than can be achieved with acute illness managed in the inpatient setting.[3]

Clinical pharmacists and pharmacotherapy specialists provide care in a wide variety of ambulatory care and primary care settings.[1,2] There are two major types of practice that are very distinct. While currently more common in structured settings such as hospitals and health maintenance organizations, primary care is increasingly being provided in many settings including community pharmacies. The first type of practice is one in which the pharmacist is *independently responsible for providing primary care*, typically between regularly scheduled physician visits. This includes conducting complete histories; obtaining objective information including physical assessment and ordering laboratory tests; starting, stopping, or changing drug therapy; and determining the appropriate timing of follow-up visits. These activities are common in pharmacist-managed clinics in the Indian Health Service, medical centers, and VA hospitals, including hypertension, diabetes, hyperlipidemia, anticoagulation, and pharmacy service clinics. These activities are in contrast to those provided by other professionals such as physician assistants or nurse practitioners who may perform functions traditionally performed by a physician.

The second type of setting is an *interdisciplinary team approach* to care of the patient where the pharmacist sees patients with physicians. Pharmacists who work in such teams assist with care at the same time other health professionals see the patient. In this setting, they may have independent patient care activities but these would not be as extensive as are generally seen in pharmacist-managed clinics. These settings would include family practice offices, general medicine clinics, or pediatric clinics.

ESTABLISHING AN AMBULATORY CARE OR PRIMARY CARE PRACTICE

Obtaining Clinical Privileges

Prior to any patient intervention, it is essential that the clinical pharmacist has in place a document that outlines specifically the practitioner's scope of practice.[5] It is important that the scope of this document be sufficient to allow the clinical pharmacist to function as a member of an interdisciplinary primary care team. This document could be in the form of clinical privileges or a scope of practice statement (Appendix 1). This approach could be used for developing a practice for a new practitioner or used for a previous clinician who has not formally obtained scope of practice privileges. If the facility is an organized health care center (hospital or managed-care organization), it would be worthwhile to review the facility's guidelines for clinical privileges granted to the physician's assistants and/or nurse practitioners if these are available. Depending on the institution, approval is required by the Chief of Pharmacy, Chief of Staff, Clinical Executive Board, and the Institutional Director. Once these privileges are established, only then can the clinical pharmacist provide primary patient care (e.g., in pharmacist-managed clinics).

In the past, formal guidelines to obtain clinical privileges were often not commonly developed in private practice or other settings such as outpatient family practice settings. However, clinical pharmacists in private practice, community pharmacies, family practice residencies, and health maintenance organizations should develop these guidelines for their clinical pharmacy practitioners to ensure quality and evaluate performance. For clinicians in these settings, it is less likely that formalized arrangements for scope of practice privileges exist with physicians or other health professionals. However, similar templates as those for institutions (Appendix 1) could be used and modified for these settings.

The application form in Appendix 1 also requests data on whether the clinical pharmacist is board certified in pharmacotherapy or another specialty. Board certification should be considered strongly desirable, if not required. At the present time, the most appropriate specialty certification process for ambulatory or primary care pharmacists would be certification in pharmacotherapy. This would be analogous to physician certification in the broad-based specialty of family practice. Board certification in pharmacy will be increasingly important and it

should be achieved by all ambulatory care/primary care pharmacy practitioners who provide the services outlined in this report.

QUALITY OF CARE PROVIDED BY PHARMACISTS IN PRIMARY CARE: EVALUATING PROCESS AND OUTCOMES OF PATIENT CARE

A comprehensive discussion of quality of care assessments is beyond the scope of this paper. This area of assessment, however, will become increasingly important in the near future. This report is intended to provide the pharmacist who practices in ambulatory care with an understanding of basic principles used to assess quality. For more in-depth reviews in this area, the reader is referred to the references and the Appendixes.

There is a great deal of interest in measuring or assessing patient outcomes. As Donabedian points out, however, outcomes can only be assessed within the overall context of health care.[6] For instance, the therapy that a pharmacist selects may have minimal influence, or perhaps even a detrimental influence on patient care, depending on the care of other practitioners, demographic factors, and the interpersonal relationship. Donabedian maintains that quality can only be assessed by examining the three components: structure, process, and outcome.[6] He suggests that there must be a knowledge of how structure and process are linked, and how outcome and process are linked before quality assessments can be made. Structure not only refers to the facility, its services and its location, but also the number and characteristics of the providers. For providers this would mean whether they are in solo or group practice and whether they are board certified.[6,7] For physicians it has been shown that board certification is a predictor of good process, but only by implication, of good outcomes. Process refers to what is done for the patient in providing care.[6,7] This includes making diagnostic and treatment decisions. Outcome refers to what happens to the patient and this may include the patients' knowledge or satisfaction with care.

Lohr and Brook have stated that quality of care is composed of both technical care and the art of care.[7] The art of care includes the practitioner's ability to provide reassurance, obvious concern of the patient's well-being, good counseling, and sensitivity to the patient. As examples, they cite whether the provider introduces himself to the patient, refers to the patient specifically by name, announces and/or explains activities before or while doing them (such as physical examination), and says goodbye to the patient. Obviously, these are all critical factors to address if patient satisfaction is being assessed. Providing these personal services is not new but it is increasingly important when patient satisfaction drives third-party contracts in managed care. Pharmacists must provide these personal levels of care if they truly are delivering pharmaceutical care.

Evaluating the Process of Delivering Care

Performing quality assurance evaluations of specific pharmacists' performance does not measure patient outcomes, but rather, the process of delivering care. However, providing an acceptable or ideal process (or standard of care) should, by implication, create an environment conducive to better patient outcomes. However, to move from evaluating process to evaluating outcome, other specific tools must be used (see below). Appendix 2 is an example of a quality assurance form that might be used in a pharmacist-managed primary care clinic.

Guidelines are being developed for a wide range of disease states and conditions. These essentially describe processes for delivering care. They can be used by the individual clinician to prospectively guide appropriate therapy. In contrast, they can be used retrospectively as a quality assurance measure. Appendix 3 lists 12 disease states that are critical to outpatient primary care, and that are currently the most common conditions cared for by pharmacists in primary care settings. While these are not all-inclusive, they provide examples that can be followed in other therapeutic areas. Where possible, nationally accepted clinical practice guidelines are provided for each of these disease states. It is imperative that clinical pharmacy practitioners be aware of nationally accepted guidelines for specific conditions they may treat in their settings. The importance of this is discussed below.

The Agency for Health Care Policy and Research (AHCPR) was created by Congress to be the successor of the National Center for Health Services Research. This agency explores medical conditions that affect large populations, have multiple therapeutic interventions, and have a large economic impact. Through the Medical Treatment Effectiveness Program (MEDTEP), the AHCPR examines variations in health care practices on patient outcomes.[8] The MEDTEP involves patient outcomes research, clinical guideline development, scientific data development, and research dissemination. The agency has supported the development of numerous guidelines such as the guidelines for depression and for angina.[9] The AHCPR is currently developing practice guidelines for the effective therapeutic management of asthma, arthritis, hypertension, and congestive heart failure. In addition to this federal agency, private groups such as the American College of Physicians, the American Medical Association, the BlueCross BlueShield Association, and other specialty societies are developing new treatment guidelines.

It is important to note that AHCPR is not a regulatory agency and is not involved with reimbursement. Application of the guidelines is not enforced by the government. Using these guidelines that were prepared by multidisciplinary panels of experts may allow primary care providers to deliver scientifically sound care to the patient.

There are also medical-legal issues pertaining to clinical practice guidelines developed by specialty societies. The general counsels who are involved with these issues, private practice attorneys, and the counsel of the American Medical Association generally believe that following established clinical practice guidelines would be a strong defense in malpractice cases. However, if a practitioner deviated widely from these guidelines, he or she would need to have a strong rationale, documented in the patient's record, to support the use of an alternate regimen.

A major issue that needs to be addressed is what standards or methodologies should be followed when guidelines are developed.[10] A structured, systematic, science-based approach should be used whenever developing these guidelines. The Institute of Medicine has identified the necessary characteristics which would enhance a guideline's effectiveness: sensitivity, specificity, patient responsiveness, readability, minimal intrusiveness, feasibility, and computer compatibility.[11,12] If guideline development followed these scientific methods, it would be difficult to criticize the process.

In contrast to good guideline development, the determination of whether guidelines are useful depends upon their readability, computer compatibility, and other factors. Outcomes management takes the results of the outcomes research and incorporates them into clinical practice guidelines to theoretically help ensure all patients receive the most effective treatment available.[11,12]

Assessing Health Outcomes

Another objective of this report is to determine the best method to measure the impact of pharmacotherapeutic decisions made by clinical pharmacists on patient outcomes. There are several approaches that can be used to assess outcomes. These include disease- or treatment-related outcomes (e.g., blood pressure, seizure frequency, medication adherence, target serum concentrations). The Task Force felt that it was not appropriate for this report to delineate specific clinical outcomes such as level of blood pressure control or serum drug concentrations. While these are important outcome measures, the Task Force wanted to highlight optimum methods for documenting positive outcomes of clinical pharmacy interventions. To keep in step with health care reform, a good method of assessing the impact of therapy on a specific chronic disease is health-related quality-of-life (HRQL) outcome measures. The pharmaceutical industry, the medical profession, and governmental agencies have shown increasing interest in assessing new measures of a drug's overall effectiveness. Quality of life (QOL) will be considered as seriously as safety and efficacy when evaluating response to therapy.

Even when primary care providers follow accepted clinical practice guidelines, there is no assurance of a favorable outcome. That is why it is important for the clinical pharmacist to understand and use appropriate, clinically relevant outcome measures to quantify the impact of their interventions.[12] Bungay and Wagner argue that HRQL outcome measures should assess physical, social, and role functioning; emotional distress and well-being; general health perceptions; and energy and fatigue.[13] They also stress that the assessment of health status must be integrated into the care of patients. HRQL measures can be used to assess a population with a specific disease, or as a research method to examine how changes in process affect outcomes.[14] The current challenge is to develop tools and operations that can be used in the office setting to evaluate care, and hopefully direct treatment for individual patients. It is critical, however, that these assessments be performed while considering the patient mix, timing of data collection (timing during the evolution of a disease process), patient characteristics, and measurement properties. The reader is referred to a more comprehensive discussion of these issues.[14,15] We will briefly discuss the importance of HRQL outcomes, the types of instruments available, and how to choose a specific instrument for a specific patient population.

Quality of life includes many issues occurring in a person's life, such as health status, job satisfaction, family issues, and overall well-being.[6,7,14,15] Since these are nonspecific, this measurement may not be the best indicator of positive or negative pharmacotherapeutic interventions made by a clinical pharmacist. Health-related quality-of-life assesses those aspects of a patient's life specifically related to physical and mental well-being. ''Hard data'' such as treadmill time in patients with heart failure may be of interest to clinicians, but is of little value to the patients. Frequently, ''hard data'' correlate poorly with the patient's actual functional status. An additional reason to add HRQL instruments to clinical outcomes measurements pertains to the phenomenon that patients with the same medical condition often respond differently to therapy. HRQL is a complementary method of meas-

uring the impact of therapy on chronic disease. Thus, HRQL might be used in tandem with explicit or implicit quality assurance review that is measuring the process of delivering care along with clinical outcome measures (e.g., blood pressure for hypertension or peak flow measures for asthma).[6]

Primary care providers, patients, and health care administrators are interested in HRQL outcomes because they are a method to measure the impact of therapy on the disease process. Hospital administrators and other policy makers have a high stake in these issues because payers are beginning to use HRQL data in their reimbursement policies.[16]

It is imperative that clinical pharmacists involved in providing primary patient care have a good working knowledge of HRQL instruments and are competent in choosing the appropriate methods to assess their interventions. Generic instruments and disease-specific instruments are the two general means by which HRQL can be measured.

The first modern health status questionnaires were very long, but their results were well validated. The Sickness Impact Profile (SIP) is an example of an early profile. It includes a physical dimension and a psychosocial dimension.[14–18] A dimension is a quality or aspect that is a component of health. The SIP also includes five independent categories including sleep, rest, eating, work, and home management, as well as recreation and pastimes. More recently, shorter profiles have been developed such as the Nottingham Health Profile and the Medical Outcomes Study (MOS) 36-Item Health Survey (SF-36) (Appendix 4).

The other approach in assessing HRQL is to focus on the aspects of health status that are specific to a particular area of interest or disease. By narrowing the area being observed, it is possible to gain increased responsiveness to changes in therapeutic interventions. Responsiveness relates to the instrument's ability to detect changes in the patient's status over time.[15–17] The instrument may be specific to a particular disease state (e.g., angina or arthritis), or to a population (e.g., the frail elderly), or to a physiologic problem (e.g., pain). In addition to responsiveness, these disease-specific instruments evaluate areas routinely addressed by primary care providers.[17]

Most generic and specific HRQL measures used today have been validated, but not in all populations. If an instrument is valid, it has been statistically determined to measure what it is intended to measure. Compendia of available measures, including critical reviews, can facilitate the choice of an instrument for a specific setting or purpose.[18] Appendix 4 contains some generic health profile instruments that can be used for various disease states, and, where possible, a disease-specific instrument was listed.

The Health Outcomes Institute has developed and validated several outcome instruments that can be used to evaluate patient outcomes following interventions by pharmacists.[19] These include hypertension/lipids, angina, asthma, chronic obstructive pulmonary disease, chronic sinusitis, hip replacement, hip fracture, depression, low back pain, osteoarthritis, alcohol abuse, stroke, rheumatoid arthritis, and prostatism (Appendix 4). The Health Outcomes Institute is located at 2001 Killebrew Drive, Suite 122, Bloomington, MN 55425; telephone (612) 858-9188.

Example

If pharmacists were providing primary care for hypertensive patients and wanted to compare the results of an intervention, they should first provide interventions based upon established therapeutic guidelines for treating hypertension such as those outlined by the Fifth Joint National Committee on Detection, Evaluation, and Treatment of Hypertension (JNC-V). With each patient encounter, they would collect the data in Appendix 2. These two procedures would ensure that the pharmacist is providing an appropriate process of care.

Prior to the intervention, the pharmacist would assess health outcome measures such as blood pressure, current medication adherence, and forms such as a general form (e.g., SF-36) and a disease-specific form (e.g., Hypertension/Lipid Form 5.1) (Appendix 4). After the pharmacist intervention, a predetermined period of time must elapse before these questionnaires can be repeated (e.g., 6–12 mo). The questionnaires and blood pressure assessments are then repeated and it is determined whether the intervention had any effect on the patient outcome.

Recommendations

1. When appropriate, generic assessment measures should be used to develop methods to evaluate overall patient outcomes after pharmacists' interventions. However, since these may not be the most appropriate techniques for specific pharmacotherapy interventions, disease-specific methods should also be considered.
2. Centers or individuals who want to evaluate patient outcomes that result from pharmacists' interventions should utilize instruments that have been developed and evaluated by experts.
3. When appropriate, patient outcomes after pharmacists' interventions and primary care activities should be assessed with disease-specific instruments that have been validated appropriately.

4. The choice of generic and/or disease-specific instrument(s), should be made by the multidisciplinary team when patient care is being assessed.

THE PROFESSIONAL RELATIONSHIP

Since primary care often involves an interdisciplinary team, many health care professionals provide care to the patient. Clinical pharmacists need to understand the legal implications of the care they provide, or of their patient interventions. Some of the medical-legal concepts that need to be addressed include: what establishes a professional relationship, how to terminate this relationship, abandonment, and harmful neglect. These issues are rooted in both tort and contract law. In actions of negligence, four legal elements must be addressed: duty, breach of this duty, damage, and causation.[20] In determining a pharmacist's duty, the central question is whether a particular conduct is a standard of pharmaceutical care. This is often quite controversial in that there may be certain activities, such as duty to warm, that are not accepted by all courts as a standard of care for pharmacists. If it is decided that the action is not a standard of care, the pharmacist cannot be held negligent. If it is, then the issues are whether the pharmacist breached that duty (standard of care), whether the patient was harmed (and to what extent), and whether the breach of duty caused the harm.

The essence of primary care is taking responsibility for the care of the patient to improve outcomes. Therefore, the following discussion is essential for the pharmacist–patient relationship in primary care.

Duty to Care

It is the pharmacist–patient relationship that gives rise to the pharmacist's duty to care.[21] The pharmacist–patient relationship usually involves an expressed or implied contractual agreement whereby the pharmacist offers to treat the patient with proper professional skill and the patient agrees to pay for such treatment. The pharmacist has the responsibility for practicing all facets of the profession competently. This could involve, for example, drug distribution, providing primary care, patient monitoring, patient and provider consultation/education, and other activities. As a result, the legal principles governing contract formation apply to the establishment of the pharmacist–patient relationship. At issue, however, is whether this contractual arrangement really exists between the pharmacist and the patient or whether it is between the pharmacist and some other entity such as the physician. The answer may depend on what the pharmacist is actually doing. If the pharmacist provides primary care functions, the contractual arrangement should be viewed as being with the patient.

Terminating the Relationship

If a pharmacist–patient relationship exists and it is to be terminated, the pharmacist must give the patient sufficient notice so that he may secure other professional care.[21] Even though the pharmacist's powers in terminating the relationship are limited, the patient has broad powers in terminating the relationship. The patient is free to unilaterally terminate the relationship at any time. From the moment the pharmacist is dismissed or discharged, he is relieved of all future professional responsibility to the patient.

Abandonment

Once established, the pharmacist–patient relationship imposes a duty of care upon the pharmacist that continues as long as attention is required, unless the pharmacist gives sufficient notice of termination or is discharged.[21] While this case law currently only applies to physicians, pharmacists who assume a caregiver role would also be subject to this duty. To recover on the theory of abandonment, the plaintiff must prove the following:

a) Existence of a pharmacist–patient relationship.
b) Unilateral severance by the pharmacist without reasonable notice and without providing an adequate substitute.
c) Necessity of continuing pharmaceutical attention.
d) Proximate cause.
e) Damages.

A pharmacist is immune from the abandonment charge when the patient voluntarily chooses not to return or discharges the pharmacist.

Abandonment may thus occur in two ways: through explicit withdrawal from a case or failure to attend the patient with due diligence. If the pharmacist fails to attend the patient with due diligence, he may also be liable under negligence principles. If he prematurely terminates the relationship despite the patient's continued need for care, he may also have abandoned the patient. The pharmacist has a definite right to withdraw from the case provided he gives the patient reasonable notice so that a patient may secure other attention. Failure by the patient to cooperate with the pharmacist may justify termination of the professional relationship by the pharmacist. The pharmacist is not justified in abandoning the patient unless the patient obstinately refuses treatment. Differences of

opinion on the factors surrounding a case may occur. Therefore, it would be prudent for the pharmacist to document carefully events and to offer to obtain a substitute clinician for the patient, and even then alternative care arrangements must be made.

Harmful Neglect

Decisions concerning frequency of patient visits are an important medical-legal issue. Pharmacists can be held liable for harmful neglect, an act of negligence involving nondiligent care of the patient. Courts have ruled, ''A physician is not chargeable with neglect on account of the intervals elapsing between visits, where the injury requires no attention during the intervals, but is negligent where attention is required.''[22]

The establishment for ''proximate'' or ''legal'' causation is the first step.[23] A factual link between the pharmacist's conduct and the patient's injury and whether the pharmacist could have foreseen the harm must be determined. The plaintiff must compare what did occur with what would have occurred if contrary-to-fact conditions existed. As an example, in a case involving a pharmacist, if the pharmacist had provided more frequent visits, would a more favorable outcome have resulted? The plaintiff would have to prove, by a preponderance of the evidence, that the infrequency of visits was the cause of damages to him. In addition, even if it is established that the pharmacist's conduct caused the patient's injury, a question of forseeability may be raised. In general, unless the pharmacist could have foreseen that harm would occur, there will be no liability. The issue of foreseeability would most likely be part of the determination of duty. Many of the consensus statements and guidelines included in this paper describe appropriate intervals of follow up that, if followed, might reduce the liability of clinical pharmacists.

The jury would be instructed not to consider a pharmacist's workload as a legitimate determination of frequency of follow-up care. Pharmacists should be aware that having more patients than time allows does not relieve them of their responsibility to provide proper follow-up care.

Pharmacists do not carry the sole burden of what happens to their patients during intervals between appointments. Patients also have responsibilities with regard to the management of their illnesses. The Supreme Court of Maine ruled ''it is the duty of a patient to follow the reasonable instructions and submit to the reasonable treatment prescribed by his physician or surgeon.''[24] If the patient fails in his duty and his conduct directly contributes to the injury, he may be precluded from or limited in seeking damages. Some state laws provide for contributory negligence where any negligence by the plaintiff completely bans recovery. Other states have comparative negligence where blame is essentially apportioned between the plaintiff and defendant.

In summary, pharmacists' decisions regarding follow-up care are subject to legal scrutiny. As standard guidelines concerning the appropriate frequency of follow-up visits for outpatient management of most diseases are not routinely available, clinicians are vulnerable to actions for harmful neglect. Lacking such standards, a jury of laypersons listens to ''expert testimony'' and decides whether appropriate care was given. Busy workloads of pharmacists who service a large number of patients are not considered a defense against harmful neglect. If a practitioner cannot provide adequate care to each patient, an equally competent substitute must be named. Finally, it is essential to remember that the patient has obligations in the management of his own health. To document appropriate pharmacists' advice to patients, written instructions should be provided that clearly and specifically outline what the patient should do during intervals between visits, and full and appropriate records of patient visits must be maintained.

SUMMARY

This Task Force report is designed to provide administrators and pharmacy practitioners with recommendations that assist them in establishing and evaluating pharmacy services and assessing patient outcomes in ambulatory/primary care. Each setting will have unique features requiring specific processes be tailored to that institution or clinic. By utilizing the outcome instruments, practice guidelines, and other materials listed in this report, the clinician should be able to establish a valuable practice in most primary care settings.

ACKNOWLEDGMENTS

This article was written in collaboration with Joseph L. Fink III, JD, and Francis B. Palumbo, PhD, JD, who assisted with the legal discussions and Kathleen M. Bungay, PharmD, and Eleanor Perfetto, PhD, who provided significant guidance and advice with the health care process and outcomes sections.

The article was written by the following ACCP Task Force on Ambulatory Care Clinical Pharmacy Practice: William Linn, Pharm.D. (Chair), Barry L. Carter, Pharm.D., FCCP, BCPS (Board Liaison); Betsy Carlisle, Pharm.D.; Allan Ellsworth, Pharm.D., BCPS; Timothy Ives, Pharm.D., BCPS; Susan Maddux, Pharm.D., BCPS; Patricia Taber, Pharm.D, BCPS. Approved by the ACCP Board of Regents on May 4, 1994.

Appendix 1 Application for scope of practice

Clinical Pharmacy Specialists

Name: ____________________
Position on hospital staff: ____________________
Pharmacy school(s): ____________________
Date(s) of graduation: ____________________
Graduate degree: ____________________
Graduation: ____________________
Board certified in pharmacotherapy?: Yes____ No____
Board certified in other pharmacy specialty?: Yes____ No____ NA____
Specialty area: ____________________
States currently licensed: ____________________

The following are the clinical scope of practices granted to you as a member of the staff of the __________ Hospital (Clinic), located in __________ (city), _____ (state). These determinations were made through a thorough review of your education, training, and experience, and demonstrated competence by the Professional Standards Board and approved by the Director. If you change positions and/or if your duties change (i.e., a geriatric clinical pharmacist moves to medical oncology), then you must reapply for practices specific to that area.

Areas of Practice:

A = Ambulatory Care

A. Routine duties: Routine duties are defined as those duties that are performed on a regular, repetitive basis.

(1) Category A-1: Routine duties that require review by the physician supervisor who will note concurrence or addendum as indicated. Countersignature of the medical record is required within 24 hours.

	Requested
• taking and recording verbal orders from physicians	A

(2) Category A-2: Routine duties that do not require review by the physician supervisor unless so indicated. These duties will be reviewed by the physician supervisor on a regular basis through a random sampling process. Results of this review will be discussed with the clinical pharmacist as appropriate.

	Requested
• provision of formal written consultations upon request in the areas of pharmacotherapy and pharmacokinetics	A
• provision of written initial assessments in the progress notes	A
• provision of follow-up notes within the progress notes	A
• taking medication/therapeutic histories	A
• measuring vital signs and performing physical examinations of relevant organ systems for the purpose of monitoring drug therapy	A
• collecting laboratory specimens (i.e., drawing blood)	A
• order the following noninvasive tests:	
(a) laboratory tests (e.g., PT, CBC)	A
(b) EKGs	A
(c) Holter monitors	A
(d) PFTs	A
(e) echocardiograms	A
(f) x-rays (e.g., CXR)	A
• order appropriate consultations from the following services:	
(a) dental	A
(b) dietetics	A
(c) medical specialties	A
(d) psychiatry	A
(e) psychology	A
(f) radiology	A
(g) social work	A
(h) surgical specialties (old problems)	A

(*Continued*)

P

Appendix 1 Application for scope of practice (*Continued*)

B. Non-Routine/Non-Emergency Duties:

	Requested	
• authority to write prescriptions for medication refills for medical problems that are stable in patients followed in outpatient clinics. The clinical pharmacist is not authorized to write prescriptions that are used to initiate any form of drug therapy.	A	
• authority to make adjustments in dosage as clinically indicated for a period of up to 3 months between physician visits using the following classes of drugs:		
1. antihistamine drugs	A	
2. antiinfective agents	A	
3. antineoplastic agents		{indicates not applicable to this ambulatory care pharmacist}
4. autonomic drugs	A	
5. blood formation and coagulation	A	
6. cardiovascular drugs	A	
7. central nervous system agents	A	
8. gastrointestinal drugs	A	
9. hormones and synthetic substitutes	A	
10. respiratory smooth muscle relaxants	A	
• limited authorization to approve the use of restricted or nonformulary medications when the use of such agents is within the established guidelines or approved criteria for use at this facility (i.e., antibiotics, chemotherapy)	A	

C. Emergency Duties: Carried out for patients in life-threatening situations where a physician is not immediately available. The clinical pharmacist initiates this activity but makes every effort to summon a physician as soon as possible (i.e., cardiopulmonary resuscitation, and, if advanced cardiac life support-certified, electrodefibrillation).

D. Miscellaneous Duties: Those duties that do not fall into the first category.

• conduct clinical research protocols ____A____

I do hereby request the above outlined scope of practices. I have read and agree to abide by the bylaws of the ____________ Hospital.

Signature of applicant ______________________ Date ______________

Signature of physician supervisor ______________________ Date ______________

Chief, Pharmacy Service ______________________ Date ______________

Chief of Staff ______________________ Date ______________

Director ______________________ Date ______________

Appendix 2 Evaluating process of care: Example quality assurance in primary care

Medical Records will be reviewed on a quarterly basis. Twenty-five charts will be randomly selected and reviewed for the following items:

1. Progress notes written in an appropriate S.O.A.P. format.
2. Determine if the subjective and objective information is consistent with the assessment and plan.
3. Past medical history and family history is obtained at least once for each patient.
4. Social, diet, and exercise history is recorded at least every 4 months.
5. Medication history recorded at least once.
6. Current prescription and nonprescription medication recorded on each visit.
7. Compliance is assessed on each visit.
8. Each visit contains thorough questioning concerning disease control, signs or symptoms of disease progression or new complications, and signs or symptoms of adverse reactions.
9. Each visit documents appropriate objective information such as laboratory, physical assessment data, vital signs, etc.
10. All patient counseling concerning drug therapy, compliance, diet, exercise, and other lifestyle factors are recorded.

(*Continued*)

Appendix 2 Evaluating process of care: Example quality assurance in primary care (*Continued*)

11. Therapeutic goals are clearly stated.
12. Appropriate recommendations and drug regimen changes are made and documented in the plan.
13. Documentation of any actions that are beyond the scope of practice that were authorized by a physician.
14. Appropriate timing of follow-up visit is included in every plan.

Appendix 3 Treatment guidelines and review articles

Hypertension guidelines

1. The fifth report of the Joint National Committee on Detection, Evaluation, and Treatment of High Blood Pressure (JNC V). Arch Intern Med 1993; 153: 154–83.
2. Frohlich ED, Apstein C, Chobanian AV, et al. The heart in hypertension. N Engl J Med 1992; 327: 998–1008.
3. National High Blood Pressure Education Program. National High Blood Pressure Education Program Working Group report on hypertension and chronic renal failure. Arch Intern Med 1991; 151: 1280–7.
4. National High Blood Pressure Education Program. National High Blood Pressure Education Program Working Group report on primary prevention of hypertension. Arch Intern Med 1993; 153: 186–208.
5. National High Blood Pressure Education Program Working Group report on the heart in hypertension. National High Blood Pressure Education Program. U.S. Department of Health and Human Services. NIH Publication No. 91-3033. September 1991.
6. National High Blood Pressure Education Program Working Group. National High Blood Pressure Education Program Working Group report on hypertension in the elderly. Hypertension 1994; 23: 275–85.
7. National High Blood Pressure Education Program Working Group. National High Blood Pressure Education Program Working Group report on hypertension in diabetes. Hypertension 1994; 23: 145–58.
8. Bussey HI, Hawkins DW. Hypertension. In: Carter B, Angaran D, Sisca T, eds. Pharmacotherapy self-assessment program, 1st edition. Kansas City: American College of Clinical Pharmacy, 1991: 1–26.
9. Bussey HI, Hawkins DW. Hypertension. In: Carter BL, Angaran DM, Lake KD, Raebel MA, eds. Pharmacotherapy self-assessment program, 2nd edition. Kansas City: American College of Clinical Pharmacy, In press.

Diabetes guidelines

10. Lebovitz HE, Clark CM, DeFronzo RA, et al., for the American Diabetes Association Clinical Education Program. Physician's guide to non-insulin-dependent (type II) diabetes. Diagnosis and treatment, 2nd edition. American Diabetes Association, 1990.
11. American Diabetes Association. Clinical practice recommendations, 1992–1993. Diabetes Care 1993; 16 (suppl 2): 1–113.
12. The Diabetes Control and Complications Trial Research Group. The effect of intensive treatment of diabetes on the development and progression of long-term complications in insulin-dependent diabetes mellitus. N Engl J Med 1993; 329: 977–86.
13. Bartels DW. Diabetes mellitus. In: Carter B, Angaran D, Sisca T, eds. Pharmacotherapy self-assessment program, 1st edition. Kansas City: American College of Clinical Pharmacy, 1993: 145–167.
14. Bartels DW. Diabetes mellitus. In: Carter BL, Angaran DM, Lake KD, Raebel MA, eds. Pharmacotherapy self-assessment program, 2nd edition. Kansas City: American College of Clinical Pharmacy, In press.

Hyperlipidemia guidelines

15. Expert Panel on Detection, Evaluation, and Treatment of High Blood Cholesterol in Adults. Summary of the second report of the National Cholesterol Education Program (NCEP) Expert Panel on Detection, Evaluation, and Treatment of High Blood Cholesterol in Adults (Adult Treatment Panel II). JAMA 1993; 269: 3015–23.
16. Israel MK, McKenney JM. Hyperlipidemias. In: Carter B, Angaran D, Sisca T, eds. Pharmacotherapy self-assessment program, 1st edition. Kansas City: American College of Clinical Pharmacy, 1991: 27–47.
17. Israel MK. Hyperlipidemias. In: Carter BL, Angaran DM, Lake KD, Raebel MA, eds. Pharmacotherapy self-assessment program, 2nd edition. Kansas City: American College of Clinical Pharmacy, In press.

Heart failure guidelines

18. Chow MSS, Scheife RT, eds. Therapeutic and research strategies for congestive heart failure. Pharmacotherapy 1993;13 (5 pt 2): 71S–99.
19. Munger MA, Stanek EJ. Heart failure. In: Carter B, Angaran D, Sisca T, eds. Pharmacotherapy self-assessment program, 1st edition. Kansas City: American College of Clinical Pharmacy, 1991: 85–99.
20. Stanek EJ, Moser LR, Munger MA. Heart failure. In: Carter BL, Angaran DM, Lake KD, Raebel MA, eds. Pharmacotherapy self-assessment program, 2nd edition. Kansas City: American College of Clinical Pharmacy, In press.

(*Continued*)

Appendix 3 Treatment guidelines and review articles (*Continued*)

21. Brutsaert DL, Sys SU, Gillebert TC. Diastolic failure: pathophysiology and therapeutic implications. J Am Coll Cardiol 1993; 22: 318–25.
22. Heart failure: evaluation and care of patients with left ventricular systolic dysfunction. Agency for Health Care Policy and Research (under development).

Coronary artery disease guidelines

23. Frishman WH. Conference on optimizing antianginal therapy. Am J Cardiol 1992; 70: 1G–76.
24. Hamilton SF. Angina pectoris. In: Carter B, Angaran D, Sisca T, eds. Pharmacotherapy self-assessment program, 1st edition. Kansas City: American College of Clinical Pharmacy, 1991: 49–64.
25. Hamilton SF. Angina pectoris. In: Carter BL, Angaran DM, Lake KD, Raebel MA, eds. Pharmacotherapy self-assessment program, 2nd edition. Kansas City: American College of Clinical Pharmacy, In press.
26. Diagnosis and management of unstable angina. Agency for Health Care Policy and Research (under development).

Asthma/chronic obstructive pulmonary disease guidelines

27. The Expert Panel on the Management of Asthma. National Asthma Education Program. Guidelines for the diagnosis and management of asthma. National Asthma Education Program. Office of Prevention, Education, and Control. National Heart, Lung, and Blood Institute, National Institutes of Health, Bethesda, Maryland 20892. NIH Publication No. 91-3042, August 1991.
28. Ferguson GT, Cherniack RM. Management of chronic obstructive pulmonary disease. N Engl J Med 1993; 328: 1017–22.
29. Kelly HW. Asthma. In: Carter B, Angaran D, Sisca T, eds. Pharmacotherapy self-assessment program, 1st edition. Kansas City: American College of Clinical Pharmacy, 1992: 3–16.
30. Kelly HW. Asthma. In: Carter BL, Angaran DM, Lake KD, Raebel MA, eds. Pharmacotherapy self-assessment program, 2nd edition. Kansas City: American College of Clinical Pharmacy, In press.
31. Stratton MA. Chronic obstructive pulmonary disease. In: Carter B, Angaran D, Sisca T, eds. Pharmacotherapy self-assessment program, 1st edition. Kansas City: American College of Clinical Pharmacy, 1992: 17–38.
32. Stratton MA, Noyes M. In: Carter BL, Angaran DM, Lake KD, Raebel MA, eds. Pharmacotherapy self-assessment program, 2nd edition. Kansas City: American College of Clinical Pharmacy, In press.

Peptic ulcer disease guidelines

33. Soll AH. Pathogenesis of peptic ulcer and implications for therapy. N Engl J Med 1990; 322: 909–16.
34. Fish DN, Siepler JK. Gastroesophageal reflux disease. In: Carter B, Angaran D, Sisca T, eds. Pharmacotherapy self-assessment program, 1st edition. Kansas City: American College of Clinical Pharmacy, 1992: 123–33.
35. Fish D. Gastroesophageal reflux disease. In: Carter BL, Angaran DM, Lake KD, Raebel MA, eds. Pharmacotherapy self-assessment program, 2nd edition. Kansas City: American College of Clinical Pharmacy, In press.
36. Piscitelli SC, Garnett WR. Peptic ulcer disease. In: Carter B, Angaran D, Sisca T, eds. Pharmacotherapy self-assessment program, 1st edition. Kansas City: American College of Clinical Pharmacy, 1992: 145–57.
37. Piscitelli SC. Zollinger–Ellison disease/NSAID-induced gastropathy/peptic ulcer disease. In: Carter BL, Angaran DM, Lake KD, Raebel MA, eds. Pharmacotherapy self-assessment program, 2nd edition. Kansas City: American College of Clinical Pharmacy, In press.
38. Hixson LJ, Kelly CL. Current trends in the pharmacotherapy for gastroesophageal reflux disease. Arch Intern Med 1992; 152: 717–23.
39. Hixson LJ, Kelly CL. Current trends in the pharmacotherapy for peptic ulcer disease. Arch Intern Med 1992; 152: 726–32.
40. National Institutes of Health Consensus Development Conferences Statement. *Helicobacter pylori* in peptic ulcer disease. February 7–9, 1994.

Seizure disorders guidelines

41. Scheuer ML, Pedley TA. The evaluation and treatment of seizures. N Engl J Med 1990; 323: 1468–74.
42. Miyahara RK. Seizure disorders. In: Carter B, Angaran D, Sisca T, eds. Pharmacotherapy self-assessment program, 1st edition. Kansas City: American College of Clinical Pharmacy, 1992: 47–61.

Thromboembolic disorders guidelines

43. Dalen JE, Hirsh J, for the Third ACCP Consensus Conference on Antithrombotic Therapy. Chest 1992; 102: 303S–549.
44. Carter BL. Therapy of acute thromboembolism with heparin and warfarin. Clin Pharm 1991; 10: 503–18.
45. Rodvold KA, Erdman SM. Pulmonary embolism. In: Carter B, Angaran D, Sisca T, eds. Pharmacotherapy self-assessment program, 1st edition. Kansas City: American College of Clinical Pharmacy, 1992: 39–53.

(*Continued*)

Appendix 3 Treatment guidelines and review articles (*Continued*)

46. Rodvold KA, Erdman SM. Thrombosis. In: Carter BL, Angaran DM, Lake KD, Raebel MA, eds. Pharmacotherapy self-assessment program, 2nd edition. Kansas City: American College of Clinical Pharmacy, In press.

Estrogen replacement therapy guidelines

47. American College of Physicians. Guidelines for counseling postmenopausal women about preventive hormone therapy. Ann Intern Med 1992; 117: 1038–41.
48. Lourwood DL. Estrogen replacement therapy. In: Carter B, Angaran D, Sisca T, eds. Pharmacotherapy self-assessment program, 1st edition. Kansas City: American College of Clinical Pharmacy, 1993: 189–202.
49. Lourwood DL. Hormone replacement therapy. In: Carter BL, Angaran DM, Lake KD, Raebel MA, eds. Pharmacotherapy self-assessment program, 2nd edition. Kansas City: American College of Clinical Pharmacy, In press.
50. Grady D, Rubin SM. Hormone therapy to prevent disease and prolong life in postmenopausal women. Ann Intern Med 1992; 117: 1016–41.
51. Wood H, Wang-Cheng R. Postmenopausal hormone replacement: are two hormones better than one? J Gen Intern Med 1993; 8: 451–8.
52. Belchetz PE. Hormonal treatment of postmenopausal women. N Engl J Med 1994; 330: 1062–71.

Rheumatologic disorders guidelines

53. Anonymous. Drugs for rheumatoid arthritis. Med Lett Drugs Ther 1991; 33: 65–70.
54. Harris ED. Rheumatoid arthritis: pathophysiology and implications for therapy. N Engl J Med 1990; 322: 1277–89.

Depression guidelines

55. Clinical practice guidelines. Depression in primary care, volume 1. Detection and diagnosis. U.S. Department of Health and Human Services, Public Health Service, Agency for Health Care Policy and Research. AHCPR Publication No. 93-0550, April 1993.
56. Clinical practice guidelines. Depression in primary care, volume 2. Treatment of major depression. U.S. Department of Health and Human Services, Public Health Service, Agency for Health Care Policy and Research. AHCPR Publication No. 93-0550, April 1993.
57. American Psychiatric Association. Practice guidelines for major depressive disorders in adults, 2nd edition. Washington, DC, 1993.
58. Grimsley SR. Depression. In: Carter B, Angaran D, Sisca T, eds. Pharmacotherapy self-assessment program, 1st edition. Kansas City: American College of Clinical Pharmacy, 1992: 127–50.

Appendix 4 Bibliography on patient outcomes measurement

Generic tools

1. Stewart AL, Ware JE Jr, eds. Measuring functioning and well-being: the medical outcomes study approach. The Rand Corporation. Durham, NC: Duke University Press, 1992.
2. Parkerson GR, Gehlbach SH, Wagner EH, et al. The Duke-UNC health profile: an adult health status instrument for primary care. Med Care 1981; 19: 806–28.
3. *Hunt SM, McEwen J, McKenna SP. Measuring health status: a tool for clinicians and epidemiologists. J R Coll Gen Pract 1985; 35: 185–8.
4. Nelson EC, Wasson JH, Kirk JQ. Assessment of function in routine clinical practice: description of the COOP chart method and preliminary findings. J Chronic Dis 1987; 40 (suppl 1): 55S–63.
5. Stewart AL, Hays RD, Ware JE. The MOS short-form general health survey: reliability and validity in a patient population. Med Care 1988; 26: 724–35.
6. Ware JE, Snow K, Kosinski M, et al. SF-36 health survey manual and interpretation guide. Boston: Nimrod Press, 1993.
7. Bergner M, Bobbitt RA, Carter WB, et al. The sickness impact profile: conceptual formulation and methodology for the development of a health status measure. Int J Health Serv 1976; 6: 393–415.

*The Nottingham Health Profile

Disease-specific tools to measure outcomes

Hypertension outcomes

1. Flack JM, Grimm RH. Hypertension/Lipid Form 5.1. Bloomington, MN: Health Outcomes Institute, 1993. (200 Killebrew Dr., Suite 122, Bloomington, MN 55425).

(*Continued*)

Appendix 4 Bibliography on patient outcomes measurement (*Continued*)

2. Bulpitt CJ, Fletcher AE. Quality of life in hypertensive patients on different antihypertensive treatments: rationale for methods employed in a multicenter randomized controlled trial. J Cardiovasc Pharm 1985; 7 (suppl 1): 137–45.
3. Chang SW, Fine R, Siegel D, et al. The impact of diuretic therapy on reported sexual function. Arch Intern Med 1991; 151: 2402–8.
4. Croog SH, Levine S, Testa MA, et al. The effects of antihypertensive therapy on the quality of life. N Engl J Med 1986; 314: 1657–64.
5. Levine S, Croog S, Sudilovsky A, et al. Effects of antihypertensive medications on vitality and well-being. J Fam Pract 1987; 25: 357–63.
6. Testa MA, Anderson RB, Nackley JF, et al. Quality of life and antihypertensive therapy in men: a comparison of captopril with enalapril. N Engl J Med 1993; 328: 907–13.
7. Testa MA, Sudilovsky A, Rippey RM, et al. A short form for clinical assessment of quality of life among hypertensive patients. Am J Prev Med 1989; 5: 82–9.
8. The Treatment of Mild Hypertension Research Group. The treatment of mild hypertension study: a randomized, placebo-controlled trial of a nutritional-hygienic regimen along with various drug monotherapies. Arch Intern Med 1991; 151: 1413–23.
9. Watters K, Campbell B. Can an antihypertensive agent be both effective and improve the quality of life? A multicenter study with indapamide. Br J Clin Prac 1986; 40: 239–44.
10. Wenger NK, Mattson ME, Furberg CD, Elinson J, eds. Assessment of quality of life in clinical trials of cardiovascular therapies. New York: Le Jacq Publishing Inc., 1984.

Diabetes outcomes

11. Given CW. Measuring the social psychological health states of ambulatory chronically ill patients: hypertension and diabetes as tracer conditions. J Comm Health 1984; 9: 179–95.
12. Hanestad BR, Polit RN. Perceived quality of life in a study of people with insulin-dependent diabetes mellitus using the Hornquist model. Qual Life Res 1994; 3: 79.
13. Hanestad BR, Hornquist JO, Albreksten G. Self-assessed quality of life and metabolic control in persons with insulin-dependent diabetes mellitus (IDDM). Scand J Soc Med 1991; 19: 57–65.
14. Kaplan SH. Patient reports of health status as predictors of physiologic health measures in chronic disease. J Chronic Dis 1987; 40: 27S–35.
15. Marquis KH, Ware JE, Johnston R, et al. Appendix B to summary report: an evaluation of published measures of diabetes self-care variables. Publication No. N-1152-HEW. Santa Monica, CA: The Rand Corporation, June 1979.
16. Nerenz DR, Repasky DP, Whitehouse FW, et al. Ongoing assessment of health status in patients with diabetes mellitus. Med Care 1992; 30 (suppl 5): MS112–24.
17. Testa MA, Simonson DC. Measuring quality of life in hypertensive patients with diabetes. Postgrad Med J 1988; 64 (suppl 3): 50–8.
18. The Diabetes Control and Complications Trial (DCCT) Research Group. Reliability and validity of a diabetes quality-of-life measure for the Diabetes Control and Complications Trial. Diabetes Care 1988; 11: 725–32.
19. Wenneker MB, Mchorney CA, Kieszak SM, et al. The impact of diabetes severity on quality of life: results from the medical outcomes study [abstr]. Clin Res 1991; 39: 612A.

Hyperlipidemia outcomes

20. Flack JM, Grimm RH. Hypertension/Lipid Form 5.1. Bloomington, MN: Health Outcomes Institute, 1993. (200 Killebrew Dr., Suite 122, Bloomington, MN 55425).

Heart failure outcomes

21. Guyatt GH, Nogradi S, Halcrow S, et al. Development and testing of a new measure of health status for clinical trials in heart failure. J Gen Intern Med 1989; 4: 101–7.
22. Bulpitt CJ, Fletcher AE. Measurement of the quality of life in congestive heart failure—influence of drug therapy. Cardiovasc Drug Ther 1988; 2: 419–24.
23. Feinstein AF, Fisher MB, Pigeon JG. Changes in dyspnea-fatigue ratings as indicators of quality of life in the treatment of congestive heart failure. Am J Cardiol 1989; 64: 50–5.
24. Gorkin L, Norvell NK, Rosen R, et al. Assessment of quality of life as observed from the baseline data of the Studies of Left Ventricular Dysfunction (SOLVD) trial quality-of-life substudy. Am J Cardiol 1993; 71: 1069–73.
25. Guyatt G. Methodological problems in clinical trials in heart failure. J Chronic Dis 1985; 38: 353–63.
26. Massie BM, Berk MR, Brozena SC, et al. Can further benefit be achieved by adding flosequinan to patients with congestive heart failure who remain symptomatic on diuretic, digoxin, and an angiotensin converting enzyme inhibitor? Results of the Flosequinan-ACE Inhibitor Trial (FACET). Circulation 1993; 88: 492–501.

(*Continued*)

Appendix 4 Bibliography on patient outcomes measurement (*Continued*)

27. Rector TS, Tschumperlin LK, Kubo SH, et al. Clinically significant improvements in the living with heart failure questionnaire score as judged by patients with heart failure. Qual Life Res 1994; 3: 60–1.
28. Rector TS. Outcomes assessment: functional status measures as therapeutic endpoints for heart failure. Top Hosp Pharm Manag 1990; 10: 37–43.
29. Rector TS, Cohn JN. Assessment of patient outcome with the Minnesota living with heart failure questionnaire: reliability and validity during a randomized, double-blind, placebo-controlled trial of pimobendan. Am Heart J 1992; 124: 1017–25.
30. Rector TS, Kubo SH, Cohn JN. Patients' self-assessment of their congestive heart failure—Part 2: content, reliability and validity of a new measure, the Minnesota living with heart failure questionnaire. Heart Fail 1987; 3: 198–209.
31. Rector TS, Kubo SH, Cohn JN. Validity of the Minnesota living with heart failure questionnaire as a measure of therapeutic response to enalapril or placebo. Am J Cardiol 1993; 71: 1106–7.
32. Retchin SM, Brown B. Elderly patients with congestive heart failure under prepaid care. Am J Med 1991; 90: 236–42.
33. Romm FJ, Hulka BS, Mayo F. Correlates of outcomes in patients with congestive heart failure. Med Care 1976; 15: 765–76.
34. Wiklund I, Lindvall K, Swedberg K, et al. Self-assessment of quality of life in severe heart failure: an instrument for clinical use. Scand J Psychol 1987; 28: 220–5.

Coronary artery disease outcomes

35. Vaisrub S. Quality of life manque. JAMA 1976; 236: 387.
36. Permanyer-Miralda G, Alonso J, Anto JM, et al. Comparison of perceived health status and conventional functional evaluation in stable patients with coronary artery disease. J Clin Epidemiol 1991; 44: 779–86.
37. Smith HC, Frey RL, Piehler JM. Does coronary bypass surgery have a favorable influence on the quality of life? Cardiovasc Clin 1983; 12: 253–64.
38. VandenBurg MJ. Measuring the quality of life of patients with angina. In: Walker SR, Rosser RM, eds. Quality of life: assessment and application. Lancaster, England: MTP Press Limited, 1988: 267–78.
39. Westaby S, Sapsford RN, Bentall HH. Return to work and quality of life after surgery for coronary artery disease. Br Med J 1979; 2: 1028–31.
40. Spertus JA, Winder JA. The Seattle angina questionnaire: validation of a new functional status measure for patients with coronary artery disease (submitted 1993).
41. Pryor DB. Angina Form 9.1. Bloomington, MN: Health Outcomes Institute, 1993. (200 Killebrew Dr., Suite 122, Bloomington, MN 55425).

Asthma/chronic obstructive pulmonary disease outcomes

42. Alonso J, Auto JM, Gonzalez M, et al. Measurement of general health status of non-oxygen dependent chronic obstructive pulmonary disease patients. Med Care 1992; 30: MS125–35.
43. Barber BL, Santanello NC, Epstein RS. Clinical meaning of change in an asthma-specific questionnaire. Qual Life Res 1994; 3: 58–9.
44. Bishop J, Carlin J, Nolan T. Evaluation of the properties and reliability of a clinical severity scale for acute asthma in children. J Clin Epidemiol 1992; 45: 71–6.
45. Eakin EG, Kaplan RM, Ries AL. Measurement of dyspnea in chronic obstructive pulmonary disease. Qual Life Res 1993; 2: 181–91.
46. Guyatt GH, Thompson PJ, Berman LB, et al. How should we measure function in patients with chronic heart and lung disease? J Chronic Dis 1985; 38: 517–24.
47. Guyatt GH, Townsend MB, Pugsley SO, et al. Quality of life in patients with chronic airflow limitations. Br J Dis Chest 1987; 81: 45–54.
48. Guyatt GH, Berman LB, Townsend M, et al. A measure of quality of life for clinical trials in chronic lung disease. Thorax 1987; 42: 773–8.
49. Hyland ME, Finnis S, Irvine SH. A scale for assessing quality of life in adult asthma sufferers. J Psychosoc Res 1991; 35: 99–110.
50. Juniper EF, Guyatt GH, Griffith LE. Determining a minimal important change in a disease-specific quality of life questionnaire. J Clin Epidemiol 1994; 47: 81–7.
51. Juniper EF, Guyatt GH, Epstein RS, et al. Evaluation of impairment of health related quality of life in asthma: development of a questionnaire for use in clinical trials. Thorax 1992; 47: 76–83.
52. Bethel RA. Asthma Form 10.1. Bloomington, MN: Health Outcomes Institute, 1993. (200 Killebrew Dr., Suite 122, Bloomington, MN 55425).
53. Bethel RA. COPD Form 15.1. Bloomington, MN: Health Outcomes Institute, 1993. (200 Killebrew Dr., Suite 122, Bloomington, MN 55425).

(*Continued*)

Appendix 4 Bibliography on patient outcomes measurement (*Continued*)

Peptic ulcer disease outcomes

54. Dimenas E, Glise H, Hallerback B, et al. Quality of life in patients with upper gastrointestinal symptoms. An improved evaluation of treatment regimens? Scand J Gastroenterol 1993; 28: 681–7.
55. Garratt AM, Ruta DA, Abdalla MI, et al. The SF-36 health survey questionnaires: an outcome measure suitable for routine use within the NHS? Br Med J 1993; 306: 1440–4.
56. Guyatt GH, Mitchell A, Irvine E, et al. A new measure of health status for clinical trials in inflammatory bowel disease. Gastroenterology 1989; 96: 804–10.
57. Pincus T, Griffin M. Gastrointestinal disease associated with non-steroidal anti-inflammatory drugs: new insights from observational studies and functional status questionnaires. Am J Med 1991; 91: 209–12.

Seizure disorder outcomes

58. Baker GA, Smith DF, Dewey M, et al. The initial development of a health-related quality of life model as an outcome measure in epilepsy. Epilepsy Res 1993; 16: 65–81.
59. Baker GA, Smith DF, Dewey M, et al. The development of a seizure severity scale as an outcome measure in epilepsy. Epilepsy Res 1991; 8: 245–51.
60. Chaplin JE, Yepez R, Shorvon S, et al. A quantitative approach to measuring the social effects of epilepsy. Neuroepidemiology 1990; 9: 151–8.
61. Jacoby A, Baker G, Smith D, et al. Measuring the impact of epilepsy: the development of a novel scale. Epilepsy Res 1993; 16: 83–8.
62. Jacoby A. Epilepsy and the quality of everyday life: findings from a study of people with well-controlled epilepsy. Soc Sci Med 1992; 34: 657–66.
63. Jacoby A, Johnson A, Chadwick D. Psychosocial outcomes of antiepileptic drug discontinuation. Epilepsia 1991; 33: 1123–31.
64. Perrine KR. A new quality-of-life inventory for epilepsy patients: interim results. Epilepsia 1993; 34 (suppl 4): S28–33.
65. Smith D, Baker G, Davies G, et al. Outcomes of assessing treatment with lamotrigine in partial epilepsy. Epilepsia 1993; 34: 312–22.
66. Smith DF, Baker GA, Dewey M, et al. Seizure frequency, patient-perceived seizure severity and the psychosocial consequences of intractable epilepsy. Epilepsy Res 1991; 9: 231–41.
67. Vickrey BG. A procedure for developing a quality-of-life measure for epilepsy surgery patients. Epilepsia 1993; 344 (suppl 4): S22–7.
68. Vikrey BG, Hays RD, Brook RH, et al. Reliability and availability of the Katz adjustment scales in an epilepsy sample. Qual Life Res 1992; 1: 63–72.
69. Wagner AK, Bungay KM, Bromfield EB, et al. Health-related quality of life of adult person with epilepsy as compared with health-related quality of life of well persons. Epilepsia 1993; 6: 5.
70. Wagner AK, Keller S, Baker GA, et al. Improving health-related quality of life measures for use with people with epilepsy. Epilepsia 1993; 2: 30.
71. French J. The long-term therapeutic management of epilepsy. Ann Intern Med 1994; 120: 411–22.

Estrogen replacement therapy outcomes

72. Daly E, Gray A, Barlow D, et al. Measuring the impact of menopausal symptoms on quality of life. Br Med J 1993; 307: 836–40.
73. Wiklund I, Holst J, Karlberg J, et al. A new methodological approach to the evaluation of quality of life in postmenopausal women. Maturitas 1992; 14: 211–24.

Rheumatologic outcomes

74. Bombardier C, Ware J, Russell J, et al. Auranofin therapy and the quality of life in patients with rheumatoid arthritis. Am J Med 1986; 81: 565–78.
75. Blair PS, Silman AJ. Can clinical trials in rheumatology be improved? Curr Opin Rheumatol 1991; 3: 272–9.
76. Bjelle A. Functional status assessment. Curr Opin Rheumatol 1991; 3: 280–5.
77. Bakker CH, Rutten-van MM, Van Doorslaer E, et al. Health related utility measurement in rheumatology: an introduction. Patient Educ Counsel 1993; 20: 145–52.
78. Meenan RF, Gertman PM, et al. Measuring health status in arthritis: the arthritis impact measurement scales. Arthritis Rheum 1980; 23: 146–52.

Depression outcomes

79. Alden D, Austin C, Sturgeon R. A correlation between the geriatric depression scale long and short forms. J Gerontol 1989; 44: 124–5.

(*Continued*)

Appendix 4 Bibliography on patient outcomes measurement (*Continued*)

80. Borson S, McDonald GJ, Gayle T, et al. Improvement in mood, physical symptoms, and function with nortriptyline for depression in patients with chronic obstructive pulmonary disease. Psychosomatics 1992; 33: 190–201.
81. Brooks W, Blair J, John S, et al. The impact of psychologic factors on measurement of functional status: assessment of the sickness impact profile. Med Care 1990; 28: 793–804.
82. Given C, Stommel M, Given B, et al. The influence of cancer patients' symptoms and functional states on patients' depression and family caregivers' reaction and depression. Health Psychol 1993; 12: 277–85.
83. Grégoire J, de Leval N, Mesters P, et al. Validation of the quality of life in depression scale in a population of adult depressive patients aged 60 years and above. Qual Life Res 1994; 3: 51–2.
84. Gurland B, Golden RR, Teresi JA, et al. The short-care: an efficient instrument for the assessment of depression, dementia and disability. J Gerontol 1984; 39: 166–9.
85. Lambert MJ, Hatch DR, Kingston MD, et al. Zung, Beck and Hamilton rating scales as measures of treatment outcome: a meta-analytic comparison. J Consult Clin Psychol 1986; 54: 54–9.
86. Light RW, Merrill EJ, Despars JA, et al. Prevalence of depression and anxiety in patients with COPD: relationship to functional capacity. Chest 1985; 87: 35–8.
87. Norris JT, Gallagher D, Wilson A, et al. Assessment of depression in geriatric medical outpatients: the validity of two screening measures. J Am Geriatr Soc 1987; 35: 989–95.
88. Prusoff BA, Klerman GL, Paykel ES. Concordance between clinical assessments and patients' self-report in depression. Arch Gen Psychiatry 1972; 26: 546–52.
89. Radloff LS. The CES-D scale: a self-report depression scale for research in the general population. J Appl Psychol Meas 1977; 1: 385–401.
90. Revicki DA, Turner R, Brown R, et al. Reliability and validity of a health-related quality of life battery for evaluating outpatient antidepressant treatment. Qual Life Res 1992; 1: 257–66.
91. Stewart AL, Sherbourne CD, Wells KB, et al. Do depressed patients in different treatment settings have different levels of well-being and functioning? J Consult Clin Psychol 1993; 6: 849–57.
92. Stoker MJ, Sunbar GC, Beaumont G. The SmithKline Beecham 'quality of life' scale: a validation and reliability study in patients with affective disorder. Qual Life Res 1992; 1: 385–95.
93. Tambs K, Moum T. How well can a few questionnaire items indicate anxiety and depression? Acta Psychiatr Scand 1993; 87: 364–7.
94. Tanaka JS, Huba GJ. Confirmatory hierarchical factor analyses of psychological distress measures. J Pers Soc Psychol 1984; 46: 621–35.
95. Tanaka-Matsumi J, Kameoka VA. Reliabilities and concurrent validates of popular self-report measures of depression, anxiety, and social desirability. J Consult Clin Psychol 1986; 54: 328–33.
96. Wells KB. Commentary: assessment of psychological morbidity in primary care. J Chronic Dis 1987; 40 (suppl 1): 81S–3.
97. Wells KB, Stewart A, Hays RD, et al. The functioning and well being of depressed patients. JAMA 1989; 262: 914–19.
98. Wells KB, Hays RD, Burnam MA, et al. Detection of depressive disorders for patients receiving prepaid for fee-for-service care: results from the medical outcomes study. JAMA 1989; 262: 3298–3302.
99. Radloff LS. The CES-D scale: a self reported depression scale for research in the general population. Appl Psychol Meas 1977; 1: 385–401.

REFERENCES

1. Carter, B.L. Ambulatory Care. In *Handbook of Institutional Pharmacy Practice,* 3rd Ed.; Brown, T.R., Ed.; American Society of Hospital Pharmacists: Bethesda, 1992; 367–373.
2. The ACCP Clinical Practice Affairs Committee, 1990–1991. Clinical pharmacy practice in the noninstitutional setting: A white paper from the American College of Clinical Pharmacy. Pharmacotherapy **1992**, *12*, 358–364.
3. Rakel, R.E. The Family Physician. In *Textbook of Family Practice,* 4th Ed.; Rakel, R.E., Ed.; WB Saunders: Philadelphia, 1990; 3–18.
4. Short, E.M. *Primary Care Education (PRIME) Program for Medical Residents and Associated Health Trainees in AY 94–95*; Department of Veterans Affairs Memorandum, 1993, July 19.
5. Hutchinson, L.C.; Wolfe, J.J.; Padilla, C.B.; Forrester, C.W. Clinical privileges program for pharmacists. Am. J. Hosp. Pharm. **1992**, *49*, 1422–1424.
6. Donabedian, A. The quality of care: How can it be assessed? JAMA, J. Am. Med. Assoc. **1988**, *260*, 1743–1748.
7. Lohr, K.N.; Brook, R.H. Quality assurance and clinical pharmacy: Lessons from medicine. Drug Intell. Clin. Pharm. **1981**, *15*, 758–765.
8. Raskin, I.E.; Maklan, C. Medical Treatment Effectiveness

Research: A View from Inside the Agency for Health Care Policy and Research. In *U.S. Department of Health and Human Services, Public Health Service, AHCPR Pub. No. 91-0025, June*; 1991.

9. Agency for Health Care Policy and Research. Clinical Practice Guideline Development. In *U.S. Department of Health and Human Services, Public Health Service, AHCPR Pub. No. 93-0023, August*; 1993.
10. Woolf, S.H. Practice guidelines: A new reality in medicine. II: Methods of developing guidelines. Arch. Intern. Med. **1992**, *152*, 946–952.
11. Ware, J.E., Jr. Standards for validating health measures: Definition and content. J. Chronic Dis. **1987**, *40*, 473–480.
12. Woolf, S.H. Practice guidelines: A new reality in medicine. III: Impact on patient care. Arch. Intern. Med. **1993**, *153*, 2646–2655.
13. Bungay, K.M.; Wagner, A.K. Comment: Assessing the quality of pharmaceutical care. Ann. Pharmacother. **1993**, *27*, 1542–1543.
14. Greenfield, S.; Nelson, E.C. Recent developments and future issues in the use of health status assessment measures in clinical settings. Med. Care **1992**, *30* (Suppl. 5), 23–41.
15. Patrick, D.L.; Erickson, P. *Health Status and Health Policy: Quality of Life in Health Care Evaluation and Resource Allocation*; Oxford University Press: New York, 1993.
16. Guyatt, G.H.; Feehey, D.H.; Patrick, D.L. Measuring health-related quality of life. Ann. Intern. Med. **1993**, *118*, 622–629.
17. Bungay, K.M.; Ware, J.E. Measuring and Monitoring Health-Related Quality of Life. In *Current Concepts*; The Upjohn Company: Kalamazoo, 1993.
18. Palmer, R.H.; Banks, N.; Edwards, J.; Fowles, J.; Necenz, D.; Zapka, J. *Interim Report: External Review Performance Measurement of Medicare HMOs/CMPs*; Delmarva Foundation for Medical Care, Inc.: New York, 1994; February.
19. Health Outcomes Institute. *Outcomes Management System (OMS), Bloomington, Minnesota*; 1993.
20. *Pharmacy Law Digest, Facts and Comparisons*; Fink, J.L., III, Marguardt, K.W., Simonsmeier, L.M., Eds.; Facts and Comparisons: St. Louis, Missouri, 1994.
21. Perdue, J.M. The Law of Texas Medical Malpractice. In *Houston Law Review, University of Houston Law Center,* 2nd Ed.; University of Houston Law Center: Houston, 1985; 2–20.
22. v. Aiken, T., et al., 101 N.W. 769 (Iowa 1904).
23. Keeton, W.P. *Prosser and Keeton on the Law of Torts,* 5th Ed.; West Publishing Company: St. Paul, 1984; 263–321.
24. v. Odiorne, M., 94 A. 753 (Maine 1915).

Professional Associations, Clinical Pharmacy Careers in

Peggy G. Kuehl
American College of Clinical Pharmacy, Kansas City, Missouri, U.S.A.

INTRODUCTION

A career in a professional association can be an extremely rewarding way to practice clinical pharmacy while serving the members of the profession. Positions may focus on many different programmatic areas (Table 1). In addition to positions in national associations, pharmacists serve important roles in state and regional associations. There are also many related biomedical associations in which pharmacists may be employed.

TYPES OF POSITIONS AVAILABLE

Many types of positions are possible within associations. Those working for large pharmacy associations may have more than 100 coworkers. Employees of large associations typically have fairly narrow responsibilities and focus. Within smaller associations, with staffs of 40 or less, areas of responsibility are broad, and many positions are combined or do not exist.

Approximately 20 of the 100 employees of the American Pharmaceutical Association are pharmacists. Of 20 employees of the Academy of Managed Care Pharmacy, 4 are pharmacists. The American College of Clinical Pharmacy, with a staff of 30, employs 6 who are pharmacists. The American Society of Health-System Pharmacists, with a staff of approximately 175 persons, has a larger proportion who are pharmacists due to their extensive publishing efforts.

Clinical pharmacists within associations are most frequently employed within areas where their backgrounds are of most benefit. Some of these areas include serving as the chief executive officer, working with chapter relations, education, government and professional affairs, marketing, membership, and publications, including the journals of the association.

Depending on the size of the association staff, clinical pharmacists may start out working with specific publications, membership and chapter activities, or educational program planning. As they gain experience and familiarity with their roles, they may take on greater managerial roles and broader areas of responsibility. Those with the greatest experience and expertise may move into director, or vice presidential, positions, or even become the chief executive officer for the association.

According to the American Society of Association Executives (ASAE), the average total compensation for association executives in 1996 ranged from $48,000 to $117,000.[1] Compensation has increased over the last few years by percentages greater than business averages (5–10% for many positions). Many associations use ASAE survey information, faculty and deans' salary data collected by the American Associations of Colleges of Pharmacy,[2] and salary data from other segments of the profession (e.g., managed care pharmacists, industry pharmacists) as benchmarks for pharmacist executive staff salaries.

Regardless of the association's scope or focus, pharmacists must be responsive to the needs and desires of the membership served. Each association employee is ultimately responsible to the members and prospective members for whom the association exists. The membership at large is generally represented by an elected governing board (Board of Directors, Board of Regents, Board of Trustees), and sometimes a house of delegates. Typically, the activities of the association are guided through a strategic planning process, which sets the direction for the association for the next few years.

CAREER PATHS AND PREPARATION

Several paths exist for entry into a career with a professional association. Many association executives have had an earlier career as a practicing member of their profession. They have worked ''in the trenches'' for a

Encyclopedia of Clinical Pharmacy
DOI: 10.1081/E-ECP 120006264

Table 1 Programmatic and management positions within professional associations

Position title(s)	Major responsibilities	Examples of primary focus
Association Foundation Executive, Director of Research Institute	Leads the not-for-profit, charitable arm of the association.	Research training and grants programs; professionwide research; fund-raising.
Attorney	Legal counsel for association.	Resource for staff, board, and officers in matters of nonprofit or antitrust law; personnel and copyright issues; legal disputes.
Chapter Relations, Manager of	Primary association liaison with chapters.	Programs and services for chapters and other affiliates.
Chief Executive Officer, Executive Director, Executive Vice President	Oversees all aspects of the association; reports to the elected governance of the association.	Strategic direction; ''the big picture.''
Computer/Management Information Systems, Manager of	Computer support systems.	Database development; membership database management; web site development and maintenance.
Deputy Chief Executive, Associate Executive Director	Responsible for major projects; often works with personnel issues.	Specific projects of importance to the association.
Education, Director/Vice President or Manager	Development and delivery of educational programming, including certificate programs and other training programs.	Scientific abstracts; meeting standards for continuing pharmaceutical education; development of program content; working with speakers; delivery of live, printed, and online programming; educational grants.
Finance, Director of/Controller	Oversees financial management of the association.	Accounting; management of short-, intermediate-, and long-term investments; financial reporting to chief executive officer and governing board.
Government Relations (often paired with Professional Affairs), Director/Vice President of	Leads association's advocacy efforts to government, legislative, and regulatory bodies.	Lobbying; tracking legislative initiatives to alert membership of important issues; writing suggested legislation; interaction with representatives of government bodies.
Human Resources, Manager of	Personnel management.	Hiring and firing; employee policies; employee benefits.
Marketing, Director/Vice President or Manager	Promotion of the association.	Promotion of membership, meetings, publications.
Meetings and Expositions, Director/Vice President or Manager	Planning for conventions of the association.	Meeting logistics; site selection; audiovisual requirements; exhibits program; vendor contracts.
Member Services, Director/Vice President or Manager	Membership recruitment and retention.	Dues notices; recruitment of new members; services for members; promotional materials; member complaints.
Professional Affairs (often paired with Government Relations), Director/Vice President	Primary contact with other professional associations.	Development of professional policies; attends meetings of other related associations; offers comment on behalf of association regarding position statements of other associations; collaboration with other associations regarding mutual interests.

(*Continued*)

Table 1 Programmatic and management positions within professional associations (*Continued*)

Position title(s)	Major responsibilities	Examples of primary focus
Public Relations Director/ Vice President or Manager	Image of the association to the rest of the world.	Press releases; quotations for publication; "official voice" of the association.
Publications Director/Vice President or Manager/Editor	Printed and online publications to meet needs of membership.	Official journal; books; home study programs; newsletters.

number of years, and were active in the association as a volunteer, through committee work, attendance at meetings, presentation of scientific abstracts, or elected office. By becoming known as a reliable volunteer, they set the stage for movement into association employment. This path is advantageous in providing the association with an employee whom they know and who understands the membership and the profession. However, a disadvantage of this path is that clinical pharmacists have rarely received the management training and experience necessary for association work.

Others who identify association work as their career objective early in their training may enter this type of work through summer internships, experiential clerkships and externships, and executive residencies, or by accepting their first postgraduate job with an association and then advancing though various positions and departments. Many of these opportunities are listed in a document prepared by the member organizations of the Interorganizational Council on Student Affairs.[3]

Regardless of the career path chosen, preparation for a career in association management should ideally include several elements. Exposure to association work can be obtained through an internship, a clerkship, or an executive residency. These experiential activities provide excellent opportunities for exposure to what it might be like to work for an association.

Experience as a practicing member of the profession can be an excellent way to prepare for association work. Through spending time in actual patient care, research, and teaching, prospective association pharmacists can learn firsthand what members do and what their struggles are. Furthermore, by becoming involved with associations on a volunteer basis, it is possible to experience how the association works. Volunteering also provides an opportunity to showcase an applicant's talents and dependability to the association. Choosing to work for employers who will allow time and support for association work is also beneficial.

Other areas of preparation that are very helpful involve skills aside from those taught in traditional pharmacy curricula. Those thinking of careers in associations should keep their eyes open to issues and events, the actions of other related groups, and the opinions of colleagues. They need to pay attention to the business and the politics of the profession and be informed about them.

Finally, training in how to manage projects, people, and finances can be very helpful for association managers. Many of the skills necessary to be an excellent clinical pharmacist are opposite those required to be an excellent manager. Training and experience in this area are crucial to excelling in an association career.

BENEFITS OF ASSOCIATION WORK

There are many benefits of association work. Working within an association provides increased opportunities to be involved with the discipline of clinical pharmacy. For someone who has practiced patient care, conducted research, or taught, this can provide an exciting new direction with new challenges and new skills to learn. Those who have been members of the association before working for it often appreciate the opportunity to work for a group having a mission they believe in and from which they have benefited as a member.

Association work necessarily involves communication with its members. Those working in this field have many more acquaintances among their peers and with representatives of the pharmaceutical industry. Establishing and maintaining contacts is an important part of this work. Those who enjoy interacting with people will especially enjoy this aspect of association life.

A clinical career within an association can be an opportunity for continued learning and professional growth. Association managers often need to enhance their skills in business management, public relations, marketing, project management, advocacy, politics, personnel management, legal issues, and many other areas, depending on the scope of their responsibilities. Adopting the attitude of a lifelong learner is essential, as the profession and the world change rapidly. It is necessary to constantly

learn new knowledge and skills to keep up with these rapid changes.

Working for an association opens many career opportunities in related fields, with other associations, with the pharmaceutical industry, in other management jobs, and in private consulting. The skills gained through association work are widely transferable to other professions, as well as back to clinical practice.

Clinical pharmacists within associations have many opportunities to work on new projects pertaining to the membership and the profession. They can be creative not only in developing a project, but also in funding, implementing, and promoting it. Association work provides a wide variety of opportunities to practice clinical pharmacy in ways that, although different from traditional roles, affect patient care and the profession.

Many of the knowledge and skills taught in clinical pharmacy curricula are applicable to association work. For example, those working with educational programs often draw on their pharmacy backgrounds to help members design better educational offerings. Also, research techniques learned in academic training are applicable to association research.

Perhaps the greatest benefit of association work is the ability to represent the association's members, seeking to improve their practices and opportunities, and to make a difference in the profession.

CHALLENGES OF ASSOCIATION WORK

Those choosing to practice clinically within a professional association must have excellent organizational and time management skills. They will experience many pressures from various directions and must be skilled at prioritizing their activities. Excellent written and verbal communications skills are also essential.

Association executives must find a balance between being a member of the association and serving on the staff. An attitude of servanthood is absolutely essential. It is also critical to find a balance between listening to members and responding to their needs versus taking the lead and establishing the direction of the association. The most responsive associations are typically ''member driven,'' meaning that the association staff take their lead from the elected officers of the organization, with a focus on implementing, rather than establishing policy.

When working with speakers, authors, and committees, who by definition are performing volunteer work, association executives must be prepared to help members meet their deadlines, through reminders, clear communication, and advance planning. Missed deadlines by members have the potential to result in periods of great stress for association staff who in turn have printing or other deadlines that cannot be changed. As this can be a source of great stress, it is essential to have the personal fortitude to deal with the unexpected and uncontrollable.

Because associations must function as businesses, serving as stewards of their members' resources, there is less freedom for association employees to function as independently as they might in an academic environment. Budgets must be developed and adhered to, and performance targets established and met. Also, because the elected officers change regularly, the direction of the association and the style of the governing board can change frequently. Clinical pharmacists working within associations must be flexible, as issues and responses can change, may not always agree with their personal opinions, and are often outside their control.

CONCLUSION

Working as a pharmacist within a professional association is a challenging, yet rewarding way to practice clinical pharmacy. This career choice should be strongly considered by those who seek to serve their colleagues and the profession by helping them manage and adapt to everchanging issues in clinical pharmacy.

REFERENCES

1. Casteuble, T. What today's association executives earn. Assoc. Manage. **1997**, *49* (4), 53–61.
2. American Association of Colleges of Pharmacy. 1999–2000 Profile of Pharmacy Faculty. In *Institutional Research Report Series*; American Association of Colleges of Pharmacy: Alexandria, Virginia, 1999.
3. Interorganizational Council on Student Affairs. *Interorganizational, Financial, and Experiential Information Document*; Interorganizational Council on Student Affairs, American Pharmaceutical Association: Washington, DC, 2000.

PROFESSIONAL DEVELOPMENT

Psychiatric Pharmacy Specialty Practice

Julie A. Dopheide
University of Southern California, Los Angeles, California, U.S.A.

INTRODUCTION

Psychiatric pharmacy practice is a rewarding specialty with diverse opportunities to have an impact on the care of patients directly through innovative practice models, quality assurance, and patient education, and indirectly through professional education, the development of treatment guidelines, clinical research, and careers in the pharmaceutical industry.[1–3]

Psychiatric illness occurs in 50% of the population over the course of a lifetime and nearly 30% over the course of a year.[4] It affects all age groups and causes significant morbidity, mortality, and diminished quality of life.[4,5] Medication management is essential to treat acute symptoms of psychiatric illness and to prevent relapse. In fact, psychotropic drugs (antidepressants, antipsychotics, anxiolytics, hypnotics mood stabilizers, stimulants) comprised 10% of the top 200 brand name and generic drugs dispensed from retail pharmacies in the year 2001.[6]

Similar to other specialty practice areas, psychiatric pharmacy requires core knowledge of disease state management and the ability to demonstrate clinical skills in applying knowledge to improve patient care. A distinguishing aspect of psychiatric pharmacy is the requirement that the practitioner have an interest and aptitude for interpersonal communication and developing professional relationships. Patient evaluation and drug therapy assessment is based on an interview that uses mental status examination, as well as standardized psychiatric rating scales, to assess symptoms and adverse effects. Clinical service involves collaboration with physicians, nurses, social workers, psychologists, and other health care professionals.[1,3]

A specialized residency is ideal to prepare a pharmacist for a career in psychiatric pharmacy practice. There are approximately 25 to 30 1-year, post-PharmD residencies in psychiatric pharmacy practice across the United States, with 17 receiving accreditation by the American Society of Health-System Pharmacists (ASHP).[7,8] Each residency may offer a unique feature such as an ambulatory care focus or teaching skills development; however, the emphasis of residency training is on specialized clinical knowledge and skill development.[9]

Given the high prevalence of psychiatric illness and specialized expertise required for successful pharmacy practice, it makes sense that psychiatric pharmacy has become one of the five specialty practice areas certified by the U.S. Board of Pharmaceutical Specialties. As of December 2001, there were 387 certified psychiatric pharmacy specialists.[10] The certification process started in 1990, when a coalition of educators and practitioners identified a need to define the specialized knowledge and skills required to function as a competent psychiatric pharmacy specialist.[2] The coalition's petition was sponsored by the ASHP and the first examination took place in December 1996.

This article presents opportunities in psychiatric pharmacy, provides examples of model practice settings, discusses the impact of psychiatric pharmacy on health outcomes, reviews the tools used by specialty practitioners, and discusses networking opportunities in psychiatric pharmacy specialty practice.

OPPORTUNITIES IN PSYCHIATRIC PHARMACY

Opportunities in psychiatric pharmacy continue to expand with specialists practicing in hospitals, clinics, long-term care facilities, developmentally disabled centers, prisons, academia, and the pharmaceutical industry.[1,3] Although acute care facilities exist to treat the most severely ill patients, primary care clinics provide service for the majority of patients. Model practice settings exist for both acute and primary care, and are discussed later in this article. Other opportunities are discussed in this section.

Encyclopedia of Clinical Pharmacy
DOI: 10.1081/E-ECP 120006238

It is estimated that 25–50% of patients in long-term care facilities suffer from neuropsychiatric disorders that are functionally impairing.[11] At least 26% of incarcerated adults[12] and 52% of children in the juvenile justice system meet criteria for a DSM-IV-TR disorder.[13] When substance abuse/dependency is included in the adult population, the incidence rises to 71%.[12] Psychiatric pharmacist specialists can provide consultation on optimizing drug therapy for patients in these settings.

Half of the homeless suffer from mental illness, and rely on community outreach, missions, and government-run clinics to provide service.[14,15] A psychiatric pharmacist is uniquely qualified to manage the coordination of medication follow-up for these patients. Developmentally disabled individuals are often treated for a combination of neurologic and psychiatric problems and, therefore, have unique drug interaction considerations and communication obstacles. Psychiatric pharmacists currently serve as consultants and members of multidisciplinary treatment teams to optimize the care of these individuals.[3,16]

Quality of care for individuals with psychiatric illness has come under scrutiny with several studies documenting a need for improvement in medication management.[17–19] Psychiatric pharmacists have the opportunity to improve care through quality assurance surveys/drug-use evaluations, patient and staff education,[3,20] and direct medication management.[1] Psychiatric pharmacist specialists are an important community resource for consumer advocacy groups such as NAMI, also known as the National Alliance for the Mentally Ill.[21]

In addition, psychiatric pharmacists have opportunities for involvement in treatment guideline development for psychiatric disorders. The American Pharmaceutical Association recruited psychiatric pharmacy specialists to develop their guidelines for psychiatric disease state management.[22]

Dr. Lynn Crismon's work in developing the Texas Medication Algorithms for treatment of depression in children and adults and the treatment of attention-deficit hyperactivity disorder in children[2,23,24] laid the groundwork for psychiatric pharmacists to work with psychiatrists, psychologists, other health care professionals, and consumer groups to develop and implement national therapeutic guidelines.

MODEL PRACTICE SETTINGS

An attractive feature of psychiatric pharmacy is that it offers creativity in developing a practice that is individualized to the setting and patient population of interest to the practitioner. Model practice settings in hospitals and several clinics are discussed in the following paragraphs.

Hospital

Psychiatric pharmacy specialists in hospitals provide a wide range of services, including participation in multidisciplinary treatment planning, medication education groups for patients, therapeutic drug monitoring, discharge counseling, and quality assurance.[3,20] Model practices exist across the United States for the acute care psychiatric pharmacist; however, the scope of practice is variable based on staffing, institutional support, and interest of the practitioner. Through the years, a patient-focused model has evolved using the principles of pharmaceutical care whereby the pharmacist develops a professional relationship with the patient in addition to staff and takes responsibility for health outcomes.

Psychiatric pharmacy in the acute care setting involves patient interviews for initial assessment and follow-up monitoring. The pharmacist obtains a medication history to facilitate treatment plan development, in addition to participating in multidisciplinary rounds for exchange of information and therapeutic decision making. The inpatient psychiatric pharmacist conducts therapeutic drug-level monitoring of lithium and anticonvulsants. Conducting medication education groups and individual medication counseling sessions are standard functions of the inpatient psychiatric pharmacist.[25]

Primary Care

In the primary care setting, there are several practice models for pharmacy-run clinics. Typically, patients are evaluated by a psychiatrist and referred to the psychiatric pharmacist for medication management and ongoing assessment.[26] The Veteran's Administration (VA) health care system was one of the first to use psychiatric pharmacy specialists in mental hygiene clinics in the 1970s. Currently, the VA health care system supports psychiatric pharmacist specialist involvement in several psychiatric clinics, including the cognitive disorders, mood disorders, psychiatry emergency, geropsychiatry, and clozapine clinics.[27,28]

Extent of involvement varies across VA systems. For example, at the VA clinic in La Jolla, California, a psychiatric pharmacist's scope of practice includes: 1) assessing clinical response to medication via mental status exam and psychiatric interviewing techniques; 2) assessing development of adverse drug reactions; 3) ordering and evaluating appropriate laboratory tests to assess clinical response, assess development of adverse drug reactions, and evaluate therapeutic drug levels; 4) making changes in psychotropic drug therapy using the physician order form or through direct order entry into the computer; 5) assessing patient compliance with medications by analyzing computer dispensing records and quantities of medications dispensed versus doses remaining; 6) documenting findings, actions, and plans in the patient's medical records on the progress notes form or through direct progress note entry into the computer; 7) providing prescriptions for all medications with enough medications to last until the patients' next appointment; 8) providing patient medication education, including methods of coping with certain side effects, recognizing symptoms of toxicity, and emphasizing the importance of compliance, and when appropriate, providing written information; and 9) rescheduling patients for follow-up appointments with an appropriate clinician. In a VA clinic in Waco, TX, a psychiatric pharmacist specialist has similar scope of practice but is assessed quarterly, in writing, by the supervising psychiatrist.

Scott and White hospital in Texas is an example of a pharmacy-run women's health clinic. In this model, the obstetrics-gynecology physician or nurse identifies patients at risk for mood disorders, including premenstrual syndrome and premenstrual dysphoric disorder, and refers them to the pharmacist for further evaluation, treatment, and drug therapy monitoring. Patient approval ratings were 96% excellent.[29]

Industry

Psychiatric pharmacy specialists are vigorously recruited by industry to serve as medical science liaisons, neuroscience managers, members of advisory boards, and resources of drug information for physicians and other health care professionals.[1]

BENEFITS OF SPECIALTY PRACTICE

Several U.S. studies report improved care, improved level of functioning, and economic savings attributed to psychiatric pharmacist interventions;[3] however, the studies are small, generally retrospective in nature, and most lack rigorous controls. One chart review of 60 clinic patients compared pharmacist prescribing with psychiatrist prescribing and determined pharmacist prescribing was as good as or better than physician prescribing.[26] The largest study was a chart review in some 4700 patients, which documented that 66% of pharmacist recommendations were implemented with an extrapolated cost savings based on decreased clinic visits and decreased prescriptions of $22,241.25 over 3 months.[30]

One prospective study from Australia analyzed clinical pharmacy interventions on an inpatient psychiatric unit over a 6-month period. Two hundred and four interventions were proposed for 69 patients, 91.7% of which were accepted. Some of the interventions (20.3%) were estimated to be of major clinical significance, with added cost savings of 24,700 based on cutting 38 days of inpatient care at $650/day.[31]

TOOLS

Psychotropic drug therapy expertise, interview technique, and the mental status exam are the most used tools of the psychiatric pharmacist. Validated psychiatric rating scales are also used and allow objective measurement of drug therapy outcomes. Psychiatric pharmacists develop expertise in using standardized rating scales such as the Hamilton-Depression Rating Scale and the Monitoring of Side Effects Scale (OSES).[32,33] American Psychiatric Association rating scales and online references, such as *Clinical Pharmacology*,[34] are available on CD-rom and make the information easily retrievable in settings with computer capabilities. Clinical psychiatric pharmacists use portable laptop or notebook computers and personal data assistants, or PDAs, to keep track of patient profiles, drug therapy recommendations, and outcomes.

NETWORKING

Networking can be accomplished by participating in national meetings where psychiatric pharmacy specialists gather to exchange ideas, participate in neuropsychiatric educational programming, and discuss professional issues such as residency training, reimbursement, and health care policy. These national meetings include the College of Psychiatric and Neurologic Pharmacists,[8] ASHP's Section of Clinical Specialists in Psychiatric and Neurologic Pharmacy,[7] and the American College of Clinical Pharmacy's (ACCP's) central nervous system practice-related network.[35] The "psypharm" listserv has approximately 400 participants (George Foose, personal communication) and is an effective tool for "rapid-response" networking with national and international psychiatric pharmacists. To become a member of psypharm and/or of the College of Psychiatric and Neurologic Pharmacists (CPNP) listserv, log on to the CPNP web site or contact Dr. George Foose via e-mail at gfoose@msn.com.[8,36]

INTERNATIONAL

Psychiatric pharmacy is an internationally recognized specialty, with organized representation in Canada, the United Kingdom, Scotland, and the Netherlands. The United Kingdom Psychiatric Pharmacy Group has its own web site,[37] and offers a postgraduate certificate in psychiatric pharmacy for practitioners in Great Britain and other countries. Currently, certification is ongoing for practitioners in Hong Kong, Canada, and New Zealand. The Scottish organization is known as The Scottish Pharmacists in Mental Health. The Dutch Association of Hospital Pharmacists has a special interest group in the field of psychiatry, with goals to elevate psychiatric pharmacy practice to include pharmaceutical care and therapeutic drug monitoring, in addition to distributive services. Practitioners from the United Kingdom and Canada report practice activities that are similar to U.S. psychiatric pharmacists with regard to multidisciplinary treatment planning, patient and staff education, and quality assurance.

REFERENCES

1. Cohen, L.J. The role of neuropsychiatric pharmacists. J. Clin. Psychiatry **1999**, *60* (Suppl. 19), 54–58.
2. Crismon, M.L.; Fankhauser, M.P.; Hinkle, G.H.; Jann, M.W.; Juni, H.; Love, R.C.; Ray, M.D.; Stimmel, G.L.; Wells, B.G. Psychiatric pharmacy practice specialty certification process. Am. J. Health-Syst. Pharm. **1998**, *55* (15), 1594–1598.
3. Jenkins, M.H.; Bond, C.A. The impact of clinical pharmacists on psychiatric patients. Pharmacotherapy **1996**, *16* (4), 7-8-714.
4. Kessler, R.C.; McGonagle, K.A.; Zhao, S.; Nelson, C.B.; Hughes, M.; Eshleman, S.; Wittchen, H.U.; Kendler, K.S. Lifetime and 12-month prevalence of DSM-III-R psychiatric disorders in the United States. Results from the national comorbidity survey. Arch. Gen. Psychiatry **1994**, *51* (1), 8–19.
5. *Diagnostic and Statistical Manual of Mental Disorders, Fourth Edition, Text Revised*; American Psychiatric Association, APA Press; 2000; copyright 2000.
6. *Top 200 Drugs of 2001 as Published in Drug Topics*; www.drugtopics.com (accessed March 4, 2002).
7. *American Society of Health System Pharmacist's (ASHP) Residency Directory*; The ASHP website at www.ashp.org (accessed July 2001).
8. College of Psychiatric and Neurologic Pharmacists (CPNP). www.cpnp.org (accessed July 2001).
9. ASHP Residency standards for psychiatric pharmacy practice. ASHP's Residency Directory **2001**, (2), 237–245.
10. *Board of Pharmaceutical Specialties (BPS) Pharmacy Specialization Newsletter*; American Pharmaceutical Association (APhA), December 2001; 11 (3).
11. Furniss, L.; Burns, A.; Craig, S.K.L.; Scobie, S.; Cooke, J.; Faragher, B. Effects of a pharmacist's medication review in nursing homes: Randomised controlled trial. Br. J. Psychiatry **2000**, (176), 563–567.
12. Birmingham, L.; Mason, D.; Grubin, D. Prevalence of mental disorder in remand prisoners: Consecutive case study. BMJ **1996**, *313* (7071), 1521–1524.
13. Garland, A.F.; Hough, R.L.; McCabe, K.M.; May, Yeh.; Wood, P.A.; Aarons, G.A. Prevalence of psychiatric disorders in youths across five sectors of care. J. Am. Acad. Child Adolesc. Psych. **2001**, *40* (4), 409–412.
14. Tolomiczenko, G.S.; Goering, P.N.; Durbin, J.F. Educating the public about mental illness and homelessness: A cautionary note. Can. J. Psychiatry **2001**, *46* (3), 253–257.
15. Hotchkiss, B.D.; Pearson, C.; Lisitano, R. Pharmacy coordination of an indigent care program in a psychiatric facility. Am. J. Health-Syst. Pharm. **1998**, *55* (12), 1293–1296.
16. Berchou, R.C. Effect of a consultant pharmacist on medication use in an institution for the mentally retarded. Am. J. Health-Syst. Pharm. **1982**, *39*, 1671–1674.
17. Young, A.S.; Sullivan, G.; Burnam, A.; Brook, R.H. Measuring the quality of outpatient treatment for schizophrenia. Arch. Gen. Psychiatry **1998**, *55*, 611–617.
18. Lehman, A.F.; Steinwachs, D.M. Translating research into practice: The schizophrenia patient outcomes research team treatment recommendations. Schizophr. Bull. **1998**, *24*, 1–10.
19. McCombs, J.S.; Nichol, M.B.; Stimmel, G.L.; Shi, J.; Smith, R.R. Use patterns for antipsychotic medications in medicaid patients with schizophrenia. J. Clin. Psychiatry **1999**, *60* (Suppl. 19), 5–11.
20. Moebius, R.E.; Jones, B.B.; Liberman, R.P. Improving pharmacotherapy in a psychiatric hospital. Psychiatr. Serv. **1999**, *50* (3), 335–338.
21. Finley, P. Ask your Psychopharmacist. In *NAMI, Advocate*; NAMI: Arlington, Virginia, Winter 2002.
22. *Treatment Guidelines for Schizophrenia, Major Depression, Bipolar Illness, Panic Disorder, Generalized Anxiety Disorder, Obsessive Compulsive Disorder and Social Anxiety Disorder*; American Pharmaceutical Association Publications: Bethesda, Maryland, copyright 1998–2000.
23. Hughes, C.W.; Emslie, G.J.; Crismon, M.L. The Texas children's medication algorithm project: Report of the Texas consensus conference panel on medication treatment of childhood major depressive disorder. J. Am. Acad. Child Adolesc. Psych. **1999**, *38* (5), 517–528.
24. Plizska, S.R.; Greenhill, L.L.; Crismon, M.L. The Texas children's medication algorithm project: Report of the Texas Consensus conference panel on medication treatment of childhood attention-deficit/hyperactivity disorder

Parts 1 and II. J. Am. Acad. Child Adolesc. Psych. **2000**, *39* (7), 908–927.

25. Stimmel, G.L.; Wincor, M.Z. Mental Health Pharmacy. In *Handbook of Institutional Pharmacy Practice*, 2nd Ed.; Smith, M.C., Brown, T.R., Eds.; Williams & Wilkins Co.: Baltimore, 1986; 535–544.
26. Stimmel, G.L.; McGhan, W.F.; Wincor, M.Z. Comparison of pharmacist and physician prescribing for psychiatric inpatients. Am. J. Hosp. Pharm. **1982**, *39*, 1483–1486.
27. Defilippi, J.L. *Personal Communication on Approved Scope of Psychiatric Pharmacy Practice in the Veteran's Administration Mental Hygiene Clinic, Waco, Texas*; May 2001.
28. Dishman, B. *Personal Communication on Approved Scope of Practice for Psychiatric Clinical Pharmacy Specialist at the Veteran's Administration Healthcare system, La Jolla, California*; July 2001.
29. Jermain, D.M.; Sulak, P.J.; Woodward, B.W.; Knight, A.G. Psychopharmacy medication clinic in a managed care women's health setting. Am. J. Health-Syst. Pharm. **1997**, *54* (23), 2717–2718.
30. Lobeck, F.; Traxler, W.T.; Bobiner, D.D. The cost-effectiveness of a clinical pharmacy service in an outpatient mental health clinic. Hosp. Commun. Psychiatry **1989**, *40*, 643–644.
31. Alderman, C.P. A prospective analysis of clinical pharmacy interventions on an acute psychiatric inpatient unit. J. Clin. Pharm. Ther. **1997**, *22* (1), 27–31.
32. Sajatovic, M.; Ramirez, L.F. *Rating Scales in Mental Health*; Lexi-comp Inc.: Hudson, Ohio, 2001, copyright 2001.
33. *American Psychiatric Association (APA) Rating Scales*; 2000, copyright 2000.
34. *Clinical Pharmacology*; www.clinicalpharmacology.com.
35. American College of Clinical Pharmacy (ACCP). www.accp.org.
36. Psypharm listserv. psypharm@listserv.unc.edu.
37. United Kingdom Psychiatric Pharmacy Group (UKPPG). www.ukppg.org.uk (accessed July 2001).

Quality Assurance of Clinical Pharmacy Practice

Pilar Mas Lombarte
Fundación Hospital-Asil de Granollers, Granollers, Spain

INTRODUCTION

Organizations exist as long as they are able to produce services that are of interest to someone—this someone being customers. Pharmacists create small organizations (e.g., hospital pharmacy services, community pharmacies, ambulatory settings) that are located within other organizations (e.g., hospitals, primary care, home care), which in turn are integrated into what we may consider the healthcare system organization.

The final customers of the healthcare service are patients and their families, and the final product is health. Health as a product is not easy to define and measure. A way to measure the product of healthcare systems is to look at their outcomes. Today, the outcomes of healthcare systems are considered to be economic, clinical, and humanistic in nature.

Pharmacists participate in the healthcare system and in its final product. Sometimes they directly access the patient, and other times, they indirectly access the patient through other health professionals (doctors, nurses), providing them with what belongs to us—our knowledge of drugs. Drugs are widely used as resources in all healthcare systems. Our objective would be to achieve a safer and effective medication-use process.

However, the healthcare system is only furnished with limited resources, facing endless new technologies and growing service demand by the population. All this translates into tremendous economic pressure on healthcare professionals.

This is the starting point in the development of the subject of quality assurance in pharmacy services.

ADVANCING TOWARD QUALITY

There are many definitions of quality;[1,3] however, we reduce the concept to the bare bones. That is, quality is doing things right, or better said, allowing our customers to receive good service because they, precisely, are the ones who rate the quality of the service (Fig. 1).

Here, we outline two basic questions:

1. *Who are the customers of the pharmacy service?* Patients, their relatives, physicians, nurses, society, managers and administrators, politicians, students, and so on are the customers. They do not always share the same values and can even have conflicting interests. This is due to agency relationships, so frequent in healthcare.

2. *Which are the services offered by the pharmacy service?* This question does not refer to what pharmacists do, how they invest their time, or what their tasks are—all that is well known. We know what is done, but it is more difficult to prove what it is for; that is, what is the usefulness for the patient? The question refers to what clients receive from pharmacy services, from their own point of view. As stated previously, what customers get is what makes sense to pharmaceutical organizations.

It is strategic for clinical pharmacy services to identify and segment customers to meet their expectancies and needs. It results in good customer management.

As a profession, the services we offer are not fully agreed upon or conceptualized, but delivery of medications, and manufacture and handling, are. These services are our baseline services. It is far more difficult to consensuate and conceptualize the pharmacists' cognitive services and their added value to health process, yet there have been some attempts.[4–6] The reason is likely to be that we are not paid to do that,[7] so we do not feel obligated to do so, unlike other professions with a great knowledge load.

We should find a way to accomplish it and offer a portfolio of services. This portfolio of pharmaceutical services should be patient oriented and customer focused. There are some works in the literature that describe the needs assessment with different customers (patients, healthcare teams, etc.).[7–11]

Encyclopedia of Clinical Pharmacy
DOI: 10.1081/E-ECP 120006379

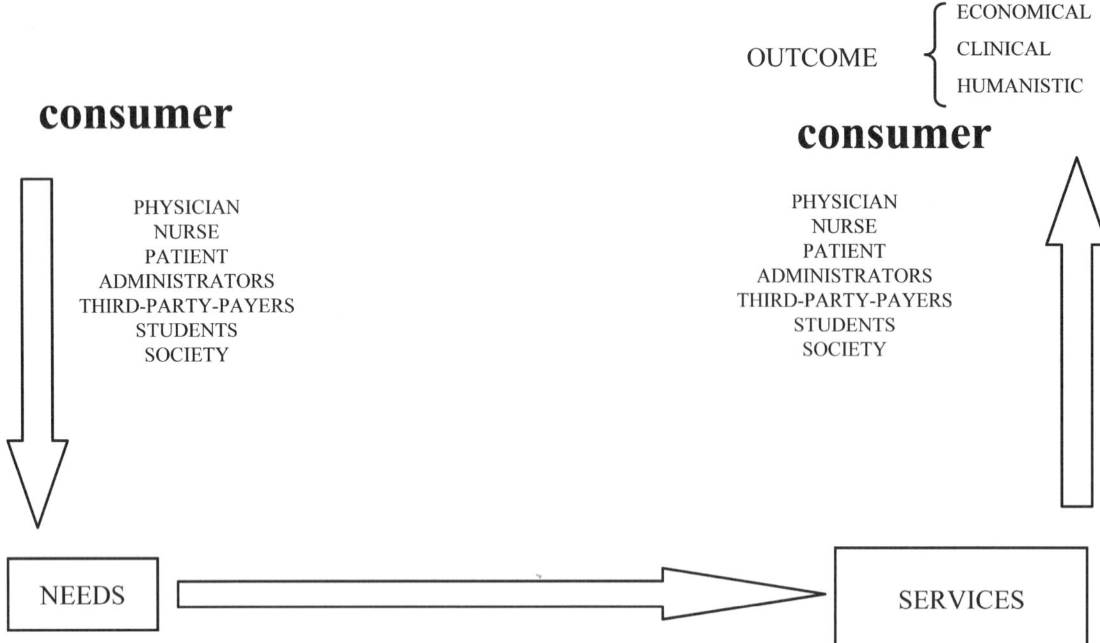

Fig. 1 Rating quality of service.

The next step is to define what the characteristics are of each service we offer in this portfolio. By these characteristics, we are defining the variables of quality itself. Quality would also include, besides technical aspects, those aspects related to communication skills, which are quite valued in customer satisfaction. The cognitive, behavioral, and emotional aspects of communication are thus considered.

In working toward quality, there are several steps and different tools:

- Certification.
- Accreditation.
- Self-evaluation.

Certification

The object of certification is to ensure that what is done is what was said would be done. In certification, we normalize both processes and procedures. A *process* is a sequence of activities performed to provide a service. The activity is what is done, whereas a *procedure* is the documentation that describes how to perform a process (a set of activities). A *service* is what the patient gets, and the *outcome* is the result of the service.

To provide our customers with these services, we are furnished with material and human resources that produce them. The production of services takes place by means of processes (Fig. 2). To obtain high-quality services, processes need to be very well defined, and to be known and accepted by those performing them. A good way to define them is using flowcharts.

Quality should never be improvised; quality is planned. Service production processes should be well defined. Those performing them should known them, and be well prepared and trained for this objective of quality. People want to work properly, so organizations should prepare the ground, clearly stating what it is expected from them, training them adequately, and commending that is done well.

In this normalization of processes, the activities to be performed should be defined, as well as their sequence and who does it and how.[12,13] This is a good time to incorporate technologies such as information system, robotics, communication systems, and so on.[14,15]

In the definition stage of this process, it is essential to be flexible, to have an open and imaginative mind, and to question everything—what is done, why, for what reason, and for whom (in these steps, some tasks are often found to be unnecessary and can be left undone because they provide nothing to the final service or are repeated; some activities are done just because it has always been that way). It is good to have a variety of people with different points of view, and it is advisable that the staff involved in the process also participate in it, whatever their train-

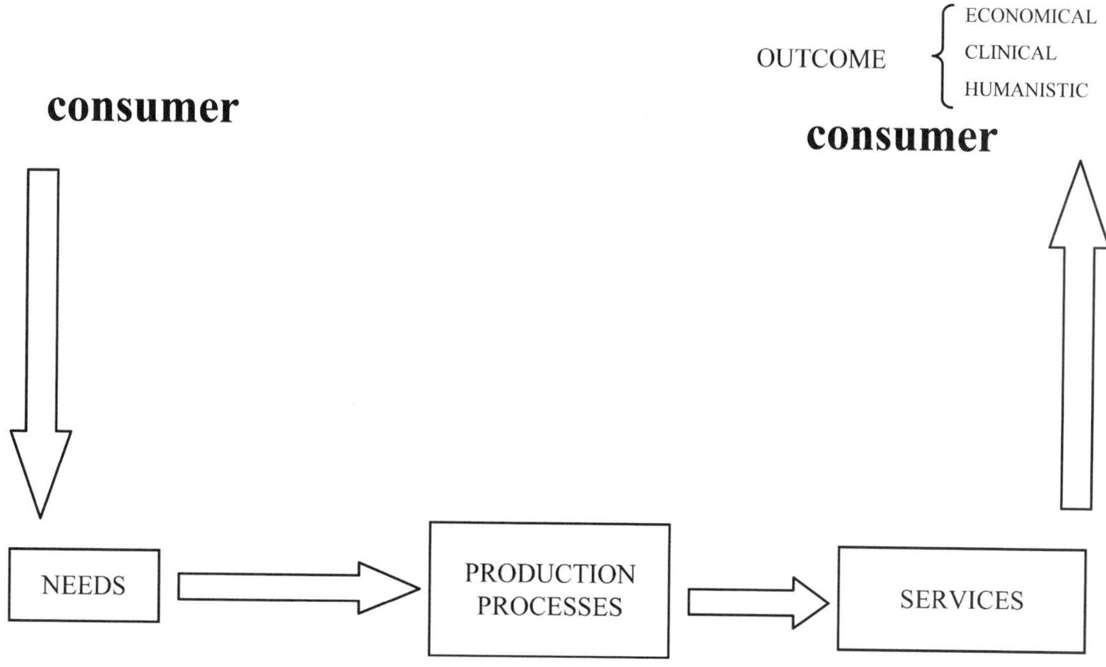

Fig. 2 Production of service by means of a process.

ing, because they are the ones who know their work best and know how to improve it. It has the advantage of facilitating empowerment, homogenizing criteria, and forcing the culture of quality to develop.

Once the process is defined, it simply has to be performed as described. The objective of this is to avoid variability in its execution and thus in its quality, depending on the person who performs it.

Everything should be normalized: process controls, equipment servicing, staff training, quality control (few indicators on critical points), frequency of the process review, and definition of responsibilities. The documentation of these activities should also be normalized.[16–18]

Processes normalization is distinct in each organization, because each organization is unique. This should be the second step toward quality. Defining our customers and our service portfolio is the first step. Each service should have a well-defined and normalized elaboration process, with a clear beginning and end.

So far we have discussed the operative processes—that is, the service production processes—but there are other equally relevant processes, such as support services (not perceived by the customer, but also essential, e.g., maintenance, purchases, etc.) and strategic services (they orient the whole organization; Fig. 3). The quality of all these processes is susceptible to evaluation.

Accreditation

Although certification ensures the homogeneity in the quality of organizations, accreditation is based on the creation of quality standards in service quality and in the comparison among several organizations.[19,20] Accreditation is granted by external organizations, which set the criteria and standards that are used as indicators of health. Before undergoing external accreditation, it is essential to know the requirements and to specifically prepare for them. An external accreditation is an acknowledgment that the quality requirements established by that organization are fulfilled. Actually, quality is predefined by means of indicators and standards. Accreditations may include structure, processes, and results standards. The current trend is toward the assessment of results, whenever possible.

Self-Evaluation

In self-evaluation, organizations enter a constant circle of questioning what they are doing, how, for whom, and how they can improve it. It involves an important degree of dynamism and maturity throughout the organization, with a clear, decided focus on the customer and society. It is a path toward excellence: the continuous culture of

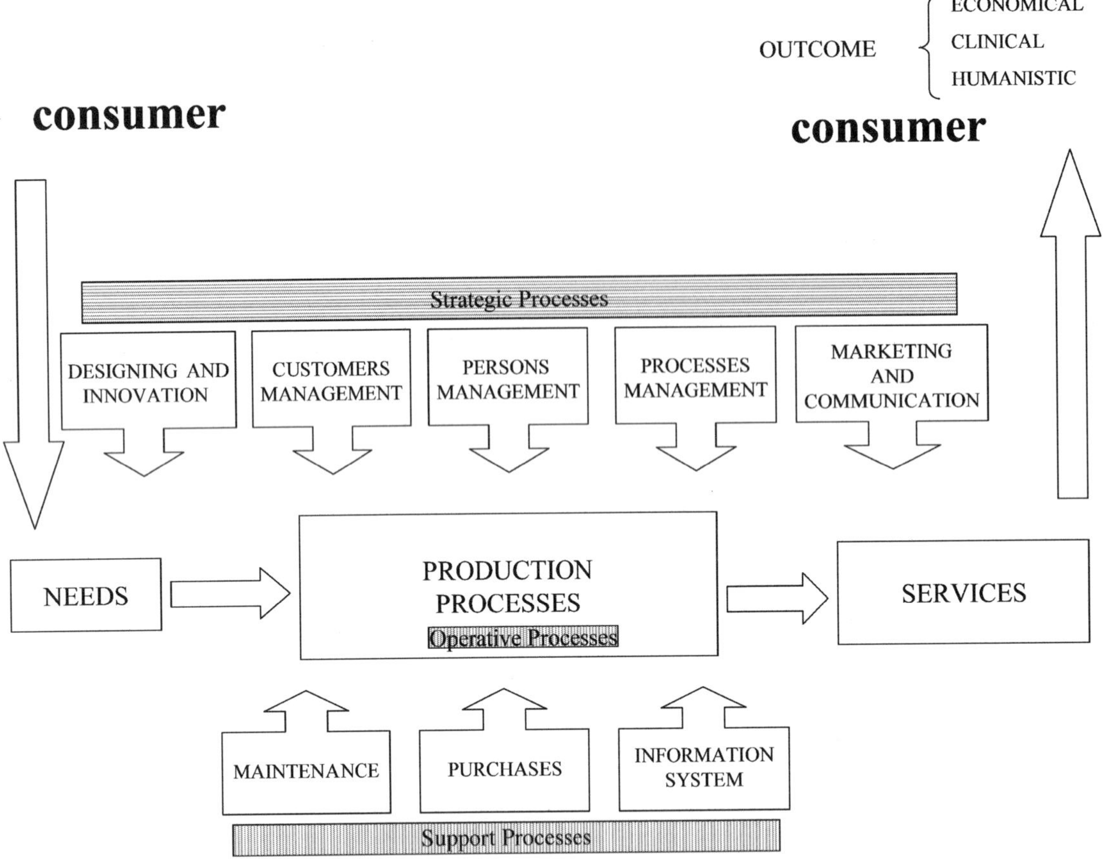

Fig. 3 Strategic, operative, and support processes.

quality. Their common feature is a proactive approach to quality. It deals with double-checking that what the organization designs is what customers really want and need, and if that it is precisely what the organization is actually providing.

At this level, we introduce concepts that are in vogue such as benchmarking and innovation. Both look for improvements in services and/or processes. The first concept does so by searching for better ways of working in other organizations; the second concept does so by means of imagination and creativity.

Benchmarking was first introduced in business science as an efficient and effective way of optimizing companies. It deals with looking for practices that proved successful in other companies, and implementing them into one's own company, or in plain words, copying the best ones.[21–24]

Innovation is the key to the future. It means to be alert as to how society, the healthcare system, and the macro- and micro-context in which we live, will evolve. An example of adaptation to the environment is the very concept of pharmaceutical care meant a great innovation for the profession,[25] and the contribution of the pharmacist in the detection and prevention of therapeutical errors, designing specific programs for it.[26,27]

In the healthcare sector, innovation is closely related to research and scientific evidence. We should increase our research on quality and also make it better.[28–32] Then, it has to be published and disseminated.[33] This should be done with high scientific level works, producing scientific evidence of quality. The study question should be less "how am I doing it?" (pharmacist oriented) and more "how do I improve the care offered to the patient?" (patient oriented).

KEY POINTS

Conceptualization and measurement are the key points. Quality management is achieved by the management of customers, people, and processes. It is essential to follow a strategy because quality is not improvised. Conceptualization is what gives congruity to the whole system. In practice, this implies the following tactic.

Defining the Portfolio

Defining the portfolio of services and the level of quality of each service. Knowing the needs before designing and performing the service. Besides expressed needs (demand), we have to explore the hidden, or not expressed, needs. We should know what services and with what characteristics of quality are of interest to our customers, avoiding unnecessary efforts in services that will not be valued or that will be useless to us (although our unit may consider them interesting). In this section, qualitative research is a tool to consider. This is a strategic step.

Designing Process

It is important to design, or redesign the process so it fits the needs of each service.[34] Many of our pharmaceutical services are related to the production of clinical services. Considered with products, these services have many special characteristics to consider:

- They have a high content in customer care.
- Their main element is the human factor.
- Part of the quality of the service depends on the customer's perception.
- The satisfaction of the service provider is transmitted to the customer.
- Many services waste time while being produced, and the customer participates in its production (e.g., interviews with the customers). This means that the services cannot be stored or repeated. This is the concept of *servuction*, or production of services.
- Services can be personalized.
- Demand is not stable and is often unpredictable.

These conditions are especially met by services related to knowledge. This is why the human factor is so important in the quality of pharmaceutical services with high added value. It is essential to reach an agreement on criteria, continuous education, and documentation. The degree of outcomes we observe in our patients depends in great measure on this quality. Human quality potenciate the technical quality.

These services are of an intangible nature, so it is sometimes positive to introduce elements to make invisible operations visible (e.g., to develop a physical support allowing the customer to see it clearly). Ours is a multidisciplinary job, and this must be reflected not only in our research, but also in our attitude and daily work, as full patient-focused members of the healthcare team.[35,36]

We have to be readily accessible, both in terms of space and time. We have to leave pharmacy offices and go to the patient and the healthcare team.[37] We must clearly and firmly lead all those activities related to drugs because this is the knowledge field of our profession. Leadership has a lot to do with effectively communicating to influence others' actions, attitudes, beliefs, and so on. Pharmacy services must relocate themselves strategically as proactive agents in the healthcare team.

Dissemination

Disseminate by notifying the customer about our work, and any improvement of it (marketing). Pharmaceutical services should notify the value of the processes in which they are leaders, and it should be loud and clear.

Due to its very nature, the work of pharmacists is multidisciplinary, as they are full, patient-focused members of the healthcare team.[38,39] To the patient, integration of the different areas or specialists who provide care is desirable. It is also true for the healthcare system. It is fundamental, of course, not only to have extensive clinical and drug knowledge, but also communication, negotiation, and leadership skills, as well as specific training in clinical interviews, pertaining to the customer and to the specific situation within the exercise of the profession. The best way to make ourselves known is with our daily work. It is then that we should let the healthcare staff know it, participating in other professions' forums, in consumers' associations, teaching other professionals, and using different channels, including mass media.

A multidisciplinary team works in coordination, sharing both patients and responsibility, both in the patient's direct care and in planning such care. Quality is defined in care planning, such as clinical sessions, elaboration, implementation, follow-up and evaluation of protocols, clinical paths and clinical guidelines, participation in different clinical commissions, and so on. Clinical pharmacists must be recognized as leaders in all drug-related aspects. It is not a right, it must be earned.

In an automated, natural manner, but in different settings from the strictly pharmaceutical, clinical pharmacists integrate clinical research and publications, together with other professionals. This involves disseminating the services that clinical pharmacy has to offer.

Actually, all healthcare professionals should learn that the center of the healthcare system is the patient, and rather than classifying the patients' functions, what is important are those processes that provide an added value to the patients' health, and those who lead them. It means establishing alliances between members of the healthcare team. Perhaps we should start viewing the members of health teams as being coresponsible for care, and even incorporate patients themselves into the team, with a first-line role, taking active part in decisions. The clinical pharmacists' goal is for patients to see them as an ally, someone on whom they can rely.

Measure

It is important to have a discussion with the customer after delivering our service to ensure that, once started, this service effectively fulfills its mission (considering its three dimensions: economic, clinical, and humanistic, along with patient satisfaction). In practice, this means to measure. Pharmaceutical services have to be measurable to be self-evaluated, corrected, and improved.

We can measure quality in our activities and in the services we produce,[40,41] as well as the impact of our contribution to the outcomes of the healthcare system in terms of quality[42] (Fig. 4).

For economic outcomes, we can use pharmacoeconomy as the main tool.[43] There is a great variety of clinical indicators that we can relate to drugs efficacy and safety. These clinical indicators, obtained with designs from clinical epidemiology (observational and experimental),[44] are excellent to measure clinical outcomes. To obtain indicators for humanistic outcomes, such as satisfaction and quality of life, we have different tools, such as surveys, and different qualitative research methods (interviews, focus groups, etc.).[45–48]

Measuring wastes some resources, and the concept of cost-opportunity makes us cautious when tempted to measure whatever is at hand. We should measure little, and just what is critical (even monitoring and

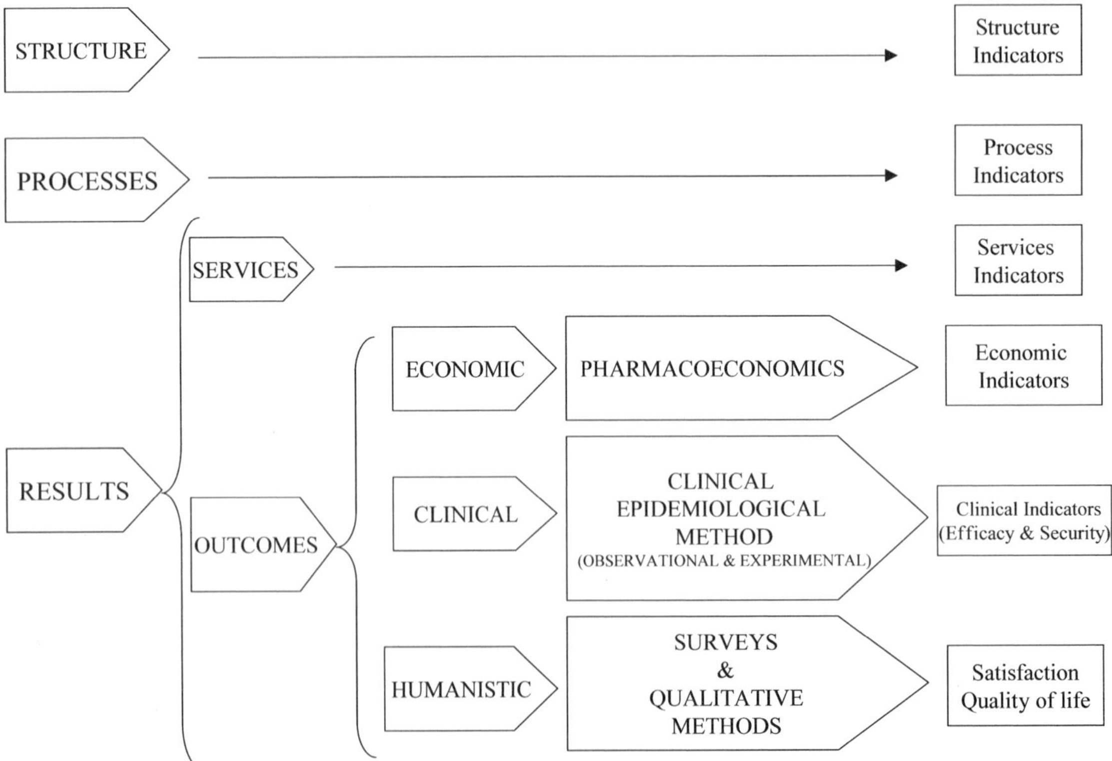

Fig. 4 Quality research.

measuring at a preestablished frequency) and/or important and significant. Before deciding to measure, we must be sure that we are using the best indicator as possible to reach our objective. We must know why and for what we are measuring—it is not the same to obtain a certification, make a decision, or research and publish.

The same criteria should be applied for technical questions of internal interest and for the clinical aspects of interest to the patient and the rest of healthcare staff. In this case, the language should be common to the rest of the healthcare staff, and indicators should be clinical indicators. This information, both positive or negative, has to be disseminated, clarified, and shared. We must insist on the use of clinical and satisfaction indicators, as well as of economic ones.

Feedback

Quality is something one pursues in an active manner. Once we start the path toward quality, we cannot abandon it: it just becomes part of our daily routine.

CONCLUSION

In this article, it was our intent to present a perspective on this topic, providing different tools to approach the quality of pharmaceutical services. These have been placed in a context where the concepts of limited resources, cost-opportunity, and efficacy are implicit. The concept has been abstracted so it can be applied to the public or private sectors, to hospital or ambulatory settings, and to different societies, with different values.

As for the external quality systems, we deliberately avoided the description of official quality systems, such as the Joint Commission on Accreditation of Healthcare Organizations (accreditation), International Standarization Organization (certification), and European Federation for Quality Management (self-evaluation), because it would have been impossible to cover them all. Instead, we abstracted the topic of quality, and all these external quality systems will probably assume most of the things discussed in this article.

Quality is planned and evaluated. In other words, quality is managed; it cannot be detached from management in its broader sense. Quality management is achieved by quality customers, people and process management. Quality varies according to the context and also with time. Quality is dynamic because the environment is dynamic. We must continuously adapt ourselves to the environment and innovate. What now is a gold standard will perhaps be a minimum tomorrow.

Quality is evaluated by measuring relevant indicators. Outcomes indicators (economic, clinical, and humanistic) will be the major importance in the future for clinical pharmacist services. Clinical pharmacist services must relocate themselves strategically as a proactive agent and lead drug therapy in the healthcare team.

The challenge will be the economic evaluation of pharmacy services. Once the products of pharmaceutical services and its outcomes are defined, considering economic, clinical, human outcomes, and once the minimal or standard quality if the services (both basic and value-added services, with high degree of knowledge), then we must know its cost.[49] In the context of healthcare systems, where they are currently immersed, with huge economic pressure, it is necessary to prove that the resources used by pharmaceutical services are cost effective for the healthcare system and the patient. In other contexts, and maybe in the future, it will be important to price these pharmaceutical services that our patients are going to receive.[50,51] By orienting our services through quality, they will probably be more effective, because we will not incur in the cost of nonquality.

We need to develop methods for measuring and meeting customers' expectations (needs and demands) to provide both baseline and value-added services, successful in kind and in quality. It is said we are in the age of knowledge. We pharmaceutics are well versed on the medications-use process. Let this knowledge be useful for our customers and society. Let us convey it in a clear, effective way. It is just quality.

REFERENCES

1. Deming, W.E. Calidad, productividad y competitividad. In *La salida de la crisis*; Diaz de Santos S.A.: Madrid, 1989.
2. Juran, J.M. El liderazgo para la calidad. In *Manual para directivos*; Diaz de Santos S.A.: Madrid, 1990.
3. Crosby, P. *La calidad no cuesta*; Compañía Editorial Continental S.A.: Mexico DF, 1990.
4. Ellis, S.L.; Billups, S.J.; Malone, D.C.; Carte, B.L.; Covey, D.; Mason, B.; Jue, S.; Carmichael, J.; Guthrie, K.B.S.; Sintek, C.D.; Dombrowski, R.; Geraets, D.R.; Amato, M. Types of interventions made by clinical pharmacist in the IMPROVE study. Pharmacotherapy **2000**, *20* (4), 429–435.

5. Maine, L.L. Pharmacy practice activity classification. J. Am. Pharm. Assoc. **1998**, *38* (2), 139–148.
6. Rudis, M.I.; Brandl, K.M. Position paper on critical care pharmacy services. Crit. Care Med. **2000**, *28* (11), 3746–3750.
7. Williams, S.E.; Bond, C.M.; Menzies, C. A pharmaceutical needs assessment in a primary care setting. Br. J. Gen. Pract. **2000**, *50*, 95–99.
8. Ruta, D.A.; Duffy, M.C.; Farquharson, A.; Young, A.M.; Gilmour, F.B.; McElduff, S.P. Determinig priorities in primary care: The value of practice-based needs assessment. Br. J. Gen. Pract. **1997**, *47*, 353–357.
9. Murie, J.; Hanlon, P.; McEwen, J.; Rusell, E.; Moir, D.; Gregan, J. Needs assessment in primary care: General Practitioners' percepcions and implications for the future. Br. J. Gen. Pract. **2000**, *50*, 17–20.
10. Hanlon, P.; Murie, J.; McEwen, J.; Moir, D.; Russell, E. A study to determine how needs assessment is being used to improve health. Public Health **1998**, *112*, 343–346.
11. Jordan, K.; Ong, B.N.; Croft, P. Researching limiting long-term illness. Soc. Sci. Med. **2000**, *50*, 397–405.
12. Kassam, R.; Farris, K.B.; Cox, C.E.; Volume, C.I.; Cave, A.; Schopflocher, D.P.; Tessier, G. Tools used to help community pharmacist implement comprehensive pharmaceutical care. J. Am. Pharm. Assoc. **1999**, *39* (6), 843–856.
13. Holdford, D.A.; Kennedy, D.T. The service blueprint as a tool for designing innovative pharmaceutical services. J. Am. Pharm. Assoc. **1999**, *39* (4), 545–552.
14. Gulliford, S.M.; Schneider, J.K.; Jorgenson, J.A. Using telemedicine technology for pharmaceutical services to ambulatory care patients. Am. J. Health-Syst. Pharm. **1998**, *55*, 1512–1515.
15. Bates, D.W. Using information technology to reduce rates of medication errors in hospitals. BMJ **2000**, *320*, 788–791.
16. Strand, L.M.; Cipolle, R.J.; Morley, P.C. Documenting the clinical pharmacists' activities: Back to basics. Drug Intell. Clin. Pharm. **1998**, *22*, 63–67.
17. Canaday, B.R.; Yarborough, P.C. Documenting pharmaceutical care: Creating a standard. Ann. Pharmacother. **1994**, *28*, 1292–1296.
18. Overhage, J.M.; Lukes, A. Practical, reliable, comprehensive method for characterizing pharmacists' clinical activities. Am. J. Health-Syst. Pharm. **1999**, *56*, 2444–2450.
19. Godwin, H.N. Achieving best practices in health-system pharmacy: Eliminating the ''practice gap''. Am. J. Health-Syst. Pharm. **2000**, *57*, 2212–2213.
20. Rupp, M.T. Establishing realistic performance standards for community pharmacists' drug-use—review activities in managed care contracts. Am. J. Health-Syst. Pharm. **1999**, *56*, 566–567.
21. Tweet, A.G.; Gavin-Marciano, K. *The Guide to Benchmarking in Healthcare*; Quality Resources, Kraus Org: New York, 1998.
22. Knoer, S.J.; Couldry, R.J.; Folker, T. Evaluating a benchmarking database and identifying cost reduction oportunities by diagnosis-related group. Am. J. Health-Syst. Pharm. **1999**, *56*, 1102, 1107.
23. Bhavnani, S.M. Benchmarking in health-system pharmacy: Current research and practical applications. Am. J. Health-Syst. Pharm. **2000**, *57* (Suppl. 2), S13–S20.
24. Murphy, J.E. Using benchmarking data to evaluate and support pharmacy programs in health-systems. Am. J. Health-Syst. Pharm. **2000**, *57* (Suppl. 2), S28–S31.
25. Heppler, C.D.; Strand, L.M. Opportunities and responsibilities in pharmaceutical care. Am. J. Hosp. Pharm. **1990**, *47*, 533–543.
26. Khon, L.T.; Corrigan, J.M.; Donaldson, M.S. *To Err is Human. Building a Safer Health System*; National Academy Press: Washington, DC, 1999.
27. Manasse, H.R. Pharmacist and the quality-of-care imperative. Am. J. Health-Syst. Pharm. **2000**, *57*, 1170–1172.
28. Vermeulen, L.C.; Beis, S.J.; Cano, S.B. Applying outcomes research in improving the medication-use process. Am. J. Health-Syst. Pharm. **2000**, *57*, 2277–2282.
29. Indritz, M.E.S.; Artz, M.B. Value added to health by pharmacist. Soc. Sci. Med. **1999**, *48*, 647–660.
30. Rogers, S.; Humphrey, C.; Nazareth, I.; Lister, S.; Tomlin, Z.; Haines, A. Designing trials of interventions to change professional practice in primary care: Lesson from an exploratory study of two change strategies. BMJ **2000**, *320*, 1580–1583.
31. White, K.L. Fundamental research at primary care level. Lancet **2000**, *355*, 1904–1906.
32. Park, G.D.; Ball, W.D. Problems with evaluations of clinical pharmaceutical services. Am. J. Hosp. Pharm. **1980**, *37*, 1290–1291.
33. Singhal, P.K.; Raisch, D.W.; Gupchup, G.V. The impact of pharmaceutical services in community and ambulatory care settings: Evidence and recommendations for future research. Ann. Pharmacother. **1999**, *33*, 1336–1355.
34. Schumock, G.T.; Michaud, J.; Guenette, A.J. Re-engineering: An opportunity to advance clinical practice in a community hospital. Am. J. Health-Syst. Pharm. **1999**, *56*, 1945–1949.
35. Franic, D.; Pathak, D.S. Situational factors as predictors of pharmaceutical care in a community setting: An application of the adaptive decision-making model. Pharm. Pract. Manage. Q **1999**, *18* (4), 19–28.
36. Nimmo, C.R.; Jewesson, P.J. Documenting pharmacists' activities in the health record: A mechanism to provide ongoing quality assurance. Ann. Pharmacother. **1993**, *27*, 1405–1406.
37. Campbell, R.K.; Saulie, B.A. Providing pharmaceutical care in a physician office. J. Am. Pharm. Assoc. **1998**, *38*, 495–499.

38. Monk-Tutor, M.R. Pharmacist too reluctant to promote services. J. Am. Pharm. Assoc. **1998**, *38*, 389–391.
39. Spalek, V.H.; Gong, W.C. Pharmaceutical care in an integrated health system. J. Am. Pharm. Assoc. **1999**, *39*, 553–557.
40. Zimmerman, C.R.; Smolarek, R.T.; Stevenson, J.G. Peer review and continous quality amprovement of pharmacists' clinical interventions. Am. J. Health-Syst. Pharm. **1997**, *54*, 1722–1727.
41. ASPH report. Summary of the final report of the ASPH quality assurance indicators development group. Am. J. Hosp. Pharm. **1992**, *49*, 2246–2251.
42. Lipowski, E.E. Evaluating the outcomes of pharmaceutical care. J. Am. Pharm. Assoc. **1996**, *NS36* (12), 726–734.
43. Sanchez, L.A. Applied pharmacoeconomics: Evaluation and use of pharmacoeconomic data from literature. Am. J. Health-Syst. Pharm. **1999**, *56*, 1630–1638.
44. Fisher, L.R.; Scott, L.M.; Boonstra, D.M.; DeFor, T.A.; Cooper, S.; Eelkema, M.A.; Hase, K.A.; Wei, F. Pharmaceutical care for patients with chronic conditions. J. Am. Pharm. Assoc. **2000**, *40*, 174–180.
45. Schommer, J.C.; Kucukarslan, S.N. Measuring patients satisfaction with pharmaceutical services. Am. J. Health-Syst. Pharm. **1997**, *54*, 2721–2732.
46. Green, J.; Briteen, N. Qualitative research and evidence based medicine. BMJ **1998**, *316*, 1230–1232.
47. Mays, N.; Pope, C. Observational methods in healthcare settings. BMJ **1995**, *311*, 182–184.
48. Carroll, N.V.; Gagnon, J.P. Parmacists' perceptions of consumers demand for patient-oriented pharmacy services. Drug Intell. Clin. Pharm. **1984**, *18*, 640–644.
49. George, B.; Silcock, J. Economic evaluation of pharmacy services—fact or fiction? Pharm. World Sci. **1999**, *21* (4), 147, 151.
50. Hawksworth, G.M.; Chrystyn, H. Clinical pharmacy in primary care. Br. J. Clin. Pharmacol. **1998**, *46*, 415–420.
51. Susla, G.; Masur, H. Critical care pharmacist: An essential or a luxury. Crit. Care Med. **2000**, *28*, 3770–3771.

Residencies

Donald E. Letendre
American Society of Health-System Pharmacists, Bethesda, Maryland, U.S.A.

INTRODUCTION

A pharmacy residency is an organized, directed, postgraduate training program in a defined area of pharmacy practice. Residencies exist primarily to train pharmacists (called "residents" during the training program) by providing them the opportunity to accelerate their growth beyond entry-level professional competence in direct patient care and in practice management, and to further the development of leadership skills that can be applied in any position and in any practice setting. Pharmacy residents acquire substantial knowledge required for skillful problem solving, refine their problem-solving strategies, strengthen their professional values and attitudes, and advance the growth of their clinical judgment, a process begun in the clerkships of the professional school years but requiring further extensive practice, self-reflection, and shaping of decision-making skills based on feedback on performance. The residency provides a fertile environment for accelerating growth beyond entry-level professional competence through supervised practice under the guidance of model practitioners. Residents are held responsible and accountable for pursuing optimal medication therapy outcomes in patients.

The residency also provides a fertile environment for accelerating the growth of residents' leadership skills. Each residency offers the opportunity to exercise leadership under the watchful eye of effective leaders. Examples of leadership skills and traits that may be enhanced during a residency include trustworthiness and integrity, comfort with ambiguity, organizational commitment, cross-cultural sensitivity, internalization of the role of service to patients and other customers, recognizing the need for change, change management, persuasive communication, team-building, confidence in one's ability to lead, and realistic self-assessment. To ensure continuous development of future leaders in pharmacy practice, it is widely accepted that leadership skill development be an integral part of pharmacy practice residency training.

TYPES OF RESIDENCIES

There are two types of residency programs: pharmacy practice and specialized. Pharmacy practice residencies, often referred to as "general" residencies, are intended to provide training and education in the fundamentals of exemplary contemporary pharmacy practice. Areas in which the resident typically receives training are acute patient care, ambulatory patient care, drug information and drug-use policy development, and practice management. Because pharmacy practice residencies are offered in a variety of practice settings (e.g., hospitals, health systems, home care programs, long-term care facilities, managed care environments, community practice sites), the relative time that the resident might spend in each area of practice will likely vary depending on the nature of the site and the types of patients being served, as well as the particular practice interests of the resident. Fundamentally, pharmacy practice residencies are intended to produce a well-grounded general clinical pharmacy practitioner.

Specialized residencies are offered in a wide variety of specialty practice areas and are intended to build upon the practice experience gained through completion of a pharmacy practice residency. The following is a list of specialized areas of practice in which American Society of Hospital Pharmacists (ASHP)-accredited residency programs are offered:

- Critical care
- Drug information
- Emergency medicine
- Geriatric
- Infectious diseases
- Internal medicine
- Managed care
- Nuclear
- Nutritional
- Oncology
- Pediatric
- Pharmacotherapy

Encyclopedia of Clinical Pharmacy
DOI: 10.1081/E-ECP 120006359

- Pharmacy practice management
- Primary care
- Psychiatric
- Subspecialities (e.g., cardiology, pulmonary, renal)

ACCREDITATION—WHAT DOES IT MEAN?

ASHP, in cooperation with its professional association partners, administers the only process that grants accreditation status to practice sites conducting residencies. ASHP's authority to grant accreditation is recognized by the Health Care Financing Administration (HCFA). The accreditation process requires that each site demonstrate compliance with established standards of practice and offer a residency that meets the requirements for training.

The process of accreditation ensures that accredited programs are peer reviewed and that requirements for providing a state-of-the-art practice environment are fulfilled.

EVOLUTION OF RESIDENCIES

The Early Years

One of the fundamental issues that led to the founding of the ASHP in 1942 was the need to train practitioners for hospital pharmacy. This segment of practice was largely ignored by pharmaceutical education. The first proposed standard for postgraduate pharmacy training was published for comment in ASHP's *The Bulletin* in 1948.[1] Since then, the requirements for training, as reflected in the accreditation standards that have evolved throughout the years, have always attempted to challenge programs to be at the vanguard of practice.

The initial pharmacy residencies—or ''internships'' as they were called—were established in the 1930s, first at the University of Michigan Hospital in Ann Arbor under the direction of Harvey A. K. Whitney and Edward C. Watts. Other early programs were at the University of California Hospital in San Francisco, Duke University Hospital, and St. Luke's in Cleveland. By the early 1950s, it was estimated that at least a dozen hospitals offered hospital pharmacy internship training.[2] In December 1962, preceptors of the programs in existence at the time and other hospital pharmacy leaders convened a meeting that in retrospect helped set the framework for the accreditation process that is in use today. At that meeting, the term ''residency'' was first adopted (replacing the term ''internship'') and participants agreed on a set of standards that would be used to conduct the first accreditation site surveys in the spring of 1963.

Throughout the early years, postgraduate pharmacy training, as reflected in the standards, focused primarily on the manufacture and preparation of pharmaceutical products and on systems that could be implemented to help ensure the integrity of those products up to the point of administration. Moreover, substantial effort was placed on providing trainees with the skills needed to pursue leadership roles in the hospital pharmacy community. The standards used to guide postgraduate training in institutional settings during the early years were reflective of practice at that time in that they focused heavily on the ''product'' side of the profession with little mention of the end user, the patient.

Clinical Pharmacy Emerges

In the late 1960s and early 1970s, a few residency programs began to place greater emphasis on patient care. Clearly, these programs provided a philosophical shift away from product-focused training and toward a greater emphasis on pharmacist participation in patients' drug therapy management. They also helped facilitate the ''clinical pharmacy'' movement that was emerging by providing training to many of the early would-be clinical pharmacy pioneers. The rapidity with which change was occurring is perhaps best reflected in the *Clinical Services* segment of the Qualifications of the Pharmacy Service section of the revised Accreditation Standard for Pharmacy Residency in a Hospital that was approved in November 1974:[3]

The functions which comprise clinical services are difficult to identify, partly because there is no common agreement among practitioners as to the definition of a clinical *service*, partly because there are ''clinical'' components associated with most, if not all, of the service functions of the hospital pharmacy department, and partly because, in current practice, no clear distinction has been made between clinical *teaching* activities and clinical *service* activities. What are frequently purported to be service activities are, more often than not, teaching (or learning) activities. For this reason, in evaluating clinical *service* activities, only those services...which are continuously performed even in the absence of students and trainees, are considered.

This document marked the first time that requirements for clinical pharmacy services in postgraduate training programs were addressed. Nonetheless, over the

next few years, it became increasingly more difficult to evaluate the growing number of ''clinical residencies'' that were emerging against the hospital pharmacy accreditation standard. In those programs, the scope of clinical services provided to patients grew substantially and the attendant requirements for residency training to assist in the delivery of those services were not addressed adequately in the accreditation standard—it was like trying to fit the proverbial square peg in a round hole. Hence, in 1980, ASHP approved the first Accreditation Standard for Residency Training in Clinical Pharmacy.[4]

Maintaining accreditation standards for both hospital and clinical residencies throughout the 1980s was beneficial for a variety of reasons. Among them were the following:

1. *The profession's differentiated workforce needs.* Many staff pharmacist positions in organized health care settings required a blend of the administrative and clinical skills that graduates of general hospital residencies received, whereas most management positions required graduates to complete hospital residencies that focused primarily on honing a resident's administrative skills (most often these were tied to MS degree programs). However, graduates of clinical residencies typically pursued faculty appointments and clinical practice positions that provided greater opportunities to deal directly with patients' drug therapy issues.
2. *An insufficient level of residency candidates holding advanced degrees.* A prerequisite to pursuing a clinical residency was the completion of the PharmD degree or a commensurate level of life experience to satisfy this requirement. Although the number of pharmacy school graduates holding the PharmD degree continued to increase throughout the 1980s, several clinical residencies struggled early on to obtain only candidates who had satisfied this prerequisite. This situation was particularly noteworthy for programs that were in regions where there was a paucity of PharmD graduates.
3. *Qualifications of the pharmacy service.* Virtually all departments of pharmacy that offered general hospital residency training in the 1980s provided some level of clinical pharmacy services. However, for the majority of these programs, the level of clinical services provided—particularly during the early 1980s—was inadequate to meet the requirements of the clinical residency accreditation standard. Hence, providing programs with sufficient time to establish these services under the direction of clinically competent practitioners was a key factor in maintaining both the hospital and the clinical tracks in residency training.

Specialization

The training of clinical specialists and pharmaceutical scientists took hold in the 1970s and progressed throughout the 1980s, paralleling advances in drugs and drug delivery systems. Outgrowths of this movement included the birth of the Board of Pharmaceutical Specialties and the American College of Clinical Pharmacy (ACCP), and the establishment of the ASHP Special Interest Groups (SIGs). Throughout this period, it became increasingly more accepted in the profession that pursuing optimal drug therapy in patients, particularly when extremely complex drug regimens were involved, often required a more sophisticated level of service than was typically provided by a general pharmacy practitioner. Spurred on by the increasing need for more highly trained individuals, in 1980, ASHP approved the first Accreditation Standard for Specialized Pharmacy Residency Training[5] (frequently referred to as the specialized ''umbrella'' standard) and the Supplementary Standard and Learning Objectives for Residency Training in Psychiatric Pharmacy Practice,[6] which had been developed by the SIG on psychopharmacy practice. Since then, supplemental standards have been approved for 16 specialized areas of pharmacy practice. Two of these standards, pharmacotherapy and infectious diseases, were developed jointly with the ACCP and the Society of Infectious Disease Pharmacists, respectively.

Pharmacy Practice Residencies

ASHP approved the Long-Range Position Statement on Pharmacy Manpower Needs and Residency Training[7] at the same time as the clinical residency standard. The position statement intended to guide the thinking of members about the categories of professional and technical pharmacy workforce needed in organized health care settings and the types of training programs required to meet that need. It also acknowledged that over time the distinction between a ''generalist'' and a ''clinical practitioner'' would diminish and that, at some point in the future, the need to maintain both clinical and hospital residency standards would no longer exist.

Consistent with this thinking, practitioners at the 1985 Hilton Head Conference on ''Directions for Clinical Practice in Pharmacy''[8] and the 1989 National Residency Preceptors Conference[9] offered several recommendations to merge the hospital and clinical residency standards. In particular, participants at these conferences noted that, due to revisions in the hospital and clinical residency standards, major segments of both documents, especially those that relate to requirements in departmental services, were now virtually identical. Moreover, it was also noted that portions of the requirements in the general hospital standard—notably, the training objectives for the clinical and drug information services areas—approached the clinical standard more closely than in the past. Hence, in 1991, following more than 2 years of development that included a series of open forums to receive input from residency preceptors on all proposed segments of the document, the first Accreditation Standard for Residency in Pharmacy Practice was approved.[10]

Adoption of the term ''pharmacy practice'' in the new standard was significant. Clear consensus was established that the time had come to no longer distinguish between ''clinical pharmacy practice'' and ''pharmacy practice'' as the practice of pharmacy is inherently clinical.[11] As of July 1, 1992, all ASHP-accredited general hospital and clinical residencies were converted to residencies in pharmacy practice. Since then, three separate standards governing pharmacy practice residencies have emerged: ASHP Accreditation Standard for Residencies in Pharmacy Practice (approved April 25, 2001); Accreditation Standard and Learning Objectives for Residency Training in Pharmacy Practice (with Emphasis in Community Care), prepared jointly by ASHP and the American Pharmaceutical Association; and Accreditation Standard and Learning Objectives for Residency Training in Managed Care Pharmacy Practice, prepared jointly by ASHP and the Academy of Managed Care Pharmacy.

PREREQUISITES FOR TRAINING

In all instances, the applicant to a residency must be a highly motivated pharmacist who desires to obtain advanced education and training, leading to an enhanced level of professional practice. The applicant must be a graduate of a college of pharmacy accredited by the American Council on Pharmaceutical Education or be otherwise eligible for licensure.

A residency in pharmacy practice is predicated on prior clerkship and externship experiences. For this reason, applicants to this type of program should have completed a comprehensive clerkship and externship program such as is required in contemporary clinically based pharmacy curricula. For pharmacy practice residencies, there is no absolute requirement concerning the type of pharmacy degree the applicant must possess. However, it is clear that the PharmD degree provides the applicant with the level of knowledge and skills needed to meet the requirements for training in a pharmacy practice residency and that, over time, this degree is expected to become an absolute prerequisite for applicants. In all cases, it is the residency program director's responsibility to assess each applicant's baseline knowledge and skills, and to ascertain overall qualifications for admission. The applicant should bear in mind that it is permissible for programs to establish more stringent entry-level requirements; hence, applicants must take responsibility for contacting programs directly to determine specific entry requirements.

Prerequisites for entry into a specialized residency may vary depending on the type of program. In all instances, however, completion of the PharmD degree (or other advanced degree in pharmacy) is required, except in unusual cases (e.g., BS degree graduate who has been in practice several years). Although completion of a pharmacy practice residency is required typically for admission to a specialized program, some programs may accept an applicant without residency training provided that the individual has a comparable level of professional practical experience. The reason for this requirement is that a specialized program requires the applicant to have a sound foundation in general practice skills that the pharmacy practice residency provides. Applicants interested in pursuing specialized residency training are urged to contact programs directly to determine the specific requirements for application.

INFORMATION ON RESIDENCIES

There are a number of ways to learn about residencies. However, the avenues that will likely provide potential applicants with the most information are the *ASHP Residency Directory*, the *ACCP Directory of Residencies and Fellowships*, and the ASHP Residency Showcase. The *ASHP Residency Directory* is a two-volume series: Volume I covers pharmacy practice residencies and volume II covers specialized residency programs. This series is available through most colleges of pharmacy's departments of pharmacy practice and can be obtained directly following application to the ASHP Resident Match Program. The *ACCP Directory of Residencies and*

Fellowships is likewise available through most colleges of pharmacy and can be obtained directly by contacting ACCP. Both directories provide interested individuals with a wealth of pertinent program information. Individuals interested in learning more about the residency resources offered by these associations are encouraged to visit the following web sites: www.ashp.org and www.accp.com. Moreover, the following web sites offer specific information about community-based and managed care residencies, respectively: www.aphanet.org and www.amcp.org.

All students who are in their last year of pharmacy school and are interested in residency training are encouraged to attend the ASHP midyear clinical meeting in December of each year to participate in the Residency showcase. The showcase is an area set up in booth format that enables representatives of the various programs to meet personally with prospective residency candidates to address any questions they might have about a program. Among other things, it provides applicants with the best opportunity to meet preceptors and residents from each program, learn first hand about specific elements of each program, and gain insights into programs that are often difficult to obtain through the directories. Because virtually all programs require candidates to participate in on site interviews, participation in the showcase can help a prospective candidate better identify those programs of greatest interest, which will help minimize unnecessary travel expenses that might have otherwise been incurred traveling to a less desirable program. It is important to note that more than 98% of all ASHP-accredited residencies typically participate in the showcase.

Before searching for a residency, the pharmacist graduate must determine his or her career objectives. Because programs vary in terms of their relative degree of emphasis in the areas in which training is provided, it is particularly important for the applicant to determine, to the extent possible, the area(s) that are most suited to achieving career objectives. For example, an individual interested in some dimension of ambulatory care would be wise to pursue programs that offer a broad range of experiences in this area of practice.

It is not uncommon for career objectives to change following experience in the many areas of practice typically provided in residencies. Hence, for applicants who are unsure about future professional goals, the best advise is to pursue training in a practice site that provides a solid grounding in patient care because many of the skills a resident will acquire are transferable across most areas of pharmacy practice. To this end, it is essential that potential applicants allow adequate time to review program information, as well as ask program representatives about the overall strengths and weaknesses of their program. Frequently, the best assessment of what a program has to offer is through communication with residents currently in the program.

RESIDENT MATCHING PROGRAM

Virtually all pharmacy practice residencies are required to participate in the Resident Matching Program (RMP). The only exception is those programs offered through the military and public health services. Hence, it is essential that potential applicants to these programs register for the match. Applicants to specialized programs are not required to participate in the RMP.

ASHP contracts with the National Matching Service (NMS) to operate the RMP. The RMP ensures that each pharmacy practice residency program is matched with the preferred individuals who have applied and who have selected the program as an acceptable site in which to train. To apply for the match, applicants must contact NMS directly (595 Bay Street, Suite 300, Toronto, Ontario, Canada; telephone: 416-977-3431 or fax: 416-977-5020). There is a modest application fee, and the applicant agreement form must be received by January 15. Once NMS has received the agreement form, the applicant will be sent a personalized match number (this number will be required by all pharmacy practice residencies to which the student graduate applies) and, under separate cover, a copy of the *ASHP Residency Directory*. As noted previously, the Directory provides descriptions of all accredited programs, including important contact information. After consulting the Directory and identifying programs of interest, the student graduate must request applications forms directly from the individual residency program directors, not from ASHP. Most programs require the applicant to participate in an on-site interview; hence, student graduates are advised to check directly with programs about application procedures as policies governing application do vary among programs.

REFERENCES

1. Fiske, R., et al. Standards for internships in hospital pharmacies. The Bulletin **September/October 1948**, *5* (5), 233–234.
2. Zellmer, W.A. Twenty-Five Years of Pharmacy Residency Accreditation—Part I. Early Efforts to Establish the Program (Unpublished manuscript).

3. Accreditation standard for pharmacy residency in a hospital. Am. J. Hosp. Pharm. **February 1975**, *32*, 192–198.
4. ASHP accreditation standard for residency training in clinical pharmacy. Am. J. Hosp. Pharm. **September 1980**, *37*, 1223–1228.
5. ASHP accreditation standard for specialized pharmacy residency training. Am. J. Hosp. Pharm. **September 1980**, *37*, 1229–1232.
6. ASHP supplementary standard and learning objectives for residency training in psychiatric pharmacy practice. Am. J. Hosp. Pharm. **September 1980**, *37*, 1232–1234.
7. ASHP long-range position statement on pharmacy manpower needs and residency training. Am. J. Hosp. Pharm. **September 1980**, *37*, 1220.
8. Directions for clinical practice in pharmacy. Proceedings of an invitational conference conducted by the ASHP Research and Education Foundation and the American Society of Hospital Pharmacists. Am. J. Hosp. Pharm. **June 1985**, *42*, 1287–1342.
9. Directions for postgraduate pharmacy residency training. Proceedings of the 1989 National Residency Preceptors Conference conducted by the American Society of Hospital Pharmacists. Am. J. Hosp. Pharm. **January 1990**, *47*, 85–126.
10. ASHP accreditation standard for residency in pharmacy practice. Am. J. Hosp. Pharm. **January 1992**, *49*, 146–153.
11. Letendre, D.E. Reflections on the future of pharmacy residency programs: An ASHP perspective. Am. J. Pharm. Educ. **Fall 1992**, *56*, 298–300.

Role of the Clinical Pharmacist in Clinical Trials (Spain)

Josi-Bruno Montoro-Ronsano
Hospital General Vall d'Hebron, Barcelona, Spain

INTRODUCTION

The development of a new drug, from its initial synthesis until its approval and registration by the regulatory agency, goes through different stages in which clinical trials are of paramount importance. The use of any drug in humans requires the fulfillment of previous valid clinical trials, which ensure the efficacy and safety of the drug and guarantee that the basic ethical rights concerning the patients are respected.[1]

The pharmaceutical industry supports much of the drug research in Spain, as it occurs in other countries.[2] The main goal of most clinical trials initiated is to achieve data on the efficacy and safety of a research drug, to complete the corresponding file, and to present it to the regulatory agency for its approval. Clinical research represents a major component of investment and development activity and takes an appreciable 36% of the research and development budget. This amount is similar to those corresponding to the European Community (39%) and to the world as a whole (36%).[3] It is important to notice that since the 1980s the activity in clinical research, and consequently the number of clinical trials that have been developed, has increased continuously. Concretely, the number of clinical trials developed in Spain has increased from 88 during the period 1987–1989 to 520 in 2000 (until November).[1,4,5]

From the profile of clinical trials developed in Spain, different descriptive studies allow us to obtain a precise picture. Most of the protocols evaluated are phase III (i.e., designed to fulfill the registration schedule and get the approval of the regulatory agency) (42% and 49%, depending on the descriptive study), multicenter (79% and 98%), controlled (76% and 82%), and double-blind (42% and 52%).[6,7] According to the therapeutic activity of the investigational drug, the most frequent groups were anti-infective and antineoplastic drugs.

Currently, as pointed out by Lunik,[8] clinical trials are a matter of great importance in the world of health care. There are many sophisticated compounds under development, with specific properties and unique storage, preparation and monitoring requirements. The need for clinical sites that can undertake quality research and the need for health professionals who are able to manage the more complex protocols is outstanding. Moreover, clinical trials have been put on aggressive time-to-development schedules. In addition, the research process has become globalized, and efforts have been made to standardize good clinical practices (GCP) among the developed communities.

Within this context, clinical pharmacists face effective participation in the research environment. Protocol development and execution, adherence to GCP and ethical principles, together with the balance between revenues and expenses, draw a specific working scenario that requires additional education and training, and represents an emerging challenge for clinical pharmacists. Conceptually, pharmacists are responsible for the safe and effective use of all medications, and this is especially important for medications used in clinical trials.[9] Different pharmacy associations have been pointed out and have defined how far drug development and its attendant activities are a core function of the pharmacy profession.[10–12]

Although regulations and recommendations currently in force in Spain assign the responsibility for studied drugs to the investigator and the promoter–sponsor, rather than the pharmacy service, it is clearly established that the management of the samples for investigation must be performed by the pharmacy. From this point of view, the development of pharmacy-based investigational drug services (the so-called Unidades de Ensayos Clínicos) has been extensive, and more interestingly, the concept that clinical trials represent a clear opportunity to expand the role of the clinical pharmacist, has been widely understood. Nevertheless, the attitude and participation of pharmacists in clinical trials is not homogeneous within the different hospitals. Passive reactions to innovate in this field can still be detected, and the pharmacists' training in many cases is not optimal, especially with a lack of standardized guidelines for training residents.

Encyclopedia of Clinical Pharmacy
DOI: 10.1081/E-ECP 120006378

REGULATORY ISSUES

The development and achievement of clinical trials is based on ethical postulates, specific legislation in use, and international guidelines for GCP research.

The assessment of health technologies is placed under the Ley General de Sanidad.[13] This evaluation is guided to determine the health, social, and economic impacts of the new technology under scrutiny, and is compared with other alternative interventions. According to the Ley del Medicamento,[14] a clinical trial is defined as experimental evaluation of a substance or a drug, by means of its administration or application to human, oriented to any of the following targets:

- To point out its pharmacodynamic or pharmacokinetic effect.
- To establish its efficacy for a given therapeutic, prophylactic, or diagnostic indication.
- To define its adverse effects profile and establish its safety.

The term ''experimental evaluation'' refers to studies that test substances nonauthorized as proprietary medicine products or proprietary medicine products used in nonauthorized conditions.

Clinical trials may only be carried out when all the following principles have been complied with:

- The preclinic data about the product under study are reasonably sufficient to guarantee that risks to the subjects on whom the trial is conducted are admissible.
- The study is based on current available data, and the research represents, or may represent, an improvement of scientific knowledge about humans or an improvement of human health, and by its design should minimize the risk to the subjects.
- The interest of the search for information justifies the risks to which the subjects taking part in the clinical trial are exposed.

The clinical trial development of drugs is regulated by Royal Decree 561/1993,[15] whereas clinical trials referring to health devices are regulated by Royal Decree 414/1996.[16] Clinical trials must be authorized by the Spanish Medicine Agency with previous assessment by the ethics commitees. The ethics commitees of clinical research will be accredited by the Regional Health Authority in each autonomous community, which has to communicate with the Ministry of Health and Consumer Affairs.[14] Clinical trials will be performed respectfully in relation to fundamental human rights and ethical postulates affecting biomedical research in humans, following the contents of the declaration of Helsinki and its succesive updates.[17] It will be necessary to obtain and document informed consent, freely stated, from each subject of the study before the patient is included.

Furthermore, it is strongly recommended that investigators, sponsors, monitors, and all others involved in clinical trials adhere to international standards for the proper management of clinical trials (i.e., GCP released by the International Conference on Harmonisation, involving the European Medicines Evaluation Agency [EMEA], and the agency's counterparts in other countries [United States and Japan]).[18]

CLINICAL TRIALS AND THE CLINICAL PHARMACIST

The role of the clinical pharmacist in clinical trials is going through a remarkable expansion process. Currently, there are different ways to integrate pharmacists into the clinical research environment.[4,19,20] Among these, it is important to mention the contribution of clinical pharmacists, as staff members, to the ethics commitee of clinical research, their participation in the development of protocols by managing the dispensing and use of investigational drugs, and their direct role as investigators, reviewing protocols and performing educational activities.

According to Royal Decree 561/93 (article 41),[15] on the minimum requirements for the accreditation of an ethics commitee of clinical research, it is stated that at least one hospital pharmacist must belong necessarily to the commitee. In this way, clinical pharmacists are capable of evaluating methodological and logistical aspects of each protocol and are competent to consider the ethical issues of clinical trials.

As mentioned previously, Royal Decree 561/93 (article 18),[15] on the distribution of investigational drug samples, it is stated that pharmacy service is responsible for written acknowledgment of receipt, custody, and dispensing of the drug being evaluated. Together with this recognition are the latest guidelines for GCP.[18] The result is easy to see. Years ago the problem for the pharmacy was to take care of the investigational drug, whereas today the problem is to find resources to facilitate the work load, which includes control of the entire process, as stated by Ferrer.[22]

Although regulations do not initially consider the role of the clinical pharmacist as an investigator,[1] it has been stated that hospital pharmacists who are experienced in research are fully prepared to participate in all processes within the research environment.[23] In our understanding, sponsors are still reticent to ask the pharmacist to play the role of investigator in clinical trials. Nevertheless, because of the pharmacist knowledge of the principles governing therapeutics, together with the intense relationship that they maintain with all the departments within the hospitals, the hospital pharmacist is a valuable asset for the design, development, review, and collaboration for preparation of the protocol, as stated before.[24]

Different surveys on the status of the role of the clinical pharmacist in clinical trials have been presented.[3,21,25] The average number of clinical trials started per year and hospital is 9.6, and there is a wide range of variation (up to 70 trials per year in large hospitals). However, the latest data indicate a dramatic increase in activity (around 100–150 trials per year). To summarize the results, it can be stated that, in 76.2% of hospitals, the pharmacy is the receiving site for investigational drugs, and 90.4% give written acknowledgment of receipt. Direct dispensing to the patient is executed in 59.5% of centers, whereas dispensing to the investigator was executed in 29.2% of hospitals.[3] Inventory control of samples for investigation is intensively observed in 85.7% of institutions. If the point of view of the industry is taken into account,[25] similar results are obtained; that is, distribution of the investigational drug is performed by the pharmacy service in 82.8% of centers. In addition, the pharmacy's role of dispensing drugs facilitates the monitor's task in 71% of cases. Interestingly, sponsors recognize that pharmacists are investigators in 12% of clinical trials, and that almost all of them (94% of sponsors) consider that the presence of pharmacist on the ethics commitee of clinical research is essential for evaluating protocols. The value of the pharmacists' activity in clinical trials is essentially not different from those obtained in surveys developed in other countries.[26]

Regarding human resources in this field, most centers manage clinical trials by assigning one or more part-time pharmacists or technicians who also work on traditional activities such as drug delivery and drug information. Therefore, it is difficult to estimate the number of protocols that can be managed by one full-time employee. Only centers developing more than 70 clinical trials per year are able to have one full-time employee dedicated for clinical research.[3] This fact is linked to another extremely important issue: The revenue for pharmacy costs is associated with research. In the current context of hospital operating costs, the structural resources and time spent by the hospital pharmacy service for clinical trial-related activities are difficult to justify. As it has been pointed out,[27] only by carefully quantifying the costs for providing this service and demonstrating that the benefits outweight the costs, can the pharmacy justify a new expansion program. In this way, a model for estimating and evaluating the cost of pharmacy activity for any given clinical trial was described. The average time spent per trial was 91 hours, the average cost per trial was $1766, and the average cost per patient was $174.[27] Although it is clear that clinical research activities should be funded by specific economic resources, this consideration is not always taken into account. Likewise, costs other than those of the investigator, such as the costs of pharmacist activities, are usually not taken into account when the total cost of a clinical trial is calculated. Even, if these costs are considered, they are fixed by people who are not familiar with the pharmacy. Consequently, the current sources for funding used to support the pharmacists' participation in clinical trials are not homogeneous, ranging from fees charged to investigators (usually reimbursed through the funds the investigator receives from the sponsor) to the covering of the costs assumed by the institution as part of the cost of participating in a research program. This picture does not differ essentially from that described by Rockwell et al. in their survey.[26]

The economic impact of drug studies on the pharmacy budget should be considered. Spanish regulations[15] give responsability to the sponsor for providing evaluated drugs free of charge. Along these lines, two reports measured the savings resulting from the development of clinical trials. In one case, 23 clinical trials saved almost 0,30 million €, during the period 1994–1996.[28] In a second study, of the 105 clinical trials conducted in our hospital during January 2000, 36 were evaluated, and the results indicated a total savings of 1,65 million €; from the rest of the studies, 55 therapies were evaluated without an established treatment, 2 were sponsored by the own hospital, and 9 added, not substituted, the investigational drug to the conventional treatment.[29] These savings are consistent with those corresponding to two pharmacy-based investigational drug services in the United States for fiscal period 1996–1997.[30]

From the institutions point of view, the Spanish Society of Hospital Pharmacy [Sociedad Española de Farmacia Hospitalaria (SEFH)] considers the clinical trials a matter of interest. As proof of this fact, there was a meeting organized in the early 1990s in which more than 50 experts described and analyzed the role of the hospital pharmacists in the growing world of clinical trials.[31] Furthermore, the reference text Farmacia Hospitalaria dedicates an entire chapter to the clinical trials issue.[32] Nevertheless, whereas different associations (e.g. American Society for Health-System Pharmacists, Joint Committee on Accreditation of Healthcare Organizations)[8] have edited guidelines oriented to establish and define the rules of the clinical pharmacist in clinical trials, the SEFH has still not taken the step. Regarding educational programs for residents, there are no standardized guidelines for rotations in investigational drug services, and each hospital acts according to its own criterion. Moreover, the recent edition of the resident's manual[33] does not consider specifically the issue of clinical trials.

As mentioned previously, the role of the clinical pharmacist in clinical trials in Spain must be considered in three main ways: the participation in the development of protocols by managing the dispensing and drug accountability by means of the pharmacy-based investigational drug services; the contribution of the clinical pharmacist, according to its condition of permanent member, to the ethics commitee of clinical research; and the direct role as investigators, conducting clinical trials.

PHARMACY-BASED INVESTIGATIONAL DRUG SERVICES

The application of the current regulations[15] and GCP recommendations[18] leads to the constitution of the investigational drug services within the pharmacy department in medium-size and large hospitals, which are involved in the developing process of a great number of protocols. The need for a specialized area in which to manage clinical trials is widely recognized among pharmacists.[34] In addition, the investigational drug services allow hospital pharmacists to specialize in the concrete area of clinical research, which is continuously evolving. Moreover, the workload that clinical trials represent can be accurately measured in the context of pharmacy service, and strategies for figuring costs and reimbursements can be adequately carried out. In some cases, a clinical trials agency is created in an attempt to integrate all clinical procedures in the hospital setting.[35]

In our hospital (a typical third-level hospital having more than 1500 beds), more than 150 clinical trials are under development. Years ago, the pharmaceutical support unit for clinical trials was created; at the present time, it involves a clinical pharmacist and a technician, both full-time employees dedicated to clinical trials-related activity.

The main goals of the pharmacy-based investigational drug services are as follows:

- To offer sponsors, investigators, and patients a guaranteed quality in the clinical trial process, by executing a correct reception, storage, and dispensation of the investigational drug.
- To collaborate with investigators and sponsors in the design of clinical trials and their follow-up.
- To collaborate with the ethics commitee of clinical research by offering information periodically on the number of patients included in each protocol and reporting incidences considered serious for the clinical trial follow-up.

The description of the basic activities (in chronological order) are as follows:

- Reception and study of protocol

  Opening protocol file
  Checking preceptive documentation
  Protocol study
  Meetings with monitors and/or investigators
  Evaluating pharmacy participation
  Opening computerized record
  Establishing dispensing procedures
  Information activities

- Drug management

  Drug reception
  Dispensing procedure
  Storage
  Stock management
  Returned drugs management
  Returning unused drug to sponsor
  Feeding computer records of activity

- Finishing the study

  Final inventory
  Balance of drug accountability (receptioning/dispensing)
  Closing computerized record
  Meetings with monitors and/or investigators
  Closing the protocol file
  Institutional or sponsor audits

As a function of the clinical trial or the nature of the investigational drug, there are some other specific activities:

- Keep the randomization codes and opening them, if necessary
- Assigning patients to treatment groups, according to randomization tables
- Packing or acconditioning samples for suitable use and administration
- Preparing solutions or mixtures and diluting the drug for parenteral use under aseptic conditions
- Preparing or acconditioning samples that require specific handling conditions (cytotoxics, biologics, gene therapy)

- Patient information and training (especially in the case of outpatients)
- Destruction of the returned or unused drug samples through a suitable procedure
- Evaluation of the patient's and investigator's adherence to protocol

For the adequate management of the investigational samples in the context of the activity of the pharmacy, it is necessary to consider the following:

- Integrating the pharmacy-based investigational drug activity within the pharmacy service, making use of common settings and facilities
- Establishing a differentiated flow to manage samples for investigation within the pharmacy service
- Storing the investigational drugs in a specific, unique place, far from the conventional drugs
- Establishing measures to keep the samples safe and guaranteeing their integrity (i.e., locks in shelves, temperature alarms)
- Relaying on trained and specialized pharmacists and technicians, even full-time employees, if the number and complexity of clinical trials requires it
- Defining an individual and concrete prescription sheet, essential for dispensing the investigational drug
- Keeping all documents and activity records in separate files for each clinical trial
- Computerizing all the clinical trial development, using available software programs (purchased software or developed in-house systems)
- Establishing safety measures to guarantee the confidentiality of the protocol and the patients, to include protecting investigational samples, files, and computer programs
- Elaborating suitable standard operating procedures for all activities related to the clinical trial performance

It is important to define the reimbursment process for providing the investigational drug services. There are two main procedures: a fixed reimbursement per trial for traditional services and a variable reimbursement estimated per clinical trial for other services;[36] and a fee structure (fixed and variable) in which each activity involved in a clinical trial is detailed.[27,37] In our case, we use the first procedure because of its simplicity. Nevertheless, a new fee questionnaire describing the cost of each pharmacy-related activity is under evaluation.

As stated previously, a specifically defined computer program is essential for maintaining inventory control and accountability. There are different software programs available, but beside conventional databases, there are two main programs: TRIALS[38] designed by a working party of the SEFH; and the recently updated GECOS, designed in our pharmacy service.[39,40]

ETHICS COMMITEE OF CLINICAL RESEARCH

The ethics commitee of clinical research is the institutional entity at the local institution that is responsible for protecting the rights of human.[15] It plays a similar role to the institutional review board in other countries.[41] The main functions of this entity are to examine the methodological, ethical, and regulatory issues of each protocol proposed for development in the hospital.

In detail, the commitee must evaluate, among other things, the following:

- The appropriateness of the protocol related to the aim of the study, its scientific efficiency, and the balance between the expected risks and benefits of the new therapy
- The appropriateness of the investigator and the research team responsible for developing the clinical trial within the hospital
- The informed consent document to be given to the patients, from the point of view of the respect for human rights
- The clinical trial follow-up, from the beginning of the study to the receiving of the final report

The ethics commitee is composed of members with or without a background in health care and research. At least one clinical pharmacist must be a member of the ethics commitee. According to a survey,[3] the pharmacist is president of the commitee in 4.8% of cases, secretary in 36.5% of cases, adviser in 12.1% of cases, or voting member in 46.6% of cases. Thus, it is understood that clinical pharmacists have the knowledge and the responsability to evaluate and monitor a diversity of clinical, methodological, and ethical aspects relating to clinical trials.

For this reason, it is remarkable that, in many cases, clinical pharmacists lead the follow-up process for the clinical trials. It was in a pharmacy-promoted audit that a serious deviation in GCP recommendations was detected. In this study, it was concluded that compliance was adequate in 50% of protocols, so-so in 25% of cases, and poor in the remaining 25%. Surprisingly, in 13% of the patients, the informed consent document was not signed; in 15% of the cases, the patient's clinical records did not reflect that the patients were included in a clinical trial.[42] In a different way, a long-term follow-up detected that only 32% of clinical trials previously evaluated by the local ethics commitee was finally published at the end of the protocol. Moreover, the sponsor sent the final report of the clinical trial to the ethics commitee in only 33 of 135 finished clinical trials.[43]

The role of the clinical pharmacist in the surveillance of ethics principles and the maintenance of the rights of human can also be notorious, especially when referring to written information and the patient's informed consent.[44] The most relevant study undertaken in Spain that evaluated the quality of the written information provided to patients was carried out by pharmacists involved in quality assurance and/or ethics commitees.[45] The main outcome of this study was to confirm that most of the written information to patients (65.3% of clinical trials) required high-level studies to be completely understood by the patients. As far as we know, a more ambitious, multicentric study is currently under development on the true comprehension and awareness that patients have of the clinical trials in which they are involved.

As a whole, the pharmacist's presence in ethics commitees allows the pharmacy service to solve practical problems related to the execution of the clinical trial, prior to the approval of the protocol. Moreover, by evaluating methodological aspects of protocols, the clinical pharmacist is asked to develop future protocols, linking with pharmacist's role as clinical researcher.

THE CLINICAL PHARMACIST AS CLINICAL RESEARCHER

It is clearly understood that pharmacy services and clinical pharmacists should play an important role in clinical research.[46] In this sense, a survey noticed that a clinical pharmacist was the investigator in 12% of clinical trials.[3] Nevertheless, the percentage of protocols in which a pharmacist is integrated into the research team is higher.

In fact, it is not rare that pharmacists collaborate in activities that are not considered to be routine drug management. The pharmacists may be responsible for randomization of the patients in treatment groups. If the study is double-blind, pharmacists can calculate dosages, keeping the investigator unaware of treatment assignments. Nevertheless, the main field in which a clinical pharmacist can expand is with the institutionally sponsored research of off-label treatments. Such research may involve an existing drug product, a new formulation, a new method or route of administration, or any combination of these that is not covered by the Spanish Medicine Agency-approved labeling; as well as clinical trials that involve activities familiar to pharmacists according to their background (i.e., pharmacokinetics, pharmacoeconomics).

In the near future, within the context of our changing health care environment and research environment, there is a need for multidisciplinary approaches. The clinical pharmacist should demonstrate, according to their expertise in the principles governing therapeutics, that they are able to participate fully in this expanding process within the research environment. Thus, a hospital pharmacist would be valuable in the design, development, review, and preparation of the protocol and, finally, in evaluating and reporting results.

REFERENCES

1. *Ministerio de Sanidad y Consumo. Ensayos Clínicos en España (1982–1988)*; Monografías Técnicas: Madrid, 1990; Vol. 17.
2. Vlasses, P.H. Clinical research: Trends affecting the pharmaceutical industry and the pharmacy profession. Am. J. Health-Syst. Pharm. **1999**, *56*, 171–174.
3. Quevedo, A.; Pérez, L.; Fernández, A. Participación del farmacéutico de hospital en la investigación clínica. Farm. Hosp. **1999**, *23*, 24–41.
4. Idoate, A.; Giráldez, J.; Idoipe, A. Trials: Aplicación informática para la dispensación y control de medicamentos en fase de investigación clínica. Farm. Hosp. **1995**, *19*, 24–30.
5. Data not published. Pesonal communication. Agencia Española del Medicamento. Ministerio de Sanidad y Consumo.
6. Vázquez, J.R.; Moncín, C.; Huarte, R.; Idoipe, A.; Palomo, P.; Marco, R. Ensayos clínicos: Participación del servicio de farmacia. Farm. Clín. **1995**, *12*, 178–188.
7. Escoms, M.C.; Novoa, C.; Jódar, R.J.; Suñé, J.M. Perfil de los ensayos clinicos evaluados en un hospital general durante quince años: Utilidad de un programa informático. Farm. Clín. **1997**, *14*, 341–347.
8. Lunik, M.C. Clinical research: Managing the issues. Am. J. Health-Syst. Pharm. **1999**, *56*, 170–171.
9. Phillips, M.S. Clinical research: ASHP guidelines and future directions for pharmacists. Am. J. Health-Syst. Pharm. **1999**, *56*, 344–346.
10. Cipywnyk, D.M. A survey of pharmacy-coordinated investigational drug services. Can. J. Hosp. Pharm. **1991**, *44*, 183–188.
11. Giraldez, J.; Idoate, A.; Jimenez-Torres, N.V.; Ribas, J.; Rodríguez-Sasiain, J.M. Aspectos Prácticos de la Participación del Servicio de Farmacia en los Ensayos Clínicos del Hospital. In *Sociedad Española de Farmacia Hospitalaria. Jornadas de Actualización Sobre Ensayos Clínicos*: Madrid, 1990; 65–118.
12. American Society of Health-System Pharmacists. ASHP guidelines on clinical drug research. Am. J. Health-Syst. Pharm. **1998**, *55*, 369–376.

13. Ley 14/1986, de 25 de abril, general de sanidad. Ministerio de Sanidad y Consumo. B.O.E. no. 78, de 27 de abril de 1986.
14. Ley 25/1990, de 20 de noviembre, del medicamento. Ministerio de Sanidad y Consumo. B.O.E. no. 306, de 22 de diciembre de 1990.
15. Real Decreto 561/1993, de 16 de abril, por el que se establecen los requisitos para la realización de ensayos clínicos con medicamentos. Ministerio de Sanidad y Consumo. B.O.E. no. 114, de 13 de marzo de 1993.
16. Real Decreto 414/1996, de 1 de mayo, por el que se regulan los productos sanitarios. Ministerio de Sanidad y Consumo. B.O.E. no. 99, de 24 de abril de 1996.
17. Declaración de Helsinki: recomendaciones para orientar a los médicos en los trabajos de investigación biomédica con sujetos humanos. Adoptada por la 18ª Asamblea Médica Mundial (Tokio, Japón, octubre de 1975), por la 35ª Asamblea Médica Mundial (Venecia, Italia, 1983) y por la 41ª Asamblea Médica Mundial, Somerset West, República de Sudáfrica, octubre 1996.
18. Organización Mundial de la Salud (International Conference on Harmonization of technical requirements for registration of pharmaceuticals for human use) Good Clinical Practice: Consolidated Guideline. Documento no. E6GCPD12WP6. ICH Secretary. Ginebra, may 1996.
19. Ordovás, J.P.; Jimenez-Torres, N.V. El Servicio de Farmacia en el Àmbito de la Investigación Clínica de Medicamentos en el Hospital. In *El Ensayo Clínico Como Tarea Cooperativa*; Barlett, A., Serrano, M.A., Torrent, J., Eds.; Monografías Dr. Esteve, Fundación Dr. Esteve: Barcelona, 1992; Vol. 13; 27–35.
20. Rodilla, F.; Magraner, J.; Fuentes, M.D.; Ferriols, F. Ensayos clínicos y su repercusión en un servicio de farmacia hospitalaria. Farm. Hosp. **1994**, *18*, 249–254.
21. Sacristan, J.A.; Bolaños, E. Papel de los servicios de farmacia en la realización de ensayos clínicos: Punto de vista de la industria farmacéutica. Farm. Hosp. **1995**, *19*, 364–367.
22. Ferrer, M.I. El control de los medicamentos de ensayo clínico: Perspectivas y necesidades actuales. El Farmacéutico Hospitales **1994**, *52*, 38–39.
23. York, J.M.; Alexander, J.G.; Barone, J.A. The value of the clinical pharmacist in the drug development process. Clin. Res. Pract. Drug Regul. Aff. **1998**, *6*, 23–39.
24. Gouveia, W.A. Regulatory Authority affecting american drug trials: Role of the hospital pharmacist. Drug Inf. J. **1993**, *27*, 129–134.
25. Gallastegui, C.; Ascunce, P.; Dávila, C.; Boado, A. Servicios de farmacia y ensayos clínicos: Encuesta a la industria farmacéutica. Farm. Hosp. **1999**, *23*, 231–237.
26. Rockwell, K.; Bockheim-McGee, C.; Jones, E.; Kwon, I.W.G. Clinical research: National survey of U.S. pharmacy-based investigational drug services-1997. Am. J. Health-Syst. Pharm. **1999**, *56*, 337–344.
27. Idoate, A.; Ortega, A.; Carrera, F.J.; Aldaz, A.; Giráldez, J. Cost-evaluation model for clinical trials in a hospital pharmacy service. Pharm. World Sci. **1995**, *17*, 172–176.
28. Barroso, E.; Ferrer, M.I.; Jódar, R. Direct Economical Impact of Clinical Trials 1994–1996. In *3rd Congress of the European Association Hospital Pharmacists*; Edimburgh: UK, 1998.
29. Valdés, G.; Suñé, M.P.; Montoro, J.B. Impacto económico de los ensayos clínicos en un hospital general: Ahorro en costes farmacéuticos. Farm. Hosp. **2000**, *24*, 93–94, (Esp. Congr.).
30. McDonagh, M.S.; Miller, S.A.; Naden, E. Costs and savings of investigational drug services. Am. J. Health-Syst. Pharm. **2000**, *57*, 40–43.
31. *Sociedad Española de Farmacia Hospitalaria. Jornadas de Actualización Sobre Ensayos Clínicos*: Madrid, 1990; 300 pp.
32. Idoate, A.J.; Caínzos, M.D.; Idoipe, A. Ensayos Clínicos. In *Farmacia Hospitalaria*; Bonal, J., Domínguez-Gil, A., Eds.; Editorial Médica Internacional SA: Madrid, 1993; 645–691.
33. Bermejo, T.; Cuña, B.; Napal, V.; Valverde, E. *Manual del Residente de Farmacia Hospitalaria*; Sociedad Española de Farmacia Hospitalaria, 1999; 917 pp.
34. Jódar, R. Los ensayos clinicos constituyen una sección con personalidad propia en el Servicio de Farmacia (entrevista). El Farmacéutico Hospitales **1994**, *52*, 6–8.
35. Gómez, B.; Codina, C.; Ribas, J. In *Clinical Trials Agency*, ASHP Annual Meeting, Orlando, Florida, 1999, poster presentation, Intl-51.
36. Swartz, A.J. Reimbursement for involvement in investigational drug protocols. Am. J. Hosp. Pharm. **1989**, *46*, 1334.
37. Anandan, J.V.; Isopi, M.J.; Warren, A.J. Fee structure for investigational drug studies. Am. J. Hosp. Pharm. **1993**, *50*, 2339–2343.
38. Idoate, A.; Giráldez, J.; Idoipe, A.; Garrido-Lestache, S. Trials®:Aplicación informática para dispensación y control de medicamentos en investigación clínica. Farm. Hosp. **1995**, *19*, 24–30.
39. *Gecos v1.1. Programa Multiusuario de Gestión de Ensayos Clínicos Desarrollado para Servicios de Farmacia Hospitalaria*; Grifols SA: Barcelona, Spain, 2000.
40. Oliveras, J.; Jódar, R.J.; Suñé, J.M. Programa informático aplicado al control de ensayos clínicos. Farm. Hosp. **1990**, *14*, 83–87.
41. Wermelling, D.P. Clinical research: Regulatory issues. Am. J. Health-Syst. Pharm. **1999**, *56*, 252–256.
42. De la Llama, F.; Gutierrez, P. Auditoría sobre ensayos clínicos. Farm. Hosp. **1996**, *20*, 114–117.
43. Monzó, M.; Pla, R. Seguimiento de los ensayos clínicos finalizados y sus posteriores publicaciones. Farm. Hosp. **2000**, *24*, 107–108, (Esp. Congr.).
44. Ordovás, J.P. Consentimiento informado del paciente en los ensayos clínicos (editorial). Farm. Hosp. **1999**, *23*, 267–270.
45. Ordovás, J.P.; López, E.; Urbieta, E.; Torregrosa, R.; Jiménez-Torres, N.V. Análisis de las hojas de información al paciente para la obtención de su consentimiento informado en ensayos clínicos. Med. Clín. (Barc.) **1999**, *112*, 90–94.
46. Ordovás, J.P. Investigación clínica en España y servicios de farmacia hospitalaria. Farm. Hosp. **1996**, *20*, 189–191.

Society of Hospital Pharmacists of Australia, The

Naomi Burgess
Society of Hospital Pharmacists of Australia,
South Melbourne, Australia

INTRODUCTION

The Society of Hospital Pharmacists of Australia (SHPA) is the professional body that represents pharmacists practicing in hospitals and similar institutions throughout Australia.

Established in 1941, the Society has a long-standing commitment to the profession of hospital pharmacy and to its role in ensuring optimal health outcomes for Australians. Fundamental to its success is the Society's culture of cooperation and contribution, which is reflected in the high level of membership involvement.

In November 1999, SHPA members voted in a new Constitution based upon revised Memorandum and Articles of Association, developed to incorporate changes in Corporation Law and to ensure that SHPA keeps abreast of professional, technology, and social changes.

New membership categories were introduced as a consequence of the new Constitution. Membership has now expanded to include pharmacy technicians, inactive members, and overseas members.

ORGANIZATIONAL STRUCTURE AND GOVERNANCE

SHPA is governed by a Federal Council which is supported by a divisional structure. There also local Branch Committees in each State and Councilors are elected by these Committees to represent the States at the federal level. All Councilors, Branch representatives, and Division/Committee members work in a voluntary capacity.

Federal Council (As of 1 March 2001)

Executive

Helen Dowling, Federal President
Naomi Burgess, Federal Vice President
Helen Matthews, Federal Treasurer
Sue Kirsa

Federal councilors

Neil Keen
George Taylor
Christine Maclean
Andrew Matthews
Paul Muir

Executive director

Yvonne Allinson

The Federal Secretariat plays a critical role in serving the needs of members and in supporting the activities of the Society. It is based in the Society's offices in Melbourne, Victoria, Australia.

The Divisions and Committees, which report to the Federal Council, develop and implement policy in the areas central to SHPA's goals. They include the Publications Division; Research, Grants and Development Committee; Division of Specialty Practice; Liaison and Promotion Unit, and the Division of Education. Each is responsible for advising the Federal Executive of new developments and opportunities that will allow SHPA to continue its mission of promoting and developing the practice of pharmacy in hospitals and other healthcare settings.

There are 15 Committees of Speciality Practice, which reflect a wide range of professional interests and are responsible for contributing to the professional development of SHPA members as well as maintaining and enhancing standards throughout hospital pharmacy practice.

MEMBERSHIP

SHPA currently represents over 1,500 pharmacists and pharmacy technicians working in hospitals and related institutions. This represents over 80% of Australian hospital pharmacists.

Encyclopedia of Clinical Pharmacy
DOI: 10.1081/E-ECP 120006304

MISSION AND KEY OBJECTIVES OF SHPA

Mission Statement

SHPA is committed to promoting the quality use of medicines through the ongoing development, application, and implementation of leading-edge pharmacy practice in hospitals and other healthcare organizations.

Goals

- To increase the status, influence, and profile of the profession of hospital pharmacy.
- To provide appropriate services to SHPA members.
- To expand the career opportunities for members of SHPA.
- To foster strong and positive corporate unity within SHPA.
- To generate resources to maintain and expand SHPA activities and to facilitate research.
- To establish and promulgate standards and guidelines in all relevant areas of pharmacy practice.
- To ensure the availability and accessibility of comprehensive, relevant education.
- To develop an effective network between SHPA and other relevant organizations.
- To provide appropriate services to organizations and individuals external to the membership.

MEMBER BENEFITS

Members have access to a range of services and programs aimed at enhancing professional standards and contributing to the ongoing development of hospital pharmacy practice. These include:

The Australian Journal of Hospital Pharmacy

This renowned publication provides a forum for the exchange of knowledge and ideas about new initiatives and developments in hospital pharmacy. It publishes the research of new and established authors.

Publications and Textbooks

SHPA publishes an impressive range of publications including:

- SHPA Practice Standards and Definitions.[1]
- Australian Injectable Drugs Handbook.[2]
- Clinical Pharmacy—A Practical Approach.[3]
- Australian Drug Information Procedure Manual.[4]
- Drug Usage Evaluation—Starter Kit.[5]
- Directory of Hospital Pharmacy and Pharmaceutical Organisations.[6]

SHPA Bulletin and Branch Newsletters

These newsletters keep members up-to-date with news and views at a national and local state level.

Education Programs

SHPA co-ordinates a comprehensive range of education services for its members and other pharmacists and healthcare workers interested in pharmacy and medication use. Continuing Professional Development is delivered throughout Australia, with branches in each state facilitating regular meetings on topical issues. SHPA works in conjunction with universities to deliver postgraduate diplomas designed to give hospital pharmacists the skills for pharmacy leadership. Postgraduate studies are recognized by SHPA through the Society's Fellowship program.

Research, Development, and Grants

SHPA administers a grants program aimed at supporting research and development in the practice of hospital pharmacy. During 1999, 103 grants were awarded with just under A$140,000 being distributed. SHPA also actively supports research and development by commissioning studies in areas of major importance to the profession including the impact of clinical pharmacy services[7,8] and casemixbased funding and management for hospital pharmacy.[9,10]

Advocacy

A key role of SHPA is advocacy for the profession and its expanding role in provision of quality healthcare services. SHPA represents the members, at both state and national levels, on many important issues in relation to public health policy, funding, and education.

CURRENT INITIATIVES AND FUTURE DIRECTIONS

The recently completed Clinical Pharmacy Intervention Study[8] was commissioned by SHPA to quantify the benefits of clinical pharmacist's interventions. Eight teach-

ing hospitals were involved in the study, the first multisite cost-benefit analysis of clinical pharmacy interventions in Australia. The results showed that over A$4 million per annum were saved by the hospitals because of the direct intervention of clinical pharmacists in drug treatments. The study will form a key component of a campaign to increase awareness of the important role hospital pharmacists have in ensuring safe, high quality, and cost-effective healthcare and of maintaining adequate levels of clinical pharmacy services.

With the rapidly aging Australian population in mind, SHPA signed an agreement with the internationally recognized Commission for Certification in Geriatric Pharmacy in 2000. The agreement sees the two organizations teaming up to offer the first examination-based competency assessment for Australian pharmacists. SHPA has negotiated with the Australian Government for CCGP-credentialed pharmacists to access funding for clinical services provided to patients in both hospitals and community settings.

Provision of quality educational services in pharmacy and related areas is one of the core businesses of SHPA and an area of ongoing expansion and development. In recent times, the focus of these services has broadened from predominantly addressing the needs of hospital pharmacists, to incorporating nursing and community pharmacist education. With the inclusion of pharmacy technicians in our membership starting in 2000, educational needs of this group will be more formally addressed.

SHPA's most recent strategic planning cycle began in February 2001.

MAJOR MEETINGS

Biennial Clinical Pharmacy Conference

October 2000: Gold Coast, Queensland
October 2002: Sydney, New South Wales

Biennial Federal Conference

November 2001: Hobart, Tasmania
November 2003: Canberra, Australian Capital Territory

SHPA State Branch Conferences

These meetings are annual or biennial, usually held in alternating years with Biennial Federal Conference.

Annual general meeting

November 2001: Hobart, Tasmania

REFERENCES

1. *Practice Standards and Definitions*; Johnstone, J.M., Vienet, M.D., Eds.; The Society of Hospital Pharmacists of Australia: Melbourne, Australia, 1996, ISBN 0 9588944 9 3.
2. *Australian Injectable Drugs Handbook,* 2nd Ed.; Morris, N., Ed.; The Society of Hospital Pharmacists of Australia: South Melbourne, Australia, 1999, ISBN 0 9586881 17.
3. *Clinical Pharmacy—A Practical Approach*; Hughes, J., Donnelly, R., James-Chatgilaou, Eds.; Macmillan Education Australia Pty Ltd.: South Yarra, Australia, 1998, ISBN 0 7329 4719 7.
4. *Australian Drug Information Procedure Manual*; Foran, S., Ed.; The Society of Hospital Pharmacists of Australia: South Melbourne, Australia, 1996.
5. *Australian Drug Usage Evaluation Starter Kit*; SHPA Drug Usage Evaluation Committee of Specialty Practice, Ed.; The Society of Hospital Pharmacists of Australia: South Melbourne, Australia, 1998.
6. *Directory of Hospital Pharmacy and Pharmaceutical Organisations*; Vernon, G., Thomson, W.A., Eds.; The Society of Hospital Pharmacists of Australia: South Melbourne, Australia, 2001.
7. Allen, K.M.; Dooley, M.J.; Doecke, C.J.; Galbraith, K.J.; Taylor, G.R.; Bright, J.; Carey, D.L. A prospective multicentre study of pharmacist initiated changes to drug therapy and patient management in acute care government funded hospitals, *in press*.
8. Dooley, M.J.; McLennan, D.N.; Galbraith, K.J.; Burgess, N.G. Multicentre pilot study of a standard approach to document clinical pharmacy activity. Aust. J. Hosp. Pharm. **2000**, *30* (4), 150–157.
9. Burgess, N. *The National Pharmacy Casemix Bridging Study—Report to the Commonwealth Department of Health and Family Services*; The Society of Hospital Pharmacists of Australia: Adelaide, Australia, 1998.
10. Howitt, K.; Burgess, N.; Brooks, C. *The Refinement and Application of PharmGroup Classifications as a Management Tool for Competitive Neutrality in the Health Industry*; Report to The Society of Hospital Pharmacists of Australia: Melbourne, Australia, December 2000.

Spanish Society of Hospital Pharmacy

Ma. Isabel Crespo Gonzalez
Urbanización Monteclaro, Madrid, Spain

INTRODUCTION

The Spanish Society of Hospital Pharmacy (SSHP) is the only national association of its kind to group together hospital pharmacists in Spain. The society is very well established within the profession due to its long tradition of protecting the interests of the sector. It has not always been easy to maintain the importance and representative of the society. However, due to the efforts of its presidents and the pharmacists themselves, it is now one of the most relevant hospital pharmacy societies in Spain and in Europe.

BACKGROUND

The growth of the society is directly related to the incorporation of newly trained specialists who tend to join the society once they have completed their studies and participate in the development of new work groups. This coupled with the experience and scientific knowledge of the older members makes the SSHP one of the most advanced hospital pharmacy societies in Europe. For more information about this organization, please visit our web site at www.sefh.es.

With targets very similar to its present-day objectives and members drawn from pharmacists working for charity, healthcare, and social service institutions, the Asociación Nacional de Farmacéuticos de Hospitales Civiles (National Civil Hospital Pharmacy Association) was founded in Madrid in 1955. The promotion of the hospital pharmacist's scientific, technical, and teaching activities and, in general, of all issues relating to health system pharmacy, remains enshrined in the bylaws today. After changing its name to Asociación Española de Farmacéuticas Hospitalarios (the Spanish Hospital Pharmacists Association), the original association finally became the Sociedad Española de Farmacia Hospitalaria (the Spanish Society of Hospital Pharmacists) in 1988.

Two landmarks in particular are closely linked to the development of hospital pharmacy in Spain. The first was the official recognition of the speciality; the second, the creation of the Comisión Nacional de la Especialidad Española de Farmacia Hospitalaria (National Commission for the Spanish Hospital Pharmacy Speciality), on which the SSHP is represented. These two developments led to the first Specialists in Hospital Pharmacy qualifications being awarded in 1986.

In 1988, the foundations were laid for the modernization of the structure of what was then the association, with the creation of a Madrid-based General Technical Secretariat. The General Technical Secretariat rapidly became integrated into the new structure, which was definitively introduced in 1992, after the relevant changes to the bylaws, and under which the society's activities were separated into a number of well-defined areas. Projects currently in progress could then be included in, for instance, the economic, educational, publishing, or institutional relations areas. Projects such as Pharmacovigilance, GRDs, DDDs, and Information on Medicines and Drugs were among the priority concerns.

Below is a list of the commissions and task forces set up over the years. These bodies generate specific projects on issues and subjects on which to focus within their areas:

(Therapeutic Evaluation Commission)
Comisión de Bioética (Bioethics Commission)
Comisión de Nutrición Artificial (Artificial Nutrition Commission)
Comisión de Normas de Procedimientos (Procedure Regulations Commission)
Grupo Español de Farmacia Pediátrica (Spanish Paediatric Pharmacy Task Force)
Comité de Acreditación (Accreditation Committee)
Grupo de Estudio de Utilización de Medicamentos (Drug Use Task Force)
Grupo de Trabajo sobre GRDs (GRD Task Force)

In line with the spirit of the bylaws, particular emphasis was placed on the areas of education and publications. In fact, so much work was done in these two areas that, in 1995, they were moved to a separate

Encyclopedia of Clinical Pharmacy
DOI: 10.1081/E-ECP 120006380

new headquarters also in Madrid and more suited to their requirements.

Created in 1997, the Fundación Española de Farmacia Hospitalaria (Spanish Foundation for Hospital Pharmacy) is also closely linked to the SSHP.

Over the years, the SSHP has given top priority to its educational and publications areas. With regard to the educational area, this special interest has resulted in a substantial number of courses, grants, prizes, and aids to research. Efforts have been made to diversify so that courses oriented toward residents have coexisted alongside others directed toward the staff, in an attempt to cover the needs and requirements of the whole group.

Substantial aids are also available to ensure research projects lead to complete, fully defined developments in highly contemporary areas and subjects, often achieved by multidisciplinary and multicentric teams chosen with a view to encouraging scientific and professional exchanges. Aid is also available to outstanding young trainees, giving them access to specific knowledge in other countries in lines of work that may or may not be directly linked with our areas, but that are certain to provide a wealth of new knowledge that can then be passed on to the group as a whole.

Among the more than 130 publications issued by the SSHP's publications area are a series of regularly revised and updated monographs, general compendia, and regular journal publications. The most outstanding of these is undoubtedly the SSHP journal *Farmacia Hospitalaria* (*Hospital Pharmacy*), which, over 20 years or so, has provided a platform for some of the major landmarks in the group's story, collaborating closely in diffusion and the training of Spanish and foreign professionals. The *SSHP Information Bulletin* is another regular publication that has also accompanied us in our efforts.

Activities undertaken by the area of institutional relations intensified in 1988, after the society's 1987 entry as full member in the European Association of Hospital Pharmacists, to which it had previously belonged as an associate member. This was then accompanied by a series of agreements with our institutions, including the 1989 Pharmacovigilance Agreement, the approval given in 1993 of the SSHP's Narcotics Computer Control Programme, and the more recent official blessing for the calculation of specific pharmacy weights in GRD, and the Spanish Medicines Agency itself, with which we have collaborated in a wide-ranging series of task forces in our acknowledged role as experts in medicine and drugs.

Fully aware of this role, the society promotes and develops a range of studies, reports, and projects over a range of contemporary and future health issues via the foundation of the same name.

Finally, a brief mention of our annual Congress (first held in Madrid in 1955) is in order. The Congress has accompanied and evolved together with the society itself. This year we held the 45th Congress, the latest in a virtually uninterrupted series that provides a meeting place for Spanish hospital pharmacists and that has now become a highlight on the international health care circuit. This year, the 47th meeting of Congress is to take place in Barcelona from October 1 to October 4 2002.

Besides the Congress, approximately 20 zoned scientific meetings are held every year, although the venues for the coming year have not yet been decided. These events keep the society alive and have helped it to grow from the 50 or so associates who attended the first General Meeting in 1955 to the current membership of 1850.

GOVERNING BODY (ELECTED POSTS)

Presidents of Honor
Felipe Gracia; Juan Manuel Reol; Manuel Ruiz-Jarabo; Joaquim Bonal de Falgas

President
Eduardo Echarri Arrieta

Vice-President
Ma. Cinta Gamundi Planas

Secretary
David García Marco

Treasurer
Rafael Molero Gómez

Member Zone 1
Ma. José Martínez Vázquez

Member Zone 2
Felipe de la Llama Vázquez

Member Zone 3
Carmen Lacasa Díaz

Member Zone 4
Manuel Alós Almiñana

Member Zone 5
Luis de la Morena del Valle

Member Zone 6
Esperanza Quintero Pichardo

Member Zone 7
Antonio López Pastor

Member Zone 8
Miguel Angel Wood Wood

Member Residents
Cristina L. Crespo Martínez

PERMANENT BOARD

Eduardo Echarri Arrieta
Ma. Cinta Gamundi Planas
Rafael Molero Gómez
David García Marco
Felipe de la Llama Vázquez
Luis de la Morena del Valle

DIRECTORS OF PUBLICATIONS

Farmacia Hospitalaria: Esteban Valverde Molina
SSHP Information Bulletin: David García Marco

OTHER POSTS

General Technical Secretary: Isabel Crespo González
Clerks: Ma. Luisa Avila Fernández; Manuela Florencio Serrano
Technical Documentation Officer: Henar Martínez Hernández
GRD Project Officer: Marc Montero Pardillo

CONCLUSION

Due to its long existence and experience, the SSHP now has a very well-developed structure and is considered as a solid reference within the health field. The society has eight commissions, work groups, and task forces, more than 130 publications, a bimonthly scientific journal, an information bulletin, and educational activities (prizes, courses, grants). This and the high degree of membership among the pharmacist's professionals make this society an important and representative organization within the medical field.

BIBLIOGRAPHY

Boletín Informativo Extraordinario. Memoria 1991–1999; Sociedad Española de Farmacia Hospitalaria, 1999.

Historia de la Sociedad Española de Farmacia Hospitalaria. Tomo I; Sociedad Española de Farmacia Hospitalaria, 1995.

Historia de la Sociedad Española de Farmacia Hospitalaria. Tomo II; Sociedad Española de Farmacia Hospitalaria, 1995.

Sociedad Española de Farmacia Hospitalaria. *Anuario*; Meditex, 1997.

PHARMACY PRACTICE ISSUES

Therapeutic Guidelines Australia

T

Mary Hemming
Therapeutic Guidelines Limited, North Melbourne, Australia

INTRODUCTION

Therapeutic Guidelines Limited (TGL) is an independent, not-for-profit enterprise that focuses on the writing, publication, and sale of prescribing guidelines for health professionals. This article presents the major initiatives of TGL, its goals and organizational structure, and publishing process.

HISTORY

TGL had its genesis in the late 1970s, when a group of enthusiastic individuals joined forces to improve prescribing in an Australian public hospital.

At that time, Melbourne hospitals were experiencing an upsurge in the isolation of antibiotic-resistant organisms. There was concern among health professionals that this was the result of the inappropriate use of antibiotics. A multidisciplinary group from the major Melbourne hospitals responded to the problem by developing an agreed position on what constituted appropriate and cost-effective therapy for a range of common conditions. In 1979, the first edition of *Antibiotic Guidelines* was launched.

In 1985, the Therapeutics Committee of the Victorian Medical Postgraduate Foundation (VMPF-TC) was established to take responsibility for the distribution, promotion, and evaluation of the *Antibiotic Guidelines*, which at that time was the only guideline booklet available.

In 1986, in recognition of the importance of appropriate drug use, a ministerial advisory committee, the Victorian Drug Usage Advisory Committee (VDUAC) was established. The VDUAC believed the availability of independent information regarding drug therapy was integral to informed decision making by prescribers. In light of the increasing acceptance and use of the *Antibiotic Guidelines*, the Committee resolved to build on this concept and to produce additional prescribing guideline booklets.

In conjunction with this expansion, a number of innovative educational marketing and outreach projects were undertaken. The result was increased implementation of the Guidelines in both hospital and community practice.

By 1996, the Guidelines were firmly entrenched and growing rapidly. The VDUAC was responsible for the production of the manuscripts, and the VMPF-TC for publication, distribution, promotion, and evaluation of the books. The number of titles was increased; the target group was extended to include community practice; and authorship was widened to ensure input from the most eminent Australian experts.

With the expansion, however, administration became difficult under the complex committee structure described above; therefore, in 1996 a separate entity, TGL, was established under which all Guidelines activities were consolidated.

ORGANIZATIONAL STRUCTURE

TGL is a not-for-profit company limited by guarantee. It is independent of government and the pharmaceutical industry and is funded solely through sales of its Guidelines.

It is governed by a board of nine directors, four of whom are nominated by organizations that reflect the genesis of the company: the Victorian Medical Postgraduate Foundation; the Victorian Drug Usage Advisory Committee; The Royal Australian College of General Practitioners; and the Commonwealth Department of Health and Ageing.

The current directors of TGL are medically and pharmaceutically qualified professionals working in the areas of health education, medicine, and pharmacy. They are:

- Professor D. Birkett (Clinical Pharmacology), South Australia
- Dr. J.S. Dowden (Medical Publishing), Australian Capital Territory

Encyclopedia of Clinical Pharmacy
DOI: 10.1081/E-ECP 120006375

- Dr. J.E. Marley (General Practice), New South Wales
- Dr. M.L. Mashford (Clinical Pharmacology), Victoria
- Associate Professor F.W. May (Pharmacy), South Australia
- Dr. J.G. Primrose (International Medicinal Drug Policy) Australian Capital Territory
- Professor J.W.G. Tiller (Postgraduate Medical Education), Victoria

The organization employs the equivalent of approximately ten full-time staff members, including the Chief Executive Officer, Mrs. Mary Hemming, B Pharm., Grad Dip Epi Biostat.

Staff are employed in the areas of production (both print and electronic), information technology, marketing, sales and administration, and evaluation. To supplement the expertise and skills within the organization, consultants in finance, law, and marketing are utilized.

The Board appoints expert writing groups as ad hoc committees, who work with editors to develop content for the titles. Each writing group is chaired by a Director.

MISSION

TGL's mission statement is promoting quality use of medicines through the preparation, publication, and sale of independent, authoritative, up-to-date, problem-orientated information.

OBJECTIVES

The key objectives of TGL are:

- To encourage the use of Therapeutic Guidelines by health professionals to assist them in making therapeutic decisions to facilitate optimum healthcare.
- To encourage the use of Therapeutic Guidelines internationally through licensing agreements with appropriate organizations.
- To preserve the integrity of the organization through strict policies for Directors, staff, and members of the writing groups, on issues such as conflict of interest.
- To maintain financial independence.
- To establish strategic alliances with other organizations to further the objectives of the organization.
- To increase the range of products, e.g., information for consumers.
- To allocate funds for research and development.

CURRENT INITIATIVES

Guideline Development

The core activity of TGL is the writing, publication, and sale of Therapeutic Guidelines. These are published as a series of pocket-sized books, and also in electronic formats suitable for both stand-alone and networked computers. All titles are regularly updated in iterative cycles, thus allowing for shifts with evidence, response to feedback, and increased input over time.

The intention of the Guidelines is to provide prescribers with clear, practical, succinct, and up-to-date therapeutic information for a range of diseases. For experienced prescribers, the recommendations provide a valuable "second opinion" against which they can compare their own prescribing. For those less experienced, the recommendations form an acceptable basis for the management of patients. For other health professionals such as pharmacists and nurses, Therapeutic Guidelines provide information to support their contribution to the pharmaceutical care of their patients.

Basis of Recommendations

The Guidelines are derived from the best available scientific evidence, but the experience, insight and opinions of Australian experts are an essential element of the writing process, with the final text reflecting independent and expert interpretation.

Because the Guidelines cover all common disorders and not just those for which there is a body of evidence, there are many instances where trial data are not available, where published data fail to answer questions relevant to prescribers, or where research findings may not be relevant to local practice. To resolve gaps in the evidence, recommendations for reasonable therapy are developed, with criteria such as a drug's adverse effect profile, long-term safety data, and cost being taken into consideration.

Review and Endorsement

Critical steps in the development of the manuscripts are an extensive external review process, reconciliation of text with that of other relevant therapeutic guidelines, and endorsement by national or peak bodies such as The Royal Australian College of General Practitioners and the National Prescribing Service.

Postpublication Evaluation

The evaluation unit liaises with a network of approximately 250 users (general practitioners, specialists, pharmacists, and students) to actively solicit feedback on the various texts. Participants in the network are provided with all titles free-of-charge, and staff visit these users regularly to discuss and record comments.

Before any new edition is commenced, accrued feedback on the previous edition is collated and passed on to the writing group for consideration in the revision of the text.

MAJOR DIRECTIONS

International

Internationally, TGL has been active. A model for an international publishing agreement has been established that is working extremely well. A Japanese not-for-profit organization (comprising doctors and pharmacists) is licensed to translate and modify the texts to suit the Japanese market. The translated books are printed and sold in Japan, with a royalty returning to TGL. Similar licenses are in place with groups in China, Spain, and Russia.

Electronic

The impact of written clinical guidelines is limited by their accessibility at the time of decision making. Therefore, the long-term goal of TGL is to optimize clinical practice by integrating Therapeutic Guidelines into decision support systems, free of commercial bias, so they can assist health practitioners at the time of consultation.

As an initial step toward this goal, all current titles were converted into electronic versions suitable for both stand-alone computers networks. The next step was the aggregation and integration of all ten Therapeutic Guidelines to produce a single comprehensive prescribing resource.

Alternate modes of delivery of the information are now being investigated. For mobile prescribers in institutional settings, the use of handheld computers is being explored. For prescribers in office settings, information embedded in prescribing software is being developed.

Research

Research is an integral part of the organization and the major focus of TGL research is on improving access to the Guidelines at the point of clinical decision making and increasing integration of the Guidelines with other information sources, prescribing modules, and patient databases.

TGL is an industry partner in a major research project, which provides one postdoctoral and two doctoral scholarships to further develop and optimize the production of electronic Therapeutic Guidelines, including the integration of Therapeutic Guidelines with prescribing, dispensing, and medical record management software.

GUIDELINES AVAILABLE

The series includes ten titles in the Therapeutic Guidelines series and one Management Guidelines title:

- Therapeutic Guidelines: Analgesic, Version 4
- Therapeutic Guidelines: Antibiotic, Version 11
- Therapeutic Guidelines: Cardiovascular, Version 3
- Therapeutic Guidelines: Dermatology, Version 1
- Therapeutic Guidelines: Endocrinology, Version 2
- Therapeutic Guidelines: Gastrointestinal, Version 3
- Therapeutic Guidelines: Neurology, Version 2
- Therapeutic Guidelines: Palliative Care, Version 1
- Therapeutic Guidelines: Psychotropic, Version 4
- Therapeutic Guidelines: Respiratory, Version 2
- Management Guidelines: People with Developmental and Intellectual Disabilities

The Guidelines are extensively used in public teaching hospitals, general practice, community pharmacies, government instrumentalities, and medical and pharmacy schools.

ADDITIONAL READINGS

Hemming, M. Therapeutic Guidelines: An Australian Experience. J. Pharm. Med. **2000**, *14*, 259–264.

http://www.tg.com.au.

Therapeutic Interchange

Anthony Compton
St. Joseph's Hospital of Atlanta, Atlanta, Georgia, U.S.A.

INTRODUCTION

In the 1980s, hospital pharmacies were important revenue generators for the institution. However, the impact of managed care coupled with reductions in public programs has significantly changed how budgets of pharmacy departments are viewed. Today, pharmacy has moved from being a revenue generator to a cost center for hospitals, and pharmacy expenses have increasingly become the focus of cost reduction strategies. In general, medication costs represent only 4–8% of total hospital expenses. However, this percentage is increasing due to the influx of more costly medications, increased acuity of hospitalized patients, and the increasing number of medications derived from technology.

Since the late 1990s, considerable literature has been published on methods to manage drug expenditures in the hospital setting. The method that is recognized as being the most effective is a well-managed drug formulary system. Institutions with well-managed formulary systems demonstrate significant decreases in medication costs per patient day compared with institutions without these systems.[1] The formulary system, as defined by the ''ASHP Statement on the Formulary System,'' is a method for evaluating and selecting suitable drug products for the formulary of an organized health care setting.[2] Formulary management is a process employing various techniques to monitor the therapeutic, economic, and clinical outcomes of medication use in the organization.

Therapeutic interchange is the cornerstone of an effective formulary system. Therapeutic interchange is defined as the exchange of one therapeutic equivalent therapy for another with the intent of improving patient outcomes.[2] Reductions in medication expenditures resulting from therapeutic interchange programs are realized by the reduction in the number of medications routinely stocked and, in most instances, contractual-based reductions in drug costs. The primary intent of this article is to provide an overview of the therapeutic interchange process.

In this article, the following definitions are used:

- *Drug Formulary System*: An ongoing process whereby a health care organization, through its physicians, pharmacists, and other health care professionals, establishes policies on the use of drug products and therapies, and identifies drug products and therapies that are the most medically appropriate and cost effective to best serve the health interests of a given population.[3]
- *Drug Formulary*: A continually updated list of medications and related information, representing the clinical judgment of physicians, pharmacists, and other experts in the diagnosis and/or treatment of disease and promotion of health.[3]
- *Therapeutic Substitution*: The act of dispensing a therapeutic alternative for the drug product prescribed without prior authorization of the physician. This is a controversial procedure.[3]
- *Therapeutic Interchange*: The authorized exchange of therapeutic alternatives in accordance with previously established and approved written guidelines or protocols within a formulary system.[3]
- *Generic Substitution*: The substitution of drug products that contain the same active ingredients and are chemically identical in strength, concentration, dosage form, and route of administration to the drug product prescribed.[3]
- *Pharmacy and Therapeutics (P&T) Committee*: An advisory committee that is responsible for developing, managing, updating, and administering the drug formulary system.[3] This committee should be composed of a variety of health care providers such as physicians, pharmacists, nurses, microbiologists, administrators, discharge planning, quality assurance, and risk managers.

HISTORY OF THE FORMULARY SYSTEM

In the United States, most health care systems establish formularies based on the recommendations of the

Encyclopedia of Clinical Pharmacy
DOI: 10.1081/E-ECP 120006336

Pharmacy and Therapeutics (P&T) Committee. The first guidelines describing a formulary system were published jointly in the early 1930s by the *Journal of the American Medical Association* and the *Journal of the American Pharmaceutical Association*.[4] In 1965, the Joint Commission on Accreditation of Hospitals (JCAHO) mandated that hospitals develop formularies.[5] In 1959, efforts by the American Society of Hospital Pharmacists (ASHP) and the American Hospital Association (AHA) were directed at providing a description of the role of the P&T Committee. Based on the recommendations of the two organizations, the P&T Committee was established as the major committee within the health care organization for determining which medications should be routinely used in the hospital. Data from the American Medical Association (AMA) suggest that 90% of hospitals have some type of formulary.[6] Historically, P&T Committees have determined which medications were selected to the formulary based on the safety, efficacy, and acquisition cost of the medications.

The function of the P&T Committee has evolved secondary to greater pressures caused by managed care, costly medications, and expedited Food and Drug Administration (FDA) review processes. Modern P&T Committees must consider the cost of therapy instead of simplistically evaluating solely the acquisition cost of the medication when deciding formulary status. Until recently, cost effectiveness information was not a routine component of data submitted to the FDA during the drug approval process. In addition, shorter FDA evaluations of new drugs mandate that hospitals, through P&T Committees, constantly review information regarding clinical outcomes, medication errors, and adverse drug effects long after the medication has been approved for use.

THERAPEUTIC INTERCHANGE PROCESS

The term *therapeutic interchange* is often used interchangeably with *therapeutic substitution*. There are important differences between the two methods of formulary management; therefore, the terms should not be used in this manner. Based on the list of definitions previously presented, the major difference between these methods of formulary management is whether the prescribing physician has given approval of the interchange of one medication for another or received notification that the interchange occurred. In an effective therapeutic interchange system, physicians are informed before the alternative medication is dispensed and administered. A great deal of controversy surrounds therapeutic interchange and substitution in the outpatient setting. The Public Advocate for the City of New York contended that the use of formularies and therapeutic interchange in ambulatory care settings resulted in inappropriate, or less appropriate, drug therapy.[7] Furthermore, this inappropriateness in drug therapy occurs in great part because financial considerations—much more than clinical considerations—drive formulary decisions in managed care organizations.[7] In a published comparison evaluating the utilization of ambulatory services for patients in a health maintenance organization (HMO), the use of formularies and therapeutic interchange was found to increase the utilization of health care resources.[8] Proponents of formularies and therapeutic interchange in the outpatient settings argue that these criticisms are unfounded. The Academy of Managed Care Pharmacy (AMCP) states that a well-developed and maintained medication formulary decreases patients costs and improves patient care.[9]

In the inpatient setting, therapeutic interchange has been accepted by a number of medical and pharmacy organizations. These organizations include the AMA American Society of Health-Systems Pharmacists, and the American College of Physicians.[2] These organizations support therapeutic interchange within the established guidelines of a formulary system to meet the therapeutic and institutional goals for the ultimate benefit of patient care.

In general, there are three mechanisms by which therapeutic interchange can occur.[10] The first mechanism consists of a system whereby pharmacists independently substitute medications on the basis of their

Table 1 Therapeutic interchanges at SJHA

Low molecular weight heparins
Extended spectrum penicillins
First-generation cephalosporins
Second-generation cephalosporins
Third-generation cephalosporins
Proton pump inhibitors
H^2 blockers
5HT3 inhibitors
Fluroquinolones
Oxazolidinones/streptagramins
Parenteral amino acids
Sedative hypnotics
Enteral feedings
Insulins
IV immunoglobulins
Narcotic analgesics
Multivitamins

own professional judgment, without the consent of the prescribing physician. This mechanism is highly discouraged by the majority of professional organizations and is considered illegal in several states. The second mechanism is the most commonly accepted. In this mechanism, medications are automatically interchanged with prior physician approval based on established institutional protocols. These protocols are established through the hospital's P&T Committee or through HMO policies. In most instances, this mechanism relies on the institution's computer system to activate the therapeutic interchange. The last mechanism involves obtaining physician approval prior to each therapeutic interchange. Although effective in some situations, this mechanism can be time consuming for pharmacists, physicians, nurses, and patients.

Table 2 Guidelines for implementing a therapeutic interchange program

Step 1: Identify the feasibility of a therapeutic interchange based on medication class usage trends, patient outcomes, clinical trial data, medication error trends/potential, adverse reaction information, cost benefits, or resistance patterns.
Step 2: Determine the purpose of the therapeutic interchange. If after determining that two agents are therapeutically equivalent, cost savings can become an important factor of the therapeutic interchange.
Step 3: Obtain support for the therapeutic interchange from the P&T Committee, Antibiotic Subcommittee, and any medical staff sections that would be involved in the therapeutic interchange.
Step 4: Communicate the proposed therapeutic interchange to members of the medical and hospital staffs. In general, input should be obtained from members of the laboratory, nursing staff, administration, risk management, key physician groups, and quality improvement staffs.
Step 5: Once approved, advertise the therapeutic interchange to the medical, nursing, administrative, and pharmacy staffs. Communication can be in the form of medical staff newsletters, pharmacy newsletters, mailings, inservices, and posters.
Step 6: Implement the interchange. Physicians are generally more accepting of therapeutic interchange programs if they know that nonformulary medications can be obtained if indicated. At SJHA, a nonformulary medication will not be interchanged if "brand necessary" or "do not substitute" is designated by the physician on the medication order.
Step 7: Monitor the program for positive or negative impacts. This is perhaps the most important aspect of developing a therapeutic interchange program. Feedback regarding the program should be continuously obtained from members of the medical, nursing, discharge planning, and laboratory staffs and, in some instances, from patients. Be prepared to modify the program, if necessary.

A survey published in 1992 by the American Society of Health-Systems Pharmacists suggested that only 61% of the hospitals participating in the survey had well-controlled formulary systems.[11] Of these hospitals, 69% had formulary systems based on the concepts of therapeutic interchange. Common medication classes involved in therapeutic interchanges were antimicrobials, antacids, and multivitamins. The medication class of antimicrobials is particularly suitable for therapeutic interchange. Antimicrobials represent a pharmacologic class with multiple therapeutic redundancies.[2] This pharmacologic class represents approximately 10–40% of a typical hospital's drug expenditures.[2]

Saint Joseph's Hospital of Atlanta (SJHA) is a 348-bed, tertiary care, nonprofit hospital that specializes in interventional cardiology, oncology, cardiac surgery, vascular surgery, orthopedics, and neurology. The hospital currently has a drug budget of approximately $10 million. Therapeutic interchange has been used extensively to manage the pharmacy drug budget. It is estimated that a well-managed formulary system based on therapeutic interchange avoids $800,000 to $1 million annually based on medication acquisition costs. Table 1 contains a listing of therapeutic interchanges by medication class.

One of the most recent medication classes reviewed for potential therapeutic interchange was the fluroquinolone antibiotics. A listing of the procedures involved in the process is provided in Table 2. This listing should serve as a guide for implementing a therapeutic interchange. Other guidelines have also been published.[12,13]

CONCLUSION

When used appropriately, therapeutic interchange has proven to be an extremely effective method of medication cost management. An active and well-organized P&T Committee is essential to the success of any therapeutic interchange program. As the focus on the costs involved with medication therapy in health care organizations increases, P&T Committees will have to become even more creative in methods used to promote, evaluate, and monitor the effects of therapeutic interchange. The evaluation of quality of life issues, pharmacoeconomics,

clinical outcomes, and humanistic/behavioral outcomes will become increasingly important.[5]

REFERENCES

1. Massoomi, F. Formulary management: Antibiotics and therapeutic interchange. Pharm. Pract. Manage. Q **1996**, *16* (3), 11–18.
2. American Society of Hospital Pharmacists. ASHP guidelines on formulary system management. Am. J. Hosp. Pharm. **1992**, *49*, 648–652.
3. Carroll, N.V. Formularies and therapeutic interchange: The health care setting makes a difference. Am. J. Health-Syst. Pharm. **1999**, *56*, 467–471.
4. Dillon, M.J. *Drug Formulary Management in Managed Care Pharmacy Practice*; Navarro, R.P., Ed.; Aspen Publishers, Inc.: Gaithersburg, Maryland, 1999; 145–165.
5. Wade, W.E.; Spruill, W.J.; Taylor, A.T.; Longe, R.L.; Hawkins, D.W. The expanding role of Pharmacy and Therapeutics Committees, the 1990s and beyond. PharmacoEconomics **1996**, *10* (2), 123–128.
6. American Medical Association. AMA policy on drug formularies and therapeutic interchange in inpatient and ambulatory care settings. Am. J. Hosp. Pharm. **1994**, *51*, 1808–1810.
7. Green, M.; Vasisht, R.; Martin, J. *Compromising Your Drug Choice: How HMOs are Dictating Your Next Prescription*; Public Advocate: New York, 1996.
8. Horn, S.D.; Sharkey, P.D.; Tracy, D.M.; Horn, C.E.; James, B.; Goodwin, F. Intended and unintended consequences of HMO cost-containment strategies: Results from managed care outcomes project. Am. J. Manage. Care **1996**, *II* (3), 253–264.
9. *Managed Care Pharmacy Factsheet-Therapeutic Substitution*; Academy of Managed Care Pharmacy: Alexandria, Virginia, 1997.
10. Abood, R.R. Legal issues confronting the therapeutic substitution of drug products. Clin. Trends Pharm. Practice **1995**, *9*, 1–11.
11. Crawford, S.Y.; Myers, C.E. ASHP national survey of hospital-based pharmaceutical services 1992. Am. J. Hosp. Pharm. **1993**, *50*, 1371–1404.
12. Wall, D.S.; Abel, S. Therapeutic-interchange algorithm for multiple drug classes. Am. J. Health-Syst. Pharm. **1996**, *53*, 1295–1296.
13. Sesin, G.P. Therapeutic decision-making: A model for formulary evaluation. Drug Intell. Clin. Pharm. **1986**, *29*, 581–583.

Therapeutic Interchange, Guidelines (ACCP)

American College of Clinical Pharmacy
Kansas City, Missouri, U.S.A.

INTRODUCTION

Therapeutic interchange involves the dispensing of chemically different drugs that are considered to be therapeutically equivalent. Therapeutically equivalent drugs are chemically dissimilar but produce essentially the same therapeutic outcome and have similar toxicity profiles. Usually these drugs are within the same pharmacologic class. They frequently differ in chemistry, mechanism of action, and pharmacokinetic properties, and may possess different adverse reaction, toxicity, and drug interaction profiles.

Interest in therapeutic interchange has risen as a result of two primary influences: rapid expansion in numbers of drugs within the same therapeutic class, and the need to contain medication and health care costs while maintaining rational drug therapy. Therapeutic interchange policies grant pharmacists the authority to interchange one drug for another without prior consent from a physician, according to procedures outlined in a specific policy.

The policies are usually developed and guided by an advisory group such as the Pharmacy and Therapeutics Committee. This committee is composed of physicians, pharmacists, and other health professionals who combine their expertise, knowledge, and experience to recommend policies to the medical staff and administration of an organization on matters related to the therapeutic use of drugs. Among other duties, the committee 1) serves in an advisory capacity to the medical staff and administration in all matters pertaining to the use of drugs, including therapeutic interchange; 2) establishes programs and procedures that help ensure cost-effective drug therapy; 3) establishes or plans suitable educational programs for the professional staff on matters related to drug use; 4) participates in quality assurance activities; and 5) initiates or directs drug use evaluation programs and reviews their results.[1–7] Thus, a successful therapeutic interchange policy is directly related to the effectiveness of the Pharmacy and Therapeutics Committee in performing its functions.

The concept of therapeutic interchange has aroused much controversy and debate. Several definitions or interpretations exist, which have fueled this debate and resulted in considerable confusion.

Concerns associated with responsibility, liability, communication, and monitoring influence an organization's statement on therapeutic interchange. This publication presents the American College of Clinical Pharmacy's (ACCP's) definition of and position on therapeutic interchange. It also reviews and discusses current views on therapeutic interchange held by the American College of Physicians (ACP), the American Medical Association (AMA), the American Society of Hospital Pharmacists (ASHP), the American Pharmaceutical Association (APhA), the American Association of Colleges of Pharmacy (AACP), the Generic Pharmaceutical Industry Association (GPIA), and the Pharmaceutical Manufacturers Association (PMA).

OTHER ORGANIZATIONS' VIEWS ON THERAPEUTIC INTERCHANGE

Organizations representing the medical, pharmacy, and drug manufacturing communities have position statements or policies regarding therapeutic interchange. These statements differ in their definitions and support of the concept. For example, the ACP published a position paper in which therapeutic interchange in institutional and ambulatory settings is supported when a functioning Pharmacy and Therapeutics Committee and formulary system are in place.[8] In addition, the ACP recommends that ''immediate prior consent'' be obtained from the authorized prescriber with ''appropriate documentation of the substitution in a timely and proper manner.'' Although this document supports therapeutic interchange, the requirement of ''immediate prior consent from the authorized prescriber'' may defeat the purpose. If a pharmacist is required to call the physician to gain per-

Encyclopedia of Clinical Pharmacy
DOI: 10.1081/E-ECP 120006409

mission to interchange medications, the physician can either give or withhold permission, or provide a new prescription order (verbal order). This new prescription order can override the therapeutic interchange policy and therefore negate its effectiveness.

Although it has not published a position statement or paper on the topic, the AMA does not support the use of therapeutic interchange in any setting. Their position is evident when reviewing policies 23.023, 23.033, 23.035, 23.057, and 23.060 from their current Policy Compendium.[9] These policies state firm opposition to the interchange of ''(1) ... a drug product that is administered in the same route and which contains the same pharmaceutical moiety and strength, but which differs in the salt or dosage form; and (2) ... a drug product containing a different pharmaceutical moiety but which is of the same therapeutic and/or pharmacological class.'' At their June 1990 annual meeting, the AMA House of Delegates debated and adopted Resolution 161, which states opposition to the establishment of a system at the federal or state level for therapeutic interchange. The resolution states that this ''will inevitably interfere with the ability of the patient's physician to assure that the medication prescribed is dispensed to the patient,'' and thus individuality of patient care can be lost.

Of note is the difference between therapeutic interchange policies endorsed by individual health care organizations, and regulations required by federal or state laws. Federal or state regulations could be misinterpreted as being blanket policies that grant pharmacists authority to interchange medications across all classes and categories. Such regulations might also inappropriately suggest that pharmacists be given the authority to override a physician's prescription without their knowledge.

Therapeutic interchange policies should not grant prescribing authority to pharmacists. Patients can be assured of getting the best care possible only when a pharmacist acts in collaboration with a physician to provide an optimal drug product.

Thus, both parties must endorse therapeutic interchange policies before such policies can be used effectively. Some states have enacted or proposed legislation that requires physicians to be involved in the development and management of therapeutic interchange guidelines, whereas other states have proposed legislation that seeks to prohibit pharmacists from enacting a unilateral interchange. The proposed or enacted legislation agrees with ACCP's definition in that physician involvement is paramount to the successful development of guidelines for therapeutic interchange, and pharmacists should not make unilateral substitution of a drug.

The ASHP supports ''therapeutic interchange by pharmacists when a pharmacist and a physician interrelate on behalf of the patient.''[10] They define therapeutic interchange as ''the interchange of various therapeutically equivalent drug products by pharmacists under arrangements between pharmacists and authorized prescribers who have previously established and jointly agreed upon conditions for interchanges.''[11] They believe that an advisory committee should develop these policies, that the policies should be reviewed and revised over time, and that prescribers should have the prerogative to override therapeutic interchange on behalf of patients. In addition, ASHP recommends that pharmacists enacting therapeutic interchange monitor affected patients to ''identify and prevent any unexpected or untoward patient response.''[11] Similarly, physicians should be notified when a therapeutic interchange policy has been enacted and be provided with educational materials supporting it when appropriate.

The APhA House of Delegates and the AACP support ''the concept of therapeutic interchange of various drug products by pharmacists under arrangements in which pharmacists and authorized prescribers interrelate on behalf of the care of patients.''[12,13] The AACP further ''views initiatives to prohibit therapeutic interchange to be counterproductive and confusing because this criticism strikes at an important element of the clinical practice of pharmacy.''[13] They believe that pharmacy students are adequately trained to participate in the clinical environment, and to assist in the development and process of therapeutic interchange. To this end, AACP is committed to training pharmacists who are competent to perform these activities.

The GPIA ''supports the practice of therapeutic interchange in institutional or ambulatory settings where a functioning P and T committee and a formulary system are in place and where there is active consultation on behalf of the patient between physicians and pharmacists.''[14] The GPIA also ''supports an ongoing drug utilization evaluation process for regular review of therapeutic interchange policies and formularies, and for providing information and education to prescribers.''[14]

The PMA (representing pharmaceutical manufacturers marketing brand name drugs) opposes therapeutic interchange. Its policy statement concerning drug selection states that it ''supports those public policies which retain the authority and responsibility for prescription drug selection exclusively with the individual attending physician, dentist or podiatrist.... Specifically, PMA believes that inappropriate drug selection policies such as pharmaceutical or therapeutic substitution: (1) may adversely affect a patient's health, (2) raises serious legal questions, (3) may result in increased cost for medical services and (4) are highly inefficient and intrusive.''[15] The policy states, however, that it ''is not intended to address the manner in which hospital pharmacy practices are implemented in an inpatient care situation. Direct physician involvement in determining the hospital formulary distinguishes such formulary practices from therapeutic and pharmaceutical substitution. Additionally, it is clear that the inpatient care environment is a highly controlled setting where patients are constantly being monitored and, consequently, is much different than the outpatient care situation.''[15]

The view of many research-based pharmaceutical manufacturers is based on the economic impact therapeutic interchange may have on their organizations. Although this is an important consideration for their welfare, we believe that the manufacturers should join physicians and pharmacists in the quest to provide excellent medical care at reasonable costs, especially in light of the federal, state, and local regulations directed at achieving this goal. Such a commitment would foster a better relationship among the represented communities.

The ACCP believes that if physicians and pharmacists communicate, collaborate, and jointly agree on the purpose for and appropriate monitoring of therapeutic interchange policies, many of the opposing viewpoints outlined above can be brought together in an effort to provide state-of-the-art therapeutics and optimal patient outcomes.

ACCP GUIDELINES FOR THERAPEUTIC INTERCHANGE

The ACCP supports therapeutic interchange policies when pharmacists and physicians collaborate to develop policies designed to provide patients with the best possible care at the best overall price. These policies should result from a synergistic combination of the expertise and knowledge of pharmacists and physicians whose common goal is to ensure optimum patient care. They should not be interpreted as ''bestowing prescribing authority on pharmacists.'' Although the policies may vary in complexity, most involve the interchange of one drug for another that is therapeutically equivalent. Thus, they should not be viewed as or become blanket policies allowing pharmacists to choose an alternative agent from an entire class or category of drugs.

The ACCP supports the following guidelines for implementing therapeutic interchange policies within health care organizations.

Guideline I

Therapeutic interchange is appropriate in institutional and ambulatory settings that have a functioning formulary system and Pharmacy and Therapeutics Committee or equivalent advisory committee.

Rationale

The success of any therapeutic interchange program is related to the effectiveness of the Pharmacy and Therapeutics Committee or its equivalent body, and the drug formulary system. This committee, representing both the pharmacy and medical staffs, must develop, implement, review, and change policies and procedures to ensure optimum patient care while containing costs. Similarly, it should recommend and assist in educating professional staff regarding current therapeutic interchange policies, the success of such policies, and any approved exceptions to them.

The Pharmacy and Therapeutics Committee should establish or assign a committee or department to monitor the effectiveness of the interchange policies. Audits or reviews should be conducted according to set policies. Criteria should be developed and used to determine when and why the therapeutic interchange policies may be ineffective (see Guideline II). The issues identified should be addressed and the professional staff notified of resulting changes to policies and procedures.

The specific types of institutional or ambulatory settings for which therapeutic interchange policies are most likely to be effective are those that have drug formularies with a functioning Pharmacy and Therapeutics Committee or its equivalent. This may include, but is not limited to, hospice settings, health maintenance organizations, community hospitals, university teaching hospitals, and ambulatory clinics affiliated with a hospital. If the policies dictate that laboratory or medical record data should be readily available to pharmacists prior to dispensing a therapeutic alternative, the setting must provide access to this information. Regardless of the setting, pharmacists and physicians must have a good working relationship and a mutual goal to provide excellent medical care while containing costs appropriately.

Guideline II

A continuous drug use evaluation process must be in place for regular review of endorsed therapeutic interchange policies and procedures.

Rationale

Drug use evaluation (DUE) reviews the types of medications prescribed within an institution. This process could identify prescription orders to which therapeutic

interchange policies apply. Once these orders are identified, the success of the policies could be further evaluated, reviewed, and reported. For example, one would have to identify whether an approved alternative agent was dispensed in place of the one originally prescribed. The DUE could also determine if appropriate patient monitoring occurred following the interchange, and whether the interchange resulted in an altered response. Results of these evaluations should be presented to the Pharmacy and Therapeutics Committee or its equivalent to determine if changes or exceptions to existing policies are indicated, or if additional educational endeavors should be made available to participating professional staff.

Guideline III

Therapeutic interchange, as defined herein, may be executed by pharmacists if the authorized prescriber is notified either verbally or in writing within a reasonable time frame, and if the pharmacists have access to medical records and appropriate laboratory or other test results as required by the therapeutic interchange policy. Exceptions to this procedure must be stated clearly in the policy.

Rationale

Therapeutic interchange policies and procedures should describe in detail the conditions and processes for interchanging medications. These should include who has authority to enact the interchange, special exceptions to a policy or procedure, criteria to be evaluated before and after the interchange occurs, and the definition of "a reasonable time frame" for notifying the physician. For example, the policy may not require that a physician be notified for interchange of certain drugs, such as multivitamins. As another example, a reasonable time frame for notification of the physician when therapeutic interchange has occurred may be defined as within 24–72 hours when the interchange involves antibiotics.

Guideline IV

The Pharmacy and Therapeutics Committee or its equivalent should ensure that professional staff are educated regarding the rationale, policies, and procedures for therapeutic interchange.

Rationale

Proper educational methods should be developed, implemented, reviewed, and revised as necessary to inform the professional staff regarding the rationale, procedures, and monitoring criteria for therapeutic interchange. These methods may include, but should not be limited to, newsletters, fliers, departmental meetings, meeting minutes, and publication of therapeutic interchange policies, such as in an organization's formulary. Physicians should be informed of the policies prospectively, and a list of active policies should be readily available to the professional staff. Concerns or problems regarding the policies should be addressed to the Pharmacy and Therapeutics Committee.

Guideline V

The therapeutic interchange policies should define a mechanism that enables authorized prescribers to disallow therapeutic interchange.

Rationale

Therapeutic interchange may not be applicable to all patients. For example, a patient's preference may play a role in a physician's decision to override a policy. An acceptable method of overriding must be made available to authorized prescribers. Ideally, it would allow one to capture information regarding decisions to override policy so that the data can be reviewed easily by an advisory committee. This could, for example, be accomplished by asking the physician to complete a brief survey or request form for disallowing therapeutic interchange. The physician would have to notify the pharmacist either in writing or verbally regarding the desire to override the existing policy.

ACKNOWLEDGMENTS

The guidelines were written by the following subcommittee of the 1990–1991 ACCP Clinical Practice Affairs Committee: Terri Graves Davidson, Pharm.D.; Mary Beth O'Connell, Pharm.D.; Ryon Adams, Pharm.D.; Veronica Moriarty, Pharm.D.; Anthony Ranno, Pharm.D.; Nathan Schultz, Pharm.D.; Barry Carter, Pharm.D., FCCP, Chair; and Marsha Raebel, Pharm.D., FCCP. Other members of the committee were Richard Berchou, Pharm.D., Dennis Clifton, Pharm.D., Joseph F. Dasta, M.S., FCCP; Carl Hemstrom, Pharm.D.; Donald Kendzierski, Pharm.D.; Bruce Kreter, Pharm.D.; Louis Pagliaro, Pharm.D.; Richard Ptachcinski, Pharm.D.; Christine Rudd, Pharm.D., FCCP; and Dominic Solimando, Jr., Pharm.D. Staff editor is Toni Sumpter, Pharm.D. Approved by the Board of Regents on November 3, 1992.

From *Pharmacotherapy* 1993, 13(6):252–256, with permission of the American College of Clinical Pharmacy.

REFERENCES

1. American Society of Hospital Pharmacists. Statement on the pharmacy and therapeutics committee. Am. J. Hosp. Pharm. **1984**, *41*, 1621.
2. Green, J.; Chawla, A.K.; Fong, P.A. Evaluating a restrictive formulary system by assessing nonformulary-drug requests. Am. J. Hosp. Pharm. **1985**, *42*, 1537–1541.
3. Sesin, G.P. Therapeutic decision-making: A model for formulary evaluation. Drug Intell. Clin. Pharm. **1986**, *20*, 581–583.
4. Butler, C.D.; Manchester, R. The P&T committee: descriptive survey of activities and time requirements. Hosp. Formul. **1986**, *21*, 90–99.
5. Weintraub, M. Effective functioning of the P&T committee: One chairperson's view. Hosp. Formul. **1977**, *12*, 260–263.
6. Martin, L.A.; Watkins, J.B.; Green, S.A., et al. Procedural compliance and clinical outcome associated with therapeutic interchange of extended-spectrum penicillins. Am. J. Hosp. Pharm. **1990**, *47*, 1551–1554.
7. Oh, T.; Franko, T.G. Implementing therapeutic interchange of intravenous famotidine for cimetidine and ranitidine. Am. J. Hosp. Pharm. **1990**, *47*, 1547–1551.
8. American College of Physicians. Therapeutic substitution and formulary systems. Ann. Intern. Med. **1990**, *113* (2), 160–163.
9. American Medical Association. AMA Policy Compendium. In *Current Policies of the AMA House of Delegates Through the 1989 Interim Meeting*; Chicago, 1990; 63–64, 67.
10. American Society of Hospital Pharmacists. *Therapeutic Interchange*; Bethesda, Maryland, 1982.
11. American Society of Hospital Pharmacists. Guidelines on formulary system management. Am. J. Hosp. Pharm. **1992**, *49*, 648–651.
12. American Pharmaceutical Association. *Position of the American Pharmaceutical Association on the Issue of Therapeutic Interchange*; Washington, DC, 1987.
13. American Association of Colleges of Pharmacy. *Perspective on Therapeutic Interchange*; Alexandria, Virginia, 1987.
14. Generic Pharmaceutical Industry Association. *Policy on Therapeutic Interchange*; New York, 1992.
15. Pharmaceutical Manufacturers Association. *Policy Statement Concerning Drug Selection*; Washington, DC, 1991; 1–6.

PROFESSIONAL DEVELOPMENT

Transplantation Pharmacy Practice

Iman E. Bajjoka
Marwan S. Abouljoud
Henry Ford Hospital, Detroit, Michigan, U.S.A.

INTRODUCTION

Transplantation of human tissues is one of the most important medical achievements of the 20th century. Transplantation began in 1902, with work of vascular surgeon Alexis Carrel, who established many of the vascular anastomosis techniques that led to effective transplantation. A century later, transplantation has become a life-saving procedure for a variety of irreversible acute and chronic diseases for which no other therapy is available. Nearly every thoracic and abdominal organ may now be successfully transplanted. In 1999, a total of 21,516 solid organ transplants were performed at centers throughout the United States, of which 16,802 were cadaveric and 4,714 were from a living donor.[1]

Increased experience and advances in surgical techniques, tissue preservation and posttransplant care have helped to improve the overall success of transplantation. In 1988, the 1-year graft survival of renal transplants using cadaver grafts was 76%, but by 1999, the 1-year cadaver graft survival rose to 89%. Results of living donors also improved from 89% to 94% for the same time period. Improvement in survival have also been achieved for other solid organ transplants with approximately 70% to 80% of grafts functioning at 1-year after transplantation.[1]

The success of any organ transplantation is due largely to control of the immune system by immunosuppressive therapy to avert rejection of the allograft. Immunosuppressive treatment strives to prevent allorecognition and subsequent destruction of the transplanted tissues. As a result, transplant recipients rely on lifelong immunosuppressive drug therapy for preventing rejection and maintaining their graft. Pharmacists, as experts on immunosuppressive drugs have a vital role in optimizing pharmacological therapy to enhance transplant outcomes and minimize drug-related complications.

OPPORTUNITIES

Since the mid 1990s, the Food and Drug Administration (FDA) has approved various potent immunosuppressive drugs for use in transplantation. Currently, there are four major classes of immunosuppressive drugs available: corticosteroids, antilymphocyte, antimetabolites, and calcineurin inhibitors (Table 1). These immunosuppressants are known to cause short-and long-term complications, including infections, cardiovascular disease, and malignancy (Table 2).[2–6] In addition, solid organ transplant recipients may have their care further complicated by preoperative conditions. The majority of patients have chronic end-stage illnesses that are inevitably fatal and necessitate their transplant. As a result, post transplantation most patients are on complex prophylactic and therapeutic multidrug regimens, leading to many drug-related complications that may compromise transplant outcomes and significantly increase the cost of transplantation.

From its inception, organ transplantation has been approached in an extraordinarily multidisciplinary manner. The transplant team has historically included surgeons, immunologists, pathologists, internists, nurses, dietitians, social workers, and spiritual representatives. With continued success of transplantation and the immunsuppression-related complications, opportunities have been created for many of the pharmacists that are now members of various transplant teams throughout the United States. As part of the team, transplant pharmacists provide direct patient care in a variety of settings, especially because they have the understanding of transplant immunology and the immunosuppressants used to preserve the graft.

Appropriately trained transplant pharmacotherapy specialists may practice in traditional clinical roles or they may fill positions outside the usual sphere of pharmacy practice. Many positions include a clinical

DOI: 10.1081/E-ECP 120006239

Table 1 Immunosuppressive drugs

Antimetabolites	Azathioprine, mycophenolate mofetil, sirolimus
Calcineurin inhibitors	Cyclosporine, tacrolimus
Corticosteroids	Methylprednisolone, prednisolone
Monoclonal antibodies	Muromonab-CD3, basiliximab, daclizumab
Polyclonal antibodies	Antilymphocyte gamma globulin—horse Antilymphocyte gamma globulin—rabbit

practice site in addition to teaching and research opportunities.[7] These individuals may provide inpatient or ambulatory patient services. The ideal role would involve both of these areas and, in so doing, ensure continuity of care. Although the need to guarantee appropriate follow-up of treatment plans and monitoring is not unique to the field of transplantation, the vast number of specialists and subspecialists involved in the care of an organ transplant recipient provides the pharmacist with many opportunities to facilitate communications between members of the transplant and nontransplant teams. Having pharmacists coordinate the patient's pharmacotherapeutic plan when the patient is hospitalized and following discharge would increase the consistency in treatment. This would have the obvious benefit of improving patients' quality of life while reducing the cost of medical resources. The pharmacists' focus adjusts as the patients' clinical issues shift from postoperative concerns immediately following the transplant procedure to more chronic medical matters.

With the expansion of investigational immunosuppressants and clinical drug trials, many pharmaceutical drug companies working in the area of transplantation have created nontraditional roles for transplant pharmacists. Pharmacists can use their skills as researchers (clinical or basic science), educators, or medical information liaisons to assist transplant centers to achieve their ultimate goal of superior patient care.

The cost of transplant medications accounts for approximately 25% of the total cost of care in the first year after transplantation and up to 90% in subsequent years.[8] The absolute dollar amount of immunosuppressants and adjunct therapy is also very high. The average wholesale price of one immunosuppressive agent may range between $6,000 to $12,000 per patient per year. In most situations, long-term immunosuppression consists of double- or triple-combination therapy. Furthermore, in the period immediately following transplantation, many centers use induction therapy that involves expensive agents such as monoclonal or polyclonal antibody preparations.[9,10] With the introduction of more induction therapies and other expensive immunosuppressive agents, the potential for cost savings by increasing participation of clinical pharmacists is undoubtedly considerable and will continue to increase. Some literature has demonstrated this in solid organ transplant patients, but more rigorous studies will be needed before this concept is widely accepted.[11]

Table 2 Possible complications of immunosuppressive drugs

Bone disease	Osteoporosis
Cardiovascular	Hyperlipidemia, hypertension
Cosmetic	Acne, sun sensitivity, hirsutism, weight gain
Gastrointestinal	Peptic ulcer disease, diarrhea, nausea, vomiting
Infections	Bacterial, fungal, viral
Hematologic	Leukopenia, thrombocytopenia
Malignancies	Solid tumors, post-transplant lymphoproliferative disorders
Metabolic	Diabetes mellitus
Neurological	Headache, tremor, neuralgia
Ophthalmological	Cataracts, glaucoma
Psychological	Depression, mood changes
Renal	Acute or chronic dysfunction

CLINICAL PRACTICE

The clinical role of the transplant pharmacist is variable, depending on the practice of the transplant center. Although each clinician may be unique in his or her position, all share the common goals and objectives of improving patient care. Inpatient responsibilities may include daily participation in medical and/or surgical rounds. The transplant pharmacist offers recommendations to optimize drug regimens for patients, based on their own history of illness. The use of immunosuppression therapy is often considered a balancing act between

Table 3 Drugs that interact with calcineurin inhibitors

Decrease concentrations of CNIs
Carbamazepine, cholestyramine, dexamethasone, isoniazid, nafcillin, octreotide, phenobarbital, phenytoin, primidone, rifampin, sulfadimidine, sulfinpyrazone, ticlopidine
Increase concentrations of CNIs
Clarithromycin, danazol, diltiazem, erythromycin, fluconazole, fluvoxamine, grapefruit juice, itraconazole, ketoconazole, methylprednisolone, methyltestosterone, nicardipine, norethisterone, protease inhibitors, verapamil

CNIs, calcineurin inhibitors.

suppression of rejection and immunosuppressive drug toxicities. The fulcrum should always be adjusted to the requirements of the individualized patient. For example, with older transplant recipients in whom the immune response is dampened, it may be more appropriate to select a lower-intensity immunosuppressive regimen to decrease side effects. In contrast, African Americans are considered immunologic high-risk recipients with increased risk for rejection and, therefore, will require more intense immunosuppression to ensure successful therapeutic outcomes.[12,13] Certain patients may be predisposed to abnormalities that are part of their disease or a result of the transplant, and a particular immunosuppressive drug may exacerbate or worsen them. To prevent jeopardizing the allograft or further complications in these patients, it may be necessary to modify the doses of certain immunosuppressants or avoid them entirely.

Once a particular regimen is chosen, pharmacists will frequently perform pharmacokinetic and pharmacodynamic evaluations for immunospppressants, antibiotics, and other agents. Many of the immunosuppressants such as cyclosporine, tacrolimus, and sirolimus have significant intra- and interpatient differences that may lead to suboptimal blood concentrations.[14,15] Therapeutic drug monitoring of immunosuppressive drugs is essential to make accurate predictions for appropriate drug therapy. Furthermore, many of the drugs used in the management of transplant recipients are extensively metabolized by the cytochrome P-450 enzyme system. Drugs, foods, or nutrients that induce, inhibit, or compete for the same isoenzyme could effect blood concentrations of these immunosuppressive agents (Table 3).[16,17] Drug interactions that increase blood concentrations could expose transplant recipients to serious adverse effects, whereas drug interactions that decrease blood concentrations may cause rejection episodes and possible loss of the graft as a result of inadequate immunosuppression. Pharmacists can prospectively review the patient's medication profile including over-the-counter (OTC) medications, herbs, and dietary supplements to predict such interactions and to help reduce adverse outcomes.

PROTOCOLS AND GUIDELINES

It has been suggested that there are as many transplant drug protocols for managing patients as there are transplant centers. This situation has been created by the fact that in the area of transplantation, there is an absence of consensus statements and guidelines in the management of immunosuppressive therapy. This is not to imply that adequate literature addressing transplant issues is lacking, only that it has not reached the standardization and review procedures achieved in other specialties. Guidelines and consensus statements regarding the optimal management of transplant recipients have not been established because there are few circumstances in which the support of one treatment over another is overwhelming. This is probably due to the large sample size required to show statistically significant differences and the variability of patients in regard to demographics, socioeconomics, and other risk factors. Most of the resources currently available to aid in the selection of specific medications are publications from the primary literature. National and international conferences play a significant role in disseminating information concerning advances in therapy and new ideas. Symposia provide another way to learn of contemporary practices, usually in a setting that fosters networking while addressing individual practice issues. They also provide more opportunity to discuss cases with colleagues, draw upon the experience of others, and learn innovative ways that other centers address common problems. This puts a greater onus on the transplant pharmacist to continuously review the current literature, as well as to evaluate it based on its merit and relevance to one's own practice setting and specific population. Again, this provides an opportunity for pharmacist involvement with the transplant team. Evaluation of the literature from transplant and nontransplant sources becomes important. At times, one will need to extrapolate information from nontransplant patients to overcome certain voids of information concerning transplant patients. Of course, experience and clinical training will determine when this extrapolation is appropriate and when it is not. Herein lies part of the attraction of transplant pharmacy; one must possess very specific knowledge on how to manage a unique group of patients, namely organ transplant recipients, while being able to deal with a multitude of medical issues, each of which also requires specialist knowledge.

Table 4 Useful web resources

http://thedrugmonitor.com
http://transweb.org
http://fujisawa.com/medinfo/cont_educ/tran_trnd/
http://transplantation.medscape.com/Home/Topics/transplantation/transplantation.html
http://tpis.upmc.edu/tpis/immuno/compre.htm
http://ntpr.registry@mail.tju.edu
http://stadtlander.com/transplant/
http://mdconsult.com
http://clinicaltrials.gov
http://unos.org

Table 4 outlines various transplant web resources that a pharmacist specializing in transplantation may find useful. Although some of these web sites include information for patients, others are for health care providers working with the transplant recipients.

EDUCATION

Transplant pharmacists are considered unbiased resources for drug information that promotes better patient care. This service may be provided to a variety of individuals and disciplines, such as patients, caregivers, and health care providers, as well as medical residents, new pharmacists, and pharmacy students.

During the first year after transplant surgery, the average drug regimen of a transplant recipient consists of at least 10 different medications with variable time schedules for administration. For many patients, the significant number of medications and their complex schedules can be overwhelming. This creates the potential for errors and drug noncompliance at a time when the patient's condition makes them especially vulnerable to the results of subtherapeutic pharmacologic treatment.

Pharmacists can educate patients on indications for immunosuppressive therapy, appropriate use of drugs, anticipated adverse effects and expected outcomes (Table 5). Drug regimens must be simplified and made relevant to a patient's individual needs. As time passes after transplantation, the risk of graft rejection will decrease. As a consequence, patients have frequent alterations in therapy as their risk for infections and other complications decreases with the reduction of their immunosuppressive therapy. Patients must be kept aware of these changes, and teaching opportunities during each admission and clinic visit should not be missed. Because of the potential for drug interactions with immunosuppressants, pharmacists usually counsel patients not to treat themselves with any of the numerous OTC medications. The pharmacists may also develop a variety of educational tools to enhance patient learning.[18] This could take the form of recording audio instructions for visually impaired diabetic patients, creating daily medication logs, and providing written information in language appropriate for patient understanding.

Table 5 Common educational needs transplant pharmacists address with patients

Signs and symptoms of infection
Medication adverse effects
Management of adverse events
Drug–drug and drug–food interactions
Blood glucose control and insulin requirements
Pain management
Over-the-counter medication use
Alternative medicine use
Nutritional requirements
Drug compliance

Patient noncompliance to immunosuppressive therapy is a significant cause of graft failure.[19–22] Some investigators have reported rates ranging from 2% to a high of 43% in pediatric renal transplant recipients.[23] In addition to the complexity of the drug regimen, other factors may lead to drug noncompliance. These include concerns in physical appearance due to the side effects of immunosuppressants, poor provider–patient communication (e.g., dosage change that is not fully explained), depression or anxiety, dissatisfaction with medical care, cost of drugs, and misconceptions that missing doses will not cause rejection.[24,25] Potential health and economic consequences of noncompliance to immunosuppressive therapy include cost of initial and additional prescriptions, physician visits, clinic or emergency department visits, hospitalization, diagnostic costs and additional care such as dialysis in case of renal transplant recipients.[26]

Pharmacists can provide various strategies for maintaining drug compliance. They can help patients to select a reminder or ''cue'' such as clock times, meal times, or daily rituals or activities to assist with medication administration. Meeting with patients more frequently to assess compliance also reinforces the drug regimen. Intervention programs using support groups and partnerships between the recipient, family, and transplant team have also been shown to be effective strategies to increase compliance.[27] Increased emphasis on patient counseling and education about the drug regimen can lead to improved compliance.

A major role of all pharmacists in hospitals and outpatient clinics encompasses education for medical and nursing staff, as well as other colleagues. Understanding the various immunologic pathways that can be modified is essential to understanding the role of the various immunosuppressive medications and their subsequent effects. Knowledge regarding identification and treatment of complications, such as infections, hypertension, diabetes, hyperlipidemia, osteoporosis, renal insufficiency, depression, and cancer, just to name a few, is crucial when caring for transplant recipients. As these patients live longer and become healthier, new issues arise as they contemplate such possibilities as conceiving children, concepts that would not have been entertained previously. Although at one time management of transplant recipients may have revolved around surgical care

and immunosuppression, it now includes every discipline from internal medicine to obstetrics and gynecology and all the pharmacologic issues that go along with each specialty. Each of the professionals contributing to the care of the patient needs to have some understanding of the transplant pharmacotherapy treatment used. This will undoubtedly affect their differential diagnosis, how they treat the patient, and what kind of problems need to be brought to the attention of the transplant team, who often continues to care long-term for the patient in some manner.[28]

Education may be provided through informal in-service programs about new drugs, new drug indications, and countless other topics. Tailoring these educational programs to each discipline may pose a challenge to the presenter because of the varied background and needs of the many people involved in the care of a transplant patient. For example, educational needs of a surgeon are often different from those of a medical specialist such as a nephrologist or hepatologist or those of a nurse or pharmacist.

Pharmacists may also perform formal presentations at professional meetings and symposia. Drug information can be regularly disseminated in a form of drug monographs, newsletters, and continuing education articles. In addition, many transplant pharmacy specialists have didactic teaching responsibilities, provide clinical rotations for pharmacy students in transplantation, and serve as preceptors.

RESEARCH

The development of newer immunosuppressive drugs has promoted a need for investigating their safety and efficacy to ensure appropriate use. Research on various approved or investigational immunosuppressants continues to be executed at an accelerated rate. Many transplant pharmacists have established precedents for successful drug research practices.[12,29,30] Most work with clinically relevant trials in areas such as drug pharmacokinetics, pharmacodynamics, and patient outcomes. This provides the transplant pharmacist with additional avenues to demonstrate to the team his or her unique suitability at coordinating and conducting such research. As is recognized by the FDA, the role of a person with a doctorate degree in pharmacy can be to serve as primary investigator in collaboration with physicians. This is due to the recognition that pharmacists have the appropriate training and the clinical perspective to participate or manage clinical drug research. Pharmacists may be involved in all aspects of the research project, including protocol development, approval from the institution's Investigational Review Board, budget analysis, patient recruitment, obtaining informed consent, data gathering, data analysis, drawing conclusions, and finally submission for presentations and publication.

Since the infancy of transplantation, the goal of drug therapy has been to create an agent that can promote true immunologic tolerance, defined as graft acceptance without immunosuppression. As a result, the National Institutes of Health, various pharmaceutical companies, and many independent scientists have all taken this into developments to foster investigation into strategies that might help to discover such a compound. Some pharmacists are currently performing such translational research in the hope of eventually crossing paths with clinical practice.

Emphasizing cost containment while achieving rational pharmacotherapy has become an important feature of many formulary decisions. A transplant pharmacist may also develop pharmacoeconomic studies that examine the financial impact associated with different immunosuppressive regimens. Such analysis extends beyond comparison of the acquisition costs of the individualized immunosuppressive agent. Other charges, such as the costs of organ procurement, room and board, laboratory tests, operating room, nuclear medicine, and physician and surgeon professional fees, must also be included because they may substantially add to the total cost of transplantation. Feasibly, a more expensive immunosuppressant agent or regimen may prove to be more cost effective than cheaper agents if graft survival is increased or the incidence of morbidity is decreased.[31] A panel of investigators drawn from major transplant centers and representatives of the pharmaceutical industry drafted standards for economic and quality-of-life studies. This panel recommended specific guidelines for the design and conduct of trials for economic analysis of transplantation. These recommendations serve as a template for better design of pharmacoeconomic trials to evaluate the cost effectiveness of transplantation in the future.

The field of transplantation, therefore, is fertile ground for numerous pharmacist-driven research projects of significant clinical and experimental relevance. The results of such research can serve to guide protocol decisions to improve patient outcomes and, ultimately, improve patient's quality of life.

INDUSTRY PRACTICE

Although the available number of transplant-related industry practice positions is relatively small, some pharmaceutical companies offer medical or scientific liaison positions well suited to pharmacists with prior transplant experience. These positions require the individual to interact with physicians, nurse coordinators, and other pharmacists regarding issues of research and education. Coordination of multicenter studies and facilitating communication among centers through symposia, continuing education programs, and advisory boards are especially vital in the transplant discipline. Many transplant practitioners rely heavily on these sources for up-to-date clinical information. Advances in transplantation pharmacotherapy are happening with increasing frequency each year to the point where there is no standard of practice. Much reliance is placed on ''experts in the field,'' often identified by industry professionals who travel between centers.

CONCLUSION

Pharmacists working with transplant patients are continuing to define their practice model to meet the pharmaceutical and primary care needs of all transplant patients throughout the spectrum of care. The role for transplant pharmacists will continue to expand as the field of organ transplantation becomes a viable option for more patients. In a relatively short period of time, for example, kidney transplantation has gone from being an experimental procedure in the 1950s to being a cost-effective treatment for end-stage renal failure. The type of practice one chooses does not need to be limited to any one of the possibilities listed here. A transplant pharmacist position can involve different combinations of these responsibilities and others, as necessary, tailored to the preferences of the practitioner and the needs of the center. As transplantation of other organs becomes more attractive, the need for qualified pharmacists as part of the transplant team will continue to increase. The field of transplantation provides numerous opportunities for pharmacists, many of which can still be individualized to suit the strengths and preferences of the practitioner. Of all the reasons that demonstrate the need for transplant pharmacists, none is as convincing as the needs of the patient. The patient's needs define what roles are available to pharmacists and how valuable those services are to the transplant community and society in general.

REFERENCES

1. 2000 Annual Report. In *U.S. Scientific Registry of Transplant Recipients and the Organ Procurement and Transplantation Network: Transplant Data*; U.S. Department of Health and Human Resources and Services Administration: Richmond, Virginia, 2000.
2. Jindal, R.M.; Sidner, R.A.; Hughes, D.; Pescovitz, M.D.; Leapman, S.B.; Milgrom, M.L.; Lumeng, L.; Filo, R.S. Metabolic problems in recipients of liver transplants. Clin. Transplant. **1996**, *10*, 213–217.
3. Kasiske, B.L.; Guijarra, C.; Massy, Z.A.; Wiederkehr, M.R.; Ma, J.Z. Cardiovascular disease after renal transplantation. J. Am. Soc. Nephrol. **1996**, *7*, 158–165.
4. Epstein, S.; Shane, E.; Bilezikian, J.P. Organ transplantation and osteoporosis. Curr. Opin. Rheumatol. **1995**, *7*, 255–261.
5. Carson, K.L.; Christine, M.H. Medical problems occurring after orthotopic liver transplantation. Dig. Dis. Sci. **1997**, *42*, 1666–1674.
6. Bennett, W. The nephrotoxicity of immunosuppressive drugs. Clin. Nephrol. **1995**, *43* (1), S3–S7.
7. Shaefer, M.S. Current topics in immunotherapy and the role of the pharmacist on solid organ transplant service. Pharmacotherapy **1991**, *11* (6), 136S–141S.
8. Hilbrands, L.B.; Hoitsma, A.J.; Koene, R.A. Costs of drugs used after renal transplantation. Transplant Int. **1996**, *9* (1), S399–S402.
9. Barlow, C.W.; Moon, M.R.; Green, G.R.; Gamberg, P.; Theodore, J.; Reitz, B.A.; Robbins, R.C. Rabbit antilymphocyte globulin versus OKT3 induction therapy after heart–lung and lung transplantation: effect on survival, rejection, infection, and obliterative bronchiolitis. Transplant Int. **2001**, *14*, 234–239.
10. Nashan, B.; Moore, R.; Amlot, P.; Schmidt, A.G.; Abeywickrama, K.; Soulillou, J.P. Randomised trial of basiliximab versus placebo for control of acute cellular rejection in renal allograft recipients. Lancet **1997**, *350*, 1193–1198.
11. Brethauer, B.; Devine, B.; Jue, M.; Quan, D.; Louie, C. Cost-benefit analysis of a clinical pharmacist's presence on a post-liver transplant service. Hosp. Pharm. **2000**, *35* (11), 1197–1202.
12. Vasquez, E.M.; Benedetti, E.; Pollak, R. Ethnic differences in clinical response to corticosteroid treatment of acute renal allograft rejection. Transplantation **2001**, *71*, 229–233.
13. Zetterman, B.K.; Belle, S.H.; Hoofnagle, J.H.; Lawlor, S.; Wei, Y.; Everhart, J.; Wiesner, R.H.; Lake, J.R. Age and liver transplantation. Transplantation **1998**, *66*, 500–506.
14. Johnston, A.; Holt, D.W. Therapeutic drug monitoring of immunosuppressant drugs. Br. J. Clin. Pharmacol. **1999**, *47*, 339–350.
15. Yatscott, R.W.; Aspeslet, L.J. The monitoring of immunosuppressive drugs: A pharmacodynamic approach. Ther. Drug Monit. **1998**, *20*, 459–463.
16. Anaizi, N. Drug interactions involving immunosuppressive agents. Graft **2001**, *4*, 232–247.
17. Levy, G.A. Long-term immunosuppression and drug interactions. Liver Transplant **2001**, *7*, S53–S59.

18. Partovi, N.; Chan, W.; Nimmo, C.R. Evaluation of a patient education program for solid organ transplant patients. Can. J. Hosp. Pharm. **1995**, *48* (2), 72–78.
19. Bittar, A.E.; Keitel, E.; Garcia, C.D.; Bruno, R.M.; Silviera, A.E.; Messias, A.; Garcia, V.D. Patient noncompliance as a cause of late kidney graft failure. Transplant. Proc. **1992**, *24*, 2720–2721.
20. Dunn, J.; Golden, D.; Ban Buren, C.T.; Lewis, R.M.; Lawen, J.; Kahan, B.D. Causes of graft loss beyond two years in the cyclosporine era. Transplantation **1990**, *49*, 349–353.
21. Matas, A. Chronic rejection in renal transplants—risk factors and correlates. Clin. Transplant. **1994**, *8*, 332–335.
22. Douglas, S.; Blixen, C.; Bartucci, M.R. Relationship between pretransplant noncompliance and posttransplant outcomes in renal transplant recipients. J. Transplant Coord. **1996**, *6*, 53–58.
23. Beck, D.E.; Fennell, R.S.; Yost, R.L.; Robinson, J.D.; Geary, D.; Richards, G.A. Evaluation of an educational programme on compliance with medication regimens in paediatric patients with renal transplants. J. Pediatr. **1980**, *96*, 1094.
24. Rodin, G.; Abbey, S. Kidney Transplantation. In *Psychiatric Aspects of Organ Transplantation*; Craven, J., Rodin, G.M., Eds.; Oxford University Press: Oxford, 1992; 145–163.
25. Swanson, M.A.; Palmeri, P.; Vossler, E.D.; Bartus, S.A.; Hull, D.; Schweizer, R.T. Noncompliance in organ transplant recipients. Pharmacotherapy **1991**, *11* (6), 173S–174S.
26. Whiting, J.F.; Martin, J.; Zavala, E.; Hanto, D. The influence of clinical variables on hospital costs after orthotopic liver transplantation. Surgery **1999**, *125*, 217–222.
27. Kiedar, R.; Katz, P.; Nakache, R. "Living again": Heterogeneous support group for transplant patients and their families. Transplant. Proc. **2001**, *33*, 2930–2931.
28. McCashland, T.M. Posttransplantation care: Role of the primary care physicians versus transplant center. Liver Transplant **2001**, *7*, S2–S12.
29. Alloway, R.; Kotb, M.; Hathaway, D.K.; Gaber, L.W.; Gaber, A.O. Randomized double-blind study of standard versus low-dose OKT3 induction therapy in renal allograft recipients. Am. J. Kidney Dis. **1993**, *22* (1), 36–43.
30. Hebert, M.F.; Ascher, N.L.; Lake, J.R.; Emond, J.; Nikolai, B.; Linna, J.; Roberts, J.P. Four-year follow-up of mycophenolate mofetil for graft rescue in liver allograft recipients. Transplantation **1999**, *67*, 707–712.
31. Sullivan, S.D.; Garrison, L.P.; Best, J.H. The cost effectiveness of mycophenolate mofetil in the first year after primary cadaveric transplant. U.S. renal transplant mycophenolate mofetil study group. J. Am. Soc. Nephrol. **1997**, *8* (10), 1592–1598.

Tri-Council Policy Statement: Ethical Conduct for Research Involving Humans

Yasmin Khaliq
Rochester, Minnesota, U.S.A.

INTRODUCTION

The Tri-Council Policy Statement is a Canadian document that has been adopted, in 1998, as the standard of ethical conduct in Canada when performing research involving human subjects. It reflects the desire to promote research conducted according to the highest ethical standards. The Policy was developed by three groups: the Medical Research Council of Canada, the Natural Sciences and Engineering Research Council of Canada, and the Social Sciences and Humanities Research Council of Canada in an effort to aid and support research in Canada. The three councils are aware that issues regarding ethical conduct are complex and continually evolving and therefore intend to update this document regularly.

GOALS AND RATIONALE

The Policy attempts to address the responsibilities of researchers, institutions, and Research Ethics Boards (REBs) to their subjects. Ethical principles are shared across disciplines such that subjects should expect equal rights as well as benefits and risks across all fields. The Policy provides a framework of common procedures by which the ethics review process may be standardized. The Policy continues to recognize the diversity of various fields, but promotes the sharing of general ethical principles. Thus, this document can lend itself to many areas. The Policy is not intended to address specific ethical dilemmas but to provide guiding principles and standards as well as promote thought regarding areas of controversy.

The Policy describes ethics as using morally acceptable means to attain morally acceptable ends. Guiding principles are found in Table 1. It is critical that the subjects' interests are primary when conducting research, and that it is understood that benefit and harm are not viewed by the subject in the same manner as the researcher. Trust by the subject or hope for cure places further emphasis on the need for accuracy and full objective disclosure regarding the proposed research. It is important that the researcher maintain the right to pursue knowledge freely but with responsibility. This means ensuring the highest scientific and ethical standards including honesty and accountability. The law regulates research standards with respect to areas such as privacy, equality, property, and competence. With the continuing advances in knowledge, technology, and areas of controversy, ethics will likely play a role in defining the law in the future.

The Policy acknowledges that principles and guidelines require flexibility and exceptions to the rule. There can be many approaches to an ethical problem and debate regarding the answers will likely always occur. This will ensure continued thought and evolution in applying ethical principles to research with humans.

This summary of the Policy is divided into ten sections similar to the original document: ethics review; free and informed consent; privacy and confidentiality; conflict of interest; inclusion/exclusion of populations; aboriginal peoples; clinical trials; human genetics research; research involving human gametes, embryos, or fetuses; and human tissue.

ETHICS REVIEW

Research requiring ethics review is that which includes living human subjects as well as human remains, cadavers, tissues, biological fluids, embryos, or fetuses. The REB is mandated by its institution to approve, reject, modify, or terminate submitted research proposals to be conducted within the institution or by its members. The REB is to have five members: two with knowledge in research, one in ethics, one in law (specifically biomedical research), and one without affiliation to the institution who is from the community.

The exposure of a subject to minimal risk is thought to be equivalent to that encountered in everyday life. Greater

Encyclopedia of Clinical Pharmacy
DOI: 10.1081/E-ECP 120006350

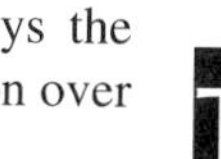

Table 1 Guiding principles for ethics

- Respect for human dignity
- Respect for free and informed consent
- Respect for vulnerable persons
- Respect for privacy and confidentiality
- Respect for justice and inclusiveness
- Balancing harms and benefits
- Minimizing harm
- Maximizing benefits

risk requires increased scrutiny of the project. Risk must also be considered in terms of interventions the patient would have been exposed to outside of the clinical trial setting. Risks can be categorized as therapeutic or not, the latter requiring recognition and consideration by the REB. The REB must ascertain that any protocol that poses more than minimal risk to the subject is properly designed to answer the research question. Compensation considered greater than that usually available to people is also worth higher scrutiny to ensure undue incentive is not offered.

Ethics Review Procedures

The proportionate approach to review is taken, which is geared toward the principle that the greater the invasiveness or potential risk, the greater the care that should be taken to review the proposal. Three levels of review are recommended: 1) full REB review, 2) expedited review for projects of minimal risk, annual renewals, chart reviews, or projects where REB conditions have been met, and 3) departmental review. The scope of research that requires REB review is found in Table 2.

Regular meetings are essential for full REB reviews with records of minutes documenting all decisions and dissents with explanations. Decisions must be made impartially. If desired, an investigator may participate in the discussion with an opportunity to reply to comments but cannot be present for the decision process. Protocols may be resubmitted if rejected. If a decision cannot be reached initially an appeal board may be utilized; however, appeals following REB decisions are not considered.

When a member of the REB has a personal interest in the project he or she cannot be present during the decision-making process, although disclosure of conflict and opportunity to rebut are allowed. All ongoing research must have a follow-up review proportionate to the level of risk. Such review often consists of evaluating the submission at an annual update. Ethics review must be performed by the appropriate REB of all involved institutions or jurisdictions. Ethics review must be performed by the REB of the institution that employs the researcher(s) as well as by any REB with jurisdiction over the location of the research.

FREE AND INFORMED CONSENT

Requirements

Free and informed consent is essential prior to participation as well as throughout a study and is usually documented in writing. The need for consent may be altered or waived when there is minimal risk or low probability that the action will have negative implications for the subject, when the research is not possible without alteration or waiver, or when such action does not involve therapeutic intervention. After participation, additional information should be provided if possible and appropriate, and consent may be obtained following subject debriefing. For subjects not proficient in the language of the consent form, consent may still be obtained provided an objective competent translator has fully informed the subject as to the contents of the consent form and has assisted the subject in participating in discussion of the study.

Consent must be given fully and with the knowledge that it can be withdrawn at any time without consequence. Power and undue influence are of concern particularly in

Table 2 Scope of research that requires REB review

Funding	Available or not Internal or external source
Subjects	Recruitment internal or external to institution Compensation awarded or not Is subject focus of research?
Location	Within Canada or not Inside or outside institution In person or from a distance (e.g., by mail)
Researchers	Staff or students
Data collection	Direct from subjects or indirect (e.g., chart review)
Publication	Planned or not
Study design	Observational, experimental, correlational, or descriptive Pilot study or full project
Purpose	For basic or applied knowledge For teaching or for acquisition of knowledge
Competition	Is a similar project approved elsewhere?

restricted or dependent subjects and under such circumstances must be judged in context.

REB review is generally necessary for studies of naturalistic observation. However, studies observing participants in rallies, demonstrations, or meetings, for example, do not require REB approval, although issues of privacy must be considered.

Informing Potential Subjects

Consent may not be obtained without full disclosure of all information that may affect consent and an opportunity for discussion. Obtaining consent includes:

- Invitation.
- Indication of the research purpose.
- Identity of the investigator(s).
- Nature of the study and time commitment.
- Procedures.
- Foreseeable harms and benefits.
- Alternatives to the research or consequences of nonparticipation.
- Assurance that one is free to not participate or may decide at a later time or date.

The likelihood of publication of the findings and any conflicts of interest by the investigators, institution, or sponsors must be indicated. Other information may be necessary to include for select projects, such as subject responsibilities, confidentiality as to subject identity, and anticipated use of the data.

Competence

Competence refers to a prospective subject's ability to understand the information provided and its consequences, and make a free and informed decision in accordance with personal values. If a subject is not considered competent, a balance must be sought as to the subject's vulnerability versus the injustice of exclusion from potentially beneficial research. There is a moral preference to use competent subjects. Subjects not legally competent should only be asked to participate when the research question can only be addressed using the identified group, and the risk is minimal when there are no direct benefits. The researcher must also demonstrate how the subject's best interests are protected and the method of obtaining free and informed consent from an objective third party. If the subject should become competent during the study, consent must be obtained for continued participation. Of note, some subjects even if not legally competent may be able to express their wishes in a meaningful way, e.g., children, patients with Alzheimer's disease. Their dissent precludes participation.

Research in Emergency Health Situations

Research that is to be performed on an individual without his or her consent is subject to all legislation and regulations as pertinent, and may be allowed by the REB if:

- A serious threat exists requiring immediate intervention and research offers a possibility of benefit when no standard therapy exists.
- Risk is not greater than with standard care or that benefits justify risk.
- Third party authorization cannot be secured despite diligent efforts.
- No prior directive by the subject is known to exist.

This is an uncommon situation that may occur if a subject has lost consciousness or competence. Additional safeguards to protect the subject's rights and interests may include additional scientific/medical/REB consultation, procedures to identify subjects in advance to obtain consent prior to occurrence of an emergent situation, monitoring procedures by safety boards, and careful REB review for assessment of harms and benefits of participation.

PRIVACY AND CONFIDENTIALITY

The need for privacy when participating in research is based upon a subject's dignity and autonomy and thus warrants respect. As a result, all information regarding a research subject must be kept confidential. The researcher has a duty not to share any information without a subject's free and informed consent. Respect for privacy is an internationally recognized ethical standard and is reflected in Canadian law as a constitutional right. However, there are some circumstances where public health and safety require protection and such privacy may be lost, e.g., child abuse, sexually transmitted diseases. The REB can play an important role in assessing the balance between the need for research versus invasion of privacy, thus protecting individuals from harm as a result of unauthorized use of personal information.

Personal Interviews and Data Collection

REB approval is required for interviews and documentation of personal information by research subjects, and free and informed consent is necessary. REB approval is not necessary for access to public information or materials. REB approval requires consideration of the purpose and type of data, limits on its use, safeguards for confidentiality, modes of observation that may identify the subject, anticipated secondary uses of the data, and anticipated linkage with other data. Subjects must be informed who will have access to any identifying information, e.g., government or agencies, research sponsors, or the REB, and must provide consent for such access to occur.

Secondary Use of Data

When data are used for purposes other than the research for which they were collected, REB approval must be sought if individuals are identifiable. Researchers must demonstrate that knowledge of the subjects' identities is necessary for the current use, that confidentiality will be maintained and that the involved individuals have not objected. If deemed appropriate, the REB may require either informed consent from those who contributed the data, an appropriate strategy for informing subjects, or consultation with representatives of those who contributed the data. REB approval is also required for situations when a researcher wishes to contact individuals to whom data refers. Linkage of databases where research subjects may be identifiable requires REB approval.

CONFLICT OF INTEREST

With increasing concern for conflicts of interest, such issues must be identified by the researcher, institution, and REB for professionalism, to maintain public trust, and for accountability. All conflicts, whether actual, perceived, or potential, must be disclosed to the REB and addressed. They must also be disclosed to the subject during consent. For the REB to identify and best address conflicts, it should be provided with details of the research project, budgets, commercial interests, consultative relationships, and other information as relevant. REB members must withdraw from the discussion and decision-making when their own research is under review. Methods to address and resolve conflicts must be developed. REBs must be given the authority and financial and administrative independence from the institution to fulfill their duties.

INCLUSION/EXCLUSION OF POPULATIONS

Choice of inclusion into research studies should follow the principle of distributive justice, i.e., no member in society should bear an unfair share, nor be unfairly excluded in research participation. Vulnerable subjects have been excluded to avoid the issue of exploitation yet have thus also been denied the potential benefits of participation. Researchers shall not exclude potential research subjects based on culture, religion, race, mental or physical disability, sexual orientation, ethnicity, gender, or age, unless a valid reason applies. Women must not be excluded on the basis of gender or reproductive capacity although harms and benefits must be considered. Incompetent subjects must not be excluded from research that may be beneficial to themselves or the group they represent.

ABORIGINAL PEOPLES

The aboriginal peoples are recognized as having a unique culture and history and perspective in life. Research regarding their customs and community has in some cases been respectful; however, there has also been inaccurate, insensitive research conducted causing stigmatization and thus apprehension with respect to future research proposals. Thought must be given to language differences and different ideas about public and private life. Involvement of academic or community members from this group is essential for appropriate ethics review. The needs and concerns of the people along with respect for their property, culture, traditions, and unique viewpoints must be considered by investigators. The community must also be given the opportunity to respond to findings prior to completion of research reports.

CLINICAL TRIALS

Clinical trials are addressed in the context of biomedical research. Research can include case studies, cohort studies, randomized controlled trials, or multicenter trials. While the methodology of these studies is greatly varied, the guiding ethical principles and procedures remain the same. To begin research there must be clinical equipoise, or a reasonable uncertainty warranting pursuit of an answer. Such an approach should ensure bias is minimized in all therapeutic arms and that participation is not considered disadvantageous. Research is generally categorized into four phases as found in Table 3.

Research with new medical devices must be carefully evaluated to ensure free and informed consent can be obtained. REBs must aid researchers to prevent conflicts of interest. This could be concerning subject selection or

Table 3 Phases of research

Phase	Description	Comments
I	Dose-finding studies of new medications to determine acute toxicity. Healthy subjects or patients.	These studies warrant close REB review with continuous monitoring independent of the trial sponsor. This is especially important as more of these trials include patients refractory to standard therapy, and with the increasing number of new medications. Unexpected adverse events are a major concern.
II	Determination of short-term pharmacologic toxicity, some degree of efficacy. Patients with specific diseases.	Phase I and II studies must be carefully examined for free and informed consent procedures. An independent continuous review may be necessary.
III	Determination of pharmacologic efficacy to increase survival or quality of life, some toxicity. Patients with specific diseases.	Phase II and III often include placebos to assess toxicity of a new agent. Use of placebo must be justified and minimization of harm ensured.
IV	Determination of long-term efficacy and toxicity of marketed drugs. Postmarketing surveillance studies.	These studies are often conducted in private practice for postmarketing with a per-capita fee paid to assess side effects and patient acceptance. Such payments can create bias. REBs must assess the science and ethics of these studies as much as with the other phases.

payment by sponsors to the investigators. Safety standards of new devices must be assured. Continued provision of therapy beyond the trial termination must also be examined.

If research is to be used for regulatory approval of a drug, the International Conference on Harmonization (ICH) guidelines that have been adopted by Canada must be followed. Budgets must be evaluated to ensure no conflicts of interest exist. Placebo-controlled studies are not considered acceptable when effective standard therapies or interventions are available. Investigators must inform subjects as to why placebos are necessary when used, and, if any treatment is to be withdrawn and the associated impact. Although sponsors may obtain preliminary data for analysis, final analysis and interpretation should be by the researchers to ensure the accuracy and integrity of the work.

HUMAN GENETICS RESEARCH

The study of genes that determine human traits is very topical and not without controversy. Identification of genes that comprise the human genome, their function, and their ability to predict disease or a predisposition to disease is central to this research. However, because genetic material in one individual is shared by biological relatives, privacy and confidentiality affect not just the individual who may wish to consent to research participation. The effects of such research and how information will be used and interpreted are unknown, requiring prudence in pursuing research in genetics. An individual may feel his/her identity and self-worth are threatened, not to mention any associated stigma that may occur toward the group to which the individual belongs.

Important considerations are as follows:

- Consent is critical in genetics research, and concern that the individual's family may be coercive and that information may affect other members in the family should be considered.
- Results of genetic tests must be protected from any third party access unless free and informed consent is given. DNA banking allows family and clinical information as well as genetic material to be a resource for other researchers; however, removal of the subject's and family's identity is important. Any potential harm must be disclosed to the REB as well as how it will be dealt with. For example, information about an individual's susceptibility to disease may provoke anxiety, particularly if effective therapy or prevention is not available.
- Genetic counseling must be available for subjects as needed. The issue of cultural differences in the perception of this research must be accounted for.
- Gene alteration, including gene therapy, using human germline cells or embryos, is not considered ethically acceptable. Gene alteration done for therapeutic reasons and using human somatic cells may be considered.

Irreversibility and lack of long-term follow-up suggest the use of this technology should be with care.
- Genetic research should be performed for the pursuit of knowledge or therapeutic benefits in disease, not to enhance or improve a population by cosmetic manipulation.
- Banking or storing of genetic material may yield benefit or harm. A defined term of storage and provisions for confidentiality and anonymity including for the future must be addressed.
- The likelihood of commercial use of genetic material and information must be discussed and consent obtained.

RESEARCH INVOLVING HUMAN GAMETES, EMBRYOS, OR FETUSES

There are complex concerns as to the possible harm to the embryo or fetus as well as the respect they deserve, and a cautious and controlled approach to such research is essential.

The human reproductive cells (ova and sperm) may be used for research provided free and informed consent is given. Use of gametes derived from cadavers is prohibited and use of gametes from fetuses or individuals unable to consent for themselves is deemed unacceptable. The use of gametes acquired by commercial transaction is considered unethical. Creation of hybrid individuals (e.g., a mix of human and animal gametes or transfer of somatic or germ cell nuclei between cells of humans and other species) is unethical and does not show consideration for human life and dignity.

It is not ethical to create human embryos specifically for research. If an embryo was created for reproduction but is no longer needed, it may be used for research if 1) consent is given, 2) no commercial transaction took place, 3) no genetic alteration of the gametes or embryo occurred, 4) manipulations not directed to normal development are not used for continued pregnancy, 5) that the research occurs within 14 days of the creation of the embryo by joining the gametes. It is not ethical to participate in research cloning human beings (ectogenesis).

With maternal consent, research may be undertaken in utero for fetuses with genetic or congenital disorders. Use of fetal tissue cannot affect a woman's decision regarding continuation of her pregnancy, nor can fetal tissue be directed for use in specific individuals as the fetus deserves respect as a potential person not just a source of tissue.

HUMAN TISSUE

Human tissue should be respected to maintain the dignity of the donor. Canadian law allows competent persons to donate, although not sell, human tissue for research. Concerns for confidentiality of tissue identity has led to the following categorization:

1. Identifiable tissue: donor known.
2. Traceable tissue: by database.
3. Anonymous or anonymized tissue: donor identity stripped.

The REB must consider the likelihood of the traceability of tissue in a research project with respect to privacy and confidentiality.

Collection of human tissue for use in research requires the researcher to demonstrate to the REB that free and informed consent will be obtained from competent donors, or from an authorized third party for incompetent donors or deceased donors without advance directive regarding the topic. Minimally, researchers must provide the research purpose; the type, location, and amount of tissue to be taken; the manner of acquiring the tissue with respect to safety and invasiveness of the procedure; the uses of the tissue; and how its use could affect privacy. Consent is also necessary for use of previously collected tissues when identification is possible.

CONCLUSION

The Tri-Council Policy strongly promotes research and attempts to provide guidance regarding ethical principles that ensure an individual's rights and interests are protected. Rapidly advancing knowledge and technology present us with new research issues and controversies that will require continual review of ethical guidelines and much opportunity for thought and debate regarding future research.

BIBLIOGRAPHY

Tri-Council Policy Statement: Ethical Conduct for Research Involving Human Subjects; Public Works and Government Services, Canada, 1998, Catalogue No: MR21-18/1998E. ISBN 0-662-27121-1.

Also found on the World Wide Web at: http://www.nserc.ca/programs/ethics/english/policy.htm (accessed October 2000).

PROFESSIONAL ORGANIZATIONS

UK Clinical Pharmacy Association

Pat Murray
Royal Edinburgh Hospital, Edinburgh, U.K.

INTRODUCTION

The UK Clinical Pharmacy Association (UKCPA) was launched in 1981 and since then membership has risen to over 2000 members. The success of UKCPA reflects the fact that the organization has been a fundamental part of the recognition of clinical pharmacy as part of the mainstream of pharmacy practice. As envisaged by the founders of UKCPA, clinical pharmacy is not a specialty within pharmacy but a discipline that is practiced by all pharmacists working with patients. Pharmaceutical care developments in all areas, including Primary Care, are now showing the relevance of the examples of practice that UKCPA has promoted. UKCPA members' activities have also had an influence on the Schools of Pharmacy to embrace clinical pharmacy within the core curriculum and to teach the skills by problem-based learning. The UKCPA workshops have been a major part of the well-known high quality conferences of which the Association is proud. The UKCPA has developed an organizational structure and a team approach in its work. Access to UKCPA has been further enhanced following the launch of our Website last year. (www.ukcpa.org).

Following is a description of the organization.

BUSINESS MANAGEMENT GROUP

The Business Group (BMG) focuses on ensuring the Association runs smoothly and efficiently. UKCPA is regularly invited to comment on new professional development proposals and it is the role of the BMG to respond on behalf of UKCPA, taking into account comments received from consultation.

The success of recruiting new members is assigned to the Membership and Marketing Committee.

This committee has set three main goals for the forthcoming year:

- To maintain and increase the current membership by highlighting the opportunities that UKCPA can offer.
- To identify and co-operate with other pharmacy groups in order for mutually beneficial collaborations to take place.
- To strengthen and increase the corporate membership.

UKCPA can only benefit if enthusiastic pharmacists are attracted to the organization. In return it can offer an exciting arena for such people to learn, participate and meet other pharmacists contributing to the profession. Clinical pharmacy now envelops all strands of the profession and it is the role of the Marketing and Membership Committee to publicize this information.

UKCPA publishes a newsletter 'In Practice' a range of practice guides and workshop support material. The two annual symposia attract over one hundred submissions of research and practice development work and the conferences provide a highly interactive format and polished performances. Conference presentations (oral communications and posters) are published as short papers or abstracts in the UKCPA official journal *Pharmacy World and Science*. All this is the remit of the Publications Committee.

Handling UKCPA conference communications and the adjudication process requires a scrupulous process to make it work for members. UKCPA has developed a housestyle and is a transparent process of adjudication and feedback to assure the final quality of work presented and published. This applies equally to the adjudication process for the four industry sponsored awards, AstraZeneca Travelling Fellowship Award, the four GlaxoSmithKline Poster Awards, the Pharmacia Pharmacoeconomics Award and the Wyeth Education and Training Award. New awards for 2002 are Unichem Community/Primary Care Pharmacy Award, Napp Palliative Care Award and the Merck Pharmaceuticals for Medicines Management Award.

The UKCPA Office plays an important role in helping the association organize many of the UKCPA events.

It is the Education Programming Committee who have the responsibility of identifying and implementing the educational content.

Encyclopedia of Clinical Pharmacy
DOI: 10.1081/E-ECP 120006297

The committee, which is made up of the Chairs of six Practice Interest Groups and four General Committee members, oversees the main educational events described in the programme. It has a challenging job in co-ordinating the main symposia while seeking to be visionary, to ensure the content of future symposia will attract participants some 12 months in advance of the planned event. There is much to do with the planning, implementation and review of each programme, which attracts up to 350 participants.

Practice Interest Groups reflect the broad range of interest and experience among members and provide a network of support among members by maintaining a data base of members and their area of expertise. Each group regularly contributes to 'In Practice,' reviewing current research/developments in their field and training opportunities. The groups comprise:

Care of the Elderly
Critical Care
Education and Training
Medicines Management in Primary Care
Quality Assurance
Surgery and Theatres
Infection Management
Each group sends a message.

Care of the Elderly

The aim of the group is to enhance the pharmaceutical care of elderly people, within both primary and secondary care. It does this by facilitating the continuing education and networking of its members.

The Group organizes study days, mini-symposia and workshops. Topics have included the pharmaceutical care of patients with stroke, dementia and osteoporosis plus a day on ophthalmological problems.

Critical Care Group

The Group provides support for all pharmacists who have an interest in the field of critical care, which includes intensive care, high dependency and coronary care. The groups core objectives are:

- To continue to provide regular, objective and high quality educational meetings.
- To provide supportive documentation, especially for those new to the field.
- To promote the role of the critical care pharmacist within a multi-disciplinary setting through a closer liaison with the Intensive Care Society.
- To push forward the debate on specialist pharmacists.
- To promote and support research within the field of critical care.

Education and Training Group

Training is a component of most jobs: many people are expected to teach others, but many have had little guidance on how to perform this key activity successfully. The Education and Training Group of UKCPA aims to share and promote good practice in the field, and to support these individuals in their role.

Training sessions are organized, either as part of the full UKCPA symposia, or as 'stand alone' days. For example, a recent study day on the topical issue of competency-based training was held. The group has also written a guide for good practice in education and training, as a short practical aide memoir for consideration when planning an educational event. Commissioned resource guides in key areas such as teaching communication skills and assessment have also been published.

Medicines Management in Primary Care Group

The Medicines Management in Primary Care Group was formed four years ago to provide support for the increasing number of pharmacists developing a clinical role in Primary Health Care Teams. The objectives are to:

- Provide education and training for pharmacists to develop clinical pharmacy in primary care.
- Facilitate liaison between primary and secondary care pharmacists.
- Encourage communication between all pharmacists working in this area by providing a forum for exchange of ideas.
- Promote innovation and encourage practically applicable research.

Recent workshops have included patient perspectives, therapeutic topics including cardiovascular and gastrointestinal systems, medicine resource management, medication review and pharmacist-led clinics.

Quality Assurance Group

The QA group was formed in the early days of UKCPA when intervention monitoring was the flavor. Since then, the Group's focus has broadened to address a whole range

of issues which are general to all clinical pharmacists, managers including those in Primary Care.

Recent examples of workshop topics at our two annual UKCPA symposia included sessions on compliance/concordance, PILs, off-label use of medicines, medication errors and ADR reporting. For the future, clinical governance and all its implications are high on the agenda and, to paraphrase the definition, we intend to help....create an environment in which pharmaceutical care will flourish.

Surgery and Theatres Pharmacy Group

The Surgery and Theatres Pharmacists Group aims to enable pharmacists world-wide to network and share best practice thereby furthering the role of pharmacy in these fields of medical practice. A system has been developed to allow pharmacists to share guidelines and protocols through a Resource Centre. Access to the Group's services are via the internet. The Group has its own website at www.users.globalnet.co.uk/~phoward/satg.htm.

Other services provided are an interactive Newsgroup that allows members to canvas views of other members, a bibliography of key references, an on-line newsletter, links to over 180 related sites, and a network of subgroups. The Newsletter reviews initiatives that pharmacists have introduced successfully. It also reviews the surgical and anaesthetic literature.

The Group also has a directory of members that allows pharmacists to contact specialists in a particular area for advice or information.

The Group would like to see a member from all 700+ hospitals within the UK ensuring all hospitals have access to best practice.

Infection Management Group

The Infection Management Group is a new group formed this year.

UKCPA has links with the European Society of Clinical Pharmacy. This has afforded opportunities for our members to network beyond the UK with opportunities for some industry sponsored award winners to present posters at an ESCP meeting on an annual basis. This link has enriched our symposia by ESCP member attendance or ESCP contributed workshops.

In early 2000 the General Committee met to prepare a UKCPA business plan for the next few years. Use this opportunity to influence future activities of the Association through sending your comments/suggestions to: UKCPA, Alpha House, Countesthorpe Road, Wigston, Leicestershire, LE18 4PJ. E-mail: pkennedy@ukcpa.u-net.com. Web site: www.ukcpa.org.

United States Pharmacopeia

Roger L. Williams
U.S. Pharmacopeia, Rockville, Maryland, U.S.A.

INTRODUCTION

Standards established by the *United States Pharmacopeia* (USP) are accepted worldwide as the highest assurance of the quality of medicines and other products used to improve the health of the public. We discuss the development of the publication and the organization, and its current and future initiatives toward promotion of worldwide health.

A BRIEF HISTORY

The first *Pharmacopoeia of the United States*, published in 1820 by USP, was an important milestone in American medicine and pharmacy. It was a compendium of about 217 of the best understood drugs and drug preparations and a lexicon of standard drug names. Three medical visionaries—Lyman Spalding and Samuel Mitchill from New York and Jacob Bigelow from Boston—were largely responsible for this work. They intended the *Pharmacopeia* to accomplish a degree of drug standardization and to facilitate communication between physicians and pharmacists. Today, USP publishes more than 3400 official standards monographs for drugs and other healthcare products.

MISSION AND FOCUS

According to Roger L. Williams, M.D., USP's Executive Vice-President and Chief Executive Officer, USP pursues its mission to promote public health through multifaceted activities:

- Establishing and disseminating officially recognized standards of quality for therapeutic products.
- Collecting information from practitioners to help reduce and prevent medication errors.
- Educating healthcare professionals and patients on the appropriate use of therapeutics and improving the health of special populations worldwide.

DISTINCTIVE STRENGTHS

As a pharmacopeial organization, USP has certain unique characteristics:

1. *Non-governmental with statutory recognition*: USP is the world's leading nongovernmental pharmacopeia. According to the Federal Food, Drug, and Cosmetic Act, USP standards are enforceable by the Food and Drug Administration (FDA) for all drugs manufactured and sold in or imported into the United States. The standards are also officially recognized in many other countries.
2. *Volunteers lead the way*: USP achieves its goals through the contributions of volunteer experts representing pharmacy, medicine, and other healthcare professions, as well as science, academia, the U.S. government, the pharmaceutical industry, and consumer organizations. Volunteers constitute USP's membership, Board of Trustees, and scientific decision-making committees. They are supported by a staff of over 350.
3. *Standards set by a participatory process*: USP standards are developed by an exceptional process of public involvement. The process gives those who manufacture, administer, regulate, and use therapeutic products the opportunity to comment on the development and revision of standards used to measure the quality of these products.
4. *Role that expands beyond standards*: Unlike most of the world's major pharmacopeias, USP's role is not limited to the establishment of written and chemical pharmacopeial standards. USP also helps to monitor and prevent medication error problems through reporting programs for healthcare professionals. The valuable information and feedback obtained through these programs enables USP to enhance the accuracy and utility of its standards.

Encyclopedia of Clinical Pharmacy
DOI: 10.1081/E-ECP 120006298

ORGANIZATIONAL STRUCTURE

USP's direction and priorities are determined by more than 400 credentialed members through resolutions that they vote to adopt at quinquennial meetings of the USP Convention. Members also elect the USP Convention officers, Board of Trustees, and Council of Experts—the scientific decision-making body.

USP membership is on a 5-year cycle. Within each membership category, eligible organizations (Fig. 1) are invited to appoint their representatives. Membership composition is determined to ensure suitable representation of those sections of the healthcare system that are impacted by, and in turn impact, USP's activities. D. Craig Brater, M.D., Dean of the Indiana University School of Medicine, is the USP Convention President for 2000–2005.

USP's Board of Trustees is elected for a 5-year term and has fiduciary responsibility for the organization. Larry L. Braden, R.Ph., is the Chairperson of the 2002–2003 USP Board. A complete list of Board members is available at www.usp.org.

Sixty-two scientific experts serve on the Council of Experts and chair expert committees, which are responsible for making scientific decisions. Visit www.usp.org for a list of USP's 2000–2005 expert committees and chairpersons. The Council of Experts is chaired by Roger L. Williams, M.D., USP's Executive Vice-President and CEO.

USP PRODUCTS

USP-NF

Official pharmaceutical standards are published in a single volume combination of the *United States Pharmacopeia* (*USP*) and the *National Formulary* (*NF*), commonly referred to as *USP-NF*. This authoritative reference provides pharmaceutical standards of identity, strength, quality, purity, packaging, storage, and labeling. Monographs and general chapters relating to drugs for human as well as veterinary use are found in *USP*, and those relating to excipients are found in the *NF*. Monographs for bota-

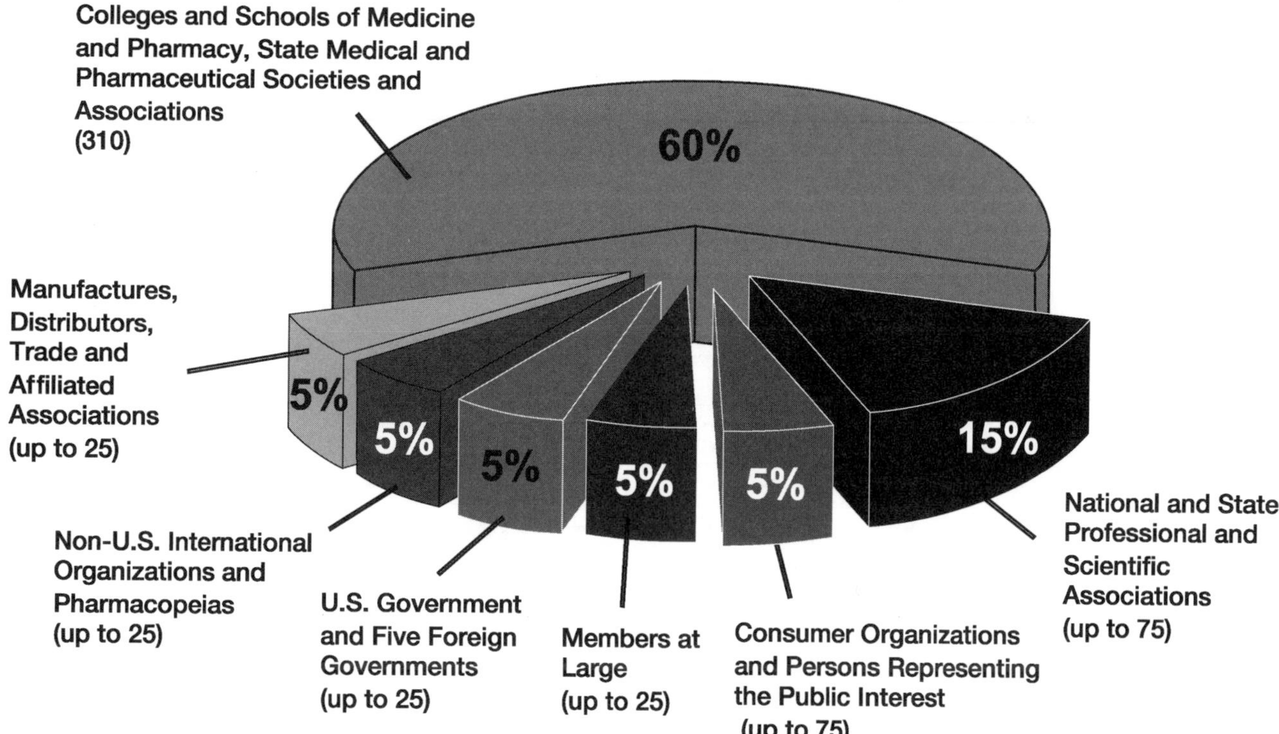

Fig. 1 USP convention membership.

nicals used as dietary supplements and having a FDA-approved or USP-accepted use are found in the *USP*. Monographs for botanicals recommended for official adoption by the USP Council of Experts and for other botanicals with no accepted or approved use, but with no safety concerns, are found in the *NF*. Monographs for nutritional supplements—vitamins and minerals—appear in a separate section.

USP-NF monographs include assays and various analytical methods—identification, dissolution, content uniformity, etc. *USP-NF* also provides guidance and standards on biotechnology, radiopharmaceuticals, pharmacy compounding, and pharmaceutical waters. General chapters outline requirements for microbiological, biological, chemical, and physical tests and assays.

USP-NF is available in four formats—a thumb-indexed hardcover, online, a CD for Windows®, and an Intra net version.

Statutory recognition

In the United States, the Federal Food, Drug, and Cosmetic (FD&C) Act affords recognition to *USP* and *NF* as official compendia and to the articles contained therein. According to sections 201(g) and (h) of this Act, the terms "drug" and "device" include articles recognized in the official *USP* or *NF* or any supplement to any of them. Section 501(b) provides that a drug is adulterated if it does not conform to *USP* standards for strength, quality, and purity. Section 502(g) requires a drug to conform to packaging and labeling requirements in the official compendium.

The Dietary Supplement Health and Education Act of 1994 (DSHEA) and the Food and Drug Administration Modernization Act of 1997 (FDAMA) extended the utilization of the *USP* and *NF* by amending the FD&C Act. Section 403(s)(2)(D) of the FD&C Act provides

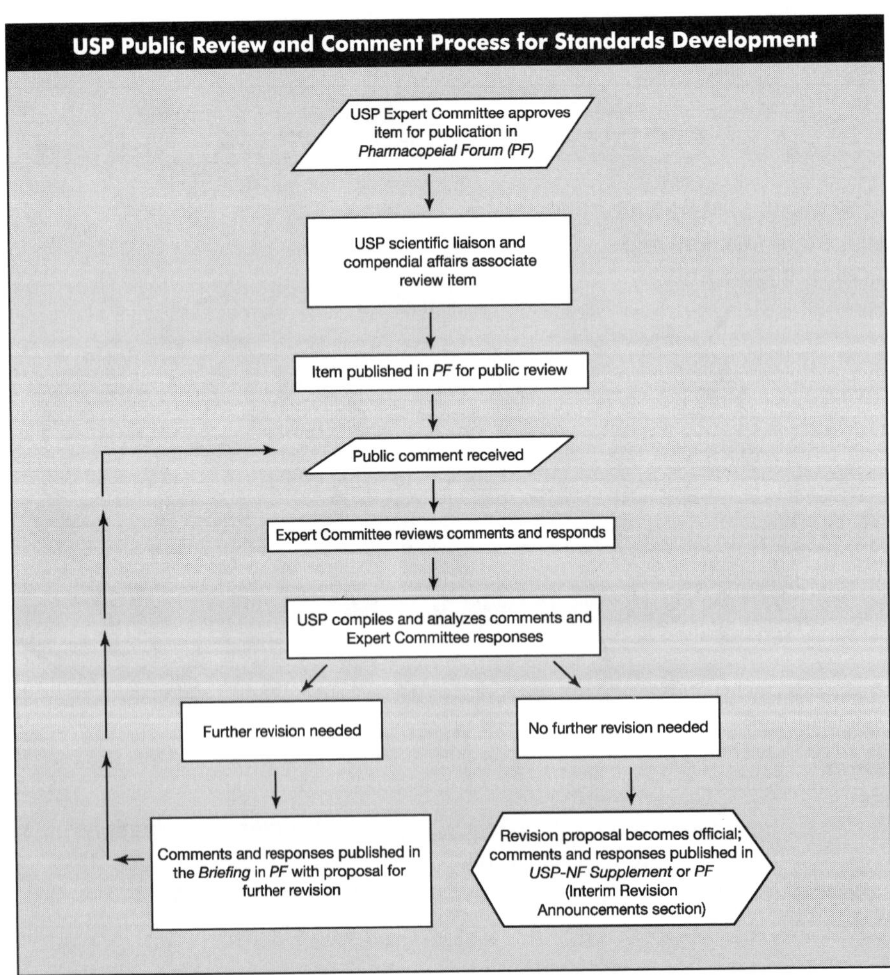

Fig. 2 USP standards development. © 2000 The United States Pharmacopeial Convention, Inc. All rights reserved. Used with permission.

that a dietary supplement represented as conforming to the specifications of an official compendium shall be deemed misbranded if it fails to do so.

Pharmacopeial Forum

Pharmacopeial Forum (*PF*) is the journal through which USP institutes the public review and comment process for standards development and revision (see Fig. 2 for a process overview). Proposed changes to official *USP-NF* standards are published in *PF* so that all interested parties may review them, offer their opinions, and prepare for their impact. *PF* also provides notice of binding revisions to the official standards. It includes authoritative scientific articles and comments on current pharmaceutical industry issues.

Reference Standards

USP establishes and validates chemical Reference Standards that help to ensure pharmaceutical quality. USP Reference Standards are used to conduct the official tests for product identity, strength, quality, and purity specified in *USP-NF*. Potential Reference Standards are rigorously and collaboratively tested in USP, FDA, and industry laboratories (see Fig. 3). USP currently has more than 1250 pharmaceutical Reference Standards—one of the world's largest collections—and about 500 new items under development.

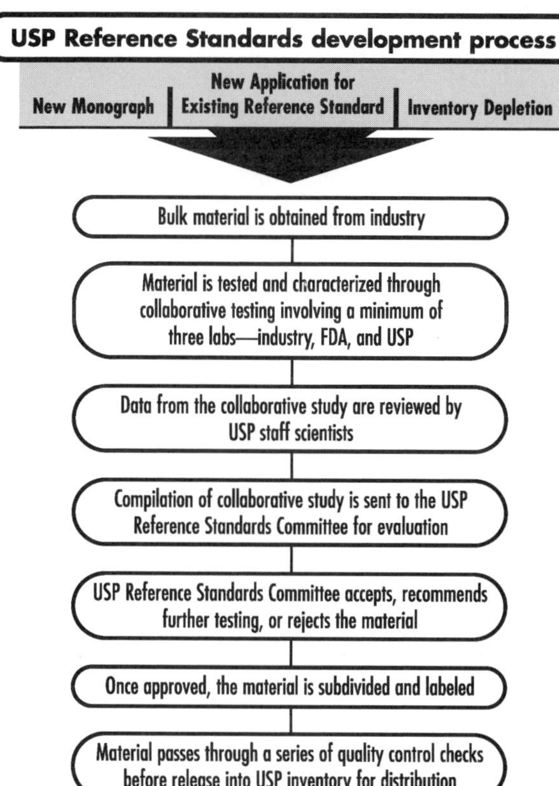

Fig. 3 Reference standards process. © The United States Pharmacopeial Convention, Inc. All rights reserved. Used with permission.

MedMARx^SM

MedMARx^SM is an effective, proactive solution to the serious and costly problem of medication errors in the United States. It is an Internet-accessible, anonymous, national database through which hospitals can report, track, and analyze medication errors to help prevent their recurrence. Errors are reported to MedMARx in a standardized format and categorized according to a nationally recognized index. This allows facilities to share information anonymously and learn from others' errors and preventive strategies. Participating facilities also receive national medication error alerts and reports on key error trends. MedMARx is backed by more than 30 years of USP experience in operating voluntary medication error and drug product problem reporting programs for healthcare professionals.

CURRENT INITIATIVES

Dietary Supplement Verification Program

In 2001, USP created the Dietary Supplement Verification Program (DSVP) to help inform and safeguard the growing number of consumers who use dietary supplements. The program responds to the need to assure the public that dietary supplement products contain the ingredients stated on the product label. If a product submitted to DSVP meets USP's rigorous standards, it will be awarded the DSVP verification mark. The mark helps assure consumers, healthcare professionals and supplement retailers that a product:

- Contains the declared ingredients on the product label.
- Contains the amount or strength of ingredients declared on the product label.
- Meets requirements for limits on potential contaminants.
- Has been manufactured properly by complying with USP and proposed FDA standards for ''good manufacturing practices'' (GMPs).

New Monograph Development

Currently, there are no public standards for about 35% of drugs in the U.S. market. USP seeks to develop public quality standards for all therapeutic products in the market, especially for the top 200 prescribed drug products and active pharmaceutical ingredients. Increased cooperation is being obtained from the pharmaceutical industry for a program to prepare and publish proposed standards monographs for newly approved drugs and drug products six years before patent expires.

International

USP, through its participation in the Pharmacopeial Discussion Group (PDG), is working closely with the European and Japanese Pharmacopoeias to explore the harmonization of pharmacopeial standards for excipients, microbiological testing, general methods of analysis, and methods of analysis for biotechnology-derived products. Several general test methods have already been harmonized. USP serves on Quality Expert Working Groups at the International Conference on Harmonization (ICH). It also works through the Pan American Health Organization and with individual countries around the world to ensure drug quality.

New Priorities

The USP Convention membership, at its April 2000 Quinquennial Meeting, adopted resolutions directing USP to explore in the next 5 years:

- The impact of applied genomics on drug discovery and development and therapeutic applications.
- Approaches to assure equivalence of complex active ingredients, including botanicals and dietary supplements.
- The application of modern biopharmaceutic principles to assure the equivalent performance of immediate- and modified-release drug products.
- Compounded drug formulations for special populations.
- Packaging, labeling, nomenclature, and dosage form characteristics for medicines to reduce medication errors.
- Standardized imprint coding for all solid oral dosage forms.
- Methods research on botanical ingredients.
- Compounding guidelines for veterinary extra-label use of medications.
- Education and training programs for health professionals to support the appropriate use of the *USP-NF* and expand its use.

MEETINGS

The USP Convention membership meets once every 5 years to discuss the areas of healthcare in which USP should be involved. These quinquennial meetings help to define USP's strategic focus and priorities for a 5-year period.

USP conducts open meetings as needed to invite the views of interested parties on issues of current significance in the pharmaceutical industry and the nature of USP's involvement with these issues. Notices of open meetings are posted on USP's web site at www.usp.org.

Universal Precautions and Post-exposure Prophylaxis for HIV

Brent M. Booker
Patrick F. Smith
Gene D. Morse
University at Buffalo, Buffalo, New York, U.S.A.

INTRODUCTION

As early as 1982, the Centers for Disease Control and Prevention (CDC) began issuing precautions for the handling and care of laboratory specimens from patients infected with the human immunodeficiency virus (HIV-1). The goal was to minimize the risk of exposure to infectious blood and blood-containing body fluids. Although it was previously recognized that blood and hypodermic needles posed a significant risk to the healthcare worker, this issue moved to the forefront of clinical practice with the recognition of the HIV/AIDS epidemic, when it became clear that healthcare workers would be at increased risk for exposure to HIV-1 and other blood-borne pathogens. In 1987, the CDC issued guidelines that were designed to minimize the risk of HIV-1 transmission in the healthcare setting. These guidelines have since become known as the ''universal precautions.''[1] Universal precautions became mandatory in 1991 with the passage of the Occupational Safety and Health Administration (OSHA) Blood-Borne Pathogens Standard, which required healthcare employers to establish, at minimum, an exposure control plan and offer training to employees.[2–4] Since their introduction, there have been numerous modifications of these guidelines by the CDC, Public Health Service (PHS), and OSHA. These modifications incorporate new scientific knowledge and address additional concerns that have arisen in the area of occupational exposures.[5–10] Recently, the Public Health Service interagency working group, comprised of members of the CDC, Food and Drug Administration (FDA), and NIH, issued updated guidelines that consolidate recommendations regarding the management of healthcare workers who have experienced occupational exposure to blood and other body fluids that may contain HIV.[10]

WORKING DEFINITIONS

Universal precautions are intended to supplement standard infection control procedures and rely on the use of protective barriers such as gloves, goggles, facemasks, as well as proper safety conduct, such as hand washing, avoiding recapping of needles, and proper disposal of medical waste. The concept of universal precautions is based on the underlying assumption that blood and certain body fluids from all patients is potentially infectious for HIV, hepatitis B virus (HBV), hepatitis C virus (HCV), methicillin-resistant *Staphylococcus aureus* (MRSA), or any other blood-borne pathogen. In this context, potentially infectious body fluids include: cerebrospinal fluid, vaginal secretions, semen, pleural fluid, peritoneal fluid, pericardial fluid, and amniotic fluid.[1,10] Universal precautions do not apply to body substances such as feces, sweat, tears, urine, human breast milk, vomitus, nasal secretions, sputum, or saliva unless visibly bloody; but this in no way lessens the need for practicing standard infection control procedures.[1,10] The term ''healthcare worker'' is very broad and refers to any person whose activities involve contact with patients or with blood or other body fluids from patients in a healthcare setting.[3,4,8,10] An ''exposure'' that places a healthcare worker at risk for HIV or other blood-borne pathogens and requires consideration for post-exposure prophylaxis is defined as any percutaneous injury, contact with mucous membrane or nonintact skin (i.e., chapped, abraded, or skin with dermatitis), or contact with intact skin when the duration of contact is prolonged (i.e., several minutes or greater) or extensive with blood, tissue, or body fluids.[10]

RATIONALE FOR UNIVERSAL PRECAUTIONS

Given the HIV/AIDS epidemic, in addition to the increase in other viral infections such as HBV and HCV, the need

Encyclopedia of Clinical Pharmacy
DOI: 10.1081/E-ECP 120006246

to adopt some form of protective measures, and thus the rationale for universal precautions, is clear. It has been suggested that there are three main reasons for applying universal precautions.[11] The first stems from the inability to test and identify every patient to discern if they are infectious. Therefore, it is most prudent to assume that all patients are infectious and thus act accordingly. The second reason is psychological in nature and takes into account human behavior and performance regarding identification and labeling practices. Lastly, there are administrative and legal reasons that embrace OSHA's mandate for employers to provide a safe working environment for employees.

Table 1 Factors associated with an increased or decreased risk of occupational HIV transmission following an exposure

Increased risk of transmission	Decreased risk of transmission
Exposure to large quantity of blood	Use of gloves
Exposure occured during placement of needle into artery or vein	Mucous membrane exposure
Exposure occurred during invasive procedure	Prompt initiation of post-exposure prophylaxis
Deep injury	Intact skin exposure
Visible blood on needle/object	
Source with increased viral load	

TRANSMISSION RISK AND EPIDEMIOLOGY

The true risk of transmission following occupational exposure is difficult to determine due to the high rate of underreporting of exposures and the underlying risk of healthcare workers being infected with HIV-1 outside of the workplace. However, one prospective study determined that a percutaneous exposure to HIV-infected blood is associated with a 0.3% risk of transmission.[12] The risk is lower after a mucous membrane exposure at 0.09%.[12] The risk associated with HIV transmission after skin exposure has not been precisely defined, because presently, no HCW enrolled in any prospective study has seroconverted following an isolated skin exposure; nonetheless, the risk is thought to be less than for mucous membrane exposure.[10,13] The risk of HIV transmission following exposure to bodily fluids other than blood, such as cerebrospinal fluid, vaginal secretions, semen, pleural fluid, peritoneal fluid, pericardial fluid, and amniotic fluid is not presently known. Factors that increase or decrease the risk of HIV transmission following an occupational exposure are listed in Table 1. The increased risk resulting from exposure to blood from terminally ill AIDS patients likely reflects an increased viral load.[14] In these instances, it has been suggested that the risk of HIV transmission exceeds 0.3%, but the practical utility of using HIV titers from source patients as a marker for transmission is unknown, and a low HIV titer does not rule out the possibility of HIV transmission.[14]

As of June 1997, documented HIV seroconversion temporally associated with occupational HIV exposure was reported in 52 healthcare workers.[10] Forty-seven cases involved exposure to HIV-infected blood, 45 exposures were percutaneous in nature, and most involved a hollow-bore needle (91%).[10] The remaining five cases involved exposure to HIV-infected bloody body fluid, concentrated virus in laboratory, and one case was unspecified. This data may not represent the true number of occupational HIV infections, because not all workers are evaluated for HIV infection following occupational exposure, not all workers with occupational exposure/infection are reported, and an estimated 40% or more of needlestick injuries may go unreported.[15–17] Needlestick injuries, whether a consequence of recapping or improper disposal techniques, continue to be a frequent and important mode of HIV exposure.[15] A large prospective surveillance study, the CDC Cooperative Needlestick Study, found more than 80% of all exposures were related to needlestick injuries.[18] Surveillance data indicates laboratory technicians, followed by nurses and physicians, represent the most common occupations associated with occupational HIV transmission.[19] Some have argued that universal precautions may not significantly decrease exposure, as a glove cannot prevent a needlestick injury. However, there is evidence that despite the ability of needles to pass through latex gloves, there may be a substantially lower blood innocula associated with the use of latex gloves, perhaps suggestive of a reduced likelihood of occupational infection.[20,21]

POST-EXPOSURE PROPHYLAXIS FOR HIV

Post-exposure prophylaxis, or the administration of pharmacological agents with the intent of preventing occupational HIV transmission after exposure, may be useful in certain circumstances. The decision to use post-exposure prophylaxis is complex and depends on a number of important factors, which should be evaluated on an individual, case-by-case basis. This decision should take

into account the nature of the exposure, including volume of blood or body fluid involved, source of exposure and risk of transmission, time between exposure and start of post-exposure prophylaxis, and documented efficacy of post-exposure prophylaxis. The potential benefits of post-exposure prophylaxis must be weighed against the potential hazards in deciding treatment duration, potential adverse effects of antiretroviral agents, viral resistance, and pregnancy. Fig. 1 outlines general considerations and procedures that should be followed after an occupational exposure. Note that the algorithm is a suggested approach and does not represent all possible scenarios; each ex-

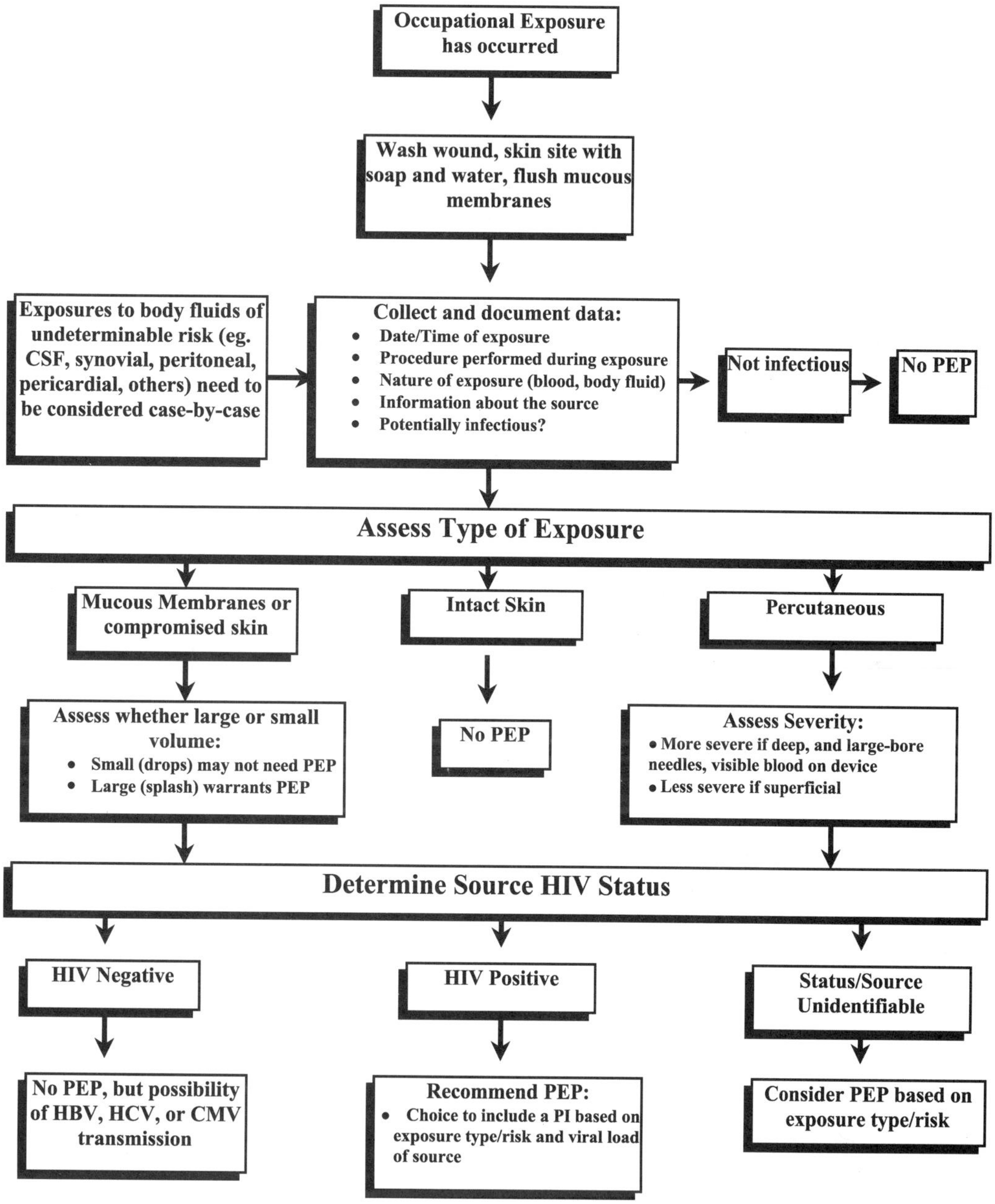

Fig. 1 General schematic of the considerations and procedures following an occupational exposure. This algorithm serves only as a suggested approach; although it reflects the PHS guidelines, it is not meant to replace published guidelines. In addition, this algorithm does not apply to other types of infectious exposures, such as sexual exposures. For specifics, please refer to the text and to the Public Health Service guidelines.[10]

posure needs to be individualized on a case-by-case basis. Additionally, this algorithm only pertains to occupational exposures and not to exposures differing in nature, such as sexual exposures.

Because evidence suggests that systemic infection does not occur immediately following primary HIV exposure, there appears to be a period of time in which initiation of post-exposure prophylaxis promptly after occupational exposure may prevent or inhibit systemic infection by limiting viral replication and proliferation. The consensus is that post-exposure prophylaxis should, therefore, be initiated as soon as possible following exposure, ideally within a few hours of exposure, because some animal data show a reduction in post-exposure prophylaxis efficacy with delayed initiation.[22,23] Post-exposure prophylaxis may be beneficial even up to 36 hour after exposure, but the efficacy of post-exposure prophylaxis given this late is largely undetermined. The optimal duration of post-exposure prophylaxis is unknown, but presently, the Public Health Service guidelines recommend a 4-week regimen of zidovudine and lamivudine with or without a protease inhibitor.[10]

Following an occupational exposure, it is vital that healthcare workers are cognizant of institutional policies and procedures to allow for the timely and organized collection of data and initiation of post-exposure prophylaxis if indicated. Institutions must have policies and procedures in place to react quickly to occupational exposure to avoid unnecessary delays in therapy. The date and time, details pertaining to the type of activity being performed, nature of the exposure (type, amount, severity, percutaneous, mucous membrane, time of contact, condition of skin), and details about the source (HIV infected, viral load, history of antiretroviral therapy) should be recorded in the healthcare worker's medical record. It is recommended that skin sites or wounds that are contaminated should be washed with soap and water.[10] The use of antiseptics may be considered, but application of caustic substances such as bleach is not recommended, as this would compromise the integrity of the skin barrier. Mucous membranes should be flushed extensively with water.[10]

Following each occupational exposure, proper assessment of the risk for HIV transmission should be performed by qualified personnel. If at all possible, the source patient should be evaluated for HIV, HBV, and HCV status to help rule out viral infection.[10] Information pertaining to intravenous drug use and other source risk factors relevant for consideration of post-exposure prophylaxis should be sought. If this information is unavailable, the source person should be notified of the incident and consent sought to facilitate testing for serologic evidence for viral infection. If the source is seronegative for virus at the time of exposure, with no clinical manifestations of HIV infection, then no further testing of the source is needed.[10] Inevitably, there will be occasions where the source is unknown. In these instances, the Public Health Service recommends assessing the risk for HIV transmission and the need for post-exposure prophylaxis based on integration of the nature of exposure and epidemiological information.[10]

EFFICACY OF POST-EXPOSURE PROPHYLAXIS

Although there are some data in animals, there is limited information available that can be used to assess the efficacy of post-exposure prophylaxis in humans. This is a manifestation of the infrequent rate of seroconversion of healthcare workers following exposure to HIV-infected blood and makes it doubtful that a prospective study with adequate statistical power could be performed. The notion of using zidovudine as post-exposure prophylaxis was addressed by the CDC in 1990 and has been shown to be beneficial by a retrospective case-controlled study, which documented the risk for HIV infection among healthcare workers who used zidovudine as post-exposure prophylaxis reduced by 81%.[6,14] However, there are at least 14 instances of zidovudine post-exposure prophylaxis failure, and there is growing concern regarding transmission of zidovudine-resistant virus, especially if the source patient's HIV treatment regimen included zidovudine.[10,24] Despite the fact that zidovudine has been shown to be effective in reducing perinatal transmission, it cannot be concluded that it is effective in reducing occupational exposure HIV transmission, because the occupational exposure route is not similar to the mother-to-infant route of transmission.[25]

OPTIONS FOR THERAPY

Because the treatment of HIV infection is rapidly evolving and takes into consideration multiple issues such as viral resistance and optimal drug combination regimens, there is growing concern over which drug or drugs should be used as the standard for post-exposure prophylaxis. Presently, there are no data available to assess whether the addition of other antiretrovirals to zidovudine enhances the efficacy of post-exposure prophylaxis regimens, or whether the increased prevalence of zidovudine resistance is an important factor for post-exposure prophylaxis regimen selection and efficacy. The CDC states that zidovudine should be considered for all post-exposure prophylaxis regimens, because it is the only agent that data supports efficacy for

post-exposure prophylaxis. Lamivudine should usually be added for increased antiretroviral activity and activity versus zidovudine-resistant viral strains.[8] The current Public Health Service guidelines maintain that it is reasonable to continue zidovudine as the drug of choice in post-exposure prophylaxis regimens, and there has been no additional information to suggest altering lamivudine as the second agent for post-exposure prophylaxis.[10] Table 2 lists the commonly used antiretroviral agents in post-exposure prophylaxis regimens, along with the usual dose and expected adverse effects. Other nucleoside reverse transcriptase inhibitors (NRTIs) that may be used with zidovudine for post-exposure prophylaxis are didanosine and zalcitabine, resulting largely from documented efficacy and use in combination with zidovudine for treatment of HIV.[26] The addition of protease inhibitor drugs such as indinavir, nelfinavir, and efavirenz are generally reserved for an expanded post-exposure prophylaxis regimen following high-risk exposures in lieu of their additional toxicity and drug interactions. Nonnucleoside reverse transcriptase inhibitors such as delavirdine and nevirapine are generally not included in post-exposure prophylaxis regimens.

The use of any drug, even as a prophylactic measure, may be associated with adverse effects. Adverse effects of antiretroviral medications are well-described in the HIV population; however, there is sparse data about adverse effects in the non-HIV-infected individuals. Between October of 1996 and December of 1998, an HIV post-exposure prophylaxis registry prospectively followed

Table 2 Post-exposure prophylaxis regimens

Drug(s)[a]	Indication	Drug regimen	Adverse effects	Specifics
Zidovudine	In combination with lamivudine for occupational exposures that have recognized risk for HIV transmission	*Basic regimen*: Zidovudine 300 mg BID, given with lamivudine for 28 days	Nausea, vomiting, headache, fatigue, anemia, neutropenia	Renal excretion of metabolite Food may affect peak plasma concentrations but not overall exposure
Lamivudine	In combination with zidovudine for occupational exposures that have recognized risk for HIV transmission	*Basic regimen*: Lamivudine 150 mg BID given with zidovudine for 28 days	Occasional nausea, headache, diarrhea, rash	Renal excretion, requires dose reduction based on creatinine clearance Food does not affect bioavailability
Indinavir	For use with zidovudine and lamivudine in cases where there is an increased risk for HIV transmission (i.e., source has high viral titer or large volume exposure)	*Expanded regimen*: Zidovudine 300 mg BID, lamivudine 150 mg BID, and indinavir 800 mg TID for 28 days	Dose-related hyperbilirubinemia, nephrolithiasis, metallic taste, rash, dry mouth/ mucous membranes	Hepatic elimination Give with water at least 1 hr prior, or 2 hr after a meal; food will substantially reduce bioavailability Minimum of 48 oz fluids daily to reduce nephrolithiasis Numerous drug interactions
Nelfinavir	For use with zidovudine and lamivudine in cases where there is an increased risk for HIV transmission (i.e., source has high viral titer or large volume exposure)	*Expanded regimen*: Zidovudine 300 mg BID, lamivudine 150 mg BID, and nelfinavir 750 mg TID for 28 days	Diarrhea, rash, asthenia, anemia	Hepatic elimination Food will increase bioavailability; take with light snack

[a]Consider potential drug interactions with other concurrent medications.
(From Refs. [10] and [27].)

healthcare workers receiving post-exposure prophylaxis following occupational exposure and monitored them for adverse effects.[27] Of the 250 healthcare workers who completed a post-exposure prophylaxis regimen, and for whom follow-up data was available, 76% reported some symptoms or adverse events.[27] The most frequently reported symptoms were nausea (57%), fatigue/malaise (38%), headache (18%), vomiting (16%), and diarrhea (14%). Minor laboratory abnormalities were reported in 8% of healthcare workers.[27] Of those healthcare workers who discontinued all post-exposure prophylaxis drugs, 50% cited symptoms or adverse effects as the reason for discontinuation.[27]

Little is known regarding the potential effects of antiretroviral drugs on the developing fetus. Although carcinogencity or mutagenicty is evident in screening tests for zidovudine and all other FDA-licensed NRTIs, limited data on zidovudine use in the second or third trimesters of pregnancy was not associated with serious adverse effects.[10,25] Whether lamivudine is teratogenic is unknown.[8] This complicates the decision to provide post-exposure prophylaxis to pregnant women; however, pregnancy should not preclude the use of optimal post-exposure prophylaxis, and post-exposure prophylaxis should not be denied to a healthcare worker solely on the basis of pregnancy.[10] In these situations, the patient shall reserve the right to decline post-exposure prophylaxis after they have been informed about the risk of HIV transmission given the nature of the exposure, limited data regarding teratogenicity in various trimesters of pregnancy, and the risks versus benefits of post-exposure prohylaxis therapy.

The impact that an occupational exposure can inflict on a healthcare worker and their family is enormous. Healthcare workers should receive post-exposure counseling and education, along with a detailed synopsis of what events the healthcare worker may expect, including the need for follow-up HIV antibody testing. Additionally, facts about the risk for transmission and other issues such as the choice, efficacy, adverse effects, and adherence of post-exposure prophylaxis medications should be discussed. Additionally, the possible need for sexual abstinence or use of a condom to prevent secondary transmission to their partners should be addressed in a manner allowing for questioning and feedback. Organ dysfunction, concurrent disease states, and other medications the healthcare worker was previously taking on a regular basis, and the potential for drug interactions with post-exposure prophylaxis medications must be considered. Exposed healthcare workers should be reminded to refrain from donation of blood, plasma, organs, tissue, and semen donation, and to temporarily discontinue breastfeeding.[10]

Table 3 Available resources and registries for HIV post-exposure prophylaxis

Resource or registry	Contact information[a]
Centers for Disease Control	Telephone: (404) 639-6425 (For reporting HIV seroconversion in HCWs that received postexposure prophylaxis) or www.cdc.gov/hiv (for general information)
National Clinicians' Postexposure Hotline	Telephone: (888) 448-4911; (888) 737-4448; (888) PEP4HIV
Food and Drug Administration	Telephone: (800) 322-1088 (For reporting unusual or severe toxicity from PEP regimens)
Antiretroviral Pregnancy Registry	Telephone: (800) 258-4263
UCSF On-line information	http://arvdb.ucsf.edu (For information about antiretroviral drugs)
National HIV/AIDS Clinician's Consultation Center	http://www.ucsf.edu/hivcntr
HIV/AIDS Treatment Information Service (ATIS)	http://www.hivatis.org Telephone: (800) 448-0440

[a]Contact information is subject to change.

There are numerous factors to consider before and after initiation of a post-exposure prophylaxis regimen, which may lead to providers and educators feeling overwhelmed and unsure about some choices that need to be made. To facilitate this process, Table 3 lists some available resources and registries that one may contact to aid in the decision process.

CONCLUSION

The practice of universal precautions is Federal Law in the United States, and it is the responsibility of every employer or institution that healthcare workers have the resources and training necessary to adhere to these safety precautions.[2] Additionally, support for continued practice of universal precautions needs to come from all levels of administration. Observations by Gershon et al. indicate that one of the strongest correlates with compliance is the institutional ''safety climate.''[15] This implies that if healthcare workers perceive their work environment to be conducive to practicing universal precautions, then they will be more likely to do so.

Each occupational exposure needs to be considered on a case-by-case basis. The risk of occupational HIV transmission appears to be greatest if the exposure is percutaneous in nature. In order to optimize post-exposure prophylaxis, the duration between exposure and initiation of therapy must be kept at a minimum; preferably within a few hours. Despite the concern of viral resistance, post-exposure prophylaxis regimens should include 4 weeks of zidovudine plus lamivudine, and under certain circumstances, a protease inhibitor.

Although the focus of this discussion emphasizes HIV infection, healthcare workers must be aware that other blood-borne pathogens such as HBV and HCV may also be of significant concern. The healthcare worker is encouraged to refer to alternative references for more detailed discussions pertaining to these infections.[28,29]

Unfortunately, healthcare worker exposure to infectious blood and body fluids will continue to be a reality. Hopefully, with the practice of basic infection control in conjunction with universal precautions, this can be kept at a minimum. Universal precautions will likely be an evolving process considering the increased prevalence of viral resistance, and the continued emergence of new, and more difficult to treat infectious entities.

REFERENCES

1. Centers for Disease Control. Recommendations for prevention of HIV transmission in healthcare settings. Morb. Mort. Wkly. Rep. **1987**, *36* (Suppl. 2S), 1S–18S.
2. Occupational Safety and Health Administration. Occupational exposure to bloodborne pathogen: Final rule. Fed. Regist. **1991**, *56*, 64004–640182.
3. DeJoy, D.M.; Gershon, R.R.M.; Murphy, L.R.; Wilson, M.G. A work-systems analysis of compliance with universal precautions among health care workers. Health Educ. Q. **1996**, *23* (2), 159–174.
4. Center for Disease Control. Update: Universal precautions for prevention of transmission of HIV, HBV, and other blood-borne pathogens in health care workers. Morb. Mort. Wkly. Rep. **1988**, *37*, 229–234.
5. Centers for Disease Control. Guidelines of prevention of transmission of human immunodeficiency virus and hepatitis B virus to healthcare and public-safety workers. A Response to P.L. 100-607. The Health Omnibus Programs Extension Act of 1988. Morb. Mort. Wkly. Rep. **1989**, *38* (S-6), 3–37.
6. Centers for Disease Control. Public Health Service statement on management of occupational exposure to human immunodeficiency virus, including considerations regarding zidovudine post-exposure use. Morb. Mort. Wkly. Rep. **1990**, *39* (RR-1), 1–14.
7. Centers for Disease Control. Recommendations for preventing transmission of human immunodeficiency virus and hepatitis B virus to patients during exposure-prone invasive procedure. Morb. Mort. Wkly. Rep. **1991**, *40* (RR08), 1–9.
8. Center for Disease Control. Provisional public health service recommendations for chemoprophylaxis after occupational exposurre to HIV. Morb. Mort. Wkly. Rep. **1996**, *45* (22), 468–472.
9. Centers for Disease Control. Guideline for infection control in health care personnel, 1998. Am. J. Infect. Control **1998**, *26*, 289–354.
10. Centers for Disease Control. Public health service guidelines for the management of healthcare worker exposures to HIV and recommendations for post-exposure prophylaxis. Morb. Mort. Wkly. Rep. **1998**, *47* (RR-7), 1–28.
11. Hopkins, C.C. Implementation of universal blood and body fluid precautions. Infect. Dis. Clin. North Am. **1989**, *3* (4), 747–762.
12. Ippolito, G.; Puro, V.; De Carli, G. The Italian study group on occupational risk of HIV infection. The risk of occupational human immunodeficiency virus infection in health care workers. Arch. Intern. Med. **1993**, *153*, 1451–1458.
13. Fahey, B.J.; Koziol, D.E.; Banks, S.M.; Henderson, D.K. Frequency of nonparenteral occupational exposures to blood and body fluids before and after universal precautions training. Am. J. Med. **1991**, *90*, 145–153.
14. Cardo, D.M.; Culver, D.H.; Ciesielski, C.A., et al. A case-control study of HIV seroconversion in health care workers after percutaneous exposure. N. Engl. J. Med. **1997**, *337*, 1485–1490.
15. Gershon, R.R.M.; Karkashian, C.; Felknor, S. Universal precautions: An update. Heart Lung **1994**, *23* (4), 352–358.
16. Mangione, C.M.; Gerberding, J.L.; Cummings, S.R. Occupational exposure to HIV: Frequency and rates of underreporting of percutaneous and mucocutaneous exposures by medical house staff. Am. J. Med. **1991**, *90*, 85–90.
17. Hamory, B. Underreporting of needlestick injuries in a university hospital. Am. J. Infect. Control **1983**, *11* (3), 174–177, Centers for Disease Control and Prevention. HIV/AIDS Surveill. Rep. **1993**, *5* (no.3), 13.
18. Centers for Disease Control Cooperative Needlestick Surveillance Group. Surveillance of health care workers exposed to blood from patients infected with human immunodeficiency virus. N. Engl. J. Med. **1988**, *319*, 1118–1123.
19. Centers for Disease Control and Prevention. HIV/AIDS Surveill. Rep. **1993**, *5* (3), 13.
20. Wollowine, J.; Mast, S.; Gerberding, J. Factors Influencing Needlestick Infectivity and Decontamination Efficacy: An ex vivo Model. In *32nd Interscience Conference on Antimicrobial Agents and Chemotherapy,*

Anaheim, CA, American Society for Microbiology, 1992, Abstract 1188.
21. Mast, S.; Gerberding, J. Factors predicting infectivity following needlestick exposure to HIV: An in vitro model. Clin. Res. **1991**, *39*, 58A.
22. Martin, L.N.; Murphy-Corb, M.; Soike, K.F.; Davison-Fairburn, B.; Baskin, G.B. Effects of initiation of 3′-azido, 3′deoxythmidine (zidovudine) treatment at different times after infetion of rhesus monkeys with simian immunodeficiency virus. J. Infect. Dis. **1993**, *168*, 825–835.
23. Shih, C.-C.; Kaneshima, H.; Rabin, L., et al. Post-exposure prophylaxis with zidovudine suppresses human immunodeficiency virus type 1 infection in SCID-hu mice in a time-dependent manner. J. Infect. Dis. **1991**, *163*, 625–627.
24. Tokars, J.I.; Marcus, R.; Culver, D.H., et al. Surveillance of HIV infection and zidovudine use among health care workers after occupational exposure to HIV-infected blood. Arch. Intern. Med. **1993**, *118*, 913–919.
25. Connor, E.M.; Sperling, R.S.; Gelber, R., et al. Reduction of maternal-infant transmission of human immunodeficiency virus type 1 with zidovudine treatment. N. Engl. J. Med. **1994**, *331*, 1173–1180.
26. Center for Disease Control. Guidelines for the use of antiretroviral agents in HIV-infected adults and adolescents. Morb. Mort. Wkly. Rep. **1997**, *175*, 1051–1055.
27. http://www.cdc.gov/ncido/hip/Blood/PEPRegistry.pdf (accessed Dec. 2000).
28. Center for Disease Control. Immunization of healthcare workers. Recommendations of the Advisory Committee on Immunization Practices (ACIP) and the Hospital Infection Control Practices Advising Committee (HICPAC). Morb. Mort. Wkly. Rep. **1997**, *46*, RR-18.
29. Centers for Disease Control. Recommendation for prevention and control of hepatitis C virus (HCV) infection and HCV-related chronic disease. Morb. Mort. Wkly. Rep. **1998**, *47* (RR19), 1–39.

DISTINGUISHED PERSONALITIES

Walton, Charles

David Hawkins
Mercer University, Atlanta, Georgia, U.S.A.

INTRODUCTION

Charles Walton has contributed substantially to the development and maturation of clinical pharmacy. Among his most important contributions were the development of drug information, the introduction of rational therapeutics into the pharmacy curriculum, and his strong advocacy for a sustained clinical practicum as an essential component in the training of competent clinical pharmacists. He is regarded by many in the discipline as the father of drug information training and one of the founding fathers of the American College of Clinical Pharmacy.

BIOGRAPHY AND ACCOMPLISHMENTS

Charles Anthony Walton was born on April 3, 1926, in Auburn, Alabama, but was raised in Tallassee, which is 35 miles northeast of Montgomery. He recalls how people passing through his hometown accused the city fathers of leaving out the ''ha'' in spelling the name of their town. Tallasseeans, according to Walton, are a noble people and would tell such critics that they needed to travel further south into the panhandle of Florida if it was laughter they wanted.

Walton's interest in pharmacy began when he was a young lad and an avid listener of radio. Television had

Encyclopedia of Clinical Pharmacy
DOI: 10.1081/E-ECP 120006257

not yet been invented. Walton was particularly interested in radio commercials and some of their dramatic claims. He remembers one mellifluous radio announcer saying "Take Sal Hapitica for the smile of health and brush with Ipana for the smile of beauty." But what really piqued his interest in drugs was the radio commercial that advertised Carter's Little Liver Pills. He found it incredible that such a small, round pill could produce such extraordinary and wonderful outcomes. In fact, he was so intrigued with commercials that he decided that when he grew up he would pursue either a career in radio announcing or in the study of drugs. His Southern drawl, however, pretty much excluded a career in radio.

After graduating from Tallassee High School, Walton enlisted in the United States Navy and served two years during the Second World War. After completing his tour of duty in the Navy, he enrolled as a freshman at Auburn University, where he would later study pharmacy. He was a good student of pharmacy and his interest and curiosity in pharmacology continued to grow. However, the faculty did not readily encourage his enthusiasm and desire to do postgraduate work in pharmacology and suggested instead that he enter pharmacy practice where he could earn a decent living. But there was one faculty member who learned of Walton's desire to continue his education and pursue a graduate degree in pharmacology. She had a master's degree in hospital pharmacy from Purdue University, and she alone urged Walton to follow his dreams and enter graduate school. He did just that. After one year at Purdue, Walton was awarded a master of science degree in pharmacology, and in 1950 he accepted a faculty position at the University of Kentucky School of Pharmacy in Louisville.

Later on Walton took advantage of a part-time sabbatical to begin work on a Ph.D. degree in pharmacology at Purdue. He completed the program in 1956. Shortly after that, the University of Kentucky School of Pharmacy moved to Lexington, and, in 1960, opened a medical school on this new Lexington campus. The presence of a medical school on the same campus as the pharmacy school led Walton to dream about training pharmacy and medical students in the same educational environment. He presented his idea to the two deans. The dean of the medical school supported the idea, but the dean of the pharmacy school had some reservations, and so Walton's dream was temporarily suspended.

Walton may have been dismayed but he was not distracted. He soon became acquainted with the Director of Hospital Pharmacy at the University of Kentucky Medical Center—a very bright and energetic young man who had some novel ideas of his own. Charles Walton and Paul Parker rapidly developed a strong collegial relationship. They were so enthusiastic and passionate about their ideas that often they would work well into the early morning hours generating new concepts and crafting plans to expand the mission of pharmacy. These late-night meetings were charged with so much creative energy that, the next days, neither man felt tired, but instead longed to get back together to continue their stimulating discussions. Walton recalls one occasion when Parker invited him to his Lexington home to explore the idea of training pharmacists to provide rational drug information to hospital formulary committees and to physicians taking care of patients in the hospital. They worked tirelessly during the night, drafting a concept paper that outlined the establishment of a hospital pharmacy service that would be dedicated to the advancement of rational therapeutics. They called the new service initiative drug information. They knew that their idea would be popular with the medical school's chairman of medicine, Ed Pellegrino, because Pellegrino was a pioneer and national leader in the teaching and practice of rational drug therapy. A key development in the concept came about with the hiring of David Burkholder as the first director of the drug information center. Working together, Burkholder and Pellegrino developed the first hospital formulary that was based entirely on the principles of rational therapeutics. (Today, we refer to rational therapeutics as evidenced-based medicine.) The major underlying principle is that no drug should be used in the hospital setting that has not been proven in proper studies to be safe and more effective than alternative therapies.

The idea of melding drug information with pharmacists monitoring drug therapy at the patient's bedside came from a drug information conference held in the early 1960s at the Carnahan House, located on a horse farm near Lexington. The conference was sponsored by the American Society of Hospital Pharmacists (ASHP). At that conference, William Smith presented the so-called "9th floor project" being conducted at the University of California at San Francisco. The marriage of drug information to hospital pharmacists working at the bedside gave birth to a new concept in pharmacy that the conference delegates referred to as clinical pharmacy. Those representing ASHP at the conference expressed some concern that a new competing organization might emerge. The conference attendees toyed with the idea of establishing an association of clinical therapologists. A delegate from Canada quickly moved that the organization be named the International Association of Clinical Therapologists (IACT). It was all in jest.

Walton became the director of the hospital pharmacy's Drug Information Center in 1967. He also started working more closely with the hospital pharmacy residency pro-

gram directed by Paul Parker. He continued his teaching of pharmacology in the pharmacy school but with a different slant, incorporating in his lectures the principles of rational therapeutics. At that time, James Doluisio, who had recently joined the pharmacy school faculty at the University of Kentucky, was serving as chairman of the curriculum committee. Doluisio had been a key player in the development of the Pharm.D. Program at the Philadelphia College of Pharmacy and Science. Walton was asked to chair a committee to develop a doctor of pharmacy program at Kentucky. Working closely with Doluisio and Parker, Walton's committee successfully established a strong post-baccalaureate Pharm.D. program deeply rooted in rational therapeutics and including a sustained clinical practicum through the hospital pharmacy residency program. Walton's dream of pharmacy students being educated and trained alongside medical students had finally become a reality.

In 1973, Doluisio became Dean at the University of Texas College of Pharmacy in Austin. One of his first administrative decisions was to hire Walton to develop a Pharm.D. program in Texas. Because there was no medical school in Austin, Doluisio negotiated an agreement with the University of Texas Health Science Center in San Antonio to establish a jointly administered post-baccalaureate Pharm.D. degree program. And, because there was no existing hospital pharmacy residency program at the health science center in San Antonio, a one-year sustained clinical practicum was incorporated as part of the academic program. Walton was appointed associate dean for clinical programs and recruited a young, dedicated faculty to help build the Pharm.D. program at Texas. The new program contained a strong didactic background in the biomedical sciences and provided experiential training in medicine, pediatrics, psychiatry, ambulatory care, and drug information.

Walton strongly insisted that pharmacists engaged in therapeutic decision making and pharmacotherapy consultations had to be well-trained clinicians who were directly involved in the management of patients. The founders of the American College of Clinical Pharmacy (ACCP) were strongly influenced by Walton's philosophy on what constitutes a credible clinical pharmacy practitioner as they developed the mission, goals, and purpose of the organization. In fact, Walton played a prominent role in the early development of the College not only as a source of encouragement, wisdom, and guidance but also by serving ACCP as an advocate before the pharmacy academy, particularly the American Association of Colleges of Pharmacy. The major impetus of ACCP was to strive for excellence. The new organization was criticized by some as being composed of elitists. When questioned about this, Walton said, "If striving for excellence makes me an elitist, then I am an elitist." Because of his important contributions to the development of the ACCP, Walton was inducted as the first honorary fellow of the College.

Charles Walton will forever be admired for his intellect, his humor, his wisdom, and his passion for excellence. To the pioneers in clinical pharmacy and to many of those who followed in their footsteps, Charles Anthony Walton is a hero and beloved friend.

DISTINGUISHED PERSONALITIES

Weaver, Lawrence

Robert J. Cipolle
University of Minnesota, Minneapolis, Minnesota, U.S.A.

INTRODUCTION

Lawrence C. Weaver is one of the true patriarchs of clinical pharmacy. He was Dean of The College of Pharmacy at the University of Minnesota from 1966–1984 and 1994–1995. He initiated the post-B.S. Pharm.D. program in 1971. His international efforts have resulted in the establishment of clinical pharmacy educational programs in Europe, Africa, the Middle East, and the Pan-Pacific Rim. He led efforts to facilitate the worldwide access and distribution of orphan drugs required to treat patients with rare diseases while working for the Pharmaceutical Manufacturers Association from 1984–1989. In 1997, the University of Minnesota recognized his life-long contributions and leadership by naming the Pharmacy and Nursing building, Weaver–Densford Hall, in his honor.

PERSONAL BACKGROUND

Born January 23, 1924 in Bloomfield, Iowa to Wilbur and Faye Weaver, Lawrence C. Weaver was the eldest of three sons. He spent his childhood living and learning on the family farm and one-room school in southeastern Iowa. He joined the U.S. Air Force in 1942, served as a pilot, and rose to the rank of Captain. He flew in the China–Burma–India theatre and earned the Distinguished Flying Cross and an Air Medal. He returned to the States and earned a Bachelor's degree in Pharmacy from Drake University in 1949.

Larry married Delores (Dee) Hillman on September 9, 1949 in Oakland, California. Dee and Larry adopted and raised four children. Their extended family includes the thousands of pharmacy students throughout the world who have benefited from their personal support and guidance.

PROFESSIONAL ACHIEVEMENTS

After earning his Ph.D. in pharmacology from the University of Utah in 1953, Weaver joined the Pitman–Moore Company. He was the Head of Biomedical Research that included both human and veterinary research in pharmacology, microbiology, and parisitology. In 1964 while with Pitman–Moore (now a division of the Dow Chemical Company), Weaver organized and directed the Biohazards Department where he pioneered efforts in the emerging field of biological hazard (environmental) control.

Weaver's research interests in Pharmacy Education, Healthcare Delivery Systems, and Drug Combinations in Therapy, among others, led him to become the fourth Dean of the College of Pharmacy at the University of Minnesota in 1966. As the Dean of Pharmacy and Professor of Pharmacology in the School of Medicine, Weaver was far-sighted enough to bring the pharmacy program at the University of Minnesota into the newly organized Health Sciences Center in order to foster interdisciplinary education and the clinical role of pharmacists. He introduced the two-year post-Baccalaureate program that became the basis for the Pharm.D. program. His vision of clinical pharmacy faculty and Pharm.D. students practicing and learning side-by-side with physician faculty and medical students became the platform for the future expansion and growth of the Pharm.D. program. His 10-year effort to join the Medical, Dental, Nursing, and Public Health programs on the campus of the University Hospital culminated with approval of funding for the Health Sciences Unit F building, which was built in 1981. He lead the effort to secure the over $20,000,000 required to build the nine-story building that houses the College of Pharmacy and the School of Nursing. The building was dedicated in 1996 when it was renamed Weaver–Densford Hall to honor Lawrence C. Weaver and Katherine Densford, the first Dean of the School of Nursing at the University of Minnesota.

Dean Weaver initiated the two-year post-Baccalaureate Pharm.D. program at the University of Minnesota in 1971. As the Dean of one of the first clinical pharmacy programs in the United States, Weaver collaborated with the Medical School to have his Pharm.D. students take pathophysiology classes with the medical students in order to both bring essential disease-related content to the

Encyclopedia of Clinical Pharmacy
DOI: 10.1081/E-ECP 120006258

pharmacy curriculum and to introduce medical students and faculty to clinical pharmacists. Dean Weaver was relentless in his belief that clinically prepared pharmacists could substantially improve the quality of health care. Under his leadership, in 1981 the Pharm.D. curriculum expanded from a two-year post B.S. program to a six-year Pharm.D. degree. Weaver felt that the curriculum should be designed to increase the clinical skills of all entry-level pharmacists. The new curriculum featured courses in pharmacokinetics and biopharmaceutics, as well as a pathophysiology and therapeutics course sequence taught for the first time entirely by clinical pharmacy faculty. Larry, as he is known by the thousands of students he taught and mentored, brought a vision of improved patient care through clinical pharmacy service that inspired students and clinical faculty for over three decades.

After 18 years as the Pharmacy Dean, Weaver left academia to become the Vice President of Professional Relations for the Pharmaceutical Manufacturers Association (PMA). In his role as the Executive Director of the PMA Commission for Rare Diseases, he championed the goal of making orphan drugs available to patients throughout the world who suffer from rare diseases. His efforts have resulted in over 800 drugs, serums, and vaccines being designated as ''orphans'' by the Food and Drug Administration. His continued efforts to help those in need of orphan drugs required interaction with federal agencies, pharmaceutical firms, researchers, practitioners, and patients and their families. He has been described as a peacemaker and a matchmaker who can bring industry, academia, and families together.

WORLD RECOGNITION

Larry Weaver is recognized around the world for his leadership and advocacy for clinical pharmacy. He assisted international pharmacy leaders from several countries in the design and implementation of clinical pharmacy programs. Throughout the late 1970's and 1980's, Weaver served as an educational consultant and advisor to numerous countries including Dalhousie University, Royal Danish College of Pharmacy, Welsh School of Pharmacy, Universities in Cairo, Tanta, Nairobi, Zimbabwe, Republic of South Africa, Amman Jordan, and the University of Sains Malaysia. In his role as Education Advisor to the University of Riyadh, Saudi Arabia, Weaver assisted in the conversion of the curriculum from the European System to the credit system and to the use of English as the language for teaching in pharmacy. He worked with Saudi faculty to establish clinical pharmacy courses and initiated the first continuing education program in pharmacy.

Larry Weaver is one of pharmacy's most well respected leaders due to his endless efforts to establish clinical pharmacy as the standard for American pharmacy education. He served in leadership roles including President and Vice President of both the Academy of Pharmaceutical Sciences of the APhA and the American Association of Colleges of Pharmacy. He also served as a member of first Board of Trustees of the Research Institute for the American College of Clinical Pharmacy.

RECENT SERVICE

In 1994, Weaver was summoned back to the Deanship of the College of Pharmacy at the University of Minnesota. He served in this capacity for two years and directed the implementation of the new entry level Pharm.D. curriculum. Since 1996, Dean Emeritus Lawrence C. Weaver has been a member of the Peters Institute of Pharmaceutical Care, developing educational and research programs in pharmaceutical care practice. Larry Weaver has had a distinguished career in pharmacy.

World Health Organization

Patrice Trouiller
University Hospital of Grenoble, Grenoble, France

INTRODUCTION

The World Health Organization (WHO) is a United Nations specialized agency, which has 193 member countries. Its objective is the attainment by all people of the highest possible level of health. WHO has four main constitutional functions: to act as the directing and cooperating authority on international health issues; to provide assistance including maintaining epidemiological and statistic services; to promote research; and to develop and promote international norms or standards. The work of WHO is carried out by the World Health Assembly, the Executive Board, the Secretariat, and six regional offices.

HISTORY

The origin of WHO dates from the need for international preventive and control measures against epidemics such as cholera and plague, particularly in the European region, during the nineteenth century. Later, different cooperation arrangements were adopted and the International Agency for Public Hygiene was established in Paris (1907) to control epidemics and communicable diseases. Later the Health Organization of the League of Nations was set up in Geneva (1919). During the San Francisco Conference, which laid the foundations for the United Nations Organization (1945), it was decided to establish a permanent and autonomous international organization for health issues. In June 1946, an international health conference convened in New York by the UN Secretariat approved the creation of WHO. Its specific Constitution, following a ratification by 61 states, was set into place on April 7, 1948 (now marked as World Health Day). WHO, once the main player, is now one of many UN and other organizations or institutions concerned with health. The World Bank plays an increased financial and technical role, bilateral agencies and the private sector as well, such as the non governmental organizations, make significant contributions to international health.[1]

MISSIONS

With 193 member states currently, WHO has a constitutional mandate to direct and coordinate international efforts in relation to health, to promote technical cooperation among nations, to develop and transfer appropriate health technology, and to set global standards for health. Finally its overall mission is the ''attainment by all people of the highest possible level of health,'' with special emphasis on closing the gaps within and among countries. According to WHO, health is defined as ''a state of complete physical, mental, and social well-being and not merely the absence of disease or infirmity.'' This ambitious and idealistic objective was summarized by the slogan ''Health for All by the Year 2000.''[5] In spite of major achievements (e.g., smallpox eradication declared in 1980; polio and guinea-worm disease on the threshold of eradication; leprosy, lymphatic filariasis and neonatal tetanus targeted for elimination), this ''health for all'' objective is currently challenged by the spread of the HIV/AIDS pandemic; by the persistence of malaria and many other parasitic diseases; by tuberculosis, which is still a major health priority since 1948; and increasing gap between least developing and developed countries (e.g., average life expectancy is 38.5 in Zambia and 79.5 in Sweden; 2 million children die each year from diseases for which vaccines exist; etc.), and the growing impact of globalization on health issues.[2]

To carry out its missions, WHO is endowed with a decentralized system including a central office (the Geneva headquarters) and six regional offices. At the central level, the World Health Assembly (WHA), a deliberating body which represents the 193 member states, sets the policy, approves the budget, and passes agreements or conventions (to be adapted by the member states). In addition, it can issue international health regulations for technical matters (WHO normative functions) directly and compulsorily applicable to member states. The WHA elects the Executive Board composed of 32 members and appoints a Director-General (DG) to give effect to the decisions and policies of the Health Assembly and to

Encyclopedia of Clinical Pharmacy
DOI: 10.1081/E-ECP 120006305

advise it. At the decentralized level, six regional offices have been created by the WHA to take local specificities into account.

The funding for WHO comes from member countries and dues are based on their national economies and as well on extra-budgetary funds allocated by countries or institutions earmarked for specific purposes or projects. The total budget for WHO was about $1.3 billion in 2000, and this amount has not increased in recent years despite increased demands on the organization. One problem has been the persistent failure of some major funders to fully pay their assessed dues and the fluctuation in member states' voluntary contributions.

FUNCTIONS AND ACTIVITIES

Normative and Harmonization Functions

According to articles 2 and 21 of its Constitution, WHO has a regulatory power for developing, establishing, and promoting standards and nomenclature (International Health Regulations). These regulations are limited to technical matters: health and international quarantine regulations; health codification for international travels; international nomenclature of diseases; nomenclature and standards with respect to the safety, quality, and efficacy for pharmaceutical, biological, and similar substances, including their advertising and labeling; international standards for biological; international cooperation for health statistics; food standards (through the joint WHO/FAO ''Codex Alimentarius'' Commission); and dependence and misuse of drugs regulations.

Numerous matters related to public health and biomedical sciences can be the subject of Recommendations but without compulsory nature contrary to international regulations. WHO Recommendations may be endorsed by the WHA under article 23 or by the Executive Board and issued as resolutions (e.g., WHA 52.19 Resolution on the revised drug strategy, 1999).

They may be issued by an individual WHO Expert Committee and included in their reports (e.g., WHO Expert Committee on specifications for pharmaceutical preparations, technical report number 863, 1996), or issued as special WHO publications under the authority of the DG (e.g., Globalization and access to drugs, health economics and drugs, 1998). Most of the WHO Recommendations are directed at developing countries (e.g., the ''International Pharmacopoeia'').

In practice, WHO standards, guidelines, and recommendations serve as advice to member states. They can however, be adopted as legally binding national regulations if a national authority so desires, or they may form the basis of national standards and technical regulations.

Operational Activities

In support of its objectives, WHO develops a wide range of operational activities to provide appropriate technical assistance, to stimulate and advance work on prevention and control of epidemic, endemic, and other diseases, and to cooperate with governments for strengthening health services. WHO works closely with other UN organizations or programs [e.g., Food and Agricultural Organization (FAO), United Nations Children's Fund (UNICEF), United Nations Development Program (UNDP), and the World Bank], and maintains working relationships with bilateral agencies and intergovernmental and nongovernmental organizations (NGOs). In addition, nearly 1200 health-related institutions are officially designated as WHO collaborating centers. More than 50 affect the pharmaceutical sector, such as the Uppsala Monitoring Center for pharmacovigilance issues.

Communicable diseases

Approximately, 56 million people died in 1999, of which 14 million died from infectious and parasitic diseases with more than 90% occurring in developing countries. (In rank order: acute lower respiratory infections, HIV/AIDS, diarrhea, tuberculosis, malaria, measles.) To prevent and control them, WHO developed and launched activities and programs in the field such as the Expanded Programme on Immunization (EPI) in collaboration with UNICEF (EPI targets are poliomyelitis, measles, diphtheria, whooping cough, tetanus and tuberculosis); the Onchocerciasis Control Programme; the global programs on AIDS; the ''Roll Back Malaria'' program; the Africa 2000 Initiative on water supply and sanitation issues; and others.

Noncommunicable diseases

Chronic diseases such as cardiovascular diseases, cancers, and respiratory diseases, affect both developed and developing countries. WHO's priorities are an integrated and coordinated approach to prevent, treat, and cure through disease-specific interventions, global campaigns to encourage healthy lifestyles (e.g., worldwide no-tobacco day), and healthy public policies promotion.

Emergency and humanitarian action

The emergency and humanitarian division of WHO/HQ plays an active role in assisting the member states in tackling health emergencies. It relies on its emergency preparedness and response programs, providing countries timely support to tackle acute emergency and supporting them during the reconstruction phase. WHO can assure coordination with donors (through consolidated appeals), UN agencies (e.g., UNHCR, UNICEF, WFP) and other entities involved (e.g., ICRC, NGOs). To face large movements or sudden influxes of refugees which create

Table 1 WHO Core functions in pharmaceuticals

Core functions	Objectives	Components	Tools
National drug policy	• Help countries formulate and implement a national drug policy (NDP) and integrate the work into their national health system in ensuring commitment of all stakeholders.	• Implementation and monitoring of NDP. • Health system development supported by essential drugs policies and programs.	Guidelines for developing a NDP. Indicators for monitoring NDP, 1999. Essential Drug List (EDL), 11th list, Nov 1999.
Access to essential drugs	• Ensure equitable availability and affordability of essential drugs.	• Access strategy and monitoring for essential drugs (in particular for priority health policies and newly developed drugs). • Financing mechanisms and affordable essential drugs. • National and local public sector drug supply systems and supply capacity.	Standard indicators to measure equitable access. Drug price information. Operational principles for good pharmaceutical procurement, 1999. Good drug donation practices, 1999. The new emergency health kit, 1998.
Quality and safety of drugs	• Ensure the quality and safety and efficacy of all medicines. • Put into practice regulatory and quality assurance (QA) standards.	• Norms, standards, and guidelines for pharmaceuticals. • Drug regulation and QA systems. • Information support for pharmaceutical regulation. • Guidance for control and use of psychotropics and narcotics.	Norms, standards and guidelines developed and updated. Quality control specifications. Drug nomenclature and classification. WHO certification scheme on the quality of pharmaceuticals. Good manufacturing practices (GMP). Good laboratory practices (GLP). Good clinical practices (GCP), 1995. The International Pharmacopoeia, 1994.
Rational use of drugs	• Ensure therapeutically sound and cost-effective use of drugs by health professionals (prescribers, nurses, and dispensers) and consumers.	• Rational drug use strategy and monitoring through national strategies. • National standard treatment guidelines, EDLs, educational programs. • Independent and unbiased drug information system.	Guide to good prescribing, 1994. Medical products and the internet: a guide to finding reliable information, 1999. The use of essential drugs. 8th report of the WHO expert committee.

(From Ref. [5].)

immediate needs for basic health services (e.g., Rwanda in 1994 and Kosovo in 1999), WHO has collaborated with most international agencies and NGOs to developed a standard kit of essential medicines, medical supplies, and basic equipment called the ''New Emergency Health Kit.'' The same interagency group has developed guidelines for obtaining the maximum benefit from drug donations through the ''Guidelines for Drug Donations'' development.[3]

Health Research

Research activities range from epidemiological surveillance for new and re-emerging diseases, the tropical disease research program (WHO/TDR in Geneva, Switzerland) tackling epidemiological research, medicines research and development (on malaria, trypanosomiasis, leishmaniasis, Chagas disease, etc.), cancer research (e.g., International Centre for Cancer Research in Lyon, France), and monitoring of the progress of genetic engineering laboratories.

WHO-Specific Activities in the Pharmaceutical Sector

While medicines alone are not sufficient to provide adequate healthcare, their health impact is remarkable. The economic incidence of medicines is substantial, representing less than 20% of total public health expenditure in developed countries ($138 per capita), but 25–66% in developing countries ($8–12 per capita). It is clear that priorities need to be set in order to maximize the health benefits from limited public spending.[4] The rational use of medicines, their accessibility and quality and the development of national drug policies are therefore critical and constitute the four main areas of work for WHO.[5] This is the mission of the ''Department of Essential Drugs and Medicines Policy'' created within the Health Technology and Pharmaceuticals cluster (EDM/HTP, Geneva Headquarters), in close collaboration with the six regional offices and other stakeholders (Table 1).

Despite the obvious medical and economic importance of pharmaceuticals, there are still widespread and persistent problems:

1. *Lack of access*: Over one-third of the world population still has no regular access to essential medicines.
2. *Poor quality of medicines*: Between 10 and 20% of sampled medicines fail quality control tests in developing countries.
3. *Irrational prescribing and use*: Up to 75% of antibiotics are prescribed inappropriately.
4. *Lack of pharmaceutical research and development* for tropical diseases.[6,7]

Reasons are complex and go beyond simple financial constraints. (In most developed countries, virtually 100% of the population has health insurance; median coverage is 35% in Latin America and less than 10% in Africa and Asia.) Changes in the patterns of disease (e.g., rise of HIV/AIDS, increasing drug resistance for malaria and tuberculosis, increase in chronic diseases and diseases of the elderly) and drug demand also represent major challenges and contribute to increased spending on drugs and growing pressure on health resources. The potential impact of international trade agreements, such as the World Trade Organization Agreement on trade-related aspects of intellectual property rights (WTO/TRIPS), is also a matter of growing concern with relation to access to new essential medicines (e.g., antiretroviral drugs, antibiotics, second-generation vaccines).[8] It highlights the need for strengthened international cooperation and a stronger public health response.

REFERENCES

1. WHO. About WHO. URL: http://www.who.int/aboutwho/en/healthforall.htm (accessed on Oct. 2000).
2. WHO. The World Health Report 2000. In *Health System: Improving Performance*; World Health Organization: Geneva, 2000; 215 pp.
3. WHO. Guidelines for Drug Donations, Revised 1999. In *Interagency Guidelines*; World Health Organization: Geneva, 1999, (WHO/EDM/PAR/99.4).
4. WHO/DAP. Global Comparative Pharmaceutical Expenditures. In *Health Economics and Drugs*; WHO/DAP Series, World Health Organization: Geneva, 1997; Vol. 346.
5. WHO. WHO and medicines: Role and mission of EDM, URL: http://www.who.int/medicines/edm_about.html (accessed on Oct. 2000).
6. WHO. WHO Medicines Strategy. In *Framework for Action in Essential Drugs and Medicines Policy, 2000_2003*; World Health Organization: Geneva, 2000; 70 pp.
7. Trouiller, et al. Drug development for neglected diseases: A deficient market and a public-health policy failure. Lancet **2002**, *359*, 2188–2194.
8. WHO. *Globalization and Access to Drugs, Perspectives on the WTO/TRIPS Agreement*; WHO/DAP Series, World Health Organization: Geneva, 1999; Vol. 7.

World Health Organization Essential Drug List

Patrice Trouiller
University Hospital of Grenoble, Grenoble, France

INTRODUCTION

Health is a fundamental human right. Access to healthcare, which includes access to essential medicines, is a prerequisite for realizing that right. First introduced in 1975, the concept of essential drugs (called ''essential medicines'' since 2001) is now widely accepted as a pragmatic approach to providing the best of evidence-based and cost-effective healthcare. This is a global concept that can be applied in any country, in private and public sectors, and at different levels of the healthcare system. The first model list of essential drugs (EDL) was prepared by an Expert Committee and published by WHO in 1977 and now includes 306 active ingredients (11th edition of November 1999).[1] The list does not exclude all other medicines but rather focuses on therapeutic decisions, professional training, public information, and financial resources. The essential medicines represent the best balance of quality, safety, efficacy, and cost.

THE ESSENTIAL MEDICINES CONCEPT

Virtually no system in the world offers unlimited access to all medicines and vaccines, even in the wealthiest countries. Careful selection of priority medicines is indispensable and should be based on considerations of quality, safety, efficacy, cost-effectiveness, and local appropriateness. The concept of essential medicines (a term that describes all pharmaceutical preparations used in clinical healthcare practice) is that a limited number of carefully selected medicines based on agreed standard treatment guidelines leads obviously to a better supply of medicines, to more rational prescribing and dispensing, and to the most cost-effective use of health resources. In developing countries, the essential medicines concept has also provided a means by which governments and donor agencies can plan and monitor drug procurement. It is within this framework that in 1978 a conference on primary healthcare (PHC) was jointly convened by WHO and UNICEF at Alma-Ata, Kazakhstan to confirm that the provision of essential medicines was one of the eight components of PHC. Essential medicines are defined as those that ''satisfy the healthcare needs of the majority of the population and they should therefore be available at all times in adequate amounts and in appropriate dosage forms and at a price that individuals and the community can afford.'' Their safety and relative cost-effectiveness are also major considerations in the choice.[2]

Since the first WHO EDL was published 25 years ago, more than 146 countries have adopted national EDLs, the WHO list serving as a model and not as a global standard. The list is biennially updated by a WHO Expert Committee, and the current eleventh revision (November 1999) includes 306 active ingredients (medicines, vaccines, contraceptives, insect repellents, and diagnostic agents) specified by international nonproprietary names (INN) or generic names—50% more drugs than in 1977 and of which 250 are included in WHO clinical guidelines. The list is divided into a main list, a complementary list, and there is a separate category of ''reserve antimicrobials.'' However, selection and dissemination of the list are not by themselves sufficient for ensuring rational drug use. Guidelines (e.g., clinical therapeutic guidelines, guide to good prescribing, WHO ethical criteria for medicinal drug promotion, drugs and therapeutic committee), professional training (e.g., training courses organized by the International Network for rational use of drugs, INRUD, Washington, D.C.), formulary manuals and independent drug information centers (e.g., drug bulletin publications with the aid of the International Society of Drug Bulletins, ISDB) are needed for health professionals to put lists into clinical practice.

Over the past 25 years, the model list has led to a global acceptance of the concept of essential medicines as a powerful means to promote health equity. By the end of 1999, 156 WHO member states had official essential medicines lists, of which 127 had been updated in the previous five years.

SELECTION OF ESSENTIAL MEDICINES

The selection of essential medicines is one of the seven key components of a national drug policy (the other com-

Encyclopedia of Clinical Pharmacy
DOI: 10.1081/E-ECP 120014641

ponents being legislation, regulation and guidelines, access to essential medicines, quality assurance, rational use of medicines, research, and human resources development). Usually, market approval of a pharmaceutical product is granted by drug regulatory authorities on the basis of efficacy, safety, and quality, and rarely on the basis of comparison with other products already on the market or cost. Different criteria are used for the selection of essential medicines; e.g., medicines with adequate evidence of efficacy and safety; relative cost-effectiveness with comparisons between medicines (comparative efficacy and safety, meta-analysis) and considerations of the total cost of the treatment; pharmacological criteria (e.g., pharmacokinetic properties, bioavailability, galenic stability); and formulations as single compounds, with fixed-ratio combination products being acceptable only when the combination has a proven advantage (e.g., therapeutic effect, adherence to treatment improvement).

There may still be oppositions to the use of the essential medicines list. Physicians may see it as questioning their prescription freedom, pharmacists may be worried about the financial implications, while manufacturers may fear a market erosion, and consumers may think that they are being offered second-rate cheap medicines. These concerns must be considered and addressed, and this is why the selection process should be consultative, and why education plays an important part. In fact, an essential medicines policy is nothing but an extension of the selective exercise carried out by the state, on behalf of the rights a community has to useful and safe products, to identify medicines that deserve marketing approval. The principle of convenience is under consideration in an increasing number of countries, especially as the pharmaceutical industry becomes more prolific, more complex, and uses products that are increasingly powerful and, consequently, more hazardous.

ACCESS TO ESSENTIAL MEDICINES

Despite important achievements, the lack of regular access to essential medicines still remains a major health problem, recently highlighted by the magnitude of the HIV/AIDS pandemic [e.g., in the 11th EDL, while 15 drugs of importance to HIV-related opportunistic infections are present, only two antiretroviral drugs (ARVs) are listed for the prevention of mother-to-child infection, and no ARVs are listed for the HIV treatment itself because of their high costs and prerequisites for treatment implementation and compliance]. These problems are greatest in low-income countries, but middle- and high-income countries are also increasingly facing difficult therapeutic and economic decisions.

To ensure access to essential medicines, WHO has defined a strategy by focusing on four key objectives: rational selection and use (through a strengthening of links between WHO EDL and standard therapeutic guidelines for priority diseases); affordable prices (through good procurement practices and indicative cost information policies, and TRIPS safeguards as parallel import and compulsory licensing for new essential medicines); sustainable financing, national spending on pharmaceuticals varying from $2 to $400 per capita and per year (through public funding and a health insurance scheme to maximize risk-pooling and equity shifts); and reliable health systems ensuring a proper diagnosis and treatment and a responsible supply system.

Despite the obvious medical and economic importance of essential medicines, there are still problems with lack of access, poor quality, and irrational use. In many health facilities in developing and in developed countries, essential medicines are not used to their full potential.[3] While the EDL is regarded as a key aspect of global health and economics, the WHO EDL is, for the past few years, at the center of debates with respect to access issues (such as new essential medicines for HIV/AIDS, tuberculosis, and bacterial infections) and availability issues (such as the lack of pharmaceutical research and development for tropical diseases, most of the current tropical pharmacopoeia having been driven by colonization requirements during the first part of the 20th century).[4] In 2001, still one-third of the world's population (over 50% in the poorest part of Africa and Asia) does not have regular access to the most vital essential medicines.

REFERENCES

1. WHO. The WHO essential drugs list. URL: www.who.int/medicines/edl.html (accessed on Oct. 2000).
2. WHO. The Use of Essential Drugs. In *6th Report of the Expert Committee*; WHO Technical Report Series, World Health Organization: Geneva, 1995; Vol. 850.
3. WHO. *WHO Medicines Strategy: Framework for Action in Essential Drugs and Medicines Policy, 2000–2003*; World Health Organization: Geneva, 2000, (WHO/EDM/2000.1).
4. Pécoul, B.; Chirac, P.; Trouiller, P.; Pinel, J. Access to essential drugs in poor countries. A lost battle? JAMA, J. Am. Med. Assoc. **1999**, *281*, 361–367.

Index